MW00488958

ANTPITTAS
AND
GNATEATERS

HELM IDENTIFICATION GUIDES

ANTPITTAS AND GNATEATERS

Harold F. Greeney

Illustrated by David Beadle

H E L M

LONDON · OXFORD · NEW YORK · NEW DELHI · SYDNEY

HELM
Bloomsbury Publishing Plc
50 Bedford Square, London, WC1B 3DP, UK

BLOOMSBURY, HELM and the Helm logo are trademarks of
Bloomsbury Publishing Plc

First published in the United Kingdom 2018

A catalogue record for this book is available from the British Library

Library of Congress Cataloguing-in-Publication data has been applied for.

ISBN: HB: 978-1-4729-1964-9; ePDF: 978-1-4729-4254-8; ePub: 978-1-4729-1965-6

2 4 6 8 10 9 7 5 3 1

Maps by Julian Baker
Design by Susan McIntyre
Printed and bound in India by Replika Press Pvt. Ltd

To find out more about our authors and books visit www.bloomsbury.com and sign up for our newsletters

CONTENTS

PREFACE

If memory serves, the journey that culminated in the writing of this volume began in the year 2000, on a drizzly afternoon when Paul Martin and Rob Dobbs walked into the Yanayacu Biological Station and introduced me to the wonderful world of avian breeding biology. Until that day, my 'study' of bird nests had been largely restricted to brief field notes on nests encountered only by happenstance while searching for caterpillars, my passion since childhood. These two amazing natural historians, however, introduced me to birds at a level that, for the first time, inspired the passion and excitement I had so frequently seen in my 'birder' friends. I eagerly followed them about and soaked up their considerable knowledge of birds and their nests. In the weeks that followed, during long, soaking wet cloud forest excursions, and while gathered at night around the light of a candle and the warmth of a bottle of cheap rum, I heard several things that sparked my interest in nests and antpittas. First, I was shocked to learn that the nests of many birds had not been described, particularly those inhabiting the steep slopes of my east Andean home. This seemed shocking to me, given that it seemed every other researcher I met was working on birds, and given the seeming ease with which Paul and Rob were able to find nests. Indeed, at the time, more than a third of the species nesting in Ecuador lacked descriptions of their nests. Among those species with undescribed nests, at that time, were 18 of the 23 species of Ecuadorian antpittas! Caught up in the bravado of our youth, we decided it was time to rectify this dearth of knowledge on antpitta breeding biology. Nevertheless, given the recognised difficulty of even seeing an antpitta, let alone finding their nests, we prudently chose to start with the 'low-hanging' fruit offered by two of the most common species in eastern Ecuador. It took less than a morning of tunneling through large stands of *Chusquea* bamboo to find our first Chestnut-crowned Antpitta nest, and less than an hour of searching isolated shrubs in the paramo to find our first Tawny Antpitta nest. Though these were encouraging results, my continued search for the nests of antpittas proved somewhat more challenging, and I feel it is only fair that I publicly admit my next antpitta nest (Peruvian Antpitta in 2003) was found by accident while searching for caterpillars. Regardless, I was hooked on nests, particularly of antpittas, and my passion quickly grew. I will not soon forget the satisfaction I felt when, while stalking a pair of Jocotoco Antpittas, I avoided the need to relinquish my hidden position in the bamboo to a swarm of army ants by urinating on my legs (and thus discouraging the ants from climbing them). Though that brief story is somewhat of an over-share, and will undoubtedly raise questions about my sanity for most of my readers, it perfectly illustrates how much of an antpitta-addict I had become.

It was around the time I was first introduced to antpittas that I first read about the breeding of gnateaters. At that time I lived at an elevation where no gnateaters occurred, and I still walked largely in the world of entomology. If truth be told, at the time I don't think I would have recognised a gnateater in life. Apart from their similarities to antpittas, their English name was well-suited for arousing curiosity in an entomologist, so I was pleased to stumble upon one of the rare papers that included information on their reproductive habits. In an offhand comment, Oniki (1981b) commented that Rufous Gnateaters 'do not peer much in the nest', a detail that, at the time, seemed such a curious thing to say that I just had to know more. Alas, in that particular paper, she had little else to say on the subject. Now, however, after watching thousands of hours of behavioural nest videos, Oniki's (1981b) observation seems like a perfectly relevant detail to comment on. Luckily, because I read that paper in 2001, I did not have to wait two years for the follow-up (see Willis *et al.* 1983). In fact, I was able to dive straight into this wonderful summary of Rufous Gnateater behaviour, full of incredible details. After this intriguing introduction to gnateaters, I could not help but feel compelled go look for my first gnateater nest!

Now, some 20 years after antpittas and gnateaters were brought to my attention, I have grown even more passionate about studying them. Despite the fact that now only two species of 'Ecuadorian' antpittas lack nest descriptions (Ochre-striped and Bicoloured), we still have a long way to go before they will be considered a well-studied group. Indeed, the species treated here are, arguably, some of the least-studied members of the Neotropical avifauna, and this book represents the first volume dedicated solely to them. I can, therefore, confidently boast that this book is the most complete! Nevertheless, I offer this book as a first draft, and it is my sincere hope that my fellow avian mavens and natural historians, both present and future, will work hard to correct its flaws, fill its shortcomings, and round out its completeness. Indeed, the enthusiasm I encountered in discussing this work with others, and the near-constant influx of new and exciting information that poured in as I compiled it, have already foreshadowed a second edition that will be even more exciting to write.

ACKNOWLEDGEMENTS

Although it goes without saying, any attempt at compiling such a large volume of published and unpublished information can hardly be credited to the author alone. This book would never have been possible without the enthusiastic support of a prestigious list of mentors, students, friends and peers. To all of these, who gave generously of their time, knowledge and support, I am forever indebted.

Every name mentioned here is fully deserving of more personal thanks, but two friends and fellow ornithologists truly went above and beyond. Not only have the following people contributed in most or all of the other ways detailed below, they have also shared their valuable time by carefully reviewing content, offering insightful suggestions, unpublished thoughts and sympathetic comments, thereby greatly improving the final product in just about every way possible. Thank you to Juan Freile and Fernando Angulo Pratolongo: you are about to read something made better by their efforts. Matt E. Kaplan also deserves a special mention. He continues to be a much-loved friend and deeply appreciated benefactor. His support of my fieldwork over the years, emotionally and financially, has been crucial for my continued professional and personal improvement.

I am especially grateful to the following museum personnel and their respective institutions for generously hosting my visits, providing access to their databases, and providing me with photos or answering my questions about specimens that I was unable to examine in person: Linnea S. Hall and René Corado (Western Foundation of Vertebrate Zoology, Camarillo), Ben Marks and John Bates (Field Museum of Natural History, Chicago), Luis Sandoval (Museo de Zoología de la Universidad de Costa Rica, San José), Christopher Milensky (Smithsonian Institution, US National Museum of Natural History, Washington DC), Christopher Witt (Museum of Southwestern Biology, University of New Mexico, Albuquerque), Paul R. Sweet (American Museum of Natural History, New York), Alexandre Aleixo and Maria de Fatima Cunha Lima (Museu Paraense Emílio Goeldi, Belém), Jonathan C. Schmitt (Museum of Comparative Zoology, Cambridge, MA), Manuel V. Sánchez N. (Museo del Instituto Nacional de la Biodiversidad del Ecuador, Quito), Fernando Forero and Claudia Medina (Instituto Alexander von Humboldt, Villa de Leyva), Margarita Martínez (Colección Ornitológica Phelps, Caracas), Gabriela G. Araujo (Museu Nacional, Universidade Federal de Rio de Janeiro), Miguel Angel Aponte Justiniano (Museo de Historia Natural Noel Kempff Mercado, Santa Cruz), L. Mauricio Ugarte-Lewis (Museo de Historia Natural de la Universidad Nacional San Agustín), Isabel Gómez (Colección Boliviana de Fauna, Museo Nacional de Historia Natural, La Paz), Ghisselle Alvarado (Museo Nacional de Costa Rica, San José), J. Van Remsen and Stephen W. Cardiff (Louisiana State University Museum of Zoology, Baton Rouge), Sara Bertelli and Walter Sebastian Aveldaño (Fundación-Instituto Miguel Lillo, Tucumán), Stephen P. Rogers (Carnegie Museum of Natural History, Pittsburgh), B. Patricia Escalante Pliego and Marco Antonio Gurrola Hidalgo (Colección Nacional de Aves, Instituto de Biología de la Universidad Nacional Autónoma de México, Mexico City), A. Alfredo Bueno Hernández (Museo de Zoología de la Universidad Nacional Autónoma de México, Mexico City), Luis Germán Gómez (Museo de Historia Natural de la Universidad del Cauca, Popayán), and Gary A. Voelker (Collection of Birds, Biodiversity Research and Teaching Collections, Texas A&M University, Austin). Additional thanks are owed to Matthew L. Brady, Jonathan C. Schmitt, Alán Palacios Vázquez, Manuel V. Sánchez N. and Samuel S. Snow for extra time spent measuring and photographing specimens for me.

This book was greatly improved by thoughtful reviews of the text by various regional or taxonomic experts. For their time and help, I thank Luis Sandoval (Costa Rica), Juan Freile (Ecuador), Fernando Angulo Pratolongo (Peru), Alejandro Bodrati (Argentina), Robert C. Dobbs (Grallariidae natural history), Peter A. Hosner (Peruvian *Grallaria*), Thomas Donegan (*Grallaricula nana*), Terry Chesser and Mort Isler (*Grallaria rufula*), Benjamin M. Winger (*Grallaria hypoleuca*), Patrick O'Donnell (*Pittasoma michleri*) and Alejandro Solano-Ugalde (*Pittasoma rufopileatum*). For help identifying reptiles and amphibians mentioned in the text, I thank W. Christopher Funk, Alejandro Arteaga and Omar Machado.

For generously sharing their unpublished observations, data and photographs I thank the following students, ornithologists, guides and institutions: Roger Ahlman, Almir Cândido de Almeida, Pedro Xavier Astudillo W., Javier Barrio, Dusti Becker (Life Net Nature and Earthwatch Institute sponsored data), Jürgen Beckers, Alejandro Bodrati, Michael Braun, Galo Buitrón-Jurado, Carlos Daniel Cadena, Diego Carantón, Hugo Fernando del Castillo, Luis Fernando Castillo C. (Asociación Calidris, www.calidris.org.co), Santiago David, Caroline Dingle, Tulio Dornas, Sandra M. Durán, Jon Fjeldså, Luis A. Florit, Benjamin G. Freeman,

Antonio García-Bravo, Rudy A. Gelis, Felipe Bittioli Rodrigues Gomes, Caleb E. Gordon, J. Berton C. Harris, Boris Herrera, Steve Hilty, Richard C. Hoyer, José B. L. Hualinga, Niels Krabbe, Daniel Lebbin, Gustavo A. Londoño, Jane Lyons, Rogério Machado, Oswaldo Maillard Z., Oscar Humberto Marín-Gómez, Paul R. Martin, Borja Mila, Jhonathan Eduardo Miranda T., Ángel Paz, Felipe de Oliveira Passos, Julia Patiño, J. Van Remsen, Forrest Rowland, Gianpaolo Ruzzante, Luis A. Salagaje M., César Sánchez, Luis Sandoval, Tatiana Santander, Marcos Pérsio Dantas Santos, Glenn F. Seeholzer, José Simbaña, Thomas Smith, Alejandro Solano-Ugalde, Andrew Spencer, F. Gary Stiles, Luis Eduardo Urueña, Jarol Fernando Vaca B., John van Dort, Juan Diego Vargas-Jiménez, Cesar Villamil, Catherine Vits and Boudewijn de Roover (Copalinga Ecolodge), Venicio Wilson and Julián Andrés Zuleta-Marín. In addition, the following individuals were very helpful in providing access to references or other information and sharing insightful thoughts on the same: Edward Dickinson, Guy M. Kirwan, Luke Powell, Thomas. S. Schulenberg and Edson Guilherme da Silva. In particular, I am grateful for the knowledge and companionship shared during fieldwork of Rudy A. Gelis and José Simbaña.

For financial support during the writing of this book and, most importantly during the past 20 years of fieldwork on antpittas and other Neotropical organisms, I am especially grateful to the following organisations (in chronological order): PBNHS, Population Biology Foundation, Neotropical Bird Club, Rufford Foundation, Association of Field Ornithologists, Field Guides Inc., Chicago Board of Trade Endangered Species Fund, Hartford Foundation, and the John Simon Guggenheim Memorial Foundation. Of particular importance to me over the years, John V. and the late Ruth Ann Moore have been incredibly generous in their support. Matt E. Kaplan has also never failed to help.

For their help throughout the long process of creating this book, I would like to thank Jim Martin and Jenny Campbell at Bloomsbury for their patience with the endless stream of concerns and questions of a first-time book author. I am particularly grateful for their willingness to allow me to write this book, with virtually no abridgement, the way that I envisioned that a work like this should be.

Last, and not at all least, thank you to my wife, Lindsey Cohen and to our three amazing children, Phoenix Francis, River Wren and Tamia Brooks. I am forever thankful for their love, as well as for their patience during, and support of, my long hours of writing and trips to the field.

Crescent-faced Antpitta, *Grallaricula lineifrons*. Adult attending a single, mid-aged nestling in its shallow mossy open-cup nest, 10 February 2012, Papallacta, Napo, Ecuador (*Harold F. Greeney*).

LAYOUT OF THE BOOK

I have made every effort to maintain a similar organisation and flow within each account, but the myriad of sources, original ways that data were presented, and the various languages in which they were printed, all make precise replication of the account format rather challenging. Below is a general outline of the organisation of species accounts with a description of the order and manner in which information is presented within each that, hopefully, will allow the interested reader to find desired information with the minimum of hassle. In recognition of the somewhat broad audience that I hope will find this book of interest, I have made every attempt to structure the species accounts such that those of you who are curious or wish to quickly reference straightforward information on range, appearance or behaviour are easily appeased by reading or scanning the first several sections of each account. Conversely, those of you seeking data, facts or further reading will likely wish to jump to the latter portions where you will find the 'data-heavy' sections. In broad-brush terms, this means that accounts flow from creatively and lightly written prose, to dry and data-rich text, a format that I hope will facilitate both the enjoyment and scientific value of their content, and appeal to both the aficionado and the scientist.

The photographs

The wonderful photographs included within this book are meant to complement both the text and the beautiful paintings of David Beadle. I have selected, where possible, images that show interesting facets of species' life histories or plumage variation. Sadly, there are several species treated here for which there were no *in situ* photos to be found. For these species I have chosen to include specimen photos, and hope the reader will also enjoy Beadle's paintings. Even though space limitations prohibited inclusion of all of their wonderful images, I am grateful to the following photographers who offered their images for consideration: Anselmo d'Affonseca, Arvind Agrawal, Roger Ahlman, Ciro Albano (www.nebrazilbirding.com), Almir Cândido de Almeida, Bilal Al-Shahwany, Gordon Appleton, Juan José Arango, David Ascanio, Nick Athanas, Ian Ausprey, José Daniel Avendaño, Guto Balieiro, Jean-Luc Baron, David Beadle, Josh Beck, Arlei Bertani, Joseph Blowers, Roger Boyd, Dušan M. Brinkhuizen, Caio Brito, Demis Bucci, Chris Burney, Linda Bushman, Carlos Calle, Diego Carantón, Ramiro Ramírez Cardona, Ernesto M. Carman Jr, Thiago Carneiro, Brandon Caswell, Karll Cavalcante, Murray Cooper, George Cruz (www.sanjorgeecolodges), Andrés M. Cuervo, Robson Czaban, Santiago David, Stephen Davies, Luis Vargas Durán, Susan Ellison, Francisco Enríquez (www.penttours.com), Jeisson Andrés Zamudio Espinoza, Don Faulkner, Marcelo Jordani Feliti, Caio Feltrin, Francisco Valdicélio Ferreira (www.lattes.cnpq.br/0242658467815633), Fernando Moreira Flores, Johan Flórez, Tom Friedel, Isabel Garavito, Rudy A. Gelis, George Golumbeski, Felipe Bittioli Rodrigues Gomes, Robert Gowan, Thomas Grim, Arthur Grosset, Alexandre Gualhanone, Marco Guedes (www.wikiaves.com.br/perfil_maguede), Jorge Obando Gutiérrez (www.flickr.com/photos/jobando), Christian Hagenlocher, Beth Hamel, Peter Hawrylyshyn (www.pahphoto.com), Ron Hoff, Jon Hornbuckle, Richard C. Hoyer, Paul B. Jones, Leif Jönsson, Mery E. Juiña J., Eric Kofoed, Yann Kolbeinsson, Martjan Lammertink, Daniel F. Lane, Daniel J. Lebbin, Eli Lichter-Marck, Jefferson Luis Gonçalves de Lima, Jorge Eduardo Lizarazo B., Gustavo A. Londoño, Thomas Love, Juan José Chalco Luna (www.facebook.com/chalcobirdwatching/?ref=br_rs), Oswaldo Maillard Z., Stuart Malcom, Neil Orlando Díaz Martínez, Miguel Felipe López Martínez, Ian Maton, Tomaz Melo, Eliot T. Miller, Kurt Miller, Jorge Montejo, Graham Montgomery, Luciano Morães, Anderson Muñoz, Andrea Narvaez, Juan Ochoa, Júlio César Vaz de Oliveira, Scott Olmstead, Angelica Hernandez Palma, Jean Paul Perret, João Quental (www.flickr.com/photos/jquental), Alonso Quevedo, Ramiro Ramírez, Galo Real, Stuart Reeds, Martin Reid, J. Van Remsen, Luiz Carlos da Costa Ribenboim, Zigmar Riedtmann, Joel N. Rosenthall, Laval Roy, Mauricio Rueda, Marco Saborio, Luis A. Salagaje M., Sidnei Sampaio, Manuel Andrés Sánchez, Håkan Sandin, Luis Sandoval, Fabrice Schmitt, Floyd Schrock, Luke Seitz, Thiago Tolêdo e Silva, Kleber Silveira, José Simbaña, Aldo Sornoza, Francisco (Pancho) Sornoza, Adam Spencer, Kristian Svensson, Andrés Terán, Paulo Tinoco (www.avesdemacae.com), Joseph A. Tobias, Roberto Torrubia, Olivier Tostain, Thomas J. Trombone, Daniel Hernández Ugarte, Daniel Uribe, Jarol Fernando Vaca B., Marcio R. Varchaki, Juan Diego Vargas-Jiménez, Júlio César Vaz de Oliveira, Carlos Verea, Ela Villanueva, Lee Vining, Deborah Visco, Nigel Voaden, Sam Woods, Alan Woodward and Glauco Zeferino.

The accounts

Below the species name and plate number, I have given the full citation for the species description, using the scientific name under which it was first described. In some cases, this formal reference is followed by some additional information such as the origin or meaning of the name, details about the location of type specimens or collecting localities, or details of particularly confusing changes or additions to the species taxonomy which may prove informative for someone wishing to search the literature for further information. The introductory paragraph for each species is the most variable section of the accounts. In general, it begins with a several-sentence broad-brush summary of the account, perhaps highlighting something of particular interest or opening with an informative, imaginative or amusing quote concerning the species. In some cases I have included short anecdotes or personal opinions meant to entertain and provoke thought. The opening remarks are usually followed by a brief summary of general information that is *missing* from our knowledge of the species, often highlighting particularly rewarding or useful directions for future research or conservation action. Lastly, where relevant, I have pointed out any significant contributions that the account itself makes with regards to previously unpublished (or unnoticed) information.

Identification

This section is designed to give a brief overview of the species' appearance and size, with remarks on similar species or important plumage variation between the sexes. For each species, where our knowledge permits, the size range is an approximation of the 'usual' size (length), from the tip of the bill to the tip of the tail. In most cases, particularly from the older literature or for species with few data, these measurements are taken or estimated from preserved skins that can vary greatly in this measurement simply due to how they were prepared. Usually this results in an overestimation of the true size of the bird when seen in life. The size range provided here is designed to represent a rough guide for someone viewing the bird in nature and should by no means be considered a scientifically accurate data point. In this identification section I have often dropped the extreme upper and lower size ranges found in the literature. For a complete range of reported sizes and associated references see **Morphometric Data**.

Streak-chested Antpitta, *Hylopezus perspicillatus*. Adult attending two mid-aged nestlings in its untidy, shallow open-cup nest of sticks and dead leaves. Drawn from photos courtesy of Ian Ausprey, 21 June 2013, Quebrada Gonzalez, Heredia, Costa Rica (*Harold F. Greeney*).

Distribution

In each case I open with the broadest and most inclusive statement that can be made about a species' geographical range, generally just mentioning the latitudinal and longitudinal extremes by country and, in many cases, by smaller geopolitical subdivisions such as state or province. If relevant I point the reader to particular gaps in our knowledge that would be useful to fill, or discuss particularly disjunct distributions in more detail. This section is designed to directly and simply supplement the visualisations of species' ranges portrayed by the distribution maps. For monotypic species, further detail on the localities used to generate the maps and known distributions, as well as further discussion of the subject, see **Distribution Data**. For species where more than one putative taxon is involved, these data can be found under Taxonomy and Variation. After some debate, I have included descriptions and details of altitudinal distributions in the **Habitat** section, as elevation and habitat are so strongly correlated. If known, with any degree of certainty, the holotypical locations for each taxon are indicated with a red dot on the distribution maps (see example below). In addition, distributional records of questionable validity are indicated with a black ×.

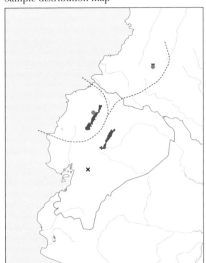

Sample distribution map

Key to the rivers

a	Río Cauca	k	Río Ucayali
b	Río Magdalena	l	Rio Juruá
c	Río Orinico	m	Rio Purus
d	Río Putumayo	n	Río Apurimac
e	Río Napo	o	Río Madre
f	Río Pastaza	p	Rio Tapajós
g	Rio Negro	q	Rio Teles Pires
h	Rio Branco	r	Rio Xingu
i	Río Marañon	s	Rio Araguaia
j	Río Huallaga	t	Rio Tocantins

Key to the distribution

Distribution area

● Holotype

× Disputed record

Habitat

This section seeks to provide the reader with a quick sense of the 'habitat type' occupied by a species (i.e., humid forest, seasonal forest, *páramo*, etc.) and, if possible, to suggest 'microhabitat' preferences within these broader categories (i.e., treefalls, riparian zones, swamps, etc.). Like many other aspects of the ecology and behaviour of a species, there is much subjectivity when discussing habitat, despite (or due to) the many divisions, subdivisions and proposed category labels available in the literature, and categorically assigning a species to a habitat is frequently challenging or impossible. As a simplistic example, forest edges can occur along waterbodies, at man-made habitat disruptions of various types ranging from linear (roads) to complete (forest removal), or even at naturally occurring habitat breaks such as gaps created by windblown trees, swamps or other natural heterogeneity. A species considered to prefer 'forest edges' will, almost certainly, have fairly strong preferences among all of these various forms of 'edge' habitats. Rather than trying to unify the habitat designations of many authors under one system, or to lump them into broad categories such as 'forest edge', therefore, I have tried to be as descriptive as possible. If applicable I have summarised the findings of previous habitat studies (relatively few for the species involved here) and have attempted to avoid unnecessary generalisations, unless there is strong evidence of habitat preference. Habitat, as defined herein, should be considered a description of where a species is most frequently encountered and, unless I have referenced specific studies, is not meant to imply dependency or specialisation. Generally at the end of this section, I review altitudinal range descriptions and specific localities that help to better define both the altitudinal range and habitat use of the species.

Riparian Central American rainforest, Costa Rica (*Harold F. Greeney*).

Bunch grass paramo, 4,100m, Papallacta Pass, Napo, Ecuador (*Harold F. Greeney*).

Riparian zone within gallery forest in Brazilian cerrado, near Brazilia, Distrito Federal, Brazil (*Harold F. Greeney*).

Humid rainforest, Ecuador (*Harold F. Greeney*).

Logging destruction of lowland rainforest in the Choco Bioregion, north-west Ecuador (*Harold F. Greeney*).

Napo River, Ecuador (*Harold F. Greeney*).

Tropical deciduous forest of the Tumbesian bioregion during the rainy season, Reserva Privada Jorupe, Loja, Ecuador (*Harold F. Greeney*).

Movements

Few of the taxa treated here are thought to show migratory or altitudinal shifts of any kind, and none are *known* to do so. As such, if there is nothing to report in this section I have omitted it from the account entirely. That said, for some species I have included some interesting observations or information under this heading.

Voice

I have, for the most part, relied entirely on published sources for descriptions of vocalisations. If there is known geographic or racial variation, this is presented here. A word of caution: I have used the terms 'call' and 'song' rather loosely as, in most cases, the distinction is not well known for the species treated here. I have, except in specific instances, removed all reference to the sex of the vocalising bird as previous authors have relied mostly on (possibly unreliable) inferences based on behaviour to assign vocalisations to one sex or the other. Wherever possible, to facilitate recognition in the field, I have tried to include more than one onomatopoeic transcription of at least the most frequently heard vocalisations.

Natural History

This is a catch-all section of miscellaneous notes covering a variety of traits such as postures or movements while vocalising, roost locations or other aspects of general natural history not specifically covered elsewhere. Information on foraging mode and behaviour is included here, while a description of prey consumed can be found under **Diet**. This section also includes, in some cases, information on territoriality, predation and parasitism.

Diet

This section includes all items known or inferred to be eaten by the species in question. For foraging behaviours refer to **Natural History**. The diets of most antpittas and gnateaters are, sadly, known only in a general sense (i.e., small invertebrates). Wherever possible I have referenced both previously published descriptors of diets as well as original information from specimen labels.

Reproduction

Near and dear to my heart, this section begins with a broad-brush look at our level of knowledge on breeding. Where applicable I have presented introductory remarks about the sources of information and attempted to give a brief 'history' of the progress of published knowledge. Also found here is a two- or three-sentence description of the nest and egg for those readers not in need of quantified details and sample sizes. Following this are up to four detailed, rather self-explanatory, subsections: (1) **Nest & Building** – detailed description of nest architecture, composition, nest site and habitat, behaviours associated with site-selection and construction; (2) **Egg, Laying & Incubation** – detailed description of egg colour, size and mass, timing of laying, and behaviours associated with laying and incubation; (3) **Nestling & Parental Care** – detailed description of hatching, nestling appearance, growth and plumage development, and behaviours associated with the care of young both in and outside of the nest; and (4) **Seasonality** – there is almost nothing known about the duration of post-fledging care or development for any species discussed here. Because of this, in most cases I have refrained from more than a cursory suggestion as to when reproductive peaks may or may not occur throughout the year. In as many cases as possible I have tried to suggest when interested fieldworkers should concentrate their efforts and in most cases my interpretation of seasonality should be viewed as a starting point, not a statement of fact. CAUTION (pet peeve follows): many authors, myself included in the folly of my youth, have taken reports of reproduction during certain months and implicitly interpreted these as a description of a species' reproductive seasonality *across its range*. This often leads to statements along the lines of: 'with the inclusion of our records, it appears that reproduction occurs throughout the year.' I would venture to say, however, that with the possible exception of species with extremely limited ranges, *most species* will eventually be found breeding during all months of the year. This is a fact that is ripe for misinterpretation and is unlikely true for any given *population*, each of which is undoubtedly responding to locally experienced seasonal changes in temperature, rainfall, prey abundance etc. My message here is that anyone wishing to improve the impact of their data on reproduction should make it a point to include local or regional descriptions of climatic variables, and use these data when interpreting the reports of others. As a subsection, at the end I include a list of specific data indicating reproductive activity under **Breeding Data**. These are ordered primarily by reproductive

Adult Thrush-like Antpitta, *Myrmothera campanisona signata*, incubating in its bulky stick nest, Reserva Privada Yankuam, Zamora-Chinchipe, Ecuador (*Harold F. Greeney*).

An adult White-bellied Antpitta, *Grallaria haplonota castanea*, bringing prey to its young at the Estación Biológica Yanayacu, Napo, Ecuador, 2,050m (*Harold F. Greeney*).

stage (from building to fledging) and to a lesser degree by the 'precision' with which they can be used to calculate breeding seasonality. By precision, I mean that a nest with eggs clearly indicates that nesting is occurring on that date at that location, whereas a bird in subadult plumage (see **Technical Description**) indicates an active nest at some (usually) poorly defined date in the past.

Technical Description

This section represents an expansion of the **Identification** section, including additional details and specifics that are not necessarily characters useful for identification in the field. This section also includes detailed descriptions of age-related plumage changes. In most cases the adult plumage characters are those of the nominate race or for the most widespread taxon. Plumage variation is described and discussed under **Taxonomy and Variation.** For immature plumages, however, any known racial or geographic variation is included in this section. More so than any other part of this work, my categorisation and descriptions of immatures and ontogenetic changes in plumage should be viewed as developing hypotheses in need of rigorous re-examination and further refinement. For all species, I have ventured as far as I dared into largely unexplored territory in this regard, fully accepting that I may have, at least in some cases, misinterpreted or misrepresented the true plumage maturation process. For the purposes of this 'first pass', I have categorically lumped plumage phases into four stages: adult, subadult, juvenile and fledgling. My definition of these latter three plumages varies to some degree by genus, and a description of my definition of each can be found immediately prior to the first species in each. In all cases, due to the incompleteness of available information, my descriptions of plumage ontogeny are, to the best of my knowledge, accurate with respect to order. I feel they provide a fairly accurate description of the *relative* age (time since fledging) of an individual based on plumage, but should *not yet* be considered to convey information as to the reproductive status of an individual. This is especially true of my subadult plumage category. Although birds in this category retain plumage characters that are clearly not those of adults, it remains to be seen if the ability to reproduce is delayed until full adult plumage is acquired.

Morphometric Data

This was, perhaps, the most difficult section of the text to maintain continuity between accounts. Data compiled here (linear measurements and mass) are presented in so many ways and derived from so many methods of data collection, that maintaining a standard presentation has proved essentially impossible. So, in an effort to keep each section as concise as possible, while still making data extraction of interest as painless as possible, I have simply followed the most logical path of organisation dictated by the data. For species with little or no geographic variation or taxonomic subdivisions, I generally organise the information by sex. For those with complicated phylogeographic divisions I have followed a more taxonomic organisation and, in some cases, location. Wherever possible I have included information on the amount of fat, in brackets, along with mass, using the terminology on the specimen's labels: (tr) = trace; (no) = none; (lt) = light; (vl) = very light; (m) = moderate; (h) = heavy; and (vh) = very heavy. Bill length presented a particular challenge as there are at least three ways commonly used to describe this: (1) from the base of the culmen where it meets the skull (hidden by the feathers), herein = total culmen; (2) from the basal-most portion not covered by skin, = exposed culmen; and (3) from the anterior edge of the nostril, = front of nares. In cases where the precise nature of measurements reported in the literature is unclear, for example when simply 'bill length' is given, I have indicated the presumed type of measurement in square brackets, based largely on comparing reported measurements with those taken by myself. For example, Hellmayr (1905a), for *Conopophaga roberti*, reports simply 'bill 15mm'. This will be presented herein as '*Bill* [total culmen] 15mm.'

Distribution Data

This section is included only for monotypic species, with distributional data provided for polytypic species under **Taxonomy and Variation**. Data are organised, roughly from north to south, first by political region and then by latitude. GPS coordinates are given to the nearest 0.5 miles. As it was obviously impossible to include all 30,000 antpitta and gnateater records, I gave preference first to all specimens with viable data that I was able to verify or found no reason to question. I omit valid records only in cases where many collecting localities were in close proximity, with the exception of taxa with small ranges or very few records. Once available specimens had been plotted, I evaluated other sources of distributional data and chose to include those records that best described a taxon's distribution. I prioritised my use of records, by first including verifiable observations (vocal recordings, photos, videos), for the most part using sight records only in cases where they were the only available data to bridge gaps in records or define distributional limits. For unvouchered records I attempted to first include localities with many reports from at least several observers. Data extracted from eBird were, in as many cases as possible, verified by correspondence with at least one observer per location, and these are generally the names credited in the text. This was particularly true for 'extralimital' or unusual records. Specific cases of doubt, confusion or error regarding locations are discussed individually. In many cases, without further discussion, I have taken the small liberty of slightly (0.5–1.5°) adjusting coordinates previously given in gazetteers (e.g. Paynter 1982, 1993, Paynter & Traylor 1991, Stephens & Traylor 1983, etc.) to more accurately approximate the location of imprecise collecting localities. These adjustments were made based on the known or likely elevational range of the species in question, or to more accurately place a point within the appropriate watershed or on the correct side of a river. I made the decision to include these data after considerable debate, and their inclusion in the final version was (thankfully!) accepted by the publisher despite the obvious augmentation to the length of the manuscript. As I wrote the descriptive text and drew range maps, two key points became quickly apparent to me. (1) This was going to be one of the most time-consuming aspects of this work if I wished it to be as accurate as possible. (2) While writing the many regional or faunal treatments that have included antpittas and gnateaters in the past, my predecessors had already performed at least a significant portion of the task at hand. The first of these assertions proved to be, unquestionably, true. As to the second point, my re-evaluation of each and every available distributional data point revealed that some authors had, indeed, waded through this enormous task, while others had probably not done so. This latter revelation was evident in the (relatively few) distributional inaccuracies that have been perpetuated, at various times, via the secondary 'field guide' literature. For those authors who probably did tackle this arduous task, what made it to the pages of their work (i.e., what was available for me to use) was a boiled-down, albeit correct, description of range that was usually further 'watered down' by a lack of discussion of subspecific ranges. I mean to imply no fault in either approach by previous authors,

Thrush-like Antpitta, *Myrmothera campanisona*. Adult adding a thin fibre to the the lining of its nest, 25 February 2012, Reserva Privada Yankuam, Zamora-Chinchipe, Ecuador (*Harold F. Greeney*).

but instead intend to convey that I became painfully aware that, had I access to their *data*, I would have saved myself *countless* hours of work. That is to say I would not have had to *re*-wade through the literary miasma of taxonomic name changes, despairingly vague and/or cryptically misspelled locality names, sadly faded and blurry handwritten original specimen labels, and other complex issues that, all told, are spread across more than two centuries, a dozen languages, and most of the world's continents. On countless occasions I spent hours following the literary trail back to a specimen's collecting locality only to arrive at the same obvious conclusion as my predecessor: the location was incorrect, or the specimen was misidentified, or there was some other obvious reason to discount the record. Even with the invaluable Biodiversity Heritage Library placing many of these resources within easier reach for the modern 'digital researcher', the thought of making the next student of antpittas do the same seemed cruel and unusual, especially when it was so easily rectified by simply providing him or her with the actual data used to define a species' known range. I am fully aware that many of you reading this will still fail to find value in the space used to include these data. For the rest of you: you're welcome. At the very least you will now find it much easier to challenge or accept *my* interpretation of the data at hand.

Taxonomy and Variation

This section opens with an overall view of geographic variation including, if applicable, possible unrecognised subspecies, areas of confusion and potentially needed changes. Following this, each recognised subspecies is discussed in as much detail as possible. They are ordered, as far as possible, by distribution from north to south. Each subspecies is discussed in the following order: (1) comments on the holotype and collecting locality; (2) a broad-brush description of its range; (3) descriptions or observations on plumage; and (4) a listing of specific literary, specimen, photographic or other record types that best describe the distribution of the subspecies (see **Distribution Data**). These data are separated by political divisions (country, state/province/department) and ordered roughly latitudinally, from north to south. Keep in mind, however, that subspecific ranges are (likely) subject to revision and re-interpretation. In other words, my organisation of records and literary sources by taxon may also require future revision.

Status

This section covers the details of current conservation status (level of threat), the distribution of protected populations, and comments or suggestions specific to conservation. In some cases I have elaborated on suggestions for future research given previously in the account. At the end of this section I include as many '**Protected Populations**' as I was able to identify, and these protected areas are organised by subspecies, where applicable.

Other names

This section could, potentially, be expanded *ad nauseum* to include a translation of, or suggested name, in all of the world's many languages. Instead, I have made an effort to provide a useful list of synonyms, alternate common names and misspellings that I have come across in the published literature. With the exception of key synonymies useful for facilitating interpretation of older literature, little more than a decade ago the names listed here would have been little more than a curiosity. With so much of the older literature now available electronically, however, I hope that the names provided (including misspellings) will be a great tool for those conducting electronic searches. It became apparent during my research for this book that, as just one example, if I had searched only for '*Myrmothera campanisona*' I would have missed a significant portion of available information on this species. Once I became aware of the frequency with which the specific epitaph was spelled '*campanisoma*', I uncovered a large body of 'digitally invisible' literature. In some cases, either as a curiosity or as an important scientific aid, I have discussed the use of some names in detail. This section has that added benefit of citing additional references of potential worth that may not otherwise have been cited in the preceding text.

Acronyms used in the text

Where applicable, the following abbreviations are used in reference to a specific specimen, photograph or recording number (i.e., AMNH 88391, XC 78757, WA 1883423, etc.). In the case of digital collections that contain various types of media, all records pertain to either photographs in the case of WA and IBC, or to audio recordings in the case of ML (unless otherwise stated). Records from my own unpublished observations are indicated with my initials (HFG).

AMNH = American Museum of Natural History (New York)
ANSP = Academy of Natural Sciences of Philadelphia (Philadelphia)
BDFFP = Biological Dynamics of Forest Fragments Project (Manaus), previously the Minimum Critical Size of Ecosystems Project (www.stri.si.edu/english/research/facilities/affiliated_stations/bdffp/)
BRTC = Texas A&M Biodiversity Research and Teaching Collections, formerly the Texas Cooperative Wildlife Collection (College Station, TX)
CAMUUA = Colección de Aves Museo Universitario de la Universidad de Antioquia (Medellín)
CM = Carnegie Museum of Natural History (Pittsburgh, PA)
CNAV = Colección Nacional de Aves, Instituto de Biología, Universidad Nacional Autónoma de México (Ciudad de de México)
COMB = Coleção Ornitológica Marcelo Bagno, Universidade de Brasília (Brasília)
COP = Colección Ornitológica Phelps (Caracas)
CUMV = Cornell University Museum of Vertebrates (Ithaca, NY)
DMNH = Delaware Museum of Natural History (Greenville, DE)
eBird = https://www.ebird.org
EcoReg = EcoRegistros, Registros Ecológicos de la Comunidad (http://www.ecoregistros.org)
ENCB = Colección de Aves, Escuela Nacional de Ciencias Biológicas, Instituto Politécnico Nacional (Ciudad de México)
FIML = Fundación-Instituto Miguel Lillo (Tucumán, Argentina)
Flickr = https://www.flickr.com
FLMNH = Florida Museum of Natural History (Gainesville, FL)
FMNH = Field Museum of Natural History (Chicago, IL)
GPDDB = Guyra Paraguay Distribution Database. A collection of Paraguayan distribution records made by various observers and organized by Guyra Paraguay (http://guyra.org.py/); communicated by Hugo Fernando del Castillo.

IAvH = Instituto de Investigación de Recursos Biológicos Alexander von Humboldt (Villa de Leyva, Colombia)

IBC = Internet Bird Collection (http://www.hbw.com/ibc)

IBUNAM = Colección Nacional de Aves, Instituto de Biología de la Universidad Nacional Autónoma de México (Ciudad de México)

ICN = Instituto de Ciencias Naturales, Universidad Nacional de Colombia (Bogotá)

INPA = Instituto Nacional de Pesquisas da Amazônia (Manaus)

KUNHM = Kansas University Natural History Museum (Lawrence, KA)

LACM = Los Angeles County Museum of Natural History (Los Angeles, CA)

LSUMZ = Louisiana State University Museum of Zoology (Baton Rouge, LA)

MACN = Museo Argentino de Ciencias Naturales "Bernardino Rivadavia" (Buenos Aires)

MBML = Museu de Biologia Professor Mello Leitão (Santa Teresa, Espírito Santo)

MCZ = Museum of Comparative Zoology of Harvard (Cambridge, MA)

MCNLS = Museo de Ciencias Naturales de La Salle (San José, Costa Rica)

MEBRG = Museo de la Estación Biológica de Rancho Grande (Parque Nacional Henri Pittier, Aragua, Venezuela)

MECN = Museo del Instituto Nacional de la Biodiversidad del Ecuador, formerly Museo Ecuatoriano de Ciencias Naturales (Quito)

MHNCI = Museu de História Natural Capão da Imbuia de Curitiba/Prefeitura Municipal de Curitiba (Curitiba)

MHNG = Museum d'Histoire Naturelle de Genève (Geneva)

MHNNKM = Museo de Historia Natural Noel Kempff Mercado (Santa Cruz)

MHNSM = Museo de Historia Natural de San Marcos, formerly Museo de Historia Natural "Javier Prado" (Lima)

ML = Macaulay Library, Cornell Lab of Ornithology (http://www.macaulaylibrary.org)

MLS = Museo de Historia Natural, Universidad de la Salle (Bogotá)

MLZ = Moore Lab of Zoology, Occidental College (Los Angeles, CA)

MNCR = Museo Nacional de Costa Rica (San José)

MNHN = Muséum national d'Histoire naturelle (Paris)

MNHUC = Museo de Historia Natural de la Universidad del Cauca (Popayán)

MNRJ = Museu Nacional, Universidade Federal de Rio de Janeiro (Rio de Janeiro)

MPEG = Museu Paraense Emílio Goeldi (Belém)

MPUSP = Museu Paulista da Universidade de São Paulo (São Paulo)

MSB = Museum of Southwestern Biology, University of New Mexico (Albuquerque)

MUSA = Museo de Historia Natural de la Universidad Nacional San Agustín (Arequipa)

MVZB = Museum of Vertebrate Zoology Berkley (Berkley, CA)

MZFC = Colección Ornitológica, Museo de Zoología, Facultad de Ciencias, Universidad Nacional Autónoma de México (Ciudad de México)

MZFES = Colección Ornitológica, Museo de Zoología, Facultad de Estudios Superiores Zaragoza, Universidad Nacional Autónoma de México (Ciudad de México)

MZPW = Museum and Institute of Zoology of the Polish Academy of Sciences (Warsaw)

MZUCR = Museo de Zoología de la Universidad de Costa Rica (San José)

MZUNAM = Museo de Zoología de la Universidad Nacional Autónoma de México (Ciudad de México)

MZUSP = Museu de Zoologia da Universidade de São Paulo (São Paulo)

NHM = Naturhistorisches Museum Wien (Vienna)

NHMUK = Natural History Museum (Tring, UK), formerly British Museum of Natural History

NR = Naturhistoriska Riksmuseet/Swedish Museum of Natural History (Stockholm)

PSOCZ = Universidad de Nariño, (Pasto)

QCAZ = Museo de Zoología de la Pontificia Universidad Católica del Ecuador (Quito)

RMNH = Nationaal Natuurhistorisch Museum (Leiden, Netherlands), formerly Rijksmuseum van Natuurlijke Historie

ROM = Royal Ontario Museum (Toronto)

SBMNH = Santa Barbara Museum of Natural History (Santa Barbara, CA)

SNOMNH = Sam Noble Oklahoma Museum of Natural History (Norman, OK)

UAZ = University of Arizona Ornithology Collection (Tucson, AZ)

UFAC = Laboratório de Ornitologia, Universidade Federal do Acre (Rio Branco, AC)
UFMG = Museu de Zoologia, Universidade Federal de Minas Gerais (Belo Horizonte)
UMMZ = University of Michigan Museum of Zoology (Ann Arbor, MI)
USNM = United States National Museum of Natural History (Washington DC), also known as the Smithsonian Institution
UV = Universidad del Valle (Cali)
UWBM = Burke Museum of Natural History, University of Washington (Seattle, WA)
WA = Wiki Aves (http://www.wikiaves.com.br)
WFVZ = Western Foundation of Vertebrate Zoology (Camarillo, CA)
XC = Xeno-canto (http://www.xeno-canto.org)
YPM = Yale Peabody Museum of Natural History, Yale University (New Haven, CT)
ZISP = Zoological Museum of the Zoological Institute of the Russian Academy of Sciences (St. Petersburg)
ZMUC = Zoological Museum University of Copenhagen (Copenhagen)
ZSM = Zoologische Staatssammlung München (Munich)

Protected areas and parks
ACM = Área de Conservación Municipal (Ecuador, Peru)
ACP = Área de Conservación Privada (Peru)
ACR = Área de Conservación Regional (Peru)
ANMIN = Área Natural de Manejo Integrado Nacional (Bolivia)
ANU = Área Natural Única (Colombia)
APA = Área de Proteção Ambiental (Brazil)
BP = Bosque Protector (Ecuador, Panama), Bosque de Protección (Peru)
CC = Concesión Para la Conservación (Peru)
DMI = Distrito de Manejo Integrado (Colombia)
EPDA = Estação de Preservação e Desenvolvimento Ambiental (Brazil)
ESEC = Estação Ecológica (Brazil)
FLONA = Floresta Nacional (Brazil)
FLOTA = Floresta Estadual (Brazil)
MN = Monumento Natural (Brazil, Mexico, Venezuela)
PE = Parque Estadual (Brazil)
PN = Parque Nacional (all countries except Colombia, see PNN)
PNM = Parque Natural Municipal (Brazil)
PNN = Parque Nacional Natural (Colombia)
PP = Parque Provincial (Argentina)
PRN = Parque Regional Natural (Colombia)
RB = Reserva Biológica (Costa Rica, Honduras)
RBP = Reserva Biológica Privada (Ecuador)
RC = Reserva Comunal (Peru)
RDS = Reserva de Desenvolvimento Sustentável (Brazil)
REBIO = Reserva Biológica (Brazil)
RESEX = Reserva Extrativista (Brazil)
RFP = Reserva Forestal Protectora (Colombia)
RFPR = Reserva Forestal Protectora Regional (Colombia)
RE = Reserva Ecológica (Ecuador, Brazil)
RMN = Reserva Municipal Natural (Colombia)
RN = Reserva Nacional (Peru), Reserva Natural (Colombia)
RNA = Reserva Natural de las Aves (Colombia)
RNFF = Reserva Nacional de Flora y Fauna (Bolivia)
RNP = Reserva Natural Provincial (Argentina)
RNSC = Reserva Natural de la Sociedad Civil (Colombia)
RP = Reserva Privada. A general term that I have used to refer to lands protected by foundations or individuals that have no particular political designation.

RPF = Reserva de Producción de Fauna (Ecuador)

RPN = Reserva Privada Natural (Paraguay)

RPPN = Reserva Particular do Patrimônio Natural (Brazil). Land owned privately and may be used for research, education, ecotourism and recreation. However, this use must be compatible with the goals of preserving the environment and maintaining biodiversity.

RVS = Refugio de Vida Silvestre (Brazil, Ecuador, Honduras, Peru)

SFF = Santuario de Flora y Fauna (Colombia)

SH = Santuario Histórico (Peru)

SN = Santuario Nacional (Peru)

UHE = Usina Hidroeletrica (Brazil)

ZR = Zona Reservada (Peru)

Miscellaneous abbreviations

SVL = snout–vent length (standard way of measuring the size of amphibians and reptiles)

DBH = diameter at breast height (standard way of measuring tree trunk size)

EBA = Endemic Bird Area. Regions identified by BirdLife International as encompassing the overlapping breeding ranges of restricted-range bird species, such that the complete ranges of two or more restricted-range species are entirely included within the boundary (Stattersfield *et al.* 1998).

n/m = not measured or data unavailable

♂? = sex unknown or unreported

Peruvian Antpitta, *Conopophaga peruviana*. Adult male on nest incubating, 3 October 2012, Shiripuno Research Station, Pastaza, Ecuador (*Harold F. Greeney*).

PLATE 1: GENUS *CONOPOPHAGA* I

Black-bellied Gnateater *Conopophaga melanogaster*   Map and text page 73

13.0–15.8cm. Sexes differ. Males nearly unmistakable, females similar to females of the much smaller Hooded Gnateater but have more prominent postocular brow tuft, and more extensive grey on the face and forecrown. Brazil, south of the Amazon from east bank of Rio Madeira, east to lower Rio Tapajós and south to northern Rondônia and northern Mato Grosso, disjunctly from the eastern bank of Rio Xingu to the west bank of the lower Rio Tocantins. Dense undergrowth, generally below 700m. The nest and eggs are undescribed.

a   **Adult female**.

b   **Adult male** with postocular tuft flared.

c   **Adult male**.

Black-cheeked Gnateater *Conopophaga melanops*   Map and text page 78

10–12cm. Sexes differ. Males distinctive, females or immatures may resemble slightly larger Rufous Gnateater, but are separated by their spotted wing coverts and all-black bill. Brazilian endemic, from south-eastern Rio Grande do Norte, to central-eastern Santa Catarina. Mature second-growth forest and forest edge, also drier forest remnants of Rio de Janeiro and the semi-deciduous forests of São Paulo, mostly below 800m, occasionally to 1,100m; replaced at higher elevations by Rufous Gnateater. Nesting biology known almost exclusively from studies of nominate subspecies.

d   **Adult female** *melanops* South-eastern Brazil Espírito Santo and eastern Minas Gerais, south to Santa Catarina.

e   **Adult male** *melanops*.

f   **Adult male** *nigrifrons* Restricted to the coast of north-east Brazil, from Rio Grande do Norte through Pernambuco to south-east Alagoas.

Black-bellied Gnateater

Black-cheeked Gnateater

PLATE 2: GENUS *CONOPOPHAGA* II

Chestnut-belted Gnateater *Conopophaga aurita* Map and text page 86

11.5–13.0cm. Sexes differ. Males unmistakable, females see Ash-throated Gnateater. May involve at least two species. The Guianas to south-east Amazonian Brazil, and west to eastern Colombia, north-east Ecuador and eastern Peru. Interior of tropical lowland forest, also scrubby *campina* habitats, generally below 300m, occasionally to 1,000–1,300m. Nesting poorly known, mostly of nominate subspecies.

a **Adult female** *aurita* Guyana east to French Guiana, south to northern Brazil from Manaus east to northern Pará.

b **Adult male** *aurita.*

c **Adult male** *australis* South of the Amazon from north-east Peru (south to Ucayali) to western Brazil, east to the Rio Madeira and south to northern Acre and northern Rondônia.

d **Juvenile male** *snethlageae.*

e **Adult female** *snethlageae* Brazil, south of the Amazon, from both banks of the lower Rio Tapajós and east bank of Rio Teles Pires, east to central Pará.

f **Adult male** *snethlageae.*

g **Adult male** *pallida* Brazil, north-west Pará, east of the lower Rio Xingu.

Ash-throated Gnateater *Conopophaga peruviana* Map and text page 94

11–12cm. Sexes differ. Males resemble male Slaty Gnateater, but have buff-tipped wing-coverts and a greyer back with more distinct scaling, also lack bicoloured bill. Females separated from female Chestnut-crowned Gnateater by distinctly scaled back and prominent spotting on the upperwing-coverts. Eastern Ecuador, west and south of the Río Napo, through eastern Peru to northern Bolivia (northern La Paz), and east to central Amazonian Brazil, south of the Rio Amazonas. Amazonian *terra firme*, *varzea* forests and foothills below 400–600m, occasionally to 1,000m. Nesting only recently described in detail.

h **Adult female.**

i **Adult male.**

j **Fledgling.**

k **Subadult male.**

Chestnut-belted Gnateater

Ash-throated Gnateater

Hooded Gnateater *Conopophaga roberti*

Map and text page 103

12–13cm. Sexes differ. Males superficially similar to male Black-bellied Gnateaters, but significantly smaller with bicoloured bill, duller brown back and lesser extent of black below. Female distinguished from females of Black-bellied by pale russet-brown cheeks and ear-coverts (not blackish). North-east Brazil, south of the Amazon from north-east Pará to northern Ceará. Evergreen and gallery forests in west, also seasonally dry forests in the east, to *c.*300m. Nest and egg described, but reproductive biology poorly known.

a   **Adult female.**

b   **Adult male.**

Ceara Gnateater *Conopophaga cearae*

Map and text page 99

11–12cm. Sexes separable only by minor differences. Unmistakable within its range, replaced to the south by similar Rufous Gnateater. North-eastern Brazil from northern Ceará to central Bahia, at least to Ibicoara, Iramaia and Ituaçu. Mostly upland evergreen forest, also deciduous forest and scrubby restinga habitats within the narrow coastal sand dunes, from near sea level, usually above 300m, probably up to *c.*2,000m. Nesting unknown.

c   **Adult female.**

d   **Adult male.**

Rufous Gnateater *Conopophaga lineata*

Map and text page 108

12–13cm. Sexes similar. Presence of white crescent on upper breast variable in most populations; separated from Black-cheeked Gnateaters by lack of strong black mask and grey underparts (males), and lack of distinctly spotted wing coverts (females). See also Ceara Gnateater. South-east Brazil, from southern Bahia south to northern Uruguay, and west to central Mato Grosso, north-eastern Argentina (Misiones, Corrientes), and eastern Paraguay. Understorey of semi-humid to humid evergreen montane and lowland forests, also semi-deciduous and gallery forests within *cerrado*, 300–2,400m, mostly above 500m. Nesting biology fairly well studied, but mostly for race *vulgaris*.

e   **Adult *vulgaris*** Southernmost Bahia to extreme north-east Uruguay, west to Argentina and Paraguay.

f   **Adult *lineata*** South-central Bahia, north-east Goiás, northern Minas Gerais, and disjunctly in southern Mato Grosso.

g   **Adult *lineata*** (South-central Bahia, see Taxonomy and Variation, p114).

h   **Juvenile *vulgaris*.**

i   **Fledgling *vulgaris*.**

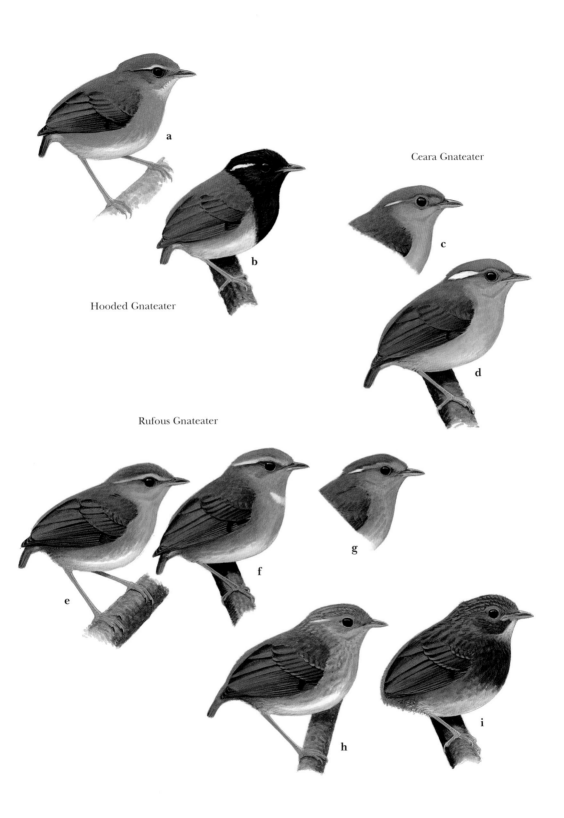

Ceara Gnateater

Hooded Gnateater

Rufous Gnateater

Chestnut-crowned Gnateater *Conopophaga castaneiceps*   Map and text page 121

13.0–13.5cm. Sexually dimorphic and somewhat variable across its range. Females distinguished from both sexes of Slaty Gnateater by orange-rufous throat and breast, males by brighter and richer rufous crown, and from Chestnut-belted Gnateater by lack of black face contrasting with the reddish breast. North-west Colombia, through all three Andean chains, but confined to eastern slope of East Andes in Ecuador to south-east Peru. Upper tropical and lower subtropical montane forest, 500–2,500m, most frequently 1,200–2,000m. Few nests described, all of race *chocoensis*.

a   **Adult female *castaneiceps*** Central and East Andes of Colombia, south to eastern Ecuador (Napo).

b   **Adult male *castaneiceps*.**

c   **Adult female *brunneinucha*** East slopes of the Andes in central Peru (Junín, Huánuco), south at least as far as the Manu road in Cuzco.

d   **Juvenile *castaneiceps*.**

Slaty Gnateater *Conopophaga ardesiaca*   Map and text page 131

12–14cm. Sexes similar but distinguishable. See Chestnut-crowned Gnateater and Ash-throated Gnateater. East slope of the Andes from south-east Peru to south-east Bolivia. Humid subtropical and upper tropical foothills on the east slope of the Andes, 400–2,450m, mostly 1,000–1,700m in Bolivia and 850–2,000m in Peru. Nesting biology only recently studied in detail, further behavioural data desirable.

e   **Adult female *ardesiaca*** Bolivia and southern Peru (Puno).

f   **Adult male *ardesiaca*.**

g   **Adult male *saturata*** Peru, at least as far south as Cuzco.

h   **Fledgling *saturata*.**

Chestnut-crowned Gnateater

Slaty Gnateater

Black-crowned Antpitta *Pittasoma michleri*

Map and text page 137

16.5–17.5cm. Sexes similar but separable by the largely blackish throat of males. Striking black-and-white scalloped underparts and heavy pale bill make species nearly unmistakable. Caribbean slope of Costa Rica, south through Panama to extreme north-west Colombia, understorey of primary forest and tall second growth, mostly at 300–1,000m. Breeding known only from a single nest and clutch of the nominate subspecies.

a   **Adult female** *zeledoni* Costa Rica and western Panama, precise distribution in Panama uncertain.

b   **Adult male** *zeledoni.*

c   **Adult female** *michleri* Both slopes in Panama, east and south of Veraguas to north-west Colombia.

d   **Adult male** *michleri.*

Rufous-crowned Antpitta *Pittasoma rufopileatum*

Map and text page 142

15–18cm. Gender differences vary slightly between subspecies. Unmistakable. West of the Andes from west-central Colombia (northern Chocó), south through Nariño to north-west Ecuador, understorey of humid lowland and foothill forests, generally below 1,100m. Nest and egg undescribed.

e   **Adult female** *rufopileatum* Ecuador, west of Andes in Pichincha, Esmeraldas and Imbabura, historical records south to Los Ríos, possibly extreme south-west Colombia.

f   **Adult male** *rufopileatum.*

g   **Adult female** *harterti* Colombia, Pacific slope of the West Andes in Cauca and Nariño.

h   **Adult female** *rosenbergi* Colombia (most records from Chocó), as far south as Valle del Cauca and as far north as north-east Antioquia.

i   **Adult male** *rosenbergi.*

j   **Adult male** *harterti.*

Black-crowned Antpitta

Rufous-crowned Antpitta

PLATE 6:   GENUS *GRALLARIA* I

Undulated Antpitta *Grallaria squamigera*
Map and text page 149

21–23cm. Sexes similar. Distinguished from larger Giant Antpitta by thinner bill, distinct malar streak and paler underparts. Andes, from Venezuela to Bolivia (Cochabamba and La Paz), humid montane forest understorey and adjacent second growth, 1,700–3,800m, but mostly above 2,600m. Only one nest and one egg described, breeding biology unstudied.

a   **Adult *canicauda*** Eastern Andes of Peru to central Bolivia, possibly also extreme south-east Ecuador (Cordillera de Cutucú).

b   **Adult *squamigera*** Western Venezuela, all three ranges of Colombia to northern Peru, precise southern limit uncertain.

c   **Subadult *squamigera*.**

d   **Fledgling *squamigera*.**

Giant Antpitta *Grallaria gigantea*
Map and text page 158

22.5–26.7cm. Sexes similar. Possibly two species involved, one race perhaps invalid and with no recent records. Separable from Undulated Antpitta by stouter bill, rufous feathering on crown and more deeply coloured, orange-rufous ground colour of underparts. Andean slopes of Colombia and Ecuador, humid and wet montane forest, 1,200–3,000m, infrequently above 2,400m, generally below range of Undulated Antpitta. Few found, eggs incompletely described, all data pertain to *hylodroma*, behavioural data scarce. Classified as Vulnerable in the IUCN Red List of Threatened Species.

e   **Adult *hylodroma*** Western Andes of Colombia and Ecuador from Nariño to Cotopaxi (possibly south to Cañar).

f   **Adult *gigantea*** Apparently endemic to the east slope of the Ecuadorian Andes, from eastern Carchi and Napo, south to the slopes of Volcán Tungurahua.

g   **Subadult *hylodroma*.**

Great Antpitta *Grallaria excelsa*
Map and text page 165

23–25cm. Sexes similar. Superficially resembles Undulated Antpitta, but noticeably larger, with a stouter bill and less contrasting crown and no malar stripe. Mostly eastern slope of Venezuelan Andes, from eastern Táchira, north through Mérida, to south-east Trujillo and south-east Lara, historical records from Sierra de Perijá (north-west Zulia), Cordillera de la Costa in Aragua and one locality on the west slope of the Venezuelan Andes. Understorey of humid cloud forest, generally 1,600–2,300m, mostly above 2,000m. Nest and eggs described (one study, nominate race), breeding biology very poorly known. Classified as Vulnerable in the IUCN Red List of Threatened Species.

h   **Adult *excelsa*** Andes (Lara to Táchira) and Sierra de Perijá (Zulia) in western Venezuela, not yet reported from adjacent Colombia.

i   **Adult *phelpsi*** Apparently endemic to the Cordillera de la Costa in Aragua (no recent records).

Undulated Antpitta

Giant Antpitta

Great Antpitta

PLATE 7: GENUS *GRALLARIA* II

Variegated Antpitta *Grallaria varia* Map and text page 169

18.0–20.5cm. Sexes similar. Possibly two species involved. Similar antpittas are all considerably smaller, Scaled and Tachira Antpittas lack pale shaft-streaks on back, and usually have a clearly visible white submalar streak and broken white necklace, Plain-backed Antpitta lacks scaling on back and heavy streaking on breast. Suriname and the Guianas and lower Amazonian Brazil, west and north to south-west Venezuela and north-east Peru, south to eastern Paraguay and north-east Argentina (Misiones and north-east Corrientes). Mature humid lowland and hilly forest, usually below 1,000m, occasionally as high as 1,800m. Several descriptions of nests and eggs (*varia* and *imperator* only) but nesting behaviour poorly known.

a **Adult** *varia* Suriname and the Guianas to Brazil, north of the Amazon and west of the lower Rio Negro.

b **Adult** *cinereiceps* Extreme southern Venezuela and immediately adjacent Brazil, along the upper Rio Negro in western Amazonas to north-east Peru (north of Río Napo).

c **Adult** *imperator* South-east Brazil (from southern Minas Gerais) to eastern Paraguay and north-east Argentina.

d **Fledgling** *imperator.*

Moustached Antpitta *Grallaria alleni* Map and text page 178

16–18cm. Sexes similar. Confusingly similar to sympatric populations of Scaled Antpitta (race *regulus*), most reliably separated by subtle differences in breast pattern, its more prominent 'moustache' and by thinner and straighter bill (see p185). Andes of Colombia and northern Ecuador, understorey of humid montane forests and mature second growth, 1,800–2,500m; though some overlap, generally replaces Scaled Antpitta at higher elevations, recently recorded at 3,100m in north-west Colombia. Nests and eggs of both races described, but no information from east of the Andes. Classified as Vulnerable in the IUCN Red List of Threatened Species.

e **Subadult** *andaquiensis* Southern Colombia from upper Magdalena Valley south on both Andean slopes in Huila to central Ecuador.

f **Adult** *andaquiensis.*

g **Juvenile** *andaquiensis.*

h **Fledgling** *andaquiensis.*

Variegated Antpitta

Moustached Antpitta

Scaled Antpitta *Grallaria guatimalensis*   Map and text page 185

17–19cm. Sexes similar. Compare with similarly sized Moustached and Plain-backed Antpittas, much larger Variegated Antpitta (little overlap) has a smaller white breast patch, flanks more streaked or spotted. South-central Mexico (Jalisco, Hidalgo), south through Central America and the tropical Andes to central Bolivia, also east through Venezuela and Trinidad to the Guianas. Wide variety of habitats including semi-deciduous forest, humid evergreen forest, pine-oak and fir forest, and lowland rainforest, mostly 500–2,500m. Numerous reports of nests and eggs, but details of nesting behaviour studied only for *regulus*.

a   **Adult *ochraceiventris*** Endemic to southern Mexico, precise range unclear, presumably western Jalisco, south-west along the Sierra Madre del Sur in Michoacán, southern México, and Distrito Federal, Morelos and Guerrero, to coastal south-west Oaxaca.

b   **Adult *guatimalensis*** From southern Mexico (north-east Hidalgo along Pacific slope to both slopes in Chiapas), south through northern Nicaragua (Matagalpa).

c   **Adult *princeps*** Costa Rica and western Panama.

d   **Older fledgling *regulus*.**

e   **Juvenile *regulus*.**

f   **Adult *regulus*** Most of species' Venezuelan range (except Zulia, Bolívar and Amazonas), in Colombia along Central and Eastern Andes, in Western Andes on east slope and on west slope south of southern Risaralda, through Ecuador to central Peru (Ucayali, northern Junín).

g   **Adult *sororia*** South-east Peru (south of Apurímac Valley) to Bolivia.

PLATE 9:   GENUS *GRALLARIA* IV

Tachira Antpitta *Grallaria chthonia* Map and text page 201

16–18cm. Sexes similar. See Scaled Antpitta. Endemic to north-west Venezuela (south-west Táchira). Undergrowth of dense cloud forest, 1,800–2,100m. Only recently rediscovered, nest and eggs unknown. Classified as Critically Endangered in the IUCN Red List of Threatened Species.

Grey-naped Antpitta *Grallaria griseonucha* Map and text page 249

16cm. Sexes similar. Fairly unique within its range, easily separated from boldly streaked Chestnut-crowned Antpitta of lower elevations, no overlap with similarly plain-coloured Rufous Antpitta. Endemic to Venezuelan Andes in Mérida, eastern Trujillo, and north-east Táchira (*tachirae*). Dense vegetation around treefalls and landslides in humid montane forests, 2,300–2,800m. Nest and eggs unknown.

a **Adult *griseonucha*.**

b **Fledgling *griseonucha*.**

Plain-backed Antpitta *Grallaria haplonota* Map and text page 204

16–17cm. Sexes similar. Distinguished from sympatric races of similarly sized Scaled Antpitta by brighter and more uniformly ochraceous underparts, and lack of a white chest crescent. Fragmented range includes coastal mountains of northern Venezuela, western Andes of Colombia and Ecuador, and east slope of the Andes from southern Colombia to central Peru (Ucayali). Understorey of humid and wet montane forest and forest borders, 700–1,950m. Only one reported nest (*chaplinae*), breeding behaviour unstudied.

c **Adult *chaplinae*** Amazonian slope of the Eastern Andes in Colombia, Ecuador and Peru.

d **Juvenile *haplonota*** Northern Venezuela.

Ochre-striped Antpitta *Grallaria dignissima* Map and text page 211

18–19cm. Sexes similar. Smaller, drabber, Thrush-like Antpitta lacks orange on the breast and bold, contrasting streaks below, see Elusive Antpitta (no overlap). South-east Colombia, Amazonian Ecuador to north-east Peru, forest floor of tall, mature Amazon forests, 100–450m. Nest and eggs not described.

Elusive Antpitta *Grallaria eludens* Map and text page 215

18–19cm. Sexes similar. Allopatric Ochre-striped Antpitta is similarly sized and patterned, but is immediately distinguished by the orange-ochre wash on the throat and chest; Thrush-like Antpitta considerably smaller and less boldly patterned. Western Amazon Basin in eastern Peru and adjacent western Brazil. Mature, bottomland Amazon forests, especially in low-lying areas and thickets, 120–500m. Nest and eggs not described.

Tachira Antpitta

a

Grey-naped Antpitta

b

c

Plain-backed Antpitta

d

Ochre-striped Antpitta

Elusive Antpitta

Chestnut-crowned Antpitta *Grallaria ruficapilla*   Map and text page 217

18–19cm. Sexes similar. Precise distribution and validity of some races unclear. Extensive streaking below combined with a contrasting chestnut hood separate this species from occasionally sympatric Watkins's Antpitta, by bright chestnut head, bolder streaking on the breast, and blue-grey tarsi (not pale pinkish). North-east Venezuela, south through Andean ranges to northern Peru (southern Cajamarca and northern San Martín). Naturally and artificially disturbed habitat, especially thickets of *Chusquea* bamboo, in dry inter-Andean valleys, seasonal cloud forests and humid montane forests, generally 1,200–3,100m. Nest and egg described only for nominate subspecies, breeding behaviour poorly known.

a   **Adult *ruficapilla*** Most of the Colombian Andes except the extreme north-east (north of Río Chicamocha), through most of the Ecuadorian Andes except the south-west, to north-east Peru (north of Río Marañon).

b   **Juvenile *ruficapilla*.**

c   **Fledgling *ruficapilla*.**

d   **Adult *perijana*** Perijá Mountains in extreme northern Colombia, at the border with western Venezuela.

e   **Adult *connectens*** Arid forests of south-west Ecuador (southern El Oro and southern Loja) and adjacent north-west Peru (north of upper Río Piura drainage).

f   **Adult *albiloris*** Both slopes of the West Andes in north-west Peru, west of Río Huancabamba and north Río Marañón, north at least to south-central Piura (Abra Cruz Blanca) and south to extreme south-west Cajamarca (Bosque Cachil).

Watkins's Antpitta *Grallaria watkinsi*   Map and text page 231

18–19cm. Sexes similar. Smaller and paler overall than similarly patterned Chestnut-crowned Antpitta, also with more extensive white on the sides of the face, narrow streaking on the back and pinkish (not bluish) legs. South-west Ecuador and adjacent north-west Peru. Arid, deciduous and semi-deciduous forests and second growth, sea level to 1,800m. Few nests described, breeding behaviour poorly known. Classified as Near Threatened in the IUCN Red List of Threatened Species.

Chestnut-crowned Antpitta

Watkins's Antpitta

PLATE 11: GENUS *GRALLARIA* VI

Santa Marta Antpitta *Grallaria bangsi*
Map and text page 236

17–18cm. Sexes similar. Unmistakable within its range. Endemic to the Sierra Nevada de Santa Marta in north-east Colombia. Humid montane cloud forest, mature second growth, and tangled forest borders, mostly 1,600–2,400m. Nest and eggs unknown. Classified as Vulnerable in the IUCN Red List of Threatened Species.

a   **Adult.**

b   **Juvenile.**

c   **Fledgling.**

Cundinamarca Antpitta *Grallaria kaestneri*
Map and text page 240

15–16cm. Sexes similar. Dull, mottled, olive-brownish plumage separate this species from sympatric antpittas. Colombian endemic known only from the east slope of the East Andes, from south-east of Bogotá. Undergrowth of humid, montane forest and mature secondary woodland, 1,700–2,500m. Nest and eggs unknown. Classified as Endangered in the IUCN Red List of Threatened Species.

d   **Adult.**

e   **Juvenile.**

Stripe-headed Antpitta *Grallaria andicolus*
Map and text page 243

16.0–16.5cm. Sexes similar. Bold streaking on head and body make this species unmistakable. High Andes of Peru and extreme north-west Bolivia. *Polylepis–Gynoxys* woodland and arid *puna* grasslands, generally above 3,500m. Nest and eggs known from only one study (*punensis*).

f   **Adult *andicolus*** Most of Peru from southern Amazonas (Atuén) south to Apurímac.

g   **Adult *punensis*** South-east Peru (Cuzco, Puno) and adjacent north-west Bolivia (La Paz).

h   **Juvenile *punensis*.**

Santa Marta Antpitta

a

b

c

d

e

Cundinamarca Antpitta

f

g

h

Stripe-headed Antpitta

Chestnut-naped Antpitta *Grallaria nuchalis*
Map and text page 260

19–21cm. Sexes similar, races vary considerably, including immature plumages. Much smaller Bicoloured Antpitta has a similar hooded appearance but lacks bright chestnut on the crown, smaller White-bellied Antpitta is white below, see also distinctive Jocotoco Antpitta. Northern Colombia to north-east Peru (north of the Río Marañón). Bamboo thickets, tangled *Alder* forests and adjacent humid cloud forest undergrowth, 1,900–3,400m [in Colombia], mostly 2,400–2,900m. Breeding poorly known, one nest (*nuchalis*) and one clutch (*ruficeps*) described.

a   **Adult *nuchalis*** East Andes of Ecuador and Peru (probably extreme south-east Colombia).

b   **Adult *ruficeps*** Confined to Colombia (all three Andean ranges).

c   **Adult *obsoleta*** West Andes of northern Ecuador (possibly extreme south-west Colombia).

d   **Fledgling *nuchalis*.**

e   **Juvenile *obsoleta*.**

f   **Fledgling *ruficeps*.**

g   **Juvenile *ruficeps*.**

Bicoloured Antpitta *Grallaria rufocinerea*
Map and text page 251

15–16cm. Sexes similar. Sympatric race of Chestnut-naped Antpitta (*ruficeps*) is significantly larger, has a paler bill, a pale iris and prominent white postocular spot. Colombia on both slopes of the Central Andes, from southern Antioquia and Caldas to southern Quindio and Tolima (*rufocinerea*) and south-east Colombia (Narino, Putumayo) to Sucumbíos in extreme northern Ecuador (*romeroana*). Understorey of dense, humid montane cloud forest, mostly 2,200–3,150m. Nest and eggs not described. Classified as Vulnerable in the IUCN Red List of Threatened Species.

h   **Adult *rufocinerea*.**

i   **Juvenile *rufocinerea*.**

Chestnut-naped Antpitta

Bicoloured Antpitta

Pale-billed Antpitta *Grallaria carrikeri*   Map and text page 267

19–20cm. Sexes similar. The heavy pale bill of Pale-billed Antpitta alone should preclude confusion with other species. East slope of the northern Peruvian Andes, south of the Río Marañón from the Cordillera Colán south at least to La Libertad (Cumpang). Broken-canopy, epiphyte-laden cloud forests, association with bamboo habitats uncertain, 2,350–2,900m. Only one nest described, eggs unknown. Classified as Near Threatened in the IUCN Red List of Threatened Species.

a   **Adult**.

b   **Juvenile**.

Jocotoco Antpitta *Grallaria ridgelyi*   Map and text page 255

20–23cm. Sexes similar. Unmistakable. Known only from the Eastern Andes in extreme south-east Ecuador and adjacent Peru. Understorey epiphyte-laden cloud forests on steep mountain slopes, 2,300–2,650m. Nest and eggs described but breeding behaviour poorly documented. Classified as Endangered in the IUCN Red List of Threatened Species.

c   **Adult**.

d   **Older juvenile**.

e   **Young juvenile**.

f   **Fledgling**.

a

b

Pale-billed Antpitta

c

d

Jocotoco Antpitta

e

f

White-bellied Antpitta *Grallaria hypoleuca* Map and text page 275

16–17cm. Sexes similar. Fairly distinct within range, Bicoloured and Chestnut-naped Antpittas are larger, considerably darker below and generally occur at higher elevations. Central and East Andes of Colombia, and on the Amazonian slope in Ecuador and northern Peru. Interior and borders of humid montane forests and secondary forests in advanced stages of regeneration, 1,400m–2,700m. Nesting poorly known, one published description of nest and eggs (*castanea*).

a **Adult** *castanea* East slope of the East Andes in Colombia (upper Río Cusiana watershed in Boyacá), and at the head of the Magdalena Valley to northern Peru (north and west of the Marañón).

b **Juvenile** *castanea.*

c **Fledgling** *castanea.*

White-throated Antpitta *Grallaria albigula* Map and text page 271

18.5–20.0cm. Sexes similar. White throat and prominent ocular ring diagnostic in most of range, Red-and-white Antpitta has chestnut-brown upperparts and distinct rufous markings on the breast, Rufous-faced Antpitta has dark crown contrasting with a rufous face. East slope of the Andes from southern Peru (Cuzco) to north-west Argentina (Jujuy, Salta). Understorey of humid montane forest, 600–2,700m, most frequently at 1,200–2,100m. Nest and eggs not described.

d **Adult.**

e **Subadult.**

White-bellied Antpitta

White-throated Antpitta

Rusty-tinged Antpitta *Grallaria przewalskii*   Map and text page 285

16–17cm. Sexes similar. Confusion unlikely within limited range, see White-bellied and Bay Antpittas. Peru, south and east of the Río Marañón in Amazonas, San Martín and La Libertad. Humid montane forests, 1,700–2,750m, replaced southward by Bay Antpitta, northward by White-bellied Antpitta. Nest and egg not described. Classified as Vulnerable in the IUCN Red List of Threatened Species.

a   Adult.

b   Fledgling.

Bay Antpitta *Grallaria capitalis*   Map and text page 288

16–17cm. Sexes similar. Monotypic, but newly discovered populations in south likely represent an undescribed taxon. With a clear view, the larger size, bluish-grey bill, and bright chestnut plumage make confusion unlikely with duller-plumaged, black-billed Chestnut Antpitta and sympatric races of Rufous Antpitta. Peruvian endemic confined to east slope of the Andes from Huánuco to Ayacucho (north of the Río Apurimac). Humid subtropical forests and adjacent second growth, primarily 1,800–3,000m. Nest and egg not described.

c   Adult (south).

d   Adult (north).

e   Juvenile (north).

f   Subadult (north).

Red-and-white Antpitta *Grallaria erythroleuca*   Map and text page 281

17–18cm. Sexes similar. Simple colour pattern fairly distinct within range, see White-throated Antpitta. East slope of the Andes in southern Peru (Junín, Cuzco). Humid subtropical montane forest, forest borders and tall secondary forest, generally 2,100m–3,000m. Nest and egg not described.

g   Adult (typical).

h   Adult (Cordillera Vilcabamba).

i   Juvenile (typical).

Rusty-tinged Antpitta

a

b

c

Bay Antpitta

d

e

f

g

h

i

Red-and-white Antpitta

Rufous Antpitta *Grallaria rufula*

Map and text page 294

14.5–15.0cm. Sexes similar. Racial limits, distribution and taxonomic status currently under revision, with multiple species involved. Largely unmistakable in most of range based on small size, dark, relatively narrow bill, and nearly uniform rufous plumage. Similar Chestnut Antpitta lacks the eye-ring of the somewhat duller, browner-plumaged races of nearby populations of Rufous Antpitta. Western Venezuela, northern Colombia to central Bolivia. Humid, mossy, epiphyte-laden montane forests, forest borders and adjacent *páramo*, 1,900–3,650m, mostly 2,500–3,300m. Nest and egg fully described only for nominate race, breeding behaviour poorly known.

a   **Adult *rufula*** Táchira and Apure in south-west Venezuela, south and west patchily through most of the Colombian Andes and Ecuador to northern Peru.

b   **Fledgling *rufula*.**

c   **Adult *spatiator*** Sierra Nevada de Santa Marta of northern Colombia.

d   **Adult *obscura*** Northern Peru, on east Andean slope, Amazonas south of the Río Marañon, to Junín.

e   **Adult *saltuensis*** Perijá Mountains on the Venezuela (Zulia) / Colombia (César/La Guajira) border.

f   **Adult *cajamarcae*** North-west Peru, in southern Piura, Cajamarca and eastern Lambayeque.

g   **Adult *cochabambae*** Bolivian endemic, from central La Paz, in the Cordillera Real, south-east to Santa Cruz.

h   **Adult *occabambae*** South-east Peru, south of the Río Apurímac, including the Cordillera de Vilcabamba, to north-west Bolivia (northern La Paz).

i   **Fledgling *occabambae*.**

Rufous-faced Antpitta *Grallaria erythrotis*

Map and text page 291

17–18cm. Sexes similar. Within small range unlikely to be confused, except with more strongly patterned White-throated Antpitta, which occurs at lower elevations and has the entire head bright rufous and a duller, greyer belly. Endemic to Bolivia, known from La Paz, Cochabamba and Santa Cruz. Montane forest edges and second growth, mostly 2,000–3,000m. Nest and eggs known only from captivity.

Rufous Antpitta

Rufous-faced Antpitta

PLATE 17: GENUS *GRALLARIA* XII

Tawny Antpitta *Grallaria quitensis*
Map and text page 312

16–18cm. Sexes similar. The three races differ in voice and plumage (including immature), and are probably deserving of species rank. At elevations where it occurs, confusion unlikely except with the smaller, much more uniformly rufous or rufescent Rufous Antpitta. East and Central Andes of Colombia south to northern Peru. Humid montane forests and adjacent *páramo*, mostly above 2,800m. Nest and egg described only for nominate race.

a   **Adult** *quitensis* Central Andes of Colombia, throughout the Andes of Ecuador, to extreme northern Peru in Cajamarca and Piura, north of the Rio Marañón.

b   **Juvenile** *quitensis.*

c   **Fledgling** *quitensis.*

d   **Adult** *alticola* East Andes of Colombia in Santander, Boyacá, Cundinamarca and Bogotá.

e   **Juvenile** *alticola.*

Chestnut Antpitta *Grallaria blakei*
Map and text page 308

14.5–15.0cm. Sexes similar. Near-uniform rufous plumage confusingly similar to nominate race of Rufous Antpitta (see p299), but generally separable from the other three Peruvian subspecies of Rufous Antpitta (*cajamarcae, obscura* and *occabambae*) by darker chestnut plumage, the lack of contrasting feather tips on the back and breast, thicker bill and legs, and lack of a relatively distinct eye-ring. Range poorly understood, known only from a few locations south of the Río Marañón in northern Peru, from the Cordillera Colán, south to Apurímac Valley in Ayacucho. Bamboo and dense undergrowth of humid montane forests, mostly 1,700–2,500m. Nest and eggs unknown. Classified as Near Threatened in the IUCN Red List of Threatened Species.

f   **Adult.**

g   **Juvenile.**

Tawny Antpitta

Chestnut Antpitta

Urrao Antpitta *Grallaria fenwickorum*

Map and text page 323

16–17cm. Sexes similar. Most similar to Brown-banded Antpitta, but distinguished by more olivaceous upperparts, brownish-olive throat and lack of a brown breast-band. Similarly uniform plumage of nominate Rufous Antpitta is distinctly bright rufescent. Extremely small range confined to the type locality on the south-east slope of the Páramo del Sol massif, at the northern tip of the West Andes of Colombia. Epiphyte-laden montane cloud forest, *c*.3,100m. Nest and egg unknown. Classified as Critically Endangered in the IUCN Red List of Threatened Species.

a   **Adult.**

b   **Juvenile.**

c   **Fledgling.**

Brown-banded Antpitta *Grallaria milleri*

Map and text page 327

16.5–18.0cm. Sexes similar. Generally separable from sympatric *Grallaria* by the distinctive band on the breast. Confined to the Central Andes of Colombia, from the west slope in Caldas to Quindío and on the east slope in Tolima. Humid second growth and forest fragments in montane regions where little natural habitat remains, 1,800–3,150m. Nest and egg unknown. Classified as Vulnerable in the IUCN Red List of Threatened Species.

d   **Adult.**

e   **Juvenile.**

Yellow-breasted Antpitta *Grallaria flavotincta*

Map and text page 331

17–18cm. Sexes similar. Confusion unlikely, most similar to allopatric White-bellied Antpitta, which occurs only east of the Andes and lacks yellow wash to the underparts. Pacific slope of the West Andes in Colombia and Ecuador from Antioquia to Pichincha and Santo Domingo de las Tsáchilas. Shady understorey of humid montane forests, especially on steep slopes and in riparian areas, 1,170–2,350m. Few nests described, eggs not properly described.

f   **Subadult.**

g   **Adult.**

h   **Juvenile.**

Urrao Antpitta

a

b

c

Brown-banded Antpitta

d

e

f

g

h

Yellow-breasted Antpitta

Streak-chested Antpitta *Hylopezus perspicillatus*　　Map and text page 336

13–14cm. Sexes similar. Only sympatric congener, Thicket Antpitta, has a distinctly plainer face (no eye-ring), lacks spotted wing-coverts (when fully adult), is more extensively washed with colour below, and has much fainter breast streaking. South-east Honduras through Panama, eastward from southern Panama, across the northern end of the Colombian Andes to Santander, and southward west of the Andes to northern Ecuador. Wet lowland evergreen forest, sea level to 1,250m, usually below 900m. Nest and eggs described in detail only for race *lizanoi*.

a　**Adult *intermedius*** Caribbean slope of Central America, from north-east Honduras (Gracias a Dios) to western Panama.

b　**Adult *lizanoi*** Pacific slope of Costa Rica.

c　**Adult *perspicillatus*** Central Panama (Veraguas) to north-west Colombia (Chocó, north-west Antioquia).

d　**Juvenile *lizanoi*.**

e　**Fledgling *lizanoi*.**

Spotted Antpitta *Hylopezus macularius*　　Map and text page 344

13.5–14.0cm. Sexes similar. Similarly sized Thrush-like Antpitta lacks malar stripe, prominent buffy eye-ring and boldly marked underparts. In western part of range, similar White-lored Antpitta has only a weak eye-ring, distinctly more diffuse streaking below and a darker crown. Eastern and southern Venezuela, east through Guyana, Suriname and French Guiana, and from there south and west across Brazil (north of the Amazon) to south-east Colombia and north-east Peru. Mature tropical lowland forests, to 500m. Only one nest described (*macularius*), breeding behaviour poorly known.

f　**Adult *macularius*** North-east Venezuela through the Guianas to north-east Brazil, north of the Amazon and east of the Rios Branco and Negro.

g　**Juvenile *macularius*.**

Alta Floresta Antpitta *Hylopezus whittakeri*　　Map and text page 349

14cm. Sexes similar. Very similar to Spotted and Snethlage's Antpittas and not safely separated from either in the field except by range (no overlap) and by voice. Endemic to south-central Brazil, confined to the Rio Madeira–Rio Xingu interfluvium, swampy or flooded areas in upland *terra firme* forest, below 500m. Nest and egg not described.

Masked Antpitta *Hylopezus auricularis*　　Map and text page 353

14cm. Sexes similar. Unique within its range, allopatric Amazonian Antpitta lacks the blackish 'mask', has a darker crown and a buffier breast. Endemic to northern Bolivia, known from only half a dozen localities in Beni and Pando. Dense, tangled, wet second growth, 100–200m. Few nests described, breeding behaviour largely undescribed. Classified as Vulnerable in the IUCN Red List of Threatened Species.

Snethlage's Antpitta *Hylopezus paraensis*　　Map and text page 351

13–14cm. Sexes similar. Very similar to Spotted and Alta Floresta Antpittas, and not safely separated from either in the field except by range (no overlap) and by voice. Endemic to north-east Brazil, south of the Amazon from the Rio Xingu east to western Maranhão. Undergrowth of mature *terra firme* forest, below 500m. Nest and eggs not described.

Thicket Antpitta *Hylopezus dives*　　Map and text page 356

13–15cm. Sexes similar, and two additional races differ little from nominate (illustrated). Sympatric with only one other *Hylopezus*, see Streak-chested Antpitta. Eastern Honduras south, through Panama, and along the Pacific slope of Colombia to western Nariño. Dense, tangled thickets and forest edges, sea level to 900m, occasionally 1,100m. Nest only recently described (nominate race), eggs unknown.

Streak-chested Antpitta

Masked Antpitta

Alta Floresta Antpitta

Spotted Antpitta

Thicket Antpitta

Snethlage's Antpitta

White-lored Antpitta *Hylopezus fulviventris* Map and text page 361

14.5–15.0cm. Sexes similar, race *caquetae* of southern Colombia differs little from nominate (illustrated). Only known *Hylopezus* within most of range. Thrush-like Antpitta lacks grey crown, strong facial pattern and fulvous wash to the underparts, see also Amazonian Antpitta. South-east Colombia to northern Peru. Dense undergrowth and edge of Amazonian forests, generally below 600m. Nest and egg not described.

Amazonian Antpitta *Hylopezus berlepschi* Map and text page 364

14–15cm. Sexes similar, race *yessupi* of the Río Ucayali drainage in north-eastern Peru differs little from nominate. Thrush-like Antpitta is similarly plainly patterned, but usually has more diffuse breast streaking, White-lored Antpitta has a distinctly grey crown and nape, a more distinctly patterned face and is somewhat paler overall below. Spotted Antpitta has a bold ochraceous eye-ring and a spotted (not streaked) breast. East-central Peru and Amazonian Brazil, south of the Amazon, south to northern Bolivia. Forest edges, overgrown clearings, riparian gallery forests within *cerrado*, to 700m, usually below 500m. Nest and eggs not described.

a   **Adult** *berlepschi* Vocalising.

b   **Adult** *berlepschi*.

White-browed Antpitta *Hylopezus ochroleucus* Map and text page 368

12–14cm. Sexes similar. Unmistakable. Allopatric Speckle-breasted Antpitta lacks postocular stripe and has a distinctly more speckled breast. Brazil, from northern Ceará to northern Minas Gerais (Serra do Espinhaço). Dense tangles in both tall and shrubby *caatinga*, mostly 400–1,000m. Only one nest described, eggs unknown. Classified as Near Threatened in the IUCN Red List of Threatened Species.

Speckle-breasted Antpitta *Hylopezus nattereri* Map and text page 372

13–14cm. Sexes similar. Unmistakable in range, see White-browed Antpitta. Endemic to the Atlantic Forest of south-east Brazil, easternmost Paraguay and north-east Argentina. Humid and montane forest, mostly 1,200–1,900m, considerably lower in some areas and to 2,450m in the Serra de Itatiaia. Nest and eggs only recently described.

c   **Adult (north)**.

d   **Adult (south)**.

White-lored Antpitta

a

Amazonian Antpitta

b

Speckle-breasted
Antpitta

c

White-browed Antpitta

d

Thrush-like Antpitta *Myrmothera campanisona* Map and text page 377

14.5–15.0cm. Sexes similar, perhaps more than one species involved. Confusion with other antpittas unlikely, distinctly browner above and less marked below than sympatric *Hylopezus* antpittas. May be sympatric with similar Tepui Antpitta, but separable by its streaked underparts. South-west Venezuela and eastern Colombia to Suriname and the Guianas, south through Brazil to Mato Grosso and along the base of the Andes to north-west Bolivia. Lowland rainforest and foothills, to 1,200m. Nest and eggs described but data scarce for most subspecies.

a   **Adult *campanisona*** Eastern Venezuela, the Guianas and Suriname, south to the north bank of the Amazon and west to Rios Negro and Branco.

b   **Adult *signata*** Base of the East Andes in southern Colombia south to northern Peru north of Río Marañón and eastward to Río Napo.

c   **Adult *minor*** Peruvian Amazon south of Río Marañón, east into western Brazil at least to Rio Purus, and south to north-west Bolivia.

Tepui Antpitta *Myrmothera simplex* Map and text page 390

15–16cm. Sexes similar and races not well defined. Generally occurs at higher elevations than Thrush-like Antpitta which has distinctly streaked underparts. Tepui region of south-east Venezuela, adjacent Guyana and northern Brazil. Dense, humid montane forests, 600–2,400m, mostly above 1,200m. Single nest reported (*simplex*).

d   **Adult *guaiquinimae*** South-east Venezuela in central and south-east Bolívar.

e   **Adult *simplex*** Sierra de Lema highlands and tepuis surrounding the Gran Sabana in Venezuela (south-east Bolívar) and immediately adjacent Guyana.

f   **Adult *duidae*** Southern Venezuela (Amazonas) and adjacent Brazil.

Thrush-like Antpitta

Tepui Antpitta

a

b

c

d

e

f

PLATE 22: GENUS *GRALLARICULA* I

Hooded Antpitta *Grallaricula cucullata* Map and text page 410

10–12cm. Sexes similar, two described races possibly synonymous. Orange-rufous hood and lack of bold markings make it unmistakable within its limited range. Range poorly understood; known from scattered localities in all three ranges of the Colombian Andes (*cucullata*, figured) and western Venezuela (*venezuelana*, Táchira and Apure). Undergrowth of humid montane forests, mostly 1,800–2,150m. One clutch described, nest not described. Classified as Vulnerable in the IUCN Red List of Threatened Species.

Ochre-breasted Antpitta *Grallaricula flavirostris* Map and text page 395

10.0–10.2cm. Sexes similar. Separable from sympatric congeners by yellowish bill, generally at elevations below Rusty-breasted and Slate-crowned Antpittas, both generally lacking streaked underparts (but see Slate-crowned Antpitta account for possible confusion in south-east Ecuador), and from most races of Rusty-breasted Antpitta by lack of a distinct pale eye-ring or postocular crescent. Widely distributed from northern Costa Rica to central Bolivia. Humid and wet montane forests, mostly 900–2,200m. Nest and eggs described for most races, but sample sizes low.

a   **Adult** *costaricensis* Costa Rica and Panama east to northern Ngäbe-Buglé, precise distribution in Panama uncertain.

b   **Adult** *flavirostris* Amazonian slope from central Colombia to south-east Ecuador, probably northern Peru.

c   **Adult** *mindoensis* South-west Colombia (Nariño) and north-west Ecuador.

d   **Juvenile** *mindoensis*.

e   **Adult** *boliviana* Central Peru (Pasco, eastern Junín), south-east to Bolivia.

f   **Adult** *ochraceiventris* Colombian endemic, western foothills of the West Andes, south to central Cauca, provisionally includes populations on the east slope of the West Andes in central Antioquia and on the west slope of the East Andes in Santander (Serranía de los Yariguíes).

Scallop-breasted Antpitta *Grallaricula loricata* Map and text page 406

10–11cm. Sexes similar. Bold markings make it unmistakable within its limited range. Endemic to the coastal mountains of northern Venezuela from north-west Falcón to Distrito Federal. Understorey of humid montane cloud forests, 800–2,200m. Nest and egg only recently described, breeding behaviour poorly studied. Classified as Near Threatened in the IUCN Red List of Threatened Species.

g   **Subadult.**

h   **Adult.**

Hooded Antpitta

a

f

Ochre-breasted Antpitta

c

d

e

b

g

h

Scallop-breasted Antpitta

PLATE 23: GENUS *GRALLARICULA* II

Crescent-faced Antpitta *Grallaricula lineifrons*   Map and text page 439

11.5–12.0cm. Sexes similar. Unmistakable and apparently does not occur with any other *Grallaricula*. Known from scattered localities from Colombia (Caldas) to southern Ecuador (Loja). Undergrowth of humid, montane cloud and elfin forest, mostly 2,900m to treeline. Few nests found, egg not formally described, breeding biology unstudied. Classified as Near Threatened in the IUCN Red List of Threatened Species.

a   Adult.

b   Juvenile.

Ochre-fronted Antpitta *Grallaricula ochraceifrons*   Map and text page 418

11.5cm. Sexes differ, largely by the brightness of the rufous crown. Distinctive. Ochre-breasted Antpitta, of generally lower elevations, lacks the distinctive ochraceous-buff forecrown and has a bold buffy eye-ring, see allopatric Peruvian Antpitta. Endemic to small area on east Andean slope in north-east Peru, south of Río Marañón. Understorey of humid montane forests, 1,850–2,400m. Nest and egg not described. Classified as Endangered in the IUCN Red List of Threatened Species.

Peruvian Antpitta *Grallaricula peruviana*   Map and text page 413

10–11cm. Sexes differ, largely by the brightness of the rufous crown and boldness of breast scalloping. Rufous crown and strongly scalloped underparts make it unmistakable in range, see allopatric Ochre-fronted Antpitta. Known from only a few localities along east slope of the Andes, north-east Ecuador to northern Peru (north of the Río Marañon), altitudinal range uncertain, reported at 1,650–2,100m. Nesting studied in only one population. Classified as Near Threatened in the IUCN Red List of Threatened Species.

c   Adult male.

d   Adult female.

e   Fledgling.

a Crescent-faced Antpitta

b

Ochre-fronted Antpitta

c

Peruvian Antpitta

d

e

Rusty-breasted Antpitta *Grallaricula ferrugineipectus*   Map and text page 421

10–12cm. Sexes similar. Two or three species probably involved. Similar to Slate-crowned Antpitta of generally higher elevations, but lacks distinctly slate-grey crown, and from Ochre-breasted Antpitta of lower elevations by lack of scalloped underparts. Patchily distributed from the coastal mountains of Venezuela south through Andes to Bolivia. Undergrowth of humid and semi-humid montane forests, especially near thickets and vine tangles, mostly 2,500–3,200m. Nesting biology studied only in Venezuela.

a   **Adult** *ferrugineipectus* North-west Colombia (Sierra Nevada de Santa Marta), Venezuela in coastal mountains from Lara and Yaracuy to Distrito Federal and Miranda, south in Andes to Mérida and Barinas.

b   **Adult** *leymebambae* Extreme south-west Ecuador (Loja) and north-west Peru (Piura), eastern Peru south of the Río Marañón to western Bolivia (La Paz), disjunct records in north-west Ecuador (Imbabura, Pichincha) and western Colombia (Valle del Cauca) provisionally included.

c   **Adult** *rara* Western Venezuela (Sierra de Perijá), Colombia on the east slope of the East Andes in Norte de Santander, and on the west slope of the East Andes in Cundinamarca.

d   **Fledgling** *leymebambae*.

Slate-crowned Antpitta *Grallaricula nana*   Map and text page 428

10–12cm. Sexes similar. Distinguished from Rusty-breasted and Ochre-breasted Antpittas by distinctive slate-grey crown; some individuals in south-east Ecuador have dusky-edged breast feathers giving scaled look, similar to Ochre-breasted Antpitta which appears to be confined to lower elevations in this region. See also allopatric Sucre Antpitta. From the coastal mountains of northern Venezuela, south through the Andes to northern Peru, disjunctly in the Tepui region (*kukenamensis*). Montane forests near dense thickets of *Chusquea* bamboo, 700m–3,500m, mostly above 1,800m. Nesting biology of most races remains unknown.

e   **Adult** *nana* East Andes of Colombia, in southern Santander, Boyacá and Cundinamarca.

f   **Adult** *occidentalis* Scaled form.

g   **Juvenile** *nanitaea* Venezuelan Andes, south from north-east Trujillo to north-east Colombia, north of the Río Chicamocha.

h   **Fledgling** *occidentalis* North-west Colombia in West and Central Andes, south through the Central Andes, and along the east slope of the Andes in Ecuador, to northern Peru.

Sucre Antpitta *Grallaricula cumanensis*   Map and text page 436

11.0–11.5cm. Sexes similar, races weakly differentiated. See allopatric Slate-crowned Antpitta. Venezuelan endemic, on Paria Peninsula in Sucre (*pariae*, figured) and disjunctly on the Turimiquire Massif at the borders of the states of Sucre, Anzoategui and Monagas (*cumanensis*). Understorey of humid, epiphyte-laden, montane forests at 600–1,850m. Eggs known from one clutch, nest not described. Classified as Vulnerable by the IUCN Red List of Threatened Species.

Rusty-breasted Antpitta

Slate-crowned Antpitta

Sucre Antpitta

FAMILY CONOPOPHAGIDAE

Genus *Conopophaga*: the 'gnateaters'

Conopophaga L. P. Vieillot, 1816, Analyse d'une nouvelle ornithologie élémentaire, p. 39. Type, by subsequent designation: *Conopophaga aurita* (J. F. Gmelin, 1789) *sensu* Gray (1840) based on 'Fourmillier à ailes (= oreilles) blanches' of Buffon (1771–86).

'I observed a display that apparently intended to divert my attention from the nest [ ]. A female crawled on the ground very slowly, flapping her wings, her feathers on her back flared, moving with occasional pauses for a distance of 5–6ft for 3–4 min.' – **William Henry Belton, 1972, Gramado, Rio Grande do Sul (Rufous Gnateater)**

Remarks on the immature plumages of *Conopophaga* While researching this book I examined more than 150 photographs or specimens that I judged to be in immature plumage. While this still leaves me with an incomplete picture for most species, a few generalities appear to apply to most or all *Conopophaga*. Gnateaters fledge while still covered in dark chocolate-brown or blackish-brown, fluffy nestling down on the upperparts, chest and (usually) sides. Above and across the chest the down is coarsely spotted paler, usually buffy or chestnut- to rufescent-brown. Markings tend to be smallest and densest on the crown, becoming larger and sparser on the back and breast, and gradually merging to form indistinct pale striping on the flanks and rear body. Their bellies are covered in similarly fluffy plumage, but are usually dusky or buffy-white and either unmarked or faintly barred. The ocular area may be bare, or only sparsely feathered, exposing dusky-orange or dull pinkish skin. Individuals fresh from the nest have (in general) paler and shorter bills than adults, with obviously inflated rictal flanges that vary from white to yellow. The young of all species appear to have bright pale markings at the tips of the greater and (usually?) lesser upperwing-coverts. These markings are generally similar in coloration to the species' other plumage spots, but distinctly brighter and ranging from rich buff to bright chestnut. They vary somewhat in shape from broad crescents spanning the width of the feather, to indistinct triangles or half-crescents in some species, where the marking is largely confined to the leading vane of the feather. Though not always obvious from photographs or in the field, it appears that the tertial and inner secondary flight feathers are similarly marked, though less brightly so, and with thinner and paler terminal crescents. Most covert spots are cleanly subtended by a terminal black margin on the feather, something that is rarely (never) seen in fully adult birds. This black margin tends to wear off and/or fade to become less obvious with age, but I believe these immature coverts are replaced with similarly patterned feathers at least once prior to a final moult to adult coverts. The transition into adult-like contour feathers appears to occur relatively evenly across most regions of the body, unlike the often patchy replacement patterns seen in some antpittas (see *Grallaria*). My general sense is that non-plumose contour feathers develop first on the sides, breast, back and crown, closely followed by the flanks and rump. In most of the older individuals I have seen the last vestiges of plumose fledgling feathers were found on the vent and nape. In most species this first covering of adult-like contour feathers is, with the exception of the hindcrown and nape, patterned similarly to that of fully adult birds, making most species separable by sex at this stage. While in this plumage phase, signs of immaturity are most apparent on the nape and back of the head, with these regions usually still distinctly spotted but the spots often approximate the coloration of the adult plumage in this region and may create a more scaled look (as opposed to distinct spots). For now I will lead with the hypothesis that the upperwing-coverts are replaced with a new set of immature-patterned coverts sometime around the age when that last evidence of spotting or scaling is lost on the nape. I base this on the observation that most individuals with spotted napes have worn, faded-looking coverts, while most individuals with no (or few) spotted nape feathers seem to have fresh, brightly patterned coverts. Like almost all genera treated herein, the last signs of immaturity are seen in the wing-coverts. In males of some species that have deep-black plumage tracts (i.e., *melanops, roberti, melanogaster*), these black areas may also remain dull or dingy black for some time before developing their adult sheen. The recognition of immatures in this last phase is, in those species of *Conopophaga* that lack patterned coverts as adults, relatively easy. The situation is more complex in those species with spotted adult coverts in one (e.g., *melanops*) or both (e.g., *peruviana*) sexes. In these species the difficulty is compounded by my feeling that this set of immature coverts is retained for a significant amount of time, gradually wearing and fading, being replaced only around the time an individual reaches sexual maturity. In fact, gonadal condition from specimen labels of some individuals with well-worn immature coverts suggests the *possibility* that gonads may begin developing prior to moulting into definitive adult plumage, at least in some species. For the purposes of the following accounts I tentatively define three principal categories of immatures as follows: Fledgling Mostly still covered in plumose, fluffy, nestling-like feathers, rictal flanges still obviously inflated and contrastingly coloured, sexes similar, probably still heavily dependent on adults for food and largely remaining stationary while awaiting adult food delivery; Juvenile Little or no plumose feathering, rictal flanges still brightly coloured but not (or only slightly) inflated, gender usually fairly easily recognised, probably fairly mobile, moving behind adults and actively soliciting food, may attempt increasing amounts of foraging for themselves; Subadult Very close to adults in general coloration, usually only separable by their immature wing-coverts, perhaps slightly different bill coloration near gape, this stage may not be readily distinguishable from adults in the field and is probably fully independent, but may remain near their natal territories (see *C. lineata*). Overall, plumage ontogeny is fairly gradual, rather than step-wise, and I use the terms '**fledgling transitioning**' and '**juvenile transitioning**' for birds that are somewhere between the above-defined stages. The term '**subadult transitioning**' is reserved for individuals with heavily worn and faded (but still not adult-patterned) upperwing-coverts that are otherwise fully adult in appearance. In a few species (see *melanogaster* for further discussion) I found it a challenge to distinguish between faded immature coverts and the very faintly marked coverts of full adults.

BLACK-BELLIED GNATEATER
Conopophaga melanogaster   Plate 1

Conopophaga melanogaster Ménétriés, 1835, Mémoires de l'Academie Imperiale des Sciences de Saint Petersbourg, 6th series, vol. 3, p. 537, plate 15, fig. 2, 'Cuyaba' [= Cuiabá, Mato Grosso, 15°35'S, 56°05'W; Paynter & Traylor 1991a – probably in error, see Distribution]. The holotype, an adult male, was collected by Georg Heinrich von Langsdorff and is at ZMRAS in Saint Petersburg.

The species is named for the distinctive all-black underparts of males. The strikingly plumaged Black-bellied Gnateater is, by a considerable measure, the largest *Conopophaga*. The large white postocular tufts of the male can be flared 90° from the head and offer stark contrast with the velvety black head and underparts. Black-bellied Gnateater is probably endemic to lower Amazonian Brazil; although there is an old specimen from northern Bolivia, there are some doubts as to whether the locality is valid. Generally, the species is considered to be fairly common in most of its range, but without knowledge of its voice and microhabitat requirements, Black-bellied Gnateater is easily overlooked. Like most of its congeners, detection difficulties are compounded by its shy and retiring nature, and affinity for areas of dense undergrowth in *terra firme* forest. The breeding biology of Black-bellied Gnateater remains completely undescribed. As currently described, its range is disjunct, and surveys should be undertaken in the gaps. All aspects of its behaviour, ecology and phylogenetic relationships require further study. As non-vocal sound production is rare amongst antpittas, antthrushes and gnateaters (Krabbe & Schulenberg 2003, Whitney 2003), anyone with the opportunity to make careful observations on the non-vocal display noises (reportedly produced by wing rattling) would find this a rewarding line of enquiry. Indeed, it seems likely that the courtship rituals of this species would be fantastic to watch, as they probably also involve flaring of their large postocular tufts.

IDENTIFICATION 13.0–15.8cm. Male Black-bellied Gnateaters are distinctive, with a black head and underparts, the head marked with a contrasting white postocular tuft. The back, wings and tail are bright chestnut. Females have the back and wings similar to males, but are otherwise different in appearance. Females have a dark grey forecrown with the hindcrown and nape dark brown, and a pale grey supercilium. The throat and centre of the belly are whitish with the remainder of the underparts pale grey. The strikingly patterned male Black-bellied Gnateater is nearly unmistakable within its range. Females are somewhat similar to females of the much smaller Hooded Gnateater of north-eastern Brazil, but have a more prominent postocular brow tuft and more extensive grey on the face and forecrown.

DISTRIBUTION The distribution of Black-bellied Gnateater is poorly understood and shows several unexplained irregularities. It occurs only south of the Rio Amazonas from the east bank of the Rio Madeira east to both banks of the lower Rio Tapajós and south to northern Rondônia and northern Mato Grosso, in Brazil. Apparently somewhat disjunct, with no records in the Rio Madeira–Rio Xingu interfluvium, it occurs from the eastern bank of Rio Xingu east to the west bank of the lower Rio Tocantins (Ridgely & Tudor 1994, 2009, Whitney 2003). Despite being almost universally considered part the Bolivian avifauna (Meyer de Schauensee 1966, Ruschi 1979, Kempff-Mercado 1985, Remsen & Traylor 1989, Ridgely & Tudor 2009, Greeney 2013c), so far as is *known*, Black-bellied Gnateater is endemic to southern Amazonian Brazil, with only a single record from adjacent Bolivia (Allen 1889a, Bond & Meyer de Schauensee 1942). This record, a female specimen that formed the basis for Allen's (1889) description of *Conopophga rusbyi*, was supposedly collected at Reyes (14°19'S, 67°23'W; Paynter 1992; AMNH 30701; LeCroy & Sloss 2000), a small town in north-west Beni surrounded by extensive, level-ground, seasonally inundated grassy savanna (Whitney 2003). To date, however, there has been no subsequent confirmation of its presence there. This specimen was collected by Henry Hurd Rusby, a botanist who travelled north along the Río Beni (= Rio Madeira where it enters Brazil). Based on geographic distance from known populations and the seemingly inappropriate habitat around the collecting locality, it seems most likely that it was actually collected somewhere substantially north of Reyes, perhaps even along the Rio Madeira in Brazil. This seems especially likely given a recent record from Guajará-Mirim, just across the border from Riberalta. Two additional uncertainties surround the distribution of Black-bellied Gnateater. The type locality, given simply as 'Cuyaba' by Ménétriés (1835), was later amended to 'Cuyaba, Mattogrosso' (Hellmayr 1907). With apparently

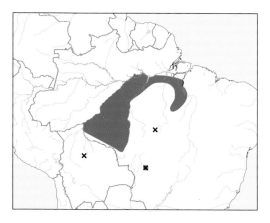

no explanation for his opinion, Hellmayr (1910) stated this location 'requires confirmation,' later upgrading his scepticism to 'no doubt erroneous' and proposed the unfortunately vague type locality of 'Rio Madeira' (Cory & Hellmayr 1924). This decision seems at odds with the fact that he never questioned Allen's (1989) Bolivian record which, at the time, was even more disjunct from other known populations. Nevertheless, this decision is probably a sound one, given the lack of further records from that region and the notoriously inaccurate localities used by Ménétriés (Pacheco 2004) when publishing information on Langsdorff's poorly labelled collections (G.M. Kirwan *in litt.* 2017). Finally, there is an as-yet unconfirmed sight record from the Alta Foresta region, on the left bank of Rio Teles Pires, *c.*450–750km from any confirmed locality ('Teles Pires Trail', *c.*09°38'S, 55°56'W; Zimmer *et al.* 1997). The record refers to an adult male 'seen clearly' by T.A. Parker in October 1989 (Zimmer *et al.* 1997) and its presence there seems unlikely given the subsequent lack of records from this well-inventoried area (Whitney 1997, 2003, Lees *et al.* 2013b). Indeed, Bret Whitney (*in litt.* 2017), informs me that after discussing the record with Parker, they both agreed that his 'lifer' Black-bellied Gnateater was probably a male Chestnut-belted Gnateater. The local race (*snethlagae*) has considerably more black below than races with which Parker was familiar with at that time. Nevertheless, in my depiction of its range I have included all three of the above-mentioned disputed localities. As a final point of interest, the collecting locality 'Arumatheua' (Snethlage 1914a: = Arumatéua, 03°55'S, 49°35'W; Paynter & Traylor 1991a), can apparently now be considered a 'historical record', as the area has been underwater since September 1984, when the Tucuruí Reservoir on the Rio Tocantins was filled.

HABITAT Black-bellied Gnateater inhabits areas of thick, tangled vegetation in *terra firme* forest in southern Amazonia, apparently often near streams (Ridgely & Tudor 1994, 2009, Whitney 2003, Pacheco *et al.* 2007). It favours vine tangles and dense successional growth in old treefalls and other light gaps, and stands of homogeneous vegetation such as understorey palms and bamboo (Whitney 2003). In areas of anthropogenic disturbance it can be found at road edges, but seems not to venture into areas that have lost most or all of their canopy-sized trees. This apparent preference (requirement?) for very dense growth usually leads to widely-separated patches of suitable habitat within areas of intact forest, and thus relatively low population densities in well-preserved areas (Whitney 2003). So far as is known, it is restricted to elevations below 700m (Whitney 2003, Ridgely & Tudor 2009).

VOICE The song of Black-bellied Gnateater lasts *c.*2.5s and consists of an evenly paced (0.6s intervals) series of up to five short, dry, rattles, each lasting less than 1s. It is produced at a frequency of 2–2.5kHz and recalls the alarm calls of *Pittasoma* (Whitney 2003). Call notes are abrupt, low grunts, barks or short rattles (e.g., *chidt-chit-it-it!*) (Whitney 2003). Whitney (2003) has recorded and described males, in courtship or aggressive displays, making a loud whirring sound with their outer primaries while flying low and rapidly. This behaviour has not been studied in detail, however, and requires thorough documentation.

NATURAL HISTORY Black-bellied Gnateaters usually are solitary, or in pairs. Published data on its behaviour are almost non-existent. Like other species of *Conopophaga*, it

forages low to the ground, and scans surrounding foliage and leaf litter for prey (Whitney 2003). There are no published data on territorial defence, maintenance or fidelity. Whitney (2003) suggested that the Black-bellied Gnateater's predilection for small patches of dense understorey in otherwise comparatively open forests, may limit its territory size to 1ha or less, but no firm data have been published on territory size. Although the bright white postocular tufts of the male can be flared out dramatically (see photos by K.J. Zimmer *in* Greeney 2013c) and Whitney (2003) described mechanical wing sounds made by males in 'aggression or courtship', there appear to be no detailed accounts of such behaviour aside from responses to recording playback.

DIET The diet of Black-bellied Gnateater is very poorly documented but, like congeners, presumably largely comprises invertebrates and small invertebrates. Stomach contents of an adult male from near Altamira (USNM 572604) included insect parts and a 4.5cm-long larval lepidopteran.

REPRODUCTION The nest and eggs are undescribed. Bret Whitney (*in litt.* 2017) found two nests along the banks of the Rio Aripuanã, Amazonas, Brazil, in July 2015. He found one nest after returning to the area where Micah Riegner and Albert Burgas had recently observed a juvenile Black-bellied Gnateater (see below). Only 10m from where the young bird had been seen, he found a fresh, empty nest that in all ways matched those of other *Conopophaga* he had seen. It was fairly exposed, *c.*1m above ground and supported by a network of thin-trunked (<1cm) saplings. The nest was platform-like, composed of sticks and large, dead leaves, overall resembling a mass of fallen detritus trapped by the saplings. He found a second nest, nearly identical in form to the first, also near where the fledgling had been seen. This one was clearly older, and not recently used. It was only slightly higher above ground and was partially concealed by surrounding vegetation, but was overall similar in placement to the first nest. **BREEDING DATA** I was able to find a few additional records of immatures: juvenile, 30 January 1906, Itaituba (AMNH 488920); juvenile, 30 July 2015, upper Rio Aripuanã (A. Burgas *in litt.* 2017); subadult, 30 April 1930, Borba (AMNH 280533); subadult, 3 October 1907, Santa Isabel (AMNH 488923); subadult, 24 January 1920, Vila Braga (YPM 29711, S.S. Snow photos); subadult, 6 August 2016, Canaã dos Carajás (WA 2226218, L.F. Matos); subadult, 24 September 2016, Anapu (WA 2294932, D. Mota); subadult, 7 June 1969, Urucurituba (FMNH 344548); subadult, 3 October 1907, Santa Isabel (AMNH 488923).

TECHNICAL DESCRIPTION Sexes similar but distinguishable by plumage. *Adult Male* In males the entire head, including nape, face, throat and sides of the neck, is black. The underparts are also uniform black, grading only slightly into dark grey on the lower belly and vent. The flanks and undertail coverts are light rufescent-brown mottled ash-grey. The upperparts are a rich rufous-chestnut that forms a clean line where it meets the black of the head. The most striking feature is a long, bright white postocular tuft of feathers. Although these tufts reportedly can be flared out laterally, there are no detailed accounts of such behaviour. *Adult Female* Female Black-bellied Gnateaters have rufous-chestnut upperparts, similar to males, but the chestnut extends onto the nape and crown, becoming duller and browner anteriorly, before fading to a grey forecrown. The face, ear-coverts and sides of the neck are grey. Below, areas that are black on the male are grey in females, broken only by a white throat and variably sized (and inconspicuous) white belly patch. The postocular tufts of females, though still present, are greatly reduced. **Adult Bare Parts** (sexes similar) *Iris* dark brown; *Bill* black; *Tarsi & Toes* grey to blue-grey or plumbeous. *Juvenile Male* I examined the juvenile male (AMNH 488920) described by Hellmayr (1907) as having the crown 'dull black, with irregular, pale rufous bars.' The individual feathers are black, broadly tipped dull chestnut, creating a *weakly* barred (more spotted in my view) pattern. To my eyes there are nearly equal amounts of chestnut and black across most of the crown, with a gradual reduction in the amount of chestnut on the sides of the upper neck and the forecrown (pure black at base of bill). In life, the crown undoubtedly looks blackish with coarse chestnut spotting that becomes less evident as immature feathers are replaced by pure black ones (probably beginning with those on the forecrown and working backwards). Hellmayr (1907) described the back as 'clearer chestnut than in the adult, and crossed by some blackish bands'. This specimen appears to be in the process of acquiring adult plumage on the back (rich chestnut). Approximately half of the feathers have yet to be replaced (mostly on the central back and rump), and these are distinctly more yellowish-chestnut with a broad black band across the centre and a thinner black margin at the tip. These create irregular, thin black bars, probably decidedly more apparent in younger individuals. The upper breast and throat are largely black, somewhat washed cinereous on the upper throat and mixed with fulvous-barred immature feathers on the upper breast. The breast, sides and flanks are largely covered with black- and fulvous-barred feathering, the barring fading to buffy on the lower breast and central belly. In these areas the bases of the feathers are cinereous, creating a somewhat sloppy mixture of grey, black, buff and fulvous. The upperwing-coverts are largely dusky-brown with a broad black apical margin, and a broad, subterminal, buffy-cinnamon band (reduced to a spot on inner coverts). **Juvenile Male Bare Parts** (from AMNH 488920 label) *Iris* brown: *Bill* black; *Tarsi & Toes* grey. *Subadult Male* I examined the 'immature' male (AMNH 488923) described by Hellmayr (1910), an individual clearly in the last phases of acquiring adult plumage. It is largely as adult males (especially above) but the breast feathers are narrowly edged grey giving an indistinct scaled appearance. The black extends only as far as the upper belly. The rest of the belly is a messy mixture of blackish, grey and white. A slightly younger male (AMNH 280533) is similar below but shows very little black on the belly and there are still a few fulvous-barred immature feathers admixed, especially on the flanks. The upperwing-coverts in both individuals are spotted as described below for juveniles. Males at this stage of plumage maturation might be easily confused with Hooded Gnateater males, being best separated by the brighter upperparts, larger size, all-black bill and messy belly plumage. **Subadult Male Bare Parts** (from AMNH 88923 label) *Iris* dark brown; *Bill* black; *Tarsi & Toes* plumbeous. *Juvenile Female* The precise appearance and ontogeny of immature female plumages is not fully apparent to me. The following description is based on a single, side-view image of an immature that was still attended by its parents (courtesy of Albert Burgas; Fig. 4).

It is clearly not in fluffy plumage and I thus categorically consider it to be a juvenile. At first glance, it is overall very similar in pattern, but duller in hue, to an adult female, including a prominent white (though somewhat dull) superciliary stripe. The face and forecrown are slate-grey as in adult females, but are concolorous with the grey throat (not contrastingly white as in adults). The crown is not fully visible but is clearly washed with brown and appears to be indistinctly scaled (brown and buffy-brown) from the centre of the crown to at least the hindcrown. Similarly, my view of the upper back and nape is incomplete but these areas do not appear to be distinctly spotted or scaled as in other *Conopophaga*. At most these regions seem only weakly marked by thin dusky margins on otherwise brownish feathers. The remainder of the back and the flight feathers appear similar in coloration to adults (tail not visible). Below, the grey of the throat extends at least as far as the upper belly with little or no white, except perhaps in the very centre. Lower on the belly, and higher up on the sides, the grey blends into the buffy-brown flanks and vent, with just a hint of dusky barring. Unfortunately, photo quality prohibits close inspection of the upperwing-coverts, but the greater and lesser primary-coverts appear only weakly patterned. The greater coverts are broadly tipped pale brown and finely margined black, while the lesser coverts appear more distinctly marked, tipped buffy-brown and (possibly) also finely edged with black. **Juvenile Female Bare Parts** *Iris* dark; *Bill* clearly not fully developed, being shorter, overall dusky (not shiny) black, slightly paler at the tip, with contrasting pinkish rictal flanges that seem to be still slightly enlarged; *Tarsi & Toes* similar to adult but seemingly darker.

MORPHOMETRIC DATA *Wing* 79mm, 77mm; *Tail* 39mm, 38mm; *Bill* exposed culmen 15mm, 15mm (*n* = 2, ♂,♀, Novaes 1947). *Wing* 81mm, 80mm, 80mm, 79mm; *Tail* 40mm, 40mm, 42mm, 42mm; *Bill* [total?] culmen 18mm, 18mm, 18mm, 17.5mm; *Tarsus* 33mm, 33mm, 32mm, 32mm (*n* = 4, ♂,♂,♂,♀, Hellmayr 1910). *Wing* 78mm; *Tail* 42mm; *Bill* 17mm (*n* = 1, 'immature', Hellmayr 1910). *Wing* 84mm; *Tail* 44mm; *Bill* [total culmen?] 20mm; *Tarsus* 33mm (*n* = 1, ♂, Hellmayr 1907). *Bill* from nares 9.7mm; *Tarsus* 33.2mm (*n* = 1, ♂, USNM 572604). *Wing* 79mm; *Tail* 41.2mm; *Bill* [exposed?] culmen 14mm, from nares 6.1mm; *Tarsus* 37mm; (*n* = 1, ♂, holotype, Chrostowski 1921). *Wing* 81mm, 82mm; *Tail* 43mm, 44mm; *Bill* [total?] culmen 19mm, 20mm (*n* = 2, ♂♂, Hellmayr 1905b). *Wing* 77mm; *Tail* 45mm; *Bill* exposed culmen 16mm; *Tarsus* 34mm (*n* = 1, ♀, type of *rusbyi*, Allen 1889a). *Wing* 85mm, n/m; *Tail* 47mm, n/m; *Bill* from nares 10.0mm, 12.0mm, total culmen 16.5mm, 17.9mm, depth at nares 5.0mm, 5.0mm, width at nares 6.0mm, 6.1mm, width at base of mandible 8.5mm, 8.1mm; *Tarsus* 35.0mm, 38.1mm (*n* = 2, ♂♂, FMNH 344087/254878). *Wing* 84mm, 81.5mm, 77.5mm; *Bill* from nares 11.3mm, 10.7mm, 10.5mm, exposed culmen 12.9mm, 12.0mm, 11.6mm; *Tarsus* 35.1mm, 33.0mm, 32.5mm (*n* = 3, ♂,♂,♀, MCZ, Jonathan C. Schmitt *in litt.* 2017). **Sex unspecified** *Wing* 73.7mm; *Tail* 38.1mm (n = ?, Sclater 1890). *Wing* 83mm; *Tail* 38mm; *Bill* 17mm; *Tarsus* 36mm (n = ?, Snethlage 1914). *Wing* 83mm; *Tail* 38mm; *Bill* 17mm; *Tarsus* 36mm (n = ?, Ruschi 1979; although Snethlage (1914a) is not cited, Ruschi probably took these measurements directly from that source). **Mass** 42g (n = 1, ♂, Graves & Zusi 1990; USNM 572604); mean 42.5g, range 42–43g (n = 2, ♂♂, Dunning 1993).

Sex unspecified 37.0–43.5g (n = ?, Whitney 2003). **Total Length** 12.7–15.8cm (Burmeister 1856, Sclater 1890, Meyer de Schauensee 1970, Ruschi 1979, Sick 1993, Ridgely & Tudor 1994, 2009, Whitney 2003, van Perlo 2009).

DISTRIBUTION DATA Endemic to **BRAZIL:** *Amazonas* Parintins, 02°38.5'S, 56°44'W (Pinto 1938, 1978); Borba, 04°24'S, 59°35'W (von Pelzeln 1871, Sclater 1890, Hellmayr 1907, AMNH 280533); Maués, 05°18.5'S, 58°11.5'W (WA 152262/69, A, Whittaker); Igarapé Arauazinho, 06°18'S, 60°21.5'W (J. VanderGaast *in litt.* 2017); Rio Bararati, *c.*07°25.5'S, 58°13'W (B. Whitney recording July 2011); Terra Indígena Nove de Janeiro, *c.*07°28'S, 62°56'W (Santos & da Silva 2005); Igarapé Nove de Janeiro, 07°38'S, 62°31'W (MPEG 47852); Pousada Rio Roosevelt, 08°29.5'S, 60°58'W (Whittaker 2009, XC 32013, B. Davis); Rodovia do Estanho, km126, 08°39'S, 61°25'W (Aleixo & Poletto 2007, MPEG 57697). *Pará* Vicinity of Juruti, *c.*02°09'S, 56°05'W (Santos *et al.* 2011a); Platô Capiranga, 02°28'S, 56°00'W (MPEG 61644); Acampamento Capiranga, 02°30'S, 56°13'W (MPEG 60972); upper Rio Arapiuns, 02°34'S, 55°22'W (Haugaasen *et al.* 2003); Alto Mentai, 02°47'S, 55°36'W (MPEG 72356); Boim, 02°50'S, 55°11'W (Griscom & Greenway 1941, AMNH, MCZ, MPEG); 'Apaçy,' near Aveiro, 03°15'S, 55°10'W (Griscom & Greenway 1941, CM P77727/28); Anapu, 03°28.5'S, 51°12'W (WA 1604717, V. Castro); Urucurituba, 03°47.5'S, 55°31.5'W (FMNH 344548); Vila Braga, 03°32'S, 55°30'W (Griscom & Greenway 1941, Novaes 1947, CM, YPM, ANSP, MCZ, FMNH, UMMZ, MPEG, MNRJ); 52km SSW of Altamira, 03°39'S, 52°22'W (Graves & Zusi 1990, USNM 572604); Novo Repartimento, 04°15'S, 49°57'W (WA 2218142, A. Grassi); Itaituba, 04°16'S, 56°00'W (Hellmayr 1907, Griscom & Greenway 1941, AMNH, ANSP, CM, YPM); *c.*7.5km W of Moloca, 05°30'S, 56°59'W (XC 147398, G. Leite); *c.*3.7km ENE of Cedro, 05°35'S, 49°12'W (MPEG 38078); FLONA Serra dos Carajás, multiple localities W of Parauapebas, *c.*06°00'S, 50°30'W (Pacheco *et al.* 2007, LSUMZ 184629, MPEG, ML); Manganês, 06°10'S, 50°24'W (MPEG 37242–245); São Félix do Xingu, 06°28.5'S, 51°08'W (MPEG 76043); Canaã dos Carajás, 06°32'S, 49°51'W (WA 2226218, L.F. Matos); Rio Riozinho, Aldeia Aukre, 07°46'S, 51°57'W (MPEG 52315/16). *Mato Grosso* Upper Rio Aripuanã, *c.*09°06'S, 59°23'W (B. Whitney recording July 2015); Comodoro, 13°49'S, 59°40'W (XC 171002, G. Leite); Chapada dos Parecis, 13°58'S, 59°45'W (Whittaker 2009). *Rondônia* Calama, 08°03'S, 62°53'W (Helmayr 1910, AMNH 488921/22); Santa Isabel, 08°03'S, 62°54'W (AMNH 488923); N of Machadinho d'Oeste, 08°57'S, 61°58'W (XC 167250; G. Leite); Machadinho d'Oeste, 09°14'S, 61°55'W (WA 2175478, S.F. Linhares); Cachoeira de Nazaré, 09°45'S, 61°55'W (Stotz *et al.* 1997, FMNH 389958/59, MPEG 39936–938); Pedra Branca, 10°03'S, 62°07'W (Stotz *et al.* 1997, FMNH 344087); 'Maruins' [= 'Milho' Paynter & Traylor 1991b; = Maroins?], 09°10'S, 61°56'W (Hellmayr 1910); left bank of Rio Ouro Preto, 10°50'S, 64°45'W (MPEG 55031); 10km SE of Guajará-Mirim, 10°51'S, 65°16'W (Whittaker 2009); BR-429, km87, 11°24'S, 62°24'W (MPEG 38807); Vilhena, 12°43'S, 60°07'W (MNRJ 3976).

TAXONOMY AND VARIATION Monotypic. Whitney (2003) suggested, primarily based on vocal similarities, that Black-bellied Gnateater is most closely related to the 'lineata group', comprising Rufous, Hooded, Ash-throated,

Slaty and Chestnut-crowned Gnateaters. Currently, as perhaps the most divergent (morphologically) member of the genus, I have followed Remsen *et al.* (2017) in placing it at the front of the present linear arrangement. Allen (1889) described *Conopophaga rusbyi* from a single specimen (female by plumage) from Reyes, El Beni (*contra* Peters 1951), Bolivia, which, if valid, would presumably be a subspecies of Black-bellied Gnateater, but *rusbyi* has not been considered a valid taxon by most subsequent authors (Hellmayr 1907, 1910, Brabourne & Chubb 1912, Cory & Hellmayr 1924, Whitney 2003).

STATUS Black-bellied Gnateater has a fairly large range and, despite the fact that its populations appear to be declining, BirdLife International (2017) has evaluated this species as one of Least Concern. Though, to some degree, all tropical organisms face some degree of habitat threat due to anthropogenic impact, a recent analysis by Bird *et al.* (2012) suggested that habitat within the species' range is not under sufficient immediate threat to upgrade the species to one of conservation concern. No specific threats have been studied in detail, but Haugaasen *et al.* (2003) found Black-bellied Gnateaters absent from areas of habitat that had been partially burned less than two years previously. Although relatively few protected areas exist within the species' range, it is particularly difficult to detect for those unfamiliar with its voice (Whitney 2003). It seems likely, therefore, that additional populations will be discovered, hopefully within areas offered at least some degree of protection. It can be locally fairly common (e.g., in the Serra dos Carajás; Whitney 2003), so hopefully substantial populations remain within protected areas other than those listed below. Somewhat ironically, as suggested by Whitney (2003), the Black-bellied Gnateater undoubtedly benefits to some degree from 'less invasive' forms of anthropogenic habitat alteration such as selective logging and road construction. Its long-term persistence in such areas, however, is probably dependent upon these areas of disturbance being connected to more natural mosaics of intact and naturally disturbed habitat. Apart from identifying additional areas of occurrence within its known range, determination of its status in Bolivia is a priority. Whitney (2003) suggested that it should be searched for in humid mosaics of intact and naturally disturbed forest east of the Río Beni, e.g. in the Riberalta area. **Protected Populations** FLONA Tapajós (Oren & Parker 1997); FLONA Serra dos Carajás (Pacheco *et al.* 2007; MPEG); FLONA Tapirapé-Aquiri (MPEG); PN Amazônia (XC 343509, G. Leite); REBIO Ouro Preto (MPEG 55031); RESEX Tapajós-Arapiuns (MPEG 72356); Reserva Indígena Kaiapó (MPEG 52315/16); RPPN Vale do Rio Doce, Rio Sororó (MPEG 38078).

OTHER NAMES *Pseudoconopophaga melanogaster* (Chubb 1918c, 1921); *Conopophaga rusbyi* (Allen 1889a, Ogilvie-Grant 1912); *Conopophaga rushyi* (Dubois 1900). Some authors have misspelled the specific epithet 'melanogastra' (e.g., Sclater 1858c, 1890, Dubois 1900, Hagmann 1904, Hellmayr 1905b, 1907 [in text]). **English** Black-breasted Gnat-eater/Gnat Eater (Brabourne & Chubb 1912, Cory & Hellmayr 1924). **Portuguese** chupa-dente-grande (Willis & Oniki 1991b, van Perlo 2009); chupa-dente-vermelho-preto (Ruschi 1979). **Spanish** Jejenero Ventrinegro (Whitney 2003). **French** Conophage à ventre noir. **German** Westlicher Schwarzkopf-Mückenfresser (Whitney 2003).

Black-bellied Gnateater, adult male, Parauapebas, Pará, Brazil, 19 July 2014 (*Ciro Albano*).

Black-bellied Gnateater, adult male, Parauapebas, Pará, Brazil, 19 July 2014 (*Caio Brito*).

Black-bellied Gnateater, juvenile female, upper Rio Aripuanã, Mato Grosso, Brazil, 30 June 2015 (*Albert Burgas*).

Black-bellied Gnateater, adult female, Maués, Amazonas, Brazil, 30 May 2010 (*Andrew Whittaker/Birding Brazil Tours*).

BLACK-CHEEKED GNATEATER
Conopophaga melanops   Plate 1

Platyrhynchos melanops Vieillot, 1818, Nouveau dictionnaire d'histoire naturelle, vol. 27, p. 14, 'South America'. The holotype is a male collected by Delalande, Jr., in the vicinity of Rio de Janeiro and is at the Paris Museum (MNHN).

Endemic to the Atlantic Forest of eastern Brazil, between Paraíba in the north and Santa Catarina in the south, the Black-cheeked Gnateater inhabits wet forest and dense, older second growth in the lowlands and foothills, as well as ranging locally into drier forest. Adults generally forage low in the undergrowth, sometimes on the ground, darting rapidly between perches. Males are stunning birds, best detected by their trilling song, with a bright rufous crown, black mask, white throat, grey underparts and largely brown upperparts. Females are predominantly orange-rufous below, with a brownish face and darker upperparts. Three subspecies have been recognised, but there is substantial individual variation, and it is unclear how many of these might be considered valid.

IDENTIFICATION 10–12cm. Sexes differ. Males have bright orange-rufous crowns that extend onto the nape, and broad black masks that extend onto the sides of neck. The back, wings and tail are brown with broad blackish streaking and indistinct scaling. The upperwing-coverts are plain brown or (infrequently) show small, inconspicuous white or pale buff spots near the tips of the greater coverts, especially those of the primaries. No distinct postocular tuft as in other species, but may have a few long white feathers in this location. The throat is bright white with the remaining underparts grey except for variable amounts of white on the central belly. Females are rufescent-brown to grey-brown above, including the crown. They lack the black mask but have a long, thin, white or off-white superciliary. The back is marked as in males but both the greater and lesser wing-coverts are distinctly tipped with buffy spots. Below, females are mostly orange-rufous. They resemble slightly larger Rufous Gnateater but that has a pale lower mandible (bill all black in Black-cheeked) and no dots on the wing-coverts or back scaling. Behaviour as in other gnateaters, but seems more terrestrial. Rufous Gnateaters are somewhat similar to females of the present species, but Black-cheeked Gnateaters are quickly separated by their more distinctly 'capped' appearance, their all-black bills and the black markings on the back (van Perlo 2009). Subadult and juvenile Rufous Gnateaters are slightly more similar, as they too have pale tips on the wing-coverts. In these cases the distinctly bicoloured bill (or generally paler, dirty yellow in immatures) of Rufous Gnateater is probably the best field character.

DISTRIBUTION This Brazilian endemic is found from coastal regions of north-east Brazil, from south-eastern Rio Grande do Norte, to central-eastern Santa Catarina in the south. It is also found on at least two of the larger inshore islands, Ilha Santa Catarina (Naka *et al.* 2002) and Ilha Grande (Marsden *et al.* 2003). Mention of this species occurring inland as far as Distrito Federal (Ribeiro *et al.* 2013) is presumably due to typographical errors.

HABITAT Black-cheeked Gnateater is generally a bird of primary and mature, tall second-growth forest and forest edge (Höfling & Lencioni 1992, Aleixo & Galetti 1997, Goerck 1999), preferring the interior and edge

of dark, often mossy Atlantic Forest (van Perlo 2009). It frequents stream margins and ravines, and patches of dense second growth such as regenerating treefalls. It is also found in the drier forest remnants of Rio de Janeiro (Whitney 2003) and the semi-deciduous forests of São Paulo (Antunes 2007). Most records are from below 800m (Machado & da Fonesca 2000, van Perlo 2009), it tending to be replaced at higher elevations by Rufous Gnateater (Machado & da Fonesca 2000). In south-east Brazil (*melanops*) is sometimes found as high as 1100m (Mallet-Rodrigues *et al.* 2015).

VOICE The trilling song of adult Black-cheeked Gnateater is a long series of tightly spaced notes, lasting up to 5–10 seconds, and becoming slightly louder as it rises in pitch and slows fractionally at the end (Whitney 2003, Lima & Roper 2013). In overall effect, Whitney (2003) likened the song to the sound made by dragging your thumbnail along teeth of a comb. In a general sense, Whitney (2003) gives the range of frequencies included in the song as 1.5–3.5kHz, while the more detailed analysis of Lima & Roper (2013) revealed 2–4 (usually 3) diagnosable harmonics, with the lower frequency harmonic ranging around 2.7–4.2kHz and the two succeeding harmonics delivered at 4.7–5.7kHz and 6.2–7.7kHz. In the same study, of nominate *melanops* in Paraná, Lima & Roper (2013) reported a mean 12.2 notes per song ($n = 10$). The precise functions of other vocalisations are not fully documented. Whitney (2003) reported regularly heard, ringing *zhink!* or *zhweenk!* calls, described by Sick (1984, 1993) as *xchit, schist,* or *kssrr* notes, by Ridgely & Tudor (2009) as *bzheeyk!* and by Willis (1985) as sneeze-like. Sick (1993) also reported a *psEEeh* vocalisation he considered an alarm call. The alarm call was quantitatively described by Lima & Roper (2013) as a single, high-frequency, short pulse, lasting *c.*0.2s, with a broad frequency range (3.2–16.0kHz). In south-east Brazil (nominate *melanops*), Lima & Roper (2013) noticed that the vocalisations of

nestlings closely resembled those of adults, specifically the adult song. In their examination of age-specific variation in song structure and the ontogenetic development of song in nestlings, they found that nestling vocalisations and adult songs differed primarily in their maximum frequency and total length of the song, and suggested that, although vocalisations may be innate, there also appears to be a developmental stage during which individuals also develop capabilities through learning.

NATURAL HISTORY Black-cheeked Gnateater forages in the understorey, hunting predominantly on low foliage and in the leaf litter, and capturing insects by making short reaches and gleans (Willis 1985a, Alves & Duarte 1996). Although it may join mixed-species foraging flocks (Stotz 1993, Brandt *et al.* 2009) and follow ant swarms (Willis 1985a, Roda *et al.* 2003, Pizo & Melo 2010), most sources seem to agree that it most frequently forages alone or in pairs. A pair observed by Willis (1985) while over ants foraged exclusively below 6m, spending the great majority of their time at or below 2m. They perched on thin (<6cm) branches, at varying angles. They captured prey by sallying to the ground, tree trunks and leaves, once capturing a moth suspended in a spider web. A few studies have examined the prevalence of blood parasites (Sebaio *et al.* 2012) and ectoparasites (Storni *et al.* 2001, Nogueira *et al.* 2005) of *C. melanops*, but overall this aspect of their biology is poorly understood. Black-cheeked Gnateaters are territorial, at least during the breeding season. In Paraná, Lima & Roper (2009a) recorded the apparent annual survival of adults as 0.44. This relatively low survival rate suggests that territories are not very permanent, and these authors suggested that they may be abandoned following nest failure.

DIET Like other members of the genus, the diet of Black-cheeked Gnateater generally consists of small arthropods, but few studies have examined prey items in detail. Lopes *et al.* (2005b) examined the stomach contents of two individuals (*melanops*) from Fazenda Santana, Salto da Divisa, Minas Gerais, identifying 69 individual arthropods which included: 32 ants (Formicidae), 20 unidentified Hymenoptera, seven beetles (Coleoptera), five larval flies (Diptera), one larval and one adult Lepidoptera, one Hemiptera, one harvestman (Opiliones) and one millipede (Diplopoda). Additional records (all *melanops*) include a small (1.5–2.0cm SVL; *cf.* Dendrobatidae) frog captured by a female (WA 1787788, H. Duarte), a large (4–5cm) lepidopteran larva captured by a male (WA 1648641, I. Epifani), a small (1.0–1.5cm) spider captured by a subadult male (WA 455681, J. Mattos), a large (3–4cm) insect (*cf.* Orthoptera) captured by a female (WA 504031, M. Nema), a large (3–4cm) insect (*cf.* Orthoptera) carried to a nest by an adult female (WA 1599315, T. Carneiro), a *c.*2cm adult Trichoptera (WA 1215237, R. von Mühlen) and a small (1.0–1.5cm) lepidopteran larva captured by a female (WA 2077426, V.E. Florencio).

REPRODUCTION Until recently, the breeding biology of Black-cheeked Gnateater was almost unknown, and many aspects of reproduction have yet to be studied. Sick (1957) mentioned two nests of Black-cheeked Gnateater (*melanops*?) with no specifics other than stating that one of them included *Marasmius* fungi (horse-hair fungus, presumably as lining). Alves *et al.* (2001, 2002) provided the first formal description of the nest and eggs, reporting a nest of nominate *melanops* containing two eggs that hatched on

7 November 2000 on Ilha Grande in the Atlantic Forest of Rio de Janeiro. In Paraná, André Magnani Xavier de Lima (Lima 2008, Lima & Roper 2009a) studied 15 nests, reporting many great details of nesting behaviour. The reported duration of the nesting cycle, from laying to fledging, was 38–44 days (*n* = 3 nests), with pairs attempting up to three nests per season. More recently, Lima & Roper (2009a) reported an intensive study of 18 nests in Paraná. **NEST & BUILDING** Nestbuilding is performed by both sexes, and lasted 14 and 20 days at two nests followed by Lima (2008). Alves *et al.* (2001, 2002) described the nest of Black-cheeked Gnateater as 'bowl-shaped' and provided measurements: external diameter 10–12cm; external height 7.5–9.5mm; internal diameter 6.5cm; internal depth 4.3cm. The total fresh weight of the nest was 12.5g. It was composed of small sticks and dead leaves, 87cm above the ground, and built into the rosette of a large bromeliad (*Neoregelia johannis*). A nest in southern Bahia was 0.8m above ground in small tree, a Paraná nest was 0.6m above ground among Heliconia stems (Whitney 2003). Of 13 nests in Rio de Janeiro, most were in Dracaena bushes, 20–80cm above the ground (Stenzel & Souza 2014). **EGG, LAYING & INCUBATION** Clutch size was two eggs in 14 nests, and one egg in one nest (Lima 2008), and given as two eggs in an unspecified number by Stenzel & de Souza (2014). Eggs are laid on subsequent days, with regular incubation beginning only after clutch completion and the eggs hatching after 18 days (Alves *et al.* 2002, Stenzel & Souza 2014). The eggs of Black-cheeked Gnateater are buffy-white or pale brown with dark brownish speckling, fairly sparse but often concentrated near the larger pole (Stenzel & Souza 2014). The egg of nominate *melanops* described by Alves *et al.* (2002) as 'salmon color' probably did not differ much from the description above, and the pinkish-orange look to the eggs was probably a result of them being undeveloped. These authors provided the following measurements for this clutch of two eggs: 23.0–23.1mm × 16.8–16.9mm; 3.4g and 3.6g. Two eggs of nominate *melanops* measured 22 × 17mm and 21 × 17mm (von Ihering 1900). Based on 30h of focal observations, both sexes participated in diurnal incubation, with the male spending more time at the nest than the female. Only the female, however, spends the night on the nest (Alves *et al.* 2002, Stenzel & Souza 2014). **NESTLING & PARENTAL CARE** Stenzel & Souza (2014) reported a nestling period of 14–15 days, most commonly 15 days. Adults brought food to the nest with roughly equal frequency, but the male spent twice as much time brooding. Only the female, however, brooded at night. While brooding, adults generally remain until an observer is almost at the nest. In response to human observers near a nest, adults perform exaggerated distraction displays by fluttering theatrically across the ground. These displays become more intense as nestlings approach fledging age (Whitney 2003). **Fledgling Behaviour and Care** Two pairs of fledglings remained on their home territories for at least 65 days after hatching (Lima 2008). **Nest Success** In Paraná, Lima (Lima 2008, Lima & Roper 2009a) reported daily survival rates of nests as 0.94–0.96, calculating that 11% of nesting attempts produced young and that the annual nesting success of adults was 22%, or 0.36 young per year, per adult in the population. For 15 nests followed closely, ten were predated, four apparently failed due to torrential rain, and one was abandoned. Following six of the predation events, the nest remained intact, with no outward signs of disturbance, possibly suggesting snakes as important nest predators in the area. The remaining four nests were clearly torn up by

predators, and at one of these with freshly hatched nestlings, an opossum (*Didelphis aurita*) was the known culprit. They also calculated that 75% of fledglings (*n* = 6) survived at least two months after fledging. At a nest in Paraná, both nestlings were infested with botflies (*cf.* Philornis), one with six larvae and the other with 12 (Whitney 2003). SEASONALITY In Paraná (*melanops*), nestbuilding commences in the middle of October, peaks in November, and may continue until the middle of January (Lima 2008), suggesting that the last nests of the season fledge sometime in late February. This period corresponds with the onset of the wet season, increased daily photoperiod, and a reduction of extreme minimum temperatures at night. Thirteen nests found at the Jardim Botânico do Rio de Janeiro were active between August and December (Stenzel & Souza 2014) and in the Serra dos Órgãos active brood patches were found only in September–February (Mallet-Rodrigues 2005). Almost all of the information available on breeding pertains to the nominate subspecies. BREEDING DATA *C. m. nigrifrons* Juvenile transitioning, 26 December 2013, Murici (WA 1194543, S. Leal). *C. m. perspicillata* ♀ sitting on nest, 3 January 2011, Valença (WA 399457, S.S. Santos). *C. m. melanops* Eggs, 26 October 2000, Ilha Grande (Alves *et al.* 2002); eggs, 18 September 2015, Biguaçu (WA 1845779, F. Llanos); fresh eggs, 2–5 November 2009, São Pedro da Aldeia (WA, L. Freire & M. Carvalho); eggs, 15 November 2008, near Rio de Janeiro (WA 327571, L.B. Junior); two nests with eggs, 24 October 2013 and 17 January 2015, São Pedro da Aldeia (WA 1131176/1708432, D. Steinwender); nest with two eggs, 12 January 1995, RPPN da Vale do Rio Doce (eBird: R. Laps); eggs, 11 November 2012, Natividade da Serra (WA 833905, E. Kaseker); incubation, 28 October 1986, Porto de Cima (Straube 1989); hatching, 20 January 2014, São Pedro da Aldeia (WA 1221410, D. Steinwender); young nestlings, 16 November 2014, Santa Leopoldina (WA 1518536, L. Loureiro); active nest, 15 November 2008, FLONA Tijuca (L.A. Florit *in litt.* 2016); fledgling, 21 December 1929, Baixo Grandu (AMNH 317477, = 'juv?' in Naumburg 1937); two fledglings, Caraguatatuba, 22 November 2011 and 31 November 2015 (WA 504031/1969372, M. Nema); fledgling, 22 November 2014, Florianópolis (WA 1523587, L. Motta); fledgling, 29 December 2010, Teresópolis (WA 270413, P. Tinoco); fledgling, 28 December 2012, Guaraqueçaba (WA 546879, E. Kaseker); adult carrying food, 20 December 2014, Caraguatatuba (T. Carneiro *in litt.* 2015); immature fed by adult [fledgling or juvenile], 23 February 2014, Serra dos Tucanos Lodge (eBird: J. Faragher); juvenile, 12 October 2016, Salesópolis (WA 2339664, L. Cardim); juvenile, 7 February 2015, Miracatu (WA 1604159, E. Kaseker); juvenile, 31 January 2014, Angra dos Reis (WA 1235604, C. Kleske); juvenile, 18 December 2011, Paraty (WA 525929, F. Arantes); three juveniles at Ubatuba, 30 December 2011 (WA 534192, W. Coppede), 4 January 2013 (WA 853239, J.H. Nuñez), 25 January 2015 (WA 1596140, L. Amaral); juvenile transitioning, 10 March 2015, Blumenau (WA 1633610, N.M. Pereira); juvenile transitioning, 6 March 2017, Angra dos Reis (WA 2488954, C.E. Blanco); juvenile transitioning, 22 December 2016, Guapimirim (WA 2422598, C. Belleza); juvenile transitioning, 10 March 2016, Blumenau (WA 2064887, S. Oliveira); juvenile transitioning, 11 January 2017, Peruibe (WA 2434686, T. Martins); juvenile transitioning, 21 February 2016, Caraguatatuba (WA 2034485, A. Gomes); juvenile transitioning, 2 May 2016, Florianópolis (WA 2106424, L.L. Nunes); juvenile transitioning, 11 February 2017, Santo Amaro da Imperatriz

(WA 2464384, S. Leal); juvenile transitioning, 23 February 2010, Florianópolis (WA 113303, F. Farias); juvenile transitioning, 21 January 2012, São Sebastião (WA 566794, J. Casoni); juvenile transitioning, 26 January 2013, Santo Amaro da Imperatriz (WA 1157667, S. Leal); juvenile transitioning, 20 March 2016, Ilha de Santa Catarina (WA 2059278, M. Moreto); juvenile transitioning, 21 February 2015, Blumenau (WA 1618622, E. Godoz); juvenile transitioning, 24 February 2014, Rio de Janeiro (WA 1255939, A. Gaertner); immature, 23 February 2001, PNM Lagoa do Peri (eBird: A.L. Roos); immature, 17 November 2003, Fazenda Angelim near Ubatuba (eBird: P. Bono); immature, 10 April 2016, PE Serra do Mar (eBird: A. Ferrari); subadult, 20 April 2014, Blumenau (WA 1303141, C.E. Zimmermann); subadult, 26 March 2014, Ribeirão Grande (WA 1321270, R.G. Lebowski); subadult, 28 March 2014, RE Guapiaçu (ML 45317101, photo N. Voaden); subadult, 14 July 2016, Ubatuba (WA 2215467, C. Goulart); subadult, 13 February 2017, Morretes (WA 2485734, A. Gabriel); subadult, 15 March 2016, Guarujá (WA 2060680, M. De Castro); subadult, 29 September 2016, Salesópolis (WA 2311199, E. 'Japão'); subadult, 31 December 2015, Caraguatatuba (WA 1969102, J. do Egito Maracajá Junior); four subadults (photos V.E. Florencio), Joinville, 8 August 2015 (WA 1790445), 17 August 2016 (WA 2241187), 23 December 2016 (WA 2410028), 8 January 2017 (WA 2463935); subadult, 30 September 2008, Serra dos Tucanos (ML 51601501, photo M. Hopkins); two subadults, FLONA Tijuca, 27 January and 11 February 1967 (AMNH 803048/49); subadult, 18 February 2017, *c.*5.5km NNW of São João (ML 49620481, photo I. Thompson); subadult, 20 January 2013, Forte Marechal Luz (ML 33228111, photo F. Gorleri); subadult, 22 March 2012, Antonina (WA 590419, M. Cruz); subadult, 29 December 2011, Guaraqueçaba (WA 546878, E. Kaseker); subadult, 18 February 2017, Morretes (WA 2477894, I. Thompson); subadult, 13 August 1956, Tinguá (LACM 27813); subadult, 20 April 1957, Iguape (LACM 28990); subadult, 11 March 1957, Porto Estrada (YPM 100007, S.S. Snow photos); subadult, 2 May 1969, Rio Guaratuba (YPM 91701, S.S. Snow photos); subadult, 12 November 1960, Pousinho (FMNH 265226); subadult, 28 August 1966, Barra do Icapara (FMNH 344553); subadult transitioning, 11 July 1969, Iguape (FMNH 344554); two subadults transitioning, 13 August 1956, Tinguá (LACM 27814/15); subadult transitioning, 23 September 2011, Caraguatatuba (WA 455681, J. Mattos).

TECHNICAL DESCRIPTION The genders of all races of Black-cheeked Gnateater are distinguishable by plumage. Whitney (2003) pointed out that, despite statements to the contrary, *all* Black-cheeked Gnateater populations show a distinct black frontal band in males. He also, correctly, noted that the described variation in dorsal plumage coloration used to distinguish subspecies does not agree well with the presumed geographic limits of the respective races. The following is a 'general' descripton for the species, roughly matching most individuals, irrespective of race. *Adult Male* Bright rufous crown, usually with a pale orange edge, especially along the sides and front. They have a well-defined, glossy black mask that extends across the lores but varies in width and extends rearward to include the ear-coverts. It also extends from the malar line across the sides of the neck. Some males (older?) have a few of the typical *Conopophaga*, elongated, postocular white feathers. I have not seen any individuals with more than a couple of these and have seen no photographs of individuals with these flared out, as in other species. They have a bright, clean white throat, bordered above by the mask and below by a contrastingly dark grey breast. The back and rump are brown, with the feathers of the upper back thickly edged along the outer margins with black, creating a coarse, irregularly striped or scale-like pattern. The scapulars and lesser wing-coverts are warmer brown, sometimes somewhat rufous-brown, while the greater wing-coverts are bicoloured, dark brown on the posterior vane and warmer brown on the leading vane. Some adults have small whitish spots at the tips of a few of the greater primary-coverts and, in those individuals I have seen, some of these white areas have a small black spot centred near the tip of the feather shaft. Below, males are dark grey on the upper breast and sides, becoming warmer lower down and tinged orange-buff on the flanks. The central belly and vent are white, the amount varying considerably; a large patch in some individuals and a narrow stripe in others. **Adult Male Bare Parts** *Iris* dark brown; *Bill* black; *Tarsi & Toes* dusky-pinkish. **Adult Female** Lacks the broad black mask of males, it being replaced by a similarly shaped but less prominent mask of dusky brown. This is bordered above by a narrow grey postocular stripe, becoming paler and less defined above the eyes and on the lores. Just below this, the feathers of the lores and face are darkest, forming an indistinct eyestripe, which is more apparent in certain lights. The back and rump are brown, with some black patterning on the upper back as described for males. The crown is somewhat variable, being fairly bright orange-brown in some individuals, duller brown and similar in coloration to the upperparts in others, and even dull grey-brown in some. Below, females are largely orange-rufous, varying in saturation, darkest on the breast and sides, paler on the throat, central belly and vent. Sides and flanks washed dusky. There appears to be some variation in saturation between individuals, with the more pigmented individuals having brighter and less extensive pale areas. **Adult Female Bare Parts** Similar to males, legs perhaps somewhat browner. **Immature** A few key details of post-fledging development were provided by Lima (2008). By 40 days post-fledging, the sex of the young is recognisable, and by 60 days the young were nearly indistinguishable from adults. This latter observation suggests that immatures make the final transition from juvenile to subadult plumage sometime around 7–8 weeks after leaving the nest. As mentioned above, some males, which we have no reason to suspect are not fully adult, have white-spotted wing-coverts. I am fairly certain that these do represent natural variation and are not a sign of immaturity. Immature covert markings are distinctly buffy, while these other markings are clearly white or at most very pale buff or off-white. Interestingly, there appears to be a correlation with the presence of white spots and of white feathers behind the eye. The number of adult males with this plumage variation appears to be relatively small. Given the apparently low annual survival rates, it seems possible that these marked wing-coverts and white tufts might be acquired by older males. **Fledgling** Not, as far as I can tell, distinguishable as to sex, although most of the following description is based on photographs. I have not yet seen any fledglings in transition to juvenile plumage. The crown is rufous-brown, similar in coloration to some adult females, breaking into fine spotting posteriorly. It is separated by a distinct buffy to buffy-orange postocular stripe of variable extent. My guess is that this may be more prominent and more extensive in males, but this is not

certain. The spotting gradually merges into indistinct black barring on the back and rump. Greater and lesser upperwing-coverts and tertials distinctly and broadly tipped with buffy bars that are subtended by thin black margins. **Fledgling Bare Parts** *Iris* brown; *Bill* maxilla mostly black, paler along tomia, around nares, and at tip, the mandible is pale grey, yellow along the tomia and bright orange on the inflated rictal flanges; *Tarsi and Toes* pinkish to pinkish yellow. **Juvenile Male** These have the features that distinguish subadults more exaggerated (e.g., more mottling on face mask). Pale orange feathering on the crown is noticeably more extensive, especially on the lores and base of bill. These pale feathers often form an indistinct buffy postocular stripe that separates the crown from the mask. Some areas of the mask may be spotted with brown-rufous feathers. The upperwing-coverts and tertials, especially the greater coverts, are distinctly but thinly edged black. The hindcrown, nape, and sometimes the neck-sides are finely spotted brownish-rufous. The black striping of the back is generally reduced compared to adults. **Juvenile Male Bare Parts** *Iris* dark brown; *Bill* black, the maxilla usually with a pale tip, rictal flanges slightly inflated and bright orange or yellow-orange; *Tarsi and Toes* grey-pink, seemingly paler in younger individuals. **Juvenile Female** Similar to adults but have the face pattern 'messy' looking and retain varying amounts of fine spotting on the back of the head as described for males. In the few juvenile specimens I have examined, as described for males, the wing-coverts show more extensive buff and are narrowly edged black. This latter trait is perhaps the easiest way to distinguish juveniles from adults and subadults. **Subadult Male** Similar in overall pattern to adult males but *may* show a reduction in the number of back feathers edged in black. It is still unclear how much of this variation is age-related. The black mask is distinctly present but may still be slightly mottled with paler feathers. Incidentally, this is not a reliable character in specimens as ruffling of the feathers may create a mottled look, even in fully adult birds. The extent of paler orange feathers on the crown tends to be less than in full adults. The most marked differences are in the upperwing-coverts. The lesser coverts may be narrowly edged buff or already coloured as in adults. The greater wing-coverts and tertials are more broadly edged at their tips with buff, broadest on the primary-coverts. These tend to fade and be reduced to narrower margins with age, probably due to wear, prior to their replacement with adult feathers (see Immatures, below). **Subadult Male Bare Parts** Similar to adults. From *in situ* photographs of birds matching this description it appears that the rictal flanges may retain some orange coloration, but generally this appears to be lost by this age. **Subadult Female** Very closely resemble adults. Because adult females have buff-tipped wing-coverts, their distinction from subadults is not yet clear, although one good indicator is the continued presence of a thin black margin subtending each spot. This appears to disappear with wear and it is unclear how long these coverts are retained into adulthood. I believe that they are eventually replaced with fully adult feathers that have buff areas reduced to well-defined central spots rather than bands, and lack black margins. Most birds in this plumage also show some degree of 'messiness' to the face pattern **Subadult Female Bare Parts** As adults.

MORPHOMETRIC DATA *C. m. nigrifrons* *Bill* from nares 9.5mm, 8.8mm, 9.0mm, 10.0mm; *Tarsus* 27.2mm, 25.3mm, 27.0mm, 26.7mm (*n* = 4, ♀,♂,♂,♂, LACM). See also Berla

(1946). ***C. m. perspicillata*** *Wing* 65mm, 65mm; *Bill*, culmen 15mm, 15mm; *Tarsus* 27mm, 26mm (*n* = 2, ♂♀, *perspicillata*, Lima & Grantsau 2005). ***C. m. melanops*** *Bill* from nares 9.4mm, 9.9mm; *Tarsus* 27.6mm, 27.2mm (*n* = 2♀♀; LACM). *Bill* from nares 9.4 ± 0.2mm; *Tarsus* 28.2 ± 0.5mm (*n* = 6♂♂, 4 adult, 2 subadult; LACM). *Bill* from nares 9.0mm; *Tarsus* 27.0mm (*n* = 1, ♂, WFVZ 32609). *Bill* from nares 10.0mm, 9.7mm, 9.0mm, n/m, exposed culmen, 13.4mm, 13.5mm, 12.4mm, n/m, depth at nares 4.5mm, 4.3mm, n/m, 4.5mm, width at nares 5.3mm, 5.9mm, 5.5mm, 5.5mm, width at base of mandible 6.1mm, 6.2mm, 6.7mm, 6.7mm; *Tarsus* 25.8mm, 29.2mm, 28.2mm, 25.9mm (*n* = 4, 2♀♀, 2 subadult ♂♂, FMNH). *Bill* from nares 9.5 ± 0.5mm (*n* = 6), exposed culmen 12.8 ± 0.4mm (*n* = 6), depth at nares 4.7 ± 0.3mm (*n* = 4), width at nares 5.7 ± 0.4mm (*n* = 4), width at base of mandible 7.1 ± 0.2mm; *Tarsus* 27.7 ± 0.4 (*n* = 6) (adult ♂♂; FMNH). See also Novaes (1947). **Mass** Mean 20.1g (*n* = 3, race and gender not given, Whitney 2003); 20g, 27g (*n* = 2, ♂♀, *perspicillata*, Lima & Grantsau 2005). **Total Length** 10.5–12.0cm (Sick 1984, Whitney 2003, Lima & Grantsau 2005, van Perlo 2009, Ridgely & Tudor 2009).

TAXONOMY AND VARIATION The three subspecies provisionally recognised by Whitney (2003) are maintained here. Note, however, in the following text which summarises Whitney's (2003) observations, it is abundantly clear that the status and distributions of all races of *C. melanops* are in dire need of revision, especially given the high degree of genetic differentiation in some portions of their range (Lunardi 2004, Lunardi *et al.* 2007). Among the characters used to distinguish the various races of *C. melanops*, the width of the black frontal band on adult males has played a prominent role. Although there does appear to be a tendency for birds in the north of the range to have broader frontal bands than southern birds, this character is so variable, even among individuals from a single locality, that its importance for taxonomy is doubtful. Similarly, the coloration of the upperparts (both genders), the presence or absence of spots on the upperwing-coverts, and the extent of grey or white on the underparts, are all characters which do not vary predictably with geography, and thus are also of questionable taxonomic validity. Whitney (2003) considered that *perspicillata*, in particular, was of dubious validity, pointing out that Lichtenstein (1823) probably lacked a strong series of nominate *melanops* for comparison when describing *perspicillata*, making its distinction from the nominate race unclear. Race *nigrifrons* (northeast), given currently understood subspecific ranges, is somewhat geographically isolated from nominate *melanops* (south-east), potentially lending strength to its status as a separate taxon. Still, the morphological diagnosability of this population has not been tested, and Whitney (2003) pointed out that Pinto (1954) described *nigrifrons* based on comparisons with *perspicillata* from 'Baía,' a locality now known as Salvador on the north coast of Bahia. The type locality for *nigrifrons* is in south-eastern Alagoas, not so very far north of Salvador, and recent records from the intervening area in Sergipe may belong to either *nigrifrons* or *perspicillata*/*melanops*. To the south, the southern limit of putative *perspicillata* has never been explicitly defined. Indeed, Ruschi (1953) listed both *perspicillata* and *melanops* from Espírito Santo. Currently (and herein), records from Bahia are referred to *perspicillata* and those from adjacent Espírito Santo to the south as nominate *melanops*. Race

perspicillata may intergrade with *melanops* somewhere in this area or, perhaps, is best considered synonymous with nominate. In sum, neither *perspicillata* nor *nigrifrons* is clearly defined morphologically or geographically. Given the individual variability in plumage within populations, and the potentially continuous distribution of the species from *nigrifrons* in the north to nominate in the south, Whitney (2003) concluded, correctly, that only via careful study of long series of specimens, from the entire range of the species, will this confusing situation be resolved. Although not yet formally published, it appears that there is little support for continuing to recognise *perspicillata* (Pessoa & Silva 2003), and the following summaries of subspecific distributions should be considered provisional.

Conopophaga melanops nigrifrons Pinto, 1954a, Papéis Avulsos do Departamento de Zoologia, vol. 12, no. 1, p. 55, 'Mangabeira (Usina Sinimbu)' [Sinimbu, sea level, *c.*09°55'S, 36°08'W], south-eastern Alagoas, Brazil. Race *nigrifrons* was described from two adult females collected 29 September and 15 October 1951 at São Miguel dos Campos and Canoas (Rio Largo), respectively, as well as two adult males collected 22 October and 7 November 1952 at Mangabeira. The latter of these specimens (MZUSP 36417) in the Coleção Ornitológica do Departamento de Zoologia da Secretaria da Agricultura de São Paulo, is the designated holotype (Pinto 1954). The currently understood range of *nigrifrons* is restricted to the coast of north-east Brazil, from Rio Grande do Norte and Paraíba, south through Pernambuco to south-east Alagoas, making it endemic to the 'Atlantic slope of Alagoas and Pernambuco' EBA 071 (Stattersfield *et al.* 1998). This distribution should, however, be considered provisional. Apart from the confusing variation and conflicting accounts discussed above, Berla (1946) collected *C. melanops* at Dois Irmãos Recife (Pernambuco) that he considered to belong to *perspicillata*. More recently (Lima & Grantsau 2005, Lima 2006) specimens collected by Rolf Grantsau at Sauipe (northern Bahia) appear to be clearly assigned to *nigrifrons*. Pinto (1954) and most authors distinguish *nigrifrons* from other races by the greater extent of black at the base of the bill that connects the two sides of the mask. As observed by Whitney (2003), however, this appears to vary greatly, both within populations of *nigrifrons* as well as between populations of other races. However, the three topotypical male specimens I examined (LACM 26942–944) did all appear to have more black on the forecrown than most others I have seen. **Specimens & Records** ***Rio Grande do Norte*** RPPN Mata Estrela, 06°22.5'S, 35°00.5'W (Olmos 2003, Junior *et al.* 2008). ***Paraíba*** Mamanguape, 06°50'S, 35°07.5'W (WA 673685, M. Holderbaum); Cruz do Espírito Santo, 07°08'S, 35°05'W (WA 1446937, recording C. Brito); Santa Rita, 07°08'S, 34°58'W (WA 675540, F.A. Sonntag). ***Pernambuco*** Timbaúba, 07°31'S, 35°19'W (FMNH 392378/79); Mata do Estado, *c.*07°36'S, 35°29'W (Pereira 2009, FMNH 399263–268); 13.5km SW of Timbaúba, 07°36.5'S, 35°23.5'W (XC 126727, M. Braun); Paulista, 07°56'S, 34°52'W (WA 1017644, M. Menêzes); Dois Irmãos, 08°01'S, 34°56.5'W (Berla 1946, MNRJ 24689); RE Gurjaú, 08°14.5'S, 35°03'W (Lyra-Neves *et al.* 2004); Lagoa dos Gatos, 08°40'S, 35°54'W (WA 628615, S.J. Jones); RPPN Pedra D'Anta, 08°42'S, 35°51.5'W (ML 63352871, photo C. Gussoni); Água Preta, 08°42.5'S, 35°31'W (WA 1686127, recording F. Pacheco); RPPN Frei Caneca, 08°43.5'S, 35°50'W (XC 115156, A. Lees); Engenho Cachoeira Linda, 08°49'S, 35°29'W (XC 5611, G.A. Pereira). ***Alagoas*** Ibateouara, 08°58.5'S, 35°54.5'W

(FMNH 399269–278); RPPN Mata do Engenho Coimbra, 09°00'S, 35°52'W (Silveira *et al.* 2003); Usina Serra Grande, Engenho Coimbra, 09°01.5'S, 35°55.5'W (MPEG 70465/66, ML 127982/83, C.A. Marantz); Grotão do Brás, 09°13'S, 35°31'W (Silveira *et al.* 2003); Fazenda Bananeiras, 09°14'S, 35°52.5'W (XC 5389, R.C. Hoyer); Quebrangulo, 09°17.5'S, 36°27.5'W (WA 1127231, S.J. Jones); Murici and ESEC Murici, *c.*09°18'S, 35°56'W (Albano 2010a, XC, WA); Usina Santo Antônio, 09°23'S, 35°35'W (Silveira *et al.* 2003); Canoas, 09°30'S, 35°45'W (Pinto 1954a); Marechal Deodoro, 09°42'S, 35°53.5'W (WA 855237, F.B.R. Gomes); Usina Porto Rico, 09°46'S, 36°14'W (Araújo *et al.* 2008, Lobo-Araújo *et al.* 2013); São Miguel dos Campos, 09°47'S, 36°05.5'W (Pinto 1954a); Usina Coruripe, 10°02'S, 36°16.5'W (ML 128071, C.A. Marantz); Mata do Riachão, 10°03'S, 36°16'W (Silveira *et al.* 2003).

Conopophaga melanops perspicillata (M. H. C. Lichtenstein, 1823), Verzeichniss der Doubletten des Zoologischen Museums der Königl, p. 43, 'Bahia.' Described as *Myiothera perspicillata*, this race, if valid, is restricted to eastern Bahia (Naumburg 1937, Freitas *et al.* 2007, Freitas 2011) with the recent records from adjacent Sergipe provisionally included following Whitney (2003). SPECIMENS & RECORDS *Sergipe* Aracaju area, 10°56.5'S, 37°04'W (eBird: G. Ziarno); Santa Luzia do Itanhy, 11°21'S, 37°27.5'W (WA 506908, M.C. Sousa). *Bahia* Baixio, 12°08'S, 37°42.5'W (Naumburg 1937); Cajazeiras, 12°54'S, 38°24.5'W (Naumburg 1937, AMNH 242773); Valença, 13°22'S, 39°04'W (WA 399457, S.S. Santos); near Camamú, *c.*13°56.5'S, 39°06.5'W (MNRJ 38199); Ibirataia, 13°57'S, 39°40'W (WA 2307060, recording A. Cafeseiro); Fazenda Agua Boa, 14°15.5'S, 39°11'W (C. Albano *in litt.* 2017); Boa Nova, 14°22'S, 40°12'W (WA 2425998, J. Filho); Uruçuca, 14°31'S, 39°18'W (WA 2085268, I. Camacho); near Ilhéus, 14°47.5'S, 39°03'W (MNRJ 22321/25167); Fazenda Fausto, *c.*15°00'S, 39°06'W (XC 67982, F. Lambert); Una, 15°12.5'S, 39°12'W (WA 149895, recording B. Rennó); Rio Verruga, *c.*15°16'S, 40°37'W (AMNH 178150); Macarani, 15°34'S, 40°25.5'W (XC 339546/50, R. Souza); RPPN Estação Veracel, 16°23.5'S, 39°10'W (C. Albano *in litt.* 2017); *Minas Gerais* Bandeira, 15°53'S, 40°34'W (WA 1594644, C. Albano); Jequitinhonha, 16°27'S, 41°02'W (WA 2380786, M.B. Lima).

Conopophaga melanops melanops (Vieillot, 1818), Nouveau dictionnaire d'histoire naturelle, vol. 27, p. 14, 'l'Amérique méridionale' [= South America]. Although Vieillot (1818) was not specific about the type locality, Cory & Hellmayr (1924) reported that the type, examined by Hellmayr in the Paris Museum, is a male collected by Delalande, Jr., in the vicinity of Rio de Janeiro. The nominate race is confined to south-eastern Brazil, from Espírito Santo and eastern Minas Gerais, south through coastal Rio de Janeiro, São Paulo, Paraná and Santa Catarina, including Ilha Grande and Ilha do Santa Catarina. SPECIMENS & RECORDS *Espírito Santo* REBIO Sooretama, *c.*19°01'S, 40°05'W (ML 113350, C.A. Marantz, ML 63749931, photo L. Merçon); Córrego do Cupido, 19°03'S, 39°57'W (MNRJ 39595); RPPN Vale do Rio Doce (Linhares), 19°09'S, 40°04'W (Stotz 1993, ML 115358/59, C.A. Marantz); Lagoa Juparanã, 19°17'S, 40°08'W (Naumburg 1937, AMNH 317473/74, MNRJ 16477–480); Linhares and vicinity, *c.*19°23.5'S, 40°04'W (MNRJ 27280, MNRJ 26456/60, XC 128921/24, N. Eiterer); Baixo Guandu, 19°31'S 41°01'W (Naumburg 1937, AMNH 317475–477); 'Pau Gigante' [= Ibiraçu; Paynter & Traylor 1991b], 19°50'S, 40°22'W (USNM

368292/93, MNRJ 19229/30); *c.*7.5km E of Laranja da Terra, 19°53.5'S, 40°59.5'W (Raton & Gomes 2015); near Santa Cruz, 19°57.5'S, 40°09.5'W (MNRJ 28104); ESEC Santa Lúcia, *c.*19°58'S, 40°32'W (MNRJ, XC 336119, O. Vieira da Fonseca); Jatiboca, 20°05'S, 40°55'W (MNRJ 25928); Santa Leopoldina, 20°06'S, 40°32'W (MCZ 273674); Domingos Martins, 20°22'S, 40°39.5'W (MNRJ MNA5625/5599); Marechal Floriano, 20°26.5'S, 40°46.5'W (WA 2075928, E.S. Neves); near Cachoeira de Itapemirim, 20°49'S, 41°10'W (ML 34487681, photo D. Czaplak); Serra das Torres, 21°00'S, 41°13'W (Simon *et al.* 2008). *Minas Gerais* São Benedito, 19°30'S, 41°16'W (Naumburg 1937); Caratinga, 19°47'S, 42°08.5'W (WA 438143, C. Zaparoli); Raul Soares, 20°06'S, 42°27'W (LACM 30228); Rio Doce, 20°14.5'S, 42°54'W (ML 43301, T.A. Parker); Tombos, 20°54'S, 42°02'W (WA 2012928/31, E. Almeida); Itamarati de Minas, 21°25'S, 42°50'W (WA 420383, F. Guimarães); REBIO Municipal Poço D'Anta, 21°45.5'S, 43°18.5'W (XC 164115, M. Manhães); Além Paraíba, 21°50'S, 42°44'W (WA 1884023, F. Medeiros). *Rio de Janeiro* RE Guapiaçu, *c.*22°25'S, 42°44'W (XC, ML); Teresópolis, 22°25'S, 42°58.5'W (WA 270413, P. Tinoco); Serra dos Tucanos Lodge, 22°26.5'S, 42°38.5'W (ML 45410401/47385401, photos N. Voaden); Hotel Donati, PN Itatiaia, 22°26.5'S, 44°36'W (XC 108739, A. Silveira); PN Serra dos Órgãos, 22°28'S, 43°00'W (ML 63897, D.W. Finch); Guapimirim, 22°32'S, 42°59.5'W (WA 2422598, C. Belleza); BR-40, km90 near Petrópolis, 22°33'S, 43°14.5'W (MNRJ MNA8661); Tinguá, 22°36'S, 43°26'W (LACM 27813–815, MNRJ 33161); São Pedro da Aldeia, 22°50'S, 42°06'W (WA 1708432, D. Steinwender); Lídice, 22°50'S, 44°11.5'W (MNRJ MNA5057); Mazomba, *c.*22°51.5'S, 43°14'W (MPEG 26649); PE Cunhambebe, 22°55.5'S, 44°00'W (ML 64045861, photo A. Mesquita); Vila Muriqui, 22°55'S, 43°57'W (LACM 59515–517); Mangaratiba, 22°56.5'S, 44°02.5'W (MNRJ MNA8410); FLONA Tijuca, *c.*22°57'S, 43°17'W (USNM, AMNH, CM, ANSP, MPEG, MNRJ); Horto Florestal, 22°58'S, 43°14.5'W (MNRJ, ML 33344481, photo A. Lees); Angra dos Reis, 23°00.5'S, 44°19'W (WA 1235604, C. Kleske); Ilha Grande, 23°11'S, 44°12'W (Alves *et al.* 2002, Vecchi & Alves 2015); near Paraty, 23°19.5'S, 44°41'W (MNRJ, XC 347518, G. Leite). *São Paulo* Fazenda Bacury, 22°41'S, 48°07'W (ML 25851671, T.V.V. Costa); Anhembi, 22°47'S, 48°08'W (WA 1637742, G. Panucci); near Tremembé, 22°57'S, 45°41'W (XC 118072, R. Dela Rosa de Souza); Taubaté, 23°01'S, 45°33'W (WA 256882/87, R. Valério); Natividade da Serra, 23°23.5'S, 45°26.5'W (WA 833905, E. Kaseker); Fazenda Angelim, *c.*23°23.5'S, 45°04'W (XC 252847, J. Minns, ML 47875601/611, K. Hansen); Itamambuca Eco Resort, 23°24'S, 45°01'W (ML 27202391/511, photos A. Lees); Ubatuba, 23°26'S, 45°05'W (USNM 561302, WA); Sítio Folha Seca, *c.*23°28'S, 45°10'W (ML 20122241, photo D. Slager, ML 113396/97, C.A. Marantz); Salesópolis, 23°32'S, 45°51'W (WA 2339664, L. Cardim); Caraguatatuba, 23°37'S, 45°25'W (T. Carneiro *in litt.* 2015); Ibiúna, 23°39.5'S, 47°13.5'W (WA 2102790, L. Frare); Condomínio Morada da Praia, 23°43'S, 45°52.5'W (XC 20342, G.R.R. Brito); Rio Guaratuba, 23°45'S, 45°55'W (LSUMZ, KUNHM, YPM); São Sebastião, 23°48'S, 45°27'W (AMNH, ROM, USNM); Ilhabela, 23°50'S, 45°21'W (Olmos 1996, XC 163562/663, E.D. Schultz); São Miguel Arcanjo, 23°53'S, 48°00'W (WA 400615, S. Salvador Jr); Guarujá, 24°00'S, 46°15'W (WA 2060680, M. De Castro); RPPN Parque do Zizo, 24°01'S, 47°49'W (XC 4314/15, B. Planqué); Mongaguá, 24°06.5'S, 46°41'W (WA 2115619, L. Souto); Ribeirão Grande, 24°12.5'S, 48°21.5'W (WA

1321270, R.G. Lebowski); Miracatu, 24°16.5'S, 47°28'W (WA 1604159, E. Kaseker); Ana Dias, 24°18'S, 47°03'W (XC 347852, R. Silva e Silva); Porto Estrada, 24°19'S, 47°51'W (YPM 100007); Pousinho (Barra do Rio Juquía), 24°22'S, 47°49'W (FMNH 265225/26); Guaraú, 24°22'S, 47°01.5'W (ML 30002311, photo F. Barata); Boa Vista, 24°35'S, 47°38'W (FMNH 265221); Iporanga, 24°35'S, 48°36'W (WA 659093, J. Petar); Embu, 24°38'S, 47°25'W (FMNH 344555); Icapara (Barra de Icapara), 24°40.5'S, 47°28'W (FMNH 344549–553); Iguape, 24°43'S, 47°33'W (LACM 28990, FMNH 344554); Cananéia, 25°01'S, 47°56'W (WA 1741938, V. Wruck). *Paraná* RPPN Salto Morato, 25°11'S, 48°18'W (XC 172519, J.A.B. Vitto); Guaraqueçaba, 25°14.5'S, 48°19'W (WA 546879, E. Kaseker); *c*.5.5km NNW of São João, 25°21.5'S, 48°53'W (ML 49620481, photo I. Thompson); Quatro Barras, 25°22'S, 48°59'W (WA 947714, H. Neto); Antonina, 25°26'S, 48°43'W (WA 590419, M. Cruz); Porto de Cima, 25°28'S, 48°50'W (Straube 1989); Morretes, 25°29'S, 48°50'W (WA 2477894, I. Thompson); Paranaguá, 25°31'S, 48°31'W (WA 2264291, E. Modesto); RPPN Bicudinho-do-brejo, 25°45.5'S, 48°43.5'W (ML 46110571, photo C. Gussoni); Guaratuba, 25°53'S, 48°34'W (MNRJ 38413, WA 602520, L. Milano). *Santa Catarina* Garuva, 26°02'S, 48°51'W (WA 760327, L. Moraes); Itapoá, 26°05'S, 48°39.5'W (WA 450224, S. Licco); Forte Marechal Luz, 26°10'S, 48°31.5'W (ML 33228111, photo F. Gorleri); Joinville, *c*.26°18.5'S, 48°51'W (WA: V.E. Florencio); Guaramirim, 26°29'S, 48°56'W (WA 1328161, I. Ghizoni-Jr.); *c*.3km NW of Timbó, 26°48'S, 49°17'W (XC 174269, D. Meyer); Blumenau area, *c*.26°55'S, 49°04.5'W (Müller *et al.* 2003, WA: many photos); Balneário Camboriú, 26°59'S, 48°39'W (XC 192729, R.B. Stringari); Ibirama, 27°01'S, 49°33'W (WA 2204962, A. Bernardi); Lajeado Alto, 27°07'S, 49°05'W (XC 30568, A.E. Rupp); RPPN Prima Luna, 27°15.5'S, 49°01'W (XC 179692/93, G. Köhler); Biguaçu, 27°30'S, 48°39'W (WA 1845779, F. Llanos); Ilha de Santa Catarina, *c*.27°35'S, 48°28.5'W (WA 2059278, M. Moreto); Florianópolis, 27°35.5'S, 48°33'W (WA 1523587, L. Motta); Santo Amaro da Imperatriz, 27°41'S, 48°47'W (WA 1157667, S. Leal); Matadeiro, 27°45.5'S, 48°29.5'W (ML 43506991, photo J. Wioneczak); Garopaba, 28°02'S, 48°37'W (WA 676102, J. Battistella); São Martinho, 28°06.5'S, 49°01'W (WA 1575240, A. Bianco); Imaruí, 28°21'S, 48°49'W (WA 1262492, E. Inacio).

STATUS Black-cheeked Gnateater is usually common to fairly common at well-sampled locations (Silva *et al.* 2008, Dario & Vincenzo 2011), especially so in the Ubatuba region of north coastal São Paulo and the Guaraqueçaba area of Paraná (Whitney 2003). As a Brazilian Atlantic Forest endemic, however, it is most certainly threatened to some degree by deforestation (Brooks *et al.* 1999). Black-cheeked Gnateater is currently evaluated as Least Concern (BirdLife International 2017) but subspecies *nigrifrons* is considered Vulnerable in Brazil (Roda 2008). Black-cheeked Gnateater, however, not only tolerates second growth (Aleixo 1999), but appears to flourish and nest in it, provided it is fairly dense. This has allowed it to persist in locations that other Atlantic Forest endemics have been expatriated from, even within residential neighborhoods inside the city of Rio de Janeiro (Nogueira *et al.* 2005). PROTECTED POPULATIONS *C. m. nigrifrons* ESEC Murici (XC 80425, J. Minns); RPPN Frei Caneca (Mazar Barnett *et al.* 2005, A. Lees *in litt.* 2017); RPPN Jaqueira (XC 80426, J. Minns); RE Murici (Whitney 2003); RPPN Mata Estrela (Olmos 2003, Junior *et al.* 2008); RE Gurjaú

(Lyra-Neves *et al.* 2004, Telino-Júnior *et al.* 2005); RPPN Pedra D'Anta (ML 63352871, photo C. Gussoni); RPPN Engenho Gargaú (eBird: A.L. Roos); RPPN Mata do Engenho Coimbra (Silveira *et al.* 2003). *C. m. perspicillata* PN Boa Nova (D. Beadle *in litt.* 2017); RPPN Serra Bonita (C. Albano *in litt.* 2017); REBIO de Una (Gonzaga *et al.* 1987, MPEG 70732/33); RPPN Serra do Teimoso (eBird: R. Laps); RPPN Mata do Passarinho and RPPN Estação Veracel (C. Albano *in litt.* 2017); probably occurs in RVS Mata dos Muriquis. *C. m. melanops* PN Serra dos Órgãos (Mallet-Rodrigues *et al.* 2010, ML 63897, D.W. Finch). PN Superagüi (eBird: F. Farias); PN Serra do Itajaí (XC 30568, A.E. Rupp); PN Itatiaia (XC 62230, H. van Oosten); PN Serra da Bocaina (eBird: B. Rennó); FLONA Tijuca (USNM, AMNH, CM, ANSP, MPEG); PE Intervales (Aleixo & Galetti 1997, Pizo & Melo 2010); PE Sete Barras (Willis & Oniki 1981); PE Serra do Mar (XC 34121, M. Melo); PE Ilhabela (XC 163562/663, E.D. Schultz); PE Pedra Branca (eBird: J. Faragher); PE Ilha Grande (Marsden *et al.* 2003); PE Três Picos (Mallet-Rodrigues & Noronha 2009); PE Intervales (XC 256275, P. van Els); PE Serra do Tabuleiro (eBird: F. Farias); PE Cunhambebe (ML 64045861, photo A. Mesquita); PE Jurupará (eBird: F. Arantes); PE Alto Ribeira (Petar) (eBird: A.E. Rupp); PE Xixová-Japuí (eBird: M.H. Achado); RPPN Vale do Rio Doce, Linhares (Stotz 1993); RPPN Prima Luna (XC 179692/93, G. Köhler); RPPN Salto Morato (Straube & Urben-Filho 2005a, Lima 2007, Lima & Roper 2009b, Scherer-Neto *et al.* 2011); RPPN Parque do Zizo (XC 4314/15, B. Planqué); RPPN Feliciano Miguel Abdala (eBird: F. Tavares); RPPN Cafundo (eBird: N.P. Dreyer); RE Guapiaçu (XC, ML); RPPN Bicudinho-do-brejo (ML 46110571, photo C. Gussoni); PNM São Francisco de Assis (Müller *et al.* 2001, 2003); PNM Lagoa do Peri (eBird: A.L. Roos); PNM Nascentes de Paranapiacaba (eBird: I. Marques); ESEC Juréia-Itatins (Lima 2011, IBC, XC); ESEC Santa Lúcia (XC 336119, O. Vieira da Fonseca); REBIO Guaribas (eBird: A.L. Roos); REBIO Sooretama (Scott & Brooke 1985, ML); REBIO Tinguá (XC 317194, O. Vieira da Fonseca); REBIO Municipal Poço D'Anta (Scott & Brooke 1985, Manhães & Loures-Ribeiro 2011); APA Anhatomirim (XC 45586, A.E. Rupp).

OTHER NAMES There are quite a few confusing synonyms in the literature, both of scientific and common names. *Conopophaga dorsalis* [*melanops*] (Ménétriés 1835, Burmeister 1856, Ogilvie-Grant 1912). According to Sclater (1858c), *C. dorsalis* was described on the basis of a single female specimen, and the two authors cited were unaware that it was the female of previously described *C. melanops* (Lichtenstein 1823). Interestingly, Sclater (1890) later reversed his decision and considered it a full species. Cory & Hellmary (1924), however, found that Sclater's (1858c) first supposition was correct, discovering that specimens labelled male *C. dorsalis* were actually *C. m. perspicillata*, and those labelled as females were nominate *melanops*. Additional synonyms: *Conopophaga Maximiliani* [*melanops*] (von Pelzeln 1871, Koenigswald 1896); *Conopophaga nigrigenys* [*melanops*] (Nehrkorn 1910, Ogilvie-Grant 1912); *Conopophaga nigrogenys* [*melanops*] (Lesson 1831, Ménétriés 1835, von Ihering 1900); *Conopophaga melanops nigrigenys* [*melanops*] (von Ihering 1902); *Conopophaga ruficeps* [*melanops*] (Swainson 1941); *Conopophaga melanops perspicillata* [*nigrifrons*] (Berla 1946); *Conopophaga melanops perspicillata* [*melanops*] (Ruschi 1967); *Conopophaga perspicillata* [*perspicillata*] (Brabourne & Chubb

1912, Cory & Hellmary 1924); **English** Black-fronted Gnat-Eater [*melanops*] (Brabourne & Chubb 1912); Spectacled Gnat-Eater [*perspicillata*] (Brabourne & Chubb 1912, Cory & Hellmary 1924); Rufous-crowned Flatbill [*melanops*] (Swainson 1941). **Portuguese** chupa-dente-de-máscara (Willis & Oniki 1991b, Rosário 1996); cuspidor-de-máscara preta (Sick 1984); chupa-dente enegrescido [*melanops*], Chupa-dentre (*sic*) do Mucurí [*perspicillata*] (Ruschi 1967). **Spanish** Jejenero carinegro (Whitney 2003). **French** Conophage à joues noires (Whitney 2003); Le Platyrhynque a joues noires 'The Black-cheeked Flycatcher' (Vieillot 1818). **German** Rotscheitel-Mückenfresser (Whitney 2003).

Black-cheeked Gnateater, male (*melanops*) consumed by snake (*Tropidodryas serra*, Colubridae), Morretes, Paraná, Brazil, 18 April 2015 (*Marcio R. Varchaki*).

Black-cheeked Gnateater, female (*melanops*) carrying prey for nestlings, Caraguatatuba, São Paulo, Brazil, 20 December 2014 (*Thiago Carneiro*).

Black-cheeked Gnateater, male (*perspicillata*), Bandeira, Minas Gerais, Brazil, 28 January 2015 (*Ciro Albano*).

Black-cheeked Gnateater, fledgling (*melanops*), Teresópolis, Rio de Janeiro, Brazil, 29 December 2015 (*Paulo Tinoco*).

Black-cheeked Gnateater, adult female (*melanops*), Macaé, Rio de Janiero, Brazil, 18 August 2013 (*Paulo Tinoco*).

Black-cheeked Gnateater, adult male (*melanops*), Macaé, Rio de Janiero, Brazil, 18 August 2013 (*Paulo Tinoco*).

CHESTNUT-BELTED GNATEATER
Conopophaga aurita   Plate 2

Turdus auritus J. F. Gmelin, 1789, Systema Naturae, vol. 1, pt. 2, p. 827, 'Cayenne' [French Guiana]. The specific name should be spelt *aurita*, not *auritus*, to conform to ICZN gender-agreement rule.

'One male shot in April at Archidona, Eastern Ecuador, among the thick undergrowth on the river-bank. It was disputing with, or seemed to be disturbed by, the presence of a male *Chiromachaeris gutturosa*, and both were shot together. With lowered head, it kept opening and shutting its wings very rapidly, making a curious whirring sound all the time.' – **Walter Goodfellow, 1901, Archidona, Ecuador**

Chestnut-belted Gnateater occupies the rainforest understorey in north-west and central Amazonia, from east of the Andes in Colombia, Ecuador and Peru, east through central Brazil to the Guainas. The English name for the species refers to the belt-like chestnut feathering separating the breast from the belly (reduced or absent in races *snethlageae* and *pallida*). Males of all subspecies have a black throat and mask, and a bright white postocular tuft that flares out in agitation or excitement, and to which the specific name refers (from Latin: *auris* = ear; Jobling 2010). Females are quite different, lacking the black mask and throat. In Ecuador and north-east Peru, Chestnut-belted Gnateater is one of two allopatric lowland Amazonian gnateaters, with Chestnut-belted restricted to the east and north banks of the Río Napo, and Ash-throated Gnateater found on the west and south banks. Further south, in Peru south of the Río Amazonas, however, the ranges of these two species apparently overlap. Chestnut-belted Gnateaters forage on insects, mainly on the ground, and are apparently highly territorial. The vocalisations of southern populations also vary markedly, with songs exhibiting abrupt changes in pace and structure across the Rio Madeira/Tapajós interfluvium (Whitney 2003). In addition to the taxonomic complexities, the natural history of Chestnut-belted Gnateater remains largely unstudied. Although some information has been published on its reproductive biology, there have been no detailed studies of breeding behaviour, parental care or habitat requirements.

IDENTIFICATION 11.5–13.0cm. The following generally applies to nominate *aurita* and follows Whitney (2003) and Greeney (2014). Male Chestnut-belted Gnateaters are distinctive, with a dark, deep chestnut crown and a long white postocular tuft. When flared, this tuft contrasts strikingly with the black forecrown, face and throat. The back, wings and tail are olive-brown, with the feathers of the lower back (somewhat variably) finely edged in black to give a coarsely scaled appearance. A bright orange-chestnut 'belt' crosses the upper breast above their whitish belly, giving rise to their English name. Females are similar on the back and lower underparts, but the lores and ocular area are grey and blend posteriorly into a less-prominent postocular plume. They lack the black on the head and throat, with the head-sides being brown and the throat pale rufous to buffy-white. Male Chestnut-belted Gnateaters are unmistakable. Females also are unlikely to be confused, but in the south-west of their range female Ash-throated Gnateater is similar.

Ash-throated females, however, have pronounced black scalloping on their backs and prominent buffy spotting on the wing-coverts.

DISTRIBUTION Chestnut-belted Gnateater is resident from the Guianas (Donahue & Pierson 1982) south to south-east Amazonian Brazil and west to eastern Colombia, north-east Ecuador and eastern Peru (Peters 1951, Pinto 1978, Whitney 2003, Ridgely & Tudor 2009). Although still unreported, it seems possible that it may also occur in adjacent Venezuela (Meyer de Schauensee & Phelps 1978, Hilty 2003). In Peru there are apparently no records west (left bank) of the Río Ucayali, and in eastern Ecuador it is considered to be absent from south (right bank) of the Río Napo (Ridgely & Greenfield 2001). This creates a large gap in distribution in western Amazonia, between subspecies *occidentalis* and *australis*. There are, however, two specimens labelled as having been collected from this region: MECN 2634, ♀, Río Bobonaza (02°35'S, 76°38'W); FMNH 99579, ♂, Río Conambo (02°07'S, 76°03'W). I confirmed the identity of these two skins and examined the original labels, and must conclude that they are either mislabelled or that Chestnut-belted Gnateater does occur there, albeit at very low density. I was unable to conclusively assign these two records to subspecies, but they seem to represent either *occidentalis* or *inexpectata*. A recent record of Chestnut-belted Gnateater (presumably *australis*) from extreme south-west Acre, Brazil (Aleixo & Guilherme 2010), suggests that the species potentially occurs in northern Bolivia. There is also an unexplained gap in records between the lower reaches of the Rios Tapajós and Madeira, a region that might potentially be inhabited by either subspecies *australis* or *snethlageae*. A single record in this region, from the vicinity of Juruti (*c.*02°09'S, 56°05'W; Santos *et al.* 2011a), is based on vocalisations (not recorded) and best considered hypothetical (Alexandre Aleixo *in litt.* 2017).

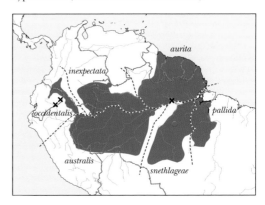

HABITAT Chestnut-belted Gnateater inhabits the interior of tropical lowland forest, most frequently in well-drained, tall *terra firme* forest with fairly dense understorey plants (Cohn-Haft *et al.* 1997, Reynaud 1998, Borges *et al.* 2001, Alvarez *et al.* 2003, Cintra *et al.* 2007, Beja *et al.* 2010), but occasionally in flooded forests (Braun *et al.* 2000). Whitney (2003) felt that this species avoids the densest thickets, or is not especially attracted to them, but Guilherme & Borges (2011) found *australis* in western Brazil (Acre) using tall *campinarana* habitats within the low, scrubby, tangled *campina*. In one study in eastern Pará (aurita), surveys across a disturbed gradient failed to find them outside primary forest (Barlow *et al.* 2007). The species is generally found below 300m, but may occur to 700m (Clements & Shany 2001, Ridgely & Tudor 2009) or 1,000m (Estación de Bombeo Guamues, FMNH 292917, race *occidentalis*), and nominate *aurita* was recently found at 1,300m in Guyana (Barnett *et al.* 2002).

MOVEMENTS Chestnut-belted Gnateaters are probably largely sedentary, but may display small-scale shifts of breeding territories within a somewhat larger home range (Stouffer 2007).

VOICE The song is a brief, 2–3s series of rapid, evenly paced, near-identical notes delivered at *c.*3kHz (Whitney 2003). Also described as a dry, chattery rattle or slow trill of 30–40 notes (Hilty 2003) or a liquid trill of even-paced notes at the same pitch (Spaans *et al.* 2016). Whitney (2003) observed that the songs of subspecies *snethlageae* and *pallida* in eastern Brazil are delivered at a significantly slower rate, the individual notes being about twice as long in duration as those produced by other taxa. Hilty (2003) described call notes of birds in north-east Peru (*occidentalis?*) as a weak, rustling chief and also a two-part, flat chat'up, chat'up given irregularly. Willis (1985) described a female *australis* giving sneeze-like kiff and cough sounds while performing a broken-wing distraction display. Fledglings apparently make soft, tinny *li-i-I* chitters and faint, descending *wee we we weed* begging calls (Willis 1985a). Male Chestnut-belted Gnateaters, apparently only in flight, produce a fairly loud whirring (Goodfellow 1901). This behaviour has not been studied in detail, but is presumably generated by fluttering the remiges. Whitney (2003) suggested that it is associated with aggression or courtship.

NATURAL HISTORY Overall, the general behaviour is poorly documented. The species forages alone or in pairs, near or, very occasionally, on the ground (Davis 1953, Peres & Whittaker 1991, Thiollay & Jullien 1998, Cintra *et al.* 2007), capturing prey with short reaches and gleans from foliage and leaf litter (Willis 1985a, Whitney 2003). It opportunistically follows army ant swarms in Colombia, Ecuador and Brazil, usually alone or in pairs (Willis 1985a). While following ants, adults perch briefly crosswise on vertical stems or vertically for longer periods on horizontal branches or logs, frequently dropping to the ground among the ants to seize prey (Willis & Oniki 1972, Willis 1985a). Willis (1985a) noted adults periodically flicking their wings while perched and searching for prey. The only mention of the species foraging in mixed-species flocks, other than with ants, is Willis' statement that they do occasionally join flocks (Hilty 1975), but it is unclear if he was referring to army ant flocks or not. Like other congeners, Chestnut-belted Gnateaters are territorial, and studies of the nominate race suggest that 41% of territories occupied in a given year are occupied the following year (Stouffer 2007). Territory size of nominate *aurita* in northern Brazil is estimated at 6.3ha (Stouffer 2007, Johnson *et al.* 2011). A study in French Guiana (*aurita*) suggests that Chestnut-belted Gnateater rarely joins mixed-species flocks (Thiollay & Jullien 1998), as noted by other authors (Peres & Whittaker 1991, Cintra *et al.* 2007). Willis (1985) once observed a Rufous-capped Antthrush *F. colma* displace a Chestnut-belted Gnateater at an antswarm, the gnateater in turn displacing a Common Scale-backed Antbird *Willisornis poecilinotus*. There are no published data on interactions with predators, but Glowska

& Schmidt (2014) described a new species of feather mite from *aurita* in Guyana.

DIET Chestnut-belted Gnateaters, like congenerics, probably mostly consume small arthropods (Schubart *et al.* 1965, Whitney 2003). The contents of the stomach of one individual included centipedes (Chilopoda) and ants (Formicidae) (Schubart *et al.* 1965).

REPRODUCTION Several nests have been described and recent papers have added significant details concerning adult behaviour but, overall, the nesting of Chestnut-belted Gnateater is poorly known. **NEST & BUILDING** There is no published information on the behaviour or process of nestbuilding. Whitney (2003) described nests in French Guiana (*aurita*) as: 'situated among the multiple thin trunks of a small understorey palm; another was open cup of blackish-violaceous fungus fibres 7.5cm wide, 2.8cm deep, built into pile of dead leaves accumulated on a large leaf of the palm Attalea attaleoides, *c.*80cm above ground.' Most recently, a nest in Brazil (Leite *et al.* 2012), was 56cm above ground, supported by epiphytic ferns growing on the trunk of a small, understorey tree. This nest was lined with thin, dark rootlets, with the external portion made of dead leaves and thin sticks. Internally, the egg cup was 5.8cm in diameter and 3.2cm deep. **EGG, LAYING & INCUBATION** Whitney (2003) also described a clutch of two eggs as 'pale ochre with pinkish-brown ring at large end' and provided measurements: 22.2 × 17.5mm, 22.9 × 17.5mm. **NESTLING & PARENTAL CARE** Both sexes have been seen attending fledglings (Willis 1985a, Tostain *et al.* 1992). The broken-wing display of adults at nests can be elaborate and insistent. Both sexes moved back and forth across the ground, dragging their wings and vocalising, near a nest with an older nestling (Leite *et al.* 2012). The male apparently was somewhat less bold than the female. During this injury-feigning display, the photographs in Leite *et al.* (2012) clearly show the female flaring her white postocular feathers, suggesting another function for these. **SEASONALITY** The breeding season of *aurita* in French Guiana is said to be March to April (Tostain *et al.* 1992). Data from north of Manaus (below) suggest that most nesting occurs October–March. A male and a female were collected with enlarged gonads on 5 and 11 1976 in French Guiana (*aurita*; Dick *et al.* 1984). Hellmayr (1910) reported an 'immature' male collected 6 November 1907 from the Rio Madeira (*australis*). **BREEDING DATA** *C. a. aurita* Laying (Tostain *et al.* 1992), 7 April 1992, Savane du Galion; nest with eggs (Olivier Tostain *in litt.* 2015), 8 September 2007, Montagne Cottica; older nestling (Leite *et al.* 2012), 10 December 2010, BDFFP site; pair performing distraction display (Oniki & Willis 1982, Willis 1985a), 8 February 1974, Reserva Florestal Adolpho Ducke; fledgling (Tostain *et al.* 1992), 15 April 1986, Montagne Bellevue de l'Inini; fledgling (Oniki & Willis 1982, Willis 1985a), 13 July 1973, Reserva Florestal Adolpho Ducke; juvenile (Tostain *et al.* 1992), 16 April 1986, Montagne Bellevue de l'Inini; juvenile (USNM 515629), 6 June 1966, Serra do Navio; nine juveniles (BDFFP database; Phillip Stouffer *in litt.* 2015), 17–18 January 1991 (*n* = 3), 3 March 1982, 9 April 1987, 4 June 1986, 14 October 1987, 17 October and 21 November 2000, BDFFP site; subadult (USNM 625231), 25 August 1998, Sipu River. *C. a. occidentalis* ♀ with two ruptured follicles (LSUMZ 83360), 4 March 1976, Limoncocha; ♀ with two developing eggs, 11 × 11mm, 8 × 8mm (LSUMZ 83365), 11 October 1975, Limoncocha; ♀ with two developing eggs, 9 × 9mm, 4 × 4mm (LSUMZ

83362), 14 August 1976, Limoncocha; nest with eggs (José Hualinga *in litt.* 2003), 15 August 2003, La Selva Lodge; juvenile transitioning (LSUMZ 83357, Matthew L. Brady *in litt.* 2017), 2 December 1975, Limoncocha; subadult (Caroline Dingle *in litt.* 2015), 25 January 2000, Pañacocha; subadult (FMNH 185274), 9 December 1946, Río Cotapino; subadult (MLZ 7602), 12 April 1927, lower Río Suno; subadult (FMNH 287191), 23 October 1969, San Antonio del Guamues; two subadults (LSUMZ 83359/61, Matthew L. Brady *in litt.* 2017), 11 February and 20 May 1976, Limoncocha; subadult transitioning (FMNH 293242), 26 March 1971, San Antonio del Guamues. **C. a. inexpectata** Subadult (FMNH 183851), 15 July 1935, Codajás; subadult (FMNH 457235), 30 July 2007, Maraã: **C. a. australis** Laying female (Robbins *et al.* 1991, LSUMZ 115605), 26 July 1983, Quebrada Vainilla; 'juvenile' roosting with adult (O'Shea *et al.* 2015), 14–20 October 2014, Quebrada Yanayacu; subadult (YPM 29683, S.S. Snow *in litt.* 2017), 21 December 1921, 'Amazon River,' Brazil; subadult (YPM 29689, S.S. Snow *in litt.* 2017), 16 April 1923, São Paulo de Olivença; subadult (FMNH 344086), 17 February 1988, Pedra Branca. **C. a. snethlageae** Fledgling (WA 1354687, Breno Vitorino), 28 May 2013, Brasil Novo; subadult (FMNH 183853), 4 February 1934, Maraí.

TECHNICAL DESCRIPTION Sexes very similar, with minor differences. The following description refers to race *aurita* (see Taxonomy and Variation). **Adult Male** Chubb (1921) described the male as: 'General colour above golden brown with dark edges to the feathers of the back; upperwing-coverts more rufous-brown than the back, with pale edges to the feathers like the innermost secondaries; bastard-wing [= alula] and primary-coverts dark brown; quills brown, rufous-brown on the outer edges and pale margins to the inner webs; tail-feathers brown; crown of head and nape deep chestnut-brown; a broad patch of silky-white feathers on the sides of the hind crown and nape; base of forehead, lores, a line above the eye, entire sides of face, and throat deep black; breast bright chestnut; middle of abdomen, underwing-coverts, and undertail-coverts white; sides of the body and flanks greyish brown.' **Adult Female** Differs from male chiefly in the absence of the black on the throat and head, which are rufous. The brow stripe is slightly reduced and blends into grey feathering around the eye and onto the lores. Most individuals have paler throat. Otherwise, females are similar in plumage to males in the coloration of the lower breast, belly and upperparts. **Adult Bare Parts** Iris dark brown; *Bill* black to brown, sometimes with paler tip; *Tarsi & Toes* grey to blue-grey (or brownish; Haverschmidt 1968). **Juvenile** Whitney (2003) described the juvenile plumage of races *inexpectata* and *snethlageae* as having extensive dark barring and chestnut-tipped feathers on crown. The back feathers have dark centres and the upperwing-coverts are dark brown with broad chestnut tips and fine black margins. The flanks are weakly barred. Hellmayr (1910) gave the bare-part coloration of a 'juvenile' male *australis* as: *Iris* brown; *Bill* black; *Tarsi & Toes* greyish-black.

MORPHOMETRIC DATA *C. a. aurita Bill* culmen 17mm; *Wing* 72mm; *Tail* 29mm; *Tarsus* 27mm (n = 1♂, Chubb 1921). *Wing* 65mm (*n* = 1, ♀?, Haverschmidt 1968). *Wing* mean 65.7 ± 2.3mm, range 56–70mm (*n* = 59, ♂♂, BDFFP). *Tail* mean 30.0 ± 2.1mm, range = 24–38mm (*n* = 54, ♂♂, BDFFP). *Bill* from nares mean 9.4 ± 0.6mm, range = 8.5–9.9mm; *Tarsus* mean 30.4 ± 1.4mm, range

28.5–31.9mm (*n* = 4, ♂♂, BDFFP). *Wing* 64mm, 63mm, 66mm, 61mm; *Tail* 29mm, 30mm, 30mm, 27mm (*n* = 4, ♂♂, juveniles, BDFFP). *Wing* mean 65.8 ± 1.3mm, range 62.5–68.0mm; *Tail* mean 31.3 ± 1.4mm, range = 29–35mm (*n* = 20, age?, ♂♂, BDFFP database). *Bill* from nares mean 9.8 ± 0.2mm, range = 9.6–10.0mm; *Tarsus* mean 31.3 ± 0.3mm, range 31.0–31.7mm (*n* = 4, age?, ♂♂, BDFFP). *Wing* 68mm (n = 1, ♀, Chubb 1921). *Wing* mean 63.4 ± 2.6mm, range 56–68mm (*n* = 17, ♀♀, BDFFP). *Bill* from nares 9.8mm (*n* = 1, ♀, BDFFP). *Tail* mean 29.7 ± 1.9mm, range 26–33mm (*n* = 16, ♀♀, BDFFP). *Tarsus* 29.9mm (*n* = 1, ♀, BDFFP database). *Wing* mean 64.1 ± 1.7mm, range 61–68mm; *Tail* mean 28.6 ± 2.4mm, range 23–32mm (*n* = 19, ♀♀, age not determined, BDFFP). *Bill* from nares mean 9.3 ± 0.4mm, range 8.9–9.6mm (*n* = 3, ♀♀, immatures, BDFFP). *Tarsus* mean 30.4 ± 0.6mm, range 28.9–31.0mm (*n* = 3, ♀♀, immatures, BDFFP). *Wing* 62mm; *Tail* 26mm (*n* = 1, ♀, juvenile, BDFFP database). *Wing* mean 65.02 ± 2.44mm (*n* = 54); *Tail* mean 30.37 ± 1.78mm (*n* = 52) (genders combined, Bierregaard 1988). *Bill* from nares 8.9mm; *Tarsus* 24.9mm (*n* = 1, ♂, USNM 515631). *Bill* from nares 9.2mm, 9.8mm; *Tarsus* 25.9mm, 26.5mm (*n* = 2, ♂♂, subadults, USNM). *Wing* 71mm (*n* = 1, ♂, Haverschmidt & Mees 1994). **C. a. occidentalis** *Bill* exposed culmen 14mm; *Wing* 67mm; *Tail* 30mm; *Tarsus* 28mm (n = 1, ♂, Chubb 1917). *Wing* 66mm (n = 1, ♀, Chubb 1917). *Wing* 64.5mm, 64.5mm, 66mm, 64mm, 67mm, 65mm; *Tail* 33mm, 32.3mm, 33.8mm, 32.5mm, 37mm, 34.4mm; *Bill* from nares 9.3mm, 9.6mm, 9.0mm, 9.9mm, 10.0mm, 10.0mm, exposed culmen 13.0mm, 15.5mm, 14.2mm, 14.2mm, 12.7mm, 13.6mm, depth at front of nares 4.4mm, 4.1mm, 4.3mm, 4.6mm, 4.5mm, 3.9mm, width at front of nares 6.3mm, 5.4mm, 6.1mm, 5.9mm, 6.0mm, 5.9mm; *Tarsus* 28.1mm, 25.0mm, 28.0mm, 26.7mm, 26.0mm, 27.7mm (*n* = 6, 1♀,5♂♂, Tom B. Smith *in litt.* 2017). **C. a. inexpectata** *Wing* 67mm; *Tail* 28mm; *Bill* exposed culmen 12.25mm, total culmen 17.75mm; *Tarsus* 28mm (n = 1 ♂, Zimmer 1931). **C. a. snethlageae** *Wing* 71.5mm, 71.0mm, 69.0mm, 67.5mm; *Tail* 34mm, 32mm, 32mm, 30.5mm; *Bill* [total?] culmen 14.5mm, 13.5mm, 12.75mm, 12.75mm; *Tarsus* 26.75mm, 26.75mm, 25.75mm, 24.75mm (*n* = 4 ♂♂, type series, von Berlepsch 1912). *Wing* 66mm, 63mm, 64mm; *Tail* 26mm, 27mm, 27mm; *Bill* exposed culmen 12mm, 12mm, 11mm (*n* = 3, ♂,♀,♀, Novaes 1947). *Bill* 9.7mm, 10.2mm, 9.9mm, 9.8mm; *Tarsus* 27.9mm 28.2mm, 28.3mm, 27.6mm (*n* = 4, ♂,♂,♂,♀, LACM). *Wing* 66.5mm; *Tail* 31.5mm; *Bill* culmen 12.75mm; *Tarsus* 26.75mm (n = 1♀, von Berlepsch 1912). **C. a. australis** *Wing* 72mm, 69mm; *Tail* 35mm, 35mm; *Bill* culmen ?mm, 13mm; *Tarsus* 24mm, 23mm (n = 2, ♂,♀, Gyldenstolpe 1945a). *Wing* 72mm; *Tail* 33mm; *Bill* culmen 14mm; *Tarsus* 26mm (*n* = 1♂, Gyldenstolpe 1951). *Wing* 69mm; *Tail* 34mm; *Bill* 13.5mm; *Tarsus* 26mm (n = 1 'immature' ♂, Hellmayr 1910). **C. a. pallida** *Bill* from nares 9.2mm, exposed culmen 12.2mm, depth at nares 4.4mm, width at nares 6.1mm, width at base of mandible 7.5mm; *Tarsus* 26.6mm (*n* = 1♀, adult; FMNH). *Bill* from nares 9.6mm; *Tarsus* 27.1mm (n = 1♂, adult; USNM 572605). *Wing* 64mm, 68mm, 63mm, 65mm; *Tail* 27mm, 31mm, 29mm, 27mm; *Bill* exposed culmen 13mm, 12.5mm, 11mm, n/m (n = 4, 2♂♂, 2♀♀; Novaes 1947). **Mass Males** 22g, 25g (n = 2, pallida; Graves & Zusi 1990); 23–28 g (n = 14, *australis*, Robbins *et al.* 1991); 20g, 24g, 25g (n = 3, *snethlageae*, Silva *et al.* 1990); mean 24.1 ± 1.5g, range 20.5–27.0g (n = 75, *aurita*, BDFFP database); mean 23.9 ± 1.6g, range 21.8–28.0g (*n* = 30, *aurita*, age undetermined, BDFFP database); 21g, 21g, 22g, 24g (*n* = 4, *aurita*, juveniles, BDFFP database); 23.6g (*n* = 1, *aurita*, subadult, USNM 625231); 26.4g, 24.9g (*n* = 2, *occidentalis*, Borja Mila *in litt.* 2017); 23.9g (*n* = 1, *aurita*, Haverschmit & Mees 1994); 22g (*n* = 1, *occidentalis*, LSUMZ, Matthew L. Brady *in litt.* 2017). *Females* 20g, 24 g (n = 2, pallida; Graves & Zusi 1990); 24–31g (n = 7, australis; Robbins *et al.* 1991); 13.2g, 12.0g (*n* = 2, ♂, ♀, respectively, *aurita*, Dick *et al.* 1984); mean 23.4 ± 1.7g, range 20.8–27.0g (*n* = 17, *aurita*, BDFFP database); mean 23.2 ± 1.9g, range 19–27g (*n* = 29, *aurita*, age not determined, BDFFP database); 22.0g, 22.5g (*n* = 2, *aurita*, juveniles, BDFFP database); 20.6g (*n* = 1, *occidentalis*, MECN 2633, Manuel V. Sánchez N. *in litt.* 2017); 24.7g, 29g(m), 24g(l) (*n* = 3, *occidentalis*, LSUMZ, Matthew L. Brady *in litt.* 2017). *Sexes combined* 23.71 ± 1.45g (*n* = 109, mean ± SD, *aurita*, Bierregaard 1988). *Gender unspecified* 24g (n = ?, occidentalis; Canaday 1997); 23.7g (n = ?, aurita; Stouffer 2007). **TOTAL LENGTH** 10.5–13.4cm (Burmeister 1856, Chubb 1917, 1921, Meyer de Schauensee 1964, Hilty & Brown 1986, Hilty 2003, Whitney 2003, Spaans *et al.* 2016).

TAXONOMY AND VARIATION The intrageneric affinities of Chestnut-belted Gnateater are unclear, but Whitney (2003) suggested a relationship with Black-cheeked Gnateater. Six subspecies are generally recognised (Novaes 1947, Howard & Moore 1984, 1991, Whitney 2003, Greeney 2014b). Across the species' range, however, there is much confusing variation, especially in populations south of the Amazon. In the north-west of its range, from eastern Ecuador and south-east Colombia to north-west Brazil and north-east Peru, races *inexpectata* and *occidentalis* are generally distinguished from other races by their whitish bellies. Even in this area, however, some adults occasionally have the buffy wash to the underparts more typical of males south of the Amazon, but this character does seem to vary slightly even in southern populations. Most markedly, southern males generally have more extensive black on the throat and reduced or nearly absent chestnut on the breast. Subspecies *snethlageae*, found along both banks of the lower Rio Tapajós east to central Pará, has most of the breast black with the remaining underparts distinctly buffy. Immediately east of the range of *snethlageae*, individuals currently assigned to the race *pallida*, are very similar but with much paler bellies and generally have little or no chestnut on the sides of the black bib. The black scaling of the back greatly reduced or even absent in *snethlageae*, and is similarly reduced in *pallida*, but to a lesser degree. Whitney (2003) suggested that *pallida* represents the pale extreme of a gradual cline east from the range of *snethlageae*, noting that their ranges may meet between the Rios Xingu and Tocantins, an area from which intermediate specimens are known (Serra dos Carajás). In the south-west of the species' range, race *australis* also shows rather confusing plumage variation, with some specimens from the Rio Purús (near the type locality) sharing plumage characters with *snethlageae*. Despite this, however, the fairly consistent plumage differences between *snethlageae* and northern and western races, along with reported vocal differences, led Whitney (2003) to suggest that *snethlageae* may deserve species rank. Subspecies *pallida*, then, would be either a subspecies or a synonym of *snethlageae*. Some authors, however, have already chosen to recognise *snethlageae/*

pallida as a separate species, Black-breasted Gnateater (del Hoyo & Collar 2016).

Conopophaga aurita aurita (J. F. Gmelin, 1789). The nominate subspecies is found from Guyana east to French Guiana (Blake 1963), south to northern Brazil from Manaus east to northern Pará (Aleixo *et al.* 2011) and Amapá (Cory & Hellmayr 1924, Whitney 2003). **SPECIMENS & RECORDS GUYANA**: *Upper Demerara-Berbice* Rockstone area, *c.*05°58.5'N, 58°31.4'W (ANSP, AMNH); Ororo Marali [= Great Falls; Stephens & Traylor 1985], 05°19'N, 58°32'W (AMNH 230184). *Upper Takatu-Upper Essiquibo* Upper Rewa River, 02°37'N, 58°32'W (Glowska & Schmidt 2014; USNM 637100); Sipu River, 01°25'N, 58°57'W (Robinson *et al.* 2007; USNM 625231). **SURINAME**: *Marowijne* Paloemeu, 03°20'N, 55°23'W (FMNH 262325). *Sipaliwini* Bakhuis Mountains, 04°44'N, 56°46'W (XC 6099, O. Ottema), Voltzberg, 04°41'N, 56°11.5'W (ML 25397, T.H. Davis); Bakhuis Gebergte, 04°29'N, 55°03'W (LSUMZ 178487); 'Kasikasima camp,' 02°58.5'N, 55°23'W (O'Shea & Ramcharan 2013a). Three locations from O'Shea & Ramcharan (2013b): Werehpai, 02°22'N, 56°42'W; Sipaliwini River, 02°17.5'N, 56°36.5'W; Kutari River, 02°10.5'N, 56°47'W. Sipaliwini Airstrip, 02°05'N, 56°10'W (Haverschmidt & Mees 1994). **FRENCH GUIANA**: *Saint Laurent Du Maroni* Tamanoir, 05°09'N, 53°45'W (CM); Saül, 03°51.5'N, 53°18.5'W (ROM 125858/59); Montagne Cottica, 03°57.5'N, 54°12.5'W (O. Tostain *in litt.* 2015); Montagne Bellevue de l'Inini, 03°32.5'N, 53°34.5'W (Tostain *et al.* 1992). *Cayenne* Savane du Galion, 04°53.5'N, 52°29.5'W (Tostain *et al.* 1992); Nouragues Field Station, 04°05'N, 52°41'W (Renaud 1998, Thiollay & Jullien 1998). **BRAZIL**: *Amapá* Rio Anotaie, 03°13'N, 52°06'W (Coltro 2008); Serra do Navio, 01°38'N, 52°16.5'W (Oniki & Willis 1972, USNM 515629–631, MNRJ 29738/899); upper Rio Tajauí, 01°31'N, 52°00'W (Novaes 1974; MPEG 21198/205); Rio Mutum, 01°23'N, 51°55.5'W (Coltro 2008); upper Rio Araguari, *c.*01°06'N, 50°30'W (Novaes 1974); Areia Vermelha, 01°00'N, 51°40'W (Novaes 1974, MPEG 20176); Foz do Rio Falcino, 00°56'N, 51°35'W (Novaes 1974, LSUMZ 67362, MPEG 20174/75); Macapá, lower Rio Amapari, 00°43'N, 51°32'W (Novaes 1974, LSUMZ 67363, MPEG 20177–179); Igarapé Novo, 00°30'N, 52°30'W (Novaes 1974, MPEG); Cachoeira Amapá/Itaboca, 00°02'N, 51°55'W (Novaes 1974, MPEG 28692/93). *Pará* FLOTA Faro, 01°42'N, 57°12'W (Aleixo *et al.* 2011); ESEC Grão-Pará Sul, 00°09'S, 55°11'W (Aleixo *et al.* 2011); FLOTA Trombetas, 00°57'S, 55°31'W (Aleixo *et al.* 2011); ESEC Jari, 01°31.5'S, 52°34'W (MPEG 51041); Monte Alegre, 01°59.5'S, 54°04.5'W (Snethlage 1907b). *Roraima* Caracaraí, *c.*01°29'N, 61°00'W (Santos 2003); near São João da Baliza, 00°59'N, 59°52'W (Santos 2004a, MPEG 56879). *Amazonas* BDFFP site, *c.*02°20'S, 60°00'W (Bierregaard & Lovejoy 1989, Ferraz *et al.* 2007, Stouffer & Bierregaard 2007, Leite *et al.* 2012); Rio Itabani, 02°47'S, 58°14'W (Pinto 1938); Reserva Florestal Adolpho Ducke, 02°57.5'S, 59°55.5'W (Oniki & Wills 1982); near Itacoatiara, 03°08'S, 58°25'W (MNRJ 31912/13).

Conopophaga aurita inexpectata J. T. Zimmer, 1931, American Museum Novitates, no. 500, p. 8, Tabocal, Rio Negro, Brazil [Amazonas, 00°48'N, 67°15'W, *c.*100m]. The type specimen is an adult male collected 11 September 1929 by Ramón Olalla and, although Zimmer (1931) gave the registration number as AMNH 301500, the correct number is AMNH 310500 (LeCroy & Sloss 2000).

Subspecies *inexpectata* is found from south-east Colombia in Putumayo and Caquetá (Meyer de Schauensee 1950b, Hilty & Brown 1986) and north-west Brazil, south and east along the west bank of the Rio Negro. Overall this race is similar to nominate, but the belly and lower parts are duller, more buffy-white. Zimmer (1931) stated that *inexpectata* is most similar to *occidentalis*, but has generally brighter upperparts, including a brighter rufescent crown (especially in females) and a warmer ochraceous-olive (less greenish-olive) back. He compared the present race with *aurita* by noting that the underparts tend towards buffy rather than white, and the crown is overall darker. The following plumage details are extracted from Zimmer's (1931) description (the holotype male) as follows. The feathers of the central back bear well-marked blackish margins preceded by an indistinct submarginal band of tawny-olive, but these markings become less conspicuous forward on the upper back, as well as rearward on the rump and uppertail-coverts. The bright chestnut coloration of the breast blends to buffy-cinnamon on the upper flanks, with this colour extending onto the undertail-coverts. The thighs are dull brown to olivaceous-brown. Zimmer (1931) stated that most of the upperwing-coverts and tertials are brown, narrowly edged with dusky and have a 'pronounced submarginal lunule of bright Hazel or Amber Brown'. The greater primary-coverts are dull blackish with greyish-olive outer margins. The feathers of the alula have their inner webs dull blackish and the inner half of the outer web greyish-olive. Half of their outer webs are white. The primary and secondary flight feathers are dusky, with brown outer margins. The underwing-coverts are mostly white, buffier at their tips, and there is a large patch of dusky near the leading edge of the wing at the base of the primaries. The rectrices are sepia-brown with slightly brighter outer margins. **SPECIMENS & RECORDS COLOMBIA:** *Vaupés* Senda Cerro Guacamaya, 01°13.5'N, 70°14'W (N. Athanas *in litt.* 2017); *c.*17km SSE of Urania, 01°08.5'N, 70°07'W (F. Rowland *in litt.* 2017). *Caquetá* Four locations from Álvarez-R. *et al.* (2003): upper Río Mesay, 00°15'N, 72°55'W; Río Cuñaré/Río Amú confluence, 00°13'N, 72°25'W; Río Sararamano, 00°08'N, 72°36'W; Puerto Abeja, 00°04.5'N, 72°27'W. *Amazonas* Near Leticia, 04°10.5'S, 69°57'W (eBird: A.M. Diaz). **BRAZIL:** *Amazonas* Tauá, 00°37'N, 69°06'W (AMNH 434362–364); 'São Gabriel' [= Uaupés; Paynter & Traylor 1991b], 00°08'S, 67°05'W (Pinto 1938); Arabo, 00°14'S, 66°51'W (MPEG); Barcelos, 00°47'S, 63°10'W (AMNH 14326); Maraã, 01°33.5'S, 65°53'W (FMNH 457234–236); Maguari, 01°51'S, 65°24'W (MPEG 42751–757); near Lago Amanã, 02°28.5'S, 64°36.5'W (ML 38620791/911, photos P. Beja); Novo Airão, 02°38.5'S, 60°57'W (WA 1970458, A. d'Affonseca); Rodovia AM-352, km75, *c.*02°51'S, 60°52'W (AMNH 14140, MPEG 59613); Tonantins, 02°52.5'S, 67°48'W (long series CM, YPM 29667–672); Rio Manacapuru, 02°56.5'S, 61°44.5'W (CM, CUMV); Lago Acajutuba, 03°06'S, 60°30'W (MPEG 12470); Manacapuru area, *c.*03°14'S, 60°41'W (Pinto 1938, long series CM, YPM 29673–677, MNRJ 16469–471); Codajás, 03°50.5'S, 62°03.5'W (FMNH 183851). **PERU:** *Loreto* Quebrada Bufeo, 02°20'S, 71°36.5'W (Stotz *et al.* 2016); Campamento Choro, 02°36.5'S, 71°29'W (Stotz & Alván 2011).

Conopophaga aurita occidentalis Chubb, 1917, Bulletin of the British Ornithologists' Club, vol. 38, p. 34, 'Río Napo,' Ecuador. Distributed from north-east Ecuador

(Berlioz 1927, 1932, Butler 1979), possibly only north of the Río Napo (Ridgely & Greenfield 2001, Hollamby 2012) east to north-east Peru east of the Río Napo (Cory & Hellmayr 1924), race *occidentalis* most resembles *inexpectata*, but has reduced scaling on the back. Chubb (1917) provided the following plumage comparisons. From nominate *aurita*, *occidentalis* differs in having the crown darker, the back 'olive-brown instead of golden-brown', the breast darker chestnut, the flanks darker overall, and a greatly reduced amount of white on the central belly. Chubb (1917) further stated that female *occidentalis* differs from female *aurita* by being darker on the head, back, wings and uppertail, and by the absence of white on the belly. **SPECIMENS & RECORDS COLOMBIA:** *Caquetá* Florencia, 01°36.5'N, 75°36'W (AMNH 116182); La Morelia, 01°29'N, 75°43.5'W (ANSP, AMNH). *Cauca* Río Guayuyaco, *c.*01°00'N, 76°22'W (ROM 103550). *Putumayo* Puerto Umbría, 00°51.5'N, 76°35'W (ANSP 159949–952); San Antonio del Guamues (00°33.5'N, 76°50'W (FMNH 287189–192); Nuevo San Miguél, 00°15.5'N, 76°33.5'W (ROM 98600). **ECUADOR:** *Sucumbíos c.*14km N of Tigre Playa, 00°14.5'N, 76°16'W (ANSP, MECN); 'Santa Cecilia,' 00°03'N, 76°58'W (KUNHM, LSUMZ); Tarapoa, 00°07.5'S, 76°20.5'W (Canaday & Rivadeneyra 2001); Güeppicillo, 00°10.5'S, 75°40.5'W (Stotz & Valenzuela 2008); Pañacocha, 00°23.5'S, 76°07.5'W (B. Mila *in litt.* 2011); Limoncocha, 00°24'S, 76°37'W (Pearson *et al.* 1972, Tallman & Tallman 1994, LSUMZ, MECN); La Selva Lodge, 00°30'S; 76°22.5'W (XC 264647/48; P. Coopmans). *Orellana* 'San José de Sumaco' [= San José Nuevo, Paynter 1993], 00°26'S, 77°20'W (MLZ 7207, AMNH 179209); Zancudococha, 00°36'S, 75°29'W (ANSP 183355); Ávila area, *c.*00°38'S, 77°26'W (AMNH, USNM, MLZ, SBMNH); Loreto area, *c.*00°41'S, 77°18'W (MLZ 7604/7751); lower Río Suno, *c.*00°43'S, 77°13.5'W (MCZ 138238/39, MLZ 7602); Guaticocha, 00°45'S, 77°24'W (MCZ 299255); Río Pucuno, *c.*00°46'S, 77°28.5'W (MCZ 299256–258); Río Cotapino, 00°48'S, 77°26'W (FMNH, UMMZ). *Napo c.*17km W of Lago Agrio, 00°05'N, 77°06.5'W (DMNH 59460/64954); Eugenio, 00°46'S, 77°42'W (MCZ 299254); Archidona, 00°54.5'S, 77°48.5'W (AMNH 488906). **PERU:** *Loreto* Río Ere headwaters, 01°41'S, 73°43'W (Stotz & Inzunza 2013); mouth of Río Curaráy, 02°22'S, 74°04'W (AMNH 255655–658, MCZ 138237); 'Campamento Medio Algodón,' 02°35.5'S, 72°53.5'W (Stotz *et al.* 2016); 'Campamento Piedras,' Río Algodoncillo, 02°47.5'S, 72°55'W (Stotz & Alván 2010); 'Campamento Curupa', Río Yanayacu, 02°53'S, 73°01'W (Stotz & Alván 2010); 'Quebrada Yanayacu,' 03°05'S, 73°08'W (LSUMZ); near Sucusari, 03°15.5'S, 72°55.5'W (LSUMZ, ML, XC); Quebrada Orán, 03°25'S, 72°35'W (LSUMZ 119924–933).

Conopophaga aurita australis Todd, 1927, Proceedings of the Biological Society of Washington, vol. 40, p. 150, Nova Olinda, Rio Purús, Brazil [Nova Olinda do Norte, 03°45.5'S, 59°05'W]. The holotype (CM P91917) is an adult male collected 14 July 1922 by Samuel M. Klages (Todd 1927, 1928). This subspecies occurs south of the Amazon from north-east Peru (south to Ucayali) to western Brazil, east to the Rio Madeira (Gyldenstolpe 1945a) and south to northern Acre (Aleixo & Guilherme 2010) and northern Rondônia (Olmos *et al.* 2011). The population east of the Madeira in north-west Rondônia was provisionally assigned to this subspecies by Whitney (2003), but its subspecific affinities remain poorly known. I examined three specimens (FMNH) collected at Cachoeira de Nazaré and Pedra Branca, *c*.220km south-east of the upper Rio Madeira. These, along with several photographs taken at Machadinho d'Oeste, just north-west of these localities (WA: A Grassi, F. Arantes), suggest to me that Whitney (2003) was correct in assigning these populations to *australis*, and this is reflected in my range map. Photographs taken *c*.290km directly east of the above-mentioned FMNH localities (near Aripuanã, Mato Grosso) all appear to be best assigned to *snethlageae* (WA: C. Veronese, M.P. Cena Neto, D. Zenere). Similarly, *c*.290km south of Machadinho d'Oeste, at Vilhena in extreme eastern Rondônia, a male photographed from below clearly shows the extended black bib of *snethlageae* (WA 1285740, V. Cordasso), making this the southernmost record of both *snethlageae* and the species. The specimen record (MPEG 49695) of Chestnut-belted Gnateater at Porongaba, Acre (Whittaker & Oren 1999), was examined by Guilherme (2009, 2016) and refers to a female Ash-breasted Gnateater. Subspecies *australis* tends to have more black on the throat than *occidentalis*, extending onto the upper breast to form a *slight* bib, albeit not as extensive as *snethlageae* and *pallida*. Overall, however, Todd (1927) described *australis* as most similar to *occidentalis*, but with the 'upper parts averaging more rufescent, and under parts more decidedly buffy, the black of the throat in the male more extended, reaching the upper breast'. He further stated that 'the characters on which [the description of *occidentalis* are] based are carried a step further in the present form, and by just that much tend toward those of the lower Amazonian *C. snethlageae*, without, however, showing actual intergradation.' In males, the greater extent to which the black feathering of the throat extends rearward, compared at least with *occidentalis*, is also characteristic of this subspecies. The greater extent of black results in a narrowing of the chestnut colouring on the breast, which is generally also paler than in *occidentalis*. Todd (1927), comparing a series of specimens with *occidentalis*, stated that *australis* averages more rufescent, less olivaceous, above. **SPECIMENS & RECORDS BRAZIL:** *Amazonas* Comunidade São João do Acurau, 03°16.5'S, 67°19.5'W (MPEG); RESEX Rio Jutaí, 03°17.5'S, 67°19.5'W (MPEG 79648); Caviana, 03°25'S, 60°38'W (CM); São Paulo de Olivença, 03°29'S, 68°58'W (CM, ANSP, YPM); Caitaú/Uará, 03°29'S, 66°04'W (MPEG 50034–037); Lagoa Tefé, 03°32'S, 64°58.5'W (Johns 1991); RPPN Palmarí, 04°17.5'S, 70°17.5'W (XC 16575, N. Athanas); RDS Cujubim, 04°39'S, 68°19.5'W (MPEG 60224); Base Petrobras Urucu, 04°52'S, 65°18'W (MPEG 57151); near Bauana, 05°26'S, 67°17'W (XC 88027, A. Lees); *c*.7km SSW of Tapauá, 05°41.5'S, 63°12.5'W (XC 119606, G. Leite); Aliança, 06°34'S, 64°24'W (AMNH 488907); Prainha, 07°16'S, 60°23'W (Novaes 1976); Huitanaã area, *c*.07°40'S, 65°46.5'W (CM, KUNHM, FMNH, MCZ). *Acre* Mâncio Lima, 07°25'S, 73°39'W (Whitney *et al*. 1997 in Guilherme 2009); São Domingos, 07°33'S, 72°59'W (Guilherme 2009, MPEG 62105); Colônia Dois Portos, 08°20'S, 72°36'W (Guilherme 2009, MPEG 62103–105); ESEC Rio Acre, *c*.10°59'S, 70°24'W (Aleixo & Guilherme 2010). *Rondônia* Rio Jamari, Jusante, 08°45'S, 63°27.5'W (MPEG 46838/39); Candeias do Jamari, 08°48'S, 63°29'W (WA 519553, D. Mota); Fazenda Rio Candeias, 08°57'S, 63°41.5'W (MPEG 35129–132); Cachoeira de Nazaré, 09°44'S, 61°53'W (FMNH 389956/57, MPEG 39930–935); Pedra Branca, 10°03'S, 62°07'W (FMNH 344086); Terra Indígena Igarapé Lourdes, 10°25.5'S, 61°39.5'W (Santos *et al*. 2011b); REBIO Ouro Preto, 10°50'S, 64°45'W (MPEG 55032); Rodovia BR-429, km87, 11°24'S, 62°24'W (MPEG 38804–806). **PERU:** *Loreto* 'Santa Cecilia' on Río Maniti, *c*.03°32'S, 72°52'W (ANSP 176218/19); Quebrada Vainilla, 03°32.5'S, 72°44.5'W (Robbins *et al*. 1991, long series LSUMZ); Centro de Investigación Jenaro Herrera, 04°54'S, 73°38.5'W (eBird: D.F. Stotz); Choncó, 05°33.5'S, 73°36.5'W (Stotz & Pequeño 2006); Quebrada Yanayacu, 06°16'S, 73°54.5'W (O'Shea *et al*. 2015); Ojo de Contaya, 07°07'S, 74°35.5'W (Schulenberg *et al*. 2006); upper Río Tapiche, 07°12.5'S, 73°56'W (Schulenberg *et al*. 2006). *Ucayali* Otorongo, 10°23'S, 73°43'W (LSUMZ 189039–043). **Notes** Santos *et al*. (2011a) reported Chestnut-belted Gnateater from the vicinity of Juruti (*c*.02°09'S, 56°05'W), the only record of this species from the lower Rio Madeira/ Rio Tapajos interfluvium. However, this is based on a single vocalisation that was not heard a second time and, despite the authors' experience with the species, its presence in the region is best considered hypothetical until further evidence surfaces (M.P.D. Santos & A. Aleixo *in litt*. 2017). If the species does occur there, its seems probable that the race concerned will be *australis*.

Conopophaga aurita snethlageae Berlepsch, 1912, Ornithologische Monatsberichte, vol. 20, p. 17, Tucumare, Rio Jamauchim, Brazil [Tucunaré, 05°18'S, 55°51'W, *c*.50m]. Endemic to Brazil, Whitney (2003) gives the range of *snethlageae* as south of the Amazon, from both banks of the lower Rio Tapajós (Griscom & Greenway 1941) and the east bank of the Rio Teles Pires east to central Pará. Birds from the west bank of the Rio Tapajós (Vila Braga; CM) do appear to represent *snethlageae*, but it is unclear where (or if) its range abuts that of *australis* in the central Tapajós–Madeira interfluvium. I was unable to examine sufficient material to resolve the issue of clinal variation (Whitney 2003) between *snethlageae* and *pallida*, but should they prove to be distinct taxa, it appears that their ranges are separated by the lower Rio Xingu. This subspecies has the black scaling on the back much reduced compared to other races, and sometimes lacks scaling altogether. The black plumage of males is considerably more extensive below, reaching onto the chest and almost eliminating the 'belt'. **SPECIMENS & RECORDS** *Pará* Cuçari, 01°53.5'S, 53°21'W (Snethlage 1914a, MPEG 3779); Porto de Moz Rio Xingu, 02°05'S, 52°32'W (WA 1348519, L. Faria); Aramanai, 02°40.5'S, 54°59'W (AMNH 286337); Mojuí dos Campos, 02°41'S, 54°38.5'W (CM, KUNHM, MCZ, YPM); Caxiricatuba, 02°50.5'S, 55°02.5'W (Pinto 1938, AMNH 286343/44, MCZ 175133–135); Vitória do Xingu, 02°53'S, 52°01'W (WA 1193540, recording B. Vitorino); Maraí, 03°04'S, 55°06'W (Pinto 1938, FMNH 183853); Patauá, 03°05'S, 55°03'W (MCZ 175139–141); Tauari, 03°05.5'S, 55°07'W (AMNH 286338/39, MCZ 175136/37); Pinhy, 03°06'S, 55°06'W (MCZ 175138); Itapoama/Aveiro, 03°15'S, 55°10'W (Pinto 1939, FMNH 183852); Brasil Novo, 03°15.5'S, 52°40'W (WA 1354687, B. Vitorino); Agrovila União, 03°17'S, 52°22.5'W (MPEG 30144/45); Rodovia BR-163, km117, 03°21.5'S, 54°57'W (MPEG 53922/23, MPEG 56105); Piquiatuba, 03°23'S, 55°09.5'W (AMNH 286340–342); Paquiçama, 03°23'S, 51°45'W (MPEG 55069); UHE Belo Monte, 03°23'S, 51°56'W (MPEG 55493/94); Lago Arauepá, *c*.03°36.5'S, 55°20'W (LACM 32121–125); Santa Júlia, 04°14.5'S, 53°28'W (MPEG 10699, MNRJ 16472); Mirituba, 04°16'S, 55°54'W (ANSP, YPM, MCZ, CM, UMMZ); Vila Braga, 04°25.5'S, 56°17.5'W (CM, YPM, ANSP, MPEG, MNRJ); Rio Ratão near Curuçá, 05°25.5'S, 56°54.5'W (MPEG

75917/25); Itaituba, 05°30'S, 56°59'W (XC 147419, G. Leite); Comunidade Terra Preta, 05°37.5'S, 57°17'W (MPEG 76473); Vila Mamãe-anã, 05°45.5'S, 57°25'W (MPEG 76238); two locations on right bank of Rio Crepori, 05°46'S, 57°17'W (MPEG 76422), 05°51'S, 57°12'W (MPEG 75977); PN Serra do Pardo, 'Base 1,' 05°52.5'S, 52°47.5'W (MPEG 74251); Comunidade São Martins, 06°06.5'S, 57°37'W (MPEG 75612); Mina de Ouro Palito, 06°18.5'S, 55°47'W (MPEG 72156); Cotovelo, 06°31'S, 57°26.5'W (MPEG 65219); Rio Jamanxim, 06°45'S, 55°23'W (MPEG 66033); Novo Progresso, 07°02.5'S, 55°25'W (WA 16056, B. Davis); Sítio Nardino, 07°08'S, 55°43'W (MPEG 69687); 20km SW of Novo Progresso, 07°11.5'S, 55°29.5'W (MPEG 59187–189); two sites on Rio São Benedito, 09°07'S, 56°56.5'W (MPEG 54693) and 09°08'S, 57°03'W (MPEG 54709–711); Rio Azul Jungle Lodge, 09°15'S, 55°59.5'W (ML 20125111, photo B. Davis). *Mato Grosso* Alta Floresta, 09°53'S, 56°28'W (Lees & Peres 2006, 2009); RPPN Rio Cristalino, *c.*09°36'S, 55°56'W (XC, ML); Aripuanã, 10°10.5'S, 59°27'W (WA 1625077, D. Zenere); Fazenda Rio do Ouro, 11°20'S, 54°09'W (MPEG 74716); Sinop, 11°51.5'S, 55°30.5'W (WA 2156306, D. Almeida). *Rondônia* Vilhena, 12°44.5'S, 60°08'W (WA 1285740, V. Cordasso).

Conopophaga aurita pallida Snethlage, 1914b, *Ornithologische Monatsberichte,* vol. 22, p. 39, Cametá, Rio Tocantins, Brazil [02°14.5'S, 49°30'W]. The holotype (MPEG 7998) is an adult male collected 18 January 1911. Given the discussion above, it appears that the range of *pallida* is restricted to north-west Pará, east of the lower Rio Xingu, and does not extend as far south as Mato Grosso, or as far east as the Rio Tocantins (*contra* Cory & Hellmayr 1924, Whitney 2003). Race *snethlageae* is most similar, but *pallida* is said to differ by usually showing slightly more extensive dark dorsal scaling (but still less than other races) and having an even more extensive black bib, frequently completely eliminating the chestnut 'belt.' There seems to be much individual variation, some of which may be age-related, but which is possibly clinal. Overall, however, I concur with Whitney (2003) and suggest that *pallida* might best be synonymised with *snethlageae*. SPECIMENS & RECORDS *Pará* Portel, 01°57'S, 51°36'W (FMNH 457237–241, MPEG 61874–878, XC); Senador José Porfírio, 02°35'S, 51°57'W (WA 2262713, E. Yosheno); Anapu, 03°28.5'S, 51°12'W (WA 2295852, K. Lara); *c.*7.8km NW of Três Palmeiras, 03°32'S, 51°44'W (MPEG 55697/98); Itapuama, 'área 1', 03°36.5'S, 52°20.5'W (MPEG 63447); 52km SSW of Altamira, 03°39'S, 52°22'W (USNM 572605/06); Vale do Caripé, 03°45'S, 49°41'W (MPEG 36242–244); Novo Repartimento, 04°15'S, 49°57'W (WA 2218143, A. Grassi); left bank of Rio Tocantins, 04°35'S, 49°34'W (MPEG 36055–057); near Marabá, 05°21'S, 49°06.5'W (MPEG 47850, MNRJ MNA3279/3411); *c.*3.7km NW of Cedro, 05°35'S, 49°12'W (MPEG 38079–081); FLONA Tapirapé-Aquiri, 05°40.5'S, 50°18.5'W (MPEG 67712); 'Projeto Salobo/Vale', 05°48'S, 50°30'W (MPEG 53802/03); Ponta Nova, 05°49'S, 54°31'W (MPEG 10698, MNRJ 16474); Serra dos Carajás and Parauapebas area, *c.*06°00'S, 50°20'W (FMNH 391447/48, MPEG 70413/14, XC); Fazenda Bitoca Grande, 06°02'S, 51°58'W (ML 49686971, photo I. Thompson); Manganês, 06°10'S, 50°24'W (MPEG); Serra das Andorinhas, *c.*06°19'S, 48°34'W (MPEG 52408/09); São Félix do Xingu, 06°38.5'S, 51°59'W (WA 2359451, I. Thompson); Luzilândia, 07°05.5'S, 49°57'W (MPEG 34678); *c.*24km NNW of Conceição do Araguaia, 08°04'S, 49°21.5'W (ML

63942691, A. Aleixo); Fazenda Fartura, 09°40'S, 50°23'W (Somenzari *et al.* 2011); Fazenda Barra das Princesas, 09°40'S, 50°11'W (MPEG 48824–827).

STATUS Largely due to its fairly extensive range, Chestnut-belted Gnateater is not considered globally threatened. This species is uncommon through much of its range (Hilty & Brown 1986, Ridgely & Greenfield 2001a, Schulenberg *et al.* 2007, Ridgely & Tudor 2009), with only one pair/100ha in one study in French Guiana (Jullien & Thiollay 1998). Chestnut-belted Gnateaters are more common in parts of Brazil (Whitney 2003, Stouffer 2007) and less so in others (Borges 2006). They appear to be fairly intolerant of anthropogenic disturbance (Stouffer & Bierregaard 1995, Canaday 1997, Laurance 2004, Lees & Peres 2006, Wunderle *et al.* 2006, Powell *et al.* 2016), apparently 'dropping out' of fragments smaller than 10 ha (Stratford & Stouffer 1999) and showing reduced abundance following selective logging (Henriques *et al.* 2003, 2008). PROTECTED POPULATIONS *C. a. aurita* **Suriname** Sipaliwini NP (Haverschmidt & Mees 1994); Raleigh Vallen NP (ML: T.H. Davis); **Guyana** PN Montanhas do Tumucumaque (Coltro 2008). **French Guiana** Reserve Naturelle des Nouragues (Renaud 1998, Thiollay & Jullien 1998). **Brazil** PN Viruá (Santos 2003); Reserva Florestal Adolpho Ducke (Oniki & Wills 1982); FLOTA Trombetas, FLOTA Faro, and ESEC Grão Pará (Aleixo *et al.* 2011). *C. a. occidentalis* **Ecuador** RE Limoncocha (Pearson *et al.* 1972, Tallman & Tallman 1994, LSUMZ, MECN); Reserva de Producción de Fauna Cuyabeno (Canaday & Rivadeneyra 2001). *C. a. inexpectata* **Colombia** PNN Serranía de Chiribiquete (Álvarez *et al.* 2003). **Brazil** PN Jaú (Borges & Carvalhães 2000, Borges *et al.* 2001, Borges 2006, 2007, Borges & de Almeida 2011). *C. a. australis* **Peru** ZR Sierra del Divisor (Schulenberg *et al.* 2006); Reserva Comunal Matsés (Stotz & Pequeño 2006). **Brazil** ESEC Bauana (XC 88027, A. Lees); PN Serra do Divisor (Whitney *et al.* 1997); ESEC Rio Acre (Aleixo & Guilherme 2010); RESEX Rio Jutaí (MPEG 79648); RDS Cujubim (MPEG 60224); RPPN Palmarí (XC 16575, N. Athanas). *C. a. snethlageae* PN Serra do Pardo (MPEG 74251); PN da Amazônia (eBird: B. Davis); FLONA Tapajós (Oren & Parker 1997, Henriques *et al.* 2003, 2008, MECN, XC); FLONA Altamira (MPEG); FLONA Jamanxim (MPEG 69687); RPPN Rio Cristalino (MPEG 51640–642, XC, ML). *C. a. pallida* FLONA Serra dos Carajás (Pacheco *et al.* 2007, FMNH, MPEG, XC); FLONA Caxiuanã (FMNH 457237–241); FLONA Tapirapé-Aquiri (MPEG 67712).

OTHER NAMES *Turdus auritus* (Linné *et al.* 1788, Latham 1790, Lesson 1831); *Pipra leucotis* (J.F. Gmelin *in* Linné *et al.* 1789); *Conopophaga aurita* (*occidentalis*; Sclater 1854); *Conopophaga snethlageae conspec nov.* (*pallida*; Snethlage 1914b); *Conopophaga leucotis* (Vieillot 1823, 1834, Ménétriés 1835); *Myiothera aurita* (Lichtenstein 1823); *Conopophaga snethlageae* (Snethlage 1914a). **English** Black-breasted Gnateater (*snethlageae*; del Hoyo *et al.* 2017a); Snethlage's Gnat-eater (*snethlageae*; Brabourne & Chubb 2012); Chestnut-belted Gnat Eater (*aurita*), Western Gnat Eater (*occidentalis*), Snethlage's Gnat Eater (*snethlageae*), Pallid Gnat Eater (*pallida*; Cory & Hellmayr 1924); White-eared Manakin (Latham 1785). **Spanish** Jejenero orejudo (Whitney & Kirwan 2017); Zumbador Pechirrufo (Salaman *et al.* 2007a); Jejenero de Snethlage (*snethlageae*; del Hoyo *et al.* 2017a); Jejenero de Faja Castaña (O'Shea *et al.* 2015); Jejenero faijicastaño (Valarezo-Delgado 1984,

Ortiz-Crespo *et al.* 1990, Ridgely & Greenfield 2001). **Portuguese** chupa-dente-de-cinta (Guilherme 2016); pula-pula pequeño (FMNH 183853, Pará; A. M. Olalla label). **French** Conophage à oreilles blanches (Vieillot 1823, 1834, Haverschmidt & Mees 1994, Whitney & Kirwan 2017); Fourmillier à oreilles blanches (Buffon 1771–86);

Conophage de Snethlage (*snethlageae*; del Hoyo *et al.* 2017a). **Dutch** Cayenne-muggeneter (Spaans *et al.* 2016). **German** Rostbrust-Mückenfresser (Whitney & Kirwan 2017); Schwarzbrust-Mückenfresser (*snethlageae*; del Hoyo *et al.* 2017a). **Locally:** Mirafowru (French Guiana; Haverschmidt & Mees 1994).

Chestnut-belted Gnateater, female (*aurita*), Manaus, Amazonas, Brazil, 17 August 2014 (*Robson Czaban*).

Chestnut-belted Gnateater, male (*australis*), Novo Airão, Amazonas, Brazil, 5 September 2013 (*Marco Guedes*).

Chestnut-belted Gnateater, nest with eggs (*aurita*) Cottica Mountains, French Guiana, 8 September 2007 (*Olivier Tostain*).

Chestnut-belted Gnateater, male (*occidentalis*), Zancudococha, Orellana, Ecuador, 3 May 2014 (*Nick Athanas*).

Chestnut-belted Gnateater, male (*pallida*), Parauapebas, Pará, Brazil, 19 June 2015 (*Ciro Albano*).

Chestnut-belted Gnateater, male (*pallida*), Parauapebas, Pará, Brazil, 24 July 2010 (*João Quental*).

ASH-THROATED GNATEATER
Conopophaga peruviana   Plate 2

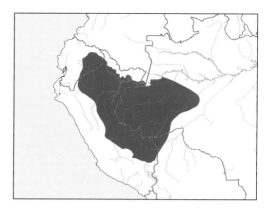

Conopophaga peruviana Des Murs, 1856, in Castelnau: Expédition dans les parties centrales de l'Amérique du Sud, Oiseaux, vol. 18, p. 50, Pl. 16, fig. 1, 'à Pebaz et à Nauta (Haut-Amazone).' The designated type locality is Pebas, on the left bank of the Río Marañón, Loreto, Peru [03°19'S, 71°51.5W] (Gyldenstolpe 1945a). The type specimen is housed at the Museum national d'Histoire naturelle in Paris.

A generally uncommon and not very vocal member of the genus, Ash-throated Gnateater is confined to upper Amazonia in eastern Ecuador (Butler 1979) and Peru (Parker *et al.* 1982), south-west Amazonian Brazil (Meyer de Schauensee 1966, Pinto 1978), and south to northern Bolivia (Kempff-Mercado 1985). It inhabits the understorey of tall *terra firme* forest, apparently preferring treefalls and denser undergrowth. Like its congeners, Ash-throated Gnateater forages on or near the ground, but there are no detailed studies of its diet. A relatively small gnateater, males have grey underparts and faces, with a small white postocular tuft and a brown crown. Both sexes have dark brown wings with prominent pale white or buffy spotting, a character that, prior to the collection of many specimens, probably led Sclater (1858a) to suggest the possibility that the present species was simply an immature Slaty Gnateater.

IDENTIFICATION 11–12cm. Sexes dimorphic. Males are mostly grey, with whitish throat, a contrasting white postocular tuft and a paler grey line separating the brownish crown from the dark grey face. The back is darker grey than the underparts, with a brownish wash and has thick, irregular black scalloping. The flight feathers and tail are brown, the upperwing-coverts and tertials with buffy tips. Females have a rufous face and breast, and dark brown upperparts. Like males, the wings have buff or buffy-rufous spots at the tips of the tertials and wing-coverts. The iris, bill and legs are dark in both sexes. Female Ash-throated Gnateater can be separated from female Chestnut-crowned Gnateater by the distinctly scaled back and prominent spotting on the upperwing-coverts (Ridgely & Greenfield 2001).

DISTRIBUTION Ash-throated Gnateater occurs in eastern Ecuador, west and south of the Río Napo (Ridgely & Greenfield 2001, Hollamby 2012), through most of eastern Peru and east into central Amazonian Brazil, south of the Rio Amazonas (Schulenberg *et al.* 2010). Apart from some continued ambiguity as to the easternmost extent of its range in Brazil (see below), the distribution of Ash-thoated Gnateater is arguably one of the best documented, with specimens, photographs or recordings available from nearly every part of the proposed range. In the north, its range is clearly limited by the entire length of the Río Napo and by the Río Amazonas east of its mouth. At the western edge of its range, specimen records document its presence fairly continuously as far south as Astillero and the upper Río Távara, which flows into the Río Tambopata at Astillero. Eastward, the closest records to the west bank of the Rio Madeira are near Bom Lugar and Arimã, both on the east bank of the Río Purús in the Purús–Madeira interfluvium. Closer to the base of the Andes, however, the range of *peruviana* does extend slightly into northern

La Paz, Bolivia (Kempff-Mercado 1985, Parker *et al.* 1991, Montambault 2002), but has not yet been reported from Pando. Bangs & Noble (1918) mentioned a record of *C. peruviana* from Cajamarca, referred to *C. castaneiceps brunneinucha* by Cory & Hellmayr (1924), which is here considered to represent *C. castaneiceps chapmani* (see that species' account). As a final note, Taczanowski (1884) mentioned records from Pebas. This small town (as well as the adjacent location Pévas) is on the north (left) bank of the Río Amazonas in Loreto (*c.*03°20'S, 71°49'W; Stephens & Traylor 1983). This is presumably in error or the actual collecting locality was on the opposite bank of the river.

HABITAT Generally found at elevations below 600m (Ridgely & Greenfield 2001), occasionally to 900m or 1,000m in Peru (Walker *et al.* 2006, Ridgely & Tudor 2009), but generally below 400m in Bolivia (Hennessey *et al.* 2003b). Ash-throated Gnateaters inhabit and nest in (Hillman & Hogan 2002) the relatively open understorey of tall *terra firme* forest, appearing to have a proclivity for slight disturbed areas such as vine tangles and treefalls (English 1998), but rarely in *Guadua* bamboo (Servat 1996, Lebbin 2007). They are also occasionally found in *várzea* habitats (Parker & Bailey 1991, Alonso *et al.* 2012). English (1998) also found them more common on slopes and ridgetops. Perhaps entirely replaced by *C. aurita australis* in blackwater drainages (Whitney 2003), especially south of the Amazon, but quite possibly just overlooked due to its retiring nature. Altitudinally, it is generally replaced at higher elevations by Chestnut-crowned Gnateater but occurs sympatrically at some sites in southern Ecuador and eastern Peru (Whitney 2003, Schulenberg *et al.* 2010).

VOICE The somewhat frog-like, weakly disyllabic notes of the Ash-throated Gnateater's song were described by Whitney (2003) as hollow-sounding *hwrickik!* notes strung together, and by Schulenberg *et al.* (2007) as a bisyllabic 'sneezing sound,' the first syllable louder and rising and the second lower pitched (*shrEE'dit*). The principal portion of notes rise sharply from around 1.0–3.5kHz in <0.25s, are loudest at the top end, and are delivered at irregular intervals (but usually every 5–6s), occasionally continuing for >30s. Ridgely & Greenfield (2001), however, describe a loud and inflected *zhweeik* repeated at well-spaced but regular intervals. Described calls include a dry chattering (Whitney 2003), a low, harsh, monosyllabic *shreff* (Ridgely & Greenfield 2001) or a coughing *tchk* (Schulenberg *et al.* 2007). Parker 1982) reported an immature male making a loud 'chink' note, to which it responded aggressively after playback (see Natural History). According to Whitney

(2003), males occasionally produce a fairly loud whirring sound with the outer remiges, which seems to be associated with aggression or courtship. This behaviour has not been studied in detail, however, and requires further attention.

NATURAL HISTORY Generally found alone or in pairs (Parker 1982, English 1998) and rarely associates with mixed-species flocks (Ridgely & Tudor 1994), adults are shy and difficult to see (O'Neill 1974, Parker 1982). Between short, rapid hops and flights, adults perch on small horizontal branches in the understorey, sometimes on vertical stems (Parker 1982), taking prey by gleaning from nearby foliage (Whitney 2003, HFG). They may also occasionally capture prey in the leaf litter. In response to playback of its loud 'chink' call, an immature male hopped about in the open undergrowth, 1m above ground, and held its white postocular tufts flared out (Parker 1982). This is, however, the only mention of use of these tufts by *C. peruviana*. There have been no quantified studies of territorial behaviour in Ash-throated Gnateaters, but Terborgh *et al.* (1990) estimated 3 pairs/100ha in eastern Peru. A few studies have examined the prevalence of ectoparasites of *C. peruviana*, including avian chewing lice (Phthiraptera; Clayton *et al.* 1992, Price & Clayton 1996, Clayton & Walther 2001).

DIET The diet of Ash-throated Gnateater is poorly known, presumably consisting primarily of a variety of small arthropods (Whitney 2003).

REPRODUCTION The first reported nest was found with incubation underway at the Reserva Nacional Pacaya-Samiria (Begazo & Valqui 1998), and two nests have been described from Peru that contained nestlings when found (Dreyer 2002, Hillman & Hogan 2002). These provided nest descriptions and only minimal behavioural data, but very recently our knowledge has been substantially increased by the unpublished observations of Gustavo A. Londoño and colleagues, who have studied 43 nests in south-east Peru (data not included here). NEST & BUILDING There are no available data on pre-laying behaviour and building. My own observations showed that, at least during the first few days after the clutch is completed, adults of both sexes continue to add lining to the nest. Almost certainly both also participate in all stages of nest construction. The nest is a fairly broad, shallow cup composed externally of loosely piled leaves, twigs and leaf petioles. Internally, they are sparsely lined with thin dark fibres, flexible leaf petioles and fungal rhizomorphs. The only previously reported nest dimensions were: external diameter 6.8cm; internal diameter 5.8cm; internal depth 3.7cm (Hillman & Hogan 2002). Two nests in Ecuador (HFG) and one in Peru (Daniel Lebbin *in litt.* 2016) measured, respectively: external diameter 9.5 × 11cm (nest oblong), 10.0cm, 8.6cm; external height 6.5cm, 7.0cm, 6.2cm; internal diameter 5.0 × 7.5cm, 5.5cm, 6.2cm; internal depth 4.0cm, 3.5cm, 4.0cm. Most nests that have been described or that I have found or examined pictures of (*n* = 15) have been fairly loosely placed on the substrate, not tightly woven into it. Supporting substrates are, in general, fairly flimsy and include the base of rosettes of leaf bases (i.e., palm, aeroid), loosely crisscrossed thin stems (i.e., ferns) and, most frequently, small clumps of epiphytes attached to thin (>5cm) saplings. They appear to favour locations where detritus might naturally accumulate, or where it already has. This has the effect of making nests very well camouflaged, appearing as little

more than random clumps of dead vegetation and twigs. The heights of six nests were 84cm (Begazo & Valqui 1998), 35cm (Dreyer 2002), 70cm (Hillman & Hogan 2002), 51cm (Daniel Lebbin *in litt.* 2016), 45cm and 30cm (HFG). EGG, LAYING & INCUBATION Laying behaviour is largely unknown, but at one nest the eggs were laid before noon, roughly 48h apart. The first documented nest (Begazo & Valqui 1998) was found with two eggs, and all completed clutches I have observed (*n* = 12) have been of two eggs. Eggs vary somewhat in shape from short subelliptical to short-oval. They also vary in coloration, from very pale buff or nearly white, and nearly unmarked, to rich buff with a mix of cinnamon, vinaceous and pale brown blotches and flecks that can form a distinct cap or ring at the larger end. I have seen eggs (*n* = 23) of nearly all variations between these two extremes, with apparently no geographic component to this variation. Within-clutch variation in colour, however, is minimal except in one clutch photographed in SE Peru (Gustavo Londoño *in litt.* 2017). In this clutch the ground colour of both eggs was pale buff, but one was only very sparsely spotted with pale cinnamon spots forming an indistinct ring, while the other was washed with pale cinnamon across the entire shell, with a slightly darker ring of merged blotches at the larger end. Overall the effect was of a bicoloured clutch like that described at one nest of Chestnut-belted Gnateater (which see). Egg dimensions were previously unreported. The eggs of one clutch in Ecuador measured 21.2 × 16.3mm and 21.3 × 16.2mm, and weighed 2.7g and 2.8g, respectively, 2–3 days prior to hatching. The first egg of a second Ecuadorian clutch measured 21.9 × 16.7mm and weighed 3.3g the day after it was laid (HFG). Two eggs in a clutch from Madre de Dios (Daniel Lebbin *in litt.* 2016) measured 20.8 × 16.2mm and 20.5 × 16.1mm. Incubation behaviour and rhythm are undocumented, but at one nest I watched on the day following clutch initiation the adults spent very little time there during daylight. The female, however, returned to spend the night on the nest prior to laying the second egg the following morning. At one nest nearing the end of incubation I found the adults extremely wary, usually flushing from the nest before I could see it, generally while I was 6–10m away. On the few occasions I managed to observe their departure, they dropped over the edge and made a rather weak, fluttery descent into the nearest dense vegetation. I have not observed any other, more exaggerated, distraction displays. NESTLING & PARENTAL CARE Both adults feed and brood the young, but only the female spends the night on the nest. Little else is known of parental care, but a female flushed from the nest in Peru during the late nestling stage performed a broken-wing distraction display (Hillman & Hogan 2002). Nestlings hatch completely devoid of natal down and almost entirely dark, sooty-grey or black, paler and pinkish-orange below. The maxilla is shiny black along the culmen and at the tip, paler dusky on the sides, with a bright white egg tooth. The mandible is dull yellow, fading to pale yellow-white at the base and darkening to black at the tip. The inflated rictal flanges are bright white on the outer margins and pale yellowish internally. The mouth lining is uniform yellow. Within two days their coloration remains unaltered but feather pins have begun to break the skin on all tracts, longest on the posterior portion of the spinal tract and the femoral tracts. Contour pins begin breaking their sheaths at 4–5 days of age, those of the capital tract emerging last. By day 8 or 9 they are fairly well covered in overall downy plumage, secondary flight feathers have broken their sheaths 1–2mm, secondary coverts are emerged 2–3mm, primary pins are just breaking open, only slightly ahead of most primary-coverts. All wing feathers show dark brown feather tips while those of the greater and lesser secondary-coverts are bright fulvous-buff. Feathers on the femoral tracts are dull olive-buff and extremely long and downy. Feathers of the underparts are also downy, but slightly shorter (less fluffy) than on the flanks. Those of the ventral (belly) tracts are bright white centrally and pale olive-buff laterally, while those of the ventral sternal and cervical tracts are dark brown at the ends and many (but not all) are dull olive-buff basally. Above, the capital tract is dark chestnut-brown, the back is brownish and mottled dull ochre. A row of longer fulvous-buff pin feathers are emerging in the region that will eventually bear the bright white postocular tufts. The bill coloration is little changed and the maxilla still bears its egg tooth. The eyes are slitted and just opening. By day 10 or 11 the nestlings are dark blackish-brown above with dark chestnut spotting on the crown that becomes thick transverse barring on the upper back. Coloration gradually changes posteriorly to dark brown with indistinct olive-buff barring on the rump. The throat and breast are dark brown, fading to dull ochre on the sides and flanks. The belly is white, tinged ochre laterally. Secondary wing-coverts are distinctly tipped with broad, rich buffy bars that are narrowly edged black along their distal margins. The markings on the coverts of the innermost secondaries and on the tertials, especially the latter, tend to buffy-chestnut rather than pure buff. The primaries have emerged 4–6mm from their sheaths. Their eyes are not yet fully open and their bill retains the white egg tooth. Overall the bill is dusky-black with bright orange-yellow rictal flanges. At 14 or 15 days of age the nestlings are very alert but can still be easily handled and returned to the nest. They remain similar in coloration to the previous description, but the head is now spotted or scaled chestnut and blackish-brown, becoming slightly buffier on the nape, transitioning to coarse ochre-buff spotting and striping on the upper back. An ochre-buff stripe extends from just above and behind the eye to the postocular area. The tarsi and toes are dusky-pink to dark vinaceous. Apart from their undeveloped flight feathers, they are now largely as they will appear on fledging. **Seasonality** Birds with enlarged gonads collected in northern Peru (Loreto) in July and September, indicating that breeding may begin at end of the dry season (O'Neill 1974), and the records below suggest that nesting may occur year-round in eastern Ecuador. An immature male was observed in January in south-east Peru (Parker 1982). **Breeding Data** Clutch initiation, 6 March 2013, Gareno Lodge (HFG); laying female (gonads), 16 August 1987, 54km SE of Macas (WFVZ 42681); nest with eggs, 30 July, RN Pacaya-Samiria (Begazo & Valqui 1998); nest with eggs, 3 October 2012, Shiripuno Research Centre (HFG); nest with eggs, 17 August 2008, Gareno Lodge (R. Ahlman *in litt.* 2015); nest with eggs, 2 October 2004, Centro de Investigación y Capacitación Río Los Amigos (D. Lebbin *in litt.* 2016); developing brood patch, 30 April 2005, Estación de Biodiversidad Tiputini (J. Freile *in litt.* 2016); nestlings, 29 November 2000, Cocha Cashu (Hillman 2000); nestlings, 16 August 1999, Estación de Biodiversidad Tiputini (Dreyer 2002); active nest, 27 March 2014, Arajuno (S. Davies *in litt.* 2016); active nest, 20 July 2015, Centro de Investigación y Capacitación Río Los Amigos (eBird: S. Williams); active brood patch, 12 February 2000,

Estación de Biodiversidad Tiputini (B. Mila *in litt.* 2015); juvenile, 7 October 2007, Centro de Investigación y Capacitación Río Los Amigos (Tobias 2009); subadult, 8 April 2001, San Pedro (FMNH 429997); subadult, 1 September 1974, Villa Gonzálo (FMNH 299217); subadult, 17 July 1980, Cerro de Pantiacolla (FMNH 310703); subadult, 13 March 1923, Puerto Bermudéz (FMNH 65780); subadult, 29 September 2001, 13.4km NNW of Atalaya (FMNH 433387); immature with adults, 14 January 1978, Concesión de Ecoturismo Inka Terra (ML 13322, T.A. Parker).

TECHNICAL DESCRIPTION Sexes easily separable by plumage. *Adult male* Dark brown or grey-brown crown with a small silvery white postocular tuft that usually is hidden while foraging. This is one of the few species bearing such tufts that I have yet to see any photos of an individual flaring them out in an obvious display of some sort. A rather broad, but indistinct pale grey stripe extends forward from the postocular tuft, over the eye, sometimes as far as the lores. This separates the brownish crown from the clean grey underparts. The face, sides of the head and neck, and the breast are clear, slate-grey. The upperparts are mostly warm brown or olivaceous-brown. The primary upperwing-coverts are uniform dark brown, while both the greater and lesser secondary coverts, as well as the tertials, bear distinct, pale buffy markings at the tips, these usually somewhat trapezoidal or diamond-shaped. The throat is white, either entirely or narrowly in the centre, the remainder being heavily washed grey. The grey of the breast and sides blends smoothly into fulvous-brown on the flanks and then more orange-brown on the vent. The mid-belly is white, varying in extent as described for the throat. **Adult Male Bare Parts** *Iris* dark brown; *Bill* black; *Tarsi & Toes* apparently somewhat variable from dark grey to pinkish-grey, horn-coloured or grey-brown (Whitney 2003, Restall *et al.* 2006). *Adult Female* Crown more distinctly rufous than in males, but still dark, with the sides of the head and chest orange-rufous rather than grey. The orange-rufous is separated from the crown by an indistinct pale buffy-orange superciliary that extends from the front of the eye to the base of the small (usually hidden) postocular tuft of white feathers. The upperparts are brown, warmer and less greyish than in males, but also show irregular black markings across the upper back. The orange-rufous breast varies somewhat in intensity, inclining to ferruginous-red in some individuals, but slowly fades onto the sides and flanks, gradually blending to a warm olivaceous-brown on the vent. The belly and throat are white, as in males, and the wings and flight feather coverts are as described for males. **Adult Female Bare Parts** Similar to male. *Subadult* I still do not feel that I can adequately describe the plumage differences that consistently separate subadults from those in fully adult plumage. I tentatively suggest that subadults of both sexes will still have narrow black edges to the tips of the wing-coverts and tertials, and will have these markings inclining towards deep, rich buff that gradually becomes more reddish or rufous-buff on the feathers of the inner wing, especially on the tertials. Comparatively, the spotting on the wings of adults tends to be paler, has little or no reddish coloration, even on the tertials, and be more 'spot-shaped' than bar-like. Retained immature features on the contour feathers are not readily apparent, but it seems that the last areas to moult into adult plumage are the upper back and nape, and rump. The former remain indistinctly spotted or barred, and the latter retain a few indistinct dusky-grey-barred feathers. The extent of dark markings on the back may also increase with age. *Juvenile* Can be sexed at some point during transition from fledgling plumage. In general, in both sexes, juveniles are fairly similar to adults but for both sexes the following characters distinguish them from full adults. The nape and variable parts of the hindcrown retain their spotted or scaled appearance. A few immature feathers (pale ochraceous with dark bars near their tips) may remain on the rump or lower back. Below, the feathers are largely adult in coloration but scattered immature feathers remain, especially just below the throat, creating an overall 'messy' appearance to the breast and belly. It appears that birds of this age may just be gaining independence from their parents, and may still show indications of a paler, orange or yellow gape (although it is no longer inflated). Otherwise, bare-part coloration appears to be as in adults. *Fledgling* Not, apparently, possible to sex. They vary little from the description given for 14–15-day-old nestlings (see Reproduction). Their tail remains very short for some time after leaving the nest, and is possibly not even visible for several weeks after fledging due to the downy plumage of the rump and vent.

MORPHOMETRIC DATA *Wing* 67mm, 65mm; *Tail* 31mm, 31mm; *Bill* [exposed] culmen 14mm, 11mm; *Tarsus* 22mm, 23mm (*n* = 2, ♂,♀, Gyldenstolpe 1945a). *Wing* 68mm (*n* = 1, ♂, Gyldenstolpe 1951). *Wing* 72mm; *Tail* 32mm; *Bill* [total culmen?] 15mm; *Tarsus* 27mm (*n* = ?, ♀?, Snethlage 1914a, apparently repeated by Ruschi 1979). *Wing* 68mm, 68mm, 64mm; *Tail* 34mm, 35mm, 32mm; *Bill* 19mm, 20mm, 20mm; *Tarsus* 27mm, 24mm, 26mm (*n* = 3, ♂,♀,♀, the latter female was the holotype of Sclater's *C. torridus*, Taczanowski 1884). *Bill* from nares 9.4mm, 10.0mm, 9.6mm, 9.3mm; *Tarsus* 26.1mm, 25.0mm, 24.9mm, 24.7mm (*n* = 4, ♀,♂,♂,♂, WFVZ and AMNH). *Wing* 64mm, 66mm, 63mm, 65.5mm; *Tail* 28.5mm, 31.4mm, 29.8mm, 31.4mm; *Bill* from nares 8.8mm, 9.5mm, 9.6mm, 9.0mm, exposed culmen 12.9mm, 12.8mm, 12.7mm, 12.7mm, depth at front of nares 4.5mm, 4.1mm, 3.9mm, 4.0mm, width at front of nares 6.7mm, 5.6mm, 6.1mm, 5.5mm; *Tarsus* 26.3mm, 25.4mm, 25.0mm, 24.9mm (*n* = 4, ♀♂,?,♂, B. Mila *in litt.* 2017). **Mass** 23–26g (Whitney 2003); 23.9g (*n* = 5♀?, Clayton & Walther 2001); 23g (*n* = ?, ♀?, Terborgh *et al.* 1990); mean 24.5g (*n* = 4, ♀?, Londoño *et al.* 2015). 23.7g(lt), 24.4g(lt), 21.2g(lt), 28.5g(lt) (*n* = 4, ♂,?,♂,♀, B. Mila *in litt.* 2017); 25.1g, 22.4g, 24.3g, 22.7g(tr), 23.1g(m) (*n* = 5, ♀,♂,♂,?, WFVZ and MSB); 18g(lt), 21.5g(no), 22g, 24.5g, 21.5g(tr) (*n* = 5, ♂,♂,♂,♀, and ♂; MECN, M.V. Sánchez-Nivicela *in litt.* 2017); Londoño *et al.* (2015) also provide a mean basic metabolic rate of 0.34 watts from southern Peru. **Total Length** 11–12cm (Des Murs 1856, Whitney 2003, Restall *et al.* 2006, van Perlo 2009).

DISTRIBUTION DATA ECUADOR: *Orellana* Sunka, *c.*00°30'S, 76°59'W (MECN 2635); Laguna del Yuturi, 00°33'S, 76°02.5'W (XC 258626, J.V. Moore); Estación de Biodiversidad Tiputini, 00°38.5'S, 76°09'W (Svensson-Coelho *et al.* 2013, Blake & Loiselle 2015, XC); km 37 on Maxus road, 00°38.5'S, 76°27.5'W (MECN 6969, XC 249170/71, N. Krabbe); Estación de Investigación Yasuní, 00°40.5'S, 76°24'W (XC 61298/99, A. Spencer); Daimi, 00°59.5'S, 76°12'W (ANSP 183356); Shiripuno Research Centre, 01°06'S, 76°43'W (XC 165551, J.F.Vaca B. , ML 51551041, photo S. Woods); Boanamo, 01°16'S, 76°23'W (Greeney *et al.* 2018). *Napo* 40km SE of Coca, *c.*00°44'S,

76°44'W (WFVZ 45750–752); Yachana Lodge, 00°52.5'S, 77°16'W (HFG); Gareno Lodge, 01°02'S, 77°23.5'W (HFG). *Pastaza* Tzapino, 01°11'S, 77°14'W (LSUMZ 83355/56); Arajuno, 01°14'S, 77°41'W (S. Davies *in litt.* 2016); Campamento Villano B., 01°27'S, 77°26.5'W (J. Freile *in litt.* 2017, QCAZ); Canelos, 01°32.5'S, 77°45'W (XC 249585, N. Krabbe); Sarayacu, 01°44'S, 77°29'W (MECN 2643); Territorio Achuar, *c.*01°45'S, 76°30'W (J. Freile *in litt.* 2017); Río Rutuno, 01°55'S, 77°14'W (SBMNH 8630–631); Montalvo, 02°03'S, 77°00.5'W (MECN); Río Tigre, 02°03'S, 76°04'W (J. Freile *in litt.* 2017); Río Copatza, 02°07'S, 77°27'W (ANSP 169664); Río Conambo, 02°07'S, 76°03'W (MECN 2637); W of Andoas, *c.*02°34'S, 76°38'W (ANSP 169665). *Morona-Santiago* 5km SW of Taisha, *c.*02°20'S, 77°27.5'W (ANSP 182532–536, MECN 2642); Kapawi Lodge, 02°43'S, 77°01.5'W (J. Freile *in litt.* 2017); 54km SE of Macas, *c.*02°46.5'S, 77°52'17.24'W (WFVZ 42681); Yaupi, 02°49.5'S, 77°59'W (J. Freile *in litt.* 2017; QCAZ); Santiago, 03°03'S, 78°00.5'W (ANSP 181678, MECN 2636). **PERU:** *Loreto* Panguana, 02°08'S, 75°09'W (Stotz & Alván 2007); Río Curaray, *c.*02°23'S, 74°06'W (MECN 2638); Teniente López, 02°34'S, 76°07'W (KUNHM 87589); Alto Nanay, 02°48.5'S, 74°49.5'W (Stotz & Alván 2007); 1.5km S of Libertad, 03°02'S, 73°22'W (LSUMZ 111367–369, 110254–259); Puerto Indiana, 03°28'S, 73°03'W (AMNH 231621); RN Allpahuayo-Mishana, 03°55'S 73°33'W (Alonso *et al.* 2012); Belén de Juda, 03°58.5'S, 73°23'W (MUSA 3714); Nauta, 04°30.5'S, 73°35'W (Taczanowski 1884); Quebrada Buenavista, 04°50'S, 72°23.5'W (Lane *et al.* 2003); RN Pacaya-Samiria, 05°02'S, 74°59'W (Begazo & Valqui 1998); Chamicuros, 05°30'S, 75°30'W (Taczanowski 1884); Esperanza, 05°32.5'S, 75°44'W (LSUMZ); Santa Cruz, 05°33'S, 75°48'W (Taczanowski 1884); 'Campamento Alto Cahuapanas', 05°40'S, 76°50.5'W (Stotz *et al.* 2014); 'Campamento Alto Cachiyacu', 05°51.5'S, 76°43'W (Stotz *et al.* 2014); NE bank of upper Río Cushabatay, 07°08'S, 75°41'W (LSUMZ 161805–807); *c.*85km SE of Juanjui, 07°36'S, 75°56'W (LSUMZ 170919–921). *Amazonas* Caterpiza, 04°01'S, 77°35'W (LSUMZ 92502); Villa Gonzálo, Río Santiago, 04°08'S, 77°45'W (FMNH 299217/18); Pagat, Río Cenepa, 04°20'S, 78°14'W (LSUMZ 84966); Quebrada Achunts/Kus, 9.5km WNW of Chávez Valdivia, *c.*04°27'S, 78°16.5'W (LSUMZ 88094/95, 34242). *San Martín* Sianbal, 06°38.5'S, 76°05'W (Merkord *et al.* 2009, MSB 27837). *Ucayali* SE slope of Cerro Tahuayo, 08°08'S, 74°01'W (LSUMZ 156624/25); Quebrada Sipiria, 09°28.5'S, 74°26.5'W (AMNH); Río La Novia, 09°56'S, 70°42'W (Angulo *et al.* 2016); Balta, 10°06'S, 71°14'W (LSUMZ). *Huánuco* Mirador Pintuyacu, 09°22.5'S, 74°48.5'W (eBird: J.A. Uribe); *c.*26km NE of Yuyapichis, 09°29'S, 74°47'W (Socolar *et al.* 2013). *Pasco* Puerto Bermúdez, 10°17.5'S, 74°56.5'W (Zimmer 1930, FMNH 65780); Puerto de Yesup, 10°27'S, 74°54'W (Bond 1953, ANSP 92300); San Juan, 10°30'S, 74°53'W (KUNHM 69032/33). *Cuzco* Heliopuerto Pagoreni A, 11°43'S, 72°54'W (MUSA 3531); *c.*8km SW of Comunidad Nativa Camisea, 11°46'S, 73°00'W (LSUMZ 64231); Las Malvinas, 11°53.5'S, 72°57'W (Angehr *et al.* 2001); Río Tono, 13°00'S, 71°11'W (FMNH 322361); San Pedro, 13°03.5'S, 71°33'W (FMNH 429997–999). *Madre de Dios* 13.4km NNW of Atalaya, 10°38.5'S, 73°49'W (FMNH 433384–388); PN Manú, 11°55'S, 71°18'W (ML 53505, S. Connop); Estación Biológica de Cocha Cashu, 11°59'S, 71°23'W (Robinson *et al.* 1990, Terborgh *et al.* 1990, Hillman 2000, AMNH 824074); Concessión ArBio, 12°10.5'S, 69°23.5'W (F.

Angulo P. *in litt.* 2017); Romero Rainforest Lodge, 12°13.5'S, 70°59'W (R.C. Hoyer *in litt.* 2017); Río Chilive– Río Madre de Dios confluence, 12°20'S, 70°58'W (FMNH 260923); Concesión de Ecoturismo Inka Terra, 12°30'S, 69°00'W (ML 13322, T.A. Parker); Centro de Investigación y Capacitación Río Los Amigos, 12°34'S, 70°06'W (Tobias 2009, MSB 35671); Albergue, Cuzco Amazónico, 12°34.5'S, 69°04.5'W (MVZB 169631); Río Palotoa, 12km from mouth, 12°34.5'S, 71°26'W (Walker *et al.* 2006, FMNH 322350–353); Pantiacolla Lodge, 12°39'S, 71°14'W (XC 221516, P. Boesman); mouth of Río Colorado, 12°39'S, 70°20'W (FMNH 251973/74); Quebrada Aguas Calientes, 12°40'S, 71°16'W (FMNH 398029); Cerro de Pantiacolla, 12°40'S, 71°13'W (FMNH 322354–358); Amazonia Lodge, 12°52'S, 71°22'W (XC 6440, N. Athanas); RC Amarakaeri, 12°59.5'S, 71°00.5'W (MUSA); RN Tambopata, 13°04.5'S, 69°35'W (Donahue 1994). *Puno* Astillero, 13°22'S, 69°37'W (AMNH 146081); Río Tavára, 13°25'S, 70°20'W (AMNH 147660); La Pampa, 13°40'S, 69°35.5'W (AMNH 146080). **BRAZIL**: *Amazonas* São Paulo de Olivença, 03°39'S, 69°06'W (CM); Estirão do Equador, 04°32'S, 71°37'W (MPEG 73006/92); vicinity of Arimã, 05°47'S, 63°38'W (Gyldenstolpe 1945a, 1951, YPM, CM, MVZB, USNM, UMMZ); Igarapé do Gordão, 06°18'S, 69°30'W (Gyldenstolpe 1945a); Eirunepé, 06°50.5'S, 70°14.5'W (Gyldenstolpe 1945a); Pôrto Alegre, 08°34'S, 67°29.5'W (Snethlage 1907a, 1914a, MPEG 3654); *c.*8km NE of Boca do Acre, 08°43'S, 67°20'W (MPEG 3653). *Acre* Igarapé Ramon, 07°27.5'S, 73°46.5'W (Whitney *et al.* 1997, MPEG 52776); Feijó, 08°10'S, 70°21'W (WA 1699355, T. Melo); near Porongaba, *c.*08°45'S, 72°49'W (Whittaker & Oren 1999, Guilherme 2009, MPEG 49695); Seringal Oriente, near Vila Taumaturgo, *c.*08°48.5'S, 72°46'W (Novaes 1957, Guilherme 2009, MPEG 26650/51); RESEX do Alto Juruá, 09°00'S, 72°32'W (Whittaker *et al.* 2002); Ramal 'Oco do Mundo,' km16, 09°50'S, 67°10'W (Guilherme 2009, MPEG 64385); near Rio Branco, *c.*10°00'S, 67°49'W (MNRJ 31091); Ramal Novo Horizonte, km09, 10°07.5'S, 675'W (Guilherme 2009, MPEG 60885); RESEX Chico Mendes, 10°22'S, 68°43'W (Mestre *et al.* 2010); Cabeceira do Rio Acre, 10°51.5'S, 69°58'W (Aleixo & Guilherme 2010); Foz do Igarapé dos Patos, 10°56'S, 69°55'W (Guilherme 2009, UFAC 0036/37). **BOLIVIA**: *La Paz* 25km NW of Santa Ana de Madidi, 13°10'S, 68°45'W (ML 52340, T.A. Parker).

TAXONOMY AND VARIATION Monotypic. *Conopophaga peruviana* was, for many years, thought to be allied to *C. ardesiaca* and *C. castaneiceps*, as well as at least some of the *C. lineata* species group (including *C. roberti* and, perhaps, *C. melanogaster*) (Whitney 2003). It was thus most frequently placed between *C. roberti* and *C. ardesiaca* in linear sequences (Cory & Hellmayr 1924, Peters 1951, Sibley & Monroe 1990, Howard & Moore 1991, Altman & Swift 1993, Whitney 2003, Clements 2007) or, alternatively, between *C. castaneiceps* and *C. ardesiaca* (Sclater 1958c); *C. melanops* and *C. ardesiaca* (Meyer de Schauensee 1982, Altman & Swift 1986); *C. aurita* and *C. castaneiceps* (Restall *et al.* 2006), or *C. melanops* and *C. aurita* (Ridgely & Tudor 1994, 2009). The molecular study by Batalha-Filho *et al.* (2014), however, has recently led to the placement of *C. peruviana* between *C. aurita* and *C. cearae* (Remsen *et al.* 2017), as I have elected to do here.

STATUS Not globally threatened. Although rarely observed and generally uncommon, Ash-throated Gnateater is not considered at risk (BirdLife International 2017). Seems to sing less frequently than other conopophagids, making it more difficult to detect. Rare to uncommon in Peru, at least where it is known to occur, but its distribution becomes rather poorly documented near the southern end of its range. Nevertheless, it has been reported from a fair number of protected areas. **PROTECTED POPULATIONS Ecuador** PN Yasuní (MECN, ANSP, XC); RP Yachana Lodge (HFG). **Peru** PN Manú (Terborgh *et al.* 1984, Robinson *et al.* 1990, Hillman & Hogan 2002); PN Otishi (eBird: D.G. Olaechea); RN Pacaya-Samiria (Begazo & Valqui 1998); RN Allpahuayo-Mishana (Alonso *et al.* 2012); RN Tambopata (Parker *et al.* 1994a); RC Amarakaeri (MUSA); Reserva Cuzco-Amazónico (Davis *et al.* 1991, KUNHM); PN Cordillera Azul (LSUMZ 161805–807); Centro de Investigación y Capacitación Río Los Amigos (Tobias 2009). **Brazil** PN Serra do Divisor (Whitney *et al.* 1997, MPEG); RESEX Chico Mendes (Mestre *et al.* 2010); ESEC do Rio Acre (Aleixo & Guilherme 2010); RESEX do Alto Juruá (Whittaker *et al.* 2002). **Bolivia** PN y ÁNMI Madidi (Hennessey *et al.* 2003b).

OTHER NAMES *Conopophaga torrida* was described by Sclater (1858c) based on an adult female collected at Chamicuros, Loreto, Peru (NHMUK 1857.11.28.237) (Warren & Harrison 1971). This name appeared infrequently in the literature (Sclater 1862, Gray 1869, Taczanowski 1884) before it was synonymised by Sclater himself (1890), when he concluded that *C. torrida* was, in fact, a female *C. peruviana*. This synonymy was quickly adopted (e.g., Dubois 1900). Quite a few common names are recored in the literature. **English** Fulvous-bellied Gnat Eater/ Gnat-eater (Brabourne & Chubb 1912, Cory & Hellmayr 1924, Gyldenstolpe 1945a, 1951); Ash-breasted Gnateater (Parker 1982). **French** Conopophage du Pèrou (Des Murs 1856). **Portuguese** chupa-dente-do-peru (Willis & Oniki 1991b, Guilherme 2016); chupa-dente-barriga-cinzenta (Ruschi 1979). **Spanish** Jejenero gorgicenizo, Pájaro cuchi, Pájaro zancudo (Valarezo-Delgado 1984); Jejenero golicenizo (Ortiz-Crespo *et al.* 1990); Jejenero golicinéreo (Ridgely & Greenfield 2001).

Ash-throated Gnateater, adult female, Shiripuno Research Centre, Pastaza, Ecuador, 5 October 2012 (*Harold F. Greeney*).

Ash-throated Gnateater, adult female incubating, Shiripuno Research Centre, Pastaza, Ecuador, 5 October 2012 (*Harold F. Greeney*).

Ash-throated Gnateater, nest and complete clutch, Shiripuno Research Centre, Pastaza, Ecuador, 3 October 2012 (*Harold F. Greeney*).

Ash-throated Gnateater, older nestlings in nest, Pantiacolla, Madre de Dios, Peru, 20 October 2013 (*Gustavo A. Londoño*).

Ash-throated Gnateater, young nestlings in nest, Pantiacolla, Madre de Dios, Peru, 29 October 2013 (*Gustavo A. Londoño*).

CEARA GNATEATER
Conopophaga cearae   Plate 3

Conopophaga lineata cearae Cory, 1916, Field Museum of Natural History (Zoological Series), vol. 1, issue 10, p. 337, Serra de Baturité, Município de Pacotí, Ceará, Brazil [*c.*04°14'0.24'S, 38°55'0.12'W]. The holotype, an adult male (FMNH 47264), was collected by R. H. Becker, on 18 July 1913. Topotypical skins are held at USNM and FMNH, and recordings made at or very near the type locality are available on Xeno-canto (D.F. Lane, H. Matheve, N. Athanas) and WikiAves (L. Caranha, J.A. Alves).

Ceara Gnateater was recently split from the wide-ranging Rufous Gnateater, having long been considered a subspecies of it (Remsen *et al.* 2017). As a full species, the range-restricted Ceara Gnateater was, for more than a decade after its description, known only from the type (Hellmayr 1929). It is best known from the Serra de Baturité, in northern Ceará, where it favours montane evergreen forest, especially near thickets of bamboo. In other parts of its distribution, now known to extend as far south as Chapada Diamantina in northern Bahia, it can be found in a broader range of forest types, but is still generally a bird of more humid regions. Ceara Gnateater is very poorly studied, and all aspects of its biology and natural history are urgently in need of attention, especially in light of the need for a careful evaluation of its global conservation status.

IDENTIFICATION 11–12cm. The sexes are separable only by minor differences. Males are generally similar to the nominate race of Rufous Gnateater, but differ in having the crown brighter, more rufous-brown, than the back. The lores and a narrow superciliary are clear grey to near-white, and this streak extends into a bright white postocular tuft. The throat, the sides of the face and neck, and also the breast and sides are clean, bright peachy or tangerine-orange, usually considerably brighter than in Rufous Gnateater. The mid-belly is pure white, grading to pale orange or orange-buff on the flanks and vent. The bill is bicoloured, the maxilla black and the mandible dusky-yellow or pinkish. Females are very similar, perhaps not reliably separated in the field without considerable experience. They have the white postocular tuft reduced, with the bases of the elongated feathers usually tinged pale orange to some degree. They also tend to have the grey of the lores paler and more extensive. The range of Ceara Gnateater overlaps with only one congener, Black-cheeked

Gnateater, males of which are immediately distinguished by their strong black mask and grey underparts. Female Black-cheeked Gnateaters are more similar but are browner above, less orange below, and have distinctly spotted wing-coverts. In general, no other gnateaters are so plainly patterned, except Rufous Gnateater, which replaces the present species to the south.

DISTRIBUTION Ceara Gnateater is endemic to, and somewhat patchily distributed through, north-eastern Brazil in northern Ceará, along the coast from Rio Grande do Norte to Alagoas, and in the Chapada Diamantina region of northern and central Bahia. Whitney (2003) pointed to the fact that there is some question as to the racial affinities of birds from Pernambuco and northern Bahia. Indeed, Naumburg (1937) assigned specimens collected at Brejó da Madre de Deus and Quipapá (Pernambuco) to nominate *lineata*, but this was because he was basing his definition of *lineata* on two recently collected skins from Ituaçu in the southern Chapada Diamantina. As he felt these were 'practically topopypes' of *lineata* from Vitória da Conquista, despite *c.*130km separating the two locations, he may have only cursorily examined the holotype of *lineata*. I find this somewhat odd, as this skin should have been available to him, but perhaps its poor condition hindered proper comparision, particularly with respect to the characters that I feel are key for separating *lineata* from *cearae* (see Rufous Gnateater). My examination of the two Ituaçu skins (AMNH 242781/82) revealed them to clearly represent *cearae*. Pinto (1978) felt that specimens from north-east Bahia (Senhor do Bonfim) are representative of *cearae*, while Novaes (1947), examining a specimen from eastern Pernambuco (Recife), felt that it showed plumage intermediate between *cearae* and nominate *lineata* (but was possibly immature). In the face of these, and other conflicting opinions, Whitney (2003) decided to include Pernambuco and northern Bahia populations in the nominate (Rufous Gnateater). After examining numerous specimens and photographs of birds from Ceará, Paraíba, Pernambuco, Alagoas, Paraiba and Bahia, however, in plumage I feel that populations assignable to Ceara Gnateater extend as far south as central Bahia, at least to Ibicoara, Iramaia and Ituaçu. Records from Bahia, south and east of these locations, are here considered to represent Rufous Gnateater (which see). Birds from Ibiquera, mentioned by Whitney (2003) as differing

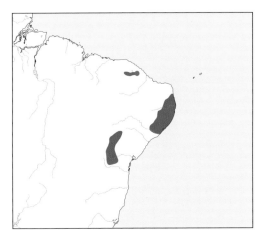

somewhat in plumage, are provisionally included within the range of Ceara Gnateater. Furthermore, until the vocalisations of the species involved are better known, a single recording from Maracás is provisionally considered to represent Ceara Gnateater. More information from this area (east of the Chapada Diamantina) is desirable, however, as populations in this region are potentially better assigned to Rufous Gnateater.

HABITAT Ceara Gnateater inhabits upland evergreen forest, known locally as *brejo* (Silveira *et al.* 2003, Roda & Carlos 2004, Albano & Girão 2008, Pereira *et al.* 2016), to a lesser degree deciduous forest (Brooks *et al.* 1999), but in Rio Grande do Norte it also inhabits denser scrubby *restinga* habitats within the narrow coastal sand dunes (Lobo-Araújo *et al.* 2013). It is replaced by *C. melanops* inside taller, more humid forest just inland of these dunes (del Hoyo *et al.* 2017b). Although it occurs near sea level in Rio Grande do Norte and Pernambuco, Ceara Gnateater is usually found above 300m in most parts of its range and is apparently found above 600m only in the Serra de Baturité (del Hoyo *et al.* 2017b).

VOICE The vocalisations of Ceara Gnateater are greatly in need of review, especially with respect to how they may or may not differ from those of Rufous Gnateater. The song is a series of *c.*10–13 sweet, whistled notes, rising slightly and usually ending in slightly louder and higher pitched notes, the whole lasting 2.16–3.0s, at 2.7–3.0kHz. Individual notes are *c.*0.14s in duration and, when illustrated in a sonogram, are shaped like the letter 'M' (van Perlo 2009, Boesman 2016). Their call notes have not been specifically described, but are probably similar to descriptions of the calls of Rufous Gnateater, and include a piercing *tchief!* or harder *tcheek!* (Whitney 2003), also described as a sharp, thin *tsiew* (van Perlo 2009).

NATURAL HISTORY There has been nothing published on the behaviour of Ceara Gnateater. It is probably similar to that of Rufous Gnateater, foraging low in small saplings and shrubs, occasionally descending to the ground (Bertoni 1901, Chubb 1910, Erickson & Mumford 1976).

DIET The diet of Ceara Gnateater has not been investigated, but is probably similar to other gnateaters in eastern Brazil.

REPRODUCTION The nest, eggs and reproductive behaviour of Ceara Gnateater remain undescribed. Lamm (1948) reported four 'fully-grown immature birds' [juveniles transitioning?] found in the jungle near Recife on 30 April 1944. **BREEDING DATA** Subadult, 3 December 2016, Areia (WA 2388020, R. Hoppen); subadult transitioning, 24 September 2016, Jequié (WA 2294045, D. Campos).

TECHNICAL DESCRIPTION See Identification for description of female. *Adult Male* Rufous-chestnut to brownish-orange crown, with a narrow grey or greyish-orange supercilium that extends from the prominent grey lores into a silvery-white postocular tuft that is composed of *c.*5–7 elongated feathers. These are entirely pure, silver-white, without any trace of greyish at the base (as in most Rufous Gnateaters). The upperparts are usually slightly browner or greyish-brown than the crown, with the wings and tail browner still. Flight feathers usually have thin, indistinct dull rust-brown anterior margins. Alula and first 1–2 wing-coverts dark brown, remaining coverts similar to flight feathers, sometimes with similar fringes (though this is really only visible in close inspection). The sides of head and neck, chin and throat, and breast and sides are clear, orange-chestnut, matching or slightly brighter than the crown. Orange of sides becomes duller, more ochraceous posteriorly. Flanks and undertail-coverts approach dusky-orange to grey-brown or orange-buff, leaving a distinctive bright white to pale buff-white belly patch. Occasionally, especially in eastern Bahia, adults may have an indistinct whitish crescent-shaped patch on upper breast. *Adult Female* The white postocular tuft is greatly reduced, tinged pale orange, or nearly absent. They also tend to have the grey of the lores paler than in males and more extensive, sometimes extending over the eye and across the base of the bill. **Adult Bare Parts** *Iris* dark brown; *Bill* maxilla black, mandible dusky-yellow or yellowish-pink, gape, even in adults, often appears slightly inflated and pinkish; *Tarsi & Toes* dusky-yellow, tending to brighter yellow or even orange-yellow at the posterior margin of the tarsi and the toe-pads. *Immature* I have not yet identified characters which permit separation of the sexes in immature plumage. I have also yet to see any examples of recently fledged birds or those likely to still be in the care of their parents. As the extent of plumage variation in adults is unknown, therefore the following description of birds that I consider to be not fully adult should be used with caution. *Subadult* Similar to adult in overall plumage, somewhat duskier on the crown. The most notable difference is the upperwing-coverts which are broadly tipped with a buffy bar above a thin black margin. **Subadult Bare Parts** As in adults, gape possibly somewhat yellower, but rictal flanges not inflated.

MORPHOMETRIC DATA *Wing* 72–75mm (*n* = 3, Serra de Baturité), 66mm (*n* = 1, Dois Irmãos, Pernambuco) (♂♂, Novaes 1947). *Wing* 66mm; *Tail* 35mm; *Bill* 14mm (*n* = 1, ♀?, Dois Irmãos, Berla 1946). *Wing* 73mm; *Bill* [exposed culmen] 13mm (*n* = 1, ♂, holotype, Cory & Hellmayr 1924). *Bill* from nares 9.9mm, 9.4mm, exposed culmen 13.5mm, 12.7mm, depth 4.4mm, 4.4mm, width at nares 5.4mm, 5.3mm, width at base 6.7mm, 6.9mm; *Tarsus* 29.7mm, 28.8mm (*n* = 2, ♂,♀, FMNH 122749/50). **Mass** 16–27g, mean of 11 birds, 22.1g (for *C. lineata*, may or may not include some individuals of *C. cearae*; Whitney 2003).

DISTRIBUTION DATA Endemic to **BRAZIL**: *Ceará* Mulungu, 04°18'S, 38°59.5'W (WA 1788703, A. Adeodato); Santa Quitéria, 04°19.5'S, 40°09'W (WA 1061211, M. Holderbaum); Baturité, 04°20'S, 38°52.5'W (WA 925959, L. Cruz); Canindé, 04°21'S, 39°18.5'W (WA 1045071, M. Holderbaum); Serra do Machado, 04°31'S, 39°39'W (Girão *et al.* 2007, Girão & Albano 2008); Itatira, 04°37'S, 39°34'W (Girão & Albano 2008). *Rio Grande do Norte* Baía Formosa, 06°22'S, 35°00'W (Olmos 2003, WA 2159276, D. Gurgel). *Paraíba* Mamanguape, 06°50'S, 35°08'W (Girão & Albano 2008, WA 914403, M. Holderbaum); Areia, 06°58'S, 35°42'W (WA 2234336, C. Brito); Conde, 07°15.5'S, 34°55'W (WA 1423564, J. Medcraft). *Pernambuco* Refúgio Ecológico Charles Darwin, 07°48.5'S, 34°56.5'W (Magalhães *et al.* 2007); Vertentes, 07°54'S, 35°59'W (WA 2302872, A. Gomes); Paulista, 07°56'S, 34°52'W (Girão & Albano 2008); Dois Irmãos, 08°01'S, 34°56.5'W (Berla 1946, Novaes 1947); Recife, 08°03'S, 34°54.5'W (Novaes 1947, Lamm 1948, USNM 383287); Brejó da Madre de Deus, 08°09'S, 36°22'W (Naumburg 1937, Novaes 1947); Reserva Estadual Gurjaú, 08°14.5'S, 35°03'W (Lyra-Neves *et al.* 2004); Caruaru, 08°17'S, 35°58'W (Girão & Albano 2008); Lagoa dos Gatos, 08°39.5'S, 35°54'W (WA 137168,

S.J. Jones); Caetés, 08°47.5'S, 36°40.5'W (Girão & Albano 2008); Quipapá, 08°48.5'S, 36°01'W (Naumburg 1937); Brejão, 09°02'S, 36°34'W (Girão & Albano 2008, AMNH 242775–780); Francês/Barra de São Miguel area, 09°49'S, 36°00.5'W (Lobo-Araújo *et al.* 2013). *Alagoas* Murici, 09°18.5'S, 35°56.5'W (Girão & Albano 2008); São Luís do Quitunde, 09°19'S, 35°33.5'W (Girão & Albano 2008); Quebrangulo, 09°19'S, 36°28'W (WA 57396, C. Albano); Pedra Talhada, 09°19'S, 36°54'W (Girão & Albano 2008); Mata da Sálvia, 09°32'S, 35°50'W (Silveira *et al.* 2003); Traipu, 09°58'S, 37°00'W (WA 400517, S. Leal); Pontal do Coruripe, 10°09'S, 36°08'W (Girão & Albano 2008). *Bahia* Senhor do Bonfim, 10°27'S, 40°11'W (Pinto 1938, 1978); Morro do Chapéu, 11°33'S, 41°09.5'W (WA 2130395, C.E. Agne); Palmeiras, 12°30.5'S, 41°35'W (WA 1677274, C.S. Oliveira); Lençóis, 12°34'S, 41°23.5'W (WA 715946, A. Fraga, WA 1242931, F. Pacheco, WA 126391, C. Albano); *c.*1.75km W of Ouricuri, 12°35'S, 41°22'W (XC 46257/80415, C. Albano/J. Minns); Fazenda Bananeira, 12°35'S, 40°50'W (MPEG 51160/61); Ibiquera, 12°39'S, 40°56'W (Whitney 2003); Iramaia, 13°17.5'S, 40°57.5'W (WA 1513982, S. Sampaio); Maracás, 13°26'S, 40°26'W (WA 1923928, recording S. Sampaio); Ibicoara, 13°25'S, 41°17'W (WA 987873, M. Barreiros; WA 2440328, T. Moura; recordings WA 981805, M. Barreiros, WA 1297087, F. Pacheco); Ituaçu, 13°49'S, 41°21.5'W (Naumburg 1937, AMNH 242781/82); Jequié, 13°51.5'S, 40°05'W (WA 2294045, D. Campos).

TAXONOMY AND VARIATION Monotypic. Ceara Gnateater has previously been recognised as a full species (Cory & Hellmayr 1924, Hellmayr 1929, Morony *et al.* 1975, Roda & Carlos 2003, Silva *et al.* 2003, Gill & Wright 2006). The last authoritative treatment of the group (Whitney 2003), however, included it as a subspecies of Rufous Gnateater, but recognised that the morphological and vocal differences probably merited elevation to species rank. This assertion was supported by the findings of Batalha-Filho *et al.* (2014) who, using DNA sequence data (two mitochondrial genes, two nuclear introns, 3,300 bp), recovered Ash-breasted Gnateater as the probable sister species to Ceara Gnateater. In light of this, Remsen *et al.* (2017) decided to elevate *cearae* to species rank (Remsen 2015). Whitney (2003), while treating Ceara Gnateater as conspecific with Rufous Gnateater, mentioned differences in specimens from Ibiquera, suggesting the possibility of an unnamed taxon in that area. I was unable to investigate this further, but that population is here included as Ceara Gnateater.

STATUS Ceara Gnateater has a restricted and patchy distribution, with all populations facing severe and ongoing habitat fragmentation. It appears common at only a few of its known sites and appears to be genuinely rare at most sites. It is, however, quite common throughout the Serra de Baturité (Bret Whitney *in litt.* 2017). In Alagoas, where only 2% of Atlantic Forest habitat remains (Brown & Brown 1992), only three of 15 fragments were found to contain Ceara Gnateater (Silveira *et al.* 2003). Overall, the situation appears rather dire, as it does for many range-restricted Atlantic Forest endemics. Indeed, Ceara Gnateater may be among the most threatened species in the genus, as it has the most restricted range, most of which is severely fragmented, leading to the species being considered Vulnerable in Brazil (Girão & Albano 2008), even before it was split from Rufous Gnateater. More recently, its status has been upgraded to Endangered by the Brazilian Ministério do Meio Ambiente (Pereira *et al.* 2016). BirdLife International (2017) has recognised the split of the present species from Rufous Gnateater, but has maintained its global threat level at Least Concern. It seems likely that, once there has been time to re-evaluate, Ceara Gnateater will merit an upgrade in threat level. More information on the distribution of Ceara Gnateater is certainly a priority, if sound conservation plans are to be implemented. Girão & Albano (2008) have suggested that the hills around Maranguape, Uruburetama and Meruoca would be likely areas to search. In this region its range may overlap with Hooded Gnateater. If my inference is correct regarding the distribution of *cearae*, however, then hope is not lost, as it appears the species has a fairly healthy population within Chapada Diamantina National Park, especially around Lençóis. Studies into all aspects of Ceara Gnateater's biology are needed, as most of what was previously considered to apply to its conservation needs was derived from studies of Rufous Gnateater in the south. Rufous Gnateater is thought to be tolerant of at least mild anthropogenic habitat alteration and forest fragmentation (Antunes 2005), being slightly more prone to cross open areas than other understorey species (Oliveira *et al.* 2011, 2012) and is even able to persist in small (19ha) fragments in urban areas (Silva 2001). Given what we know to date, this may not apply to Ceara Gnateater, which is suspected to quickly disappear from fragments <70ha (Girão & Albano 2008). Although studies are lacking, the creation of more protected areas and conservation corridors is of utmost priority for the species (Girão & Albano 2008). These authors also suggest that investigations into captive

breeding and reintroduction might be useful in efforts to reduce the effects of reproductive isolation caused by forest fragmentation. **Protected Populations** ÁPA do Maciço do Baturité (Rodrigues *et al.* 2003; FMNH, USNM); REBIO Federal Guaribas (Girão & Albano 2008); ESEC Tapacurá, ESEC Caetés, REBIO Federal de Saltinho (Girão & Albano 2008); Parque Ecológico Municipal Vasconcelos Sobrinho (Girão & Albano 2008); Parque Ecológico Dois Irmãos (Girão & Albano 2008); REBIO Federal de Pedra Talhada (Girão & Albano 2008); ESEC Murici (Girão & Albano 2008); RPPN Mata Estrela (Olmos 2003); RPPN Pedra D'Anta (ML 63352811, photo C. Gussoni); Refúgio Ecológico Charles Darwin (Magalhães *et al.* 2007).

OTHER NAMES As Ceara Gnateater was, for most of its history, considered a subspecies of Rufous Gnateater, there are many uses of *Conopophaga lineata cearae* or simply *Conopophaga lineata* in the old literature (not listed), with far fewer in the past two centuries (e.g., Silveira *et al.* 2003, Albano & Girão 2008, Albano 2010, Girão & Albano 2008); *Conopophaga lineata lineata* (Pinto 1940, Berla 1946); *Conopophaga* (*lineata*) *cearae* (Brooks *et al.* 1999, Olmos 2003); *Conopophaga cearae lineata* (Lobo-Araújo *et al.* 2013). **English** Ceará Gnat Eater (Cory & Hellmayr 1924); 'Caatinga' Gnateater (Forrester 1993); Silvery-tufted Gnat-eater (for Rufous Gnateater, which see; Brabourne & Chubb 1912). **Portuguese** sabiazinha, chupa-dente, cuspidor-do-nordeste (Girão & Albano 2008). **Spanish** Jejenero de Ceará (del Hoyo *et al.* 2017b). **French** Conophage du Ceara. **German** Cearamückenfresser (del Hoyo *et al.* 2017b).

Ceara Gnateater, Chapada Diamantina, Bahia, Brazil, 24 July 2012 (*Ciro Albano*).

Ceara Gnateater, Guaramiranga, Céara, Brazil, 8 July 2015 (*Ciro Albano*).

Ceara Gnateater, Guaramiranga, Céara, Brazil, 28 May 2016 (*Caio Brito*).

Ceara Gnateater, Pacoti, Céara, Brazil, 21 July 2009 (*João Quental*).

Ceara Gnateater, Serra Do Baturité, Céara, Brazil, 4 December 2014 (*Ciro Albano*).

HOODED GNATEATER
Conopophaga roberti   Plate 3

Conopophaga roberti Hellmayr, 1905, Bulletin of the British Ornithologists' Club, vol. 15, p. 54, 'Igarapé-Attú, near Pará, Brazil' (= Igarapé-Açu, *c.*100km NE of Belém, Pará, 01°08.5'S, 47°37'W; coordinates more accurate than those of Paynter & Traylor 1991a). The scientific name honours the collector of the type specimen, Monsignor Alphonse Robert (Hellmayr 1905a). The holotype, an adult male collected 4 April 1904, was deposited at Tring Museum (Hartert 1922) but later sent to the AMNH as part of the Rothschild Collection and is currently housed there (AMNH 488908) (LeCroy & Sloss 2000). Monotypic.

One of three gnateaters endemic to Brazil, Hooded Gnateater is restricted to the area between the east bank of the lower Rio Tocantins in Pará, east to north-west Ceará (Peters 1951, Pinto 1978) where, based on the number of eBird records, it is apparently fairly easy to find in the area north and west of Sobral. In the west of its distribution, the species seems to be solely reliant on thickets at the edge of humid forest, but further east

Hooded Gnateater inhabits increasingly xeric wooded habitats, although it always prefers densely vegetated areas near the ground (Whitney 2003). Like all of its congeners, this is a beautiful bird when seen well, but it can be hard to get a glimpse of in the species' tangled habitat. Males are strikingly patterned, largely black below and on the head, with narrow white postocular tufts, while females are much more rufous-brown above and on the crown, with a grey face and underparts. At the time of its description, Hellmayr (1905a) considered Hooded Gnateater most closely related to Black-bellied Gnateater, a thought shared by subsequent authors (Novaes 1947). It appears, however, that Hooded Gnateater may, in fact, be more related to Rufous, Ash-throated, Slaty and Chestnut-crowned Gnateaters (Whitney 2003). There are no aspects of the biology, ecology, evolution or taxonomy of Hooded Gnateaters which are not in need of further research. Until recently, it was one of the few *Conopophaga* with a published description of the nest and eggs (although this was previously overlooked; Whitney 2003). To date, however, only two nests and one clutch have been described and there is nothing known of the species' behaviour, nestling diet or other aspects of their reproductive biology, details of which may prove useful for comparative or conservation purposes in the future.

IDENTIFICATION 12–13cm. Hooded Gnateater is distinctive, with a black head and neck. The following is based on Greeney (2013g). The black 'hood' extends to the lower breast, ceasing abruptly above the grey flanks and white belly. Males have a strongly contrasting white postocular tuft, which is often partially flared and usually quite visible. The upperparts are somewhat uniform chestnut-brown, sometimes with slight dusky edges to the feathers of the upper back. Females are rather different, with a warm rufous-brown crown and (duller) rufous-brown face and ear-coverts. Females are grey below except for their white throat and belly. The extensive amount of black on male Hooded Gnateaters is superficially quite similar to male Black-bellied Gnateater and, in fact, the figure of '*Conopophaga melanogaster*' in Goeldi's *Album de aves amazonicas* (1901–06) is a depiction of Hooded Gnateater. Cory & Hellmayr (1924) summarise the differences between the two species as follows. Overall, Hooded Gnateater is significantly smaller and has a weaker bill than Black-bellied. The mandible of Hooded is also distinctly pale rather than dark. Hooded has a much browner back (not chestnut) than Black-bellied. In adult male Hooded Gnateaters the black markings do not extend below the chest, leaving an extensive area of white centrally, while in Black-bellied the entire underparts are black (except the lower flanks and undertail-coverts, which are grey). Female Hooded is most easily distinguished from females of Black-bellied by its pale russet-brown (instead of sooty-black) cheeks and ear-coverts. Male Chestnut-belted Gnateater is similar in size to Hooded Gnateater, but easily separated by its chestnut-brown crown and nape and by having the lower underparts buffy. Female Chestnut-belted Gnateater is distinctive from female Hooded Gnateater in having bright rufescent lower parts.

DISTRIBUTION Hooded Gnateater is endemic to north-east Brazil, south of the Amazon from north-east Pará to northern Ceará. Well known from around Belém (Griscom & Greenway 1941, Novaes 1947, Pinto 1953, Oniki 1974, Novaes & Lima 1998), its range extends north and east through the coastal 'Salgado Paraense' region (Lees *et al.* 2014) and south along the eastern (right) bank of the Rio Tocantins, as far south in Pará as the Tucuruí Dam. From there, specimen records extend east and south across most of Maranhão (Oren 1991, Almeida *et al.* 2003), south to extreme southern Maranhão (Alto Parnaíba area; Morrinhos). Records from adjacent areas of extreme southern Piauí (São Gonçalo do Gurguéia) are the species' southernmost. The range of Hooded

Gnateater in Ceará is apparently limited to the extreme north-west corner, north and west of Sobral. Although its western range is commonly described as being limited by the east bank of the Rio Tocantins (Whitney 2003, Greeney 2013g), I have been unable to document its presence in the Rio Araguaia–Rio Tocantins interfluvium (= northern Tocantins), suggesting that the latter river limits its westernmost distribution in the south.

HABITAT Hooded Gnateater inhabits both primary forest and mature second growth, but appears to prefer densely vegetated and tangled parts of forest, often near the edge (Whitney 2003). While found in evergreen and gallery forests in the west of its range (Snethlage 1928), it also inhabits more seasonally dry forests at the eastern extremity (Ridgely & Tudor 1994, Whitney 2003, van Perlo 2009). Altitudinally, it ranges from sea level to around 300m (Ridgely & Tudor 2009).

VOICE Like Black-bellied Gnateater, Whitney (2003) recorded and described males producing a loud whirring sound in the context of courtship or aggressive displays. This behaviour has not been examined in detail, however, and merits further study. Lasting only *c.*1.5s, the song of Hooded Gnateater is a rapid, slightly musical ascending series of notes delivered at *c.*3.5kHz (Whitney 2003). The song is sometimes described as slightly longer and upslurred at the end (van Perlo 2009). The call notes are described as a piercing *tchief!* or a hard *tcheek!* They occasionally produce a one-second-long chatter of scratchy notes (Whitney 2003).

NATURAL HISTORY Published data of the species' behaviour are almost non-existent. Like other *Conopophaga*, it is said to forage low above ground and scan surrounding foliage and leaf litter for prey (Whitney 2003). Hellmayr (1929) noted that adults give the appearance of being sluggish or 'lazy,' pausing frequently in the undergrowth to perch quietly. Though the species' association with ant-following flocks is largely undocumented, apparently it will occasionally forage in the wake of ant swarms (Willis *in* Hilty 1975).

DIET The Hooded Gnateater's diet is apparently undocumented but, like congeners, is presumably composed primarily of small arthropods.

REPRODUCTION Snethlage (1935) provided the first description of a nest based on one found at 'Pará' [= Belém; Paynter & Traylor 1991a], on an unspecified date. The only other information published on the reproductive habits appears to be the description of an incomplete nest found in eastern Pará in mid-April (Whitney 2003). **NEST & BUILDING** The nest found by Snethlage (1935) was low to the ground in a bush near a thicket. In form, it was an open cup of coarse plant materials constructed on a platform of fine twigs and leaves. The nest described by Whitney (2003) was built by both sexes and was 29cm above ground 'atop some saplings' in a relatively open area beside a dense thicket of vines and small palm saplings. Excluding longer twigs and material protruding out, the outer portion of the open-cup nest was 10cm wide and 10.5cm tall. Internally, the cup measured 7cm wide by 4.5cm deep. **EGG, LAYING & INCUBATION** The two eggs in the nest found by Snethlage (1935) presumably constituted a full clutch and were described as pale ochre-yellow with a few small reddish flecks near the larger end. Although Snethlage (1935) did not provide measurements, two eggs described by Velho

(1932) and collected by Snethlage in Pará, are almost certainly these eggs. Velho (1932) described the eggs as being yellowish-white with red-brown flecking forming a ring at the larger end and gave measurements of 20 × 17mm for both. Schönwetter (1979) cited Snethlage's (1935) qualitative description of the eggs but did not provide any measurements. NESTLING & PARENTAL CARE There are no published data on any aspect of nestling growth, appearance or parental care. SEASONALITY A 'juvenile' male [= juvenile in transition?] was collected 10 May 1910 at Peixe-Boi, east of Belém (Hellmayr 1912), and I examined two possibly similarly aged juvenile males in transition to subadult plumage, collected 9 and 23 May 1960 along the Rodovia Belém–Brasília (LACM 43006/10). I have little doubt that these last two males are immatures, but the following records of subadults should be used with caution until immature plumages are better understood (see Technical Description). In the event that the characters I have used to identify immatures vary between the sexes, I indicate gender with each record. BREEDING DATA Juvenile, 24 August 1918, Bragança (YPM 29698, S.S. Snow *in litt.* 2017); juvenile ♂ transitioning, 28 July 1925, Alto Parnaíba (FMNH 63888); juvenile ♂ transitioning, 2 March 2017, Altos (WA 2490370, F. Fernandex); two juvenile ♂♂ transitioning, Rodovia Belém–Brasília (BR-10), 9 and 23 May 1960 (LACM 43010/06); subadult ♂, 22 November 1923, Turiaçu (FMNH 63882); subadult ♀, 11 February 2016, Altos (WA 2019877, L. Gaspar); subadult ♂, 25 September 2016, Caxias (WA 2325361, E. Loreto); two subadult ♂♂ near Belém, 27 March 1963, and 5 August 1964 (USNM 513349/514291); two subadults, 13 August 2016, Meruoca; (♀ WA 2270522, ♂ WA 2238727, R. Czaban); subadult ♂ transitioning, 23 July 2016, Meruoca (WA 2208454, M. Sidrim); subadult ♀ transitioning, 24 April 2016, Caxias (WA 2109706, O. Villela); three subadults transitioning, 25, 27 and 28 July 1925, Alto Parnaíba (♀ FMNH 63886, ♀ FMNH 63887, ♂ FMNH 63890); two subadult ♂♂ transitioning, 29 April and 11 May 1924, Rosário (FMNH 63883/85); two subadults transitioning, 1 February and 25 March 1963, near Belém (♂ USNM 513347; ♀ USNM 513348). I cannot be sure, but I believe that the subadult ♂, photographed 21 September 2016 at Caxias (IBC 1388931, P. Tizzard), is the same individual photographed four days later at the same location (see above; WA 2325361).

TECHNICAL DESCRIPTION *Adult Male* Hellmayr (1905b) described the holotype: 'Top and sides of the head, throat and foreneck black; postocular pencil of elongated feathers silky white; back and upperwing-coverts pale rufous-brown; quills dusky, outer webs and tertiaries pale rufous-brown, rather lighter than the back; tail rather more olive-brown. Sides of the body pale greyish with a slight olivaceous brown admixture on the flanks; middle of the breast and abdomen white; undertail-coverts whitish. Axillaries pale grey with white margins; underwing-coverts whitish, those near the edge of the wing black. Inner edge of the quills very indistinctly dirty greyish white; thighs dark grey with paler tips.' *Adult female* Following Snethlage (1906) and Hellmayr (1906), the forecrown and crown are dull brown to russet-brown while the lores and superciliary are pale cinereous (not black). The white postocular stripe is narrower than in the male and is bordered below by a fine blackish line, not always visible in life. The cheeks, malar region and foreneck are all pale cinereous. The throat is white and the ear-coverts rufescent-brown. The flanks are more distinctly washed brownish than in the male. **Adult Bare Parts** (sexes similar) *Iris* dark brown; *Bill* black with mandible pinkish-dusky, from USNM specimen label: mouth lining yellow-orange, gape blackish-brown; *Tarsi & Toes* grey to bluish-grey. *Immature* Below I describe what I consider is the subadult plumage and, in both sexes, it is differentiated from adult plumage by the presence of rufous-brown-tipped upperwing-coverts that are additionally finely edged black. I encountered a relatively large number of specimens, and found a fair number of photographs, of both sexes, that showed distinctly rufous-tipped coverts. Many, but not all, of these also had black fringes, sometimes only on some of the coverts. Some individuals clearly lacked both markings on their coverts, and I am left uncertain concerning the meaning of patterned coverts in otherwise adult-plumaged birds. It seems clear that immatures have patterned coverts, but the length of time these are retained into reproductive age (adult) is unclear. Given the seeming utility of patterned coverts for recognising immatures in most other *Conopophaga*, I have provisionally considered all birds with upperwing-covert markings as immatures, assigning those with black-margined tips to the subadult stage and those with distinctly rufous-brown tips, but largely or entirely lacking black margins, to the subadult transition phase. *Subadult* Hellmayr (1912) described a 'juvenile' from Pará (see Breeding Seasonality) as being similar in general coloration to ♀♀females, having a white throat and belly, greyish chest and brownish crown. The basal half of the mandible, however, is dark brown (from specimen?) and the upperwing-coverts are fringed narrowly blackish and a broader, submarginal band of rufescent-brown. This description probably refers to what I would consider a subadult (female), no longer dependent on its parents and beginning the final transition to adult plumage. This supposition is borne out by my direct observation of two juvenile males in the final transition to subadult plumage (LACM 43006/10). They have nearly full adult male plumage, but have a few brown feathers scattered on the crown and nape and have the greater secondary wing-coverts and, to a lesser degree, the tertials, patterned as above. **Subadult Bare Parts** Apparently similar to adults. *Juvenile/Fledgling* Undescribed.

MORPHOMETRIC DATA Cory & Hellmayr (1924) provide the following summary measurements for ♂ and ♀ *C. roberti*, respectively (no sample size given): *Wing* 69–72mm, 65–69mm; *Tail* 35–39mm, 32–36m; *Bill* [total culmen; ♂♀ combined] 15–16mm. Data from FMNH, LACM, USNM (HFG) and MCZ (Jonathan C. Schmitt *in litt.* 2017), for adults and subadults combined. Adult ♀♀: *Wing* 73.5 ± 1.5mm (*n* = 5); *Bill* from nares 9.1 ± 0.2mm (*n* = 6), exposed culmen 12.3 ± 1.1mm (*n* = 2), depth at nares 4.1 ± 0.2mm (*n* = 2), width at nares 5.3 ± 0.3mm (*n* = 2), width at base of mandible 6.2 ± 0.3mm (*n* = 2); *Tarsus* 26.9 ± 1.2mm (*n* = 8). Adult ♂♂: *Wing* 70.1 ± 2.8mm (*n* = 9); *Bill* from nares 9.5 ± 0.3mm (*n* = 19), exposed culmen 13.1 ± 0.4mm (*n* = 6), depth at nares 4.3 ± 0.2mm (*n* = 6), width at nares 5.7 ± 0.3mm (*n* = 6), width at base of mandible 6.8 ± 0.3mm (*n* = 6); *Tarsus* 27.7 ± 1.2mm (*n* = 18). Juvenile ♂♂: *Wing* 75.0mm (*n* = 1); *Bill* from nares 9.5 ± 0.4mm (*n* = 3), exposed culmen 12.3mm (*n* = 1), depth at nares 4.1mm (*n* = 1), width at nares 5.4mm (*n* = 1), width at base of mandible 6.6mm (*n* = 1); *Tarsus* 27.5mm (*n* = 3). See also Hellmayr (1905a,b, 1906, 1912), Naumburg (1937), Novaes (1947), Novaes & Lima (1998). **MASS Males** 21g, 25g (*n* = 2, Silva *et al.* 1990); 27g (USNM 514291). *Females* 24g

(n = 1, Silva *et al.* 1990). **Sex unspecified** 23.0g (n = 1, Silva *et al.* 1990); 20.8g (n = 1, Oniki 1974); 23.0g, 25.0g (n = 2, Novaes & Lima 1998). **Other** Cloacal temperature 39.8°C (n = 1♀?, Oniki 1974). **TOTAL LENGTH** 11–14cm (Meyer de Schauensee 1970, 1982, Ruschi 1979, Sick 1993, Ridgely & Tudor 1994, Whitney 2003, van Perlo 2009).

DISTRIBUTION DATA Endemic to **BRAZIL**: *Pará* *c.*6.5km N of Vigia, 00°48'S, 48°08'W (LSUMZ 71710/11, MPEG 30466); Fazenda Jaburu, 00°54'S, 47°24'W (MPEG 49275); Estudo Impacto Ambiental Votorantim, 00°59'S, 47°07'W (MPEG 72478/79); Terra Alta, 01°02.5'S, 47°54.5'W (MPEG 35190–192); Bragança, 01°04'S, 46°47'W (YPM 29696–698); Baía do Sol, 01°04'S, 48°21'W (LSUMZ 67365); 'Flor do Prado', 01°08'S, 47°00.5'W (MPEG 5804–06, MPEG 12749–52, MNRJ 17567/72, AMNH 128387); Peixe-Boi, 01°12'S, 47°19.5'W (Hellmayr 1912, Griscom & Greenway 1941); 'Prata' [= Santo Antônio do Prata], 01°18'S, 47°36'W (Snethlage 1907c, AMNH 488912–918, MPEG 4099/100); Castanhal, 01°18'S, 47°55'W (Stone 1928, ANSP 80525/26); Benevides, 01°21.5'S, 48°14.5'W (CM, MCZ, USNM, MPEG); Belém and vicinity, 01°27'S, 48°29'W (Novaes 1970, YPM, MCZ, USNM, LSUMZ, MPEG); Sítio Fé em Deus, 01°33'S, 47°06'W (MPEG 32217/18); Guajará-Açu, 01°38'S, 48°07'W (MPEG 33310); Bacaba, 01°51'S, 47°03.5'W (MPEG 31505/06); Acará, 02°01'S, 48°19'W (MCZ 175146–148); Rodovia Belém–Brasília (BR-10), km92/93, 02°12.5'S, 47°33.5'W (MPEG 15763); Tomé-Açu, 02°25'S, 48°09'W (LSUMZ 67364, MPEG 70272); Mina de Caulim, 02°27.5'S, 47°39.5'W (MPEG 51989); Reserva Florestal Grupo Agropalma, 02°36'S, 48°44'W (Portes *et al.* 2011); Mocajuba, 02°38'S, 49°24'W (AMNH 430691–694); Baião, 02°41'S, 49°41'W (AMNH 430686–689, MPEG 7697–700); Paragominas area, *c.*03°00'S, 47°21'W (Lees *et al.* 2012); Fazenda Vitória, 03°00'S, 47°30'W (Silva *et al.* 1990, MPEG 39308–310); Tailândia, Fazenda Rio Capim, 03°40'S, 48°33.5'W (Portes *et al.* 2011, MPEG 58983); Cauaxi, 03°44'S, 48°17.5'W (Portes *et al.* 2011, MPEG 58810/11); Rodovia Belém–Brasília (BR-10), km307, 04°07.5'S, 47°33.5'W (MPEG 18088); Tucuruí Dam, 04°14'S, 49°25'W (XC 27205, S. Dantas); Jacundàzinho, 04°27.5'S, 49°27.5'W (MPEG 36240/41); just N of Rondon do Pará, 04°46'S, 48°04'W (MPEG 72272). *Maranhão* Pedra Chata, 01°13'S, 46°06'W (MPEG 34812); Fazenda Santa Bárbara, 01°22'S, 46°02'W (MPEG 36924/27); Turiaçu, 01°40'S, 45°22'W (FMNH 63881/82, MNRJ 16463/64); Bom Jesus da Mata, 02°22'S, 45°54'W (MPEG 34980); Miritiba, *c.*02°37'S, 43°27'W (AMNH 488909–911); Rosário, 02°56'S, 44°15'W (Hellmayr 1929, FMNH 63883–885); Aldeia Zé Gurupi, 02°59'S, 45°46'W (MPEG 38652–654); REBIO Gurupi, 03°49'S, 46°48'W (XC 146938, J. Engel); Fazenda São Francisco, 04°03'S, 44°56.5'W (MPEG 36866); Fazenda do Caximbo, 04°04'S, 44°08'W (LSUMZ 71706–709); *c.*11.7km ESE of Roça Grande, 04°07.5'S, 46°16'W (MPEG 37360/61); Buriticupu, 04°29'S, 46°19'W (WA 692773, D. Machado); Caxias, 04°52'S, 43°21.5'W (WA 2109706, O. Villela); Povoado Canaã, 05°00.5'S, 45°08.5'W (MPEG 68129); Vila Nova dos Martírios, 05°13.5'S, 48°12.5'W (MPEG 70183); Fazenda Castiça, 05°28.5'S, 43°13'W (MPEG 70201); Santa Filomena do Maranhão, 05°32'S, 44°35'W (AMNH 242769); near Amarante do Maranhão, *c.*05°34'S, 46°45'W (MPEG 37873-78); Grajaú, 05°49.5'S, 46°10'W (WA 864948, G. Serpa); Mancha Verde, 06°36'S, 43°37'W (Olmos & Brito 2007, XC 1411, G.R.R. Brito); Povoado Jatobá dos Noletos,

06°36.5'S, 43°37'W (MPEG 68130–133); Balsas, 07°32'S, 46°02'W (WA 1316198, S. Almeida); Povoado Feira Nova, 07°37'S, 47°02'W (MPEG 42324/25); Inhumas, 08°45'S, 45°58'W (Hellmayr 1929); Alto Parnaíba, near Morrinhos, 09°24.5'S, 46°15.5'W (FMNH 63886–890). *Piauí* José de Freitas, 04°40'S, 42°28'W (Silva *et al.* 2004); Eco Resort Nazareth, 04°47'S, 42°37'W (MPEG 68123–126); Altos, 05°03' S, 42°28'W (WA 2076425, E. Kaseker); Belo Horizonte, 06°00'S, 43°00'W (Naumburg 1937, Silva *et al.* 2003, AMNH 242771); Uruçuí, 07°44'S, 44°33'W (Naumburg 1937, AMNH 242770); Baixão do João Carlos, 08°52.5'S, 43°53.5'W (MPEG 76777/80); Santa Filomena, 09°02'S, 45°41'W (Naumburg 1937, WA 948885, E. Patrial); Jurema, 09°02'S, 43°13'W (WA 1617119, C. Rasta); Caracol, 09°04.5'S, 43°22'W (WA 1970439, R. Rizzaro); Gilbués, 09°32'S, 45°22'W (WA 1153775, P.C. Lima); Chapadão do Gurguéia region in Serra Vermelha, 09°41'S, 44°14'W and 09°38'S, 44°10'W (Santos *et al.* 2012, MPEG 68127/70345); near São Gonçalo do Gurguéia, 10°02'S, 45°18' W (Lima & Lima 2005). *Ceará* Granja, 03°07'S, 40°50'W (WA 127008, recording C. Albano); Viçosa do Ceará, 03°34'S, 41°05.5'W (Naumburg 1937; WA 1558155, C. Ferreira); Meruoca, 03°35'S, 40°27'W (WA 2208454, M. Sidrim); Ponto do Adeo, 03°38'S, 40°24.5'W (MPEG 82883); Sobral, 03°42'S, 40°21'W (XC: C. Brito, R.S.L. Junior); Tianguá, 03°44'S, 40°59.5'W (XC 294536, R.S.L. Junior); Ubajara, 03°51'S, 40°55.5'W (XC 197206/213, C. Brito); Ipu, 04°19'S, 40°43'W (WA: P. Lira); São Paulo, Serra do Ibiapaba, 04°19.5'S, 40°48.5'W (MPEG 7221–225, MNRJ 16465). **Notes** For a discussion of this final location (São Paulo), see Distribution Data for White-browed Antpitta. Paynter & Traylor (1991a) were unable to locate Flor do Prado, placing it somewhere south of Quatipuru, between Peixe-Boi and Bragança. The coordinates above reflect this approximate location and place it at or near Igarapé Castanho (MPEG 31550). Finally, a fair number of collecting localities for Hooded Gnateater are within 15km of the GPS given above for Belém. Those not listed above include: Ananindeua (MPEG 9775/76); Marco da Légua (Bairro do Marco) (MPEG 1960); Marituba (MPEG 47851); Tanaquará (= Benjamin) (MPEG 11279).

STATUS Originally, Hooded Gnateater was the only species in the family to be considered 'potentially threatened' (Collar & Andrew 1988, Mountfort & Arlott 1988) and was considered 'insufficiently known' by the IUCN (1988, 1990). Collar *et al.* (1994) later specified the threat status of Hooded Gnateater as Near Threatened, but in subsequent evaluations it has been downgraded to Least Concern (BirdLife International 2017). To date, there have been no studies of population trends, but extensive and ongoing deforestation east of the Rio Tocantins, especially around Belém, including within protected areas (Forrester 1993, Stotz *et al.* 1996, BirdLife International 2017), is likely to be having a negative impact. The species is apparently tolerant of, and perhaps even prefers, mature second growth (Ridgely & Tudor 1994). Nevertheless, understorey degradation, even in areas where forest is left largely intact, may have undocumented detrimental effects on the foraging and nesting of Hooded Gnateaters. Similarly, poor dispersal ability probably hinders movements between isolated forest patches, and the long-term effects of widespread fragmentation within its range should form an important aspect of future conservation research (Collar & Andrew 1988). Although thought to possibly occur within just one protected area

(BirdLife International 2017), populations persist within several forests under varying levels of formal protection. **Protected Populations** PN Sete Cidades (Santos *et al.* 2013); PN Serra das Confusões (MPEG); PN de Ubajara (MPEG 82878–82880); Eco Resort Nazareth (Silva *et al.* 2004); Reserva Forestal Grupo Agropalma (Portes *et al.* 2011); FLONA Palmares (MPEG 74362); REBIO Gurupi (MPEG 76963); Parque Ecológica Gunma (Portes *et al.* 2011); PE Utinga (MPEG 11122); Reserva Indígena Alto Turiaçu (MPEG 38652–654).

OTHER NAMES English Robert's Gnat-eater (Brabourne & Chubb 1912); Robert's Gnat Eater (Cory & Hellmayr 1924). **Portugese** chupa-dente-de-capuz (Novaes & Lima 1998, Willis & Oniki 1991b, van Perlo 2009); chupa-dente-cabeça-negra (Ruschi 1979). **Spanish** Jejenero Encapuchado (Whitney 2003). **French** Conophage capuchin. **German** Östlicher Schwarzkopf-Mückenfresser (Whitney 2003).

Hooded Gnateater, adult male, Tianguá, Ceará, Brazil, 12 July 2014 (*Caio Brito*).

Hooded Gnateater, adult male, Meruoca, Ceará, Brazil, 16 February 2016 (*Ciro Albano*).

Hooded Gnateater, adult female, Cristino Castro, Piauí, Brazil, 13 June 2013 (*Ciro Albano*).

Hooded Gnateater, adult male, Caxias, Maranhão, Brazil, 19 June 2012 (*João Quental*).

Hooded Gnateater, adult female, Caxias, Maranhão, Brazil, 17 June 2012 (*João Quental*).

RUFOUS GNATEATER
Conopophaga lineata   Plate 3

Myiagrus lineatus Wied, 1831, Beiträge zur Naturgeschichte von Brasilien, vol. 3, part 2, p. 1046, 'Arrayal da Conquista' [= Vitória da Conquista, 14°51'S, 40°51'W], Bahia, Brazil. Note that Cory & Hellmayr (1924) incorrectly gave the page as 1064. The holotype is an adult female (AMNH 6777) (Allen 1889c). Rufous Gnateater was described by Prince Alexander Philipp Maximilian zu Wied-Neuwied, or the Principe de Wied (Hagmann 1904) during his pioneering explorations of Brazil.

'Inhabits the most intricate of tangled growth within the forest. It is never found far from the ground, rarely more than a meter. There they flit calmly from stick on stick, stopping every few minutes. In this manner they detect insects in the soil and drop down to snatch them up. They also follow the armies of ants, eating many (*Oeciton* [= *Eciton*]). Occasionally utters a strange call.' – **Arnoldo de Winkelried Bertoni, 1901 [translation from Spanish]**

While reviewing the literature, it was immediately apparent that Rufous Gnateater leads its relatives with respect to the number of studies that involve it, outstripping even the Scaled Antpitta, which holds several other related records. Among other ways in which Rufous Gnateater deserves top marks, it must surely be considered as tied with Scaled Antpitta in having been reported as suffering from a most unexpected source of mortality for an understorey insectivore. That is, at least, if one considers large panes of glass and speeding vehicles to be unusual components of the tropical forest understorey. If I have succeeded in arousing your curiosity, please proceed to the Status section. The account that follows is, at best, an attempt to summarise the 350+ publications that involve this common species. The vast majority of these, however, contribute only details of distribution, habitat and status. For those interested in the general natural history and breeding biology of Rufous Gnateater, I urge you to read the wonderfully detailed account of their lives published (1983) by Edwin O. Willis, Yoshika Oniki and Wesley R. Silva. Rufous Gnateater, as currently defined, is generally considered to be an Atlantic Forest endemic, but its range extends well into the *Cerrado* biome and, at least partially, into the southern part of the *Caatinga* biome. As implied by its monochromatic English name, Rufous Gnateater is the most plainly marked member of the genus, being mostly orange-rufous to rufous-brown, with a fairly variable amount of grey or whitish feathering forming a superciliary stripe that ends in the typical *Conopophaga* postocular tuft. Despite its uncomplicated plumage and (relatively) small amount of clearly defined variation in appearance, complex and poorly understood vocal variation and genetic diversity have already hinted at the need for further study. Undoubtedly, the distributional and taxonomic limits of the two subspecies considered here will change, once a much-needed revision occurs. Vocalisations vary in tempo, in general terms from north to south, and there is also significant variation in note structure that hints at five or more different song types (Whitney 2003). My preliminary investigations into the matter suggest that at least three taxa are involved and that future work will reveal at least one of these to be extremely range-restricted, prompting the need for immediate re-evaluation of conservation status.

IDENTIFICATION 12–13cm. Despite dissention in the literature, the sexes are not reliably recognised in the field. One of the most plainly coloured gnateaters, Rufous Gnateater is brown above with a (usually) prominent grey eyebrow that ends in a silvery-white postocular tuft of several elongated feathers. The brown crown is slightly warmer than the remaining upperparts and grades to rufous on the sides of the head, throat and breast. Most populations have a distinct white, crescent-shaped patch separating the throat from the breast. The sides and flanks are duller, washed brownish or olive, surrounding a large, whitish belly patch. The bill is bicoloured, dark brownish or black above, with the mandible mostly or entirely whitish or yellowish, sometimes darker on the distal half. Legs dusky-greyish or greenish-yellow. Females are said to have the postocular tuft greatly reduced or absent, but this is probably more age-related than specific to females. The range of Rufous Gnateater overlaps with just one congener, Black-cheeked Gnateater, males of which are immediately distinguished by their strong black mask and grey underparts. Female Black-cheeked Gnateaters are more similar, but are browner above, less orange below and have distinctly spotted wing-coverts. The most likely confusion, at least until taxonomy and distribution are re-evaluated, is with the recently split Ceara Gnateater. It is unclear if, or where, the ranges of these two species meet, but they are not known to be sympatric anywhere. Ceara Gnateaters are brighter, more orange-rufous overall, have a grey spot on the lores that extends to the postocular tuft only as a very thin line, being much less prominent than in Rufous Gnateater. However, see Taxonomy & Variation with respect to the south-eastern Bahia population.

DISTRIBUTION Rufous Gnateater probably has one of the largest ranges of any of the genus and is almost certainly the commonest. It lives in south-east Brazil, from southern Bahia along the coast to Rio Grande do Sul, fairly recently being found in Uruguay. Its range also extends inland to eastern Mato Grosso and the Distrito Federal and then, somewhat disjunctly, in central Mato Grosso. In the south it ranges inland into north-eastern Argentina (Misiones, Corrientes) and eastern Paraguay. Overall, Rufous Gnateater appears to be less common further from the coast, with the exception of populations in Misiones,

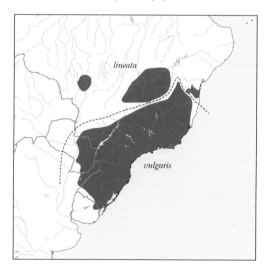

being largely unknown from the Pantanal but for several records in its northernmost part. **See also** Zotta (1944), Contreras (1987), Narosky & Yzurieta (1989), Belton (1985), Rosário (1996), Bencke (2001), Guyra Paraguay (2004), Gwynne *et al.* (2010), Scherer-Neto *et al.* (2011).

HABITAT Rufous Gnateater predominantly inhabits the understorey of semi-humid to humid evergreen montane and (to a lesser degree) lowland forests in the Atlantic Forest region (Brooks *et al.* 1999). It is also found in seasonal, semi-deciduous forest (Maldonado-Coelho & Marini 2003, Willis 2003b, Pacheco & Olmos 2006, Lopes *et al.* 2008) and in gallery forests within *cerrado* at the eastern edge of the Planalto Central (Willis & Oniki 1991b, Silva 1995, 1996, Willis 2004, Telles & Dias 2010, dos Anjos *et al.* 2011, Valadão 2012). Although best considered a forest understorey insectivore, with perhaps some preference for riparian areas in at least parts of its range (Willis & Oniki 2002b, dos Anjos *et al.* 2007, Valadão 2012), the species can persist in second growth and forest fragments in agricultural landscapes (Sousa *et al.* 2004, Willis 2006, Cavarzere *et al.* 2011, 2012), provided they are fairly mature (Willis & Oniki 2002a, see Status). Willis & Oniki (2002a) observed Rufous Gnateaters leaving the cover of a woodlot to forage in adjacent sugarcane fields. In the south-east of its range Rufous Gnateater occupies more humid areas, including *Araucaria* forests (dos Anjos & Graf 1993, Bispo & Scherer-Neto 2010, Scherer-Neto & Toledo 2012), within which it favours dense undergrowth such as bamboo thickets (Holt 1928, Santana & dos Anjos 2010, Buckingham 2011). Across its range, Rufous Gnateater is most often found above 500m, but in the far south is regularly found as low as 300m. The altitudinal range, however, is rather broad, and in some areas the species occurs to the treeline (2,400m) on some of the highest peaks in south-east Brazil (Sick 1993), as well as below 100m elsewhere (Buzzetti 2000, Anciães & Marini 2000), including the Serra do Mar (Höfling & Lencioni 1992, Goerck 1999). The distribution of Rufous Gnateater overlaps with just one other *Conopophaga*, Black-cheeked Gnateater, which the present species replaces at higher altitudes (with some overlap), in the higher mountain ranges such as the Serra do Mar (Sick 1993, Mallet-Rodrigues & Noronha 2003, Mallet-Rodrigues *et al.* 2010).

VOICE The presumed song of Rufous Gnateater is a series of *c*.5–15 whistled notes rising in tempo and pitch into a trill; *too too to to ta ta te ti ti tititititititee* (de la Peña & Rumboll 1998). Songs last *c*.1.5–4s, are delivered at 2.5–3.5kHz and frequently end with several slightly louder notes (Whitney 2003). In São Paulo this is described as a slow *tree-tree-tree-tree-tree-tree* similar to the vocalisation of Crescent-chested Puffbird *Malacoptila striata* (Willis & Oniki 2002b). According to Whitney (2003), in the southernmost parts of its range the song is slightly longer and includes more (*c*.30) notes. In Argentina it is described as a rapid *Chüchüchüchichichihhhh* (de la Peña 1988). Also in the south (*vulgaris*), Brooks *et al.* (1993) likened the ascending trill to the vocalisation of Russet-winged Spadebill *Platyrinchus leucoryphus*. Rufous Gnateaters appear to have several types of calls, although their various functions are not well known. The calls which give them their vernacular names in Portuguese and Spanish are the 'snap-like sucking of teeth' 'sneezing' or 'spitting' noises (Willis *et al.* 1983, Narosky & Yzurieta 1987, Brooks *et al.* 1993). These calls are abrupt and sneeze-like, variously described as: a piercing *tchief!* (Whitney

2003), a hard thick *tcheek!* (Whitney 2003), *chuk* (Brooks *et al.* 1993), a short series of similar high notes, *tze tze tze* (Brooks *et al.* 1993) and a 'chittering' of rapidly delivered 'sneezes' (Willis *et al.* 1983). Several vocalisations are associated with chases or disputes, and the details of their various uses would no doubt represent a fascinating line of investigation. Sneeze calls, even delivered by distant birds, often lead to one or more individuals converging to give rapid 'chatters' (Willis *et al.* 1983). These chatters frequently terminate in loud whirring or buzzing sounds created by the modified outer remiges of males (Sick 1965, Willis *et al.* 1983, Canevari *et al.* 1991). Songs are frequently given by males once disputes such as these are resolved.

NATURAL HISTORY Several authors have remarked or insinuated that Rufous Gnateater forages a good deal on the ground (Bertoni 1901, Erickson & Mumford 1976), with Chubb (1910) going so far as to say 'can generally be found by the rustling of the dead leaves on the ground.' Generally they forage low, perching on small branches and vertical stems, and making sally strikes and sally gleans, often for ground-based prey (Canevari *et al.* 1991, Gwynne *et al.* 2010). They seem to usually forage alone or in pairs (Willis *et al.* 1983), but also join mixed-species flocks (Gwynne *et al.* 2010). They appear to join flocks at ant swarms, at least occasionally (*c*.30% of swarms in one study) (Willis 1985a, Faria & Rodrigues 2005, 2009, Pizo & Melo 2010). Although apparently active throughout the day, vocalisations are mostly heard at dusk and dawn (Antunes 2008, Esquivel-M. & Peris 2008), with these crepuscular songs sometimes not delivered at full volume (Willis *et al.* 1983). Details of territory acquisition, maintenance and fidelity are largely unknown, but in the Paraguayan Atlantic Forest (*vulgaris*), Cockle *et al.* (2005) estimated 42 pairs per 100ha. Willis *et al.* (1983) found that near São Paulo (*vulgaris*), home ranges were *c*.150m in diameter and territories *c*.100m across. In this population there appeared to be a fair number of 'floaters' during the non-breeding season that wandered through the year-round territories of other adults. Adults who lost their mates quickly acquired a new one. Wandering appeared to cease or was greatly reduced in the breeding season, when territorial disputes intensified. Although little is known of the effect of parasites, the species is a known host to ectoparasites such as *Formicophagus* bird lice (Phthiraptera, Philopteridae; Cicchino & Valim 2008), *Amblyomma* and *Ixodes* ticks (Acari, Ixodidae; Arzua & Barros-Battesti 1999, Rojas *et al.* 1999, Arzua *et al.* 2003, González-Acuña *et al.* 2005, Ogrezewalska *et al.* 2008, 2009), *Analloptes* feather mites (Astigmata, Xolalgidae; Mironov & Hernandes 2014), and subcutaneous *Pelecitus* filaroid worms (Nematoda, Onchocercidae; Oniki *et al.* 2002). They are also know to suffer from blood parasites, including *Plasmodium*, *Haemoproteus* and *Trypanosoma* protozoans (Woodworth-Lynas *et al.* 1989, Ribeiro *et al.* 2005, Sebaio *et al.* 2012). Although little is known of moult from most parts of their range, Rufous Gnateaters in São Paulo (Willis *et al.* 1983) began to moult shortly after their young fledged and were in heavy moult around six weeks later (just as the young became independent). These authors also found that, in this region, adult moult occurred from November to March, and that the moult cycle lasted *c*.2.3 months. A male in São Paulo (*vulgaris*) was recaptured several times over the course of *c*.4 years (Lopes *et al.* 1980), but little else has been published on longevity.

DIET Rufous Gnateater feeds on invertebrates, based on scattered reports of stomach contents or direct observations (Rudge *et al.* 2005, Krabbe 2007), with only a single report of a vertebrate (frog) that was fed to nestlings (Willis *et al.* 1983, Lopes *et al.* 2005a). Two studies using emetics (Lopes *et al.* 2005b, Lima *et al.* 2010) and one using faecal samples (Lima *et al.* 2011), all in Minas Gerais (*vulgaris*), quantified dietary items from 63 individuals. Food (*n*) = Formicidae (508); Coleoptera (87); Aranae (41); 1.2–4.0mm seeds (28); larval Coleoptera/ Lepidoptera (22); Hymenoptera (14); Isoptera (11); Hemiptera (10); Hemiptera (9); Pseudoscorpiones (4); Orthoptera (4); Blattodea oothecal (3); Neuroptera larva (1); Diptera (1); Diplopoda (1); Scorpiones (1); Acari (3); Opilliones (1); Trichoptera (1); unknown invertebrate (2); and unknown plant tissue (1). In these, and other dietary studies (Durães & Marini 2005, Manhães 2007, Manhães *et al.* 2010), ants (Formicidae) and beetles (Coleoptera) are consistently among the most numerous prey items. Seeds/fruits are usually detected, but in small quantities and it is unclear how often fruit is consumed. Whitney (2003) reported an individual eating a blue berry of the rubiaceous genus *Coccocypsellum*.

REPRODUCTION There is a fair amount of breeding information for Rufous Gnateater, but it almost entirely concerns subspecies *vulgaris*, to which all of the following pertains unless otherwise stated. The first description of a nest is that of von Ihering (1900), who examined a previously collected nest. He described it as a carefully crafted bowl of broad leaves, well lined with flexible fibres, the shape and form of which led him to suggest that it was likely sited on a rock or the ground. The second mention of a nest appears to be that of Bertoni (1901). From his description, however, it seems clear that its owner was misidentified. He described a nest woven of fungal rhizomorphs and moss, 60cm above ground, on a small branch in plain view! This nest apparently held three white eggs. The egg colour and clutch size do not necessarily preclude Rufous Gnateater as the owner (see below) but I feel these data should be used with caution. To date, despite the discovery of many nests, reported cursorily here and there, the work of Willis *et al.* (1983) has not been surpassed in its completeness and detail. All of the wonderful behavioural details in the following are from this study, unless otherwise indicated. **NEST & BUILDING** The following general description is taken from my own examination of photographs as well as from Willis *et al.* (1983), Fraga & Narosky (1985), de la Peña (1988) and Buzzetti & Silva (2008). Nests are proportionately broad, shallow, open cups composed of loosely layered sticks, leaf petioles and dead leaves, lined somewhat sparsely (but apparently variably) with coiled and interwoven dark (reddish-brown or black) fibres, rootlets and fungal rhizomorps. It may be suspended among small branches, wedged into the 'Y' formed by small vertical trunks, or built into the rosette of leaf petioles emerging from a bamboo shoot. *Nest height*: 0.25m, (*n* = 1; Lopes *et al.* 2013); 0.5m (*n* = 1; Frisch & Frisch 1964); range 0.2–0.5m, 0.35 ± 0.11m, (*n* = 7; Marini *et al.* 2007); 0.4m, 0.5m, 0.6m, 0.6m (*n* = 4; Willis *et al.* 1983). *Nest Dimensions*: external diameter 11cm, 13cm, 12cm, 12cm, 10–14cm; external height ?, 7cm, 5cm, 10cm, 7–9cm; internal diameter 6cm, 6.5cm, 5.5cm, 6.3cm, 6.5–7.0cm; internal depth 3.5cm, 3.5cm, 3.0cm, 5.5cm, 3.5–4.0cm (from von Ihering 1900, *n* = 1; Fraga & Narosky 1985, *n* = 2; Lopes *et al.* 2013, *n* = 1; Willis *et al.*

1983, *n* = 3). Marini *et al.* (2007) gave means and SD for several nests: external diameter 12.0 ± 1.1cm (*n* = 4); internal diameter 7.4 ± 0.4cm (*n* = 4); internal depth 5.1 ± 0.3cm (*n* = 3). *Nest Dry Weight*: 21g, 31g, 37g (*n* = 3, Willis *et al.* 1983). Near the onset of a nesting attempt, adults remain in close association with their mates. Males vocalise softly, but frequently, often with their crests raised. During this period males frequently signal the capture of prey with a soft 'chitter' call, uttered when their mate approaches. Females also solicit food with 'faint calls.' Apart from these apparent courtship and mate-guarding behaviours, nothing has been written concerning nest-site selection or nest construction. **EGG, LAYING & INCUBATION** The eggs are variously described as buff, rusty, yellowish or 'flesh-colored' with an indistinct wreath of markings on the larger pole (Nehrkorn 1899, 1910, Willis *et al.* 1983, de la Peña 1988), or 'spheroidal' to 'pointed oval' in shape, slightly glossy, cream or pinkish-cream, smeared here and there with pale brownish-pink or clouded, chiefly at the broad end, with pale rufous-brown (Iguapé, *n* = 2, presumably a clutch as a single date is given, 20 October, Oates & Reid 1903). Kreuger (1968) described two eggs from 'Igvapé' [Iguape] and three eggs from 'Bahia Prov.' as 'reddish-cream with very few and very small pale speckles mostly on the larger end.' Clutch size is 1–3 eggs, but appears to be most frequently two. *Clutch Size Records* (*vulgaris*): two eggs (von Ihering 1900, *n* = 1; Willis *et al.* 1983, *n* = 3; Fraga & Narosky 1985, *n* = 1; Lopes *et al.* 2013, *n* = 1; Maurício *et al.* 2013, *n* = 2); one egg (Marini *et al.* 2007, *n* = 1; Maurício *et al.* 2013, *n* = 1); three eggs (Marini *et al.* 2007, *n* = 1; Frisch & Frisch 1964, *n* = 1). *Egg Measurements* (all *vulgaris*): 23–24 × 17–18mm, 3.4–4.0g (*n* = 4), 20 × 18mm (2.8g), 21 × 18mm (2.9g) (*n* = 2) (Willis *et al.* 1983); 23 × 18mm, 24 × 18mm (Oates & Reid 1903), 22 × 18mm, 24 × 18mm (*n* = 2; Ihering 1900; range 22–23 × 18mm (*n* = 4; von Ihering 1900); range 21.0–23.6 × 17.0–18.5mm, estimated fresh weight 3.9g (*n* = 24; Schönwetter 1967); 22.5–22.6 × 16.8–17.1mm (*n* = 2; Fraga & Narosky 1985, MACN); *c.*21–23.5 × 17.0–18.5mm (de la Peña 1988); 23.3 × 18.0mm, 22.1 × 17.6mm (*n* = 2, Lopes *et al.* 2013); 20.5–21.0 × 17.0–17.5mm (Nehrkorn 1899, 1910); 22.9 × 18.25mm, 21.4 × 16.85mm (*n* = 2; Kreuger 1968); 23.0 × 17.5mm, 22.65 × 17.85mm, 22.15 × 17.2mm (*n* = 3, from 'Bahia Prov'; Kreuger 1968). These latter three eggs potentially belong to Ceara Gnateater, but without more precise information it is impossible to know for certain. Kreuger (1968) assigned them to nominate *lineata*. The eggs described by Nehrkorn were suggested by von Ihering (1900) to belong to Black-cheeked Gnateater. He based this, however, on size, but I see no reason to doubt their origin. As expected, both sexes warm the eggs, taking turns during the day but leaving nocturnal duties to the female alone. Incubation bouts are usually long, with Willis *et al.* (1983) reporting one bout that lasted six hours! While covering the eggs, adults remain largely motionless except for slow head movements to scan for danger. Off-bouts are short, usually just a few minutes and occasionally only a few seconds, with one adult arriving as the first abandons its post. Changeovers are silent and so unceremonious that Willis and colleagues reported occasionally missing them even while the nest was under direct observation. Frequently the incubating adult drops from the nest simultaneously with the arrival at the rim of the second adult, combining the adults' movement at the nest into one fluid motion. The authors suggested that this 'cooperative switch' was likely to create the illusion of a single bird passing the nest site, with the departure of one adult drawing the eye of potential observers away from the nest, while the arriving adult glanced briefly at its contents before settling immediately and freezing in position to incubate. Males (at three nests) averaged more diurnal hours at the nest than females, usually dedicating 1–4 hours per day more than their mates. Females depart the nest at or before first light and are quickly replaced by the male beginning their first on-bout of the day. The female usually returns after a 'breakfast break' of 1–2h, remaining for an hour or so before the male returns to assume most of the afternoon duties. Generally, it appears that each adult contributes 2–6 on-bouts per day. Males occasionally leave the nest in the middle of their turn to briefly engage in territorial disputes. NESTLING & PARENTAL CARE Rufous Gnateaters are dark-skinned on hatching and devoid of natal down. The 'gape' (= mouth lining) is bright orange and is enthusiastically, but silently, displayed on the arrival of adults or if the nestling is touched by an observer. Developing pin feathers become evident three days after hatching. At around four days of age the nestlings grip the nest lining when they are removed by an observer and shortly after this they begin to make audible 'peeping' sounds while begging. By day 6 the contour feathers of the spinal tract begin to break their sheaths but their eyes remain closed (but well-slit) until at least day 7, opening around day 8. Flight feather pins begin to break their sheaths on day 9, while they are still regularly brooded by adults. At 10–11 days of age the nestlings are quite active while being brooded, frequently sticking their heads out through the adult's flank feathers. By this time feathers are well developed on the crown, but the face and neck remain bare. The bill is now largely dark but is orange at the base and on the inner surface, and the bare skin of the throat is also orange. Young begin to protest with loud vocalisations when removed from the nest at around day 12, and by day 13 they preen and stretch their wings and legs in the absence of adults, sometimes even making aborted movements towards the rim of the nest as if to follow the departing adults. Both sexes are devoted parents, and Rufous Gnateaters are, of the species treated here, the 'Oscar winners' of distraction displays. Their theatrical attempts to draw observers from the nest have been described by various authors (Belton 1985), and involve the spasmodic fluttering of their wings through the leaf litter, creating as much noise and movement as is possible for such a small bird. Nestlings are brooded at night by the female during the entire nestling period. By day the sexes feed and brood somewhat equally but, at one of three nests followed closely, which contained only a single nestling, the male was notably less attentive and his participation dropped off sharply once brooding ended. Prior to this he had brought more food to the nest than the female, who often returned to brood without bringing food. At a nest with two seven-day-old nestlings, feeds were brought equally by the sexes at a rate of 2.2 feeds/nestling/h. At one nest where the male reduced his participation after the cessation of brooding, the female apparently compensated by increasing her own visits four-fold. During this intense feeding by the female, she frequently foraged near the nest, but otherwise both adults appear to search for most prey at greater distances from the nest. On the whole, feeding rates peak at dusk and dawn, with well-lit periods of the day coinciding with periods of relative inactivity at the nest, apart from a brief increase around midday. Across the day, the relative contributions of the sexes to nestling care appear to follow roughly the same pattern as described for incubation. Food items vary in size but are predominantly small, less than half of mean adult culmen length. In 586 feeds for which food size was estimated by Willis *et al.* (1983), 69% of prey were >5mm and almost 75% were <1cm in length. Across the nestling period, the diets of young nestlings are almost entirely very small items, with the majority of large items delivered in the days prior to fledging. Though not explicitly stated by Willis *et al.* (1983) prey appear to be delivered almost exclusively singly and, of these nearly 600 prey deliveries, all but one were invertebrates, the exception being a small frog. The latter was apparently a costly item to deliver, both in terms of handling time and increased movement at the nest, as the adults repeatedly dropped to the ground to thrash the hapless frog against it. Even after successful delivery, the young were forced to remain with the frog's legs protruding from its gape for some time, before it was finally able to swallow it. Small items were usually brought in the tip of the adults' bills and delivered immediately on arrival. Larger items usually arrived already thrashed beyond recognition. If nestlings voided a faecal sac after being fed, adults consumed them at the nest early in the nestling period, gradually increasing the frequency with which they carried them away. Nestlings make a rapid peeping sound before being fed, changing the quality of their vocalisation once they had been (sonograms in Willis *et al.* 1983). If the young failed to respond on the adults' arrival, the parents solicited the nestlings to gape for food with faint 'chittering notes,' reminiscent of a brief wing flutter or bill chatter. If this elicited no response, which tended to happen more frequently when the nestlings were very young, the adults consumed the item themselves and settled over the young to brood. By age 13–14 days the young reply with similar 'chitter' sounds, whenever the female chittered nearby. **Post-Fledging Care and Behaviour** See Technical Description for appearance at fledging. Departure from the nest by a single nestling was described in detail by Willis *et al.* (1983). After about a dozen morning feeds, 2.0–2.5h after sunrise, the young followed the female from the nest, leaving behind an egg that had failed to hatch. It hopped to a nearby twig, peeped softly, then to the ground in response to the nearby female's muted song. The female gradually led it along the ground, away from the nest, softly singing and 'sneezing' while hopping back and forth just ahead of the nestling. It appeared that she purposefully led the fledgling to a treefall tangle *c.*30m from the nest. About an hour after the nestling's departure, the male arrived at the nest with food, made the gape-soliciting chitter, peered into the nest and then ate the prey. He pecked several times at the abandoned egg and settled briefly as if to incubate it. Overall his actions suggested that he had not anticipated fledging as, after sitting on the egg he hopped a short distance but returned immediately to again peer at, and then peck at, the egg. After hesitant, uncertain movements on the nest rim and several pecks, he left the area. Each time that observers approached the treefall within which the nestling was hiding the female appeared nearby, nervously flitting her wings, chittering and singing softly. Not until the young was located and captured, at which time it screeched loudly, did she attempt to more actively distract the observers. Remaining 10m away she fluttered about and made 'loud *stauf* variants' of the sneezing call. While the young was still being held she repeatedly wandered closer, producing chitters and soft songs. In the hand, the fledgling continued to make raucous vocalisations and would gape at and raise its crown feathers towards an approaching hand. When left on a low perch the young sang softly but, instead of approaching, the female went to the position within the vegetation from where the young had been removed and sneezed repeatedly until the young left its perch to join her. She then led it away from the area as described above. The male, who had been foraging just 30m away when the fledgling was captured, did not make its presence known during its offspring's ordeal. At two weeks post-fledging, the young watched by Willis *et al.* (1983) was awkwardly following the adult, clinging to low perches and generally not moving until hearing the soft song. It moved to the adult to receive food in response to chittering calls and would freeze in response to alarm vocalisations. The fledgling was still largely fed by the female at five weeks, but sometimes foraged on its own, apparently usually by dropping to the ground to capture prey in the leaf litter. Six weeks post-fledging it still remained with its parents but was more independent, foraging on its own most of the time, singing softly and only occasionally received food. Once it was harassed by another gnateater and responded in an apparently submissive-aggressive manner by gaping and spreading its wings (but not begging). The fledgling in this case (cared for by female) gained full independence at 7–8 weeks post-fledging. It remained, however, in its parents' territory for another *c*.1 month. Although apparently relegated to a small part of the territory, it was largely ignored by its parents. **SEASONALITY** Available information suggests that breeding is seasonal, but that timing varies greatly across the range. In southeast Brazil (*vulgaris*) there may be an increase in vocal activity during the wet season, August–January (Antunes 2008) and nesting apparently occurs late September–November in Argentina (Whitney 2003). Marini *et al.* (2007) reported that, around Belo Horizonte (*vulgaris*), 12 clutches were laid October–January, most in the first month. They found three of nine males and one of four females captured in August to have brood patches. Between October and December, 11 of 13 males and nine of 11 females had brood patches. Further data, see Breeding Data below. Nestling period *c*.14 days (Willis *et al.* 1983, Skutch 1996). **Nest Success** Of nine nests followed from laying, three fledged, five were predated, one abandoned and three had unknown outcomes (Marini *et al.* 2007). Four nests produced just one fledgling near São Paulo, with one predated, one abandoned and three of the remaining four eggs were infertile (Willis *et al.* 1983). **BREEDING DATA** *C. l. lineata* Subadult, 3 April 1963, Brasília (FMNH 295749); subadult, 16 October 2016, Poções (WA 2397712, P. Barros); subadult, 23 March 2016, Vale dos Sonhos, PN Cavernas do Peruaçu (ML 27021421, photo H. Peixoto); subadult, 24 December 2008, Boa Nova (IBC 1322587, photo L. Petersson). *C. l. vulgaris* Carrying nest material, 23 October 2015, Araras (WA 1883423, G. Muniz); carrying nest material, 26 October 2012, Mairiporã (WA 788522, M.C. Zumkeller); carrying nest material, 4 December 2010, Jundiaí (WA 254676, R. Gallacci); nearly complete nest, 25 October 2014, Pouso Alegre (WA: several photos M. Fujihara); two nests with eggs, 8 November 1941 and 17 November 1942, near Santa Teresa (Willis & Oniki 2002b, MBML); three nests with eggs, October and two in December 1976, Fazenda Santa Genebra (Willis *et al.* 1983); four eggs [two nests?],

October 1896, Piquete (von Ihering 1900); two eggs [single clutch?], 20 October, Iguape (Oates & Reid 1903); nest with eggs, 29 December 1992, Santo Amaro (Maurício *et al.* 2013); nest with eggs, 27 November 1997, General Câmara (Maurício *et al.* 2013); nest with eggs, January 2004, Arroio Cadeia (Maurício *et al.* 2013); nest with eggs, 28 September 1953, Tobuna (Fraga & Narosky 1985, MACN); nest with eggs, 10 October 2013, Conchas (WA 1115181/183, H. Neto); nest with eggs, 8 October 2004, *c*.6km E of Vargina (Lopes *et al.* 2013); nest with eggs, 30 October 2012, Joaçaba (WA 789130, C. Geuster); nest with young, 5 December 2014, Timburi (WA 1537676, F. Zurdo); nest with half-grown nestlings, 23 September 1976, Fazenda Santa Genebra (Willis *et al.* 1983); nest with young, 21 October 1960, Arroyo Urugua-í (Fraga & Narosky 1985, MACN); nest with young, 16 January 2015, Poços de Caldas (WA 1729525, E. Godoy); nest with young, 22 November 2009, Ribeirão Grande (WA 82785/83723, L. Monferrari); nest with young, 8 November 2013, Salesópolis (WA 1151683, E.J. Japão); nestlings, 27 November 2000, Polo Petroquímico (Maurício *et al.* 2013); active nest, 26 November 2015, São Paulo (WA 1926376, C. Ferreira); nest collected, contents not noted, 3 November 1960, Arroyo Urugua-í (Fraga & Narosky 1985, MACN); active nest, 18 September 2012, San Sebastián de la Selva (EcoReg 15882, J.L. Merlo); active nest, 10 January 2017, Monte Sião (WA 2486699, R. Silva); two pairs with fledglings, January 1977, Fazenda Santa Genebra (Willis *et al.* 1983); fledgling, 8 February 2015, Mairiporã (WA 1612774/1609311, E. & C. Bolochio); fledgling, 14 December 2014, Marau (WA 1546419, C. Longo); 'fledgling', 10 November 1996, PN San Rafael (Madroño *et al.* 1997a); fledgling transitioning, 14 January 2017, Sabinópolis (WA 2434759, L.H. Pinto); nine 'juveniles', two in February, four in March, two in November, one in December, São Paulo (state) (Willis *et al.* 1983, MZUSP not examined); three 'juveniles', January, 'Paraná and Santa Catarina' (Willis *et al.* 1983, MZUSP not examined); 'juvenile', 23 December 2001, Fazenda do Sr. João Ribas (Lopes *et al.* 2013, COMB not examined); juvenile, 16 January 2010, Campos do Jordão (WA 100107, F. Kallen); juvenile, 17 January 2014, Resende (WA 1249139, R. Czaban); juvenile, 13 March 2012, Ilhabela (WA 602445, F. Yanes); juvenile, 31 December 2012, Luís Antônio (WA 842579, C. Frateschi); juvenile, 23 February 2014, Monte Verde (Flickr: C. Henrique); two juveniles, 15 and 20 December 1959, Arroyo Urugua-í (LACM 48919/95398); juvenile, 28 February 2013, São Paulo (WA 920343, P. Marcelli); juvenile, 2 March 2013, São Paulo (WA 900539, R.G. Lebowski); juvenile, 24 January 2016, Tapiraí (WA 1998377, D. Rodrigues); juvenile, 18 January 2015, São Valério do Sul (WA 01596547, F. Delgiovo); juvenile, 5 February 2015, Avaré (WA 1607082, M. Camacho); three 'young' skins, January, 2 February, March, near Santa Teresa (Willis & Oniki 2002b, UFMG not examined); subadult, 4 August 1960, Boa Vista (FMNH 265219); subadult, 25 July 1966, Salesópolis (FMNH 295747); subadult, 11 January 2011, Saltos del Moconá (F. Schmitt photos); subadult, 29 October 2016, Ibiúna (WA 2359515, L. Cardim); subadult, 13 February 2017, Curitiba (WA 2473167, C. Menezes); subadult, 16 April 2016, Inconfidentes (WA 2176200, R. Garcia); subadult, 6 March 2016, São José dos Campos (WA 2047566, D. Sala); subadult, 22 January 2017, Caxias do Sul (WA 2442756, F. Hofmann); subadult, 23 March 2016, Erval Seco (WA 2062207, L.E. Fritsch); subadult, 28 May 1969, Embu

(FMNH 344556); subadult, 18 June 2016, Petrópolis (WA 2167766, S. Adalberto); subadult, 16 November 1960, Barra do Rio Juquía (FMNH 265222); subadult, 12 June 2016, Pelotas (WA 2154166, D. Berbare); subadult, 21 May 1917, Puerto Segundo (FMNH 57721); subadult, 27 June 2016, PN Itatiaia (IBC 1280796, D.M. Brinkhuizen); subadult, 21 August 2016, Nova Lima (WA 2244995, G. Brandão); subadult transitioning, 10 July 2016, Pindamonhangaba (WA 2189199, J. Gomes); subadult transitioning, 16 June 2016, Santa Rosa de Lima (WA 2159321, J.C. Smith); subadult transitioning, 24 June 2016, Itariri (IBC 1251087, photo L. Souto); greatly enlarged testes, 7 October 1972, Rio Grande do Sul (Belton 1994); distraction display, 20 December 1974, Gramado (Belton 1994); two immatures [juveniles?] (eBird: A.L. Roos), 4 and 7 February 2015, PN São Joaquim.

TECHNICAL DESCRIPTION Sexes similar. Previously proposed plumage differences between them do not appear reliable for in-the-field separation (Dantas *et al.* 2009), especially in light of the dearth of information on geographic variation. The following applies to the most widespread plumage type (*cf. vulgaris*), in which adults possess a distinct superciliary. *Adult* Warm brown above with a fairly prominent grey superciliary stripe that begins just before the eye and ends in a bright silvery-white postocular tuft of elongated feathers. The brown crown is slightly warmer than the remaining upperparts and grades to rufous on the sides of the head, throat and breast. Most populations have a distinct white, crescent-shaped patch separating the throat from the breast. The sides and flanks are duller, washed brownish or olive, surrounding a large, whitish belly patch. The undertail-coverts are slightly more rufescent than the flanks. **Adult Bare Parts** *Iris* brown; *Bill* bicoloured, largely black on maxilla, mandible pale, usually distinctly yellow; *Tarsi & Toes* brownish-green to greyish-yellow. *Immature* There are few previous descriptions of immatures, and none are complete or specific as to the age to which they apply. Bertoni (1901) described immatures as similar to adults but with each feather with a cinnamon spot. He likened their appearance to juvenile *Turdus* and, because he did not describe their plumage as in any way fluffy, this leads me to believe he was describing subadult plumage (as defined here). Other descriptions (Belton 1994, Whitney 2003) also likely refer to juvenile plumage and are largely in agreement in describing 'immatures' or 'juveniles' as lightly streaked ferruginous or buffy on the crown and back, with buffy-white postocular tufts, pale 'orange' spots on lesser and median wing-coverts ('occasionally also on greater coverts'), a mottled chest and faintly barred flanks. Unlike the post-fledging plumages for most other species treated here, based largely on descriptions of individual young seen at a specific age, for Rufous Gnateater we are fortunate to have the description of a young (*vulgaris*, São Paulo), which was followed for several months post-fledging by Willis *et al.* (1983). Much of the following is based on their text, with my own examination of photographs used to complete some details. *Fledgling* At fledging (Willis *et al.* 1983), young are well covered with contour feathers, but their 'dark face and pink cere', as well as the dusky-orange throat are still bare. The feathers of the crown and back are brown with blackish fringes, creating a finely streaked effect. The rump is brown with fine black speckling created by black feather tips. The developing

postocular plume feathers are already visible as small, buff-tipped, whitish tufts behind the eyes. The fluffy, brownish breast feathers are similarly edged blackish, and those of the belly are white with grey margins. Two weeks after leaving the nest, fledglings are nearly adult size, but still have a distinctly short tail and small bill. The ocular area remains bare, the skin blackish, but freshly emerging brown feathers start to replace the streaked forecrown feathers. The remaining upperparts are as previously described, but the postocular tuft is now more evident, albeit buffy. The chest is mottled with emerging adult contour feathers, the belly whitish washed grey, with indistinct barring on the flanks. The lesser and median wing-coverts are tipped with pale orange spots, the greater coverts and remiges are dark. At 2–4 weeks after leaving the nest, young transition into what I here consider juvenile plumage. **Fledgling Bare Parts** (HFG, from photos) *Iris* brown; *Bill* largely dark with inflated, pale orange-white rictal flanges; *Tarsi & Toes* orange-pink to yellow-orange. *Juvenile* At 4–5 weeks post-fledging, young have full-length tail and the frontal half to third of the crown is brown with adult-type contour feathers that lie flat. The hindcrown and nape, possibly extending to the scapulars, are still fluffy, with the face and neck-sides transitioning. The grey superciliary line of adults is now apparent, the back is largely as in adults, but probably has scattered patches of downy feathers on the lower back and rump. The underparts are almost fully adult, tending to appear duskier or 'messy' at the sides, with the flanks and sides of the belly still probably showing some faint dusky barring. The wings are as adults, except the prominent orange-spotted wing-coverts. By 7–8 weeks post-fledging the back is no longer noticeably streaked. Juveniles continue moulting into adult-like plumage, but probably still retain a fair amount of streaking on the hindcrown and nape until 8–9 weeks after fledging, finally losing the last vestiges of nape streaking over next two weeks. Sometime around 10–11 weeks after leaving the nest, juveniles have probably now lost sufficient of their immature plumage to be considered in transition to subadult plumage. At first glance they might be mistaken for adults but probably retain a few spotted or streaked feathers across the nape and possibly upper back. They complete the transition to subadult 11–12 weeks post-fledging. **Juvenile Bare Parts** *Iris* dark brown; *Bill* maxilla dark, similar to adults, rictal flanges still slightly inflated and pale yellowish or white, mandible pale pinkish-white, yellowish mostly at the base; *Tarsi & Toes* dusky-orange to dull yellowish-pink. *Subadult* Largely separable from adult only by the retained immature wing-coverts, which appear to be kept over most of their post-fledging development. Wear removes the terminal black markings and appears to lead to some fading of the rufous-orange spots, which appear pale buff during the transition to subadult plumage. I have seen several examples of subadults that appeared to have fresh, bright orange-buff-spotted wing-coverts. These had clean black edging at the tips that broadened slightly at the centre to create a small black spot. My guess is that most subadults moult into a second set of immature wing-coverts while they are losing the last of their original contour feathers (i.e. during juvenile–subadult transition). Considering that there is certainly some individual and geographic variation, my guess is that most subadults will also be overall duller, specifically in the saturation or brightness of the orange throat and breast.

MORPHOMETRIC DATA Data for adults and subadults combined. *C. lineata lineata* *Wing* 75mm; *Bill* [exposed culmen] 14mm (*n* = 1, ♂, Hellmayr 1908). *Wing* 70mm, 78mm, 73mm, 75mm; *Tail* 50mm, 48mm, 45mm, 48mm; *Bill* exposed culmen 13mm, 12mm, 12mm, 12mm. (*n* = 4, 3♂,1♀?, Novaes 1947). Data from LACM, FMNH. Adult ♀♀: *Bill* from nares 8.8 ± 0.1mm (*n* = 4), exposed culmen 13.2mm (*n* = 1), depth at nares 4.4mm (*n* = 1), width at nares 5.4mm (n = 1); *Tarsus* 28.8 ± 0.6mm (n = 4). Adult ♂♂: *Bill* from nares 9.0 ± 0.2mm (*n* = 5), exposed culmen 12.5 ± 1.3mm (*n* = 3), depth at nares 4.3 ± 0.2mm (*n* = 2), width at nares 5.6 ± 0.2mm (*n* = 2), width at base of mandible 7.8 ± 0.2mm (*n* = 3); *Tarsus* 29.8 ± 0.2mm (*n* = 5). *C. lineata vulgaris* Data from LACM, FMNH. Adult ♀♀: *Bill* from nares 8.5 ± 0.5mm (*n* = 18), exposed culmen 11.2 ± 0.9mm (*n* = 6), depth at nares 4.1 ± 0.3mm (n = 4), width at nares 5.1 ± 0.3mm (*n* = 4), width at base of mandible 6.3 ± 0.2mm (*n* = 4); *Tarsus* 29.6 ± 0.8mm (*n* = 22). Adult ♂♂: *Bill* from nares 8.8 ± 0.4mm (*n* = 33), exposed culmen 11.7 ± 0.9mm (*n* = 13), depth at nares 4.2 ± 0.3mm (*n* = 10), width at nares 5.3 ± 0.3mm (*n* = 10), width at base of mandible 6.3 ± 03mm (*n* = 10); *Tarsus* 30.1 ± 0.9mm (*n* = 33). Adult ♂?: *Wing* 73.7 ± 8.6mm, range 66–99mm (*n* = 12); *Tail* 48.1 ± 3.5mm, range 42–53mm (*n* = 9); *Bill* [exposed] culmen 12.7 ± 1.1mm, range 11.0–14.5mm (*n* = 12); *Tarsus* 30.7 ± 1.2mm, range 28.7–32.6mm (*n* = 12; Bugoni *et al.* 2002). Data from FMNH, LACM, MLZ. *Bill* from nares 8.5 ± 0.5mm (*n* = 3), exposed culmen 11.3 ± 0.7mm (*n* = 3), depth at nares 3.8 ± 0.9mm (*n* = 2), width at nares 5.1 ± 0.1mm (*n* = 2), width at base of mandible 6.3 ± 0.0mm (*n* = 2); *Tarsus* 29.2 ± 0.6mm (*n* = 8). Juveniles (*n* = 2, ♀,♂, LACM): *Bill* from nares 9.0mm, 7.9mm; *Tarsus* 29.8mm, 29.0mm. See also Bertoni (1901), Dabbene (1919). **MASS** Range 16–27g, mean of 11 birds, 22.1g (race unspecified, may include *C. cearae*, Whitney 2003); 22.0g, 24.0g, 19.0g, 21.0g, 24.0g, 20.0g, 21.0g, 24.0g, 19.0g, 20.0g, 22.0g, 23.0g, 22.0g, 21.0g, 21.0g, 22.0g, 23.0g, 18.0g, 21.0g, 21.0g, 21.0g, 22.0g, 23.0g, 19.0g, 20.0g (*n* = 25, 5♂♂,1♀,19♀?, *vulgaris*, Faria & Paula 2008); 22g, 25g, 21g (*n* = 3, ♂,♂,♀, *vulgaris*, Belton 1994); 23.0 ± 1.3g, range 20–25g (*n* = 11,♂?, *vulgaris*, Bugoni *et al.* 2002); 21.1 ± 1.6g (*n* = 11 ♂♂), 20.5g (*n* = 1, ♀), 20.0g, 20.0g (*n* = 2 immatures, ♂?), 19.0g, 21.0g, 22.0g, 23.0g, 23.0g (*n* = 5, ♂?) (*vulgaris*, Reinert *et al.* 1996); 19.0g, 20.0g, 18.0g (*n* = 3, ♂?, immatures, *vulgaris*, Faria & Paula 2008): 20.7g (*n* = ?, ♂?, *vulgaris*, Faria & Rodrigues 2009); mean 21.0g, range 20.0–22.5g (*n* = 4, ♂?, *vulgaris*, Oniki 1981a); 21g (*n* = 1, ♂?, *vulgaris*, Salvador & di Giacomo 2014); 20g (*n* = 1, ♂?, *vulgaris*, Alderete & Capllonch 2010); 21.5 ± 0.7g, range 21–23mm (*n* = 11), also three additional individuals 21.0g, 22.0g, 25.0g (♂?, *vulgaris*, Marini *et al.* 1997); 27.0g, 23.0g, 24.5g, 28.0g, 23.0g, 24.5g, 25.0g (*n* = 7, 1♂,3♀,3♂?, *lineata*, Marini *et al.* 1997); 20g, 21.5g, 22g, 20g, 20g (*n* = 5, 3♂,.2♀; *vulgaris*, Paraguay, Storer 1989). **TOTAL LENGTH** 11.0–14.4cm (von Berlepsch & von Ihering 1885, von Ihering 1898, Bertoni 1901, Ruschi 1979, de la Peña 1988, Narosky & Yzurieta 1989, Dubs 1992, Sick 1993, Belton 1994, de la Peña & Rumboll 1998, Gwynne *et al.* 2010, Esquivel-M. & Peris 2011). **Other** Cloacal temperature: mean 39.6°C, range 38.0–41.4°C (*n* = 4♂?, Oniki 1981a).

TAXONOMY AND VARIATION Three subspecies were provisionally recognised by Whitney (2003) prior to the recent elevation of *cearae* to species rank (Ceara Gnateater; which see). Our understanding of the taxonomic and distribution limits of the other two subspecies are exceedingly poor. During my attempts to sort out some of these uncertainties, it became clear that the situation is far too complex to be resolved without significant additional time and effort. The complexities of the situation mentioned by Whitney (2003), especially the complex vocal variation but rather minimal plumage variation, are but two of the challenges to be overcome if we are to arrive at a clear picture. The literature is littered with misunderstandings and misinterpretations by authors who rarely or never had sufficient specimens to gain a full understanding. Add to this the challenges of transportation and communication faced by early workers, combined with the rather poor condition of the species' holotype, and the result was the description of three additional taxa, two of which are currently subsumed within *lineata*, and one within *vulgaris* (see below). Genetic data also point to a complex history of population expansion and contraction, probably resulting in multiple regions of secondary contact (Pessoa 2001, Pessoa *et al.* 2006, Dantas *et al.* 2007) and leaving us with a complex genetic landscape to navigate if we are to clearly define the number of taxa involved and their precise distributions (Pessoa *et al.* 2004, 2005, Sari & Santos 2004, Sari *et al.* 2005, 2006). This will probably require fairly dense sampling, of vocal, genetic and morphological characters, from across the entire range. The available records suggest that Rufous Gnateater occurs nearly continuously from northern Uruguay, eastern Paraguay and north-east Argentina, north to south-central Bahia, where it quite possibly comes into contact with Ceara Gnateater somewhere east of the Chapada Diamantina. The status and taxonomic affinities of the only (apparently) isolated population in Mato Grosso (Serra das Araras) have not been investigated. It is provisionally included in the nominate race. Sadly, I am left with little recourse but to continue to use the two most widely used names and to propose subspecific ranges that are unlikely to be wholly correct. I hope, however, the suggestions below will at least provide future workers with some helpful lines of inquiry and a starting point to develop testable hypotheses.

Conopophaga lineata lineata (Wied, 1831) The distribution of the nominate race is usually described as including east-central and south-central Bahia, north-east Goiás, northern Minas Gerais, and the disjunct population in southern Mato Grosso, in the Serra das Araras. Whitney (2003) suggested that if populations from the interior of the species' range, usually included within the nominate race (as here), prove separable, the name *rubecula* would likely pertain to them. Neumann (1931) described *rubecula* from Veadeiros, in north-east Goiás, and soon afterwards, Pinto (1936) described *hellmayri* from Fazenda Thomé Pinto, *c.*250km south-east of Veadeiros, and the Serra do Espinhaço. There seems to be relatively little plumage variation in this region and, even if *rubecula* is resurrected, *hellmayri* would be its junior synonym. Unfortunately, the holotype of nominate *lineata* was originally prepared as a display mount, and is in rather poor condition, which may have added to the confusing assignment of names by early authors. Much confusion with respect to distribution and taxonomy stems from misinterpretations of populations in Bahia, where it is now clear that Ceara Gnateater *and* at least one race of Rufous Gnateater occur. Indeed, few authors were in agreement with respect to the racial affinities of populations in Bahia and coastal *cearae*. At least in one case (Naumburg 1937), perhaps due to the poor condition of the *lineata* holotype, specimens of Ceara Gnateater were used to represent nominate *lineata* for comparisons (see Ceara Gnateater). It appears that Naumburg's (1937) flawed inferences as to racial boundaries, which were disseminated in the literature, may have been the basis for much of the confusion during the past century. I have yet to see any truly topotypical specimens or photographs of nominate *lineata* and, in fact, have found just one sight record that might pertain to topotypical *lineata*, by Helberth Peixoto, from west of Barra do Choça (see below), just 10–15 km east of Vitória da Conquista. The nearest locations with vouchered material, however, are the humid hilly forests around Poções and Boa Nova (WikiAves), *c.*60–70 km east-north-east of Vitória da Conquista. I feel that, without exception, all of the photographs available to me from that area are distinguishable by plumage from all other populations of Rufous and Ceara Gnateaters. At first glance this population is similar to both *cearae* and accepted descriptions of *lineata* + *vulgaris*. These birds share the brighter, orange-rufous breast and head of *cearae*, but are consistently distinguishable from both *cearae* and *lineata* + *vulgaris* by having the face, lores and ocular area the same rufous as the surrounding plumage. They lack the distinctive grey loral patch and ocular area of *cearae*, and also lack the distinctive, broad pale superciliary of '*lineata*' which extends in front of the eyes from the base of the postocular tuft. In fact, these characters are so distinctive and consistent that from photos I was easily able to correctly recognise individuals from this population even as 'thumbnails', without needing to open the files. Based on careful examination of all available photographs, this population extends from Jequié south at least as far as Poções and south-east as far as Camacan. Once material is available for comparison, I propose that these birds will prove indistinguishable from topotypical records from the Barra do Choça–Vitória da Conquista area. Should this be true, and should this population be as taxonomically unique as I suggest, it would then be the only one to actually represent nominate *lineata*. Birds elsewhere that are herein treated as nominate *lineata* based on convention would either require another name (*rubecula?*) or would perhaps be best allied with *vulgaris*. Re-examination of photographs of the *lineata* holotype (courtesy of Paul R. Sweet) reveal that it is largely congruent with the above-mentioned characters, but unfortunately its condition precludes certain conclusions. To the south, south-west and west, from the nearest locations with photo documentation (Macarani, Caetité, Taiobeiras) birds clearly possess the distinct greyish superciliary that has, perhaps erroneously, become one of the characters used to define Rufous Gnateater. I believe that the need for this somewhat radical subspecific overhaul has been overlooked because the perpetuation of past misconceptions has led to incorrect character descriptions. Furthermore, only recently have photographs (WikiAves, Flickr) become available for comparison. In the list of localities below, those representing this distinctive population are marked [†]. I have attempted, where possible, to include vocal, photographic and specimen records from pertinent localities, to facilitate future studies. **SPECIMENS & RECORDS** **Tocantins** Serra Traíras, 13°24'S, 47°42.5'W (Pacheco & Olmos 2006). **Bahia** *c.*4.5 km N of São Desidério, 12°19'S, 44°58.5'W (C. Albano *in litt.* 2017); Colônia do Formoso,

13°45'S, 44°28'W (MPEG 45279–281); Jequié, 13°51.5'S, 40°05'W (WA 2294045, D. Campos)[†]; *c.*3.5 km SW of Água Suja, 14°03'S, 44°28.5'W (MPEG 80832); Itacaré, 14°16.5'S, 38°59.5'W (WA 2517648, M.E. Salgado)[†]; vicinity of Boa Nova, *c.*14°22'S, 40°12.5'W (XC: H. Matheve, R.C. Hoyer, J. Minns; see also WA 106064, C. Albano, and many WA photos)[†]; Poções, 14°31.5'S, 40°22'W (WA 2501961/2471352, M.G. Santos)[†]; 10.5 km ESE of Poções, 14°32.5'S, 40°16'W (XC 340981, M.G. Santos)[†]; *c.*11 km W of Barra do Choça, 14°51.5'S, 40°41'W (eBird: H. Peixoto)[†]; REBIO Una, 15°11.5'S, 39°03.5'W (eBird: R. Laps)[†]; Arataca, 15°13'S, 39°25.5'W (WA 161637/640; recordings B. Rennó)[†]; RPPN Serra Bonita, 15°23.5'S, 39°34'W (XC 284728/329401, M. Braun)[†]; Camacan, 15°25.5'S, 39°29'W (WA 2293405, A.L. Briso, many other WA photos)[†]. **Distrito Federal** Planaltina, 15°37'S, 47°40'W (Novaes 1947, MNRJ 14415); Brasília, 15°48'S, 47°52'W (FMNH 295749, LSUMZ 31765, ML, WA); Fazenda Água Limpa, 15°57'S, 47°56'W (Marini *et al.* 1997). **Goiás** Serra Geral de Goiás, 13°00'S, 46°15'W (Novaes 1947, MNRJ 16372); Veadeiros, 14°07'S, 47°31'W (Neumann 1931, FMNH 75124–126); São João d'Aliança, 14°26'S, 47°22'W (LACM 40007–011); Barro Alto, 14°58.5'S, 48°55'W (WA 2123336, T. Junqueira); Fazenda Esperança, 15°25'S, 49°38'W (Hellmayr 1908, AMNH 488930); Fazenda São Francisco, 15°27'S, 48°00'W (MPEG 80836/37); Fazenda Thomé Pinto, 15°43'S, 49°20'W (Pinto 1938); Ferreiro, 15°52'S, 50°04'W (Novaes 1947, MNRJ 14406); Alexânia, 16°05'S, 48°30.5'W (WA 319432, R. D'Alessandro); Nova Veneza, 16°21.5'S, 49°18.5'W (Novaes 1947, MNRJ 14408); Inhúmas, 16°22'S, 49°30'W (Pinto 1936); Cristalina, 16°46'S, 47°37'W (WA 2527143, A. Alves); Fazenda Transvaal, *c.*17°47.5'S, 50°55'W (MCZ 269922–224). **Minas Gerais** PN Cavernas do Peruaçu, 15°09'S, 44°15'W (Kirwan *et al.* 2001); Januária, 15°29'S, 44°22'W (WA 2070540, F. Olmos); Fazenda São Miguel, 15°50'S, 46°30'W (Lopes *et al.* 2008); 'Sangrador,' 16°02'S, 57°15'W (Naumburg 1937). **Mato Grosso** Serra das Araras, 15°03'S, 57°12'W (Oniki *et al.* 2002); ESEC Serra das Araras, 15°38'S, 57°14'W (Valadão 2012); Santo Antônio do Leverger, 15°52'S, 56°05'W (WA 1663662, recording D. Meyer); between Poconé and Poto Jofre, *c.*16°48'S, 56°43.5'W (Cintra & Yamashita 1990, Tubelis & Tomas 2003).

Conopophaga lineata vulgaris Ménétriés, 1835, Mémoires de l'Academie Imperiale des Sciences de Saint Petersbourg 6th series, vol. 3, p. 534 (Pl. 14, fig. 1). The description is based on specimens from 'near Rio de Janeiro and in State of Minas Gerais', with at least some of the type series lost (Chrostowski 1921, Cory & Hellmayr 1924, Peters 1951). As a point of reference I have mapped Rio de Janeiro as the type locality. Subspecies *vulgaris* was not recognised, by most authors for many years (Burmeister 1856, Gray 1869, Cory & Hellmayr 1924). Perhaps as a result, *Ceraphanes anomalus* was described by Bertoni (1901) from 'Puerto Bertoni,' Alto Paraná. Chubb (1910) rejected Bertoni's (1901) description of a new genus for *C. lineata*, but elected to maintain *anomalus* as a taxon distinct from nominate *lineata*. This name, however, was rarely used in the literature before being synonymised with *vulgaris*. As noted by Whitney (2003) should some of the vocal diversity lead to the recognition of other taxa, *anomalus* may well be applied to one of these. Past descriptions of the range of *vulgaris* are vague, and often conflicting, depending on how they are interpreted. As here defined, the distribution of *vulgaris* extends from extreme north-east Uruguay, north through all of the Brazilian coastal states to southernmost Bahia. West from Rio Grande do Sul it ranges through Misiones and northern Corrientes across most of south-east Paraguay. Further north, there are a few records from south-east Mato Grosso do Sul and western São Paulo. North of São Paulo, the western limit of the range probably adjoins or even overlaps with that of nominate *lineata* somewhere in western or central Minas Gerais. In fact, records from western São Paulo may prove to be better allied with the population of *lineata* in Goiás and Distrito Federal. For discussion of the northern limit of *vulgaris*, see above under *lineata*. Subspecies *vulgaris* is usually defined as being less rufous-brown above, more olivaceous-brown. The rufous throat and chest tend to be duller, paler and less extensive, with the white belly patch more extensive. The undertail-coverts show little or no rufous, which seems to be conspicuous in northern birds (Chubb 1910). Overall, descriptions of *vulgaris* make it out to be little more than a 'faded' version of *lineata*. To my eyes, this description best fits the majority of specimens from the interior of its range in Argentina and Paraguay, particularly with respect to the reduced saturation and extent of rufous on the breast, which does not extend to the flanks. There appear to be some individuals further north that seem intermediate between this description and the more saturated coloration of northern birds. The change, however, is not distinctly clinal, but rather it appears that the percentage of 'pale *vulgaris*' declines with increasing latitude, with few or none matching this description in Rio de Janeiro and eastern Minas Gerais (from where *vulgaris* was described!). Birds here are much more similar to nominate *lineata*, especially below, than Argentina/Paraguay populations. The most consistent difference from *lineata* seems to be the reduced brightness of the crown, with 'Rio de Janeiro *vulgaris*' tending to have a dull chestnut-brown crown that gradually becomes greyish or olivaceous-brown on the back. This description also appears to fit most individuals in the entire coastal region, at least as far south as Paraná. The situation is far from clear but, at first glance, plumage variation in *vulgaris* appears to follow Gloger's rule (Gloger 1883, Delhey 2017), and may include a pale, Cerrado/Chaco-distributed form and a darker, more saturated Atlantic Forest form.

SPECIMENS & RECORDS BRAZIL: Bahia Caetité, 14°04'S, 42°29'W (WA 432707, J. Ministro); 5km S of Brejinho das Ametistas, 14°18.5'S, 42°32.5'W (XC 39993, C. Albano; see also WA 2338800, R. Souza, WA 2327382, R. Prates; and XC, R. Souza); RPPN Mata do Passarinho, 15°47.5'S, 40°31.5'W (C. Albano *in litt.* 2017). **Minas Gerais** Fazenda Americana, 16°16'S, 42°55'W (Vasconcelos & Neto 2007); Santa Maria do Salto, 16°19.5'S, 40°06.5'W (WA 1577200, W. Nogueira); W of Montes Claros, 16°44'S, 43°54'W (Vasconcelos & Neto 2007); Mata do Lobo, 16°47'S, 43°01'W (Vasconcelos & Neto 2007); Campo Limpo area, 17°12'S, 42°51'W (Vasconcelos & Neto 2007); Paracatu, 17°13'S, 46°52'W (Krabbe 2007); Pirapora, 17°23'S, 44°48'W (Kirwan *et al.* 2001); João Pinheiro, 17°44.5'S, 46°10.5'W (WA 2218310, S. Pacheco); Poté, 17°49.5'S, 41°48'W (WA 1274260, recording E. Luiz); São Gonçalo do Abaeté, 18°20.5'S, 45°50'W (WA 1775235, recording W. Nogueira); Sabinópolis, 18°37.5'S, 43°03.5'W (WA 2434759, L.H. Pinto); Agua Suja, 18°53'S, 47°38'W (Laubmann 1940); Uberlândia, 18°55'S, 48°16.5'W (Marini 2001); Patrocínio, 18°56'S, 46°59.5'W (WA

356424, V. Castro); 5km ESE of Conceição do Mato Dentro, 19°03'S, 43°22'W (Willis & Oniki 1991a); Antônio Dias, 19°08'S, 42°56'W (Machado & Fonseca 2000); Iapu, 19°17.5'S, 42°15.5'W (WA 1274186, L. Moreira A.); Cabeça de Boi, 19°24'S, 43°23.5'W (Freitas & Costa 2007); Cauaia, 19°28'S, 44°03'W (Christinsen & Pitter 1997); Ibiá, 19°29'S, 46°325'W (WA 642444, F. Rage); Pitangui, 19°34.5'S, 44°53'W (WA 1422145, T. Silva); Cauê, 19°37'S, 44°01'W (Christinsen & Pitter 1997); Nova Era, 19°42'S, 43°04'W (Machado & Fonseca 2000); São Gonçalo do Rio Abaixo, 19°49'S, 43°18.5'W (WA 1683110, E. Franco); Pará de Minas, 19°51.5'S, 44°36.5'W (WA 1902455, L. Faria); Sacramento, 19°52'S, 47°27'W (WA 1607618, L. Amui); Rio Piracicaba, 19°58'S, 43°08'W (WA 1315629, J.S. Barros); Nova Lima, 19°59.5'S, 43°51'W (WA 2244995, G. Brandão); Santo Antônio do Monte, 20°05'S, 45°17.5'W (WA 1176893, P. Couto); Matão, 20°09'11'S, 44°22'35'W (Lopes *et al.* 2012); Itatiaiuçu, 20°13'S, 44°27.5'W (WA 1262800, W. Nogueira); Reduto, 20°13.5'S, 41°56.5'W (WA 1830828, G. Adams); Serra da Canastra, 20°15'S, 46°37'W (XC 221490/91, P. Boesman); Mariana, 20°22.5'S, 43°25'W (Lima *et al.* 2010); Serra do Caparaó, c.20°27'S, 41°51'W (USNM 499043, MNRJ 26095); Casa Queimada, c.20°28'S, 41°48'W (MNRJ 14414/15); c.38km SSW of Manhuaçu, 20°34'S, 42°12'W (Faria & Paula 2008); Fazenda do Sr. João Ribas, 20°40'S, 42°30'W (Lopes *et al.* 2013); near Viçosa, 20°45.5'S, 42°52.5'W (Ribon *et al.* 2001, 2003, ANSP 175875, XC); Jacuí, 21°01'S, 46°44.5'W (WA 2268209, J.M.C. Monteiro); São João Del Rei, 21°08'S, 44°15'W (Lombardi *et al.* 2007); near Muriaé, c.21°08'S, 42°22'W (MNRJ 16359/67/69); Ijaci, 21°11'S, 44°55'W (Ribon 2000); Lavras, 21°15'S, 45°00'W (Vasconcelos *et al.* 2002, XC); Guaxupé, 21°18.5'S, 46°42.5'W (WA 1115209, A. Carosia); Mata dos Bentes, 21°30.5'S, 43°10'W (Ribon *et al.* 2004); São João Nepomuceno, 21°32'S, 43°01'W (WA 2199425, J.L. Grazia); c.6km E of Vargina, 21°35'S, 45°30'W (Lopes *et al.* 2013); São Thomé das Letras, 21°43'S, 44°59'W (WA 1464208, C. Souza); Poços de Caldas, 21°47'S 46°33.5'W (MNRJ 30256); Lambari, 21°58'S, 45°21'W (WA 1273220, J.A. Marins); Bocaina de Minas, 22°10'S, 44°23.5'W (MNRJ MNA4560); Pouso Alegre, 22°13.5'S, 45°56'W (WA 1492769, M. Fujihara); Inconfidentes, 22°20.5'S, 46°16'W (WA 2176200, R. Garcia); Monte Sião, 22°26'S, 46°34.5'W (WA 2486699, R. Silva); Marmelópolis, 22°27'S, 45°10'W (WA 353813, E. Kaseker); near Sebastião Lauriano, 22°41'S, 45°54.5'W (Vasconcelos & Neto 2009); Monte Verde, 22°51.5'S, 46°03'W (Flickr: C. Henrique). **Espírito Santo** Alto Rio Novo, 19°01.5'S, 40°58.5'W (WA 2335768, J. Rezende); Baixo Guandú, 19°30.5'S, 41°01'W (Novaes 1947, MNRJ 14413); Itarana, 19°52.5'S, 40°52.5'W (Venturini *et al.* 2001); Santa Teresa, 19°56'S, 40°36'W (Willis & Oniki 2002a, YPM, MCZ, MBML, MZUSP, UFMG, MNRJ); ESEC Santa Lúcia, 19°57'S, 40°32'W (Ruschi 1977, Simon 2000, MNRJ); Jatiboca, 20°05'S, 40°55'W (MNRJ 26309); PE Pedra Azul, 20°23.5'S, 41°01.5'W (ML 64200261, photo L. Merçon); Vargem Alta, 20°40'S, 41°00.5'W (XC 80130, J. Minns). **Rio de Janeiro** Pedra da Elefantina, c.20°59'S, 42°06'W (Pacheco *et al.* 1996); c.7.5km NNW of Carmo, 21°52'S, 42°35'W (XC 291976, J. Fischer); near Comendador Levy Gasparian, 22°02'S, 43°11'W (MNRJ MNA3924); 'Morro Queimado' [= Nova Friburgo] and vicinity, 22°17'S, 42°32'W (Krabbe 2007, MNRJ 35579–582, ZMUC); c.6km ENE of Barra Mansa, 22°18'S, 43°04.5'W (XC 8462/33296, R. Gagliardi); Visconde de Mauá, 22°20'S, 44°32'W (Bauer & Pacheco 2000); Pico das Agulhas Negras, 22°23'S, 44°38'W (AMNH

188971–978); Serra dos Tucanos Lodge, Trilho Bamboo, 22°23′S, 42°33′W (ML 45406751, photo N. Voaden); Rio Bonito de Cima, 22°23.5′S, 42°22.5′W (XC 20309, R. Gagliardi); Ingá, 22°24.5′S, 42°51′W (MNRJ); Teresópolis and vicinity, c.22°25′S, 42°58.5′W (Novaes 1947, Sousa et al. 2004, LACM 27812, MNRJ, XC); PN Itatiaia, c.22°26.5′S, 44°36′W (LACM 66620–623); Miguel Pereira, 22°27.5′S, 43°29′W (LACM 66624); Resende, 22°28′S 44°27′W (WA 1249139, R. Czaban); Petrópolis, c.22°30.5′S, 43°10.5′W (Novaes 1947, MNRJ 14100–112); Fazenda Bela Aliança, c.22°36′S, 43°54.5′W (Pacheco et al. 1997b); Perequê, 23°00.5′S, 44°32′W (XC 80421, J. Minns). **São Paulo** Ituverava, 20°20′S, 47°47.5′W (AMNH 140017/18); 'Foz do Rio Tietê', c.20°39.5′S, 51°20′W (CM P143493); Fazenda Moro Chato, 21°22′S, 49°41′W (Almeida et al. 1999); Luís Antônio, 21°33.5′S, 47°42′W (WA 842579, C. Frateschi); Lins, 21°40.5′S, 49°45′W (FMNH 123489); Universidade Federal de São Carlos, 21°58′S, 47°52′W (Motta-Junior 1990, Marini et al. 1997); Fazenda Água Branca, 22°10′S, 48°10′W (Marini et al. 1997); Fazenda Santa Elisa, 22°17′S, 48°08′W (Almeida et al. 1999); Araras, 22°21.5′S, 47°23′W (WA 1883423, G. Muniz); Rio Claro–Araras area, c.22°22′S, 47°28′W (Cândido 2000, Willis & Oniki 2002b, Gussoni 2007, Mironov & Hernandes 2014); trail to Gruta do Fazendão, 22°25.5′S, 47°47.5′W (ML 34225891, photo L.C. Ramassotti); Fazenda Rio Claro, 22°27′S, 48°51′W (Donatelli et al. 2004); PE Morro Do Diabo, 22°31′44.22′S, 52°18′W (ML, C.A. Marantz); Piquete, 22°37′S, 45°10.5′W (von Ihering 1898, von Ihering & von Ihering 1907, AMNH 488931–935); Fazenda São João, 22°44′S, 46°54′W (Anciães & Marini 2000); Campos do Jordão, 22°44.5′S, 45°35.5′W (WA 100107, F. Kallen); Serra de Bananal, c.22°45′S, 44°20′W (Pacheco et al. 1997a, MZUSP 27271); Fazenda Barreiro Rico, 22°45′S, 48°09′W (Willis 1979, ML); Victoria [= Vitoriana, Paynter & Traylor 1991b], 22°46.5′S, 48°24′W (AMNH 488936/37, MCZ 120063); Fazenda Santa Genebra, 22°49′S, 47°07′W (Willis 1979, Willis et al. 1983, Aleixo & Vielliard 1995); Pindamonhangaba, 22°50.5′S, 45°29.5′W (WA 2189199, J. Gomes); Casa da Bocaina, 22°52′S, 44°27′W (XC 344738/39, J. Minns); Caiuá near Salto Grande, 22°53.5′S, 49°59′W (FMNH 49013, UMMZ 89434; see Paynter & Traylor 1991a); Conchas, 23°01′S, 48°00.5′W (WA 1115181/183, H. Neto); c.8km SSE of Taubaté, 23°06′S, 45°31′W (Faria & Paula 2008); Avaré, 23°06.5′S, 48°55.5′W (WA 1607082, M. Camacho); Tietê, 23°07′S, 47°43′W (von Ihering 1898, von Ihering & von Ihering 1907); Atibaia, 23°07′S, 46°33.5′W (XC 144006, W. Zaca); Jundiaí, 23°11′S 46°53′W (USNM 177688, WA); Nazaré Paulista, 23°12′S, 46°21.5′W (Ogrzewalska et al. 2008); Timburi, 23°12.5′S, 49°36.5′W (WA 1537676, F. Zurdo); São José dos Campos, 23°13.5′S, 45°54′W (WA 2047566, D. Sala, XC); Itu, 23°16′S, 47°19′W (Krabbe 2007, ZMUC, XC); c.17.3km ESE of Catuçaba, 23°16′S, 45°01.5′W (XC 177590, A. Silveira); Mairiporã, 23°19′S 46°35′W (WA 1612774/1609311, E. & C. Bolochio); Fazenda Rio das Pedras, 23°23′S, 48°36′W (Donatelli et al. 2007); Ipanema, 23°26′S, 47°36′W (von Pelzeln 1871, AMNH 78279); Pico do Corcovado, 23°28′S, 45°12′W (Goerck 1999); Mogi das Cruzes, 23°31.5′S, 46°11′W (von Pelzeln 1871, von Ihering 1898, MCZ 169105–110); Salesópolis, 23°32′S, 45°51′W (WA 1151683, E.J. Japão, LSUMZ, FMNH, KUNHM, UAZ); near Itapetininga, 23°35.5′S, 48°03′W (LSUMZ, KUNHM, UMMZ); Estação Biológica Boracéia, 23°38′S, 45°52′W (Cavarzere et al. 2010, XC); Ibiúna, 23°39.5′S, 47°13.5′W (WA 2359515, L. Cardim); São Bernardo do Campo, 23°42′S, 46°33′W (LSUMZ 63374); Campo Grande,

23°46′S, 46°21′W (ANSP, FMNH, MPEG); Paranapiacaba, 23°46.5′S, 46°18′W (FMNH 107077–088, LSUMZ 70450); Ilhabela, c.23°49.5′S, 45°20′W (von Ihering & von Ihering 1907, Olmos 1996); c.5km E of Vila Élvio, 23°50′S, 47°20′W (Banks Leite et al. 2012); c.7.5km SW of Marsilac, 23°57′S, 46°46′W (XC 186186–188, F. I. de Godoy); Tapiraí, 23°58′S, 47°30.5′W (WA 1998377, D. Rodrigues); RPPN Parque do Zizo, 24°01′S, 47°48.5′W (XC 4650, B. Planqué); Ribeirão Grande, 24°06.5′S, 48°22.5′W (WA 82785/83723, L. Monferrari); near Itanhaém, 24°11′S, 46°47′W (XC 330796, M.J. Feliti); Itariri, 24°17.5′S, 47°10.5′W (IBC 1251087, photo L. Souto); 'Ribeirão Onça Parda' (Figueiro Rasa), c.24°20.5′S, 47°53′W (CM P143610; see Paynter & Traylor 1991b); Barra do Rio Juquiá (Pousinho), 24°22′S, 47°49′W (FMNH 265220/222); near Guaraú, 24°22′S, 47°01.5′W (ML 48259361, photo F. Barata); Boa Vista, 24°35′S, 47°38′W (FMNH 265218/19); Embu, 24°38′S, 47°25′W (FMNH 344556, ML); Vila Barra de Icapara/Icapara area, c.24°41′S, 47°28′W (FMNH 344557–559); Iguape, c.24°42′4.65′S, 47°32′59.15′W (Oates & Reid 1903, Kreuger 1968, LACM 28706); Baía do Trapandé, 25°04.5′S, 47°56′W (FMNH 258118). **Paraná** Marilena, 22°44.5′S, 53°04′W (Straube & Urben-Filho 2005b); Fazenda Caiuá, 22°53.5′S, 49°59′W (FMNH 49013, UMMZ 89434, loc. contra Paynter & Traylor 1991a); Londrina, 23°18′S, 51°09′W (Lopes et al. 2001, XC); Pôrto Camargo, 23°22.5′S, 53°45′W (Pinto & de Camargo 1955); São Pedro do Ivaí, 23°51.5′S, 51°52.5′W (Straube & Urben-Filho 2005b); Guaíra, 24°05′S, 54°15.5′W (AMNH 318386–389); Fazenda Barra Mansa, 24°05.5′S, 49°50′W (Straube 2008); Campo Mourão, 24°06′S, 52°19′W (XC 313527/28, L.C. Silva); near Tibagi, 24°31′S, 50°24.5′W (AMNH 318378/79); Cândido de Abreu, 24°33′S, 51°20′W (Hinkelmann & Fiebig 2001, Straube et al. 2005); FLONA Piraí do Sul, 24°34′S, 49°55′W (Carvalho et al. 2016); Pôrto Britânia, 24°38′S, 54°17′W (AMNH 318383–385); Rio Cantu, 24°43′S, 52°24′W (Straube & Urben-Filho 2008); Fazenda Lira, 24°47′34′S, 53°17′49′W (Straube & Urben-Filho 2008); Vermelho, 24°55′S, 51°25′W (Sztolcman 1926b); Corvo, 25°13′S, 48°32′W (Straube 2003, AMNH 318390–398, MNRJ 11404/07); Fazenda Thá, 25°17′S, 48°47′W (Straube 2003); Curitiba, 25°25.5′S, 49°16′W (WA 2473167, C. Menezes); Roça Nova, 25°28.5′S, 49°01′W (AMNH 488926/27); Foz do Iguassú, 25°31′S, 54°39′W (AMNH 318381/82); Fazenda Arapongas, 25°41′S, 49°57′W (Mestre 2004); Rio Azul, 25°45′S, 50°47′W (Pichorim & Bócon 1996, Straube et al. 2005); Fazenda Concórdia, 25°45′S, 51°05′W (Sztolcman 1926b, Straube et al. 2005); Foz do Rio Jordão, 25°45′S, 52°15′W (Straube et al. 1996, 2005, MHNCI); UHE-Segredo, 25°55′S, 52°25′W (Straube et al. 2005); Mallet, 25°55′S, 50°50′W (Pichorim & Bócon 1996); c.2.8km SW of Lagoinha, 25°58.5′S, 49°14′W (Scherer-Neto & Toledo 2012); Solais, 26°02′S, 51°58′W (Straube 1988, Straube et al. 2005); UHE-Foz do Areia em Bituruna, 26°05′S, 51°35′W (Straube et al. 2005). **Mato Grosso do Sul** Rio Amambaí, 23°22′S, 53°56′W (AMNH 319405). **Santa Catarina** Fazenda Naderer, c.26°15′S, 49°23′W (MNRJ 16360); c.14km WNW of Joinville, 26°16′S, 49°00′W (MCZ 147923/24); c.8km NW of Dedo Grosso, 26°20′S, 48°58′W (XC 268550, F. Lambert); c.15km NE of Abelardo Luz, 26°29.5′S, 52°12′W (XC 172124/27, A.A.C. Junior); Doutor Pedrinho, 26°43′S, 49°29′W (XC 294703/333058, D. Meyer); Salete, 26°58′S, 50°00′W (Meyer 2016); Rio Itajaí drainage, c.27°00′S, 49°00′W (Brandt et al. 2009); Joaçaba, 27°10.5′S, 51°30.5′W (WA 789130, C. Geuster); Praia Vermelha, 27°11.5′S, 48°35′W (Marenzi et al. 2006); 4km

NNE of Vidal Ramos, 27°21.5'S, 49°20.5'W (XC 173698, D. Meyer); *c.*2.6km WNW of Angelina, 27°34'S, 49°00.5'W (XC 43767, E. Legal); Santa Rosa de Lima, 28°00.5'S, 49°11'W (WA 2159321, J.C. Smith); Fazenda Santa Rita 28°18'S, 49°48'W (dos Anjos & Graf 1993); Siderópolis, 28°36'S, 49°26'W (Just *et al.* 2015). *Rio Grande do Sul* FLOTA Turvo, 27°08'S, 53°53'W (ML 68468, D.W. Finch); Iraí, 27°11.5'S, 53°15.5'W (ML 25429, W. Belton); 15km E of Planalto, *c.*27°19'S, 52°54.5'W (ML 68437, D.W. Finch); Erval Seco, 27°30.5'S, 53°32.5'W (WA 2062207, L.E. Fritsch); São Valério do Sul, 27°47'S, 53°56'W (WA 01596547, F. Delgiovo); Urupema, Morro das Antenas, 27°55.5'S, 49°51.5'W (ML 59590291, photo I. Thompson); Santo Ângelo, 28°18.5'S, 54°16'W (WA 1113388, P.B. Rodrigues); Marau, 28°27'S, 52°12'W (WA 1546419, C. Longo); Cruz Alta, 28°39'S, 53°36'W (WA 2472263, C. Boufleur); Maçambará, 29°02'S, 55°53.5'W (Accordi 2003); Itaqui, 29°08'S, 56°33'W (Accordi 2003); Caxias do Sul, 29°10'S, 51°11'W (WA 2442756, F. Hofmann); *c.*2.7km S of Eletra, 29°20'S, 50°41'W (ML 19272, W. Belton); SE corner of Lagoa do Jacaré, 29°20.5'S, 49°48.5'W (AMNH 813066); *c.*1.5km W of Serafim Schmidt, 29°22'S, 52°24'W (ML 19279, W. Belton); Gramado, 29°22.5'S, 50°52.5'W (Belton 1994, MNRJ 32189); São Francisco de Paula, 29°27'S, 50°35'W (USNM 503769/70); *c.*3km NW of Monte Alverne, 29°33'S, 52°21.5'W (Bencke 1996, ML 91099, G. Bencke); Novo Hamburgo, 29°41'S, 51°08'W (von Ihering & von Ihering 1907); Santa Cruz do Sul, 29°43'S, 52°26'W (USNM 503771); Polo Petroquímico, 29°52.5'S, 51°23.5'W and General Câmara, 29°52.5'S, 51°53.5'W (Maurício *et al.* 2013); near Santo Amaro, *c.*29°55.5'S, 51°56'W (Accordi & Barcellos 2006, Maurício *et al.* 2013); Foz do Rio Vacacaí, 29°57'S, 53°05'W (Accordi & Barcellos 2006); Sans Souci, 30°03'S, 51°18'W (Repenning & Fontana 2011); Banhado dos Pacheros, 30°05'S, 50°53'W (Accordi & Barcellos 2006); Turuçu, 31°25'S, 52°10.5'W (XC 22091, F. Jacobs); Candiota, *c.*31°32.5'S, 53°38'W (WA 611102, R. De Matos); Arroio Cadeia, *c.*31°33'S, 51°33'W (Maurício *et al.* 2013); Estação Experimental Cascata, 31°37'S, 52°31.5'W (Bergmann *et al.* 2015); Pelotas, 31°46'S, 52°20.5'W (Sclater 1890, Maurício & Dias 1998); Rio Santa Maria, 31°53'S, 53°05'W (Maurício & Dias 2000). **PARAGUAY:** *Amambay* 40km WSW of Capitán Bado, 23°27'S, 55°56'W (UMMZ 101795–800). *San Pedro* Estancia Laguna Blanca, 23°49'S, 56°17.5'W (Smith *et al.* 2005); Nueva Germania, 23°54'S, 56°34'W (Laubmann 1940, Hayes 1995); Estancia Caaguyrory, 24°46'S, 56°26'W (Lowen *et al.* 1996). *Cordillera* RPN Sombrero, 25°00'S, 56°38'14'W (Lowen *et al.* 1996). *Canendiyú* Puesto de Guardaparques Lagunita, 24°07'S, 55°26'W (Brooks *et al.* 1993); 13.3km N by road of Curuguaty, 24°22'S, 55°42'W (UMMZ 200818–821); Estancia Itabó, 24°27'S, 54°38'W (Brooks *et al.* 1993, Lowen *et al.* 1995, Cockle 2003). *Caaguazú* E of Caaguazú, *c.*25°28'1.07'S, 55°58.5'W (AMNH 320399). *Alto Paraná* Estancia San Antonio, 25°18'S, 55°18'W (Brooks *et al.* 1993); *c.*3.7km S of Presidente Franco, 25°35.5'S, 54°36.5'W (UMMZ 108836–841); Puerto Bertoni, *c.*25°38'S, 54°40'W [see Paynter 1989] (Bertoni 1901). *Guairá* Independencia, 25°41.5'S, 56°16'W (AMNH 320043–045). *Caazapá* Estancia La Golondrina, 25°33'S, 55°29.5'W (Brooks *et al.* 1993); 7.5km E of San Carlos, 26°07'S, 55°44'W (KUNHM 87952/53); PN San Rafael, *c.*26°20.5'S, 55°31.5'W (KUNHM 88444–445). *Paraguarí* Sapucái, 25°40'S, 56°55'W (Hayes 2014); PN Ybycuí, *c.*26°01'S, 57°03'W (Lowen *et al.* 1996, UMMZ 202173).

Itapúa *c.*14.5km WNW of Itapúa Poty, 26°36.5'S, 55°40'W (Esquivel-M. & Peris 2008); Santa Inés, 26°23'S, 55°46'W (XC 55579, M. Velazquez); 8km N of San Rafael del Paraná, 26°35'S, 54°56'W (UMMZ 200820); 19.5km by road NNE of Encarnación, 27°10'S, 55°49.5'W (USNM, MVZB).
ARGENTINA: *Misiones* San Sebastián de la Selva, 25°51.5'S, 53°58.5'W (EcoReg 15882, J.L. Merlo); Arroyo Uruguá-í, 10km from mouth, 25°52'S, 54°30.5'W (YPM, LSUMZ, UMMZ, SNOMNH; location probably underwater now, after river was dammed in 1981); Puerto Bossetti, 25°52'S, 54°34'W (CM P141177, YPM 66936); Arroyo Uruguá-í, 25°54'S, 54°36'W (Fraga & Narosky 1985, MACN, LACM, AMNH); Arroyo Uruguá-í, 30km from mouth, *c.*25°55'S, 54°19'W (LSUMZ 55425–428); Puerto Segundo, 25°59.5'S, 54°38'W (Dabbene 1919, FMNH); Eldorado, 26°24'S, 54°37.5'W (FMNH 57896–98, CM P138567/68); Tobuna, 26°28'S, 53°54'W (Fraga & Narosky 1985, LACM, MLZ, BMNH, SDMNH, MACN); 8km SW of San Pedro, *c.*26°41'S, 54°10.5'W (KUNHM 104888); 48km SE of San Pedro, *c.*26°55'S, 53°47'W (FIML 14510); near Puerto Mineral, *c.*26°57'S, 55°05'W (UWBM 70298/303); near Dos de Mayo, *c.*27°02'S, 54°40.5'W (FIML 13417–420); Aristóbulo del Valle, 27°06'S, 54°53.5'W (Giraudo *et al.* 1993, 2008); Corpus Christi, 27°08'S, 55°28'W (Krauczuk 2008); Saltos del Moconá, 27°08.5'S 53°53.5'W (Giraudo *et al.* 1993); Campo Grande, 27°12.5'S, 54°58.5'W (FIML 6194/6277); *c.*20km SE of San Ignacio, 27°22'S, 55°24.5'W (AMNH); Parada Leis, 27°22'S, 55°30'W (AMNH 795294–296); Santa Ana, 27°22'S, 55°34'W (Dabbene 1919); Yacaratía Lodge, 27°23.5'S, 54°14'W (XC 60479, M. Castelino); Bonpland, 27°29'S, 55°28.5'W (FIML 525); Cerro Corá, 27°30.5'S, 55°36'W (XC 337876, L. Pradier); San José, 27°46'S, 55°47'W (AMNH 795293); Gobernador Lanusse, 27°59'S, 55°31.5'W (FIML 13593/94); Barra Concepción, 28°07'S, 55°35'W (AMNH, LSUMZ, UCLA). *Corrientes* Garruchos, 28°10.5'S, 55°39'W (YPM, CM); Colonia Garabí, 28°14'S, 55°47'W (AMNH, LSUMZ, UMMZ); Estancia La Blanca, 28°29'S, 55°57'W (Capllonch *et al.* 2005, Alderete & Capllonch 2010). **URUGUAY:** *Cerro Largo* Paso del Centurión, *c.*32°08.5'S, 53°44'W (Maurício & Dias 2000, ML photos).

STATUS The widespread and locally common Rufous Gnateater is not considered of conservation concern, despite its enormously reduced Atlantic Forest habitat (Brooks *et al.* 1999, Jenkins *et al.* 2010, BirdLife International 2017). Rufous Gnateater does not appear to shy away from forest edges (Cândido 2000) and can be fairly tolerant of anthropogenic habitat alteration and fragmentation (Antunes 2005, dos Anjos *et al.* 2009, 2010, Martensen *et al.* 2012). It can even persist in very small (<10ha) fragments (Motta-Junior 1990, Santos 2004) including, in some cases, those in urban areas (Silva 2001, Santos & Cademartori 2010, Fontana *et al.* 2011, Alexandrino *et al.* 2013), and has been reported nesting in fragments as tiny as 21ha. In fact, Oliveira *et al.* (2011, 2012) found that, although somewhat reluctant to cross open areas, like roads, Rufous Gnateater does cross roads more than other understorey birds like Short-tailed Antthrush and White-shouldered Fire-eye *Pyriglena leucoptera*. Marini (2010) also found evidence to suggest the ability to cross considerable gaps between isolated fragments, an assertion supported by molecular studies suggesting that Atlantic Forest fragmentation has had little effect on gene flow (Dantas *et al.* 2007). Sadly, this 'bravery' may not always work to its benefit, as evidenced by Rufous Gnateaters being among the casualties of car collisions in several studies (Becker *et al.* 2007, Coelho *et al.* 2008). Its ability and proclivity to inhabit fragments and plantations appears to vary somewhat, but may ultimately depend on the proximity of more botanically diverse habitat (Brandt *et al.* 2009). Rufous Gnateaters were not found in *yerba mate* (Cockle 2003) or coffee plantations (Moura *et al.* 2015) with adjacent forest, and Motta-Junior (1990) noted their absence from *Eucalyptus* plantations despite presence in an adjacent forest patch. Conversely, Jacoboski *et al.* (2016) and Straube (2008) did, however, find them in *Eucalyptus* monocultures of varying ages, and Willis (2003a) noted long-term persistence (with small declines) in a *Eucalyptus* plantation allowed to become overgrown. Overall, compared with most species treated herein, I could continue nearly indefinitely with examples of how tolerant the Rufous Gnateater is of anthropogenic disturbance. Nevertheless, it has likely already suffered considerable reductions in numbers, especially in landscapes converted almost entirely to agriculture such as western Rio Grande do Sul (*vulgaris*; Belton 1994). As another word of caution, in a small woodlot in São Paulo (Willis *et al.* 1983), half of the six eggs observed (three nests) failed to hatch. The authors suggested this low hatching rate may have resulted from inbreeding or the effects of insecticides on a nearby cotton field, either of which seems quite possible. Statements that Rufous Gnateater may 'benefit' from habitat disturbance (Develey & Metzger 2006) should certainly be interpreted with caution. Rufous Gnateater occurs in many parks, apparently with proportionately fewer populations of the nominate race afforded protection, even when its smaller range is considered. Our knowledge of the protection of individual taxa or populations will require immediate re-evaluation once we have a better understanding of the true diversity involved. **PROTECTED POPULATIONS** *C. l. lineata* PN Cavernas do Peruaçu (Kirwan *et al.* 2001, ML 27021421, photo H. Peixoto); PN de Boa Nova (D. Beadle *in litt.* 2017); FLONA Brasília (Braz & Cavalcanti 2001, ML); ESEC Serra das Araras (Valadão 2012); REBIO de Una (eBird: R. Laps); PE dos Pirineus (eBird: C. Gussoni); RPPN Fazenda VagaFogo (eBird: A. Aguiar); RPPN Serra Bonita (XC, MPEG, ML). *C. l. vulgaris* **Brazil** PN Aparados da Serra (ML 19976, W. Belton); PN Serra do Gandarela (eBird: D. Avelar); PN Caparaó (eBird: T. Pongiluppi); PN Itatiaia (Parker & Goerck 1997, Anciães & Marini 2000, Maia-Gouvêa *et al.* 2005, Mallet-Rodrigues *et al.* 2015, AMNH, LACM, ML, XC); PN Superagüi (eBird: F. Olmos); PN Lagoa do Peixe (eBird: K. Riding); PN São Joaquim (eBird: A.L. Roos); PN Serra da Bocaina (Buzzetti 2000, Mallet-Rodrigues *et al.* 2015); PN Serra da Canastra (Silveira 1998); PN Serra do Cipó (eBird: F. Murphy); PN Serra do Itajaí (eBird: A.E. Rupp); PN Serra do Mar (XC 177590, A. Silveira); PN Serra dos Órgãos (Scott & Brooke 1985, Mallet-Rodrigues *et al.* 2007, 2010, 2015); PN Tijuca (Krabbe 2007, ZMUC); FLONA Canela (Franz *et al.* 2014); FLONA Piraí do Sul (Carvalho *et al.* 2016); FLONA Três Barras (Corrêa *et al.* 2008); FLONA Nonoai (ML 68437, D.W. Finch); FLONA São Francisco de Paula (XC 212147, J.G. Just); ÁPA Macaé de Cima (Pacheco *et al.* 2014); ÁPA Jacarandá (Mallet-Rodrigues *et al.* 2010); ÁPA Manancial do Barreiro (Durães & Marini 2003); ÁPA Ilha Comprida (Avanzo & Sanfilippo 2000); ÁPA Municipal do Capivari-Monos (XC 33233, M.A. Melo); ÁPA Reserva de Taboões (Anciães & Marini 2000); EPDA Peti (Carnevalli *et al.* 1989, Amaral *et al.* 2003, Faria & Rodrigues 2005, 2009,

Faria *et al.* 2006); ESEC Assis (Willis & Oniki 1981); ESEC Itirapina (Telles & Dias 2010, Cavarzere & Arantes 2017); ESEC São Carlos (Pozza & Pires 2003); ESEC Caetetus (Donatelli & Ferreira 2009); ESEC Juréia-Itatins (Lima 2011); ESEC Mata Preta (XC 172124/27, A.A.C. Junior); ESEC Santa Lúcia (Ruschi 1977, Simon 2000); RE Guapiaçu (Pimentel & Olmos 2011); RE Gália (Anciães & Marini 2000); RE Samuel Klabin (dos Anjos & Schuchmann 1997, dos Anjos 2002); RE Santa Genebra (Anciães & Marini 2000); REBIO Aguaí (Just *et al.* 2015); REBIO Augusto Ruschi (Parker & Goerck 1997, Willis & Oniki 2002a); REBIO Mata Escura (Lopes *et al.* 2005b); REBIO Lami José Lutzenberger (eBird: C. Gussoni); REBIO Estadual Sassafrás (XC 24494, A.E. Rupp); REBIO/ESEC Mogi Guaçu (Willis & Oniki 1981); Reserva Ambiental Fazenda Santa Cecília (Pozza & Pires 2003); Reserva Florestal Morro Grande (Develey & Martensen 2006); Parque Florestal Pioneiros (Krügel & dos Anjos 2000); RVS Mata dos Muriquis (eBird: H. Peixoto); PE Campos do Jordão (Oniki 1981a); PE Carlos Botelho (Oniki 1981a, Willis & Oniki 1981, Oliveira *et al.* 2011); PE Cantareira (eBird list S24516537, photos G. Bravo); PE Ilha do Cardoso (FMNH 258118); PE Lapa Grande (eBird: F. Olmos); PE Pedra Azul (ML 64200261, photo L. Merçon); PE Pedra Branca (Pacheco & Maciel 2005); PE Pedra Selada (eBird: B. Rennó); PE Serra da Concórdia (eBird: L. Bianquini); PE Serra do Brigadeiro (Simon *et al.* 1999, Anciães & Marini 2000, Lopes *et al.* 2005b); PE Serra do Tabuleiro (eBird: A.L. Roos); PE Caxambu (dos Anjos & Schuchmann 1997); PE Espigão Alto (eBird: R.D. Agnol); PE Porto Ferreira (eBird: C. Gussoni); PE Ibitipoca (Pacheco *et al.* 2008, Manhães *et al.* 2010, Manhães & Dias 2011); PE Itacolomi (eBird: C.S. de Oliveira); PE Jaraguá (photos ML 25463431, C. Gussoni, ML 66236021, U. Rêgo); PE Três Picos (Mallet-Rodrigues & do Noronha 2009, Mallet-Rodrigues *et al.* 2010); PE Fazenda Paraíso (Willis & Oniki 1981); PE Ilhabela (eBird: N. Strycker); PE Intervales (Willis & Schuchmann 1993, Pizo & Melo 2010, XC, ML); PE Jacupiranga (Willis & Oniki 1981, Willis 1985a); PE Jurupará (Banks-Leite *et al.* 2012); PE Mata dos Godoy (dos Anjos & Schuchmann 1997, dos Anjos *et al.* 1997, dos Anjos 2001, 2002, Santana & dos Anjos 2010, Marques & dos Anjos 2014); PE Mata São Francisco (Bornschein & Reinert 2000); PE Morro do Diabo (Oniki 1981a, Willis & Oniki 1981, ML); PE Serra do Mar (Goerck 1999, XC 177590, A. Silveira); PE Serra Furada (ML 21274251/24345921, recording/photo J.G. Just); PE Vassununga (Willis & Oniki 1981); PE Vila Rica do Espírito Santo (Scherer-Neto & Bispo 2011); PE Vila Velha (Scherer-Neto *et al.* 1994, dos Anjos & Baçon 1999, dos Anjos & Schuchmann 1997); FLOTA Edmundo Navarro de Andrade (Mironov & Hernandes 2014, ML 27464671, photo L.C. Ramassotti); FLOTA Turvo, 27°08′S, 53°53′W (ML 68468, D.W. Finch); FLOTA Itapetininga (LSUMZ, KUNHM, UMMZ); FLOTA São Bárbara do Rio Pardo (Willis & Oniki 1981); Bosque Municipal Manoel Júlio (Bornschein & Reinert 2000); PNM Arthur Thomas (XC 37217, R. Campos de Oliveira); PNM Ronda (Franz *et al.* 2014); PNM Mangabeiras (eBird: H. Peixoto); PNM Açude de Concordia (eBird: J. Faragher); PNM Morro Azul (eBird: A.E. Rupp); PNM Nascentes de Paranapiacaba (eBird: O. Prioli); PNM Saint'Hilaire (Efe *et al.* 2001); PNM São Francisco de Assis (eBird: A.E. Rupp); RPPN Bugerkopf (eBird: A.E. Rupp); RPPN Caraça (Vasconcelos & Melo-Júnior 2001); RPPN Caraguatá (eBird: A.L. Roos); RPPN Estância Hermínio e Maria (Straube & Urben-Filho

2008); RPPN Fazenda Campo Alto (Straube & Urben-Filho 2008); RPPN Feliciano Miguel Abdala (Lopes *et al.* 2005b); RPPN François Robert Arthur (eBird: E. Franco); RPPN Guainumbi (eBird: C. Bell); RPPN Mata do Passarinho (C. Albano *in litt.* 2017); RPPN Mata do Sossego (Lopes *et al.* 2005b); RPPN Mata Samuel de Paula (Ferreira *et al.* 2009); RPPN Monte Sinai (eBird: D. Lorin); RPPN Papagaios de Altitude (eBird: F. Olmos); RPPN Parque do Zizo (XC 4650, B. Planqué); RPPN Parque Ecológica da Raposa (eBird: D. Lorin); RPPN Parque Ecológica Quedas do Rio Bonito (eBird: V. Torga); RPPN Rio das Furnas (eBird: G. Carvalho); RPPN Salto Morato (Anciães & Marini 2000; Straube & Urben-Filho 2005a); RPPN Santo António (eBird: B. Rennó); RPPN Santuário do Caraça (ML 47826741/751, photos K. Hansen, IBC 1123309, recording A. Davis); RPPN São Judas Tadeu (eBird: E. Kaseker); RPPN Sítio Curucutu (eBird: Z.E. Camargo); RPPN Tarumã (dos Anjos 2002); RPPN Universidade de Santa Cruz do Sul (Oliveira & Köhler 2010); RPPN Volta Velha (eBird: A.E. Rupp). **Argentina** PN Iguaçu/Iguazú (Parker & Goerck 1997, Straube & Urben-Filho 2004, Buckingham 2011, ML, XC); PP Cañadon Profundidad (ML 35964501 photo, XC 336043, 'Aves del NEA'); PP Araucaria (M. Lammertink *in litt.* 2017); PP La Sierra Martínez Crovetto (eBird: 'Aves del NEA'); PP El Piñalito (eBird: D. Almiron); PP Esmeralda (eBird: R. Ramírez); PP Moconá (Giraudo *et al.* 1993); PP Puerto Esperanza (eBird: G. Martínez); PP Salto Encantado del Valle del Arroyo Cuñá Pirú (eBird: G. Pugnali); PP Teyú Cuaré (eBird: F.J. Castía); PP Urugua-í, (Fraga & Narosky 1985; MACN, LACM, AMNH. LSUMZ). **Paraguay** PN Caaguazú (Madroño *et al.* 1997b); PN Ybycuí (Hayes & Scharf 1995, Lowen *et al.* 1996, UMMZ 202173); PN San Rafael (Smith *et al.* 2006, Esquivel-M. & Peris 2008, 2011, 2012, Oosterbaan 2008, Esquivel-M. *et al.* 2007, KUNHM); Monumento Científico Moisés Bertoni (XC 55580, M. Velázquez); Refugio Biológico Carapá, Reserva Forestal Agropeco, and RPN Kuri'y (GPDDB); RPN Bosque Mbaracayú, RPN Estancia La Golondrina, RPN Estancia San Antonio (Brooks *et al.* 1993); RPN Estancia Itabó (Brooks *et al.* 1993, Lowen *et al.* 1995, Cockle 2003); RPN Sombrero and RPN Ypetí (Lowen *et al.* 1996).

OTHER NAMES *C. l. vulgaris Ceraphanes anomalus* (Bertoni 1901); *Conopophaga anomala* (Chubb 1910), *Conopophaga lineata anomala* (Dabbene 1919); *Conopophaga* (*lineata*) *lineata* (Parker & Goerck 1997), *Conopophaga lineata lineata* (Pinto 1944, Ruschi 1953, 1967, 1977, Sick & Pabst 1968). *C. l. lineata Myiagrus lineatus* (Wied 1831); *Conopophaga lineata hellmayri* (Pinto 1936). **Portuguese** chupa-dente (Brabourne & Chubb 1912; Pinto 1954b, Frisch & Frisch 1964, Belton 1994, Bencke 2001, CBRO 2011, Scherer-Neto *et al.* 2011); cuspidor (von Ihering & von Ihering 1907, Pinto 1938, Ruschi 1953, Sick & Pabst 1968); guspidor [*sic*] (von Ihering 1898, Brabourne & Chubb 1912); chupa-dente-marrom (Willis & Oniki 1991b, Dubs 1992, Pacheco *et al.* 1996, 1997a,b); chupa-dente-rufa (Ruschi 1979); limpa-dente (Sick 1993); curumanço (*lineata*; Wied 1831). Allen (1889c) was curious as to the significance of curumanço, as it appears on the original labels. Wied (1831), however, stated that it was a local name for the bird. **English** Silvery-tufted Gnat Eater (*lineata*, Cory & Hellmayr 1924, Whitney 2003); Lineated Gnat-eater (Oates & Reid 1903). **Spanish** Jejenero Rojizo (Whitney 2003); Tokotoko (de la Peña & Rumboll 1998, Guyra Paraguay 2004); Toco-toco (Zotta 1944, Olrog 1963, Frisch & Frisch 1964); Chupadientes (Narosky & Yzurieta 1989, Mazar Barnett & Pearman 2001, Guyra Paraguay 2004, Esquivel-M. & Peris 2011); Mosquitero castano (Brooks *et al.* 1993); Tocotoco Castaño (Fraga & Narosky 1985); Tocotoco rojizo (Olrog 1979); Batará Castaño (de la Peña 1988, Canevari *et al.* 1991). **French** Conophage roux (Whitney 2003). **German** Rotkehl-Mückenfresser (Whitney 2003, Hoffmann & Geller-Grimm 2013); Ameisenschnäpper mit grauem Schläfenstreifen (Wied 1831).

1

Rufous Gnateater, adult (*vulgaris*), São José do Barreiro, São Paulo, Brazil, 22 October 2013 (*Luiz Carlos da Costa Ribenboim*).

Rufous Gnateater, adult (*lineata*), Boa Nova, Bahia, Brazil, 2 February 2010 (*Ciro Albano*).

Rufous Gnateater, subadult (*lineata*), Brasília, Distrito Federal, Brazil, 8 July 2014 (*Thiago Tolêdo e Silva*).

Rufous Gnateater, adult (*vulgaris*), Macarani, Bahia, 31 December 2015 (*Caio Brito*).

Rufous Gnateater, adult (*vulgaris*), Petrópolis, Rio de Janeiro, 4 March 2014 (*João Quental*).

CHESTNUT-CROWNED GNATEATER
Conopophaga castaneiceps   Plate 4

Conopophaga castaneiceps P. L. Sclater, 1857, Proceedings of the Zoological Society of London, vol. 25, p. 47, 'Bogota'. There is, unfortunately, no way of determining the type locality; the species having been described from one of the numerous Bogotá trade skins.

'These birds kept to the seclusion of overgrown thickets in dense forest where their dark plumage blended with the shadows on the ground, only the silvery white occipital tufts coming startlingly into view when displayed. The only note I heard from the species was a sharp '*sheep.*'' – **John T. Zimmer, 1930, Huánuco, Peru**

Chestnut-crowned Gnateater is one of only two Andean *Conopophaga*, and occurs on the slopes of all three Andean chains in Colombia, south along the Amazonian slope of the East Andes through Ecuador to south-central Peru. With only a small amount of overlap, it is replaced southward by its putative sister species, Slaty Gnateater. Four subspecies of Chestnut-crowned Gnateater are generally recognised, although a fifth might remain to be described from central Colombia (Whitney

2003). In general, males are very dark below, with a bright rufous-orange forehead, long white ear-tufts and brownish upperparts, whereas females have less prominent ear-tufts, and largely deep orange-rufous head and underparts. Despite some confusing variation, in general adults of the east Ecuadorian and north Peruvian race, *chapmani*, differ from nominate adults by having a brighter, less brown-tinged crown. The Peruvian endemic, *brunneinucha*, is darker overall, has bright orange-rufous restricted to the forehead, and a larger white belly patch. The west Colombian endemic, *chocoensis*, is most similar to *brunneinucha*, but is smaller, more olivaceous above and has more extensive chestnut on the crown. Behaviour and natural history are poorly known. Only recently, nearly half a century after its nest was described, have detailed data on its reproductive biology been published (Lizarazo-B. & Londoño in review). The following account provides the first detailed description of immature plumages and post-fledging development, and adds more than 25 indirect records of reproductive activity.

IDENTIFICATION 13.0–13.5cm. Sexually dimorphic and somewhat variable across its range. Males are dark greyish-olive above with a bright orange-rufous forecrown and chestnut crown, and dark slate-grey face and underparts. Central belly slightly paler. Females tend to be warmer brown above with the entire head and breast dark orange-rufous, and a whitish belly and throat. The postocular tuft is less obvious in females. Both sexes have brown eyes, bicoloured bills (dark maxilla, pale whitish mandible) and blue-grey or vinaceous legs. In parts of Ecuador, it overlaps with the noticeably smaller Ash-throated Gnateater. Both sexes of the latter, however, have distinct spotting on the wing-coverts and lack the bright chestnut crown (Ridgely & Greenfield 2001). At the southern end of its range, Chestnut-crowned Gnateater is replaced by the only Andean congener, Slaty Gnateater, from which it is generally distinguishable by its far brighter and richer rufous crown (see Identification under Slaty Gnateater). Chestnut-belted Gnateater occurs allopatrically in the Amazonian lowlands below some of Chestnut-crowned Gnateater's range. Chestnut-belted Gnateater, however, is separable at a glance by its black face contrasting with the reddish breast. Though confusion is unlikely, several authors (Sclater 1868, Hellmayr 1911a) likened female Chestnut-crowned Gnateater to the smaller, but similarly shaped Hooded Antpitta, which has an entire hood of much brighter chestnut and completely lacks the postocular white tuft.

DISTRIBUTION Chestnut-crowned Gnateater occurs from north-west Colombia to south-east Peru, although our understanding of its range is especially fragmentary on the Pacific slope of the Colombian West Andes, where the southernmost record is currently a geographically isolated report from central Cauca (Donegan & Dávalos 1999). Unconfirmed reports exist from as far south as Río Ñambí

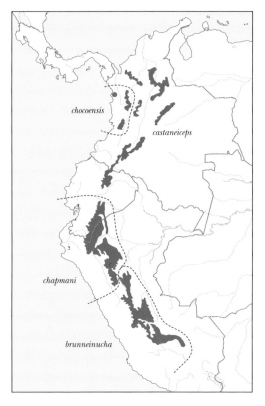

in west-central Nariño that presumably represent *chocoensis*, but these require confirmation (Diego Carantón & Oscar Humberto Marín-Gómez *in litt.* 2017). In Colombia, it occurs in all three Andean chains, but is only found east of the East Andes in Ecuador and Peru. In the latter countries it is also known from several outlying ranges such as Volcán Sumaco, Cordillera Napo-Galeras, Cordillera del Cóndor, Cordillera de Cutucú, Cerros del Sira and Cordillera Vilcabamba (Hilty & Brown 1986, Ridgely & Tudor 1994, Sibley & Monroe 1990, Ridgely & Greenfield 2001, Ayerbe-Quiñones *et al.* 2008, McMullan *et al.* 2010, Schulenberg *et al.* 2010, Calderón-Leytón *et al.* 2011).

MOVEMENTS None described. Hilty (1997) who made careful observations on seasonal, altitudinal movements in an avian community in western Colombia that included *C. c. chocoensis*, did not notice any shifts in populations of this species.

HABITAT Inhabits upper tropical and lower subtropical montane forest. It apparently prefers dense undergrowth in humid montane forests, often near treefalls, landslides and edges, but also occurs in fairly open understorey (Whitney 2003, Ridgely & Tudor 2009). Nevertheless, like many congeners, indeed like most species treated herein, *C. castaneiceps* appears to avoid crossing relatively open areas. Hilty (1975) noted that adults 'are hesitant to cross trails if the clearings are more than 8–10m'. Across its range, altitudinal records span 500–2,500m, but preferred and most frequently reported elevations are 1,200–2,000m (Sibley & Monroe 1990, Ridgely & Tudor 1994). In Colombia, it is found mostly at 1,000–1,800m (McMullan *et al.* 2010), but as low as 500m in western Meta (Hilty & Brown 1986) and 680m on the west slope of the Andes in the Río Anchicayá Valley of Dept. Valle (Hilty 1975). In eastern Ecuador *C. castaneiceps* is best known in Morona-Santiago and Zamora-Chinchipe (where it occurs as far south as Panguri), and generally appears to prefer elevations of 800–2,000m, locally as low as 600m (Ridgely & Greenfield 2001). At lower elevations it may occur sympatrically with *C. peruviana*. In Peru, *C. castaneiceps* inhabits montane forests at 1,000–2,200m, becoming somewhat scarcer further south (Schulenberg *et al.* 2007) and with a somewhat narrower altitudinal range on the outlying slopes of Cerros del Sira in central Peru (Terborgh & Weske 1975, Harvey *et al.* 2011, Socolar *et al.* 2013).

VOICE The song is similar to that of Slaty Gnateater but delivered at a slightly higher frequency (Whitney 2003). It has been described as a raspy, chipping rattle, *chit, chit-it, chit-it-it-it* (Hilty 1975, Hilty & Brown 1986) or a slightly rising, grinding phrase which stutters at the start and end, *grrrew-grr'grr'grr'grr'grree-grrew* (Schulenberg *et al.* 2007). Whitney (2003) described the acoustics as 'a series of frog-like, slightly disyllabic notes at 3–4kHz, pace accelerating and amplitude increasing after first 1–2 notes, the whole lasting 3–6 seconds, with longer series returning to more widely spaced and slightly quieter notes at end'. Calls are variously described as: a quiet, gravelly *grew* (Schulenberg *et al.* 2007), a harsh *zhiek!* (Whitney 2003), a low *schek* (Hilty & Brown 1986) or a sharp '*PSEW!*' (Schulenberg *et al.* 2007). Apparently in aggression or courtship (Whitney 2003), males occasionally produce a fairly loud, low whirring sound in flight, sometimes introduced by a *PSEW!* call (Hilty & Brown 1986, Schulenberg *et al.* 2007).

NATURAL HISTORY Despite the fact that Chestnut-crowned Gnateaters are infrequently seen, and probably overlooked in many parts of their range, Ridgely & Tudor (1994) remarked that they are not particulary shy. Most of what follows is taken from Hilty (1975), who provided the most complete description of general behaviour based on observations of *chocoensis* in western Colombia. They are usually found singly or in pairs, perching low on narrow horizontal branches or the thin stems of vertical saplings, and moving by hopping or jumping short distances in pursuit of prey (Hilty 1975, Ridgely & Tudor 1994, Whitney 2003), only rarely dropping to the ground to capture a prey item, but rarely or never searching leaf litter by overturning or flicking leaves (Hilty 1975). Hilty (1975) provided the only quantified foraging data, recording 48 prey-capture heights as follows: six on the ground; ten at 0.1–0.8m up; 20 at 0.8–1.5m; nine at 1.5–2.5m; three at 2.5–4.0m; and just one above 4 m. For 27 foraging-perch diameters he recorded: nine perches <1.0cm; 12 perches 1–5cm; and six perches >5cm. Typically, individuals observed by Hilty (1975) chose thin sloping stems of aroids or heliconias, or the tops of mossy logs to perch and peer about, cocking their heads sideways or upwards to peer at overhanging vegetation or the forest floor. When prey was located, they made a quick lunge or bound to glean small items from trunks, limbs or leaves. They often flutter-perched briefly from a movable leaf or stem to capture prey, but hovering manoeuvres were clumsily executed. He observed them occasionally make a loud snapping sound with the bill while taking prey in flight. As noted above, Hilty (1975) usually found *C. castaneiceps* alone or in pairs. He suggested that, while not prone to participate in mixed-species foraging flocks, they appear to be 'stimulated' by them. As the flock passes, *C. castaneiceps* increase their own activity, wing-flicking, rattle-calling and foraging more actively. They apparently remain in their preferred microhabitat of dense undergrowth and vine tangles, seemingly more responsive to flocks of noisy canopy species than smaller flocks foraging lower. Hilty (1975) noted both an increase in foraging rate, as well as increased vocalisation and wing-whirring, which were infrequently heard when gnateaters were foraging on their own. When alarmed, *C. castaneiceps* responds by wing-flicking and making a raspy call, apparently never flicking or pounding its tail as described for many Thamnophilidae (Willis 1972). When startled, Hilty (1975) states that 'they may bound up from a low concealed perch to cling adroitly to nearly vertical branches in full view', often approaching an observer and inspecting them at close range. Despite that *C. castaneiceps* does not appear to follow swarms of army ants (Hilty 1975), these somewhat curious and oddly bold behaviours of an otherwise skulking bird in the presence of human observers sounds rather similar to my own observations of some species of antpittas, particularly *Grallaricula* species. This leads me to suggest that more extensive observations may reveal at least a facultative association between *C. castaneiceps* and foraging mammals as suggested for antpittas (Greeney 2012b). There are no concrete observations on territoriality or territory size. Songs, presumably functioning in territorial establishment and/or defence, are apparently most frequently heard at dusk and dawn (Schulenberg *et al.* 2007). Hilty (1975) suggested that pairs are fairly sedentary but may range up to *c.*250m while foraging. In northern Ecuador, estimated to occur at densities of 1 pair/ha at appropriate elevations on the slopes of the Sumaco Volcano (Whitney 2003).

DIET There are no concrete data concerning the diet, but it presumably consists largely of small arthropods (Hilty 1975, Whitney 2003). On several occasions Hilty (1975) observed a female gleaning small, ripe, berries while in flight, once from *Hedyosmum brenesii* (Chloranthaceae) and once from an unidentified shrub in the understorey.

REPRODUCTION The first nest description was of a nest of *chocoensis* found by Hilty (1975) at 1,000m in the Río Anchicayá Valley on the Pacific Andean slope of Colombia. The two nestlings were estimated to be *c.*2 days old, but were not described in detail. Only recently have the eggs been described (Lizarazo-B. & Londoño in review). These authors also considerably augmented our understanding of the species' reproductive habits, based on three nests in western Colombia (also *chocoensis*). NEST & BUILDING The nest found by Hilty (1975) was on a steep, west-facing slope and was *c.*1.5m from the trunk of a large buttress-root canopy tree. It was *c.*5–6m uphill from an old landslide, densely regrown with Melastomataceae and *Cecropia* sp. saplings, vines and small aroids. The nest was sited 0.65m above ground in a thicket, wedged into a supporting platform of ferns, aroids and a fallen limb. In general aspect, it resembled a slightly oval, naturally collected pile of debris. External dimensions were not given, but the slightly oblong inner cup measured 6 × 4cm. The external portion comprised loosely assembled small twigs, dicot leaves and leaf pieces, and flexible rootlets and fibres. The internal cup was lined with fine rootlets and small pieces of dry leaves. Above, the nest was completely sheltered by a curled *Heliconia* leaf and by other vegetation at the sides, making it almost invisible from all angles. The three nests found by Lizarazo-B. & Londoño (in review) were lined with dead dicot leaves, small pieces of dead fern leaves and large dried bamboo leaves (*Guadua* spp.). The leaves were interwoven with fine rootlets and black fungal rhizomorphs. Externally they comprised predominantly bamboo leaves and small sticks, with a relatively small amount of moss intermixed. The inner cups of two nests, measured at perpendicular angles, were 7.1 ± 0.01 × 6.8 ± 0.3cm wide and 4.4 ± 0.7cm deep. Externally they measured 10.1 ± 3.4 × 10.3 ± 2.5cm wide by 7.0 ± 0.5cm tall. Two qualitatively similar nests were found by C. Vits and B. de Roover at Copalinga Ecolodge (*chapmani*), one 50cm above ground in low vegetation and one built 75cm up, nestled into the top of the very loose leaf rosette of a terrestrial *Pitcairnia* bromeliad. EGG, LAYING & INCUBATION Two of the described nests held two eggs and two nests held two nestlings when found (Hilty 1985, Lizarazo-B. & Londoño in review). One of the nests at Copalinga Ecolodge (C. Vits *in litt.* 2017) contained two eggs, suggesting that two is the regular clutch. The eggs (photos courtesy of Jorge Lizarazo-B. & Gustavo Londoño) are qualitatively similar to most *Conopophaga* eggs I have observed and, even with such a small sample, appear to show a fair degree of variability in coloration (as in Ash-breasted Gnateater). Eggs (*n* = 3 eggs, two clutches) are short sub-elliptical, creamy-beige or off-white, marked with various shades of cinnamon blotches and flecks. On all three eggs markings were concentrated in a distinct cap at the larger end, with the rest of the shell lightly and sparsely flecked. Markings on one egg were significantly paler overall. Three eggs (Lizarazo-B. & Londoño in review) measured 22.5 ± 1.1 × 17.9 ± 0.9mm and weighed 3.9 ± 1.5g. NESTLING & PARENTAL CARE When first observed, Hilty's (1975) nest contained two blind, 'nearly naked' young, which Hilty (1975) estimated to be *c*.2 days old. In light of my own examination of photographs of nestlings (of known age: provided by Jorge Lizarazo-B. & Gustavo Londoño), however, it seems more likely that they were already 4–5 days old (see below). They weighed 6.0g and 6.3g, and one chick weighed 21.5g, 11 days later. At this nest, the second chick apparently disappeared several days after the nest was discovered. When estimated to be *c*.12–14 days old (14–15 days; my estimate), Hilty (1975) described the remaining nestling as 'largely feathered', with a considerable covering of down and only half-emerged primaries. It fledged when 16–18 days old (my estimate). From my examination of photographs, nestlings hatch entirely devoid of natal down and almost all dark, sooty-grey or black, paler and pinkish-orange below. The maxilla is shiny black on the culmen and tip, paler dusky-white on the sides and base, with a small, bright white egg tooth. The mandible is dull yellow, or pale yellow-white at the base, darkening to black at the tip. The inflated rictal flanges are bright white on the outer margins and pale yellowish internally. Only a day after hatching the developing contour feather tracts are visible through the skin. Contour pins begin breaking their sheaths at 6–7 days, those of the capital tract slightly delayed. Primary pin feathers begin to break their sheaths around day 8 or 9, and the nestlings are fairly well covered in downy plumage by day 10. Hilty (1975) was only able to make brief observations of adult behaviour at the nest, but documented both sexes provisioning the young. The adults repeatedly favoured the same approach to the nest while bringing food, from behind the nearby tree trunk before moving to the nest. Hilty (1975) noted that adults did not engage in any noticeably elaborate distraction displays when an observer approached. Both adults would apparently make a soft *chek* call, flick their wings and flutter between low perches, always remaining near vegetation. **SEASONALITY** Apart from Hilty's (1975) nest, the recent study by Lizarazo-B. & Londoño (in review), and breeding-condition adults reported between March and June in the Central Andes of Colombia (*n* = 6; Hilty & Brown 1986), I located a handful of other breeding records. The reproductive period almost certainly varies across the species' large range. Too few data are available to accurately assess seasonality, but the records below suggest that they may prefer the drier months. **BREEDING DATA** *C. c. chocoensis* Two nests with eggs, 22 March 2014 and 9 June 2015, Cerro Montezuma (Lizarazo-B. & Londoño in review); young nestlings, 12 February 1973, Alto Yunda, Río Anchicayá (Hilty 1975); nestling, 28 February 2014, Cerro Montezuma (Lizarazo-B. & Londoño in review). *C. c. castaneiceps* Adult carrying material, 25 May 2016, Bosque Nuevo Mundo (ML 29948271, photo B.C. Jaramillo); two fledglings, 2 and 15 December 1970, El Carmen (FMNH 292597/602); fledgling, 14 January 2010, Cordillera Guacamayos (ML 44033611, photo R. Batie); fledgling, 10 May 1912, La Candela (AMNH 116179); juvenile, 28 June 2013, RNA Arrierito Anioqueño (D. Calderón-F. *in litt.* 2017); juvenile transitioning, 10 January 2001, Río Hollín (C. Dingle *in litt.* 2015); immature with adult female [fledgling or young juvenile], 5 September 2013, RNA Arrierito Antioqueño (eBird: G. Ewing); subadult, 21 July 2006, Finca Bodega Vieja (ML 28372531, photo A.M. Cuervo); subadult, 15 December 1970, El Carmen (FMNH 292598); two subadults transitioning, 14 and 15 December 1970, El Carmen (FMNH 292596/601); subadult transitioning, 2 January 1915, La Frijolera

(FMNH 50992 = AMNH 133370); distraction display, 9 April 2014, WildSumaco Lodge (F. Rowland *in litt.* 2017); active gonads, 26 April 1997, La Grilla (Stiles *et al.* 1999); immature, 10 July 2015, RNA Arrierito Antioqueño (eBird: J. Beck). *C. c. chapmani* Nest with eggs, 31 October 2012, Copalinga Ecolodge (C. Vits *in litt.* 2017); active nest (incubation based on behaviour), 17 July 1999, Copalinga Ecolodge (C. Vits *in litt.* 2017); fledgling transitioning, 7 November 1983, 15km NE of Jirillo (LSUMZ 116980); juvenile, 8 September 1975, Bosque Udina (FMNH 299608); juvenile, 14 April 1923, Chaupe (AMNH 181292); subadult, 19 June 2010, Quebrada Huancabamba (FMNH 474175); subadult, 22 June 2003, La Canela (MECN 7955); two subadults transitioning, 13 June 2012, *c*.1.5km N of Gosen (MSB 41795/41801); subadult transitioning, 28 June 2010, Quebrada Huancabamba (FMNH 474174); two subadults transitioning, 1 March 1999, Nuevo Peru and 16 November 1983, *c*.15km NE of Jirillo (LSUMZ 172192/116981, photos M.L. Brady). *C. c. brunneinucha* Immature [juvenile?; skull 30% ossified], 22 June 2011, 2.7km S of Plataforma (MSB 36310, not examined); subadult, 29 June 1985, Cushi (LSUMZ 128569, photos M.L. Brady); subadult, 22 September 1922, Huachipa (FMNH 65782); subadult, 10 April 2007, 'Andy's Bamboo' (MSB 27452); subadult transitioning, 22 September 1922, Huachipa (FMNH 65784); subadult transitioning, 14 August 1968, Cerro Conchapén (FMNH 285112); subadult transitioning, 4 January 1971, Río Cacazú headwaters (FMNH 296715); subadult transitioning, 14 June 2011, 2.7km S of Plataforma (MSB 36159). **Additional Remarks** A single fledgling (*chocoensis*) stayed with its parents for at least seven weeks, perhaps longer (Hilty 1975). Hilty (1975) did not report how frequently it was fed or exactly how independent it was at this age, but noted that the family group still spent a fair amount of time just 5m from the nest, suggesting that it may not have been fully independent. His cursory description of the fledgling at seven weeks (see Technical Description) suggests that I would have considered it a juvenile in transition to subadult or perhaps fully subadult.

TECHNICAL DESCRIPTION Sexes differ. The following is most applicable to nominate *castaneiceps*. **Adult Male** Bright chestnut-orange forecrown that blends to duller chestnut on hindcrown and to brownish-chestnut on the nape. The sides of the head, throat and underparts are dark, ashy grey, with only the central belly whitish (sometimes invisible in the field). The flanks and vent are fulvous-brown to olive-brown. The ear-coverts, head-sides and throat are usually slightly darker than the other grey areas, and sharply contrast with the elongated, bright silvery-white feathers of the postocular tuft. Unlike some species of gnateaters, they lack a superciliary stripe extending forward from the base of the white tuft. The back is warm brown to grey- or olivaceous-brown, or greyish-olive, the feathers variably edged dusky, creating an indistinct mottled or scaled appearance (see Taxonomy and Variation). The wings and tail are uniformly dark brown or olive-brown, usually similar to the back. The upperwing-coverts are similar to the flight feathers and are unmarked (*contra* Whitney 2003). **Adult Male Bare Parts** *Iris* brown; *Bill* black, mandible pale whitish at base, sometimes for most of its length; *Tarsi & Toes* blue-grey or vinaceous. **Adult Female** Orange-rufous head and breast, brightest on lores, ear-coverts and breast, dullest or darkest on crown. Also has a white postocular tuft, though it tends to be slightly less prominent, or is at least held less exposed in life. The sides of the breast gradually shade to dull orange-brown and to brown on the flanks and vent. The throat and belly are white, the throat somewhat variable, but sometimes heavily washed orange-rufous. The extent of white on the belly varies more than in males, sometimes being largely white. Upperparts are similar to males but tend to be warmer brown. The wings, tail and bare parts are as described for males. *Immature* Previously only partially described. An unsexed 'juvenile' (*chocoensis*), about seven weeks out of the nest, was described by Hilty (1975) as differing from adults in having 'a duller coloration, pale horn-colored legs, and only a trace of the broad silvery eye stripe'. Caroline Dingle mist-netted an immature (*castaneiceps*) that she described as having the crown mottled rufous-chestnut and black, a white postocular stripe, a dark grey chest, mottled white and paler grey lower on the belly. This description fits that of a juvenile male, one I might roughly estimate was 1–2 months out of the nest, perhaps starting to transition to subadult plumage. These field observations are supplemented here with my observations of specimens and mist-netted individuals in various plumages. It appears that birds in juvenile plumage may still accompany their parents and beg, probably still being fed at least occasionally. By this stage they are easily separable by sex. I have not yet examined sufficient immatures to determine if racial differences are apparent. *Fledgling* Short-winged, with little or no visible tail, and mostly covered with dusky-brown, fluffy fledgling feathers. The entire crown, nape and neck-sides are mottled with varying numbers of emerging orange-to orange-chestnut-tipped feathers. The back is dull brown with indistinct, broad dusky barring. The feathers gradually become brighter brown to chestnut-buff towards the rump, making the barring more evident. Uppertail-coverts are bright orange-chestnut to chestnut-buff and distinctly barred dusky-black. The breast and throat are dark chocolate-brown becoming coarsely striped brown and dull buff on the sides and flanks, mixed with grey and white on the lower breast to white mottled with grey on the central belly. The vent and undertail-coverts are dusky-brown. Flight feathers appear nearly fully developed and are overall similar to adults. The greater wing-coverts are marked with variable amounts of orange-chestnut that often form an indistinct crescent near the tip, usually restricted to the anterior vane. The secondary greater and lesser wing-coverts are broadly tipped bright orange-chestnut and thinly margined in black. The tertials are usually marked at their tips with orange- or buff-chestnut, with small black spots or edges at the tips, but these appear to be somewhat variable. It appears that fledglings become possible to sex sometime during the transition to juvenile plumage. In females the entire head and upper breast gradually become bright orange-chestnut with grey and black markings, while in males the head and breast become largely dark grey with dark chestnut markings. **Fledgling Bare Parts** *Iris* dark brown; *Bill* maxilla dusky-yellow, brighter around nares and orange-yellow along tomia, young fledglings may retain their egg tooth, mandible yellow, orange-yellow along tomia and at base, dusky near tip, rictal flanges inflated and bright yellow to orange-yellow; *Tarsi and Toes* pinkish-grey, yellowish at posterior margin and on toepads. *Juvenile Male* Usually have already replaced the feathers of the forecrown, which is bright chestnut and blends to coarse chestnut with black spotting on crown and nape. The postocular tuft is already fairly prominent in juveniles but the face and ear-coverts are a mixture of dusky and rufous-brown feathers that give it a 'messy' appearance. Above, the central back feathers are the first to be replaced, and a variable amount of chestnut and black barring remains in the interscapular area, as well as a few dull chestnut immature feathers irregularly scattered across the back. Some or all of the uppertail-coverts are chestnut-buff with indistinct dusky striping. Below, overall pattern as adults but breast browner and belly mottled white, brown and grey. Vent pale, dusky-brown with coarse, indistinct dusky barring. Flight feathers similar to adults but the greater and lesser secondary wing-coverts are broadly tipped bright orange-chestnut, somewhat more than in other species, and narrowly edged black. Tertials have varying amounts of chestnut-buff combined with indistinct black spots or edges at their tips. *Juvenile Female* Entire head, throat and breast orange-brown with variable amounts of indistinct dusky spotting on the crown, nape and neck-sides. The ear-coverts are mottled brown and grey, the lores have already developed the bright orange-rufous of adults. The upperparts, and the underparts other than the breast, are similar to males, perhaps with slightly less white on the belly. Wings as described for males. **Juvenile Bare Parts** *Iris* dark brown; *Bill* maxilla black, mandible yellow-pinkish, rictal flanges yellow and at most slightly inflated; *Tarsi & Toes* similar to adults, sometimes with varying pinkish or yellowish at posterior margin. *Subadult Male* Usually appear fully adult, perhaps slightly browner on the breast and face. The most apparent difference is the upperwing-coverts, which are as described for juveniles. *Subadult Female* Also largely as adults, but also retain immature upperwing-coverts. **Subadult Bare Parts** *Iris* may still be slightly darker than adults; *Bill* similar to adults; *Tarsi & Toes* like adults.

MORPHOMETRIC DATA Data for adults and subadults combined. **C. c. chocoensis** Data from MCZ (J.C. Schmitt *in litt.* 2017) and live birds (G.A. Londoño *in litt.* 2017). Adult ♀♀ (*n* = 5): *Wing* 67.0 ± 3.4mm; *Tail* 38.5 ± 4.5mm; *Bill* exposed culmen 13.4 ± 1.1mm, total culmen 18.6 ± 1.0mm, depth at nares 4.7 ± 0.2mm, width at nares 7.0 ± 0.5mm, width at gape 11.6 ± 1.8mm; *Tarsus* 34.1 ± 4.0mm. Adult ♂♂ (n = 3): *Wing* 74.2 ± 2.4mm; *Bill* from nares 9.8 ± 0.5mm, exposed culmen 12.8 ± 0.5mm; *Tarsus* 27.6 ± 0.7mm (*n* = 2 for tarsus alone). See also Hellmayr (1911a), Chapman (1915), Hilty (1975). **C. c. castaneiceps** Data from USNM, FMNH, AMNH. Adult ♀♀: *Bill* from nares 9.9 ± 0.4mm (*n* = 12), exposed culmen 12.5 ± 0.3mm (*n* = 4), depth at nares 4.7 ± 0.1mm (*n* = 3), width at nares 6.6 ± 0.1mm (*n* = 4), width at base of mandible 8.1 ± 0.6mm (*n* = 3); *Tarsus* 28.6 ± 0.6mm (*n* = 12). Adult ♂♂: *Bill* from nares 10.2 ± 0.6mm (*n* = 9), exposed culmen 12.9 ± 0.7mm (*n* = 3), depth at nares 4.6 ± 0.0mm, width at nares 6.4 ± 0.5mm, width at base of mandible 7.9 ± 0.4mm; *Tarsus* 29.9 ± 1.2mm (*n* = 9). Data from live birds (J. Freile; *n* = 1♀, 1♂): *Wing* 66.0mm, 70.5mm; *Tail* n/m, 43.8mm; *Bill* from nares 9.4mm, 10.0mm, exposed culmen 12.5mm, 14.0mm, depth at nares 4.2mm, 5.8mm, width at nares 6.0mm, 6.3mm; *Tarsus* 29.6mm, 28.5mm. Fledglings (*n* = 3; AMNH, FMNH): *Bill* from nares 7.0mm, 8.9mm, 7.9mm, exposed culmen n/m, 12.1mm, 10.4mm, depth at nares n/m, 4.4mm, 5.4mm, width at nares n/m, 4.9mm, 5.4mm, width at base of mandible n/m, 6.7mm, 6.6mm; *Tarsus* 28.3mm, 27.5mm, 28.4mm. See also Sclater (1857b),

Chapman (1915). *C. c. chapmani* Data from MECN (M.V. Sánchez-Nivicela *in litt.* 2017), MCZ (J.C. Schmitt *in litt.* 2017), FMNH, AMNH, MSB (HFG). Adult ♀♀: *Wing* 71.3 ± 4.0mm (*n* = 3); *Tail* 38.0 ± 1.4mm (*n* = 2); *Bill* from nares 9.9 ± 0.6mm (*n* = 5), exposed culmen 12.9 ± 1.1mm (*n* = 4); total culmen 17.6 ± 1.2mm (*n* = 2), depth at nares 5.4 ± 1.0mm (*n* = 3), width at nares 6.9 ± 0.8mm (*n* = 3), width at base of mandible 8.1 ± 0.8mm (*n* = 3), width at gape 11.3 ± 0.8mm (*n* = 2); *Tarsus* 30.0 ± 1.7mm (*n* = 4). Adult ♂♂: *Wing* 70.4 ± 2.4mm (*n* = 5); *Tail* 32.6 ± 1.3mm (*n* = 4); *Bill* from nares 9.8 ± 0.4mm (*n* = 8), exposed culmen 12.9 ± 0.9mm (*n* = 6), total culmen 18.4 ± 1.2mm (*n* = 4), depth at nares 5.6 ± 0.5mm (*n* = 5), width at nares 7.2 ± 0.3mm (*n* = 5), width at base of mandible 8.2 ± 0.7mm (*n* = 3), width at gape 12.0 ± 2.1mm (*n* = 4); *Tarsus* 30.1 ± 2.2mm (*n* = 8). Two juvenile ♂♂: *Bill* from nares 8.3mm, 9.5mm; *Tarsus* 27.9mm, 29.3mm. *C. c. brunneinucha* Data from FMNH, MSB (HFG), MCZ (J.C. Schmitt *in litt.* 2017). Adult ♀♀: *Wing* 75mm (*n* = 1); *Bill* from nares 9.3 ± 0.4mm (*n* = 4), exposed culmen 12.0mm (*n* = 1), depth at nares 4.9mm (*n* = 1), width at nares 5.6mm (*n* = 1), width at base of mandible 7.2mm (*n* = 1); *Tarsus* 28.9 ± 1.2mm (*n* = 4). Adult ♂♂: *Wing* 73.2 ± 1.8mm (*n* = 7); *Bill* from nares 9.4 ± 0.5mm (*n* = 8), exposed culmen 12.0 ± 1.2mm (*n* = 7), depth at nares 4.9 ± 0.5mm (*n* = 3), width at nares 6.0 ± 0.5mm (*n* = 4), width at base of mandible 7.5 ± 0.4mm (*n* = 4); *Tarsus* 28.5 ± 1.1mm (*n* = 8). See also Taczanowski (1884), Berlepsch & Sztolcman (1896). **Mass** 27.6g (*n* = 10♀?, multiple subspecies, Dunning 2008, including data from Weske 1972 and Rahbek *et al.* 1993). *C. c. chocoensis* Range 24–30g, mean 26.8g (*n* = 6♀?, Hilty 1975); 25.5 ± 1.7g (*n* = 3 ♀♀; G.A. Londoño *in litt.* 2017). *C. c. castaneiceps* 28.2g(lt), 29.1g(m), 26.3g(lt) (*n* = 3, ♀,♀,♂, B. Mila *in litt.* 2015); 38.3g(h) (*n* = 1, ♂, J. Freile *in litt.* 2017). *C. c. chapmani* 25.5g (*n* = 1♀; Rahbek *et al.* 1993). Adult ♀♀: 21.3g(no); 21.5g, 21.7g(lt), 22.0g, 22.0g(no), 23.0g(vl), 23.0g(lt), 23.5g, 24.0g(tr), 24.0(no), 24.2g(tr), 24.4g, 24.4g(lt), 25.1g(no), 26.0g, 26.4g, 27.0g. Adult ♂♂: 20.5g(no), 21.0g(no), 22.0g, 22.0g(no), 22.0g(tr), 22.1g(lt), 22.2g(no), 22.5g(no), 22.5g, 22.7g(no), 23.0g(no), 23.0g(lt), 23.5g(tr), 23.7g(no), 24.0g(lt), 24.0g, 24.5g, 24.8g, 25.0g(no), 25.4g(lt), 25.5g(no), 25.5g(no), 27.1g(no), 27.9g(no), 28.0g, 28.0g, 30.1g(lt). Fledgling ♂: 25.0g (LSUMZ 116980). *C. c. brunneinucha* Mean 27.6 ± 2.5g (*n* = 4♀♀), 27.6g (*n* = 3♂♂ (Weske 1972). Adult ♀♀: 21.1g(no), 22.4g(tr), 23.0g(no), 24.7g(lt), 25.0g(no), 26.9g(lt), 28g(no) (LSUMZ, MSB, FMNH). Adult ♂♂: 22.8g, 23.5g, 25.7g(no), 26.2g(tr), 26.7g(lt), 27.0g, 27.2g(lt), 30.0g(no) (LSUMZ, MSB); 27.5g(no) (juvenile ♀; MSB 36310). **Total Length** 12.5–14.6cm (Meyer de Schauensee 1964, 1970, Clements & Shany 2001, Whitney 2003, Restall *et al.* 2006).

TAXONOMY AND VARIATION The taxonomic affinities of Chestnut-crowned Gnateater are not well known but it appears to belong to the 'Rufous Gnateater lineage' (Whitney 2003), with recent molecular evidence supporting a sister relationship with Slaty Gnateater (Batalha-Filho *et al.* 2014). Four subspecies are currently recognised (Whitney 2003). Following careful review of published descriptions, but with admittedly little direct comparisons of skins, I have the impression that the variation described below, supposedly delineating the four races, may reflect individual variation more than geographic. Chapman (1915) and Carriker (1933) suggested that *chapmani* is, plumage-wise,

somewhat intermediate between nominate *castaneiceps* and *brunneinucha*. Even so, to distinguish *chapmani* from other races Carriker (1933) cited several characters, including the amount of scaling and coloration of the upperparts, that he noted were somewhat variable, 'perhaps' based on age. Chapman (1915) in his description of *chocoensis* mentioned several specimens that were intermediate between *chocoensis* and *castaneiceps*. Overall, there appears to be a gradual clinal variation with latitude in many traits. The ground colour of the back, in particular, appears to change from brown (*chocoensis*) to greyer (*castaneiceps*) to olive-washed (*chapmani*) and browner again in the south (*brunneinucha*). Other authors have been similarly confused. For example, Berlepsch & Sztolcman (1896), despite using belly colour as a character to diagnose *brunneinucha*, admitted that this character does appear to be quite variable. They cited Taczanowski's (1884) description of a bird from the Huayabamba Valley (by range supposedly *chapmani*) as lacking any white on the belly. They also noted that the colour of the throat, and the size, shape and colour of the bill of the Huayabamba bird is identical to Colombian specimens of nominate *castaneiceps*. They concluded that birds in northern Peru (subsequently described as *chapmani* by Carriker 1933) may be indistinguishable from the nominate race. As just one example pointing towards a need for reconsideration of racial divisions are the descriptions of the upperparts of the various taxa, by multiple authors, many of them examining the same specimens. The descriptions are as follows: *chocoensis* back 'mummy-brown with an olivaceous cast' (Chapman 1915), back 'darker grey (than nominate) with more olive brownish wash' (Todd 1932); *chapmani* back 'brownish-olive' (Bond 1953), 'dark olive' (Carriker 1933); *brunneinucha* back 'rufescent' Bond (1953), 'rich seal brown' (Carriker 1933), 'decidedly mummy-brown' (Cory & Hellmayr 1924); *castaneiceps* back 'slate grey' (Carriker 1933), 'deep neutral grey' (Cory & Hellmayr 1924), 'dark greyish-olive' (Meyer de Schauensee 1964), 'olivaceous brown' (Sclater 1890), 'dark mouse grey with little olive brownish wash' (Todd 1932).

Conopophaga castaneiceps chocoensis Chapman, 1915, Bulletin of the American Museum of Natural History, vol. 34, p. 641, Baudó Mts, Dept. Chocó, Colombia, 3500ft [Serranía de Baudó, 1,065m, probably near Alto del Buey, *c.*06°07'N, 77°15.5'W]. The holotype, an adult male (AMNH 123321), was collected on 18 July 1912 by Elizabeth L. Kerr (Chapman 1915, LeCroy & Sloss 2000). The Serranía de Baudó is a coastal mountain range extending from the Panamanian border to central Chocó, with peaks in most areas reaching only 500–700m but some of the higher mountains reach *c.*1,800m around Alto del Buey, the slopes of which seem the most likely origin of the holotype. This mountain range is separated from the West Andes Río Atrato and Río San Juan valleys and populations there, given the known altitudinal range of Chestnut-crowned Gnateater, may be isolated to some degree from populations in the Andes. Interestingly, however, this species has remained unreported from these mountains since the holotype's collection. Although the species' presence there is not unlikely, the lack of further records suggests that perhaps the question of the type locality requires reconsideration. This Chocó bioregion endemic is found west of the West Andes from Chocó, south to Cauca. Chapman (1915) described the back of male *chocoensis* as 'mummy-brown' with an olivaceous cast.

Race *chocoensis*, following Chapman (1915), is similar to nominate *castaneiceps* but in males, at least, the back is much browner and less grey. Todd (1932), however, stated that the back of *chocoenis* was 'darker grey with more olive brownish wash'. Both these authors agreed that *chocoensis* differs from nominate and *brunneinucha* by having more extensive and richer chestnut on the crown. Bare-part colours for *chocoensis* provided by Hellmayr (1911a) are: *Iris* dark brown; *Bill*, maxilla black, mandible grey; *Toes and tarsi* grey. A fifth race, *subtorridus*, was synonymised with *chocoensis* soon after its description (Peters 1951; see Other Names). Todd (1932) described *subtorridus* from a single pair of birds from the 'heights above Caldas'. In his comparisons of these birds with *brunneinucha*, he felt they 'had nothing to do' with that race. He described the male as being much more similar to nominate *castaneiceps*, but darker above, with both the grey ground colour and olive wash being deeper. He described the male's forecrown as chestnut, darkening rapidly into 'deep Prout's brown' on the crown and nape, adding to the overall darker appearance of the upperparts. In his opinion, in nominate *castaneiceps* the pileum is almost uniform 'deep hazel' and the back is 'dark mouse grey' with little olive-brown wash, except towards the rump. Todd (1932) felt that the underparts of his male *subtorridus* were no different than those of males of the nominate race. He stated, however, that the female holotype of *subtorridus* 'differs sharply from the same sex of *castaneiceps* and *brunneinucha* in having the breast grey (like the male, but not so deep), contrasting with the rusty throat and malar region', and in having the upperparts and pileum darker than in female *castaneiceps*, just as in the other sex. Todd (1932), in comparing the male with Chapman's (1915) holotype of *chocoensis*, was convinced that the two were distinct. He felt that his *subtorridus* was distinctly dark greyish above, not the dark brownish or 'mummy-brown' of *chocoensis* (Chapman 1915). Todd (1932) felt that the flanks, central belly and vent were uniform dark grey, compared to the brown flanks and vent of *chocoensis* which contrast with the grey central belly. He concluded that *subtorridus* was not only a valid taxon, but could well merit species rank, as might *chocoensis*. Cory & Hellmayr (1924) were not so convinced, but upheld Todd's (1932) distinction while implying that *subtorridus* might be synonymous with *brunneinucha*. Peters (1951), however, followed Meyer de Schauensee's (1950) suggestion, and synonymised *subtorridus* with *chocoensis*. Although I have not examined the specimens in question, I suspect that this is another example of individual variation within *C. castaneiceps* leading to confusing and unnecessary subspecific divisions. **Specimens & Records** *Chocó* La Selva, 04°55'N, 76°09'W (Hellmayr 1911a, ANSP 157973). *Risaralda* Cerro Montezuma, 05°14'N, 76°05.5'W (Lizarazo-B. & Londoño in review). *Valle del Cauca* RFP Bosque de Yotoco, 03°52'N, 76°26'W (ML 22262761/771, photos J. McGowan); near Pavas, 03°41'N, 76°35'W (MCZ 124480–482); above Dagua (= 'Heights of Caldas', see Paynter 1997), *c.*03°40'N, 76°41'W (Meyer de Schauensee 1950, CM P67150/80); Alto Yunda, Río Anchicayá, 03°32'N, 76°48'W (Hilty 1975, 1997, ANSP 173497). *Cauca* RNSC Tambito, 02°30'N, 77°00'W (Donegan & Dávalos 1999).

Conopophaga castaneiceps castaneiceps P. L. Sclater, 1857, Proceedings of the Zoological Society of London, vol. 25, p. 47, 'Bogota.' An adult male syntype (NHMUK 1854.1.25.83) and another syntype (data not provided) are held in Tring (Warren & Harrison 1971). These authors point out that, although Sclater (1890) designated a type, he used a specimen that was not included in Sclater's (1862) summary of his collection, and therefore must have been received by him well after the description was published. The nominate race occurs, somewhat patchily, in the subtropical zone of the Central and East Andes of Colombia (Peters 1951, Bohórquez 2002). South into Ecuador, it is known, again patchily, along the Amazonian slope of the Andes to around Volcán Sumaco in Napo (Ridgely & Greenfield 2001). In general coloration females of *C. c. castaneiceps* are similar to Hooded Antpitta, but are easily recognised by the white postocular stripe and generally duller 'hood' (Sclater 1857b, Hellmayr 1911a). In *castaneiceps* the crown is uniform chestnut while the back is slate-grey, becoming more olivaceous posteriorly. Feathers of the central back are marginally tipped dusky, giving a slightly scaled appearance (Carriker 1933). **Specimens & Records COLOMBIA:** *Bolívar* Vereda Guacima, 08°17.5'N, 74°06.5'W (ML 59275461, photo C.M.W. Wagner); Near La Punta, 08°09'N, 74°12.5'W (Salaman *et al.* 2002b); Santa Cecilia, 07°58.5'N, 74°13'W (XC 99273/74, T. Donegan). *Cordoba* Cerro Murrucucú, 07°59'N, 76°00'W (Meyer de Schauensee 1950b, ANSP 160866). *Antioquia* Quimarí, 08°07'N, 76°23'W (Meyer de Schauensee 1950a); above Valdivia, 07°11'N, 75°27'W (USNM 402980–987); La Frijolera, 07°10'N, 75°25'W (Chapman 1917, AMNH 133369, FMNH 50992); Vereda Santa Gertrudis, 07°08'N, 75°09'W (Cuervo *et al.* 2008, ICN 34551); Alto Combate, 07°05'N, 75°09'W (ICN 33798); Mampuestos, 07°04'N, 75°10'W (Cuervo *et al.* 2008); Finca Bodega Vieja, 06°58'N, 75°03'W (Cuervo *et al.* 2008, ICN 34422, ML 28372531, photo A.M. Cuervo); RNA Arrierito Anioqueño, 06°59'N, 75°06.5'W (Salaman *et al.* 2007a, XC 154854, H. Matheve); Bosque Las Ánimas, 06°56'N, 75°01'W (Cuervo *et al.* 2008, ICN 34596). *Santander* NE of Bucaramanga, *c.*07°08.5'N, 73°06'W (Donegan *et al.* 2007); RNA Reinita Azul, 06°52.5'N, 73°23'W (Salaman *et al.* 2007a); near Vereda Honduras, *c.*06°37.5'N, 73°29.5'W (Donegan *et al.* 2007, ML 59199111, photo C.M.W. Wagner). *Quindío* Bosque El Silencio, 04°39'N, 75°38.5'W (O.H. Marín-Gómez *in litt.* 2017). *Tolima* Juntas, 04°33'N, 75°19.5'W (O. Cortes-Herrera *in litt.* 2017). *Caldas* La Sofía, 05°38'N, 75°04'W (USNM 436487–489); RFP Río Blanco, 05°05'N, 75°25'W (eBird: P. Kaestner). *Boyacá* Vereda La Grilla, 05°49'N, 74°18'W (Stiles *et al.* 1999, Stiles & Bohórquez 2000, ICN 32833/33766); Pajarito, 05°17.5'N, 72°42'W (Salaman & Donegan 2001, Salaman *et al.* 2002a); Finca Guayabal, 05°23.5'N, 72°42.5'W (Borhóquez 2002); RNSC La Almenara, 04°52.5'N, 73°15'W (P.C. Pulgarín-R. *in litt.* 2017). *Meta* RNA Halcón Colorado, *c.*04°11'N, 73°39'W (Salaman *et al.* 2009a); Buena Vista, 04°10'N, 73°41'W (Chapman 1917, AMNH 121830–833); Acacías, 03°59'N, 73°46'W (Iafrancesco *et al.* 1987, MLS 3976B); Cubarral, Vereda Aguas Claras, 03°46'N, 73°50'W (ICN 32601/623). *Cauca* Camino Pitalito, 01°25'N, 76°29.5'W (J. Beckers *in litt.* 2017). *Huila* El Isno, 02°14'N, 75°55'W (ANSP 155856–859); 45km SW of La Plata, 02°06.5'N, 76°11'W (USNM 446760); Andalucia, 01°54'N, 75°41'W (Chapman 1917, AMNH 116175–177, USNM 256095/96); La Candela, 01°50'N, 76°20'W (Chapman 1917, AMNH 116172–174/178–179); PNN Cueva de los Guácharos, 01°36.5'N, 76°06.5'W (O. Cortes-Herrera *in litt.* 2017). *Nariño* El Carmen, 00°40'N, 77°10'W (FMNH 292595–602, ICN 29464); Estación de Bombeo Guamués, 00°39'N, 77°04'W (J. Beckers *in litt.* 2017); Río Rumiyaco, 00°22.5'N,

77°16.5'W (Borhóquez 2002). ***Caquetá*** *c.*1.9km SE of Guacamaya, 02°48'N, 74°51'W (ML 89203/04, M. Álvarez); Vereda La Esperanza, 02°44.5'N, 74°53.5'W (Borhóquez 2002); Paraíso, 01°45'N, 75°41.5'W (Willis 1988). ***Putumayo*** Bosque Nuevo Mundo, 00°41'N, 76°57.5'W (ML 29948271, photo B.C. Jaramillo). **ECUADOR**: ***Sucumbíos*** Mirador Bermejo, 00°18'N, 77°24.5'W (Schulenberg 2002); Río Verde, 00°14'N, 77°34.5'W (Stotz & Valenzuela 2009); Sinangoe, 00°11'N, 77°30'W (Schulenberg 2002); Ccuccono, 00°08'N, 77°33'W (Schulenberg 2002). ***Napo*** Cordillera Guacamayos, *c.*00°37.5'S, 77°50.5'W (ML 44033611, photo R. Batie); head of Río Guataraco, *c.*00°40'S, 77°35'W (MCZ 299266–272); RP WildSumaco, 00°41'S, 77°36'W (XC 20790, R. Ahlman); Río Chontayacu, *c.*00°43.5'S, 77°45.5'W (HFG); Río Hollín, 00°45.5'S, 77°41.5'W (B. Mila *in litt.* 2016); *c.*5.5km NW of Muyuna, 00°56.5'S, 77°53.5'W (R.A. Gelis *in litt.* 2016). ***Orellana*** upper Río Pucuno, 00°31.5'S, 77°33.5'W (MCZ 299261–263); 'Palm Peak', S of Volcán Sumaco, *c.*00°39'S, 77°36'W (MCZ); near Mushullacta, 00°50'S, 77°34'W (XC 250709, N. Krabbe).

Conopophaga castaneiceps chapmani Carriker, 1933, Proceedings of the Academy of Natural Sciences of Philadelphia, vol. 85, p. 14, Rio Jelashte, San Martín, Peru, alt. 5,000ft [Río Jelache, 06°48'S, 77°12'W, 1,525m]. The adult male holotype (ANSP 108121) was collected 16 August 1932 by M. A. Carriker. Subspecies *chapmani* is found in the Amazonian foothills of south-east Ecuador in Morona-Santiago and Zamora-Chinchipe, south to the Río Huallaga valley in north-east Peru, in Cajamarca and San Martín (Carriker 1933, Peters 1951, Bond 1953, Ridgely & Greenfield 2001, Schulenberg *et al.* 2007). Apparently, in Ecuador, the northernmost records are from the Cordillera de Cutucú in Morona-Santiago, but there is a fairly significant gap in records between there and north-central Ecuador (Napo, where *C. c. castaneiceps* occurs), presumably due to incomplete sampling. If the species does occur in the intervening region, the race is unknown. From Carriker's (1933) description of the holotype of *chapmani*, the forecrown and crown, as far as the eye, are bright chestnut-rufous, gradually darkening posteriorly into chestnut-brown on the hindcrown and nape. The nape is washed blackish. The head bears a long postocular, thin plume of white feathers. The rest of the upperparts, including the wings and tail, are dark olive to olive-brown, more brownish on the median and greater wing-coverts, remiges and rectrices. The feathers of the upper back are edged black and have black shafts. The ear-coverts are blackish. Below, there is an indistinct blackish band on the chest, while the rest of the underparts are slate-grey, becoming whitish centrally on the breast and belly. The flanks and vent are olive-brown. In females, the chestnut-rufous throat and chest are paler and brighter, and the flanks somewhat more olivaceous. Carriker (1933) suggested that *chapmani* is somewhat intermediate between nominate *castaneiceps* and *brunneinucha*, and that in *chapmani* the crown is similar to nominate *castaneiceps*, while the upper- and underparts are nearer *brunneinucha*. The transition from the dark, chestnut nape to the olive upperparts is very abrupt in *chapmani*, while in *brunneinucha* the transition is much less defined, simply grading to an overall browner back. In *brunneinucha* the sides of the head, lower throat and chest are much blacker than in *chapmani*, while the flanks and vent are dull cinnamon-brown versus olive-brown in *chapmani*. Female *chapmani* differs from

brunneinucha in having the crown darker chestnut and the back darker and more olivaceous-brown, with more visible dark fringes to the feathers of the central back. They also have the same marked contrast between the crown and back, which is absent in female *brunneinucha*. Females of the nominate race differ from *chapmani* in being darker, more olivaceous-brown above with back feathers faintly edged blackish (Ridgely & Greenfield 2001). SPECIMENS & RECORDS **ECUADOR**: ***Morona-Santiago*** Nueva Alianza, 02°05'S, 78°09'W (J. Freile *in litt.* 2017); Nuevo Israel, 02°10.5'S, 77°52.5'W (J. Freile *in litt.* 2017, QCAZ); Colimba, 02°15'S, 78°13'W (AMNH 408362–364); W slope of Cordillera de Cutucú, 02°26'S, 78°03'W (XC 249780–784, N. Krabbe); Unnsuants, 02°33'S, 77°54'W (XC 249901, N. Krabbe); E of Logroño, 02°37.5'S, 78°12'W (ANSP 176899–900); Tayuntza, 02°43'S, 77°535'W (WFVZ 42680); *c.*19km SE of Logroño, 02°43.5'S, 78°04.5'W (Robbins *et al.* 1987, USNM 559994); Mirador Condor, 03°38.5'S, 78°24'W (MECN 8056, XC 86248, A. Spencer). ***Zamora-Chinchipe*** Zamora, 04°04'S, 78°57'W (AMNH 129641); Sabanilla, 04°02'S, 79°01'W (AMNH 156216/17); above Chinapinza, 04°02'S, 78°35'W (ANSP 186183); Copalinga Ecolodge, 04°05.5'S, 78°57.5'W (XC 209297, J. Fischer); Río Bombuscaro, 04°07'S, 78°58'W (XC 5514/15, N. Athanas, ZMUC *in* Bloch *et al.* 1991); Cabañas Yankuam, 04°15'S, 78°39.5'W (Freile *et al.* 2013b, XC 86311/13, A. Spencer); Filo de Chumbiriatza, 04°16'S, 78°41'W (XC 250710–712, N. Krabbe); Shaime, 04°19'S, 78°40'W (Juan Freile *in litt.* 2016); Valladolid, 04°33'S, 79°08'W (ML 68268, P. Coopmans); Cerro Panguri, 04°36'S, 78°58'W (ANSP 185512/13); La Canela, 04°37'S, 78°57'W (MECN); San Francisco del Vergel, 04°40'S, 79°02'W (MECN 6383). **PERU**: ***Amazonas*** Upper Río Comaina, 03°54'S, 78°25'W (Schulenberg & Wust 1997, ML 92479/882, T.S. Schlenberg); Quebrada Katerpiza, 04°01'S, 77°35'W (Inzunza *et al.* 2012); *c.*1.5km N of Gosen, 05°12'S, 78°38.5'W (MSB); 12km E of La Peca, 05°36'S, 78°19'W (LSUMZ); Bosque Udina, *c.*05°45.5'S, 78°26.5'W (FMNH 299607/08, see Stephens & Traylor 1983); Huambo, 06°22S, 77°28'W (Taczanowski 1884); Quebrada Huancabamba, 06°35.5'S, 77°33'W (FMNH 474174–178). ***Loreto*** 'Campamento Alto Cachiyacu', 05°51.5'S, 76°43'W (Stotz *et al.* 2014). ***Cajamarca*** *c.*2km WSW of Hito Jesús, 04°53'S, 78°53'W (LSUMZ 179020/21); ENE of San José de Lourdes, 05°04.5'S, 78°51.5'W (LSUMZ 82002/03, MCZ 330978/79); San Ignacio, *c.*05°08.5'S, 79°00.5'W (ANSP 117234/35, CM P117966/68); Chaupe, 05°10'S, 79°10'W (Bangs & Noble 1916, Bond 1953, AMNH 181285–292, ANSP 117232); Perico, 05°15'S, 78°45'W (Bangs & Noble 1916, MCZ 79962); Nuevo Peru, 05°16'S, 78°40'W (LSUMZ 172189–193); below Bosque Laurel, *c.*28km WSW of Huarango, 05°19.5'S, 79°01.5'W (A. García-Bravo *in litt.* 2017). ***San Martín*** *c.*24km ENE of Florida, 05°41'S, 77°45'W (LSUMZ 174023–026); Afluente, 05°44'S, 77°31'W (Hornbuckle 1999, LSUMZ 84968); *c.*15–20km by trail NE of Jirillo, 06°03'S, 76°44'W (Davis 1986, LSUMZ 116976–982, ML); Río Blanco, 06°46'S, 77°33'W (J. Barrio *in litt.* 2017). ***La Libertad*** Utcubamba, 08°12'S, 77°10'W (LSUMZ 92503–505).

Conopophaga castaneiceps brunneinucha Berlepsch & Sztolcman, 1896, Proceedings of the Zoological Society of London, vol. 1896, p. 385, La Gloria, Chanchamayo [*c.*11°04'S, 75°19'W] and Vitoc, Garita del Sol [11°17'S, 75°21'W], Junín, Peru. Berlepsch & Sztolcman (1896) based their description of *brunneinucha* on a male (La Garita del Sol) and two females (La Gloria) collected by Jan Kalinowski in August 1890 and August and October 1891, respectively, but did not designate a type. Subsequently, Sztolcman & Domaniewski (1927) designated the male (MZPW 33977) as the type (Mlíkovský 2009), while Cory & Hellmayr (1924) referred to La Gloria as the type locality. The male from La Garita del Sol, however, should be considered the lectotype. One of the other syntypes (MZPW 33970) is a female collected 16 August 1890 at La Garita del Sol (Mlíkovský 2009). I was unable to determine the whereabouts of the third syntype (a female collected in October). The southernmost race, *brunneinucha*, is restricted to the east slopes of the Andes in central Peru, in Junín (Bond 1953) and Huánuco (Zimmer 1930) south through Pasco (Eneñas; Bond 1953) at least as far as the Manu road in Cuzco (Peters 1951, Schulenberg *et al.* 2007, Whitney 2003). Males of *brunneinucha* are mostly chestnut-brown above, paler and more rufous on the forecrown, with a distinctive silvery-white postocular tuft. Below, males are dark slate-grey, becoming paler to whitish on the lower midbelly. Females are similar, but overall paler, with a more orange-rufous forecrown similar to the throat and breast, which fades to greyish on the belly and whitish on the central belly (Ridgely & Tudor 1994). Males of this Peruvian subspecies differ from those of nominate *castaneiceps* by having the posterior crown duller, more admixed brownish, and uniform brown without any blackish fringes to the mantle feathers (Berlepsch & Sztolcman 1896). Additionally, they have the central belly more extensively white than the nominate race, which tends to be duller, more greyish-white. Female *brunneinucha* are much more similar to *castaneiceps*, but also have more extensive white on the lower underparts (Berlepsch & Sztolcman 1896; Cory & Hellmayr 1924). Berlepsch & Sztolcman (1896) described the bare parts as: *Iris* dark brown, *Bill*, mandible black, maxilla whitish, *Tarsi and Toes* 'bluish ash.' They suggested that the pale maxilla is another character separating this race from nominate *castaneiceps*, which they claimed tends to have the apical third of the maxilla dusky. **SPECIMENS & RECORDS** *Loreto* Payua, 07°33.5'S, 75°55'W (Schulenberg *et al.* 2001b); *c.*86km SE of Juanjui, 07°34'S, 75°54'W (LSUMZ 170916–918). *San Martín* 2.7km S of Plataforma, 07°25'S, 76°17.5'W (MSB). *Huánuco* Divisoria en Cordillera Azul, 09°05'S, 75°46'W (LSUMZ 62323–325); Trocha Garza, 09°26'S, 74°41.5'W (S.J. Socolar *in litt.* 2017); Cerros del Sira, *c.*09°27'S, 74°45'W (Mee *et al.* 2013; Bosque Río Abra, 09°29'S, 75°55'W (LSUMZ 81999–2000); Huachipa, 09°30'S, 75°52'W (Zimmer 1930, FMNH 65781–784); Sariapampa, 09°43'S, 75°54'W (FMNH 296714); Cushi, 09°58.5'S, 75°41.5'W (LSUMZ 128567–569, ML 40090, T.S. Schulenberg). *Pasco* 27km E of Yuyapachis, 09°38.5'S, 74°44'W (AMNH 820982); Lanturachi, 10°23'S, 75°35'W (MUSA); Río Cacazú headwaters, 10°38'S, 75°07'W (FMNH 296715); Eneñas, 10°45'S, 75°14'W (ANSP 92297/99, CM P117792); 4km SW of Vila Rica, 10°44.5'S, 75°15'W (ML 98838, M.J. & W.P. Widdowson). *Ucayali* 3km by road NE of Abra Divisoria, 09°03'S, 75°43'W (LSUMZ 84967); 22.8km SW of Río Cohengae (mouth), 10°26.5'S, 74°06'W (LSUMZ 189047); upper Chipani Valley, 10°41'S, 74°05.5'W and Santeni River valley, 10°42'S, 74°09.5'W (G.F. Seeholzer *in litt.* 2017). *Junín* Cerro Conchapén, 10°50'S, 75°07'W (FMNH 285112); Yapaz, 10°50'S, 75°12.5'W (ANSP 92295, MCZ 179657); Chanchamayo, 11°03'S, 75°19'W (AMNH 408615); Puya Sacha, 11°05.5'S, 75°25.5'W (MUSA 4253); Soriano, *c.*11°10'S, 75°16'W and nearby

Masayacu, *c.*11°10'S, 75°15'W (Taczanowski 1884); bridge at Pan de Azucar, 11°11'S, 75°27'W (LSUMZ 72515–516); along Río Satipo, 11°27'S, 74°47'W (KUNHM 113753–756). *Cuzco* near Alto Manguriari, 12°34'S, 73°05'W (Robbins *et al.* 2011; KUNHM); Cordillera Vilcabamba, 12°38'S, 73°39'W (AMNH, USNM); near Alto Materiato, 12°42'S, 72°52.5'W (Robbins *et al.* 2011, KUNHM); near Llactahuaman, 12°52'S, 73°31'W (Pequeño *et al.* 2001b, MUSA 4095/96); 'Andy's Bamboo', 13°02'S, 71°31.5'W (MSB 27452). *Ayacucho* Monterrico, 12°28'S, 73°54'W (Taczanowski 1884); Tutumbaro, 12°44'S, 73°57.5'W (KUNHM 112722/882/895); Huanhuachayo, 12°44'S, 73°47'W (Berlepsch & Sztolcman 1896, AMNH 820668, USNM 512025). **Notes** I have adjusted the location of Cushi fairly significantly from that given by Stephens & Traylor (1983), placing it in Huánuco, not Pasco. Google Earth places Cushi *c.*35km ESE of Panao, not 40km E as stated by Stephens & Traylor (1983) and reflected by their coordinates (09°51'S, 75°37'W). The coordinates given above are also closely aligned with the location and elevation (1,800m) accompanying T.S. Schulenberg's recording (ML 40090), as well as those derived from the Map of Hispanic America cited in Stephens & Traylor (1983).

STATUS Considered fairly common to uncommon and local (Stattersfield *et al.* 1998), and is not currently considered to be of conservation concern (BirdLife International 2017). As it is rather quiet and easily overlooked (Whitney 2003), the species may well occur at more localities than are currently documented. It is apparently somewhat tolerant of disturbance, sometimes persisting in only remnants of intact forest (Bloch *et al.* 1991), leading Whitney (2003) to suggest that it may benefit from natural or artificial openings in forest, allowing denser growth. All of the subspecies are known from protected areas. **PROTECTED POPULATIONS** *C. c. castaneiceps* Colombia PNN Los Nevados (Pfeifer *et al.* 2001); PNN Cueva de los Guácharos (O. Cortes-Herrera *in litt.* 2017); PNN Serranía de los Yariguíes (Donegan *et al.* 2007a); PRN Ucumarí (Negret 1994); RNA Arrierito Antioqueño, and RNA Reinita Cielo Azul (Salaman *et al.* 2007a); RNA Halcón Colorado (Salaman *et al.* 2009a); RNSC La Almenara (P.C. Pulgarín-R. *in litt.* 2017); RFP Río Blanco (eBird: P. Kaestner); RFP La Forzosa (Cuervo *et al.* 2008, ICN 33798). **Ecuador** PN Sumaco-Galeras (Ridgely & Greenfield 2001, XC, ML); PN Cayambe-Coca (Schulenberg 2002); RP WildSumaco (many recordings in XC). *C. c. chocoensis* RNSC Tambito (Donegan & Dávalos 1999); PNN Tatamá (Lizarazo-B. & Londoño in review); RFP Bosque de Yotoco (ML, photos J. McGowan). If the the type locality of *chocoensis* is correct, this subspecies probably occurs in PNN Utría. *C. c. chapmani* Ecuador PN Podocarpus (Bloch *et al.* 1991, XC); PN Cordillera del Condor (ANSP, MECN, XC); BP Alto Nangaritza (Freile *et al.* 2013b, XC); **Peru** BP Alto Mayo (Hornbuckle 1999, Rosas 2003, ANSP); SN Tabaconas-Namballe (B.M. Winger *in litt.* 2017); ÁCR Cordillera Escalera (Davis 1986, LSUMZ 116976–982, ML). *C. c. brunneinucha* PN Manu (Whitney 2003); PN Cordillera Azul (Schulenberg *et al.* 2001b); Reserva Comunal El Sira (Forero-Medin *et al.* 2011).

OTHER NAMES *Conopophaga gutturalis* described by Sclater (1868) from a female specimen (NHMUK 1889.9.20.665) labelled only 'Bogotá' was synonymised with nominate *castaneiceps* (Sclater 1890, Dubois 1900,

Hellmayr 1911a, Cory & Hellmayr 1924) before it could be used regularly (but see Gray 1869). *Conopophaga castaneiceps subtorridus* (Todd, 1932; holotype CM 67160) is considered synonymous with *chocoensis* (Peters 1951); *Conopophaga ardesiaca* [*castaneiceps*] (Sclater 1855c), [*brunneinucha*] (von Tschudi 1845, Taczanowski 1874) [*chapmani*] (Taczanowski 1882); *Conopophaga c. castaneiceps* [*chocoensis*] (Hellmayr 1911a); *Conopophaga peruviana* [*chapmani*] (Bangs & Noble 1916); *Conopophaga brunneinucha* (Brabourne & Chubb 1912, Ogilvie-Grant 1912); *Conopophaga castaneiceps* var. *brunneinucha* (Dubois 1900). **English** Chestnut-headed Gnat Eater/Gnat-eater [*castaneiceps*] (Brabourne & Chubb 1912, Cory & Hellmayr 1924, Meyer de Schauensee 1950a); Brown-naped Gnat Eater/Gnat-eater [*brunneinucha*] (Brabourne & Chubb 1912, Cory & Hellmayr 1924); Western Chestnut-headed Gnat Eater/Gnat-eater [*chocoensis*] (Cory & Hellmayr 1924, Meyer de Schauensee 1950a). **Spanish** Jejenero Coronicastaño (Valarezo-Delgado 1984, Clements & Shany 2001, Whitney 2003, Granizo 2009); Zumbador Pechigrís (Salaman *et al.* 2007a); Jenero de Corona Castaña (Schulenberg *et al.* 2001b). **French** Conophage à couronne rousse. **German** Roststirn-Mückenfresser (Whitney 2003).

Chestnut-crowned Gnateater, adult male (*castaneiceps*), RNA Arrierito Antioqueño, Antioquia, Colombia, 22 October 2014 (*Daniel Uribe*).

Chestnut-crowned Gnateater, eggs (*chocoensis*) in nest, Cerro Montezuma, Risaralda, Colombia, 12 September 2015 (*Jorge Eduardo Lizarazo-B.*).

Chestnut-crowned Gnateater, eight-day-old nestlings (*chocoensis*) in nest, Cerro Montezuma, Risaralda, Colombia, 29 March 2014 (*Isabel Garavito*).

Chestnut-crowned Gnateater, adult female (*chocoensis*) in nest, Cerro Montezuma, Risaralda, Colombia, 28 March 2014 (*Isabel Garavito*).

Chestnut-crowned Gnateater, adult male (*castaneiceps*), RNA Arrierito Antioqueño, 11 April 2009 (*Alonso Quevedo*).

SLATY GNATEATER
Conopophaga ardesiaca   Plate 4

Conopophaga ardesiaca d'Orbigny & Lafresnaye, 1837, Magasin de Zoologie, d'Anatomie Comparee et de Palaeontologie, vol. 7, p. 13, 'Yungas (Bolivia).'

'We encountered this species [...] on the wooded mountains of the eastern slope of the Andes, in the province of Yungas, department of La Paz. It remains in the midst of the damp forests of the steep hills, where it is very rare, hiding in the densest thickets of the woods, sometimes hopping among trees and bushes, but often descending to the earth in order to run and seek there the insects on which it feeds.' – **Alcide d'Orbigny, 1838, Bolivian Andes, translated from French**

As the English name suggests, Slaty Gnateater is among the more drab-plumaged of the genus, although males do bear a well-developed postocular plume. Its English name is derived from the specific epithet, *ardesiaca*, which, in turn, is derived from the Latin *ardesiacus*, meaning slate-coloured or slatey (Jobling 2010). It is one of just two montane *Conopophaga*, being restricted to lower and middle foothill forests south of the Amazon in south-east Peru and Bolivia. The species is said to be particularly challenging to see, due to its propensity to remain in dense vegetation, but also because of the steeply sloping terrain it inhabits. There is much plumage variation within the species, especially females in the south of its range, and it has been suggested that undescribed taxa are involved (Whitney 2003). Slaty Gnateater is considered closely allied to Ash-throated and Chestnut-crowned Gnateaters, all three of which belong to the Rufous Gnateater group (Whitney 2003). Other members of the group include Hooded Gnateater and (possibly) *Black-bellied Gnateater*. To date, however, no taxonomic work has addressed the specifics of these relationships. Although one nest has been described, the species' general natural history is poorly known. In the account I expand on previously published breeding data as well as correct and attempt to resolve past errors and conflicts concerning range limits and collecting localities.

IDENTIFICATION 12–14cm. The following description is from Whitney (2003). Sexes similar but distinguishable. Males are brown above with dark scaling and grey below, with brownish on belly and flanks. The forecrown, face and neck-sides are a similar grey to the underparts and a white postocular plume contrasts with its surrounds. Females are similar to males but the forecrown and lores are rufous or tawny-brown, with white on the central belly, and the postocular tuft is greatly reduced in size and brightness. The bill of both sexes is dark above and pale below. Male Chestnut-crowned Gnateater can be distinguished from male Slaty Gnateater by the orange-rufous forecrown, more rufescent upperparts without dark scaling, and considerably darker throat and breast (Ridgely & Tudor 1994). Female Chestnut-crowned Gnateater is immediately distinguished from both sexes of Slaty Gnateater by its orange-rufous throat and breast. Male Ash-throated Gnateater somewhat resembles male Slaty Gnateater, but has buff-tipped wing-coverts and a greyer back with more distinct scaling. Ash-throated Gnateater also lacks the bicoloured bill of Slaty Gnateater.

DISTRIBUTION The range of Slaty Gnateater extends over the east slope of the Andes from south-east Peru to south-east Bolivia. Previous reports from Ecuador (Dubois 1900, Brabourne & Chubb 1912, Valarezo-Delgado 1984) are traceable to the mention of a female specimen from Napo (Sclater 1890), an error noted by Hellmayr (1921) and Cory & Hellmayr (1924). Although I have not examined this specimen, it is likely that it was a female Chestnut-crowned Gnateater. At the northern extent of its known range (Cuzco), Slaty Gnateater is known from many sites along the Cuzco–Manu road (Walker *et al.* 2006, Jankowski 2010). In Bolivia, its range is frequently said to extend as far as Tarija (Hennessey *et al.* 2003b, Whitney 2003, Ridgely & Tudor 2009). This appears, however, to be based entirely on a specimen (CM P80655) cited by Remsen *et al.* (1986) as being collected at Bermejo, Tarija (22°10'S, 64°42'W; Paynter 1992). Steve Rogers provided photos of the specimen labels for three specimens of Slaty Gnateater collected by J. Steinbach on 2 November 1919: all three were collected at Bermejo, La Paz. Discarding the Tarija record, the southernmost record is a specimen (FMNH 293994) taken 72km ESE of Monteagudo, Santa Cruz. Our knowledge of the species' precise distribution is, however, far from complete, and recent records have shown it to be more widespread than was believed for almost a century following its description (Bond & Meyer de Schauensee 1942, Meyer de Schauensee 1966, Remsen 1984, Remsen & Traylor 1983, Perry *et al.* 1997, Macleod *et al.* 2005, Martinez & Rechberger 2007, Aben *et al.* 2008).

HABITAT Slaty Gnateater is one of only two predominantly Andean *Conopophaga*, found in humid subtropical and upper tropical foothills on the east slope of the Andes (Parker *et al.* 1982, Whitney 2003). Interestingly, although in most parts of its range the species occurs in montane evergreen habitats (Hennessey & Gomez 2003a, Jankowski 2010), in the south Slaty Gnateater also ranges into drier deciduous and semi-deciduous forests (Perry *et al.* 1997, Aben *et al.* 2008). In general, at least in the more humid parts of its range, Slaty Gnateater appears to prefer the 'densest growth available' (Whitney 2003), including treefalls, old clearings, road edges and similar areas of disturbance (Whitney 2003, Macleod *et al.* 2005, Martinez & Rechberger 2007). In the drier parts of its range, however, there is some evidence to suggest that it prefers more intact habitat (Aben *et al.* 2008). Across its range, altitudinal records span 400–2,450m (Fjeldså & Krabbe 1990, Whitney 2003, Hennessey *et al.* 2003b), with most at 1,000–1,700m in Bolivia (Remsen 1984) and 850–2,000m in Peru (Clements & Shany 2001, Schulenberg *et al.* 2010). There is, however, a fair degree of variability between localities (Hennessey *et al.* 2003a, Walker *et al.* 2006, Jankowski 2010), with sight records to 2,530m in the Serranía de Siberia (eBird: Julián Quillén Vidoz) not implausible, but requiring confirmation. In south-east Cuzco, along the Cuzco–Manu road, the range (*saturata*) overlaps very slightly with Chestnut-crowned Gnateater (*brunneinucha*), where the two taxa apparently segregate by using structurally different habitats (Whitney 2003).

VOICE The song is described by Whitney (2003) as 'a series of slightly drawn-out, frog-like, weakly disyllabic notes at 2.5–3.5kHz, sounding like "g-reeep" or "w-reeep", with warbled quality owing to modulation.' He described each note as being around 0.25s in duration but with the precise number of notes and their spacing varying. In general, most songs include 3–6 notes and last 2–8s. Several vocalisations described as calls were considered by Remsen (1984) to resemble sounds produced by tyrant-flycatchers (Tyrannidae) rather than more closely allied species such as antbirds (Thamnophilidae). The most commonly heard is a piercing *tseet* (Whitney 2003) or *psii* (Ridgely & Tudor 2009). Another call, described by Ridgely & Tudor (2009), is a repeated, sometimes querulous *jereeé*. Whitney (2003) also mentioned harsher notes and 'dry chatters'. Although the precise mechanical details have not been investigated, during aggressive or courtship encounters, males apparently produce loud bursts of sound in flight, possibly with their wing feathers (Whitney 2003).

NATURAL HISTORY Despite our increasing knowledge of distribution and habitat use, there are few data concerning the species' basic natural history and behaviour. Their foraging behaviour is apparently similar to other *Conopophaga* (Hilty 1975, Willis *et al.* 1983), with most time spent 0.5–2.0m above ground, capturing prey with upward-directed sallies to glean the undersides of leaves (Remsen 1984, Fjeldså & Krabbe 1990). They apparently will also drop to grab prey from the leaf litter (d'Orbigny 1849) or perch-glean insects from nearby foliage (Whitney 2003). They are most frequently encountered alone and were noted by Remsen (1984) to rarely or never join mixed-species flocks. There is no published information on interactions of Slaty Gnateater with predators but, due largely to the work of entomologists interested in avian chewing lice (Phthiraptera, Amblycera and Ischnocera) and feather mites (Acari), several papers have explored the taxonomy and ecology of these ectoparasites collected from Slaty Gnateater (race *saturata*; Clayton *et al.* 1992, Price & Clayton 1996, Byers 2013). In fact, the known louse fauna of Slaty Gnateater includes several genera of lice (Clayton & Walther 2001). This last paper, using a large survey of chewing lice collected from many species in south-east Peru (including Slaty Gnateater), found a significant correlation between bill morphology and the mean abundance of lice on different species of birds. They proposed the idea, which to my knowledge has been largely overlooked by the ornithological community, that bill shape, especially the degree to which the maxilla overlaps the mandible, may be an important trait influencing birds' ability to remove parasites. Their study indicated that birds with just the right amount of maxillary overhang were perhaps more adept at capturing and crushing lice, while those with little or no overhang were *louse-y* at it.

DIET The diet is poorly known, but presumably consists primarily of small arthropods (Whitney 2003) as suggested by Remsen's (1984) examination of stomach contents. Remsen (1984) did, however, mention finding small seeds, and seeds are included in the stomach contents of at least two other specimens (MSB), but the degree to which Slaty Gnateater may be frugivorous is unknown.

REPRODUCTION Apart from scattered reports of fledglings and juveniles (see below), the breeding biology remained completely unknown until the first nest was described by Sánchez & Aponte (2006). It was found with incubation underway on 16 October 2004 near San Onofre, in the Río Paracti valley, Chapare, Cochabamba (nominate). The following details are taken from the paper, to date the only detailed information available. Photographs of the nest and eggs of *saturata* (Londoño 2014), however, suggest that at least these characteristics are similar in the two races. Behaviour around the nest and the specifics of nest-site selection, habitat use and nestling care are presently unreported. Fortunately, details are forthcoming based on the study of 44 nests of *saturata* as part of the recent cornucopia of excellent natural history studies published by Gustavo A. Londoño and his team (e.g., Sánchez-Martínez & Londoño 2016a,b, Valdez-Juarez & Londoño 2011, 2016). **NEST & BUILDING** There are no data concerning construction. The described nest was in a small (4.5ha) fragment of second growth largely dominated by *Guadua* bamboo and surrounded by agriculture. It was not stated how isolated this fragment was from the nearest intact habitat. The nest was 1.1m above ground and supported by several small, angled shoots emerging from the main stalk of a bamboo. From photographs it seems likely that there was already a fair amount of material collected naturally by these supports, creating a solid platform for the nest. The bulk of the shallow-cup nest comprised various bamboo parts, including dead leaves and interlaced leaf petioles, but also included a few woody sticks and dicot leaves, some of the sticks being considerably longer than the width of the nest (22cm). Overall, due to its location atop naturally collected detritus and the use of materials common in the area, the nest was very well camouflaged, appearing as little more than a natural clump of detritus stuck to the side of a *Guadua* shoot. The nest was lined with thin, unbranched, flexible fern-leaf rachises, which were apparently relatively uniformly brown, smooth and somewhat shiny. Only one, rather long (20.5cm) fungal rhizomorph was found in the lining, presumably coiled among the other materials. Measurements of the nest were: external width 9.0cm; external height 8.5cm; internal diameter 7.5cm; internal depth 6.0cm. **EGG, LAYING & INCUBATION** On first discovering the nest, the authors noted the eggs were being incubated by the male, but no further behavioural data were reported. The two subelliptical eggs were pale buffy-brown. One of them was completely unmarked while the second had small spots and flecks of various shades of brown and grey forming an indistinct ring at the larger end. On collection, the eggs were determined to be in the early stages of development, but embryonic size was not stated. They measured 22.9 × 16.6mm and 21.3 × 15.4mm. A nest of *saturata* in south-east Peru held two almost completely unmarked eggs (Londoño 2014). **NESTLING & PARENTAL CARE** Nothing published to date, but apparently similar to other gnateaters (Gustavo A. Londoño *in litt.* 2017).

SEASONALITY Remsen (1984) described three 'fledglings' collected at Abra de Maruncunca on 9–10 November 1980, each of which is included in the breeding data below and probably correspond to the 'fledglings Nov (Puno)' in Fjeldså & Krabbe (1990). Remsen (1984) also reported that, of 52 specimens collected in Bolivia in June, July and November, 'most were in breeding condition'. From label data at LSUMZ I have provided specific information from the same specimens reported by Remsen (1984). In addition to these records, summarised by Whitney (2003), Fjeldså & Krabbe (1989) describe what I interpret to be a bird in subadult plumage. This supposition is supported by the fact that it had a significantly shorter wing than adults. Their description stated that was brown above with only weak scaling, relatively pale greyish on the sides of the head, neck, throat breast and flanks, and some of the greater wing-coverts had narrow buffy tips on an otherwise brown wing. Unfortunately, with so few data available on post-fledging plumage development for *Conopophaga*, it is impossible to accurately estimate this individual's fledging date. Based on the evidence above, it seems that the breeding period of *C. a. ardesiaca* in Bolivia spans at least June–November, possibly starting somewhat earlier, depending on how future data allow us to interpret the age of described immatures. This, according to Sánchez & Aponte (2006) spans the end of the dry season and the start of the wetter period. BREEDING DATA *C. a. saturata* Nest with eggs, 5 November 2013, San Pedro (I. Ausprey *in litt.* 2015); nestlings, 1 October 2013, San Pedro (I. Ausprey *in litt.* 2015); two juveniles, 19 November 1981, Consuelo (FMNH 311468/69); juvenile, 29 December 2009, Cock-of-the-Rock Lodge (IBC 1366758, D. Shapiro); two subadults at Cadena, 15 June 2011 (MSB 36768) and 22 June 1958 (YPM 81717, S.S. Snow photos); subadult, 11 November 1981, Consuelo (FMNH 311460); three subadults, 31 August, 1 and 3 September 1985, 4km ENE of Shintuya (FMNH 322368/69/75); two subadults transitioning, 25 August 1985, 4km ENE of Shintuya (FMNH 322365/66); four subadults transitioning, 19 October, 5, 17 and 19 November 1981, Consuelo (FMNH 311454/57/65/66); subadult transitioning, 30 November 1985, Río Tono (FMNH 322362); subadult transitioning, 13 June 2011, 16.5km WNW of Cadena (MSB 36721). *C. a. ardesiaca* Two fledglings, 10 November 1980, Abra de Maruncunca (Remsen 1984, LSUMZ 98382/84); fledgling transitioning, 9 November 1980, Abra de Maruncunca (Remsen 1984, LSUMZ 98381); juvenile, 1 January 2005, near Zongo (IBC 983145, J.A. Tobias); juvenile transitioning, 25 April 1954, Alto Palmar (LSUMZ 36126); subadult, 1 April 2005, Padilla (IBC 983147, J.A. Tobias); subadult, 13–23 October 1938, Finca Silala (Fjeldså & Krabbe 1989); subadult, 25 June 1927, Incachaca (FMNH 180576); subadult transitioning, 25 June 1927, Incachaca (FMNH 110365); female with post-laying gonadal condition, 10 November 1976 (LSUMZ 82751), and two males with enlarged testes, 12 November 1976, Río Huari-Huari (LSUMZ 82752/53).

TECHNICAL DESCRIPTION Sexes differ. The following refers to nominate *ardesiaca* and is taken from Meyer de Schauensee (1970) and Fjeldså & Krabbe (1990). *Adult Male* Above near-uniform dark olive-brown with feathers inconspicuously fringed black or dark grey giving a slight scaled appearance. The forecrown is slate-grey, while the sides of the head and neck are grey to dark grey, contrasting with a long silvery-white postocular tuft. Below

dark grey to grey, sometimes with whitish on the central belly. *Adult Female* Similar to male but has the upperparts more reddish-brown instead of olivaceous, and the vent and flanks ochraceous-brown rather than grey. Most notably, they have a tawny-buff or orange-rufous forecrown and lack the postocular white tuft. *Adult Bare Parts* *Iris* dark brown; *Bill* maxilla black, mandible pale, dull pink (Remsen 1984); *Tarsi & Toes* pale grey. *Subadult* Fjeldså & Krabbe (1990), presumably based on the plumage of a specimen from La Paz which they had described previously (Fjeldså & Krabbe 1989), suggested that 'immature' plumage is overall browner above, and that the greater wing-coverts are narrowly tipped buff. *Fledgling* Fjeldså & Krabbe (1990) described the plumage of 'juveniles' as brown above with small pale spots on the crown and larger pale spots on the back. Below, the feathers have pale buff centres creating a scaled appearance, except on the central belly, which is white. This description appears similar to, and is perhaps based on, three 'fledglings' collected in Bolivia (*ardesiaca*) described as follows by Remsen (1984): 'Except for the remiges and rectrices, which are very dark brown to golden brown, these birds are speckled and scalloped throughout; each contour feather has a tawny-brown V-shaped, crescent-shaped, or round centre bordered proximally and distally by blackish-brown. In the centre of the belly, the tawny brown is replaced by white, and feathers in the vent area are downy and white. The postocular stripe can be faintly discerned as a row of nearly all tawny-brown feathers that are basally white.'

MORPHOMETRIC DATA Data for adults and subadults combined. *C. a. saturata* Data from FMNH and MSB. Adult ♀♀: *Wing* 70.5 ± 1.8mm (n = 10); *Bill* from nares 9.0 ± 0.4mm (n = 10), exposed culmen 12.2 ± 0.5mm (n = 10), depth at nares 4.4 ± 0.5mm (n = 8), width at nares 5.5 ± 0.2mm (n = 8), width at base of mandible 7.1 ± 0.4mm (n = 8); *Tarsus* 29.9 ± 0.5mm (n = 10). Adult ♂♂: *Wing* 74.7 ± 2.8mm (n = 14); *Bill* from nares 9.3 ± 0.4mm (n = 17), exposed culmen 12.4 ± 0.7mm (n = 14), depth at nares 4.6 ± 0.4mm (n = 7), width at nares 5.7 ± 0.2mm (n = 7), width at base of mandible 7.4 ± 0.4mm (n = 7); *Tarsus* 30.0 ± 0.8mm (n = 17). Two juvenile ♂♂ (FMNH 311468/69): *Wing* 72mm, 72mm; *Bill* from nares 9.0mm, 8.9mm, exposed culmen 12.2mm, 12.1mm, depth at nares 4.3mm, 4.1mm, width at nares 5.4mm, 5.5mm width at base of mandible 6.8mm, 7.3mm; *Tarsus* 29.5mm, 30.5mm. See also Berlepsch & Sztolcman (1906). *C. a. ardesiaca* *Wing* 66mm, 74mm (two specimens of *ardesiaca* from La Paz and Cochabamba, respectively; Fjeldså & Krabbe 1989, former possibly subadult (see Reproduction Seasonality). Adult ♀ FMNH 180577: *Bill* from nares 8.5mm, exposed culmen 11.5mm, depth at nares 4.1mm, width at nares 5.0mm, width at base of mandible 6.3mm; *Tarsus* 27.0mm. Adult ♂♂: *Wing* 77.5mm; *Tail* 50mm; *Bill* [exposed culmen] 14mm (n = 1, holotype, Hellmayr 1921). Data from FMNH. *Wing* 69mm, n/m, 75mm, 72mm; *Bill* from nares 8.2mm, 8.9mm, 9.3mm, 7.9mm, exposed culmen 11.0mm, 12.6mm, 11.8mm, 10.8mm, depth at nares 4.3mm, 4.2mm, 4.3mm, 4.2mm, width at nares 5.3mm, 5.1mm, 5.4mm, 4.8mm, width at base of mandible 6.7mm, 6.7mm, 6.9mm, 6.3mm; *Tarsus* 29.5mm, 30.3mm, 30.1mm, 29.9mm. MASS *C. a. ardesiaca* Mean 23.2g, range 20–27g (n = 27♂♂), mean 21.9g, range 17.5–27.0g (n = 24♀♀) (Remsen 1984). *C. a. saturata* mean 26.8g (n = 10, 5♂♂, 5♀♀, Clayton & Walther 2001). Data from MSB, FMNH. Adult ♀♀: 24.9g(lt), 25.3g(no), 25.7g(no), 26.0g(no), 26.0g(m), 26.4g(no), 27.0g(m),

27.5g(m), 30.2g(no), 30.5g(m). Adult ♂♂: 24.0g(lt), 24.0g(lt), 24.4g(lt), 24.9g(m), 25.0g(m), 25.0g, 26.0g(lt), 26.0g(lt), 26.0g(lt), 26.1g(no), 26.8g, 26.8g(m), 26.8g(tr), 28.0g(m), 28.5g(tr), 28.6g(no), 28.6g, 29.3g(m), 31.0g, 31.2g. Juvenile ♂♂: 23.2g, 26.3g. **Total Length** 12–14cm (d'Orbigny & de Lafresnaye 1837, Meyer de Schauensee 1970, Fjeldså & Krabbe 1990, Clements & Shany 2001, Ridgely & Tudor 2009, Schulenberg *et al.* 2010).

TAXONOMY & VARIATION The precise range limits of the two subspecies of *C. ardesiaca* are not well established (Whitney 2003), although many sources simply refer to Bolivian birds as *C. a. ardesiaca* and Peruvian birds as *C. a. saturata* (Peters 1951, Howard & Moore 1980, 1984, 1991). Plumage variation within Bolivia is especially confusing, 'with seemingly both dark and light specimens from Cochabamba, and three geographically defined types of back colour in females' (Whitney 2003). The upperpart variation in females apparently includes dark brownish backs in Cochabamba specimens, olivacous-brown backs in Santa Cruz, and distinctly rufous-brown backs in some females from La Paz. Some females from Santa Cruz, however, are somewhat more olivaceous above, and were suggested by Whitney (2003) to possibly represent an undescribed taxon. Whitney (2003) stated that the range of the nominate subspecies 'probably' includes part of Puno in south-east Peru. Bond (1953), however, based on specimens from Bella Pampa, stated that 'These birds differ from 4 males of *C. a. ardesiaca* from the Department of La Paz, Bolivia, chiefly in being decidedly darker and browner above without a pronounced olivaceous tinge. The grey of the under parts, including the sides of the head, merely averages darker, but the flanks and undertail coverts are definitely browner. The skins agree fairly well with the description of *saturata*.' Stephens & Traylor (1983) were unable to locate Bella Pampa but noted that it was equivalent to 'La Oroya' of Chapman. Using their approximate location of Bella Pampa (in the Quitún Valley above the Río Inambari), I located a small town named Oroya, almost certainly the correct location of this site. A fifth specimen, from Río Limbani (NHMUK), was probably collected near Oroya. Also not located, Río Limbani was posited to be near the town of Limbani (14°08'S, 69°42'W) by Stephens & Traylor (1983). As Limbani (3,350m) is above the altitudinal range of *C. ardesiaca*, the true origin of the NHMUK specimen is probably east of (and below) Limbani, near Oroya. These localities are separated from southern Puno localities (here considered to harbour *ardesiaca*) by just *c.*65km, but from the southernmost records of *saturata* (Cuzco) by at least 120km. Here, Oroya is included in the range of *ardesiaca*, but the true affinities of this population are unknown.

Conopophaga ardesiaca saturata von Berlepsch & Sztolcman, 1906, Ornis, vol. 13, pt. 2, p. 119, 'Huaynapata, Marcapata', south-eastern Peru [= ? near Mandor, 13°22'S, 70°54'W, *c.*1,000m, Dept. Cuzco; see below]. *C. a. saturata* was based on three males and one female collected by Jan Kalinowski. Berlepsch & Sztolcman (1906) failed, however, to describe the female, despite presumably having it before them as part of the Kalinowski collection. They based their diagnosis entirely on the males, leaving the female supposedly 'unknown' (Cory & Hellmayr 1924). The three syntypes were deposited in the privately owned Muzeum Zoologiczne Branickich (Branicki Zoological Museum), a collection that in 1919 was donated to the Polish

government and incorporated into the newly founded National Museum of Natural History (MZPW) (Mlíkovský 2007). One of the males and the female are apparently lost, but two male syntypes remain: MZPW 33972, 13 September 1897; MZPW 33981, 19 July 1897 (Mlíkovský 2009). Stephens & Traylor (1983) stated that Berlepsch & Sztolcman (1906) placed Huaynapata in the Marcapata Valley, and suggested the location to be near 13°13'S, 70°24'N, Vaurie (1972) noted that Zimmer felt the location was synonymous with 'Río Cadena' (*c.*13°16'S, 70°36'W; Stephens & Traylor 1983). The location given by Berlepsch & Sztolcman (1906), however, is as quoted above, suggesting perhaps a closer association with Marcapata (13°30'S, 70°55'W) than previously acknowledged. As Marcapata (*c.*3,150m) is probably above the range of *C. ardesiaca*, a location below this, along the Río Marcapata, perhaps near Mandor, seems a more likely location for Huaynapata. The limits of the range of *saturata* are not entirely clear, as populations in Puno, though usually assigned to this race (Meyer de Schauensee 1966), are possibly best included within *ardesiaca* (see above). Most authors agree, however, that populations from Cuzco involve *saturata* (Peters 1951). On the upperparts, male *saturata* is generally darker, deeper rufous-brown rather than olivaceous-brown. Below they are also darker, with the throat, upper chest and head-sides being blackish rather than paler grey. The undertail-coverts are also more rufescent than in *ardesiaca* (Berlepsch & Sztolcman 1906, Cory & Hellmayr 1924, Whitney 2003). Berlepsch & Sztolcman (1906) stated that the bill of *saturata* averages larger, in length and width. It is their overall darker plumage which gives them their subspecific name, derived from the Latin words *satur* (rich), *satura* (copious) or *satis* (enough) (Jobling 2010). Records from Soriano, Masayacu (Taczanowski 1874), actually refer to *C. castaneiceps brunneinucha* (Cory & Hellmayr 1924), and those from Huamba (Taczanowski 1882) to *C. castaneiceps chapmani* (Peters 1951). **Specimens & Records PERU:** *Madre de Dios* Cordillera de Pantiacolla, *c.*4km ENE of Shintuya, 12°40'S, 71°14'W (FMNH 322364–406); ridge above Hacienda Amazonia (= Amazonia Lodge), 12°52'S, 71°23.5'W (FMNH 322407). *Cuzco* Katarompanaki, 12°11.5'S, 72°28'W (Lane & Pequeño 2004); Río San Pedro, 12°58'S, 71°34'W (Gustavo A. Londoño *in litt.* 2017); Río Tono, 13°00'S, 71°11'W (FMNH 322362/63); Trocha Bamboo, 13°03'S, 71°32'W (MSB); San Pedro, 13°03'S, 71°33'W (MUSA, FMNH, XC); Cock-of-the-Rock Lodge, 13°05'S, 71°25.5'W (XC 75356, F. Schmitt, IBC 1126690, recording N. Athanas); Consuelo, 13°08'S, 71°15'W (very long series FMNH); *c.*9.75 WSW of Pilahuata, 13°10'S, 71°30'W (XC 221481/82, P. Boesman); *c.*16.5km WNW of Cadena, 13°21'S, 70°52'W (MSB 36721); Cadena, 13°24'S, 70°43'W (YPM, FMNH, MSB).

Conopophaga ardesiaca ardesiaca d'Orbigny & Lafresnaye, 1837, Magasin de Zoologie, d'Anatomie Comparee et de Palaeontologie, vol. 7, p. 13, 'Yungas (Bolivia)' [= La Paz Dept., Río Choqueyapu drainage, near Circuata, *c.*16°38'S, 67°15'W, *c.*1,600–1,800m, collected between 31 August and 4 September 1930; see below]. Several uncertainties surround the type of *C. ardesiaca*, which was mounted for display at MNHN in Paris, with its original label apparently affixed to the base of the pedestal (Hellmayr 1921). According to Hellmayr (1921) the original label states that it was collected by M. Alcide d'Orbigny in 1834. The type was obtained, however, during d'Orbigny's travels in southern South America, the inclusive dates of which (1826–33), as well as his specific itinerary (see below) exclude 1834 as a potential year of collection. The full account of his travels appeared in a multi-volume series (*Voyage dans l'Amérique méridionale*), published in Paris in 1835–47. The account of the ornithological collections made during his explorations is included in the third livraison of the fourth volume but, as is the case with many such ambitious works of the era, there is some confusion as to the date of issue of the various parts (Dickinson 2017, Dickinson & Lebossé 2017). Although never completed, the livraison that includes *C. ardesiaca* (pp. 159–232) was apparently published in 1838 (Sherborn & Woodward 1901, Sherborn & Griffin 1934, Dickinson & Lebossé 2017). During the same period that these sections were issued, d'Orbigny published with Lafresnaye a series of papers detailing the collections and describing new taxa. This series was published under the title Synopsis Avium and issued via the *Magasin de Zoologie*, but was similarly never finished (Hellmayr 1921). Nevertheless, the first part (d'Orbigny & Lafresnaye 1837) included the description of *C. ardesiaca*. Somewhat confusingly, various references alternate between providing either Lafresnaye or d'Orbigny as the first author of the species name, either correctly citing d'Orbigny as first author (Sclater 1858, Sclater & Salvin 1879, Taczanowski 1882, Peters 1951, Meyer de Schauensee 1966) or erroneously crediting Lafresnaye (Sclater 1855, 1862, 1890, Taczanowski 1874, Dubois 1900, Ménégaux & Hellmayr 1906, von Berlepsch & Sztolcman 1906, Brabourne & Chubb 1912, Hellmayr 1921, Cory & Hellmayr 1924, Bond & Meyer de Schauensee 1942). Gray (1869) even credited the name to Lafresnaye alone. Hellmayr (1921), who reproduced the text of the handwritten original label on the type, noted that it read 'Conopophaga ardesiaca Lafr. Et d'Orb. Type,' which perhaps explains his own mistake. Conversely, this label may have been written by d'Orbigny or Lafresnaye and, perhaps, although not explicit in their description, they had agreed upon Lafresnaye as first author of *ardesiaca*. Having examined the original description, I find no reason that d'Orbigny should not be credited as first author, as is now commonly the case (Whitney 2003). With respect to the type locality, it is unfortunate that neither the label nor the original description provides any more precision (d'Orbigny & Lafresnaye 1837, Hellmayr 1921). In the brief account provided by d'Orbigny (1838), he mentioned it being common at Cajuata (16°49'S, 67°15'W) and around the Río Miguilla (16°16'S, 67°12'W), both in the middle Río Choqueyapu (Río La Paz) valley as it flows north-east through the Cordillera Real. From 30 August to 9 September 1830 d'Orbigny (1843) explored the region between these localities. Undoubtedly, therefore, the type locality was somewhere in this region, perhaps near Circuata, where he spent several nights. The sex of the holotype was not stated by d'Orbigny & Lafresnaye (1837), but is apparently male (Hellmayr 1921). The nominate race, as currently defined, is largely confined to Bolivia (Remsen & Traylor 1989, Hennessey *et al.* 2003b). Following previous authors (Whitney 2003; see discussion above), however, here I consider records from southern Puno as belonging to *ardesiaca*. I have also, provisionally, included records from Oroya and 'Rio Limbani,' although this requires careful scrutiny (see Distribution, also Bond 1953). **SPECIMENS & RECORDS PERU:** *Puno* PN Bahuaja Sonene, 13°30'S, 69°22.5'W (ML 65736701, J.Q. Vidoz); Oroya [= Bella Pampa],

13°51'S, 69°41'W (Bond 1953, ANSP 103291–294); Río Huari-Huari, 5km NE of San José, *c.*13°58.5'S, 69°27.5'W (LSUMZ, DMNH); road to Pampa Yanamayo, 14°01.5'S, 69°12'W (ML 43888581, G.F. Seeholzer); *c.*5.5km NNE of San Juan del Oro, 14°14'S, 69°12'W (KUNHM); near Porompata, 14°17'S, 69°11'W (XC 82780, D. Geale); Abra de Maruncunca, 14°18'S, 69°14'W (Remsen 1984, LSUMZ 98379–390, ML). **BOLIVIA:** *La Paz* Middle Río Tuichi, 14°30'S, 68°35'W (Perry *et al.* 1997); Sumpulo, 14°35'S, 68°47'W (S.K. Herzog *in litt.* 2017); Fuertecillo, 14°35'S, 68°56'W (Hennessey & Gomez 2003); Tokoaque, 14°37'S, 68°57'W (Hennessey & Gomez 2003); 'Torcillo-Sarayoj', 14°39'S, 68°37'W (ML 120907, A.B. Hennessey); Cerro Asunta Pata, 15°03'S, 68°29'W (LSUMZ 162694–707); Río Yuyo, 15°02'S, 68°27'W (S.K. Herzog *in litt.* 2017); *c.*52km WNW of Yucumo, 15°06'S, 67°31'W (ML 101647/95, A.B. Hennessey); Calabatea (on Río Coroico), *c.*15°27'S, 67°50'W (Bond & Meyer de Schauensee 1942, ANSP 121244–246); *c.*19km NNE of Caranavi, 15°41'S, 67°30'W (long series LSUMZ, DMNH, ML); near Zongo, *c.*16°06'S, 68°03'W (Hellmayr 1921, IBC 983145, J.A. Tobias); Padilla, 16°06'S, 67°42.5'W (IBC 983147, J.A. Tobias); Sandillani, 16°12'S, 67°54'W (Hellmayr 1921, Bond & Meyer de Schauensee 1942, ANSP 121241–243); San Antonio, 16°15'S, 67°47'W (Martínez & Rechberger 2007); *c.*3.2km NNW of Apa Apa, 16°21'S, 67°30'W (ML 10147, A.B. Hennessey); 'Pitiaguaya' [= La Florida; Paynter 1992], 16°21'S, 67°46'W (AMNH 229201/02); Finca Silala, 16°30'S, 67°25'W (Fjeldså & Krabbe 1989). *Cochabamba* Laguna Carachupa, 16°14'S, 66°25'W (S.K. Herzog *in litt.* 2017); Río Pampa Grande Valley, 16°40'S, 66°29'W (Macleod *et al.* 2005); 'Quebrada Honda' [near Cocapata; Paynter 1992], *c.*16°57'S, 66°43'W (Hellmayr 1921, AMNH 99205); Miguelito, 17°00'S, 65°45'W (Fjeldså & Krabbe 1989); Río Mascota Mayu, 17°04.5'S, 65°46.5'W (Brumfield & Maillard 2007); Palmar, 17°06'S, 65°29'W (Bond & Meyer de Schauensee 1942, ANSP, UMMZ); Alto Palmar, 17°08'S, 65°35'W (LSUMZ); old Cochabamba–Villa Tunari road, 17°10'S, 65°35'W (XC 3579, S.K. Herzog); San Onófre, 17°09'S, 65°47'W (Sánchez & Aponte 2006, LSUMZ, BMNH, MHNNKM); Incachaca, 17°14'S, 65°49'W (CM P120363/64, FMNH); 30km N of Monte Punco, *c.*17°20'S, 65°16'W (Olrog 1984, AMNH 823800–803); near Sehuencas, 17°27.5'S, 65°16.5'W (eBird: J.Q. Vidoz). *Chuquisaca* Río Azuero, *c.*19°18'S, 64°00'W (Bond & Meyer de Schauensee 1942, ANSP 141713); Serrania de Iñau, 19°30'S, 63°57''W (MHNNKM 4862). *Santa Cruz* Serranía de Siberia, 17°50.5'S, 64°41.5'W (eBird: J.Q. Vidoz); Refugio Los Volcanes, 18°06'S, 63°36'W (XC 144956, D.F. Lane); Achiras, 18°09'S, 63°48'W (LSUMZ 37771); Bermejo, 18°10'S, 63°36'W (CM); 23.2km E of Samaipata, 18°16.5'S, 63°40.5'W (MHNNKM 2213, LSUMZ 171327); El Chacheal, 18°22'S, 63°38'W (LSUMZ, MHNNKM); Río Parabano, 18°25.5'S, 63°30'W (MHNNKM 3898); Loma Larga, 18°47'S, 63°53'W (S.K. Herzog *in litt.* 2017); 72km ESE of Monteagudo, 19°57'S, 63°18'W (FMNH 293994).

STATUS Slaty Gnateater is evaluated as Least Concern (BirdLife International 2017). Across its entire range it is described as fairly common (Stotz *et al.* 1996), but is uncommon in Peru (Parker *et al.* 1982, Clements & Shany 2001), and seems to be generally thought of as uncommon (Hennessey *et al.* 2003a, Martinez & Rechberger 2007) or rare (Hennessey & Gomez 2003) locally. Within its fairly large range (from a conservation standpoint), both subspecies occur within formally protected areas.

BirdLife International (2017) states that the population trend appears to be increasing, apparently basing this on Whitney's (2003) suggestion that Slaty Gnateater may actually benefit to some degree from mild forest disturbance such as 'selective logging, road construction, and from landslides and other consequences of such activities.' That anthropogenic habitat alteration might lead to increasing populations is, perhaps, overly optimistic. Although I agree that a species like Slaty Gnateater may be capable of tolerating habitat alteration that might preclude other, more sensitive, species, at least one study has suggested that populations are negatively impacted by disturbance (Aben *et al.* 2008). **Protected Populations** *C. a. saturata* PN Manú (Walker *et al.* 2006); SN Megantoni (Lane & Pequeño 2004). *C. a. ardesiaca* **Peru** PN Bahuaja Sonene (ML 65736701, J.Q. Vidoz). **Bolivia** PN y ANMI Amboró (Gemuseus & Sagot 1996); PN

Carrasco (Hennessey *et al.* 2003b); PN y ANMI Cotapata (Martinez & Rechberger 2007); PN Madidi (Perry *et al.* 1997, Hennessey & Gomez 2003); Reserva de la Biósfera Pilón Lajas (Hennessey *et al.* 2003a); ANMI Apolobamba (Hennessey *et al.* 2003b); RNFF Tariquía (Hennessey *et al.* 2003b); RP Refugio Los Volcanes (Aben *et al.* 2008, XC).

OTHER NAMES *Conopophaga saturata* (Brabourne & Chubb 1912). **English** d'Orbigny's Gnat-eater/Gnat Eater [*ardesiaca*], Peruvian Gnat-eater/Gnat Eater [*saturata*] (Brabourne & Chubb 1912, Cory & Hellmayr 1924); Slaty Gnataeter [misspelling] (Valarezo-Delgado 1984). **Spanish** Jejenero Pizarroso (Clements & Shany 2001, Whitney 2003); Jejenero apizarrado (Valarezo-Delgado 1984). **French** Conophage ardoisé. **German** Olivgrauer Mückenfresser (Whitney 2003).

Slaty Gnateater, adult male (*saturata*), Manu road, Cuzco, Peru, 22 October 2014 (*David Beadle*).

Slaty Gnateater, older nestling (*saturata*) in nest, Cock-of-the-Rock Lodge, Cuzco, Peru, 16 November 2013 (*Gustavo A. Londoño*).

Slaty Gnateater, eggs (*saturata*) in nest, Cock-of-the-Rock Lodge, Cuzco, Peru, 20 August 2008 (*Gustavo A. Londoño*).

Slaty Gnateater, mid-aged nestlings (*saturata*) in nest, Cock-of-the-Rock Lodge, Cuzco, Peru, 6 October 2011 (*Gustavo A. Londoño*).

Genus *Pittasoma*: the 'gnatpittas'

Pittasoma Cabanis, 1860, Proceedings of the Academy of Natural Sciences of Philadelphia, vol. 12, p. 189. Type, by monotypy: *Pittasoma michleri* (J. F. Gmelin, 1789)

Remarks on the immature plumages of *Pittasoma* Of the six genera treated herein, *Pittasoma* remains the least understood with respect to the appearance and ontogeny of immatures. There is no description of the nestling or fledgling for either species. Based on the youngest specimen I have examined (*P. r. rufopileatum*), the fledglings of *Pittasoma* are probably similar to those of *Conopophaga* in having dark fluffy plumage with coarsely spotted upperparts and distinctly rufescent (not buffy) markings on the upperwing-coverts. Much work remains to be done, however, and any observations of immatures, especially in nature where their plumage can be placed into a behavioural context, should be reported in detail. In the following, I have discussed, in as much detail as possible, most of the complexity that presently impedes our understanding of the immature plumages of *Pittasoma*. Until our understanding of these are improved, I err on the side of caution in assigning individuals to one of three age classes, as follows. **FLEDGLING** (Presumably) mostly still covered in plumose, fluffy, nestling-like feathers, rictal flanges still obviously inflated and contrastingly coloured, sexes similar, probably still heavily dependent on adults for food and largely stationary while awaiting its delivery. **JUVENILE** Little or no plumose feathering, rictal flanges (presumably) still brightly coloured but not obviously inflated, probably fairly mobile and sufficiently developed to assign gender. **SUBADULT** Very close to adults in general coloration, only separable by their immature wing-coverts, this stage may not be readily distinguishable from adults in the field, especially once most of the immature upperwing-coverts are replaced. For both species, it appears that the lesser wing-coverts are replaced before the greater coverts, generally in sequence from the outer to inner wing. As for other genera, '**fledgling transitioning**' and '**juvenile transitioning**' refer to birds somewhere between the above-defined stages.

BLACK-CROWNED ANTPITTA
Pittasoma michleri   Plate 5

Pittasoma Michleri Cassin, 1860, Proceedings of the Academy of Natural Sciences of Philadelphia, vol. 12, p. 189, 'Río Truandó, above its junction with the [Río] Atrato, but before reaching the Cordilleras' [07°17'N, 77°13'W, Chocó, Colombia]. Although the type locality is commonly truncated to 'Río Truandó, Colombia', Cassin (1860) provided the more detailed locality provided above, used to generate the suggested GPS. The holotype is an adult male collected 22 January 1858 by the US Topographical Engineer Corps expedition led by First Lieutenant Nathaniel Michler. Naming the species after the expedition's leader, Cassin (1860) credited the collection of the holotype to William S. Wood Jr. and Charles J. Wood, but the original label bears the name of another collector, Arthur C. von Schott (HFG). Cassin (1864) illustrated the male, and the female was described a few years later from a bird collected in central Panama (Lawrence 1862a,b; MCZ 46435).

'April 14, 1947, on the upper Río Jaqué, as I followed a hunting trail in heavy forest, one of these birds appeared suddenly on the ground only 3 or 4 meters distant, and ran, scolding with chattering calls, through the heavy undergrowth. I followed it slowly for nearly a quarter of an hour while it called regularly but always remained hidden. Finally, its calls ceased and I returned to the original point to continue along the trail. Then, by chance, as I looked about, the reason for the actions described became evident as in the crown of a low tagua palm I saw a large cup-shaped nest with two beautiful eggs.' – **Alexander Wetmore, 1972, Darién, Panama**

The large, boldly patterned Black-crowned Antpitta is fairly unique within its range, and unlikely to be misidentified even by the most 'wet behind the ears' of us.

As noted by O'Donnell (2014), however, this doesn't make the species any easier to see as it skulks in its dimly lit tropical home. Inhabiting the humid forest understorey in eastern Costa Rica, western Panama and north-west Colombia, the species appears to require primary habitat. As its common name implies, this species and its congener were long considered part of the Formicariidae and, more recently, within the more narrowly defined Grallariidae. Nevertheless, its *Grallaria*-like appearance and behaviour certainly make antpitta a fitting common name. Black-crowned Antpitta is well known to forage with army ants, either alone or in pairs, confining its movements to the understorey. Like most species treated here, it is infrequently seen, and the species' presence is most often revealed by its explosive and distinctive alarm call (O'Donnell 2014). The species is rather poorly known within its Colombian range but is seen with some regularity at several areas in Costa Rica and Panama. Almost nothing is known of its natural history, and much remains to be learned of even basic details such as the meaning of fairly significant variation in wing colour and patterning.

IDENTIFICATION 16.5–17.5cm. The sexes are somewhat similar, but usually separable in the field given a good look at the underparts. Equally, the two subspecies are easily distinguished with a good view of the head. Both sexes and both subspecies have a large, predominantly ivory-coloured or silvery-grey bill, brownish upperparts with indistinct, coarse black streaking, brown wings with distinctly spotted coverts, and pale underparts with broad, black, transverse striping and scalloping. In the north, *zeledoni* has a largely black head, appearing 'hooded', while the black of nominate *michleri* is largely confined to the crown, creating a "capped" look. In both subspecies, males are largely blackish on the throat, while females usually appear to have a largely chestnut or buffy throat, not as dark as in males. Females are typically washed buffy or chestnut below, not as clean black and white as males. The striking black-and-white scalloped underparts, heavy pale bill, and upright, tail-less "antpitta" shape of Black-crowned Antpitta make it nearly unmistakable. The smaller Black-headed Antthrush *Formicarius nigricapillus*, which is frequently sympatric, also has a black head, but is shorter-legged, less upright with a visibly cocked tail, and lacks scalloped underparts. The other similar-shaped antpittas that overlap in range with the present species, Streak-chested, Thicket and Scaled Antpittas, both lack the bold horizontal barring on the underparts and the bright, pale bill of Black-crowned Antpitta.

DISTRIBUTION Black-crowned Antpitta occurs from the Caribbean slope of Costa Rica south through western and central Panama, and on both the Caribbean and Pacific slopes in southern Panama. Its range just reaches north-west Colombia. Although it may frequently be overlooked, it appears that the species has been extirpated from parts of its range, especially at lower elevations. Several locations with historical records (Karr 1971) lack any modern reports for more than a decade. These include La Selva Biological Station in Costa Rica, as well as the Achiote and Pipeline roads in Panama (O'Donnell 2014).

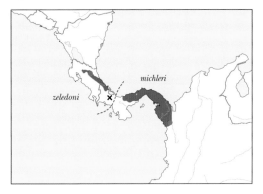

HABITAT Black-crowned Antpitta is a denizen of humid or wet forests, being confined to the understorey of primary forest and tall second growth (Stiles & Skutch 1989, O'Donnell 2014). It occurs mostly in foothill forests at 300–1,000m in Central America (Stiles & Skutch 1989, Angehr *et al.* 2004, Garrigues & Dean 2007), but appears to range to lower elevations in Colombia (Hilty & Brown 1986) and has been reported as high as 1,240m (Angehr & Christian 2000). In north-east Costa Rica, specimen records extend almost to sea level (MZUCR).

VOICE The song is a long (>1min) series of "whistled *piw* or "pulsating *pinking* notes, gradually increasing in pitch and slowing in tempo (Stiles & Skutch 1989; Angehr & Dean 2010). The ten or so final notes are of the same pitch and slightly longer in duration. What is apparently the alarm call is described as 10–16 descending, hoarse, guttural, coughing, squirrel-like notes (lasting 1.0–1.5s) (Karr 1971b, Hilty & Brown 1986).

NATURAL HISTORY Black-crowned Antpitta is largely terrestrial, moving across the forest floor by running or hopping, but sometimes climbs onto low limbs and vine tangles (Wetmore 1972). Probably, the majority of its foraging is performed on the ground and, although it will vocalize from the forest floor, especially in alarm, it appears

to more frequently choose a horizontal branch, 1–2m above ground, from which to sing (Hilty & Brown 1986, Stiles & Skutch 1989, O'Donnell 2014). While dashing or bounding across the forest floor, the species snatches prey from the leaf litter and low branches, occasionally digging into the soil with its large bill (Stiles & Skutch 1989). It is said to typically forage alone, but is said to also favour foraging at army ant swarms (Karr 1971b, Angehr & Dean 2010, O'Donnell 2014). While at swarms, Black-crowned Antpitta appears to be aggressive and largely dominant over other birds in attendance (Stiles & Skutch 1989). A perhaps spurious observation, but nonetheless interesting, was made by O'Donnell (2014) who observed, on several occasions, Stripe-throated Hermits *Phaethornis striigularis* harassing singing males at Quebrada González (*zeledoni*). While a male antpitta sang from an understorey perch, the hummingbird apparently hovered in front of it and made 'sputtering' vocalisations, being likened to the aggressive behaviour hummingbirds sometimes show towards owls or other potential predators. Given the large size of *Pittasoma* antpittas, perhaps they may be mistaken as a potential predator by a hummingbird. It is also possible that the long legs and powerful bill of *P. michleri* make it a regular nest predator of low-hanging *Phaethornis* nests! Known to be infested by *Formicaphagus* bird lice (Mallophaga) (Carriker 1957).

DIET Black-crowned Antpitta is said to consume a variety of invertebrates, as well as small vertebrates such as frogs and reptiles (Stiles & Skutch 1989). There are, however, almost no first-hand, quantified descriptions of its diet. The stomach contents of an adult male *zeledoni* collected while following an ant swarm included amblypygid(s?), Dermaptera and Coleoptera (MZUCR 2531; F.G. Stiles) and the stomach of an adult female contained ants, snails and other insect parts (WFVZ 35492; M. Marín).

REPRODUCTION There is virtually nothing known of the breeding biology of Black-crowned Antpitta. A single nest and its clutch of two eggs have been described (*michleri*; Wetmore 1972). The nest described by Wetmore (1972) was just over 1m above ground, built atop a rosette of leaf bases of a small understorey palm in primary forest. It was described as nothing more than a thin-walled cup of interwoven dark rootlets, *c.*9cm in diameter, but my examination of the nest (USNM 40975) reveals that this description refers only to the inner lining. This was constructed within an outer structure of twigs, leaf petioles, dead leaves and other detritus that doubtless appeared to have collected naturally in the top of the palm, but was probably at least partially constructed by the birds themselves. I estimate that the nest's outer diameter *in situ* was 20–24cm, its total thickness was *c.*10–12cm and the inner cup depth was *c.*6cm. The two eggs are nearly white, perhaps slightly buffy before being collected, but I feel that 'pinkish buff' (Wetmore 1972) is a misleading description of their true colour. They are marked with cinnamon and lavender splotches and flecks, concentrated into a cap at the larger end. They are sub-elliptical and measure 31.3 × 23.5mm and 32.2 × 23.0mm. Overall, they do not resemble the eggs of *Conopophaga* as much as Wetmore's (1972) description suggests, and they are quite dissimilar from those of any antpitta genera. The quote describing adult behaviour at the beginning of this account, this nest and egg description, and the report of a dependent fledgling in the Canal Zone (Karr 1971b) are the only previously published data on the reproduction of Black-crowned

Antpitta. **Breeding Data** *P. m. zeledoni* Juvenile, 6 November 1978, 5km S of Puerto Vargas (MZUCR 1952); subadult, 28 October 1898, Carillo (MCZ 117057); subadult, 20 April 1925, Hacienda La Iberia (AMNH 390388); subadult, 17 August 1899, Cariblanco de Sarapiquí (AMNH 492121). *P. m. michleri* Lightly incubated eggs, 14 April 1947, Río Jaqué (Wetmore 1972, USNM 40975); nestling, 12 June 2014, Donoso (eBird/photo: E. Campos); fledgling, 12 July 1969, *c.*8km NE of Gamboa (Karr 1971b); young with two adults [juvenile?], 25 November 2012, PN Altos de Campana (eBird: M. Gerber); subadult, 2 May 2015, Nusagandi (photo; A. Lewis); subadult transitioning, 20 May 1915, Cituro (AMNH 136604).

TECHNICAL DESCRIPTION Sexes differ. The following refers to the nominate subspecies. *Adult Male* Deep, slightly glossy, crown, including the upper half of the lores and superciliary. The lower portion of the lores, including feathers at the base of the bill, is pale grey to white, creating a distinct pre-ocular line that extended rearwards below the eye in a narrow stripe of pale feathers that forms the lower half of an eye-ring. The upperparts are olive-brown, with the feathers of the scapulars having narrow, pale shaft-streaks and broadly edged sooty-black that creates a streaked or vaguely scalloped pattern. Tail and wings are warm brown. The upperwing-coverts have pale shaft-streaks with a whitish or buffy transverse (usually somewhat diamond-shaped) bar at the tip, which usually has a black spot in the upper portion and is bordered below by a narrow black fringe. The flight feathers are paler or brighter chestnut at the leading edge and the tertials are marked like the wing-coverts, but the terminal markings are usually less distinct and most apparent on the leading vane. Cassin (1860) described the underwing-coverts of the holotype as 'dull greenish brown, striped and spotted with white and black; quills greenish rufous, some of the shorter quills having sub-terminal spots of light rufous, edged with black ; tail greenish rufous.' The ear-coverts form a rich chestnut patch that continues onto the neck-sides and across the nape, separating the black crown from the brownish back. The ear-coverts and neck-sides are sometimes lightly speckled black and white. The chin and throat are largely black with fine white and chestnut flecks. The remaining underparts are largely white with heavy black U-shaped bars creating a wavy, barred appearance. The barring fades on the flanks and undertail-coverts, but extends onto the thighs, with the ground colour becoming warm brown to buffy-chestnut. *Adult Female* Similar to males above, with more extensive and somewhat cleaner chestnut on the face and neck-sides, extending onto the malar region and blending into the largely chestnut throat. The throat is finely flecked white and flecked and narrowly barred with black. The remaining underparts are similar to males, but the barring tends to be narrower and the white areas are often washed chestnut to varying degrees. **Adult Bare Parts** *Iris* dark brown; *Bill* overall pale ivory or silvery-white, maxilla duskier, especially on culmen; *Tarsi & Toes* pale grey to silvery grey, duskier on feet, nails yellowish-ivory. *Immature* Overall, plumage development is poorly understood. Especially with respect to coloration and pattern of the upperwing-coverts, it is unclear which characters are clear indicators of immaturity and which the result of individual and gender-related variation. Two previous descriptions of immatures have been published, neither of which appears to be entirely correct. Stiles & Skutch (1989) described this plumage (gender not specified, presumably of *zeledoni*) as being duller overall, with a sooty-grey crown, a mottled brown-and-rufous face, and faintly barred black and light brown on the upperparts. Their description states that the spotting on the upperwing is more prominent, almost forming wingbars, and the scalloping below is distinctly fainter. Ridgway (1911) described the immature plumage of *michleri* as similar to that of adult females but with the 'lower throat tawny, chin and upper throat white, with a few narrow streaks of black'. Despite having examined numerous specimens and photographs of both subspecies, I find it difficult to define immature plumage with any degree of precision and I cannot state, with certainty, that I have seen individuals matching either of these descriptions. In general, for both subspecies, however, it is clear that the degree of buffy wash to parts of the underparts is not a good predictor of age, for either sex or race. Based largely on knowledge of plumage maturation in other gnateaters and antpittas, the key to assigning age using plumage, once clear evidence from the bill and gape are no longer visible, undoubtedly lies in the coloration and pattern of the upperwing-coverts. The youngest individual that I have examined (♀ *zeledoni*; MZUCR 1952), already had the crown uniform black (*contra* Stiles & Skutch 1989) but the feathers of the upper back are broadly tipped bright rufescent, with black central spots and broad black fringes. The greater and lesser coverts, tertials and tail feathers all bear similar markings. The second youngest skin I examined (♂ *michleri*; AMNH 492119) had only a few spots on the upper back, but none had as much rufescent coloration surrounding the spots. Overall, if there had been more immature feathers in this region, the effect would have been much finer spotting compared to the large obvious spots of MZUCR 1952. The degree to which this difference reflects age, sex and/or race is unclear. AMNH 492119 also had immature markings on the tertials and tail feathers, as well as fine black and rufescent barring on the rump. The wing-coverts, however, differed significantly from MZUCR 1952 in that the lesser coverts were patterned as in adults, while the greater coverts were broadly tipped bright rufescent *but lacked* central black spots and broad black fringes. For our purposes, I have considered MZUCR 1952 a juvenile and AMNH 492119 to represent a juvenile in transition to subadult plumage.

MORPHOMETRIC DATA *P. m. zeledoni* Wing 99.5mm, 100.8mm, 115mm, 97.5mm, 98.1mm, 100.5mm; *Tail* 31.5mm, 33.5mm, 34mm, 33.5mm, 33.4mm, 36mm; *Bill* [exposed] culmen 29.0mm, 30.0mm, 31.0mm, 27.0mm, 28.5mm, 30mm; *Tarsus* 50mm, 50.6mm, 51.5mm, 46.0mm, 46.0mm, 52.0mm, middle toe 28.5mm, 31.1mm, 31.0mm, 29.0mm, 30.0mm, 31.0mm ($n = 6$, 3♂♂,3♀♀, Ridgway 1911). *Wing* 99.5mm, 101mm, 99.5mm; *Bill* from nares 19.1mm, 19.1mm, 17.5mm, exposed culmen 22.1mm, 21.7mm, 20.4mm; *Tarsus* 50.7mm, 49.3mm, 47.8mm ($n = 3$, ♂♂♀; MCZ; J.C. Schmitt *in litt.* 2016). *P. m. michleri* Data from MCZ (J.C. Schmitt *in litt.* 2016), AMNH, FMNH, USNM (HFG). Adult ♀♀: *Wing* 97.2 ± 2.5mm ($n = 8$); *Tail* 37mm ($n = 1$); *Bill* from nares 17.2 ± 0.6mm ($n = 14$), exposed culmen 21.7 ± 3.6mm ($n = 8$), depth at nares 8.8mm, 8.9mm ($n = 2$), width at nares 9.1mm, 8.8mm ($n = 2$), width at base of mandible 13.3mm, 12.8mm ($n = 2$); *Tarsus* 48.8 ± 1.9mm ($n = 16$). Adult ♂♂: *Wing* 99.2 ± 2.4mm ($n = 15$); *Tail* 36.0 ± 2.2mm ($n = 5$);

Bill from nares 17.8 ± 0.9mm (n = 21), exposed culmen 21.5 ± 2.6mm (*n* = 10), depth at nares 9.1mm, 8.6mm (*n* = 2), width at nares 9.0mm, 8.9mm (*n* = 2), width at base of mandible 12.2mm, 10.4mm (*n* = 2); *Tarsus* 50.3 ± 1.8mm (*n* = 24). See also Cassin (1860), Ridgway (1911). **Mass** 110g (Stiles & Skutch 1989); 109.3g (*n* = 1♂, Karr 1971a,b); 125g(vl), 129g (lt) (*n* = 2, ♂,♀, MZUCR 2531, WFVZ 35492); 106g(vl) (*n* = 1♀, juvenile, MZUCR 1952). **Total Length** 15–19cm (Cassin 1860, Ridgway 1911, Sturgis 1928, Hilty & Brown 1986, Dunning 1987, Ridgely & Gwynne 1989, Garrigues & Dean 2007, Angehr & Dean 2010, O'Donnell 2014).

TAXONOMY AND VARIATION Two rather distinctive subspecies. Wetmore (1972) felt that birds from Bocas del Toro showed some intermediate characteristics but, from the material I have examined, this does not seem to be obviously the case.

Pittasoma michleri zeledoni Ridgway, 1884, Proceedings of the United States National Museum, vol. 6, p. 414, Río Sucio, Costa Rica [*c.*10°12'N, 83°55'W, Heredia; see below]. The Río Sucio is a large river that flows broadly north from the northern slopes of Volcán Irazú. It is impossible to know the precise location on the river where the holotype was collected, but the suggested GPS places it east of Guapiles, a location where the species is still fairly common and one that was probably fairly accessible during the late 1800s. The holotype is an adult male (USNM 91841) collected in 1881 by J. Cooper. Subspecies *zeledoni* occurs on the Caribbean slope in Costa Rica east to western Panama (Griscom 1935). Past descriptions of the distribution of this race have generally been somewhat vague (Costa Rica to 'western Panama'; Griscom 1935), but my examination of specimens makes it clear that the easternmost definite records of *zeledoni* are from the Almirante area and the upper Río Culubre, and this limit is reflected in the range maps. Some uncertainty surrounds the racial affiliation of birds in the Reserva Forestal Fortuna, slightly east of there (see Notes), but most Panamanian populations, including those over the majority of the Caribbean coast, are not of *zeledoni* (*contra* Krabbe & Schulenberg 2003). **Specimens & Records COSTA RICA:** *Alajuela* Estación Biológica Pocosol, 10°21'N, 84°39.5'W (P. O'Donnell *in litt.* 2016); Cariblanco de Sarapiquí, 10°16'N, 84°11'W (AMNH 492120/21). *Heredia* Finca Plástico, 10°18'N, 84°02'W (WFVZ 35492). *San José* Quebrada González, 10°09.5'N, 83°56.5'W (XC 169508/09, P. O'Donnell). *Limón* Carillo, 10°09'N, 83°55'W (Ridgway 1911, MCZ 117057); Jiménez, 10°10'N, 83°46.5'W (Salvin & Godman 1892, USNM 198338); Hacienda La Iberia, 10°05'N, 83°39'W (UCLA 21883, AMNH 390388); 5km S of Puerto Vargas, 09°41'N, 82°49'W (MZUCR 1952); RB Hitoy Cerere, 09°40'N, 83°01.5'W (P. O'Donnell *in litt.* 2017); Kéköldi Hawkwatch Observatory, 09°38'N, 82°48'W (P. O'Donnell *in litt.* 2016); Suretka, 09°34'N, 82°56'W (MZUCR 2531); Río Sixaola, 09°33'N, 82°34'W (Carriker 1957, Slud 1964, CM). **PANAMA:** *Bocas del Toro* near Changuinola, *c.*09°27'N, 82°29'W (Angehr *et al.* 2008); Almirante, 09°18'N, 82°25.5'W (Griscom 1935, MCZ 141247/48); 'Changuena River, lower camp' [Río Culubre; see Siegel & Olson 2008], *c.*09°08.5'N, 82°30'W (Wetmore 1972, USNM 476071). **Notes** Black-crowned Antpitta is apparently known from Reserva Forestal La Fortuna (Angehr 2003, Garcés 2007), but I was unable locate specimens or other evidence to confirm this, and

the racial affiliation of this population is unclear. The reserve includes forests on both sides of the continental divide, at *c.*08°46'N, 82°14'W on the Ngäbe-Buglé-Chiriquí border. Irrespective of its distribution in this area, it seems probable that birds in this region are assignable to *zeledoni*. Confirmation of its presence and racial affinities in this region would certainly be worth reporting.

Pittasoma michleri michleri Cassin, 1860. Some confusion, perhaps derived from vague or poorly worded descriptions, also surrounds the distribution of the nominate subspecies. It is not confined to the Pacific coast of Panama as implied by Krabbe & Schulenberg (2003), but is instead present on both slopes, at least from Veraguas, east and south to north-west Colombia. I examined an adult male of nominate *michleri* (AMNH 492118) labelled only 'Cascajal Coclé'. Google Earth places a Cascajal in Coclé at the first GPS provided below, the second is that provided by Siegel & Olson (2008) for a Cascajal in Colón (locality in Coclé not provided). Irrespective of this specimen's true origin, I can confirm that it should correctly be assigned to nominate *michleri*. **Specimens & Records PANAMA:** *Veraguas* Calovébora, 08°47'N, 81°13'W (Salvin 1870, Cory & Hellmayr 1924); Chitra, 08°32'N, 80°55'W (AMNH 77402); Santa Fé, 08°31'N, 81°05.5'W (Sclater 1890, Cory & Hellmayr 1924, Angehr *et al.* 2008, AMNH 187269–272). *Coclé* Cascajal, Coclé, 08°43'N, 80°28'W/09°33'N, 79°37'W (see text; AMNH 492118); El Cope, 08°26.5'N, 80°41'W (XC 65230, K. Allaire). *Colón* Mouth of Río Indio, 09°19'N, 80°00'W (USNM 206684); Gatún, 09°16'N, 79°55'W (Stone 1918, ANSP 64779/80, USNM 206686) and vicinity (USNM 206683/85); Lion Hill (now an island), 09°13.5'N, 79°53.5'W (Sclater & Salvin 1864, Cory & Hellmayr 1924, USNM 41581); Alhajuela, 09°13'N, 79°37'W (USNM 232350); *c.*8km NE of Gamboa, 09°09.5'N, 79°44.5'W (Karr 1970, 1971a,b, 1977); Donoso, Río del Medio, 08°55.5'N, 80°39'W (eBird/photo: E. Campos). *Panamá* Cerro Vistamares, 09°14'N, 79°23.5'W (XC 78757/58, W. Adsett); Quebrada Carriaso, 09°14'N, 79°14'W (USNM 347719/409508); Cerro Azul, 09°09'N, 79°23.5'W (Angehr *et al.* 2008); Río Bayano, *c.*09°06'N, 79°05'W (USNM 241127); Cerro Campana, 08°41'N, 79°56'W (Karr 1971b, AMNH 786240); Altos del María, 08°40'N, 80°02'W (XC 17193, A. T. Chartier); Charco Del Toro, headwaters of Río Majé, 08°43'N, 78°37'W (loc. see Fairchild & Handley 1966, USNM 423315). *Guna Yala* near Nusagandi, *c.*09°21.5'N, 78°59.5'W (Angehr *et al.* 2008, ML 105079, L.R. Macaulay); Burbayar Lodge, 09°20'N, 78°58'W (XC 3878, N. Athanas); Permé, 08°45'N, 77°34'W (Griscom 1932a, MCZ 155541–552, ANSP 128904); Quebrada Venado, near Armila, 08°40'N, 77°29'W (Wetmore 1972, USNM 477762); Puerto Obaldía, 08°40'N, 77°25'W (Griscom 1932a, MCZ 155540). *Embera* Cerro Tacarcuna, 08°10'N, 77°18'W (AMNH 135788). *Darién* Cerro Chucantí, 08°47.5'N, 78°27'W (Angehr & Christian 2000, Angehr *et al.* 2008, USNM 423316–319); Río Subcuti, 08°44'N, 77°59'W (AMNH 272077); Laguna de Matusagratí, 08°20'N, 77°56'W (Salvadori & Festa 1898b, Cory & Hellmayr 1924); lower Río Tuira, *c.*08°13.5'N, 77°57.5'W (Haffer 1975); Capetí, 08°04'N, 77°33'W (AMNH 136603); Estación Pirre, 08°00.5'N, 77°43.5'W (ML 477350, video S. Dzielski); Cituro, 08°00'N, 77°36'W (AMNH 136604); Tapalisa, 07°59'N, 77°26'W (AMNH 135789–791); Cerro

Sapo, 07°58'N, 78°22'W (Bangs & Barbour 1922, MCZ 87363); 4.8km SW of Cerro Sapo, *c.*07°56.5'N, 78°23.5'W (ANSP 148895); Cana, 07°46'N, 77°41'W (Robbins *et al.* 1985, LSUMZ 108380, XC, ML); Cerro Antaral, 07°37'N, 78°00'W (Angehr *et al.* 2004), Río Jaqué, 07°31'N, 78°09'W (Wetmore 1972, USNM 40975); Las Peñitas, 07°28'N, 77°59'W (USNM 389793). **COLOMBIA**: *Chocó* Río Salaquí, 07°27'N, 77°07'W and Río Juradó, 07°07'N, 77°46'W (Meyer de Schauensee 1950b). **NOTES** A good number of specimens that I examined (directly or via photos J.C. Schmitt) bore collection labels too vague to plot, but were probably collected close to one of the above-mentioned locations. These include: 'Darien', MCZ 155550; line of the Panama railroad (MCZ 46434/35, loc. see Olson & Siegel 2008); 'Panama' (AMNH 43561–565, USNM 53777/78, FMNH 50764, MCZ 32718). A specimen from 'Chiriquí, Panama' (AMNH 492119) is clearly a male *zeledoni* and, if correctly (albeit vaguely) labelled, would extend the distribution of this race a significant distance to the west. However, until further information becomes available, I consider it as probably mislabelled.

STATUS Black-crowned Antpitta is currently considered Least Concern (Birdlife International 2017). Although it has a restricted range, numbers are believed to be stable. However, as it seems probable that the species requires fairly large tracts of primary forest, Black-crowned Antpitta must be considered highly susceptible to habitat alteration and fragmentation (O'Donnell 2014). Luckily, a substantial part of the range of *zeledoni* in Costa Rica lies within protected areas. In Panama, nominate *michleri* is also currently fairly well protected in parks and also by the remoteness of much of its range. Its status in Colombia is largely unknown, but it probably occurs in PNN Los Katíos. **PROTECTED POPULATIONS** *P. m. zeledoni* **Costa Rica** PN Braulio Carillo (XC 169508/09, P. O'Donnell); RB Hitoy Cerere and Bosque Eterno de los Niños (P. O'Donnell *in litt.* 2017). **Panama** BP Palo Seco (USNM 476071). *P. m. michleri* PN General Omar Torrijos (XC 65147/230, K. Allaire); PN Darién (LSUMZ, AMNH, XC, ML); PN Santa Fé (IBC 1067828; photo T. Noernberg); PN Chagres (USNM, XC); PN Altos de Campana (Angehr *et al.* 2008, AMNH); PN Soberanía (Karr 1970, 1971a, 1977, Robinson *et al.* 2004, Angehr *et al.* 2008); BP San Lorenzo (historically, USNM; possibly now absent Angehr *et al.* 2008).

OTHER NAMES *Pittasoma zeledoni* (Sclater 1890, Salvin & Godman 1892, Dubois 1900, Sharpe 1901, Ogilvie-Grant 1912); *Calobamon michleri* (Heine & Reichenow 1890); *Pittosoma michleri* (misspelling; Sclater 1868); *Pittisoma michleri* (misspelling; Sclater & Salvin 1864). **English** Michler's Antpitta/Ant-pitta (*michleri*; Ridgway 1911, Stone 1918, Cory & Hellmayr 1924, Sturgis 1928, Meyer de Schauensee 1950b); Black-faced Antpitta (Haffer 1975); Zeledon's Antpitta (*zeledoni*; Ridgway 1911, Cory & Hellmayr 1924). **Spanish** Tororoí capinegro (Krabbe & Schulenberg 2003); Tororoí Pechiescamado (Salaman *et al.* 2007a); Tororoí Pechiescamaso (van Perlo 2006); Pittasoma coroninegro (Angehr 2003, Garcés 2007). **French** Grallaire à tête noire (Krabbe & Schulenberg 2003); Pittasome à tête noire (Krabbe *et al.* 2017). **German** Schwarzscheitel-Ameisenpitta (Krabbe & Schulenberg 2003); Schwarzscheitel-Mückenfresser (Krabbe *et al.* 2017).

Black-crowned Antpitta, adult female (*zeledoni*), Quebrada González, Heredia, Costa Rica, 10 October 2016 (*Jorge Obando Gutiérrez*).

Black-crowned Antpitta, adult female (*michleri*), Los Altos del María, Cocle, Panama, 30 December 2010 (*Scott Olmstead*).

141

Black-crowned Antpitta, adult female (*zeledoni*), Quebrada González, Heredia, Costa Rica, 10 October 2016 (*Jorge Obando Gutiérrez*).

Black-crowned Antpitta, adult (*zeledoni*), Quebrada González, Heredia, Costa Rica, 3 July 2008 (*Patrick O'Donnell*).

Black-crowned Antpitta, adult male (*zeledoni*), Quebrada González, Heredia, Costa Rica, 27 August 2016 (*Hansell Rodríguez*).

Black-crowned Antpitta, adult male (*michleri*) Cerro Pirre, Darien, Panama, 21 March 2013 (*Andrew Spencer*).

RUFOUS-CROWNED ANTPITTA
Pittasoma rufopileatum   Plate 5

Pittasoma rufopileatum Hartert, 1901, vol. 8, p 370, Novitates Zoologicae, 'Salidero, Bulún, Rio Bogotá, in N.W. Ecuador. 160 to 350 ft. above the sea' [= near Pulún, 01°05'N, 78°40'W, Esmeraldas, *c*.75m; see Paynter 1993]. The type specimen was originally considered to be a female, but is, in fact, a male (Hartert 1922). It was collected on 31 December 1900 by R. Miketta and G. Flemming (Hartert 1922). In the original description, Hartert (1901) failed to specify a holotype, but later (1922) designated AMNH 492122 the lectotype. There are four paralectotypes: AMNH 492123 (♂), AMNH 492124 (♂), AMNH 492125 (♂) and AMNH 492126(♀) (LeCroy & Sloss 2000). LeCroy & Sloss stated that all are adults, except AMNH 492123, which is 'immature'. In fact, my examination revealed that, based on the criteria defined herein, all are adults.

Rufous-crowned Antpitta is the more southerly of the two species of *Pittasoma*. It is a Chocó bioregion endemic, found from north-west Colombia to north-west Ecuador (Parker *et al.* 1996, Rodner *et al.* 2000), and is confined to humid forests in the lowlands and foothills. Within its relatively small range, there is considerable sexual and racial variation, with three subspecies, each with a fairly restricted range (Clements 2007, Ridgely & Tudor 2009). Although no one has examined the intraspecific relationships of these taxa, Krabbe & Schulenberg (2003) suggested that *harterti* might be synonymous with the nominate race. Currently considered Near Threatened globally (BirdLife International 2017), with the Ecuadorian-endemic nominate considered Vulnerable nationally (Jahn & Valenzuela 2002), the species is in sore need of studies documenting its breeding, behaviour, foraging strategies and habitat use (Greeney 2013j).

IDENTIFICATION 15–18cm. The following is from Greeney (2013j). In their lowland and foothill haunts the rufous-chestnut crown and nape, demarcated by a black supercilium, render the species unmistakable. The upperparts are olive-brown with broad black longitudinal stripes. The wings are brownish with small whitish spots on the wing-coverts and tertials. The cheeks, neck, and throat are pale ochraceous, sometimes sparsely dotted with black, while the breast, belly and flanks are buffy-ochraceous with heavy black barring. Females are similar but with less barring below and white speckling on the black supercilium. The boldly patterned Rufous-crowned Antpitta is easily distinguished from its only congener, Black-crowned Antpitta, with which there is no known geographic overlap, by having the head extensively rufous, rather than mostly black.

DISTRIBUTION Confined to the Chocó centre of endemism (Cracraft 1985, Stiles 1997), where it is resident from west-central Colombia (northern Chocó) south through Nariño to north-west Ecuador (Pichincha, possibly slightly further, at least historically) (Hilty & Brown 1986, Ridgely & Greenfield 2001, Krabbe & Schulenberg 2003, Jahn 2011).

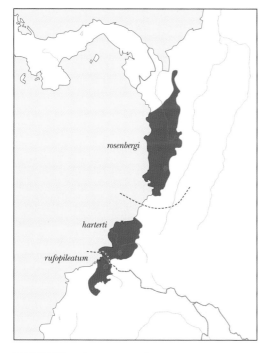

HABITAT Rufous-crowned Antpittas inhabit the understorey of humid lowland and foothill forests, generally below 1,100m (Sibley & Monroe 1990, Ridgely & Tudor 1994, Ridgely & Greenfield 2001, Jahn 2011). There have been no studies documenting specific habitat preferences and use, but these are presumably similar to the better-studied *P. michleri* of Central America (Greeney 2013j, O'Donnell 2014). According to J. Freile (*in litt.* 2017), subspecies *rufopileatum* appears to prefer mature forest on moderate to steep slopes, especially in places where the understorey includes some dense thickets or tangled undergrowth.

VOICE The acoustic details of the song were provided by Krabbe & Schulenberg (2003). In all, the whistle-like song lasts 0.4–0.8 seconds and is produced at 2.2–2.4Khz,

distinctly descending at the end and occasionally also at the start. This piercing *keeee-yurh* note is given at intervals of 1.3–1.6 seconds for up to several minutes (Krabbe & Schulenberg 2003, Restall *et al.* 2006). The alarm call, given by both sexes, is a decelerating chatter of 9–28 notes lasting 1–3 seconds. Each note in the rattle spans a wide frequency range. Also reported is a loud, emphatic, sharp *tche-tchik!* (Ridgely & Greenfield 2001, Krabbe & Schulenberg 2003). The latter's function is unknown, but it is probably an aggressive or alarm vocalisation.

NATURAL HISTORY Generally shy and difficult to observe, with a dearth of published behavioural information. While foraging it often pauses abruptly (Ridgely & Tudor 2009), freezing in a somewhat awkward-looking, spread-leg stance, termed 'spraddle-legged' by Willis (1985a). The best descriptions of general behaviour were provided by Willis (1985a), who watched several individuals (*rosenbergi*) foraging at ant swarms. Like similar-shaped *Grallaria* and *Hylopezus* antpittas, Rufous-crowned Antpittas are most frequently seen running or making long (0.4m) bounds, and rarely perching more than 0.5m above ground. Willis (1985a), while watching adults at ant swarms, reported several instances of dominance over smaller species of Thamnophilidae, and in north-west Ecuador they are sometimes aggressive towards Bicoloured Antbirds *Gymnopithys bicolor* while foraging together near the heads of swarms (A. Solano-Ugalde *in litt.* 2016). Rufous-crowned Antpitta is frequently reported to follow army ants (Willis & Oniki 1978, Willis 1985b, Ridgely & Tudor 1994). Recently, an adult male has been trained to receive food from humans (see Diet) in the fashion now well known for *Grallaria* antpittas (Woods *et al.* 2011). As has been suggested to be the case for *Grallaria* antpittas that are easily trained to receive food, in Mashpi Shungo Rufous-crowned Antpitta is known to regularly forage in the wake of mammals like White-collared Peccary *Tayassu pecari* and Nine-banded Armadillo *Dasypus novemcinctus* (A. Solano-Ugalde in *litt.* 2017). Although the ectoparasite fauna of *P. rufopileatum* has not been extensively studied (Price & Clayton 1996), Carriker (1957) described a new species of avian chewing louse (*Formicophagus pittasomae*) from specimens collected on subspecies *rosenbergi* in Antioquia. Leucism is rarely reported for any of the species treated in this book, so it is worthy of note that Olivares (1958) collected an adult female (*harterti*) in Cauca with one of the right tertials completely white. Interestingly, as noted by Hartert (1911b) some of the interscapular feathers of *rosenbergi* are white at their bases. Although it has not yet been observed, it seems possible that these feathers may be flared during displays, either intra-specific as, for example in some becards *Pachyramphus* (Miller *et al.* 2015), or as part of an interspecific warning or distraction display, as in some thamnophilids (e.g. genus *Sipia*; HFG).

DIET Almost completely unknown. The stomach of a female from western Dept. Cauca, Colombia (*harterti*), contained insects and spiders (Olivares 1958). Given their predominantly terrestrial foraging habits, they probably have preferences similar to *Grallaria* antpittas. With their heavy bills, they can forage through dense leaf litter, soil and rotting wood, and are perhaps capable of capturing small vertebrates as do *Grallaria*. Quite recently, at Mashpi Shungo, D. Chalá trained an adult male to arrive at a worm-feeding station in the same manner as many *Grallaria* species have been trained. Following the lead of the '*Grallaria* feeders', the feeding sessions were originally largely worm-based. With some trial and error, however, Sr. Chalá learned that crickets and grasshoppers (Orthoptera) were much preferred by his new dinner guest, now officially named 'Shunguito'. Naturally captured prey observed at this site (A. Solano-Ugalde *in litt.* 2017) include earthworms, crickets and beetles (both adults and larvae).

REPRODUCTION Breeding biology is virtually undocumented (Greeney 2013j). A female *harterti* with 'developed ovaries' was collected near Guapí, Dept. Cauca, in western Colombia, in late November (Olivares 1958) and Hilty & Brown (1986) reported a male *rosenbergi* in breeding condition in mid-February in Colombia's Dept. Chocó. The nest, eggs and young are undescribed, but presumably similar to those of Black-crowned Antpitta (O'Donnell 2014). Nests are probably broad, relatively shallow, open cups, well supported from below and nestled into low vegetation, vine tangles or large leaf bases (i.e., palms, cyclanths). They are likely to be composed of a loose, messy aggregation of leaves and sticks, with a sparse, but neatly woven lining of fine fibres. BREEDING DATA *P. r. rosenbergi* Juvenile, 23 December 1911, Nóvita (Chapman 1917, AMNH 111961); subadult ♀ [supercilium, coverts], 13 April 2017, Finca La Alborada (ML 54768841, photo D.M. Montoya). *P. r. rufopileatum* Fledgling transitioning, 25 February 1968, Guembí (MECN 2591, M.V. Sánchez N. photos); adult feeding young at ant swarm [juvenile?], 9 April 2014, RP Mashpi Shungo (eBird: G. Zambrano); subadult ♂ [coverts], 18 October 2012, Playa del Oro (D.M. Brinkhuizen photos).

TECHNICAL DESCRIPTION The following refers to the nominate race. *Adult male* Bright rufous crown separated from buffy or ochraceous-buff sides of neck and face by a bold black band from lores to nape. Upperparts olive-brown with broad black scaling on back. Wings and tail slightly browner than back and rump, with tips of wing-coverts and tertials bearing small whitish to buffy-white spots. Underparts whitish, buffier on throat and vent, and more ochraceous-buff on flanks. From lower throat, underparts prominently marked with wavy black bars. *Adult female* Similar to male, but crown duller, ear-coverts and neck-sides buffy-rufous. Black supercilium less pronounced and bears small white spots. Below, more creamy-whitish with weaker barring. Wing-covert spotting buffier, less white, and not as pronounced as in male. *Immature* Not well described, and the only published description possibly refers to a juvenile (as defined here) female *rosenbergi* (Chapman 1917). Chapman (1917) stated that the belly is less fulvous and has less of an ochraceous tinge than adult females, and the supercilium is barely evident. This part of the head is, instead, similar to the crown, which is dull chestnut bordered blackish. The tips of the upperwing-coverts are ochraceous, similar in colour to the throat. In this specimen, a few soft, downy, blackish feathers of an earlier plumage stage remain on the flanks. The whitish feathers at the sides of the abdomen are very faintly barred. Although I undoubtedly examined the same specimen on which Chapman's description was based, my analysis suggests that we disagree on at least some of the characters that indicate immaturity (see below). *Fledgling/Juvenile* The following description is based on an immature male (MECN 2591), which is clearly the youngest examined specimen examined for either species of *Pittasoma*, and I would consider a fledgling

transitioning to juvenile. Collected in 1968, it was clearly unavailable to Chapman (1917), but its appearance suggests that it may be similar in age to the bird described by him, perhaps slightly younger. This is the only specimen I examined that lacked an obvious supercilium, the feathers in this region being still in pins, having only emerged (asymmetrically) to begin forming a black line on one side of the hindcrown. The bill is noticeably smaller than that of adults, but there are no clear signs of an inflated gape. The crown and nape are uniform deep chestnut, similar to that of adults, the face and neck-sides are paler, rufescent orange, the throat paler still, slightly flecked blackish and mottled whitish. Below, the plumage is a mix of fluffy, down-like, brown feathers, and strongly barred buffy-white feathers clearly similar to an adult male. The vent and flanks are nearly entirely covered in fluffy brown feathers, somewhat paler, brownish-buff in the central belly. The tips of the upperwing-coverts, as well as the feathers of most of the upperparts, are tipped rufescent orange, similar in colour to the face. The feathers of the back are, in fact, rufescent over at least the distal half, broken only by a broad black bar. These feathers create an overall thickly spotted appearance that, in life, may have been similar to the immature plumages of many *Turdus* thrushes, but with the addition of an orange-chestnut 'hood'. Apart from the adult-like barring over some of the underparts, a difference to be expected based on subspecies, the obviously spotted upperparts is the most striking difference from the description by Chapman (1917) of immature *rosenbergi*, and suggests a possibly important difference between immature plumages of the two subspecies. Indeed, if the fledgling plumage of *harterti* proves to lack spotting, this would appear to be a divergence unique within the entire family. *Subadult* Based on the rufescent-spotted wing-coverts of the above-described juvenile, I define subadults as being those individuals with retained immature coverts, but otherwise adult-like. It appears that, as in Black-crowned Antpitta, the lesser wing-coverts are all replaced before the greater coverts. Some females, however, have the pale speckling in the superciliary stripe buff or rich buffy-orange. Although I found no females with the 'typical' white-speckled supercilium *and* immature wing-coverts, several had buff-speckled superciliary stripes but apparently fully adult wing-coverts. It appears that in females buff or orange-buff speckling may be the final vestige of immaturity to be lost. As it is possible that this character is simply variable in adult females, I have indicated the method (coverts or supercilium) by which I have aged immature females in the Breeding Data section.

MORPHOMETRIC DATA *P. r. rosenbergi* Wing 97mm, *Tail* 34mm; *Bill* [exposed culmen] 24.5mm; *Tarsus* 45mm ($n = 1$, ♂, holotype, Hellmayr 1911b). *Wing* 93mm ($n = 1$, ♀, Hellmayr 1911b); *Bill* from nares 17.0mm, 16.6mm, 16.0mm; *Tarsus* 46.8mm, 48.2mm, 46.8mm ($n = 3$, ♀,♂,♂, USNM). *P. r. harterti* Data from USNM, LACM, AMNH, FMNH, WFVZ. Adult ♀♀: *Wing* 92.5mm, 92.0mm ($n = 2$); *Tail* 29.5mm, 32.0mm ($n = 2$); *Bill* from nares 16.9 ± 1.0mm ($n = 4$), exposed culmen 26.2 ± 2.6mm ($n = 3$), depth at nares 8.5mm ($n = 1$), width at nares 8.4mm ($n = 1$), width at base of mandible 10.9mm ($n = 1$); *Tarsus* 46.6 ± 1.2mm ($n = 6$). Adult ♂♂: *Wing* 89.0mm ($n = 1$); *Tail* 30mm ($n = 1$); *Bill* from nares 17.3 ± 0.4mm ($n = 7$), exposed culmen 28.0mm, 26.0mm ($n = 2$), depth at nares 7.6mm ($n = 1$), width at nares 7.5mm ($n = 1$), width at base of mandible

10.3mm (*n* = 1); *Tarsus* 47.4 ± 1.4mm (*n* = 8). See also Chapman (1917), Olivares (1958). **P. r. rufopileatum** *Wing* 101mm; *Tail c.*30mm; *Bill* exposed culmen 28mm; *Tarsus* 47–50mm (*n* = 1♂ [holotype], except tarsus measurements which include an adult ♀ and an immature, type series; Hartert 1901). *Wing* 95mm, 97mm (*n* = 2, ♀♀, type series, Hartert 1901).*Wing* 96–99mm (♂♂), 93mm, 93mm (♀♀); *Tail* 30–34mm; *Bill* 25.0–26.5mm (*n* = 10, 8♂♂, 2♀♀, Cory & Hellmayr 1924). **Mass** 96g, 97g (*n* = 2, ♂,♀, Krabbe & Schulenberg 2003). **Total Length** 15.0–17.8cm (Meyer de Schauensee 1964, 1970, Dunning 1982, 1987, Hilty & Brown 1986, Ridgely & Greenfield 2001, Krabbe & Schulenberg 2003, Restall *et al.* 2006, McMullan *et al.* 2010).

TAXONOMY AND VARIATION Each of the three subspecies recognised by most authors (e.g. Krabbe & Schulenberg 2003, Greeney 2013j) was originally described as a species (Hartert 1901, Hellmayr 1911b, Chapman 1926), but subsequently considered to be subspecifically related (Cory & Hellmayr 1924, Peters 1951). As there is considerable variability in adult plumage, even within subspecies, Krabbe & Schulenberg (2003) suggested that *harterti* is a synonym of *rufopileatum*. Some of this variability, however, may reflect incompletely described immature plumages, and until true variation can be separated from ontogenetic changes, I treat three subspecies here. Admittedly, after examining a fair number of specimens I tend to agree with Krabbe & Schulenberg (2003) in questioning the validity of *harterti*, especially given the additional observation that there is no clear separation of their ranges (see discussion under *harterti*).

Pittasoma rufopileatum rosenbergi Hellmayr, 1911, Revue Française d'Ornithologie, vol. 2, p. 51, Sipí, Río Sipí, 150ft, Chocó, western Colombia [= 04°39'N, 76°36'W, 46m]. The subspecies was named in honour of W.F.H. Rosenberg of London, who organised the M. Palmer expedition on which the adult male holotype (ZSM 09.5759) was collected by Mervyn G. Palmer on 25 September 1908. Subspecies *rosenbergi* is endemic to Colombia, with most localities in Dept. Chocó, but records range as far south as Valle del Cauca. Very recently, photographic evidence has documented its presence as far north as north-east Antioquia (ML 54768841/58156841; photos D.O. Montoya & W.B. Castrillón). The following details are from Hellmayr (1911b). Adult males have the crown and nape deep cinnamon-rufous, the back dull olive, becoming more brownish towards the rump. Some of the interscapular feathers have white bases. Each feather of the back is broadly fringed black on either side, giving the upperparts a striped appearance. The secondary wing-coverts are sepia-brown, washed russet, especially on the outer webs, each feather with a small, but distinct, whitish apical spot, and a distinct black margin at the tip. The greater primary-coverts are uniform blackish-brown. The flight feathers are blackish with dull, russet-brown outer webs. The rectrices are mostly dusky, the central ones washed russet. The lores and broad superciliary stripe, which extends onto the neck-sides, are uniform deep black. The cheeks, sub-ocular and malar region, ear-coverts, chin and throat are ochraceous, somewhat darker on the head-sides. The foreneck, breast-sides and flanks are dull olive-brown, while the central breast and abdomen are uniform buffy-white. This buffy coloration is separated from the ochraceous throat by a dull brown crescent.

The undertail-coverts are 'hair-brown' and thinly edged whitish. The underwing-coverts are dusky-brown, with those of the primaries tipped white. **Bare Parts** *Iris* dark brown; *Bill* black; *Tarsi & Toes* black. It is overall similar to the nominate race but generally duller. Adult female *rosenbergi* differs from male *rosenbergi* most notably in having the broad, black supercilium finely striped with white (Chapman 1917). Both sexes of *rosenbergi* are significantly smaller than the nominate subspecies, and they lack markings on the buffy underparts. Most notably, both sexes have a mostly rufous head (rather than just a rufous crown like nominate). Female *rosenbergi* is additionally separated from females of the nominate subspecies by having a browner back and bright buff (rather than white) apical spots on the wing-coverts. **Specimens & Records** *Antioquia* Finca La Alborada, 07°50.5'N 76°28.5'W (ML 54768841, photo D.O. Montoya); Villa Arteaga, 07°20'N, 76°26'W (USNM 426436); Reserva La Bonga, 07°11'N, 76°22.5'W (ML 59829191, photo E. Munera). *Chocó* Vereda Avejeros, El Carmen de Atrato, *c.*05°44'N, 76°19.5'W (Flickr: J.A. Muñoz); Andagoya, 05°06'N, 76°41'W (CM P66291/370); near Bahía Solano, 06°11'N, 77°22'W (B. G. Freeman *in litt.* 2017); Río Jurubidá, *c.*05°52'N, 77°11'W (Meyer de Schauensee 1950b); Noanamá, 04°42'N, 76°56'W (AMNH); Nóvita, 04°57'N, 76°34'W (AMNH 111961/62); Río Nuquí, 05°38'N, 77°14'W (USNM 443344/45); 'Serranía de Baudó', near Pizarro (AMNH 123350; loc. see Paynter 1997). *Valle del Cauca* Córdoba, 03°53'N, 76°56'W (CM P66769).

Pittasoma rufopileatum harterti Chapman, 1917, Bulletin of the American Museum of Natural History, vol. 36, p. 392, Barbacoas, Nariño, Colombia [35m, 01°41'N, 78°09'W; Paynter 1997]. The holotype, an adult male (AMNH 117876), was collected on 25 August 1912, by William B. Richardson (LeCroy & Sloss 2000). Named in honour of Ernst Hartert, the subspecies is endemic to Colombia, being known only from the Pacific slope of the West Andes, in Cauca and Nariño (Ayerbe-Quiñones *et al.* 2008, Calderón-Leytón *et al.* 2011). The southernmost records traditionally treated as *harterti* are separated from the type locality of *rufopileatum* by scarcely 50km, with no obvious geographic barrier separating them. Until the issue is investigated further, I have provisionally extended the range of *harterti* to the right (north) bank of the Río Mira. Male (from Chapman 1917) has crown and nape bright rufous-chestnut, only slightly paler at the sides. The lores and broad supercilium, which extends to the nape, are black. The rest of the upperparts are generally light brown-olive (slightly browner than *rosenbergi*), with each feather widely fringed black. The rump is browner and unstriped, and the feathers of the rump are 'much elongated and 'fluffy'.' The tail is 'raw umber'. The flight feathers are black and have brownish outer margins, which are increasingly broad inwardly until the entire outer web of the inner secondaries and both webs of the tertials are brownish. The tertials have a rounded, buffy terminal shaft-spot, and slightly blackish edges. Greater primary-coverts are uniform blackish and unmarked. The remaining wing-coverts are a similar brown to the exposed surface of the wing, but bear conspicuous buffy-white terminal spots that cover most of the feather tip, except for a narrow black margin. The underwing-coverts are blackish, some washed rust-brown, with those at the base of the outer primaries broadly tipped white, forming a conspicuous patch. The throat and head-sides are deep, clear orange-rufous (somewhat richer than *rosenbergi*). The remaining underparts are the same colour as the throat, but duller, especially on the sides and flanks. The thighs are brownish-olive while the vent and undertail-coverts are buffy-brown. The underparts, from the posterior margin of the throat to, and including, the upper part of the tibiae and undertail-coverts, but excluding the thighs and flanks, are evenly barred black. The extent of the barring varies individually, but the vent and undertail-coverts seem to be almost always barred. **Bare Parts** (from skins) *Bill*, maxilla black, mandible black but sometimes with a paler tip; *Tarsi and Toes* brownish-black. Female (from Chapman 1917) resembles the male but the lores are blackish with a whitish supraloral stripe. The supercilium is speckled white. The underwing-coverts and the white patch at the base of the primaries are tinged rufous. The throat is similar to male's, and the rest of the underparts show only a hint of pale, irregularly broken barring. Overall, this subspecies is somewhat intermediate between *rosenbergi* and nominate *rufopileatum*, with the black markings of the male varying between spotting and barring. The head, however, including the face and nape, is entirely bright rufous. Back and rump are slightly browner than *rosenbergi*, but a similar brownish-olive to the upperparts of *rufopileatum*. **Specimens & Records** *Cauca* Guapí, 02°34'N, 77°53'W (Olivares 1958). *Nariño* Guayacana, 01°26'N, 78°27'W (LACM, ANSP, FMNH, LSUMZ, WFVZ, USNM); Barbacoas, 01°41'N, 78°09'W (AMNH).

Pittasoma rufopileatum rufopileatum Hartert, 1901, Novitates Zoologicae, vol. 8, p. 370, near Pulún, 01°05'N, 78°40'W, Esmeraldas, *c.*75m. This taxon was named for its distinctive rufous crown, which separated it from both races of *P. michleri*, the only other *Pittasoma* known at the time of its description. The nominate race, so far as is known, is endemic to Ecuador and is currently known only from the provinces of Pichincha, Esmeraldas and Imbabura, apparently having been extirpated from the only known site in Los Ríos (Leck 1979, Leck *et al.* 1980, Pearson *et al.* 2010). Nevertheless, as discussed under *harterti*, I have provisionally depicted its range as extending north to the left (south) bank of the Río Mira. It is the only subspecies to have rufous head markings restricted to the crown and differs from *rosenbergi* in having the underparts barred, especially in males. Female *harterti* also usually lacks ventral barring, and male *harterti*, although barred, tends to be buffier rather than white below. Soon after its description, there was some confusion over plumage differences between males and females, in part due to an imprecise understanding of immature plumages and adult variation. The following characters were included in Hartert's (1901) description of the male holotype (see Technical Description). In adult males, the crown is bright rufous, edged paler rusty-rufous. From the base of the bill, through the eye to the neck-sides, is a broad black line. The head-sides are pale rufous, and the throat feathers rusty-rufous with white bases, partially black shafts and tiny black spots near the tip and on the lateral margins; together these characters produce a finely speckled appearance. The underparts are finely barred black and white, becoming more olive- or grey-brown on the sides and chest, with the flanks uniform brownish-olive. The upperparts are pale olive, the inter-scapular feathers are widely fringed black. The flight feathers are blackish-brown, exteriorly fringed rufous olive-brown. The upperwing-coverts have buffy-white tips and narrow

black terminal margins. Underwing-coverts are deep brown, with two white patches. Thighs are olive, with pale buff shaft-streaks, and sometimes slightly barred in front. **Bare Parts** *Iris* brown; *Bill* blackish-slate; *Tarsi and Toes* dark bluish-slate. Hartert (1901) described a second individual of which he was unsure of the age or the gender. His description: 'Upperside like the adult male, the superciliary black stripe spotted with white along the shafts of the feathers. Under-surface bright rusty-buff, olive on the flanks and sides, sparsely spotted with black, uniform on throat and along the middle of the abdomen.' He examined two other skins that he felt had intermediate plumages between the two birds described, concluding that those with rusty underparts were immatures. He was apparently confused by the labels of these specimens, stating that 'the sexes noted on the labels would indicate that the sexes are alike, but I do not know if they are reliable.' Shortly afterwards, Hartert (1902) described young as resembling adult females. Again, however, he was still unsure, qualifying the previous statement with 'if the specimens are correctly sexed, but there is some discrepancy in the sex-notes on the labels.' Indeed, Hellmayr (1911b) pointed out that the 'juvenile' figured in Hartert (1902) was, in fact, the female. **Specimens & Records** *Esmeraldas* Río Durango, 01°05'N, 78°42'W (ANSP 62016); Río Bogotá, 01°03'N, 78°50'W (AMNH 492128); Guembí, 00°57'N, 78°47'W (MECN 2590–592); 20km by road NNW of Alto Tambo, 00°55'N, 78°34'W (ANSP 182526); *c.*2.5km SSE of Alto Tambo, 00°53'N, 78°32.5'W (XC 112529, D.M. Brinkhuizen); *c.*7km E of Playa del Oro, 00°52'N, 78°44'W (XC 262128–132, O. Jahn); El Placer, 00°51'N, 78°34'W (ANSP, USNM MECN); Río Negro Chico, 00°50'N, 78°33'W (Jahn & Valenzuela 2006); Ónzole, 00°42'N, 79°08'W (J. Freile *in litt.* 2017); Estero Pindiupi, 00°40.5'N, 78°52.5'W (Jahn & Valenzuela 2006); RP Río Canandé, 00°31.5'N, 79°13'W (HFG). *Pichincha* RP Mashpi Shungo, 00°11'N, 78°54.5'W (XC 342241, N. Krabbe); RP Mangaloma, 00°07.5'N, 78°59.5''W (XC 53754–756, D.M. Brinkhuizen). *Los Ríos* RP Estación Científica Río Palenque [possibly extirpated], 00°35'S, 79°23'W (Pearson *et al.* 2010, LSUMZ, ANSP).

STATUS As a range-restricted endemic of the Chocó EBA (Stotz *et al.* 1996, Stattersfield *et al.* 1998), inhabiting areas facing continued and increasing threats of deforestation, Rufous-crowned Antpitta is suspected to be experiencing a fairly rapid population decline, and is afforded Near Threatened status (Stattersfield & Capper 2000, BirdLife International 2015). Habitat within the species' range has already experienced considerable fragmentation due to the construction of roads for colonisation, agriculture and logging (Salaman 1994, Stattersfield *et al.* 1998, Sierra *et al.* 1999, Conservation International 2001, Jahn 2011). The nominate subspecies, apparently endemic to Ecuador, is considered Vulnerable there (Jahn & Valenzuela 2002), and several authors (Ridgely & Greenfield 2001, Krabbe & Schulenberg 2003) have suggested that the species may be more threatened globally than is currently recognised. Nonetheless, no specific threats to the species have been identified and more information on specific habitat requirements, home-range size and movements is sorely needed for a proper assessment. Historically, the nominate race occurred at the Río Palenque Reserve in western Ecuador but, ominously, has not been observed there since 1975 (Leck 1979). Subspecies *harterti* is apparently not known from any protected areas. **Protected Populations**

P. r. rosenbergi PNN Utría (Meyer de Schauensee 1950b); PRN Santa Emilia (eBird: A. Sánchez). *P. r. harterti* RNA El Pangan (Salaman *et al.* 2007a). *P. r. rufopileatum* RE Cotacachi-Cayapas (Jahn & Valenzuela 2002, 2006); RE Cayapas-Mataje (Jahn & Valenzuela 2002); RP Mangaloma (XC 53754–756, D.M. Brinkhuizen); RP Mashpi Shungo (XC 342241, N. Krabbe); RP Río Canandé (HFG); BP Mashpi, within the Área de Conservación Mashpi-Guaycuyacu-Sahuangal (J. Freile *in litt.* 2017).

OTHER NAMES *Pittasoma rosenbergi*, *Pittasoma harterti* (Chapman 1917). **English** Rufous-crowned Gnatpitta (McMullan *et al.* 2010); Rufous-crowned Ant-Thrush [nominate], Rosenberg's Ant-Thrush [*rosenbergi*] (Brabourne & Chubb 1912); Rosenberg's Antpitta [*rosenbergi*] Cory & Hellmayr 1924; Rosenberg's Ant-pitta [*rosenbergi*] (Meyer de Schauensee 1950b); Hartert's Antpitta [*harterti*] (Cory & Hellmayr 1924); Hartert's Ant-pitta [*harterti*] (Meyer de Schauensee 1950b). **Spanish** Tororoí Capirrufo (Krabbe & Schulenberg 2003); Tororoí Cejinegro (Salaman *et al.* 2007a); *Pitasoma coronirrufa* (Ridgely & Greenfield 2001, Granizo *et al.* 2002, Granizo 2009); *Pittasoma coronirrufa* (Ortiz-Crespo *et al.* 1990); *Gralaria coronirrufa* (Valarezo-Delgado 1984). **French** Grallaire à sourcils noirs. **German** Rostscheitel-Ameisenpitta (Krabbe & Schulenberg 2003).

Rufous-crowned Antpitta, adult male (*rufopileatum*), Playa del Oro, Esmeraldas, Ecuador, 4 August 2011 (*Dušan M. Brinkhuizen*).

Rufous-crowned Antpitta, adult male (*rufopileatum*), Playa del Oro, Esmeraldas, Ecuador, 2 August 2011 (*Dušan M. Brinkhuizen*).

Rufous-crowned Antpitta, adult male (*rufopileatum*), Playa del Oro, Esmeraldas, Ecuador, 24 February 2012 (*Luke Seitz*).

Rufous-crowned Antpitta, adult male (*rosenbergi*), Carmen de Atrato, Choco, Colombia, 8 February 2017 (*Anderson Muñoz*).

Genus *Grallaria*: large montane antpittas

Grallaria, Vieillot, 1816, Analyse d'une nouvelle ornithologie élémentaire, p. 43. Type, by original designation: *Grallaria varia* (Boddaert, 1783) based on 'Roi des Fourmilliers' of Buffon (1771–86).

'The bird inhabits the floor of the thickets in the low temperate zone, where it is never an obtrusive resident. Of those I found, one seemed to materialize out of the shadows as my eyes became accustomed to the gloom on entering a shadowed ravine, and was quietly disappearing behind an adjacent tree when I stopped it with a shot. Another was in a sheltered pathway as I rounded a bend and, rail-like, made instant use of its long legs to carry it into concealment among the grasses.' – **J.T. Zimmer, 1930, from** *Birds of the Marshall Field Expedition* **(Undulated Antpitta)**

Remarks on the immature plumages of *Grallaria* While researching this book I examined more than 250 photographs or specimens of various *Grallaria* species that I judged to be in immature plumage. Although I still have not seen examples of all species, some generalities appear to apply to most or all *Grallaria* and there are a few emerging patterns of similarities among species that bear mention. *Grallaria* fledge while still covered in dense, fluffy nestling plumage. For those species where I have observed fledging, the young are incapable of flight at this age and do not even appear capable of sustained or coordinated ambulatory movement. It is my guess that most species leave the nest and move immediately to a sheltered location on or very near the ground, remaining there for at least a few days while they are tended by their parents. With respect to ground colour, fledgling plumage varies from pale (buff to yellow-buff) to dark (deep chocolate-brown to black). In terms of patterns, fledglings, at least while still very young, may be nearly unmarked (e.g. *G. nuchalis ruficeps*, *G. albigula*, *G. hypoleuca* and at least some members of the *G. rufula* complex), bear fine spotting or streaking on the upperparts and breast (e.g. *G. guatimalensis*, *G. alleni*, *G. haplonota*, *G. varia*, *G. capitalis*, *G. blakei*, *G. andicolus*) or have barred upperparts (e.g. *G. squamigera*, *G. gigantea*, *G. ruficapilla*, *G. quitensis*, *G. bangsi*, *G. n. nuchalis*, *G. carrikeri*, *G. ridgelyi*, *G. kaestneri*, *G. przewalskii*, *G. urraoensis*, *G. milleri*, *G. erythroleuca*). So far as is known, fledglings of all species have obviously inflated rictal flanges that vary from white to yellow or orange, but are most frequently predominantly or entirely bright orange. Their bills are usually notably shorter than those of adults and almost always show a good deal of orange, especially on the tomia and basally. At least some species may leave the nest with their egg-tooth still attached! Immatures of all species may be recognised by their patterned upperwing-coverts. Like the pattern of fledgling plumage, immature coverts appear to fall into at least three broad categories. The commonest marking appears to be fine barring across the feather tip, from one to four thin bars, sometimes alternating between pale (buff or cinnamon) and dark (blackish). Examples of species with barred immature upperwing-coverts include: *G. squamigera*, *G. gigantea*, *G. ruficapilla*, *G. bangsi*, *G. watkinsi*, *G. capitalis*, *G. carrikeri*, *G. ridgelyi*, *G. hypoleuca*, *G. przewalskii*, *G. blakei* and *G. erythroleuca*. Several species, including *G. guatimalensis*, *G. alleni*, *G. haplonota* and *G. varia* have distinctly spotted coverts, each bearing a single pale, diamond-shaped or roughly triangular marking at the tip. In this regard, *G. andicolus* stands out in having a broken bar with a small central shaft-spot. Finally, there appears to be a group with immature coverts that bear only a thin, contrasting margin at their tips (pale or dark). Those species which apparently have such weakly marked coverts (e.g., *G. urraoensis* and at least some members of the *G. rufula* complex) are not entirely clear, and further evidence may uncover that at least some of the species I have described in this way are best considered of the 'fine-barred' type described above and are either weakly pigmented versions or, alternatively, are species where the only examples I examined were of individuals with coverts too worn or faded for me to accurately access their original markings. Though not always obvious from photos or in the field, it appears that the tertials and inner secondaries of immatures have variably patterned tips. The transition to adult-like contour feathers appears to almost always occur in an irregular, patchy manner, such that often an individual may appear to be in slightly different plumage stages, depending on the angle from which it is viewed! For those species that I have examined the largest numbers of immatures (*G. guatimalensis*, *G. quitensis*, *G. ruficapilla*), it was not uncommon to find individuals with immature plumage on one side of their nape and not the other, or to have only one half of their breast still in fluffy immature plumage, with nearly fully adult-like plumage on the opposite side. Nevertheless, on the whole, the last tracts from which *Grallaria* lose fully immature plumage are the central breast, rump and nape. This latter region appears to usually be the last to retain signs of immaturity, except in *G. squamigera* and *G. gigantea* (and probably *G. excelsa*). In these species it appears that older immatures retain an (often asymmetrical) scattering of barred immature feathers on the back. Other than these examples, the last indications of immature plumage to be lost appear to be the upperwing-coverts and, as with other genera treated here, it seems most likely that these are replaced with a new set of immature coverts at least once after fledging. For the purposes of the following accounts I define three principal categories of immatures as follows. FLEDGLING Mostly still covered in plumose, fluffy, nestling-like feathers, rictal flanges still obviously inflated and contrastingly coloured, still heavily dependent on adults for food and largely remaining stationary while awaiting prey delivery. JUVENILE Little or no plumose feathering, though the loss of their 'fluffiness' may be largely due to the loss of fine filaments from the feather tips, the rictal flanges are still brightly coloured and may still be noticeably inflated, the bill overall still smaller than in adults and probably still washed orange at least on the base of the mandible and often at the tip of the maxilla, probably fairly mobile, moving with adults and actively soliciting food, but may attempt increasing amounts of foraging for themselves. **SUBADULT** Very close to adults in general coloration, usually only separable by their immature wing-coverts, perhaps slightly different bill coloration near gape, and may not be readily distinguishable from adults in the field. Overall, as with *Conopophaga* in particular, plumage ontogeny is fairly gradual, rather than step-wise, and I use the terms '**fledgling transitioning**' and '**juvenile transitioning**' for birds that are somewhere between the above-defined stages. The term '**subadult transitioning**' is used for individuals that are essentially fully adult with respect to plumage but retain heavily worn and faded (but still not adult-patterned) wing-coverts. In some cases, subadults in transition may retain an (often asymmetrical) scattering of little-worn immature coverts.

UNDULATED ANTPITTA
Grallaria squamigera   Plate 6

Grallaria squammigera Prévost & Des Murs, 1846, *Voyage autour du monde sur la frégate la Vénus commandée par Abel de Petit-Thouars. Oiseaux*, p. 198, pl. 3. No locality was given by the describers and the type locality was subsequently given as 'Santa Fé de Bogotá' by some authors (Lafresnaye 1842, Brabourne & Chubb 1912) until Hellmayr (Cory & Hellmayr 1924) used published records of the collectors' travels to approximate the locality as Bogotá, setting this as the official type locality (Peters 1951). There is, however, a great deal of confusion concerning the year of publication and the correct authorities for the description, in all likelihood due to the protracted period over which the above work was published. Many authors simply omitted the year when referencing *G. squamigera* (Sclater 1855c, 1858b, c, 1877, 1890, Gray 1869, Sharpe 1901), while others gave the year of description as 1849 (Brabourne & Chubb 1912, Chapman 1917, 1926) or 1842 (Meyer de Schauensee 1966). Peters (1951) gave the description date as '1846 (1842)' while Cory & Hellmayr (1924) appeared to give two dates, '1846 = 1842' and as '1855 = 1849.' The former apparently refers to the publication of the plate illustrating *G. squamigera*, while the latter refers to the descriptive text.

'*Grallaria squamigera* was to me the most interesting species. It is a huge, heavy-bodied bird, olive above and tawny barred with black below. From a distance the coloration reminds one of a large immature robin, but the tail is very short and protrudes only about half an inch beyond the lower coverts, and the long legs measure fully five inches. The plumage is long and full. Occasionally we saw the shy creatures as we worked in front of our tent in the afternoons; we always made it a point to be very quiet and the reward came in the way of shadowy forms that unconcernedly pursued their lives among the logs and brush without suspecting our presence. This shows the advantage of camping in the midst of the wilderness, where one is sure to see and hear wild things at the most unexpected times—experiences that are lost if one does not spend his entire time in the very heart of their environs.' – **Alden H. Miller, 1918,** *In the Wilds of South America*

Over the years, Undulated Antpitta has been called 'enormous', 'huge' and other similar descriptors, by a variety of authors. It is, indeed, rivaled in size only by its two putative relatives, Giant Antpitta and Great Antpitta. Although it is not especially scarce in parts of its range, it is definitely one of the trickiest antpittas to see. It occurs from western Venezuela, through the Andes of Colombia and Ecuador, and along the Amazonian slope and central Andes of Peru, as far south as Cochabamba and La Paz in Bolivia. Like other large *Grallaria*, Undulated Antpitta seems largely terrestrial and rather tied to closed-canopy habitats. Occasionally, especially when occurring near

the treeline, it ventures into the open, apparently more frequently on overcast days and in twilight. Two subspecies are generally recognised, but these are questionably separable on the basis of plumage, at least as currently defined. See Taxonomy and Variation for my thoughts on this. On the whole, behaviour and ecology remain poorly understood, its diet is virtually unknown, and the species' nest was only recently described. Although there are no specifically documented effects of human activity on the species, and BirdLife International (2017) currently considers it of Least Concern, I found it amusing that several authors have remarked that the species may be hunted in parts of its range. Indeed, Fraser (in Sclater 1858b) stated that, in eastern Ecuador, 'the flesh is much prized for eating', while Olivares (1969) was told that the taste of Undulated Antpittas is 'exquisite'. Undoubtedly, however, the threat posed by hunters is far overshadowed by continuing anthropogenic habitat alteration.

IDENTIFICATION 21–23cm. A very large, unmistakable antpitta in most parts of its range. The upperparts are uniform grey or olive-grey, with the crown and nape greyer, especially in the north of its range. The throat, lores and area immediately behind the eye are buffy-white to white, contrasting with the black malar streak. Sides of head and neck and most underparts ochraceous-buff with thick, wavy black barring, sometimes broken into a more scalloped-like pattern. Below, only the central lower belly and vent are unmarked, the former pale and the latter bright ochraceous. Both Giant Antpitta and Great Antpitta are, in fact, larger than Undulated Antpitta, and both have heavier bills. Both of these species, although superficially similar in plumage, lack a distinct malar stripe, which is probably the best field character. Giant Antpitta, which may be sympatric in some areas, has the ground colour of the underparts much more cinnamon-buff or orange-rufescent (not yellowish), and has no white on the throat. Great Antpitta differs by having a brown back in sharp contrast to the grey crown and nape, and lacks a moustachial streak.

DISTRIBUTION Found from the Venezuelan Andes, south through all three ranges of the Colombian Andes, both Andean slopes in Ecuador, and along the Amazonian slope and central Andes in Peru. It reaches its southern limit in Bolivia, in Cochabamba and La Paz (Meyer de Schauensee & Phelps 1978, Butler 1979, Parker *et al.* 1982, Hilty & Brown 1986, Remsen & Traylor 1989, Ridgely & Greenfield 2001, Hennessey *et al.* 2003b, Hilty 2003, McMullan *et al.* 2010, 2013, Schulenberg *et al.* 2010).

MOVEMENTS No movements documented, but the possibility that some populations engage in short-distance altitudinal movements (like Giant Antpitta, which see) should be explored. In addition, a subadult was collected in Parque La Carolina, in downtown Quito (MECN 7434; photos M.V. Sánchez N.). This suggests that Undulated

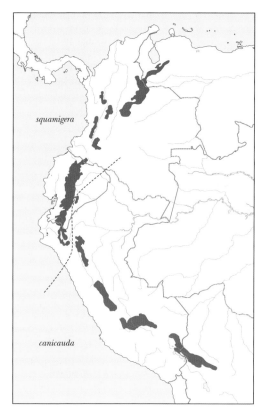

squamigera

canicauda

Antpitta may show long-distance post-fledging dispersal as is suspected for Scaled Antpitta (which see).

HABITAT Inhabits humid, wet, epiphyte-laden montane forest, mossy *Polylepis* woodland and high-elevation elfin forest (Hilty & Brown 1986, Fjeldså & Krabbe 1989, Arbeláez-Cortés *et al.* 2011a). In these habitats it is apparently often found in areas of second growth (both natural and man-influenced) such as thickets of *Chusquea* bamboo or *Neurolepis* cane (Fjeldså & Krabbe 1990, Krabbe & Schulenberg 2003) as well as the nearly monotypic stands of alder (*Alnus*: Betulaceae) (Kattan & Beltrán 1999) that frequently appear in Andean riparian zones following flooding. It also uses disturbed semi-humid shrubbery and second growth, especially in western Ecuador and the inter-Andean valley, as illustrated by a nest built very close to an exotic pine plantation, even including needles in its structure (Greeney & Juiña 2011). At dawn and dusk, and during foggy weather, Undulated Antpitta may venture into pastures and other open areas adjacent to forest, but it is rarely (never?) seen away from cover in strong sunshine. It rarely leaves the ground except occasionally when vocalising or if flushed by an observer, when the species frequently perches immobile on a low branch (Meyer de Schauensee & Phelps 1978). Across its entire range, I have found altitudinal records ranging from 1,700m to 3,800m (Hilty & Brown 1986, Fjeldså & Krabbe 1990, Clements & Shany 2001), with the majority of records from the band 2,600–3,800m (Meyer de Schauensee 1950b, Giner & Bosque 1998, Hennessey & Gomez 2003, Hennessey *et al.* 2003b, Hilty 2003, Walker & Fjeldså 2005, Ridgely & Tudor 2009). At any given locality,

however, this seems to vary, perhaps due to competition with other *Grallaria* or ground-foraging species (Weske 1972). In the Cordillera Cocapata, Cochabamba, Bolivia, Scaled Antpitta (ssp. *sororia*) ranges unusually high, up to 2,800m, at which point it is replaced by Undulated Antpitta (Herzog *et al.* 1999). Andrade *et al.* (1994) reported a range of 2,000–3,000m in the East Andes of Cundinamarca, while it apparently reaches 3,300m at Páramo de Frontino (Arbeláez-Cortés *et al.* 2011b) and 3500m at some sites in Tolima (Parra-Hernández *et al.* 2007) in the Central Andes of Colombia. In some areas of Risaralda and Antioquia, Undulated Antpitta tends to range somewhat lower (1,800–2,550m; Naranjo 1994, Kattan & Beltrán 1999).

VOICE Walker & Fjeldså (2005) described the song as a series of quavering notes, 'quite haunting in quality', the series lasting 4–5s and repeated every 20s or so: *hohohohohoho-ha-ha-ho-ho*. The specifics of the song were provided by Krabbe & Schulenberg (2003). The full song lasts 4–6s and is repeated every 4–15s. Their description was of a rolling series of notes delivered at 0.6–0.8kHz and evenly paced, 14–16 notes/s. The song gradually increases in volume and pitch, often with a slight jump and fall in pitch at the very end. Krabbe & Schulenberg (2003) also described a call, of unknown function and rarely heard, as a hollow *rrhooh-rrhooh-rrhooh*.

NATURAL HISTORY Like most *Grallaria*, Undulated Antpitta is generally encountered alone and spends most time on the ground (Zimmer 1930, Krabbe & Schulenberg 2003). The first-hand accounts of many authors and my own observations suggest that, even among 'ground-antbirds', Undulated Antpitta is particularly terrestrial. Peters & Griswold (1943), in fact, failed to detect the species via numerous hours of mist-netting effort, capturing their only specimen in a steel trap set to capture tinamous. It bounds or runs along the ground or on low, fallen logs, pausing frequently, often with a tilt of the head while searching for prey. Upon spotting a prey item, it lunges forward to swiftly snatch it up with its heavy bill. Apparently, often after swallowing food, the species fluffs out its plumage and simultaneously flicks both wings, bobbing up and down a few times (Krabbe & Schulenberg 2003). More so than any other antpitta I have researched for this book, many authors have noted that Undulated Antpitta may be found along trails (Zimmer 1930, Olivares 1969). Given that, at least before the advent of worm-feeding stations, this species was no more frequently observed than other montane *Grallaria*, its apparent preference for foraging along trails *may* indicate that it is more dependent on foraging in association with large mammals than other montane antpittas (Greeney 2012a). Perhaps, however, this is simply an indication of the readiness with which Undulated Antpitta will occupy marginal or edge habitats. There are no published data on territorial defence, maintenance or fidelity, but in Peru (*canicauda*) one study estimated population densities of 13 pairs/100ha, with a mean (± SD) territory size of 3.2 ± 0.56ha (Kikuchi 2009). Kattan & Beltrán (2002) approximated territory sizes for two pairs of Undulated Antpittas as 2ha and 4.5ha, observing that one banded bird (sex not specified) held its territory over two years. In the same study, the territory of a radio-tracked Bicoloured Antpitta had 90% overlap with a territory of Undulated Antpitta, while a territory of Brown-banded Antpitta overlapped around 50% with the same Undulated Antpitta territory (but the Brown-banded Antpitta and Bicoloured Antpitta territories did not overlap). Singing, perhaps especially for territorial purposes, is from an elevated but concealed perch, up to 2–3m above ground (Asociación Bogotana de Ornitología 2000).

DIET Undulated Antpitta is among the numerous species that has been 'trained' to visit worm-feeding stations (Woods *et al.* 2011), and earthworms are thought to also form a large portion of the 'natural' diet (Poulsen 1993, Ridgely & Tudor 2009). Other items reported are large arthropods (Schulenberg & Williams 1982) and lizards (Asociación Bogotana de Ornitología 2000). In Colombia, the stomach of one adult included Coleoptera, Dermaptera, Orthoptera, Opiliones and Phasmida (Fierro-Calderón *et al.* 2006).

REPRODUCTION Only one nest (race *squamigera*) has been described: a nest found at Cerro Guachaurco near Huachanama, Loja, Ecuador, at an elevation of 3,020m (Greeney & Juiña 2011). When discovered, the nest contained a single, newly hatched young and an infertile egg. The following account is taken entirely from this paper. **NEST & BUILDING** The only described nest was in scrubby second growth, with a largely broken canopy <15m high. The nest was a bulky cup composed mostly of fresh green moss intermixed with a few sticks and dead leaves. It was fairly densely lined with dark, flexible fibres and rootlets, but also included pine needles. The only (introduced) pines in the immediate area were more than 30m away, indicating that the adults had actively sought such materials. The nest was 2.5m above ground and supported from below by several small (>5cm) and two large (10cm) horizontal branches. Overall, especially from below, the nest was well camouflaged, appearing to be a naturally accumulated clump of debris. The authors suggested that it was probably built on a pre-existing accumulation of vegetable material. The nest had the following measurements: external diameter (measured at the widest and narrowest points) 27cm × 25cm; external height 22cm tall; internal diameter 13cm × 14cm; internal depth 6–7cm. **EGG, LAYING & INCUBATION** The single described egg (addled) was turquoise with brownish and pale lavender spots, heaviest at both ends but more so at the larger end. It measured 33.5mm × 28.0mm. There are no data available on any aspect of incubation. **NESTLING & PARENTAL CARE** The nestling described by Greeney & Juiña (2011) was estimated to be no more than 1–2 days old and had dark greyish-pink skin, darkest on the legs and feet. The bill was dusky-orange with an orange mouth lining and slightly brighter (more yellowish) inflated rictal flanges. Qualitatively, the authors suggested that the bill and mouth lining were not quite as intensely coloured as other *Grallaria*. While this is very subjective, it is interesting in light of the rather pale rictus of fledgling Giant Antpitta (see that species), perhaps indicating that these two putative relatives may have bill and gape characters which, so far as is known, appear to differ from other *Grallaria*. The natal down was fairly long and dense, and very dark grey or black. At this age it weighed 16.5g, doubling its weight to 36g during the next three days. When estimated to be 4–5 days old, the nestling's appearance was similar, but contour pin-feathers were starting to break through the skin on all feather tracts and the (unbroken) primary pin-feathers were roughly 5mm long. Few behavioural data were gathered at this nest other than the observation that the adult brooding the 0–1 day-old nestling allowed the observer to approach to within 30cm before abandoning the nest, and when approached while brooding the 4–5 day-old nestling it remained frozen until the observer approached to within 1m. Upon leaving the nest on both occasions the adult dropped to the ground and disappeared silently into dense vegetation. **SEASONALITY** The duration and sequence of plumage maturation is not well known, and it is thus early to speculate on breeding seasonality. From the following records, however, I tentatively suggest that nesting in north-west Ecuador occurs largely in December–January. The fledgling AMNH 492139 listed below is the basis for the 'fledgling' in November from 'NW Ecuador' reported by Fjeldså & Krabbe (1990) (J. Fjeldså *in litt.* 2015). **BREEDING DATA** *G. s. squamigera* Hatching, 25 February 2010, Cerro Guachaurco (Greeney & Juiña 2011); '*pullus*' [fledgling], March 1859, 'Pichincha' (Sclater 1860c, 1890); fledgling, 12 April 2016, RP Yanacocha (F. Enríquez photos); fledgling, 23 June 2002, Cajanuma (R.C. Dobbs *in litt.* 2017); fledgling, 8 July 1914, Peñón (NHMUK 1916.9.21.38, photos G.M. Kirwan); fledgling, 19 March 2009, San Jorge Ecolodge (Quito) (G. Cruz photos); fledgling transitioning, 1 November 1898, Volcán Pichincha (AMNH 492139); fledgling transitioning, 17 June 1909, Valle (FMNH 57173); fledgling transitioning, 18 June 1909, Páramo de La Culata (FMNH 57168); juvenile, 11 May 2001, RP Bellavista (B. Mila *in litt.* 2017); juvenile transitioning, 15 October 1905, Valle (FMNH 57172); juvenile, 10 July 1931, Volcán Sangay (MLZ 7766); juvenile, 3 April 2014, RP Yanacocha (R. Ahlman *in litt.* 2017); juvenile, 8 July 2017, RP El Corazón (ML 63016361, photo P. Molino); two subadults, 14 June 1904 and 18 June 1911, Cacute (FMNH 57166/170); subadult, 20 January 1988, Mérida (FMNH 57165); subadult, 27 August 2012, RFP Río Blanco (ML 20284381, photo P. Hawrylyshyn); subadult, 27 March 1920, Río Mucujún (FMNH 53436); subadult, 3 September 2002, BP Cashca Totoras (MECN 7684); subadult, 20 June 1955, Cerro Munchique (YPM 32136, photos S.S. Snow); subadult, 19 June 1980, Batan (LSUMZ 97677, photos M.L. Brady); subadult transitioning, 1 October 1916, Choachí (AMNH 176532, holotype); subadult transitioning, 28 May 1974, BP Tinajillas (MECN 2593, photos M.V. Sánchez N.); subadult transitioning, 18 July 1998, Parque La Carolina (MECN 7434; photos M.V. Sánchez N.); subadult transitioning, 18 October 1910, Valle (FMNH 57171); subadult transitioning, 30 April 1904, Páramo de La Culata (FMNH 57176); subadult transitioning, 8 November 1992, above San Andrés (MECN 6492, photos M.V. Sánchez N.); 'juvenile', December 1898, 'Pichincha' (Goodfellow 1901, 1902); 'juveniles', June, 'Venezuela' (Krabbe & Schulenberg 2003); 'immature', November, 'NW Ecuador' (Fjeldså & Krabbe 1990); 'large gonads', August, Antioquia, W Andes and ♂ in breeding condition in August, N end of W Andes in Colombia (Hilty & Brown 1986). I also examined the following specimens with no collection date: fledgling, 1873, Sierra Nevada de Mérida (NHMUK 1889.7.10.839, photos G.M. Kirwan); fledgling transitioning, 'Bogotá' (AMNH 492136); juvenile, 'Colombia' (FMNH 11330); juvenile, Mérida (AMNH 492152); subadult, 'Mérida' (USNM 147348); subadult, 1880, 'Ecuador' (USNM 147346); subadult, 'Bogotá' (FMNH 57163); subadult, 1936, Ecuador (FMNH 99572). *G. s. canicauda* Juvenile transitioning, 26 April 1954, Alto Palmar (LSUMZ 37761); two subadults, 19 March 1974 and 10 May 1975, Bosque de Ampay (FMNH 299213/603); subadult, 14 June 1980, Copata (LSUMZ 96067); subadult,

22 May 1926, Nequejahuira (AMNH 229211); subadult, 1 July 1903, Cushi (AMNH 492132); subadult, 15 March 1973, Papayajo (FMNH 296712); subadult, 1 June 1922, 'Huánuco Mts' (FMNH 66255); subadult, 29 June 1967, Cordillera de Vilcabamba (AMNH 820282); subadult, 17 June 1985, Playa Pampa (LSUMZ 128519); subadult, 6 August 2002, *c.*22km ENE of Florida (LSUMZ 174007); subadult, 1 August 1953, Cachupata (FMNH 222146); subadult transitioning, 26 July 1953, Amacho (FMNH 222147); subadult transitioning, 10 November 2011, Abra Patricia (IBC 989192, D. Shapiro); subadult transitioning, 4 June 1966, Bosque Taprag (FMNH 281498); subadult transitioning, 2 July 1970, Bosque Zapatogocha (FMNH 287804); two records of 'fledglings', Apurímac in March and La Paz in December (Fjeldså & Krabbe 1990).

TECHNICAL DESCRIPTION Sexes similar. *Adult* The following describes the nominate race. Above, ash-brown variably washed olivaceous, with most feathers having darker margins, giving a scaled appearance, especially in the interscapular region. Wings and tail more brownish, less grey. Lores and central throat buffy-whitish or buffy-ochraceous, and throat bordered with black moustache-like streaks. Head-sides and underparts, including underwing-coverts, yellowish-fulvous (varying from richly coloured to somewhat pale) and crossed by numerous, thick black undulations. These wavy bars are somewhat evenly spread across the underparts except on the lower belly and vent, which are usually densely marked, the central belly being pale and the vent a deeper fulvous or ochraceous-orange. **Adult Bare Parts** *Iris* dark brown; *Bill* maxilla dark grey to blackish, mandible pinkish-grey, duskier distally and paler basally; *Tarsi & Toes* pinkish blue-grey. *Immature* Previous descriptions of immature plumages range from Chapman's (1926b) statement that 'immature birds are dark olive above', to the more detailed descriptions below, all of which I have interpreted in light of the present categorical age classes used here. *Fledgling* Carriker (1935) described a female in 'downy juvenal plumage' of *canicauda* as having the 'entire upper parts and sides of head and neck and breast slaty black, with the feathers all broadly tipped with cinnamon ochraceous, and the abdomen, flanks and crissum plain cinnamon ochraceous.' Overall, the coloration of the fluffy fledgling plumage might best be described as dark olive-brown with 'blonde highlights' that are buffer on the head and gradually become more cinnamon towards the rump. The breast coloration is dominated by the buffy down with a gradual reduction of dark plumage lower until the central belly is clean buff and the flanks and vent are clean cinnamon-buff or cinnamon-ochraceous. Tail not yet visible, with all rectrices still largely sheathed. Upperwing-coverts broadly tipped buffy-ochraceous to cinnamon-ochraceous with a thick subterminal black bar. **Fledgling Bare Parts** (from photographs) *Iris* brown, perhaps slightly paler than in adult; *Bill* mostly orange-grey, greyest on culmen and brightest orange at nares, along tomia and at base, inflated rictal flanges pale yellow to pale orange; *Tarsi & Toes* dull grey to blue-grey, washed orange along posterior margin and yellowish on feet, nails dull yellow. *Juvenile* Taczanowski (1884) described 'young' birds (presumably *canicauda*) as differing from adults by having the olive-brown feathers of the nape, neck-sides and upper back with rusty-brown spotting. He described the upperwing-coverts as being thinly fringed the same colour, and the underparts as being washed rusty-brown,

less so on the flanks and undertail-coverts. Carriker (1935) described an 'immature male' *canicauda* as having the head and back dark and the tips of the feathers edged with light reddish-brown. The breast was speckled black and brown, much darker near the throat. Ménégaux & Hellmayr (1906) described an 'immature' male *squamigera* from Pichincha (unfortunately no date) as having tawny spots on the central back, nape and neck-sides forming a partial collar. What is apparently the same specimen, also lacking a date, is mentioned in Ménégaux (1911). Overall, unlike the spotted or finely barred plumage of other *Grallaria*, the juvenile plumage of Undulated Antpitta is more coarsely barred and appears to retain a 'fuzzy' appearance on the nape and crown for longer than in other species. **Juvenile Bare Parts** (from photographs) *Iris* like adult; *Bill* maxilla mostly dusky, may retain some orange around nares and on tomia, mandible dusky at tip, fading to pinkish white or orange-white basally, rictal flanges not apparent, at most only slightly inflated; *Tarsi & Toes* largely as adult, may be indistinctly washed yellowish or orange on posterior margin and around feet, nails dusky-yellow. *Subadult* Overall, coloration rather similar to adult, but subadults retain their immature wing-coverts and usually have scattered immature feathers on the back. Most feathers of the underparts are fully adult.

MORPHOMETRIC DATA *G. s. squamigera* Wing chord 143.7 ± 3.1mm; *Bill* culmen 30.0 ± 2.0mm, width at gape 17.6 ± 1.9mm; *Tarsus* 60.7 ± 0.8mm (*n* = 4 unsexed adults, Kattan & Beltrán 1999). *Wing* 139mm, 145mm; *Tail* 62mm, 62mm; *Bill* [total?] culmen 36.5mm, 35mm; *Tarsus* 63.5mm, -?- (*n* = 2♀♀, Chapman 1926a). *Wing* 142mm, 145mm; *Tail* 63mm, 64mm; *Bill* [exposed? culmen] 30mm, 30mm (*n* = 2, unsexed adults, Ménégaux & Hellmayr 1906). *Wing* 139mm, 145mm; *Tail* 63mm, 59mm; *Bill* [total?] culmen 35.5mm, 35.5mm; *Tarsus* 59mm, 61mm (*n* = 2♂♂, Chapman 1926a). *Wing* 148.0mm, 142.0mm, 150.0mm; *Bill* from nares 17.3mm, 18.9mm, 19.3mm, depth at nares 10.5mm, 11.6mm, 12.0mm, width at nares 8.0mm, 7.7mm, 8.2mm; *Tarsus* 60.9mm, 60.8mm, 60.3mm (*n* = 3, ♂, FMNH, E.T. Miller *in litt.* 2015). *Wing* 141.0mm, 136.2mm, 140.0mm; *Bill* from nares 18.3mm, 18.7mm, 19.3mm, depth at nares 10.5mm, 11.2mm, 10.5mm, width at nares 7.5mm, 7.9mm, 7.6mm; *Tarsus* 58.6mm, 60.8mm, 58.5mm (*n* = 3♂♂, FMNH, E.T. Miller *in litt.* 2015). *Wing* 140.2mm, 141.0mm; *Bill* from nares 18.4mm, 19.5mm, depth at nares 10.6mm, 11.1mm, width at nares 7.7mm, 7.8mm; *Tarsus* 58.6mm, 57.5mm (*n* = 2♀♀, FMNH, E.T. Miller *in litt.* 2015). *Bill* from nares 21.5mm, exposed culmen 30.6mm, depth at nares 12.0mm, width at nares 8.3mm, width at base of lower mandible 11.7mm; *Tarsus* 62.1mm (*n* = 1♂, USNM 436481). *Bill* 19.1mm, exposed culmen 26.7mm, width at gape 17.0mm; *Tarsus* 56.0mm (*n* = 1♂, juvenile, MLZ 7766, see Reproduction). *Bill* from nares 19.6mm, 20.8mm, 22.3mm, 22.6mm; *Tarsus* 57.3mm, 60.4mm, 61.4mm, 60.4mm (*n* = 4, ♂,?,♂,♂, WFVZ, USNM). *Bill* from nares 20.0mm, 21.3mm; *Tarsus* 55.9mm, 60.1mm (*n* = 2, ♀?, subadults, USNM). *Wing* 144.5mm, 151mm, 146mm; *Bill* from nares 19.3mm, -?-, 21.0mm, exposed culmen 21.8mm, -?-, 24.1mm; *Tarsus* 54.9mm, 57.5mm, 59.3mm (*n* = 3, ♂,?,?, MCZ, J.C. Schmitt *in litt.* 2017). *G. s. canicauda* Wing 136mm; *Tail* 60mm; *Bill* [total? culmen] 39mm; *Tarsus* 58mm (*n* = ?♂, Taczanowski 1884). *Wing* 142mm, 141mm; *Tail* 69mm, 68mm; *Bill* [total?] culmen 35mm, 34mm; *Tarsus* 60mm, 61mm (*n* = 2♂♂, Chapman 1926a). *Bill* 18.7mm, 19.2mm, 20.1mm; *Tarsus* 59.1mm,

57.2mm, 58.8mm (*n* = 3♂♂, subadults, AMNH). *Wing* 132.2mm, 136.0mm, 137.2mm; *Bill* from nares 18.1mm, 19.1mm, 19.6mm, exposed culmen 24.2mm, 24.7mm, 25.3mm, depth at nares 10.3mm, 10.2mm, 10.7mm, width at nares 6.9mm, 7.5mm, 7.8mm; *Tarsus* 57.0mm, 54.9mm, 58.3mm (*n* = 3, ♂♀♂, FMNH, E.T. Miller *in litt.* 2015). *Bill* from nares 19.5mm; *Tarsus* 56.4mm (*n* = 1, ♀?, AMNH 810750). *Wing* 137mm; *Bill* from nares 19.2mm, exposed culmen 22.3mm; *Tarsus* 57.0mm (*n* = 1♀, MCZ 46074, J.C. Schmitt *in litt.* 2017). **Mass** 131.6g (*n* = 1♀, *squamigera*, Krabbe & Schulenberg 2003), 166g(vh) (*n* = 1♀, *canicauda*, Remsen 1985, LSUMZ 96067); 129.1g (*n* = 1♀, *squamigera*, Echeverry-Galvis *et al.* 2006); 116g(lt), 86.9g(m) (*n* = 2♀ *canicauda*, ♂ *squamigera*, MSB 33988/42610); range 116–174g (♂ *canicauda*), range 105–166g (♀ *canicauda*) (Krabbe & Schulenberg 2003); 122g, 125g, 131g, 150g(vh), 132g(m), 125g(lt), 110g(m), 118g, 126g, 116g/ 120g(m), 134g(h), 130g(h), 144g(lt), 118g, 174g(h), 138g(lt), 130g, 120g(m), 116g(m), 144g, 161g, 156g (*n* = 23, 10♀♀/13♂♂, *canicauda*, LSUMZ); 127g(m) (*n* = 1♀ *squamigera*, LSUMZ 78551); 120g, 120g(m), 112g(lt) (*n* = 3, subadult ♂♂, *squamigera*, MECN); 129g(m) (*n* = 1♂, *squamigera*, MECN 6609). **Sex unspecified** 112g, 129g, 149g (*n* = 3, race?, Krabbe *et al.* 1994); mean = 144.3 g (*n* = 4, *canicauda*, Schulenberg & Williams 1982); mean = 115g (*n* = 3, *canicauda*, Parker & O'Neill 1980); 112–149g (*n* = ?, *squamigera* [presumed], Hilty 2003); 131.6 ± 3.1g (*n* = 4, *squamigera*, Kattan & Beltrán 1999). **TOTAL LENGTH** 20–24cm (Lafresnaye 1842, Meyer de Schauensee & Phelps 1978, Hilty & Brown 1986, Dunning 1987, Iafrancesco *et al.* 1987, Fjeldså & Krabbe 1990, Clements & Shany 2001, Hilty 2003, Walker & Fjeldså 2005, Ridgely & Tudor 2009).

TAXONOMY AND VARIATION Two subspecies currently recognised. Birds from Cordillera de Cutucú are somewhat isolated from Andean populations and may represent an undescribed taxon (Krabbe & Schulenberg 2003). Like these authors, I have tentatively assigned them to *canicauda* until a proper analysis is made. However, it is my belief, strengthened by the apparent lack of strong variation in vocalisations across its range, that *G. squamigera* might best be considered monotypic, something that numerous authors have hinted at. The subspecific differences illustrated on the plate are representative of 'typical' plumages for the two subspecies, but my own observations of a long series of specimens shows substantial intra-population variation and numerous intermediates. Chapman (1926a) while deriving informative characters to distinguish *canicauda* from nominate, noted that 'two Colombian specimens [of *squamigera*], while darker in tone, are nearly as pure grey above as are both our specimens of *canicauda* [from Bolivia].' In searching for characters, Chapman (1926a) stated that immature *squamigera* has variable amounts of, and sometimes the entire back, olive-washed. He stated that only fully adult *squamigera* are 'pure grey above'. This immediately appears to contradict his distinction of *canicauda* being greyer above (less brownish) than *squamigera*. He continued to state that 'one of our specimens of *canicauda*, although exhibiting traces of immaturity, has no olivaceous wash above', interpreting this as evidence that *canicauda* lacks (or has very slight) olivaceous on the upperparts. Zimmer (1930), comparing two males and a female from Huánuco with a series of 22 Colombian birds, felt that size and general coloration were rather similar to some Colombian specimens, but that

the tails were a closer match to *canicauda*, being grey like the back and not at all brownish. He felt that the three Huánuco birds were intermediate between *squamigera* and *canicauda* and, despite the fact that they also had tail lengths closer to those of nominate, assigned them to *canicauda*. Peters & Griswold (1943), discussing a female collected in Junín, somewhat confusingly stated that the 'underparts [presumably meaning upperparts] of this bird are a clearer bluish grey than a skin from Colombia and another from Ecuador', a character which, following Chapman (1926a), should assign it to *canicauda*. They went on to mention that the Junín specimen 'can hardly be referable to *canicauda* Chapman, even though it possesses a grayish rather than brown tail.' They then referenced Carriker's (1935) assertion that *canicauda*, in addition to having wings and tail coloured the same as the back, has a pure white throat and a white subocular region. They did not mention the colour of the lores and throat of the Junín bird. Despite having described characters that would suggest it was best assigned to *canicauda*, they concluded that it represented the nominate race. Perhaps most telling are Chapman's (1917) own comments on Colombian specimens from Quindío. When comparing those specimens to other examples of nominate *squamigera*, he found them to be 'much deeper plumbeous above'. The differences were apparently somewhat striking, as he then suggested that they might represent a 'well-marked race'. Of the Quindío birds, he further noted that their throats were whiter than nominate. The two specified differences are both characters he later (1926a) used to name Bolivian specimens *canicauda*. At the time, Chapman (1917) concluded that the grey back and white throat represented individual variation. These examples point to the confusion that existed during the peak of collecting within the range *G. squamigera*, but other authors expressed doubts as to the validity of *canicauda*. Hellmayr (Cory & Hellmayr 1924) said 'examination of a considerable series [of *G. squamigera*] from Colombia, Ecuador and Peru, reveals no racial variation. Birds from the Andes of Merida are not different either'. This statement actually mirrors Sclater's (1877) observation that a series of nine specimens, including Venezuelan, Colombian, Ecuadorian and Bolivian skins, showed 'no great amount of variation'. At the time, Hellmayr also had before him an Ecuadorian skin with the throat 'nearly white'. In addition, the observations of Carriker (1935) on material from Amazonas, San Martín and Cajamarca, suggest that, even should the two races be valid, their currently understood distributions may require re-evaluation. He felt that the Cajamarca skin was clearly referrable to the nominate, while those from Amazonas and San Martín represented *canicauda*. The birds in extreme southern Loja (Utuana) are considered to be of nominate *squamigera* (Krabbe 1991, Ridgely & Tudor 2001), making it possible that Carriker (1935) was correct, and the distribution of nominate *squamigera* extends to Cajamarca. Nevertheless, after examining photos and specimens from Venezuela, Colombia, Ecuador and Peru, I fail to see a clear separation between the two taxa, at least as currently defined. Some individuals in extreme south-east Ecuador in Zamora-Chinchipe show plumage characters suggesting *canicauda*, others nominate. An individual from Abra Patricia, not far south of the Ecuadorian border, is nearly a classic example of the nominate race. Immature Undulated Antpittas appear to retain their immature upperwing-coverts (with buffy-ochraceous or reddish-buff terminal fringes) for a significant period. This leads me to the observation that, as part of what appears to be a slow change to full adult plumage (including loss of the buff-tipped coverts), individuals slowly acquire greyer (less olive or brownish) upperparts. Similarly, there is a slow progression from pale ochraceous to buffy-white, to pure white feathers on the lores and throat. Considering that the holotype of *squamigera* is clearly described as having immature upperwing-coverts, I conclude that the proposed differences between the two forms, already known to vary within populations, represent nothing more than differences in plumage ontogeny. Individuals with *canicauda*-like plumage are in full adult plumage and those matching characteristics of *squamigera* are subadults still showing the browner or more olivaceous upperparts and buffier facial and throat feathering of juveniles.

Grallaria squamigera squamigera Prévost & Des Murs, 1846, *Voyage autour du monde sur la frégate la Vénus commandée par Abel de Petit-Thouars. Oiseaux*, p. 198, pl. 3, 'Bogotá', Colombia. The original description was in French and is supposedly based on an adult male. However, the authors stated '*les grandes couvertures lisérées finement de fauve à leur extrémité*', which I translate as 'greater coverts finely margined fawn at their tips'. It appears that this is yet another case where species or subspecies-level diagnostic plumage characters may require formal revision (see Giant Antpitta) as a result of descriptions based on immature plumage. As currently defined, nominate *squamigera* occurs in the Andes of western Venezuela (Meyer de Schauensee & Phelps 1978, Hilty 2003, Pelayo & Soriano 2010), all three ranges of Colombia (López-Lanús *et al.* 2000, Calderón-Leytón *et al.* 2001, Losada-Prado *et al.* 2005, Ayerbe-Quiñones *et al.* 2008, Arbeláez-Cortés *et al.* 2011a) and the Ecuadorian Andes to northern Peru (Chapman 1926, Ridgely & Greenfield 2001, Schulenberg *et al.* 2010). SPECIMENS & RECORDS VENEZUELA: *Trujillo* PN Guaramacal, *c*.09°14'N, 70°11'W (Hilty 2003); Teta de Niquitao, 09°04.5'N, 70°23.5'W (CM P89355). *Mérida* Llano Rucio, 08°58'N, 71°05'W (COP 14213/14); Conejos, 08°50'N, 71°15'W (USNM 263853); Laguna Negra, 08°47'N, 70°48'W (COP 4802); Páramo de La Culata, 08°45'N, 71°05'W (AMNH, FMNH, USNM, COP, CUMV, UMMZ); El Valle, 08°41.5'N, 71°06'W (FMNH 57171–174, AMNH 492145/46, CUMV 7197, COP 14365); Cacute, 08°40'N, 71°02'W (FMNH 57166/70); Tabay, 08°40'N, 71°04'W (CM P89798/821); Río Mucujún, 08°38'N, 71°10'W (FMNH 53436); Páramo Escorial, *c*.08°38'N, 71°06'W (AMNH 492147–149); Páramo La Fría, 08°36'N, 71°02'W (CM); Caserío Arangurén, 08°33'N, 70°53'W (COP 71524); Los Nevados, 08°28'N, 71°04'W (FMNH 57175); Cerro El Muerto, Páramo Aricagua, 08°16'N, 71°11'W (COP 45375–377); Páramo Aricagua, 08°13'N, 71°08'W (Phelps & Phelps 1963). *Táchira* Páramo Zumbador, 08°00'N, 72°05'W (FMNH 288323); Páramo La Negra, 08°15'N, 71°52'W (COP 64254–256). *Apure* Cerro Las Copas, 07°26'N, 72°21'W (COP 70135/36); Páramo Tamá near Colombian border, 07°24'N, 72°22.5'W (J.E. Miranda T. *in litt.* 2017). **COLOMBIA:** *Magdalena* El Dorado Lodge, 11°06'N, 74°04'W (Osborn & Olson 2015). *Norte de Santander* Ramírez, 07°48'N, 73°05'W (CM P58131); Páramo Frontibón, 07°21'N, 72°39'W (Iafrancesco *et al.* 1987; MLS 3962); *Santander* Las Picotas, 07°23'N, 72°53'W (XC 282395, J.E. Avendaño). *Boyacá* Alto de Onzaga, 06°20.5'N, 72°43.5'W (O. Cortes-Herrera *in litt.* 2017); Páramo de Onzaga, 06°11'N, 72°46'W (D. Uribe *in*

litt. 2017); PNN Pisba, 05°52'N, 72°34'W (D. Carantón *in litt.* 2017); *c*.9.5km N of Paipa, 05°52'N, 73°07.5'W (eBird: O. Acevedo Charry); SFF Iguaque, 05°38'N, 73°29'W (XC 119727, D. Edwards). *Bogotá* Quebrada La Vieja, 04°37.5'N, 74°03.5'W (XC 11222, D. Knapp). *Cundinamarca* Páramo de Guasca, 04°54.5'N, 73°51.5'W (Olivares 1969, Iafrancesco *et al.* 1987, ICN 12715–718, MLS 3960); Páramo La Calera, 04°42.5'N, 73°57.5'W (Olivares 1969, Iafrancesco *et al.* 1987, MLS 3961); Páramo de Choachí, 04°33'N, 73°57.5'W (Olivares 1969, Iafrancesco *et al.* 1987, AMNH 176532, ICN 12719, MLS 3958/59); Vereda Ferralarada, 04°33'N, 73°53'W (ICN 22185); Laguna Chingaza, 04°31'N, 73°45'W (XC 12561/62, O. Laverde); Peñón, 04°26'N, 74°18'W (NHMUK 1916.9.21.38). *Meta* 1.8km S of San Juanito, 04°26.5'N, 73°40.5'W (eBird: R. Coffman). *Antioquia* Páramo de Frontino, 06°28'N, 76°04'W (Hilty & Brown 1983, Flórez *et al.* 2004, Krabbe *et al.* 2006, Arbeláez-Cortés *et al.* 2011b, USNM 436481); RNA El Colibrí del Sol, 06°26.5'N, 76°05.5'W (XC 154920, H. Matheve); RNSC La Esperanza, 05°36.5'N, 75°50'W (J.B.C. Harris *in litt.* 2016); Jardín–Riosucio road, 05°35'N, 75°46.5'W (D. Uribe *in litt.* 2017). *Caldas* RFP Río Blanco, *c*.05°05'N, 75°25'W (Verhelst *et al.* 2001). *Tolima* Palomar, 04°39'N, 75°13'W (Martínez 2014); Guaimaral, 04°41.5'N, 75°06.5'W (Martínez 2014); Finca Indostán, *c*.04°51'N, 75°12.5'W (XC 54276, B. López-Lanús); Finca Yerbabuena, 04°32'N, 75°25.5'W (XC 51450/51, B. López-Lanús). *Risaralda* Near Potreros, 04°51.5'N, 75°29'W (F. Schmitt *in litt.* 2017); Santa Isabel, 04°47'N, 75°28'W (AMNH 111965/66); La Pastora, 04°42.5'N, 75°29'W (Naranjo 1994; UV 6169); El Cedral, 04°42'N, 75°32'W (Naranjo 1994). *Valle del Cauca* RFP Bosque Yotoco, 03°52.5'N, 76°26'W (eBird: C. Downing); Chicoral, 03°34'N, 76°35'W (eBird: L.G. Naranjo); above Cali, *c*.03°28'N, 76°33'W (Hilty & Brown 1986). *Cauca* Cerro Munchique, 02°32'N, 76°57'W (Bond & de Schauensee 1940, ROM, FMNH, WFVZ, LACM, YPM); Coconúco, 02°20.5'N, 76°30'W (Bond & de Schauensee 1940). *Quindío* Finca Bengala, *c*.04°39'N, 75°28.5'W (UV 4840); Laguneta, 04°35'N, 75°30'W (AMNH 111963/64). *Huila* 12.8km SW of San Agustín, 01°47.5'N, 76°21'W (XC 147984, D. Calderón-F.); near Palestina, 01°39.5'N, 76°11'W (XC 298650, D. Bradley). **ECUADOR:** *Carchi* RP Estación Biológica Guandera, 00°36'N, 77°42'W (Creswell *et al.* 1999a); Cerro Mongus, 00°27'N, 77°52'W (Robbins *et al.* 1994b, Buitrón-Jurado 2008). *Sucumbíos c*.3km SE of Santa Barbara, 00°37.5'N, 77°30'W (R. Ahlman *in litt.* 2017); Alto La Bonita, 00°29.5'N, 77°35'W (Stotz & Valenzuela 2009). *Orellana* Pavayacu, *c*.00°33.5'S, 77°37.5'W (Vogt 2007). *Imbabura* Yaguarcocha, 00°22'N, 78°06'W (AMNH 810746); Intag, 00°20'N, 78°24'W (Poulsen & Krabbe 1998); Laguna de Cuicocha, 00°18'N, 78°21'W (HFG); BP Cerro Blanco, 00°12.5'N, 78°20.5'W (XC 345253/346296, C. Vogt). *Pichincha* Cerro Mojanda, 00°08'N, 78°17'W (AMNH 810748); RP Santa Lucía, 00°07'N, 78°36'W (HFG); Lago San Marcos, 00°06'N, 77°57'W (J. Freile *in litt.* 2017); RP Bellavista, 00°01'S, 78°41'W (HFG); Verdecocha, 00°06.5'S, 78°37'W (MECN 2594); San Jorge Ecolodge (Quito), 00°07'S, 78°31.5'W (G. Cruz photos); RP Yanacocha, 00°06.5'S, 78°35'W (XC, ML); Loma Yanayacu, 00°08S, 78°35'W (Krabbe 1991, XC); 'Volcán Pichincha', *c*.00°10'S, 78°35'W (Goodfellow 1902, AMNH; see below); Parque La Carolina (central Quito), 00°11'S, 78°29'W (MECN 7434); Loma Guarumos, 00°03'S, 78°38'W (ANSP 169696/97); above Lloa, *c*.00°13.5'S, 78°35.5'W (MCZ 138467); Cerro Ilaló, 00°14.5'S, 78°24.5'W (J. Freile *in litt.* 2017); RVS Pasochoa, 00°26'S, 78°30.5'W

(XC 9325, A.T. Chartier); Volcán El Corazón, 00°33'S, 78°43'W (Poulsen & Krabbe 1997a, 1998). *Cotopaxi* Pilaló, 00°59'S, 79°01'W (ANSP 169695); Cerro Parcato, 00°44'S, 78°58'W (Krabbe 1991). *Bolívar* 10km NW of Salinas, 01°21'S, 79°05'W (Poulsen & Krabbe 1998, XC 249508, N. Krabbe; BP Cashca Totoras, 01°45'S, 78°58'W (Bonaccorso 2004, MECN 7684). *Chimborazo* Matus, 01°33'S, 78°30'W (Sclater 1858b); Galeras de Sangay, *c.*02°03'S, 78°29'W (MLZ 7766); RP El Corazón, 02°04'S, 78°55'W (ML 63016361, photo P. Molino). *Cañar* 11km by road S of Zhud, 02°24'S, 78°59'W (Krabbe 1991); Llavircay, 02°33.5'S, 78°37.5'W (P.X. Astudillo W. *in litt.* 2017). *Napo* Río Anatenorio, 00°59'S, 78°17'W (Krabbe 1991); Cordillera Guacamayos, 00°38'S, 77°50.5'W (Krabbe 1991). *Azuay* 5.5km SW of Molleturo, 02°48'S, 79°26'W (XC 250035, N. Krabbe); BP Mazán, 02°52'S, 79°07'W (King 1989, Poulsen & Krabbe 1998); 3.7km S of Hacienda Cancan, 02°55'S, 79°15'W (XC 223715, P. Boesman); Nero, 02°57.5'S, 79°06.5'W (B. Tinoco *in litt.* 2017); Yunga Totorillas, 03°00'S, 79°01'W (J. Freile *in litt.* 2017); San Gerardo, 03°08.5'S, 79°12'W (B. Tinoco *in litt.* 2017); upper Yunguilla Valley, 03°14'S, 79°17'W (XC 250903, N. Krabbe). *Tungurahua* El Triunfo, 01°15.5'S, 78°22'W (Benítez *et al.* 2001); Machay, 01°23'S, 78°17'W (Benítez *et al.* 2001); 'Runtun', near Baños *c.*01°26'S, 78°24'W (loc. see Paynter 1993, SBMNH 8605); previous location near San Antonio de las Montañas, 01°25.5'S, 78°24.5'W (eBird: A. Spencer); Cañayacu, 02°51'S, 79°23'W (Poulsen & Krabbe 1998). *Loja* Lagunas de Maní, 03°34.5'S, 79°25.5'W (XC 155458, L. Ordóñez-Delgado); Cerro Guachahurco, 04°02'S, 79°52'W (Greeney & Juiña 2011); *c.*2.5km N of Celica, 04°05'S, 79°57'W (XC 250296, N. Krabbe); Cajanuma, 04°20'S, 79°10'W (XC 223713/14; P. Boesman); 3km SE of Papaca, 04°21'S, 79°43'W (Krabbe 1991); Bosque Hanne, 04°22'S, 79°45'W (XC 210286/89, J. Fischer); 9km E of Jimbura by road, 04°42'S, 79°27'W (ANSP 184709); Cordillera Las Lagunillas, 04°43.5'S, 79°26'W (Krabbe & Coopmans 2000, MECN 6609, XC). *Morona-Santiago* BP Tinajillas, 02°58'S, 78°29.5'W (MECN 2593). *Zamora-Chinchipe* Río Isimanchi above San Andrés, 04°47'S, 79°20'W (MECN 6492, KUNHM 86823); RP Tapichalaca, 04°29.5'S, 79°07.5'W (HFG). **PERU:** *Piura* Cerro Aypate, 04°42.5'S, 79°34.5'W (eBird: S. Crespo); Bosque de Cuyas, 04°36'S, 79°44'W (Vellinga *et al.* 2004); Batan, 05°06'S, 79°21'W (LSUMZ 97677); 33km SW of Huancabamba, 05°20'S, 79°32'W (LSUMZ 78551); above Limón de Porculla, 05°52.5'S, 79°32'W (F. Angulo P. *in litt.* 2017). *Cajamarca* 2.8km NW of Chontalí, 05°35'S, 79°10'W (MSB 42610); near El Edén, 05°51.5'S, 78°54'W (A. García-Bravo *in litt.* 2017); Chira, 06°16'S, 78°42'W (ANSP 117492); Bosque Paja Blanca, 06°23'S, 79°07'W (XC 132124–125, D.F. Lane). **NOTES** There are a large number of Undulated Antpitta specimens labelled simply 'Bogotá' (FMNH, NHMUK, AMNH, MCZ, USNM, ANSP, ICN). Although it is quite possible that some of these did, in fact, come from the Bogotá area, many are trade skins, making their origin uncertain and unfortunately casting some doubt on all specimens so labelled. There are also many speciems labelled simply 'Mérida' (AMNH 492150–160, FMNH 57165, USNM 147348), for which it is unclear if this refers to the city of Mérida (08°36'N, 71°09'W) or to the state of Mérida in general. The coordinates given above for 'Volcán Pichincha' are general ones for the many collecting localities on the slopes of this volcano; see Specimens & Records for *G. q. quitensis* for further discussion. Uncertain locations on the slopes of Volcán Pichincha for Undulated

Antpitta include: Pichincha Abajo (AMNH 810745/51), Bosques de Pichincha (AMNH 810747–750), Rucu Pichincha (AMNH 148098).

Grallaria squamigera canicauda Chapman, 1926a, American Museum Novitates, vol. 231, p. 1, 'Cocopunco, 10,000 ft., Dept. Larecaja, Bolivia.' The holotype is an adult male collected by G.H. Tate on 7 March 1926 (AMNH 211009) and the type locality is actually in Dept. La Paz (Paynter 1992, LeCroy & Sloss 2000). This subspecies is distributed from south-east Ecuador in the Cordillera de Cutucú through the eastern Andes of Peru to central Bolivia. In general terms, *canicauda* is said to be greyer above and to have whiter (less buffy) lores and throat. In his original description of *canicauda*, Chapman (1926a) provided the following differences from nominate *squamigera*: *canicauda* averages paler below and greyer above; the wings and (especially) tail always greyer and closer to the same colour as the back; tail averages longer. He stated that in *squamigera* the tail is 'olivaceous fuscous or brownish and quite different from the back', while in *canicauda* it is rather grey and almost concolorous with the back. SPECIMENS & RECORDS **ECUADOR:** *Morona-Santiago* E of Logroño, *c.*02°38'S, 78°04'W (ANSP 176859); Cordillera Cutucú, 02°43'S, 78°05'W (see discussion under *G. s. squamigera* Robbins *et al.* 1987; ML). **PERU:** *Amazonas* Abra Patricia, 05°42'S, 77°48.5'W (IBC 989192, D. Shapiro); above San Lorenzo, 05°48'S, 78°02'W (F. Angulo P. *in litt.* 2016); Atumpampa, 05°58'S, 77°53'W (F. Angulo P. *in litt.* 2016); Bagazán, 06°08'S 77°26'W (ANSP 117490); Cerro Shucahuala, 06°14'S, 78°06.5'W (A. García-Bravo *in litt.* 2017); Abra Barro Negro, 06°42.5'S, 77°52'W (F. Angulo P. *in litt.* 2017); Lluy, 06°45'S, 77°49'W (ANSP 117491); km 404 on Balsas–Leymebamba road, 06°47'S, 77°55'W (LSUMZ 80567); La Muralla, 06°47'S, 77°46'W (XC 181, W.-P. Vellinga). *San Martín* *c.*22km ENE Florida, 05°41'S, 77°45'W (LSUMZ 114007); Laurel, *c.*06°42'S, 77°41'W (FMNH 480819); Puerta del Monte, 07°34'S, 77°09'W (LSUMZ 104486). *La Libertad* Utcubamba, 08°08'S, 77°05'W (CM P133921); Mashua, 08°12'S, 77°14'W (LSUMZ 92436/37); Cumpang, 08°12'S, 77°10'W (Ménégaux 1910, Schulenberg & Williams 1982, LSUMZ 92438/39, ML). *Huánuco* Cordillera Carpish, *c.*09°34'S, 76°08'W (Parker & O'Neill 1976b, FMNH 66253/54); Bosque Zapatogocha, 09°40'S, 76°03'W (FMNH 287804/296711, LSUMZ 74092); Quilluacocha, 09°42'S, 76°07'W (LSUMZ 75240); 2km NW of Punta de Saria, 09°43'S, 75°54'W (LSUMZ 128521); Bosque Taprag, 09°43'S, 76°04'W (FMNH 281498/LSUMZ 74091); Bosque Chinchinga, 09°44'S, 76°00'W (FMNH 281499); Bosque Unchog, 09°44'S, 76°10'W (LSUMZ, XC); Papayajo, 09°45'S, 76°03'W (FMNH 296712); Rincón, 09°52.5'S, 76°25.5'W (G. Seeholzer *in litt.* 2017); Millpo, 09°54'S, 75°44'W (LSUMZ 128522–526); Playa Pampa, 09°57'S, 75°42'W (LSUMZ 128519/20); Bosque Potrero, 09°59'S, 76°05'W (LSUMZ 113574, XC 34351, D. Geale). Cushi, 09°58'S, 75°42'W (AMNH 492132). *Pasco* Cueva Blanca–Santa Barbara road, 10°22'S, 75°36'W (A. García-Bravo *in litt.* 2017); Cumbre de Ollón, 10°39'S, 75°17.5'W (Schulenberg *et al.* 1984, loc. see Distribution Data for *G. blakei*); Santa Cruz, 10°37'S, 75°20'W (Schulenberg *et al.* 1984). *Junín* Maraynioc, 11°22'S, 75°24'W (Peters & Griswold 1943, MCZ 266698); below Toldopampa, 11°29'S, 74°54'W (KUNHM); Pampa Huasi, 11°31'S, 74°48'W (LSUMZ 127644); Chilifruta, 12°01'S, 74°54'W (F. Schmitt *in litt.* 2017). *Ayacucho* 2km S of Ccano, 12°47'S, 74°00'W

(KUNHM 112907); Chupón, 13°14.5'S, 73°30'W (P.A. Hosner *in litt.* 2017). **Apurímac** 'Pomayaco', Río Pampas valley, *c.*13°21.5'S, 73°49.5'W (Morrison 1948); Huanipaca, 13°29'S, 72°59'W (MSB 33988); Cerro Turronmocco, 13°35'S, 72°53'W (XC 47651, N. Krabbe); Bosque de Ampay, 13°38'S, 72°57'W (FMNH 299213/603, MUSA 1377, XC); *c.*9km E of Abancay, 13°39'S, 72°48'W (Fjeldså & Kessler 1996). **Cuzco** Río Timpia headwaters, *c.*12°38'S, 72°41'W (MUSA 3443); Quebrada Lorohuachana near Echarate, *c.*12°47'S, 72°35'W (MUSA 2067); Cordillera Vilcabamba, *c.*13°00'S, 73°00'W (AMNH 820257/82); below San Luís, 13°04'S, 72°23'W (LSUMZ); Canchaillo, 13°08'S, 72°19'W (LSUMZ 78552, ML); 3km NW of Pillahuata, 13°08'S, 71°28'W (ML 100945, S. Connop); upper Manu road, *c.*13°11.5'S, 71°36'W (R. Ahlman *in litt.* 2017); *c.*5km N of Urubamba, 13°15.5'S, 72°06.5'W (Fjeldså & Kessler 1996); 'Cchachupata, Bolivia' [= Cachupata, see Stephens & Traylor 1983 and Paynter 1992], 13°17'S, 71°22'W (Sclater & Salvin 1874, FMNH 222146); Paucartambo, 13°19'S, 71°36'W (ML 135319, B.J. O'Shea); Chochocca, 13°21'S, 72°39'W (Fjeldså & Kessler 1996); above Mollepata, 13°27.5'S, 72°32.5'W (T.S. Schulenberg *in litt.* 2017); 'Amacho', near Marcapata, 13°30'S, 70°55'W (see Stephens & Traylor 1983, FMNH 222147). **Puno** Abra de Maruncunca, 14°12'S, 69°13'W (A. Spencer *in litt.* 2016); above Sandia, 14°24'S, 69°29'W (D. Beadle *in litt.* 2016); Cuyocuyo–Sandia road, 14°26.5'S, 69°30'W (ML 43794051, G. Seeholzer). **BOLIVIA: La Paz** Tokoaque, 14°37'S, 68°57'W (Hennessey & Gomez 2003); Muñamachay, 14°42'S, 69°01'W (XC 73606, J. Tobias & N. Seddon); Chulina, 15°09'S, 68°58'W (S.K. Herzog *in litt.* 2017); Cocopunco, 15°30'S, 68°35'W (AMNH 211011); near Sorata, 15°46.5'S, 68°38.5'W (eBird: S. Hampton); Cotapata, 16°17'S, 67°51'W (XC 73607/08, J. Tobias & N. Seddon, LSUMZ 96067); *c.*15km SW of Coroico, 16°17.5'S, 67°49.5'W (ML 65359701, photo S. Walker); Hichuloma, 16°18'S, 67°54'W (ANSP 120372/73); Nequejahuira, 16°20'S, 67°50'W (AMNH 229211); near Titi Amaya, 16°54'S, 67°12'W (XC 1989, S. Mayer). **Cochabamba** Pampa Grande, 16°42'S, 66°29'W (S.K. Herzog *in litt.* 2017); Incachaca, 17°14'S, 65°49'W (ANSP, CM, FMNH, LSUMZ); Alto Palmar, 17°08'S, 65°35'W (LSUMZ 37761); 'Campamento el Limbo', 17°10.5'S, 65°39.5'W (MHNNKM 4312); Represa Corani, 17°14'S, 65°53'W (D.F. Lane *in litt.* 2017); Jatun Potrero, 17°31'S, 65°12'W (S.K. Herzog *in litt.* 2017). **NOTES** A specimen labelled 'Huánuco Mts' (FMNH 66255) probably emanates from the mountains E of Huánuco (*c.*09°56'S, 76°07.5'W), not far from Bosque Potrero and ÁCP San Marcos (*c.*09°54.5'S, 76°05.5'W). **SEE ALSO** Sclater & Salvin (1874, 1879b), Taczanowski (1884), Carriker (1935), Bond & Meyer de Schauensee (1942), Remsen (1985), Macleod *et al.* (2005), Mark *et al.* (2008), Jankowski (2010).

STATUS Undulated Antpitta is fairly widely distributed and not uncommon in suitable habitat. It is thus not considered a conservation priority at a global level and is rated as Least Concern (BirdLife International 2016). Nevertheless, this species is dependent on montane Andean forests, which are being cleared at alarming rates across its range. It is already thought to have faced serious decline in Colombia (Hilty 1985). However, Undulated Antpitta is at least somewhat tolerant of minor disturbance as it has been reported from disturbed habitats (Macleod *et al.* 2005) and can be found in introduced pine and *Eucalyptus* woodland (Latta *et al.* 2011). It is even known

to nest in such habitats (Greeney & Juiña 2011), but its ability to persist there for long periods is uncertain. In addition, both races occur in a variety of protected areas across their range, including several national parks. **PROTECTED POPULATIONS G. s. squamigera Venezuela** PN Tamá (COP 70135/36); PN Tapo-Caparo (Phelps & Phelps 1963); PN Guaramacal (Hilty 2003); PN Sierra Nevada (FMNH, CM, COP); PN Páramos El Batallón y La Negra (COP 64254–256). **Colombia** PNN Munchique (Bond & de Schaunsee 1940, ROM, FMNH, WFVZ, LACM, YPM); PNN Los Nevados (Pfeifer *et al.* 2001); PNN Chingaza (XC 12561/62, O. Laverde); PNN Sumapaz (O. Cortes-Herrera *in litt.* 2017); PNN Puracé (eBird: N. Ocampo-Peñuela); PNN Pisba (D. Carantón *in litt.* 2017); RN Chicaque (eBird: C. Hesse); SFF Iguaque (XC 119727, D. Edwards); SFF Otún-Quimbaya (eBird: L.G. Naranjo); RMN El Mirador (eBird: A. Bartels); RFP Río Blanco (ML 20284381, photo P. Hawrylyshyn); RFP Los Yalcones (XC 147984, D. Calderón-F.); RFP Bosque Yotoco (eBird: C. Downing); PRN Ucumarí (Naranjo 1994, UV); RNA Loros Andinos (Woods *et al.* 2011); RNA El Dorado (Osborn & Olson 2015); RNA Loro Orejiamarillo (Salaman *et al.* 2007a); RNA El Colibrí del Sol (XC 154920, H. Matheve); RNSC La Riviera (XC 298650, D. Bradley); RNSC La Esperanza (J.B.C. Harris *in litt.* 2016); RNSC La Sonadora (eBird: V. Marín). **Ecuador** PN Podocarpus (Bloch *et al.* 1991, Rasmussen *et al.* 1994); PN Cajas (Tinoco & Astudillo 2007, Tinoco & Webster 2009, Astudillo *et al.* 2015); PN Sumaco-Galeras (Krabbe 1991, XC); PN Sangay (MLZ 7766); PN Llanganates (Benítez *et al.* 2001); PN Yacuri (MECN 6609, ANSP 186010, KUNHM 86823); BP Cashca Totoras (Bonaccorso 2004, MECN); PN Cayambe Coca (HFG); RP Bosque Hanne/Utuana (XC 210286/89, J. Fischer), RP Tapichalaca, RP Bellavista, RP Santa Lucía (HFG); RP Yanacocha (XC, ML); RVS Pasochoa (ML 83132/134, T.A. Parker); RE El Ángel, RE Cotacachi Cayapas, RE Los Illinizas, RBP Maquipucuna (J. Freile *in litt.* 2017); RBP Guandera (Creswell *et al.* 1999a); RP Verdecocha (MECN 2594); BP Mazán (King 1989, Poulsen & Krabbe 1998); BP Cerro Blanco (Imbabura) (XC 345253/346296. C. Vogt); BP Tinajillas (MECN 2593); RP San Jorge de Quito (G. Cruz photos); RP El Corazón (ML 63016361, photo P. Molino). **G. s. canicauda Peru** PN Yanachaga–Chemillén (Schulenberg *et al.* 1984); SH Machu Picchu (Walker & Fjeldså 2005); SN Ampay (FMNH, MUSA); ÁCP Abra Málaga (HFG); ÁCP Abra Patricia–Alto Nieva (IBC 989192, D. Shapiro); ÁCP Hierba Buena-Allpayacu (eBird: O. Janni). Probably occurs in ÁCP San Marcos. **Bolivia**: ANMIN Apolobamba, PN Carrasco (Hennessey *et al.* 2003b), PN Madidi (Hennessey & Gomez 2003).

OTHER NAMES *Grallaria squammigera* (= race *squamigera*; Prévost & Des Murs, 1846); *Myiotrichas squamigera* (= race *squamigera*, Cabanis & Heine 1859, Heine & Reichenow 1890). **English** Undulated Ant-Thrush (Brabourne & Chubb 1912). **Spanish** Tororoí ondoso (Krabbe & Schulenberg 2003); Tororoí Ondulado (Phelps & Phelps 1950, Salaman *et al.* 2007a); Hormiguerito ondulado (Kempff-Mercado 1985); Gralaria ondulada, Tóbalo (Valarezo-Delgado 1984); Hormiguero Tororoí Ondulado (Meyer de Schauensee & Phelps 1978, Phelps & Meyer de Schauensee 1979, 1994); Shumpo (Quechuan name?; Sclater 1858b). **French** Grallaire squammigère (Prévost & Des Murs, 1846); Grallaire ondée (Krabbe & Schulenberg 2003). **German** Ockerbauch-Ameisenpitta (Krabbe & Schulenberg 2003).

Undulated Antpitta, subadult (*squamigera*), RN Rio Blanco, Caldas, Colombia, 27 August 2012 (*Peter Hawrylyshyn*).

Undulated Antpitta, egg and one-day-old nestling (*squamigera*), Cerro Guachaurco, Loja, Ecuador, 26 February 2010 (*Mery E. Juiña J.*).

Undulated Antpitta, adult (*squamigera*), RP Tapichalaca, Zamora-Chinchipe, Ecuador, 30 November 2013 (*Roger Ahlman*).

Undulated Antpitta, adult (*canicauda*), Abra Patricia, Amazonas, Peru, 19 October 2011 (*Nick Athanas*).

Undulated Antpitta, subadult (*squamigera*), RN Rio Blanco, Caldas, Colombia, 17 September 2012 (*Stewart Elsom*).

GIANT ANTPITTA
Grallaria gigantea   Plate 6

Grallaria gigantea Lawrence, 1866, Annals of the Lyceum of Natural History of New York, vol. 8, p. 345, 'Ecuador'. The type specimen (USNM 35101) is housed in the Smithsonian Collection in Washington D.C.

Among the largest of the *Grallaria*, Giant Antpitta is aptly named. Until the present century it was rarely seen and extremely little was known of its ecology. Almost overnight, however, the now-famous Angel Paz, a rural man living outside Mindo (north-west Ecuador), trained a pair of Giant Antpittas to appear on demand when called and to eagerly gobble earthworms (and pieces of giant earthworms) proffered along one of the trails traversing a forested portion of his property. He gave his new star birdwatching attractions names and started what is, today, a widespread and highly successful type of ecotourist attraction: worm-feeding stations (Collins 2006, Woods *et al.* 2011). Like most of its relatives, Giant Antpitta is shy and retiring, to this day still rarely seen except where feeding stations have been established. Currently, Giant Antpitta can be seen regularly at Sr. Paz's 'ground-zero', as well as several other reserves in the Mindo area. A montane antpitta, found mostly between 1,200m and 3,000m, most of the species' range lies in Ecuador, extending north only to southernmost Colombia. It occurs on both Andean slopes and is replaced only much further north by Great Antpitta, but is often sympatric with the equally impressive Undulated Antpitta. All three are similar in size and plumage, and are thought to be closely related (Fjeldså 1992, Krabbe & Schulenberg 2003). Undoubtedly this group merits further taxonomic scrutiny. Overall, the biology, ecology, evolution, taxonomy, behaviour and conservation needs of Giant Antpitta are poorly understood and also in need of further research. Its breeding biology has only recently been investigated, with the first nest being described only in 2009 at Angel Paz's finca (Solano-Ugalde *et al.* 2009). Importantly, the distribution and habitat needs of race *lehmanni* are essentially unknown, data which are sorely needed to understand the just-mentioned taxonomic conundrum. Collar & Andrew (1988) suggested that nominate *gigantea* may also be particularly threatened, and it is much less studied than the (now) often photographed and observed *hylodroma* (Woods *et al.* 2011). Throughout its range, Giant Anptitta is in need of further surveys to determine if it still inhabits historic locations and to find new ones that may be in need of formal protection from the ever-looming threat of deforestation (Greeney 2015b, BirdLife International 2017).

IDENTIFICATION 22.5–26.7cm. Sexes similar. A particularly plump-looking *Grallaria* with a heavy bill and strongly hooked tip. Lores deep tawny-buff or buffy-chestnut, becoming olivaceous-brown on forecrown and grading to grey on hindcrown and nape. Rest of upperparts olivaceous-brown, with wings and tail browner than back. Head-sides and entire underparts a distinctive, deep rusty-orange, traversed by broken black bars and scales. Central belly and undertail-coverts are deeper coloured and unbarred. Giant Antpitta is among the largest of the genus and is most similar to Undulated and Great Antpittas, the latter of which does not overlap geographically. Above, Giant Antpitta is similar to the

slightly smaller Undulated Antpitta but is stouter, with a heavier bill. Above it is darker and browner with a more rufous forecrown and a slaty crown and nape, fairly different from the nearly uniform slaty-brown upperparts of Undulated Antpitta. The underparts of Giant Antpitta are a much deeper orange-rufous overall, barred black, compared to the more 'washed-out', pale buffy underparts of Undulated Antpitta. The black barring on the underparts of Undulated is also noticeably thicker and somewhat more 'scaled'. Overall, Giant Antpitta is stouter with a heavier bill.

DISTRIBUTION Colombia and Ecuador. For many years, known from Colombia by just two specimens of the mysterious race *lehmanni* collected on the west slope of the Central Andes in Cauca (Meyer de Schauensee 1950b, 1970). Note that the type locality for *lehmanni* is given (incorrectly) by many authors as being in Huila, as this region was apparently the focus of a governmental dispute for some time (Lehmann-V. 1957). Hilty & Brown (1986) predicted that it would likely be found in the south-west (*hylodroma*), where it is now known from one locality in Nariño, with apparently a large gap in observatations between there and the area west of Quito. Similarly, in the East Andes, early records include a specimen from eastern Carchi (Salvadori & Festa 1899), with no modern confirmation of its presence there or in adjacent Colombia, and a similarly large gap in observations south to the Río Cosanga Valley. The south-east limit (*gigantea*) appears to be the upper Río Pastaza Valley (Tungurahua), while in the west it is known as far south as Cotopaxi (Ridgely & Greenfield 2001) and possibly Cañar (See Taxonomy and Variation).

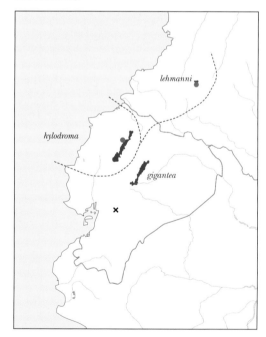

MOVEMENTS There are no studies documenting seasonal movements. In eastern Ecuador (*gigantea*), however, other field biologists (R.A. Gelis *in litt.* 2014) and I have witnessed distinct seasonal changes in the species' abundance at its lower altitudinal limits, most notably at Hacienda SierrAzul, 14km west of Cosanga. During some, but apparently not all, years, Giant Antpitta is fairly abundant in the broad floodplain at the confluence of the Ríos Cosanga and Aragón (*c*.2,300m), always between September and January, i.e. the drier season. It is absent or extremely rare during other months. This area is largely cloaked in various types of second growth including both artificial (pasture) and natural habitats (*Alnus* forests, *Chusquea* bamboo). While present here, Giant Antpittas are relatively silent, rarely singing and only reluctantly responding to playback. They are most frequently encountered alone at dusk and dawn, foraging along trails and at the edges of pastures, with casual observations suggesting that they are not especially territorial during this period. My feeling is that these seasonal fluctuations in abundance represent one (or both) of the following. These individuals may be far-dispersing young (though plumage characters have not been noted), as may occur with Scaled Antpitta (which see). This phenomenon may also represent some degree of altitudinal movement during the non-breeding season, perhaps in response to distribution of resources. This season is when giant earthworms are most frequently seen above ground in this area. The reasons for this are unclear, but their increased abundance is marked and is also capitalised on by overwintering Swainson's Hawks *Buteo swainsoni* in December and January (HFG). Over the past 20 years Giant Antpitta has also made sporadic appearances at Cabañas San Isidro (M. Lysinger *in litt*. 2003), *c*.10km further down the Cosanga Valley (1,950m), but it is unlikely to breed there. I suggest that seasonal movements may be widespread but are infrequently noted due to the relatively silent nature of individuals during these times. The reason this behaviour is quite marked at SierrAzul is simply a product of the 'dead-end' nature of this locality, where further dispersal is hindered by two rivers and results in an artificial concentration of individuals.

HABITAT Inhabits intact, humid and wet montane forest, apparently preferring level ground and swampy areas, but is also occasionally found on steep slopes (Fjeldså & Krabbe 1990, Krabbe & Schulenberg 2003). Sometimes seen foraging in open pasture adjacent to intact forest, especially just before dawn and at dusk, or during foggy weather (de Soye *et al*. 1997, Krabbe & Schulenberg 2003), overall Giant Antpitta is probably a species of relatively undisturbed forest, at least for breeding (see Movements). It is found at elevations of 1,200–3,000m (Chapman 1926a, Meyer de Schauensee 1950b, de Soye *et al*. 1997, Ridgely & Tudor 2009). In north-west Ecuador it occurs mostly at 1,400–2,300m. Records as high as 3,350m are thought to be erroneous (Ridgely & Greenfield 2001) and, in fact, at least in north-west Ecuador, I suggest that records above 2,400–2,500m should be treated cautiously.

VOICE In Ecuador, the song of *G. g. hylodroma* is a fast, quavering, hollow trill lasting *c*.5s, slightly increasing in strength before ending in a crescendo, while that of nominate *gigantea* is described as 'slightly higher pitched and typically lacking a strongly accented ending' (Ridgely & Greenfield 2001). Following Krabbe & Schulenberg (2003), the song of *hylodroma* lasts 4–6s (rarely 7s) and is repeated at intervals of 5–15s. It consists of a rolling series of notes at 0.6–0.9kHz, evenly paced, with 16–18 notes/s. The song gradually increases in volume and pitch (up to *c*.0.1kHz). Frequently, the song ends with a sudden rise in pitch. Overall, it is extremely similar to that of Undulated

Antpitta, but is slightly faster paced. The song of *gigantea* is usually longer than that of *hylodroma* (around 8s) and is slightly faster (19–20 notes/s). The vocalisations of subspecies *lehmanni* are completely undocumented.

NATURAL HISTORY Largely terrestrial and forages as it hops on forest floor, often pausing to peer around and at the ground for long intervals. They frequently pound their heavy bill into the soil, quickly grabbing food items from the surface (de Soye *et al*. 1997, Krabbe & Schulenberg 2003). Singing usually occurs just before dawn (de Soye *et al*. 1997) and around sunset, and is usually performed by adults perched 2–8 m above ground. Foraging activity apparently may continue all day (de Soye *et al*. 1997). Territories in north-west Ecuador (*hylodroma*) are estimated to be *c*.1.0–1.5ha (de Soye *et al*. 1997).

DIET Fjeldså & Krabbe (1990) suggested tadpoles and frogs as food and, although this has not been confirmed, seems likely given the species' massive bill, which would surely have no trouble dismembering small vertebrates. Prey items include large beetles, grubs, slugs, giant earthworms (*Rhynodrylus*) and other large earthworms (Krabbe *et al*. 1994). Giant earthworms are quickly cut into 8cm-long pieces and gobbled down (Krabbe & Schulenberg 2003). As mentioned, worm-feeding stations are regularly visited by *G. gigantea* in several parts of their range (Woods *et al*. 2011) and it seems likely that these are an important part of their natural diet, as suggested for other large antpittas (Greeney *et al*. 2008a, Greeney 2012a).

REPRODUCTION The first (and only) description of the nest and breeding biology comes from a nest studied in north-west Ecuador (*hylodroma*) in November 2007 (Solano-Ugalde *et al*. 2009), from which the following is derived. NEST & BUILDING Over the course of 13 days, a pair of Giant Antpittas was observed carrying nesting material (mainly small dead branches, mosses, lianas, ferns, *Sellaginella* sp., dark rootlets, fibres from petioles and leaflets of *Ceroxylon* palms) to a nest built *c*.7m above ground atop a fairly dense tangle of thin vines and partially supported and attached to a small (18cm dbh) tree. Bulkier matter was placed first (dead branches and larger fibres), with medium-sized materials (ferns, *Sellaginella* sp. and rootlets) added subsequently. These finer materials began to shape the cup, which was firmed up with the final addition of moss and thin vines. Casual observations suggested that one individual gathered most of the material, while the other spent more time arranging it. Adults apparently used their feet and bills to arrange materials and shaped the cup by pressing their breast into the inner depression, with wings pushed back in the manner described for other antpittas (Greeney & Sornoza 2005, Greeney *et al*. 2006) and passerines during nest-cup shaping (Greeney *et al*. 2008b). At this nest, the process of construction lasted at least 13–14 days, but possibly not much longer than this and potentially less if the eggs were laid earlier than was confirmed (see below). The nest was partially obscured by overhanging epiphytes and difficult to detect at most angles. The deep, bulky cup had the following dimensions: outer diameter 27.5cm; outer height 24cm, with an additional 3cm of fibres and moss draped below it in a loose tail; inner diameter 13cm; inner depth 8.2cm. The habitat surrounding the nest was seasonally humid, montane cloud forest, with a fairly dense understorey and subcanopy, and a canopy height of *c*.25m. The nest tree was on fairly steep terrain and located *c.*15m from a small stream. **Egg, Laying & Incubation** At this nest, by 20 December, adult behaviour suggested that incubation was underway, an assumption confirmed two days later using video monitoring. At this nest, clutch size was two eggs, subelliptical and uniform turquoise, although they were not inspected closely. Because the observers were unable to approach the nest frequently, only a very rough estimate of incubation period (11–18 days) was possible. Solano-Ugalde *et al.* (2009) reported adults vocalising at the nest, documenting both full song and shorter modifications of typical song, but did not confirm if both sexes vocalised. Both sexes did participate in incubation, as confirmed by observed changeovers. While incubating, adults occasionally 'napped', arranged material in the nest and engaged in rapid probing (Haftorn 1994, Greeney 2004). Although details were not reported, the authors also suggested that adults occasionally fed one another. Apparently, only one of the two eggs hatched, the other disappearing from the nest several days later. **Nestling & Parental Care** On hatching, the nestling's eyes were closed, the skin was dark grey, and it was covered in fine blackish-grey down. Roughly two weeks after hatching the nestling retained some tufts of natal down on the head, but most contour feathers were fully exposed. The contour feathers, however, differed from adult plumage in having buffy edges and, especially on the head, were overall browner. The mandible was mostly dusky with a paler base and tomia, while the maxilla was dusky with a paler tip. The rictal flanges were enlarged and pinkish-orange, while the mouth lining was bright orange. The back feathers had broad dark fringes while the underparts were slightly more rufescent, with narrower black edges, already showing some scaling. The chin and throat were slightly paler. Flight feathers were almost fully emerged from their sheaths by *c.*18 days after hatching, differing little from the adults, and the wings bore only a few wisps of natal down at this point. Several days after it was last seen in the nest, the chick was found in the dense understorey nearby, where both adults brought it large pieces of earthworms. These observations suggested a nestling period of *c.*22 days. **Seasonality** Krabbe & Schulenberg (2003) reported spontaneous singing from October to January in north-west Ecuador (*hylodroma*) and in October and March in eastern Ecuador (*gigantea*). Until recently, the only direct observations of breeding, apart from one 'immature' and one 'juvenile' in November in Pichincha (*hylodroma*, Fjeldså & Krabbe 1990, Collar *et al.* 1992, Knox & Walters 1994) were those of Greeney & Nunnery (2006), who watched an adult (*hylodroma*) feeding a fledgling pieces of an earthworm in north-west Ecuador. **Breeding Data** *G. g. lehmanni* Subadult, San Marcos, 7 November 1941 (USNM 376732, holotype). *G. g. hylodroma* Building, 28 November 2007, Refugio Paz de las Aves (Solano-Ugalde *et al.* 2007); two nests during incubation, Refugio Paz de las Aves, 12 November 2008 (R.A. Gelis *in litt.* 2008) and 4 January 2010 (R. Ahlman *in litt.* 2016); nestling, 22 March 2015, Refugio Paz de las Aves (J. Patiño *in litt.* 2015); fledgling, 19 April 2001, RP Refugio Natural Pacha Quinde (Greeney & Nunnery 2006); juvenile, 22 July 2010, Tandayapa Bird Lodge (R. Ahlman photos); juvenile, 10 July 1931, Campanario (MLZ 5203); juvenile, 24 September 2011, Refugio Paz de las Aves (R. Ahlman photos); juvenile, 31 May 1955, Río Blanco (= near Mindo, see Paynter 1993) (MECN 2596); two subadults, both 2 August 1936, 'Cerro Castillo' (AMNH 447368/369); subadult, 13 October 2015, RP Tandayapa Bird Lodge (P.W. Wendelken photos); six

subadults, RP Refugio Paz de las Aves, 5 February 2000 (L. Frid photos), 9 February 2006 (M. Reid photos), 25 October 2011 (IBC 987879, G. Poisson), 22 June 2012 (S. Walker photos), 31 May 2016 (ML 33391391, photo D. Summers, IBC 1255452, G. Baker) and 15 November 2016 (ML 43362671/681, photos N. Wingert); three subadults, Gualea, 28 September 2010 (USNM 376961), 8 August 1921 (AMNH 173010) and 14 June 2013 (AMNH 124426, holotype); subadult, 22 July 1998, RP Bellavista (MECN 2475); subadult, 11 August 2017, El Refugio de Intag Lodge (ML 65802921, photo H. Wolf); two records of adults carrying food (= nestling or fledgling), RP Refugio Paz de las Aves, 15 September 2010 (D. Faulkner *in litt.* 2015), and 28 November 2014 (Flickr: R. Gowan); and the following specimens with incomplete data, subadult, near Mindo (MECN 2595); subadult, 1909, Gualea (AMNH 173012). *G. g. gigantea* Juvenile transitioning, 10 September 2011, La Guatemala (S. Woods *in litt.* 2015); juvenile transitioning, 23 November 2013, RP Río Zuñac (photos M. Tuston, L. Jost *in litt.* 2017); subadult, 12 October 1993, RP SierrAzul (MECN 6841).

TECHNICAL DESCRIPTION Sexes similar. The following description refers to subspecies *hylodroma*. *Adult* Forecrown unmarked rufous-chestnut, very slightly paler on lores, mixed with olivaceous and grading to slate-grey on crown and upper nape, but not extending to upper back as sometimes illustrated (e.g. Ridgely & Tudor 1994, Ridgely & Greenfield 2001, Krabbe & Schulenberg 2003). Rest of upperparts olivaceous-brown. Wings overall similar to back, except leading vanes of primaries, which are rusty-brown to chestnut. According to Ménégaux & Hellmayr (1906) and Wetmore (1945), the underwing-coverts are 'russet', with a few small scattered sooty-black spots. Face, ear-coverts and underparts rufous-chestnut, narrowly barred black. This rufous-chestnut coloration is deepest (darkest) on the face and ear-coverts such that the black barring is barely visible. Conversely, palest on central throat, approaching rufescent-tawny. Below, the black barring is broadest on upper chest and, in many individuals especially in certain postures, the broad barring of the upper breast forms a fairly distinct black collar just below the pale throat. The barring becomes finer on the rear underparts and, in full adults, it extends across the belly, vent and most of the throat. This is contrary to the descriptions of previous authors, like Wetmore (1945), who stated that barring in these areas is 'nearly absent' in some individuals, a trait that I consider to be a reflection of subadult plumage. This inference is further supported by the illustration in Dubois (1900), which clearly shows other immature plumage characters such as rufescent edges to the upperwing-coverts or scattered feathers of the back, as well as a finely barred rump. In reference to this figure, Wetmore (1945) stated that it 'differs from any specimen seen in the restriction of the black barring on the under surface', which is illustrated as completely unbarred on the flanks, belly and vent. The rufous-chestnut coloration of the underparts is richest on the upper breast, fading lower and along flanks, with the central belly and vent palest, only slightly darker than throat. **Adult Bare Parts** *Iris* warm brown; *Bill* dusky pinkish, maxilla darker, approaching black on culmen and paler on tomia and at tip, mandible paler overall, palest at base and tip; *Tarsi & Toes* dark flesh-brown to dusky-pinkish. There is a somewhat variable amount of bare postocular skin, which is pale pink to bluish-pink. This bare skin behind the eye, despite being illustrated by Ménégaux (1911), appears to have remained unmentioned and unillustrated in subsequent works, although it is present in all individuals I have observed either directly or in photos. *Juvenile* Not previously well described. Wetmore (1945) noted that the 'plumage of the first year' of nominate *gigantea* bears rufescent and black feather tips and bars on the wing-coverts, back and rump. He further pointed out that the plate in Dubois (1900), apparently of only the second-known specimen of Giant Antpitta (Sclater 1890), and frustratingly bearing a locality label of 'Ecuador' is, in fact, a juvenile of race *hylodroma*. Hilty & Brown (1986) also note that immatures 'show a few tawny rufous and black bars on mantle'. Juveniles are said to have 'narrow ochraceous tipping' on the upperwing-coverts (Ridgely & Tudor 2009). Restall *et al.* (2006) illustrated race *lehmanni* in subadult plumage, stating that juveniles are paler brown from forehead to tail, with no grey on head and irregular spotting on back and wings formed by black subterminal spots and orange fringes. They indicate that, prior to moulting into full adult plumage, young birds (presumably also of *lehmanni*) develop the grey crown and nape of adults, but retain 'spotting on back and wings'. The youngest individual I have examined, via photos provided by R. Ahlman of race *hylodroma* (see Reproduction Seasonality), represent a juvenile-plumaged bird presumably somewhere in between the descriptions of Restall *et al.* (2006) and those of the other authors mentioned above. This individual was generally similar to adults, but the forecrown was duller, more rusty- or orange-brown, the crown was already slate-grey but the hindcrown and nape were finely barred ochraceous-orange and blackish, with only a few adult feathers. The remaining upperparts were similar in coloration to adults, but some black-barred immature plumage was retained on the scapular region and upper back, and at least one-third of the back feathers were broadly tipped tawny-ochraceous and had a thin, subterminal blackish band. The feathers of the face were brownish-ochraceous, similar to adults, but duller, and the ear-coverts appeared rather mottled. Below, this individual was also largely similar to adult plumage, but overall was paler ochraceous. The wings and tail were as adults, but the upperwing-coverts were broadly tipped ochraceous with thin, black subterminal bands. The rictus was noticeably paler than in adults, pinkish-white, but the bill was otherwise similar to that of adults. *Subadult* I examined what was presumably the same immature bird photographed as a juvenile by R. Ahlman, arriving at the same worm-feeding station, *c.*4 months later. It was now almost completely in adult plumage but the upperwing-coverts were still distinctly banded as described above. The back still bore a scattering of feathers with immature barring, but much fewer than previously, and these could easily be overlooked in the field. Bare-parts coloration was largely as in adults.

MORPHOMETRIC DATA *G. g. hylodroma* The holotype (USNM 376961) male *hylodroma* measured by Wetmore (1945): *Wing* 152.6mm; *Tail* 61.8mm; *Bill* culmen from base 38.7mm; *Tarsus* 70.6mm. I took the following measurements from the same specimen: *Bill* from nares 24.1mm, exposed culmen 39.0mm, depth at front of nares 14.4mm, width at gape 21.5mm, width at front of nares 10.1mm, width at base of mandible 19.0mm; *Tarsus*

60.4mm. Additional measurements: *Wing* 144.2–152.6mm, mean 148.3mm; *Tail* range 54.5–61.8mm, mean 57.0mm; *Bill* culmen from base range 38.7–40.0mm, mean 39.4mm; *Tarsus* range 66.3–70.6mm, mean 69.7mm (n = 4♂♂, but only three for tarsus, type series, Wetmore 1945). *Wing* 153mm; *Tail* 67mm; *Bill* 39mm; *Tarsus* 58.5mm (n = 1♀, Ménégaux & Hellmayr 1906). *Wing* 147.3mm; *Tail* 61.3mm; *Bill* culmen from base 39mm, (n = 1♀, Wetmore 1945). *Wing* 165mm; *Tail* 80mm; *Bill* total culmen 32.7mm, 'width at base' 14.8mm, from nares 25.5mm, width at nares 11.6mm (n = 1, ♀?, de Soye *et al.* 1997). *G. g. gigantea* Wetmore (1945) provided the following measurements for Lawrence's holotype (USNM 35101): *Wing* 142.2mm; *Tail* 54.3mm; *Bill* [total culmen] 39.4mm; *Tarsus* 69.8mm. I measured the same specimen: *Bill* from nares 23.3mm, exposed culmen 35.9mm, depth at front of nares 14.5mm, width at gape 19.4mm, width at front of nares 10.3mm, width at base of mandible 18.4mm; *Tarsus* left 71.0mm, right 70.6mm. Additional measurements include: *Wing* 155mm; *Bill* from nares 22.8mm, depth at front of nares 15.3mm, width at front of nares 10.0mm; *Tarsus* 68.3mm (n = 1♂, FMNH 73911, E.T. Miller *in litt.* 2013). *Bill* from nares 23.9mm, exposed culmen 34.6mm, width at gape 20.2mm; *Tarsus* 71.0mm (n = 1, ♀?, MLZ 5204). *Bill* exposed culmen 36.6mm, width at gape 19.0mm (n = 1♀, juvenile, MLZ 5203). *Bill* from nares 23.9mm, 24.5mm, 24.1mm, 21.2mm; *Tarsus* -?-, 70.7mm, 66.7mm, 65.3mm (n = 4, ♀,♂,♂,?, subadults, AMNH). *G. g. lehmanni* *Wing* 154.1mm; *Tail* 60.7mm; *Bill* culmen from base 38mm; *Tarsus* 73mm (n = 1, ♀?, holotype USNM 376732, Wetmore 1945). *Bill* from nares 21.9mm, exposed culmen 35.0mm, depth at front of nares 15.0mm, width at gape 20.7mm, width at front of nares 11.2mm, width at base of mandible 17.9mm; *Tarsus* left 72.2mm, right 72.0mm (n = 1, holotype, my measurements). **MASS** 204–266g (n = 3, ♂♂, *gigantea*, Krabbe & Schulenberg 2003); 266g (n = 1♂, *gigantea*, Krabbe *et al.* 1994); 218g (n = 1♂, *hylodroma*, Krabbe *et al.* 1994); 254g (n = 1, ♀?, *hylodroma*, de Soye *et al.* 1997). **TOTAL LENGTH** 22.5–26.7cm (Hilty & Brown 1986, Dunning 1987, Fjeldså & Krabbe 1990, Ridgely & Greenfield 2001, Krabbe & Schulenberg 2003, Ridgely & Tudor 2009, McMullan *et al.* 2010).

TAXONOMY AND VARIATION Three subspecies are currently recognised but a revision is needed. Together, Great Antpitta, Undulated Antpitta and Giant Antpitta form a fairly uniform group of closely related species that are considered members of the subgenus *Grallaria* (Lowery & O'Neill 1969), the remaining members of which are somewhat smaller. Giant Antpitta and Great Antpitta have been suggested to be conspecific (Krabbe & Schulenberg 2003). Alternatively, nominate *gigantea* and *hylodroma* Giant Antpittas might be best considered as a species together, while the more northern subspecies, *lehmanni*, may be better treated as a race of Great Antpitta as it shows more similarities to the latter than do the other two subspecies of Giant Antpitta (Wetmore 1945). All things considered, and putting aside for a moment the matter of *lehmanni*, it seems most likely to me that *hylodroma* should be elevated to species status, while maintaining both *gigantea* and *excelsa* as valid species, and subsuming *lehmanni* within nominate *gigantea*.

Grallaria gigantea lehmanni Wetmore, 1945, Proceedings of the Biological Society of Washington, vol. 58, pp 19–20. Race *lehmanni* was described from a single, unsexed specimen (USNM 376732, Deignan 1961) collected 7 November 1941 at 3,000m, San Marcos, Moscopán, Dept. Cauca, in south-west Colombia at the head of Magdalena Valley on the east slope of the Central Andes (*c*.02°22'N, 76°12'W). Dr F. Carlos Lehmann-V., the collector of the holotype, indicated at the time of its description that another specimen was at MNHUC in Popayán, Colombia. This specimen was said to be a female (age not indicated) collected 8 March 1944 at Tijeras (02°22'N, 76°16'W), in the same area as the holotype, but at 2,300m (Wetmore 1945). Apparently in reference to the same specimen, Lehmann-V. (1957) stated that it was an adult male and the elevation of the collecting locality was 2,800m. No one, apparently, published a description or any further details concerning this second specimen. A third specimen of *lehmanni* (WFVZ 11892) was reported from Cerro Munchique on the east slope of the West Andes in Cauca (Collar *et al.* 1992, Negret 1994), but this was later determined to be a misidentified adult male *G. s. squamigera* (Collar *et al.* 1994, Krabbe *et al.* 1994). To date, no other specimens or published records of *lehmanni* are known (Hilty 1985, Krabbe *et al.* 1994) and it is possible that this taxon is now extinct. While the holotype is still on deposit at the Smithsonian, I was sadly informed by Dr Luis Germán Gómez, currently at MNHUC, that the above-mentioned female specimen is no longer extant, having apparently been destroyed in the earthquake that hit Popayán on 31 March 1983 (Hoyos 2013), along with many other undoubtedly valuable specimens. Overall, *lehmanni* is apparently most similar to nominate *gigantea*. Published descriptions, however, are somewhat conflicting. Fjeldså & Krabbe (1990) and Krabbe *et al.* (1994) stated that *lehmanni* has even more heavily barred underparts than *gigantea*. Conversely, the text and figure in Restall *et al.* (2006) suggested that *lehmanni* is the least barred of the three subspecies, both in terms of the number of bars as well as their thickness. In his description of the type, Wetmore (1945) described *lehmanni* as differing from *gigantea* by having the barring on the central underparts 'definitely heavier'. Interpretation of these different comparative remarks, however, is complicated (or perhaps explained) by the fact that Wetmore (1945) stated of *lehmanni* that 'part of wing coverts, and a few feathers on lower back tipped with tawny and barred subterminally with dull black; upper tail coverts tipped also with tawny, with subterminal barring dark to deep neutral gray.' He was, then, apparently basing his description of *lehmanni* on immature plumage. Wetmore (1945) described *lehmanni* as follows: 'forehead tawny basally, with tips of olive-brown, becoming olive brown that extends back over pileum past posterior angle of eyes; back of pileum and upper hindneck rather dull neutral gray; lower hindneck, back, wings and tail olive brown; primaries edged narrowly externally with cinnamon-brown, this changing on the outermost to a narrow outer margin of ochraceous-tawny; lores tawny; rictal bristles black; sides of head and neck tawny, barred with dull black; center of throat and upper foreneck ivory yellow, barred with black; chin, foreneck (except for ivory yellow area), upper breast, sides and flanks ochraceous-tawny, heavily barred with black; center of breast and abdomen light ochraceous buff heavily barred with black; undertail coverts ochraceous-tawny without bars or spots; under wing coverts tawny to ochraceous-tawny heavily barred and spotted with black. Maxilla dusky neutral gray; mandible benzo brown; tarsus benzo brown; toes fuscous (from dried skin).' Wetmore (1945) further compared the

specimen with *hylodroma*: 'From *Grallaria g. hylodroma* the type of this race differs in being decidedly less rufescent on the lower surface, with the dark bars much heavier, these being heavy and distinct on the flanks where they are weak or absent in the west Ecuadorian birds. The under wing coverts are heavily barred and spotted with black as in typical *gigantea*. The series of *hylodroma* is so uniform in maintaining these differences that I do not hesitate to describe *lehmanni*, even though only a single specimen is available.' Given the paucity of comparative material for *lehmanni* and because it is based on immature plumage, the taxon's validity remains in question. Based on written descriptions of *lehmanni* and my own observations of *gigantea* and *hylodroma*, I suggest that *lehmanni* may best be considered synonymous with nominate *gigantea*.

Grallaria gigantea hylodroma Wetmore, 1945, Proceedings of the Biological Society of Washington, vol. 58, pp 18–19. Race *hylodroma* was described from Gualea, 6,000ft, Prov. Pichincha, north-western Ecuador [2,000m, 00°05'N, 78°44.5'W; corrected slightly from Paynter 1993]. The holotype is a male (AMNH 124426) collected 14 June 1913 by William B. Richardson. The distribution of *hylodroma* is poorly known, but it is confined to the West Andes from Nariño to Cotopaxi (and possibly Cañar, see below). Ridgely & Greenfield (2001) excluded records from Lloa, because the town lies outside the presumed altitudinal range of Giant Antpitta. Less than 10km west of there, however, just below Chinguil, there appears to be acceptable Giant Antpitta habitat. Records of specimens from 'El Tambo', Loja (Meyer de Schauensee 1966, Collar *et al.* 1992) were discussed in the literature (Paynter 1993, Krabbe *et al.* 1994), but are now thought to actually be from Cañar, further north (Ridgely & Greenfied 2001). Its continued presence this far south in the West Andes has not been confirmed nor, for that matter, has the true location of 'El Tambo' been uncovered. In general, most authors agree that the most apparent difference in plumage from *gigantea* is the much deeper, rufous-chestnut coloration of the underparts of *hylodroma* (Wetmore 1945, Ridgely & Greenfield 2001, Krabbe & Schulenberg 2003, Ridgely & Tudor 2009). Interestingly, although this fact should not alter currently defined taxon limits, the holotype is in subadult plumage. Wetmore (1945) stated that the 'wing coverts [have] a slight margin of russet, the inner greater coverts [have] a narrow subterminal bar of dull black and a narrow tip of russet' and that the 'back, scapulars and rump [have] a few feathers tipped with russet crossbarred narrowly with one or two narrow bands of dull black, [with the] rectrices very narrowly tipped with russet.' My examination of the holotype confirmed the presence of these characters, which are clearly remnants of immature plumage. SPECIMENS & RECORDS COLOMBIA: *Nariño* RNSC Planada, 01°09'N, 78°00'W (Willis & Schuchmann 1993, de Soye *et al.* 1997; GPS incorrect in publications). ECUADOR *Imbabura* El Refugio de Intag Lodge, 00°22.5'N, 78°28.5'W (ML 65802921, photo H. Wolf). *Pichincha* RP Santa Lucía, 00°07'N, 78°36.5'W (HFG); RP Tandayapa Bird Lodge, 00°00'N, 78°40.5'W (MECN 6850); RP Refugio Natural Pacha Quinde, 00°00.5'S, 78°40.5'W (Greeney & Nunnery 2006); 'Cerro Castillo' [= near Mindo; Paynter 1993] (AMNH 447368/69); Mindo and vicinity, 00°03'S, 78°46'W (de Soye *et al.* 1997); Miraflores, 00°03'N, 78°43'W (C. Hesse *in litt.* 2017); Guarumos, 00°03'S, 78°38'W (Berlioz 1937;

coordinates adjusted slightly from Paynter 1993 based on elevation); RP Refugio Paz de las Aves, 00°00.5'N, 78°43'W (Solano-Ugalde *et al.* 2009, Woods *et al.* 2011, XC, ML, IBC); RP Las Gralarias, 00°01'S, 78°45'W (XC 5585, N. Athanas); RP Bellavista Cloud Forest Lodge, 00°01'S, 78°41'W (Welford 2000, Welford & Nunnery 2001, MECN 2475); San Tadeo, 00°02'S, 78°44'W (Berlioz 1937, MECN 2597); Campanario, 00°02'S, 78°40'W (MLZ 5203/04, loc. T. Santander *in litt.* 2016); Lloa [below?], *c.*2km W of Chinguil, *c.*00°12.5'S, 78°40'W (NHMUK 1938.12.20.1, Knox & Walters 1994 see discussion above). *Santo Domingo de las Tsáchilas* BP Río Guajalito, 00°14'S, 78°49'W (J. Freile *in litt.* 2017); BP Río Lelia, *c.*00°20.5'S, 78°59'W (XC 212925, J.C.C. Arteaga). *Cotopaxi* Reserva Integral Otonga, 00°25'S, 79°00'W (Freile & Chaves 2004); Las Palmas, 00°35'S, 79°01'W (XC 248862/63, N. Krabbe); Las Palmas–La Chala road, 00°36'S, 79°04'W (XC 332918/19, F. Sornoza); Caripero, 00°36'S, 79°00'W (Ridgely & Greenfield 2001). *Cañar* 'El Tambo' (see above), 02°30'S, 78°54'W (ANSP 163658/9700).

Grallaria gigantea gigantea Lawrence, 1866, Annals of the Lyceum of Natural History of New York, vol. 8, p. 345. The type specimen of the nominate race (USNM 35101) bears only 'Ecuador' as the locality (Deignan 1961). With respect to the collection locality, Wetmore (1945) stated that the specimen 'came to Baird with a small lot of skins from John Akhurst, a dealer in natural history material of Brooklyn with a letter dated November 25, 1864.' He felt that it was most likely from the east slope of the Andes, noting that birds delivered by the 'dealer' included Bare-necked Fruitcrow *Gymnoderus foetidus* and Northern Mountain Cacique *Cacicus l. leucoramphus*, two taxa which do not occur on the west slopes of the Ecuadorian Andes. Race *gigantea* is, so far as known, endemic to the east slope of the Ecuadorian Andes, from eastern Carchi (Salvadori & Festa 1899) and Napo (Krabbe 1991) south to Volcán Tungurahua (Meyer de Schauensee 1966, Collar *et al.* 1992). Nominate *gigantea* is larger than *hylodroma* and differs by being distinctly barred instead of spotted on the underwing-coverts. It is also paler overall below (Krabbe *et al.* 1994). The following description is based on Lawrence (1866) and Wetmore (1945). Above, the forecrown is dark, rich rufous, the hindcrown and nape are dark lead-grey, grading on the upper back to rich olive-brown which extends to the rump. The head, lores, face, neck-sides and ear-coverts are dark rufous. The wings are brown, slightly warmer than the upperparts. Underparts are described as mostly dark rufous, paler (more pale ochraceous-buff) at the rear and approaching tawny-white on the central belly. The neck-sides and rufous underparts are finely barred black, more prominently on the flanks, but the undertail-coverts are apparently unbarred. Lawrence (1866) stated that the type specimen bore 'a few minute black spots' on the undertail-coverts. The wording of Lawrence's (1866) account implies that the belly and vent are also barred, but other authors described the vent and lower belly as unbarred (Restall *et al.* 2006), a character perhaps reflective of age (see Description of *hylodroma* above). The tail is brown, somewhat warmer than the brownish-olive back, and similar to the wings. The rectrices have blackish-brown shafts and indistinct olive-brown margins. The underwing-coverts are rufous, heavily banded and spotted dark grey-brown. The bare-part colours given by Salvadori & Festa (1899) of an adult male were: iris red-brown; bill brown; tarsi plumbeous. My own observations,

however, suggest that they are nearly identical to the bare parts of *hylodroma*, including the bare postocular patch. SPECIMENS & RECORDS *Carchi* 'El Pun', 00°40'N, 77°37'W (Salvadori & Festa 1899). *Napo* RP SierrAzul (= Hacienda Aragón in older literature), 00°40'S, 77°55'W (ANSP 185501/6178, MECN 6841, XC); Cordillera Guacamayos, 00°37.5'S, 77°50.5'W (ML 57097/74695, M.B. Robbins). *Pastaza* RP Río Zuñac, 01°22.5'S, 78°08'.5W (L. Jost *in litt.* 2016). *Tungurahua* RP Viscaya, 01°21'S, 78°23'W (L. Jost *in litt.* 2016); La Guatemala, 01°23'S, 78°10'W (S. Woods *in litt.* 2015); RP Machay, 01°23'S, 78°16'W (L. Jost *in litt.* 2016); 'Runtun Hills' near Baños, 01°24'S, 78°25'W (ANSP 169698); RP Cerro Candelaria, 01°28.5'S, 78°19.5'W (L. Jost *in litt.* 2016).

STATUS Found in both the Northern Central Andes EBA (*gigantea*) and Chocó EBA (*hylodroma*), with the historical records of *lehmanni* from the Colombian inter-Andean slope EBA (Stattersfield *et al.* 1998). For many years there was almost no information available on Giant Antpitta, and it was considered Data Deficient (Groombridge 1993). More recently the species is currently considered Vulnerable globally (Collar *et al.* 1994, BirdLife International 2017) and at a national level in Colombia (Renjifo *et al.* 2014). Previously afforded Vulnerable status in Ecuador (Granizo *et al.* 2002), Freile *et al.* (2010b) suggested its status there be raised to Endangered. Given that the healthiest known populations are those of *hylodroma* in north-west Ecuador, perhaps its status should be upgraded at a global level as well? Certainly, like other montane species, habitat loss is a large and ongoing threat (Stattersfield & Capper 2000). Nonetheless, in eastern Ecuador, where the species can occasionally be locally common, one large ecological reserve and four national parks cover nearly 60% of the species' potential range in Ecuador (Freile 2000). PROTECTED POPULATIONS At least historically, race *lehmanni* occurred in PNN Puracé (Negret 1994, Wege & Long 1995, Ayerbe-Quiñones *et al.* 2008). *G. g. hylodroma* Colombia PNN Munchique (Negret 1994); RNSC La Planada (de Soye *et al.* 1997). Ecuador BP Mindo-Nambillo (Kirwan & Marlow 1996); BP Río Lelia (XC: J.C.C. Arteaga); BP Río Guajalito (BirdLife International 2017); Reserva Integral Otonga (Freile & Chaves 2004), RP Refugio Paz de las Aves (Collins 2006, Solano-Ugalde *et al.* 2009, XC, ML, IBC); RP Bellavista (Welford 2000, Welford & Nunnery 2001, MECN), RP Las Gralarias (XC: N. Athanas, J. King); RP Santa Lucía, RP San Jorge de Tandayapa (HFG); RP Tandayapa Bird Lodge (MECN 6850); RVS Pasochoa (J. Freile *in litt.* 2017); RBP Maquipucuna (HFG); RP Refugio Natural Pacha Quinde (Greeney & Nunnery 2006). *G. g. gigantea* PN Sumaco-Galeras (ML: M.B. Robbins); PN Llanganates (L. Jost *in litt.* 2017); RE Antisana (HFG); RP Río Zuñac, RP Viscaya, RP Machay, RP Cerro Candelaria (L. Jost *in litt.* 2017); RP Cabañas San Isidro, and RP Estación Biológica Yanayacu (HFG); RP SierrAzul (ANSP, MECN, XC).

OTHER NAMES English Greater Undulated Ant-Thrush (Brabourne & Chubb 1912); Pichincha Antpitta [*hylodroma*] (Monroe & Sibley 1993, Ridgely & Greenfield 2001). Spanish Tororoí Gigante (Krabbe & Schulenberg 2003); Gralaria gigante (Valarezo-Delgado 1984, Ortiz-Crespo *et al.* 1990, Granizo 2009); Tóbalo grande (Valarezo-Delgado 1984); Peón Gigante (Negret 2001); Licuango Grande, Tóbalo de Montaña (Granizo *et al.* 2002). French Grallaire géante. German Riesenameisenpitta (Krabbe & Schulenberg 2003).

Giant Antpitta, adult (*hylodroma*), Refugio Paz de las Aves, Pichincha, Ecuador, 16 October 2012 (*Dušan M. Brinkhuizen*).

Giant Antpitta, adult gathering prey for its young (*hylodroma*), Refugio Paz de las Aves, Pichincha, Ecuador, 16 January 2009 (*Jon Hornbuckle*).

Giant Antpitta, subadult (*gigantea*), La Guatamala Trail, Tungurahua, Ecuador, 10 September 2011 (*Sam Woods*).

Giant Antpitta, juvenile (*hylodroma*), Tandayapa Bird Lodge, Pichincha, Ecuador, 22 July 2010 (*Roger Ahlman*).

Giant Antpitta, older nestling in nest (*hylodroma*), RP Paz de las Aves, Pichincha, Ecuador, 16 December 2014 (*Murray Cooper*).

Giant Antpitta, adult with single nestling (*hylodroma*), RP Paz de las Aves, Pichincha, Ecuador, 16 December 2014 (*Murray Cooper*).

GREAT ANTPITTA
Grallaria excelsa   Plate 6

Grallaria excelsa Berlepsch, 1893, Ornithologische Monatsberichte, vol. 1, p. 11, Montaña Aricagua, Andes of Mérida, western Venezuela [08°13'N, 71°08'W; Paynter 1982].

'When *Grallaria excelsa phelpsi* was brought into camp the residents of Colonia Tovar were as fascinated as we were. At least twenty natives viewed it but, with two exceptions, none had seen the bird before. One young man claimed to have killed a *Grallaria* several years previously but had never seen another. The second, Florencio Ruthman, who collected our specimen in a dense belt of forest, stated that, although he had never seen such a bird before, he had imitated its strange whistle and lured it out of the deep forest debris. At our request he imitated the whistle a number of times. It sounded very much like the low vibrant note created by blowing over the neck of an empty bottle.' **E.T. Gilliard, 1939, Colonia Tovar, Estado Aragua, Venezuela**

Great Antpitta is thought to form a superspecies with Giant Antpitta, which generally replaces it further south in the Andes. The two are so closely related that some authors believe them to be conspecific or, alternatively, that at least one subspecies of Giant Antpitta is actually better placed with Great Antpitta. Unlike its sister species, Great Antpitta is extremely poorly known, with the only detailed field study pertaining to its nesting habits. Currently, most aspects of its biology and distribution remain a mystery. Great Antpitta is ranked as globally Vulnerable. Until recently, it was virtually unknown in life, and it is thought that many early records probably involved misidentified Undulated or Plain-backed Antpittas. It is endemic to humid cloud forests at 1,600–2,300m in the Andes and north-coastal ranges of Venezuela, where the species has been observed at just a handful of localities. There is no definite evidence that Great Antpitta occurs in immediately adjacent Colombia, although its presence has been suggested. Proposed conservation measures include the use of playback to assess its current distribution and ecological requirements, especially at upper elevations of the Coastal Cordillera, where *phelpsi* has not been found in recent years (Sharpe & Ascanio 2008). A general understanding of its vocalisations is also needed, to aid in these surveys, as the species may prove to be more widely distributed once its song is better known (Krabbe & Schulenberg 2003). For a proper understanding of its conservation needs, detailed studies are needed to determine whether either subspecies is, in fact, valid.

IDENTIFICATION 23–25cm. One of the largest *Grallaria*, Great Antpitta is similar in overall form to other putative relatives. The following description is based on Krabbe & Schudulenberg (2003). Sexes similar. Adult has whitish lores, throat and eye-ring, but most of the crown is brown, becoming grey on the nape. Rest of upperparts olive-brown (contrasting with grey hindcrown). Cheeks and most of underparts buff, finely barred black, with buff of underparts palest, approaching white, just below throat, forming a crescent. Superficially resembles Undulated Antpitta, but is noticeably larger, with less contrasting crown and no malar stripe. Upperparts coloration appears more brownish-olive and slate-grey feathers on head are restricted to the hindcrown and nape. In overall shape, both species present the typical, stout, long-legged appearance of a large *Grallaria*, but the bill is noticeably stouter and (although difficult to see in the field) the tarsi are thicker as well (Cory & Hellmayr 1924, Meyer de Schauensee 1970, Hilty 2003).

DISTRIBUTION Endemic to Venezuela, with most of its range along the east slope of the Venezuelan Andes (Cordillera de Mérida) from eastern Táchira, north through Mérida, to south-east Trujillo and south-east Lara (Meyer de Schauensee & Phelps 1978, Ridgely & Tudor 1994). Historically, there are specimens from the Sierra de Perijá in north-west Zulia, the Cordillera de la Costa in Aragua (race *phelpsi*) and one locality on the west slope of the Venezuelan Ande. The only modern records, however, are from the east slope of the Venezuelan Andes.

MOVEMENTS None described, but if the ecology of Great Antpitta is similar to that of Giant Antpitta, this species may show seasonal movements across elevations.

HABITAT An inhabitant of the understorey of humid cloud forest, especially along creeks, at the edges of treefall gaps and near large, fallen trees (Schäfer & Phelps 1954, Kofoed & Auer 2004, Ridgely & Tudor 2009). Few specific or quantified data are available, however its habitat preferences are probably similar to Giant Antpitta. They may tolerate some level of disturbance, but this is not well documented (Manne & Pimm 2001). The altitudinal range is generally given as 1,600–2,300m (Meyer de Schauensee & Phelps 1978, Giner & Bosque 1998, Pelayo & Soriano

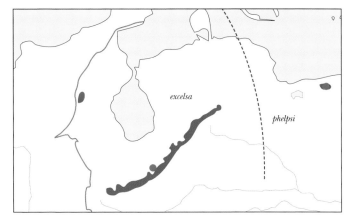

2010), while Krabbe & Schulenberg (2003) noted that it is most common above 2,000m.

VOICE The dawn song is a series of low-frequency, vibrating, hollow notes, lasting on average 4.3s (Krabbe & Schulenberg 2003). It is very similar to that of Undulated Antpitta, but differs by accelerating slightly and ending abruptly. This song is also similar to the voice of Vermiculated Screech-owl *Megascops roraimae*, although the latter's vocalisation is softer, slower and ends smoothly rather than abruptly. Kofoed & Auer (2004) reported hearing this vocalisation, noting that this 4–5s phrase is repeated at regular 10s intervals and is delivered at 7.1–8.2kHz and a rate of 22 notes/s.

NATURAL HISTORY Probably monogamous, the species appears to forage alone or in pairs, rarely or never found with flocks (Kofoed & Auer 2004). Scant observations suggest that it is highly terrestrial, at least while foraging, hopping and pausing while searching for food on the ground with head slightly tilted, darting quickly forward to grab prey (Krabbe & Schulenberg 2003, Kofoed & Auer 2004). After swallowing food, they often bob up and down several times while puffing out their tail feathers and flicking their wings (Krabbe & Schulenberg 2003). Like other antpittas (Greeney 2012b), they apparently forage, at least opportunistically, with army ant swarms (Kofoed & Auer 2004). While not foraging, adults may climb as high as 4–5m and song perches are generally *c*.4m above ground (Kofoed & Auer 2004). As they deliver their trilled song, the throat appears to gradually inflate, the feathers puffing outward suddenly, simultaneous with the song's abrupt ending (IBC 1157203, video D. Ascanio).

DIET No information is available, although with a species this large and with such a heavy bill, it is undoubtedly capable of taking small vertebrates, along with the likely staple diet of invertebrates. Kofoed & Auer (2004) noted that earthworms made up a large percentage of prey items delivered to nestlings, as is the case in other *Grallaria* (Greeney *et al.* 2008a).

REPRODUCTION Our knowledge of the nesting biology is, to date, based on one study of three nests found in PN Yacambú (race *excelsa*; Kofoed & Auer 2004). The following details are taken from this study. **NEST & BUILDING** The nest is a large, bulky, open cup, similar to the putative relatives, Giant Antpitta (Solano-Ugalde *et al.* 2009) and Undulated Antpitta (Greeney & Juiña 2011). It is composed externally of densely packed mosses, rootlets, leaves and small twigs, and is lined with a thick matt of black rootlets and fungal rhizomorphs. Nests are supported by dense clusters of epiphytes and branches on sides of trees. Mean nest height is 3.8 m above ground. Both adults participated in all aspects of nest construction. **EGG, LAYING & INCUBATION** Eggs are subelliptical, turquoise-blue and unmarked. Incubation duties are shared by both adults. During incubation the adults warm the eggs for a mean 98.8% of daylight hours. **NESTLING & PARENTAL CARE** Nestlings are fed at seemingly rather low rates, despite both adults contributing, with less than one food delivery per hour. Kofoed & Auer (2004) described nestling plumage on the day that the primary pin-feathers broke through their sheaths as 'similar to that of the adults [with] the breast feathers [...] scalloped, and its down and emerging tail and wing feathers [...] light olive-brown.' With respect to bare-parts coloration they described the bill as bright orange, and the gape as

being a 'striking, bright-red hue.' Using photos provided by the authors, Greeney (2015d) amplified this description as follows. 'Remnants of the natal down remain, mostly around the head and upper parts, indicating that it was black or dark gray. Contour feathers, however, are already emerging from their sheaths. The head, back, and breast are tawny or yellowish-buff with fine black barring. The upper breast and sides of the underparts appear to be somewhat darker tawny-orange, while the lower breast and belly are paler. The sides of the head are overall black or dark gray, showing less buffy barring than other areas. The tarsi and toes are pinkish gray, approaching dusky violet. The maxilla is dusky, black along the culmen, becoming pink-orange along the tomia and near the base posterior to the nares. It has a small whitish tip and a bright white egg-tooth still attached. The mandible is almost entirely pink-orange, paler near the tip, and the rictal flanges are inflated and bright orange to orange-red.' Given what we know as to the condition at fledging of other *Grallaria*, it is likely that this description will be roughly accurate for young fresh from the nest. Similarly, using the progression of feather development known for related species post-fledging (see Giant Antpitta and Undulated Antpitta), young probably begin to lose this black-barred, yellowish-buff plumage gradually, starting on the rump and lower back, followed by the underparts and crown. By 3–4 months post-fledging the plumage will probably be similar overall to that of adults, with the exception of remnants of nestling plumage on the nape and buffy-orange scaling on some feathers of the upper back and wing-coverts. It is not yet clear how long subadults retain this immature plumage. **SEASONALITY** All available reproductive evidence suggests that Great Antpitta breeds during the wet season in Venezuela, probably initiating most clutches in May and June (Schäfer 1969, Kofoed & Auer 2004). **BREEDING DATA** *G. e. phelpsi* Subadult, 14 November 1937, Colonia Tovar (AMNH 322974; holotype). *G. e. excelsa* Only three active nests found, at PN Yacambú in 2003, with eggs on 25 May, nestlings 5 June, under construction 18 June (Kofoed & Auer 2004); two subadults, Guárico, 30 January (USNM 313771) and 6 February 1911 (AMNH 428768).

TECHNICAL DESCRIPTION Sexes similar. The following description of plumage and bare parts is derived from Restall *et al.* (2006) and Hilty (2003). *Adult* The forecrown is brown or olive-ochre, becoming grey with faint black scaling on midcrown and nape. Rest of upperparts olive-brown. Wings and tail similiar in colour to back, with inner web of remiges conspicuously fringed ochraceous near their bases. Central throat clean white to buffy-white, bordered by a blackish submalar stripe. In *excelsa* (see Geographic Variation) these throat markings extend somewhat lower, onto upper breast. Head- and neck-sides, and rest of underparts ochraceous to pale rufous, boldly scalloped with wavy black lines. Lores can be white (*excelsa*) or similar in coloration to underparts (See Geographic Variation). Undertail-coverts bright tawny-ochre. **Adult Bare Parts** *Iris* dark brown; *Bill* mandible black, maxilla paler, greyish-horn or dusky-pinkish; *Tarsi & Toes* dark grey to greenish-brown. *Immature* Undescribed, but I agree with Restall *et al.* (2006) that plumage characteristics and moult progression are likely very similar to Undulated and Giant Antpittas. Indeed, based on photos of nestling Great Antpittas (Kofoed & Auer 2004) and my own observations of fledgling Undulated Antpittas, these two species at this age may be nigh-on indistinguishable. **Subadult** I examined three subadults (one *phelpsi*, two *excelsa*; See Reproduction), all in roughly the same stage: the final moult to full adult plumage. The most notable differences from adult plumage were noted by Gilliard (1939) for *phelpsi*. All had scattered feathers across the back and rump that were tipped with two narrow blackish lines separated by a broader bar of buffy-orange. Some bore just the buffy-orange bar with a single dark bar above that, but the lower dark bar appeared to have worn off. The uppertail-coverts were indistinctly tipped rufescent and the upperwing-coverts were broadly tipped buffy-orange below a thin dark line. In one specimen, grey adult feathers on the nape were mixed with scattered buffy immature feathers faintly barred black.

MORPHOMETRIC DATA Very few published data available and none for females. For males: *Wing* 159mm; *Tail* 68.5mm; *Bill* exposed culmen 31mm, culmen from base 37.5mm; *Tarsus* 59.5mm ($n = 1$, holotype *phelpsi*, AMNH 322974, Gilliard 1939). *Bill* from nares 20.7mm, exposed culmen 29.8mm; *Tarsus* 60.7mm (holotype *phelpsi*, my measurements). *Bill* from nares 23.5mm, 22.0mm, exposed culmen 33.6mm, 32.0mm, depth at front of nares 13.4mm, 12.2mm, width at gape 20.0mm, -?-, width at front of nares 10.0mm, 9.4mm, width at base of mandible 14.4mm, 14.1mm; *Tarsus* 64.0mm, 61.2mm ($n = 2$, ♂,♀ *excelsa*, subadults, AMNH 428768, USNM 313771). *Wing* 159mm; *Tail* 70mm; *Bill* [exposed culmen?] 35mm; *Tarsus* 68mm ($n = 1$, holotype? *excelsa*, Cory & Hellmayr 1924). A single unsexed *excelsa* was measured by von Berlepsch (1893) and similarities between his measurements and those of Cory & Hellmayr (1924) suggest they are of the same individual: *Wing* 158mm; *Tail* 70.5mm; *Bill* [exposed?] culmen 34.75mm; *Tarsus* 68mm. **MASS** No data. Dunning (2008) provided a range of 218–266g, citing Hilty (2003) as source. Hilty (2003), however, under the account for *G. excelsa*, gave this range but clearly stated that these data refer to *G. gigantea*. Hilty (2003) did not cite his source, but it seems clear that it was Krabbe *et al.* (1994) who mentioned these figures for two individuals of Giant Antpitta. **TOTAL LENGTH** Overall size varies considerably in the literature; 22.5–28.0cm (von Berlepsch 1893, Meyer de Schauensee & Phelps 1978, Hilty 2003, Krabbe & Schulenberg 2003, Ridgely & Tudor 2009).

TAXONOMY AND VARIATION Undoubtedly forms a superspecies with Giant Antpitta (Lowery & O'Neill 1969, Krabbe & Schulenberg 2003). Some authors have even suggested they might be conspecific, but Krabbe & Schulenberg (2003) alternatively proposed that race *lehmanni* of Giant Antpitta should actually be considered a race of Great Antpitta. Fjeldså (1992) proposed that Undulated and Giant Antpittas are sister species, based on plumage similarities. Rice (2005) found that Undulated and Variegated Antpittas are sister taxa, but his sampling did not include Great or Giant Antpittas. It seems likely that Great Antpitta, Giant Antpitta and Undulated Antpitta are closely related, and all of them were included in Lowery & O'Neill's (1969) subgenus *Grallaria*. Two subspecies are currently recognised, but given the paucity of available specimens and the fact that *phelpsi* was clearly described based on a subadult (see below), Great Antpitta may best be considered monotypic.

Grallaria excelsa excelsa Berlepsch, 1893, Ornithologische Monatsberichte, vol. 1, p. 11, Montaña Aricagua, Andes of Mérida, Venezuela. The nominate subspecies occurs in the Sierra de Perijá in western Venezuela, possibly extending into adjacent Colombia (Krabbe & Schulenberg 2003), on the east slope the Venezuelan Andes from Lara to Táchira and at a single locality on the west slope in Trujillo. There are no recent records from west of the Cordillera de Mérida (Hilty 2003), however, and its status in this region is unclear. Hilty (2003) doubted records from Páramo Aricagua, but I have included this locality given that P. Boesman (XC) has fairly recently recorded the species from nearby (but at lower elevation). The GPS given by Kofoed & Auer (2004) (09°24'N, 69°30'W), supposedly within PN Yacambú, appears to be incorrect. **RECORDS AND SPECIMENS** *Zulia* Cerro Pejochaina, 10°02'N, 72°49'W (COP-54843/44). *Trujillo* Los Potreros, 09°21'N, 70°43'W (COP 82162/63); 2.5km SE of La Vega de Guaramacal, 09°10'N, 70°11'W (XC 223619, P. Boesman). *Lara* Laguna El Blanquito, 09°42.5'N, 69°34.5'W (COP 82161, IBC 1157203, video D. Ascanio); PN Yacambú, 09°42'N, 69°36.5'W (XC 6855, N. Athanas); Guárico, 09°37.5'N, 69°48'W (CM P36806/07, CM P36684/831, AMNH 428768, USNM 313771). *Barinas* Túnel de San Isidro road, 08°50'N, 70°35'W (eBird: J.G. León). *Mérida* La Cuchilla, *c.*08°54'N, 70°38'W (CM); El Morro–Aricagua road, 08°17'N, 71°09'W (XC 223616/17, P. Boesman); Páramo Aricagua, 08°13'N, 71°08'W (Phelps & Phelps 1963). *Táchira* Boca de Monte, 08°04'N, 71°50'W (COP 24538); Páramo Zumbador, 08°00'N, 72°05'W (ML 62421–426, P.A. Schwartz); 3.75km S of La Honda, 07°55.5'N, 71°57.5'W (J.E. Miranda T. *in litt.* 2017); Queniquea, 07°54.5'N, 72°01'W (COP 9092).

Grallaria excelsa phelpsi Gilliard, 1939, American Museum Novitates, vol. 1016, p. 1, Colonia Tovar, Estado Aragua, Venezuela, alt. 5,900 ft [1,800m, 10°25'N, 67°17'W; GPS from topotypical specimens COP 1466/3441]. The type specimen (AMNH 322974) is a male collected 14 November 1937 by a local guide, Florencio Ruthman, who was working with the AMNH's 'W. H. Phelps Expedition'. This subspecies, if valid, is known from just three specimens from the type locality. There are no modern records. It is named in honour of William H. Phelps. Sr. Gilliard (1939) stated that the holotype is an adult but then described it as having what are clearly immature characters (see below). Race *phelpsi* is similar to the nominate race, but greener above, with a darker forecrown and with brown extending only to the mid-crown. Additionally, the white is less extensive on the chin, and the throat and upper breast are paler with broader terminal and subterminal bars. Comparing the holotype to nominate *excelsa*, Gilliard (1939) stated that the 'remiges and rectrices above and below not so warm a brownish olive; [...] anterior breast less cinnamomeous, [...] a more pronounced crescentic patch; exposed culmen at least 3.5mm shorter, tarso-metatarsus at least 4mm shorter.' Gilliard (1939) described the back, rump and uppertail-coverts as 'dull Olive-Citrine', the nape and hindcrown as 'Deep Neutral Gray' and the forecrown as 'Orange-Citrine'. There is an indistinct buffy line above the eye, slightly speckled black. The lores and face are light cinnamon-grey. The ear-coverts are bicoloured, with the anterior portion reddish-brown with indistinct blackish barring, and the posterior part dark yellowish-buff. The central throat is dull white washed buffy-yellow and the feathers of the chin and throat have blackish hair-like extensions at their tips. The breast, flanks and abdomen are golden-buff, warmer on the breast-sides and flanks, with the vent distinctly warmer. The feathers of the upper breast have broad black tips forming an irregular dark band. Below this, the blackish feather tips become narrower than the subterminal bands, with more golden-buff colour in between, creating more regular barring. These bars become more widely spaced and narrower lower down, fading to the unbarred vent. The undertail-coverts and thighs are richer yellow-buff than the rest, except the vent, with the thighs slightly barred. The wings are 'Dark Mouse Gray' except for a 'light Cinnamon-Buff patch at inner webs of the primaries (seen best from below).' The leading margins of the outer primaries are pale brownish-olive, paler on the inner primaries and secondaries. The upperwing-coverts are dull citrine and the underwing-coverts are 'Deep Chrome flecked laterally with black.' The tail is dark reddish-brown above and rich brown below. From the skin, Gilliard (1939) described the mandible as sooty-black, the maxilla as horn-coloured, and the feet as 'Dark Quaker Drab'. In his description, Gilliard (1939) described scattered feathers on the back and rump as bearing 'two narrow dusky lines separated by a broader bar of Raw Sienna' and a few feathers on the rump as being terminally barred with pale yellowish-orange. In addition, he noted that a few feathers on the nape and hindcrown are terminally and subterminally tinged buffy and faintly barred black. Although Gilliard (1939) acknowledged the likelihood that these feathers represent immature plumage, he considered the rest of the plumage to be that of a full adult. Given our (still) imperfect understanding of adult, juvenile and transitional plumages in all antpittas, that *phelpsi* was described based on an immature calls into question other characters that supposedly separate it from nominate *excelsa*.

STATUS Until the late 1990s, Great Antpitta was afforded a conservation status of Near Threatened (Collar *et al.* 1994, Stattersfield *et al.* 1998). Currently, however, because of the dearth of recent records and the likelihood that its range is quite small, this poorly studied species is considered Vulnerable (BirdLife International 2017). It is a restricted-range species (Stattersfield *et al.* 1998) found in the Colombian East Andes EBA, Cordillera de la Costa Central EBA and Cordillera de Mérida EBA (Krabbe & Schulenberg 2003). Great Antpitta until recently was essentially unknown in life. Earlier sight records probably referred to misidentified Undulated and Plain-backed Antpittas (Sharpe & Ascanio 2008). In particular, there are apparently no recent records of the (possibly) distinctive

Coastal Cordillera race *phelpsi*. Though scarce and local across its small range, Great Antpitta is now recorded with some regularity in several national parks (Boesman 1998). Although some have suggested there may be an extant population in the PN Sierra de Perijá, these mountains have been extensively deforested due to the cultivation of narcotics, cattle-ranching and mineral exploitation, with the last known records there in 1952 (Rodríguez & Rojas-Suárez 1995, Hilty 2003, Krabbe & Schulenberg 2003, BirdLife International 2017). Although some large areas of forest remain in Mérida and Aragua, severe and extensive deforestation has occurred throughout its range, and ongoing clearance for agriculture is a major threat in the Mérida chain (Stattersfield *et al.* 1998, Krabbe & Schulenberg 2003). The MN Pico Codazzi in Aragua is potentially within the range of race *phelpsi*, but there have been no records since 1939, despite considerable search efforts in recent years (J.E. Miranda T. *in litt.* 2017). Similarly, the single record from the west slope of the Andes (Betijoque) has not been repeated since 1971. It is considered nationally Vulnerable in Venezuela (Sharpe & Ascanio 2008). Although there are a few patches of intact forest in the Cordillera de Mérida and Cordillera de la Costa portions of the species' range, deforestation has been severe in many parts of these ranges (Huber & Alarcón 1988, BirdLife International 2017). Agricultural clearance is an ongoing threat in the Sierra de Perijá, Cordillera de Mérida and Cordillera de la Costa (Stattersfield *et al.* 1998, Sharpe & Ascanio 2008). **PROTECTED POPULATIONS** *G. e. phelpsi* PN Henri Pittier (Schäfer & Phelps 1954). *G. e. excelsa* PN Guaramacal (Boesman 1998); PN Yacambú (Kofoed & Auer 2004, COP, XC); PN Sierra Nevada (eBird; J.G. León); PN Tapo-Caparo (Phelps & Phelps 1963, XC); PN Chorro El Indio (J.E. Miranda T. *in litt.* 2017); PN Páramos Batallón y La Negra (COP 9092); PN Perijá (COP 54843/44). Probably also occurs in PN Dinira (Trujillo) and possibly at higher elevations in PN El Gauche (Portuguesa).

OTHER NAMES English Aricagua Antpitta (*excelsa*, Cory & Hellmary 1924). **Spanish** Tororoí Excelso (Schäfer & Phelps 1954, Krabbe & Schulenberg 2003); Hormiguero tororoí excelso (Phelps & Phelps 1963, Hilty 2003, Sharpe & Ascanio 2008, Verea et al. 2017); Tororoí Excelso Merideño [*excelsa*], Tororoí Excelso de Phelps [*phelpsi*] (Phelps & Phelps 1950). **French** Grande Grallaire. **German** Fahlbauch-Ameisenpitta (Krabbe & Schulenberg 2003).

Great Antpitta, an 8–10-day-old nestling (*excelsa*), PN Yacambú, Lara, Venezuela, 6 June 2003 (*Eric M. Kofoed*).

Great Antpitta, adult (*excelsa*), PN Yacambú, Lara, Venezuela, 23 October 2011 (*David Ascanio*).

VARIEGATED ANTPITTA
Grallaria varia   Plate 7

Formicarius varius Boddaert, 1783, Table des planches enluminées d'histoire naturelle de M. D'Aubenton, p. 44, 'Cayenne' (= Suriname) (Brabourne & Chubb 1912, Cory & Hellmayr 1924).

'I killed one inside the vast forest, far away from the shore of the Paraná River. He walked below dense thickets and among the ferns, with tail raised, shaking it as do rails, whose manner it imitated, yet it was far from their damp, swampy habitat. From what I've seen, their customs are like those of *Chamaeza tshororo* [Short-tailed Antthrush *C. campanisona*]. It must be very rare, because during six years at Djaguarasapá, where I killed one, I have not seen another, or elsewhere. In truth, the forest is so thick that one might pass you by without ever being. In July, however, finding myself in the woods with my brother, we heard a scream, similar to that emitted by the feared indigenous peoples of Guayaquí as they call to each other in the forest. Finally there appeared a bird, much like the present species, that I glimpsed for no more than a few seconds before it cried out and disappeared into the weeds. From this brief view I could not be sure it was not an unknown species of tinamou. Another, that may also be this species, was a bird which sang out a mu mu mu mu, audible at a great distance. This bird is heard always on rainy days and I've yet to see him, despite much striving to do so.' – **A. de Winkelried Bertoni, 1901, Paraguay**

I translated the above quote with the creative liberty needed to match English-speaking authors of the time and, regardless of whether or not all of Bertoni's observations pertain to the present species, the passage captures well the mystery that surrounded Variegated Antpitta for this natural historian. The species is among the larger and more robust *Grallaria*. It has a brownish back, slate-grey crown, pale lores and a prominent white or buffy malar. Across its geographic range, the underparts vary from finely barred to coarsely streaked. The five recognised subspecies fall into two groups. Taxa in the northern and western Amazon (the nominate group) range from southern Venezuela and the Guianas through northern Brazil to southern Amazonia, as far west as north-east Peru. The '*imperator* group' is distributed in southern and eastern Amazonia, and the Atlantic Forest region of south-east Brazil to Rio Grande do Sul, eastern Paraguay and north-east Argentina (Mazar Barnett & Pearman 2001). Throughout the species' range it is most easily detected by virtue of its slightly melancholy sounding, monotone, low hooting song, given most frequently at dawn and dusk. Variegated Antpitta is fairly well known, yet there is probably no aspect of its biology that does not deserve closer attention. Despite its prevalence in collections, there are few published morphometric data. Taxonomic work is needed to verify the status of the two (northern and southern) populations.

IDENTIFICATION 18.0–20.5cm. Sexes similar. A fairly large (the largest of the lowland antpittas), stout-bodied *Grallaria*, with the typical long-legged, upright posture of the genus. Overall, it is brownish, with a grey crown, prominent whitish malar stripe and (of course) a distinctly variegated breast pattern. More precisely, the crown and nape are grey, usually with a more olive-brown forecrown.

The cheeks are paler buffy above a white or creamy-white malar. The back and wings are olive-brown while the flight feathers and tail are warmer brown. The throat and breast are dark brown or rufous-brown streaked white, with the underparts becoming gradually paler with various amounts of fine black barring, and white spots and streaks. Birds from the east and south of the species' range (*imperator* group) are so much larger than other antpittas with similar ranges that confusion is unlikely. The smaller, more northerly forms (nominate group) might be confused with the often sympatric Scaled Antpitta *G. guatimalensis* (Restall *et al.* 2006). In general, Variegated Antpitta is a true lowland Amazonian species, while Scaled Antpitta is found in foothill and lower montane forests, with no currently known area of sympatry (Ridgely & Tudor 1994). When comparing the two species, Scaled Antpitta is even smaller than the smallest races of Variegated Antpitta and always lacks pale shaft-streaks on the back, but shows streaking on the breast-sides (Schulenberg *et al.* 2007). Additionally, Scaled Antpitta's belly is more tawny or rufous (not pale buff) and lacks streaking on the breast-sides (Ridgely & Tudor 1994, Restall *et al.* 2006). In western Venezuela it might be confused with the smaller Tachira Antpitta, but this species is darker and has a clearly visible white submalar streak and broken white necklace, and lacks streaking on the flanks like Scaled Antpitta (Restall *et al.* 2006). Plain-backed Antpitta is also smaller, and lacks scaling on the back and heavy streaking on the breast (Restall *et al.* 2006).

DISTRIBUTION Found in humid lowland forests from the Guianas and lower Amazonian Brazil (north of the Amazon and west to the upper Rio Negro), west and north in Brazil and into adjacent south-west Venezuela (Amazonas) and north-east Peru (Loreto). In eastern Brazil the species is found from Bahia to northern Rio Grande do Sul, and into eastern Paraguay and north-east Argentina (Misiones and north-east Corrientes) (Contreras 1983, Bencke 2001).

HABITAT Appears to favour humid forest with a high canopy and a relatively dense understorey (Volpato *et al.* 2006). Less commonly, it inhabits mature secondary woodland (Bencke 1996, Aleixo & Galetti 1997, Aleixo 1999, Goerck 1999, Restall *et al.* 2006) and forest edges

(Quintela 1986), perhaps because of the dense tangled understorey in such habitats. Across most of the Amazon Basin portion of its range the species occurs in both floodplain (Willard *et al.* 1991) and *terra firme* forests (Cohn-Haft *et al.* 1997, Whittaker 2009). Southward, it can be found in understorey dominated by bamboo (Buckingham 2011), but appears to avoid such areas if dense, broadleaf understorey areas are available (Rother *et al.* 2013). Southern populations of *imperator* occur in *Araucaria* woodland up to *c.*1,000m (Bencke & Kindel 1999), and occasionally as high as 1,800m in south-east Brazil (Ridgely *et al.* 2016). Donahue (1985) noted that the species 'prefers elevated forest regions' in Suriname, but detailed studies of specific habitat requirements are lacking. Its clear preference for less habitat disturbance than other *Grallaria* (e.g. Tawny Antpitta), has led Variegated Antpitta to often serve as an indicator species in conservation assessments (Belton 2003, Oliveira & Köhler 2010).

VOICE The songs of two individuals of race *imperator* were analysed by Krabbe & Schulenberg (2003) and described as a series of 11–14 notes at *c.*0.5kHz, beginning with 4–6 long notes, increasing in volume, then rising slightly in pitch and pace to a series of 6–8 loud, shorter notes, with the last note weaker; lasting 2.2–3.1s. They described the song of one individual of race *varia* as lasting 3s and noted that it is similar to the song of *imperator*, but steady in pitch and comprising only eight notes, with the first four drawn-out and without the weaker final note. Schulenberg *et al.* (2007) described the song as a fairly slow, decelerating then accelerating, monotone series of hoots, loudest in the middle *pu'pu-pu-poo POO POO poo-pu'pu*. Donahue (1985) described the vocalisation of *G. v. varia* from Suriname as 'a series of about six low-pitched, wooden-sounding hooting notes, similar to the measured hooting song of the Thrush-like Antpitta but lower-pitched and increasing in tempo towards the end.' Around the nest, apparently in response to observer presence, Érard (1982) described a one-note *hou* which he likened to a small owl. In addition, in response to the distress calls of nestlings while being handled, he described a harsh, metallic *tchkrrik-krrik* interspersed with the single-note *hou* and a doubled *Houou*. Protomastro (2000) similarly described this alarm call, *ooha*, as being given by adults around the nest.

NATURAL HISTORY Possibly somewhat more crepuscular than congeners, and most active pre-dawn and at dusk (Restall *et al.* 2006). Although it prefers dense tangles of forest understorey, the species may hop up to an exposed log or branch several metres above ground to sing (de la Peña & Rumboll 1998, Hilty 2003). While singing, their throat enlarges (WA 553216, C. Albano) and undulates with each note. It is unclear if this is due to the throat itself being inflated or simply ruffled feathers, although no bare skin is exposed as in several *Hylopezus* (see Amazonian Antpitta account, and Plate 20; Fig. a). Like others in the genus, Variegated Antpitta hops or runs rapidly on the forest floor (Berla 1944), pausing to flick aside leaves with its bill, probing into soft, often damp ground (Ridgely & Tudor 1994, Krabbe & Schulenberg 2003, Restall *et al.* 2006). Startled birds may flush abruptly and perch 3–4m above ground before diving back into heavy cover (Ridgely & Tudor 1994). More frequently, however, they prefer to run from danger rather than fly (Berla 1944). Though infrequently reported, it appears that Variegated Antpittas will occasionally follow swarms

of army ants (Willis 1985a). Johnson *et al.* (2011) gave a territory size of 9.2ha and suggested that territories are held year-round, but few other data are available. Stouffer (2007) gave territory size as 8ha and estimated 7.46 pairs per 100ha. Longevity is poorly known for any species treated herein, but one unsexed individual in São Paulo (*imperator*) was recaptured several times over the course of *c.*4.5 years (Lopes *et al.* 1980).

DIET Despite that the first information on diet was published in the early 1800s, we know very little. The stomach contents of one individual (*imperator*) were given by zu Weid (1831) as 'insects, especially green beetles' (my translation from German). Érard (1982) gave the food brought to nestlings as large spiders, cockroaches, orthopterans and myriapods. It is generally assumed that, like congeners, the diet of *G. varia* consists largely of invertebrates such as earthworms, spiders, cockroaches (Blattodea), grasshoppers (Acrididae) and centipedes (Chilopoda) (zu Weid 1831, Berla 1944, Krabbe & Schulenberg 2003). Although seeds are sometimes considered part of the diet (Canevari *et al.* 1991), it is not known if they are actively sought or ingested incidentally.

REPRODUCTION To the best of my knowledge, this species holds the honour of being the first of all taxa treated in this book to have its nest described. Based on information from local people, zu Weid (1831) provided the first nest description from Rio de Janeiro (*imperator*), albeit rather cursory and (probably) partially incorrect, as it was said to be on the ground but to contain unmarked blue-green eggs. As egg size was unmentioned, the description might also pertain to a tinamou. Euler (1900) repeated this description of a nest on the ground, apparently believing it to be true in light of its location matching the terrestrial habits of the bird. The first formal description of the nest, eggs and nestlings, and indeed the report from which the bulk of our knowledge of breeding is derived, came from a nest found in 1980 in French Guiana (Érard 1982; race *varia*). Donahue (1985) described a second nest (*varia*) from Suriname found in 1981 and Quintela (1987) reported on a third nest (*varia*) from Amazonas, Brazil, in 1984. Some years later, Protomastro (2000) described two nests (*imperator*) found in PN Iguazú, Misiones, Argentina. **NEST & BUILDING** Despite the previous suggestion that the species nests on the ground (zu Weid 1831, Euler 1900, Canevari *et al.* 1991), there are no confirmed reports of this. Érard (1982) gave only a somewhat cursory description of his nest, stating that it was a large, untidy *Turdus*-like nest built into damp, rotting leaves and of twigs, dead leaves, skeletonised leaves and leaf petioles. The cup measured 9cm wide by 3.5cm deep. Quintela (1987) described the nest as 'a shallow cup 20cm in diameter, lined with a thick mat of very small brown twigs and rootlets'. Donahue (1985) described the nest only as a 'shallow cup constructed of damp rootlets and mosses'. Protomastro (2000) gave the internal cup measurements of one nest as 14cm wide by 5cm deep, stating that it was constructed of dead leaves, twigs, fern petioles and thin rootlets. Érard's (1982) nest was built into the hollow depression atop a rotting stump, while Donahue's (1985) nest was built into a crevice on the trunk of a tree *c.*1m above ground and Quintela's (1987) nest was on an upright rotting stump 40–50cm in diameter and 'well concealed' by a palm frond. Both nests found by Protomastro (2000) were atop broken stumps. Fraga & Narosky (1985) described a nest found by C. Olrog in Misiones (*imperator*) which was apparently in (on?) the broken trunk of a tree. Additional published records of nest height are 1.5m (Quintela 1987), 2.6m, 2.4m (Protomastro 2000) and 1m (Érard 1982). **Egg, Laying & Incubation** Egg colour is rather subjective, but the eggs of *G. varia* are universally described as unmarked and blue-green (zu Weid 1831, Euler 1900, de la Peña 1988), turquoise (Érard 1982; faint brownish markings were attributed to staining from the nest), turquoise-blue or blue-turquoise (Donahue 1985, Protomastro 2000). With respect to size, the following data are available: de la Peña (1988) gave egg measurements (source and sample size not provided) as 32.5–36.0 × 26.7–29.5mm; Schönwetter (1979) gave measurements of 32.3–36.4 × 26.7–29.7mm, and the weight of shells is 0.85–1.04g (*n* = 6), estimating fresh egg mass to be 15.2g; and von Ihering (1900) presented measurements for two eggs, one in Nehrkorn's collection (36 × 30mm) and another one from Iguape (*imperator*; 35 × 28mm). Other published measurements include 33.5 × 26.5mm (*n* = 1, Donahue 1985), 40 × 28mm (*n* = 1, Érard 1982, but apparently estimated) and 34.6–35.8 × 28.8–29.4mm (*n* = 2, Protomastro 2000). The measurements in Fraga & Narosky (1985) are those given by von Ihering (1900) and Schönwetter (1979). Their description of a nest with one blue egg is unclear as to whether this was a complete clutch. The incubation behaviour of *G. varia* is rather poorly known. Despite that Krabbe & Schulenberg (2003) suggested that incubation is performed only by the female, my interpretation of Érard (1982) is that both adults incubated, which agrees better with other antpittas. Reported on-bouts lasted 1.0–2.5h while absences were of 40–75min (Érard 1982). He also stated that, when flushed from the nest, adults would drop silently to the ground and move to a location *c*.15m away to wait until the observer had departed. He made no mention of vocalisations during this event. The only report of incubation period is that of Érard (1982) who stated that both eggs hatched 17 days after clutch completion, one in the morning and one in the afternoon. **Nestling & Parental Care** The most complete, clear description of newly hatched young is that of Érard (1982). He described them (my translation from French) as dark-skinned with grey-brown down on the capital, cervical, spinal, humeral and femoral feather tracts, with black legs and white nails, and black bills and very bright orange-red rictal flanges. The nestlings found by Quintela (1987) were described as having their eyes open and 'covered in black down, and spotted with brown juvenal feathers.' He described their gapes as 'yellowish' and their mouth linings as bright red. Although Protomastro (2000) apparently observed nestlings a few days after hatching, his description apparently refers to nestlings about one week old, based on the published photo. He described them as having their 'wing feathers [ ] growing. They had black skin and dense black downy plumage from head to back, pink skin on throat, breast and sides, black with sparse brown down on upper legs but black lower legs, and black tarsus and toes. The bills were black with red-orange gapes, mouth-linings and lower mandible bases.' From the photo in Protomastro (2000), it appears that the 'downy plumage' he described is the same natal down described by Érard (1982) as grey-brown. On hatching two nestlings weighed 13.3g and 12.0g, apparently gaining around 5.7g daily (Érard 1982). This coarse estimation of growth, however, is of little utility given that nestling growth would not have been linear. Érard (1982) noted

that both adults brought food to the nest and that nestlings were brooded for long periods during the first six days after hatching. Prey were apparently large and delivered approximately every 30–40min, although it was not specified if items were brought singly or not. One interesting observation, which demands confirmation, is the assertion by Érard (1982) that one adult, which he thought to be the female, move to *c*.12m from the nest to receive prey brought by her mate. These were then taken to the nestling. I am unaware of such behaviour by any other antpitta, but such behaviour would be easily missed, especially given that most of what we know of nesting behaviour was documented by cameras pointed only at the nest. In general, Érard (1982) described the adults as relatively silent while attending the nest and reported hearing an adult singing only once, *c*.100m away. He also mentioned several types of vocalisations, apparently given in alarm (see Voice). Érard (1982) appears to be the first observer to describe the behaviour of adult *Grallaria* during the approach of an observer to a nest, something that I have witnessed at the nests of several species and which was also described by Dobbs *et al.* (2003). As the observer approaches, the incubating or brooding adult aligns its body with the observer while lifting the tip of the bill, raising the body slightly and drooping the wingtips. This display appears to serve two purposes. It makes the adult appear somewhat larger and more threatening, presenting the heavy bill towards the perceived threat while also showing their rather large and (hopefully) threatening eyes. Simultaneously, however, in a case where the approaching threat has not yet seen the nest, the drooped wings serve to better conceal the contents of the nest (bright orange bills or bright blue eggs!). Also, while in this position, the submalar stripe and contrasting throat pattern are exposed and presented vertically to an observer. The effect of this, when observed directly, is one of increased camouflage, breaking the outline of the nest and adult. Indeed, in his description, Donahue (1985) reported that 'the long white moustache marks and large spot on the upper breast were shown to best advantage, broken up by the dark throat and bill. The large areas of white resembled the patches of whitish lichen growing on the tree trunk.' This behaviour was also witnessed by Protomastro (2000), who also perceived it as possessing a camouflaging function. Finally, although it has not been my experience that broken-wing displays are prevalent or elaborate in *Grallaria*, Érard (1982) describes *G. varia* as performing an 'artful parade of broken-wing' display in the presence of observers at the nest. Most authors, however, describe adults flushed from the nest as dropping immediately to the ground or flying a short distance and then running silently from view (Donahue 1985, Protomastro 2000). **Seasonality** Despite quite a few breeding records and active nests, the broad range of Variegated Antpitta precludes making predictions as to breeding seasonality. Quintela's (1987) nest found 80km north of Manaus (Amazonas) contained two young on 19 June 1984. Based on his description of the young and my own observations of related species (*guatimalensis, alleni*), his observation of the nestlings disappearing from the nest on 22 June represents a successful fledging, placing clutch initiation somewhere around 16 May. Willis & Oniki (2002a) mentioned a label at MBML that reads 'on nest with 2 eggs' attached to a specimen collected 5 November 1941 in Santa Teresa, Espírito Santo (race *intercedens*), but no additional details are known. Protomastro (2000) reported a nest with eggs on 21 November 1998 that hatched a few days prior to 6 December. **BREEDING DATA** *G. v. varia* Clutch completion, 24 December 1980, Saut Pararé (Érard 1982); incubation, 22 December 1981, Brownsberg (Donahue 1985); nest with young, 19 June 1984, *c.*80km N of Manaus (Quintella 1987); three 'immatures', 13 June 1984, 20 July 2000, 7 October 2008, *c.*80km N of Manaus (BDFFP database); testes enlarged, 18 December 1960, Kayserberg Airstrip (FMNH 260493). *G. v. distincta* Incubation, 4 February 2015, Anapu (WA 1608375, J.C. Vaz de Oliveira); nestling, 12 February 2007, Novo Aripuanã (WA 1153536, M. Padua); two subadults, 15 June 1907, Calamá (Hellmayr 1910, AMNH 492193/194). *G. v. intercedens* Incubation, 5 November 1941, Santa Teresa (Willis & Oniki 2002). *G. v. imperator* Building nest, 22 October 2009, near Ribeirão Grande (WA 89389, B. Intervales); clutch completion, 12 November 2011, PP Cruce Caballero (J.M. Lammertink *in litt.* 2017); nest with single egg [laying?], 24 October 1949, Fracrán (C. Olrog in Fraga & Narosky 1985); incubation, 9 January 2016, Ribeirão Grande (WA 1983428, M. Martins); incubation, 12 December 2011, Cachoeiras de Macacu (WA 523423, E. Legal); incubation, 21 November 1998, PN Iguazú (Protomastro 2000); incubation, 9 January 2014, Tremembé (WA 1393341, F.B.R. Gomes); two nests with eggs, Salesópolis, 9 February 2013 (WA 1234784, F. Passos) and 9 December 2013 (WA 1179646; E. Jesus); incubation, 23 October 2001, *c.*6.75km SW of Ribeirão Grande (eBird: L. Gardella); incubation, 20 October 2015, Ribeirão Grande (WA 2226716, V. Moller, WA 2098583, V. Wruck); nestling, 7 November 1998, PN Iguazú (Protomastro 2000); fledgling, 2 March 2012, São Luiz do Paraitinga (WA 823649, M. Singer & WA 588999, V. Luccia); juvenile, 2 February 2009, São Luiz do Paraitinga (WA 60878, D. Bucci); juvenile, 2 February 2014, Tapiraí (WA 1238035, R. Machado).

TECHNICAL DESCRIPTION Sexes similar. The following description of nominate *varia* is based on Krabbe & Schulenberg (2003) (see also Geographic Variation). *Adult* Slate-grey crown grades to dark olive-brown on forecrown. Feathers of crown and nape tipped black with narrow pale shaft-streaks. Lores white or buffy and extend into a prominent malar stripe, often tipped black at its lower end. Ear-coverts dark rufescent-olive with very thin buff shaft-streaks and exposed orbital skin dull blue-grey. Above, olive-brown with variable black scaling across most of the dorsal surface and, in some races, most feathers (especially on scapulars) have narrow tawny shaft-streaks. Wing-coverts have narrow buff or tawny streaks and small terminal buff spots, while primaries are tawny-brown, secondaries darker, more rufescent brown and similar to the rectrices. Throat dark rufescent-olive, some feathers with narrow whitish shaft-streaks, grading to brown on breast, which bears a few black-bordered white or very pale buff feathers in centre forming a small throat/upper-breast patch. Lower breast striped pale buff, merging into pale buff or ochraceous belly and tawny-buff vent. Flanks somewhat variable, either spotted or streaked blackish or dusky-brown. **Adult Bare Parts** *Iris* dark brown; *Orbital Skin* bluish-grey; *Bill* maxilla dark grey, mandible grey with base pinkish or creamy; *Tarsi & Toes* greyish. *Immature* Previously, the only published observations on immature plumages were those by Chubb (1921) who described a 'nestling' of nominate *varia* as cinnamon-rufous above, brighter rufous on tips of greater wing-coverts and tips

of flight-quills. He described the underparts as overall darker, but browner (rather than black) on upper chest, with the sides of neck and throat much paler cinnamon than in adults, and 'almost naked'. This description lacks crucial details, but seems to be describing a bird with developed flight feathers, and there is no mention of a nest. Furthermore, Chubb (1921) did not refer to downy plumage and made no mention of any of the spotting or streaking usual in fledgling *Grallaria*. This description then, is hard to interpret, but I suggest that refers to a juvenile. *Fledgling* The following description and the illustration of a fledgling on Plate 7 (Fig. d) are based largely on a photo of fledgling *imperator* (WA 823649, M. Singer). The fledgling is largely covered in fluffy, sooty-black down that fades to off-white on the lower breast and central belly and becomes increasingly buffy then ochraceous on the flanks and vent. Visible parts of the crown, nape, upper back and sides are marked with short, white, spot-like streaks. These gradually increase in size on the breast, lower on the sides, and rump until they create a more spotted rather than finely speckled appearance, grading into the paler central belly and vent. Lores are pale whitish, and large pale feathers in the malar already hint at the adult's creamy malar. The wings are similar in colour to those of adults but the upperwing-coverts each bear a terminal, cleanly formed triangle bordered by a distinct black margin at the tip. **Fledgling Bare Parts** *Iris* dark; *Bill* partially obscured by foliage but maxilla appears mainly black, mandible largely pale, palest at the gape, with no clear indication of a bright orange gape, though it is likely present given its clear presence on older individuals (see below); *Tarsi & Toes* dusky-pink to dark vinaceous. *Juvenile* The following is based on several *imperator* juveniles photographed *in situ* (see Breeding Data). Unfortunately, I have seen no clear examples of juveniles in collections and the underparts of the individuals photographed were largely obscured. The upperparts now approach those of adults in coloration, with the nape and crown retaining downy, streaked plumage (as in the fledgling) longer than the rest of the upperparts. Although the buffy malar streak is well developed in both, in what appears to be the youngest individual the lores appear messy and not as cleanly marked buffy-white as in adults. In both juveniles, the bill is similar to that of adults but is clearly bright orange at the gape. No further details available. *Subadult* Characters that separate subadults from full adults remain unclear, especially in those subspecies that retain spotted wing-coverts as adults (*imperator, distincta*). My feeling is that older individuals lack black fringes at the tips of the wing-coverts, but whether this is due to wear or a difference in the pattern of adult coverts is unclear.

MORPHOMETRIC DATA ♂♂ *Wing* 110mm; *Tail* 38mm; *Bill* depth 10mm, exposed culmen 21mm; *Tarsus* 48mm (*n* = 1, *varia*, Chubb 1921). *Bill* from nares 17.3mm, *Tarsus* 52.6mm (*n* = 1, *imperator*, LACM). *Wing* 115.2mm; *Tail* 39mm; *Tarsus* 50mm (*n* = 1, *cinereiceps*, Friedmann 1948). *Bill* from nares 17.1mm (*n* = 1, *cinereiceps*, AMNH 310744). ♀♀ *Wing* 110mm; *Tail* 45mm; *Bill* 28mm (*n* = 1, *cinereiceps*, holotype, Hellmayr 1903). *Bill* from nares 17.8mm, 18.4mm; *Tarsus* 50.0mm, 47.4mm (*n* = 2, *cinereiceps*, AMNH). *Bill* from nares 17.5mm, 17.6mm, 17.9mm, 18.0mm, 18.9mm, 19.1mm, -?-; *Tarsus* 60.1mm, 54.2mm, 53.9mm, 54.7mm, 563mm, 54.4mm, 56.2mm (*n* = 7, *imperator*, LACM). **Sex unspecified** *Wing* 122mm; *Tail* 52mm; *Bill* culmen 25.5mm; *Tarsus* 53mm (*n* = 1, *intercedens*, holotype, von Berlepsch & Leverkühn 1890). *Wing* 125.3mm; *Tail* 57.2mm; *Bill* [from nares] 18.8mm; *Tarsus* 61.8mm (*n* = 1, *imperator*, Reinert *et al.* 1996). *Wing* 123–125mm; *Tail* 50mm; *Bill* exposed culmen 28mm; *Tarsus* 58mm (*n* = 4, ♂,♂,♂,♀, *imperator*, Dabbene 1919). *Wing* 112mm; *Tail* 41mm (*n* = ?, *varia*, Ruschi 1979). *Wing* 114.58 ± 3.81mm; *Tail* 42.29 ± 2.17mm (*n* = 19 gender unknown, ± SD, *varia*, Bierregaard 1988). **Mass** ♂♂ 90g(no) (*n* = 1; *cinereiceps*, Cardiff 1983, LSUMZ 110245); 113.0g(lt) (*n* = 1, *cinereiceps*, Willard *et al.* 1991, FMNH 319334); 121g (*n* = 1, *distincta*, Graves & Zusi 1990); 129.7g(m) (*n* = 1, *varia*, USNM 625538); 90–121g (sample size not specified, '*varia* group', Krabbe & Schulenberg 2003). ♀♀ 134.7g (*n* = 1, *imperator*, Contreras 1983); 125–130 g (*n* = 2, *intercedens*, Willis & Oniki 2002a); 98g (*n* = 1, *distincta*, Graves & Zusi 1990); 122g, 126g (*n* = 2, '*varia* group', Krabbe & Schulenberg 2003); 122.0g (*n* = 1, *cinereiceps*, Willard *et al.* 1991); 134g (*n* = 1, '*imperator* group', Krabbe & Schulenberg 2003); 125g (*n* = ?, ssp. ?, Sick 1984); 121.87 ± 10.23g (*n* = 16, *varia*, Bierregaard 1988); 130g (*n* = 1, *intercedens*, Lima 2006). **Sex unspecified** 134g (*n* = 1, *imperator*, Reinert *et al.* 1996); 119.2 ± 8.9g (*n* = 26, *varia*, BDFFP database). **Total Length** 16.6–20.5cm (von Berlepsch & von Ihering 1885, von Berlepsch & Leverkühn 1890, Chubb 1921, Snyder 1966, Meyer de Schauensee & Phelps 1978, de la Peña & Rumboll 1998, Krabbe & Schulenberg 2003, Lima 2006, Restall *et al.* 2006).

TAXONOMY AND VARIATION Variegated Antpitta is thought to be closely related to Scaled Antpitta (Lowery & O'Neill 1969, Krabbe *et al.* 1999), with which it shares many plumage characters, and which is traditionally placed alongside it in linear sequences (Cory & Hellmayr 1924, Peters 1951, Remsen *et al.* 2017). Indeed, *G. guatimalensis carmelitae* was originally described as a subspecies of the present species (Todd 1915). It was suggested by Hernández-Camacho & Rodríguez-M. (1979) that Variegated Antpitta might be conspecific with Moustached and Scaled Antpittas. Based on plumage these three species might be closely related, strong differences in their vocalisations make them easily diagnosable species (Krabbe & Coopmans 2000). Most authors recognise five subspecies of Variegated Antpitta (Krabbe & Schulenberg 2003) and several have suggested that these subspecies form two geographically well-separated groups perhaps better treated as separate species, based principally on vocal differences (Ridgely & Tudor 1994, Krabbe & Coopmans 2000). In the north and west of the species' range, the '*varia* group' includes the nominate subspecies plus *cinereiceps* and *distincta*, and in the east and south the remaining two subspecies form the '*imperator* group'. Krabbe & Schulenberg (2003) noted that the racial affiliation of birds in south-east Minas Gerais is uncertain, but included them in *intercedens* pending further study.

Grallaria varia cinereiceps Hellmayr, 1903. Verhandlungen der Kaiserlich-Königlichen Zoologisch-Botanischen Gesellschaft in Wien, vol. 53, p. 218. The type is an adult female from the Natterer collection, procured 4 May 1831 at Marabitanas [= *c.*00°58'N, 66°51'W] along the upper Rio Negro, Amazonas, Brazil. Race *cinereiceps* occurs from extreme southern Venezuela and immediately adjacent northern Brazil along the upper Rio Negro in western Amazonas (Borges 2006), to north-east Peru on the northern bank of the Rio Napo. The following comparison with the nominate subspecies is based on Krabbe & Schulenberg (2003), Restall *et al.* (2006), with

additions taken from Hellmayr's (1903) original remarks (in German). Compared to nominate *varia*, *cinereiceps* has a brighter ochraceous or cinnamon ground colour below, particularly on the throat and upper breast (blackish-brown in *varia*) while the lores and malar stripe are distinctly buffier. The wing-coverts lack pale markings, the dark brown primaries have the leading edge brighter rufous, while the secondaries tend towards rusty olive-brown. Also compared to *varia*, the upperparts of *cinereiceps* tend to have heavier scaling formed by dusky black margins to otherwise olive-brown feathers. The Río Siapa specimen listed below was originally misidentified as *G. guatimalensis roraimae* (which see) but is now confirmed to represent *cinereiceps* (Meyer de Schauensee 1966, Mallet-Rodrigues & Pacheco 2003). **Specimens & Records VENEZUELA**: *Amazonas* Río Siapa, Poste de la Frontera #2, 02°30'N, 64°03.5'W (Phelps & Phelps 1948, COP 34902); mouth of Río Ocamo, 02°47'N, 65°13'W (AMNH 432817); Caño Atamoni, 02°10'N, 66°28'W (COP 66105); Cerro de la Neblina, Río Mawarinuma and vicinity, *c.*00°51'N, 66°10'W (AMNH 816752, FMNH 319334). **BRAZIL**: *Amazonas* Serra Curicuriari, 00°20'N, 66°50'W (AMNH 310744); Salto do Hua, 00°40'N, 66°08'W (USNM 326451); Lago Cumapi, 01°33.5'S, 65°53'W (FMNH 457232/MPEG 62965); Novo Airão, 02°38'S, 60°57'W (WA 2471769, recording J.F. Pacheco); Campamento Cachimbo, 02°43'S, 70°31.5'W (Stotz & Alván 2011); São Francisco do Tonantins, 02°52.5'S. 67°48'W (CM P97204); Manacapuru, 03°13.5'S, 60°41'W (CM P98715). **PERU**: *Loreto* Quebrada Bufeo, 02°20'S, 71°36.5'W (Stotz *et al.* 2016); 'Campamento Medio Algodón', 02°35.5'S, 72°53.5'W (Stotz *et al.* 2016); 'Campamento Choro', 02°36.5'S, 71°29'W (Stotz & Alván 2011); 'Campamento Curupa', Río Yanayacu, 02°53'S, 73°01'W (Stotz & Alván 2010); 'Campamento Alto Cotuhé', 03°12'S, 70°54'W (Stotz & Alván 2011); Sucusari, 03°15'S, 72°54'W (ML, LSUMZ); ÁCP Sabalillo, 03°20'S, 72°18'W (S.J. Socolar *in litt.* 2016).

Grallaria varia varia (Boddaert, 1783), Table des planches enluminées d'histoire naturelle de M. D'Aubenton, p. 44. Nominate *varia* was originally described as *Formicarius varius* from 'Cayenne' (= French Guiana). It is known from Suriname and the Guianas to Brazil, north of the Amazon and west of the lower Rio Negro. West of the lower Rio Negro it is replaced by *cinereiceps*. It is unclear, however, the precise situation in the upper Rio Negro/ Branco watersheds, and it remains to be seen if, and where, the ranges of these taxa approach each other. Although the original description by Boddaert (1783) is somewhat cursory and written in Latin, the following description by Chubb (1921) described the nominate race as having the upperparts 'fulvous-brown' with black feather edges. Lores, forecrown and cheeks buffy-white, with ear-coverts, neck-sides and throat chestnut-brown with pale shaft-streaks that widen and become white on the central lower throat. Nape and hindneck are grey, with similar black fringes to the feathers. The upperwing-coverts and leading edge of the flight feathers are similar to the upperparts, but paler. Median and greater coverts have pale buffy tips, slightly darker on primary-coverts. The inner vanes of the flight feathers are dark brown broadly fringed pale ferruginous on the basal portion while the underwing-coverts and inner edges of quills are pale cinnamon-rufous, contrasting with the pale brownish undersides of the wing vanes. Tail cinnamon-rufous above, more brownish below. Breast, belly and undertail-coverts pale ferruginous, somewhat darker ochraceous on boy-sides. SPECIMENS & RECORDS **VENEZUELA**: *Bolívar c*.4km E of Mata Verde, 07°58'N, 61°53'W (ML 62453–457, P.A. Schwartz). **GUYANA**: *Barima-Waini* Baramita, 07°22'N, 60°29W (USNM 586402). *Cuyuni-Mazaruni* 'Kamakusa' (loc. Stephens & Traylor 1985), 05°57'N, 59°54'W (Chubb 1921, AMNH 176880). *Potaro-Siparuni* Iwokrama, 04°38'N, 58°39'W (A. Farnsworth *in litt.* 2016); Quonga, 04°10'N, 59°20'W (Chubb 1921). *Upper Takutu-Upper Essequibo c*.85km WNW of Biloku, 02°02'N, 59°15'W (ML 115734, D.W. Finch); upper Essequibo River, 01°35'N, 58°38'W (Robbins *et al.* 2007, USNM 625538); Kamoa River, 01°32'N, 58°50'W (Robinson *et al.* 2007); Acari Mountains, 01°23'N, 58°57'W (O'Shea 2013). *East Berbice-Corentyne* Itabu Creek head, Boundary Camp, 01°33.5'N, 58°10'W (Blake 1950, FMNH 120233-36). **SURINAME**: *Brokopondo* Brownsberg Natuurpark, 04°56.5'N, 55°10'W (Donahue 1985, ML 25209/10, T.H. Davis). *Sipaliwini* Bakhuis Gebergte, 04°21'N, 56°45'W (ML 134657/68, C.K. Hanks); Lely Gebergte, 04°16'N, 54°44.5'W (O'Shea 2007); Kayserberg Airstrip, 03°07'N, 56°27'W (Blake 1961, 1963, FMNH 260493); 'Kasikasima camp', 02°58.5'N, 55°23'W (O'Shea & Ramcharan 2013a); Juuru Camp, Upper Palumeu River, 02°28.5'N, 55°38'W (O'Shea & Ramcharan 2013a); Grensgebergte Rock, 02°28'N, 55°46'W (O'Shea & Ramcharan 2013a); 7.5km ENE of Kwamalasamutu, 02°22'N, 56°43.5'W (eBird: J.C. Mittermeier); Sipaliwini River, 02°17.5'N, 56°36.5'W (O'Shea & Ramcharan 2013b). **FRENCH GUIANA**: *Saint Laurent Du Maroni* Tamanoir, 05°09'N, 53°45'W (CM, YPM, AMNH); Säul, 03°52'N, 53°18'W (Renaudier 2009); Sauts du Litany, 03°19'N, 54°03'W (Berlioz 1962). *Cayenne* Saut Pararé, Rivière Arataye, 04°03'N, 52°42'W (Érard 1982). **BRAZIL**: *Roraima* PN do Viruá, 01°29.5'N, 61°00.5'W (eBird: D. Pioli). *Amapá* Igarapé Capivara, 01°02'N, 51°44'W (MPEG 21169); PN Montanhas do Tumucumaque, 02°11.5'N, 54°35.5'W (Coltro 2008); Rio Anacuí, 01°50.5'N, 52°44.5'W (Coltro 2008); Calçoene, Rio Mutum, *c*.01°23'N, 51°55.5'W (Novaes 1974, Coltro 2008, MPEG 21230); FLOTA Paru, 00°56'N, 53°14'W (Aleixo *et al.* 2011); Igarapé Rio Branco, Boa Fortuna, 00°33'N, 52°16'W (Novaes 1974, LSUMZ 67359, MPEG 16138); Areia Vermelha, 01°00.5'N, 51°40'W (Novaes 1974, ANSP 170092, MPEG 20228/30); Rio Falsino, 00°56'N, 51°35'W (Novaes 1974, MPEG 20229); Cachoeira Itaboca, 00°02'N, 51°55'W (Novaes 1974; MPEG 28747); RESEX Rio Cajari, 00°35'S, 52°16'W (Schunck *et al.* 2011); Igarapé Novo, 00°30'S, 52°30'W (Novaes 1974, MPEG 16607–611, LSUMZ 67360). *Pará* FLOTA Faro, 01°42'N, 57°12'W (Aleixo *et al.* 2011); Oriximiná, 01°17'N, 58°41'W (Aleixo *et al.* 2011, MPEG 65849); REBIO Maicuru, 00°49'N, 53°55'W (Aleixo *et al.* 2011). *Amazonas* Presidente Figueiredo, 01°27'S, 60°02'W (WA 462806, F.B.R. Gomes); BDFFP site, 02°20'S, 60°00'W (Cohn-Haft *et al.* 1997, Stouffer *et al.* 2006, Stouffer 2007, ML); *c*.11km SSW of Efigênio Sales, 02°55'S, 59°58.5'W (ML 115269/70, C.A. Marantz). SEE ALSO Brabourne & Chubb (1912), Cory & Hellmayr (1924), Donahue & Pierson (1982), Braun *et al.* (2000), Bierregaard & Lovejoy (1989).

Grallaria varia distincta Todd, 1927. Proceedings of the Biological Society of Washington, vol. 40, p. 176. The type is an adult male collected at Villa Braga, Rio Tapajós, Brazil (04°25.5'S, 56°17.5'W), on 6 December 1919 by Samuel M. Klages. It occurs in Brazil, south of

the Amazon from the Rio Madeira east to the Rio Tapajós (Todd 1927), and south to Rondônia and northern Mato Grosso (Krabbe & Schulenberg 2003). Subspecies *distincta* is most similar to the nominate but has more obvious buffy shaft-streaks on the back than other races, the brown of the underparts is overall duller and less rufescent except the underwing- and tail-coverts, which are brighter and more ochraceous. SPECIMENS & RECORDS **BRAZIL**: *Amazonas* Novo Aripuanã, 05°08'S, 60°22.5'W (WA 1153536, M. Padua); Maués, 05°18.5'S, 58°11.5'W (Dantas *et al.* 2011); Trilha Castanheira, 06°56.5'S, 60°31'W (J. VanderGaast *in litt.* 2017). *Pará* Vicinity of Juruti, *c*.02°09'S, 56°05'W (Santos *et al.* 2011a); Santarém, 02°27'S, 54°42'W (CM P72857/58, YPM 29654); Igarapé Mutum, 02°37.5'S, 56°12'W (MPEG 61642); 'Apacy', near Aveiro (loc. Paynter & Traylor 1991a), *c*.03°15'S, 55°10'W (Griscom & Greenway 1941, CM); 'Bacia 363', (loc. see Gardner *et al.* 2013), 03°18'S, 55°00'W (XC 94645, N. Moura); Anapu, 03°28.5'S, 51°12'W (J.C. Vaz de Oliveira *in litt.* 2015); 'Bacia 549', (loc. see Gardner *et al.* 2013), 03°49'S, 48°23'W (XC 84118, A. Lees); above Tucuruí Dam, right bank, 04°14'S, 49°24.5'W (XC 170618, S. Dantas); Itaituba, 04°16'S, 55°59.5'W (CM P77207); Vila Braga, 04°25.5'S, 56°17.5'W (Griscom & Greenway 1941, CM P75444); Trairão, 04°30'S, 55°40'W (Pacheco & Olmos 2005); BR-163 W of Trairão, 04°44'S, 56°12'W (MPEG 76032); FLONA Serra dos Carajás, Sector Salobo, 05°55'S, 50°30'W (XC 223718–721, P. Boesman); Igarapé Preto, 06°30'S, 57°10'W (MPEG 65699). *Mato Grosso* RPPN Rio Cristalino, 09°35'S, 55°55'W (ML 88590/885, C.A. Marantz); Alta Floresta area, *c*.09°51'S, 56°05'W (Zimmer *et al.* 1997); Rondolândia, 10°27'S, 61°08.5'W (WA 1995245, recording G. Correa). *Rondônia* Calamá, 08°01.5'S, 62°52'W (Hellmayr 1910, Griscom & Greenway 1941, AMNH 492193/94); Machadinho d'Oeste, 08°45.5'S, 62°15'W (XC 167347, G. Leite); Cujubim, 08°59'S, 62°35'W (XC 320487, F. Igor de Godoy); ESEC Antonio Mujica Nava, 09°25'S, 64°56.5'W (Olmos *et al.* 2011); Cachoeira Nazaré, 09°44'S, 61°53'W (Stotz *et al.* 1997); Pedra Branca, 10°02'S, 62°06'W (Stotz *et al.* 1997); Aldeia Gaviões, Terra Indígena Igarapé Lourdes, 10°25.5'S, 61°39.5'W (Santos *et al.* 2011b, MPEG 58219).

Grallaria varia intercedens Berlepsch & Leverkühn, 1890. Ornis, vol. 6, p. 27, 'Bahia'. Described originally in Latin, as a race of '*Grallaria imperator*', the holotype is an unsexed adult collected at an unspecified location in Bahia and deposited in the Zoological Museum of Kiel University. The unfortunately vague type locality leaves its precise distribution in Bahia somewhat unclear. As mentioned by Silveira (2008), further uncertainty surrounds a specimen reportedly from 'Pernambuco' (Sclater 1890) that is frequently mentioned in the literature (Hellmayr 1903, von Ihering & von Ihering 1907, Cory & Hellmayr 1924, Pinto 1938, 1978, Peters 1951, Krabbe & Schulenberg 2003). There are no modern records from that region and, indeed no other records north of the Rio São Francisco. It seems quite possible that the specimen in question is a mislabelled trade skin (A. Lees *in litt.* 2017). Records of Variegated Antpitta in north-east Minas Gerais are usually refered to *intercedens* (Krabbe & Schulenberg 2003), although their precise taxonomic affinity remains uninvestigated, while populations in south-east Minas Gerais probably belong to *imperator* (and are included as such herein). This race resembles *imperator*, but is overall paler and more distinctly barred ventrally. SPECIMENS &

RECORDS *Bahia* Cajazeiras, 10°46'S, 38°25.5'W (Silveira 2008); Sauípe, 12°21'S, 37°54'W (Lima 2006); near Boa Nova, *c.*14°22'S, 40°12.5'W (WA 955349; recording E. Luiz); *c.*9km NW of Camacan, 15°23'S, 39°34.5'W (XC 212636, M. Braun); Macarani, 15°34'S, 40°25.5'W (XC 338818, R. Souza); near Porto Seguro, 16°23.5'S, 39°10'W (C. Albano *in litt.* 2017). *Minas Gerais* Near Balbina, 15°48.5'S, 40°31'W (XC 215652, H. Matheve); Fazenda Montes Claros, *c.*19°50'S, 41°45'W (Gonzaga *et al.* 1987); Araponga, 20°40'S, 42°31.5'W (XC 201563, V.P. Herdy). *Espírito Santo* Itarana, 19°52'S, 40°52.5'W (Venturini *et al.* 2001); Santa Teresa area, 19°56'S, 40°36'W (Ruschi 1965, 1967, Willis & Oniki 2002a); Domingos Martins, 20°22'S, 40°40'W (Silveira 2008); Alfredo Chaves, 20°38'S, 40°45'W (Silveira 2008, MZUSP).

Grallaria varia imperator Lafresnaye 1842, Revue Zoologique par la Société Cuvierienne, vol. 5, p. 333. Described as *Grallaria imperator* from 'Saint Paul' (São Paulo, south-east Brazil; Cory & Hellmayr 1924), this race is found in southern Minas Gerais (Bauer & Pacheco 2000), São Paulo (Aleixo 1999, Aleixo & Galetti 1997, Cavarzere *et al.* 2010, Banks-Leite *et al.* 2012), Rio de Janeiro (Bauer & Pacheco 2000, Alves *et al.* 2009), including the large, offshore island, Ilha Grande (Alves & Vecchi 2009), south through Paraná (Candia-Gallardo 2005), Santa Catarina (Candia-Gallardo 2005), to eastern Paraguay (Bertoni 1901), north-east Argentina (Misiones) and central Rio Grande do Sul (Belton 1994, Bencke 1996, Bencke & Kindel 1999, Bencke *et al.* 2010). Subspecies *imperator* is the largest race and has the crown extensively tinged olive with more obvious pale shaft-streaks (Krabbe & Schulenberg 2003). The majority of feathers on the back have the shaft-streaks slightly inflated near their tips into small, somewhat triangular, terminal buff spots. The upperwing-coverts bear buffy, triangular spots that are larger and more obvious than in other races, and the throat is considerably darker (with narrow buff streaks). The lower breast and upper belly are olive-buff and only lightly barred, while the lower belly is pale buff with barring or scaling reduced or absent. This, the southernmost subspecies, is among the best-studied races, with much of our general knowledge of behaviour and natural history deriving from studies of *imperator* in Argentina (Fraga & Narosky 1985, Protomastro 2000). SPECIMENS & RECORDS BRAZIL: *Rio de Janeiro* Serrinha, 22°02'S, 41°42'W (Pacheco *et al.* 1997a); Fazenda Campestre, 22°16'S, 42°32'W (MNRJ 35630); Nova Friburgo, 22°17.5'S, 42°32'W (MNRJ MNA3398/99); Maromba and vicinity, *c.*22°20'S, 44°36.5'W (Pinto 1954b, MNRJ 46110/208/557); Pico das Agulhas Negras, 22°22.5'S, 44°37.5'W (USNM 490063); Hotel Donati, 22°23'S, 44°38'W (XC 64845, R. Gagliardi); Taquaras Lodge, 22°24'S, 42°27'W (Pacheco *et al.* 2014); Teresópolis, 22°25'S, 42°58.5'W (Davis 1945, LACM 27941/42); Cachoeiras de Macacu, 22°28'S, 42°39'W (WA 523423, E. Legal); RE Guapiaçu, 22°30'S, 42°43'W (XC 61588, F. Lambert); near Guapimirim, 22°31'S, 43°01'W (Mallet-Rodrigues & Noronha 2003); Fazenda Bracuí, 22°52'S, 44°26'W (Buzzetti 2000); Rio Taquari, Parati, 23°02'S, 44°43'W (Buzzetti 2000); Pedra Branca Parati, 23°13'S, 44°43'W (MNRJ). *Minas Gerais* Passa-Vinte, 22°07'S, 44°14.5'W (Lombardi *et al.* 2012); Serra do Juncal, *c.*22°43.5'S, 45°56'W (Vasconcelos & Neto 2009). *São Paulo* Ilha Seca, 20°40.5'S, 51°15'W (MNRJ 20235); Rio Feio near Bauru, 22°19'S, 49°03.5'W (USNM 177716); Vitoriana, 22°47'S, 48°24'W (AMNH 492195); Santo Antônio do Pinhal, 22°48'S, 45°37.5'W (XC 345979; R. Dela

Rosa); Sertão do Bracuhy, 22°50'S, 44°23'W (XC 15887, R. Gagliardi); São José dos Campos, 22°52.5'S, 45°52'W (XC 87404, R. Dela Rosa); Atibaia, 23°12'S, 46°35'W (XC 146314; W. Zaca); São Luiz do Paraitinga, 23°13.5'S, 45°18.5'W (WA 60878, D. Bucci); Ubatuba, 23°26'S, 45°5'W (USNM 561301, XC); PE Serra da Cantareira, 23°28'S, 46°37'W (XC 144756, A. Silveira); Salesópolis, 23°32'S, 45°51W (WA 1179646, E. Jesus); Tremembé, 23°32'S, 45°50.5'W (WA 1393341, F.B.R. Gomes); Cotia, 23°36'S, 46°55'W (Develey & Martensen 2006); Casa Grande, 23°37'S, 45°57'W (LSUMZ); Estação Biológica Boracéia, 23°38'S, 45°52'W (XC 1310, G.R.R. Brito); Ibiúna, 23°40'S, 47°13'W (Develey & Metzger 2006); APA Capivari Monos, 23°57.5'S, 46°39'W (XC 187630, M.A. Melo); Tapiraí, 23°58'S, 47°30.5'W (Martensen *et al.* 2012, WA 1238035, R. Machado); near Ribeirão Grande, *c.*24°06.5'S, 48°22.5'W (WA, XC); PE Intervales, 24°16'S, 48°25'W (XC 102618, F. Schmitt); Peruíbe, 24°19.5'S, 47°00'W (WA 297225, B. Lima); 'Boa Vista', (loc. Stephens & Traylor 1991a), 24°35'S, 47°38'W (FMNH 265210–212); Iguape, 24°43'S, 47°33'W (LACM 28555). *Paraná* PE Mata dos Godoy, 23°07'S, 49°49'W (dos Anjos 2002); Apucarana, 23°33'S, 51°28'W (WA 1461483, recording A. Ribeiro); Telêmaco Borba, 24°19.5'S, 50°37'W (WA 591441, M. Villegas); RE Samuel Klabin, 24°20'S, 50°35'W (dos Anjos 2002); Teresina, 24°45'S, 51°05'W (Sztolcman 1926b); Turvo, 25°03'S, 51°32'W (WA 1535209; recording R. Oliveira); 13km NNE of Guaraqueçaba, 25°11'S, 48°18'W (XC 172531, J.A.B. Vitto); Irati, 25°28'S, 50°39'W (WA 2365502, H. Neto); Roça Nova, 25°28.5'S, 49°01'W (AMNH 492196); Mananciais da Serra, 25°30'S, 48°57'W (Reinert *et al.* 1996); Tijucas do Sul, 25°45'S, 49°20'W (Kaminski & Carrano 2004); Foz do Rio Jordão, 25°45'S, 52°15'W (Straube *et al.* 1996); UHE Segredo, 25°55'S, 52°05'W (Straube *et al.* 1996). *Santa Catarina* Near Joinville, 26°18.5'S, 48°51'W (MNRJ 21894); Doutor Pedrinho, 26°43'S, 49°29'W (XC 333063, D. Meyer); Fazenda Santa Rita, 27°06.5'S, 49°09.5'W (XC 42228, A.E. Rupp); near Nova Trento, 27°15.5'S, 49°01'W (XC 45069, E. Legal); Bom Retiro, 27°49'S, 49°34'W (do Rosário 1996); Campo Belo do Sul, 27°55'S, 50°45.5'W (XC 96412, C. Espínola); Grão Pará, 28°11'S, 49°13'W (do Rosário 1996); near Nova Veneza, 28°39'S, 49°40'W (XC 186423, R. Romagna). *Rio Grande do Sul* FLOTA Turvo, 27°12.5'S, 54°00'W (ML 19256, W. Belton); Sinimbu, 29°24'S, 52°33'W (WA 1990874, P. Kuester); Hotel Veraneio Hampel, 29°26.5'S, 50°36.5'W (XC 98544, A.C Lees); Taquara, 29°39'S, 50°46.5'W (von Berlepsch & von Ihering 1885); Maquiné, 29°36.5'S, 50°13.5'W (WA 461323, recording D. Machado); Santa Cruz do Sul, 29°43'S, 52°26'W (WA 1934018, recording A.R. Mohr). ARGENTINA: *Corrientes* Garruchos, 28°10.5'S, 55°39.5'W (Contreras 1983, Nores *et al.* 2005, YPM 115074). *Misiones* *c.*10.5km ESE of Puerto Iguazú, 25°39'S, 25°39'S (Buckingham 2011); Paraje María Soledad, 25°51'S, 53°59'W (XC 255928, L. Pradier); Eldorado, 26°24'S, 54°37.5'W (LSUMZ 69189); Tobuna, 26°28'S, 53°54'W (LACM 48626–633); Fracrán, 26°45'S, 54°18.5'W (C. Olrog in Fraga & Narosky 1985); PP Salto Encantado, 27°06'S, 54°59'W (XC 52795, C. Ferrari); Corpus Christi, 27°08'S, 55°28'W (Krauczuk 2008); Guarani, 27°16.5'S, 54°12'W (XC 154396/97, B. López-Lanús); Santa Ana, 27°22'S, 55°34'W Dabbene 1919, AMNH 154199, FIML 703); PP Cañadon de Profundidad, 27°33.5'S, 55°43'W (XC 336037, Aves del NEA). PARAGUAY: *Amambay* Arroyo Blanco, 22°29'S, 56°08'W (GPDDB); PN Cerro Corá, 22°39'S, 56°00'W (GPDDB); Cerro Guazú, 23°05'S, 56°01'W (GPDDB); 40km WSW of Capitán Bado, 23°38.5'S,

55°40.5'W (UMMZ 101789). *San Pedro* Estancia Alegría, 23°33'S, 56°26'W (GPDDB); Bosques de Yaguarete, 23°50'S, 56°09.5'W (GPDDB); Estancia Cerrito, 24°45'S, 56°09.5'W (GPDDB); RE Capiibary, 24°48.5'S, 55°56'W (GPDDB). *Canendiyú* Puesto de Guardaparques Jejui-mi, 24°07'S, 55°32'W (Brooks *et al.* 1993); Puesto de Guardaparques Lagunita, 24°07'S, 55°26'W (Brooks *et al.* 1993); Estancia Felicidad, 24°09'S, 55°41'W (GPDDB); Koe Tuvy, 24°15.5'S, 55°21'W (GPDDB); RPN Estancia Itabó, 24°27'S, 54°38'W (Brooks *et al.* 1993, Cockle 2003, Cockle *et al.* 2005). *Caaguazú* Morombi, 24°37'S, 55°22'W (GPDDB); Ka'aguy Rory, 24°44'S, 55°41'W (GPDDB); Estancia Itá Pytã, 25°02'S, 55°48'W (GPDDB); Estancia Quinto Potrero, 25°14.5'S, 56°10'W (GPDDB). *Alto Paraná* Limoy, 24°47'S, 54°26'W (GPDDB); RB Itabó, 25°02'S, 54°39'W (GPDDB); Estancia Don Oscar, 25°51'S, 54°51'W (GPDDB). *Guairá* Ybytyruzú, 25°50'S, 56°13'W (H.F. del Castillo C. *in litt.* 2017). *Paraguarí* PN Ybycuí, 26°04'S, 56°48'W (GPDDB). *Itapúa* Área de Reserva para PN San Rafael, 26°36'S, 55°40'W (Esquivel-M. & Peris 2011a, 2012). *Caazapá* PN Caaguazú, fraccion de Cristal, 26°07'S, 55°44'W (Madroño *et al.* 1997). NOTES There are numerous records from within a few km of Teresópolis: Varginha, Canoas (MNRJ 37203), Jacarandá (XC 340957, R.J. Mitidieri), Fazenda C. Guinle (MNRJ 24542/89), Alpina (MNRJ 8262) and Barreira (MNRJ 8261).

STATUS Though usually uncommon to rare within its range (Krabbe & Schulenberg 2003), Variegated Antpitta can be fairly common in parts of south-east Brazil (e.g., São Paulo, *imperator*). Despite that the population trend appears to be decreasing, and the species is quickly lost from isolated fragments (Bierregaard & Lovejoy 1989, dos Anjos *et al.* 2009, 2011), Variegated Antpitta has an extremely large range and the population size is not believed to approach the thresholds for Vulnerable. It is currently considered Least Concern at a global level (BirdLife International 2017). Nevertheless, in Brazil, the coastal subspecies (*intercedens*) is considered Endangered nationally and Vulnerable in the state of Minas Gerais (Fundação Biodiversitas 2008, Machado *et al.* 2008). Overall, the species is expected to lose 13.0–14.3% of suitable habitat within its distribution over three generations based on currently projected rates of deforestation (Soares-Filho *et al.* 2006, Bird *et al.* 2011, BirdLife International 2017). Given the sensitive nature of most understorey insectivores to fragmentation (e.g., Kattan *et al.* 1994, Stouffer 2007), and the seemingly high sensitivity of Variegated Antpitta (Moura *et al.* 2016), it is possible that predicted anthropogenic disturbance will be detrimental. Continued monitoring of populations and re-evaluation of its threat status will be especially important. Due largely to its broad range, Variegated Antpitta is found in a large number of reserves. PROTECTED POPULATIONS *G. v. varia* Suriname Brownsberg Natuurpark (Donahue 1985, ML); Eilerts De Haan Natuurpark (Blake 1961, 1963, FMNH 260493). Probably occurs in the Raleigh Vallen-Voltzberg Natuurpark. Brazil PN do Viruá (eBird: D. Pioli); PN Montanhas do Tumucumaque (Coltro 2008); Reserva Florestal Adolpho Ducke (Bueno *et al.* 2012); RESEX do Rio Cajari, (Schunck *et al.* 2011); ESEC Grão-Pará (Aleixo *et al.* 2011; MPEG); FLOTA Faro (Aleixo *et al.* 2011); REBIO Maicuru (Aleixo *et al.* 2011); FLOTA Paru (Aleixo *et al.* 2011). *G. v. cinereiceps* Brazil PN Jaú (Borges 2006, Borges *et al.* 2011). Peru ÁCP Sabalillo (S.J. Socolar *in litt.* 2017); ZR Yaguas (Stotz & Alván 2011). *G. v. distincta* PN dos

Campos Amazônicos (eBird: J. Dickens); FLONA Pau-Rosa (Dantas *et al.* 2011); FLONA Serra dos Carajás (Pacheco *et al.* 2007, MPEG 70402); FLONA Rio Pacu (MPEG 65699); FLONA Tapajós (Oren & Parker 1997, Henriques *et al.* 2003); RPPN Rio Cristalino (ML 88885, C.A. Marantz); ESEC Antonio Mujica Nava (Olmos *et al.* 2011). *G. v. intercedens* PN Monte Pascoal (Silveira 2008); PN Caparaó (Vasconcelos 2003b, Silveira 2008); PN Boa Nova (C. Albano *in litt.* 2017); PN Serra das Lontras (eBird: R. Laps); PN Descobrimento (Machado *et al.* 2008); PN Pau Brasil (Silveira 2008); PE da Serra do Brigadeiro (Silveira 2008); REBIO Augusto Ruschi (Parker & Goerck 1997, Silveira 2008); ESEC Juréia-Itatins (Lima 2011); RPPN Estação Veracel (C. Albano *in litt.* 2017); RPPN Serra Bonita (XC 212636, M. Braun, ML 60918351, photo C. Albano); ESEC Nova Esperança (Silveira 2008); REBIO Una (Gonzaga *et al.* 1987, Silveira 2008); RPPN Estação Veracruz and RPPN Serra do Teimoso (Silveira 2008); RPPN Mata do Passarinho (eBird: H. Peixoto); RPPN Vale do Rio Doce, Linhares (eBird: R. Laps). *G. v. imperator* Brazil PN Aparados da Serra (Parker & Goerck 1997); PN da Serra do Itajaí (XC 42228, A.E. Rupp); PN Itatiaia (Parker & Goerck 1997, XC, USNM, MNRJ); PN Serra da Bocaina (eBird: B. Rennó); PN Iguaçu/Iguazú (Parker & Goerck 1997); PN Serra dos Órgãos (Mallet-Rodrigues & Noronha 2003, ML, XC); FLONA São Francisco de Paula (eBird: R. Kurz); PE Serra da Cantareira (XC 144756, A. Silveira); PE Jurupará (eBird: J. Dickens); PE Campos do Jordão (eBird: C. Gussoni); PE Três Picos (Mallet-Rodrigues & Noronha 2009, XC 340957, R.J. Mitidieri); PE Serra do Mar (XC 252851/302746, J. Minns); PE Intervales (Aleixo & Galetti 1997, XC, ML, IBC); PE Serra do Tabuleiro and PE Serra Furada (do Rosário 1996); PE Mata dos Godoy (dos Anjos 2001, 2002, Volpato *et al.* 2006); PE Carlos Botelho (Rother *et al.* 2013); REBIO Estadual Sassafrás (do Rosário 1996); REBIO Estadual Aguaí (Just *et al.* 2015); REBIO Araras (Alves *et al.* 2009); Reserva Florestal Morro Grande (Develey & Martensen 2006); Estação Biológica Boracéia (Cavarzere *et al.* 2010, XC 1310, G.R.R. Brito); RE Samuel Klabin (dos Anjos 2002); FLOTA Turvo (Belton 1994, ML); PNM Ronda (Franz *et al.* 2014); APA Macaé de Cima (Pacheco *et al.* 2014); APA Municipal Capivari-Monos (XC 187630, M.A. Melo); Parque Botânico Morro do Baú (do Rosário 1996); RE Guapiaçu (XC 61588, F. Lambert); RPPN Parque Ecológico do Guarapiranga (XC 186350, F. Igor de Godoy); RPPN Salto Morato (XC 172531, J.A.B. Vitto); RPPN Parque Ecológico da Artex (do Rosário 1996); RPPN Bugerkopf (eBird: A.E. Rupp); RPPN Parque do Zizo (XC 333830, J. Fischer); RPPN Prima Luna (XC 45069, E. Legal); RPPN Universidade de Santa Cruz do Sul (Oliveira & Köhler 2010); RPPN Fazenda Renópolis (XC 345979, R. Dela Rosa). Paraguay PN Caaguazú (Madroño *et al.* 1997); PN San Rafael (Esquivel-M. *et al.* 2007, Esquivel-M. 2010, Esquivel-M. & Peris 2011b, 2012); PN Ybycuí (GPDDB); PN Cerro Corá (GPDDB); RE Capiibary (GPDDB); RB Itabó [Itaipú] (GPDDB); Monumento Científico Moisés Bertoni (eBird: A. Esquivel-M.); RPN Estancia Itabó (Brooks *et al.* 1993, Cockle 2003, Cockle *et al.* 2005); RPN Bosque Mbaracayú (Brooks *et al.* 1993). Argentina PP Cruce Caballero (Bodrati *et al.* 2010); PP Salto Encantado (XC 52795, C. Ferrari); PP Cañadón de Profundidad (XC 336037, Aves del NEA); PP Urugua-í (XC 255928, L. Pradier); PP Guardaparque Horacio Foerster (eBird: Aves del NEA); PP Esmeralda (R. Ramírez *in litt.* 2016); PP El Piñalito (P.A. Hosner *in litt.* 2017); PP Moconá (XC 154396/97, B. López-Lanús).

OTHER NAMES *Formicarius varius* [*varia*] (Boddaert 1783); *Grallaria rex* [*varia*] (Lafresnaye 1842); *Chamaebates rufiventris* [*imperator*] (Bertoni 1901); *Myiothera grallaria* [*imperator*] (Lichtenstein 1823); *Myiotrichas imperatrix* [*imperator*] (Cabanis & Heine 1859); *Myioturdus rex* (Ménétries 1835); *Grallaria varia rufiventris* [*imperator*] (Bertoni 1914, 1939); *Grallaria imperator* (Salvin 1882, von Berlepsch & von Ihering 1885, Koenigswald 1896, Euler 1900, von Ihering 1900, 1902); *Grallaria imperator intercedens* (von Berlepsch & Leverkühn 1890); *Grallaria varia imperator* [*intercedens*] (Ruschi 1967); *Grallaria varia varia* [*distincta*] (Hellmayr 1910); *Turdus rex* (Gmelin 1789); *Grallaria fusca* (Vieillot 1824, 1834, von Tschudi 1844); *Pitta grallaria* (Temminck & Laugier 1820–39); *Turdus grallarius* [*varia*] (Latham 1790, Stark 1828). **English** Royal Antpitta [*varia*] (Cory & Hellmayr 1924); King Thrush [*varia*] (Latham 1783); Royal Ant-thrush (Chubb 1921, Lodge (1991); Ash-headed Antpitta [*cinereiceps*] (Cory & Hellmayr 1924); Intermediate Antpitta [*intercedens*] (Cory & Hellmayr 1924); Imperial Antpitta [*imperator*] (Cory & Hellmayr 1924). **Spanish** Tororoí Pintado (Krabbe & Schulenberg 2003); Hormiguero Tororoí Cabecinegro (Hilty 2003); Chululú Pintado (de la Peña & Rumboll 1998, Mazar Barnett & Pearman 2001); Chululú Grande (Olrog 1963, Canevari *et al.* 1991); Tovacasú (Bertoni 1939); torón-torón (Zotta 1944); Urú-i (Arribalzaga 1926); Tororoí abigarrado (Olrog 1979); Pichón Cabecicenizo [*cinereiceps*] (Phelps & Phelps 1950). **Portuguese** tovacuçu/ú (Ruschi 1953, 1965, Belton 1994, do Rosário 1996, Bencke 2001, Bencke *et al.* 2010); tovacuçú-malhado (Pacheco *et al.* 1997, Lima 2006); tovaquçu (Straube *et al.* 2007); gallinhola do mato (von Berlepsch & von Ihering 1885); galinha do mato (Ruschi 1953, 1967). **French** Grallaire roi (Krabbe & Schulenberg 2003); Le Roi des Fourmilliers (Buffon 1771–86); La Grallaire Brune (Vieillot 1824, 1834); Brève roi (Temminck & Laugier 1820–1839); Grallaire variable (Tostain *et al.* 1992). **German** Große Bartameisenpitta (Krabbe & Schulenberg 2003).

Variegated Antpitta, adult (*intercedens*), Camacan, Bahia, 29 October 2016 (*Ciro Albano*).

Variegated Antpitta, nest and complete clutch (*distincta*), Anapu, Pará, Brazil, 15 February 2015 (*Júlio César Vaz de Oliveira*).

Variegated Antpitta, adult brooding two older nestlings (*imperator*), PP Cruce Caballero, Misiones, Argentina, 28 November 2011 (*Martjan Lammertink*).

Variegated Antpitta, adult (*imperator*), Corupá, Santa Catarina, Brazil, 2 December 2013 (*Luciano Moraes*).

Variegated Antpitta, adult (*imperator*), Tapiraí, Sao Paulo, Brazil, 16 May 2016 (*Caio Brito*).

MOUSTACHED ANTPITTA
Grallaria alleni   Plate 7

Grallaria alleni Chapman, 1912, Bulletin of the American Museum of Natural History, vol. 31, p. 148, Salento, 7,000ft [2,140m], Central Andes, 'Cauca', Colombia [= Quindío, 04°38'N, 75°34'W]. The holotype is an adult female (AMNH 112005) collected by (and named for) Arthur Augustus Allen, on 2 October 2011 (LeCroy & Sloss 2000).

Moustached Antpitta was, for many years, an extremely elusive bird, being known from only a single specimen until 1979 (Chapman 1917, Meyer de Schauensee 1966, Hernández & Rodríguez 1979). It remained almost equally unknown until the late 1990s (Hilty & Brown 1986, Collar & Andrew 1988, Collar *et al.* 1992) when it was rediscovered in both Colombia and Ecuador (Renjifo 1999, Krabbe & Coopmans 2000). It is now known to be fairly well represented across its range in both countries, even fairly common in areas such as the west Andean slope in Pichincha and the west slope of the Central Andes in Risaralda. Moustached Antpitta is a mid-sized *Grallaria* with a broad white malar stripe, slate-grey crown and nape, and russet throat bordered by an irregular white chest-band. It inhabits wet, mossy cloud forest and feeds on arthropods and earthworms, occasionally small vertebrates, captured on or near the ground. Moustached Antpitta is among the first of (now) many species that have become accustomed to being fed worms by humans, much to the delight of birdwatchers who no longer must agonise over getting even a brief look at these furtive and elusive birds. Although the species' breeding biology is better known than most antpittas, details remain to be elucidated. Similarly, although its distribution is better understood today, extensive surveys of additional areas where the species might occur, aided by improved knowledge of the species' vocalisations, would enable better knowledge of its range, population density and habitat requirements. In particular, it has been suggested that the east slope of the Andes in Ecuador

and Colombia, from Tungurahua north to Caquetá, and the slopes of PN Sangay in central Ecuador may hold significant populations of Moustached Antpitta (BirdLife International 2017). Krabbe *et al.* (1998) suggested focused efforts to protect montane forests and to help local communities with land management on the Pacific slope in Pichincha and northern Cotopaxi should be a special conservation priority. Other priorities include basic studies of territory size, competitive interactions, reproductive biology, lifespan, survivorship and dispersal (Schulenberg & Kirwan 2012b). Other than its more widely distributed relative, Scaled Antpitta, the nesting biology of Moustached Antpitta is one of the best-studied of the genus, with the nests of both subspecies having published descriptions. Both adults care for the eggs and young in large, bulky and relatively deep, open-cup nests of humid organic material (Freile & Renjifo 2003, Londoño *et al.* 2004, Greeney & Gelis 2006). With this account I bring the total number of described nests to ten, expand on previous descriptions of fledglings, and describe juveniles and subadults for the first time.

IDENTIFICATION 16–18cm. The following description is based on Krabbe & Schulenberg (2003) and Schulenberg & Kirwan (2012b), and refers to nominate *alleni*. **Adult** Sexes similar. The forecrown is pale olive-brown, grading to slate-grey on the crown and nape, with feathers of the crown narrowly edged blackish. Rest of upperparts dark rufescent-brown to olive-brown, and also have the feathers edged blackish, giving a scaled effect. Tail rufous-chestnut. Lores white, with the feathers narrowly dark-tipped giving a slightly speckled appearance sometimes that is not necessarily obvious in the field. A broad white submoustachial stripe gives the species its common name, while the ear-coverts are uniform olive-brown with pale feather shafts at their bases and often invisible. Chin and throat rufescent-brown, with the white lower throat forming a crescent somewhat variable in its completeness and sometimes broken into white spots. Breast olive-brown, with narrow white streaking fading to a buffy-white belly. White underparts washed cinnamon on flanks and vent. In plumage, Moustached Antpitta is confusingly similar to sympatric populations of Scaled Antpitta (race *regulus*) (Krabbe & Coopmans 2000, Schulenberg & Kirwan 2012b). The two species seem most reliably separated by differences in breast pattern, in the width and prominence of the 'moustache', and by bill shape and colour (Krabbe & Coopmans 2000). Krabbe & Coopmans (2000) further compared their plumages as follows. In both species, the breast is predominantly buff or olive-brown, with feathers in the centre of the breast contrastingly patterned. In Scaled Antpitta, ssp. *regulus*, these feathers are pale buff (rarely white) 'with a noticeable difference in the distribution of white or buff between the inner and outer web of individual feathers, and with black on each side of the basal half of the shafts.' In contrast, these feathers are white in Moustached Antpitta and have an 'equal distribution of white on the outer and inner webs of each feather'. The white of these feathers forms 'large, symmetrically elongated, narrowly black-edged spots extending onto the lower breast as thin, white to buff streaks surrounding the shafts'. With practice, it is clear that the submoustachial streak of Moustached Antpitta is much more prominent than in Scaled, usually twice the width of the latter. That of Moustached is almost always cleaner white, while that of Scaled is noticeably buffy-white in most individuals. With respect to general shape, the bill of Moustached is thinner and straighter than that of Scaled, and is all black while the outer half of the maxilla of the latter is pale.

DISTRIBUTION Restricted to the North Andean centre of endemism (Cracraft 1985) and known, rather patchily, from several localities in the Andes of Colombia and Ecuador. It occurs in the West Andes (Krabbe *et al.* 2006) and the west slope of the Central Andes (Chapman 1912, Renjifo 1999, Renjifo *et al.* 2002) in the upper Magdalena Valley of Colombia (Hernández & Rodríguez 1979, Renjifo *et al.* 2002), and on both slopes of the Andes of northern Ecuador, south to the Guacamayos Ridge in Napo in the east and to north-west Cotopaxi on the west slope (Krabbe & Coopmans 2000, Schulenberg 2002, Freile & Chaves 2004).

HABITAT Occurs in the dark understorey of very humid, humid and seasonally humid montane forests, especially in ravines and steep slopes of undisturbed mossy forest (Krabbe & Coopmans 2000, Schulenberg & Kirwan 2012b). There have been no focused studies, however, documenting the specific habitat requirements or preferences. For nesting, however, several studies have described in detail the microhabitat characteristics and general landscape surrounding nest sites, giving us some insight into the species' preferences for breeding, as well as tolerance of habitat disturbance. Subspecies *alleni*, near the type locality in Dept. Risaralda, Colombia, was found nesting in a mosaic of ash (*Fraxinus chinensis*) plantations and small patches of forest, some primary but most in various states of regeneration. At least one nest was inside a plantation (Londoño *et al.* 2004). Freile & Renjifo (2003) also found a nest in this habitat, at the base of a small, shallow valley in secondary forest with a *c.*13m-high canopy. This site was equidistant between old-growth forest and the edge of an abandoned pasture, and was close to a small creek. Nearby, Londoño *et al.* (2004) also found *alleni* nesting in forest of varying successional stages, most <25 years old, separated by ash, pine and *Eucalyptus* plantations but with some remnants of old-growth forest

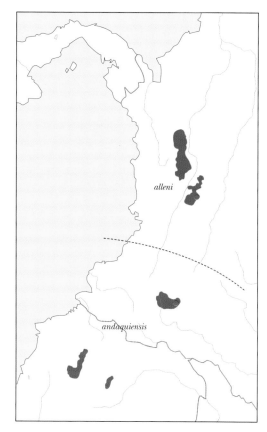

especially in riparian areas. In Ecuador, race *andaquiensis* also displays a similar ability to nest in less-than-pristine cloud forest habitats. The Ecuadorian nest described by Freile & Renjifo (2003) was in secondary forest with a fairly open understorey on a gentle slope, and was directly adjacent to a well-used path. The other *andaquiensis* nest (Greeney & Gelis 2006) was in old second growth or mature cloud forest with a well-developed canopy and a moss-and-epiphyte-laden understorey. Similarly, a previously unreported nest that I found at Reserva Mindo Loma (see below) was well inside forest with a closed canopy and shaded understorey. The altitudinal range in Colombia is generally given as 1,800–2,135m (Renjifo *et al.* 2002) and in Ecuador as 1,850–2,500m (Ridgely & Greenfield 2001a), but J. Freile (*in litt.* 2017) has found several Ecuadorian localities as low as 1,800m and Flórez *et al.* (2004) and Krabbe *et al.* (2006) recently recorded the species at 3,100m at Páramo de Frontino. In most regions, Moustached Antpitta replaces (with some overlap) the closely related Scaled Antpitta at higher elevations (Krabbe & Coopmans 2000).

VOICE Ridgely & Greenfield (2001b) described the song of *andaquiensis* as 'a slightly ascending series of hollow notes that gradually become louder'. It is apparently delivered at a slightly higher pitch and slower pace than that of Scaled Antpitta. The details provided by Krabbe & Schulenberg (2003) described it as a 1.8–3.0s series of 17–25 notes delivered at 0.8kHz and an even pace of 8–9 notes/s, with the volume increasing initially and the pitch slightly higher over the latter half. This song is produced at intervals of 7–13s (Krabbe & Schulenberg 2003). Though not well known, an additional vocalisation, given frequently at dawn, is a shorter, simpler, version of the song but with an *Otus* owl-like quality (Braun *et al.* 2003).

NATURAL HISTORY Although this species was included in the original 'family' of antpittas trained to consume worms at feeding stations (Collins 2006, Woods *et al.* 2011), we still know very little of the behaviour and general natural history in the 'wild'. Unlike many species of *Grallaria*, it usually does not approach the observer in response to playback (Schulenberg & Kirwan 2012b). Adults forage alone on or within 3m of the ground, generally within dense vegetation, climbing onto horizontal vines or small branches, 0.5–3.0m above ground to sing (Krabbe & Coopmans 2000). In Colombia, singing and (presumably) territorial behaviour appear to occur year-round, despite apparently seasonal breeding (Londoño *et al.* 2004). These authors further reported that most singing occurs at dawn and dusk, and that preferred habitats are humid areas within forest, especially along creeks and flat swampy areas. Similar crepuscular singing habits and a preference for riparian areas has been noted in north-west Ecuador (Chaves & Freile 2005, J. Lyons *in litt.* 2017). Vocalisations may continue into late morning and afternoon on overcast days (Chaves & Freile 2005).

DIET The diet of naturally foraging adults (not at worm feeding stations) is essentially unknown. Freile & Renjifo (2003) observed adults provisioning nestlings with arthropods, which were taken on the ground, both on the forest floor as well as along trails. At another nest, Greeney & Gelis (2006) documented adults bringing arthropods, including 3–5cm katydids (Tettigoniidae). They also found that at least 22% of prey deliveries (*n* = 328) included one or more earthworms (in an area without nearby feeding stations). At the same nest, they also observed the only recorded case of vertebrate predation, when adults fed a small frog to their nestlings. It seems likely, however, that this and other *Grallaria* feed on small vertebrates more often than currently documented. Nevertheless, J. Lyons (*in litt.* 2017) reports that individuals trained to feed on hand-proffered worms seem uninterested in other invertebrate prey offered.

REPRODUCTION Five nests have been described. The first known nest (*alleni*) was found in 1995 at Otún-Quimbaya Floral and Faunal Sanctuary, east of Pereira, on the west slope of the Central Andes, Risaralda, Colombia, at an elevation of 1,800m (Freile & Renjifo 2003). The second (*andaquiensis*) was found in 1999 in north-west Ecuador at the private reserve Bosque Integral Otonga, Cotopaxi province, at 1,900m (Freile & Renjifo 2003). The third and fourth nests were both found in 2003 (Londoño *et al.* 2004), one (*alleni*) in the La Nona Municipal Natural Park, on the west slope of the Central Andes, Risaralda. In the same study, they described a fourth nest, also of the nominate race, at Otún-Quimbaya Floral and Faunal Sanctuary. The fifth and final nest described in the literature (*andaquiensis*) was also found in 2003 at Tandayapa Bird Lodge, near the small town of Tandayapa, Pichincha, Ecuador, at 1,650m (Greeney & Gelis 2006). In addition to these published records, I describe a sixth nest that I found in 2007 at Mindo Loma private reserve above Mindo, Pichincha, at 1,700m, and J. Lyons described three nests to me, all found at Las Gralarias private reserve, also

above Mindo, at elevations of 1,850–2,000m. **NEST & BUILDING** No published data on nest-site selection or nest construction. Almost certainly, however, both adults play a role. The first nest described by Freile & Renjifo (2003) was built into a hanging tangle of epiphytes close to the trunk of a tree and further supported by the small branches of a shrub. This nest comprised, externally, dead leaves, sticks and other damp plant material. Internally the cup was lined with flexible, dark fibres, primarily thin rootlets. The second nest described by Freile & Renjifo (2003) was built into a tangle of small crisscrossing branches that supported the nest, as well as providing some overhead cover and camouflage. It was composed primarily of dead leaves, rootlets and small sticks, but included abundant green mosses, especially externally, which aided its camouflage. The inner cup was not described in detail, but photos provided by the authors suggest that it was rather sparsely lined with a few thin, dark, branched rootlets. The first nest described by Londoño *et al.* (2004) was supported by the branch of a small (10cm dbh) tree and was composed primarily of small sticks and moss woven together, and lined with thin, flexible, slightly branched rootlets. No measurements were taken. Their second nest was well supported between the two thick trunks of a palm. It was composed externally mostly of moss, mixed with a few dead leaves and sticks. The internal cup was thickly lined with thin black rootlets and fungal rhizomorphs. Details of the nest studied by Greeney & Gelis (2006) were not published previously, but it was supported by crisscrossing vines and small branches that formed a sturdy platform. My Mindo Loma nest was supported against a 35cm dbh tree trunk by several sturdy vines. Its external part was composed largely of humid, dead leaves and small sticks, only sparsely incorporating green moss. Internally, the cup was sparsely lined with dark flexible rootlets. Few data were collected at the three Las Gralarias nests, but one was built in a somewhat dense area of spiny-trunked palms beside a small stream. It was supported by three slanting, crossed palm trunks (*c.*10cm diameter) which appeared to have collected a substantial amount of detritus, on top of which the nest was built. Nest measurements for all of the nine nests mentioned above, in order presented, are: nest height 1.3m, 1.35m, 1.3m, 1.1m, 1m, 1.6m, 1.4m, 1m, 1.2m; external diameter 23cm, 22cm, ?cm, 11.1x16.8cm, 24.5cm, 22cm, 17.5cm, ?cm, 23cm; external height 17cm, 22.5cm, ?cm, 13.3cm, 22cm, 13cm, 15cm, 20cm, 18cm; internal diameter 10cm, 11.8cm, ?cm, 10.5cm, 10.5cm, 10cm, 9.5cm, 10cm; 11cm, internal depth 7cm, 4.5cm, ?m, 4.6cm, 6.5cm 7cm, 7.5cm, 8cm. 7.5cm. **EGG, LAYING & INCUBATION** All eggs, including those described for the first time here, are short, subelliptical and blue to blue-green with no markings. Seven nests held two eggs, one was found with an egg and a newly hatched nestling, and the other held two young nestlings, suggesting that two is the normal clutch size (Freile & Renjifo 2003, Londoño *et al.* 2004, Greeney & Gelis 2006, J. Lyons *in litt.* 2017). Egg measurements are as follows: 30.7 × 24.7mm (Freile & Renjifo 2003); 30 × 25mm, (8g), 29 × 25mm (8g), 30 × 24.3mm (8.75g), 30 × 24mm (8.5g) (Londoño *et al.* 2004). As with other antpittas (Greeney *et al.* 2008a), constancy of incubation is relatively high. At a nest (*alleni*) presumed to be somewhere mid-incubation (Londoño *et al.* 2004), total daylight coverage of the eggs on three consecutive days was 86.1%, 83.6% and 84.6%. At this nest, during *c.*53 hours of incubation monitoring during daylight hours, the eggs were incubated for a total of 80.9% of the day. Periods of absence from the nest during the day ranged from 4min to 67min, and periods of constant attendance (including quick changeovers) ranged from 11min to 240min. The incubation period is unknown. NESTLING & PARENTAL CARE Freile & Renjifo (2003) described newly hatched nestlings as having 'dark gray skin covered with dark gray down, and a prominent vitelline sac. The bill (except for a dark tip) and the gape were vermilion.' They observed that, when handled, the nestlings appeared to hide their heads under their own bodies, postulating that this might function to hide their brightly coloured gapes. My experience with this and other species of antpittas and passerines is that, just after hatching, most nestlings find it hard to hold their own head up for very long, especially when being moved from their natural position in the nest. Thus I interpret these authors' observations as simply the natural dropping of the nestlings' heads once their neck muscles are exhausted. Freile & Renjifo (2003) reported that, two days after hatching, the nestlings of one clutch weighed 12.0g and 13.5g and had tarsal measurements of 13.6mm and 14.8mm, respectively. Two nestlings, roughly eight days old, were well covered in black down with sparse buffy-chestnut spotting (Greeney & Gelis 2006). The chestnut spotting was barely visible on the hindcrown, nape and upper back, and more extensive and buffier on the lower belly and flanks. The skin was dusky-pink, blacker on the head and the legs were dark blue-black. The maxilla was black, becoming orange at the gape, while the mandible was orange, black only near the tip. The mouth lining was a striking red-orange contrasting sharply with the black plumage. Primary pin-feather sheaths were unbroken, but the secondaries had broken their sheaths 2–4mm. Nestlings c.10–11 days old were described by Greeney & Gelis (2006) as similar to eight-day-old nestlings except for a predominance of buffy on the lower belly and flanks, and having the lower breast and upper belly appearing scaled, with pale buff feather shafts and plume bases. Primaries had emerged from their sheaths 3–4mm by day 10–11, presumably having broken their sheaths c.9–10 days after hatching. By age 12–13 days (four days prior to fledging) light chestnut speckling is visible across most of the upperparts and primary pin-feathers are exposed from their sheaths 10–15mm (Greeney & Gelis 2006, see also figure in Greeney 2012a). Tarsal measurements of two 12–13-day-old nestlings were 31.3mm and 31.6mm. At the time of fledging, young appear quite similar in coloration as described for 12–13-day-old nestlings. Their primaries are still half to one-third sheathed (pers. obs.) and they are unable to fly. During the final three days of the nestling period, Greeney & Gelis (2006) reported that the nestlings often moved about in the nest, preening themselves and frequently flapping their wings, as described for older Scaled Antpitta nestlings (Dobbs et al. 2001). On the day that the nestlings fledged from this nest, they were fed ten times before 07.45 and, after the final adult visit, appeared to be watching something moving below the nest. Their behaviour was interpreted as interest rather than fear and it was presumed that the departing adult was below the nest. Several minutes after the adult departed, the excited nestlings hopped one at a time to the nest rim and jumped, rather unceremoniously, from the edge. No vocalisations were heard prior to the first nestling leaving. After the first had fledged, however, a few loud, harsh, squawking calls were heard. After the second nestling left, these calls increased in volume and number for several seconds, but it is unclear if they emanated from the adults or the young. Nestlings are brooded at night by an adult (of unknown sex, but presumably the female) right up until fledging (Greeney & Gelis 2006), a behaviour which is certainly facilitated by the voluminous nests of this species. During the final week that the young remained in the nest, they were brought food 44–65 times per day (mean± SD = 53 ± 9). This range in feeding rates is equivalent to 1.8–2.7 feeds per nestling per hour. Adults appear to usually bring more than one prey item per visit (Freile & Renjifo 2003, Greeney & Gelis 2006), and both adults received faecal sacs from the nestlings, 61% of which were carried away while the rest were eaten at the nest (n = 166) (Greeney & Gelis 2006). Faecal sacs were always produced in the presence of adults during the first half of the nestling period (prior to primaries breaking their sheaths), but were occasionally deposited on the rim of the nest in the absence of adults during the days just prior to fledging. As it appears that the integrity of the faecal sac membrane begins to weaken prior to fledging, as evidenced by them frequently falling apart when grasped by the adults, the pair of antpittas followed by Greeney & Gelis (2006) spent much time in the final two days of the nestling period perched on the rim of the nest avidly cleaning away the bright white faecal stains. Throughout the course of the observations made (via video camera) at the nest with nestlings (Greeney & Gelis 2006), adults frequently probed the nest lining with their bills, possibly removing parasites (Haftorn 1994, Greeney 2004). They noted that, unlike the rather smooth 'sewing-machine-like' probing described for Scaled Antpitta (Dobbs et al. 2003), probing by this pair of Moustached Antpittas was much more erratic and included more sideways movements of the bill in a jerky, uncoordinated fashion. In fact, this behaviour reminded the authors of the thrusting movements made by most Grallaria while foraging in leaf litter or soil. Whether Moustached Antpittas differ in their 'style' of nest probing from other Grallaria, or the observed variation was an individual characteristic, remains to be seen. The duration of the nestling period is not known with certainty. Greeney & Gelis (2006) gave an estimated nestling period of 15–17 days at one nest, and at one of the nests studied by J. Lyons hatching occurred on 4 March 2010 and fledging 20–22 days later, but was not monitored closely. For G. alleni, both the incubation and nestling periods are probably similar to Scaled Antpitta, perhaps marginally longer given its slightly higher altitudinal range. NEST SUCCESS Far too few nests have been studied to make an accurate assessment of breeding success. All three nests of alleni in Colombia, however, were unsuccessful (Freile & Renjifo 2003, Londoño et al. 2004). In Ecuador, one nest of andaquiensis was presumed to have been destroyed by a predator (Freile & Renjifo 2003), while two fledged successfully (Greeney & Gelis 2006, J. Lyons in litt. 2015) and the Mindo Loma nest was discovered during fledging (HFG). SEASONALITY Nests of the nominate subspecies on the west slope of the Central Andes in Colombia have been found with incubation underway or nearly complete in September–November (Freile & Renjifo 2003, Londoño et al. 2004), with fledglings observed there in October–February (Londoño et al. 2004, D. Beadle in litt. 2017). These records all represent breeding during the second rainy period of the year (Freile & Renjifo 2003, Londoño et al. 2004), suggesting a preference for the wetter months. The situation in north-west Ecuador is less clear. The nest at Otonga was found at the peak of the wet season while the Tandayapa nest was active during the peak of the dry season, but fledging at Tandayapa occurred close to when rains usually start (Freile & Renjifo 2003, Greeney & Gelis 2006). Fledging occurred at the nest at Mindo Loma on 14 February 2007, indicating that nesting was initiated in the very early wet season, or possibly late dry season. The three nests at RP Las Gralarias were discovered with incubation underway, two during the peak rains and one during peak dry season. The juvenile photographed by N. Voaden was first observed in early August, suggesting that it probably fledged during the previous wet season. A fledgling at Tandayapa Bird Lodge and one at the Guacamayos Ridge (R.A. Gelis *in litt.* 2015) both represent wet-season nests in their respective regions, the latter being the only breeding data for Moustached Antpitta from the east Andean slope. Overall, the peak reproductive period appears to coincide with the wetter seasons across its range, but year-round breeding at a more constant rate cannot be eliminated. **BREEDING DATA** *G. a. alleni* Nest with eggs, 8 September 2003, SFF Otún-Quimbaya (Londoño *et al.* 2004); nest with eggs, 15 October 2003 PMN La Nona (Londoño *et al.* 2004); nestling, 23 November 1995, SFF Otún-Quimbaya (Freile & Renjifo 2003); fledgling, 10 February 2010, SFF Otún-Quimbaya (D. Beadle *in litt.* 2017); fledgling, 24 October 2003, PMN La Nona (Londoño *et al.* 2004); juvenile 5 July 1996, SFF Otún-Quimbaya (Freile & Renjifo 2003). *G. a. andaquiensis* Three nests with eggs, 18 February 2010, 14 April 2010 and 18 October 2011, RP Las Gralarias (J. Lyons *in litt.* 2015); nests with eggs, 2 December 2006, RP Refugio Paz de las Aves (R. Ahlman *in litt.* 2015); nestling, fledge 14 February 2007, Mindo Loma (HFG); nestling, 8 March 1999, Reserva Integral Otonga (Freile & Renjifo 2003); nestling, 5 December 2003, RP Tandayapa Bird Lodge (Greeney & Gelis 2006); fledgling, 23 December 2003, RP Tandayapa Bird Lodge (Greeney & Gelis 2006); fledgling, 28 July 2003, Cordillera de Guacamayos (R.A. Gelis *in litt.* 2015); fledgling transitioning, 16 March 2004, RP Tandayapa Bird Lodge (R.A. Gelis *in litt.* 2015); juvenile, 11 August 2009, RP Refugio Paz de las Aves (N. Voaden photos); fledgling/juvenile, 15 September 2011, RP Refugio Paz de las Aves (eBird: A. Lewis).

TECHNICAL DESCRIPTION Sexes similar. The following pertains to the nominate race, confined to western Colombia. *Adult* From Chapman's (1912) description of nominate *alleni*. Crown and nape, down to malar stripe slate-grey, with the feathers very narrowly edged blackish and lacking evident shaft-streaks. Forecrown and lores whitish, tinged russet. Upperparts deep olivaceous-brown with the feathers conspicuously edged black giving a scaled look (but lacking shaft-streaks). Tail deep tawny, and greater and median upperwing-coverts similar to back, while primary wing-coverts are blackish. Secondaries similar to back, as is trailing edge to primaries. Leading edge (exposed when at rest) browner, with that of outer primary paler. Underwing- and undertail-coverts ochraceous (Meyer de Schauensee 1970). Ear-coverts similar to back but with a slight rusty tinge and bordered posteriorly by the grey nape. Malar streak broad and white, as is central/lower throat spot, with (at least some) feathers of these areas thinly edged black or olivaceous. Chin and upper throat mixed russet, black and olive-brown, and lower throat and breast similar to, but somewhat paler than, back. Feathers of central breast have white streaks edged with black, and these feathers often form a broken collar across upper breast. The darker breast grades into the creamy-white, unmarked belly. The slightly elongated flank feathers are

rich ochraceous. **Adult Bare Parts** *Iris* dark brown; *Bill* black, with mandible slightly paler in some individuals (Schulenberg & Kirwan 2012b); *Tarsi & Toes* purplish-grey or pinkish-grey. *Fledgling* A fledgling of unknown age (referred to as a 'juvenile' by Krabbe & Schulenberg 2003), but still dependent, was described by Londoño *et al.* (2004) as generally being dark brown above and on upper breast, and creamy-white below. Crown, nape and back streaked rufous, and dark feathers of upper breast thinly streaked cream. No evidence of the moustachial streak. Bill generally black, with distinctly reddish rictal flanges. On leaving the nest, plumage is fluffy or 'wool-like', black to dark sooty-grey on head, upperparts, breast and sides (HFG), all with somewhat sparse, evenly spaced buffy to buffy-ochraceous or buffy-chestnut streaks. Those on crown smallest, becoming slightly longer on nape, and longest on breast and rump. Streaks on back thickest (almost like spots) and most ochraceous or chestnut. There may or may not be a few blackish, white-streaked feathers (see Juvenile) already emerged on lower breast. Soft wispy plumage of belly and flanks mottled buffy-white, grey and buff, with the richest buff feathers on flanks and vent. There may or may not be a few scattered chestnut-spotted feathers on hindcrown, nape and upper back, and the lower belly and flanks may be more extensively washed buff. The ear-coverts is the only tract with non-fluffy plumage, rich brown streaked buffy to buffy-ochraceous. Primaries and secondaries only half to two-thirds emerged from their sheaths, ochraceous-brown on leading vane, blackish on trailing vane. Alula and greater primary-coverts uniformly blackish. Greater secondary-coverts blackish with an ochraceous-brown wash on leading edge, black on terminal margins, and bright ochraceous spots just above the black fringes. Median secondary-coverts similar but washed ochraceous-brown only subterminally. Rectrices only just beginning to emerge from their sheaths. **Fledgling Bare Parts** *Iris* dark brown; *Bare Ocular Skin* dark grey to bluish- or violet-grey; *Bill* blackish to dark grey, maxilla pale at very tip and still has whitish egg-tooth, dull orange tomia and base behind nares, mandible also orange on tomia, rictal flanges fully inflated and bright orange; *Tarsi & Toes* grey to violet-grey, nails blackish. *Juvenile* N. Voaden kindly shared several photos of a juvenile, on which the following is based. No remnants of downy fledgling plumage. Upperparts generally as adult but crown and nape not distinctly grey. Hindcrown and nape still bear sparse streaking similar to fledglings. Face dull olivaceous-brown, with warmer brown ear-coverts streaked ochraceous. White moustachial prominent, breaking into long white streaks across neck-sides and large spots forming a necklace on upper breast. Wings as fledgling, but feathers now fully emerged, and tail shorter than adult but coloured similarly. Throat dark blackish-brown, becoming black or dark sooty-grey on breast and upper belly. Breast-sides mottled with dark ochraceous feathers similar to adult, becoming entirely ochraceous lower and on flanks and belly (vent invisible in photos). Central belly paler ochraceous. **Juvenile Bare Parts** *Iris* dark brown; *Bill* blackish, slightly paler near tip and on tomia, possibly still with pale egg-tooth, rictal flanges orange, still slightly inflated; *Tarsi & Toes* purplish- to pinkish-grey. *Subadult* The bird just described as a juvenile was a regular visitor to feeders and was photographed by several people subsequently. G. Appleton photographed it 20 days later, in what I would consider the juvenile transitioning phase. Upperparts now generally similar to adult with only a few scattered streaks and spots. Moustachial now extends below ear-coverts, but is duller white posteriorly and no longer extends across breast in broken necklace. Wings as described above and tail similar to adult. Below, this individual was similar to adults, but central breast and lower throat still had irregular patches of blackish, white-streaked plumage described above. Rest of breast ochraceous-brown with scattered, thin white to buffy-white streaks. Bill now largely similar to adult, but rictus dull orange and no longer inflated. Legs pinkish-grey. Finally, this (presumably) same individual was photographed 64 days (S. Woods) after my description of its juvenile plumage. It was in essentially full adult plumage with at least the greater secondary-coverts still bearing ochraceous spots (median coverts invisible). The tertials, not visible in previous photos, were dark brown with pale brownish tips. **Subadult Bare Parts** As described for adult.

MORPHOMETRIC DATA Almost no linear measurements published. *G. a. alleni* Chapman (1912) provided the following for the ♀ holotype: *Wing* 113mm; *Tail* 38mm; *Bill* culmen 25mm; *Tarsus* 43mm. Londoño *et al.* (2004) provided the following: *Wing* flattened chord 96mm; *Bill* exposed culmen 24mm; *Tail* 47mm; *Tarsus* 50mm (*n* = 1, ♂?). Measurements for an unsexed fledgling (Londoño *et al.* 2004): *Wing* flattened chord 90mm; *Tail* 37mm; *Tarsus* 42mm. **G. a. andiquiensis** G. Ruzzante (*in litt.* 2015) provided measurements from adults mist-netted in Cotopaxi: *Wing* flattened chord 102.0mm, 110.0mm; *Bill* exposed culmen 26.9mm, 27.3mm; *Tarsus* 41.8mm, 44.4mm (*n* = 2, ♂?). J. Lyons (*in litt.* 2015) measured an adult (Pichincha): *Wing* 100mm; *Bill* exposed culmen 22.4mm; *Tail* 40mm (*n* = 1, ♂?). **Mass** 64g, 77g (*n* = 2, ♂, race?, Krabbe & Schulenberg 2003), 66.1g (n = 1, ♀?, *andaquiensis*, G. Ruzzante *in litt.* 2015), 71.5g (*n* = 1 ♂, *andaquiensis*, J. Lyons *in litt.* 2015), 87g (*n* = 1, ♀?, fledgling, *alleni*, Londoño *et al.* 2004). **Total Length** 15.7–18cm (Hilty & Brown 1986, Ridgely & Greenfield 2001, Krabbe & Schulenberg 2003, Restall *et al.* 2006, McMullan *et al.* 2010, J. Lyons *in litt.* 2015).

TAXONOMY AND VARIATION Two subspecies recognised (Schulenberg & Kirwan 2012b). The relationships of *G. alleni* to other members of the genus are unresolved, as it has not been sampled molecularly (Krabbe *et al.* 1999, Rice 2005). Hernández & Rodríguez (1979) suggested that *G. alleni*, *G. guatimalensis* and *G. varia* might be conspecific. However, they differ vocally, morphologically and in altitudinal distributions, and are consequently best regarded as species (Krabbe & Coopmans 2000). Indeed, in at least some areas (i.e. north-west Ecuador), *G. alleni* and *G. guatimalensis* occur sympatrically. Thus, although it is probable that *alleni* is closely related to *guatimalensis*, treating them as separate species is unequivocal.

Grallaria alleni alleni Chapman, 1912. Endemic to the west slope of the Central Andes of Colombia in Antioquia, Quindío, Caldas and Risaralda. **Specimens & Records COLOMBIA**: *Antioquia* Páramo de Frontino, 06°26'N, 76°05'W (Flórez *et al.* 2004). *Caldas* Cerro Ingruma, 05°25'N, 75°43'W (eBird: J.A. Zuleta-Marín); El Zancudo, 05°03.5'N, 75°26'W (Verhelst *et al.* 2001); RFP Río Blanco, 05°05'N, 75°25'W (Ocampo-T. 2002). *Risaralda* Cerro Montezuma, 05°17'N, 76°02'W (López-Ordóñez *et al.* 2013); PRN Santa Emilia, 05°12'N, 75°54'W (Zuleta-Marín 2012); DMI Cuchilla del San Juan, 05°11.5'N, 75°57.5'W

(XC 103650, J.A. Zuleta-Marín); Vereda Las Cumbres, 05°06.5'N, 76°00'W (XC 146170, J.A. Zuleta-Marín); RMN La Nona, 04°53'N, 75°43'W (Londoño *et al.* 2004); SFF Otún-Quimbaya, 04°43.5'N, 75°34.5'W (Freile & Renjifo 2003, Londoño *et al.* 2004, XC); La Pastora, 04°42.5'N, 75°29'W (eBird: L.G. Naranjo). *Quindío* Vereda Cruces, 04°41.5'N, 75°38'W (XC 240081, O.H. Marín-Gómez); five locations from Arbeláez-Cortés *et al.* (2011): Río Santo Domingo, Calarcá, 04°31.5'N, 75°39'W; Finca La Ofrenda, 04°35'N, 75°39.5'W; Cedro Rosado, 04°32.5'N, 75°46'W; Estrella de Agua, 04°37.5'N, 75°26'W; Alto Quindío, Salento, 04°38'N, 75°27'W.

Grallaria alleni andaquiensis Hernández-Camacho & Rodríguez-M., 1979, Caldasia, vol. 12, p. 574, vicinity of Cueva le los Guácharos, Parque Nacional Natural Cueva de los Guácharos, Huila, Colombia [01°37'N, 76°00'W]. The more widespread of the two subspecies, *andaquiensis* occurs in the upper Magdalena Valley of Colombia south on both Andean slopes in Huila to central Ecuador, in Pichincha and Cotopaxi in the west, and Sucumbíos and Napo in the east. Reports in Bolívar, 9.5km NW of Salinas (Poulsen & Krabbe 1998, Poulsen 2002) are in error (N. Krabbe *in litt.* 2016) and there are no confirmed reports of its presence south of Carmela in NW Cotopaxi. The southern end of its range in eastern Ecuador is also poorly understood. There are no records south of the Río Pastaza, but modelling suggests its range may extend to the Río Paute (Freile *et al.* 2010b). Generally similar to nominate *alleni*, *andaquiensis* is browner dorsally and has the belly ochraceous (not white). Some individuals have scattered whitish feathers below (Schulenberg & Kirwan 2012b). **Specimens & Records COLOMBIA**: *Huila* c.11.5km SW of San Agustín, 01°48.5'N, 76°21W (XC 148011, D. Calderón-F.). **ECUADOR**: *Pichincha* 4.75km E of Nanegal, 00°08N, 78°38'W (Krabbe & Coopmans 2000, MECN); RP Santa Lucía, 00°07'N, 78°36'W (HFG); RBP Maquipucuna, 00°05'N, 78°37'W (ML, XC: P. Coopmans); Gualea, 00°02.5'N, 78°44.5'W (Krabbe & Coopmans 2000, AMNH 124427); RP Refugio Paz de las Aves, 00°01.5'N, 78°42.5'W (XC 88144, D.F. Lane, ML 46027201, N. Voaden); RP Tandayapa Bird Lodge, 00°00'N, 78°40.5'W (Greeney & Gelis 2006, XC); RP Las Gralarias, 00°00.5'S, 78°45'W (XC 101029–031, J. King); 6km SE of Tandayapa, 00°01'S, 78°38'W (Krabbe & Coopmans 2000); RP Mindo Loma, 00°01.5'S, 78°44.5'W (XC 223580/81, P. Boesman); 2.8km SW of Tandapi, 00°26'S, 78°49'W (Krabbe & Coopmans 2000); Loma Alta, 00°27'S, 78°51'W (R. Ahlman *in litt.* 2017). *Santo Domingo de los Tsáchilas* BP Río Guajalito, 00°14'S, 78°49'W (Freile 2000). *Cotopaxi* Reserva Integral Otonga, 00°25'S, 79°00'W (Freile & Renjifo 2003, Freile & Chaves 2004); near Caripero, 00°34.5'S, 79°01.5'W (Krabbe & Coopmans 2000, XC 50400, B. López-Lanús). *Sucumbíos* Bermejo, 00°15'N, 77°23'W (Schulenberg 2002). *Napo* Cordillera Guacamayos, south of Cosanga, at 00°37.5'S, 77°50.5'W (ML, P. Coopmans) and at 00°39S, 77°52'W (Krabbe & Coopmans 2000).

STATUS Although it is now known from several localities each in Colombia and Ecuador, the species has been afforded Vulnerable status by BirdLife International (2017) due to its small and severely fragmented range, which is under continued threat of habitat destruction (Collar *et al.* 1992, 1997). It is considered Endangered at a national level in both Colombia (Renjifo *et al.* 2014) and Ecuador (Granizo *et al.* 2002, Freile *et al.* 2010b). Indeed, most or all of the cloud forest in the upper Magdalena

Valley has been logged and settled or converted to agriculture (Stiles 1998, BirdLife International 2017). Another potential safe haven for *G. alleni*, Cueva de los Guácharos, is increasingly encroached by human settlement and, apparently, opium production (Wege & Long 1995, Renjifo *et al.* 2002). Similarly, in Ecuador, the west slopes of the north-west Andes are, for the most part, fragmented or cleared (Krabbe *et al.* 1998). The slopes of Volcán Sumaco in Napo were still largely forested in 1990 (Krabbe 1991, BirdLife International 2015), but during the two and a half decades since, rapid human population growth and a steady increase in clearing for the farming of naranjilla has severely altered much habitat between 1,000m and 1,800m (HFG). Nevertheless, although sensitive to fragmentation (Renjifo 2001), Moustached Antpitta can tolerate disturbance (Kattan *et al.* 2006) and has even been found nesting in mature secondary forest (Greeney & Gelis 2006, BirdLife International 2017). The good news is that, across its range, the species appears to occur within a relatively large number of protected areas. On the east slope of the Andes in Ecuador potential habitat for *andaquiensis* is protected by at least five large reserves, but its true distribution there is poorly known, and the extent to which it is actually protected is uncertain (Freile 2000, Krabbe & Coopmans 2000, Freile *et al.* 2010b, BirdLife International 2017). **Protected Populations** *G. a. alleni* PNN Cueva de los Guácharos (Freile 2000, Krabbe & Coopmans 2000, Renjifo *et al.* 2002, Freile *et al.* 2010b); PNN Los Nevados (Pfeifer *et al.* 2001); PNN Tatamá (XC 146170, J.A. Zuleta-Marín), RFP Río Blanco (Ocampo-T. 2002); PRN Santa Emilia (Zuleta-Marín 2012); PRN Ucumarí (Krabbe & Coopmans 2000), SFF Otún-Quimbaya (Freile & Renjifo 2003, Londoño *et al.* 2004, Lentijo & Kattan 2005, XC); RMN La Nona (Londoño *et al.* 2004); RFP AguaBonita (J.A. Zuleta-Marín *in litt.* 2017); RNSC Alto Quindío Acaime (Krabbe & Coopmans 2000); RNA El Colibrí del Sol (Salaman *et al.* 2007a); DMI Cuchilla del San Juan (XC 103650, J.A. Zuleta-Marín); DMI Agualinda (J.A. Zuleta-Marín *in litt.* 2017). *G. a. andaquiensis* Colombia RFP Los Yalcones (XC 148011, D. Calderón-F.). **Ecuador** RE Cofán-Bermejo (Schulenberg 2002); PN Sumaco-Galeras (R.A. Gelis *in litt.* 2016); RP Refugio Paz de las Aves (Collins 2006, XC, ML); BP Río

Guajalito (Freile 2000); Reserva Integral Otonga (Chaves & Freile 2005); many reserves near Mindo including BP Mindo-Nambillo, RP Bellavista, RP Mindo Loma, RP Refugio Natural Pacha Quinde, RP Santa Lucía (HFG), RP Las Gralarias (J. Lyons photos), RP Tandayapa Bird Lodge (Greeney & Gelis 2006); RP Sachatamia (J. Freile *in litt.* 2017); RBP Maquipucuna (ML: P. Coopmans). Hope is not lost.

OTHER NAMES English Allen's Antpitta (Cory & Hellmayr 1924, Meyer de Schauensee 1950); Allen's Ant-pitta (Meyer de Schauensee 1950). **Spanish** Tororoí Bigotudo (Krabbe & Schulenberg 2003, Salaman *et al.* 2007a); Licuango Chico (Chaves & Freile 2005); Gralaria Bigotuda (Chaves & Freile 2005); Gralaria Bigotiblanca (Granizo *et al.* 2002); Peón de Bigotes, Peón del Quindío (Negret 2001); Tororó Bigotudo (Ocampo-T. 2002). **French** Grallaire à moustaches **German** Grauscheitel-Ameisenpitta (Krabbe & Schulenberg 2003).

Moustached Antpitta, nest and complete clutch (*andaquiensis*), RP Las Gralarias, Pichincha, Ecuador, 28 February 2010 (*Dušan M. Brinkhuizen*).

Moustached Antpitta, two mid-aged nestlings in nest (*andaquiensis*), Tandayapa Bird Lodge, Pichincha, Ecuador, 7 December 2003 (*Harold F. Greeney*).

Moustached Antpitta, adult incubating (*andaquiensis*), RP Las Gralarias, Pichincha, Ecuador, 28 February 2010 (*Dušan M. Brinkhuizen*).

Moustached Antpitta, two older nestlings in nest (*andaquiensis*), Tandayapa Bird Lodge, Pichincha, Ecuador, 9 December 2003 (*Harold F. Greeney*).

Moustached Antpitta, juvenile (*andaquiensis*), Refugio Paz de las Aves, Pichincha, Ecuador, 22 November 2009 (*Nigel Voaden*).

SCALED ANTPITTA
Grallaria guatimalensis   Plate 8

Grallaria Guatemalensis Prévost & Des Murs, 1846, in *Oiseaux. Voyage autour du monde sur la frégate la Vénus, commandée par Abel de Petit-Thouars. Zoologie: mammifères, oiseaux, reptiles, et poissons*, p. 199, Plate IV, 'Guatémala'.

'G.D.S[mooker]. met with a pair in deep hill-forest near the bank of the Rio Grande stream (Heights of Oropucheu) on 5 May, 1925. They were strutting along the ground in Indian file, short tails held perfectly erect and feet thrown forward in a stately, marching movement.'
– Belcher & Smooker, Trinidad, 1936

Although I sadly failed to find it mentioned in the *Guinness Book of World Records*, Scaled Antpitta is unquestionably top, in many ways, among species treated here. Not least among these distinctions, Scaled Antpitta holds the somewhat dubious honour of being the antpitta most frequently recorded dying in collisions with windows! Perhaps most notably, however, it has the widest geographic range, from central Mexico to central Bolivia; roughly 20°N to 18°S. In fact, including the somewhat disjunct population in the Pantepui of southernmost Venezuela and adjacent northern Brazil and western Guyana, the range of Scaled Antpitta also spans roughly 45° of longitude. Unsurprisingly then, this antpitta occurs in a wide range of habitats, from pine–oak forests in Mexico and seasonally humid forests in south-west Ecuador to the rain-drenched Amazonian foothills of the Andes. Numerous subspecies have been described, of which nine are recognised here, but almost all require careful revision and evaluation before being certainly accepted. Races vary most in ground colour and degree of pattern to the underparts, less so in the prominently scaled upperparts. There is some variation in vocalisations, but to date no one has examined this in detail. Behaviour has been reported by many authors, more so than perhaps any other *Grallaria*, and it is also the antpitta with the most published information on its breeding. Nevertheless, its behaviour and diet are still relatively unknown due to its reclusive and secretive nature. Many authors have also commented

on the elusive ability of Scaled Antpitta to materialise and vanish in an instant. Their often colourful accounts made choosing the opening quote a difficult task. In some cases, they admitted that specimens were only secured via accidental captures in traps set on the ground (Carriker 1910, Dickey & van Rossem 1938). One frustrated collector (Lehmann-V. 1957) was dismayed that, despite much searching, the only specimen he procured was found by his dog. Aspiring natural historians should fear not, however, as there is much still to be learned about even this, the best-studied and most published-upon antpitta!

IDENTIFICATION 17–19cm. With a fair amount of geographic variation, adults are olive-brown on the back with a slate-grey crown and nape. Feathers of the upperparts are edged black, imparting the scaled look that gives rise to the English name. Lores paler in most races, but it is the coloration and pattern of the underparts that varies most significantly. Over most of South America, the throat and chest are brownish-olive with a variably sized and (often) semi-concealed whitish crescent on the upper chest, and a distinct whitish malar stripe. Rest of underparts vary from deep orange-rufous to pale tawny-buff. Some have variable amounts of white, black, or rusty streaking and spotting on the breast. Overall, variation is considerable but the degree to which this corresponds with currently recognised races is unclear. Scaled Antpitta is similar to several other species and it overlaps locally in Colombia and Ecuador with the very similar Moustached Antpitta. These two species are most reliably distinguished by differences in breast pattern, the width of the submoustachial streak, and bill shape and colour (although the last feature requires practice). Where the two species are sympatric, the breast of both is mostly buff or olive-brown, with feathers in the central breast having a contrasting pattern. Following Krabbe & Coopmans (2000), in Scaled (race *regulus*), these feathers are pale buff (rarely white), 'with a noticeable difference in the distribution of white or buff between the inner and outer web of individual feathers, and with black on each side of the basal half of the shafts'. In Moustached, in contrast, these feathers are white, 'with equal distribution of white on the outer and inner webs of each feather', and the white forms 'large, symmetrically elongated,

185

narrowly black-edged spots extending onto the lower breast as thin, white to buff streaks surrounding the shafts' (Krabbe & Coopmans 2000). The submoustachial streak of Moustached Antpitta is much larger than that of *regulus* and is always white; the submoustachial streak of *regulus* is only half the width of that of Moustached, and only rarely is it white (Krabbe & Coopmans 2000). Scaled Antpitta is also somewhat similar to Variegated Antpitta, but there is little overlap between them. Variegated Antpitta is larger, with a smaller white breast patch, and the breast and flanks more or less streaked or spotted (flanks unmarked in Scaled Antpitta).

DISTRIBUTION Of the antpittas, only Variegated and Ochre-breasted come close to the latitudinal range of Scaled. The northernmost known localities for the species are in south-central Mexico just north of Mexico City in western Jalisco and west-central Hidalgo, south through Costa Rica into Panama and Colombia, then east through Venezuela (and on Trinidad) to the Guianas and Pantepui region, and south through all of the tropical Andes to central Bolivia. Of the presently recognised subspecies, *regulus* is among the more widely distributed, from north-west Colombia south along the west slope of the Andes through Ecuador to northern Peru. East of the Andes, *regulus* occurs from western Venezuela (Mérida) south through Colombia and eastern Ecuador to central Peru (Griscom 1950, Peters 1951, Parker *et al.* 1982, Kempff-Mercado 1985, Remsen & Traylor 1989, Howell & Webb 1995, Hilty 2003, Kenefick *et al.* 2007, Obando-Calderón *et al.* 2009, van Perlo 2009, Ridgely & Tudor 2009, Angehr & Dean 2010, McMullan *et al.* 2010, 2013).

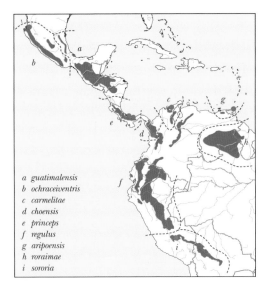

a *guatimalensis*
b *ochraceiventris*
c *carmelitae*
d *choensis*
e *princeps*
f *regulus*
g *aripoensis*
h *roraimae*
i *sororia*

MOVEMENTS There is nothing concrete known concerning potential migratory or dispersal-related movements. Stiles & Skutch (1989) stated that it 'wanders widely after breeding season, especially juveniles', and Herzog *et al.* (1999) suggested that the species shows some degree of local migration in the extreme south of its range (race *sororia*). During work on this book, based on examination of a rather surprising museum specimen in Costa Rica (MZUCR 5202) and discussions with L. Sandoval, I hypothesise that Scaled Antpittas in Costa

Rica (*princeps*) have a post-fledging dispersal period that may take them far from their natal territories. The same idea has been postulated for populations in southern Ecuador (*regulus*; Ordóñez-Delgado *et al.* 2016). The aforementioned Costa Rican specimen was a juvenile female that struck a window well within the city limits of San José (Montes de Oca, Barrio Dent) and Ordóñez-Delgado *et al.* (2016) recorded five separate immatures in the city of Loja, all a considerable distance from suitable habitat or known populations. Finally, an unusual record of Scaled Antpitta from the botanical gardens in Ciudad de Guatemala (ML 60068661, photo C.L. Burgos) also involved a subadult. Taken together, these observations are strongly suggestive of long-distance dispersal. That all such records are of just one species may be a product of the relative abundance and broad distribution of Scaled Antpita, and similar records for other species are possible. On the other hand, should such post-natal dispersal be unique, it may help to explain the species' large range.

HABITAT Inhabits a fairly wide range of habitats, including semi-deciduous forest (del Toro 1958), humid evergreen forest, pine–oak and fir forest (del Toro 1971, Navarro-Sigüenza 1992, Howell & Webb 1995) and lowland rainforest. Irrespective of habitat, they are usually associated with natural and anthropogenic second growth, thickets and other areas of dense vegetation such as along riparian corridors (Carriker 1910, Lowery & Dalquest 1951, Monroe 1968, Meyer de Schauensee 1970, O'Dea & Whittaker 2007). In north-west Ecuador the species can be found in shade-coffee agroforestry systems where the canopy is relatively complete and understorey is dense (J. Freile *in litt.* 2017). In the mountains at the north end of the Valle Central in Costa Rica, Scaled Antpitta has been found in cypress plantations (*Cupressus lusitanica*) at elevations of 1,600–2,000m (L. Sandoval *in litt.* 2017). Across the range, Scaled Antpitta usually is most numerous at elevations of 500–2,500m (Krabbe & Schulenberg 2003), but occasionally occurs as low as 200m or as high as 3,100m. Published altitudinal information is as follows: to 2,900m in Guatemala (*guatimalensis*; Land 1970); rare above 600m in Panama (*princeps* and *chocoensis*, Angher & Dean 2010); 900–2,500m in south-west Ecuador (*regulus*; Best *et al.* 1992); 700–1,600m in south-east Peru (*sororia*; Walker *et al.* 2006); and 1,400–3,100m in Guerrero, Mexico (*ochraceiventris*, Navarro-Siguenza 1992). At a few locations in north-west Ecuador, Scaled Antpitta is replaced at higher elevations by Moustached Antpitta (J. Freile *in litt.* 2017) and, at least in some areas, their altitudinal ranges overlap (e.g. RP Mindo Loma; HFG). Similarly, at other locations in western Ecuador, Scaled Antpitta appears to be replaced at lower elevations by Plain-backed Antpitta, but they also regularly exhibit some overlap (e.g., RP Buenaventura, BP Mashpi; J. Freile *in litt.* 2017). The degree to which these species compete or exclude each other at these sites is unknown, but would make a fascinating research project.

VOICE The song appears to vary relatively little across its broad range, but most populations are poorly sampled. The song of *regulus* is best known, a rapid (14–17 notes) series of 30–50 quavering, hollow notes lasting 2-7s, slowly increasing in volume and pitch and then quickly becoming quieter at the end: *cau, cau, cau-cau-caucaucaucau, cau* (Fjeldså & Krabbe 1990, Krabbe & Schulenberg 2003). The song of *sororia* differs in having the first part much slower (only *c*.6 notes), with the entire song slightly longer and composed of 18–23 notes (Krabbe & Schulenberg 2003:

see XC 39746, D. Geale, XC 1629, S. Mayer). D.F. Lane (in Schulenberg *et al.* 2007) described the song of *sororia* as a 'series of hooted notes that accelerates-decelerates at the loudest and highest-pitched notes, then accelerates again slightly: *poo-poo-pu-pu'pu'pu-pu-POO-POO-POO-pu'pu.*' The song of *roraimae* (XC 114371, A. Crease) is very similar to that of *regulus* and was described by Braun *et al.* (2003) as a long series of low hooting notes with three longer, more emphatic notes in the middle. The songs of *carmelitae* and *chocoensis* have not previously been described. Songs recorded on the west slope of Colombia's West Andes, on the Cerro Montezuma road (XC 308320/21, D. Calderon-F.) are presumably of *chocoensis* and are distinctive in being significantly shorter and lacking a distinctive up-down inflection of pitch. A very similar song has also been recorded in the Sierra de Santa Marta population (*carmelitae*; XC 17256, F. Lambert). Stiles & Skutch (1989) suggested that an infrequently heard 'low piglike grunt or croak' is given in alarm. Race *regulus*, in south-west Ecuador, also has a muted, shorter version of its song that is given around dawn, and is perhaps similar to the third song-type described above by Braun *et al.* (2003). I suspect that this particular vocalisation may facilitate the reuniting of pairs at dawn if they have not roosted together (i.e. if the female was incubating). Howell & Webb (1995) described the song of nominate *guatimalensis* in Mexico as 'a low resonant series of notes that starts as a trill, rises in pitch, and slows to distinct, pulsating notes, stop[ing] abruptly.' They also noted that birds in the highlands north of the Isthmus of Tehuantepec appear to differ somewhat. This statement would presumably refer to *ochraceiventris* or, possibly, to birds that were at one time split from the nominate race as race *mexicana* (see Taxonomy and Variation).

NATURAL HISTORY Like most congeners, Scaled Antpittas forage largely on the ground, tossing or flicking leaves aside with the bill and overturning detritus with their feet (del Toro 1971, Peterson & Chalif 1973, Stiles & Skutch 1989). Often, after displacing litter, they will pause with head cocked to one side, intently examining the exposed area for potential food (Dobbs *et al.* 2001). They may stop frequently to probe the leaf litter or soft mud (Slud 1964, Stiles & Skutch 1989). The species forages in both the darkness of the forest and (perhaps less frequently) broad open trails through forest. In the latter situation, they are rarely far from dense cover (Slud 1964). When alarmed, adults fly low and straight for short distances, usually into the nearest area of thick vegetation, sometimes perching up to 3m above ground first, before diving out of sight into the undergrowth (Stiles & Skutch 1989). Unlike most other *Grallaria*, which generally sing below 3m above ground, Scaled Antpitta has been reported singing from perches as high as 10m (Krabbe & Coopmans 2000, Braun *et al.* 2003), usually at dusk or in early morning. Though it has only rarely been reported (Coates-Estrada & Estrada 1989), it seems probable that Scaled Antpittas follow army ant swarms reasonably frequently, especially at lower elevations. Coates-Estrada & Estrada (1989) found nominate *guatimalensis* in Veracruz foraging with ants only rarely, and only with larger *Eciton* swarms, although *Labidus* swarms were also surveyed. Researchers have reported being followed by adult Scaled Antpittas in western Ecuador (*regulus*), observing them forage in leaf litter disturbed by their passage (Loaiza B. in Greeney 2012b). In support of the idea presented in this

reference, that *G. guatimalensis* may forage in the wake of large mammals, Woods *et al.* (2011) report that they are now among the growing number of *Grallaria* attracted to 'worm-feeding stations'. Overall, most aspects of the behaviour and basic natural history of Scaled Antpitta are as poorly known as most congeners, but are likewise thought to be similar (Greeney *et al.* 2013). Interestingly, however, there are several accounts of this species 'strutting' along the ground with their tails held erect (Belcher & Smooker 1936, Dickey & van Rossem 1938). I have never observed this posture or behaviour, which seems much more descriptive of *Formicarius* behaviour. It is plausible that this is a form of rarely observed courtship, particularly as Belcher & Smooker (1936) reported this behaviour by a pair in Trinidad which were then associated with an active nest thought to be of this species. Alternatively, as Dickey & van Rossem (1938) subsequently collected a fledgling in the area where they observed this behaviour, perhaps it is an odd type of threat or even distraction display, a theory that might also pertain to the Trinidad record. There are no solid data available on territorial behaviour or territory defence. Kikuchi (2009) reported a density of 0.56 pairs per ha, with a mean territory size of 0.83 ± 0.13ha in Peru (*regulus*). Nothing is known of interactions of adults with natural predators, but Dobbs *et al.* (2001) reported a nest being depredated by a Tyra *Eira barbara* (Mustelidae) in south-west Ecuador (*regulus*). The adult, which was brooding the two nestlings as the Tyra approached, flushed from the nest only when the mammal was <1m away. The adults did not return to the nest during 45min of observation following the consumption of the nestlings. There is a similar dearth of information on the relationship of the species with parasites, but Dietsch (2005) reported that an individual captured in Chiapas, Mexico, was heavily infested with chigger mites (Acarina: Trombiculidae).

DIET Adults are reported to prey predominantly on arthropods and other invertebrates (del Toro 1971), but surprisingly few data are available despite the relative frequency with which the species has been collected. Nestlings in southern Ecuador (Dobbs *et al.* 2001, 2003) were fed a variety of arthropods, including insects from several orders, and millipedes (Diplopoda). One large millipede identified more precisely (Polydesmida, Platyrhacidae, *Nyssodesmus* sp.) is apparently chemically defended by cyanide and benzaldehyde compounds (Heisler 1983). At the same nests in southern Ecuador, *c.*40% of adult provisioning visits included at least one earthworm. The diet perhaps also includes small vertebrates, such as small frogs (Stiles & Skutch 1989). The stomachs of three adult males reportedly held arthropod remains (*sororia*, Cardiff & Remsen 1981), fragments of medium-sized beetles (*chocoensis*) and piecies of a millipede (*chocoensis*) (Wetmore 1972). The stomach of a female *chocoensis* from Darién held insect remains (Robbins *et al.* 1985) and the stomach of a male *guatimalensis* from El Salvador contained 'large insects' (Dickey & van Rossem 1938).

REPRODUCTION Perhaps due to the record-setting breadth of its geographic range, Scaled Antpitta also holds the record for the most previously published nest descriptions and breeding records. Griscom (1932b) appears to have provided the first nest description, describing two nests of *guatimalensis* from Guatemala, but failing to describe the eggs and nestlings they contained.

Belcher & Smooker (1936) provided a cursory description of a nest (*aripoensis*) from Trinidad, suspected to belong to Scaled Antpitta. Subsequently, Edwards & Lea (1955), often credited with the first nest description, described a nest and eggs of *guatimalensis* at Monserrate, Chiapas. Miller (1963) provided the first description of a nest in South America, describing a nest of *regulus* from Colombia. He also provided the first description of the nestlings. Rowley (1966) described a single nest and clutch of *ochraceiventris*, and later (Rowley 1984) another nest and clutch of *guatimalensis* from Mexico. In a summary work of the birds of Chiapas, del Toro (1971) described the nest and eggs of *guatimalensis*, but it is unclear if his descriptions were based on direct observations or previous descriptions. After almost 20 years without further reported nests, Dobbs *et al.* (2001) described five nests of *regulus* from Celica, Loja, south-west Ecuador; subsequently Dobbs *et al.* (2003) provided numerous additional aspects for reproductive behaviour based on observations of an additional six nests at the same locality. Following this, Valencia-Herverth *et al.* (2012) described a nest of *guatimalensis* at San Juan, Huazalingo, Hidalgo. Most recently, Greeney & Valencia-Herverth (2016) described a nest of *regulus* from Utuana, Loja, Ecuador, bringing the total number of nests reported in the literature to 20, including that of Belcher & Smooker (1936). **NEST & BUILDING** The nesting microhabitat appears to vary little from the preferred habitat of adults, dense tangled undergrowth of mature forests or regenerating second growth. Although some nests have been near streams (Rowley 1966, 1984), there does not appear to be a marked predilection for riparian sites. Previous descriptions have implied that nests are rudimentary: a 'crude saucer' (Stiles & Skutch 1989); 'a mere platform or pad of dead leaves', 'a handful of dead leaves, with little if any attempt toward structure' (Griscom 1932b); and a 'poorly constructed platform' (Land 1970). This is, however, somewhat subjective, and most descriptions view the nest-construction abilities of the species more favourably. My summation of published descriptions and directly observed nests suggests that they are bulky, deep, open cups composed externally of a variety of materials, predominantly humid, dead plant matter such as sticks, leaves and grasses. Some nests, however, reportedly include moss, especially in very humid habitats. The inner cup is sparsely lined with thin, flexible materials such as pine needles (Edwards & Lea 1955) or rootlets (Rowley 1966, 1984, Greeney & Valencia-Herverth 2016). Nests are placed in relatively well-supported locations such as tree stumps (*n* = 1; Griscom 1932b), the fork of a large tree trunk (*n* = 4; Griscom 1932b, Dobbs *et al.* 2001, Copalinga Ecolodge data), fallen logs (*n* = 4; Edwards & Lea 1955, Rowley 1966, 1984, Greeney & Valencia-Herverth 2016), vegetation tangles or epiphytes against tree trunks (*n* = 4; Belcher & Smooker 1936, Dobbs *et al.* 2001, del Toro 1971) or the centre of a tree-fern leaf whorl (*n* = 1; Valencia-Herverth *et al.* 2012), and less frequently in less stable situations such as forking or overlapping branches (Miller 1963). Published nest measurements are as follows: height above ground 1.4m (Rowley 1984), 0.5m, 2.5m (Griscom 1932b), 1m (×2) (Edwards & Lea 1955, Copalinga Ecolodge data), 0.6m (Miller 1963), 1.3m, 0.7m, 0.7m, 1.2m, 0.8m (Dobbs *et al.* 2001), 1.6m (Greeney & Valencia-Herverth 2016); external diameter 15.2cm (Rowley 1966), 16.5–19.0cm (Edwards & Lea 1955), 17.8–30.5cm (Miller 1963), 20.1cm, 22.5cm, 16.0cm, 18.7cm, 20.0cm (Dobbs *et al.* 2001), 22.0cm (Greeney & Valencia-Herverth 2016); external height 6.1cm

(Rowley 1966), 8.5–10.0cm (Edwards & Lea 1955), 17.8cm (Miller 1963), 13.5cm, 13.0cm, 21.5cm, 19.5cm, 14.0cm (Dobbs *et al.* 2001), 14cm (Greeney & Valencia-Herverth 2016); internal diameter 10.0–11.5cm (Edwards & Lea 1955), 8.9cm (Miller 1963), 10.9cm, 8.5cm, 11.0cm, 9.75cm, 9.5cm (Dobbs *et al.* 2001), 9.5cm (Greeney & Valencia-Herverth 2016); internal depth 2.5cm (Rowley 1966), 5.0–7.5cm (Edwards & Lea 1955), 5.0cm (Miller 1963), 4.1cm, 5.0cm, 6.5cm, 5.0cm, 6.7cm (Dobbs *et al.* 2001), 6.5cm (Greeney & Valencia-Herverth 2016). **EGG, LAYING & INCUBATION** The completed clutch appears to almost always be two eggs: *n* = 15 (Griscom 1932b, Rowley 1966, 1984, Thurber *et al.* 1987, Dobbs *et al.* 2001, 2003, Valencia-Herverth *et al.* 2012, Londoño 2014, Greeney & Valencia-Herverth 2016); also, previously unpublished, *n* = 5 (R.A. Gelis, N. Athanas, E.M. Carman Jr. *in litt.* 2017, Copalinga Ecolodge data; clutch MLZ N237). Also from the literature, nests found post-hatch have held two nestlings (*n* = 9) (Griscom 1932b, Miller 1963, Thurber *et al.* 1987, Dobbs *et al.* 2001, 2003). Nevertheless, del Toro (1971) gave the clutch size of *guatimalensis* in Chiapas as 2–3, but did not provide sample size or specifics. A nest with a single egg in Chiapas (Edwards & Lea 1955) was most certainly the first of a two-egg clutch, as the female was collected with a shelled egg in the oviduct and there was no mention of additional developing ova (see also Edwards 1967). Detailed documentation of three-egg or one-egg clutches is lacking. Descriptions of the eggs were, until recently, all remarkably consistent (Greeney *et al.* 2013), and it appears that the species usually lays unmarked, pale blue to turquoise-blue eggs (Nehrkorn 1914, Edwards & Lea 1955, Edwards 1967, Rowley 1966, 1984, del Toro 1971, Stiles & Skutch 1989, Howell & Webb 1995, Dobbs *et al.* 2001, 2003). Two clutches recently described by Greeney & Valencia-Herverth (2016) both comprised two immaculate white eggs. These are the first reported eggs of any *Grallaria* that are other than blue, but the significance is unknown. Eggs are laid in late afternoon, roughly 48h apart (Dobbs *et al.* 2003) and the only documentation of incubation period (*n* = 1) suggests that they hatch after 20 days (Dobbs *et al.* 2003). At one nest that was closely monitored, regular incubation did not begin until three days after clutch completion, but over subsequent days diurnal coverage ranged from 87–96% (Dobbs *et al.* 2001, 2003). During incubation, eggs lose around 0.85% of their mass/day as they develop (Dobbs *et al.* 2003). Egg dimensions: *guatimalensis* 35 × 26mm (Edwards & Lea 1955); 30.7 × 25.2mm, 29.8 × 25.5mm (Rowley 1984); 30.2 × 25.1mm (estimated 9.3–10.3g) (Nehrkorn 1914, Schonwetter 1979, Schonwetter & Meise 1988); 30mm × 25mm (MVZB 187442; fully shelled in oviduct); *ochraceiventris* 33.1 × 26.0mm (11.2g), 31.6 × 25.9mm (10.9g) (Rowley 1966); *princeps* 34.27 × 27.19mm; 15g fresh (from label), unmarked sky blue (MZUCR 429); *regulus* 29.84 × 24.15mm (9.2g), 29.05 × 24.05mm (8.9g) (Dobbs *et al.* 2001); 30.11 × 23.63mm (9.0g), 29.13 × 23.65mm (8.6g) (Dobbs *et al.* 2003); 31.6×25.0mm (10.7g), 30.7 × 25.0mm (10.2g). Several authors have described the behaviour of adults near the nest during incubation (Edwards & Lea 1955), which is universally considered secretive or furtive, with approaches being slow, usually along the ground until a series of short hops brings the adult level with the nest. My experience, like that of other authors (Edwards & Lea 1955), is that adults flush silently, dropping immediately to the ground into the nearest dense vegetation. **NESTLING & PARENTAL CARE** Thurber *et al.* (1987) reported asynchronous hatching at one nest (*c.*1 day apart), but all other observations suggest that hatching is more frequently synchronous. Miller (1963) found two young nestlings, nearly identical in weight (12g), and suggested that they were c.3–4 days old. Although this is not an unreasonable presumption, I suggest they were more likely to have been just 2–3 days old. From the same nestlings, Miller (1963) reported hearing a 'surprisingly loud, harsh note' given by the nestlings when handled. Scaled Antpittas are born with dark pink or pinkish-black skin and long, wispy tufts of sooty-grey or black down on the capital, spinal and humeral feather tracts. Like other antpittas (Greeney et al. 2008), the young have bright orange or yellowish-orange rictal flanges, and striking crimson-orange mouth linings (Miller 1963, Dobbs et al. 2001, 2003). In south-west Ecuador (regulus), nestlings fledge after 17–19 days and, while young, are brooded roughly 80% of daylight hours and fed at a rate of 0.5–1.3 times per nestling/h (Dobbs et al. 2001, 2003). SEASONALITY The following data, most previously unpublished, are available regarding breeding activity, arranged by subspecies, then by type of evidence, and by country for subspecies with numerous records and wide ranges. BREEDING DATA G. g. ochraceiventris Lightly incubated eggs, 2 June 1965, La Cima (Rowley 1966, WFVZ 21314/25971); fledgling, 1 August 1983, Cerro Teotepec (MZFC 5783, photos A. Palacios Vázquez); juvenile, 24 October 2003, c.20km NE of Oaxaca (MZFC 18014, photos A. Palacios Vázquez); young [juvenile], July, 'Sierra Madre del Sur, Guerrero' (Salvin & Godman 1904); young [juvenile], July, Omiltemi (Salvin & Godman 1904); young [juvenile], September, Volcán Ajusco (Salvin & Godman 1904); juvenile transitioning, March 1976, 7km SW of San Pedro Nejapa (ENCB 951, photos A. Palacios Vázquez); subadult, 16 March 1897, San Sebastián (holotype, USNM 156013); subadult, 19 May 1903, Omiltemi (USNM 185810); subadult, 21 June 1967, Cerro de San Felipe (WFVZ 19162); subadult, 3 May 1965, La Cima (WFVZ 25970); subadult transitioning to adult, 24 May 1903, Omiltemi (Dickerman 1990, USNM 186490); 2♀♀ with enlarged follicles, 26 April 1965, Barranca Sin Nombre (Rowley 1966, Binford 1989, AMNH 815389). G. g. guatimalensis Salvin & Godman (1904) suggested that laying occurs June and July in Guatemala, based on 'young birds in their first spotted plumage' [= fledglings?] collected in 'upland forests' in August. In addition: MEXICO Clutch initiation, 9 August 1950, El Fenix (Edwards & Lea 1955); 'slightly incubated' eggs, 2 May 1967, Cerro Bául (Rowley 1966, 1984, WFVZ 21313); nest with eggs, observed 2–9 August 2011, Cerro del Huitepec (F. Albini in litt. 2017); incubation, 4 July 1937, Santa Rosa (MLZ N237); incubation, 9 August 2008, Huazalingo (Valencia-Herverth et al. 2012, Greeney & Valencia-Herverth 2016); fledgling, 15 June 1941, Siltepec (WFVZ 15090); fledgling transitioning, 26 August 1947, Ocosingo (FMNH 186099); fledgling transitioning, 7 August 1961, c.24km E of Tonalá (LSUMZ 40761); fledgling transitioning, 8 October 1962, 21km NE of Las Margaritas (LSUMZ 40763); fledgling transitioning, 6 November 1945, Montecristo (FMNH 153227); juvenile, 5 September 1962, 19.3km SE of San Cristóbal de las Casas (WFVZ 11245); juvenile, 6 April 1943, Presidio (MLZ 36196); 'juvenile', 3 August 1937, Santa Rosa (Berlioz 1939); two subadults, Cerro Bául, 29 April (AMNH 793445) and 20 May 1967 (WFVZ 19168); two subadults, Motzorongo, 1 September 1932 (MLZ 10273) and 27 February 1894 (USNM 154678); subadult, 3 June 1962, 12km ENE of Piedra Blanca (AMNH 768810); subadult, 5 May 1940, Cerro Tuxtla (USNM 359841); subadult, 26 May 1963, Río Lacantún (WFVZ

10736); subadult, 16 May 1963, 25km NW of San Pedro Tapanatepec (WFVZ 10740); subadult, 13 June 1901, Buena Vista (USNM 177370); subadult, 8 April 1998, Zona Arqueológica Yaxchilán (MZFC 14358, photos A. Palacios Vázquez). GUATEMALA Adult ♀ with fully shelled egg in oviduct, 4 June 2011, 2.1km NW by road of Chinatzatz (MVZB 187442); fresh eggs, 18 September 1926, Lago de Atitlán (Griscom 1932b, MCZ 355934); nest with nestlings ready to fledge, 18 September 1926, Panajachel (AMNH 394671, MCZ 146107; = '2 nestlings in down' from Griscom 1932b); fledgling, 5 August 1926, Santa María (AMNH 394669); fledgling transitioning, 29 July 1926, Tecpán (AMNH 394668; = '♂ imm' from Griscom 1932b); juvenile transitioning, 10 October 1873, 'Barranco Hondo' (USNM 69838); subadult, 13 May 1925, Sepacuité (AMNH 394667); subadult, 2 June 2017, Jardín Botánico de la USAC de Guatemala (ML 60068661, photo C.L. Burgos). EL SALVADOR Three nests, incubation, 8 June 1973 (hatch 10 June), 7 June 1975, 20 June 1978, Cerro Verde (Thurber et al. 1987); incubation, 21 July 2005, Los Andes (eBird: B. Sharp); nestling, 29 June 1975, Cerro Verde (Thurber et al. 1987); juvenile transitioning, 4 August 2016, Laguna de las Ranas (ML 32205261/271, photos G. Funes); 'young bird', 16 May 1927, Cerro de Los Naranjos (Dickey & Van Rossem 1938, UCLA 19174, not examined). HONDURAS Active nest, suspected building, 9 August 2015, PN La Tigra (J. van Dort in litt. 2015); three fledglings at El Derrumbo, 16 July (MLZ 65213) and 17 July 1933 (MCZ 158207, MLZ 65212); fledgling, 26 June 1936, Montaña de La Cruz Alta (MLZ 16815); fledgling, 30 June 1936, Montaña El Chorro (MLZ 16943); fledgling transitioning, 24 October 1890, Santa Ana (Ridgway 1891, USNM 120198); juvenile, 24 July 2015, PN Cusuco (IBC 1103508, S.E. Jones); juvenile transitioning, 23 December 2012, RB Uyuca (eBird S12369939, photos J. van Dort & R. Juárez); adult carrying worms, 8 September 2014, PN La Tigra (eBird: C. Sánchez); ovary enlarged, 26 June 1936, Merendón (FMNH 32316); testes enlarged, 27 June 1935, La Libertad (FMNH 32317). NICARAGUA Subadult, 15 January 2014, Reserva El Jaguar (ML 34101861, photo G. Duriaux). I also examined the following immature specimens with incomplete data: fledgling, 1897, 'Guatemala' (AMNH 71498); juvenile, San Cristóbal Verapaz (AMNH 492162); subadult, Sierra de Santa Elena (FMNH 30395); subadult, 'Guatemala' (USNM 50544). G. g. princeps COSTA RICA Stiles & Skutch (1989) gave the breeding season in Costa Rica as May–July. In addition: fresh egg, 12 October 2003, RF Prusia (MZUCR 429); incubation, 3 July 1981, La Montura (WFVZ 141898); nest with eggs, 3 August 2012, Finca Cristina, near Paraíso de Cartago (eBird: D. Martínez); incubation, 1 May 2010, Paraíso de Cartago (E.M. Carman Jr. in litt. 2015); incubation, 20 July 2016, RVS Curi-Cancha (J.D. Vargas in litt. 2016); hatching, 7 July 1998, RP Monteverde (eBird: R. Guindón); mid-aged nestlings, 11 August 2012, near Cartago, (L.V. Durán in litt. 2015); fledgling, 12 July 1920, Navarro (AMNH 390390); fledgling, June 1902, Volcán Irazú (MNCR 20156); fledgling transitioning, 3 January 2007, Tapantí (L. Sandoval photos); fledgling transitioning, 2 November 1898, Carrillos (AMNH 492168); fledgling transitioning, 27 August 1927, Navarro (YPM 56714, S.S. Snow photos); fledgling transitioning, 25 October 1930, Cerro Santa María (YPM 56716, S.S. Snow photos); juvenile, 8 September 1894, Azahar de Cartago (AMNH 492170); juvenile, 11 May 2016, San José (MZUCR 5202); adult carrying food, 26 June 2014, Savegre Mountain Lodge (eBird: H. Venegas); subadult, 26 January 1998, Jardín

Botánico Lankester (MNCR 24633); adult with testes slightly enlarged, 22 September 1939, Finca Lerida (El Velo) (Blake 1958, FMNH 207255). **PANAMA** Fledgling transitioning, 25 September 1936, El Velo (FMNH 207257); fledgling transitioning, 17 October 1905, Boquete (FMNH 57236). I also examined the following specimens with no collection dates: subadult transitioning to adult, 1903, Boquete (FMNH 57238); juvenile, 'Costa Rica' (USNM 90385). *G. g. carmelitae* Subadult, 28 March 1942, Hiroca (USNM 373668); subadult, 20 April 1943, Cincinati (USNM 373672); two adults with enlarged gonads, early April 1942, Hiroca (Hilty & Brown 1986, Fjeldså & Krabbe 1990, USNM). *G. g. aripoensis* Nest suspected to be under construction, 5 May 1925, Heights of Oropuche (Belcher & Smooker 1936). *G. g. chocoensis* Subadult, 5 March 1964, Tacarcuna Village (USNM 484544). *G. g. roraimae* Subadult, 11 April 2001, N slope of Mt. Roraima (USNM 626834). *G. g. regulus* **VENEZUELA** 'fledglings', June, 'Mérida' (Fjeldså & Krabbe 1990); subadult, 26 April 1903, El Valle (USNM 190376); two subadults transitioning, El Valle, 11 March (FMNH 50730) and 7 August 1904 (FMNH 50731). **COLOMBIA** Newly hatched nestlings, 23 April 1958, San Antonio (Miller 1963, MVZB 141786/87); 'fledglings', April, 'Valle del Cauca' (Fjeldså & Krabbe 1990); subadult, 6 April 1959, Río Arauca (FMNH 261454); subadult, 15 March 1971, Estación de Bombeo Guamués (FMNH 292938). **ECUADOR** Five nests with eggs, 13 and 14 February, 5 and 23 March, and 24 April 2000, 4km W of Celica (Dobbs *et al.* 2001, 2003); incubation, 5 March 2014, RP Utuana (Greeney & Valencia-Herverth 2016); incubation, 20 January 2010, El Monte Sustainable Lodge (R.A. Gelis *in litt.* 2012); incubation, 10 January 2008, Tandayapa Bird Lodge (N. Athanas *in litt.* 2010); nest with eggs, 13 April 2012, Copalinga Ecolodge (Copalinga Ecolodge data); newly hatched nestlings, 23 April 1958, San Antonio (Miller 1963, MVZB 141786/87); five nests with young, 25, 26 and 29 February, 2 March and 27 May 2000, 4km W of Celica (Dobbs *et al.* 2001, 2003); fledgling, 4 November 1988, 10km SE of Archidona (WFVZ 45742); fledgling, 30 March 1991, 4km W of Celica (WFVZ 48691); juvenile, 1 August 1923, Mindo (AMNH 180251); juvenile, 24 June 1985, *c.*9.5km NW of Piñas (MECN 2601); juvenile transitioning, 6 June 2013, RP Tandayapa Bird Lodge (IBC 1027621, N. Athanas); 'immatures', August, 'northwest Ecuador' (Fjeldså & Krabbe 1990); two active nests, Tandayapa Bird Lodge, 25 November 2008 (R. Shaw *in litt.* 2015) and 27 January 2014 (eBird: D. Bree); subadult, 29 October 1913, Zamora (AMNH 129757); subadult, 4 April 1912, Mindo (USNM 305186); adult suspected to be on nest, 18 December 2005, RP Buenaventura (D.J. Lebbin *in litt.* 2014). **PERU** Subadult, 26 February 1942, Río Chanchamayo (FMNH 123302); subadult, 26 March 1974, Cayumba Grande/Chinchavito (FMNH 299214). *G. g. sororia* Subadult transitioning, 29 October 1981, above Pilcopata (FMNH 311451).

TECHNICAL DESCRIPTION Sexes similar. See Taxonomy and Variation for details of substantial variation in the general description presented here. The following is based largely on the most widely distributed race, *regulus*. *Adult* Forecrown olive-brown, grading to slate-grey on crown and nape, where the feathers are fringed blackish. Lores whitish or buffy, ear-coverts olive-brown with narrow streaking. Rest of upperparts similar to ear-coverts, dark brown, usually with faint olivaceous cast, but feathers edged black giving it a scaled appearance.

Throat ochraceous- or tawny-brown, often with narrow pale streaking, and bordered by broad buffy or whitish malar extending into crescent-shaped paler (usually white) area, sometimes spotted or mixed black giving a necklaced appearance. Rest of underparts unmarked and vary from pale tawny in some races (individuals) to bright ochraceous or rufous. Usually, white bases to feathers of central belly give this area a paler or mottled look. Underwing and vent clean rufous. Upperwing-coverts olive-brownish, while rest of feathers of wing and tail are light brown to reddish-brown. **Adult Bare Parts** *Iris* dark brown; *Bare Orbital Skin* blue-grey; *Bill* maxilla black, mandible blackish, fading to greyish or pinkish-grey basally; *Tarsi & Toes* pinkish- or bluish-grey (plumbeous). *Fledgling* Several published descriptions of fledglings, still with fluffy nestling plumage in which they left the nest (del Toro 1971, Fjeldså & Krabbe 1990, Howell & Webb 1995), and I have examined numerous additional specimens. Downy, wool-like plumage is dark grey to black on upperparts, slightly washed brownish on throat and upper breast. Lower breast and central belly white, while flanks and vent are buffy-white to tawny. Crown and nape spotted (roundish spots or short streaks) tawny to creamy-white, with these markings being larger, sparser and darker (buffy-brown to reddish-buff) on back and rump. Throat mostly unmarked but upper chest has sparse pale buffy to white streaking, somewhat teardrop-shaped. Rest of underparts unmarked or faintly barred. Wings similar to adult, with upper median and greater secondary-coverts distinctly tipped tawny or rufescent forming teardrop- or chevron-shaped spots finely fringed dark grey or black. Tail generally not visible for some time after fledging. **Fledgling Bare Parts** *Iris* brown; *Bare Orbital Skin* grey to pinkish-grey; *Bill* dark orange, duskier on culmen, brighter orange on tomia and inflated rictal flanges, and pale whitish egg-tooth appears to be retained for some time after fledging, perhaps until transition to next plumage, at least in some cases; *Tarsi & Toes* blue-grey to violaceous-grey, often with yellowish or orange wash. *Juvenile* There are several published descriptions of juveniles (Ridgway 1891, Rand & Traylor 1954, Stiles & Skutch 1989), here supplemented with my own observations. Head, including crown and nape, similar to fledgling. Rest of upperparts similar to adult, perhaps somewhat less distinctly scaled. Wings and tail also similar to adult, but the fledgling wing-coverts retained. Below, fledgling plumage retained on throat and upper breast, but lower breast and belly are generally more similar to adult, though some downy plumage may remain on vent and flanks. **Juvenile Bare Parts** *Iris* dark brown; *Bare Orbital Skin* grey to blue-grey; *Bill* blackish, orange reduced to tomia and only slightly inflated rictus; *Tarsi & Toes* dark grey to bluish-grey, similar to adult. *Subadult* There are no published descriptions of subadult plumage, but I have seen several specimens and photos that indicate it is best distinguished by the retained spotted wing-coverts. Also appears overall duller, with scaling on upperparts not as clearly defined and crown less gray, tending to be washed brownish and not as clearly differentiated from back. **Subadult Bare Parts** As adult, but in youngest individuals some dull orange or yellowish may still be visible at gape.

MORPHOMETRIC DATA *G. g. guatimalensis Wing* 114mm, 116.5mm; *Tail* 42mm, 44mm; *Bill* culmen 25.5mm, 26mm; *Tarsus* 46mm, 50.5mm, middle toe 24mm, 25.5mm (*n* = 2♀♀, Ridgway 1911). *Wing* mean 112.6mm, range 104–120mm;

Tail mean 40.9mm, range 37–44mm; *Bill* culmen, mean 26mm, range 24–28mm; *Tarsus* mean 50.1mm, range 49.0–51.5mm, middle toe mean 24.3mm, range 24.0–25.5mm (*n* = 5, ♀?, Ridgway 1911). *Wing* range 105–122mm, mean 109.5 ± 2.1mm; *Bill* culmen, range 15.7–17.9mm, mean 17.0 ± 0.6mm (*n* = 11, ♀?, 'mexicana', Dickerman 1990). *Wing* range 107–119mm, mean 113.8 ± 2.9mm; *Bill* culmen, range 16.8–19.6mm, mean 17.8 ± 0.8mm (*n* = 16, ♂?, 'mexicana', Dickerman 1990). *Wing* range 105–112mm, mean 109.5 ± 2.1mm; *Bill* culmen, range 15.7–17.9mm, mean 17.0 ± 0.6mm (*n* = 11, ♀, from Oaxaca, Dickerman 1990). *Wing* range 109–122mm, mean 114.5 ± 3.6mm; *Bill* culmen, range 15.1–18.4mm, mean 19.0 ± 0.9mm (*n* = 21, 20, respectively, gender not specified, from Chiapas, Dickerman 1990). *Wing* mean 109.1mm, range 104.5–111mm; *Tail* mean 42mm, range 28.0–44.5mm; *Bill* culmen mean 26mm, range 25.5–27.0mm; *Tarsus* mean 47.3mm, range 43.5–49.5mm, middle toe mean 24.3mm, range 23–25mm (*n* = 4♂♂, 'mexicana', Ridgway 1911). *Wing* mean 110mm, range 107–113mm; *Tail* mean 41.5mm, range 40.5–43.0mm; *Bill* culmen mean 26.7mm, range 26.5–27.0mm; *Tarsus* mean 47mm, range 46.5–47.5mm, middle toe mean 24.2mm, range 24.0–24.5mm (*n* = 3♀♀, 'mexicana', Ridgway 1911). **G. g. aripoensis** *Wing* 102–105mm; *Bill* 33–34mm; *Tail* 32–36mm (*n* = ??, ♂♂, Hellmayr & von Seilern 1912). *Wing* 101–105mm; *Bill* 32–33mm; *Tail* 34–37mm (*n* = ??, ♀♀, Hellmayr & von Seilern 1912). *Wing* range 101–105mm, 98–105mm (*n* = 17, 11♂♂, 6♀♀, respectively, ffrench 1973). **G. g. princeps** *Wing* 109mm; *Bill* 33mm; *Tail* 43mm (*n* = 1♀ syntype, Sclater & Salvin 1869). *Wing* mean 111.7mm, range 110–115mm; *Tail* mean 41.3mm, range 40–43mm; *Bill* culmen, mean 27.4mm, range 24.5–28.5mm; *Tarsus* mean 48mm, range 44.5–52.0mm, middle toe mean 25.2mm, range 23–27mm (*n* = 12♂♂, Ridgway 1911). *Wing* means 112mm, 111.8mm; *Tail* means 41.3mm, 41.3mm; *Bill* culmen, means 27.1mm, 27.6mm; *Tarsus* means 48.3mm, 47.6mm, middle toe mean 25.3mm, 25.1mm (*n* = 12♂♂, five from Costa Rica, seven from Chiriquí, same individuals as previous, Ridgway 1911). *Wing* mean 111.1mm, range 106.5–113.5mm; *Tail* mean 40mm, range 37.5–42.5mm; *Bill* culmen, mean 26.7mm, range 25.0–27.5mm; *Tarsus* mean 48.6mm, range 46.5–50.0mm, middle toe mean 25.4mm, range 25–26mm (*n* = 4♀♀, Ridgway 1911). *Wing* mean 111.5mm, range 109.8–113.2mm; *Tail* mean 38.6mm, range 35.0–41.5mm; *Bill*, culmen from base mean 28.8mm, range 26.7–30.8mm; *Tarsus* mean 48.8mm, range 44.4–52.0mm (Wetmore 1972: precise ranges and means given first for 10♂♂, then for 8♀♀). *Bill* from nares 16.9mm, 15.9mm, 18.2mm, 16.8mm, 16.9mm; *Tarsus* 50.8mm, 51.3mm, 47.8mm, 51.5mm, 49.6mm (*n* = 5, ♀♀,♂,♂,♂, MNCR). *Tarsus* 47.7mm (*n* = 1♂, LACM 16227). **G. g. chocoensis** *Wing* 89mm; *Bill* culmen 22mm; *Tail* 28mm; *Tarsus* 42mm (*n* = 1, holotype ♂, Chapman 1917). *Wing* mean 108.8mm, range 103.3–112.6mm; *Tail* mean 38.1mm, range 36.9–41.1mm; *Bill* culmen from base mean 27.5mm, range 26.7–28.0mm; *Tarsus* mean 47.6mm, range 43.8–50.6mm (*n* = 3♂♂, Wetmore 1972). **G. g. ochraceiventris** *Wing* 114mm; *Tail* 43mm; *Bill* culmen 28mm; *Tarsus* 47mm (*n* = 1♂, subadult holotype, Nelson 1898). *Wing* 111mm, 119.5mm; *Tail* 46mm, 49.5mm; *Bill* culmen 25.5mm, 26.5mm; *Tarsus* 48.5mm, 51mm, middle toe 25.5mm, 26.5mm (*n* = 2 ♂♂, Jalisco, Guerrero, respectively, Ridgway 1911); *Wing* mean 115.7mm, range 110.5–120.0mm; *Tail* mean 47.7mm, range 45–52mm; *Bill* culmen, mean 28.1mm, range 27.5–29.5mm; *Tarsus* mean 52.1mm, range 49–55mm, middle toe mean 26.7mm, range 25–28mm (*n* = 4♀♀, Ridgway 1911). *Wing* 110.5mm;

Tail 45mm; *Bill* culmen 28mm; *Tarsus* 49mm, middle toe 26mm (*n* = 1♀ from Morelos, included in previous, Ridgway 1911). *Wing* range 106–122mm, mean 114.4 ± 3.5mm; *Bill* culmen, range 17.8–22.0mm, mean 19.6 ± 1.2mm (*n* = 20, 18, respectively, gender not specified, Dickerman 1990). *Wing* range 108–123mm, mean 115.2 ± 4.6mm; *Bill* culmen, range 17.6–19.8mm, mean 19.0 ± 0.6mm (*n* = 9, ♀?, as *binfordi*, Dickerman 1990). **G. g. carmelitae** *Wing* 100mm; *Bill* exposed culmen 21.5mm; *Tail* 38mm; *Tarsus* 45mm (*n* = 1♂, holotype, Todd 1915). *Wing* 103mm; *Tail* 35mm; *Bill* 21mm; *Tarsus* 42mm (*n* = 1♀, Todd & Carriker 1922). *Wing* 99–105mm; *Tail* 35–40mm; *Bill* 21.0–23.5mm (*n* = 9, Cory & Hellmayr 1924). **G. g. sororia** *Wing* 107.5mm; *Bill* culmen 22.75mm; *Tail* 40.5mm; *Tarsus* 45.5mm (*n* = 1♂, Berlepsch & Sztolcman 1901). **G. g. roraimae** *Wing* 108mm; *Bill* exposed culmen 23mm; *Tail* 39mm; *Tarsus* 41mm (*n* = 1♀, Chubb 1921). *Wing* 112mm; *Tail* 37mm (*n* = ?, ♀?, Ruschi 1979). **G. g. regulus** *Wing* 102mm; *Tail* 30mm; *Tarsus* 41mm (*n* = 1♀?, holotype, Sclater 1860a). *Wing* 107mm; *Tail* 42mm; *Bill* culmen 22mm (*n* = 1♀, Lehmann-V. 1957). *Wing* 102mm, 100mm; *Tail* 35mm, 33m; *Bill* 28mm, 29mm; *Tarsus* 43mm, 46mm (*n* = 2, ♂, ♀, Taczanowski 1884). *Wing* 105mm; *Tail* 40mm; *Bill* culmen 23.5mm; *Tarsus* 46mm (*n* = 1♀, Berlepsch & Sztolcman 1906). *Wing* mean 102.9 ± 2.6mm, range 97–108mm; *Tail* mean 35.0 ± 2.8mm, range 29–41mm; *Bill* culmen mean 26.0 ± 0.9mm, range 23.9–28.0mm, from nares mean 15.5 ± 0.8mm, range 14.1–16.9mm; *Tarsus* mean 43.4 ± 1.8mm, range 39.6–47.5mm (*n* = 39, genders combined, Krabbe & Coopmans 2000). *Wing* 95.5mm, 96mm, 99mm, 105mm, 102mm, 99mm, 103mm, 98mm; *Tail* 31mm, 37mm, 34mm, 34mm, 34mm, 32mm, 35mm, 34mm; *Bill* from nares 17.0mm, 15.7mm, 16.2mm, 16.8mm, 16.5mm, 15.8mm, 15.7mm, 17.1mm; *Tarsus* 42.2mm, 42.9mm, 42.9mm, 45.7mm, 45.5mm, 43.8mm, 43.7mm, 43.5mm (*n* = 8♂?, Pichincha, D. Becker *in litt.* 2015). **MASS** ♂♂ 98.9g (*n* = 1, *guatimalensis*, Blake 1957), 94.6g (*n* = 1, *guatimalensis*, Krabbe & Schulenberg 2003), 86g (*n* = 1, *regulus*, Miller 1963), 69.5–86.0g (*n* = ?, *regulus/sororia*, Krabbe & Schulenberg 2003), 89g (*n* = 1 *sororia*, Cardiff & Remsen 1981), 86g, 100g, 140g (*n* = 3, *princeps*, MNCR). ♀♀ 117.1g (*n* = 1, *ochraceiventris*, Binford 1989), 90.9g (*n* = 1, *ochraceiventris*, Rowley 1966), 95g (*n* = 1, *chocoensis*, Robbins *et al.* 1985), 79.8g (*n* = 1, *regulus/sororia*, Krabbe & Schulenberg 2003); 89g, 101g (*n* = 2, *princeps*, MNCR). **Gender not specified** mean 98g (*n* = ??, *princeps*, Krabbe & Schulenberg 2003), 94.1g, range 77–116g (*n* = 6, race?, Dunning 1993); 111.5g (*n* = 1, *princeps*, Karr *et al.* 1978), 64.5g, 71.0g 71.0g, 74.0g, 75.0g, 68.5g, 74.0g, 77.0g (*n* = 8, *regulus*, C.D. Becker *in litt.* 2017). **TOTAL LENGTH** 11.5–20.3cm (Taczanowski 1884, Chubb 1921, Blake 1953, Rand & Traylor 1954, Herklots 1961, Meyer de Schauensee 1964, Land 1970, del Toro 1971, Davis 1972, Edwards 1972, Wetmore 1972, ffrench 1973, Meyer de Schauensee & Phelps 1978, Ruschi 1979, Iafrancesco *et al.* 1987, Edwards 1989, Fjeldså & Krabbe 1990, Ridgely & Tudor 1994, Clements & Shany 2001, Garrigues & Dean 2007, Kenefick *et al.* 2007, van Perlo 2009).

TAXONOMY AND VARIATION Scaled Antpitta is clearly part of Lowery & O'Neill's (1969) core *Grallaria* subgenus, within which most authors tend to place it as closely related to Moustached or Variegated Antpittas (Cory & Hellmayr 1924, Peters 1951, Krabbe *et al.* 1999, Remsen *et al.* 2017). Across its considerable range, Scaled Antpitta is quite variable in plumage, with some variability even within populations, and up to 11 subspecies have been recognised historically. Nevertheless, most of the nine currently recognised subspecies are fairly diagnosable by plumage. The northernmost populations were, at one point, split from nominate *guatimalensis* as race *mexicana* (Sclater 1861, see below). The diagnosis was based largely on this population's larger size and paler underparts, compared to nominate. Numerous authors (e.g. Carriker 1910, Griscom 1932b, Edwards & Lea 1955, Krabbe & Schulenberg 2003) and even Sclater himself (1877) questioned the validity of this distinction. The range of putative *mexicana* was considered to be eastern Mexico, in Veracruz and adjacent northern Oaxaca and eastern Tabasco (Ridgway 1911, Binford 1989, Dickerman 1990). Griscom (1932b), on comparing a series of *mexicana* with several series of nominate *guatimalensis* from Nicaragua, Guatemala and Mexico, found it to be smaller on average than nominate from some areas, paler than nominate from some areas, but also noted the reverse was true for both characters, compared to other populations. He concluded that apparent differences in size and saturation of the underparts were more correlated with altitude than region, and that *mexicana* should be subsumed within nominate. *G. g. binfordi* is the most recently described race of Scaled Antpitta (Dickerman 1990). The holotype (AMNH 805767) is an adult female, collected above Cuernavaca (see Selander & Vaurie 1962), Morelos, on 20 February 1908. The type arrived at AMNH from the collection of Austin Paul Smith via William Beebe. Smith's label states the skin is a 'juvenile' female, but it appears to be in full-adult plumage (Dickerman 1990, LeCroy & Sloss 2000; HFG examination). It was named in honour of Lawrence C. Binford. Race *binfordi* is said to be similar to *ochraceiventris* in having pale underparts, but differs by having stronger breast markings forming vertical striping, which, according to Dickerman (1990) 'extends onto the belly in the first basic plumage'. Krabbe & Schulenberg (2003) elected not to recognise this subspecies pending further analysis, but acknowledged that it might prove valid. I agree with Krabbe & Schulenberg (2003) that the possibility remains that this population deserves recognition, but acknowledge that Dickerman's (1990) description may indeed be based on juvenile *G. g. ochraceiventris* in transitional plumage. Following Krabbe & Schulenberg (2003), I recognise nine subspecies. The population that apparently exists on the Venezuelan island of Nueva Esparta (Bisbal 1983, Sanz *et al.* 2010) was referred to by Hilty (2003) as race *schwartzii*, but the origin of this name is unclear. Without any further information on this population I have provisionally included it within *regulus*, although affinities with *aripoensis* are possible. The distributions of the various subspecies are incompletely understood and there are apparently some relatively newly discovered populations that require scrutiny. I have attempted to resolve some of the confusion, but the ranges provided here are still highly hypothetical.

Grallaria guatimalensis ochraceiventris Nelson, 1898, Proceedings of the Biological Society of Washington, vol. 12, p. 62, San Sebastían, Jalisco, Mexico [20°46'N, 104°51'W]. The adult male holotype (USNM 156013) was collected on 16 March 1897 by Edward W. Nelson and Edward A. Goldman (Deignan 1961). Race *ochraceiventris* is endemic to southern Mexico, but its precise range is unclear and has been the source of some confusion based largely on a fair degree of plumage variation in this region, compounded by vague locality data and the description

of two additional subspecies from this region (now not recognised, see below). As defined here, its range extends from the type locality in western Jalisco, south-west along the Sierra Madre del Sur in Michoacán, southern México, and Distrito Federal (Peterson & Navarro-Sigüenza 2006), Morelos and Guerrero, to coastal south-west Oaxaca. Race *ochraceiventris* is significantly paler than the nominate race, especially on the underparts. It further differs by having narrow black scaling above and lacking dusky feathers on throat. I feel that *ochraceiventris* is a fairly distinctive race. Nevertheless, there appear to be quite a few intermediate examples, leading to much confusion in the early literature. This is, in part, because many examples of *ochraceiventris* were originally assigned to the race *mexicana* (e.g. Salvin & Godman 1904), which was later subsumed into the nominate (Griscom 1932b). *G. g. mexicana* Sclater, 1861 was described from 'Jalapa', Veracruz [= Xalapa, Caribbean slope]. The unsexed adult holotype (NHMUK 1889.9.20.606) was collected by Rafael Montes de Oca (Warren & Harrison 1971). Confusingly, *mexicana* was largely defined by being paler than nominate, but not as pale as *ochraceiventris*. Further confusion was introduced by incomplete knowledge of immature plumages, compounded by the fact that the holotype of *ochraceiventris* is in subadult plumage. Thus, in distinguishing the subspecies, Nelson (1898) stated that the 'feathers of crown [...] and nape [are] olive-brown with a dark ashy shade most marked on sides of crown', and that the 'wing coverts [are] dull brown with shaft lines and spots of dull tawny brown at tips'. Both distinguishing plumage traits are, however, lost in full adults. The crown of *ochraceiventris* is, in fact, slate-grey, similar to other subspecies, and remnants of brown feathers on the crown are indications of immaturity, as I observed in the holotype. This was surmised by Dickerman (1990), but its importance not stressed.

SPECIMENS & RECORDS *Michoacán* Cerro San Andres, 19°48'N, 100°31'W (UMMZ 116910). *México* Amanalco de Becerra, 19°15.5'N, 100°01'W (Dickerman 1990, CNAV P003117); Ocuilán de Arteaga–Cuernavaca road, km14, *c.*18°59.5'N, 99°20'W (MZFC 10281); *c.*5km SW of Ocuilán de Arteaga, 18°57'N, 99°23'W (KUNHM 111995). *Distrito Federal* Tlalpan, 19°17'N, 99°12.5'W (CNAV P003111–116); Volcán Ajusco, 19°13'N, 99°15.5'W (Salvin & Godman 1904). *Morelos* Lagunas de Zempoala and vicinity, 19°04'N, 99°19'W (MZFC 18580, CNAV P014040); near Fierro del Toro, *c.*19°02'N, 99°11'W (CNAV P014011); *c.*4.5km NW of San Juan Tlacotenco, 19°02.5'N, 99°07.5'W (MZFES 73); above Cuernavaca, 19°01'N, 99°18'W (Dickerman 1990, AMNH 805767; holotype of *binfordi*). *Jalisco* *c.*9km SW of Autlán, 19°42'N, 104°25'W (Dickerman 1990); 19.5km S of Ahuacapán, 19°32'N, 104°19'W (LSUMZ 40759). *Guerrero* Coapango, 18°38'N, 99°33'W (MVZB 110482); Chilpancingo, 17°33'N, 99°30'W (Griscom 1934, LSUMZ 14282/83); Tlacotepec, 17°47'N, 99°59'W (Dickerman 1990); Centro Ecoturistico La Mona, 17°35'N, 100°36'W (MZFC 29000); Omiltemi, 17°33'N, 99°41'W (Griscom 1937, MVZB, MCZ, CAS, MLZ, USNM, ANSP, DMNH); Cerro Teotepec, 17°28'N, 100°08'W (Griscom 1957, Navarro-Sigüenza 1992, MLZ 45923/26/83, MZFC 5783); Nueva Delhi, 17°25.5'N, 100°11.5'W (Navarro-Sigüenza *et al.* 1991, KUNHM 111802); Chimicotitlán, 17°23'N, 99°22.5'W (DMNH); 14km S of Puerto el Gallo, *c.*17°20.5'N, 100°11'W (MZFC 3656). *Oaxaca* *c.*20km NE of Oaxaca, *c.*17°11.5'N, 96°36'W (MZFC 18014); Barranca Sin Nombre, 16°11'N, 97°07'W (AMNH 815389); La Cima,

16°12'N, 97°07'W (WFVZ 25970/71, CAS 71993); Cerro de San Felipe, 17°08'N, 96°43'W (Hernández-Baños *et al.* 1995, WFVZ 19161/62, MVZ 94776); 1.6km N of San Andrés de Chicahuaxtla, 17°10.5'N, 97°50.5'W (LSUMZ 33177); Cerro Piedra Larga, El Aguacate, 16°37'N, 95°48.5'W (MZFC 11645); 7km SW of San Pedro Nejapa, *c.*16°32'N, 96°01.5'W (ENCB 951). **NOTES** Apart from the *c.*30 skins labelled Omiltemi or Chilpancingo (see above), there are several other specimens from locations near these two cities: Cañada Agua Fría (MZFC 4735), 'Chilpancingo Mountains' (MCZ 163538).

Grallaria guatimalensis guatimalensis, Prévost & Des Murs, 1846. As the holotype (lost?) of the nominate race was probably a trade skin from Vera Paz, there is little hope of defining the type locality any more precisely than somewhere in Guatemala (Griscom 1932b). The nominate subspecies is distributed from the Caribbean slope of south-central Mexico south fairly continuously to northern Nicaragua. The northernmost records are from a small patch of cloud forest (*c.*1,500m) just north of Tlanchinol in north-east Hidalgo (Howell & Webb 1992a, Valencia-Herverth *et al.* 2012). Records in the mountains between there and north-east Oaxaca are extremely few and include only several scattered observations posted on eBird and a handful of specimens from the forested hills around Matzoronga (MLZ, USNM) in eastern Veracruz just south of Córdoba. As with Mexican Antthrush *Formicarius moniliger*, there is an apparently an isolated coastal population in the Sierra de Tuxtla (Andrle 1967, 1968), with at least two modern records from the highly fragmented forests around Lago Catemaco (eBird). Due south of here, with a gap in records of *c.*120km, the hills north of Benito Juárez (17°08'N, 94°58'W) appear to be the westernmost locality where nominate *guatimalensis* is (or at least was) most abundant. East and south from there, historical and recent records are fairly continuous, extending as far north as Teapa in southern Tabasco. Two old specimen from north of there, 15–30km inland from the coast of Tabasco (USNM, ROM), have not been substantiated by further specimens or modern observations, and are shown as points on the map. In Chiapas, *guatimalensis* formerly occurred throughout, except in drier areas and the flat coastal regions (del Toro 1971). Almost all modern (eBird) records, however, are from south and east of San Cristóbal de las Casas, along the coast and Guatemalan border in Chiapas. Nominate *guatimalensis* occurs on both slopes in Guatemala (Carriker & Meyer de Schauensee 1935, Howell & Webb 1992b, Eisermann & Avendaño 2007, Eisermann *et al.* 2013), and I found records from nearly every department except Petén, where it may not be present but for the extreme south-east. To date, there are also no records west of Ciudad de Guatemala in the departments of Guatemala, Santa Rosa, Jutiapa, Jalapa and Chiquimla. The range of *guatimalensis* extends only slightly into western El Salvador (Dickey & van Rossem 1938) and I found no indication that it occurs further east than San Salvador. The only slight exceptions are records from north-east Chalatenango near the Honduran border (KUNHM, eBird: O. Komar). In fact, apart from one plausible but unsubstantiated eBird record (M. Ávalos) from the slopes of Volcán San Salvador, the range of *guatimalensis* in the country appears to terminate in the hills surrounding the Volcán Santa Ana complex. I see no reason why it should not be present further east, in what appears to be suitable habitat surrounding other peaks in the coastal volcanoes.

The Honduran range of *guatimalensis* is reasonably well known and relatively well represented in collections, especially from localities around Tegucigalpa. However, somewhat inexplicably, there are no records from the north-east (Colón, Gracias a Dios), and the literature treating Scaled Antpitta in Honduras is rather scarce (but see Stone 1932, Monroe 1968). The easternmost extent of nominate *guatimalensis* is Kum (Nicaragua) and the southernmost reports are from around Matagalpa (see Salvin & Godman 1904, Martínez-Sánchez & Will 2010). Originally described in Latin, retrospectively rather vaguely given what we now know concerning this species' geographic variation, nominate *guatimalensis* was described by Prévost & Des Murs (1846) as having the crown and nape iron-grey, with individual feathers evenly fringed in black. The back and wing-coverts are olive-brown, scaled black. The secondaries are brown, differing from the more red-brown primaries and reddish tail feathers. The face bears a distinctive, postocular white crescent. The underparts, from the chin to the tail-coverts, are tawny-rufous, with the throat darker and somewhat scaly. **SPECIMENS & RECORDS MEXICO**: *Hidalgo* 4.5km NNE of Tlanchinol, 21°02'N, 98°39'W (Howell & Webb 1992a); San Juan, 20°59.5'N, 98°31'W (Valencia-Herverth *et al.* 2012). *Veracruz* Xalapa, 19°32.5'N, 96°54.5'W (Sclater 1859b, 1861); Córdoba, 18°53'N, 96°55.5'W (Sclater 1857d); Presidio, 18°41'N, 96°46'W (Griscom 1957, MLZ 36196); Matzorongo, 18°39'N, 96°44'W (Griscom 1957, MLZ 10273, USNM 154678); Estación de Biología Tropical Los Tuxtlas, 18°35'N, 95°05.5'W (Coates-Estrada & Estrada 1989); 'Cerro Tuxtla', [Volcán San Martín Tuxtla], 18°34'N, 95°12'W (Wetmore 1943, USNM 359841/842); Tres Zapotes, 18°28'N, 95°26'W (Wetmore 1943, USNM 359843); NE of Catemaco, 18°29.5'N, 95°04'W (DMNH 25131); Buena Vista, San Andrés Tuxtla, 18°26'N, 95°13'W (USNM 177369/70, MCZ 102497/98); Rancho Caracól, 18°26'N, 96°38.5'W (Griscom 1957, MLZ 48880/919); Playa Vicente, 17°50'N, 95°49'W (Sclater 1859c). *Oaxaca* 30km SSE of Jesús Carranza, *c.*17°13.5'N, 94°55.5'W (KUNHM 29179); 39km N of Matías Romero, 17°13'N, 95°02'W (WFVZ 4935); Rancho San Carlos, 17°11.5'N, 94°55.5'W (Schaldach *et al.* 1997; KUNHM 104893); 12km ENE of Piedra Blanca, 17°05.5'N, 94°59'W (AMNH 768810); Cerro Bául, 16°38'N, 94°13'W (AMNH 793445/446, WFVZ); 19.3km NNE of Zanatepec, *c.*16°36'N, 94°18'W (LSUMZ 33178/79); 25km NW of San Pedro Tapanatepec, *c.*16°32'N, 94°22'W (WFVZ 10738–745). *Tabasco* 56km N and 10.8km NW of Teapa, *c.*18°09.5'N, 93°02'W (ROM 119901); La Venta, 18°06'N, 94°02.5'W (USNM 371751); Teapa, 17°34'N, 92°56'W (Brodkorb 1943, USNM 166085/86, LSUMZ 23304). *Chiapas* PN Palenque, 17°31.5'N, 92°02'W (Patten *et al.* 2011); Tumbalá, 17°17'N, 92°19'W (USNM 154679); 26km by road N of Ocozocoautla de Espinosa, *c.*16°55'N, 93°27'W (LSUMZ 167188/89); Ocosingo, 16°54.5'N, 92°06'W (FMNH 186098–100); Zona Arqueológica Yaxchilán, 16°53.5'N, 90°58.5'W (MZFC 14358); 12km N by road of Berriozábal, *c.*16°53'N, 93°17'W (LSUMZ 167190); Finca Patichuiz, *c.*16°46'N, 91°42.5'W (WFVZ 11246/47, LSUMZ 45436); Cerro del Huitepec, 16°45.5'N, 92°41.5'W (de Silva 2005); Ocozocoautla de Espinosa, 16°45.5'N, 93°22.5'W (LACM 24415); Montecristo, 16°43.5'N, 93°11.5'W, near Tuxtla Gutiérrez (FMNH 153227); El Fenix, *c.*16°39'N, 94°01'W (Edwards & Lea 1955, FMNH 208882/83, DMNH 47577); Cerro Brujo, *c.*16°36'N, 93°23'W (MLZ 27123–129); 19.3km SE of San Cristóbal de las Casas, *c.*16°36'N, 92°31'W (WFVZ 11243–245, KUNHM

104892, LSUMZ 167191); Teopisca, 16°32.5'N, 92°28.5'W (MLZ 43804); 21km NE of Las Margaritas, 16°29'N, 91°50'W (LSUMZ 40762/63); Río Lacantún, c.16°13'N, 91°20'W (WFVZ 10736); Santa Rosa, 16°09'N, 91°32'W (Berlioz 1939, DMNH, MLZ); PN Lagos de Montebello, 16°08'N, 91°43.5'W (Rangel-Salazar *et al.* 2009); c.24km E of Tonalá, 16°05'N, 93°32'W (WFVZ 5066/67, LSUMZ 40761); Siltepec, 15°33'N, 92°19'W (UMMZ 110195, MCZ 273001, WFVZ 15090); Cerro Saxchanal, 15°30.5'N, 92°36'W (UMMZ 110194); Cerro Ovando, 15°25'N, 92°37'W (MCZH 272999/3000, UMMZ 107721/22); Volcán Tacaná, 15°07.5'N, 92°09'W (UMMZ 102462–464). **GUATEMALA: Huehuetenango** c.4.3km SW of San Mateo Ixtatán, 15°48'N, 91°30'W (MVZB 184236). *Quiche* 2.1km NW (by road) of Chinatzatz, 15°05'N, 90°55'W (MVZH 187442). **Alta Verapaz** Chajbaoc, 15°29'N, 90°21.5'W (ML 53703251, photo P. Chumil); Reserva Natural Privada Orquigonia, 15°26.5'N, 90°24.5'W (ML 53487531/541, ML 64056861, photos F.A. Ordoñez); Sepacuité, 15°26'N, 89°45'W (AMNH 394666/667, MCZ 146105); 'Finca Concepción', c.15°26'N, 89°50'W (loc. Griscom 1932b) (MCZ 146108); near Chelemhá, 15°23'N, 90°04'W (Renner *et al.* 2006); Finca Aurora, 15°22.5'N, 90°26'W (ML 32051731, photo J. Cahill); San Cristóbal Verapaz, 15°22'N 90°29'W (AMNH 492161/62). **El Progreso** San Agustín Acasaguastlán, 14°57'N, 89°58'W (ANSP 121355). *Quetzaltenango* Santa María, 14°45'N, 91°33'W (AMNH 394669). **San Marcos** Near 'Unión Reforma', 15°10'N, 92°01'W (eBird: J. Cahill); Volcán Tajumulco, 15°02.5'N, 91°54'W (FMNH 93759). *Sololá* Lago de Atitlán/Panajachel, c.14°44'N 91°10'W (AMNH 394670–673; MCZ 146106/07). *Suchitepéquez* RNP Los Andes, 14°32.5'N, 91°11'W (eBird: A. Jaramillo). *Chimaltenango* 'Sierra Santa Elena', c.14°48'N, 91°01'W (FMNH 30395); Tecpán, 14°46'N 91°0'W (AMNH 394668, FMNH 36378); 'Barranco Hondo', Volcán de Fuego, c.14°28.5'N, 90°53'W (USNM 69838). *Sacatepéquez* Parque Ecológica Florencia, 14°34'N, 90°41'W (eBird: A. Sagone); Parque Municipal Corazón de Agua, 14°31'N, 90°41'W (ML 34483171/181, photos V. Cattelan). *Guatemala* Jardín Botánico de la USAC de Guatemala, Ciudad de Guatemala, 14°37'N, 90°31'W (ML 60068661, photo C.L. Burgos). **El SALVADOR: Chalatenango** Caserio La Montañona, 14°08'N, 88°55'W (KUNHM). *Santa Ana* PN Montecristo, c.14°24.5'N, 89°22.5'W (Komar 2002a, 2002b); Laguna de las Ranas, 13°54'N, 89°43'W (ML 32205261/271, photos G. Funes); Los Andes, 13°52'N, 89°38'W (eBird: B. Sharp); Volcán Santa Ana, 13°50'N, 89°38'W (UCLA 19033/132); Cerro Verde, 13°49.5'N, 89°37.5'W (Thurber *et al.* 1987, ML 33331471, photo M. Trejo). *Ahuachapán* PN El Imposible, c.13°50'N, 89°57'W (Komar & Herrera 1995, Komar 2002b). *Sonsonate* Cerro de los Naranjos and vicinity, c.13°53.5'N, 89°42'W (UCLA 19174, MVZB 85924). **HONDURAS: Copán** La Leona, 14°52'N, 88°59'W (AMNH 326924); La Libertad, 14°49'N, 89°00'W (FMNH 32317); 'Merendón' [= San Agustín], 14°49'N, 88°56'W (loc. Monroe 1968, FMNH 32316, MCZ 172202). *Ocotepeque* Montaña El Chorro, 14°26'N, 89°05'W (MLZ 15533/16943); Montaña El Sillón, 14°24'N, 89°05'W (MLZ 16942); Montaña de La Cruz Alta, 14°21'N, 88°58'W (MLZ 16815). *Santa Bárbara* E slope of Montaña de Santa Bárbara, 14°55'N, 88°07'W (LSUMZ 29517); Santa Bárbara, 14°55'N, 88°14'W (AMNH 326923, MCZ 172201). *Yoro* Subirana, 15°12'N, 87°27'W (MCZ 158567). *Atlántida* Santa Ana, 15°38'N, 87°03'W (USNM 120198). *Intibucá* El Derrumbo, 14°18'N, 88°23'W (MCZ 158205–207, MLZ 65212/213). *Lempira* Montaña Puca, 14°45'N, 88°33'W (MCZ 265258–

261). *Cortés* Cantiles camp, PN Cusuco, c.15°30.5'N, 88°14.5'W (IBC 1103508, S.E. Jones); Las Peñitas, 15°28'N, 88°03'W (MCZ 158202–204, AMNH 326922); Lago de Yojoa, 14°53'N, 87°59'W (CM P135521); Sendero El Venado, PN Azul Meámbar, 14°52.5'N, 87°54'W (ML 25984231, J. van Dort). *Francisco Morazán* Cerro Cantoral, 14°20'N, 87°24'W (AMNH, MLZ, MCZ, ANSP); San Juancito, 14°13'N, 87°04'W (CM P135238); PN La Tigra, 14°13'N, 87°07'W (ML 30115571, photo G. Flores-Walter); RB Uyuca, 14°02'N, 87°04'W (eBird S12369939, photos J. van Dort & R. Juárez); Montaña de Izopo, 13°55'N, 87°08'W (J. van Dort *in litt.* 2015). *Olancho* Catacamas, 14°51'N, 85°53'W (MCZ 158201); 'Cerro El Triunfo', c.14°56.5'N, 85°35'W (LSUMZ 31091). **NICARAGUA: Jinotenga** Reserva El Jaguar, 13°14'N, 86°03'W (ML 34101861, photo G. Duriaux); San Rafael del Norte, 13°13'N, 86°07'W (AMNH 103386); 10km N of Matagalpa, c.13°01'N, 85°55.5'W (UWBM 70131). *Matagalpa* 'Río Tuma', c.13°09'N, 84°49'W (Martínez-Sánchez & Will 2010, AMNH 103600); Matagalpa, 12°56'N, 85°55'W (Martínez-Sánchez & Will 2010, AMNH 423523). *Nueva Segovia* Quilalí, 13°34'N, 86°01.5'W (Martínez-Sánchez & Will 2010, AMNH 103385); Ocotal, 13°38'N, 86°29'W (Paynter 1957, AMNH 102849/50). *Atlántico Norte* Kum, 13°26'N, 84°56'W (UCLA 37025). **See Also** Blake (1953), del Toro (1958, 1971).

Grallaria guatimalensis princeps Sclater & Salvin, 1869, Proceedings of the Zoological Society of London, vol. 1869, p. 418, Calovébora [08°47'N, 81°13'W, sea level Ngäbe-Buglé] and Santa Fé [08°30.5'N, 81°04.5'W, 700m, Veraguas], Panama. Salvin (1867) first recognised the distinctiveness of *princeps* while examining an unsexed specimen with 'not quite adult plumage', from Santa Fé. Sclater & Salvin (1869) then described *princeps* after receiving a 'second and more adult [female] specimen' from 'Veragua.' These two skins, both collected by Enrique Arcé, were subsequently considered syntypes by Warren & Harrison (1971) after Sclater (1890) clarified the collecting locality of the second (NHMUK 1889.7.10.856 *fide* Warren & Harrison 1971) as Calovébora. In referencing the first specimen, Warren & Harrison (1971) make the confusing, and presumably incorrect, statement that the 'other syntype is also the type of *G. guatemalensis* Sclater and Salvin, (*q.v.*).' This Central American subspecies is confined to Costa Rica and western Panama. It is more richly coloured than nominate *guatimalensis*, with heavier black scaling on the upperparts. **Specimens & Records COSTA RICA: Guanacaste** Cerro Santa María, 10°49'N, 85°19'W (UMMZ 135002–004, YPM 56716/17); Tenorio, 10°36'N, 85°05'W (Carriker 1910, MCZ 123378). *Puntarenas* Monteverde, 10°18'N, 84°49'W (Buskirk 1972, MZUCR 1198/4207); Las Alturas, 08°57'N, 82°50'W (Lindell & Smith 2003); Estación Biológica Las Cruces, 08°47'N, 82°57.5'W (Daily *et al.* 2001). *Alajuela* Hotel Villa Blanca, 10°12'N, 84°29'W (MZUCR 4932); Carillos, 10°02'N, 84°16'W (Carriker 1910, AMNH 492167–169, CM P25487). *Heredia* El Plástico, 10°18'N, 84°02'W (M.L. Brady *in litt.* 2017); Volcán Brava, c.10°08'N, 84°06.5'W (Carriker 1910, USNM 198337/39); 5km N of San José de la Montaña, c.10°07.5'N, 84°06.5'W (Orians & Paulson 1969); Cerro La Hondura, 10°04'N, 84°01'W (LACM 16227, AMNH 390389). *San José* La Montura, 10°06.5'N, 83°58.5'W (WFVZ 141898); Cerros de Escazú (e.g. Cedral, Escazú), c.09°52.5'N, 84°09'W (Carriker 1910, AMNH 492171/72, MCZ 117060/61, ROM 46563/65); San Joaquín de Dota, 09°35'N, 83°59'W (CUMV 7199); Savegre Mountain Lodge,

09°33'N, 83°48.5'W (eBird: H. Venegas). *Cartago* Volcán Irazú, 09°59'N, 83°51'W (Carriker 1910, MNCR 20156, USNM 198336); Turrialba, 09°55'N, 83°41'W (Carriker 1910, USNM 198340); Juan Viñas, 09°53.5'N, 83°45'W (Carriker 1910, CM P11056); Pavones, 09°54'N, 83°38'W (UMMZ 132707); Cartago area, *c.*09°51'N, 83°55.5'W, including many localities within 15km of the city such as Navarro (YPM 56714/15, AMNH 390390), Paraíso de Cartago (MNCR 24635), La Estrella de Cartago (USNM 199802/03), Azahar de Cartago (AMNH 492170), Jardín Botánico Lankester (MNCR 24633); Tapantí, 09°46'N, 83°48'W (L. Sandoval *in litt.* 2016). **PANAMA**: *Bocas del Toro* Río Chánguena, 09°08'N, 82°30'W (USNM 476090). *Ngäbe-Buglé* Boquete Trail, *c.*08°49'N, 82°12.5'W (Peters 1931, MCZ 137276). *Chiriquí* Quebrada Alemán, 08°42.5'N, 82°14'W (Angehr *et al.* 2008); El Velo [= 'Finca Lerida', see Fairchild & Handley 1966], 08°49'N, 82°29'W (FMNH 207255–257); Volcán de Chiriquí (Volcán Barú), 08°49'N, 82°32'W (Salvin 1870, Bangs 1902, Blake 1958, MCZ 109003–005, ANSP 128905); Boquete, 08°46.5'N, 82°27'W (Bangs 1902, FMNH, AMNH, MCZ); Volcán, 08°46'N, 82°38'W (AMNH 492174); Cerro Hornito, 08°39'N, 82°11'W (Angehr *et al.* 2008); 4.3km by road S of Lago Fortuna Dam, 08°33'N, 82°18'W (LSUMZ 163611). *Veraguas* Chitra, 08°32'N, 80°55'W (AMNH 77397/400). *Coclé* Monumento Natural Cerro Gaital, 08°37.5'N, 80°07'W (Angehr *et al.* 2008). *Panamá* Cerro Azul, 09°12.5'N, 79°25'W (Angehr *et al.* 2008); Altos del María, 08°40'N, 80°03.5'W (ML 60903861, photo J. Cubilla).

Grallaria guatimalensis chocoensis Chapman 1917, Bulletin of the American Museum of Natural History, vol. 36, p. 394, 'Baudo, alt. 3,000 ft.', Chocó, Colombia. The holotype, an adult male (AMNH 123351), was collected on 13 July 1912, by Elizabeth L. Kerr. The type locality of 'Baudo' is somewhat vague, as other collectors have used this to refer to the modern-day coastal town of Pizarro (Paynter 1997). However, the stated elevation of 900m places its origin as somewhere on the Serranía de Baudó, likely somewhere near 06°00'N, 77°05'W. This range-restricted taxon is confined to the northern Chocó region, its range extending only as far north as the Serranía de Pirre and Serranía de Darién in extreme south-east Panama. It is considered similar to *princeps*, having similar, comparatively rich coloration, but *chocoensis* is darker still, with a more olive crown and wings, and somewhat rustier or darker lores (rather than whitish). Wetmore (1972) noted of the Panamanian specimens he examined 'Assignment of the three from Panama [to *chocoensis*] is tentative, based on color, as possibly they may represent a distinct larger race.' **SPECIMENS & RECORDS PANAMA**: *Darién* Cerro Malí, 08°07'N, 77°14'W (Wetmore 1972, USNM 432691/484543); Tacarcuna village 08°05'N, 77°17'W (Wetmore 1972, USNM 484544); Cana, 07°45'N, 77°42'W (Griscom 1929, 1935, Robbins *et al.* 1985, LSUMZ 108381, MCZ 140695). **COLOMBIA**: *Antioquia* Reserva La Bonga, 07°11'N, 76°22.5'W (eBird: E. Munera). *Risaralda* Río Claro bridge on Cerro Montezuma road, 05°15'N, 76°06'W (XC 308320/21, D. Calderón-F.).

Grallaria guatimalensis aripoensis Hellmayr & von Seilern, 1912, Bulletin of the British Ornithologists' Club, vol. 31, p. 13, Aripo Mts., Trinidad [= near Cerro Aripo, 10°43.5'N, 61°14.5'W]. The only subspecies not occurring on the mainland, *aripoensis*, is considered endemic to Trinidad's northern mountains on the Aripo Massif (Beebe 1952, Junge & Mees 1961, Snow 1985, ffrench 1991, 2012). Race

aripoensis shares the rich, deep coloration of *princeps*, but lacks a dusky throat patch, has a buff-coloured malar stripe and an unmarked chest. It also appears to have broader fringes to the feathers of the upperparts, making the scaling even more apparent than in other races (Ridgway 1911, ffrench 1973). In their original description, based on a 'large series' collected in 1912, Hellmayr & von Seilern (1912b), felt that it was closest to *regulus*, but smaller, with 'much brighter, deeper ferruginous' underparts. The following description of *aripoensis* is from Ridgway (1911) and ffrench (1973). The feathers of the crown and hindneck are slate-grey fringed black, the remaining upperparts olive-brown with feathers broadly edged in black. The uppertail-coverts and tail are russet-brown to chestnut; wings olive to olive-brown with the flight quills more russet-brown, paler on the primaries, which have the outer webs paler near the tips. The greater coverts are edged russet and bear tawny terminal spots. Lores dull whitish, with white on posterior halves of both eyelids. Ear-coverts dark olive with pale shaft-streaks and malar region whitish to tawny-white. Chin and upper throat olive-brown suffused tawny with buffy shaft-streaks. Lower throat tawny and bordered posteriorly by a narrow, semi-circular line of dusky. Rest of underparts bright tawny, becoming paler on abdomen. Bill brown and, unlike other races, leg coloration is given as brown. **SPECIMENS & RECORDS** *Tunapuna-Piarco* Mt. Tucuche, 10°44'N, 61°24.5'W (ffrench 1991); Asa Wright Nature Centre, 10°43'N, 61°18'W (ffrench 1985, 2012); Las Lapas Trace, 10°43.5'N, 61°18.5'W (ffrench 2012). *Sangre Grande* Heights of Oropuche, *c.*10°46'N, 61°03'W (Belcher & Smooker 1936); Mantura National Park, *c.*10°43'N, 61°05'W (ffrench 2012).

Grallaria guatimalensis carmelitae Todd 1915, Proceedings of the Biological Society of Washington, vol. 28, p. 81, Pueblo Viejo, Colombia [= El Pueblito, 10°59'N, 73°27'W, La Guajira, 610m; Paynter 1997]. The holotype (CM 44850) is an adult male collected 6 March 1914. At the request of M. A. Carriker Jr., the collector of the type specimen, the taxon was named for his wife. In trying to define the range of *carmelitae* it is obvious that previous accounts of its distribution are both confused and, in some ways, conflicted. I have re-defined its range here but, especially given the lack of well-defined characters separating it from *regulus* (see below), a detailed re-analysis is much-needed. The range of *carmelitae*, as proposed here, consists of two apparently isolated populations, in the Sierra Nevada de Santa Marta and the Sierra de Perijá. The latter population is included because of its apparent geographic isolation from other known populations of *regulus* and (marginally) closer proximity to the type locality of *carmelitae*. This population includes records from as far north as the area near Lajas in western Zulia and as far south as El Cauca in western Norte de Santander. Krabbe & Schulenberg (2003) gave the range of *carmelitae* as extending south to northern Boyacá', presumably based on the FMNH record from La Ceiba. This site, however, is actually on the east slope of the East Andes and is thus best considered within the range of *regulus* (as described by the same authors). Collections from El Cauca, erroneously placed in César by Paynter (1997), were excluded from the range of *carmelitae* by Krabbe & Schulenberg (2003), who included the 'west slope of the Eastern Andes in southern César' within the range of *regulus*. Race *carmelitae* was originally described as *G. varia carmelitae*, considered by Todd (1915) to be similar to nominate *G. varia*, but smaller, with the vent darker, more cinnamon-rufous. The currently accepted (and likely correct) affinity with Scaled Antpitta was later noted by Todd & Carriker (1922). It is said to differ from *regulus* in having darker, more brownish (less olivaceous) upperparts, and darker underparts (more brownish, less ochraceous), with more dark mottling (Todd & Carriker 1922). These authors felt that the throat and head-sides are also darker and more uniform than in *regulus*. Bare-part colours given by Carriker are: *Bill*, black, base of mandible pale pink; *Tarsi & Toes* 'bright leaden blue'. Some authors consider it doubtfully separable from *regulus* (Fjeldså & Krabbe 1990, Krabbe & Schulenberg 2003). Apparently, intermediates have been reported (Lehmann-V. 1957) and, comparing plumages and vocalisations, I see little reason to maintain *carmelitae* as separate from *regulus*. Pending a rigorous analysis, however, I have followed previous authors in recognising it. In light of the confusion over the range of *carmelitae*, should it indeed prove to be a valid taxon, a reasonable alternative hypothesis might be that the Sierra de Perijá population is of *regulus* (Phelps & Phelps 1953b), leaving the range of *carmelitae* as confined to the Santa Marta area of endemism. Alternatively, the Perijá population could represent an undescribed taxon. **SPECIMENS & RECORDS COLOMBIA**: *Magdalena* Vereda Bellavista, 11°06'N, 74°04.5'W (Durán *et al.* 2009); Cincinati, 11°06'N, 74°05'W (USNM 373672, improved accuracy from Paynter 1997); RNA El Dorado, 11°06'N, 74°04'W (XC 17256/57, F. Lambert); RNSC El Congo, 10°59'N, 74°04'W (Strewe & Navarro 2004); San Javier, 10°51'N, 74°02.5'W (Durán *et al.* 2009). *La Guajira* San Salvador Valley, *c*.11°08'N, 73°34'W (Strewe & Navarro 20003). *César* Pueblo Viejo, 10°32'N, 73°25'W (Durán *et al.* 2009); Hiroca ('Eroca'), 09°42'N, 73°06'W (USNM 373666–671, ICN 11820). *Norte de Santander* El Cauca, 08°10'N, 73°24'W (Carriker 1955, CM P54707). **VENEZUELA**: *Zulia* Lajas, 10°20'N, 72°35'W (J.E. Miranda T. *in litt.* 2017); Sierra de Perijá, 10°00'N, 72°56'W (COP 66104).

Grallaria guatimalensis roraimae Chubb 1921, Birds of British Guiana, vol. 2, p. 80, '5,000ft, Mount Roraima, British Guiana'. The holotype adult female (NHMUK 1889.7.10.862) was procured by Henry Whitley on 13 September 1883 (Warren & Harrison 1971). As noted by Phelps (1938), however, Whitley was never in Guyana, and all of his collections from there are, in fact, taken in Venezuela, on the south-west slope of Cerro Roraima. The correct type locality (*contra* Cory & Hellmayr 1924, Peters 1951), therefore, should be Cerro Roraima, Bolívar, Venezuela, 1,534m, *c*.05°10'N, 60°45'W (see Phelps 1938, Snyder 1966). This tepui endemic race occurs in northern Amazonas (Phelps & Phelps 1947) and north-west Bolívar (Phelps & Phelps 1947, Phelps & Phelps 1950) in southern Venezuela (Meyer de Schauensee & Phelps 1978), and in western Guyana (Braun *et al.* 2007, O'Shea *et al.* 2007). Phelps & Phelps (1948) reported a specimen of *roraimae* collected on the Serra de Curupira of northern Amazonas (not Roraima as reported by Sick 1997). This specimen (COP 34902), collected by Félix Cardona in 1946 at Poste de la Frontera #2 on the Brazilian side of the border, in the headwaters of the Ríos Siapa and Padauiri, was later recognised to be a misidentified *G. varia cinereiceps* (which see) and its identity corrected by Phelps in Meyer de Schauensee (1966). Apparently overlooking this correction, many subsequent authors have extended (or

inferred) the range of *roraimae* to include Brazil based on this record (Mayr & Phelps 1967, Pinto 1978, Sick 1984, 1993, Hilty & Brown 1986, Ridgely & Tudor 1994, Sick 1997, Hilty 2003, Krabbe & Schulenberg 2003, van Perlo 2009). Although known localities from adjacent Venezuela and Guyana have strongly suggested its presence in Brazil for many years (Mallet-Rodrigues & Pacheco 2003), its occurrence there has only recently been confirmed (M. Cohn-Haft *in litt.* 2016). Originally described as *G. regulus roraimae*, based on a single female, Chubb (1921) considered *roraimae* similar to *regulus*, but to have a greyer crown and nape, and a paler and more olive-brown back with less prominent scaling. The leading vane of the remiges are more 'cinnamon-rufous' than in *regulus*, the ear-coverts darker, the throat more streaked with white or ferruginous, and the foreneck ferruginous mixed with black (rather than uniform ochraceous-brown). Chubb (1921) considered the underparts to be overall paler and brighter ferruginous than in *regulus*, and he felt it was also somewhat larger. Chapman (1931) made a cursory comparison between *roraimae* and *aripoensis* and found few differences. Without a rigorous comparison of sufficient examples of the various taxa, my impression is that *roraimae* is most similar to *carmelitae* and *aripoensis*. **SPECIMENS & RECORDS VENEZUELA**: *Bolívar* Río Nichare near Cerro Tobaro, 06°30'N, 64°45'W (COP 62727–731); *c*.7.7km W of La Denta, 06°25'N, 61°28'W (ML 52914, S.L. Hilty); km123 on El Dorado–Santa Elaena Road, 06°05'N, 61°22'W (ML 34431, T.A. Parker); La Escalara, 05°57'N, 61°26'W (XC 214759/60, J. Klaiber); Cerro Guanay, 05°51'N, 66°18'W (Mayr & Phelps 1967); Ptaritepui, 05°46'N, 61°46'W (COP 24898/99); La Gran Sabana, 04°36'N, 61°21'W (XC 114371, A. Crease); Meseta del Cerro Jáua, 04°48'N, 64°26'W (COP 72419); Santa Elena de Uairén, 04°36'N, 61°07'W (eBird: L.R. Macaulay). *Amazonas* W slope of Cerro Paraque, 05°06'N, 67°27'W (Mayr & Phelps 1967, COP 33598); Cerro Sipapo, 04°57'N, 67°24'W (Hilty 2003). **GUYANA**: *Cuyuni-Mazaruni* Cerro Roraima, 05°16'N, 60°44'W (ML, KUNHM, USNM). *Potaro-Siparuni* Mt. Ayanganna, *c*.05°20.5'N, 59°57'W (Milensky *et al.* 2016). **BRAZIL**: *Roraima* Serra da Mocidade, 01°42'N, 61°47'W (M. Cohn-Haft *in litt.* 2016).

Grallaria guatimalensis regulus Sclater 1860, Proceedings of the Zoological Society of London, vol. 28, p. 66, 'Pallatanga, Ecuador' [02°01'S, 78°58'W, 1,500m, Chimborazo province]. Sclater described this taxon from a single unsexed specimen collected by Louis Fraser. Subspecies *regulus* is one of the more widely distributed races. In Venezuela it is found on both slopes of the Andes from southern Táchira, western Apure (Meyer de Schauensee & Phelps 1978) and central Mérida (Phelps & Phelps 1953a), north to central Trujilla and south-east Lara (Hilty 2003). Records of *G. guatimalensis* from Monagas and Sucre (Boesman1998, Hilty 1999) do, presumably, involve this species, but their racial affinities have not been examined (Hilty 2003). Reported occurrence of *regulus* in Bolivia (Hennessey *et al.* 2003a) is in error. Subspecies *regulus* is generally smaller than other races, usually has a buffier malar and facial crescent (only occasionally whitish), dusky throat and pale tawny (rarely white) stripes on a dark brown breast. The richly-saturated ochraceous underparts fade to tawny on the belly and vent. The records below from Durán *et al.* (2009) were georeferenced with the aid of supplemental information provided by S.M. Durán (*in litt.* 2017). **SPECIMENS & RECORDS VENEZUELA**:

Nueva Esparta Cerro El Copey, Isla Margarita, 11°01′N, 63°54′W (Hilty 2003). *Monagas* Cerro Negro, 10°14′N, 63°34′W (XC 223639–641, P. Boesman). *Lara* 37km S of Cabudare, 09°54′N, 69°16′W (COP 71729/30). *Mérida* La Azulita, 08°43′N, 71°27′W (CM A1796, ML 62435/36, P.A. Schwartz); El Valle, 08°41′N, 71°06′W (AMNH 492176/77, FMNH 50730/31, USNM 190376, COP 36578). *Táchira* Presa Las Cuevas road, 07°56′N, 71°47′W (XC 223636/38, P. Boesman); Campamento La Trampa, 07°55′N, 71°43′W (XC 223634, P. Boesman); km15 on San Cristóbal–La Florida road, *c.*07°51′N, 72°06′W (ML 62437, P.A. Schwartz); San Cristóbal, 07°46′N, 72°14′W (FMNH 289078); José del Carmen, Cerro El Tetéo, 07°25′N, 72°04′W (COP 60629–631, ML 62431–434, P.A. Schwartz). *Apure* El Nula, 07°16′N, 71°56′W (Hilty 2003). **COLOMBIA**: *Boyacá* La Ceiba, 07°00′N, 72°10′W (FMNH 261453); Santa María, 04°52′N, 73°15′W (J. Beckers *in litt.* 2017); La Almenara, Caserío San Rafael, 04°52′N, 73°17′W (Durán *et al.* 2009). *Arauca* Río Arauca, 07°00′N, 71°53′W (FMNH 261454); Palmar, 06°10′N, 72°01′W (Carriker 1955, CM P60650). *Cundinamarca* 5.5km W of San Miguel, 04°35′N, 73°25′W (ML 80868, M. Álvarez); Pipiral, 04°15′N, 73°50′W (Durán *et al.* 2009, AMNH 144808). *Antioquia* Vicinity of Medellín, 06°15′N, 75°35′W (Durán *et al.* 2009, CAMUUA 0530); Finca Zanzibar, 06°02.5′N, 75°41.5′W (Durán *et al.* 2009); Las Nubes, 05°49′N, 75°47.5′W (Durán *et al.* 2009); Támesis, 05°43.5′N, 75°42.5′W (Durán *et al.* 2009); Cerro Amarillo, 05°37′N, 75°49′W (D. Uribe *in litt.* 2017); Jardín, 05°37′N, 75°50.5′W (Durán *et al.* 2009). *Caldas* Anserma, 05°14.5′N, 75°46.5′W (Durán *et al.* 2009); Samaria, 05°14′N, 75°34.5′W (Álvarez-R. *et al.* 2002); Hotel Tinamú, 05°05′N, 75°41′W (XC 203126/27, J. Poveda); Los López, 05°05′N, 75°47.5′W (Durán *et al.* 2009); W of Manizales, *c.*05°04.5′N, 75°34′W (Durán *et al.* 2009); Morabia, 05°01.5′N, 75°46.5′W (Durán *et al.* 2009); Playa Rica, 05°01′N, 75°35.5′W (Durán *et al.* 2009); Palestina, *c.*05°01′N, 75°37′W (Lentijo & Botero 2013); Planalto, 04°57′N, 75°35.5′W (Durán *et al.* 2009). *Risaralda* Ceibal, Quinchía, 05°18′N, 75°42′W (D. Carantón *in litt.* 2017); Alto de Plumas, 05°02′N, 75°57′W (Durán *et al.* 2009); Alto del Rey, 04°57′N, 75° 56.5′W (Durán *et al.* 2009); La Nona, 04°54′N, 75°43.5′W (Durán *et al.* 2009); Jazmín, 04°44′N, 75°43′W (Durán *et al.* 2009); SFF Otún Quimbaya, 04°44′N, 75°35′W (Durán *et al.* 2009). *Quindío* Near Filandia, 04°40′N, 75°38′W (XC 130057/60, O.H. Marín-Gómez); Fachadas, 04°38.5′N, 75°42′W (Durán *et al.* 2009); Aguilas, 04°36.5′N, 75°37.5′W (Durán *et al.* 2009); Finca El Placer, 04°21.5′N, 75°43′W (O.H. Marín-Gómez *in litt.* 2017); Armenia, 04°33′N, 75°40′W (XC 129713/14, O.H. Marín-Gómez); campus of Universidad del Quindío, 04°32.5′N, 75°46′W (Marín-Gómez 2005). *Valle del Cauca* RFP Bosque de Yotoco, 03°52.5′N, 76°26′W (Durán *et al.* 2009); San Antonio and vicinity, *c.*03°30.5′N, 76°37′W (Miller 1963, Durán *et al.* 2009, MVZB 141786–789); Bichacue Yath Arte y Naturaleza, 03°27.5′N, 76°38′W (ML 52599271, photo T. Muñoz). *Cauca* La Sierra, 02°13′N, 76°43.5′W (Durán *et al.* 2009); Camino Pitalito, 01°25′N, 76°29.5′W (J. Beckers *in litt.* 2017); Vereda Diamante Alto, 01°13.5′N, 76°31.5′W (eBird: F. Ayerbe-Quiñones). *Huila* PNN Cueva de los Guácharos, *c.*01°36′N, 76°07′W (Durán *et al.* 2009). *Caquetá* Florencia, 01°37′N, 75°36.5′W (Iafrancesco *et al.* 1987, MLS 3974). *Putumayo* RFP Cuenca Alta Río Mocoa, Sector Cristales, 01°12.5′N, 76°42′W (D. Carantón *in litt.* 2017); Puerto Umbría, 00°52′N, 76°35′W (Meyer de Schauensee 1950, ANSP 160164); Orito, 00°40′N, 77°05′W (XC 333580, M. Viganò); Estación de Bombeo Guamués, 00°40′N, 77°00′W (FMNH 292938);

Puerto Leguízamo, *c.*00°11.5′S, 74°47′W (ICN 33287). **ECUADOR**: *Carchi* Chilma, 00°52′N, 78°04′W (XC 276032, J. Nilsson). *Esmeraldas* 2.7km E of Alto Tambo, 00°53.5′N, 78°32′W (XC 74758, D.F. Lane); El Placer, 00°52′N, 78°33′W (ML 48913, M.B. Robbins); RP Río Canandé, 00°31.5′N, 79°13′W (D. Lebbin *in litt.* 2016); Tesoro Esmeraldeño, 00°27.5′N, 79°09.5′W (J. Freile *in litt.* 2017); RBP Bilsa, 00°21.5′N, 79°42′W (R.C. Dobbs *in litt.* 2017). *Imbabura* Cordillera de Toisán, *c.*00°25′N, 78°47′W (J. Freile *in litt.* 2017); Hacienda La Florida, 00°22′N, 78°29′W (XC 248395, N. Krabbe); BP Los Cedros, 00°18′N, 78°46′W (HFG). *Pichincha* Río Chalpi, 00°12′N, 78°50′W (J. Freile *in litt.* 2017); RBP Maquipucuna, 00°05′N, 78°37′W (ML 58065/68/71: P. Coopmans); Reserva Geobotanica Pululahua, 00°04′N, 78°31′W (R. Ahlman *in litt.* 2017); Milpe, *c.*00°02′N, 78°52′W (MECN 2599); Reserva Ecoturística Alpa Huasi, 00°00′N, 78°45.5′W (J. Freile *in litt.* 2017); San Tadeo, 00°02′S, 78°44′W (MECN 2598); RP Tandayapa Bird Lodge, 00°01′N, 78°41′W (XC 6633, N. Athanas); Mindo area, *c.*00°03′S, 78°46.5′W (Kirwan & Marlow 1996, AMNH 180251, ANSP 180345, USNM 305186, MECN 2600/3721, XC, ML); El Monte Sustainable Lodge, 00°03′S, 78°46′W (R.A. Gelis photos). *Santo Domingo de los Tsáchilas* Tinalandia, 00°18.5′S, 79°03′W (van den Berg & Bosman 1984, ML 28497, A.B. van den Berg). *Chimborazo* Chunchi, 02°17′S, 78°55′W (ANSP 59358). *Los Ríos* Estación Científica Río Palenque, 00°35.5′S, 79°22′W (eBird: H. Brieschke). *Manabí* Río de Oro, 00°28′S, 79°36′W (Chapman 1926, AMNH 119906); Cerro San Sebastián, 01°35′S, 80°40′W (J. Freile *in litt.* 2017). *Santa Elena* Cerro La Torre, 01°44′S, 80°35′W (C.D. Becker *in litt.* 2017); RE Comunal Loma Alta, 01°51′S, 80°36′W (J. Freile *in litt.* 2017: probably = 'Cerro Manglar', Chapman 1926; see Paynter 1993). *Guayas* Naranjito, 02°13′S, 79°29′W (Chapman 1926, AMNH 124428). *El Oro* La Chonta, 03°35′S, 79°53′W (AMNH 171387); RP Buenaventura, *c.*9km NW of Piñas, 03°39′S, 79°45.5′W (Robbins & Ridgely 1990, Krabbe 1991, ANSP 177599, MECN 2601, XC 86560, D.F. Lane). *Loja* La Tagua, 03°55′S, 80°15′W (J. Freile *in litt.* 2017); Loja, *c.*03°58.5′S, 79°12.5′W (Ordóñez-Delgado *et al.* 2016, MCZ 299201); Alamor, 04°01′S, 80°01.5′W (Best *et al.* 1992); San Bartolo, 04°02′S, 79°55′W (AMNH 171388); Tierra Colorada, 04°02′S, 79°57′W (Best *et al.* 1992); Catacocha, 04°03′S, 79°40′W (Best *et al.* 1992); 4km W of Celica, 04°06′S 79°57′W (Dobbs *et al.* 2001, 2003); Celica–Alamor road, 04°09′S, 79°50′W (Bloch *et al.* 1991); near Sozoranga, 04°18′S, 79°48′W (XC 68415, F. Lambert); RP Utuana, 04°22′S, 79°45′W (Best *et al.* 1992, Greeney & Valencia-Herverth 2016). *Sucumbíos* Sinangoe, 00°11′N, 77°30′W (Schulenberg 2002); Ccuccono, 00°08′N, 77°33′19.90″W (Schulenberg 2002); Garzacocha, 00°30′S, 76°22′W (XC 258590, J.V. Moore, ML 86041–43, J.W. Wall); Limoncocha, 00°24′S, 76°37′W (van den Berg & Bosman 1984, ML 28584, A.B. van den Berg). *Orellana* San José Nuevo (= 'San José de Sumaco', see Paynter 1993; = 'San José de Mote' on Google Earth), 00°26′S, 77°20′W (AMNH 179375–377, ANSP 83336); Río Pucuno, 00°36′S, 77°35′W (ANSP 163659); Zancudococha, 00°36′S, 75°29′W (ANSP 183353, ML 78305/13/15, M.B. Robbins); Ávila, 00°38′S, 77°26′W (ANSP 163660); RBP Río Bigal, 00°38′S, 77°19′W (Freile *et al.* 2015); *c.*27km SSE of Pompeya, 00°38′S, 76°28′W (XC 249093, N. Krabbe); 10 de Agosto, 00°46′S, 77°32′W (R.A. Gelis *in litt.* 2017); Río Shiripuno, 01°00.5′S, 76°56.5′W (eBird: J. Illanes). *Napo* WildSumaco Lodge, 00°41′S, 77°36′W (R. Ahlman *in litt.* 2017); km35 on Narupa–Loreto road, 00°43′S,

77°38'W (XC 249396, N. Krabbe); Mushullacta, 00°48'S, 77°35'W (R.C. Dobbs *in litt.* 2017); Cabañas Hakuna Matata, 00°55.5'S, 77°49'W (R.A. Gelis *in litt.* 2017); 10km SE of Archidona, 00°58'S, 77°44'W (WFVZ 45741/42); RBP Jatun Sacha, 01°04'S, 77°37'W (J. Freile *in litt.* 2017). *Tungurahua* Río Blanco, 01°23'S, 78°18'W (MLZ 5199). *Pastaza* Shiripuno Research Center, 01°06.5'S, 76°44'W (J.F. Vaca B. *in litt.* 2017); Comunidad Ingaru, 01°16'S, 78°03'W (J. Freile *in litt.* 2017); *c.*13km SW of Santa Clara, 01°22'S, 77°57.5'W (R.A. Gelis *in litt.* 2017); Omaere, 01°28'S, 77°59'W (J. Freile *in litt.* 2017); Arutam, 01°47.5'S, 77°50'W (XC 216613, C. Mroczko). *Morona-Santiago c.*12km WSW of Amazonas, 01°39.5'S, 78°05'W and *c.*13.5km W of Palora, 01°41.5'S, 78°05.5'W (Guevara *et al.* 2010); Nuevo Israel, 02°10'S, 77°53'W and near Macuma, *c.*02°10'S, 77°39.5'W (J. Freile *in litt.* 2017); Colimba, 02°15'S, 78°13'W (AMNH 408360); near Macas, 02°19'S, 78°07'W (YPM 15030); *c.*5.5km SW of Taisha, 02°22'S, 77°30'W (ML 73725, M.B. Robbins); San Luís de Ininkis, 02°24'S, 78°05'W (J. Freile *in litt.* 2017); near Unnsuants, 02°33'S, 77°54'W (XC 249948, N. Krabbe); *c.*17km WNW of Teniente Ortíz, 03°01'S, 78°11'W (R.A. Gelis *in litt.* 2017). *Zamora-Chinchipe* 3km E of Paquisha, 03°56'S, 78°39'W (XC 250236, N. Krabbe); RP Estación Científica San Francisco, 03°58'S, 79°04'W (J. Freile *in litt.* 2017); Zamora, 04°04'S, 78°57'W (AMNH 129757/167277); Cabañas Copalinga, 04°06'S, 78°58'W (XC 120117/18, E. DeFonso); 13.5km SSW of Palanda, 04°46'S, 79°11'W (XC 341361, L. Ordóñez-Delgado); El Progresso, 04°48'S, 79°08'W (XC 345861, L. Ordóñez-Delgado); *c.*11km SE of Zumba, 04°55'S, 79°03'W (ML 68280, P. Coopmans); E of Chito Juntas, 04°57'S, 79°05'W (XC 250877, N. Krabbe); Cabañas Yankuam, 04°15'S, 78°40'W (XC 175036, R. Ahlman). **PERU: *Tumbes*** Campo Verde, 03°51'S, 80°12'W (Parker *et al.* 1995). *Piura* Bosque de Cuyas, 04°36'S, 79°44'W (Vellinga *et al.* 2004, Kikuchi 2009); Abra Porculla, 05°53'S, 79°33'W (F. Angullo P. *in litt.* 2017); km34 on Olmos–Bagua Chica road, 05°54.5'S, 79°31'W (Schulenberg 1987, LSUMZ 106082). *Cajamarca* Alto Ihuamaca, 05°11.5'S, 79°05'W (B.M. Winger *in litt.* 2017); Finca Agroecoturística Santa Fé, 05°45'S, 78°51.5'W (A. García-Bravo *in litt.* 2017); Llama, 06°31'S, 79°08'W (Koepcke 1961); Bosque Cachil, 07°24'S, 78°46.5'W (Schmitt *et al.* 2013). *Lambayeque* Corral Grande, 06°17'S, 79°27'W (Angullo *et al.* 2012). *Loreto* Río Curaray, 01°53'S, 75°12'W (XC 87609, J.D. Alván); Panguana, 02°08'S, 75°09'W (Stotz & Alván 2007); Quebrada Wee, 04°12.5'S, 77°32'W (Inzunza *et al.* 2012); Río Cushabatay, 07°05'S, 75°38.5'W (Schulenberg *et al.* 2001b); PN Cordilleira Azul, 07°16'S, 75°53'W (Schulenberg *et al.* 2001b, Stotz & Alván 2007); Río Pisqui, 08°28.5'S, 75°44'W (Schulenberg *et al.* 2001b). *Amazonas* Campamento Palomino, 05°36'S, 78°15'W (F. Angullo P. *in litt.* 2016); *c.*9km E of Lonya Grande, 06°06.5'S, 78°20.5'W (MSB 32081); Valle de Montealegre, 06°39'S, 77°27.5'W (B.M. Winger *in litt.* 2017). *San Martín* Afluente, 05°44'S, 77°31'W (Parker & Parker 1982); Quebrada Mishquiyacu, 06°05'S, 76°58.5'W (XC 47829, A. Spencer); *c.*3.7km ESE of Marona, 06°05'S, 76°53'W (XC 47643, N. Krabbe); 20km by road NE of Tarapoto, 06°26.5'S, 76°18'W (Davis 1986, LSUMZ 118251); between Aguacate and Caanan, 06°54.5'S, 77°29'W (F. Schmitt *in litt.* 2017); Río Verde, 07°19'S, 76°45'W (FMNH 474180); *c.*37km E of Capirona, 07°22'S, 76°17.5'W (XC 150846, H. van Oosten); S of Selva Andina, 07°25'S, 76°14'W (ML 37370031, O. Johnson); 2.7km S of Plataforma, 07°26'S, 76°16.5'W (MSB 36299). *Huánuco* S slope of Cordillera Divisoria, 09°12'S, 75°47.5'W (F. Angulo

P. *in litt.* 2017); PN Tingo Maria, 09°22'S, 76°00'W (Pratolongo *et al.* 2015); Chinchavito/Cayumba Grande, 09°30'S, 75°56.5'W (FMNH 299214/15); Jaupar, 09°35'S, 75°54'W (ANSP 176471). *Pasco* Pozuzo, 10°04'S, 75°33'W (AMNH 492185); Puellas, 10°40'S, 75°06'W (Schulenberg *et al.* 1984). *Ucayali* 3km by road NE of Abra Divisoria, 09°03'S, 75°43'W (LSUMZ 84951); Río Shinipo valley, 10°31'S, 74°07'W (Harvey *et al.* 2011, ML). *Junín* Yurinaqui Alto, Cerro Conchapen, *c.*10°46'S, 75°05'W (FMNH 296699/700); Río Chanchamayo, 10°55'S, 75°18'W (FMNH 123302); 'La Gloria', La Merced, 11°03.5'S, 75°20'W (Berlepsch & Sztolcman 1896); Puyu Sacha, 11°05.5'S, 75°25.5'W (J. Barrio *in litt.* 2017). **See Also** Taczanowski (1879), Berlepsch & Sztolcman (1896, 1906), Chapman (1926), Parker & O'Neill 1981, Best *et al.* (1992), Ayerbe-Quiñones *et al.* (2008), Arbeláez-Cortés *et al.* (2011), Calderón-Leytón *et al.* (2011).

Grallaria guatimalensis sororia Berlepsch & Sztolcman, 1901, Ornis, vol. 10, p. 194, 'Idma, Santa Ana', 4,600ft. Stephens & Traylor (1983) place Idma at 3,000m (12°53'S, 72°49'W) on the left bank of the Río Urubamba drainage south-west of [above] Quillabamba [*c.*1,050m]. Based on a stated elevation of 1,450m for the type, a more accurate approximation of the type locality would be: 1,450m, above Quillabamba, Cuzco, Peru, 12°54'S, 72°45.5'W. The adult male holotype (MIZ 34344) was collected on 19 November 1894 by Jan Kalinowski. Originally deposited in the Branicki museum in Varsovia, it is now in Warsaw (Mlíkovský 2009). Recognition of *sororia* has been the matter of some debate since its validity was questioned by Cory & Hellmayr (1924). Krabbe & Schulenberg (2003) considered *sororia* to be doubtfully separable from *regulus*, differing only in its whiter facial markings, greyer back and paler underparts. Indeed, Peters (1951) synonymised *sororia* with *regulus*, but this decision was followed in only a few subsequent works (Howard & Moore 1984, Clements 2000). The proverbial waters are further muddied by the uncertain location of several of Kalinowski's Peruvian collecting locations (Paynter & Traylor 1983, Vaurie 1972), combined with a dearth of specimens from southern Peru and an incomplete understanding of geographic variation within and between these taxa. Indeed, Berlepsch & Sztolcman (1906) considered Kalinowski specimens from 'Huaynapata', south of the type locality for *sororia* (*c.*13°13'S, 70°24'W; Stephens & Traylor 1983) to be closest to *regulus*, leading them to be first to hint at the questionable validity of *sororia*. Without doubt there is a fair degree of plumage variation (e.g. Chapman 1926) within the widely distributed *regulus*, as currently defined. Some specimens, as far north as south-east Colombia, are actually more similar to Bolivian birds than they are to most specimens I have seen from north-east Peru. Nevertheless, given potential vocal differences (see Voice), I have chosen to maintain *sororia* as distinct, provisionally suggesting that it replaces *regulus* south of the Apurímac Valley. Thus, as here defined, *sororia* is confined to southern Peru and Bolivia. **Specimens & Records PERU:** *Cuzco* Katarompanaki, 12°11.5'S, 72°28'W (Lane & Pequeño 2004); 6.0–8.5km E of Luisiana, *c.*12°38'S, 73°39'W (Weske 1972, AMNH 819915/820039); Llactahuaman, 12°52'S, 73°31'W (Pequeño *et al.* 2001b); Cock-of-the-Rock Lodge, 13°03.5'S, 71°32.5'W (XC 101065/68, J. King); above Pilcopata, *c.*13°08'S, 71°10'W (FMNH 311451); Aguas Calientes, 13°09.5'S, 72°31.5'W (R.C. Hoyer *in litt.* 2017); Cadena, 13°24'S, 70°43'W (XC 39746, D. Geale);

Paccaypata, 13°24.5'S, 73°09.W (P.A. Hosner *in litt.* 2017).
BOLIVIA: *La Paz* Serranía Tequeje, near Río Undumo, *c.*13°45'S, 68°21'W (ML 110434/447, A.B. Hennessey); Cerro Asunta Pata, 15°03'S, 68°29'W (LSUMZ 162693); Serranías Beu & Chepete, 15°06'S, 67°32'W (XC 3121, A.B. Hennessey); 47km by road N of Caranavi, 15°41'S 67°30'W (LSUMZ 96068, ML); Apa Apa Lodge, 16°21.5'S, 67°30'W (P.A. Hosner *in litt.* 2017); Piedras Blancas, 16°45.5'S, 66°47.5'W (ML 193173, G.F. Seeholzer); San Onófre, 17°09'S, 65°47'W (LSUMZ 171323). *El Beni* along Berna logging road, 15°05.5'S, 67°07'W (S.K. Herzog *in litt.* 2017). *Cochabamba* Pampa Grande, 16°41'S, 66°29'W (S.K. Herzog *in litt.* 2017); Río Mascota Mayu, 17°04.5'S, 65°46.5'W (Brumfield & Maillard 2007); Miguelito, 17°10.5'S, 65°46'W (XC 1629, S. Maijer); Río Colomelín, 17°23'S, 64°30.5'W (S.K. Herzog *in litt.* 2017). *Santa Cruz* Cajónes de Ichilo, 17°24.5'S, 64°13.5'W (R. Hoyer *in litt.* 2015); *c.*24km ENE of Samiapata, 18°08.5'S, 63°39'W (eBird: P. van Els); Estancia San Lorenzo, 18°39.5'S, 63°55.5'W (Maijer *et al.* 2000); Loma Larga, 18°47'S, 63°52.5'W (S.K. Herzog *in litt.* 2017). **See Also** Cardiff & Remsen (1981), Parker & Rowlett (1984), Sibley & Monroe (1990), Whitney *et al.* (1994), Remsen & Parker (1995), Herzog *et al.* (1999), Hennessey *et al.* (2003b), Macleod *et al.* (2005).

STATUS Due in part to its large distribution (1,160,000 km²) Scaled Antpitta is considered to be of Least Concern (BirdLife International 2017). Similarly, Freile *et al.* (2010) found this species to be under little threat in Ecuador, and it is fairly common in most of its South American range (Stotz *et al.* 1996, Ridgely & Tudor 2009). Scaled Antpitta is not, however, immune to the ever-growing threat of habitat alteration, and it is estimated that 50% or more of its population has been lost during the last 100 years in Mexico, where it is considered threatened (Berlanga *et al.* 2010, Greeney *et al.* 2013). It is also considered at risk in El Salvador (Komar 1998, 2002b), where it has a narrower distribution and does face an increasing threat from habitat modification. Additionally, due to its limited range, race *aripoensis* (Trinidad) is possibly more at risk than other populations (King 1981, IUCN 1986), being considered by Temple (2002) as among the island's most extinction-prone species and known from only on a handful of records since the early 1980s (ffrench 2012). There have been no specific studies examining the effects of human activity on Scaled Antpitta, but it is obvious that the species is tolerant of at least some habitat alteration, as it was found breeding abundantly in south-west Ecuador in relatively small, isolated patches of forest surrounded by cattle pasture (Dobbs *et al.* 2001, 2003). Across its range, a fair number of populations of most currently recognised subspecies occur within protected areas. **Protected Populations** *G. g. ochraceiventris* PN Cumbres de Ajusco (Salvin & Godman 1904); PN Laguna de Zempoala (MZFC 18580); PN Benito Juárez (WFVZ specimens near edge); PN Volcán Nevado de Colima (eBird: D. Wheeler); Reserva de la Biósfera Sierra de Manantlán (Dickerman 1990, LSUMZ 40759); RP Estación Científica Las Joyas (Santana-Castellón 2000); probably also in PN El Tepozteco and PN Nevado de Toluca. *G. g. guatimalensis* **Mexico** PN El Triunfo (Parker *et al.* 1976, de Silva *et al.* 1999, AMNH, USNM); PN Cañón del Sumidero (eBird: M. Hilchey); PN Palenque (Patten *et al.* 2011); PN Lagos de Montebello (Rangel-Salazar *et al.* 2009); MN Yaxchilán (Puebla-Olivares *et al.* 2002); Reserva de la Biosfera Los Tuxtlas (Coates-Estrada & Estrada 1989, USNM, CUMV, DMNH). **Guatemala** PN Laguna Lachuá

and PN Las Victorias (eBird: J. Cahill); Parque Municipal Corazón de Agua (ML 34483171/181, photos V. Cattelan); Reserva Natural Privada Orquigonia (ML: photos F.A. Ordoñez). **El Salvador** PN Los Volcanes (Thurber *et al.* 1987, UCLA); PN Montecristo (Komar 2002a, 2002b); PN El Imposible (Komar & Herrera 1995, Komar 2002b); PN La Tigra (J. van Dort *in litt.* 2015). **Honduras** PN Cusuco (IBC 1103508, S.E. Jones); PN Montaña de Santa Bárbara (LSUMZ 29517); PN La Tigra (ML 30115571, photo G. Flores-Walter); PN Azul Meámbar (ML 25984231, J. van Dort); PN Montaña de Celaque (eBird: C. Caballero); RVS Montaña Puca (MCZ 265258–261); RB Tawahka (LSUMZ 31091); RB Uyuca (eBird S12369939, photos J. van Dort & R. Juárez). **Nicaragua** RN Cerro Datanli-El Diablo (Fraser 2010, Fraser *et al.* 2011); RP El Jaguar (ML 34101861, photo G. Duriaux). *G. g. princeps* **Costa Rica** PN Rincón de la Vieja (UMMZ, YPM); PN Braulio Carrillo (Blake & Loiselle 2000, WFVZ); PN Volcán Arenal (eBird: K. Barton); PN Volcán Irazú (MZUCR, MNCR); PN Tapantí (Sánchez 2002, MNCR); PN Volcán Barú (AMNH, MCZ, ANSP); PN Juan Castro Blanco (eBird: J.D. Astorga); RVS Curi-Cancha (ML 46655251, photo D.A. Rodríguez-Arias); RB Alberto Manuel Brenes (MNCR 24634); RP Monteverde (MZUCR 1198/4207). **Panama** MN Cerro Gaital (Angehr *et al.* 2008); PN Chagres (Angehr *et al.* 2008); probably occurs in PN Santa Fé. *G. g. chocoensis* **Panama** PN Darién (Wetmore 1972, Robbins *et al.* 1985). **Colombia** PNN Ensenada de Utría (AMNH 123351, holotype); PNN Tatamá (XC 308320/21, D. Calderón-F.); Reserva La Bonga (eBird: E. Munera); probably also in PNN Los Katios. *G. g. carmelitae* **Colombia** PNN Sierra Nevada de Santa Marta (CM, holotype); RNA El Dorado (XC, A. Spencer, F. Lambert); RNSC El Congo (Strewe & Navarro 2004). **Venezuela** PN Perijá (COP). Hopefully also PN Catatumbo Bari. *G. g. aripoensis* This Trinidad endemic probably occurs in many forest reserves of northern range, including Blanchisseuse and Paria (ffrench 2012); Matura NP (ffrench 1977, Kenefick 2012); Asa Wright Nature Centre (ffrench 1985). *G. g. roraimae* PN Canaima (COP, XC); PN Jáua-Sarisariñama (COP 72419). *G. g. regulus* **Venezuela** PN Chorro El Indio (XC 223636/38, P. Boesman); PN Cueva del Guácharo (XC 223639–641, P. Boesman). **Colombia** SFF Otún Quimbaya (S.M. Durán *in litt.* 2017); PRN Barbas-Bremen (XC 130057/60, O.H. Marín-Gómez); DMI Agualinda (J.A. Zuleta-Marín *in litt.* 2017); RFP Cuenca Alta Río Mocoa (D. Carantón *in litt.* 2017); RFP Bosque de Yotoco (Durán *et al.* 2009); Ecoparque Los Alcázares-Arenillo (Botero *et al.* 2005); RNSC La Montaña del Ocaso (Marín-Gómez 2012); RNSC Arbol de la Cheta (O.H. Marín-Gómez *in litt.* 2017); RP Hotel Tinamú Birding Nature Reserve (XC 203126/27, J. Poveda). **Ecuador** PN Cayambe-Coca (Schulenberg 2002); PN Sangay (Guevara *et al.* 2010); PN Podocarpus (Rasmussen *et al.* 1994, XC); PN Sumaco-Galeras (AMNH, ANSP); PN Yasuní (Canaday & Rivadeneyra 2001, XC); PN Machalilla (J. Freile *in litt.* 2017); RE Comunal Loma Alta (J. Freile *in litt.* 2017); RE Limoncocha (van den Berg & Bosman 1984, ML); RP Estación Científica Río Palenque (eBird: H. Brieschke); Reserva Geobotánico Pululahua (R. Ahlman *in litt.* 2017); BP Los Cedros (HFG); RP Río Canandé (D.J. Lebbin *in litt.* 2017); RBP Bilsa (R.C. Dobbs *in litt.* 2017); RP Utuana (Best *et al.* 1992, Greeney & Valencia-Herverth 2016); RP Buenaventura (Robbins & Ridgely 1990, Best *et al.* 1992, MECN, ANSP, XC); RP Tandayapa Bird Lodge (XC 6633, N. Athanas); RBP Río Bigal (Freile *et al.* 2015); RBP Maquipucuna (MECN 6889/90); RP Santa Lucía (XC

128014, J. Fischer); RP El Monte Sustainable Lodge (R.A. Gelis photos); RP Séptimo Paraíso (XC 210783, M. Haribal); RP El Tundo (XC 50404/05, B. López-Lanús); RBP Jatun Sacha (J. Freile *in litt.* 2017; no recent records). **Peru** RN Tumbes (Parker *et al.* 1995, Walker 2002); PN Cordillera Azul (Schulenberg *et al.* 2001b, Stotz & Alván 2007); PN Yanachaga-Chemillén (J. Barrio *in litt.* 2017); PN Tingo María (Pratolongo *et al.* 2015); SN Tabaconas-Namballe (B.M. Winger *in litt.* 2017); SN Cordillera de Colán (F. Angullo P. *in litt.* 2017); RVS Laquipampa (Pratolongo *et al.* 2012). *G. g. sororia* **Peru** PN Manú (Walker *et al.* 2006); SH Machu Picchu (Walker & Fjeldså 2005); SN Megantoni (Lane & Pequeño 2004). **Bolivia** PN Madidi (Hennessey *et al.* 2003a); PN Amboro (Gemuseus & Sagot 1996); Reserva de la Biosféra Pilón Lajas (Hennessey *et al.* 2003b).

OTHER NAMES The following list of Latin synonomies is reasonably complete and is arranged by subspecies. *G. g. ochraceiventris Grallaria ochraceiventris* (Nelson 1898, Sharpe 1901); *Grallaria mexicana* (Sclater 1890, Salvin & Godman 1892, 1904). *G. g. guatimalensis Myiotrichas guatemalensis* (Heine & Reichenow 1890); *Grallaria mexicana* (Sclater 1861, 1862, 1877, 1890, Gray 1869, Sclater & Salvin 1873, Salvin & Godman 1892, 1904, Sharpe 1901); *Grallaria princeps* (Boucard 1878b); *Grallaria mexicanus* (Sumichrast 1869); *Grallaria guatimalensis mexicana* (Ridgway 1911, Brodkorb 1943, Schaldach *et al.* 1997), *Grallaria _____?* (Sumichrast 1869). *G. g. princeps Grallaria princeps* (Sclater & Salvin 1869, 1873, Sclater 1877, Salvin 1870, Zeledón 1885, Sclater 1890, Salvin & Godman 1892, 1904, Sharpe 1901, Bond 1902, Ogilvie-Grant 1912). *G. g. carmelitae Grallaria varia carmelitae* (Todd 1915); *Grallaria regulus carmelitae* (Todd & Carriker 1922). *G. g. roraimae Grallaria regulus* (Salvin 1885, Sclater 1890, Sharpe 1901, Penard & Penard 1910); *Grallaria regulus roraimae* (Chubb 1921). *G. g. regulus Grallaria fusca* (von Tschudi 1845, Taczanowski 1886); *Grallaria regulus* (Sclater 1862, 1877, Gray 1869, Sclater & Salvin 1873, Taczanowski 1879, 1884, Sclater 1890, Berlepsch & Sztolcman 1896, 1906, Sharpe 1901, Ogilvie-Grant 1912). *G. g. sororia Grallaria sororia* (Berlepsch & Sztolcman, 1901, Chapman 1921). **Misspellings** *Grallaria guatimalensis* (Sclater & Salvin 1859, 1873, Sclater 1859b,c, 1862, 1877, 1890, Salvin 1867, Lawrence 1868, Sumichrast 1869, Garrod 1877a,b, Reichenow 1884, Salvin & Godman 1892, 1904, Sharpe 1901, Carriker 1910, Ogilvie-Grant

1912, Stone 1932, del Toro 1958, Ramos 1985, Stiles 1992, Geffen & Yom-Tov 2000, Dietsch 2005, O'Shea *et al.* 2007, Ramírez-Albores 2010). **English** Guatemalan Antpitta (*guatimalensis*; Ridgway 1911, Cory & Hellmayr 1924, Stone 1932, Dickey & van Rossem 1938, Skutch 1940, Lowery & Dalquest 1951, Rand & Traylor 1954, Davis 1972); Mexican Antpitta (northern populations of *guatimalensis*, Ridgway 1911, Cory & Hellmayr 1924), Costa Rican Antpitta (*princeps*, Ridgway 1911, Cory & Hellmayr 1924), Roraima Antpitta (*roraimae*, Cory & Hellmayr 1924), Chocó Antpitta/Ant-pitta (*chocoensis*, Cory & Hellmayr 1924, Meyer de Schauensee 1950), Trinidad Antpitta (*aripoensis*, Cory & Hellmayr 1924, ffrench 2012), Nelson's Antpitta (*ochraceiventris*, Ridgway 1911, Cory & Hellmayr 1924); Trinidad Ant-pitta (*aripoensis*, Herklots 1961, ffrench 1973); Trinidad Ant-Thrush (*aripoensis*, Roberts 1934); Trinidad Scaled Antpitta (*aripoensis*; King 1981); Fulvous-breasted Ant-Thrush (*roraimae*, Chubb 1921); Fulvous-breasted Antpitta (*regulus*, Cory & Hellmayr 1924, Meyer de Schauensee 1950, Koepcke 1961); Santa Ana Antpitta (*sororia*, Cory & Hellmayr 1924); Carmelita's Antpitta/Ant-pitta (*carmelitae*, Cory & Hellmayr 1924, Meyer de Schauensee 1950). **Spanish** Tororoí cholino (Krabbe & Schulenberg 2003); Tororoí Escamoso (Clements & Shany 2001); Tororoí Escamado (Land 1970, Valarezo-Delgado 1984); Hormiguero Tororoi Escamado (Hilty 2003); Tororoí Guatemalteco de Santa Marta [*carmelitae*], Tororoí Guatemalteco del Roraima [*roraimae*] (Phelps & Phelps 1950); Gralaria Escamada (Ortiz-Crespo *et al.* 1990, Best *et al.* 1992, Rasmussen *et al.* 1994); torontorón (Cendrero 1972); Cholina (*guatimalensis*, del Toro 1964, Land 1970, Edwards 1968, 1972, 1989, Valarezo-Delgado 1984); Hormiguero Cholino (Sada *et al.* 1987); Hormiguero-cholino Escamoso (Eisermann & Avendaño 2007, Patten *et al.* 2011, Valencia-Herverth *et al.* 2012); Hormiguero Tororoí Guatemalteco (Meyer de Schauensee & Phelps 1978, Valarezo-Delgado 1984); Pájaro Vaquero (Land 1970, Valarezo-Delgado 1984); Tororoí dorsiescamado (McNish 2007); Tóbalo garganta blanca (*regulus*, Valarezo-Delgado 1984); Tororoí montañero (Valarezo-Delgado 1984). **Portugese** tovaçucu-oliva (*roraimae*, Ruschi 1979). **French** Grallarie de Guatemala (*guatimalensis*, Prévost & Des Murs 1846); Grallaire écaillée (Krabbe & Schulenberg 2003). **German** Kehlband-Ameisenpitta (Krabbe & Schulenberg 2003).

Scaled Antpitta, adult on nest (*regulus*), El Monte Lodge, Pichincha, Ecuador, 21 January 2010 (*Rudy A. Gelis*).

Scaled Antpitta, adult brooding nestlings (*princeps*), Paraiso de Cartago, Costa Rica, 11 August 2010 (*Luis Vargas Durán*).

Scaled Antpitta, nest and two 5–6-day-old nestlings (*princeps*), Paraíso de Cartago, Cartago, Costa Rica, 15 May 2010 (*Ernesto M. Carman Jr*).

Scaled Antpitta, nest and complete clutch (*princeps*), Paraíso de Cartago, Cartago, Costa Rica, 1 May 2010 (*Ernesto M. Carman Jr*).

Scaled Antpitta, adult gathering prey for its young (*princeps*), Paraíso de Cartago, Cartago, Costa Rica, 17 August 2012 (*Ela Villanueva*).

Scaled Antpitta, fledgling transitioning to juvenile plumage (*princeps*), PN Tapantí, Cartago, Costa Rica, 3 January 2007 (*Luis Sandoval*).

TACHIRA ANTPITTA
Grallaria chthonia   Plate 9

Grallaria chthonia Wetmore and Phelps, 1956, Proceedings of the Biological Society of Washington vol. 69, p. 6, Río Chiquito, 1,800–2,100m, south-western Táchira, Venezuela. The holotype (COP 61055) is a male collected 10 February 1955 by Ramón Urbano at Hacienda La Providencia (1,800m), Río Chiquito, Táchira, Venezuela [probably *c.*07°34'N, 72°18.5'W, not 07°38'N, 72°15'W as in Paynter 1982]. It is currently on deposit in Washington as USNM 461698 (Wetmore & Phelps 1956, Deignan 1961).

Arguably the most poorly known of all species treated in this book, Tachira Antpitta may also be one of the most appropriately named species of 'ground antbirds', as its Latin name means 'stilt-walker of the earth', derived from the Latin *grallarius* and the Greek *khthonios* (Jobling 2010). There are only four specimens, all taken at the same locality in southern Táchira, in westernmost Venezuela, in the mid-1950s (Hilty 2003). Rampant deforestation in the vicinity of the type locality has made Tachira Antpitta of key conservation priority (Collar *et*

al. 1997, Adams *et al.* 2003, BirdLife International 2008), and it is currently considered Critically Endangered (BirdLife International 2017). Indeed, this elusive bird was thought possibly extinct until very recently, when a targeted expedition to PN Tamá, near the type locality, found and photographed several individuals (J.E. Miranda T. *in litt.* 2017; Fig. 1). Full details of this discovery, including the first recorded vocalisations, are forthcoming (J.E. Miranda T. *et al.* in prep.). With its rediscovery at this historical locality, additional searches are needed of surrounding areas, including in adjacent Colombia. Rodríguez & Rojas-Suárez (1995) suggested that adults are most likely to vocalise in May–June in this region, and any information is crucial. There are currently no data on any aspect of its biology. Even the taxonomic status of Tachira Antpitta requires evaluation. Is it a valid species? Alternatively, perhaps it is best considered a subspecies of one of its putative relatives.

IDENTIFICATION 16–18cm. Sexes presumed to be similar. Brown forecrown grading to grey hindcrown and nape, all of these areas scaled black. Upperparts otherwise olive-brown, including the rump, tail and wings, all narrowly scaled black except the tail and wings.

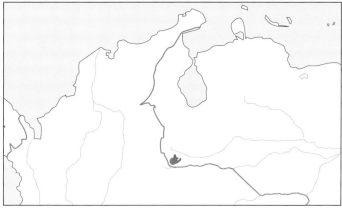

Uppertail-coverts cinnamon-brown. Chin and throat olive-brown with black-tipped feathers, while upper breast is buff-brown, becoming dull whitish with indistinct olivaceous-grey lines on lower breast. Flanks buffier and less strongly barred, and vent rich buff. Head further marked by a broad, pale buff malar stripe and central throat spot. Tachira Antpitta is fairly similar to Scaled Antpitta, but has a dark throat, a more whitish (rather than ochraceous) lower breast and belly, which bear fine black barring, and a less evident (or absent) crescent-shaped marking on the upper breast.

DISTRIBUTION To date, known only from the El Tamá Key Area (Wege & Long 1995) and the Meridian Montane centre of endemism (Cracraft 1985) in the south-west part of the state of Táchira in north-west Venezuela, where it is known only from the type locality on the Río Chiquito. Its overall geographic range is unknown, as recent records (J.E. Miranda *et al.* in prep) are the first since the type series was collected in the mid-1950s (Collar *et al.* 1992). Given the proximity of the type locality to the Colombian border, its range is considered possibly to extend into the Andes of north-east Colombia.

HABITAT No detailed information is available, but the species reportedly occupies 'the mossy undergrowth of high, dense cloud forest' (Meyer de Schauensee & Phelps 1978). All four specimens were collected between 1,800m and 2,100m. Hilty (2003) noted that the altitudinal range of Scaled Antpitta in Táchira is confined to 350–1,250m (Giner & Bosque 1998), suggesting that Tachira Antpitta may replace it at higher elevations.

VOICE There are no known descriptions or recordings of the voice, although this will soon be rectified in forthcoming publications (M. Braun *in litt.* 2017).

NATURAL HISTORY Unstudied in life, and all aspects of its behaviour are undescribed. It presumably behaves in a similar manner to other *Grallaria*, foraging on or very near the ground and remaining in dense understorey vegetation. Given its morphological and presumed taxonomic affinities, it probably shares many habits with Scaled, Moustached and Variegated Antpittas.

DIET No data.

REPRODUCTION There is no information on breeding biology; the nest and eggs are undescribed, and there is no information on seasonality (Greeney & Kirwan 2014).

Presumably the nest is like that of other *Grallaria*, i.e. a deep cup placed <5m above ground (Greeney *et al.* 2008a). Given its presumed taxonomic affinities and apparently montane distribution, the nest may be rather similar to those described for Scaled and Moustached Antpittas.

TECHNICAL DESCRIPTION With so few data available for this species, the following relies heavily on Wetmore & Phelps' (1956) description of the male holotype, supplemented by my own examination of the type and the photo of an *in situ* adult provided by J.E. Miranda T. (Fig. 1). **Adult** Sexes presumed to be similar. The forecrown is sepia-brown, becoming slightly more russet towards the base of the bill, with each feather edged narrowly with black around the exposed margin. The posterior crown and hindneck are slate-grey, the feathers also margined in black giving an indistinct scaled look to this area, except for a thin, indistinct superciliary line of unmargined grey feathers extending from the centre of the eye above the ear-coverts. This latter character varies in its visibility based on the angle. The upperparts are pale olive-brown, the feathers narrowly edged with black, similar to *G. guatimalensis*. The uppertail-coverts are warm, cinnamon-brown, the rectrices sepia, with exposed grey-brown fringes. The wings are similar in colour to the back, slightly darker overall, with the upperwing-coverts and outer webs of secondaries olive-brown. The primaries are darker, dusky brown, the leading edges margined in a paler, olive-brown. On the face, in front of the eye, the feathers are mixed buffy-brown and blackish, creating a somewhat speckled look, but overall producing a buffy loral area. The ocular area, especially the postocular region, is lightly feathered with small buffy feathers that create a thin eye-ring and indistinct pinkish-buff postocular spot of partially feathered pinkish skin. The ear-coverts and feathers below the eye are warm brown, with indistinct pale shaft-streaks. The sides of the neck are olive-brown. A pronounced buffy-white to white malar streak extends from the base of the mandible to below the posterior margin of the ear-coverts. Some of the feathers of the malar stripe are minutely tipped with black and the feathers in the rear part are thinly margined dusky-grey. The throat and upper foreneck are dark overall, the feathers olive-brown basally, grading to dull black at their tips. The slightly paler feather shafts produce scattered, thin streaks. There is a large pale central patch at the base of the throat formed by dull white feathers washed creamy-buff basally. Wetmore &

Phelps (1956) described these feathers as having buffier tips separated from the paler whitish bases by indistinct dusky lines, a character that I did not clearly observe in a meaningful number of feathers. The upper breast is buffy olivaceous-brown with mostly concealed whitish shaft-streaks fading on the lower breast and sides to dull whitish, narrowly barred with pale grey. The barring becomes less evident on the flanks, which are washed pale ochraceous-brown. The lower abdomen and vent are dull whitish washed pale olive-buff. The undertail-coverts are pale olive-brown and the underwing-coverts are ochraceous-buff, indistinctly margined olive-brown.

Adult Bare Parts *Iris* dark brown; *Exposed Ocular Skin* pinkish, especially at rear, but partially feathered; *Bill* maxilla black, perhaps slightly paler on the tomia, mandible dusky, buffy-pinkish near base; *Tarsi & Toes* vinaceous-brown. *Immature* Undescribed.

MORPHOMETRIC DATA *Wing* 97.8mm; *Tail* 38.7mm; *Bill* total culmen 26.8mm; *Tarsus* 43.7mm (*n* = 1♂, holotype, Wetmore & Phelps 1956). My own measurements taken from this specimen (AMNH 461698) are: *Bill* from nares 17.0mm, exposed culmen 23.1mm, depth at front of nares 8.4mm, width at gape 11.1mm, width at front of nares 5.9mm, width at base of lower mandible 9.9mm; *Tarsus* left 42.4mm, right 42.6mm. **Mass** No data, but probably similar to Scaled Antpitta. **Total Length** 15–19cm (Meyer de Schauensee & Phelps 1978, Hilty 2003, Krabbe & Schulenberg 2003, Restall *et al.* 2006, Ridgely & Tudor 2009).

TAXONOMY AND VARIATION Monotypic. Hilty (2003) suggested that Tachira Antpitta may be just a subspecies of the more widely distributed Scaled Antpitta, but Krabbe & Schulenberg (2003) considered it perhaps more closely related to Moustached Antpitta. Given some of the characters that apparently separate Tachira Antpitta from (most) races of both of these (e.g. dark throat, finely barred underparts), I propose that a sister relationship to Variegated Antpitta is also a viable hypothesis. Until further plumage and morphometric data can be gathered, along with molecular and vocal evidence, its true affinities will be uncertain. Nevertheless, it clearly belongs to the subgenus *Grallaria*, which also includes *G. varia*, *G. gigantea*, *G. excelsa*, *G. squamigera*, *G. guatimalensis*, *G. alleni* and *G. haplonota* (Lowery & O'Neill 1969). Within this group, its plumage suggests that Variegated Antpitta is the most likely sister species.

STATUS Searches of the type locality in 1990 and 1996 (Collar *et al.* 1992, Boesman 1998, BirdLife International 2017) failed to detect the species, making its very recent rediscovery even more exciting. Hope remains that the species may persist within other parts of the relatively large PN Tamá, which potentially includes a reasonable area of suitable habitat. There is, however, extensive deforestation along the Río Chiquito, especially below 1,600m (Collar *et al.* 1994, Sharpe & Lentino 2008). Irrespective of where else it may occur, the species undoubtedly has an extremely small population that is likely to be declining in the face of continued forest loss in the region. Initially considered to be of possibly conservation concern in 1988 (IUCN 1988, 1990), but designated 'insufficiently known', Tachira Antpitta was listed as Vulnerable by Collar & Andrews (1988) and Collar *et al.* (1992, 1994), but Endangered by Groombridge (1993), Stattersfield & Capper (2000) and Rodríguez *et al.* (2003), and finally Critically Endangered by BirdLife International (2008). It has remained at this highest threat level ever since (BirdLife International 2017). Apart from the ubiquitous assumption that the species will be detrimentally affected by habitat fragmentation, there are no specifically documented threats to Tachira Antpitta. Given that one of its putative relatives, Scaled Antpitta, can persist and even breed in areas of high disturbance (Greeney *et al.* 2008a, 2013), we can, for now, remain hopeful that additional populations of Tachira Antpitta remain to be discovered.

OTHER NAMES Spanish Tororoí de Táchira (Krabbe & Schulenberg 2003); Hormiguero Tororoí Tachirense (Meyer de Schauensee & Phelps 1978, Hilty 2003). **French** Grallaire du Tachira. **German** Táchiraameisenpitta (Krabbe & Schulenberg 2003).

Tachira Antpitta, first photograph of the species in life, PN Tamá, 19 June 2016 (*J.E. Miranda T.*).

PLAIN-BACKED ANTPITTA
Grallaria haplonota   **Plate 9**

Grallaria haplonota P. L. Sclater, 1877, Ibis, 4th ser., vol. 1, p. 442, 'vicinity of Caracas', Venezuela.

'These interesting birds were found in small numbers along a winding trail through dense, wet forest. [...] One morning as I descended this trail toward noon I had a glimpse of an alert, long legged bird that ran quickly across a little open space and then disappeared. The following day at the same place one ran down the sloping path and into the cover of leaves at the side. A moment later at a call it ran out suddenly with a quick, robinlike movement and in another moment was in my hand. On November 10 a little higher on the same trail, following a shot I heard a call that for a moment I thought came from some laborer giving me warning that he was in the forest. I wondered casually what had brought a workman into this remote woodland, when suddenly the note came more clearly and I recognised that it was a bird. It was a sound difficult to describe except to say that it was a low-pitched, rather hollow sounding whistle that I could imitate with sufficient accuracy to draw the bird up within a few feet. It answered me regularly, coming nearer and nearer, until suddenly I found the caller in one of these ant-pittas that appeared with a thrush-like flirt of its wings on a log a few feet away. It eyed me for an instant and then dropped out of sight, but presently it came up again and I secured it. [...] The breast muscles were moderate in size and light in color, indicating little use.' – **Alexander Wetmore, 1939(a), El Portachuelo, Venezuela**

Plain-backed Antpitta derives its English name from a direct translation of its scientific specific name, which is derived from the Greek words *haploos* (plain) and -*n tos* (backed) (Jobling 2010). It occurs, albeit somewhat discontinuously, from north-east Venezuela south to central Peru, inhabiting the understorey of wet montane forest at 700–1,950m, with an apparent predilection for steep slopes and creeks (Krabbe & Schulenberg 2003, Ridgely & Tudor 2009). The song of Plain-backed Antpitta is fairly easily imitated and is most similar to that of Thrush-like Antpitta, but contains more notes than the latter. The four subspecies of Plain-backed Antpitta recognised here differ only marginally in plumage. At the time of its description, Sclater (1877) considered Plain-backed Antpitta most closely related to Scale-backed Antpitta (*regulus*). When describing *G. h. parambae* from western Ecuador, however, Rothschild (1900) felt that this form was unique enough to have 'no near ally among the recorded species of the genus.' To date, its closest relative has not been determined. The known distribution of Plain-backed Antpitta is broken by large gaps within which it is only presumed to occur: between Venezuela and north-west Colombia; between north-west and south-west Ecuador; and between north-east and east-central Peru. Whether these gaps represent real breaks in the species' distribution, or simply gaps in our knowledge and/or sampling effort merits further study. Even among congeners with renowned abilities to remain hidden, Plain-backed Antpitta can be considered especially secretive (Ridgely & Tudor 2009), yet it is a fairly common voice where the species occurs. The nest was, for some time, thought to be strikingly anomalous

within the genus; a domed structure on the ground (Schäfer 2002, Krabbe & Schulenberg 2003). To date, only a single nest, confirmed to be of this species with video, has been described in the literature (Greeney *et al.* 2006), indicating that it builds a deep, open cup similar to its congeners. Published information on other basic aspects of its reproductive behaviour are also lacking, and its diet is all but undocumented. This account provides the first description of the juvenile based on specimen data for the nominate subspecies, as well as a previously overlooked description of the fledgling. A second nest description and previously unpublished breeding activities are also included.

IDENTIFICATION 16–17cm. Nominate Plain-backed Antpitta is described by Krabbe & Schulenberg (2003) as brown above with a slightly greyer crown, whitish lores, a dark rufescent tail, white throat with a dusky malar streak, and a pale buff moustachial. The remaining underparts are ochraceous, tending to be darker on the breast. Scaled Antpitta is the only species with which Plain-backed might be confused. Plain-backed Antpitta is brighter and more uniformly ochraceous below (at least compared with South American races of Scaled) and lacks a white chest crescent. Even the 'scaliest' race of Plain-backed (*chaplinae*) is much less scaled dorsally (Ridgely & Tudor 1994, Ridgely & Greenfield 2001).

DISTRIBUTION In northern Venezuela Plain-backed Antpitta is found along the interior mountain chain, in Lara, Yaracuy and southern Aragua, as well as the coastal mountains from Carabobo to Miranda, including the Paria Peninsula (*pariae*) (Phelps & Phelps 1950, Meyer de Schauensee & Phelps 1978). West of the Andes in Colombia and Ecuador, Plain-backed Antpitta has a somewhat disjunct distribution with four seemingly isolated populations. The northernmost population is in eastern Chocó and western Risaralda, the second in eastern Valle del Cauca (Stiles & Álvarez-López 1995), a third in Nariño, Carchi, Esmeraldas, Pichincha and Imbabura (Rothschild 1900, Kirwan & Marlow 1996,

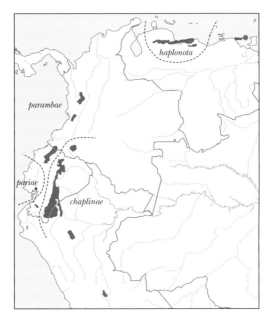

Strewe 2000a, Jahn & Valenzuela 2006, Calderón-Leytón *et al.* 2011) and, finally, the southernmost population in El Oro (Krabbe 1991, Robbins & Ridgely 1990, Freile 2000, Freile *et al.* 2010b). Its distribution east of the Andes is known to be fairly continuous, from western Putumayo to southern Ecuador and northern Peru in the Cordillera del Cóndor area (Robbins *et al.* 1987, Krabbe 1991, Krabbe & Sornoza 1994, Ridgely & Greenfield 2001, Schulenberg 2002, Freile *et al.* 2015). South of this, however, Plain-backed Antpitta is known from only two widely-separated localities, the southernmost being a population recently discovered in the Shaani Valley, Cerros del Sira (Ucayali) (Clements & Shany 2001, Schulenberg *et al.* 1997, 2010, Harvey *et al.* 2011).

HABITAT Plain-backed Antpitta inhabits the floor and lower understorey of humid and wet montane forest and forest borders, especially on steep slopes along forest streams and other more humid areas, often in dense vegetation (Wetmore 1939a, Schäfer & Phelps 1954). Like most *Grallaria*, it spends the majority of its time on the ground, but may climb 2–5m above ground to sing (Schäfer & Phelps 1954). The overall altitudinal range of *G. haplonota* extends from 700m to 1,950m (Krabbe & Schulenberg 2003). In coastal Venezuela (*haplonota*) reportedly occurs in subtropical forests at 900–1,600m, being replaced at higher elevations by *G. ruficapilla* (Schäfer 1969), presumably around the elevation where *Chusquea* bamboo begins to dominate the understorey. Subspecies *pariae* is found in similar habitat on the slopes of Cerro Azul and Cerro Humo (Paria Peninsula) at 900–1,200m (Phelps & Phelps 1949). The survey of Venezuelan races by Giner & Bosque (1998) found the species at elevations of 880–1,950m, being seemingly most abundant at 1,000–1,500m (Schäfer & Phelps 1954). Hilty & Brown (1986) gave its altitudinal range in Colombia as mostly 800–1,600m, while McMullan *et al.* (2010) reported the range as 700–1,600m. Apparently, in western Colombia, Plain-backed Antpitta (*parambae*) is most common above 1,300m, but is replaced at elevations above 1,700m by Yellow-breasted Antpitta, especially along ridges (Stiles & Alvarez-Lopez 1995). In Ecuador, *parambae* (west slope) is found at 700–1,300m (Ridgely & Greenfield 2001), apparently with a narrower altitudinal range in the more humid foothills of the south-west, where Best *et al.* (1992) recorded it at 900–1,000m. In NW Ecuador, there is probably a similar pattern of distributional displacement with Yellow-breasted Antpitta as noted for areas of sympatry in Colombia. It also overlaps in range with Scaled Antpitta in this region, where the two have been found sympatrically at several locations (e.g. BP Mashpi), but where there is some suggestion that they more frequently segregate by elevation, with Plain-backed below Scaled Antpitta (J. Freile *in litt.* 2017). I suggest that studies of habitat use and altitudinal range in antpittas might be especially rewarding in the forests around Mindo and Nanegalito where, potentially, the range of Plain-backed Antpitta overlaps both Yellow-breasted and Scaled Antpittas as well as Moustached Antpitta. Ridgely & Greenfield (2001) gave 1,100–1,700m as the range in eastern Ecuador, but it has been found nesting as low as 1,100m (Greeney *et al.* 2006) and is now known from locations at elevations down to 1,000m (Freile 2000, Guevara *et al.* 2010, Freile *et al.* 2015). In the southernmost portions of its range, Plain-backed Antpitta is rather poorly

known and its altitudinal range is given as 1,150–1,500m (Schulenberg *et al.* 2007) though it will likely be found to occur at similar elevations as east-slope populations in Ecuador.

VOICE The song is a series of mournful, hollow notes, described by Schulenberg *et al.* (2007) as *hu-hu-hu-hu-HOO-HOO-HOO-hu-hu*. Krabbe & Schulenberg (2003) described the song of Plain-backed Antpitta as lasting 4–5 seconds, delivered at 8–18 second intervals. The song consists of a series of 10–18 notes delivered at a regular pace of *c.*3 notes at a pitch of 0.6–0.7kHz, initially increasing in pitch and volume, but falling towards the end. In Venezuela, the song of *haplonota* (and *pariae?*) is apparently slightly higher pitched and does not fall in volume towards the end (Krabbe & Schulenberg 2003). Ridgely & Greenfield (2001) noted that the, *occasionally* sympatric, Thrush-like Antpitta's song is similar in quality but with fewer notes.

NATURAL HISTORY Ridgely & Greenfield (2001) described the behaviour of Plain-backed Antpitta as similar to Scaled Antpitta, but even more secretive, considering it 'the hardest antpitta to actually see.' Like other *Grallaria*, Plain-backed Antpitta rarely flies and, when it does so, stays low to the ground (Schäfer 1969). It spends most time on the ground, only occasionally climbing above the ground along angled vines and small branches, and adults are most frequently observed flitting across the forest floor in short dashes or long, bounding leaps (Schäfer 1969). Part of the difficulty in observing this species well derives from its apparently strong preference for areas (and periods) of low light. J. Freile (*in litt.* 2017), however, observed an individual in western Ecuador (*parambae*) that seemed unalarmed by his presence as it foraged along the ground in a fairly open hedgerow, albeit adjacent to darker forest. Plain-backed Antpittas sing most frequently during the crepuscular twilight (Schäfer 1969), sometimes sporadically throughout the day when skies are overcast. A pair that I observed in the foothills of eastern Ecuador (*chaplinae*) appeared to increase song rates in response to darkening skies, as small weather patterns moved through the area all day. Though not supported by specific observations, Schäfer (1969) stated that Plain-backed Antpitta roosts for the night 2–4m above ground in dense vegetation, either alone or in 'family groups'. If this proves true, it would be the only information on gregarious nocturnal roosting that I am aware of for *Grallaria* antpittas. In parts of its Venezuelan range, Plain-backed Antpitta is replaced at lower elevations by Chestnut-crowned Antpitta (Schäfer 1969) and Stiles & Alvarez-Lopez (1995) found it to be replaced at higher elevations in western Colombia by Yellow-breasted Antpitta.

DIET Schäfer (1969) described prey taken by Plain-backed Antpitta as (my translation from German) 'spiders, earthworms, myriapods, snails, and soft insects', stating that even large prey may be eaten whole. Of possible interest is Schäfer's (1969) statement that stomachs were found to contain up to 80% soil. Most likely this is due to incidental ingestion of dirt while gulping down whole prey items captured in or on the soil. It has not escaped my notice, however, that geophagy, either intentionally or incidentally, may facilitate the ingestion of toxic prey such as cyanide-defended millipedes consumed by Scaled Antpitta (Dobbs *et al.* 2003). The stomach of one specimen from Colombia (*parambae*) contained a katydid (Tettigoniidae) (Stiles & Alvarez-Lopez 1995).

REPRODUCTION There is only one confirmed nest of Plain-backed Antpitta described in the literature (Greeney *et al.* 2006), on which the following account is based unless otherwise noted. The nest account given by Krabbe & Schulenberg (2003), based on Schäfer's (2002) report of a Venezuelan nest being domed with a side entrance and placed on the ground, is almost certainly incorrect (Greeney *et al.* 2008). Greeney *et al.* (2006) found a nest of *G. h. chaplinae*, just prior to clutch initiation, on 11 April 2005. This nest was at 1,175m within the community-owned reserve of Mushullacta, contiguous with PN Sumaco-Galeras. **NEST & BUILDING** There are no data on nest construction or nest-site selection, but this is probably undertaken by both of the sexes. The nest described by Greeney *et al.* (2006) was in dense, tangled vegetation within a small drainage, surrounded by primary forest with a relatively open understorey. It was well concealed and placed 1.6m up in a 3m-wide tangle of fallen dead branches and living vines, roughly near the centre of the tangle but near the top, sheltered by several large leaves. Only a few narrow, crisscrossing vines supported the nest from below and it was not well affixed to the substrate. In form, the nest was a bulky open cup of dead, partially decaying, vegetable matter including sticks, leaves, brown moss and rootlets, sparsely lined with dark rootlets. Measurements were: internal diameter 10.5cm, internal depth 8.5cm, external diameter 23–25cm, external height 15–17cm. **EGG, LAYING & INCUBATION** Only three eggs of Plain-backed Antpitta have been described, all as turquoise blue or blue-green and unmarked (Schönwetter & Mees 1988, Greeney *et al.* 2008). The eggs from eastern Ecuador measured 31.5 × 25.0mm and 31.5 × 25.2mm, and had fresh masses of 10.8g and 11.0g, respectively. A single egg of *haplonota* from Venezuela measured 30.4 × 23.8mm, with a shell weight of 0.92g (Schönwetter & Meese 1988) and a predicted fresh weight of 9.3g (Kreuger 1968, Schönwetter 1979). Like other species of *Grallaria* for which the hour of laying has been documented (Greeney *et al.* 2008), the eggs of *G. haplonota* were laid in the afternoon. The first egg was laid after 16:30 and the second between 15:30 and 17:45, roughly 48 hours later. Both adults participated in incubation. During the laying period (on the day between eggs) adults visited the nest only twice between 10:15 and 17:30, covering the egg for just 5% of this time. On both occasions the attending adult arrived with a single dark rootlet that was added to the nest lining by dropping it into the cup. After settling over the egg, on one occasion an adult stood slightly, lifted its wings rearwards until they crossed slightly, and pressed its breast into the nest while rapidly vibrating its body. This was interpreted as a nest-shaping technique performed also by *Grallaricula nana* (Greeney & Sornoza 2005) and known for a variety of other passerines (e.g. Nolan 1978). The nest was not observed during early incubation, but ten days after the second egg was laid the nest was attended for 83.6% of daylight hours between 06:00 and 14:00. During this time, three complete on-bouts averaged 131 minutes and both adults participated, as confirmed by brief, unceremonious exchanges at the nest. On all three observed arrivals at the nest, adults brought a single rootlet and added it as

described above. While covering the eggs, adults generally sat motionless except for small, quick movements of the head, but occasionally preened or stood briefly to scratch themselves. Every 6–7 min they stood up suddenly and peered into the nest, occasionally appearing to roll the eggs with a gentle, backward movement of the bill. They also probed the nest's lining, either with sharp jabs of the bill or with a rapid, sewing-machine-like movement, behaviours thought to be important in nest sanitation and the removal of parasites (Haftorn 1994, Greeney 2004). Overall, *c.*4% of their time at the nest was spent in activities other than vigilance. In addition, while sat, the first adult to return to the nest in the morning sang the typical song of *chaplinae* (Ridgely & Greenfield 2001) almost once per minute during its 40 minute on-bout. Given that it is probably the male which takes the first turn on the nest after the female spends the night, the presumed female who replaced this individual sang just three times during its four-hour on-bout, once immediately after replacing its mate and twice immediately preceding its return. Though not remarked upon by Greeney *et al.* (2006), these observations suggest that singing on the nest may serve an important role in intra-pair communication, perhaps signalling safe conditions for change-overs. In the case of males, singing while on the nest may serve to advertise its level of parental care to the female, while also maintaining a vocal presence in their territory. **NESTLING & PARENTAL CARE** No data, but probably similar to other foothill *Grallaria* found in humid habitats. Given the presumed taxonomic affinities of *G. haplonota* (Lowery & O'Neill 1969), I suggest that nestlings will hatch with dusky-pink skin and black to dark-grey down. **NEST SUCCESS** No definite data on this aspect of reproduction for *G. haplonota*, but at the only nest described in the literature, Greeney *et al.* (2006) returned to a nest two weeks after clutch completion and found it covered in adult contour feathers. Although both eggs were intact, they were cold and obviously unattended, indicating abandonment. These authors suggested that the adult spending the night on the nest (presumably the female) was attacked and possibly killed, although at least one adult continued to sing nearby during subsequent days. **SEASONALITY** In northern Venezuela (*haplonota*) the main breeding period is said to last from April to August, beginning with the onset of the rainy season (Schäfer 1969), perhaps peaking in May and June (Schäfer & Phelps 1954), but there are no data confirming this. Presumably, seasonality in *pariae* would be similar. In western Colombia (*parambae*), Stiles & Alvarez-Lopez (1995) noted singing to be much more intense in April (compared to May and June) and suggested that reproduction occurs during the first months of the year. **BREEDING DATA** *G. h. haplonota* Fledgling, 8 October 1910, Cumbre de Valencia (Hellmayr & von Seilern 1912); juvenile, 4 November 1937, Rancho Grande (Wetmore 1939a; USNM 351951); juvenile, 14 October 1919, Cumbre de Valencia (CM P35149); subadult, 1 February 1910, Cumbre de Valencia (AMNH 492187). *G. h. parambae* Subadult, 24 July 1921, La Chonta (AMNH 171387). *G. h. chaplinae* Clutch initiation, 12 April 2005, Mushullacta (Greeney *et al.* 2006); adult gathering material, 17 February 2013, WildSumaco Lodge (G. Real photo).

TECHNICAL DESCRIPTION Sexes similar. *Adult* The following description is based on Sclater (1890) and refers to the nominate race. Upperparts nearly uniform olive-brown, with head slightly tinged greyish and showing traces of fine black fringes to the feathers. Lores are slightly paler, whitish or buffy-white, while the rest of the face is similar to the back and crown. The tail is pale brown on the dorsal side, more fulvous-brown below. The central throat is pale fulvous to whitish, bordered by an indistinct dusky submalar stripe and pale buffy malar streak. There is a broken blackish band between the lower throat and upper breast. The rest of the underparts are pale fulvous or ochraceous, tinged more olive-brown on the breast-sides and upper flanks. The underwing-coverts and vent are pale chestnut. **Adult Bare Parts** Described somewhat differently for different races, but it is unclear how much true variation there is and how much is subjective. *Iris* dark brown (agreed by most authors), pale brown (*parambae*, Rothschild 1900); *Bill* maxilla black, mandible brownish-horn (*haplonota*, Krabbe & Schulenberg 2003), maxilla brownish black, edged greyish-white, mandible horny-white, with tip dark brown (*haplonota*, Hellmayr & von Seilern 1912), mandible whitish (*parambae*, Rothschild 1900), mandible horn to dirty white (*chaplinae*, Robbins & Ridgely 1986), mandible black, greyish-purple at base (*pariae*, Phelps & Phelps 1949), mandible grey basally, flesh-coloured distally (*parambae*, Restall *et al.* 2006); *Tarsi & toes* greyish (most authors) to greyish-purple (*pariae*, Phelps & Phelps 1949). *Fledgling* Restall *et al.* (2006) illustrated and described what is here considered a fledgling. They described this plumage stage, which is depicted as being rather 'fluffy', as 'mostly dark brown, rufous on belly and undertail-coverts; streaked above, the short streaks inclining to spots on tips of feathers above; feathers of back fringed darker brown, lending a softly scaled appearance. Breast also spotted with slightly darker brown fringes to feathers.' This description is supplemented by a previous description of fledgling of nominate *haplonota* (Kreuger *in* Hellmayr & von Seilern 1912). The following description apparently refers to an individual described as still in fluffy immature plumage. The crown and nape are blackish with rusty-buff streaks and spots. The back is similarly coloured (perhaps less spotted?) with adult (olive-brown) feathers scattered across it. The throat and breast are dusky with buffy streaking. This coloration fades to dull reddish or orange-buff on the belly, duskier on the flanks and whiter near the centre. **Fledgling Bare Parts** *Iris* brown; *Bill* maxilla black, tangerine-red along the tomia and at the gape, mandible tangerine-red, grading to pinkish-orange basally and to clouded grey on terminal half; *Tarsi & Toes* light plumbeous. *Juvenile* The only previous reference to the juvenile is that of Wetmore (1939a) who described an immature as having 'a few streaks of cinnamon-buff scattered over the back and sides of the crown'. I examined a single juvenile of nominate *haplonota* (see Reproduction, Seasonality) for the following description, which applies to a juvenile in transition to subadult plumage. The upperparts are largely as described for adults, although the crown is the same olivaceous-brown (not slightly greyer) as the back, and the forecrown is slightly paler, more olivaceous and less brown. The feathers of the crown and nape are distinctly edged blackish, creating a scaled appearance, as are some feathers of the upper back. There are scattered remnants of immature plumage on the nape. Immature feathers are similar in colour to the crown but distinctly paler along the shaft, forming long streaks. These streaks are whitish near the base and broader and tawnier apically. Wings and

tail are rufescent-brown, similar to adults. The greater primary-coverts are browner than the flight feathers and thinly margined blackish, and the feathers of the alula are pale greyish-brown. The greater secondary-coverts and tertials are similar to the flight feathers but have elongated teardrop-shaped markings at the tips, which are blackish apically and contrastingly bright ochraceous proximally. Lores are dark grey and the remaining feathers of the face are largely olivaceous-grey. The ear-coverts are streaked pale ochraceous and ochraceous on the posterior margin. The throat is white, washed buffy at the sides, with an indistinct dusky malar streak (essentially a 'messy' version of the adult's pale buffy moustachial). The remaining underparts are mottled with large, irregular patches of ochraceous adult feathers and sooty-grey juvenile feathers streaked buffy and pale ochraceous, The remaining patches of juvenile feathers are largest across the lower breast and upper belly, with the sides mainly ochraceous. The streaking continues centrally to the lower belly, but the flanks are largely pale ochraceous, somewhat mottled pale grey. The vent is ochraceous-orange and the thighs similar but paler. **Juvenile Bare Parts** (from dried skin) *Iris* undescribed, probably similar to or same as adults; *Bill* maxilla dark brown, yellowish along tomia and at rictus, mandible mostly yellowish, darker near tip; *Tarsi & Toes* dark reddish-brown, nails yellowish.

MORPHOMETRIC DATA Males *G. h. parambae* Wing 100.2mm; *Tail* 33.3mm; *Bill* from nares 17.8mm, [total] culmen 29.2mm, exposed culmen 24.0mm (*n* = 1, Stiles & Alvarez-Lopez 1995). *Wing* 105mm; *Tail* 40mm; [total] culmen 28mm; *Tarsi* 43mm (*n* = 1; Chapman 1926). **G. h. pariae** *Wing* 98–101mm, mean 99.4mm; *Tail* 40–43mm, mean 41.8mm; *Bill* from base 28–30mm, mean 29.2mm (*n* = 5; Phelps & Phelps 1949). *Bill* from nares 17.8mm; *Tarsi* 45.0mm (*n* = 1, AMNH 388108, Sucre). **G. h. haplonota** *Bill* from nares 15.6mm; *Tarsi* 47.1mm (*n* = 1, subadult, AMNH 492187; see Reproduction). *Bill* from nares 16.2mm, 16.7mm; *Tarsi* 46.2mm, 46.4mm (*n* = 2, AMNH 492186, Carabobo; USNM 351950, Aragua). *Wing* 99–107mm, mean 102.2mm; *Tail* 39–42mm, mean 40.8mm; *Bill* from base 28–30mm, mean 29.2mm (*n* = 5, Phelps & Phelps 1949). *Wing* mean 100mm, longest 104–107mm; *Tail* mean 41mm, longest 43–44mm; *Bill* [exposed? culmen] 22–24mm (*n* = 7, adults and immatures, Hellmayr & von Seilern 1912). **G. h. chaplinae** *Wing* 97.2mm; *Tail* 35.8mm; *Bill* [exposed] culmen 25.3mm; *Tarsi* 41.2mm, (*n* = 1, holotype, Robbins & Ridgely 1986). **Females G. h. parambae** *Wing* 104.0mm; *Tail* 33.0mm; *Bill* from nares 17.3mm; [total] culmen 28.3mm, exposed culmen 22.1mm (*n* = 1, Stiles & Alvarez-Lopez 1995). *Wing* 100.0mm; *Tail* 38.0mm; *Bill* from nares 18.0mm, culmen 29.8mm, exposed culmen 25.0mm (*n* = 1, Stiles & Alvarez-Lopez 1995). *Wing* 107mm; *Tail* 45mm; *Bill* [total] culmen 30mm; *Tarsi* 48mm (*n* = 1; holotype [♀? see subspecific description], Rothschild 1900, Chapman 1926). *Wing* 100mm; *Tail* 33mm, culmen 25mm; *Tarsi* 45mm (*n* = 1, 'juvenile,' Chapman 1926). **G. h. haplonota** *Bill* from nares 14.4mm; *Tarsi* 46.4mm (*n* = 1, juvenile, USNM 351951; see Reproduction). *Bill* from nares 16.8mm; *Tarsi* 46.8mm (*n* = 1, AMNH 492188). *Wing* mean 99mm, maximum 104–106mm; *Tail* 40–43mm; *Bill* 22.5–24.0mm (*n* = 6, adults and immatures, Hellmayr & von Seilern 1912). *Wing* 99mm; *Tail* 33mm; *Bill* 18mm (*n* = 1, juvenile [fledgling?, note short tail and bill, see Breeding Seasonality], Hellmayr

& von Seilern 1912). *Wing* 103mm; *Tail* 41mm; *Bill* exposed culmen 22.0mm; *Tarsus* 45.0mm (*n* = 1, AMNH 805768). **G. h. pariae** *Wing* 98–103mm, mean 101.2mm; *Tail* 39–43mm, mean 41.4mm; *Bill* [total culmen] 28–30mm, mean 28.8mm (*n* = 5; Phelps & Phelps 1949). **Gender unspecified G. h. haplonota** *Wing* 108mm; *Tail* 45mm; *Bill* 24.5mm; *Tarsus* 46.6mm (*n* = 1, von Berlepsch & Leverkühn 1890). *Bill* from nares 16.4mm; *Tarsus* 45.4mm (*n* = 1, AMNH 492191, Lara). **G. h. chaplinae** *Wing* 97.2–101.3mm, mean 99.7mm; *Tail* 35.8–39.8mm, mean 38.0mm; *Bill* from base 25.0–25.3mm, mean 25.2mm; *Tarsus* 41.2–43.1mm, mean 42.1mm (*n* = 3, Robbins & Ridgely 1986). *Wing* 100mm; *Tail* 44.8mm; *Bill* from nares 16.1mm, exposed culmen 23.4mm, depth at nares 9mm, width at nares 6.8mm; *Tarsus* 44.9mm (*n* = 1, Napo, B. Mila *in litt.* 2015). **G. h. parambae** *Wing* 99.3–103.0mm, mean 101.0mm; *Tail* 33.2–39.5mm, mean 36.5mm; *Bill* [total culmen] 27.4–30.7mm, mean 28.3mm; *Tarsus* 41.7–46.44mm, mean 44.2mm (*n* = 6, Robbins & Ridgely 1986). *Wing* 93.2–104.5mm, mean 99.9mm; *Tail* 37.1–43.6mm, mean 40.6mm; *Bill* [total? culmen] 25.1–30.0mm, mean 27.1mm; *Tarsus* 37.1–46.2mm, mean 47.7mm (*n* = 13, Robbins & Ridgely 1986). **Mass Males** 83.5g (*n* = 1, *parambae*, Stiles & Alvarez-Lopez 1995), 85g (*n* = 1, *parambae*, Salaman 2001; this record cited by Dunning 2008, and possibly Hilty 2003). **Females** 89.2g (*haplonota*, AMNH 805768); 82g (*parambae*, Krabbe *in* Stiles & Alvarez-Lopez 1995). Krabbe & Schulenberg (2003) gave the mass of two males, probably the male and female individuals from Stiles & Alvarez-Lopez (1995). They also gave a range of 75–90g for 'unsexed birds in coastal Venezuela', presumably the nominate race. **Gender undetermined** 83g (*haplonota*, Faaborg 1975), 82.2g (*chaplinae*, Napo, B. Mila *in litt.* 2015). **Total Length** 16–18cm (von Berlepsch & Leverkühn 1890, Meyer de Schauensee & Phelps 1978, Dunning 1987, Ridgely & Tudor 1994, Clements & Shany 2001, Ridgely & Greenfield 2001, Hilty 2003, Schulenberg *et al.* 2007, Ridgely & Tudor 2009, McMullan *et al.* 2010).

TAXONOMY AND VARIATION Four subspecies are currently recognised (Clements 2007), although racial variation is not very pronounced and warrants further analysis. Indeed, Robbins & Ridgely (1986) in their description of race *chaplinae* noted the following. 'Both *parambae* and *pariae* do average deeper tawny underparts than either *chaplinae* or the nominate subspecies, though there are specimens of each race which can be matched with individuals of the others. For example, one *chaplinae* (ANSP 176860) is very close ventrally to a recent example of *parambae* (ANSP 177600). The amount of ochraceous on the throat is another variable character among the forms. Both *parambae* and *pariae* tend to be more ochraceous (nominate least so), with *chaplinae* intermediate. Again, as with the ventral colour, there is some overlap among the races.' The populations found (potentially) on the Amazonian slopes of the East Andes in Colombia and in northern Peru (San Martín) are of unknown race, but were tentatively placed with *chaplinae* by Krabbe & Schulenberg (2003). Plenge (2014) suggested the possibility that, at least those south of the Cordillera del Condor, may belong to an undescribed race. Stiles & Alvarez-Lopez (1995) suggested that *parambae* might warrant separation as a separate species.

Grallaria haplonota haplonota P. L. Sclater, 1877, Ibis, vol. 19, p. 442, 'vicinity of Caracas', Venezuela. The

nominate subspecies is found from northern Venezuela, in the interior, in the Andes of Lara, and in the coastal mountains other than the Paria Peninsula (Meyer de Schauensee & Phelps 1978). Warren & Harrison (1971) reported that NHMUK houses a syntype (NHMUK 1889.9.20.611) collected in 1873 by 'Spence' but (without giving a registration number) noted that there is another syntype at Tring that is apparently that on which Sclater's diagnosis was based, and as such could be regarded as the holotype. It appears, however, that the specimen number given above does, indeed, refer to the holotype. Sclater (1877), however, had previously stated 'my diagnosis of this apparently new species is from an example obtained in Venezuela by Mr. Spence.' He went on to state that 'Salvin and Godman's single specimen is likewise Venezuelan, having been procured in the wood-region of the coast near Puerto Cabello by Mr. Goering in 1873.' Along with the apparent confusion over the holotype, the precise range of the nominate race is unclear. The two localities in central Lara (see below) are both well west of the interior mountain chain and in rather arid regions. The coordinates provided below are both from Paynter (1982) and match fairly well the towns of Río Tocuyo and Bucarito, respectively. There are, however, no additional records from this area. It seems probable that either these specimens were mislabelled or, perhaps, as they are from nearly a century ago, drastic habitat alteration has lead to the species being extirpated in the region. Either way, in the accompanying map these two locations are shown outside the range of *haplonota* pending confirmation of its presence in central Lara. It is also unclear to what extent the nominate race occurs in the 'interior mountain chain' (Phelps & Phelps 1950). Restall *et al.* (2006) gave the bare-part coloration of nominate *haplonota* as: 'mandible flesh-coloured, legs and feet cool grey.' **Specimens & Records** *Lara* Tocuyo, 10°16'N, 69°56'W (AMNH 492190/91); Cerro Bucarito, 10°20'N, 69°41'W (AMNH 156294). *Yaracuy* Pico El Tigre, 10°25'N, 68°47'W (J.E. Miranda T. & K. López *in litt.* 2017); Cerro El Candelo, 10°22'N, 68°50'W (COP 64002); Sierra de Aroa, La Soledad, 10°14.5'N, 68°58.5'W and *c.*1.5km N of Rabo Frito, 10°13.5'N, 68°28.5'W (J.E. Miranda T. & K. López *in litt.* 2017). *Carabobo* Las Quiguas (= Paso Hondo, see Paynter 1982), 10°24'N, 68°00'W (CM P34708/727, AMNH 492189); Mario Briceno Iragorry, 10°21'N, 67°41'W (XC 143065/71; F. Deroussen); near Palmichal, 10°18'N, 68°13.5'W (XC 27907, D. Edwards); Cumbre de Valencia, 10°20'N, 68°00'W (AMNH, COP, CM); Hacienda Santa Clara, 10°20'N, 67°44'W (COP 24823/24). *Aragua* Rancho Grande, 10°22.5'N, 67°41'W (USNM, LSUMZ, COP, CM); Colonia Tovar–Limon Road, 10°25'N, 67°17'W (ML 62440–442; P.A. Schwartz; COP 1451–455); Maracay (probably mountains above), *c.*10°20'N, 67°36'W (ANSP 162703/04); Cerro Golfo Triste, 10°01'N, 65°55'W (Phelps & Phelps 1950; COP 19256/257). *Distrito Capital* Los Venados, 10°32'N, 66°54'W (A. Farnsworth *in litt.* 2017). *Miranda* Curupao, 10°30'N, 66°38'W (Phelps & Phelps 1950); 1.3km N of Las Marias, 10°26'N, 66°50'W (ML 62438/39, P.A. Schwartz); Urbanización Los Castores, 10°22'N, 66°58'W (J.E. Miranda T. *in litt.* 2017); PN Guatopo, *c.*10°05'N, 66°25'W (Hilty 2003).

Grallaria haplonota pariae Phelps Sr. & Phelps Jr., 1949, Proceedings of the Biological Society of Washington, vol. 62, p. 37, Cerro Azul, Cristóbal Colón, Paria Peninsula, Sucre, Venezuela [*c.*10°40'N, 61°56'W; Paynter 1982]. The holotype, an adult male collected 29 May 1948 by Ramón Urbano, was originally in the Phelps Collection (original number 44048) in Caracas, Venezuela, but is now held at AMNH. This very poorly known race is known only from the Venezuelan state of Sucre, essentially at just two locations, at opposite ends of the Paria Peninsula. There are apparently no known locations between these two points, *c.*60km apart. The following description of subspecies *pariae* is based on Phelps & Phelps (1949). Above, brown ('Dresden Brown'), slightly paler towards the rump and has the back feathers faintly edged dusky-brown, slightly more so on the nape and crown. The uppertail-coverts are reddish-chestnut. The lores are greyish, ear-coverts dusky-olivaceous, chin and centre of throat whitish bordered by dusky-olivaceous feathering and a narrow buffy-white malar stripe. Below, the breast, sides, flanks and thighs are ochraceous strongly washed with dusky. The belly is paler while the undertail-coverts are buffy orange ('Xanthine Orange'). Wings are brown, closely matching the back, with the leading edge of the first primary buffy. The underwing-coverts are similar to the undertail-coverts. The tail is brown, slightly paler below. Phelps & Phelps (1949) felt that *pariae* differed from *haplonota* in having the undertail-coverts more 'reddish' and by being darker overall below ('more ochraceous, less buffy'). Restall *et al.* (2006) described the bare parts as: *Bill* maxilla dark, mandible horn-coloured; *Tarsi & Toes* rich vinaceous. **Specimens & Records** Península de Paria, 10°41'N, 62°37'W (XC 223650, P. Boesman); Cerro Humo, 10°40'N, 62°30'W (Phelps & Phelps 1950, COP, XC); Marino, 10°37'N, 62°30'W (AMNH 388108).

Grallaria haplonota parambae Rothschild, 1900, Bulletin of the British Ornithologists' Club, vol. 11, p. 36, 'Hacienda Paramba,' Imbabura, Ecuador, 3,500ft [= Túmbez, 00°49'N, 78°21'W, 1,067m]. The holotype is an adult female (AMNH 492192) collected on 3 October 1898, by G. Flemming (LeCroy & Sloss 2000). This taxon was originally described as a species named for the type locality. Hartert (1922), however, made the suggestion that it was best considered a subspecies of *G. haplonota*, which was quickly followed by Cory & Hellmayr (1924) and subsequent authors. The type specimen (no date) was not assigned a gender by Rothschild (1900). Hartert (1901), however, referred to the type as a female. The holotype *parambae* was originally housed in the Tring Museum (Hartert 1922), but subsequently went to AMNH (Robbins & Ridgely 1986). Subspecies *parambae* is known from Risaralda and east-central Chocó in western Colombia as far south as El Oro in southern Ecuador (Robbins & Ridgely 1986), but has not yet been reported from adjacent north-west Peru. Interestingly, there is a large gap in its known range west of the Andes, creating four fairly isolated populations. Hartert (1901) described *parambae* as differing from nominate in having the upperparts dark olive-brown with a rufous tinge, as noted in Rothschild's (1900) description. Based on Rothschild (1900), subspecies *parambae* has blackish-brown remiges with rufous-brown outer webs and rusty margins to the inner webs. The tail is deep rufous and the uppertail-coverts are bright rust-coloured (but generally 'hidden by the long rump-feathers'; Hartert 1901). The feathers of the lores are pale at the base. The underparts, including the underwing-coverts, are rufous-ochraceous, with the breast-sides and flanks washed dark brown. The throat is bordered by a thin dark submalar stripe, with 'the feathers between this line and the

mandible whitish.' In Cory & Hellmayr (1924), *parambae* was said to differ from nominate *haplonota* by having the 'bill larger, crown rufescent rather than olive, underparts more tawny, with the throat hardly paler than the chest.' With the exception of stating that the underparts are 'deeper coloured, more rufous', these are nearly the same differences between *parambae* and nominate *haplonota* given by Hartert (1922). Robbins & Ridgely (1986), however, pointed out that these differences were subsequently shown to be largely incorrect once further specimens of the various subspecies were available. They noted that the bill of *parambae* is, in fact, similar to both *haplonota* and *pariae*, but is possibly larger than in *chaplinae*. Most notably, however, after re-examining the *parambae* holotype, along with several freshly collected individuals from south-west Ecuador, Robbins & Ridgely (1986) concluded that the rufescent crown ascribed to *parambae* (Hartert 1922, Cory & Hellmayr 1924, Restall *et al.* 2006) is in error, an assertion supported by the lack of this characteristic in Rothschild's (1900) original description. **Specimens & Records COLOMBIA**: *Chocó* RNA Las Tangaras, 05°48'N, 76°11'W (D. Uribe *in litt.* 2017). *Risaralda* Cerro Montezuma, 05°14.5'N, 76°06'W (XC 95939/41, D. Calderón-F.); Alto de Pisones, 05°26'N, 76°00'W (Stiles & Álvarez-López 1995); PNN Tatamá, 05°14'N, 76°05'W (P.C. Pulgarín-R. *in litt.* 2017). *Valle del Cauca* old Buenaventura road, 03°36.5'N, 76°52.5'W (J. Beckers *in litt.* 2017); Río Blanco, *c.*03°36'N, 76°50'W (Stiles & Álvarez-López 1995). *Nariño* RNA El Pangán, 01°21'N, 78°04'W (Strewe 2000a, Schulenberg *et al.* 2007a); RNSC Río Ñambí, Vereda el Barro, 01°18'N, 78°05'W (XC 241460; O.H. Marín-Gómez). **ECUADOR**: *Imbabura* BP Los Cedros, 00°18'N, 78°46'W (J. Freile *in litt.* 2017). *Esmeraldas* El Cristal Alto, Río Pachamama, 00°50'N, 78°32'W (XC 0262133, O. Jahn). *Pichincha* RP Mashpi Amagusa, *c.*00°10'N, 78°51'W (J.B.C. Harris *in litt.* 2017); near Mashpi, 00°09.5'N, 78°51'W (R. Ahlman *in litt.* 2017); near La Delicia, 00°08.5'N, 78°48'W (A. Spencer *in litt.* 2017); RBP Maquipucuna, *c.*00°07.5'N, 78°38.5'W (J. Freile *in litt.* 2017); Cabeceras del Guambuge, 00°06.5'N, 78°54'W (J. Freile *in litt.* 2017); near Milpe, 00°00'N, 78°57'W (MLZ 5197); Huila, 00°03'N, 78°53'W (AMNH 180252); *c.*11km SW of San Miguel de Los Bancos, 00°02'S, 78°59'W (SBMNH 8608). *El Oro* *c.*9km NW of Piñas, 03°39'S, 79°45'W (Robbins & Ridgely 1990, Krabbe 1991, ANSP, MECN); *c.*2.5km SW of Piñas, 03°42'S, 79°42'W (ML, M.B. Robbins); La Chonta (Satayán), 03°35'S, 79°53'W (AMNH 171387).

Grallaria haplonota chaplinae Robbins & Ridgely, 1986, Bulletin of the British Ornithologists' Club, vol. 106, p. 102, 'Yapitya', on trail from Logroño to Yaupi, west slope Cordillera de Cutucú, Morona-Santiago, Ecuador, 1,525m [02°44'S, 78°05'W]. The holotype is an adult male collected 26 June 1984 and is held at the Academy of Natural Sciences, Philadelphia (ANSP 176862). Topotypical specimens are at USNM. The specific epitaph honours Louise Chaplin Catherwood, wife of the ichthyologist Charles Clifford Gordon Chaplin. Race *chaplinae* is found on the Amazonian slope of the East Andes in Colombia, Ecuador and Peru. There are very few specimens from Peruvian populations usually assigned to this race (Krabbe & Schulenberg 2003). The underparts of *chaplinae* are very similar to those of nominate *haplonota* (Robbins & Ridgely 1986), while the back is somewhat more greenish or olive-brown and more distinctly scaled (Krabbe & Schulenberg 2003). The tail of *chaplinae* is brown with a rusty tinge and the throat is white, tinged ochraceous (Robbins & Ridgely 1986), with less white on the throat than nominate (Krabbe & Schulenberg 2003). Specifically, Robbins & Ridgely (1986) stated that *chaplinae* can be distinguished from other races by having the distal edge of the dorsal feathers reddish-brown instead of light brown (nominate *haplonota* and *pariae*; Phelps & Phelps 1949) or 'dark olive-brown with a rufous tinge' (*parambae*; Rothschild 1900), which affords this race a darker, more scaled appearance to the upperparts. As a distinguishing character for *chaplinae* Robbins & Ridgely (1986) also stated that the remiges are 'Sepia (119) with Raw Umber (123) to outer margins, [and are not the] Benzo Brown of *haplonota* and *pariae* (Phelps & Phelps 1949) or blackish-brown, with rufous-brown outer webs and rusty margins to the inner webs of *parambae* (Rothschild 1900).' SPECIMENS & RECORDS COLOMBIA: *Cauca* Serranía de la Concepción, 01°22'N, 76°24'W (J.P. López-Ordoñez *in litt.* 2017). *Putumayo* RFP Cuenca Alta Río Mocoa, Sector Cristales, 01°12.5'N, 76°42'W (D. Cantarón *in litt.* 2017). ECUADOR: **Sucumbíos** Sinangoe, 00°11'N, 77°30'W, Bermejo, 00°15'N, 77°23'W, and Ccuccono, 00°08'N, 77°33.5'W (Schulenberg 2002). *Napo* Cascada San Rafael, 00°06'S, 77°35'W (HFG); Río Malo, 00°09'S, 77°38.5'W (R.A. Gelis *in litt.* 2017); Cocodrilos, Cordillera Guacamayos, 00°39'S, 77°47'W (C.E. Gordon *in litt.* 2015); RP WildSumaco, 00°40.5'S, 77°36'W (many XC recordings, photos G. Real); Río Hollín, 00°41'S, 77°44'W (HFG); 3km NW of Guagua Sumaco, 00°42'S, 77°36'W (XC 249338, N. Krabbe); 1.8km ENE of La Y de Narupa, 00°43.5'S, 77°46'W (HFG); Río Guataracu, *c.*00°46'S, 77°15'W (loc. see Paynter 1993; MCZ 329744); *c.*9.2km S of Jondachi, 00°48'S, 77°48'W (ML 58846–847, L.R. Macaulay). *Orellana* Mushullacta, Cordillera Napo-Galeras, 00°50.5'S, 77°33.5'W (Greeney *et al.* 2006); RBP Río Bigal, 00°38'S, 77°19'W (Freile *et al.* 2015). *Morona-Santiago c.*14.5km WSW of Amazonas, 01°39.5'S 78°06.5'W (Guevara *et al.* 2010); *c.*13.5km W of Palora, 01°41.5'S, 78°05.5'W (Guevara *et al.* 2010); confluence of Ríos Huamboya and Sangay, 01°52.5'S, 78°08.5'W (MECN 7337/38); Nueva Alianza, 02°05'S, 78°09'W (J. Freile *in litt.* 2017); 2.1km W of Capadino Entza 02°07'S, 77°49'W (G. Real *in litt.* 2017); Kichikentza, 02°11'S, 77°51'W (J. Freile *in litt.* 2017); *c.*9.5km NW of Macas, 02°16'S, 78°11.5'W (R.A. Gelis *in litt.* 2017); San Luís de Ininkis, 02°24'S, 78°05'W (J. Freile *in litt.* 2017); Unnsuants, 02°33'S, 77°54'W (N. Krabbe *in litt.* 2016); *c.*9.3km ESE of Logroño, 02°39'S, 78°08'W (ANSP 176860–862, ML 36052: M.B. Robbins); Yapitya, 02°43.5'S, 78°04.5'W (USNM 560059); Limón-Santa Susana de Chiviaza road, 02°57'S, 78°25'W (XC 276035; J. Nilsson); El Pangui, 03°42.5'S, 78°38'W (XC 129136, L. Ordóñez-Delgado); Sector Colibrí, Cordillera del Cóndor, 03°45.5'S, 78°30.5'W (XC 127517/18, L. Ordóñez-Delgado). *Zamora-Chinchipe* Near Puerto Minero, 04°03'S, 78°36'W (XC 106745, L. Ordóñez-Delgado); Área de Conservación Parroquial Amuicha Entza, 04°05'S, 78°51.5'W (XC 161338, L. Ordóñez-Delgado); Copalinga Ecolodge, 04°05'S, 78°57.5'W (J. Freile *in litt.* 2017); Río Bombuscaro, 04°06.5'S, 78°58'W (XC 276033, J. Nilsson); 10km S of Zamora, *c.*04°09'S, 78°57'W (ML 90064, J. Sterling); Cabañas Yankuam, 04°15'S, 78°40'W (HFG); Quebrada Honda, 04°29'S, 79°08'W (ML 128092, R.S. Ridgely); near Shaime, 04°27'S, 78°40'W (F. Rowland *in litt.* 2017). PERU: *Loreto* Quebrada Wee, 04°12.5'S, 77°32'W (Inzunza *et al.* 2012); Alto Cahuapanas Camp, 05°40'S, 76°50.5'W

(Stotz *et al.* 2014); Alto Cachiyacu, 05°51.5'S, 76°43'W (Stotz *et al.* 2014). *Amazonas* Cordillera del Cóndor, *c.*11.3km E of Mayayacu (Ecuador), 04°00'S, 78°30'W (Krabbe 1991); Río Comainas, Cordillera del Cóndor, *c.*04°06'S, 78°23'W (Schulenberg *et al.* 1997). *San Martín c.*26.5km E of Moyobamba, 06°03'S, 76°44'W (ML 40217, T.S. Schulenberg). *Ucayali* Shaani Valley, Cerros del Sira, 10°42'S, 74°07'W (Harvey *et al.* 2011).

STATUS Plain-backed Antpitta is not considered globally threatened (BirdLife International 2016), being uncommon to locally fairly common in many parts of its range (Wetmore 1939b, Schäfer 1969, Stotz *et al.* 1996). In other parts of its range, however, its distribution is patchy and poorly understood. In western Ecuador, particularly the isolated population in the south-west, Plain-backed Antpitta is probably facing considerably more pressure due to high rates of habitat loss compared to *chaplinae* in eastern Ecuador and northern Peru (Freile *et al.* 2010b). PROTECTED POPULATIONS *G. h. pariae* PN Península de Paria (COP, XC): *G. h. haplonota* PN Henri Pittier (Schäfer & Phelps 1954); PN San Estéban (COP, AMNH, CM); PN Guatopo (Hilty 2003, Krabbe & Schulenberg 2003); PN Yurubí (J.E. Miranda T. *in litt.* 2017); PN El Ávila (A. Farnsworth *in litt.* 2017). *G. h. parambae* Colombia: RNA Las Tangaras (D. Uribe *in litt.* 2017); RNA El Pangán (Strewe 2000a, Schulenberg *et al.* 2007a); PNN Tatamá (XC 95939/41, D. Calderón-F.); RNSC Río Ñambí (XC: O.H. Marín-Gómez). **Ecuador**: BP Los Cedros (J. Freile *in litt.* 2017); RP Buenaventura (Best *et al.* 1992, Robbins & Ridgely 1986, XC); RBP Maquipucuna (J. Freile *in litt.* 2017); RP Mashpi Amagusa (J.B.C. Harris *in litt.* 2017). *G. h. chaplinae* Colombia: RFP Cuenca Alta Río Mocoa (D. Cantarón *in litt.* 2017). **Ecuador**: PN Sangay (Guevara *et al.* 2010); PN Sumaco-Galeras (Greeney *et al.* 2006); PN Podocarpus (XC 276033, J. Nilsson); RBP Río Bigal (Freile *et al.* 2015); RP WildSumaco (many XC recordings, photos G. Real); RP Narupa (HFG); BP Alto Nangaritza (F. Rowland *in litt.* 2017); RP Tapichalaca (ML: R.S. Ridgely); Área de Conservación Parroquial Amuicha Entza (XC 161338, L. Ordóñez-Delgado). **Peru**: ACR Cordillera Escalera (ML 40217, T.S. Schulenberg); ZR Santiago-Comaina (Inzunza *et al.* 2012). Few of the remaining distribution points for this species are clearly within Peruvian protected areas. Nevertheless, several reserves protect habitat at appropriate elevations within the undoubtedly exaggerated gaps in the known range of subspecies *chaplinae* in Peru, and protected populations of Plain-backed Antpitta seem likely to be found in PN Cordillera Azul, RC El Sira, and perhaps also in the BP San Matías-San Carlos and lower portions of PN Río Abiseo and BP Alto Mayo.

OTHER NAMES *Grallaria parambae* (Brabourne & Chubb 1912, Ogilvie-Grant 1912). **Spanish** Tororoí Torero (Krabbe & Schulenberg 2003, Schulenberg *et al.* 2007a); Gralaria Dorsillana (Best *et al.* 1992, Ridgely *et al.* 1998, Granizo 2009); pichón [*haplonota*] (Wetmore 1939b); Torero [*haplonota*] (Schäfer & Phelps 1954); Hormiguero Torero (Fernández-Badillo 1997, Hilty 2003); Torero de Paria [*pariae*] (Phelps & Phelps 1950); Torero del Ávila [*haplonota*] (Phelps & Phelps 1950). **English** Rothschild's Ant-Thrush [*parambae*], Grey-headed Ant-Thrush [*haplonota*], Ecuadorian Plain-backed Antpitta [*parambae*] (Cory & Hellmayr 1924); Sclater's Antpitta (Walters 1980). **French** Grallaire à dos uni. **German** Ockerbart-Ameisenpitta (Krabbe & Schulenberg 2003).

Plain-backed Antpitta, nest and complete clutch (*chaplinae*), Mushullacta, Napo, Ecuador, 15 April 2005 (*Harold F. Greeney*).

Plain-backed Antpitta, adult gathering nest material (*chaplinae*), WildSumaco Lodge, Napo, Ecuador, 17 February 2013 (*Galo Real*).

Plain-backed Antpitta, adult (*chaplinae*), WildSumaco Lodge, Napo, Ecuador, 13 July 2012 (*Roger Ahlman*).

Plain-backed Antpitta, adult (*chaplinae*), WildSumaco Lodge, Napo, Ecuador, 21 March 2014 (*Christian Hagenlocher*).

OCHRE-STRIPED ANTPITTA
Grallaria dignissima   Plate 9

Grallaria dignissima P. L. Sclater & Salvin, 1880, Proceedings of the Zoological Society of London, vol. 48, p. 160, pl. XVII, Sarayacu, Pastaza, Ecuador [on Río Bobonazo, *c*.400m, 01°44'S, 77°29'W; Paynter 1993]. Described on the basis of two unsexed syntypes (NHMUK 1889.7.10.836 and NHMUK 1889.9.20.603) collected by Clarence Buckley (Warren & Harrison 1971).

Ochre-striped Antpitta is one of just three truly lowland species in an otherwise montane genus. An infrequently seen denizen of the western Amazon's Napo EBA, occurring in Colombia, Ecuador and Peru (Parker *et al.* 1982, Altman & Swift 1986, Ridgely *et al.* 1998, Rodner *et al.* 2000), it prefers low-lying swampy areas and streambanks in *terra firme* forests. South of the Amazon, Ochre-striped Antpitta is replaced by the aptly named Elusive Antpitta. Ochre-striped Antpitta is particularly long-legged and

short-tailed, even for a *Grallaria*. Most frequently detected by its far-carrying, low, mournful two-note song, its plumage is characterised by a rusty-rufous throat and strong black and white streaking below. Ochre-striped Antpitta was, for a time, separated in the monotypic genus *Thamnocharis*. It is undoubtedly sister to allopatric Elusive Antpitta, although recent molecular analyses also suggest affinities to Chestnut-crowned and Watkins' Antpittas. Almost half of the described species of *Grallaria* still lack studies of their nesting behaviour, and Ochre-striped Antpitta is among these. When available, studies of its reproductive biology and general natural history will provide a fascinating comparison with those of its montane counterparts. More information is desirable, however, on all aspects of its taxonomy, behaviour and ecology. Of particular importance might be studies that evaluate the effects of forest fragmentation on Ochre-striped Antpitta, particularly on its dispersal abilities and territoriality. Below, I provide the first, albeit cursory, description of the species' nest and eggs.

IDENTIFICATION 18–19cm. Ochre-striped Antpitta has brown upperparts with a slightly greyer crown and pale buffy lores. The feathers of the lower back and rump have white shaft-streaks, but these are difficult to see in the field, as is its very short tail, which is usually hidden by the same feathers (Meyer de Schauensee 1970). Below, however, is 'where the party starts' and what makes this species strikingly and uniquely patterned. The throat and chest are warm, ochraceous-orange, with this 'bib' streaking downwards to the lower breast and upper belly (as if the colours were 'running'). Below this, the central underparts are contrastingly white, with the flanks boldly and broadly streaked black and white. The only species of *Grallaria* within its range, Ochre-striped Antpitta is unlikely to be confused with any species except the much drabber-plumaged Thrush-like Antpitta, with which it is sympatric over most or all of its range. The latter, however, leans towards the 'LBJ' side of the plumage spectrum, devoid of orange on the breast and lacking broad, contrasting streaks below. It is furthermore much smaller with a very different voice. The entirely allopatric Elusive Antpitta occurs only south of the Río Amazonas and, while similar, lacks the distinctive orange throat. Compare (if you must) with the only other *Grallaria* with somewhat similar amounts of orange on the throat, Santa Marta Antpitta, whose range does not even approach that of the present species.

DISTRIBUTION Ochre-striped Antpitta is found in the North Amazon or Napo centre of endemism (Cracraft 1985, Stattersfield *et al.* 1998) from south-east Colombia to north-east Peru (Ridgely & Tudor 2009). There are just two Colombian records, a historical specimen from San Miguel, in south-west Putumayo on the Ecuadorian border (00°20.5'N, 76°55'W; ROM 98961; Barlow & Dick 1969) and sight records in Amazonas (Pearman 1993), in the southern 'panhandle'. It is known from a fair number of localities across northern Amazonian Ecuador, and probably occurs in other areas of adjacent Colombia as well. South of the Colombian border, records are few but distributed evenly enough to suggest that its distribution is contiguous (Freile 2000, Ridgely & Greenfield 2001). The westernmost record I could locate is a recording made by the late P. Coopmans at Estación Biológica Jatun Sacha in 1990 (XC 264640, *c.*01°05'S, 77°37'W). I, and many others familiar with Ochre-striped Antpitta, have spent much time in this reserve during the past decade, and it seems quite likely that it is no longer found there. Similarly, historical records from *c.*35km north-north-east of there, across the Río Napo in Orellana (ANSP 169720), lack modern

confirmation of the species' continued presence, and satellite imagery suggests that habitat in this region is highly fragmented and no longer suitable. Indeed, this region of Ecuador has experienced widespread and drastic declines or disappearances of many forest-dependent species (J. Freile *in litt.* 2017). I have, however, found populations persisting in the region of Gareno Lodge (south of the Río Napo), only 35km east-north-east of Jatun Sacha (Greeney 2013b), making it the westernmost modern location in the north of its range. From there, records, modern and historical, extend north-east to the Puerto Nariño record mentioned above (all north of the Amazon) and south to the north bank of the Río Marañón at its confluence with the Río Morona, and along the Río Santiago in the Pongos Basin of northern Amazonas (Brooks *et al.* 2009), the latter being the south-westernmost record. Though records are lacking, its range presumably extends north-east from here, along the north (left) bank of the Ríos Marañón and Amazonas.

HABITAT Ochre-striped Antpitta inhabits the forest floor of tall, mature Amazon forests (Ridgely & Tudor 1994), apparently with an affinity for streams and low-lying areas (Pearman 1993, Freile 2000, Blake 2007), perhaps especially where there is tangled undergrowth (Willis 1985a, English 1998, Schulenberg *et al.* 2007). During a decade of surveys at two vegetatively similar plots of primary forest in eastern Ecuador, both of which experienced partial inundation during the wettest months, Blake & Loiselle (2015) found Ochre-striped Antpitta to be more frequent in the level-ground plot that tended to experience persistent standing water to a greater extent. My own experience with the species in eastern Ecuador (Greeney 2013b) and that of J. Freile (*in litt.* 2017) is similar, suggesting that flat, low-lying terrain is strongly favoured. I have never found Ochre-striped Antpittas on hilly or sloping terrain, with the exception of along small streams, an observation mirrored by the experience of J. Freile. Qualitatively, territories appear to be established either linearly along the margins of meandering streams or in broad, level-ground areas surrounded by hills. Preferred areas tend to be very moist or to have standing or flowing water, and usually include at least some areas of dense understorey or tangled vegetation (e.g., treefalls, vine tangles, or regenerating forest gaps). The altitudinal range is 100–450m (Hilty & Brown 1986, Clements & Shany 2001, Ridgely & Greenfield 2001).

VOICE The song is short, lasting only 1.1–1.2s and emitted at 1.3–1.5kHz. The low, mournful, two-note *whü, whaöw* or *whü, whüüw* song is delivered at intervals of 5–13s, with the second note slightly louder, distinctly longer and falling (Ridgely & Tudor 1994, Krabbe & Schulenberg 2003). The vocalisation generally thought of as a call begins with a whistle and descends into a rolling, musical trill or churr lasting about one second and consists of 25–30 notes at 0.9–1.4kHz (Krabbe & Schulenberg 2003): it was described by Schulenberg *et al.* (2007) as *heeuu'r'r'r'r.* Willis (1985a, 1988) described a vocalisation given in response to imitation of the song as a more nasal version of the latter, *hoot, hee-ont,* with an occasional faint buzzy *ruzz ruzz ruzz ruzz.* These latter two vocalisations may signal excitement or agitation.

NATURAL HISTORY There are few published details on the species' behaviour, but it is generally reported to be even more terrestrial than other members of the

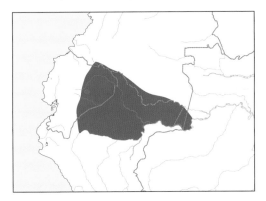

genus, running and hopping along the ground in short bursts, pausing occasionally to sing (Ridgely & Greenfield 2001, Krabbe & Schulenberg 2003). My own observations (Greeney 2013b) in eastern Ecuador confirm this. The only time I ever observed the species to leave the ground was to climb 0.5–1.0m above it in the safety of a dense branch tangle or small shrub. In all cases, this departure seemed to be for the purpose of inspecting me closer. There are no published data on territory defence, maintenance or fidelity. At Gareno Lodge (Greeney 2013b) in eastern Ecuador, a pair of Ochre-striped Antpittas was consistently found in the same area of low-lying swampy forest surrounded by small ridges over three years. I failed to detect them on several visits during the fourth year, but cannot be sure they were absent as I did not use playback. Based on their movements as I followed them on many occasions, I estimate their territory to be roughly 1.5–2.0ha. They appear not to join mixed-species flocks, or to associate with ant swarms (Willis 1985a) and generally are encountered alone or in pairs (Ridgely & Greenfield 2001). They apparently respond well to playback or a whistled imitation, but remain at a distance and are difficult to see (Willis 1985a, Pearman 1993).

DIET Absolutely nothing has been published on the species' diet and, although probably fond of invertebrates, its strong, hooked bill suggests it is quite capable of subduing small vertebrates.

REPRODUCTION There are no previously published data on any aspect of the reproductive biology of Ochre-striped Antpitta. J.B.L. Hualinga informed me that, in February 2003, at the Añangu community south of the Rio Napo in north-east Ecuador (Orellana), he found an active nest of *G. dignissima* with two 'very blue, very round' eggs. He described the nest to me as being *c.*1m above ground and supported below by the buttress-root of a large tree. It was apparently a 'large' open cup made of dead leaves and sticks. Sr. Hualingua is an extremely knowledgable local bird guide, and I have no doubt that this nest belonged to Ochre-striped Antpitta. Indeed, the details given are, in every way, expected. The nest, egg, nestling and breeding behaviour of *G. dignissima*, however, await formal description. **SEASONALITY** The above-mentioned nest was found with eggs during the early wet season. The only other information available is a subadult collected 14 March 1957, Río Morona (YPM 81714; S.S. Snow photos).

TECHNICAL DESCRIPTION Sexes similar (but see below). *Adult* Above, adults have an ochraceous-brown face, white or buffy lores, buffy-brown ear-coverts, and a grey-brown crown and nape. The greyish nape grades to brown on the back and rump, with the tail dark brown or blackish. The lower back and rump have elongated, lanceolate, white feathers fringed black. The wings are slightly warmer than the back, more rufescent-brown, with the underwing-coverts and inner margins of the wings rufous. The throat and breast are orange-ochraceous, fading to white streaked black on the lower breast and sides of the belly, with the central belly clean white. The streaking is especially pronounced on the elongated feathers of the flanks, which are mostly black with long white central streaks (Sclater 1890, Restall *et al.* 2006, Ridgely & Tudor 2009). Zimmer (1937) noted the following slight difference between male and female plumages. 'The males have the dark margins of the elongated femoral feathers quite blackish, the throat and

breast ochraceous brown, and the back about as olivaceous as the top of the head. The females have the femoral margins more brownish, the pectoral region deeper and more rufescent, and the back distinctly warmer than the crown.' **Adult Bare Parts** (Zimmer 1937, Ridgely & Greenfield 2001) *Iris* dark brown; *Bill* dark brown or grey-brown, all or basal half of mandible paler, dull yellowish or horn; *Tarsi & Toes* grey to bluish-grey. **Juvenile** The only immature I have examined was a subadult female (YPM 81714; photos S.S. Snow). It was almost entirely in fresh plumage indistinguishable from that of adults. The upperwing-coverts, however, were well worn and clearly immature. Largely similar in colour to the flight feathers, each covert was thinly fringed bright chestnut along both edges and at the tip. A faded, dusky blackish subterminal bar crossed each feather just proximal to a chestnut-buff spot (or thick bar). As the coverts were well worn, it seems probable that, when fresh, these patterned tips would have been much more visible, the pale tips highlighted on one side by the bright chestnut fringes and on the other by the black bar. The only other visible signs of immaturity were two feathers in the scapulars bordering the right wing. Each of these was similar in form to adult contour feathers, but bore a small white spot on the shaft near the tip. Though far from conclusive, this suggests that juveniles may be dark above with fine white spotting.

MORPHOMETRIC DATA Measurements for Ochre-striped Antpitta are scarce. The following (converted from inches) were provided for one of two unsexed individuals by Sclater & Salvin (1880). *Wing* 104mm; *Tail* 27.9mm; *Bill* '*rostri a rictu*' 33.0mm; *Tarsus* 55.9mm. In addition I measured the following specimens: *Bill* from nares 17.4mm, 17.5mm, exposed culmen 25.3mm, 23.4mm, depth at nares 10.6mm, 10.7mm, width at gape 15.7mm, 16.1mm, width at nares 8.5mm, 8.5mm, width at base of lower mandible 11.9mm, 13.1mm; *Tarsus* 56.2mm, 58.2mm ($n = 2$, ♂,♀, AMNH, Ecuador). *Bill* from nares 17.9mm, 18.7mm; *Tarsus* 56.1mm, 58.9mm ($n = 2$, ♀,♂, USNM 309220, Ecuador, AMNH 407151, Peru). **MASS** 110g ($n = 1$, ♂, Krabbe & Schulenberg 2003). **TOTAL LENGTH** 15.2–19.0cm (Sclater & Salvin 1880, Meyer de Schauensee 1970, Clements & Shany 2001, Krabbe & Schulenberg 2003, McMullan *et al.* 2010).

DISTRIBUTION DATA COLOMBIA: *Putumayo* San Miguel, 00°20.5'N, 76°54.5'W (Barlow & Dick 1969; ROM 98961). *Amazonas* Puerto Nariño, 03°46'S, 70°23'W (Pearman 1993). **ECUADOR**: *Sucumbíos* Tarapoa, 00°07.5'S, 76°20.5'W (Canaday & Rivadeneyra 2001); Güeppicillo, 00°10.5'S, 75°40.5'W (Stotz & Valenzuela 2008); Sani Isla, 00°26.5'S, 76°16.5'W (Hollamby 2012); vicinity of Redondococha, *c.*00°33.5'S, 75°15'W (ANSP 183352, ML: M.B. Robbins, P. Coopmans); mouth of Río Lagartococha, 00°39'S, 75°16'W (Zimmer 1937, AMNH 255987–89, USNM 309220). *Pastaza* Shiripuno Research Center, 01°06.6'S, 76°44'W (XC 26878, J.F. Vaca B.); Territorio Achuar, *c.*01°45'S, 76°30'W (J. Freile *in litt.* 2016); Río Conambo, 01°52'S, 76°47'W (LSUMZ 49044); Río Rutuno, 01°55'S, 77°14'W, (Orcés-V. 1974, MECN 2611/3113); Río Tigre, 02°03'S, 76°04.5'W (J. Freile *in litt.* 2017); Montalvo, 02°03'S, 77°00.5'W (Orcés-V. 1974, MECN 2610); Kapawi Lodge, 02°32.5'S, 76°51.5'W (XC 258613; J.V. Moore); Río Bobonaza, 02°35'S, 76°38'W (Berlioz 1932a). *Orellana* Añangu, 00°31.5'S, 76°265'W (XC 258612, J.V. Moore); Yuturi Lodge, 00°33'S, 76°02.5'W (XC 248840; N. Krabbe); Estación Científica Yasuní, 00°38'S, 76°30'W (XC 61336/37; A. Spencer); Estación de Biodiversidad

Tiputini, 00°38.5'S, 76°09'W (Dreyer 2002, Blake 2007); Sinchichicta, 00°41'S, 75°45.5'W (XC 191744; A. Naveda-Rodríguez); Cotapino, 00°48'S, 77°25'W (ANSP 169720). *Napo* Gareno Lodge, 01°02'S, 77°24'W (Greeney 2013b, XC); Estación Biológica Jatun Sacha, *c.*01°05'S, 77°37'W (XC 264640; P. Coopmans). *Morona-Santiago* 5km SW of Taisha, 02°22'S, 77°30'W (ML 73726, M.B. Robbins); Río Upano, *c.*02°31.5'S, 78°08.5'W (Berlioz 1937). **PERU**: *Amazonas* Mouth of Río Santiago, 04°26'S, 77°38.5'W (Brooks *et al.* 2009, AMNH 407151). *Loreto* Panguana, 02°08'S, 75°09'W (Stotz & Alván 2007); Quebrada Bufeo, 02°20'S, 71°36.5'W (Stotz *et al.* 2016); Río Curaráy–Río Napo confluence, 02°22'S, 74°05'W (Peters 1951, AMNH 255990); 'Campamento Choro', 02°36.5'S, 71°29'W and 'Campamento Cachimbo', 02°43'S, 70°31.5'W (Stotz & Alván 2011); 'Campamento Piedras', Río Algodoncillo, 02°47.5'S, 72°55'W (Stotz & Alván 2010); Andoas, 02°48.5'S, 76°27.5'W (Willis 1985a); 'Campamento Yaguas', 02°52'S, 71°25'W (Stotz & Alván 2011); 'Campamento Curupa', Río Yanayacu, 02°53'S, 73°01'W (Stotz & Alván 2010); Pongo Chinim, 03°07'S, 77°46.5'W (Inzunza *et al.* 2012); Quebrada Sucusari, 03°15'S, 72°55'W (ML: T.A. Parker, M. Isler); ACP Sabalillo, 03°22'S, 72°17.5'W (XC 20046, D. Edwards); Río Mazán, 03°28'S, 73°11'W (AMNH 407152/ANSP 155304); Río Morona, 04°45'S, 77°04'W (YPM 81714).

TAXONOMY & VARIATION Monotypic. Although originally described in the genus *Grallaria* (Sclater & Salvin 1880), Ochre-striped Antpitta was later separated into the monotypic genus *Thamnocharis* (Sclater 1890), a name that was subsequently reduced to subgeneric level by Lowery & O'Neill (1969). In describing the species, Sclater & Salvin (1880) remarked on its uniqueness, stating that the 'bright ferruginous red breast and strong white flammulations render this species quite unmistakable.' At the time they tentatively placed it near *G. ruficapilla*, following Sclater's (1877) previous arrangement of the genus. Ochre-striped Antpitta is undoubtedly sister to allopatric *G. eludens* (Lowery & O'Neill 1969, Krabbe & Schulenberg 2003, Rice 2005b), with which it shares many ecological and morphological similarities. More recent molecular analyses place Ochre-striped and Elusive Antpittas close to the sister-species pair of *G. ruficapilla* and *G. watkinsi* (Rice 2005a,b). Although not included in Rice's analyses, by extension, a close affinity to *G. bangsi* and *G. kaestneri* is also suggested.

STATUS Ochre-striped Antpitta has a relatively broad geographic range and, despite disheartening rates of anthropogenic habitat alteration within this region (Soares-Filho *et al.* 2006), it is currently afforded a global conservation status of Least Concern (BirdLife International 2017). In Ecuador, where most of its range lies, *G. dignissima* is also considered Least Concern (Freile *et al.* 2010). These authors noted, however, that most or all of its Ecuadorian range is under imminent threat by oil exploration (Vallejo 2003), so it is certainly a species to watch. There are no directly documented effects of human activities on the species, though it appears to be negatively impacted by the construction of roads (Canaday

& Rivadeneyra 2001). In addition, the species appears to be genuinely rare throughout its range and thus may be more severely impacted by habitat fragmentation than other small birds (Greeney 2013b). Ochre-striped Antpitta is afforded some protection in several, at least nominally protected, areas, including PNN Amacayacú (Pearman 1993) in Colombia, PN Yasuní (English 1998, Freile 2000, Canaday & Rivadeneyra 2001, Bass *et al.* 2010) and RPF Cuyabeno (Canaday & Rivadeneyra 2001) in Ecuador, and ACP Sabalillo (XC 20046; D. Edwards) in Peru. Recently, Blake & Loiselle (2015) observed a *non-significant* decline in a population of Ochre-striped Antpitta in eastern Ecuador, yet observed *significant* declines in several other sympatric terrestrial and understorey insectivores including *Formicarius colma*, *F. analis* and *Conopophaga peruviana*. The meaning of these results, however, is unclear, as their study area did not experience any anthropogenic alteration during this time.

OTHER NAMES In much of the old literature, *G. dignissima* is referred to as *Thamnocharis dignissima* (Sclater 1890, Dubois 1900, Brabourne & Chubb 1912, Chapman 1926, Zimmer 1937, Peters 1951, Meyer de Schauensee 1966). **English** Striped-sided Antpitta (Cory & Hellmayr 1924, Howard & Moore 1984); Striped Antpitta (Walters 1980, Lodge 1991); Striped-sided Ant-Thrush (Brabourne & Chubb 1912). **Spanish** Gralaria ocrelistada (Valarezo-Delgado 1984, Ridgely *et al.* 1998, Granizo 2009); Tororoí Ocre Listado (O'Neill 2003a); Tororoí del Ucayali (Krabbe & Schulenberg 2003); güicundo and lou-lou (Valarezo-Delgado 1984). **French** Grallaire secrete. **German** Fahlbrust-Ameisenpitta.

Ochre-striped Antpitta, adult in dark understorey, Yasuni National Park, Orellana, Ecuador, 21 August 2010 (*Nick Athanas/Tropical Birding*).

ELUSIVE ANTPITTA
Grallaria eludens   Plate 9

Grallaria eludens Lowery & O'Neill, 1969, Auk, vol. 86, p. 1, Balta at the point where the streams known to the local Cashinahua Indians as the Xumuya and Inuya enter the Río Curanja, 'Loreto,' Peru, *c.*300m (10°06'S, 71°14'W). The holotype is an adult male (LSUMZ 62312) collected 11 July 1967 and, along with ten paratypes, is at LSUMZ. Topotypical material can be found at several other institutions including USNM, AMNH, MPEG and ANSP. Following political reorganisation, the type locality is now in Dept. Ucayali (Cardiff & Remsen 1994).

The Elusive Antpitta is rare and local in *terra firme* and transitional forest south of the Amazon. So far as is known, it occurs only in south-east Peru and adjacent western Brazil (Parker *et al.* 1996, Clements 2007, Ridgely & Tudor 2009). Its plumage is characterised by a somewhat dull mantle, but with heavy black streaking on otherwise white underparts. Elusive Antpitta is thought to favour dense thickets near streams and ravines, but no quantified data are available. The song is very similar to that of the previous species, a hollow two-note *hoo-oooooo*, with the second note rising slightly and sounding slightly 'hoarse.' As its name suggests, Elusive Antpitta is about as well studied and as frequently seen as other infamous characters such as 'Sir Not-appearing-in-this-film' (Chapman *et al.* 1975). One of just three truly lowland *Grallaria*, Elusive Antpitta may have very different life history strategies to its montane congeners. Anyone with the opportunity to see the species would likely be richly rewarded by taking time to follow the bird, if possible.

IDENTIFICATION 18–19cm. Elusive Antpitta is strikingly and uniquely patterned. It is brown above except the lower rump, where the blackish feathers have white shaft-streaks. The lores, sides of the face and breast are buffy, and the throat is white. The belly is white, with bold, black striping, including the flanks. So far as is known, the distribution of Elusive Antpitta does not overlap with that of any other species of *Grallaria*, thus, based on size, silhouette and behaviour alone, it is unlikely confused. Elusive Antpitta does overlap with Thrush-like Antpitta, which is noticeably smaller and much duller overall. Allopatric Ochre-striped Antpitta is similarly sized and patterned, but is immediately distinguished by the orange-ochre wash on the throat and chest (Meyer de Schauensee 1970).

DISTRIBUTION Elusive Antpitta is endemic to the southern Amazon or Inambari centre of endemism (Cracraft 1985), being found only in the western Amazon Basin of eastern Peru and adjacent western Brazil. In northern Peru (Loreto) it is known only from east of the Río Ucayali (Krabbe & Schulenberg 2003), leaving an apparent gap, the Río Huallaga–Río Ucayali interfluvium, where neither Ochre-striped nor Elusive Antpittas has been found. Several unconfirmed reports from eBird, however, place *eludens* there, as expected, and I have tentatively included this region in the range of Elusive Antpitta. Overall, an understanding of the complete range of *G. eludens* remains, well, elusive. Its presence in Brazil has only fairly recently been confirmed, but it was actually first detected (by voice) south of Benjamin Constant in 1966 (Willis 1987, 1988). This was prior to its formal description, and it was assumed at the time

to have been *G. dignissima*. Now that its voice is known, additional records have expanded its known range into Acre (Guilherme 2012, 2016).

HABITAT Elusive Antpitta inhabits the floor of tall, mature, bottomland Amazon forests (Parker *et al.* 1982, Krabbe & Schulenberg 2003), especially in low-lying areas and thickets (Karr *et al.* 1990, Lebbin 2007). The habitat preferences probably are similar to those of Ochre-striped Antpitta (Greeney 2013d), although the only specifics to have been documented are of an individual singing repeatedly from streamside dense understorey bamboo (Whittaker & Oren 1999). Elusive Antpitta is assumed to be entirely confined to lowlands, with all known localities at 120–500m (Krabbe & Schulenberg 2003, Walker *et al.* 2006, Ridgely & Tudor 2009).

VOICE As described by Krabbe & Schulenberg (2003), the song lasts 1.1–1.2s and consists of two whistles, very similar to those of the closely related *G. dignissima*. The first note is longer, while the second begins with an abrupt rise in pitch: *hooHEEeoo* (Schulenberg *et al.* 2007). The song sometimes starts with a (presumably) less-audible, third note (van Perlo 2009). The low, buzzy, *churr* call, described as *ruzz ruzz ruzz ruzz* and apparently produced in alarm, is virtually identical to that of Ochre-striped Antpitta (Willis 1988).

NATURAL HISTORY To date, Elusive Antpitta has, unsurprisingly given its name, eluded ornithologists and natural historians, leaving me with little to report of its behaviour and ecology. Like previous authors, I am forced to describe its behaviour as 'similar to that of *G. dignissima*'. Our knowledge of the natural history of Ochre-striped Antpitta, however, is not much better (Krabbe & Schulenberg 2003, Restall *et al.* 2006, Ridgely & Tudor 2009). Although Willis (1985a) suggested that Ochre-striped Antpitta does not frequently join mixed-species flocks or follow ant swarms, so few data are available that I hesitate to rule such behaviours out for Elusive Antpitta. O'Neill (1974) was informed by local people that Elusive Antpitta roosts low to the ground and often continues to vocalise after arriving at its roost.

DIET Nothing has been reported, but if I were an invertebrate I would make myself elusive if crossing paths with this antpitta!

REPRODUCTION Nothing known. O'Neill (1974), however, suggested that breeding coincides with the onset of the rainy season, based on increased vocalising in February and March, as well as on the partially enlarged testes of birds collected in July.

TECHNICAL DESCRIPTION Sexes similar. The following is from the description of the holotype provided by Lowery & O'Neill (1969). **Adult** Crown and nape warm brown ('Metal Bronze'), slightly more rufescent on the mantle; feathers on the upper back have narrow white shaft-streaks bordered near their tips by black, and these shaft streaks become increasingly broad and more visible on the lower back and rump. Rectrices dark grey, slightly tipped white. Lores tawny rufous-brown, much paler than crown; ear-coverts tawny-olive; rictal bristles thin but prominent. Chin and upper throat white, with each feather tipped by a long hair, giving the throat a hairy appearance (on close inspection); shafts with prominent terminal setae; lower throat and upper breast creamy-buff, narrowly edged with black markings on some feathers at the sides; belly white; flank feathers greatly elongated with dark shafts bordered by white and each feather edged blackish-brown or dark grey. Wings similar in colour to mantle, but more rufescent, especially on outer webs of inner primaries and outer secondaries; underwing-coverts brownish to buffy-rufous with some feathers edged black. **Adult Bare Parts** *Iris* brown; *Bill* maxilla dusky, mandible pinkish-grey; *Tarsi & Toes* blue-grey. *Juvenile* Undescribed.

MORPHOMETRIC DATA Lowery & O'Neill (1969) provided the following (*n* = 10, ♂♂): *Wing* mean 104.4mm, range 100.3–111.5mm; *Tail* mean 36.9mm, range 35.0–39.9mm; *Bill* exposed culmen, mean 25.7mm, range 24.0–27.3mm, width at base of culmen, mean 12.5mm, range 11.0–13.1mm, depth at base of exposed culmen, mean 11.5mm, range 11.0–11.9mm; length from base, mean 30.3mm, range 28.9–31.2mm; *Tarsus* mean 58.0mm, range 52.3–58.9mm, length of middle toe without claw, mean 31.5mm, range 30.0–34.0mm. In addition I measured the following: *Bill* from nares 18.2mm, 18.8mm; *Tarsus* 56.8mm, 60.8mm (*n* = 2, ♂♂, AMNH 789782, USNM 533618). **Mass** 115g (*n* = 1, ♂, Lowery & O'Neill 1969); 111–115g (*n* = 3, ♂♂, presumably including the above, O'Neill 1974). **Total Length** 17–19cm (Dunning 1987, Ridgely & Tudor 1994, Clements & Shany 2001, Krabbe & Schulenberg 2003).

DISTRIBUTION DATA PERU: *Loreto* Pebas, right bank of the Río Napo, *c.*03°22'S, 71°49'W (A. Farnsworth *in litt.* 2016); Quebrada Buenavista, 04°50'S, 72°23.5'W (Lane *et al.* 2003). ***Ucayali*** Balta (type locality) and vicinity (Lowery & O'Neil 1969, O'Neill 1974, O'Neill 2003a, 2003b, LSUMZ, AMNH, USNM, ANSP, MPEG); SE slope of Cerro Tahuayo, 08°08'S, 74°01'W (LSUMZ 160493). ***Madre de Dios*** Cocha Cashu, 11°53.5'S, 71°24.5'W (Terborgh *et al.* 1984); Manu Wildlife Center, 12°24'S, 70°42'W (XC 20616/17; H. Lloyd). **BRAZIL: *Amazonas*** Atalaia do Norte, 04°29'S, 71°33'W (MPEG 72971); 13km S of Benjamim Constant, 04°30.5'S, 70°02.5'W (Willis 1987, 1988, Mazar Barnett & Kirwan 1999, ML 117228, C.A. Marantz); ***Acre*** PN Serra do Divisor, *c.*07°24'S, 73°41'W (Whitney *et al.* 1997 cited in Guilherme 2009); Tartaruga, 09°16'S, 72°41'W (Whittaker & Oren 1999); RESEX Alto Tarauacá near Marechal Thaumaturgo, *c.*09°02'S, 72°37'W (Williams 1995).

TAXONOMY AND VARIATION Monotypic. Elusive Antpitta was considered part of the subgenus *Thamnocharis* by Lowery & O'Neill (1969) in their original description, which included a revision of the Grallariidae. Prior to this, *Thamnocharis* was recognised as a monotypic genus erected for Ochre-striped Antpitta (Sclater 1890, Cory & Hellmayr 1924, Zimmer 1937, Peters 1951), which clearly is the present species' closest relative (Lowery & O'Neill 1969, Rice 2005a) and with which it might be considered to form a superspecies (Krabbe & Schulenberg 2003). Some authors have even suggested a subspecific relationship with Ochre-Striped Antpitta (Sibley & Monroe 1990). Together, Elusive and Ochre-striped Antpittas are probably sister species to the clade of Chestnut-crowned and Watkins's Antpittas (Rice 2005a). In two molecular studies with incomplete sampling, Elusive and Ochre-striped have both times proved closely related to Chestnut-crowned Antpitta (Krabbe *et al.* 1999, Moyle *et al.* 2009).

STATUS Elusive Antpitta is classified as Least Concern by BirdLife International (2017) having been recently downgraded from Near Threatened (Collar *et al.* 1994, Stattersfield & Capper 2000, BirdLife International 2012) after its range was found to be broader than previously thought (Whittaker & Oren 1999). Although small (309,168km²), the range size (Franke *et al.* 2007, Young *et al.* 2009) is not believed to approach the threshold for Vulnerable. It occurs within the South-east Peruvian Lowlands Endemic Bird Area (Stattersfield *et al.* 1998). Further research into its conservation status, however, is certainly warranted. Although no specific impacts of human activities have been documented, such a reclusive understorey bird is likely be severely affected by predicted fragmentation of its forest habitat (Bird *et al.* 2012). The species has not yet been reported from Tambopata Reserve in Madre de Dios (Lloyd 2004), but might be expected there. **Protected Populations Brazil** PN Serra do Divisor (Whitney *et al.* 1997 cited in Guilherme 2009); RESEX Alto Tarauacá (Williams 1995); RESEX Alto Juruá (Whittaker & Oren 1999, Whittaker *et al.* 2002). **Peru:** PN Manú (Terborgh *et al.* 1984).

OTHER NAMES O'Neill (1974) reported that the indigenous Cashinahua refer to Elusive Antpitta by the onomatopoeic name 'doo-shau.' **Spanish** Tororoí Evasivo (Plenge 2014); Tororoí del Ucayali (Krabbe & Schulenberg 2003). **Portuguese** tovacuçú-xodó (Willis & Oniki 1991b, Guilherme 2016). **French** Grallaire secrète. **German** Fahlbrust-Ameisenpitta (Krabbe & Schulenberg 2003).

Elusive Antpitta, adult male holotype (LSUMZ 62312), Río Curanja, Loreto, Peru, collected 11 July 1967 (*J. Van Remsen*).

CHESTNUT-CROWNED ANTPITTA
Grallaria ruficapilla   **Plate 10**

Grallaria ruficapilla Lafresnaye, 1842, Revue Zoologique par La Société Cuvierienne, p. 333, 'Habitat in Bolivia, Santa Fe de Bogota' [unknown location in Colombia, see Taxonomy and Variation, *ruficapilla*].

'In the morning, and shortly before sunset, may be heard a melancholy cry as this Ant-Thrush creeps amongst the brushwood. Many times have I followed to obtain a specimen, and after a tough scramble of an hour given it up for a bad job. At one time you seem to stand right upon it, and a moment after you hear it 4 yards off; again you reach the spot, and you hear it 20 yards behind you; you return, then it is on the right; soon after you hear it on the left. At first you imagine the bird has the power of a ventriologuist; but by dint of patience and watching you may see it creeping swiftly and silently among the grass and brushwood in places where it has to pass a rather more open spot, and the mystery is explained.' – **Thomas Knight Solomon, Antioquia, Colombia (in Sclater & Salvin 1879a)**

One other encounter with a 'Chestnut-crowned' Antpitta is worthy of mention (with a good deal of creative adaptation of my own). On 5 June 1999, in the Central Andes of Colombia, Bernabé López-Lanús had an experience that I can only imagine had his heart racing. While exploring a forested fragment near Roncesvalles, he heard the vocalisation of an antpitta he was unable to identify! Following a whistled imitation of the bird he laid eyes on what appeared to be a largely unstreaked, over-sized Chestnut-crowned Antpitta with an enormous bill. With images of this odd bird in his mind, he returned a few months later with his colleagues in the hope of capturing what appeared to be a new species of antpitta. (Bear in mind this was around the time that Krabbe *et al.* (1999) published their discovery of the Jocotoco Antpitta.) The result of this expedition was not the discovery of a new antpitta but arguably, perhaps, even more exciting. After relocating and capturing the aberrant bird, and following a detailed analysis of plumage, morphometrics, vocalisations, and mitochondrial and nuclear DNA sequences, Cadena *et al.* (2007) documented the first known occurrence of interspecific hybridisation in the Grallariidae. The mitochondrial DNA profile of the hybrid demonstrated that 'mom' was a Chestnut-naped Antpitta and 'dad' was a Chestnut-crowned Antpitta. The frequency with which such events occur in these species is unknown, but these authors tape-recorded at least two additional individuals in the area with similar vocal qualities to that of the hybrid. Anyone with the opportunity to return and explore this fascinating interaction, particularly its effects on nesting biology, would surely be richly rewarded. Apart from this, the strikingly plumaged Chestnut-crowned Antpitta is both easily identified and, despite the opening quote, one of the most easily seen among *Grallaria* species. It ranges from northern Venezuela south to central Peru, and is divided into seven subspecies, which differ mainly in small details of their head and breast markings. The species has a broad altitudinal range, at 1,200–3,600m, and is found in all manner of forest types, but favours borders, clearings, second growth, patches of bamboo and other disturbed habitats. This antpitta's song is easily imitated, and these bold, inquisitive birds will readily approach in response to human whistles. In some places, the birds even become tame, feeding in car parks and other areas subject to much human activity. Being so widespread and relatively numerous, Chestnut-crowned Antpitta is one of the most familiar of the Grallariidae. As one of the most widespread species of the genus, one that appears to be fairly adaptable to intermediate to high levels of habitat alteration, Chestnut-crowned Antpitta appears to be facing no immediate conservation threat. This, however, along with its propensity to be fairly common and easily detected, would make it an ideal study species for a number of lines of research. In particular, confusion over the taxonomic validity and distribution of its many subspecies make this an area ripe for investigation. As with all antpittas, the natural history and reproductive behaviour of Chestnut-crowned Antpitta remain nearly undocumented, but at least a few nests have been found and the specimen records included here add a substantial number of breeding records from across its range.

IDENTIFICATION 18–19cm. There are no clear field marks that distinguish the sexes, both of which have the head, including nape and sides, bright rufous or chestnut. The lores and ear-coverts are either the same or are somewhat variably washed (or entirely) white, depending on the subspecies. Overall, the head coloration gives a hooded appearance, contrasting with the bright white throat and brown or olive-brown upper back. The upperparts are uniformly dark, including the tail and wings. Below, including the belly and vent, Chestnut-crowned Antpitta is white, with bold, black-brown or olive-brown streaking, most prominent on the flanks and extending across the breast more or less, depending on the subspecies. Some show varying degrees of orange-brown wash across the breast. Throughout most of its range, Chestnut-crowned Antpitta is nearly unmistakable, and is frequently the only species of *Grallaria* with extensive streaking below combined with a contrasting chestnut hood. In south-west Ecuador, where it is occasionally sympatric with Watkins's Antpitta, it can be separated from the latter by the overall bold coloration, in particular the brightness of the head and bold streaking on the breast, the lack of fine streaking on the upperparts (but see juvenile description) and the predominantly rufous (not whitish) face. Additionally, the tarsi of Watkins's Antpitta are pale pinkish, compared to the bright blue-grey tarsi of Chestnut-crowned Antpitta. In Colombia, Santa Marta Antpitta, and to a lesser degree Cundinamarca Antpitta are somewhat reminiscent of a 'faded' Chestnut-crowned Antpitta. Though a clear view of the heavier underparts streaking and chestnut crown of the present species should immediately separate it, an over-eager birder, desperate to get a glimpse of either of the two much rarer species should beware of being fooled by a brief glimpse in low light.

DISTRIBUTION Chestnut-crowned Antpitta is among the most broadly distributed species in the genus, being resident in montane areas from the coastal mountains of north-east Venezuela, south through all three Andean ranges in Colombia, then Ecuador and as far south as southern Cajamarca in western Peru and northern San Martín in eastern Peru. Compared with other Andean taxa treated here, there are relatively few unexplained gaps in its distribution. There is an apparent gap in the West Andes of Colombia, roughly from El Limonar north at least to La Celia (see Taxonomy and Variation, *ruficapilla*). Though seemingly likely, it is unclear if populations in the north of Colombia's West Andes are contiguous with

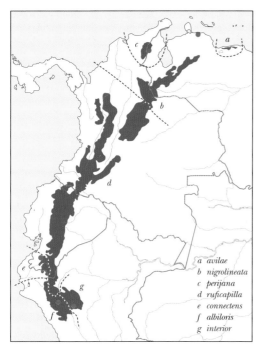

a *avilae*
b *nigrolineata*
c *perijana*
d *ruficapilla*
e *connectens*
f *albiloris*
g *interior*

those in the northern Central Andes. The *c.*160km gap in records in the East Andean cordillera in Meta and Huila probably reflects a lack of exploration. Overall, there are likely only two truly isolated populations, in the Sierra de Perijá (*perijana*) and the coastal mountains of northern Venezuela (*avilae*).

HABITAT The species inhabits a fairly wide range habitats with respect to seasonality of rainfall and humidity, from seasonally rather dry inter-Andean valleys, to extremely humid and rather aseasonal cloud forests (Greeney & Kirwan 2013). Dense thickets of *Chusquea* spp. bamboo are unquestionably the natural habitat of this Andean antpitta, but it will also use floodplains and landslides dominated by *Alnus acuminata* (Betulaceae) (Poulsen 1996, HFG). Where large areas of forest mix with patches of bamboo, the species is rarely seen far from dense tangles of *Chusquea*. Even isolated landslides and regrowing pastures, completely surrounded by intact cloud forest, are colonised (J. Freile *in litt.* 2017). This propensity to inhabit naturally disturbed habitat appears to permit the species to adapt well to more anthropomorphic habitat changes, and in many parts of its range Chestnut-crowned Antpitta is found at forest edges and in mosaics of pasture and scrubby second growth (Renjifo 1998, 1999, Kattan & Beltrán 2002, Cadena *et al.* 2007, Pulido-Santacruz & Renjifo 2011), even where little natural habitat remains such as in the Yunguillia Valley (HFG). In parts of the inter-Andean Valley of Ecuador, Chestnut-crowned Antpittas may persist at fairly high densities, confined to small fragments of tangled second growth, but occasionally emerging to forage in the open like the (sometimes) sympatric Tawny Antpitta (J. Freile *in litt.* 2017). Altitudinal reports generally range from 1,200 to 3,100m (Parker *et al.* 1996, Schulenberg *et al.* 2007, Buitrón-Jurado 2008), and reports as low as 600m (Guerrero 1996) seem unlikely. Elevation does seem to vary geographically to some degree. Venezuelan populations are reported at 1,300–3,450m

(Giner & Bosque 1998, Meyer de Schauensee & Phelps 1978, Hilty 2003), but its range is apparently narrower (1,450–2,500m) in the coastal mountains (*avilae*; Schäfer & Phelps 1954, Phelps & Phelps 1963, Schäfer 1969) and in the Serranía de Perijá (1,300–2,900m) (*perijana*; Phelps & Phelps 1963). In Colombia records span 1,300–3,000m (Hilty & Brown 1986, McMullan *et al.* 2010), in Ecuador 1,300–3,400m (Freile 2000) and in Peru 1,200–3100m (Clements & Shany 2001, Schulenberg *et al.* 2010). Although it often occurs sympatrically with other *Grallaria*, it probably often segregates by habitat. This is certainly the case around Yanayacu Biological Station east of the Andes in Napo, where the sympatric White-bellied Antpitta haunts the interior of moss-laden forest adjacent to the tangled bamboo habitat of Chestnut-crowned. Interestingly, further south in eastern Ecuador (i.e. Tungurahua), the same two species appear to replace each other elevationally, with Chestnut-crowned replacing White-bellied Antpitta above 1,800m (J. Freile *in litt.* 2017).

VOICE The most frequently heard song is a monotonously repeated, three-note *wheee, whooooo-whooo!* with the first note usually highest, the second lowest and slightly drawn-out, and the third loudest. The song is easily whistled and gives rise to its widely used local name of *com-pra-PAN* (Miller 1918, Olivares 1969). Also described as a *hee hoo-HEW* (Schulenberg *et al.* 2010). When alarmed, especially around an active nest (Greeney & Kirwan 2013), adults give a one-note, sharp *peeyu!* (e.g., XC 223677, *avilae*, Aragua, P. Boesman). The same, or similar, one-note call was described by Schulenberg *et al.* (2010) as a plaintive, whistled *clew!* This call may be similar in tone to the song, or have a distinctly 'burry' sound to it (e.g., XC 102699, *interior*, Amazonas, Peru, F. Schmitt), as if the bird were hoarse from calling (Greeney & Kirwan 2013). Birds may also deliver a harsh version of the more typical song, with the final two notes often very gravelly (e.g. XC 250077, *ruficapilla*, Azuay, N. Krabbe). In addition, the first note of the song may occasionally have a 'hitch' in it, making the song four-noted (e.g., XC 202001, *nigrolineata*, Mérida, H. Matheve). Krabbe & Schulenberg (2003) provide a quantified description: three notes, duration 1.1–1.4s, delivered at 5–12s intervals, frequency 1.5–1.8kHz. They describe the call as lasting 0.4–0.5s, delivered every 2–7s, descending from 2.6kHz to 1.4kHz. There is little obvious variation across the range. Recent recordings from Serranía del Perijá (*perijana*) suggest that, in this subspecies, the final two notes are closer to being on the same pitch than in other subspecies (e.g. XC 331811, Serranía del Perijá, D. Calderón-F.), but additional comparative material for *perijana* is needed before robust conclusions are made. Songs in Serranía de Mérida (*nigrolineata*), again based on a fairly small sample, appear to have the final note slightly closer in pitch to the first note than in the songs of other populations (e.g., XC 223676, Lara, P. Boesman). Overall, however, a quick sampling of vocalisations from across the range of *G. ruficapilla* suggests to me that there is no clear pattern of inter-population variation. Krabbe & Schulenberg (2003) suggested that the call of *interior* is usually shorter (0.2s) and lower (1.8–1.4kHz) than the typical call of *ruficapilla*.

NATURAL HISTORY Probably monogamous and territorial but, like all antpittas, the behaviour of Chestnut-crowned Antpittas is poorly documented, and few aspects have been quantified. Nevertheless, as one of the most easily observed species, we know that Chestnut-crowned Antpittas frequently forage on the ground for invertebrate prey, but also occasionally climb several metres above it in search of caterpillars and other foliage-dwelling invertebrates (Krabbe & Schulenberg 2003). Like their congeners, this species is generally shy, but may respond strongly to a whistled imitation of the song. Otherwise, they remain well hidden in thick vegetation, low to the ground (Greeney & Kirwan 2013). Chestnut-crowned Antpitta occasionally follows highland army ant swarms (Nieto-R. & Ramírez 2006, Greeney 2012b). Greeney (2012b) also reported that the species appears to actually be attracted to the sound of humans crashing through bamboo, postulating that it may regularly follow large terrestrial mammals (e.g., Spectacled Bears *Tremarctos ornatus*, Mountain Tapirs *Tapirus pinchaque*), as a facultative means of searching for prey in the earth and vegetation disturbed by the larger animal. To date, while no studies have examined or quantified this behaviour, I have observed it on several additional occasions, strengthening my opinion. Territories in Colombia are estimated to be about 1.9ha (*ruficapilla*; Kattan & Beltrán 1999, 2002) and only 0.9ha in northern Peru (*connectens*; Kikuchi 2009) but little is known about behavioural aspects of territory defence or maintenance. Longevity in the wild is unknown, but Bell & Bruning (1976) reported an adult which lived five years in captivity. I found no references to predators of Chestnut-crowned Antpittas, but they are known hosts of *Trouessartia* chewing lice (Phthiraptera; Byers 2013) and a recently described described species of *Leucocytozoon* malaria (Haemosporida, Leucocytozoidae) was described from a bird that was reportedly simultaneously infected with an unidentified parasitic nematode (Onchocercidae) and an unknown Trypanosoma protozoan (class Kinetoplastida) (Lotta *et al.* 2015).

DIET The diet is not well documented, but almost certainly dominated by invertebrates. Like others in the genus, earthworms are probably important to adults and for feeding young in the nest. In support of this, Chestnut-crowned Antpittas are among those species easily 'trained' to take earthworms proffered at feeding stations (Woods *et al.* 2011). Along with other insect fragments, the remains of a small weevil (Curculionidae) was found in the stomach of a Chestnut-crowned × Chestnut-naped Antpitta hybrid (Cadena *et al.* 2007).

REPRODUCTION Despite descriptions of the eggs from several parts of the species' range, there are only three published descriptions of the nest, all of nominate *ruficapilla*. The first description was by T. K. Salmon (*in* Sclater & Salvin 1879a), who described a nest from Antioquia in western Colombia, along with a clutch of two eggs, but provided no date, measurements or specific location. Following this, Asociación Bogotaña de Ornitología (2000) described the nest and eggs, with no specifics as to their sample size or the source of their data. The precise origin of three eggs described by Taczanowski (1880, 1884) presents a puzzle. The eggs are described first (1880) with no mention of the nest. Subsequently, Taczanowski (1884) quoted Sztolcman, who described collecting eggs from a 'nest on the ground in a sugarcane field' at Callacate (translation from French). We now know that a nest of *Grallaria* on the ground is highly improbable and, in addition, there are no reported clutches of three eggs for any antpitta. There is no specific mention of a

second nest, but the statement that the species nests in March *and* April (not through April) might be taken to imply there was a second nest from whence the 'third' egg was collected. For now, the measurements and descriptions provided by Taczanowski (1880) leave little doubt that they are, in fact, those of a *Grallaria*, and it seems safe to assume that they were collected from at least two separate nests. The only quantified, detailed description of the nest and eggs concerns a single nest studied in Napo, eastern Ecuador (Martin & Greeney 2006). **NEST & BUILDING** The description in Sclater & Salvin (1879a) was simply 'a mass of roots, dead leaves, and moss, lined with roots and fibres [ ] placed at some height from the ground.' Asociación Bogotana de Ornitología (2000) stated that the nest is a broad cup of dry leaves, roots and moss, built low to the ground. Neither of these descriptions provide any information on nesting habitat. The following details pertain to the single nest found by Martin & Greeney (2006). This nest was in an area where the canopy almost entirely comprised 20m *Alnus acuminata* (Betulaceae) trees, with an understorey of dense *Chusquea* bamboo interspersed by herbaceous Solanaceae, Urticaceae and Piperaceae. It was near the edge of the *Chusquea* patch and surrounded by relatively more open, herbaceous understorey. It was 2m above ground, apparently incorporated into a natural clump of leaves and small branches that had accumulated where 8–10 horizontal *Chusquea* stems overlapped. It was constructed primarily of sticks and twigs, with additional bamboo leaves, sparse moss, leaf petioles and a few dicot leaves mixed to form the outer cup. Internally, the cup was sparsely lined with thin, dark, flexible rootlets. Measurements were: external diameter 27 × 27cm (measured at perpendicular angles); external height 18.5cm; internal diameter 11.5 × 11.5cm; internal depth 7cm. Previously unpublished (HFG), a nest of nominate *ruficapilla* at the Jocotoco Foundation's Yunguilla Reserve in Azuay was 1.8m above ground in a 3.4m-tall *Baccaris* shrub (Asteraceae) within an area of heavily disturbed, patchy second growth and pasture. The substrate plant and surrounding vegetation formed a 2–3m in diameter patch, isolated from other plants by at least 4m of pasture in all directions, but well concealed by the leaves of the substrate. The nest was supported by three, nearly upright stems, 4cm, 6cm and 7cm in diameter. Three additional, thinner branches passed through the structure of the nest, likely providing stability, but not support. The bulk of the nest comprised long, dead twigs (15–25cm) mixed with dead leaves and grasses. Internally it was only sparsely lined with thin, generally pale-coloured, flexible rootlets and grass stems. The nest was slightly oblong, squeezed between the surrounding plant stems, and its dimensions were: external diameter 18 × 28cm; external height 10cm; internal diameter 10.5 × 9cm; internal depth 6cm. Also previously unpublished, J. Simbaña found a nest 4.5m above ground supported by multiple overlapping *Chusquea* bamboo stems, at Yanayacu Biological Station in Napo. The site was near the middle of a large (10ha+) patch of dense bamboo surrounded by primary cloud forest. The nest was well hidden by bamboo on all sides and was similar in size and composition to that described by Martin & Greeney (2006). Overall, the nests of *G. ruficapilla* appear to be similar in structure and composition to nests described for *G. watkinsi* (Martin & Dobbs 2004, Greeney *et al.* 2009b; see *watkinsi* account). **EGG, LAYING & INCUBATION** Descriptions of the eggs are all remarkably consistent and it appears that the species lays unmarked, sky-blue to blue-green eggs throughout its range. The three eggs collected by S.B. Gabaldon (AMNH 13865), which were described as 'buffy eggs with rufous blotches' by Wiedenfeld (1982), are clearly not of this species. Eggs have been described for three of the seven subspecies of *G. ruficapilla*. **G. r. ruficapilla** Oates & Reid (1903) gave the measurements of both eggs (NHMUK 1962.1.470) collected by T.K. Salmon and described by Sclater & Salvin (1879a) as 'rather round and blue': 30.0 × 25.4mm; 31.0 × 26.2mm. Schönwetter (1979) provided a mean of 30.6 × 25.6mm for three eggs. N. Heming provided me with a size-calibrated photo of a third egg in Tring (NHMUK 1901.11.30) labelled simply 'Colombia,' and apparently from the Phillip Crowley collection (Crowley 1883). My digitally calculated measurements (30.9 × 25.2mm) provide an identical mean, suggesting the three NHMUK eggs are those measured by Schönwetter (1979) and allowing me to add standard deviations (± 0.6 × ± 0.5) to Schonwetter's measurements of Colombian eggs. Martin & Greeney (2006) described two eggs as subelliptical and 'uniform turquoise or pale greenish-blue, with no flecking or spotting': 28.5 × 24.3mm; 29.9 × 24.2mm. These eggs weighed 8.8g and 9.1g, respectively, when they were found. On being re-weighed six days later, the change in weight was used to calculate a total loss of mass (water) during incubation of 11–13% of fresh egg mass. Greeney & Kirwan (2013) gave measurements of three eggs from Azuay: 31.3 × 25.0mm; 29.7 × 24.5mm; 30.5 × 25.7mm. All of the above eggs inspected by myself directly or from photographs, are short, subelliptical and unmarked pale blue. **G. r. albiloris** Taczanowski (1880) described three eggs from Cundinamarca as uniform pale bluish-green: 29.0 × 24.0mm; 29.0 × 25.0mm; 30.6 × 25.2mm. **G. r. nigrolineata** Nehrkorn (1899, 1910) gave measurements for eggs from Mérida without providing sample sizes: 30.0 × 25.0mm; 27.0 × 24.0mm. Schönwetter gave the mean measurements of three Mérida eggs as 28.4 × 24.4mm. These 16 egg measurements (counting Nehrkorn's once each) provide a mean of 29.5 ± 1.3mm × 24.8 ± 0.7mm for the species' eggs. Surprisingly, despite this relatively large sample size of measured eggs (for an antpitta), only two of these published sources explicitly provided a clutch size (both two eggs) (Sclater & Salvin 1879a, Martin & Greeney 2006). The addition of three two-egg clutches from eastern Ecuador (Greeney & Kirwan 2013) suggests that two eggs are the normal clutch (*n* = 5 clutches). There is almost nothing known of adult behaviour during incubation, other than the observations of Martin & Greeney (2006) who stated that both adults incubated and, when arriving at the nest, occasionally brought thin materials to add to the lining. While covering the eggs, adults stood periodically to probe the inside of the nest with their bills, presumably as a means of parasite removal (Haftorn 1994, Greeney 2004, Greeney *et al.* 2008a). **NESTLING & PARENTAL CARE** Chestnut-crowned Antpittas are born with pink skin and long, whispy tufts of pale grey down on the head, back and wings, but develop a dense covering of non-natal down by the time primary pin-feathers break their sheaths (Greeney 2012a, Greeney & Kirwan 2013). Previously unpublished, my observations show that contour and flight feather pins break the skin *c.*4–5 days after hatching and the contour feathers start breaking their sheaths soon after, apparently beginning with those on the alar feather tracts. By days 7–8 nestlings are well covered in a dense, buff-coloured coat of secondary nestling down. Secondary flight feather sheaths

begin to rupture around day 8 and those of the primaries around day 9. By the time they reach 15–16 days of age, nestlings are capable of walking (very awkwardly) for short distances and are already similar to the description of fledglings above. While still in the nest, however, they retain much of the wispy greyish natal down, now scattered sparsely across the dense coating of down below. Like other antpittas, from hatching until fledging the young have bright orange bills, crimson-orange mouth linings and inflated rictal flanges (Greeney *et al.* 2008a). Martin & Greeney (2006) gave a nestling period of 18 days, but provided few data on adult behaviour other than stating that both sexes participated in rearing the young. **SEASONALITY** I have previously suggested that the breeding of *G. ruficapilla* (and other Andean bamboo-inhabiting species) may coincide with periods of reduced rainfall (Greeney *et al.* 2008b, 2011a) and I maintain that this is indeed the case for populations inhabiting regions that are relatively humid year-round (e.g., eastern Ecuador and Peru). It seems likely, however, that populations in more seasonally arid regions (e.g., western Ecuador) may very well have a peak in reproduction that coincides with increased precipitation. Though still somewhat premature, the specific breeding records below somewhat support this hypothesis. The following records are scattered geographically and provide at least some information for all of the currently recognised subspecies. **BREEDING DATA** **G. r. avilae** Subadult, 17 March 1977, Pico Guacamayo (USNM 575268); subadult, 10 July 1929, Pico Naiguatá (USNM 325002). Said to breed May to July in Aragua, but evidence not provided (Schäfer & Phelps 1954). **G. r. nigrolineata** Two fledglings, 17 July and 14 September 1903, Valle (AMNH 100208/492229); fledgling, 14 June 1904, Los Nevados (FMNH 57187); fledgling, 17 August 1949, Hacienda Las Vegas (USNM 411872); two fledglings transitioning, La Culata, 25 June 1894 (AMNH 492234) and 28 August 1911 (FMNH 50738); two juveniles, La Culata, 14 May 1909 (FMNH 57181) and 15 August 1904 (FMNH 50737); juvenile transitioning, 15 May 1912, Conejos (FMNH 57188); juvenile transitioning, 18 April 1909, Valle (FMNH 57185); subadult, 28 April 1904, Quebrada La Capaz (FMNH 57189); subadult transitioning, 14 July 1909, Valle (FMNH 57183); subadult transitioning, 18 August 1912, Páramo El Escorial (FMNH 57186). **G. r. perijana** Two fledglings, 30 June 1942, Laguna De Junco (USNM 373658/59); two juveniles, 27 June 1942, Laguna De Junco (USNM 373657/63); juvenile, 19 June 1942, La Africa (USNM 373656); three subadults, 9–13 July 1942, Laguna De Junco (USNM 373660–662); testes enlarged, 17 June 1942, La Africa (USNM 373655). **G. r. ruficapilla** Nest construction almost finished, 19 September 2008, RP Estación Biológica Yanayacu (J. Simbaña *in litt.* 2008); adult with egg in oviduct, July 2000, north-west of Bogotá (Echeverry-Galvis *et al.* 2006); incubation, 11 March 2005, RP Yunguilla (HFG); incubation, 20 September 2004, RP SierrAzul (Martin & Greeney 2006); nestling, 19 September 2008, RP Estación Biológica Yanayacu (J. Simbaña *in litt.* 2008); adult carrying food, 4 May 2016, RP Bellavista (Flickr: S. Siegel); adult carrying food, 9 December 2007, Guango Lodge (ML 37462411, photo J. Stahl); adult carrying food, 7 March 2014, La Esperanza Lodge (J.B.C. Harris *in litt.* 2017); adult carrying food, 10 September 2010, RFP Río Blanco (D. Calderón-F. photos); fledgling, 29 January 1911, San Antonio (Chapman 1917, AMNH 108098); fledgling, 20 July 1947, Soatá (USNM 444550); fledgling, 9 September

1951, below Páramo de Frontino (USNM 436482); fledgling, 5 September 1957, Cerro Munchique (LACM 30925); fledgling, 10 October 2011, RFP Río Blanco (eBird: C. Downing); fledgling, 10 March 1936, Nudo de Cajanuma (MLZ 9832); fledgling, 23 August 2007, RP Cabañas San Isidro (N. Athanas *in litt.* 2016); fledgling, 7 March 2001, RP Yunguilla (MECN 8357, M.V. Sánchez N. *in litt.* 2017); fledgling, 28 February 2015, RP Bellavista (A. Agrawal photos); fledgling, 24 May 2016, RN Chicaque (ML 29787931, photo C. du Plessis); immature (fledgling/juvenile), 15 March 2015, Refugio Paz de las Aves (eBird: H. Brieschke); four 'fledglings,' January in Cauca, April in W Nariño, February and September in Pichincha (Fjeldså & Krabbe 1990); juvenile, 15 January 2017, El Mirador, Carretera 10 (ML 45680421, D. Rocha López, ML 45957151, J. Muñoz García); two juveniles, RP Cabañas San Isidro, 13 December 2012 (Flickr: S. Ellison) and 28 November 2009 (Flickr: 'dawgbirder'); juvenile, 20 August 1952, Cerro Munchique (USNM 446646); juvenile, 27 August 1920, Salvias (AMNH 167250); juvenile, 18 July 1968, Cordillera de Tandayapa [= *c.*RP Bellavista] (MECN 2607, M.V. Sánchez N. *in litt.* 2017); juvenile, 11 July 1959, El Guabo (LSUMZ 38706; photo M.L. Brady); juvenile, 19 May 2013, Quebrada el Viao, Cocorná (J. Ochoa photos); juvenile, 20 June 1936, Guarumos (FMNH 99575); juvenile, 6 April 2011, RP Bellavista (J.P. Perret photos); juvenile, 6 July 2013, RFP Río Blanco (IBC 1263136, photo D. Avendaño); juvenile transitioning, 28 October 2009, RP Yunguilla (ML 45753581, photo N. Voaden); juvenile transitioning, 6 July 1951, Páramo de Sonsón (USNM 436484); adult carrying food, 4 May 2016, RP Bellavista (Flickr: S. Siegel); four subadults at RFP Río Blanco, 27 August 2012 (P. Hawrylyshyn *in litt.* 2015), 20 May 2012 (J.J. Arango photos), 28 November 2015 (IBC 1116001, P. Edwards), 23 February 2017 (ML 50929541, photo K. Russell); subadult, 14 September 1929, 'Moya' [= near Alaspungo, see Paynter 1993] (MLZ 7210); subadult, 20 November 2012, RP Cabañas San Isidro (Flickr: A. Drewitt); subadult transitioning, 29 March 1958, El Guabo (FMNH 251118); 'immature', 15 May 2015, RP Refugio Paz de las Aves (eBird: H. Brieschke); three ♂♂ with enlarged testes (11–12mm), one at Popayán, and two at San Antonio, 14 and 30 March 1958 (Miller 1963); two ♂♂ with testes greatly enlarged, 3 April 1913, Fusagasugá area (CUMV 7218) and 2 April 1913, El Roble (CUMV 7217); testes greatly enlarged, 12 January 1956, Cerro Munchique (FMNH 226615). *G. r.* **connectens** Subadult, 9 October 1920, Guachanamá (USNM 323090); subadult, 26 March 1991, 4km W of Celica (WFVZ 48692). *G. r.* **interior** Fledgling trasitioning, 7 July 2008, 4.5km N of Tullanya (MSB 32055); juvenile, 4 February 1926, 'San Pedro' (AMNH 235530); two juveniles transitioning, 21 and 23 May 1912, E of Balsas (FMNH 44307/08); subadult, 2 October 1927, Bosque Tinyo (FMNH 299604). *G. r.* **albiloris** Juvenile, 14 June 1926, Taulis (AMNH 235884); juvenile, 13 April 1926, 64km NW of Chugur (AMNH 235893); juvenile, 2 August 1926, Seque (AMNH 235889); juvenile, 22 May 1933, Abra de Porculla (MCZ 179770); juvenile transitioning, 7 May 1912, Hacienda Limón (FMNH 44305); subadult, 1 May 1954, Canchaque (FMNH 222153); subadult, 6 July 2010, Bosque Cachil (MSB 35107); subadult, 9 May 1912, Hacienda Limón (FMNH 44306). In addition, the eggs of *albiloris* discussed above (Taczanowski 1880, 1884) almost certainly came from at least two nests collected between mid-April and late May in C Cajamarca, at least one from Callacate and the other either there or in

Cutervo. Some of the preceding records derived from specimens were also included in Fjeldså & Krabbe (1990), but only those not verified by my own direct observations are cited as theirs.

TECHNICAL DESCRIPTION Sexes similar. The following description is based on Krabbe & Schulenberg (2003) and refers to nominate *ruficapilla*. *Adult* Crown, head-sides, nape and malar rufous. Remaining upperparts olivaceous. Throat, belly and centre of lower breast clean white; rest of underparts heavily streaked olivaceous and blackish, heaviest on sides and variable across upper chest where sometimes reduced to just ochraceous feather edges forming pale streaks; feathers of flanks elongated and most heavily marked. *Adult Bare Parts Iris* brown, dark brown; *Bill* dark grey or black; base of mandible pale grey or whitish; *Tarsi & Toes* pale grey-blue. *Fledgling* Overall plumage is distinctly fluffy and less-than-upright posture usually creates a rounded or 'plump' look, distinct from that of adults. The following is based on direct observations of live fledglings. Upperparts finely barred black and tawny or tawny-rufous, forecrown slightly less barred, ear-coverts buffy and lores whitish mixed grey, with indistinct pale whitish eye-ring. Sides of neck buffy-rufous blending to reddish-olive on scapulars. Throat white, very faintly tinged buff, ending in an abrupt line on upper chest, below which is pale, buffy-chestnut with fine dusky barring, similar to back, and on upper belly becoming a mix of whitish, buffy and dusky feathers, then clean buffy-whitish, and finally white on central and lower belly, with rich buff vent. The thighs in young fledglings are indistinctly barred rufous and black, but soon become olive-brown with some white streaking, especially on upper, inner portion. Wings similar to adults but wing-coverts tipped buffy-rufous with subterminal black band. **Fledgling Soft Parts** *Iris* pale brown; *Bill* maxilla orange, dusky on culmen and yellowish at tip, may retain pale yellow egg tooth for some time after fledging; tomia and area around nares pale yellow; rictal flanges still enlarged and orange externally, yellow internally; mouth lining bright orange; mandible all yellowish-orange except dusky base; *Tarsi & Toes* former pale grey to bluish-grey at front and sides, distinctly yellow-orange posteriorly, toes similar to grey of tarsi, nails yellow to dusky-orange. *Juvenile* Forecrown and crown largely chestnut, nape, upper back and neck-sides largely barred as in fledglings, back largely like adult, patches of barring remain, sometimes asymmetrically, especially on rump; below, breast similar to adult, but traces of buffy juvenile plumage remain on belly, flanks and vent; wings as described for fledglings. **Juvenile Bare Parts** *Iris* dark brown; *Bill* maxilla black with slightly dusky-orange tip, mandible dull pinkish-orange, duskier at base and tip, only faintest hint of expanded rictus (pinkish-orange); *Tarsi & Toes* similar to adult, nails still slightly paler (see Fledgling description). *Juvenile/Subadult transition* The following is based largely on a specimen of *ruficapilla* (USNM 446646) and several other specimens and photos of birds transitioning to subadult plumage. Plumage overall approaches that of adult and signs of immaturity might be missed in the field if the nape and bill are not seen well. Crown and central nape are typical (adult) bright rufous, forecrown slightly paler; on lower nape the rufous fades gradually to ochraceous-olive and then to greyish-olive of adult plumage on the upper back and rump. Tail similar to adult, but rectrices bear very thin rufous fringes along the basal one-third to half of the vane; wings similar to adult, brown to olivaceous-brown, leading edges of primaries warmer, more tawny rufous-brown; upperwing-coverts darker and browner, tipped buffy-rufous with subterminal blackish bands; face somewhat 'messy' with transitioning plumage: lores buffy, warmer and with rufous wash just in front of and below eyes; ear-coverts tawny-buff, mottled and speckled dark grey; and neck-sides bright to olive-rufous with thin black barring, forming a thin, indistinct, 'nuchal collar' broken by a few bright rufous adult feathers on the central nape. Chin and throat white, bordered on neck-sides with mottled tawny-buff, rufous, grey and white, forming an indistinct speckled moustachial. Rest of underparts generally similar to adult plumage, but with generally reduced dark streaking, especially at sides, and some tawny streaks on the breast; vent dark grey mixed white with some buff; long plume-like feathers of flanks white with broad grey margins; undertail-coverts greyish-white; thighs brownish with a few buffy-chestnut feathers. **Subadult** The following refers to a subadult *connectens* (WFVZ 48692). Below, the white plumage is washed buff, especially on lower belly and chest. There are faded rufous streaks mixed with normal adult-like streaks. The vent is a mix of dusky and whitish feathers, appearing overall greyish-white. Above, very similar to adult, differing slightly on head. Ear-coverts a mix of buffy-white to buffy-rufous feathers and appear slightly speckled. Lores and neck-sides paler buffy-rufous than adult. The pale whitish chin and ocular area is mixed with some rufous and dusky feathers, creating a mottled or speckled appearance. The bill (in dried skin) has the mandible blackish, slightly dusky-orange on the centreline below, near the tip, and the tomia near the tip. The maxilla is similarly coloured, still dusky-orange nearer the tip, along the tomia and around the nares. Overall, in life, a bird in this final transition to adult plumage might appear simply 'messy'. The two characters to be looked for are the pale orange portions of the bill, which are likely to be much more evident in a live bird. Finally, the secondary wing-coverts still have buffy-rufous tips, but may not have subterminal black bands, as in juveniles. In this specimen, lacking black bands on the coverts, it is apparent that this is a result of wear, the terminal portion of each feather being no longer present. **Subadult Bare Parts** As adult except that the base of the mandible can be slightly pink or orange, rather than grey or whitish. *Additional observations on post-fledging plumage development* Several previous descriptions of immature plumage appear to be generally congruent with the above descriptions (e.g., Fjeldså & Krabbe 1990, Asociación Bogotana de Ornitología 2000). The most complete account is that of Restall *et al.* (2006). They described and illustrated three stages of development that I 'translate' as 'chick' (= fledgling), 'juvenile' (= juvenile) and 'immature' (= older juvenile transitioning to subadult). The only obvious inconsistency is the lack of immature wing-coverts in the transitioning juvenile. I have examined numerous juveniles and subadults, and have never seen an example where immature wing-coverts are replaced prior to acquiring fully adult plumage elsewhere. Otherwise, although my understanding of the exact progression remains incomplete, the following generalities apply (with considerable overlap in the order). As fledglings begin to lose their fluffy look, the breast feathers are replaced first, followed by those of the sides, belly and rump. The next feathers to be replaced appear to be those of the central back, face and forecrown, with a gradual

replacement of crown feathers progressing rearward. The last contour feathers to be replaced are those of the neck-sides and nape. Though I cannot be certain, based on some juveniles having apparently much fresher wing-coverts than others, I believe that fledgling wing-coverts may be replaced with similarly patterned juvenile feathers during this process. It appears that juvenile wing-coverts may be retained for some time after they otherwise achieve fully adult plumage. In at least some cases, however, the coverts become worn, losing their black and/or chestnut markings before they are replaced. Data from specimens suggests that the skull does not become fully ossified until after adult wing-coverts are acquired. The labels on three fledgling *nigrolineata* in AMNH indicate that the iris is blue. These were, however, all collected by Briceño Gabaldon, and the meaning of this remains unclear but is perhaps related to the bluish foggy appearance of the eyes of recently killed specimens.

MORPHOMETRIC DATA *G. r. ruficapilla Wing* 96mm, 96.5mm, 93.5mm, 95.5mm; *Tail* 62.0mm, 61.7mm, 57.8mm, 57.8mm; *Bill* from nares 15.7mm, 17.0mm, 17.1mm, 17.4mm, exposed culmen 24.3mm, 26.7mm, 24.9mm, 25.2mm, depth at nares 8.4mm, 8.9mm, 9.1mm, 9.6mm, width at nares 7.0mm, 7.5mm, 7.4mm, 7.0mm; *Tarsus* 52.0mm, 54.8mm, 53.2mm, 55.3mm ($n = 4$, ♀,?,?,?, B. Mila *in litt.* 2015). *Bill* from nares 15.2mm, 15.3mm, 16.2mm, exposed culmen 21.8mm, 21.0mm, 24.0mm, width at gape 13.4mm, 12.2mm, 12.3mm; *Tarsus* 54.8mm, 52.3mm, 51.9mm ($n = 3$, ♂♂, fledgling, juvenile, adult, MLZ). *Wing* 113.25 ± 4.10mm, 107.5–120.3mm ($n = 9$); *Tail* 63.26 ± 3.93, 58.1–69.8mm ($n = 9$); *Bill* height 10.43 ± 0.52, 9.9–11.4mm ($n = 6$), exposed culmen 24.13 ± 1.46mm, 22.0–26.3mm ($n = 7$), total culmen 27.55 ± 1.47mm, 25.7–30.1mm ($n = 7$), width at gape 17.30 ± 0.84mm, 16.0–18.4mm ($n = 8$); *Tarsus* 59.35 ± 2.96mm, 55.0–62.6mm ($n = 8$) (♀?, Cadena *et al.* 2007): see also Domaniewski & Sztolcman (1918), Guerrero (1996). *G. r. perijana Wing* 100.5–102.5mm; *Tail* 58-61mm; *Bill* culmen from base 29–31mm, exposed culmen 24–24.5mm; *Tarsus* 54–55mm ($n = 3$, ♂♂, Phelps & Gilliard 1940). *Wing* 98–104mm; *Tail* 54–58mm; *Bill* culmen from base 27.5–29mm, exposed culmen 22.5–23.5mm; *Tarsus* 53.5–55mm ($n = 5$, ♀♀, Phelps & Gilliard 1940). *Bill* from nares 10.9mm, 12.4mm; *Tarsus* 44.3mm, 48.6mm ($n = 2$, fledglings, ♂♂, USNM). *Bill* from nares 14.6mm, 15.0mm, 15.8mm; *Tarsus* 52.3mm, 52.9mm, 52.9mm ($n = 3$, juveniles, ♀,♀,♂, USNM). *Bill* from nares 16.6mm, 17.1mm, 17.4mm; *Tarsus* 53.9mm, 54.4mm, 56.2mm ($n = 3$, subadults, ♂,♀,♂, USNM). *Bill* from nares 17.2mm, 18.8mm, 17.8mm; *Tarsus* 55.5mm, 53.6mm, 54.1mm ($n = 3$, ♂,♂,♀, USNM). *G. r. nigrolineata Bill* from nares 15.7mm, 15.9mm, 16.8mm; *Tarsus* 52.8mm, 53.9mm, 52.9mm ($n = 3$, subadults, ♀,?,♂, USNM). *Bill* from nares 15.8mm; *Tarsus* 54.9mm ($n = 1$, ♀, USNM 190375). *G. r. avilae Wing* range 98–100mm / 97mm; *Tail* range 54–59mm / 55–57mm; *Bill* [total? culmen] range 23–24mm / 23.5–24.0mm ($n = ?$, ♂♂/♀♀, type series, Hellmayr & von Sneidern 1914). *Wing* 92–98mm; *Tail* 51.5–55.0mm; *Bill* culmen from base 26.0–28.5mm, exposed culmen 21–23mm; *Tarsus* 50–53mm ($n = 6$, ♂♂, Phelps & Gilliard 1940). *Bill* from nares 16.9mm; *Tarsus* 51.2mm ($n = 1$, ♂, USNM 605766). *G. r. connectens Bill* from nares 15.9mm, 17.6mm; *Tarsus* 53.3mm, 55.9mm ($n = 2$, ♀ adult, ♂ subadult, USNM 323091, 323090; see Reproduction). *G. r. albiloris Wing* 104mm, 99mm; *Tail* 67mm, 62mm; *Bill* from nares 17mm, 16mm; *Tarsus*

52mm, 54mm (n = 2, ♂,♀, Taczanowski 1880). *Bill* from nares 17.3mm, exposed culmen 24.2mm, depth at front of nares 8.4mm, width at gape 14.1mm, width at front of nares 7.0mm, width at base of lower mandible 10.5mm; *Tarsus* 52.7mm (n = 1, ♀, MSB 35251). *Bill* from nares 17.0mm; *Tarsus* 53.3mm (n = 1, ♀, AMNH 175282). **G. r. interior** *Wing* mean 101.5mm, range 100.0–103.5mm; *Tail* mean 58.6mm, range 53–64mm; *Bill* culmen from base mean 27mm, range 26.5–27.75mm; *Tarsus* mean 50.5mm, range 49.5–51.0mm (n= 10, ♂♂, Zimmer 1934). *Wing* mean 99mm, range 97.0–100.5mm; *Tail* mean 56.2mm, range 55–60mm; *Bill* culmen from base mean 26.5mm, range 26.0–26.75mm; *Tarsus* mean 50.5mm, range 50–51mm (n=?, ♀♀, Zimmer 1934). *Bill* from nares 14.0mm, exposed culmen 19.9mm, depth at nares 7.5mm, width at gape 15.0mm, width at nares 6.5mm, width at base of lower mandible 9.0mm; *Tarsus* 51.0mm (n = 1, ♂ fledgling, MSB 32055). **Mass** 76.3g, 79.1g, 79.1g, 79.1g, 69.9g, 85.6g, 87.6g, 89.8g (n = 8, ♂,♀,♀,♀,?,♀,?,♂, *ruficapilla*, Echeverry-Galvis *in litt.* 2017; means presented in Echeverry-Galvis *et al.* 2006); 75.6g, 83.4g, 84.1g, 89.4g, 91.5g (n = 5, ♂♂, *ruficapilla*, MVZB labels; reported by Miller 1963); ♂ range 70–98g, ♀ range 70–92g (n = ?, subspecies unspecified; Krabbe & Schulenberg 2003); 68.1g(lt) (n = 1, ♂ fledgling, *interior*, MSB 32055); 70.0g (n = 1, ♂, *interior*, MVZB 156499); 73.9g(m), 80.9g(no), 83.7g(no), 87.7g(no) (n = 4, ♀,?,?,?, *ruficapilla*, B. Mila *in litt.* 2015); 81g (n = 1, ♀?, *ruficapilla*, Guerrero 1996); 76.9g(lt) (n = 1, ♀, *albiloris*, MSB 35251); mean 88g (n = 2, sex/race unspecified, Schulenberg & Williams 1982). **Total Length** 17.8–20.0cm (Hilty & Brown 1986, Fjeldså & Krabbe 1990, Ridgely & Tudor 1994, Guerrero 1996, Ridgely & Greenfield 2001, Krabbe & Schulenberg 2003).

TAXONOMY AND VARIATION Formerly, most authors classified Watkins's Antpitta as a subspecies of the present species (e.g. Peters 1951, Meyer de Schauensee 1966). Subsequently, it was discovered that the song of *watkinsi* is very different from that of *ruficapilla*, and that *watkinsi* merited species rank (Parker *et al.* 1995). Furthermore, the distributions of *watkinsi* and *ruficapilla connectens* closely approach one another in Ecuador, with no evidence of introgression (Krabbe 1992, Ridgely & Greenfield 2001). Phylogenetic analysis of mitochondrial DNA (from the cytochrome *b* and ND2 genes), which included only 11 of the more than 30 species in the genus, confirms that, as expected, *ruficapilla* and *watkinsi* are sister species (Krabbe *et al.* 1999). I have maintained the seven currently recognised subspecies (Krabbe & Schulenberg 2003) for the purposes of this account. In researching their distributions, however, it quickly became clear that there is no general agreement as to the geographic range of several, particularly those in southern Ecuador and northern Peru. This is in large part due to the somewhat dubious plumage characters used for separating them, compounded by the lack of modern analysis of vocal differences (if any). Indeed, in his description of Watkins's Antpitta, Chapman (1919) examined specimens of what he considered to be *albiloris* from Levanto, *albiloris* from Tabaconas and nominate *ruficapilla* from Zaruma. In the present work, these specimens are considered to represent *interior*, *ruficapilla* and *connectens*, respectively. It seems likely that, at best, four taxa might stand up to rigorous analysis (*perijana*, *avilae*, *nigrolineata-ruficapilla-interior* and *connectens-albiloris*). For those interested in pursuing such questions, the ranges of *albiloris* and *connectens* appear

to meet around Balsas (Amazonas, Peru, *c.*06°50'S, 78°01'W). Similarly, the range of *albiloris* approaches that of nominate *ruficapilla* near Huancabamba (Piura, 05°14'S, 79°27'W), and the ranges of all three approach each other in the vicinity of Jaén (Cajamarca, 05°42'S, 78°48'W). Subspecies *connectens* and *ruficapilla* populations are of interest in the vicinity of the Peru/Ecuador border where Ecuador's Zamora-Chinchipe and Loja provinces meet (*c.*04°44'S, 79°25'W) and nearby localities in northern Piura (see *connectens*), as well as near Lamkapac (Loja, 03°45.5'S, 79°17'W). It remains unclear where the ranges of *nigrolineata* and *ruficapilla* meet, but the following localities (here included in *nigrolineata*) would be informative: Páramo El Zumbador (Táchira); Peña Blanca (Boyacá, 04°03'N, 73°45'W); La Pica (Santander, 06°45'N, 72°45'W); Cachiri (Santander, 07°30'N, 73°01'W); Las Ventanas (Norte de Santander, 07°48'N, 73°06'W); Buenos Aires (Norte de Santander, 08°01'N, 72°58'W); and Hacienda Las Vegas (Santander, 07°04'N, 72°56'W). The last-named locality appears (eBird) to have a particularly healthy, accessible population. Similarly, at the north end of the Venezuelan Andes, it remains unclear if populations in southern Lara and northern Trujillo are best placed with *avilae* or *nigrolineata* (as here included).

Grallaria ruficapilla avilae Hellmayr & von Sneidern 1914, Verhandlungen der Ornithologischen Gesellschaft in Bayern, vol. 12, p. 92, Galipan, 2,000m, Cerro del Avila, Venezuela [= Picacho de Galipán, Cerro El Ávila, Vargas, *c.*10°34'N, 66°54'W, near Caracas]. Described originally in German and named after the mountain on which it was discovered, the type was collected by S. M. Klages on 27 October 2013. The holotype (ZSM 13.1047) is an adult male at the Zoologische Staatssammlung München. The range of *avilae* is generally given as the northern Venezuelan Andes in southern Lara (around Cubiro, 09°47'N, 69°35'W) east in the coastal mountains to Miranda (Cory & Hellmayr 1924, Hilty 2003, Krabbe & Schulenberg 2003, Greeney & Kirwan 2013). This would seem unlikely, however, given the apparently complete distributional continuity of Andean populations of *nigrolineata* with localities in southern Lara, and the more than 200km between these records and coastal populations of *avilae*. In light of this, I consider *avilae* to be endemic to the coastal mountains of northern Venezuela (Phelps 1966), from extreme eastern Carabobo to western Vargas. This subspecies is quite similar to *nigrolineata*, but said to have the crown and nape richer chestnut, and the feathers of the back a paler greenish-olive. The streaking on the underparts is similarly heavy, as in *nigrolineata*, but the pale ochraceous fringes to the breast feathers are slightly more pronounced (Hellmayr & von Sneidern 1914, Krabbe & Schulenberg 2003). Bare-parts coloration (USNM specimen labels): bill black, legs grey, eyes brown. **Specimens & Records** *Aragua* Fundo Jeremba, 10°26'N, 67°15'W (COP, ML); 40km W of El Junquito, 10°25'N, 67°13'W (AMNH, COP); Colonia Tovar, 10°24.5'N, 67°17'W (COP, AMNH, CM, ML); Pico Guacamayo, 10°24'N, 67°35'W (USNM 575268); Puesto Guardaparques Quebrada Honda, 10°21'N, 67°09'W (ML 32434811, F. Machado-Stredel). *Vargas* Picacho de Galipán, 10°33.5'N, 66°54.5'W (YPM, AMNH, CM, ANSP); El Limón, 10°28'N, 67°17'W (CM P104110), El Junquito, 10°28'N, 67°05'W (COP, ML, AMNH). *Distrito Federal* Plan de Los Lirios, 10°33'N, 66°55'W (COP 3341–43, USNM 605766); San Antonio de Galipán, 10°33'N, 66°53'W (COP

62597–602, ROM); near Hotel Humboldt, 10°32.5'N, 66°52.5'W (USNM 504457); Estación Teleférico del Cerro El Ávila, 10°32'N, 66°52'W (COP 62601); Silla de Caracas, 10°32'N, 66°51'W (Hellmayr & von Sneidern 1914, COP, AMNH, CM, MCZ); Los Venados, 10°32'N, 66°54'W (J.E. Miranda T. *in litt.* 2017); road to El Junquito, 10°28'N, 67°03'W (COP). *Miranda* Pico Naiguatá, 10°33'N, 66°46'W (USNM 325002, CM); Hacienda Izcaragua, 10°31'N, 66°40'W (COP 18849); Laguneta de la Montaña, 10°21.5'N, 67°07.5'W (J.E. Miranda T. *in litt.* 2017). SEE ALSO Cordero-Rodríguez (1987).

Grallaria ruficapilla nigrolineata Sclater 1890, Catalogue of the Birds in the British Museum, vol. 15, p. 320, Merida [Sierra Nevada de Mérida], Venezuela. Described as *Grallaria ruficapilla nigro-lineata*, the subspecific epithet was hyphenated in much of the older literature. There are two syntypes in Tring. One of these (NHMUK 1889.9.20.634) was collected on 6 November 1886 by L. Brieño, and apparently obtained by Sclater from Dr. E. Rey of Leipzig (Warren & Harrison 1971). This subspecies is generally considered endemic to Venezuela and listed as occurring in the Andes of northern Trujillo south to north-east Táchira (Hilty & Brown 1986, Hilty 2003, Krabbe & Schulenberg 2003). At the northern end of its distribution, however, there is no clear reason why birds would, so abruptly, be referable to *avilae* (which see), and I have here extended its range to include northern Trujillo and southern Lara. Similarly, at the south end of the Venezuelan Andes, records of *G. ruficapilla* extend, rather continuously, across the Colombian border (see below) and it seems highly likely that at least the birds in this region of Colombia refer to *nigrolineata*. Therefore, I provisionally include all birds north of the Río Chicamocha (right bank) within *nigrolineata*. This arrangement was, at least formerly, considered to be the case by Meyer de Schauensee (1952a, 1964), who referred birds from Pamplona (Norte de Santander, 07°22'N, 72°39'W) to *nigrolineata*, but apparently the validity of this has not been investigated. Subspecies *nigrolineata* is similar to the nominate race, but the streaking on breast and flanks is heavier and darker. SPECIMENS & RECORDS VENEZUELA: *Lara* Cubiro, Fila Santo Domingo, 09°46'N, 69°34'W (as *avilae*, Phelps & Phelps 1950, Hilty 2003, COP). *Trujillo* Páramo Las Rosas, 09°42'N, 70°07'W (CM); Cerro Niquitaz, 09°32'N, 70°10'W (COP 20175/76); Páramo Cendé, 09°28'N, 70°05'W (Phelps & Phelps 1950, COP 19958–60); Páramo Misisí, 09°20'N, 70°18'W (as *avilae*, Phelps & Gilliard 1940, but as *nigrolineata*, Phelps & Phelps 1950; COP 4957); *c.*4km SW of Guaramacal, 09°10'N, 70°11'W (XC 223674, P. Boesman); Guamito, 09°05'N, 70°25'W (CM). *Mérida* Paramitos, *c.*09°00.5'N, 70°47'W (USNM 504458); Páramo Las Tapias, 09°00'N, 70°51'W (ML: P.A. Schwartz); Llano Rucio, 08°55'N, 71°05'W (COP 14212); La Cuchilla, *c.*08°54'N, 70°38'W (FMNH, CM); Conejos, 08°50'N, 71°15'W (FMNH 57188); La Honda, 08°49'N, 70°44'W (COP 49282/85); Hotel Páramo La Culata, 08°43.5'N, 71°05'W (ML 178440, J.W. McGowan); Quebrada La Capaz, 08°43'N, 71°24'W (FMNH 57189); El Valle, 08°41.5'N, 71°06'W (AMNH, FMNH, MCZ, COP, CUMV); Páramo El Escorial, *c.*08°38.5'N, 71°06'W (AMNH, FMNH, USNM); Tabay, 08°38'N, 71°04'W (CM, YPM); La Mucuy, 08°37.5'N, 71°04'W (Stiles 1984, XC); Quebrada La Pedregosa, 08°36'N, 71°12'W (FMNH 50736); Pico Humboldt Trail, 08°36'N, 71°03'W (many recordings XC, ML); Páramo Tambor, 08°35'N, 71°26'W

(FMNH 53437); La Culata, 08°35'N, 71°13.5'W (FMNH, AMNH, COP); Ejido, 08°33'N, 71°14'W (AMNH 492228); Caserío Arangurén, 08°32'N, 70°53'W (COP 71522/23); Los Nevados, 08°27.5'N, 71°05'W (FMNH 57187); Páramo Aricagua, 08°16'N, 71°11'W (COP 45374); 'Cordillera de Mérida' or simply 'Mérida' (Sclater & Salvin 1870, AMNH, USNM). *Barinas* Mesa de Lino, 08°51'N, 70°36'W (COP 49284). *Táchira* Pregonero, 08°04'N, 71°50'W (COP 24539–542); Páramo El Zumbador, 08°00'N, 72°05'W (Phelps & Phelps 1950, COP, CM); Queniquea, 07°54'N, 72°01'W (COP 9093); near Matamula, 07°42'N, 72°26'W (XC 223671, P. Boesman); Hacienda La Providencia, 07°33'N, 72°22'W (long series COP). *Apure* Betania, 07°25'N, 72°25.5'W (ML 29012101, J.E. Miranda T.). COLOMBIA: *Norte de Santander* RNA Hormiguero de Torcoroma, 08°12'N, 73°23'W (Salaman *et al.* 2007a); Buenos Aires, 08°01'N, 72°58'W (USNM 398001–003); Guaimaral, 08°00'N, 72°48'W (ML 62463–67: P.A. Schwartz); Las Ventanas, 07°48'N, 73°06'W (CM P57601/856); Pamplona, 07°22.5'N, 72°39'W (Meyer de Schauensee 1952a, Iafrancesco *et al.* 1987); Hacienda La Primavera, 07°00'N, 72°20'W (FMNH 261455–457). *Santander* Cachiri, 07°30'N, 73°01'W (CM); Las Picotas, 07°23'N, 72°53'W (XC 282403–405, J.E. Avendaño); Hacienda Las Vegas, 07°04'N, 72°56'W (USNM 411871–873); La Pica, 06°45'N, 72°45'W (CM P59549/571); Alto Onzaga, 06°34'N, 72°44'W (XC 12499, O. Laverde). *Boyacá* Peña Blanca, 06°33'N, 72°30'W (CM P59628/629/686).

Grallaria ruficapilla perijana Phelps & Gilliard, 1940, American Museum Novitates, no. 1100, p. 3, La Sabana, 1,500m, Río Negro, Perijá District, Zulia, Venezuela [=10°00'N, 72°45'W; Paynter 1982]. The holotype is an adult male collected 26 February 1940, by Alberto Fernández Yépez. Confined to the Perijá Mountains in extreme northern Colombia, at the border with western Venezuela. This subspecies is quite similar to *avilae*, but 'dark streakings of underparts decidedly fewer and narrower, and with a great deal more light Ochraceous-Orange suffusion especially on chest; the Ochraceous-Orange markings extending brightly to the middle abdomen and to the posterior flanks while in *avilae* it is largely confined to the chest with only subobsolete tracings on the upper abdomen and flanks' (Phelps & Gilliard 1940). They described the iris as brown and both the bill and tarsi as light grey. SPECIMENS & RECORDS COLOMBIA: *La Guajira* La Africa, 10°34'N, 72°58'W (USNM 373655/56); Laguna De Junco, 10°29N, 72°55W (USNM 373655–663). *Cesar* Above El Cinco, 10°22'N, 72°57'W (XC 331811, D. Calderón-F.); 6km ESE of La Nueva Estrella, 10°22'N, 72°58'W (ML 26154731, photo K. Fiala); RNA Chamicero del Perijá, 10°21.5'N, 72°57'W (ML 46892261, R. Gallardy); above Eroca, 09°42'N, 73°05'W (USNM 373653/54); Camp Perijá, 09°49N, 73°03W (Hilty & Brown 1983); general Sierra del Perijá (López-O. *et al.* 2014, Ellery 2015). VENEZUELA: *Zulia* Base of Nudo de Febrero, 10°14'N, 72°42'W (COP 57672/73); Cerro Tetari, 10°05'N, 72°55'W (COP 54852); La Sabana, 10°02'N, 72°45'W (COP 6422–32, AMNH 428842/43); Cerro Pejochaina area, *c.*10°02'N, 72°49'W (COP 54845–51); Fila Macoita, Campamento Avispa, 10°16'N, 72°42'W (COP 57674/75).

Grallaria ruficapilla ruficapilla Lafresnaye 1842, Revue Zoologique par La Société Cuvierienne, p. 333, Santa Fé de Bogotá 'habitat in Bolivia'. The holotype is, unfortunately, one of the many 'Bogotá trade skins,' and its true origin will likely never be known (Olivares 1969). The distribution of the nominate race is fairly well documented. It is distributed through most of the Colombian Andes except the extreme north-east at the Venezuelan border (see race *nigrolineata*) and through most of the Ecuadorian Andes except the south-west (see *connectens*). The northernmost record in the West Andes is historical, from Hacienda Portreros (Antioquia, 06°39N, 76°09W, USNM, Hilty & Brown 1983). A notable lack of records in the West Andes of Colombia, from north of El Limonar (Valle del Cauca, 03°35''N, 76°40'W) to at least around La Celia (Risaralda, 05°00'N, 76°00'W), remains unexplained. A single sight record at Páramo del Duende (eBird: Valle del Cauca, 04°06'N, 76°25'W) seems credible, but this region has been omitted from the map until further confirmation of its presence in this *c.*170km gap surfaces. With such a common and widespread taxon, the body of literature is diverse and extensive, and I include here only records that help to define its distribution, including non-literary records in geopolitical areas poorly represented in the literature. SPECIMENS & RECORDS COLOMBIA: *Antioquia* *c.*7.5km S of Angostura, 06°49'N, 75°20'W (XC 331467/68; A. Angulo); Hacienda Portreros, 06°39'N, 76°09'W (Hilty & Brown 1983; USNM 426423–428); base of Páramo de Frontino, 06°28'N, 76°04'W (USNM 436482/83); Vereda La Lana, 06°27'N, 75°35.5'W (Donegan *et al.* 2009a, XC); Vereda El Apretel, 06°25'N, 75°28.5'W (Donegan *et al.* 2009a); Caicedo, 06°24.5'N, 75°58.5'W (XC 311839; A.E.H. Guarín); Vereda Ovejas, 06°21.5'N, 75°39.5'W (Donegan *et al.* 2009a); Quebrada El Viao, *c.*06°03'N, 75°11'W (Flickr: J. Ochoa); Las Nubes, 05°48'N, 75°47.5'W (A.M. Cuervo *in litt.* 2017); Páramo de Sonsón, 05°43'N, 75°15'W (USNM 436484); RN Hacienda La Esperanza, 05°36.5'N, 75°50'W (J.B.C. Harris *in litt.* 2017); Alto de Ventanas, 05°32.5'N, 75°48'W (XC 119656/57; D. Edwards); above Finca La Primavera, 05°30'N, 75°53'W (A.M. Cuervo *in litt.* 2017). *Chocó* La 'M,' 05°59.5'N, 76°09.5'W (F. Rowland *in litt.* 2017). *Risaralda* PRN Santa Emilia, 05°12'N, 75°54'W (Zuleta-Marín 2012); DMI Agualinda, 05°07.5'N, 75°56.5'W (J.A. Zuleta-Marín *in litt.* 2017); El Pabellon, 04°51.5'N, 75°29'W (D. Calderón-F. *in litt.* 2017); La Florida, 04°46'N, 75°37'W (AMNH 109629/30); El Cedral, 04°42'N, 75°32'W (XC 299429; J. Bradley). *Quindío* El Roble, 04°40.5'N, 75°36'W (AMNH, USNM, FMNH); Salento, 04°38.5'N, 75°34'W (AMNH, ANSP, USNM); Laguneta, 04°35'N, 75°30'W (ANSP 154017–023); El Eden, 04°27'N, 75°46'W (AMNH 112006–008, FMNH 51011); general Dept. Quindío (Arbeláez-Cortés *et al.* 2011b). *Tolima* Vereda Santa Isabel, 04°45'N, 75°20'W (Cuadros 1988); Vereda Guaimaral, 04°41.5'N, 75°06.5'W (Martinez 2014); La Flor, 04°39.5'N, 75°07'W (Martinez 2014); Toche, 04°32'N, 75°25'W (ANSP 154013–016); *c.*10km NW of Cajamarca, 04°29'N, 75°29'W (XC 127872–874; O.H. Marín-Gómez); Clarito Botero, 04°28'N, 75°13.5'W (Parra-Hernández *et al.* 2007, Martinez 2014); Río Toche, 04°26'N, 75°22'W (López-Lanús *et al.* 2000, AMNH 112009). *Valle del Cauca* Miraflores, 03°35'N, 76°10'W (CUMV 7216); east of Palmira, *c.*03°32'N, 76°08'W (AMNH 108934/35); Bosque de San Antonio, 03°30'N, 76°38'W (eBird: J.V. Remsen); San Antonio and vicinity, *c.*03°12.5'N, 76°39'W (MVZB, AMNH, MCZ). *Cauca* 15km N of Popayán, *c.*02°40'N, 76°35'W (MVZB); Uribe, 02°33'N, 76°51.5'W (LACM 36437); Cerro Munchique, 02°32'N, 76°57'W (LACM, USNM, YPM, LSUMZ, CM, FMNH); near Cocal, *c.*02°31'N, 77°00'W (Donegan & Davalos 1999; AMNH 109633); Polindara,

02°29'N, 76°24'W (Ayerbe-Quiñones *et al.* 2009, YPM 26975); El Ciruelo, 02°28'N, 76°53'W (ROM 107611/12); El Tambo, 02°27'N, 76°48.5'W (ANSP 144672/673, USNM 388583); Popayán, 02°26.5'N, 76°37'W (Ayerbe-Quiñones *et al.* 2009; ANSP 8559); Páramo Puracé, 02°24'N, 76°23'W (USNM 446647/48); Timbío, 02°21'N, 76°41'W (USNM 388584); El Dorón, 01°40.5'N, 76°14.5'W (Donegan & Salaman 1999, Salaman *et al.* 1999, 2007b); Tatauí, 01°37'N, 76°16'W (Donegan & Salaman 1999). *Santander* *c.*11km W of Galán, 06°38.5'N, 73°23.5'W (XC 25437/38; T. Donegan). *Boyacá* Soatá, 06°20'N, 72°41'W (Meyer de Schauensee 1950b, USNM 444550); Arcabuco, 05°44'N, 73°23'W (XC 215991; S. Chaparro-Herrera); Tunja, 05°32.5'N, 73°21.5'W (Iafrancesco *et al.* 1987); Cerro Comijoque, 05°26'N, 72°41.5'W (Bohórquez 2002); RNSC El Secreto, 05°04.5'N, 73°21.5'W (XC 12327, O. Laverde); Santa María, 04°53'N, 73°17'W (XC 326981, D.R.R. Villamil). *Casanare* Finca Buenos Aires, 05°20'N, 72°46.5'W (XC 245584, O.H. Marín-Gómez); Finca La Garantía, 05°15'N, 72°52.5'W (XC 245523/34, O.H. Marín-Gómez). *Bogotá* Quebrada La Vieja, 04°39'N, 74°03'W (O. Cortes-Herrera *in litt.* 2017); Cerro de Montserrate, 04°36.5'N, 74°03'W (XC 345898, N. Komar). *Meta* El Calvario, 04°22'N, 73°43.5'W (XC 96234/35, O. Cortes). *Cundinamarca* Laguna de Tabacal, 05°01.5'N, 74°19.5'W (eBird: P. Holmes); Cajicá, 04°57'N, 74°03'W (ML 62502351, M.G. Gomez); Santuario Señor de la Piedra Sopó, 04°54.5'N, 73°56'W (XC 211074, J.I.G. Arango); *c.*6km E of Albán, 04°52.5'N, 74°23'W (Echeverry-Galvis 2001); Finca Santa Clara, *c.*04°51.5'N, 74°05.5'W (Echeverry-Galvis & Morales-Rozo 2007); Vereda 'Cerca de Piedra', 04°51'N, 74°03'W (Echeverry-Galvis & Morales-Rozo 2007); La Trinidad, 04°48'N, 73°54'W (XC 308984; D.R.R. Villamil); Pueblo Viejo, municipio Zipacón, *c.*04°45'N, 74°23'W (DMNH 56511); Cerro La Calera, 04°44.5'N, 73°59'W (Asociación Bogotana de Ornitología 2000); Vereda Chircal, 04°43'N, 74°22.5'W (ML 28767181, photo S.A. Collazos-González); RFP Cárpatos, 04°42'N, 73°51'W (Stiles & Rosselli 1998); Palacio, 04°41'N, 73°50'W (Lotta *et al.* 2015); RN Chicaque, 04°36.5'N, 74°19'W (XC 312397/98; G. Leite); Finca Rancho Grande, *c.*04°36'N, 74°20'W (Munves 1975); Medina, 04°35'N, 73°25'48.00'W (ML 80800/02; M. Álvarez); Choachí, 04°31.5'N, 73°55.5'W (Olivares 1969, AMNH 176533, MCZ 94037); El Peñón, 04°26'N, 74°18'W (Olivares 1969; MCZ 280109); El Roble, 04°23'N, 74°19'W (Olivares 1969; CUMV 7217); Fusagasugá area, *c.*04°20.5'N, 74°21.5'W (Olivares 1969, Iafrancesco *et al.* 1987, CUMV 7218); La Vuelta, 04°19.5'N, 74°29.5'W (eBird: L. Cardenas Ortíz). *Huila* RNSC Merenberg, 02°13'N, 76°07.5'W (Ridgely & Gaulin 1980); RFP Los Yalcones, 01°48.5'N, 76°21'W (XC: D. Calderón-F.); RNSC El Cedro, 01°40.5'N, 76°14'W (ML 60828621, photo J.L. Peña). *Caquetá* Finca Andalucía, 02°44'N, 74°51.5'W (Bohórquez 2002); Cerro Negro, 01°46'N, 75°46'W (XC 303603; J. Muñóz). *Nariño* Corregimiento de Daza, 01°16'N, 77°15'W (XC 12709/13, O. Laverde); Ricaurte, 01°13'N, 77°59.5'W (AMNH, ANSP); El Guabo, 01°07'N, 77°49'W (FMNH, LSUMZ, YPM, WFVZ); Tuquerres, 01°05'N, 77°37'W (ANSP 162060/61); Mayasquer, 00°54'N, 78°04'W (ANSP 149858); Llorente, 00°49'N, 77°15'W (FMNH 292126/27). *Caldas* Aranzazu, 05°16'N, 75°29'W (Pulido-Santacruz & Renjifo 2011); Finca Termópilas, 05°13'N, 75°29'W (Cadena *et al.* 2007); RFP Río Blanco, 05°05'N, 75°30'W (XC, IBC 1263136, photo D. Avendaño); El Zancudo, 05°03.5'N, 75°26'W (CM P70224/424); Aguas de Manizales, 05°04.5'N, 75°26.5'W (XC 100371, J. King).

Putumayo Vereda Bellavista, 01°14'N, 76°55.5'W (XC 247859; B.C. Jaramillo); El Mirador, Carretera 10, 01°04'N, 76°44.5'W (ML 45680421, D. Rocha López, ML 45957151, J. Muñoz García); Río San Miguel, *c.*00°24'N, 77°23'W (Meyer de Schauensee 1952b; ANSP 165178). **ECUADOR**: ***Carchi*** Loma Laurel, 00°50'N, 78°03'W (XC 248323, N. Krabbe); W of Tulcán, *c.*00°48'N, 78°00'W (ANSP 181200/201), NE of El Moran, 00°46'N, 78°01'W (ML: M.B. Robbins); *c.*6.3km SW of Mira, 00°31.5'N, 78°05.5'W (R. Ahlman *in litt.* 2017). ***Esmeraldas*** 4km W of Lita, *c.*00°52'N, 78°31'W (ML 86020, J.W. Wall); Cordillera de Toisán, *c.*00°27'N, 78°47'W (J. Freile *in litt.* 2017). ***Imbabura*** Sendero Bosque Agua Savia, 00°28'N, 78°14'W (eBird: M. Tellkamp); 6.5km ENE of Apuela, 00°22.5'N, 78°27'W (R.A. Gelis *in litt.* 2017); Intag, 00°20'N, 78°32.5'W (Tristram 1889, Goodfellow 1902, Poulsen & Krabbe 1998); *c.*km21 on Laguna Cuicocha–Apuela road, *c.*00°20'N, 78°25'W (LSUMZ 112586/87, MECN 2609); Laguna de Cuicocha, 00°18'N, 78°21.5'W (ML 62106621, photo 'The Wildlab'); Loma Taminanga, 00°17'N, 78°28'W (MECN 2602); Cerro Blanco Antenna road, 00°12'N, 78°20'W, (eBird: C. Vogt); La Delicia, 00°08.5'N, 78°48'W (MECN 2608). ***Pichincha*** Cerro Mojanda, 00°08'N, 78°17'W (Berlioz 1927); Reserva Geobotánica Pululahua, 00°06'N, 78°32'W (R.A. Gelis *in litt.* 2017); Gualea, 00°05'N, 78°44.5'W (AMNH, USNM, MCZ); RP Refugio Paz de las Aves, 00°00.5'N, 78°42.5'W (XC 349319, E. Hutchings); Reserva Ecoturística Alpa Huasi, 00°00'N, 78°45.5'W (J. Freile *in litt.* 2017); Alaspungo and vicinity, *c.*00°00'S, 78°36'W (Ménégaux 1911; MLZ 7210, ANSP 169710); RP Bellavista, 00°01'S, 78°41'W (Welford 2000, Becker *et al.* 2008, MECN 2607, XC, ML); Recinto 23 de Junio, 00°02'S, 78°52.5'W (XC 186919, G. Leite); Guarumos, 00°04'S, 78°38'W (AMNH 447370/FMNH 99575); 9km ENE of Mindo, 00°05'S, 78°40'W (XC 71346/47, D.F. Lane); RP Verdecocha, 00°06'S, 78°38'W (MCZ 138481); Yanacocha area, *c.*00°06.5'S, 78°35'W (MECN 2606); Loma Bahamonte, 00°17'S, 78°40'W (XC 248649, N. Krabbe); RVS Pasachoa, *c.*00°26'S, 78°29'W (J. Freile *in litt.* 2017); SW slope of Volcán El Corazón, 00°33'S, 78°43'W (Poulsen & Krabbe 1998). ***Santo Domingo de los Tsáchilas*** Old Chiriboga road, *c.*00°14'S, 78°45'W (XC 9967/68; A.T. Chartier); Estación Científica Río Guajalito, 00°14'S, 78°49'W (J. Freile *in litt.* 2017). ***Sucumbíos*** Near Santa Bárbara, 00°38.5'N, 77°31'W (R. Ahlman *in litt.* 2017); Quebrada Las Ollas, 00°33.5'N, 77°32'W (XC 248362, N. Krabbe); Alto La Bonita, 00°29.5'N, 77°35'W (Stotz & Valenzuela 2009); Bermejo, 00°14.5'N, 77°23'W (Schulenberg 2002). ***Orellana*** Upper Volcán Sumaco, *c.*00°31'S, 77°37.5'W (MCZ 138477). ***Napo*** Oyacachi, 00°12.5'S, 78°04.5'W (Poulsen & Krabbe 1997b; CAS-90677); La Cabaña, 00°16'S, 77°42'W (J. Freile *in litt.* 2017); Río Maspa Chico, 00°24'S, 78°02'W (Poulsen 1996); Cuyuja, 00°25'S, 78°01.5'W (MECN 2604/05); above Baeza, *c.*00°28'S, 77°56'W (AMNH 180270, MCZ 138478); upper Río Bermejo, 00°31.5'S, 77°56'W (HFG); RP Cabañas San Isidro, 00°35.5'S, 77°53'W (XC 250897, N. Krabbe); RP Estación Biológica Yanayacu, 00°36'S, 77°53.5'W (XC 2382, W. Halfwerk); above Cocodrilos, 00°39'S, 77°48'W (R. C. Dobbs *in litt.* 2017); RP SierraAzul (Hacienda Aragón), 00°40'S, 77°55'W (Martin & Greeney 2006; ANSP 186180). ***Chimborazo*** 25km by road Baños–Río Bamba, *c.*01°31'S, 78°31'W (FMNH 373093); 3.3km N of Pallatanga, 01°59'S, 78°58'W (Sclater 1860a; MLZ 7204); Cayandeled, 02°07'S, 78°59'W (von Berlepsch & Taczanowski 1884, Salvadori & Festa 1899); near Huigra,

*c.*02°17'S, 78°59'W (Domaniewski & Sztolcman 1918, ANSP 59362). ***Cañar*** Llavircay, 02°33.5'S, 78°37.5'W (P. Astudillo Webster *in litt.* 2017). ***Cotopaxi*** Reserva Integral Otonga, 00°25'S, 79°00'W (Freile & Chaves 2004); Las Palmas, 00°35'S, 79°01'W (Krabbe 1991); SW slope of Volcán Iliniza, Río Ratuncama, *c.*00°42'S, 78°47.5'W (MECN 2603). ***Tungurahua*** San Rafael, 01°22'S, 78°29'W (Taczanowski & von Berlepsch 1885); Machay, 01°23'S, 78°17'W (Benítez *et al.* 2001); Baños and vicinity, *c.*01°24'S, 78°25'W (AMNH, FMNH, MCZ, SBMNH); 'Yunguilla', Río Pastaza, 1900m' *c.*01°24'S, 78°18'W (MCZ 199131; loc. adjusted for elevation from Paynter 1993). ***Bolívar*** *c.*9km NW of Salinas, 01°21'S, 79°05'W (Poulsen & Krabbe 1998); BP Cashca Totoras, *c.*01°45'S, 78°58'W (Bonaccoroso 2004); Tiquibuzo, 02°01'S, 79°05'W (J. Freile *in litt.* 2017). ***Azuay*** Saucay, 02°45'S, 79°01'W (P.X. Astudillo W. *in litt.* 2017); Huasihuaico, 02°46'S, 79°20'W (Krabbe 1991); Sural, 02°47'S, 79°26'W (Krabbe 1991); Río Mazán valley, *c.*02°52'S, 79°07'W (King 1989, Guerrero 1996, Poulsen & Krabbe 1998, Latta *et al.* 2011, Astudillo *et al.* 2015); Llano Largo, 02°50.5'S, 79°23'W (P.X. Astudillo W. *in litt.* 2017); Yunga-Totorillas, 03°01'S, 78°58'W (J. Freile *in litt.* 2017); Río Jubones valley, 03°17'S, 79°15'W (Best *et al.* 1992, Krabbe 1992); RP Yunguilla, 03°14'S, 79°17'W (Greeney & Kirwan 2013, MECN 8357); Estación El Gullán, 03°20.5'S, 79°10.5'W (J. Freile *in litt.* 2017). ***El Oro*** Salvias, 03°47'S, 79°21'W (AMNH 167250). ***Loja*** Papaya, S bank of Río Paquishapa, 03°31'S, 79°16'W (XC 250142, N. Krabbe); Loja, 03°58'S, 79°12'W (AMNH 167249); Parque Universitario (PUEAR) Universidad Nacional de Loja, 04°02'S, 79°11.5'W (XC 77140/99950; L. Ordóñez-Delgado); Nudo de Cajanuma, 04°05'S, 79°12'W (MLZ-9832); Cajanuma, *c.*04°07'S, 79°10.5'W (MCZ 299197, XC). ***Morona-Santiago*** Road above Macas, 2,635m, 02°13'S, 78°23'W (background of XC 251670, N. Krabbe); upper Gualaceo-Limón road, *c.*03°00.5'S, 78°38.5'W (eBird: M. Reid). ***Zamora-Chinchipe*** Loja–Zamora pass above El Tambo, 03°59.5'S, 79°08'W (R.A. Gelis *in litt.* 2016); Quebrada Honda, 04°28.5'S, 79°07.5'W (HFG); Río Isimanchi above San Andrés, 04°47'S, 79°20'W (MECN 6496). **PERU**: ***Cajamarca*** NE of San José de Lourdes, *c.*05°05'S, 78°52.5'W (XC 132095, D.F. Lane, MCZ, ML); near Chaupe, 05°14.5'S, 79°07'W (A. García-Bravo *in litt.* 2017); Tabaconas, 05°19'S, 79°17.5'W (Zimmer 1934, Bangs & Noble 1912, as *albiloris*; MCZ 79976); Agua Azul, 05°35'S, 79°09.5'W (MSB 42184); ACM Bosque de Huamantanga, 05°40'S, 78°56.5'W (A. García-Bravo *in litt.* 2017); Quebrada Lanchal, 05°41'S, 79°15'W (LSUMZ 169899–906); San Felipe, 05°46'S, 79°19'W (Zimmer 1934, as *albiloris*); near El Edén, 05°51.5'S, 78°54'W (A. García-Bravo *in litt.* 2017). ***Piura*** E slope of Cerro Chinguela, 05°07'S, 79°23'W (ML 21739, T.A. Parker).

Grallaria ruficapilla connectens Chapman 1923, American Museum Novitates, no. 86, p. 9, Taraguacocha, 9750ft. [2,972m], Cordillera de Chilla, El Oro, Ecuador [= 03°40'S, 79°40'W; LeCroy & Sloss 2000]. The holotype (AMNH 167253) is an adult male collected by George K. Cherrie on 19 August 1920. The distribution of this subspecies is unclear, but it is generally stated to be confined to south-west Ecuador, in the provinces of El Oro and Loja (Best & Clarke 1991, Bloch *et al.* 1991, Ridgely & Greenfield 2001), where it occurs sympatrically with *G. watkinsi* in at least a few areas (Krabbe 1992). Nearly contiguous records of Chestnut-crowned Antpitta cross the Peruvian border into northern Piura south of

Cariamanga (Loja), followed by a fairly significant gap north of Cruz Blanca (see *albiloris*). Based on this, I provisionally extend the range of *connectens* as far south as the upper Río Piura drainage, *c*.15km south of Ayabaca. Subspecies *connectens* is somewhat intermediate between nominate *ruficapilla* and *albiloris*, showing characters of both, but tending towards *albiloris* of north-west Peru. Chapman (1923c) differentiated *connectens* from nominate *ruficapilla* by its overall paler upperparts, less heavily streaked, more fulvous underparts, and more olivaceous (less rufescent) external margins to the wings. From *albiloris*, Chapman (1923c, 1926a) distinguished *connectens* by its slightly darker crown, more olivaceous (less greyish) back, the greater extent of orange-rufous to the lores, malar and ear-coverts, and by having darker and broader streaking below. **SPECIMENS & RECORDS ECUADOR**: *Loja* El Cisne–Ambocas road, 03°49'S, 79°30'W (XC 81328/29; L. Ordóñez-Delgado); *c*.1.5km N of Barrio Loma Redonda, 03°55'S, 79°40'W (ML 130401, L.R. Macaulay); 7km NW of Catamayo, 03°57.5'S, 79°32'W (Krabbe 1991); Las Chinchas–Zambi road, 03°57.5'S, 79°28.5'W (N. Athanas *in litt.* 2017); Guachanamá, 04°02'S, 79°53'W (AMNH 167247/48, USNM 323090); Tierra Colorada and vicinity, *c*.04°02'S, 79°56'W (Krabbe 1991, Best *et al.* 1992, XC); San Bartolo, 04°02S, 79°55W (AMNH, ANSP, MCZ, USNM); Catacocha, 04°03'S, 79°40'W (Best *et al.* 1992, Krabbe 1991); Celica and vicinity, *c*.04°06'S, 79°58'W (Krabbe 1991, AMNH, MCZ, WFVZ, MECN, XC); Cruzpamba, 04°09.5'S, 80°00.5'W (MCZ 346608); 5km SE of Gonzanamá, 04°15.5'S, 79°23.5'W (MCZ 299199/200); Utuana, 04°20'S, 79°46'W (Krabbe 1991); Cariamanga–Colaisaca road, *c*.04°20'S, 79°37'W (MECN 8388); RP Bosque Hanne, 04°22'S, 79°45'W (XC 210285; J. Fischer), Espindola, 04°29'S, 79°23.5'W (XC 118040/41; L. Ordóñez-Delgado); Lance, 04°34'S, 79°27.5'W (P.X. Astudillo W. *in litt.* 2017); Bosque Comunal Angashcola, *c*.04°35'S, 79°22'W (J. Freile *in litt.* 2017). *El Oro* El Chiral, 03°38'S, 79°41'W (AMNH 167253); San Pablo, 03°40'S, 79°33.5'W (Freile 2000); Zaruma, 03°41'S, 79°36.5'W (AMNH 129756). **PERU**: *Piura* Bosque de Cuyas, 04°36'S, 79°44'W (Vellinga *et al.* 2004, Kikuchi 2009, XC), El Toldo, 04°40'S, 79°31'W (Vellinga *et al.* 2004); Aypate, 04°42'S, 79°35'W (Vellinga *et al.* 2004).

Grallaria ruficapilla albiloris Taczanowski 1880, Proceedings of the Zoological Society of London, p. 201, Cutervo and Callacate, eastern slope of the western Cordillera of Peru. Taczanowski (1880) based his description on four birds collected at Cutervo (06°22'S, 78°49'W) and Callacate (06°25'S, 78°56'W), but did not designate a holotype. Zimmer (1934) apparently considered one of these (a male from Cutervo) as the holotype, but stated that it had been destroyed with the collection in the Warsaw Museum. Subsequently, Warren & Harrison (1971) designated what is presumably one of these original four specimens as a syntype (NHMUK 1889.6.20.635), an unsexed skin collected 2 May 1879 by Sztolcman at Callacate. I was unable to determine if any of the other syntypes are extant, but both localities from the original series are in central Cajamarca, only *c*.13km apart, making the ambiguity surrounding the type and its collecting locality unlikely to be important. As implied by the name, this subspecies is separated from nominate *ruficapilla* chiefly by the white lores and moustachial, and white suffusion on the ear-coverts. From specimen labels (MSB; J. & D. Schmitt): maxilla black, mandible dark

grey, tarsi and toes blue-grey, toe-pads pale buff. The range of this race presents several challenges and has been the source of some confusion in the literature. This is apparently due to the suggestion of Zimmer (1934) that the ranges of *albiloris* and *connectens* meet somewhere in Ecuador and the mention of an intermediate specimen from somewhere in Loja (Ridgely & Greenfield 2001). This led subsequent authors to consider its range to include parts of southern Ecuador (e.g., Peters 1951, Fjeldså & Krabbe 1990, Krabbe 1991, Clements 2007). Although this is no longer thought to be the case, its northern limits in Peru remain poorly known. The northernmost localities from where specimens referred to *albiloris* have been collected are all in the vicinity of Abra Cruz Blanca (*c*.05°21'S, 79°32'W) in Piura (e.g., El Tambo, Palamba, ANSP), a region included in the range of *albiloris* by Zimmer (1934) but from where Chapman (1926a) considered plumage characters to be intermediate between *connectens* and *albiloris*. Unfortunately, I have not had the opportunity to carefully examine all of the available material from this region, but have done my best to reach a consensus based on the various opinions expressed in the literature. With the exception of two localities considered by Zimmer (1934) to be within the range of *albiloris*, and one included by Bangs (1950), all in the East Andes north of the Río Marañón (Tabaconas, 05°19'S, 79°18'W; San Felipe, 05°46'S, 79°19'W; Bagazán, 06°08'S, 77°26'W), I have followed the lead of previous authors in assigning records to this race. The range of *albiloris*, as hypothesized here, is confined to both slopes of the West Andes in north-west Peru. It is separated from the range of nominate *ruficapilla* in the north-east by the Río Huancabamba, and south of there, by the Río Marañón from the range of *interior*. North of the Abra Cruz Blanca area there is only one sight record of Chestnut-crowned Antpitta further south of the records I consider to represent *connectens* (eBird: San Martín de Huala, 05°00'S, 79°54'W, K. Herrera-Peralta). Though perfectly plausible, there is no way to know if this record pertains to *connectens* or *albiloris* and, until further evidence presents itself, I have abruptly truncated the range of *albiloris* only as far north as vouchered records from the mountains south-west of Huancabamba and west of the Río Huancabamba (e.g., ML recordings by T.A. Parker and specimens ANSP, LSUMZ). The southernmost vouchered records I consider to represent *albiloris* are from Seques and extreme south-west Cajamarca at Bosque Cachil (Schmitt *et al.* 2013). Both of these localities are likely at the western extremity of its range, as most habitat further west is probably too low and too arid for Chestnut-crowned Antpitta. Scattered sight records (eBird) extend the range slightly south of Bosque Cachil, but I have discounted the as-yet unconfirmed sighting of the species some 250km south in Ancash (Frimer *in* Fjeldså & Krabbe 1990). **SPECIMENS & RECORDS** *Piura* Huancabamba, *c*.05°16'S, 79°30'W (Zimmer 1934); El Tambo, 05°20'S, 79°30'W (Bond 1950, ANSP 117482, MCZ 179771); Canchaque, 05°22.5'S, 79°36.5'W (FMNH 222153) and vicinity 05°22'S, 79°34'W (LSUMZ); Palambla, 05°23.5'S, 79°36.5'W (Zimmer 1934, Bond 1950, ANSP 117477–479, AMNH 175282); Abra de Porculla, 05°51'S, 79°31'W (Bond 1950; ANSP, MCZ, CM, ML, XC); 3.5km ESE of Limón, 05°54.5'S, 79°31'W (LSUMZ 84953). *Cajamarca* 64km NW of Chugur, 06°17.5'S, 79°09'W (AMNH 235891–894); Bosque Paja Blanca, 06°23.5'S, 79°07'W (XC 37706, B. Planqué); *c*.9km W of Pencapuquio, 06°28'S, 79°03'W (XC 47644,

N. Krabbe); Casupe, 06°28.5'S, 79°23'W (R. Ahlman *in litt.* 2017); *c.*1.4km SE of La Palma, 06°30'S, 78°37'W (LSUMZ 84956); Chugur, 06°40'S, 78°44'W (Zimmer 1934); Hacienda Limón, 06°50'S, 78°05'W (Zimmer 1934, Bond 1950, FMNH, ANSP, ML); *c.*4km SE of Celendín, 06°53.5'S, 78°07'W (XC 157866, H. Matheve); Taulis, 06°54'S, 79°03'W (Zimmer 1934, AMNH 235882–884); Cruz Conga, 07°00'S, 78°12'W (D. Beadle *in litt.* 2017); Trigal, 07°00'S, 79°09.5'W (F. Angulo P. *in litt.* 2017); 0.5km NW of Sangal, 07°08'S, 78°51'W (D.F. Lane *in litt.* 2017); Lago Huanico, 07°10.5'S, 78°16.5'W (F. Angulo P. *in litt.* 2017); W side of Abra Gavilán, 07°15.5'S, 78°29'W (F. Angulo P. *in litt.* 2017); Bosque Cachil, 07°24'S, 78°46.5'W (MSB, XC); *c.*2.3km ENE of Cospán, 07°25.5'S, 78°31.5'W (F. Angulo P. *in litt.* 2017). **Lambayaque** Corral Grande, 06°21.5'S, 79°29'W (Pratolongo *et al.* 2012, XC); Seques, 06°54'S, 79°18'W (Zimmer 1934, AMNH 235885–890).

Grallaria ruficapilla interior Zimmer 1934, American Museum Novitates, no. 703, p. 16, San Pedro, 8600–9400ft., south-east of Leimebamba, Peru. There is some confusion over the precise location of the type locality (Paynter 1993), but it appears to be somewhere on the east Andean slope, south-east of Chachapoyas (LeCroy & Sloss 2000), and north-east of Leymebamba (*contra* Zimmer 1934). I located a San Pedro (Google Earth) very close (1.8km south) of the estimate given by Vaurie (1972), placing the probable location as 06°37'59.65'S, 77°41'59.95'W, south-east of Chachapoyas and north-east of Leymebamba, in Amazonas just north of the border with San Martín. The holotype (AMNH 235531) is an adult male collected 4 February 1926 by Harry Watkins. This subspecies occurs on both slopes of the central Peruvian Andes and upper Utcubamba Valley (Schulenberg *et al.* 2010), in the departments of Amazonas and San Martín. Zimmer (1934) distinguished it from the nominate subspecies as having the streaking of the breast, sides and flanks paler and more brownish-olive, without well-defined dark stripes separating the fringes from the white shaft-streaks. The centre of the breast is usually less strongly marked. The whitish shaft-streaks on the feathers of the rump are narrower, without prominent dusky borders, and are overall less apparent. The eye-ring of *interior* is broader and brighter white. Overall, the underparts are slightly buffier and the rufous of the head a little brighter. Zimmer (1934) also stated that *interior* had the breast 'more frequently washed with orange-ochraceous'. Bond (1950), however, remarked that there is as much variation in the amount of rusty wash to the breast in this subspecies as in other races, and my examination of specimens suggests that this is probably true, making this a relatively uninformative character. **Specimens & Records** **Amazonas** *c.*6km NW of Pomacochas, 05°48.5'S, 78°01.5'W (XC 296848, R. Gallardy); 33km by road NE of Ingenio, 05°52'S, 77°57'W (Schulenberg & Williams 1982, LSUMZ 81986/87); Gocta, 06°01.5'S, 77°53.5'W (XC 36242, F. Lambert); 4.5km N of Tullanya, 06°05'S, 78°19.5'W (MSB 32055); Karajia, 06°06'S, 78°04'W (A. García-Bravo *in litt.* 2017); Chachapoyas, 06°14'S, 77°52.5'W (as *albiloris*, Taczanowski 1882, 1884; as *interior* Zimmer 1934); Bagazán, 06°08'S, 77°26'W (Bond 1950, ANSP 117487); Bosque Tinyo, *c.*06°11'S, 77°35'W (FMNH 299604); Cerro Shucahuala, 06°14'S, 78°06.5'W (A. García-Bravo *in litt.* 2017); Levanto, 06°18.5'S, 77°54'W (Bond 1950, AMNH, ANSP, CM); Tamiapampa, 06°20'S, 77°52'W (Zimmer 1934); Chillo, 06°28'S, 77°45'W (Zimmer

1934); Leimebamba, 06°42.5'S, 77°48'W (Bond 1950, ANSP 66755/56, MCZ); Abra Barro Negro, 06°43.5'S, 77°52.5'W (XC 157865: H. Matheve); Atuén, 06°45'S, 77°52'W (Bond 1950, ANSP 66757); Lluy, 06°45'S, 77°49'W (Bond 1950, ANSP 117484); La Muralla, 06°48'S 77°45'W (Mark *et al.* 2008); E of Balsas, *c.*06°52'S, 77°53'W (FMNH 44307/08); Chuquibamba, 06°56'S, 77°51.5'W (A. García-Bravo *in litt.* 2017). **San Martín** *c.*25km NW of Naranjos, 05°41.5'S, 77°44'W (XC 122211, E. DeFonso); Laguna de los Cóndores, 06°51'S, 77°41.5'W (J. Barrio *in litt.* 2017).

STATUS BirdLife International (2017) and other recent evaluations of the status of Chestnut-crowned Antpitta (Freile *et al.* 2010b) agree that the broad geographic range and tolerance of disturbance (Welford 2000, O'Dea & Whittaker 2007, Castaño-Villa & Patiño-Zabala 2008, Pulido-Santacruz & Renjifo 2011) make this species one of Least Concern. It is obvious that Chestnut-crowned Antpitta can tolerate some degree of alteration of its habitat. A note to conservationists, however, that the key word in that sentence should be *tolerate*. Careful, detailed studies involving the effects of fragmentation on this species do show negative effects (Renjifo 1998, 1999). Perhaps, unlike the above-treeline Tawny Antpitta, which is also often considered tolerant of disturbance, Chestnut-crowned Antpitta is more dependent on *contiguous* habitat, even if highly altered. It should also be pointed out that the ability to persist does not necessarily equal the ability to flourish. Particularly for understorey insectivores such as Chestnut-crowned Antpitta, caution should be used when interpreting its ability to persist in fragments without long-term studies from multiple regions and habitat types. It is possible that, in some areas, human activities may even create more habitat for the species, and actually boost population levels. **Protected Populations** *G. r. avilae* PN Henri Pittier (Schäfer & Phelps 1954, Schäfer 1969, Verea & Solórzano 2011); PN El Ávila (MCZ, AMNH, CM, YPM, USNM, COP, ANSP, ROM); PN Macarao (ML 32434811, F. Machado-Stredel); MN Pico Codazzi (J.E. Miranda T. *in litt.* 2017). *G. r. perijana* RNA Chamicero del Perijá (ML 46892261, R. Gallardy). *G r. nigrolineata* **Venezuela**: PN Dinira (COP 19958–960), PN Sierra Nevada (FMNH, XC); PN Tamá (COP 62217–224); PN Guaramacal (XC 223674, P. Boesman); PN Páramos de Batallón y La Negra (J.E. Miranda T. *in litt.* 2017); PN Yacambú (eBird: L. Duque). **Colombia**: PNN Tamá (A.M. Cuervo *in litt.* 2017); RNA Hormiguero de Torcoroma (Salaman *et al.* 2007a). *G r. ruficapilla* **Colombia**: PNN Cordillera Los Picachos (Bohórquez 2002); PNN Enrique Olaya Herrera (A.M. Cuervo *in litt.* 2017); PNN Los Nevados (Pfeifer *et al.* 2001); PNN Munchique (Negret 1994); PNN Chingaza (Lotta *et al.* 2015); PNN Farallones de Cali (eBird: C. Downing); PNN Tatamá (eBird; O. Janni); RN Chicaque (XC, ML); RFP Bosque Oriental de Bogotá (Peraza 2011); RFP La Romera (XC 167963, D. Zapata-Henao); Parque Regional Ecoturistico Arví (Castaño-Villa & Patiño-Zabala 2007, 2008); PRN Santa Emilia (Zuleta-Marín 2012); PRN Ucumarí (Naranjo 1994, Kattan & Beltrán 1997, 1999, 2002); RFP Cárpatos (Stiles & Rosselli 1998); RFP Los Yalcones (XC; D. Calderón-F.); RFP Río Blanco (Ocampo-T. 2002, Nieto-R. & Ramírez 2006); SFF Otún-Quimbaya (XC 100373/374, J. King); SFF Iguaque (XC 56062–063/66–67, B. López-Lanús); SFF Volcán Galeras (eBird: R. Fernández Gómez); RMN El Mirador (Salaman *et al.* 2007a); RMN Campo Alegre (O.H. Marín-Gómez *in litt.* 2017); RNA Loro Orejiamarillo and RNA Mirabilis–Swarovski (Salaman *et al.* 2007a); RNA El Colibrí del Sol, RNA Reinita Cielo Azul, RNA Halcón Colorado and RNA Ranita Dorada (Salaman *et al.* 2009a); RNA Loro Coroniazul (eBird: A. Pinto); RNA Gorrión Andivia (eBird: J.S. Moreno); Reserva Ecológica Quimza Sopó (XC 211079, J.I.G. Arango); Parque Ecológico Pionono (XC 211072, J.I.G. Arango); DMI Agualinda (J.A. Zuleta-Marín *in litt.* 2017); RNSC Tambito (Donegan & Dávalos 1999); RNSC La Esperanza (J.B.C. Harris *in litt.* 2017); RNSC Merenberg (Ridgely & Gaulin 1980); RNSC San Sebastían de la Castellana (A.M. Cuervo *in litt.* 2017); RNSC Rogitama (D. Uribe *in litt.* 2017); RNSC El Cedro (ML 60828621, photo J.L. Peña). **Ecuador**: PN Cajas (Tinoco & Astudillo 2007, Tinoco & Webster 2009, Astudillo *et al.* 2015); PN Cayambe-Coca (Poulsen & Krabbe 1997b); PN Llanganates (Benítez *et al.* 2001); PN Podocarpus (Rasmussen *et al.* 1994); PN Sangay and PN Sumaco-Galeras (MCZ 138477, WFVZ 45743, ANSP 185508); PN LimiYacurí (R. Ahlman *in litt.* 2017); RE Cotacachi Cayapas (J. Freile *in litt.* 2017); RE Antisana (HFG); RBP Maquipucuna (O'Dea & Whittaker 2007); Reserva Geobotánica Pululahua (R.A. Gelis *in litt.* 2017); RP Alto Chocó (J. Freile *in litt.* 2017); BP Cashca Totoras (Bonaccorso 2004); BP Mazán (Guerrero 1996, Latta *et al.* 2011); Reserva Integral Otonga (Freile & Chaves 2004); RVS Pasachoa (ML: T.A. Parker); RE Los Illinizas (MECN 2603); RP Estación Biológica Yanayacu (XC 2382, W. Halfwerk); RP Cabañas San Isidro (XC 250897, N. Krabbe); RP Bellavista (XC, ML); RP Las Gralarias, RP Santa Lucía, RP Refugio Natural Pacha Quinde, RP Tapichalaca (HFG); RP Refugio Paz de las Aves (XC, ML); RP Yanacocha (MECN 2606); RP Verdecocha (MCZ 138481); RP Yunguilla (Greeney & Kirwan 2013); RP SierrAzul (Martin & Greeney 2006, ANSP 186180); in Ecuador probably occurs in PN Cotopaxi, despite dearth of records. **Peru**: SN Tabaconas-Namballe (F. Angulo P. *in litt.* 2017); ACM Bosque de Huamantanga (A. García-Bravo *in litt.* 2017). *G r. connectens* **Ecuador**: RP Utuana (HFG); RP Bosque Hanne (XC: J. Fischer, A. Spencer); RP El Tundo (eBird: S. Thompson); possibly also in extreme west of PN Yacurí. *G r. albiloris* RVS Laquipampa (Pratolongo *et al.* 2012, Schmitt *et al.* 2013); RVS Bosques Nublados de Udima (Angulo 2009); BP Pagaibamba (F. Angulo P. *in litt.* 2017). *G. r. interior* BP Alto Mayo (XC 122211; E. DeFonso); ACP Abra Patricia–Alto Nieva (eBird: A. Farnsworth); ACP Llamapampa-La Jalca (eBird: A. Sagone); ACP San Antonio (eBird: M. Leon Leon); ACP Hierba Buena-Allpayacu (A. García-Bravo *in litt.* 2017).

OTHER NAMES *Grallaria ruficapilla melanosticta* The original specimen label of USNM 147351, an example of *nigrolineata*, bears this name, but I have not been able to locate its use anywhere in the literature. *Grallaria ruficapilla taczanowskii* Domaniewski & Sztolcman (1918) based their description of *taczanowskii* on a specimen collected at Cayandeled [02°07'S, 78°59'W, 1,375m, south of Pallatanga in south-west Chimborazo (Paynter 1993)]. Both Chapman (1923c, 1926a) and Cory & Hellmayr (1924), however, agreed that is a synonym of nominate *ruficapilla*, and I found no other references to this name in the literature. *Hypsibemon ruficapillus* (Hein & Reichenow 1890); *Grallaria albiloris* (Taczanowski 1880, 1882, Sharpe 1901, Brabourne & Chubb 1912); *Grallaria nigro-lineata* (Nehrkorn 1899, Cory & Hellmayr 1924, Phelps & Gilliard 1940); *Grallaria nigrolineata* (Sharpe 1901, Nehrkorn 1910, Brabourne & Chubb 1912); *Grallaria*

ruficapilla nigro-lineata (Dubois 1900, Phelps & Phelps 1950, Peters 1951, Meyer de Schauensee 1952b, 1964, Iafrancesco *et al.* 1987); *Grallaria ruficapilla nigroliniata* (misspelling; Pelayo & Soriano 2010); *Grallaria ruficapina* (misspelling; Stiles 1984); **Spanish** Tororoí Compadre (Krabbe & Schulenberg 2003); Tororoí Comprapán (Salaman *et al.* 2007a); Compadre Merideño (*nigrolineata*), Compadre Avileño (*avilae*), Compadre Perijá (*perijana*) (Phelps & Phelps 1950); Tororoí de Corona Castaña (Schulenberg 2002); Hormiguero Comprapán (Olivares 1969); Hormiguero Compadre (Meyer de Schauensee & Phelps 1978, Phelps & Meyer de Schauensee 1979, 1994); Compadre (*avilae*; Schäfer & Phelps 1954); Gralaria Coronicastaña (Ortiz-Crespo *et al.* 1990, Granizo 2009, Stotz & Valenzuela 2009); Tororoí Coronicastaña (Clements & Shany 2001); Güicundo, Huicundo saratán, Pide pan, Viejo soy (Valarezo-Delgado 1984), Adiós Compae, Seco Estoy (Phelps & Meyer de Schauensee 1979, 1994); Comprapán (Phelps & Meyer de Schauensee 1979, 1994, Restrepo 1997); Huicuco (F. Angulo *in litt.* 2017); Hormiguerito Compadre (Ferraro & Lentino 1992); Compadre de Perijá [*perijana*], Compadre Avileño [*avilae*], Compadre Merideño [*nigrolineata*] (Phelps & Phelps 1950). **English** Chestnut-crowned Ant-Thrush (nominate *ruficapilla*), Rufous-crowned Antpitta (Mark *et al.* 2008); White-lored Ant-Thrush (*albiloris*), Venezuelan Chestnut-crowned Ant-Thrush (*nigrolineata*) (Brabourne & Chubb 1912); Caracas Chestnut-crowned Antpitta (*avilae*), Mérida Chestnut-crowned Antpitta (*nigrolineata*), South Ecuadorian Chestnut-crowned Antpitta (*connectens*) (Cory & Hellmayr 1924). **French** Grallaire à tête rousse. **German** Rostkappen-Ameisenpitta (Krabbe & Schulenberg 2003).

Chestnut-crowned Antpitta, adult vocalising (*ruficapilla*), Cabañas San Isidro, Napo, Ecuador, 25 August 2009 (*Dušan M. Brinkhuizen*).

Chestnut-crowned Antpitta, adult (*ruficapilla*), Cabañas San Isidro, Napo, Ecuador, 23 June 2014 (*Dušan M. Brinkhuizen*).

Chestnut-crowned Antpitta, older nestling in nest (*ruficapilla*), Estación Biológica Yanayacu, Napo, Ecuador, 29 September 2008 (*José Simbaña*).

Chestnut-crowned Antpitta, nest and complete clutch (*ruficapilla*), Sierra Azul Lodge, Napo, Ecuador, 20 September 2004 (*Harold F. Greeney*).

Chestnut-crowned Antpitta, two 3-day-old nestlings in nest (*ruficapilla*), Yungilla Reserve of Jocotoco Foundation, Azuay, Ecuador, 16 May 2005 (*Harold F. Greeney*).

Chestnut-crowned Antpitta, juvenile (*ruficapilla*), Manizales, Caldas, Colombia, 6 July 2013 (*José Daniel Avendaño*).

Chestnut-crowned Antpitta, newly-hatched nestling and unhatched egg in nest (*ruficapilla*), Yungilla Reserve of Jocotoco Foundation, Azuay, Ecuador, 13 March 2005 (*Harold F. Greeney*).

WATKINS'S ANTPITTA
Grallaria watkinsi   Plate 10

Grallaria watkinsi Chapman, 1919, Proceedings of the Biological Society of Washington, vol. 32, p. 255, Milagros, 2,200ft. 'Piura Dept., Peru' [= 04°06.5'S, 80°09'W, 670m, Loja Province, Ecuador; Stephens & Traylor 1983, Paynter 1993]. The type specimen (AMNH 163084) is an adult male collected 5 July 1919 by, and named for, Harry Watkins.

Watkins's Antpitta is a shy resident of dry forest and secondary scrub, with a very restricted distribution in the lowlands and foothills of south-west Ecuador and north-west Peru. This species is reclusive and occupies dense (often thorny) thickets. As it can be very vocal, like most antpittas it is more often heard than seen. Indeed, the best way to locate a Watkins's Antpitta is to listen for the distinctive song, which consists of several short, hollow, whistles that descend in pitch, followed by a rising hollow whistle. Within its restricted range, Watkins's Antpitta can be surprisingly common in appropriate habitat, and, with luck, may be seen foraging for arthropods by pecking at the leaf litter on or near the ground in dense vegetation. Most aspects of its natural history are very poorly known. Although a few brief studies have shed some light on its reproductive biology, other aspects of its natural history such as diet, foraging method, territoriality, lifespan and survivorship remain unstudied. Given the propensity of other *Grallaria* to forage on worms, and the supposition that worms might be difficult to find during the intense dry periods experienced in the species' range, diet and foraging might be a particularly interesting line of investigation. I suggest also to compare foraging and breeding biology with the closely related Chestnut-crowned Antpitta, especially where their ranges overlap and there is apparent segregation by habitat (Krabbe 1992).

IDENTIFICATION 18–19cm. Watkins's Antpitta is a medium-sized antpitta, with mostly olive-brown upperparts, except the somewhat dull, brownish-chestnut crown with faint white streaking. The underparts are white, with broad, pale dusky-grey to dusky-olive streaks on the breast and flanks. The only species with which it is liable to be confused is the much more brightly plumaged Chestnut-crowned Antpitta. Although the ranges of these two species closely approach one another in south-west Ecuador, there are no currently known areas of sympatry (Ridgely & Greenfield 2001). Watkins's Antpitta is readily separated from Chestnut-crowned Antpitta by voice, but also differs morphologically. Watkins's Antpitta is smaller and its plumage considerably paler overall, especially the crown. It can also be separated from Chestnut-crowned Antpitta by the more extensive white on the sides of the face and by being narrowly streaked on the back. Furthermore, the legs and tarsi are pinkish rather than blue-grey (Chapman 1919, Ridgely & Greenfield 2001, Schulenberg *et al.* 2007).

DISTRIBUTION Watkins's Antpitta is confined to the Tumbesian area of endemism in south-west Ecuador and adjacent north-west Peru. Most of its range falls within the Ecuadorian provinces of Loja and El Oro and the Peruvian department of Tumbes. There is an isolated population further north in Ecuador, in and around the Cordillera de Chongón Colonche in south-west Manabí and western Santa Elena. To date, it is known only from the west slope

of this isolated cordillera, being apparently absent from Cerro Blanco and Manglares-Churute (Mischler 2012, J. Freile *in litt.* 2017), despite some suggestions to the contrary (Mischler & Sheets 2007). In addition, Watkins's Antpitta was recently found in central Lambayeque (Pratolongo *et al.* 2012). Its presence in the intervening region, though probable, has not been documented. On the whole, its status and precise range is poorly known (Bond 1950, Wiedenfield *et al.* 1985, Ridgely & Tudor 1994, Ridgely & Greenfield 2001, Schulenberg *et al.* 2007).

HABITAT Watkins's Antpitta inhabits dry, semi-deciduous and, in a few areas, evergreen forests, but readily forages and nests in heavily disturbed habitats (Krabbe & Schulenberg 2003, Martin & Dobbs 2004, Schulenberg *et al.* 2013a). Parker *et al.* (1995) found Watkins's Antpitta to be 'restricted to greener portions of forest, especially the shaded, denser vegetation of narrow ravines' at one site in Peru. My experience suggests that, unlike many forest *Grallaria*, but similar to Chestnut-crowned Antpitta, Watkins's is not especially partial to shady ravines and riparian areas but, in fact, prefers hillsides and other more 'exposed' locations. The altitudinal distribution of Watkins's Antpitta extends from sea level to 1,800m in Ecuador (J. Freile *in litt.* 2017). Similarly, in Peru, it is known only at 400–900m (Schulenberg *et al.* 2007), perhaps because it lacks appropriate habitat nearer the ocean. This propensity for lower-elevation, dry habitats generally precludes much range overlap with Chestnut-crowned Antpitta, which generally occurs above 2,000m in this region. There are locations, however, where both species can be heard singing only several metres apart and, in these locations, the two species appear to segregate by habitat (Krabbe 1992).

MOVEMENTS *G. watkinsi* is not known to move altitudinally, but given the extreme differences in rainfall (and potentially in food resources) across the year within its range, small-scale altitudinal shifts seem possible.

VOICE The song is described as 'a series of 4–7 well-enunciated and emphatic whistled notes, the first set all similar but the last longer and sharply upslurred', e.g., *keeu, keew-kew-kew k-wheeeei?* (Ridgely & Greenfield 2001b) and as 'an accelerating series of descending hollow notes ending with a longer rising note': *CLEW clew-clew'clew'clew cu-HOOEE?* (D.F. Lane *in* Schulenberg *et al.* 2007). This song is 2.1–3.4s long. The initial 4–9 notes are at 1.6–1.8kHz and given at a steady pace of 2.3–2.8s; this series is followed immediately by a sharp whistle, rising from 1.4 to 2.6kHz (Krabbe & Schulenberg 2003), with songs delivered at intervals of 4–15s (Krabbe & Schulenberg 2003). Parker *et al.* (1995) and Krabbe & Schulenberg (2003) suggested that songs are usually given from the ground or low branches, primarily in the crepuscular hours. My own experience, however, suggests that, at least during the breeding season, the species can be heard nearly the entire day. Furthermore, I frequently found singing adults perched up to 4–5m above ground in areas of dense vegetation. The acoustic properties of the song, however, made discovery of the precise location of the nearly motionless vocalists extremely hard to determine. When delivered from perches on steep slopes, even individuals perched well above ground often sounded as if delivered from ground level. One frequently heard call is somewhat similar to the song, but consists of fewer introductory notes and ends with a distinctly slurred final one (Ridgely & Greenfield 2001b). My experience suggests that this note is often (but not always) given in response to playback and during conspecific encounters at edges of territories. In general, the alarm is simply the final note of this call, but louder and delivered in a more slurred or burry fashion. An additional call, described by D.F. Lane (in Schulenberg *et al.* 2007) as a 'doubled hollow whistle, rising at end: *clew-clewEE?*' also appears to serve as an alarm. While in a territory, I found that as I approached a singing individual, particularly near a nest, the full song was replaced by one of the two alarms. These frequently continued well after I had passed, usually progressing from the loud slurred single note to the querulous doubled whistle, and finally returning to the full song as my presence ceased to agitate the bird. Though I have not witnessed any direct encounters between two Watkins's Antpittas, I suspect that the doubled whistle may also function in intraspecific encounters, perhaps as aggression, while the loud single-note call is primarily used to convey agitation or alarm in the presence of potential predators.

NATURAL HISTORY Very few data are available in the literature, except the anecdotal descriptions cited here. Watkins's Antpitta tend to forage on or near the ground, preferring thickets and vine tangles (Schulenberg *et al.* 2013a). Although sometimes observed foraging in more open areas, similar to Chestnut-crowned Antpitta (Ridgely & Greenfield 2001b), Watkins's Antpitta seems less likely to emerge from dense thickets and is certainly seen in the open less frequently than many of the more montane *Grallaria* species. Like other antpittas, Watkins's Antpitta may at least opportunistically forage in the wake of larger animals that disturb the leaf litter and/or scare potential invertebrate prey (Greeney 2012b), as evidenced by their frequent appearance behind me soon after I have finished crashing through an annoyingly dense patch of undergrowth. There are no published data on territory size or maintenance, but my casual observations in south-west Ecuador suggest that territories are maintained year-round, are occupied for at least 2–4 years, and are *c*.1.0–1.5ha in size. Non-vocalising individuals are, however, extremely difficult to follow, and individuals may well range beyond my estimated territory size, as suggested by the somewhat puzzlingly large gaps between perceived territories. Watkins's Antpitta is presumed to be socially monogamous, but there are no data on extra-pair paternity. Pairs do not appear to forage together, and Watkins's Antpitta is generally encountered alone, away from mixed-species flocks. Much remains to be studied of daily movement patterns, and there is at least one published record of the species associating with a mixed flock of brush-finches and seedeaters at the edge of a clearing (Parker *et al.* 1995).

DIET Presumably, the diet consists largely of invertebrates, similar to its congeners. It seems probable that, due to the more arid habitats inhabited by Watkins's Antpitta compared to congeners, earthworms may be more difficult to extract from the soil or less abundant at the surface, leading to a paucity of such invertebrates in its diet. This would be in rather sharp contrast to the importance that earthworms appear to play in the diets of other *Grallaria*. Few concrete data are available, but the stomach contents of three individuals in Tumbes during the wet season included only unidentified insect parts and at least one grasshopper (Orthoptera) (Schulenberg *et al.* 2013a).

REPRODUCTION The first published nest, egg and nestling descriptions for Watkins's Antpitta were provided by Martin & Dobbs (2004), who studied a nest 6.5km east of Celica, Loja, Ecuador. These descriptions were supplemented by the brief observations made by Greeney *et al.* (2009) on nest architecture and parental care at a nest in RP Jorupe. These two papers remain the only available information on reproductive ecology and are summarised here. **NEST & BUILDING** No information has been published on the process of nest-site selection and nest construction, but these are probably undertaken by both sexes. The nest found by Martin & Dobbs (2004) was in a secondary patch of roadside trees in a highly fragmented landscape at 1,900m, at the ecotone between deciduous and humid evergreen forest. It was in a well-protected location, nestled in the thorny branches of an *Acacia* (Leguminosae). The nest was *c*.3m above ground and 0.5–1.0m below the top of the substrate tree that formed a canopy with adjacent foliage, concealing the nest from above. It was apparently not firmly supported, but built atop several small branches and several thin, living vines. The nest discovered by Greeney *et al.* (2009) was 4m up in a dense vine tangle (in the upper portion of the substrate) growing near the edge of a 6m bank, at the edge of a rarely used road. Its location at the edge of this incline led to it, in actuality, being nearly 10m above the ground directly below the nest. It was well concealed in the foliage of a small, spiny tree, 50cm from the edge of the vine tangle and shaded by 40cm of small, interwoven, leafy vines and branches above the nest. It was supported by several small branches, but was not well 'attached' to the substrate, with only a few small branches passing loosely through the nest's external portion. The nest was a bulky cup of dead leaves, sticks and mud, sparsely lined with thinner dark twigs and fibres. Martin & Dobbs (2004) described their nest as a 'bulky and unkempt-looking' cup. It was composed externally of predominantly woody materials such as sticks and dead vines, but also a few dead leaves and some moss, which they suggested was brought only incidentally with other materials. The lining of the cup was composed entirely of narrow black rootlets. Published measurements: outer diameter 23 × 22cm; external height 10cm (with an additional 7cm of detritus below); inner diameter 10.0 × 10.5cm; inner depth 5cm (Martin & Dobbs 2004). **EGG, LAYING & INCUBATION** The single egg measured by Martin & Dobbs (2004) was 30.1 × 25.3mm and weighed 7.9g. It was subelliptical and turquoise or greenish-blue with no markings. **NESTLING & PARENTAL CARE** A young nestling, estimated to be 1–2 days old, was described by Martin & Dobbs (2004) as being 'mostly naked [with] dark pinkish skin. Sparse blackish down was present on most, if not all, feather tracts, and pin feathers were just starting to break the skin of the wing. The eyes were closed. Except for a black nail to the tip of the upper mandible, the entire bill, gape and mouth lining were bright orange.' Interestingly, these authors further noted that the cloaca was bright orange like the bill, gape and mouth lining. The significance of this cloacal coloration is mysterious, but Martin & Dobbs (2004) proposed that it may function as an enhanced signal for emerging faecal material or to direct parental pecking of the nestling cloaca to stimulate defecation. These hypotheses are not mutually exclusive, and both would serve well in a species for which nest sanitation was a priority. The young nestling description in Martin & Dobbs (2004), viewed in the light of my more extensive experience with nestlings of *G. ruficapilla*, suggests that the above description better applies to 3–4 day-old nestlings. During 20.4 hours of observation at a nest spanning the eight days prior to fledging, Greeney *et al.* (2009) observed both adults arriving at the nest with prey 53 times to provision the two nestlings. On most visits adults arrived with multiple prey items (arthropods) and 15% of visits included earthworms. During this time, nestlings produced 37 faecal sacs, almost all of which were consumed by adults at the nest. Greeney *et al.* (2009) noted that adult Watkins's Antpittas spent much time perched on the rim of the nest, apparently shading the nestlings. Both adults frequently visited the nest simultaneously, and frequently sang at the nest, but always when alone. The fledging event of *G. watkinsi* has been observed just once (Greeney *et al.* 2009) and occurred at 06:30, with both nestlings leaving within moments of each other, just after being fed and while both adults were still present. At the nest with a single 3–4 day-old nestling (see above), Martin & Dobbs (2004) quietly observed an adult on the nest from a distance of 5m for 20–30 min. They reported that the brooding individual remained 'frozen' and looked straight out from the nest at them, with its bill held very slightly elevated, providing an excellent illustration of this posture in their paper. They remarked on the difference in 'defensive' postures between this and other species of antpittas such as *G. guatimalensis* and *G. varia* (Protomastro 2000, Dobbs *et al.* 2001), which freeze with their bills raised in the direction of the observer, revealing their strongly streaked throats (see also under *G. varia*). When the brooding adult was eventually flushed, it dropped into the thick understorey, began to sing and was immediately joined in song by its (presumed) mate. Both adults continued to sing from low, concealed perches during the remainder of their presence at the nest. **SEASONALITY** The nest found by Martin & Dobbs (2004) on 5 March 2000, contained an infertile egg and a 3–4 day-old nestling. A second nest in

south-west Ecuador (Greeney *et al.* 2009) fledged on 26 April 2006. On 4 April 2014, at RP Jorupe, L.A. Salagaje M. found a nest containing two young nestlings that I estimate were *c.*3–4 days old (Fig. 3). Unsurprisingly, given the extremely seasonal weather in the Tumbesian region, all known nesting behaviour by *G. watkinsi* coincides with the middle rainy season (January–April; Best *et al.* 1996), when avian reproductive activity peaks in south-west Ecuador and north-west Peru (Marchant 1959, 1960a,b, Best *et al.* 1992, Knowlton 2010). Although based on relatively few data, it seems highly likely that nesting of *G. watkinsi* occurs entirely in the rainy season, with nestbuilding possibly beginning in late December and clutches initiated until the end of April. My fieldwork in this region leads me to speculate that clutch initiation is probably somewhat delayed after the start of the rains, probably peaking in March, thereby fledging with the early drier period, which would surely facilitate the survival of newly fledged young. **BREEDING DATA** Near-finished nest, 6 March 2014, RP Jorupe (Greeney *et al.* in review); older nestlings, 28 April 2006, RP Jorupe (Greeney *et al.* 2009); young nestling, 5 March 2000, 3.5km W of El Empalme (Martin & Dobbs 2004); nestlings, 4 April 2014, RP Jorupe (Greeney *et al.* in review); juvenile, 20 April 1993, Pindal (MECN 6671); two juveniles transitioning, 14 and 15 June 1979, Quebrada Facial (LSUMZ 92440/41); subadult, 21 June 1979, Quebrada Facial (LSUMZ 92444); subadult, 3 September 1920, Portovelo (USNM 323088); subadult, 11 July 1919, El Alamor (USNM 323089); subadult, 15 July 2009, Cerro Los Limos (USNM 643984).

TECHNICAL DESCRIPTION The following description is based on Schulenberg *et al.* (2013a). **Adult** Sexes similar. The crown and nape are pale rufous to dull chestnut with a somewhat sun-bleached look. Though not visible in all field situations, the whitish shafts to the crown and nape feathers create thin streaks. The forecrown also bears indistinct streaking and black spotting. Lores and eye-ring white. Upper ear-coverts pale olive; lower part white, narrowly streaked olive. Back and tail pale olive, the centre of the back with a few whitish shaft-streaks. Remiges pale olive-brown or pale tawny-brown, wing-coverts pale olive. Throat white with a narrow dusky-olive lateral throat-stripe. Sides of breast olive, streaked white. Central breast white or whitish-buff, streaked dusky; a few feathers also may show ochraceous edges. Rest of underparts white, streaked dusky-olive, but central belly, vent and undertail-coverts white. **Adult Bare Parts** (Krabbe & Schulenberg 2003) *Iris* dark brown; *Bill* maxilla blackish, mandible pale pinkish-grey; *Tarsi & toes* pinkish to pale horn. **Immature** I was unable to examine sufficient immatures to provide much detail, other than to state that preliminary evidence suggests that the plumages of older juveniles and subadults are very similar to those of Chestnut-crowned Antpitta.

MORPHOMETRIC DATA *Wing* 93.5–96mm, mean 92mm; *Tail* 49–53mm, 52mm; *Bill* culmen 24–25mm, mean 24mm; *Tarsus* 50–55mm, mean 51mm (*n* = 6, 5 ♂♂, 1 ♀, Chapman 1919). *Bill* from nares 16.2mm, 17.7mm; *Tarsus* 51.9mm, 55.2mm (*n* = 2, ♀,♂, subadults (USNM 323088/89). *Wing* 95mm; *Tail* 47mm; *Bill* from nares 15.4mm, exposed culmen 23.0mm, total culmen 25.0mm, depth at nares 7.8mm, width at nares 7.7mm, width at base 8.3mm, width at gape 10.4mm; *Tarsus* 54.5mm (*n* = 1, ♂, juvenile, MECN 6671, M.V. Sánchez N. *in litt.* 2017). *Wing*

95mm; *Tail* 40mm; *Bill* from nares 17.7mm; *Tarsus* 52.9mm (*n* = 1, D. Becker *in litt.* 2015). **Mass** males 60–84g, mean 69.7g; females 58–62g, mean 60g (*n* = 12, 10 ♂♂, 2 ♀♀, Krabbe & Schulenberg 2003); 60g(mo) (*n* = 1, ♂, juvenile, MECN 6671, G. Buitrón-Jurado *in litt.* 2017). **Total Length** 18–19cm (Clements & Shany 2001, Ridgely & Greenfield 2001, Krabbe & Schulenberg 2003, Ridgely & Tudor 2009, Schulenberg *et al.* 2010).

DISTRIBUTION DATA, ECUADOR: *Manabí* RVS y Marino Costera Pacoche, 01°04.5'S, 80°53.5'W (eBird: R. Terrill); Las Goteras, 01°32'S, 80°46'W (Freile 2000); near Ayampe, 01°40'S, 80°47'W (XC 17429; A. Spencer); Cantalapiedra, 01°40'S, 80°45'W (Krabbe 1991); Cerro San Sebastián, *c.*01°35'S, 80°40'W (ANSP 184003–005, MECN 5884). *Santa Elena* Cerro La Torre, 01°41.5'S, 80°35.5'W (XC 249627, N. Krabbe); RE Comunal Loma Alta, 01°50'S, 80°39'W (J. Freile *in litt.* 2017). *El Oro* RE Arenillas, 03°32.5'S, 80°09'W (Ridgely & Greenfield 2001); La Mesa, 03°42.5'S, 79°39'W (J. Freile *in litt.* 2017); Portovelo, 03°43'S, 79°37.5'W (AMNH, USNM, MCZ, ANSP, MECN); La Victoria, 03°47'S, 80°04'W (XC 9806/07; A.T. Chartier); Río Pindo, 03°49'S, 79°45'W (AMNH 156245–247); El Tigre, 03°50'S, 80°01'W (J. Freile *in litt.* 2017); 'El Puente', probably *c.*03°55'S, 80°04.5'W (loc. see Paynter 1993, AMNH 171407/408). *Loja* El Triunfo Alto, 03°51'S, 79°40.5'W (XC 105333, L. Ordóñez-Delgado); Bosque Petrificado Puyango, 03°52.5'S, 80°02'W (R.A. Gelis *in litt.* 2017); Quebrada Zapote, 03°57'S, 80°03'W and Hacienda El Limo, 03°58'S, 80°07'W (J. Freile *in litt.* 2017); 10km E of El Limo, 04°00'S, 80°02.5'W (ANSP 186181); Alamor, 04°01'S, 80°01.5'W (AMNH, ANSP, MCZ); Catacocha area, *c.*04°03'S, 79°39'W (ANSP 184711/12, MECN 6106, XC); Quebrada Achiotes, 04°04'S, 80°17'W (Bonaccorso *et al.* 2007); Loma Socolas, 04°04'S, 80°23'W, Cuchilla Piedra Lisa, 04°05'S, 80°25'W, and El Faique, 04°07'S, 80°24'W (J. Freile *in litt.* 2017); Pindal, 04°07'S, 80°06.5'W (MECN 6671); 3.5km W of El Empalme, 04°09'S, 79°53.5'W (Martin & Dobbs 2004); 10km E of Mangahurco, *c.*04°08.5'S, 80°21'W (ANSP 185509); Cruzpamba, 04°09.5'S, 80°00.5''W (MCZ 299202); RP Laipuna, *c.*04°13'S, 79°55'W (J. Freile *in litt.* 2017); Nueva Fátima, 04°16'S, 79°49.5'W (XC 76327; L. Ordóñez-Delgado); RP Jorupe, 04°23'S, 79°53.5'W (Greeney *et al.* 2009, XC, IBC); Tambo Negro, 04°24'S, 79°52'W (Krabbe 1991, XC 250383/84, N. Krabbe). **PERU**: *Tumbes* La Laja, 03°48.5'S, 80°09.5'W (ANSP 117488/89); Campo Verde, 03°50.5'S, 80°10'35'W (LSUMZ 183152/53); Quebrada Faical, 03°50'S, 80°16'W (LSUMZ 92440–445); Cerro San Carlos, 04°00'S, 80°30'W (ML 84046; T.A. Parker); El Platano, 04°08'S, 80°37'W (LSUMZ 183155–157); Cerro Los Limos, 04°09'S, 80°38'W (USNM 643984). *Piura* Quebrada Huabal, 04°14'S, 80°45'W (eBird: A. Kratter); Quebrada Sauce Grande, 04°21'S, 80°46'W (F. Angulo P. *in litt.* 2017); Trocha Tambor, 04°26'S, 80°46'W (F. Angulo *in litt.* 2017); El Alamor, 04°28.5'S, 80°24'W (USNM 323069, AMNH 151564). *Lambayeque* RVS Laquipampa, 06°21.5'S, 79°29'W (Pratolongo *et al.* 2012).

TAXONOMY AND VARIATION Monotypic. Described as a species, subsequent authors classified *watkinsi* as a subspecies of *G. ruficapilla* (Peters 1951, Meyer de Schauensee 1966). Subsequently, the distinctive song of *watkinsi* led to its elevation to species rank (Parker *et al.* 1995). Furthermore, although the distributions of

watkinsi and *ruficapilla* closely approach one another, even overlapping at a few locations in Ecuador, there is no evidence of introgression (Ridgely & Greenfield 2001). Its fairly different vocalisations notwithstanding, it is clear that Watkins's Antpitta is sister to the similarly plumaged Chestnut-crowned Antpitta, a suggestion supported by molecular data (Rice 2005a,b).

STATUS Watkins's Antpitta is restricted to the highly threatened Tumbesian region (Wege & Long 1995, Cook 1996, Stattersfield *et al.* 1997, 1998), but overall its distribution is larger than that of many species treated herein. In suitable habitat, the relative abundance of Watkins's Antpitta is rated as fairly common in Ecuador (Ridgely & Greenfield 2001) and as locally fairly common in Peru (Schulenberg *et al.* 2007). Furthermore, population trends were, until recently, believed to be stable (BirdLife International 2013). These two factors led BirdLife International (2013) to assign Watkins's Antpitta the conservation status Least Concern. Unsurprisingly, however, this species is recognised as facing ever-increasing risks from habitat loss or degradation, primarily through clearing forest for agriculture, and perhaps also degradation from grazing by livestock (Schulenberg *et al.* 2013a). Freile & Rodas (2008) recognised Watkins's Antpitta as a species of high conservation priority and Freile *et al.* (2010b) considered it meritorious of Near Threatened status in Ecuador, where most of its range lies. Following recent re-evaluation, BirdLife International (2017), based largely on the recommendations of Freile *et al.* (2010b), currently considers Watkins's Antpitta as globally Near Threatened, believing the population size to be approaching the thresholds for Vulnerable under the range size criterion. Additionally, its range is shrinking and available habitat quality declining. Although *G. watkinsi* is currently reported from more than ten locations that are not yet considered to be severely fragmented, the global population is predicted to decline by 10–30% over the next three generations. **Protected Populations Ecuador**: PN Machalilla (ANSP 184003–005, MECN 5884); RE Arenillas (Ridgely & Greenfield 2001); RVS y Marino Costera Pacoche (eBird: R.Terrill); RP Jorupe (Greeney *et al.* 2009, XC, IBC); RP Laipuna (J. Freile *in litt.* 2017); RE Comunal Loma Alta (M.E. Juiña J. *in litt.* 2015); RP El Tundo (eBird: F. Broulik); RP Río Ayampe (HFG); BP Bosque Petrificado Puyango (R.A. Gelis *in litt.* 2017). **Peru**: PN Cerros de Amotape (LSUMZ, USNM); RN Tumbes (Walker 2002, LSUMZ); Coto de Caza El Angolo (Barrio *et al.* 2015); RVS Laquipampa (Pratolongo *et al.* 2012).

OTHER NAMES *Grallaria ruficapilla watkinsi* (Zimmer 1934, Bond 1950, Peters 1951); *Grallaria [ruficapilla] watkinsi* (Sibley & Monroe 1990). **English** Watkins's Chestnut-crowned Antpitta (Cory & Hellmayr 1924); Scrub Antpitta (Ortíz-Crespo *et al.* 1990, Sibley & Monroe 1990, Monroe & Sibley 1993, Granizo *et al.* 2002). **Spanish** Tororoí Matorralero (Krabbe & Schulenberg 2003); Gralaria Matorralera (Ortiz-Crespo *et al.* 1990); Huicundo (Granizo *et al.* 2002); Tororoí de Watkins (Clements & Shany 2001, Plenge 2014); Gralaria de Watkins (Ridgely *et al.* 1998, Ridgely & Greenfield 2001, Granizo *et al.* 2002; Granizo 2009); tunturuguay and cántaro no hay (locally; F. Angulo *in litt.* 2017). **French** Grallaire de Watkins. **German** Buschlandameisenpitta (Krabbe & Schulenberg 2003).

Watkins's Antpitta, adult brooding two young nestlings, RP Jorupe, Loja, Ecuador, 4 April 2014 (*Luis A. Salagaje M.*).

Watkins's Antpitta, adult, RP Jorupe, Loja, Ecuador, 18 October 2011 (*Galo Real*).

Watkins's Antpitta, adult vocalizing from well-hidden perch, RP Jorupe, Loja, Ecuador, 1 March 2014 (*Harold F. Greeney*).

Watkins's Antpitta, young nestlings begging for food, RP Jorupe, Loja, Ecuador, 4 April 2014 (*Luis A. Salagaje M.*).

SANTA MARTA ANTPITTA
Grallaria bangsi   Plate 11

Grallaria bangsi J. A. Allen, 1900, Bulletin of the American Museum of Natural History, vol. 13, p. 159, El Líbano, at 7,000 feet [*c*.2,140m] in the Santa Marta Mountains of Magdalena Department, Colombia [11°10'N, 74°00'W]. Named for Outram Bangs, the male holotype (AMNH 73145) was collected on 25 May 1899, by Grace H. Hull for Herbert H. Smith (LeCroy & Sloss 2000). Allen (1900) also mentioned an adult female collected at San Lorenzo on 23 May 1899, but provided no further details.

Endemic to the Santa Marta Mountains from which it derives its English name, this antpitta is largely dark olive-brown on the head and upperparts, with a tawny-ochraceous throat and heavily streaked white underparts. Although relatively common (at least by voice) within its limited range, there are virtually no aspects of the species' biology not in need of further research. Its nest, eggs and details of reproductive ecology are undescribed (Greeney 2015c). Renjifo *et al.* (2000) suggested that one priority would be determining a more accurate geographic and altitudinal range within the Sierra Nevada, calling for censuses to estimate the density and state of the population to enable a more accurate assessment of suitable habitat and to facilitate the design of appropriate conservation measures. This account provides the first description of the fledgling and first complete description of the juvenile.

IDENTIFICATION 17–18cm. Upperparts olivaceous-brown, greyer on crown and sides of the neck, with brown wings and tail. Ear-coverts and face-sides brown with whitish shaft-streaks and a narrow buffy-white ring. Throat rather bright tawny-ochraceous, but most of the rest of the underparts white with heavy olive-brown to dusky streaking. Essentially the only *Grallaria* within its restricted range, Santa Marta Antpitta possibly overlaps at lower elevations with the distinctly different Scaled Antpitta (Hilty & Brown 1986). As noted by Allen (1900), Santa Marta Antpitta bears a passing resemblance to the two *Myrmothera* antpittas, neither of which occurs close to the present species. The bold chest streaking of *G. bangsi* recalls that of *G. ruficapilla,* which likewise has no range overlap.

DISTRIBUTION Santa Marta Antpitta is endemic to the Sierra Nevada de Santa Marta, predominantly within the Colombian department of Magdalena (Cracraft 1985,

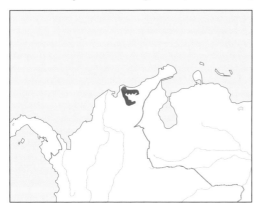

Stattersfield *et al.* 1998, Krabbe & Schulenberg 2003). Its range presumably encompasses most or all of the slopes of this isolated range, but to date there are no records from much of the eastern slope.

MOVEMENTS The species is sedentary, so far as is known, but it is possible that variations in reported altitudinal range reflect seasonal movements rather than historical changes.

HABITAT For the most part, Santa Marta Antpitta appears to favour the floor and lower understorey of humid montane cloud forest, mature second growth and tangled forest borders (Krabbe & Schulenberg 2003, Restall *et al.* 2006, Ridgely & Tudor 2009). They will, however, frequently forage in small openings, clearings or roadsides at the forest edge (Hilty & Brown 1986). Cubillos & Carantón-Ayala (2012) provided the only quantified description of habitat use, suggesting that the species prefers riparian areas on steep terrain. They found most individuals in areas with abundant leaf litter (78% coverage with *c*.6kg of litter per m²), a closed canopy (79%) of mature, epiphyte-laden trees (12–20m tall) and a dense understorey of ferns and *Chusquea* bamboo. The study did, however, find that the species is somewhat tolerant of disturbance, occasionally foraging at forest edges and adjacent pastures, provided there is ample forest nearby. The species' altitudinal range is generally given as 1,200–2,400m, with most observations above 1,600m (Meyer de Schauensee 1950b, Hilty & Brown 1986, Krabbe & Schulenberg 2003). This lower limit is borne out by modern surveys of the San Salvador Valley (Strewe & Navarro 2003) and lower records may reflect a historical distribution recently altered by anthropogenic habitat alteration. Some sources provide a maximum altitude of 2,450m (Todd & Carriker 1922, Fjeldså & Krabbe 1990), 2,500m (McMullan *et al.* 2010, Cubillos & Carantón-Ayala 2012), 2,600m (Stiles 1997) or 2,650m (Sibley & Monroe 1990), but more surveys are needed to determine its current distribution.

VOICE The song was described by Krabbe & Schulenberg (2003) as lasting 0.8–1.0s and delivered at intervals of 6–11s. It comprises two similar whistles, the first at 1.9–2.1kHz, rising slightly, and the second rising distinctly from 1.9 to 2.7kHz. Other authors have qualitatively described the song similarly. McMullan *et al.* (2010) described the song as '2.5 whistled notes, upslurred and interrogative at end', Ridgely & Tudor (2009) as a 'hollow, far-carrying *whow-whoit*', and Hilty & Brown (1986) suggested that the song recalls a bobwhite's whistle, 'a loud, flat *bob white*.' Krabbe & Schulenberg (2003) described the call as a single, squeaky, rising *queet*, 0.25s long, given at 2–5s intervals for minutes on end. The frequency of the call note ranges from 4.5 to 5.3kHz and usually includes several audible harmonics. This call note is undoubtedly, in at least some circumstances, an alarm.

NATURAL HISTORY Like many congeners, Santa Marta Antpitta is most frequently detected by voice (Hilty & Brown 1986) and forages by hopping over the ground, only occasionally assuming low horizontal perches (Krabbe & Schulenberg 2003, Cubillos & Carantón-Ayala 2012). Hilty & Brown (1986) described Santa Marta Antpitta as 'less retiring and easier to see than many *Grallaria*' and this has been repeated by most subsequent authors (Krabbe & Schulenberg 2003). Although the species seems to favour densely vegetated habitat, it sometimes feeds at the edge of trails, roads and pastures in the early morning (Ridgely & Tudor 2009, Cubillos & Carantón-Ayala 2012, Spencer 2014), occasionally until late morning (G.M. Kirwan *in litt.* 2017). There are no published data on territory defence, maintenance or fidelity, but Cubillos & Carantón-Ayala (2012) found seven individuals along a 500m transect, suggesting a somewhat greater density than the five singing individuals along 4.5km of trail by Strewe & Navarro (2004).

DIET The diet remains to be formally documented but, like many congenerics, individuals have been trained to visit worm feeding stations to entertain tourists (Woods *et al.* 2011). This suggests that it shares with other *Grallaria* the inclusion of a large percentage of earthworms in its diet (Greeney 2012b). Cubillos & Carantón-Ayala (2012), mentioned unidentified invertebrates, coleopterans (Staphylinidae, Cucurculionidae), hymenopterans (Formicidae) and myriapods in stomachs. They also found unidentified bones along with some belonging to glass frogs (Centrolenidae). Although they mentioned seeds in the stomach contents examined, it seems likely that they may have been consumed incidentally.

REPRODUCTION Although undocumented, Santa Marta Antpitta is probably socially monogamous, with both adults partaking in all aspects of reproduction. Details of the species' nests, eggs and nestlings are undescribed, but will probably prove to be similar to those of the better-known Chestnut-crowned Antpitta. **SEASONALITY** Very poorly documented. Krabbe & Schulenberg (2003) mentioned an 'immature' collected in May, possibly the same individual described by Todd & Carriker (1922) as having 'slight rufescent edging to the wing coverts' [= subadult, 25 May 1921, Cuchilla de San Lorenzo]. Two females in breeding condition were collected by Carriker, in September and January (Hilty & Brown 1986). The latter authors also mentioned a 'pin-feathered juvenile' following an adult at San Lorenzo on 7 July, citing T.B. Johnson. This observation undoubtedly refers to the following specimen. In his description of Cundinamarca Antpitta, Stiles (1992) mentioned the existence of a fledgling male Santa Marta Antpitta, but provided no further information. This specimen is now in the Instituto Alexander von Humboldt collection (IAvH 0932, see Description, Fledgling) and bears the following data: collected by T.B. Johnson on 7 July 1972 at Estación Experimental San Lorenzo, 25km SE of Santa Marta. Finally, I examined a juvenile female (USNM 387372) collected on 18 September 1945 at San Lorenzo. Taken together, the above records provide no clear indication of seasonality, with clutches potentially initiated any time between January and July.

TECHNICAL DESCRIPTION Sexes similar. *Adult* Uniform olive-brown above, slightly greyer on crown. The wings and tail are brown with the leading edge of the primaries slightly brighter. They have a whitish eye-ring and, although Allen (1900) and Krabbe & Schulenberg (2003) mentioned a whitish loral spot, I have found this to be somewhat variable, or even lacking entirely in some individuals. In those individuals with some loral markings, this generally appears rather small and buffy-ochraceous rather than whitish. Ear-coverts and face brown, with whitish shaft-streaks on the feathers below the eyes. The neck-sides are grey-brown, similar to the crown, contrasting with a tawny-ochraceous throat, which

also bears indistinct white streaks. Remaining underparts are white with heavy dusky olive-brown streaking, which becomes sparser lower down, ending in the unmarked white vent. Some individuals show a few tawny streaks on the breast (similar in colour to the throat). The breast-sides and flanks are olive-brown with sparse, thin white streaks, while the underwing-coverts are tawny-chestnut, paler on the base of the quills. **Adult Bare Parts** *Iris* dark brown; *Bill* overall bluish-grey, with maxilla often blackish along culmen, becoming paler near the tomia. Most individuals also appear to have a slightly paler tip to the maxilla, bordering on pinkish-grey. Allen (1900), working from specimens, described the bill as 'rather light horn color, very light at tip and along the edges of the commissure.' In life, the bill is certainly darker than this implies. *Tarsi & Toes* dark grey to bluish-grey or leaden-blue, though again Allen (1900) described them as 'light horn color'. *Juvenile* There is no published description of juvenile plumage (Krabbe & Schulenberg 2003), but Ryder & Wolfe (2009) indicated that pre-formative moult is partial in several species of *Grallaria*, among them *G. bangsi*, and that formative-plumaged birds retain the outer greater coverts, primary-coverts and a variable number of median and lesser coverts that are edged buff or rufous, unlike adult feathers. Thus, presumably, subadult Santa Marta Antpittas may differ from adults by having buffy or rufous fringes to the wing-coverts. This assertion is borne out by an assumed immature specimen described by Todd & Carriker (1922) as having 'slight rufescent edgings and tipping to the wing-coverts'. I examined a juvenile female in the USNM collection (see Reproduction Seasonality). Most of the upperparts are generally similar in coloration to adults, particularly the forecrown, crown and back. The hindcrown and nape, however, are distinctly barred rufous and dusky-black with a few scattered olivaceous feathers of adult plumage. This barring extends onto the neck-sides. Scattered feathers on the back, slightly more towards the rump, as well as the upperwing-coverts, are rufous-brown, brighter at the distal margin, and bear a broad, subterminal dusky-black band. These non-adult contour feathers and wing-coverts are, doubtlessly, the markings referred to by Todd & Carriker (1922). There is a fairly pronounced eye-ring of white feathers; ear-coverts buffy-rufous with pale buffy streaking, lores buffy, area below eyes similar to upperparts but speckled whitish. Below more similar to adult plumage, especially the distinctive buffy-orange or tawny-ochraceous throat. The remaining underparts differ mostly by having the amount of dusky olive-brown coloration greatly reduced and more extensive areas of white with generally thinner and paler olive-brown streaks. The specimen has more tawny-ochraceous streaking on the underparts than most adults, with the central upper breast distinctly washed with this colour. Streaking becomes sparser and fainter towards the tail, becoming almost pure white as in adults. The lower flanks and vent, however, are distinctly tawny-ochraceous, only slightly paler than the throat and breast streaks. The undertail-coverts are buffy-ochraceous. The thighs are olive-brown, washed and indistinctly barred rufous. **Juvenile Bare Parts** (from dried specimen) *Bill* dusky-black only on culmen, with the rest of the maxilla, and the mandible, dusky-yellow, brighter along the tomia of the maxilla and on the distal third of the mandible; in life, the yellowish areas were probably orange; *Tarsi & Toes* dark reddish-brown. *Fledgling* Examination of photos of the 'pin-feathered juvenile' (Hilty & Brown 1986) (see Seasonality) in the von Humboldt collection (IAvH 0932) has allowed me to describe the fledgling plumage for the first time. The fledgling, at this stage, bears little resemblance to the adult, and probably the following might well also apply to older nestlings. Above, dark brown to blackish, with broad rufous-chestnut barring on the back and rump, similar in colour on the nape and crown, but barring finer; in addition, the feathers of the nape are pale buffy-rufous on the rachis, giving a streaked appearance that partially obscures the barring. The individual bears only half a dozen 'adult' contour feathers on the upper back, these being dark olivaceous-brown. Wings brownish, with feathers not fully emerged from their sheaths; upperwing-coverts rufous-brown with paler, rufescent-brown margins. Small white feathers narrowly encircle the eye, but the face is otherwise dark olivaceous-brown with scattered pale shaft-streaks below eye and on ear-coverts. Chin and throat whitish, orangey-buff centrally. Upper and middle breast dark, chocolate-brown, only faintly barred rufescent with scattered buffy streaks, lower breast becoming more rufescent, this extending onto flanks and vent, densely barred dusky-brown throughout except somewhat paler, buffy rufous mid-belly, which appears largely unbarred. Thighs dark brown with rufous edges to some feathers. **Fledgling Bare Parts** (from specimen) *Bill*, maxilla dusky with a pale tip and pale edges to the tomia, mandible all yellowish-horn; *Tarsi & Toes* greyish-brown. The specimen label gives total length as 19.5cm, but this was presumably taken from the skin. The iris is described as very dark chocolate-brown, the legs and feet as greyish, the maxilla black with a yellow tip, and the mandible yellow. The testes of this (presumably) non-breeding individual measured 3 × 1mm. The label also bears the following measurements: wing 87mm, tail 49mm, mass 59g. Despite this description of the bill, which was taken from the museum label, it is my guess that, in life, the mandible and tomia of the maxilla are dark orange and the rictal flanges are inflated and also bright orange to yellow-orange (as illustrated, Plate 11).

MORPHOMETRIC DATA *Wing* 91mm (*n* = 1, ♂ holotype, Allen 1900), 95.6mm, 91.3mm (*n* = 2, ♂,♀ Stiles 1992). *Tail* 59mm (*n* = 1, ♂ holotype, Allen 1900), 51.3mm, 50.5mm (*n* = 2, ♂,♀, Stiles 1992). *Tarsus* 47mm (*n* = 1, ♂, holotype, Allen 1900), 50.2mm, 51.2mm (*n* = 2 ♂,♀, Stiles 1992), 48.6mm, 47.8mm, 52.7mm, 51.5mm, 51.6mm (*n* = 5, ♂,♀,♂,♂,♀,♂, USNM). *Bill* culmen 23mm (*n* = 1, ♂ holotype, Allen 1900), exposed culmen 21.8mm, 22.5mm, total culmen 25.7mm, 26.2mm, commissure width 16.6mm, 26.7mm, bill depth at nares 8.8mm, 7.9mm (*n* = 2, ♂,♀, Stiles 1992), from nares 15.0mm, 15.3mm, 15.3mm, 15.5mm, 15.7mm (*n* = 5, ♂,♀,♂,♂,♀, USNM). **Mass** 62g (*n* = 1, ♂, Stiles 1992). **Total Length** 17–18cm (Meyer de Schauensee 1970, Hilty & Brown 1986, Ridgely & Tudor 1994, Krabbe & Schulenberg 2003, McMullan *et al.* 2010).

DISTRIBUTION DATA Endemic to **COLOMBIA**: *Magdalena* Cuchilla de San Lorenzo, 3.3km ENE of La Tigrera, 11°10'N, 74°07'W (Todd 1922, FMNH, USNM, AMNH, ANSP, CM, UMMZ, XC, ML); Estación Experimental San Lorenzo, 11°06.5'N, 74°03.5'W (IAvH 0932, XC, many recordings: D. Calderon-F., F. Schmitt); RNSC La Cumbre, 11°06'N, 74°03'W (Cubillos & Carantón-Ayala 2012); Hacienda Vista Nieve, 11°05'N, 74°05'W (CM P102878/USNM 387373); 4km SSW of Palomino, 11°00'N,

73°40'W (XC 223584–588, P. Boesman); Cebolletas, 10°54'N, 73°59'5W (XC 301161/162, D. Calderon-F.); 8km SW of Hato de Simon, 10°54'N, 73°46'W (XC 18083, H. van Oosten). *Cesar* Chinchicuá, on Cesar/Magdalena border, 10°30'N, 73°35'W (USNM 387374–376). *La Guajira* Páramo de Macotama, 10°55'N, 73°30'W (O. Cortes-Herrera *in litt.* 2016); Cherua, 10°52'N, 73°23'W (ANSP 63060/061, CM P45040/041); San Miguel, 10°58'N, 73°29'W (CM, LSUMZ 127434).

TAXONOMY AND VARIATION Monotypic. The precise taxonomic affinities of Santa Marta Antpitta have yet to be investigated. It seems highly likely, however, based on morphological similarities and distribution, that Santa Marta Antpitta is sister to Cundinamarca Antpitta (Stiles 1992, Krabbe & Schulenberg 2003). These two species have been suggested to be closely allied to two other range-restricted Colombian endemics, Brown-banded and Urrao Antpittas (Carantón-Ayala *et al.* 2012). Plumage characters of both Santa Marta and Cundinamarca Antpittas, and to a lesser degree voice, seem to suggest a closer relationship with Chestnut-crowned Antpitta. Allen (1900) in his original description stated that 'this species finds its nearest ally in *Grallaria modesta* Sclater [*Myrmothera campanisona modesta*]'. This certainly erroneous taxonomic arrangement was followed by Brabourne & Chubb (1912), but subsequent authors stated or implied various affinities within the genus (Peters 1951, Clements 2007), with most including it near *G. ruficapilla* in their linear arrangements (Cory & Hellmayr 1924, Meyer de Schauensee 1964, 1966, 1970, Altman & Swift 1986, Howard & Moore 1984, Sibley & Monroe 1990, Monroe & Sibley 1993, Clements 2000). Todd & Carriker (1922) remarked on the taxonomic placement of Santa Marta Antpitta within *Grallaria*, stating that it has morphological and plumage characteristics somewhat intermediate between the subgenera *Hypsibemon* (type = *Grallaria ruficapilla*) and *Oropezus* (type = *G. rufula*). Lowery & O'Neill (1969), in their review and arrangement of the antpitta genera, definitively placed *G. bangsi* in the subgenus *Hypsibemon*, within which they also included *G. ruficapilla*, *G. watkinsi* and *G. andicolus*. The description of juvenile plumage provided above (see Technical Description, Juvenile), in particular the distinctive, fine, black-and-rufous barring on the hindcrown and nape during the final phases of acquiring adult plumage, leave me with little doubt that this species is fairly closely related to *G. ruficapilla*. Indeed, if viewed only from behind, it might be difficult to separate juveniles of the two species at this age, suggesting closer affinities to the *ruficapilla/watkinsi* lineage.

STATUS Until the 1990s, Santa Marta Antpitta was not considered at risk (IUCN 1990, Renjifo *et al.* 1997), but its status was subsequently upgraded to Near Threatened (Stiles 1992, Collar *et al.* 1994, Stattersfield *et al.* 1998). More recently, Santa Marta Antpitta was upgraded to Vulnerable (Krabbe & Schulenberg 2003, McMullan *et al.* 2010, Fundación ProAves 2014). Its current categorisation as Vulnerable (BirdLife International 2017) is based largely on its small range and rapid population decline suspected because of observed rates of habitat loss using time-series satellite imagery (Orejuela 1985, Renjifo *et al.* 2000). It occurs in the Santa Marta Mountains EBA and, despite being common within its small range, severe degradation of remaining forest habitat, despite formal protection in the PNN Sierra Nevada de Santa Marta, this species could become threatened. Krabbe & Schulenberg (2003), however, suggested that the species' relatively less shy nature (compared to other antpittas) may enable Santa Marta Antpitta to be reasonably adaptable. At least one study, however, found the species to have fairly specific habitat requirements (Cubillos & Carantón-Ayala 2012). The observation that Santa Marta Antpitta will forage in open areas and can be attracted to worm-feeding stations supports the assertion that it may be somewhat adaptable in the face of habitat alteration. Nevertheless, heavy deforestation of the preferred habitat of this antpitta is an ongoing and ever-increasing threat (Stattersfield *et al.* 1998, Renjifo *et al.* 2000). **PROTECTED POPULATIONS** PNN Sierra Nevada de Santa Marta (CM, AMNH, MCZ, XC); RNA El Dorado (Salaman *et al.* 2007a, XC); RNSC La Cumbre (Cubillos & Carantón-Ayala 2012).

OTHER NAMES Spanish Tororoí de Santa Marta (Krabbe & Schulenberg 2003, Salaman *et al.* 2007a). **French** Grallaire des Santa Marta. **German** Ockerkehl-Ameisenpitta (Krabbe & Schulenberg 2003).

Santa Marta Antpitta, adult, RNA El Dorado, Magdalena, Colombia, 11 March 2010 (*Nick Athanas/Tropical Birding*).

Santa Marta Antpitta, El Dorado Reserve, Magdalena, Colombia, 27 July 2016 (*Fabrice Schmitt*).

Santa Marta Antpitta, RNA El Dorado, Magdalena, Colombia, 3 April 2015 (*Alonso Quevedo*).

Santa Marta Antpitta, PNN Sierra Nevada de Santa Marta, Magdalena, Colombia, 24 March 2014 (*Fabrice Schmidtt*).

CUNDINAMARCA ANTPITTA
Grallaria kaestneri   Plate 11

Grallaria kaestneri Stiles, 1992, Wilson Bulletin, vol. 104, p. 389, 3km east-north-east of Monterredondo, Amazonian slope, Cundinamarca, Colombia [*c*.04°16'N, 73°48'W]. The species is named for its discoverer, Peter G. Kaestner, and its common name reflects its extremely small range, which to date is believed to be confined to the borders of Dept. Cundinamarca.

Cundinamarca Antpitta is known from just a few localities on the east slope of the East Andes in Colombia, all of them reasonably close to the country's capital, Bogotá. Given its small range and the degree of deforestation this region has suffered, the conservation status of Cundinamarca Antpitta was evaluated initially as Vulnerable, but continued threats to its habitat and apparently declining population has led to its threat status being revised to Endangered. It is a medium-sized olive-brown antpitta with a dull-white throat mottled darker, and a greyish-olive breast with very narrow white shaft-streaks. Cundinamarca Antpitta inhabits the understorey of wet primary and older secondary cloud forest, at elevations of 1,700–2,300m. It is perhaps even more terrestrial than others in the genus, foraging in the leaf litter and soil for arthropods and earthworms. There are no aspects of the biology, ecology, evolution or taxonomy of Cundinamarca Antpitta not in need of further research. Clearly, the establishment of protected areas within its very limited range, before it becomes completely deforested, is a priority. The nest, eggs and breeding behaviour are completely unknown (Greeney 2014d). A description of immature plumage, the first reproductive information of any kind, is presented here.

IDENTIFICATION 15–16cm. The sexes are similar. It is a rather dull, generally olive-brown antpitta, slightly mottled with whitish and dull olive on the face and barely noticeable pale shaft-streaks on the ear-coverts. The feathers of the back are narrowly fringed sooty-black, creating very faint barring, except on the rump and uppertail-coverts. The wings are richer brown, while the tail is slightly darker than the back. The throat is off-white and slightly mottled, similar to the face. The underparts, including the breast and flanks, are olive-brown with narrow, dull white shaft-streaks most noticeable on the sides and flanks. The central belly is dull white, the vent is dull olive-buff and the underwing feathers are buffy-cinnamon. The dull-coloured, relatively unmarked Cundinamarca Antpitta is unlikely to be confused in the field. It is similar to Santa Marta Antpitta, with which there is no overlap in range, but is generally darker overall.

DISTRIBUTION Cundinamarca Antpitta is restricted to the east slope of the East Andes in Cundinamarca, Colombia, from south-east of Bogotá near Monterredondo, above Guayabetal and at Farallones de Medina. It probably also occurs in Dept. Meta between these sites (Krabbe & Schulenberg 2003, Cortes-Herrera *et al.* 2012, BirdLife International 2017).

HABITAT Ranging at 1,700–2,500m (Stiles 1992, Ridgely & Tudor 2009, McMullan *et al.* 2010), Cundinamarca Antpitta inhabits the floor and lower undergrowth of humid montane forest and mature secondary woodland. It prefers dense understorey and, while it will occupy forest edges and gaps, the species seems to require a fairly intact canopy with heavy epiphyte loads (Stiles 1997, Cortes-Herrera *et al.* 2012).

VOICE The song is short, roughly 1s long, and consists of three similar, whistled notes increasing in pitch at 2.5–3.0kHz. There is a slight pause after the first note, and the last note sometimes is omitted (Krabbe & Schulenberg 2003, Restall *et al.* 2006). Ridgely & Tudor (2009) noted that the song is distinctly high-pitched for *Grallaria*, and described it as *whir, whee-whee*. The call is a two-noted whistle starting with a piercing note at 5.8–6.0kHz and dropping abruptly to 5kHz.

NATURAL HISTORY Stiles (1992) noted that Cundinamarca Antpitta rarely ascends to perches as high as 1m or more above ground, even to sing, suggesting that it may be even more terrestrial than most of its congeners. Stiles (1992) further noted that 'within its dense habitat, the bird is not particularly shy, and may approach a motionless observer closely', which may indicate that it is accustomed to foraging in the wake of large terrestrial mammals as has been suggested for other montane *Grallaria* (Greeney 2012b). They forage on the ground, moving in long hops, pausing periodically to displace leaf litter with their bills. They occasionally forage in low vegetation such as fallen bromeliads, and will sometimes thrash captured prey against a stem or the ground prior to consumption (Stiles 1992). There are no published data on territorial defence, maintenance or fidelity for Cundinamarca Antpitta. Although territory size has not been well described, Stiles (1992) estimated roughly four

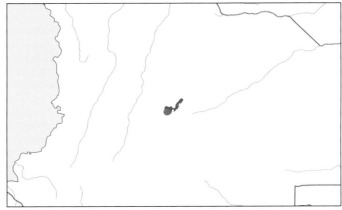

pairs of Cundinamarca Antpittas in 5ha and noted the presence of *c.*10 pairs along a transect of 1.5km.

DIET The diet is virtually unrecorded. Stomach contents of the three female specimens collected by Stiles (1992) contained mainly arthropods (beetles, roaches, katydids and spiders), and also the remains of a small earthworm.

BREEDING There are no published details of the reproductive habits or breeding seasonality. The nest, eggs and young of *G. kaestneri* are undescribed. **SEASONALITY** Based on the reproductive condition of adults and timing of singing activity, Stiles (1992) suggested that the species breeds in the latter half of the year, roughly coinciding with the end of the wetter months. M. Rueda posted an image of a juvenile (IBC 1011342) and was kind enough to share several other photographs of the same individual with me. Based on these photographs, I feel that this individual was probably still dependent (see Technical Description, Juvenile). The photos were taken on 12 October 2012 near Guayabetal. A juvenile of similar age was photographed by J. Beck on 8 June 2015, almost exactly at the type locality, and a slightly older juvenile was captured by J.V. Remsen, G. Bravo and A.M. Cuervo on 14 August 2007 along the upper Quetame–El Calvario road (series of photos including ML 28426181/221/241, A.M. Cuervo).

TECHNICAL DESCRIPTION Sexes similar, but specimens of this rare species are few and, interestingly, I was able to find more photos of juveniles than of adults! The following is adapted from Stiles' (1992) original description which used the colour names of Smithe (1975, 1981). *Adult* The upperparts, including the crown, nape, back, and scapulars are dark olive-green, with the narrow, sooty-black feather fringes creating very fine blackish barring (apparent only in the hand). Lower, the rump and uppertail-coverts are similar in colour but lack the barring, making them appear brighter olive-green. The tips of the longest tail-coverts are dark brown. The tail is blackish-brown. The greater secondary-coverts, primary-coverts and remiges are blackish on inner webs, with a broad fringe of warm brown on outer webs. The lores and face are dull, buffy-white, appearing mottled due to the dusky bases and dull olive tips of the feathers. The ear-coverts are dull olive with fine pale shaft-streaks. The throat is whitish, washed buff and lightly mottled dusky-olive and grey. The feathers of the upper chest are whitish, broadly tipped olive-grey, the underparts becoming mostly olive-grey and narrowly streaked whitish on the breast, sides and flanks. Streaking is broadest on sides and flanks. The central abdomen is dull white, tinged pale horn or buff, and the thighs more brownish. The vent and undertail-coverts are dull olive-buff, with the longest undertail-coverts tipped brown. The undertail is similar to above, but slightly washed olive-green. The underwing linings are bright ochraceous-cinnamon. **Adult Bare Parts** *Iris* dark brown; *Bill* maxilla black, mandible slate-grey, both fading to whitish on tomia and at tips; *Tarsi & Toes* greyish to light purplish-grey. *Fledgling* Previously undescribed. A fledgling photographed by J.D. Castillo Ramírez and A. Pinto (date and location not shared) was still in fluffy fledgling plumage and not yet capable of coordinated movement or full extension of its legs. The entire upperparts, throat, breast and sides were dark blackish-brown with buff barring and spotting. Only the face, lores and ear-coverts were devoid of markings, though some small spotting was visible below the eyes. The buff portions of the feathers from the

crown to the upper back tended to be more rusty-buff. The feathers of the forecrown were least marked, this area being almost entirely dark brown, increasingly barred rearwards. The feathers of the rump were most extensively marked buff and on the underparts the amount of buff slowly increased from fine buff barring on the breast to coarse barring on the sides and lower breast. A few feathers scattered across the breast-sides and under the throat bore paler buff-white, teardrop-shaped spots. The amount of buff in each feather increased gradually below the breast, with the belly and vent entirely buff, tending towards white centrally and clean buff at the base of the tail. Its tail feathers had not yet emerged sufficiently to be visible through the dense, fluffy down. The flight feathers were not fully emerged from their sheaths but were distinctly bicoloured, olivaceous-brown on the anterior webs and dark greyish-brown on the posterior webs. **Fledgling Bare Parts** *Iris* dark brown; *Bill* mostly blackish with distinctly inflated and bright orange rictal flanges and a pale orange tip on the maxilla that appeared to surround the pale yellowish egg tooth; mandible slightly browner than the maxilla and indistinctly washed orange, especially at the base and on the tomia; *Tarsi & Toes* vinaceous-pink, nails dark yellow. *Juvenile* Additional photos of a juvenile (IBC 1011342) were shared with me by M. Rueda (Fig. 1) and a series of photos of a different juvenile are attached to eBird checklist S25496050 (A.M. Cuervo on deposit at ML). The forecrown and crown are similar in colour to adults, but with a few pale shaft-streaks, the hindcrown and nape are dark brownish and narrowly streaked buffy-cinnamon, the streaks whitish on the nape but continuing onto the neck-sides and becoming buffier again. Face brownish-olive, ear-coverts browner with buffy-whitish streaking. Lores, malar, throat and sides to upper breast white with greyish 'salt-and-pepper' speckling. Back, tail and wings similar to adults, including the underwing linings which are already bright ochraceous-cinnamon (see ML 28426241). The upperwing-coverts, though not clearly visible in the photographs, appear to be edged pale rufous-brown or buffy-cinnamon, with a thin subterminal dusky band, generally similar to the immature coverts of Chestnut-crowned Antpitta. Below, similar to adult, breast and sides greenish-olive with narrow dull whitish shaft-streaks, which are broadest on sides and flanks, and central belly dull white. Lower belly, flanks and vent still largely in fluffy fledgling plumage, dull brownish, mottled buffy, grey and ochraceous. **Juvenile Bare Parts** *Iris* dark brown; *Bill* maxilla blackish-grey, with dull orange tomia and tip, mandible mostly orange, duskier at tip, rictus only slightly inflated and orange; *Tarsi & Toes* plumbeous to light purplish-grey.

MORPHOMETRIC DATA No published morphometric data except the following measurements from Stiles (1992; *n* = 3 ♀♀). *Wing* 80.3mm, 80.5mm, 78.7mm; *Tail* 41.3, 42.6, 44.6mm; *Bill* exposed culmen 20.0mm, 17.8mm, 19.5mm, total culmen 21.4mm, 22.4mm, 23.1mm, commissure width 11.5mm, 11.3mm, 11.4mm, bill depth at nostril 6.2mm, 6.1mm, 6.2mm; *Tarsus* 42.2mm, 44.7mm, 43.0mm. **MASS** 49.4g, 45.8g, 47.4.g (*n* = 3 ♀♀; Stiles 1992). **TOTAL LENGTH** 15–16cm (Restall *et al.* 2006, Ridgely & Tudor 2009, McMullan *et al.* 2010).

DISTRIBUTION DATA Endemic to **COLOMBIA**: *Cundinamarca* Upper Río Gazaunta, Miralindo, 04°35'N, 73°26'W (XC 117285–288; M.Á. Rebolledo); 5km NNE of Guayabetal, 04°16'N, 73°49'W (XC 128049, O. Cortes-Herrera); Fallarones de Medina, 04°35'N, 73°29'W and

04°30'N, 73°28'W (XC 88730–739/93011–030, O. Cortes-Herrera); 6.5km W of San Miguel, 04°35'N, 73°26'W (ML: M. Álvarez); Quetame–El Calvario road, near Meta border, 04°23'N, 73°48'W (J.V. Remsen *in litt.* 2016); 8km N of Monterredondo, 04°19.5'N, 73°49.5'W (ML 74663/64: P.G. Kaestner); 4.75km NE of Monteredondo, 04°17'N, 73°48'W (A.M. Cuervo *in litt.* 2017); 0.5km NNW of Guayabetal, 04°13'N, 73°49'W (IBC 1011342, photo M. Rueda: Fig. 1). *Meta* Near El Calvario, 04°22'N, 73°43.5'W (XC 96204/06; O. Cortes-Herrera).

TAXONOMY AND VARIATION Monotypic. When described, Cundinamarca Antpitta was thought to be the sister species of Santa Marta Antpitta (Stiles 1992). More recently, it has been suggested to be closely related to two other Colombian endemics, Brown-banded Antpitta and the newly described Urrao Antpitta (Carantón-Ayala & Certuche-Cubillos 2010). Although immature plumage is incompletely known (transition from fledgling to juvenile unknown), from what I have seen I would tend to agree more with the latter suggestion.

STATUS Cundinamarca Antpitta is currently listed as Endangered at both national and global levels (Renjifo *et al.* 2014, BirdLife International 2017). Its threat status was upgraded from Vulnerable less than a decade ago (Renjifo *et al.* 2000, Stattersfield & Capper 2000, BirdLife International 2012) as the reality of its very small range and lack of habitat became more apparent. Increasing pressures on its habitat are expected to result in as much as a 30% reduction in the species' range over the next 10–15 years, and BirdLife International (2017) predicts a moderately rapid population decline. Cundinamarca Antpitta appears to successfully occupy fairly heavily disturbed forest and old second growth, leading Stiles (1992) to suggest that it may be somewhat resistant to local extirpation, provided some tree cover remains. Ongoing habitat loss and disturbance within its range, however, seem to have led to at least a moderate population decline over the last ten years (Cortes-Herrera *et al.* 2012). The establishment of a reserve protecting at least a small part of the species' range certainly seems of utmost importance.

OTHER NAMES Spanish Tororoí de Cundinamarca (Krabbe & Schulenberg 2003, ProAves 2014). **French** Grallaire de Kaestner. **German** Cundinamarcaameisenpitta (Krabbe & Schulenberg 2003).

Cundinamarca Antpitta, juvenile, 8 June 2015, Monterredondo, Cundinamarca (*Josh Beck*).

Cundinamarca Antpitta, juvenile, 12 October 2012, Guayabetal, Cundinamarca (*Mauricio Rueda*).

Cundinamarca Antpitta, juvenile, 12 October 2012, Guayabetal, Cundinamarca (*Mauricio Rueda*).

STRIPE-HEADED ANTPITTA
Grallaria andicolus   Plate 11

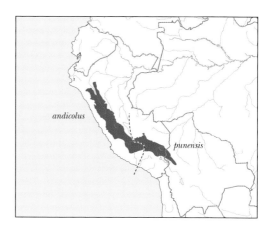

Hypsibamon andicolus Cabanis, 1873, Journal für Ornithologie, vol. 21, p. 318, Maraynioc, Junín, Peru [11°22'S, 75°24'W].

'They were very retiring in habits and required quiet stalking even to be seen. When found, the bird was usually standing large-eyed, silent, and motionless on a moss-grown root or similar perch or among the rotting leaves.' – **John T. Zimmer, 1930, Peru**

'The silent and secretive ways of this bird made them difficult to collect. They were very apt to hop around the rocks and underbrush and disappear as if by magic.' – **James L. Peters & John A. Griswold, 1943, Peru**

Stripe-headed Antpitta has rather strikingly different plumage from any other *Grallaria*, being principally brown above, with bold streaking and spotting on the face and underparts. The species is found at elevations above 3,000m from northern Peru, along the Andes, to north-west Bolivia. It is widely distributed but occurs at low densities and in high, cold habitats visited infrequently by ornithologists. Although its nest, eggs and nestlings were recently described, most other aspects of its natural history are very poorly known.

IDENTIFICATION 16.0–16.5cm. Stripe-headed Antpitta is a relatively small *Grallaria*, but nevertheless shares the distinctive long-legged form and characteristic terrestrial gait of its congeners. It is brown above, with narrow pale streaks on the crown and a more-or-less streaked back in most populations (plain brown in subspecies *punensis* in the south of the species' range). It is boldly patterned with streaks and spots on the face, breast and belly. Sclater's (1877) discussion of the nominate race described it as mouse-brown above with the head, nape, back and wing-coverts bearing 'light fulvous shaft-stripes which are more or less margined with black.' He mentioned that the tips of the rectrices and outer secondaries, as well as the underwing-coverts, are pale fulvous. The underparts are generally white, with the feathers of the throat- and breast-sides, and flanks, strongly margined with black and 'slightly varied with fulvous'. Streak-headed Antpitta is rather a distinctive bird, unlikely to be confused with any other species, especially as so few other *Grallaria* are found in the relatively patchy, high-elevation woods it inhabits. It typically occurs above elevations inhabited by the plain-coloured, unmarked Tawny Antpitta or the similarly unmarked and slightly smaller Rufous Antpitta. As mentioned by Schulenberg & Kirwan (2011), by plumage alone one might perhaps confuse Stripe-headed Antpitta with spotted juveniles of the sympatric Great Turdus fuscater and Chiguanco Thrushes *T. chiguanco*, but undoubtedly its long-legged gait and upright posture would immediately distinguish it from a fledgling thrush.

DISTRIBUTION Found in the high Andes of Peru and extreme north-west Bolivia (Schulenberg *et al.* 2007, Ridgely & Tudor 2009). In the western cordillera, its range extends from Cajamarca south to Ayacucho (Lüthi 2011), occurring in many parts of the inter-montane valleys (including the east slope of the Marañón Valley) from Dept. Amazonas south to Apurímac (*andicolus*) (Mark *et al.* 2008). South from Apurímac (*punensis*), Stripe-headed

Antpitta is found in the departments of Cuzco and Puno, and across the border into northern La Paz in Bolivia.

HABITAT Stripe-headed Antpitta occurs at higher elevations and in somewhat drier habitats than most congeners. Matched in elevation only by Tawny Antpitta and in habitat aridity only by Watkins's Antpitta, the primary habitat is *Polylepis* or *Polylepis–Gynoxys* woodland (Fjeldså 1993, Lloyd 2008), but it is common and apparently nests in patchy, rocky grasslands where it spends much time in the dense groves of stunted *Polylepis* trees, bushes and ferns, occasionally venturing into the open *puna* in search of food (Zimmer 1930, Dorst 1957, Greeney 2012a), and it is not always associated with forested areas (Fjeldså & Krabbe 1990). Although Stripe-headed Antpitta is said to tolerate heavily disturbed habitats (Krabbe & Schulenberg 2003, Schulenberg & Kirwan 2011), this may be something of a misleading statement for conservationists. Its natural high-altitude habitat consists of small patches of woodland separated by open areas of *páramo* and bunch grass, giving rise to the suggestion that it tolerates fragmentation, but to date no studies have examined the effects of anthropogenic habitat disturbance on Stripe-headed Antpitta. In Peru, the primary altitudinal range of *G. andicolus* is 3,500–4,600m, but in some areas it occurs as low as 3,000m (Schulenberg *et al.* 2007, Walker & Fjeldså 2005, Lüthi 2011).

VOICE The song is an odd, frog-like trill (Krabbe & Schulenberg 2003, Walker & Fjeldså 2005) described as 'a rolling series of wheezy notes, first slightly descending, then ascending and accelerating, last note sometimes drawn-out *ree ree … ree eeee*' (Fjeldså & Krabbe 1990) or as 'a low, grinding, froglike trill that rises and falls, often with an introductory stutter: *gr-grrrEEEEErrrr*' (D.F. Lane in Schulenberg *et al.* 2007). Other vocalisations include 'a single, somewhat wheezy, slightly descending note' (*andicolus*; Fjeldså & Krabbe 1990), '1–2, usually mellower notes alike, somewhat resembling call of Great Thrush [*Turdus fuscater*]' (*punensis*; Fjeldså & Krabbe 1990). My limited experience with these last two calls suggests that they probably play important roles in alarm and intraspecific aggression, respectively. These calls are rather far-carrying and frequently delivered from high in a bush or tree (Walker & Fjeldså 2005).

NATURAL HISTORY Few data. Although one could argue that his description might pertain to most *Grallaria*, Zimmer's (1930; see quote above) description of its habits is eloquent and appropriate of *G. andicolus*. It is primarily terrestrial, like other *Grallaria*, but do ascend thick mossy branches and horizontal trunks (Fjeldså & Krabbe 1990), especially when vocalising (Walker & Fjeldså 2005). Although most food is probably captured on the ground, Fjeldså & Krabbe (1990) described *G. andicolus* as searching for food on 'vertical cliff faces.' They can be heard vocalising or seen dashing between forest patches throughout the day, but the species is most active crepuscularly, with vocal activity beginning in the pre-dawn, even on cold foggy mornings (Lüthi 2011). While giving its putative alarm or aggression calls, but also during regular foraging, adults frequently flick their wings (Fjeldså & Krabbe 1990). As previously described (Greeney 2012a), during a brief study of *G. a. andicolus* near Cuzco, adults were generally very wary of my presence, quickly vanishing as I approached to within 10–15m, making it almost impossible to follow them through the dense, gnarled *Polylepis* woodland. Twice, however, while I sat quietly after following adults into the undergrowth, an adult reappeared within 3–6m of me, periodically melted from view then reappeared from a different direction. It is just such behaviour that I have observed in other antpittas, which leads me to believe that at least some species regularly forage behind large mammals (Greeney 2012b). A blundering Mountain Tapir *Tapirus pinchaque* or Spectacled Bear *Tremarctos ornatus* tearing apart *Puya* bromeliad bases would pose little threat to an antpitta, but could expose a plethora of otherwise inaccessible prey. While foraging, adults often freeze for very long periods, in a posture of vigilance rather than one of prey-searching (HFG). Presumably, while foraging away from cover, aerial predators are an important source of adult mortality.

DIET Despite its rather broad distribution and the ease with which it can be observed in appropriate habitat, the diet is very poorly known. Presumably it consists mostly of small to medium-sized invertebrates. Two stomachs contained unidentifiable 'larvae and insects' and several adult moths (Noctuidae) (Krabbe & Schulenberg 2003). Food items delivered to nestlings (Greeney 2012a) were predominantly earthworms, with a few larval insects similar to cranefly larvae (Tipulidae). No further information.

REPRODUCTION With the exception of a cursory description of three nests found by J. Fjeldså (Greeney *et al.* 2008), the breeding biology of Stripe-headed Antpitta was almost completely unknown until 2012. There are still no published data on nestbuilding behaviour, incubation period or rhythm, nestling period, or post-fledging parental care. **NEST** The nest is similar to those of other *Grallaria*, a bulky, open cup, composed primarily of moss intermixed with a few long sticks, especially near the base. Although not well documented, it appears that the species may 'decorate' the rim of the nest with thin twigs, an architectural feature unknown in the nests of any other antpitta. Internally they are overall pale-coloured, being sparsely lined with dried and green grass fibres and thin, flexible rootlets. Nine of the 11 described nests (Greeney *et al.* 2008, Greeney 2012a) were supported by several small tree branches, the rest were slightly better supported by larger horizontal branches. Mean nest height of eight nests in one study was 1.6m (range = 1.0–2.8m, Greeney 2012a), while mean height of three nests found by Fjeldså was 3m (Greeney *et al.* 2008). Based on the presence of both fresh

and old materials in active nests, Greeney (2012a) suggested that nests may be used for more than one breeding attempt, but this remains to be documented. Two nests measured by Greeney (2012a) had the following dimensions: maximum outer diameter, 24cm, 24cm; minimum outer diameter, 22cm, 22cm; outer height, 13cm, 15cm; inner cup diameter, 9.5cm, 12cm; inner cup depth, 6.5cm, 8.5cm. **EGG, LAYING & INCUBATION** Three eggs (Greeney 2012a) were subelliptical and pale blue. Two were sparsely blotched and flecked brown, only slightly more concentrated at the larger end. Egg measurements were 30.4 × 23.9mm, 30.0 × 22.9mm and 29.4 × 23.1mm, with the fresh mass of the last two being 8.3g. There are no data to indicate the duration of incubation, but based on altitudinal range and size, it seems likely that it is similar to the 22 days reported for *G. quitensis* (Greeney & Harms 2008). **NESTLING & PARENTAL CARE** Two nestlings, estimated to be 5–7 days old (Greeney 2012a), weighed 17.0g and 16.5g and had tarsi 18.4mm and 18.1mm long, respectively. They were pink-skinned (including legs) with sparse dark grey to blackish natal down sprouting from the dorsal feather tracts. Ventrally, they lacked natal down except a few sparse plumes on the sub-malar and ventral cervical tracts. Contour pin-feathers had emerged 1–2mm from the skin on the central back, the wings and legs. These unbroken pin-feathers were brownish basally and black near the tips. Pin-feathers on the lower feather tracts, especially the belly, were pale buffy or white, while those on the head, neck and upper breast were dark, suggesting that the first plumage would be tawny-brown with fine black barring, darker on the upper breast, nape and head, and paler on the belly and vent, much as Tawny Antpitta. Primary and secondary flight feathers were only recently emerging from the skin. Three days later most contour feathers were just breaking their sheaths. The nestlings' bills were dusky-orange with cream-coloured rictal flanges and bright orange mouth linings. They still bore bright white egg-teeth at an estimated age of 8–11 days. Both adults provisioned the nestlings, apparently with similar effort, and were relatively tolerant of observer presence less than 5m from the nest and in plain view. Once settled over the nestlings to brood, they would usually flush only if approached to within 1m. Adults approached the nest by running along the ground through dense vegetation until just below the nest, then hopping up quickly via several perches before popping onto the rim from below. Departures from the nest were quick and silent, and accomplished by dropping straight to the ground below the nest and running rapidly away. Four out of five observed prey deliveries included more than one item carried in the adult's bill. **SEASONALITY** Across its range, the breeding season appears to be fairly extended, from mid-June to mid-February, but with a relatively strong peak in September–November. Few data are available, however, with most coming from Fjeldså & Krabbe (1990) who appeared to use the term 'juveniles' to represent individuals in transitional plumage from fledglings to juveniles (as defined herein) and who used 'immature' to refer to juveniles (as defined herein). Based on their description of 'juveniles' as entirely barred and streaked, to interpret their data I consider these records as older fledglings, almost certainly still partially dependent on adults and probably 2–3(4?) months out of the nest. I interpret their use of the word fledgling as birds entirely dependent on their parents, probably less than two months out of the nest. I consider birds termed 'immature' as fully independent juveniles, but which may still associate with their parents, and which were likely 5–7 months out of the nest. My interpretation of Fjeldså & Krabbe's (1990) records is: young fledglings in February from Huánuco and Lima, and from March in La Libertad; older fledglings and/or juveniles in December from Junín, in January from Pasco and Cajamarca, in February from Junín, in March from La Libertad and Cajamarca, in April from La Libertad, and in May from Huánuco, Cajamarca and Puno. All of these records, except the juvenile from Puno, are of the nominate subspecies. Salvin (1895) reported the collection of a 'juvenile' male in March from Huamachnco and a 'juvenile' female in May from Cajamarca, both presumably of the nominate subspecies. As Salvin did not define the age of these specimens more specifically, based on other works by the same author and on use of 'juvenile' during this era, I interpret these two records to have been individuals completely independent of their parents, with few or no vestiges of subadult plumage, but likely with incomplete skull ossification and/or undeveloped reproductive organs. These individuals, by my interpretation, are 5–7 months post-fledging. Using my estimated fledgling ages, a nestling period of c.18 days and an incubation period of c.22 days, the above data suggest reproductive activity may be protracted in many areas, with most clutches initiated July–November. Specifically, from the above records and the additional information below, egg-laying dates may be estimated as: from July–November in Cajamarca; at least September–December in La Libertad; during November in Huánuco, Lima and Puno; during July in Pasco; June–August in Junín; and October–December in Cuzco. **BREEDING DATA** *G. a. andicolus* Juvenile, 2 January 2010, Ccocha (MSB 33962); two juveniles, 29 December 1921, Chipa (AMNH 174094/098); juvenile, 1 March 1894, Huamachuco (AMNH 492273); juvenile, 9 August 1974, Macachaya (FMNH 299216A); juvenile, 19 May 2011, Macate (MSB 36030); juvenile transitioning, 21 October 2012, El Molino Viejo (MSB 43060); juvenile transitioning, 10 January 2010, Laguna Anantay (MSB 34116); 'juvenile', 8 December 1954, Quebrada Victoria (Dorst 1956); 'juvenile', 18 December 1954, Huarón (Dorst 1956); subadult, 6 December 2009, Chacoche (MSB 33668); subadult, 23 July 2007, Carhuayumac (MSB 28286); subadult, 20 February 1954, Nevado Quilcayhuanca (FMNH 222154); subadult, 12 June 1974, Quilluacocha (FMNH 399869); subadult transitioning, 19 February 1954, Tullparaju (FMNH 222157); subadult transitioning, 20 February 1954, Quilluacocha (FMNH 222156); subadult transitioning, 7 June 1922, Huanuco Mountains (FMNH 66261); subadult transitioning, 20 February 1954, Nevado Quilcayhuanca (FMNH 222155); subadult transitioning, 23 July 2007, Carhuayumac (MSB 28286); subadult in last stages of transition (but skull apparently 80% ossified), 23 October 2012, El Molino Viejo (MSB 43069). *G. a. punensis* Three breeding records from Abra Málaga on 15 November 2011, a one-third complete nest, a nest with fresh eggs, and a nest with 8–10-day-old nestlings (Greeney 2012a); two separate observations of older fledglings, still with adults, 15 November 2011, Abra Málaga (HFG, G.F. Seeholzer *in litt.* 2012).

TECHNICAL DESCRIPTION Sexes similar. The following description follows Schulenberg & Kirwan (2011) and refers to nominate *andicolus*. **Adult** Crown blackish, rear crown brown or olive-brown, both streaked white or buff. Loral spot whitish-buff or buff, large and extending above eye. Prominent white eye-ring. Back and rump greyish-olive or olive-brown. Nape and back usually streaked black and white or whitish-buff; in some individuals these markings are reduced to buff shaft-streaks. Tail olive-brown or dull reddish-brown. Remiges dusky, primaries edged dull rufous; wing-coverts olive-brown with small buff apical spots. Sides of the head, neck and throat streaked whitish, buff and blackish. Centre of throat whitish or buff, unstreaked. Breast-sides olive-brown, streaked white. Breast whitish, feathers edged black and buff, creating a heavily scaled appearance; feather edgings on the flanks and belly dusky and in some individuals are incomplete, creating a spotted (not scaled) appearance. Central belly less heavily marked. **Bare parts** Iris brown; *Bill* mostly or entirely black or very dark brown (Peters & Griswold 1943), occasionally pale whitish or pinkish basally, especially on mandible; *Tarsus & toes* blue-grey. **Juvenile** Above, largely similar to adult in general coloration. Wings also similar but tips of wing-coverts, secondaries and tertials with pale-buff triangular spots. Rectrices also adult-coloured except more pointed with pale tips (Zimmer 1930, Krabbe & Schulenberg 2003) that probably disappear with wear. Carriker (1931) described a juvenile female (*punensis*) moulting into adult plumage as differing from the adult in having 'the whole of the under parts slightly suffused with rusty buff'; greater coverts, scapulars and feathers of nape tipped black except for a subterminal spot of rusty-buff; lores and eye-ring buffy-ochraceous, and upper back and scapulars with sparse buffy-ochraceous streaks bordered black (similar to adult *andicolus*). Juveniles that I have examined can also be described as having a noticeably darker breast, washed sooty-grey, that fades to 'rusty-buff' on the upper belly and flanks. **Fledgling** Not well described, but apparently 'spotted to barred throughout' (Krabbe & Schulenberg 2003). The youngest specimens I have seen appear as illustrated (Plate 11), midway between fledgling and subadult plumage.

MORPHOMETRIC DATA Measurements refer to the nominate subspecies unless indicated. *Wing* 99.0mm, 98.0mm; *Tail* 43.5mm, 46.0mm; *Bill* from nares 14.5mm, 13.2mm; *Tarsus* 45.0mm, 46.0mm (*n* = 2, ♂,♀, Domaniewski & Sztolcman 1918). *Wing* 93.5mm, 96.5mm; *Tail* 43.5mm, 45.5mm; *Bill* culmen 22.75mm, 21.75mm; *Tarsus* 45.5mm, 46.5mm (*n* = 2, ♂,♀, Berlepsch & Sztolcman 1896). *Wing* 93mm; *Tail* 37mm; *Bill* [total?] culmen 24mm; *Tarsus* 44mm (*n* = 1, ♂, Taczanowski 1884). *Wing* 96mm; *Tail* 42mm; *Bill* exposed culmen 20mm; *Tarsus* 49mm (*n* = 1, ♀, *punensis*, Chubb 1918). *Bill* from nares 14.2mm, depth at front of nares 6.6mm, width at gape 9.8mm, width at front of nares 6.0mm, width at base of lower mandible 7.5mm; *Tarsus* 49.0mm (*n* = 1, ♀, *punensis*, AMNH 229210). *Bill* from nares 12.0mm, 12.4mm, 12.8mm; *Tarsus* 42.4mm, 46.0mm, 42.3mm (*n* = 3, ♂,♂,♀, juveniles, AMNH). *Bill* from nares 13.2mm, 12.6mm, exposed culmen 18.9mm, 17.7mm, depth at front of nares 6.5mm, 6.3mm, width at gape 10.9mm, 11.1mm, width at front of nares 5.6mm, 5.6mm, width at base of lower mandible 7.3mm, 7.7mm; *Tarsus* 45.9mm, 45.3mm (*n* = 2, ♂,♀, juveniles, MSB). *Bill* from nares 14.9mm, 14.4mm, 16.0mm, exposed culmen 21.9mm, 19.3mm, 22.2mm, depth at front of nares 6.6mm, 6.8mm, 6.9mm, width at gape 9.8mm, 9.9mm, 10.5mm, width at front of nares 5.2mm, 5.8mm, 5.9mm, width at base of lower mandible 7.6mm, 8.1mm, 8.1mm; *Tarsus*

46.2mm, 47.5mm, 47.4mm (*n* = 3, ♀,♀,♂, subadults, MSB). *Bill* from nares 13.6mm, 15.3mm; *Tarsus* 44.6mm, 46.3mm (*n* = 2, ♂♂, juveniles, MSB). **Mass** 48–60g and 51–66g (*n* = ?, ♂♂ and ♀♀, respectively, race?, Krabbe & Schulenberg 2003). *G. a. punensis* 51g, 52g, 55g (*n* = 3, ♂,♀,♀, MVZB). *G. a. andicolus* 46.6g(h), 49.5g(lt), 49.9g(m), 51.6g(m), 52.7g(m), 55.2g(lt), 55.5g(tr), 57.0g(lt), 58.0g(m), 59.6g(lt) (♀♀, juveniles–adults, MSB, LSUMZ): 47.5g(tr), 49.2g(m), 52.3g(m), 52.5g(h), 52.9g(m), 56.0g, 56.0g, 57.0g(lt), 57.7g(lt), 59.1g(tr), 61.9g(lt) (♂♂, juveniles–adults, MSB, LSUMZ).

TAXONOMY AND VARIATION Two subspecies currently recognised (Schulenberg & Kirwan 2011). The intra-generic relationships of Stripe-headed Antpitta are not entirely clear, and it has not been included in the few molecular analyses of the genus (Krabbe *et al.* 1999, Rice 2005). It has been suggested previously (Krabbe & Schulenberg 2003, Walker & Fjeldså 2005) that subspecies *punensis*, which occurs in the south of the range, might be better treated as a separate species based on morphological and vocal characters. At present, however, the song of *punensis* is poorly documented and apparent vocal differences are based largely on comparisons of the calls of *andicolus* and *punensis* (Krabbe & Schulenberg 2003).

Grallaria andicolus andicolus (Cabanis 1873), Journal für Ornithologie, vol. 21, p. 318, Maraynioc, Junín, Peru. There was no official designation of a holotype for *andicolus* and a careful examination of the literature and communications between Cabanis and Taczanowski was necessary to determine that at least one syntype (ZMB 21344) remains (Mlíkovský 2009, Mlíkovský & Frahnert 2009). Topotypical material is available at Harvard (MCZ 266699/700). The nominate subspecies is the more widely distributed of the two, occupying most of the species' range. Although the East Andean population of the nominate subspecies is generally considered to range as far north as the Río Marañón valley, I was able to find just one sight record (Valle de Utcubamba, 06°08'S, 77°54'W; eBird: S. White) north of Atuén, the northernmost collecting locality. **SPECIMENS & RECORDS** *Amazonas* Atuén, 06°45'S, 77°52'W (ANSP 115257); Balsas Mountains, *c.*06°52'S, 77°53'W (FMNH 44815). *Cajamarca* Lagunas, 06°54'S, 78°37'W (XC 345775, M. Roncal-Rabanal); *c.*7.5km SSW of Sorochuco, 06°58.5'S, 78°16.5'W (XC 8503, H. van Oosten); Micuypampa, 07°02'S, 78°13.5'W (XC 120309, E. DeFonso); Cajamarca 07°09.5'S, 78°31'W (AMNH 492270–271). *San Martín* Tambo de Callangate, 07°20'S, 77°39'W (ANSP 115260). *La Libertad* Cajamarquilla, 07°18'S, 77°48'W (ANSP 115258); Pataz, 07°44'S, 77°37'W (ANSP 115263, CM P117898); El Molino Viejo, 07°46'S, 77°46'W (MSB 43060/069); Huamachuco, 07°49'S, 78°03'W (AMNH, USNM); Laguna Corneadas, 07°49.5'S, 77°28.5'W (J. Barrio *in litt.* 2017); Cochabamba, 07°50'S, 77°53'W (MCZ 179773); Quiruvilca, 08°00'S, 78°18.5'W (MCZ 179772); 24km W of Quiruvilca, 08°00'S, 78°31'W (MVZB 161373); Mashua and Quebrada La Caldera, *c.*08°12.5'S, 77°14.5'W (LSUMZ 92448–455). *Ancash* Santa Clara, 08°28'S, 77°25'W (ANSP 115264); Quebrada Tutapac, 08°37'S, 77°52'W (ML: T.A. Parker); Macate, 08°45'S, 78°02'W (MSB 36030); Quebrada Parón, 09°00'S, 77°41'W (Fjeldså & Kessler 1996); Morococha, 09°02'S, 77°33'W (FMNH 391893); Quebrada Ishinca, 09°10'S, 77°33.5'W (Fjeldså & Kessler 1996); Aquilpo, 09°21'S, 77°30'W

(FMNH 391894); Nevado Quilcayhuanca, 09°24'S, 77°23'W (FMNH 222154/55); San Marcos, 09°31'S, 77°06'W (MSB 36051/070); Tullparaju, 09°32'S, 77°32'W (FMNH 222157); 21km SW of Huaráz, 09°40'S, 77°40'W (MSB 43229); Quebrada Pucavado, 09°41'S, 77°14'W (XC 16237, N. Krabbe); Puca Jaga Patac, 09°43'S, 77°20'W (Fjeldså & Kessler 1996); Cerro Huanzala, 09°52'S, 76°58'W (LSUMZ 80571). *Huánuco* Laguna Luciacocha, 09°37'S, 77°01.5'W (A. García-Bravo *in litt.* 2017); Bosque Zapatagocha, 09°40'S, 76°03'W (LSUMZ 74107–109); Quilluacocha, 09°42'S, 76°07'W (ANSP, FMNH); Bosque Unchog, 09°44'S, 76°10'W (LSUMZ); *c.*10km E of Caramarca, 09°51'S, 76°30'W (Fjeldså & Kessler 1996); Millpo, 09°54'S, 75°44'W (LSUMZ 128528); Cayumba, 09°55'S, 75°55'W (ANSP 176467); 'Huanuco Mountains' *c.*09°56'S, 76°08'W (FMNH 66260/61). *Pasco* La Quinua, 10°37'S, 76°10'W (Parker & O'Neill 1976a, FMNH 66256–259); above Huariaca, 10°27'S, 76°07'W (LSUMZ 75241/84957); Santa Barbara, 10°20'S, 75°38'W (XC 65708, W.-P. Vellinga); Chipa, 10°42'S, 75°57'W (AMNH 174094/098/101); Huarón, 11°00'S, 76°25'W (Dorst 1956). *Junín* Quebrada Victoria, *c.*10°53.5'S, 76°01'W (Dorst 1956); Laguna Mamancocha (= Torococha), 11°07'S 75°41'W (eBird: T. Valqui); Maraynioc, 11°22'S, 75°24'W (MCZ 266699/700); Cumbre de Oroya, 11°28'S, 75°53'W (ANSP 92293); below Toldopampa, 11°32'S, 74°57'W (KUNHM); Concepción, 11°36.5'S, 74°47'W (XC 152957, H. van Oosten). *Huancavelica* Lachoc, 12°50'S, 75°06'W (Morrison 1939). *Ayacucho* 30km NE of Tambo, 12°44'S, 73°53'W (AMNH 820776); 3km S of Lirriopata, 13°00'S, 74°00'W (KUNHM); above La Quinua, 13°01'S, 74°07'W (LSUMZ 80570–573); Quebrada Ccehua, 14°38.5'S, 74°00'W (Fjeldså & Kessler 1996). *Lima* Above Oyón, 10°35'S, 76°45'W (Fjeldså & Kessler 1996); Carhuayumac, 11°46'S, 76°33'W (MSB); Contunapampa, 11°46'S, 76°31.5'W (MSB 43477); Macachaya, 12°07'S, 76°20'W (FMNH 299216A); 14km by road SW of Milloc, 11°39,5'S, 76°28'W (Plenge *et al.* 1989, LSUMZ 78559); Laraos, *c.*12°21'S, 75°47'W (MUSA 2178-80); above Hortigal, 12°57'S, 75°44'W (Fjeldså & Kessler 1996). *Apurimac* Ccocha, 13°29'S, 72°58'W (MSB 33962); SN Ampay, 13°34'S, 72°53'W (ML: P.A. Hosner); Cerro Runtacocha, 13°40.5'S, 72°48'W (Fjeldså & Kessler 1996); upper Quebrada Chua, 13°45'S, 72°41.5'W (Fjeldså & Kessler 1996); Laguna Anantay, 14°03.5'S, 73°00.5'W (MSB); Chacoche, 14°06.5'S, 73°01.5'W (MSB 33668); Mollebamba, 14°26'S, 72°55'W (MUSA 6688); 7km S of Cotaruse, 14°27'S, 73°14'W (Fjeldså & Kessler 1996).

Grallaria andicolus punensis Chubb 1918a, Bulletin of the British Ornithologists' Club, vol. 38, p. 7, Limbare, Puno, Peru [= Limbani, 14°08'S, 69°42'W, 3,000m; Carriker 1931, Stephens & Traylor 1983]. The holotype is an adult female (NHMUK 1902.3.13.1215) collected 4 July 1900 by P. O. Simons (Warren & Harrison 1971). This subspecies has a relatively limited distribution, found only in south-east Peru (eastern Cuzco, Puno) and extreme north-west Bolivia. It is quite similar to nominate *andicolus* but is, on average, slightly larger and tends to have a blacker crown streaked orange-buff rather than white. The lores and face-sides are buffier, the eye-ring tends to be cleaner white, broader and more conspicuous, and the back lacks the sparse buffy streaking of typical *andicolus* (Carriker 1931, Schulenberg & Kirwan 2011). Carriker (1931) described the crown of *punensis* as darker olive-brown, stating that the nape is the same colour as the rest of the back (but paler greyish-brown in *andicolus*). The underparts are more heavily barred in *punensis*, especially on the breast-sides and flanks, and the patch of ferruginous-buff on the lower throat tends to be slightly brighter and more prominent (Carriker 1931). Occasionally, however, *andicolus*, even from the sites as far north as Ancash, may have little or no streaking on the upperparts (Schulenberg & Kirwan 2011), suggesting the possibility of a clinal increase in the amount of streaking. SPECIMENS & RECORDS PERU: *Cuzco* Abra Málaga, 13°08'S, 72°19'W (Greeney 2012a, XC); below San Luís, 13°04'S, 72°23'W (LSUMZ 78560); 4km N of Pallata, 13°12'S, 72°13'W (FMNH 391892); Mandorcasa, 13°13'S, 72°56'W (Fjeldså & Kessler 1996); Occumare, 13°17'S, 72°43'W (Fjeldså & Kessler 1996); Bosque Choquechampi, 14°17.5'S, 72°03.5'W (XC 28664, I. Aragón). *Puno* Limbani area, *c.*14°08.5'S, 69°42'W (ANSP, CM, MVZB); Sina, 14°30'S, 69°15'W (KUNHM 21290/115474); Quebrada Metara, 15°10'S, 70°25'W (Fjeldså & Kessler 1996). BOLIVIA: *La Paz* Keara, 14°41'S, 69°06'W (background of XC 73113, J.A. Tobias); near Zongo, 16°07'S, 68°02'W (Fjeldså & Kessler 1996); 'Alaska Mine' north of Pongo, *c.*16°17'S, 67°58'W; see Paynter 1992 (AMNH 229210); Chaquetanga, 16°20'S, 67°57'W (Fjeldså & Kessler 1996).

STATUS The species' population is believed to be declining (BirdLife International 2015). Despite this, however, because of its large geographic range and relative abundance (Fjeldså & Krabbe 1990, Stotz *et al.* 1996, Schulenberg *et al.* 2007, Ridgely & Tudor 2009), it is currently assessed as Least Concern (BirdLife International 2017). Given its tolerance, even preference, for patchy habitats (Krabbe & Schulenberg 2003, Walker & Fjeldså 2005, Greeney 2012a), Stripe-headed Antpitta is probably under little immediate threat from human activities. *Polylepis* woodlands within the species' range, however, continue to be decimated by livestock grazing and harvesting for fuel (Fjeldså *et al.* 1987, Fjeldså &

Kessler 1996, Fjeldså 2002), and subspecies *punensis* has fairly recently been upgraded to Vulnerable in Bolivia (Rocha & Quiroga 1996, Balderrama 2009). Until the extent to which this species relies on forested habitats is known, irrespective of how frequently it ventures into the open, conservationists should keep a careful eye on its populations. Indeed, despite having a very large range (for an antpitta), the species is apparently found in relatively few protected areas. PROTECTED POPULATIONS *G. a. andicolus* PN Yanachaga-Chemillén (XC: W.-P. Vellinga); PN Huascarán (FMNH, XC); PN Río Abiseo (J. Barrio *in litt.* 2017); SN Huayllay (eBird); SN Ampay (Fjeldså & Kessler 1996); Reserva Paisajística Nor Yauyos-Cochas (eBird: many observers). *G. a. punensis* Peru: ÁCR Choquequirao (eBird: many observers); SH Machu Picchu (Walker & Fjeldså 2005); ÁCP Hatun Queuña-Quishuarani Ccollana and ÁCP Sele Tecse–Lares Ayllu (eBird; G. Ferro Meza); ÁCP Abra Málaga (Greeney 2012a, XC). **Bolivia**: RN Ulla Ulla, within ÁNMI Apolobamba (XC: J.A. Tobias).

OTHER NAMES Krabbe & Schulenberg (2003) note that the species' name *andicolus* is a noun and therefore invariable, so the formerly widespread use of *andicola* is incorrect (e.g. Brabourne & Chubb 1912, Domaniewski & Sztolcman 1918, Morrison 1940, Altman & Swift 1989, Fjeldså & Krabbe 1990, Sibley & Monroe 1990, Fjeldså & Kessler 1996, Stotz *et al.* 1996, Lloyd 2008). Additionally, Cabanis (1873) described Stripe-headed Antpitta as *Hypsibamon* [*sic*] *andicolus*. A few subsequent authors, however, spelled the genus *Hypsibemon* (Sclater 1877, Cory & Hellmayr 1924, Lowery & O'Neill 1969, Moyle *et al.* 2009, Plenge 2014) **Spanish** Tororoí Andino (Krabbe & Schulenberg 2003); Gralaria andina (Koepcke 1970); Gralaria de Cabeza Listada (Fjeldså & Kessler 1996). **English** Stripe-headed Antthrush (Brabourne & Chubb 1912); *Grallaria andecola* (Salvin 1895). **French** Grallaire des Andes. **German** Strichelkopf-Ameisenpitta (Krabbe & Schulenberg 2003).

Stripe-headed Antpitta, adult (*punensis*) feeding two nestlings, Abra Málaga, Cuzco, Peru, 15 November 2011 (*Harold F. Greeney*).

Stripe-headed Antpitta, adult (*andicolus*), Huaraz, Ancash, Peru, 31 October 2013 (*Carlos Calle*).

Stripe-headed Antpitta, adult (*punensis*) brooding nestlings, Abra Málaga, Cuzco, Peru, 15 November 2011 (*Harold F. Greeney*).

Stripe-headed Antpitta, eggs (*punensis*) in nest, Abra Málaga, Cuzco, Peru, 15 November 2011 (*Harold F. Greeney*).

Stripe-headed Antpitta, adult (*punensis*) preparing to brood young nestlings, Abra Málaga, Cuzco, Peru, 15 November 2011 (*Harold F. Greeney*).

Stripe-headed Antpitta, two young nestlings in nest (*punensis*), Abra Málaga, Cuzco, Peru, 12 November 2011 (*Harold F. Greeney*).

GREY-NAPED ANTPITTA
Grallaria griseonucha   Plate 9

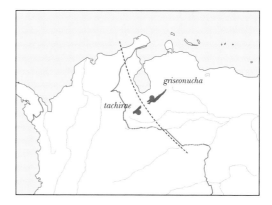

Grallaria griseonucha P. L. Sclater & Salvin, 1871, Proceedings of the Zoological Society of London, vol. 1870, p. 786, Paramo of La Culata, north of Merida [= Páramo de La Culata], 08°45'N, 71°05'W, in the Andes of Mérida, Venezuela. NHMUK 1889.9.20.626. The description of *G. griseonucha* is included in the 1870 volume of Proc. Zool. Soc., but the relevant issue was not published until 1871. This has, apparently, led to much confusion in the literature. Sclater himself (1877, 1890) incorrectly used 1870 as the date of description, as did Phelps & Phelps (1963), Meyer de Schauensee (1966) and Warren & Harrison (1971). Several authors have attempted to rectify this: Peters (1951) gave the date as '1870 (1871)' while Cory & Hellmayr (1924) used 1870, but placed it in quotation marks. There has also been some confusion as to the correct gender of the holotype, which is stated to be male in the original description (Sclater & Salvin 1871). Sclater (1890) also referred to it as a male, but Warren & Harrison (1971) listed it as a female. This confusion seems to stem from the use of an upside-down '♀' on the original label, a symbol that was previously used to denote the male gender (G.M. Kirwan *in litt.* 2017), and this was likely misinterpreted by Warren & Harrison (1971).

This Venezuelan endemic is a shy and poorly known bird of dense forest undergrowth, especially where there is bamboo. Despite its small range, two subspecies are recognised, but the differences between them are subtle. Grey-naped Antpitta has a dark brown crown and upperparts, broken by a grey eyestripe and nape patch, while the rest of its plumage is deep rufous-chestnut. It is a small- to mid-sized *Grallaria*, which appears to replace Chestnut-crowned Antpitta at higher elevations (above 2,300 m). Grey-naped Antpitta belongs to the subgenus *Oropezus* (Ridgway 1911, Lowery & O'Neill 1969), but its precise taxonomic affinities are not well known. In linear arrangements, it is usually placed next to Rufous Antpitta (Peters 1951, Sibley & Monroe 1990, Krabbe & Schulenberg 2003) or Bicoloured Antpitta (Ridgely & Tudor 2009, Remsen *et al.* 2016). All aspects of the biology, ecology, evolution, or taxonomy of Grey-naped Antpittas are in need of further research. Its breeding ecology and diet are completely undocumented, and its true habitat preferences have not been studied in detail.

IDENTIFICATION 16cm. Grey-naped Antpitta is a mid-sized *Grallaria* and is rather simply patterned, chestnut-brown above with a prominent grey band extending back from the eye to encircle the nape. Below, they are uniform bright rufous. Grey-naped Antpitta is fairly unique within its range, as the smaller and similarly plain-coloured Rufous Antpitta does not occur this far north. Chestnut-crowned Antpitta, which replaces Grey-naped Antpitta at lower elevations, is unlikely to be confused due to its boldly streaked white breast.

DISTRIBUTION Grey-naped Antpitta is endemic to Venezuela's Cordillera de Mérida EBA (Stattersfield *et al.* 1998) and Meridan centre of endemism (Cracraft 1985), being found only in Mérida, eastern Trujillo and north-east Táchira.

HABITAT The species inhabits the ground and lower undergrowth of humid montane forests (Ridgely & Tudor

1994, Krabbe & Schulenberg 2003). Generally found in dense vegetation at treefalls and landslides, but only rarely inside open forest understorey (Hilty 2003). It also appears to have something of an affinity for *Chusquea* bamboo (Restall *et al.* 2006). Most authors give an altitudinal range of 2,300–2,800m (Phelps & Phelps 1963, Meyer de Schauensee & Phelps 1978, Sibley & Monroe 1990, Stattersfield *et al.* 1998, Hilty 2003, Ridgely & Tudor 2009). Records as low as 1,800m (Giner & Bosque 1998) and as high as 3,000m (Fjeldså & Krabbe 1990) would seem to require confirmation.

VOICE The typical song is a short, fast series of hollow notes that become louder and faster *ho-ho-ho-hó-hó-hóhóhó* (Ridgely & Tudor 2009). This song is rendered by Hilty (2003) *wü, wü-wü-wúwu'wU'WU* with a hollow quality, delivered quickly and ending abruptly, with the last 2–3 notes loudest. This song lasts 1.0–1.8s and is delivered every 7–11s at a frequency of 1.0–1.2kHz (Krabbe & Schulenberg 2003). The call is a single, whistled *whüüt?* delivered singly at long intervals (Hilty 2003). An alternate song (?) is composed of the same call notes delivered in short, 2–4s bursts combined into an irregular Morse Code-like series that may continue for several minutes (Hilty 2003). This last vocalisation comprises 2–3 notes delivered at a frequency of *c.*1.1kHz. These calls last 0.7–0.8s and are given every 2–4s (Krabbe & Schulenberg 2003).

NATURAL HISTORY Very little descriptive information is available on the species' behaviour. Like others in the genus, it is usually encountered alone, hopping or running on the ground, often in dense vegetation. Apparently slightly more 'arboreal' than other species, adults occasionally climb to 4m or more to call, and may remain on the same perch for long periods. While singing, adults 'shake [their] body', sometimes squatting slightly near the song's finale (Hilty 2003).

DIET No information, but presumably the diet is similar to other congeners, and comprised predominantly of earthworms and arthropods taken on the forest floor (Greeney 2014f).

REPRODUCTION The nest, eggs and breeding behaviour are completely unknown. **SEASONALITY** The only published information concerns a 'juvenile' (*griseonucha*) in January in Mérida (Fjeldså & Krabbe 1990). I examined two immature skins labelled 'Mérida' and, unfortunately, the youngest of these (fledgling transitioning; AMNH 492259) is undated. The second (AMNH 492258), a juvenile, has the collection date '1.8.96'. Based on other labels by the same collector (Briceño), there is little doubt that this refers to 1 August 1896, and this specimen is the juvenile reported (erroneously) for January by Fjeldså & Krabbe (1990) (J. Fjeldså *in litt.* 2016).

TECHNICAL DESCRIPTION Sexes similar. *Adult* A dark brown crown is encircled by a slate-grey stripe starting at the lores, extending above the eye and across the hindcrown, where the grey extends onto the nape. The upperparts, wings and tail are dark reddish-brown, while the head-sides and most of the underparts are bright rufous-chestnut. The central belly is paler, tinged olivaceous. Underwing is reddish-brown. **Adult Bare Parts** *Iris* brown; *Bill* blackish, paler at base of mandible or over most of mandible; *Tarsi & Toes* grey to brownish-grey. *Immature* Previously, 'juveniles' were described as similar to adults but having pale shaft-streaks on the nape, back and underparts (Fjeldså & Krabbe 1990). I examined what I feel were an older fledgling (AMNH 492259) and a juvenile (AMNH 492258), both of the nominate subspecies. The following descriptions are based on these two specimens. *Fledgling* The downy plumage of fledglings is, on the whole, dull brown. The feathers of the upperparts, especially the crown, are slightly paler, buffy-brown along the shafts, creating indistinct fine speckling on the head. Towards the nape and onto the upper back the amount of buff on each feather gradually increases until much of the feather tip is slightly buffier than the base, forming very faint barring, most apparent on the nape. Below, the throat and upper breast are nearly uniform dark sooty-brown, the feathers of the breast and flanks gradually becoming paler and slightly olivaceous with distinctive pale shaft-streaks. The colour of the underparts fades to brownish-buff on the central belly and vent. Flight feathers are similar in colour to those of adults, the upperwing-coverts are slightly warmer, cinnamon-brown, and each bears a narrow subterminal blackish bar. **Fledgling Bare Parts** Undescribed in life, but the evidence from dried skins suggests that the bill is probably dusky-orange, darkest along the culmen and tip of the mandible, and brightest orange along the tomia and at the base. *Juvenile* The transition to adult plumage, similar to other *Grallaria* where our knowledge is more complete, appears to be patchy and not always symmetrical, especially with respect to the contour feathers of the breast and back. The first areas to acquire adult-like plumage appear to be the lower throat, malar region and central-lower back. In the juvenile examined this resulted in an indistinct rufous-chestnut lower throat crescent and malar stripe, and dark brown ear-coverts streaked with the same colour. Scattered adult-coloured feathers on the breast-sides and flanks create irregular patches of rufous-chestnut. The crown and nape are still as described for fledglings, but the central back and rump are largely dark reddish-brown as in adults. In this individual the undertail-coverts had already been replaced by bright rufous-chestnut adult feathers. The wings are as described for fledglings. **Juvenile Bare Parts** Undescribed in life but probably close in colour to adults, except that the bill still retains some orange or yellow-orange near the base and probably around the nares. **Subadult** Although I examined no subadult specimens, it seems likely that they are similar to full adults but still retain the thinly barred, rufescent-brown upperwing-coverts. Based on the aforementioned juvenile, it

further seems likely that, apart from the immature wing-coverts, the last area to acquire adult plumage is the nape, suggesting that individuals lacking the clean grey collar of adults may be immatures. **Subadult Bare Parts** Probably as adults.

MORPHOMETRIC DATA *G. g. tachirae Wing* 85mm, 87mm; *Tail* 36mm, 38mm; *Bill* exposed culmen 17mm, 17.5mm, culmen from base 22mm, 22.5mm; *Tarsus* 47mm, 47mm (*n* = 2, ♂♂, Zimmer & Phelps 1945). *Wing* 86mm; *Tail* 38mm; *Bill* exposed culmen 19mm, culmen from base 23mm; *Tarsus* 45mm (*n* = 1, ♀, Zimmer & Phelps 1945). *G. g. griseonucha Bill* from nares 12.3mm, 12.0mm, 12.3mm, 12.0mm, 12.5mm; *Tarsus* 47.2mm, 45.6mm, 47.0mm, 45.4mm, 46.3mm (*n* = 5, ♀,♀,♀,♂,♂, AMNH, USNM). *Bill* from nares 12.6mm, exposed culmen 17.8mm, depth at front of nares 6.8mm, width at front of nares 5.6mm, width at base of lower mandible 9.3mm; *Tarsus* 47.3mm (*n* = 1, ♂, AMNH 492253). *Bill* from nares 11.0mm, 11.3mm; *Tarsus* 47.1mm, 47.6mm (*n* = 2, unsexed older fledgling, AMNH 492259 and ♂ juvenile transitioning, AMNH 492258). **Mass** No data. **Total Length** 15–17cm (Meyer de Schauensee & Phelps 1978, Dunning 1987, Fjeldså & Krabbe 1990, Ridgely & Tudor 1994, Hilty 2003, Restall *et al.* 2006).

GEOGRAPHICAL VARIATION Two subspecies are typically recognised, but there are so few specimens that it is unclear how morphologically distinct the two forms are. Most records of the two taxa are separated by at least 100km, with nominate *griseonucha* confined to the mountains north of the Chama and Albarregas valleys (Zimmer & Phelps 1945). Records in the vicinity of Páramos de Batallón and La Negra are of uncertain racial affinity but, being south of the above-mentioned rivers, are probably best assigned to *tachirae*. As noted by Zimmer & Phelps (1945), however, the subtropical-zone habitat of Grey-naped Antpitta is fairly contiguous across the upper headwaters of these rivers.

Grallaria griseonucha griseonucha P. L. Sclater & Salvin, 1871, Proceedings of the Zoological Society of London, vol. 1870, p. 78, Páramo of La Culata, Andes of Mérida, Venezuela. This subspecies is confined to the Andes of Venezuela in eastern Trujillo (south from the vicinity of Boconó) to southern Mérida (see discussion below). The GPS given for Páramo Conejos (08°50'N, 71°15'W) by Paynter (1982) would extend the range of nominate *griseonucha* west and north from the city of Mérida to the west slope of the Western Cordillera, an area from where there are no other records. The map in Phelps & Gilliard (1941), however, clearly places this locality south and slightly west of Mérida, and this correction is reflected in the coordinates given below. The nominate subspecies is considered to differ from *tachirae* by having very little grey on the lores, an only slightly paler mandibular base, and indistinct pale shaft-streaks on the ear-coverts, breast and flanks (Restall *et al.* 2006). **Specimens & Records** *Trujillo c.*3.5km E of Boconó, 09°15.5'N, 70°13'W (eBird: J. Matheus); PN Guaramacal, *c.*09°11'N, 70°15'W (Hilty 2003). *Mérida* La Cuchilla, 08°54'N, 70°38'W (CM P89598); Quintero, 08°41'N, 71°06'W (Meyer de Schauensee 1966, COP 14525); Echicera, 08°40'N, 71°10'W (Zimmer & Phelps 1945); Páramo El Escorial, 08°40'N, 71°05'W (Phelps & Phelps 1963, COP 14578, AMNH 492257); Cacute, 08°40'N, 71°02'W (USNM 208960); Valle, 08°40'N, 71°06'W (AMNH 492253–256); Pico Humboldt Trail, 08°37.5'N, 71°02'W (XC 6856, N.

Athanas); near Mérida, c.08°37'N, 71°09'W (AMNH); La Mucuy, 08°36.5'N, 71°03.5'W (XC 43127/28, J. Klaiber); 16km ESE of Tabay, 08°35.5'N, 70°55.5'W (USNM 505782); Páramo Conejos, 08°30.5'N, 71°15.5'W (MCZ 94793).

Grallaria griseonucha tachirae J. T. Zimmer & Phelps Sr., 1945, American Museum Novitates, no. 1274, p. 3, Boca de Monte, Pregonero, Táchira, Venezuela, 2,300m [= 08°03'N, 71°51'W]. The adult male holotype (COP 24556) was collected 10 December 1943, by Fulvio Benedetti, and remains on deposit at AMNH. Subspecies *tachirae* is known with certainty only from the Andes of north-east Táchira at the type locality, south and west to around Páramo El Zumbador. A.M. Cuervo & J.E. Miranda T. (eBird) encountered Grey-naped Antpitta in the PN Páramos El Batallón y La Negra, only c.30km from the type locality (see also FMNH 288314). The subspecific identity of birds in this region is unknown, but they may represent a northward extension of the range of *tachirae* to southern Mérida, something thus far unreported. This subspecies differs in having the crown, back, wings and tail brownish-olive, rather than rufous. The underparts are less rufous as well, with the belly and vent showing slight buffy scaling and the feathers of the ear-coverts and flanks unstreaked. The throat is noticeably paler than the breast, whereas nominate *griseonucha* has more uniform underparts. The lores are paler than the grey nape (Restall *et al.* 2006). **SPECIMENS & RECORDS** *Mérida* Guaraque, 08°15.5'N, 71°43'W (J.E. Miranda T. *in litt.* 2017); Páramo La Negra, 08°15'N, 71°40'W (FMNH 288314). *Táchira* Páramo el Rosal, 08°02'N, 71°59'W (ML 62486/87, P.A. Schwartz); La Barrosa, 08°00.5'N, 71°58.5'W (A.M. Cuervo *in litt.* 2017); Páramo El Zumbador, 07°58.5'N, 72°05'W (Krabbe & Coopmans 2000; CM, FMNH, ML, XC).

STATUS Despite its small range, the population is suspected to be stable in the absence of noticeable declines or substantial threats, and is therefore not considered globally threatened (BirdLife International 2015). Within its range the species is generally uncommon, but can be quite common locally (Stotz *et al.* 1996, Hilty 2003, Greeney 2014f). The nominate race occurs in PN Guaramacal (Boesman 1998, Hilty 2003) and PN Sierra Nevada (USNM 505782, XC), while the range of *tachirae* apparently includes both PN Chorro El Indio (COP, FMNH) and PN Páramos Batallón y La Negra (XC 223628/629; P. Boesman).

OTHER NAMES *Grallaria griseinucha* (Sharpe 1901). **Spanish** Tororoí Nuquigrís (Krabbe & Schulenberg 2003); Seco-Estoy Merideno [*griseonucha*] (Phelps & Phelps 1950); Seco-Estoy Tachirense [*tachirae*] (Phelps & Phelps 1950); Hormiguero Seco Estoy (Meyer de Schauensee & Phelps 1978); Hormiguero Seco-Estoy (Phelps & Phelps 1963); Hormiguero de Nuca Gris (Hilty 2003). **French** Grallaire à nuque grise. **German** Graunacken-Ameisenpitta (Krabbe & Schulenberg 2003).

Grey-naped Antpitta, *Grallaria griseonucha tachirae*, adult male holotype (COP 24556) on deposit at the AMNH (*Harold F. Greeney*).

BICOLOURED ANTPITTA
Grallaria rufocinerea   Plate 12

Grallaria rufo-cinerea P. L. Sclater & Salvin, 1879, Proceedings of the Zoological Society of London, vol. 1879, p. 526, Santa Elena, Antioquia, Colombia [06°13'N, 75°10'W].

Long considered to be endemic to Colombia, and restricted to just one or two localities in Antioquia and Caldas (Peters 1951, Meyer de Schauensee 1970), new locations have been discovered in the past few decades which extend its range as far south as north-east Ecuador (Nilsson *et al.* 2001, Greeney 2014h). As its English name suggests, this antpitta is essentially two-toned, rufous above and grey below, with a slightly brighter reddish head. It is a globally threatened species of dense, humid montane forest and second growth at 1,900–3,150m. Despite being considered Vulnerable (BirdLife International 2017), Bicoloured Antpitta appears to tolerate considerable disturbance provided forest cover is maintained. As with so many members of the Grallariidae, most aspects of the species' biology remain poorly studied and, along with surveys for additional populations, should be considered a primary focus for future research. Bicoloured Antpitta was assigned to the subgenus *Hypsibemon* by Lowery & O'Neill (1969), but its taxonomic affinities are still

poorly established. In linear arrangements, it is generally placed near Grey-naped Antpitta *G. griseonucha* (Restall *et al.* 2006, Remsen *et al.* 2016). To properly assess its conservation status and afford it proper protection, Bicoloured Antpitta needs additional surveys to determine population levels and the true extent of its occurrence within its possible range (Stattersfield & Capper 2000). Any information on body mass, diet, foraging behaviour and nesting biology merits publication.

IDENTIFICATION 15–16cm. A plump, long-legged bird, typical of the genus. It is essentially two-toned, with a reddish or rufous-brown head and back contrasting with grey underparts, and has a hooded appearance, not unlike sympatric Chestnut-naped Antpitta (race *ruficeps*). Chestnut-naped Antpitta, however, is readily distinguished from Bicoloured by being significantly larger and more 'slender-looking', with a paler bill, a pale iris and white postocular spot, and a contrasting chestnut crown with a reddish-brown back and blackish throat. Hooded Antpitta is also somewhat similarly patterned, but is so much smaller that confusion is unlikely.

DISTRIBUTION Bicoloured Antpitta was, until recently, thought to be endemic to Colombia. It is now known from quite a few locations on both slopes of the Central Andes, from southern Antioquia and Caldas to southern Quindío and, somewhat disjunctly, the Central Andes of eastern Cauca to the East Andean slope in northernmost

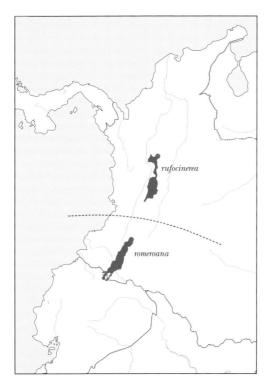

rufocinerea

romeroana

Ecuador (Sclater & Salvin 1879, Chapman 1917, Stiles 1997, Renjifo *et al.* 2000, Rodner *et al.* 2000, Negret 2001, Nilsson *et al.* 2001, Ridgely & Greenfield 2001, Losada-Prado *et al.* 2005, Ayerbe-Quiñones *et al.* 2008, Calderón-Leytón *et al.* 2011).

HABITAT Inhabits the floor and lower understorey of dense, humid montane cloud forest to the treeline (Fjeldså & Krabbe 1990, Krabbe & Schulenberg 2003). While predominantly associated with mature habitat, the species appears reasonably tolerant of disturbance and smaller numbers occur in second growth and edges (Ridgely & Tudor 1994, Restall *et al.* 2006). Bicoloured Antpitta is most common at 2,200–3,150m (Ridgely & Tudor 2009), but locally to around 1,900m (Stattersfield & Capper 2000, Parra-Hernandez *et al.* 2007) and in some areas as high as 3,300m (Fjeldså & Krabbe 1990).

VOICE The song is a high, clear, whistled *treeeee* or *treeeeeuh* (Hilty & Brown 1986, Fjeldså & Krabbe 1990). Apparently, the female sometimes follows this with a *kree-kree-kree-kree-kree-kree* (Nilsson *et al.* 2001, J. Nilsson in Ridgely & Tudor 2009). The typical song lasts only *c*.1s and is delivered at intervals of 3–4s, at a frequency of *c*.2kHz (Krabbe & Schulenberg 2003). The end of the song is usually upslurred and then downslurred (or the reverse). The presumed 'reply' by the female is described by Krabbe & Schulenberg (2003) as a 5–6s-long delivery of 6–7 whistles, descending from *c*.2.5kHz to 2kHz. It remains to be seen if the reply is gender-specific.

NATURAL HISTORY The foraging behaviour has not been described in detail, but is presumably similar to other montane *Grallaria* (Greeney 2014h). This species is among the many species of *Grallaria* that have been readily trained to receive food at worm-feeding stations (Woods *et al.* 2011,

Greeney 2012b). Recently, it has been observed following ant swarms (Nieto-R. & Ramírez 2006) and, based on the suggestion of Greeney (2012b), it appears likely that the species at least occasionally also follows large mammals such as tapirs and bears. There are no published data on territory defence, maintenance or fidelity. Territory size has not been well established either, but Renjifo (1991) estimated 1.6–5.0 birds along a 10km transect, and Kattan & Beltran (1999) estimated 3.7–5.7 birds per 1km transect.

DIET The stomach contents of the type specimen were described as including insects (Sclater & Salvin 1879), and presumably invertebrates make up most of their diet. No further data.

REPRODUCTION The nest, eggs and breeding biology await description. **SEASONALITY** The literature mentions a single breeding record: a male (nominate *rufocinerea*) in breeding condition was collected in June in south-east Antioquia, Colombia (Hilty & Brown 1986, Fjeldså & Krabbe 1990). I am able to add only a few records to this, largely from photos of immatures visiting worm feeders. **BREEDING DATA** Two juveniles, 28–29 August 1911, Laguneta (AMNH 111976/77); two juveniles, RFP Río Blanco, 28 August (ML 33536021, photo J. Blowers) and 12 October 2014 (A. Muñoz; photos). What is presumably the latter of these two juveniles at Río Blanco was videotaped on 15 November 2014 (IBC 1230335, D. Jimenez).

TECHNICAL DESCRIPTION Sexes similar. The following description refers to nominate *rufocinerea* (see Geographical Variation). *Adult* Sexes similar. Adults have the head, throat, upperparts, wings and tail near-uniform rufous-brown or ferruginous-red, and feathering around the eye slightly buffier, forming an indistinct eye-ring. The throat feathers are somewhat variable, but are very slightly more rufous than the upperparts, with grey bases giving an indistinct mottled or speckled appearance. The underparts vary from grey to sooty-grey, with the central belly tending towards whitish. Grey feathers of the breast, sides and flanks are narrowly edged whitish, giving a slightly scaled appearance, especially in certain lights. **Adult Bare Parts** *Iris* dark brown; *Bill* black; *Tarsi & Toes* grey to brownish-grey. **Juvenile** There are no previously published descriptions of the immature plumages. The following are based on two immature specimens and photographs shared by M.F.L. Martínez, A. Muñoz, S. Davis, J. Blowers and S. Reed, all of the nominate subspecies (see Breeding Seasonality). Overall, appearance is like a very 'messy' adult. The crown and hindcrown have chestnut adult feathers, the back with some chestnut feathers but largely splotched dark grey; scapulars blotched olivaceous and dark grey; face, throat and neck-sides largely dark grey with sparse chestnut feathering; fairly distinct white crescent at posterior edge of eye. Most dark grey feathers, especially on the nape, throat and breast, have narrow white shaft-streaks; upperwing-coverts browner than flight feathers and broadly edged cinereous at tips; breast and belly similar to adult in overall colour, except the white streaks and a few irregularly sized and positioned dark grey immature feathers. **Juvenile Bare Parts** *Iris* reddish-brown; *Bill* largely black but paler at tip and around nares, rictus pale orange and only very slightly inflated; *Tarsi & Toes* vinaceous-grey to pinkish-grey. **Subadult** It appears that the last areas to moult into adult plumage are the throat and face, which appear to remain mottled or 'messy'-looking with grey, cinereous and dull chestnut feathers. Adults also retain the brownish, dark-edged upperwing-coverts for some time. The bare parts are generally as in adults but the rictus may be paler. The eye-ring is not yet present and they either still have the white feathers behind the eye or this area appears somewhat bluish, while it is partially bare during moult.

MORPHOMETRIC DATA Males *Wing* chord 89.6mm, flattened wing 95.0mm; *Tail* 50.5mm; *Bill* culmen from base 23.0mm, exposed culmen 18.6mm, from nares 13.8mm; *Tarsus* 45.3mm (*n* = 1, *romeroana*, Hernández & Rodríguez 1979). *Bill* from nares 11.9mm, 13.0mm, exposed culmen 16.2mm, 17.4mm, depth at nares n/m, 6.7mm, width at nares 4.5mm, 5.6mm, width at base of lower mandible 8.1mm, 7.7mm; *Tarsus* 47.0mm, 45.0mm (*n* = 2, *rufocinerea*, AMNH, Cauca). *Bill* from nares 12.5mm, 12.1mm; *Tarsus* 44.4mm, 46.3mm (*n* = 2, *rufocinerea*, AMNH and USNM). **Females** *Bill* from nares 11.2mm, 12.9mm; *Tarsus* 43.4mm, 44.0mm (*n* = 2, *rufocinerea*, juveniles, AMNH, Cauca). **Sex unspecified** *Wing* 86.4mm (3.4 inches); *Tail* 45.7mm (1.8 inches); *Tarsus* 43.2mm (1.7 inches) (*n* = 1, *rufocinerea*, Sclater & Salvin 1879). *Wing* chord 83.0mm, flattened wing 87.0mm; *Tail* 46.0mm; *Bill* from nares 12.0mm; *Tarsus* 40.7mm (*n* = 1, *rufocinerea*, Hernández & Rodríguez 1979). *Wing* chord 83.2mm, flattened wing 86.0mm; *Tail* 40.0mm; *Bill* culmen from base 21.9mm, exposed culmen 16.5mm, from nares 11.2mm; *Tarsus* 47.2mm (*n* = 1, *romeroana*, Hernández & Rodríguez 1979). **Mass** 44.8g (*n* = ?, ♀?, Krabbe & Schulenberg 2003). **Total Length** 15.0–17.8cm (Sclater 1890, Meyer de Schauensee 1970, Hilty & Brown 1986, Dunning 1987, Fjeldså & Krabbe 1990, Restall *et al.* 2006, Ridgely & Tudor 2009, McMullan *et al.* 2010).

TAXONOMY AND VARIATION Two subspecies tentatively recognised, but the validity of *romeroana* seems questionable (see below).

Grallaria rufocinerea rufocinerea P. L. Sclater & Salvin, 1879, Proceedings of the Zoological Society of London, vol. 1879, p. 526, Santa Elena, Antioquia, Colombia [06°13'N, 75°10'W]. The holotype is an unsexed adult housed at Tring (NHMUK 1889.9.20.618) (Warren & Harrison 1971, Knox & Walters 1994). The nominate subspecies is endemic to Colombia, from Antioquia, on both slopes of the Central Andes to southern Quindío and central Tolima. **Specimens & Records** *Antioquia* Páramo de Sonsón, 05°43'N, 75°15'W (USNM 436486, XC 56350, P. Florez); Alto de El Escobero, 06°14'N, 75°30'W (Ramírez 2006). *Risaralda* Santa Rosa de Cabal, 04°50.5'N, 75°33'W (XC 302235, J. Holmes); Finca El Cortaderal, 04°46.5'N, 75°28'W (ML 63944151, E. Munera); PRN Ucumarí, 04°44'N, 75°36'W (Kattan & Beltrán 1999, 2002). *Quindío* 13.5km E of Salento, 04°38'N, 75°27'W (Arbeláez-Cortés *et al.* 2011); Finca La Judea, 04°09.5'N, 75°43'W (XC 244777; O.H. Marín Gómez); near Génova, 04°08'N, 75°44'W (XC 10773/74, N. Athanas); Laguneta, 04°35'N, 75°30'W (AMNH 111975–978); Salento, 04°40'N, 75°30'W (AMNH 111979). *Caldas* RFP Río Blanco, 05°05'N, 75°28'W (Nieto-R. & Ramírez 2006, XC, ML); above Manizales, 05°00'N, 75°20'W (XC 316985, G.M. Kirwan); *c.*2km ENE of La Siberia, 04°52.5'N, 75°28.5'W (D. Uribe *in litt.* 2017). *Tolima* near Ibagué, *c.*04°27'N, 75°15'W (Parra-Hernández *et al.* 2007); Hacienda La Carbonera, 04°32'N, 75°28'W (López-Lanús *et al.* 2000, XC); Hacienda Las Cruces, 04°31'N, 75°26'W (López-Lanús *et al.* 2000).

Grallaria rufocinerea romeroana Hernández-Camacho & Rodríguez-M., 1979, Caldasia, vol. 12, p. 577, vicinity (right bank) of the waterfall of the Río Bedó, PNN de Puracé, Municipio La Plata, Departamento Huila, 3,000m, 02°20'N, 76°17'W [now = Municipio Puracé, Dept. Cauca]. The holotype is an adult male in fresh plumage collected 15 November 1970 by Carlos Arturo León, and was named in honour of Hernando Romero Zambrano, then curator of the ornithological collection at the Instituto de Ciencias Naturales, Universidad Nacional de Colombia, in Bogotá. The holotype is currently at Tring (NHMUK 1889.9.20.618). Subspecies *romeroana* is said to occur from the head of Magdalena Valley in Huila and western Putumayo, south to Sucumbíos in north-east Ecuador (Krabbe & Schulenberg 2003), but there are large gaps in the known distribution. Most modern records are concentrated in the mountains between Mocoa and Colón in extreme western Putumayo or in the Santa Bárbara–La Bonita area just across the Colombian border in Ecuador. Its presence in the intervening region, as well as north to the type locality is presumed, as suitable habitat appears to remain. Nevertheless, surveys of these regions are recommended before its conservation status can be evaluated. Bicoloured Antpitta was, for 100 years, considered monotypic until *romeroana* was described. Indeed, the two subspecies appear very similar, with *romeroana* generally described as having a solid rufous-brown chin and throat compared to the 'black chin and throat' (Restall *et al.* 2006) or 'feathers of the throat grey basally' (Fjeldså & Krabbe 1990) seen in nominate *rufocinerea*. Hilty & Brown (1986) postulated that some individuals with a mix of grey on their throat might be females, and Ridgely & Tudor (1994, 2009) that the 'messy admixture' of grey feathers on the face and throat of some females might indicate immaturity. I suggest the latter, but have not been able to examine sufficient material to be confident of this. **Specimens & Records COLOMBIA**: *Nariño c.*20km E of San Francisco, 01°12'N, 77°05'W (Hilty & Brown 1986). *Putumayo* RFP Cuenca Alta Río Mocoa, Sector Cristales, 01°12.5'N, 76°42'W (D. Carantón *in litt.* 2017). **ECUADOR**: *Sucumbíos* 3km ENE of Santa Bárbara, 00°39'N, 77°30'W (Nilsson *et al.* 2001, XC); Santa Bárbara, 00°37'N, 77°35'W (XC 78768, R. Ahlman); 10km S of Santa Bárbara, 00°33.5'N, 77°31.5'W (R.A. Gelis *in litt.* 2017); Alto La Bonita, 00°29.5'N, 77°35'W (IBC 984691, 'sam'); *c.*9.5km NE of La Sofía, 00°26'N, 77°35.5'W (XC 236551, R.A. Gelis).

STATUS Bicoloured Antpitta previously was considered Endangered (Stattersfield *et al.* 1998), but as it subsequently became known from additional localities was downgraded to Vulnerable (Stattersfield & Capper 2000, BirdLife International 2017). Nevertheless, it is known to have been extirpated in at least some areas near the type locality in Antioquia (Castaño-Villa & Patiño-Zabala 2000, 2008), and almost its entire range has experienced heavy deforestation (Collar *et al.* 1992, Salaman & Gandy 1993). The only known locality in Ecuador (subspecies *romeroana*) is unprotected, but its threat status in Ecuador has not been evaluated. Ridgely & Greenfield (2001) suggested that Bicoloured Antpitta should be considered Endangered in Ecuador. No anthropogenic effects have been documented specifically, but presumably its habitat is threatened by human expansion and grazing, which are thought to have caused population declines since the early 1900s (Stattersfield *et* *al.* 1998). **Protected Populations** *G. r. rufocinerea* RMN El Mirador (Salaman *et al.* 2007a); RNSC Alto Quindío Acaime (Renjifo *et al.* 2002, BirdLife International 2012); SFF Otún-Quimbaya (XC 54155, B. López-Lanús); PRN Ucumarí (Kattan & Beltran 1997, 1999, 2002); RFP Río Blanco (Wege & Long 1995, Ocampo-T. 2002, Nieto-R. & Ramírez 2006); PNN Los Nevados (Pfeifer *et al.* 2001, XC); RFPR Cañón del Quindío (Rojas-R. 2012). *G. r. romeroana* PNN Puracé (Hernández-Camacho & Rodríguez-M. 1979, Negret 2001); RFP Cuenca Alta Río Mocoa (D. Carantón *in litt.* 2017).

OTHER NAMES *Grallaria rufo-cinerea* (Sclater 1890, Dubois 1900, Cory & Hellmayr 1924, Meyer de Schauensee 1950, 1964, Peters 1951). **Spanish** Tororoí Bicolor (Krabbe & Schulenberg 2003); Gralaria Bicolor (Ridgely & Greenfield 2001); Peón de pecho gris (Negret 2001); Grallaria bicolor (Granizo 2009); Tororoi Rufocenizo (Salaman *et al.* 2007a, Rojas-R. 2012); Cholongo, Cocona, Hornero bicolor (Rojas-R. 2012). **English** Ferruginous Ant-Pitta (Meyer de Schauensee 1950); Ferrugineous Ant-Thrush (Brabourne & Chubb 1912); Ferruginous Antpitta (Cory & Hellmayr 1924); Bicoloured Antpitta (Walters 1980, Howard & Moore 1984, Krabbe & Schulenberg 2003). **French** Grallaire bicolore. **German** Zweifarben-Ameisenpitta.

Bicoloured Antpitta, juvenile (*rufocinerea*), RNA Río Blanco, Caldas, Colombia, 12 October 2014 (*Anderson Muñoz*).

Bicoloured Antpitta, adult (*rufocinerea*), RNA Río Blanco, Caldas, Colombia, 30 July 2011 (*Daniel Uribe*).

Bicoloured Antpitta, adult (*rufocinerea*), RNA Río Blanco, Caldas, Colombia, 13 March 2017 (*Ramíro Ramírez Cardona*).

Bicoloured Antpitta, adult (*rufocinerea*), RNA Río Blanco, Caldas, Colombia, 7 September 2014 (*Dušan M. Brinkhuizen*).

JOCOTOCO ANTPITTA
Grallaria ridgelyi   **Plate 13**

Grallaria ridgelyi Krabbe *et al.* 1999, Auk, vol. 116, p. 883, 2,520m, Quebrada Honda, 04°29'S, 79°08'W, Zamora-Chinchipe Province, south-eastern Ecuador. The adult male holotype (MECN 7199) was collected by Francisco Sornoza M. on 28 December 1997. The scientific name honours Robert S. Ridgely.

The Jocotoco Antpitta is a striking species, not only because of its size, but also by virtue of the broad snowy-white malar stripe that contrasts sharply with the glossy black crown. Amazingly, this large and very distinctive species was not discovered until 1997 (Sornoza 2000, Ridgely 2012). Following this, it has been found at just a handful of locations, including one in extreme northern Peru *c.*50km SSE of the type locality. With an extremely small known range and its montane cloud forest habitat under continued threat, Jocotoco Antpitta is considered Endangered. Very little is known of its natural history, but its nest was recently discovered by park guards working in the Jocotoco Foundation reserve at Tapichalaca, created specifically to preserve the type locality of this amazing antpitta. As noted by BirdLife International (2017), further survey work is desperately needed in potentially suitable habitat within the species' potential range. New populations might then benefit from immediate conservation attention. In addition, like most antpittas, the behaviour, reproduction, diet and habitat requirements of Jocotoco Antpitta are in need of further study. Only with such basic natural history information will future conservation action be effective should additional populations be discovered. Greeney & Juiña (2010) suggested that the year-round surplus of food provided by the feeders in their study area may have encouraged that pair of Jocotoco Antpittas to breed year-round, a pattern that is unlikely to be common for antpittas without dietary supplementation. Given the now widespread practice of feeding antpittas (Woods *et al.* 2011) the importance of this and the long-term repercussions on adult longevity and health merit study. Tapichalaca Biological Reserve, which is the type locality and centre for current conservation efforts, is sited near a road frequently used for commercial transport, and a recent road-widening project probably affected at least two territories (R.S. Ridgely in BirdLife International 2013).

Much of the species' range is threatened by logging and gold mining, even within Podocarpus National Park, and forest degradation is ongoing throughout the known range (Snyder *et al.* 2000, BirdLife International 2017).

IDENTIFICATION 20–23cm. Sexes are similar, both having a black cap and the rest of the upperparts brownish. The throat is bright white as are most of the underparts, becoming washed pale grey posteriorly. The flanks and breast-sides are washed olive-brown. The most striking feature is the large, contrasting white patch in front of the eye that extends from the pre-loral area to below the eye and the anterior portion of the ear-coverts. It is striking and unmistakable. The somewhat similar Pale-billed Antpitta is slightly smaller, with a large pale bill, black throat above a darker grey belly, and lacks the white facial marking of Jocotoco Antpitta. Sympatric Chestnut-naped Antpitta is immediately distinguished by lacking white on the face or throat, and by having a rufous crown.

DISTRIBUTION Confined to a very small part of the eastern Andes in extreme south-east Ecuador and far northern Peru (Krabbe & Schulenberg 2003, Heinz *et al.* 2005, O'Neill 2006, Freile *et al.* 2010).

HABITAT Jocotoco Antpitta occurs on steep, humid slopes, densely covered with epiphyte-laden trees. It favours areas where the 10–25m canopy is broken by dense areas of *Chusquea* bamboo, but is perhaps less closely associated with bamboo than the similarly large, sympatric, Chestnut-naped Antpitta. Indeed, the two species do appear to segregate within the habitat mosaic, as Chestnut-naped Antpitta nests in the bamboo (Juiña *et al.* 2009), while Jocotoco Antpitta appears to prefer mossier areas with a more intact canopy (Greeney & Juiña 2010). The habitat on these steep slopes is, however, extremely complex and varies due to the frequent landslides that maintain dynamic habitat regrowth of varying ages; elucidating the microhabitat segregation of the two species will undoubtedly prove very challenging. Within this range Jocotoco Antpitta inhabits elevations of 2,300–2,650m (Ridgely & Tudor 2009).

VOICE The song is a slow series of 6–10 (or more) identical notes produced at 0.5–0.6kHz, and delivered at 1–2s intervals (Krabbe & Schulenberg 2003). These authors suggested that the song is reminiscent of a dog, but my own description would be more like the distant 'moo' of a cow. The call is a similarly low, but much softer, two-noted

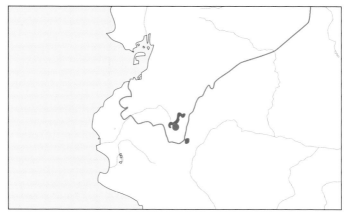

ho-co. In alarm both sexes give a similar call, but louder, and the second part may be more emphatic and almost double-noted, sometimes ending in a harsh churr and described as a *hoó-krrr* or a staccato *hoó-có-kurr* (Krabbe *et al.* 1999, Heinz 2002, Ridgely & Tudor 2009). These calls inspire the species' common name, which is the local onomatopoeic rendition 'jóco-tó-cuorrr'. Juveniles make what is apparently a contact call while foraging near adults, as suggested by Heinz (2002), a slightly drawn-out, single *woooo* note, easily imitated by a whistle, similar in quality to a single note of the common song of Ferruginous Pygmy-owl *Glaucidium brasilianum* (Greeney & Gelis 2005). **Non-vocal sounds** Agitated adults, while bobbing their heads and running back and forth, make a whooshing sound or subdued clap while simultaneously flicking their wings (Greeney 2013f).

NATURAL HISTORY The behaviour of Jocotoco Antpitta is, unsurprisingly, poorly studied. The species is generally encountered alone or in pairs (Ridgely & Greenfield 2001, Ridgely & Tudor 2009). Limited observations suggest that its behaviour is similar to other large *Grallaria*, with most of its time spent on or close to the ground. It forages by probing its heavy bill into the leaf litter, overturning dead leaves and vegetation with its bill and feet (Greeney 2013f). Jocotoco Antpittas are known to, at least opportunistically, follow swarms of montane *Labidus* sp. army ants (Greeney & Gelis 2005, Greeney 2012a). Greeney (2012a) also reported that Jocotoco Antpittas appear to be attracted to the loud crashing sounds of an observer moving through dense vegetation. This attraction to the sounds of large mammals, propensity to follow army ants and readiness with which the species' can be trained to take food from humans (Woods *et al.* 2011) are behaviours that led Greeney (2012a) to suggest that Jocotoco Antpittas may regularly associate with large mammals foraging in the understorey (ie. tapirs, bears). A commensal foraging interaction has yet to be sufficiently documented, but undoubtedly the disturbed leaf litter, broken logs and torn-open bromeliads left in the wake of a foraging bear would provide lucrative hunting grounds for a Jocotoco Antpitta. Spontaneously produced song is infrequently heard, but is usually given during the mid-morning or any time of day when light levels are low (Heinz 2002). While singing, adults throw their head back, fluffing the white throat feathers and pumping their short tails in synchrony with each hoot (Krabbe *et al.* 1999, Heinz 2002, Greeney 2013f). Jocotoco Antpittas sometimes respond so aggressively to playback that they can be heard crashing through the undergrowth before emerging into view to run agitatedly back and forth, sometimes leaning forward and bobbing their head (Krabbe *et al.* 1999, Ridgely & Greenfield 2001). Jocotoco Antpitta appears to be sedentary and to maintain territories for many years (Greeney 2013f). Most known territories overlap with one or more territories of Chestnut-naped Antpitta, yet it appears that neither species responds aggressively to playback of songs of the other (Heinz 2002). As mentioned previously, however, Jocotoco Antpitta responds very aggressively to conspecific playback, suggesting that intraspecific territorial defence is aggressive. Sometimes emerging from the undergrowth into full view, an agitated adult responding to playback will often flick its wings and fluff out its breast and throat feathers, making the bird appear larger, sometimes pausing and stretching its neck up and appearing taller. Agitated adults often perform displacement activities during this display, preening their

underwing or breast, and wiping their sizeable bill on nearby branches or the ground (Heinz 2002, Greeney 2013f). Just one aggressive interaction between two adult Jocotoco Antpittas has been described, presumably of two individuals disputing territorial boundaries. This interaction (F. Mendoza in Heinz 2002) was described as similar to a domestic cock-fight, with their heavy bills being used to peck at one another. The interaction ended with the victor chasing the loser into the undergrowth while continuing to deliver fierce blows to its back and head. Using response to playback at the type locality in south-east Ecuador, Heinz (2002) estimated mean territory size to be 13.58ha ($n = 9$, range = 11.68–15.34ha). The species does not appear to follow mixed-species flocks, however, except perhaps when attending army ant swarms (Greeney & Gelis 2005, Greeney 2012a). At worm-feeders, Snow *et al.* (2015) observed very aggressive interactions between Jocotoco and Chestnut-naped Antpitts, wherein Jocotoco Attpitta seemed competitively dominant. However, these interactions were unaccompanied by vocalisations, suggesting to the authors that there may be competition between the species that goes undetected by playback experiments such as those of Heinz (2002). They further speculated that the amount of resource competition might change through the year, and only become relevant during reproductive periods. Unfortunately, most of what we know of interspecific interations and breeding for Jocotoco Antpittas derives from observations at worm feeders, or of pairs that regularly visit them. Although the possible influence of human-inflated food availability on both of these aspects of the species' ecology cannot be ignored, the brief observations of Snow *et al.* (2015) are nevertheless a good example of how worm-feeding stations may be useful tools in studying otherwise 'inaccessible' aspects of antpitta biology. Apart from the aforementioned work of Heinz (2002), little is known of territory size in Jocotoco Antpitta, although estimated population densities at the type locality are 6.0–7.5 individuals per km² (Krabbe *et al.* 1999, Heinz 2002). Furthermore, it is unknown how far juveniles may disperse, but recent radio-telemetry studies suggest that individuals may range considerably further than this territory estimate (R.S. Ridgely in BirdLife International 2013).

DIET Jocotoco Antpittas are well-known visitors to worm-feeding stations at the type locality (Woods *et al.* 2011), where they feed predominantly on earthworms and other invertebrates. Very little, however, is known of the natural diet. The stomach contents of the five individuals in the type series contained earthworms, beetles, ants, millipedes and 'larvae' (Krabbe *et al.* 1999).

REPRODUCTION The nest has only recently been described (Greeney & Juiña 2010) from the type locality in south-east Ecuador. **NEST & BUILDING** The nest described by Greeney & Juiña (2010) was 3.6m above ground on the side of a rotting tree trunk (4.1m tall; 56cm dbh). It was built against the trunk and supported primarily by a dense tuft of *Tillansia* sp. bromeliads and also by a 4cm-diameter, horizontal branch under the nest. The nest was a deep, open cup predominantly of dead and decaying plant materials (dicot and bromeliad leaves). The inner portion was lined with a dense layer of fine black rootlets and flexible brown fern stems. Measured at perpendicular angles, the internal width of the cup was 15.5 × 14.0cm. The cup was significantly deeper in the front (17cm) and just 12.5cm deep at the back. Externally, the nest was 23.5

× 26cm in diameter and 18cm tall. The area around the nest had a dense understorey of *Chusquea* bamboo. Canopy height was *c.*20m and, along with surrounding bamboo shoots and epiphytes on the substrate, provided nearly 90% shade for the nest. The nest was on the north-east side of the trunk, on a north-east-facing slope where it received most sun in the mornings. To this I can add (previously unpublished) descriptions of three additional nests, built by the same pair in the same territory, all found by F. Mendoza Armijos. The process of nest-site selection and building is poorly known. All three of these nests were well supported by bromeliads and other large epiphytic plants, built against the trunks of large trees, and were partially hidden from above by vegetation or collected organic detritus caught by epiphytes. In one case, the density of surrounding epiphytic vegetation was such that the nest was virtually within a vegetative cavity, only accessible from two sides and completely covered from above, *c.*30cm above the nest rim. Nest heights were 5m, 3.2m, 5.5m. Measurements of these three additional nests: external diameter, 24cm, 25cm, 27cm; external height, 12–14cm, 15cm, 12–16cm; internal diameter, 14.5cm, 16cm, 15cm; internal depth, 9cm, 11cm, 9cm. One nest was built and laid in 15–17 days after the disappearance of a previous clutch. The amount of decomposing material included in the walls of the nests, and their tight fixture to surrounding vegetation, strongly suggests that old nests are refurbished and re-used. It is likely that a nest built completely from scratch would take longer than two weeks to complete.

EGG, LAYING & INCUBATION Clutch size at two nests was one egg. Two additional nests contained single nestlings when found, further suggesting that one-egg clutches are the norm for Jocotoco Antpittas. Previously unpublished, two eggs were fairly typical of other *Grallaria* in shape and ground coloration, being subelliptical and clear, bright, sky-blue. Both were fairly sparsely patterned with coarse, heavy brownish and blackish blotching and fine speckling, intermixed with paler, lavender blotching. The markings, especially larger blotches, were slightly more concentrated at the larger end. Dimensions: 38.4 × 30.5mm (17.1g); 39.7 × 30.5mm (20.64g, fresh weight). At one nest the single egg was laid sometime between 17:00 and 05:30 the following morning, the female spending this entire period on the nest. Remarkably (see generic introduction), the incubation period at this nest was 29–30 days (696–708h).

NESTLING & PARENTAL CARE On 8 November 2008, the Greeney & Juiña (2010) nest contained a single, older nestling (see below) that fledged seven days later. They videotaped the nest on several occasions prior to fledging, revealing that both adults provisioned the nestling. The nestling was brooded just 7% of the observation period. They were unable to determine if both adults participated, but unpublished observations at another nest (HFG) show that both sexes brood. The mean (± SD) duration of feeding visits was 2.2 ± 7.5min, during which adults frequently probed sharply or rapidly into the nest lining, which is presumed to be a means of parasite removal as in other antpittas (Greeney *et al.* 2008). The nestling was provisioned at a rate of 1.96 feeds/h. Close to the day of fledging, the nestling often followed the adults to the rim of the nest as they departed. Newly hatched nestlings have not been formally described, but the day-of-hatch nestling photographed in Greeney (2012b) clearly shows that they have orange-pinkish skin and legs, an orange bill (including rictal flanges) with a bright white egg-tooth, and dense tufts of blackish down on the back. Greeney &

Juiña (2010) described an older nestling, three days prior to fledging, as cinnamon-brown on the back, chestnut on the crown and upper breast, with all of these areas being finely barred black. Most of the ocular area was still bare, with only a hint of the white facial pattern of adults. The maxilla was dark blackish with a pale orange tip, while the mandible was mostly dull orange. The rictal flanges were slightly brighter orange. At this age the nestling retained only a few sparse wisps of its original natal down, mostly on the head and lower back. At one nest the single nestling was weighed one, four and eight days after hatching: 18.2g, 33.0g, 57.9g. At this nest the nestling fledged 31 days after hatching. **SEASONALITY** From the records to date (see below), it is evident that Jocotoco Antpitta is capable of nesting twice per year, apparently with most attempts coinciding with *either* the drier *or* the wetter months. All of the following data are from RP Tapichalaca and refer to the same pair of antpittas that visit the worm-feeding stations there. To the best of my knowledge, all of the following records represent different breeding events. **BREEDING DATA** Laying 2 February 2010 / fledge 6 April (HFG); fledge 13 November 2008 (Greeney & Juiña 2010); fledge 11 August 2009 (HFG); mid-incubation 14 January 2010 (F. Sornoza M. *in litt.* 2010); juvenile 30 November 2003 (Greeney & Gelis 2005; date not provided in original publication!). Records of eight additional juveniles: 2 August 2007 and 17 July 2008 (F. Sornoza M. *in litt.* 2014); 18 September 2008 (A. Sornoza M. *in litt.* 2016); 6 October 2010 (Flickr: R. Lewis); 18 March 2012 and 1 March 2014 (G. Real *in litt.* 2015); 11 April 2015 (Flickr: A. Sorokin); 16 August 2016 (F. Angulo P. *in litt.* 2016). I also believe that a subadult photographed 22 September 2013 (D. Weidenfeld *in litt.* 2015) represents a reproductive event distinct from the above.

TECHNICAL DESCRIPTION Sexes similar. **Adult** The original diagnosis of Jocotoco Antpitta by Krabbe *et al.* (1999) can be paraphrased as follows. The crown, broad malar, anterior ear-coverts and a 1–2mm-wide feathered ocular ring are jet black. The face bears a large, bright white, fan-shaped patch over the loral and subocular region. A poorly defined postocular streak, the posterior ear-coverts and neck-sides are pale neutral grey. Nape mostly black but slightly suffused with back, which is brownish-olive. The upper mantle is washed black, less so posteriorly. All back feathers have blackish shafts. The tail, of 12 rectrices, is somewhere between raw umber and Mars Brown. The wings are largely cinnamon-brown with the three innermost secondaries and outer webs of the other nine slightly paler chestnut. The distal half of the outer webs of the ten primaries are cinnamon, the inner webs of the remiges are blackish and the underwing primary-coverts are mostly black. The upperwing-coverts are brownish-olive with a very faint 2mm-wide blackish band. (Not all adults show this and its presence in the holotype description may be the result of this individual retaining subadult coverts.) Alula is brownish-olive like back, outer webs nearly uniformly this colour. Breast is white, lightly suffused pale neutral grey, especially on sides. Throat is white, feathers fairly stiff, glossy, elongated and recurved shafts which on some chin feathers are distinctly dark-tipped. Feathers of upper breast show faint, 1mm-wide, blackish fringes producing an indistinctly scaled appearance. (These fringes appears to be lost in some or all full adults.) Vent and extreme lower belly washed pale pinkish-buff. Breast-sides are brownish-olive. Sides and flanks are pale neutral grey suffused brownish-olive. Thighs are brownish-olive admixed pale neutral grey and heavily washed black. Undertail-coverts and underwing-coverts are brownish-olive, with fine pale black bars. Most of the olive-brown feathers, especially on the upper and central back have faint blackish fringes. The white cheek patch of adults is composed of rigid loose-barbed feathers with glossy shafts extending from the subocular and posterior part of the lower lores to the bill base. These throat feathers are longer and denser than in congeners and cover the base of the bill and most of the malar. Rictal bristles are absent (unlike all other congeners), but a few of the white loral feathers have elongated bare shafts, forming bristle-like extensions. The feathers of the throat are white, stiff and loose-barbed, as in Ochre-striped Antpitta, but are denser, shorter and more recurved at their tips, similar in this respect to Chestnut-naped Antpitta. The grey of the underparts is interspersed with white in a pattern resembling pale-bellied members of the *G. hypoleuca* allospecies group. The bill is shaped most like that of Chestnut-naped Antpitta and Pale-billed Antpitta, but is proportionately deeper at the base, with a straighter culmen. The tarsi and toes are proportionately larger (thicker) than in most other antpittas, approaching or equalling those of *G. gigantea*. They have relatively few tarsal scutes, and each is distinctly ridged, most like Chestnut-naped Antpitta. **Adult Bare Parts** *Iris* dark ruby-red; *Bill* black; *Tarsi & Toes* blue- to vinaceous-grey. *Juvenile* The following is expanded from Greeney & Gelis (2005), who described a mid-aged juvenile, probably just beginning to gain independence from its parents and still foraging with them. Similar in pattern to adults, but with throat white, upper to lower breast pale slate-grey and fading to whitish on belly. The distinctive malar pattern of adults is evident, but more subdued, variously washed grey and becoming brighter with age. The most distinctive difference is the chestnut (rather than black) crown, bearing fine black vermiculations that reach onto the hindcrown and fade gradually on the nape, just reaching the upper back. The black adult crown feathers appear first on the forecrown, slowly replacing immature feathers rearwards, sometimes leaving patches of barred feathers surrounded by black. In most cases, it appears that the nuchal area is last to retain vestiges of chestnut feathering. The primaries are chestnut, as in the adult, and the tertials bear indistinct, slightly wavy, subterminal black bands. The upperwing-coverts are similar in colour to those of adults but each bears a thin, subterminal blackish bar. Most are double-barred, with the upper bar somewhat less distinct, and almost absent on some. **Juvenile Bare Parts** (adapted and expanded on from Greeney & Gelis 2005) *Iris* dark brown to dull reddish-brown, becoming redder with age; *Bill* mostly black, except basal two-thirds of the mandible, which is dusky-orange, slightly paler basally, little or no remnants of inflated rictus; *Tarsi & Toes* similar to adults. Tobias (2009) includes a photograph of a juvenile similar to this description. *Fledgling* Similar to older nestling described by Greeney & Juiña (2010) (see Reproduction). Plumage still 'fluffy', especially on rump and underparts, and may retain a few wisps of dark grey or blackish nestling down. Back cinnamon-brown, brighter chestnut on crown and upper breast, all of these areas finely barred black, especially on upper breast, making it appear overall darker. Wings similar to juveniles, tertials distinctly bearing slightly V-shaped subterminal bars, the terminus of each feather having an indistinct pale central spot,

upperwing-coverts with two thin black subterminal bars, less rufescent than flight feathers. Tail not fully emerged but similar to adults. Feathers of ocular area just emerging (whitish-grey), distinctive face pattern of adults replaced by mottled grey, forming indistinct crescent. **Fledgling Bare Parts** *Iris* clear, fairly pale brown (compared with juveniles); *Bill* maxilla dark blackish with a pale orange tip and orange wash around nares and along tomia, mandible mostly dull, dusky-orange, darkening towards tip, rictal flanges only slightly inflated, thin, bright orange, and mouth lining deep yellowish-orange; *Tarsi & Toes* mostly vinaceous or bluish-grey, not significantly different from adults other than faint orange or yellowish tint to posterior edge of tarsus.

MORPHOMETRIC DATA Males *Wing* unflattened chord 130mm, flattened chord 138mm; *Tail* 59.6mm; *Bill* at nares 19.3mm, total culmen 33.5mm; *Tarsus* 67.1mm (*n* = holotype; Krabbe *et al.* 1999). *Wing* unflattened chord 131 ± 3.61mm, range 128–135mm, flattened chord 138 ± 2.52mm, range 135–140mm; *Tail* 58.0 ± 2.43mm range 55–66mm; *Bill* exposed culmen 28.2 ± 1.0mm range 27.1–29.0mm, from front of nares 19.5 ± 0.85mm range 18.9–20.5mm, total culmen 34.8 ± 1.63mm range 33.5–36.6mm, width at base 16.1 ± 1.01mm range 15–17mm; *Tarsus* 71.1 ± 3.59mm range 67.1–74.1mm, middle toe 47.1 ± 1.91mm range 46.0–49.3mm, hindclaw 13.9 ± 0.25mm range 13.6–14.1mm (*n* = 3 type series, including holotype; Krabbe *et al.* 1999). **Females** *Wing* unflattened chord 124mm, 128mm, flattened chord 131mm, 133mm; *Tail* 50mm, 57mm; *Bill* exposed culmen 27.0mm, 28.5mm, from front of nares 18.1mm, 18.6mm, total culmen 32.9mm, 33.1mm, width at base 15.8mm, 16.2mm; *Tarsus* 67.4mm, 73.4mm, middle toe 45.4mm, n/m, hindclaw 13.62mm, 14.0mm (*n* = 2, type series; Krabbe *et al.* 1999). **Mass** Krabbe *et al.* (1999) provide the only data: 176g (♂ holotype); 192 ± 14.3g, range 176–204g (*n* = 3 ♂♂ type series, including holotype); 152g, 182g (*n* = 2, ♀♀, type series). **Total Length** 20–23cm (Ridgely & Greenfield 2001, Krabbe & Schulenberg 2003, Restall *et al.* 2006, Ridgely & Tudor 2009).

DISTRIBUTION DATA Near-endemic to **ECUADOR**: *Zamora-Chinchipe* N slope of Cordillera de Tzunantza, 04°17'S, 78°57'W (Heinz *et al.* 2005); Romerillos Alto, above San Luís, 04°18'S, 78°57'W (N. Krabbe in Freile *et al.* 2010); NE slope of Cerro Toledo, 04°23'S, 79°06'W (Heinz *et al.* 2005, MECN 7284); SE slope of Cerro Toledo, *c.*04°26'S, 79°07'W (Krabbe *et al.* 1999, ANSP); Ventanilla, RP Tapichala, 04°29'S, 79°09.5'W (XC 250886–88, N. Krabbe); RP Tapichalaca, worm-feeding station, 04°29.5'S, 79°07.5'W (Greeney & Gelis 2005, Greeney & Juiña 2010, Snow *et al.* 2015, XC, ML); above Río Blanco, 2,580m, 04°35'S, 79°17'W (XC 250889, N. Krabbe). **PERU**: *Cajamarca c.*21km W of Chingozales, 04°53'S, 78°53'W (LSUMZ 179012); Hito Jesús, 04°54'S, 78°54'W (XC 6974, T. Mark). **Notes** Survey localities with presumably appropriate habitat where Jocotoco Antpitta has not been found (Heinz *et al.* 2005) include: Estación Científica San Francisco, 03°59'S, 79°04'W; Quebrada Las Palmas, 04°01'S, 78°55'W; Quebrada del León, 04°06'S, 78°59'W; Cajanuma, 04°07'S, 79°10'W; Quebrada Las Dantas, 04°15'S, 78°56'W; along the Palanda–Fatima trail, 04°36'S, 79°11'W. In addition, N. Krabbe (*in litt.* 2017) failed to find the species during targeted searches at two sites on Cerro Plateado in August 2012: 04°36'S, 78°49.5'W, 2,093m (40h); 04°37'S, 78°47.5'W, 2,600m (18h).

TAXONOMY AND VARIATION Monotypic. Thought to be most closely related to Pale-billed and Chestnut-naped Antpittas, with all three species possibly forming a sister clade to Chestnut-crowned and Watkins's Antpittas (Krabbe *et al.* 1999, Krabbe & Schulenberg 2003).

STATUS The entire extent of known range is only *c.*180km², and possibly just 25–36km² (Freile *et al.* 2010). Recent efforts to locate new populations in Ecuador met with only limited success (Heinz *et al.* 2005; see Distribution Data), although it has recently been reported from Peru (O'Neill 2006, LSUMZ, XC). Alarmingly, at least one of the northernmost localities, in the Cordillera de Tzunantza, was about to be deforested (N. Krabbe in Freile *et al.* 2010). The good news is that two of the four known Ecuadorian populations are currently protected within RP Tapichalaca and PN Podocarpus (Heinz *et al.* 2005), and new sites might yet be discovered in Podocarpus and neighbouring PN Yacurí (Freile & Santander 2005). Jocotoco Antpitta is currently listed as Endangered at global and national levels (Stattersfield & Capper 2000, Granizo *et al.* 2002, Freile *et al.* 2010, BirdLife International 2017).

Jocotoco Antpitta, adult stretching, Tapichalaca, 13 November 2008 (*Aldo Sornoza*).

Jocotoco Antpitta, nest and complete clutch, Tapichalaca, 23 February 2010 (*Eli Lichter-Marck*).

Jocotoco Antpitta, adult feeding older nestling, Tapichalaca, 1 August 2009 (*José Simbaña*).

Jocotoco Antpitta, adult, Tapichalaca, 8 December 2011 (*Dušan M. Brinkhuizen*).

Jocotoco Antpitta, adult feeding older fledgling, Tapichalaca, 1 March 2014 (*Galo Real*).

CHESTNUT-NAPED ANTPITTA
Grallaria nuchalis **Plate 12**

Grallaria nuchalis P. L. Sclater, 1859, Proceedings of the Zoological Society of London, vol. 27, p. 441, 'Río Napo', Ecuador.

Among the largest of the genus, Chestnut-naped Antpitta can be rather common in parts of its range and seemingly inexplicably rare in others. It is frequently associated with stands of *Chusquea* bamboo, seemingly more so in the south-east of its range, perhaps because it is partially sympatric there with the larger and closely related Jocotoco Antpitta. The largest of the three subspecies (*ruficeps*) is apparently endemic to Colombia, the smallest (*obsoleta*) to north-west Ecuador, and the nominate race probably occurs in Colombia, Ecuador and Peru, but has not yet been confirmed to occur in the first-named. The plumage of this handsome species is fairly simple, brownish on the back, wings and tail, reddish-chestnut on the head with a pale ocular spot, and unmarked grey below. The three subspecies differ principally in the extent and intensity of chestnut on the head and the saturation of the underparts. Along with some differences in their vocalisations, this fairly consistent variation suggests that three species are involved. Only one nest has been described and there have been no detailed studies of any aspect of its natural history or diet.

IDENTIFICATION 19–21cm. The following description applies to nominate *nuchalis*, but see Description and Geographical Variation for extensive notes on plumage variation. The adult has dark grey lores and a bare, bluish-white postocular spot. The crown and ear-coverts are dark chestnut, becoming brighter rufous-chestnut on the rear crown, nape and neck-sides. The remaining upperparts are brown, tending towards reddish-brown on the sides and rump, and the wings and tail are distinctly reddish-brown. The underparts are nearly uniform dark grey, paler on the belly and vent, and darker on the throat. The iris is grey, possibly becoming paler with age, the bill black, and the tarsi and toes pale blue-grey. In parts of its range in Colombia, Chestnut-naped may overlap with the much smaller Bicoloured Antpitta, which has a similar hooded appearance. Bicoloured Antpitta lacks bright chestnut on the crown and nape and contrastingly coloured ear-coverts. In addition, White-bellied Antpitta is smaller and white below. In the south of its range, might be confused with the larger, heavier-billed Jocotoco Antpitta, adults of which may be immediately separated by their bright white cheek patch. Juvenile Jocotoco Antpitta without a distinct cheek patch will lack the white postocular spot of Chestnut-naped Antpitta and has a heavier bill and dark red-brown eye (not pale grey).

DISTRIBUTION Found from northern Colombia, south on both slopes of the Central Andes, the east slope of the West Andes south to at least Tuluá and the west slope of the East Andes in Cundinamarca (Meyer de Schauensee 1952, Hilty & Brown 1986, Ridgely & Greenfield 2001, Schulenberg *et al.* 2010). So far as is known, subspecies *ruficeps* is confined to Colombia (all three ranges), nominate *nuchalis* occurs in the East Andes of Ecuador and Peru (and probably extreme south-east Colombia), and *obsoleta* is confined to the West Andes of northern Ecuador (and possibly extreme south-west Colombia). Not

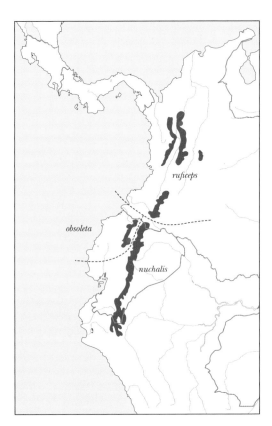

recorded in Peru until 1985 (Meyer de Schauensee 1970, Parker *et al.* 1985), its range is now known to extend as far south as extreme western Piura and northern Cajamarca (Jaén), being replaced south of the Marañón by the closely related Pale-billed Antpitta (Schulenberg & Williams 1982, Parker *et al.* 1985). It is fairly widespread and well known in Ecuador (Ridgely & Greenfield 2001). Records from Colombia, to date, suggest a fragmented and poorly known range, with the first records from the West Andes reported only within the past two decades (Strewe 2000b, Cuervo *et al.* 2003).

HABITAT Most frequently found in the dark recesses of bamboo thickets and adjacent humid cloud forest undergrowth, frequently on steep slopes and in stream ravines (Parker *et al.* 1985, Renjifo & Andrade 1987, Fjeldså & Krabbe 1990 Ridgely & Tudor 2009). In parts of the Central Andes of Colombia, adults maintain territories in secondary and alder forests (*Alnus* sp., Betulaceae) (Poulsen 1996, Kattan & Beltrán 1999). In the early morning, foraging adults are occasionally observed on forest trails (Ridgely & Tudor 2009). Some of the variation in altitudinal limits across the species' range may be due to incomplete sampling, but published records show that Chestnut-naped Antpitta inhabits montane forests at elevations above 1,900m (Krabbe & Schulenberg 2003) in north-east Ecuador (*nuchalis*) to 3,400m (Verhelst *et al.* 2001, McMullan *et al.* 2010) in Colombia (*ruficeps*). At any given locality, most work has uncovered a range of 2,400–2,900m (Parker *et al.* 1985, Bloch *et al.* 1991, Baez *et al.* 1997, Verhelst *et al.* 2001, Buitrón-Jurado 2008). In the Cosanga Valley of eastern

Ecuador (*nuchalis*), however, I have found the species fairly regularly as low as 2,150m, and *obsoleta* is known at 2,100–2,200m in RP Río Guajalito (J. Freile *in litt.* 2017). But, records as low as 1,500m (Arbalaez-Cortes *et al.* 2011b) are likely in error.

VOICE The songs of the three subspecies apparently show some variation. Hilty & Brown (1986) described the song of nominate *nuchalis* from Peru as 'a clear series of '*kook*' notes followed by a gradually rising series of notes that accelerate slightly and become very high and thin at end, '*kook, kook, kook-who-who-hu-hu-hu-hu-e-e-e-e*'.' It was described by Schulenberg *et al.* (2007) as a steeply accelerating series of hollow whistles, rising at the end, usually with a quiet introductory note or stutter: *clew TUR TUR TUR-TUR'TUR't't't't'tr'ee'ee?* and by Ridgely & Tudor (2009) as a 'far-carrying series of musical but somewhat metallic notes that hesitate, then accelerate before ending in a short series of rising tinkling notes, *tew, tew, tew, tew-tew-tew-teh- te-ti-ti - titititi?*'. This song (*nuchalis*) was acoustically quantified by Krabbe & Schulenberg (2003) as lasting 2.3–4.1s and delivered at 6–15s intervals for several minutes at a time. The song comprises a series of 7–13 notes, mostly at constant pitch of 1.9–2.2kHz, which slowly accelerates. The final notes accelerate quickly and rise to 2.1–2.4kHz. Krabbe & Schulenberg (2003) suggested that a frequently included 'introductory note' delivered at a similar pitch *c.*1s before the above-described series, may be a 'well-synchronized female vocalization'. They also mentioned another vocalisation which they considered is 'possibly [made] by female', a descending series, lasting 4–7s and comprising 6–9 wheezy notes at 5.5–4.9kHz and delivered at a steady pace. Given that both sexes sing similar or identical songs in most *Grallaria*, these proposed sexual differences in vocalisations bear close scrutiny. The latter vocalisation was described by Ridgely & Tudor (2009) as 'a repetition of a much higher-pitched note (almost like a hummingbird)', *tsi-tsi-tsi-tsew-tsew- tsi-tsi-tsi*. In northern Peru, Schulenberg & Williams (1982) described a strident *chee-chee-chee-chee-chee* call given in response to playback, suggesting that it may be delivered by the female alone. The song of *obsoleta* is said to be similar in quality to nominate *nuchalis* but lacking the initial hesitation and starting with a fast phrase *tew-te-te-tew, tew-tew-tew- tew-tew-titititi?* (Ridgely & Tudor 2009). Restall *et al.* (2006) described this as 'a distinctive and far-carrying series of somewhat metallic notes, at first hesitating, gradually accelerating, ending in short series of rapidly rising tinkling notes'. According to Krabbe & Schulenberg (2003), it lasts 3–4s and is delivered at a constant pitch of 1.8–1.9kHz, with the last note occasionally slightly higher. This song comprises 1–3 trisyllabic stutters that become a steady-paced series. They stated that this song is often preceded by a single note that might be given by the female. The alternate vocalisation delivered by *obsoleta* (presumably by females; Krabbe & Schulenberg 2003) is a 12s, steady-paced, descending series of 14 notes at 6.0–6.4kHz. Restall *et al.* (2006) described the song of *ruficeps* as consisting of two phrases separated by up to 15s. The first phrase is short, the second longer. Both phrases begin with a single (easily overlooked) 'hiccup', followed by a series of even notes, then 2–3 rising, slightly faster and sharper notes. These authors suggested that birds from around Medellín sound slightly different, with a more pronounced 'hiccup note', with the series gently rising over the duration of the song.

NATURAL HISTORY Chestnut-naped Antpitta forages predominantly on the ground, occasionally on logs or low perches. They hop rapidly forward, pausing with tilted head, then lunging forward to probe into leaf litter or soft earth, and have been observed turning over debris or flicking it aside with their bills. Known to forage with ant swarms (*nuchalis*, Greeney 2012b), their behaviour while doing so was described by Nieto-R. & Ramírez (2006). They found the species foraging at army ant swarms in Colombia (*ruficeps*), recording them foraging at all portions of the swarm, walking among the ants as well as hopping between low perches above them. Chestnut-naped Antpitta is among those species suggested to forage in association with large mammals such as tapirs (Greeney 2012b). This secretive species usually keeps to dense vegetation and bamboo thickets, but may venture onto trails and a short way into clearings at dusk and dawn or on overcast or foggy days (Poulsen 1993). In Peru, Parker *et al.* (1985) found at least four territorial pairs (*nuchalis*) along *c.*0.7km of trail (*c.*5.7 birds/km) just below the *páramo* (2,750–2,900m). Similarly, in the Central Andes of Colombia (*ruficeps*), Kattan & Beltrán (1999) estimated 16.7 birds/km at 2,600m and 6.5/km at 3,000m. The latter authors estimated densities of 0.5 birds/ha and territory sizes of *c.*9.3ha (Kattan & Beltrán (1999, 2002).

DIET Reported prey includes earthworms, ants (Formicidae), cockroaches (Blattodea), grasshoppers (Acrititae), centipedes (Chilopoda), millipedes (Diplopoda) and spiders (Araneae) (Sclater & Salvin 1879, Parker *et al.* 1985, Nieto-R. & Ramírez 2006).

REPRODUCTION The first concrete nesting data were those of Sclater & Salvin (1879), who described the eggs from Antioquia (*ruficeps*) collected on an unknown date. They did not mention the nest, suggesting that the eggs were extracted from females just prior to laying. The nest of *G. nuchalis* remained unknown until Juiña *et al.* (2009) described a nest (*nuchalis*) with young nestlings on 11 November 2006 at Tapichalaca reserve in Zamora-Chinchipe. Most of the details below are taken from this paper, which also reported observations made with a video camera. **NEST & BUILDING** The only described nest was in an area of forest with a *Chusquea* bamboo-dominated understorey, but with scattered, tall trees forming a fragmented canopy. The nest was placed 3m above ground in dense bamboo. It was supported by three, roughly horizontal, bamboo stems that crossed below the nest, 3–4cm in diameter. The nest itself was a large, irregular cup primarily of *Chusquea* bamboo leaves, but also incorporating a few small sticks and moss. The inner cup was very sparsely lined with dark, flexible rootlets. It appeared that it had been built onto a pre-existing clump of naturally accumulated detritus (mostly bamboo leaves). Overall, the supporting detritus was *c.*25–35cm in diameter. Juiña *et al.* (2009) gave the following nest dimensions: external diameter 17–20cm; external height 17cm; inner diameter 14cm; inner cup depth 10cm. **EGG, LAYING & INCUBATION** The only descriptions of the eggs appear to be those referred to by Sclater & Salvin (1879). These authors described the eggs as 'rich dark greenish-blue'. Schönwetter (1979) gave the dimensions of these eggs as 36.3–36.8 × 26.9–27.9mm. **NESTLING & PARENTAL CARE** When discovered, the two nestlings described by Juina *et al.* (2009) had dark greyish-pink skin with sparse, dark blackish natal down. Their eyes were closed and they were estimated to be *c.*3 days old based on details of other

Grallaria nestlings (Dobbs *et al.* 2001, 2003). With time, the nestlings developed a dense post-natal down, buffy below and palest on the belly. The breast bore distinct, thin, black barring. Their upperparts were dark grey to black (especially on the crown) with buff-tipped feathers giving a barred appearance on the upper back. Both nestlings had bright orange bills, gapes and mouth linings. Juiña *et al.* (2009) filmed the nest for 21.3 hours while the nestlings were estimated to be 6–12 days old. During the first two days (6.9h; nestlings 6–9 days old), the young were brooded for just 9% of the observation period and not at all thereafter (*n* = 14.4h, nestlings estimated 10–12 days old). During approach to the nest, adults hopped through the dense bamboo around it, from stem to stem, generally making one longer, final movement to the nest from below. On arrival, they frequently paused 5–7s before delivering food, appearing to survey the area around the nest. Most brooding bouts occurred during mist-like rain. While incubating, adults would occasionally (*n* = 13; 0.6 times/h) lean into the nest cup and rapidly vibrate their body, performing what has been termed 'rapid probing' in other antpittas (Greeney *et al.* 2008), and which may help rid the nest of parasites (Halftorn 1994, Dobbs *et al.* 2001). While the nestlings were estimated to be 6–9 days old, adults brought food at a (mean) rate of 0.9 times per nestling/h. During the latter part of the period they were fed 1.9 times per nestling/h. While unable to quantify faecal sac production due to the nestlings being hidden below the nest rim, of the five faecal sacs that were seen, adults ate three and carried two away. Prey appeared to be delivered singly or only a few at a time. Most were unidentifiable arthropods. On 47% of prey deliveries (*n* = 47), however, they delivered at least one earthworm (Oligochaeta). P. Hosner relayed the following observations of a juvenile being fed by adults (Greeney *et al.* 2010, see below). The adults delivered food *c*.6 times over the course of two hours. He observed the juvenile foraging tentatively, and once saw it clumsily find and eat a small worm. Although he was unable to directly observe the delivery of food, once he heard very faint *peep-peep-peep...* begging calls emanating from where the birds were at the time of feeding. The juvenile remained in the same place for several hours, while the adult came and went. **Seasonality** A synthesis of the reproductive activity of *G. nuchalis* (below) suggests that the species regularly breeds in the drier months, but this must be treated cautiously until more data are available. Nevertheless, this species is one of many Andean birds that nest within and/or specialise on foraging in *Chusquea* bamboo, and also appear to have reproductive peaks during the drier months. I have previously suggested (Greeney & Miller 2008, Juiña *et al.* 2009) that the growth rate of *Chusquea* bamboo is greatest during the wetter months, favouring dry-season nesting by birds that regularly use bamboo as a nesting substrate. Fast-growing bamboo, periodically weighed down by heavy rain, is unstable, and I have often witnessed large sections of interwoven stems collapse. Specific evidence of reproduction, other than the active nest of Juiña *et al.* (2009), is not well documented in the literature. **Breeding Data *G. n. ruficeps*** Fledgling, 25 June 1951, Páramo de Sonsón (USNM 436478); juvenile transitioning, 27 January 1907, 'Cauca Valley' (= along Río Cauca in Antioquia or Valle del Cauca; see Paynter 1997) (AMNH 492197); juvenile transitioning, 21 September 2009, RNA Colibri del Sol (A. Quevedo photos); two records from Páramo de Sonsón on 25 June

1951, a subadult (USNM 436480) and a female with very enlarged ovaries (USNM 436479). ***G. n. obsoleta*** Juvenile, 20 September 1939, Guarumos (USNM 371254); juvenile transitioning, 21 August 1923, Verdecocha (AMNH 180255). ***G. n. nuchalis*** Nestlings, 11 November 2006, RP Tapichalaca (Juiña *et al.* 2009); fledgling, 18 October 2007, RP Tapichalaca (Greeney *et al.* 2010); juvenile, 31 October 2007, RP Tapichalaca (Greeney *et al.* 2010); juvenile, 21 April 2011, RP Tapichalaca (ML 36592181, photo L. Seitz); juvenile transitioning, 9 August 1992, Río Isimanchi (MECN 6497); adult carrying worms from feeder, 12 November 2014, RP Tapichalaca (IBC 1092807, J. del Hoyo). The field notes of B.M. Winger, concerning the October 2007 records (Greeney *et al.* 2010), viewed in light of the plumage phases defined here, reveal that the bird on 18th was a fledgling, still with fluffy, barred plumage below. The young observed on 31st was a juvenile, retaining barred plumage on the hindcrown and nape. I examined two immatures from western Ecuador at AMNH and four specimens at USNM collected by Carriker in south-east Antioquia (see above). The Carriker specimens include two adult females (June, July, 1951), the label of the former indicates that the ovary was enlarged. One is an older immature (subadult) and one a fledgling, both collected June 1951. Hilty & Brown (1986) cited Carriker as collecting a 'BC ♀ and juv., Jan, se Antioquia' and Fjeldså & Krabbe (1990) reported a 'fledgling' in July in Antioquia, a bird with 'enlarged gonads' in January in 'se Antioquia' and 'juv./imm.' in August and September in 'western Ecuador'. I believe all of these refer to the USNM/AMNH specimens above. The older juvenile photographed by A. Quevedo was quite likely the same individual as the subadult he photographed at the same location on 16 January 2010 (Flickr).

TECHNICAL DESCRIPTION Sexes similar. The following adult description refers to the most widespread subspecies, *ruficeps*, and is based on Cadena *et al.* (2007). **Adult** Upperparts unpatterned, warm cinnamon-brown, becoming brighter and more rufescent on the tail and secondaries. The outer secondaries and all of the primaries are fringed dark rufous to pale cinnamon-brown. The crown, forehead and head-sides are nearly chestnut, with the crown darker and postocular area brighter. The lores are blackish, peppered white due to the pale feather bases. The malar is mixed olive-brown and dark grey, while the chin is mottled blackish and dull greyish-white, similar to the lores. The lower throat, breast, sides and flanks are neutral grey, with the abdomen slightly paler grey with an olive tinge. Thighs are olive-brown. **Adult Bare Parts** *Iris* reddish-grey; *Bill* black; *Tarsi & Toes* grey. **Fledgling (*nuchalis*)** Fjeldså & Krabbe (1990) described a 'juvenile' as 'plain light brown above, darkest on crown, light buffy brown below, darkest on breast, with bare parts of face bright orange'. This description largely matches my own made from a skin of *G. n. ruficeps* (see below). There appears to be some fairly significant differences, however, in the fledgling plumages of *ruficeps* and nominate *nuchalis*. The following description of a fresh-out-the-nest *G. n. nuchalis* was made from photographs taken by F. Sornoza (see Reproduction Seasonality). Overall, the bird appears downy or fluffy. Above, the forecrown and crown are nearly uniform black to dark charcoal-grey. The hindcrown has very faint buffy spotting and barring, becoming more apparent lower down until the nape and upper back are finely barred buffy-chestnut and blackish. The fine barring continues to the rump, but becomes much more buff and less chestnut. Below, throat dark charcoal-grey, with fine buffy barring on upper breast continuing lower, slowly transitioning to fine blackish barring on buffy, and finally to nearly unbarred buffy to buffy-white on flanks, and vent. **Fledgling (*nuchalis*) Bare Parts** *Iris* dark greyish-brown; *Bill* bright orange with maxilla dusky along culmen and still bearing a small white egg tooth, rictus deeper, crimson-orange, as is mouth lining, mandible entirely bright orange. *Tarsi & toes* pale whitish-pink. **Fledgling (*ruficeps*)** A fledgling *ruficeps* (USNM 436478) is very close in age to the above-described fledgling, and still has an overall fluffy appearance. It is, however, significantly differently coloured. The forecrown is dark grey mottled dark rufous, slowly becoming less grey and more rufous on the nape, neck-sides and upper back, which are nearly uniform rufous. This slowly fades lower down, becoming first buffy-rufous on the back then nearly uniform buffy on the rump. The lores and malar are dark charcoal-grey lightly peppered whitish. This charcoal feathering mixes with deep rufous on the ear-coverts. The superciliary is deep rufous. The throat is dark charcoal-grey, lightly peppered whitish on the chin due to pale feather bases. The breast is rufous, similar to the nape, slowly grading to buffy-rufous on the lower breast and upper belly, to buffy and then pale greyish-white washed with buff on the belly and flanks. The vent is mostly greyish-white. The bill, tarsi and toes (in the dried specimen) are all pale horn. There is no indication of barring or spotting on any area. The fluffy plumage covers most of the upperwing-coverts but they do not appear to be edged in a different colour. *Juvenile/Subadult transition (nuchalis)* I have not seen an older fledgling or juvenile of nominate *nuchalis*. An individual that was apparently transitioning from juvenile to subadult plumage, but still being fed by adults, was photographed by A. Quevedo and can be described as follows. The upperparts are largely the same as an adult's, but the crown and forecrown are washed blackish and somewhat 'unkempt', and the hindcrown and nape are more buffy-chestnut. Overall, the upperparts are somewhat duller and there are a few buffy-chestnut feathers scattered across the upper back. The upperwing-coverts are broadly edged rufous-chestnut. Below, also similar to adults, nearly uniform dark grey, but the photographs I examined do not permit further details. I did, however, examine a skin that closely matches the above-described characteristics. The following is derived from that skin. Below, similar to adults, nearly uniform dark grey, but chin and upper throat washed charcoal and lower throat faintly washed olivaceous-chestnut. Several irregular, small patches of rufous remain on the breast and several patches of buffy-chestnut on the lower breast and belly. *Juvenile/Subadult transition (nuchalis)* **Bare Parts** (from photographs by A. Quevedo) *Iris* grey to dark grey, bare postocular crescent dull grey; *Bill* maxilla almost entirely dusky-blackish, mandible almost entirely bright orange, yellower towards tip and more reddish-orange basally, rictus still slightly inflated and deep crimson-orange, as is mouth lining; *Tarsi & toes* dusky-grey, only very subtly bluish. *Juvenile/Subadult transition (obsoleta)* I examined an *obsoleta* from western Ecuador (AMNH 180255) that appeared to be transitioning from juvenile plumage. Head and face mostly dull blackish-brown, with a few pale feathers peppering the forecrown and rufous feathering creating an indistinct superciliary contiguous with the finely barred rufous and blackish-brown nape, neck-sides and upper back. Back brownish-olive with a few irregular tawny-rufous to rufous feathers with indistinct subterminal dusky bars.

Lower, the back and rump are olive-rufous. Hindcrown has only very sparse and indistinct rufous spots. Chin and throat dull blackish-brown with pale feathers peppering the malar and sides of throat. Chest heavily mottled grey-brown, rufous and tawny-rufous. The mottling continues onto the lower chest, sides and belly, but there is little or no rufous feathering lower, where predominantly mixed tawny and brownish-grey. Vent and flanks mostly buffy to tawny-buff mixed with brownish-grey. Undertail-coverts brownish-olive. *Juvenile/Subadult transition (obsoleta)* **Bare Parts** (from specimen) *Bill*, maxilla dull reddish-black with the very tip pale yellow, mandible reddish-brown with the distal third yellow.

MORPHOMETRIC DATA Chapman (1926) provided ranges of wing measurements for all three subspecies (sex and sample size not noted): *nuchalis* 102–108mm; *obsoleta* 107–111mm; *ruficeps* 115–117mm. Additional measurements as follows. ***G. n. obsoleta*** *Wing* 118mm; *Bill* [total] culmen 27mm; *Tail* 61mm; *Tarsus* 58mm (*n* = 1 ♂, holotype, Chubb 1916). *Wing* 110.4mm, 113.4mm, 108.9mm, 110.9mm, 109.9mm; *Bill* total culmen 27.7mm, 27.9mm, 27.1mm, 27.6mm, 27.7mm; *Tail* 60.1mm, 65.8mm, 61.6mm, 61.7mm, 57.8mm; *Tarsus* 56.7mm, 57.9mm, 52.9mm, 54.4mm, 51.8mm (*n* = 5, ♂♂,♀,♀,?, Schulenberg & Williams 1982). *Wing* 115.5mm; *Bill* from nares 16.0mm, exposed culmen 18.4mm; *Tarsus* 54.1mm (*n* = 1, ♂, MCZ 138468, J.C. Schmitt *in litt.* 2017). *Bill* from nares 15.3mm; *Tarsus* 57.9mm (*n* = 1, ♂ juvenile, USNM 371254). ***G. n. ruficeps*** *Wing* 117.5 ± 3.3mm; *Bill* [exposed] culmen 23.3 ± 3.1mm, width at gape 17.1 ± 2.1mm; *Tarsus* 60.5 ± 2.3mm (n = 4, ♀?, Kattan & Beltrán 1999). Schulenberg & Williams (1982) provided the following: *Wing* ♂♂ (*n* = 4) mean 115.4mm, range 113.8–120.3mm, ♀♀ (*n* = 4) mean 117.7mm, range 116.0–118.6mm; *Bill* culmen from base, ♂♂ (*n* = 2) mean 31.4mm, ♀♀ (*n* = 3) mean 30.0mm, range 29.0–30.9mm; *Tail* ♂♂ (*n* = 3) mean 67.5mm, range 65.7–70.7mm, ♀♀ (*n* = 3) mean 67.9mm, range 65.7–70.4mm; *Tarsus* ♂♂ (*n* = 3) mean 57.4mm, range 55.0–59.0mm, ♀♀ (*n* = 4) mean 59.4mm, range 57.3–60.9mm. From Cadena *et al.* (2007), means (± SD) and ranges, genders not specified: *Wing* 113.25 ± 4.10mm, 107.5–120.3mm (*n* = 9); *Tail* 63.26 ± 3.93mm, 58.1–69.8mm (*n* = 9); *Bill* exposed culmen 24.13 ± 1.46mm, 22.0–26.3mm (*n* = 7), total culmen 27.55 ± 1.47mm, 25.7–30.1mm (*n* = 7), commissure width 17.30 ± 0.84mm, 16.0–18.4mm (*n* = 8), bill height 10.43 ± 0.52mm, 9.9–11.4mm (*n* = 6); *Tarsus* 59.35 ± 2.96mm, 55.0–62.6mm (*n* = 8). ***G. n. nuchalis*** Schulenberg & Williams (1982) provided the following: *Wing* ♂♂ (*n* = 4) mean 107.2mm, range 99.6–112.7mm, ♀♀ (*n* = 5) mean 105.5mm, range 100.8–109.9mm; *Bill* culmen from base, ♂♂ (*n* = 3) mean 28.9mm, range 25.9–30.8mm, ♀♀ (*n* = 5) mean 28.9mm, range 27.0–31.2mm; *Tail* ♂♂ (*n* = 3) mean 64.2mm, range 59.8–67.7mm, ♀♀ (*n* = 4) mean 61.4mm, range 57.5–67.0mm; *Tarsus* ♂♂ (*n* = 3) mean 53.4mm, range 48.8–59.3mm, ♀♀ (*n* = 5) mean 54.3mm, range 51.2–57.4mm. **MASS** Krabbe & Schulenberg (2003) gave the range for 3 ♂♂ as 111–122g, and of 3 ♀♀ as 104–122g, but did not specify to which race they pertain. 101.2±4.3g (*n* = 4, ♀?, *ruficeps*, Kattan & Beltrán 1999). Parker *et al.* (1985) provided weights for Peruvian nominate *nuchalis*: 2 ♂♂, 115g, 122g; 2 ♀♀, 110g, 122g. **TOTAL LENGTH** 17–22cm (Chubb 1918, Hilty & Brown 1986, Fjeldså & Krabbe 1990, Ridgely & Greenfield 2001, Restall *et al.* 2006, Schulenberg *et al.* 2007, Ridgely & Tudor 2009).

TAXONOMY AND VARIATION Three subspecies are currently recognised and I have followed this arrangement here. Based on differences in adult coloration, immature plumage and voice, however, I suggest that three species may be involved, as opined by some previous authors (Ridgely & Greenfield 2001, Krabbe & Schulenberg 2003, Restall *et al.* 2006). The precise distributions of these three taxa have largely remained poorly defined, with most authors treating all Colombian populations as *ruficeps*, north-west Ecuadorian populations as *obsoleta*, and giving the distribution of nominate *nuchalis* as eastern Ecuador to north-east Peru (Krabbe & Schulenberg 2003, Ridgely & Tudor 2009). It seems likely that the ranges of *obsoleta* and nominate *nuchalis* will eventually be shown to extend at least marginally into Colombia.

Grallaria nuchalis ruficeps P. L. Sclater, 1874, Proceedings of the Zoological Society of London, vol. 41, p. 729, 'Antioquia reipubl. Columbianae.' Subsequently, Sclater (1877) referred to the type specimen with the same vague locality but then stated that the type was from Medellín (06°15'N, 75°35'W) and referenced two additional topotypical skins in his collection, without providing dates or collection numbers. Cory & Hellmayr (1923) amended the original vague type locality to Medellín, which is an unlikely location given that city's elevation (*c.*1,650m). A more precise location might be obtained if we were able to ascertain the itinerary of its collector (T. K. Salmon), but a more likely location would be the mountains east of Medellín, at or near Santa Elena (2,750m), where T.K. Salmon later collected an additional specimen (Sclater 1890; NHMUK 1889.7.10.871). (Note that the coordinates for Santa Elena given by Paynter 1997 are incorrect.) The pages containing the description of *ruficeps* are dated 18 November 1873 but, apparently, this part of the journal was not published until 1874, resulting in the confusingly dated species description. Warren & Harrison (1971) dated it as 1873, as did Sclater himself (1877, 1890), and gave NHMUK 1889.9.20.617 as one of four syntypes held in Tring, apparently as a result of Sclater's (1890) unfortunately vague designation. Described as a species, this race was considered by many authors as a valid species (Chapman 1917). Peters (1951), however, considered *ruficeps* to merit only subspecific rank, an arrangement followed to date, albeit with some reservations (see above). So far as is known, *ruficeps* is endemic to Colombia, found on the east slope of the West Andes from Paramo del Frontino in Antioquia, south to central Valle del Cauca (near Tuluá). This race is also found in the Central Andes from central Antioquia (Abriaquí) at least as far south as western Huila and eastern Cauca. In the northern Central Andes, however, records are fairly scarce (see below). Between Alto del Escobero (Antioquia) and Finca La Estrella (Caldas), there is just one historical record (specimens AMNH), I have included this region on the map, but its continued presence within this *c.*100km region requires modern documentation. Vocalisations from PNN Puracé in Huila (XC 203515, G.M. Kirwan) and specimens from Almaguer, in Cauca (AMNH) appear to represent *ruficeps*. Its distribution in the East Andes is, so far as is known, restricted to the west slope in Cundinamarca and, after an apparent gap in records of *c.*190km, in western Putumayo and eastern Nariño. The two southernmost records in Colombia are from the Eást Andes in eastern Nariño: La Cocha and Reserva Natural Herederos del Planeta El Encano. **SPECIMENS & RECORDS**

Antioquia Angostura, 06°54.5'N, 75°19.5'W (XC 77195, P. Flórez); Abriaquí, 06°34'N, 76°07'W (XC 189421, S. Chaparro-Herrera); San Pedro de los Milagros, 06°27.5'N, 75°33.5'W (Donegan *et al.* 2009); Páramo de Frontino, 06°26'N, 76°05'W (Flórez *et al.* 2004, Krabbe *et al.* 2006); RE San Sebastián La Castellana, 06°06.5'N, 75°33'W (XC 96100/116305, D. Calderón-F.); Alto del Escobero, 06°06'N 75°32'W (Cadena *et al.* 2007); Farallones del Citará, 05°46'N, 76°04'W (Pulgarín-R. & Múnera-P. 2006); Páramo de Sonsón, 05°43'N, 75°15'W (USNM 436477–780); RNA Loro Orejiamarillo, 05°37'N, 75°47'W (D. Calderón-F. *in litt.* 2017); Altos de Ventanas, 05°32.5'N, 75°48'W (XC 119660/661, D. Edwards). *Risaralda* Santa Rosa de Cabal, 04°50.5'N, 75°33'W (XC 302276, J. Holmes). *Caldas* Finca La Estrella, 05°13'N, 75°24'W (Cadena *et al.* 2007); El Zancudo, 05°03.5'N, 75°26'W (CM P70300). *Quindío* Finca La Betulia, 04°40.5'N, 75°32.5'W (O.H. Marín-Gómez *in litt.* 2017); Laguneta, 04°35'N, 75°30'W (Chapman 1917, AMNH, ANSP, MCZ); Páramo Navarco, 04°28.5'N, 75°33.5'W (XC 273787, P. Boesman). *Tolima* Anzoategui, Palomar, 04°39'N, 75°13'W (XC 96504, Y.G. Molina-Martínez); Juntas, 04°33.5'N, 75°19.5'W (D. Carantón *in litt.* 2017); Toche, 04°32'N, 75°25'W (ANSP 154005/06); Vereda San Julian, 04°29'N, 75°29'W (XC 127805/07, O.H. Marín-Gómez); RNA Giles–Fuertesi, 04°17.5'N, 75°33'W (XC 326761/62, A. Pinto-Gómez); Finca San Miguel, 04°02'N, 75°38'W (Cadena *et al.* 2007); 10km NW of Roncesvalles, 04°01'N, 75°38'W (XC 29515, N. Krabbe). *Cauca* Almaguer, 01°55'N, 76°50'W (AMNH, USNM, MCZ); Coconuco, 02°20.5'N, 76°30'W (ANSP); Sector San Nicolás of PNN Puracé, 02°19.5'N, 76°17'W (eBird: J. Muñoz García); Los Milagros, 01°41'N, 76°53'W (eBird: F. Ayerbe-Quiñones). *Huila* PNN Puracé, 02°12'N, 76°21'W (XC 203515, G.M. Kirwan). *Cundinamarca* El Peñón, 04°26'N, 74°18'W (Chapman 1917, Olivares 1969). *Nariño* Daza, 01°16.5'N, 77°15'W (eBird: C. Downing); Laguna La Cocha, 01°05'N, 77°09'W (Strewe 2000b, Calderón-Leytón 2002); RNSC Herederos del Planeta El Encano, 01°03.5'N, 77°07'W (XC 179109, A. Mendoza). *Putumayo* Torres San Francisco, 01°09'N, 76°50.5'W (XC 100382/83, J. King).

Grallaria nuchalis obsoleta Chubb, 1916, Bulletin of the British Ornithologists' Club, vol. 36, p. 47, west side of Pichincha, Ecuador, 12,000ft. [= western slopes of Volcán Pichincha, *c.*00°11'S, 78°38.5'W, 3,660m. The type (NHMUK 1916.8.24.6), an adult male, was collected by Walter Goodfellow in November of 1914 and is held at Tring (Chubb 1916, Warren & Harrison 1971). This subspecies is, so far as is known, endemic to north-west Ecuador, where it is known south from the Colombian border to southern Pichincha. The altitudinal distribution of *obsoleta* is no different from other races, leading me to question a specimen (MCZ 138468) labelled as being collected at Gualea (1,500m). Undoubtedly, it has been mislabelled or was more likely collected on the Andean slopes east of there. Unlike most authors, Sibley & Monroe (1990) and Monroe & Sibley (1993) did not recognise *obsoleta*, subsuming it within the nominate race, but I agree with previous authors (Ridgely & Greenfield 2001, Krabbe & Schulenberg 2003) that *obsoleta* should, in fact, be elevated to species rank. It is the least known of the three taxa, and differs from the other two in having the chestnut-rufous of the head restricted to the nape, while the crown is brownish. The remaining upperparts are olivaceous-brown (versus reddish-brown in nominate *nuchalis*), while the lores and eye-ring are blackish, contrasting with chestnut ear-coverts. The underparts are blackish-grey, most like *ruficeps* but noticeably darker. Interestingly, in his original description, Chubb (1916) quoted Goodfellow as reporting the bare-part colours as: 'Bill black; feet dark brown; iris dark ruby-red'. Apart from the black bill, these would seem to differ significantly from those of either of the other two subspecies. I have been unable to determine what the true iris colour is of *obsoleta*. **Specimens & Records** *Carchi* 5.9km WSW of Maldonado, 00°53'N, 78°10'W (J. Freile *in litt.* 2017); Chilmá Alto, 00°52'N, 78°04'W (R. Ahlman *in litt.* 2017). *Imbabura* Cayapachupa, 00°33'N, 78°28'W (XC 262134/35, O. Jahn); RP Siempreverde, 00°22'N, 78°25'W (J. Freile *in litt.* 2017); Las Delicias, 00°21'N, 78°26'W (XC 250885, N. Krabbe); km13 on Laguna Cuicocha–Apuela road, 00°19'N, 78°26.5'W (LSUMZ 112588); Loma Taminanga, 00°17N, 78°28W (Krabbe 1991). *Pichincha* RBP Maquipucuna, 00°05'N, 78°35'W (J. Freile *in litt.* 2016); Reserva Geobotánica Pululahua, 00°01'N, 78°30'W (R.A. Gelis *in litt.* 2016); *c.*2.8km SSW of Tandayapa, 00°01'S, 78°41.5'W (HFG); Chiquilpe, 00°01.5'S, 78°36'W (T. Santander *in litt.* 2017); Guarumos, 00°04'S, 78°38'W (USNM 371254); Verdecocha, 00°06.5'S, 78°37'W (AMNH); Yanacocha, 00°08'S, 78°35'W (Krabbe 1991); upper Chiriboga Road, 00°17'S, 78°41.5'W (S. Olmstead *in litt.* 2016); RVS Pasachoa, *c.*00°26'S, 78°29'W (ML 83175, T.A. Parker); SW slope of Volcán El Corazón, 00°33'S, 78°43'W (XC 248852, N. Krabbe). *Santo Domingo de las Tsáchilas* BP Río Guajalito, *c.*00°14'S, 78°48'W (J. Freile *in litt.* 2017).

Grallaria nuchalis nuchalis P. L. Sclater, 1859a, Proceedings of the Zoological Society of London, vol. 27, p. 441, 'Río Napo', Ecuador. Elevations along the Río Napo are far too low for this species, but there is virtually no way of knowing the true provenance of the holotype. Somewhere in the vicinity of Papallacta or Baeza, however, is a likely guess given probable travel routes through appropriate habitat en route to the Río Napo from Quito. Somewhat confusingly, Sclater (1860d) published a second, near-identical description of apparently the same specimen in a different journal. That cited above, however, takes precedence. The nominate form is confined to the East Andes where it is known from just south of the Colombian border to northern Cajamarca and Piura, north of the Marañón drainage. It seems likely that it will be eventually found in extreme southern Colombia. A specimen reportedly from Ambato (MCZ 149840) has been omitted from the map. Ambato is located in the East Andean watershed, in the upper Río Pastaza watershed and at a plausible elevation for Chestnut-naped Antpitta (*c.*2,600m), but is 'tucked behind' the main mountain chain and, in climate, resembles the warm, arid inter-Andean valley. Given what we suspect about the occurrence of long-distance dispersal in Scaled Antpitta, it would not be unreasonable to propose that this record represents an individual from the more humid slopes east of the town. Nevertheless 'Ambato' was frequently used in a very general sense by early collectors (Paynter 1993), casting further doubt as to the label's accuracy. It remains very unlikely that the arid habitat around Ambato would support a resident population of this humid-forest species. **Specimens & Records** **ECUADOR**: *Sucumbíos* Santa Bárbara, 00°38.5'N, 77°31.5'W (J. Freile *in litt.* 2017); Alto La Bonita, 00°29.5'N, 77°35'W (Stotz & Valenzuela 2009).

Napo Oyacachi, 00°13'S, 78°05'W (Baez *et al.* 2000, AMNH 180253); Baeza, 00°28'S, 77°53'W (AMNH 173023); Cordillera de Guacamayos, 00°40'S, 77°51.5'W (Krabbe 1991); Río Maspa Chica, 00°24'S, 78°02'W (Poulsen 1996) and nearby Maspa (NRM); *c.*1km WNW of Estación Biológica Yanayacu, 00°35.5'S, 77°54'W (HFG). *Orellana* Volcán Sumaco, 00°32'S, 77°36.5'W (AMNH, ANSP). *Tungurahua* vicinity of Baños (e.g., Lamas, San Nicolás; see Paynter 1993), *c.*01°24'S, 78°25'W (MCZ, SBMNH, ANSP); RP Machay, 01°23'S, 78°16'W (L. Jost *in litt.* 2017). *Chimborazo* Arenales, right bank of Río Paute, 02°34.5'S, 78°34'W (XC 250021, N. Krabbe). *Loja* Cajanuma, 04°07'S, 79°10.5'W (XC 258601, J. V. Moore, ANSP); 'Finca de Dr. Espinosa', 03°58'S, 79°09'W (Krabbe 1991). *Morona-Santiago c.*7.5km W of Zúñac, 02°12.5'S, 78°26.5'W (R.A. Gelis *in litt.* 2016); Gualaceo–Limón road, 03°02'S, 78°39'W (eBird: M. Reid). *Zamora-Chinchipe* RE Cerro Plateado, 04°36.5'S, 78°48'W (XC 250884, N. Krabbe); RP Estación Científica San Francisco, 03°58'S, 79°05'W (XC 4177, W. Halfwerk); Río Isimanchi above San Andrés, 04°47'S, 79°20'W (MECN 6497, XC); RP Tapichalaca/ Quebrada Honda, *c.*04°29.5'S, 79°07.5'W (Juiña *et al.* 2009, Greeney *et al.* 2010, Heinz 2010, Snow *et al.* 2015, MECN, XC, ML). **PERU**: *Cajamarca* Hito Jesús, 04°54'S, 78°54'W (XC 6940, T. Mark); Picorana, 05°02'S, 78°51'W (LSUMZ 172182/83); Miraflores (Tabaconas), *c.*05°15'S, 79°01.5'W (MUSA 3184); 14.5km SE of Huancabamba, 05°19.5'S, 79°21'W (XC 184222, D.F. Lane); near Laguna Azul, 05°32'S, 79°08'W (A. García-Bravo *in litt.* 2017); Agua Azul, 05°35'S, 79°10'W (MSB 42225); ACM Bosque de Huamantanga, 05°40'S, 78°58'W (A. García-Bravo *in litt.* 2017). *Piura* Majaz Mine, 04°54'S, 79°20'W (XC 7092, T. Mark); Cerro Chinguela, 05°07'S, 79°23'W (Parker *et al.* 1985, LSUMZ, ML); 3km NNE of Lechuga, 05°20.5'S, 79°31.5'W (D. Beadle *in litt.* 2017).

STATUS The global population size has not been quantified, but it has an estimated range of 40,500km^2 (BirdLife International 2017). Across its range the species is generally considered uncommon, but it can be locally fairly common (Fjeldså & Krabbe 1990, Stotz *et al.* 1996, Krabbe & Schulenberg 2003, Ridgely & Tudor 2009). As it appears that each of the three subspecific populations may deserve species rank, it should be noted that these estimations of abundance may be true for subspecies *ruficeps* (McMullan *et al.* 2010) and nominate *nuchalis* (Ridgely & Greenfield 2001, Schulenberg *et al.* 2007), but that the Choco endemic, *obsoleta*, is somewhat rarer and considerably more poorly known (Ridgely & Greenfield 2001). As emphasised by Freile *et al.* (2010), *obsoleta* might easily fulfill the criteria for threatened if it was considered a full species but, even should this not be true, isolated and potentially genetically divergent populations should be considered when designing conservation plans (Moritz 2002, Freile *et al.* 2010). **Protected Populations** *G. n. ruficeps* PNN Puracé (XC 203515, G.M. Kirwan); PNN Los Nevados (Kattan & Beltrán 1999, 2002, Pfeifer *et al.* 2001); RN Chicaque (eBird: F. Salgado); PNN Tatamá (Echeverry-Galvis & Córdoba-Córdoba 2007); PRN Ucumarí (Kattan & Beltrán 1999, 2002); RFP Río Blanco (Nieto-R. & Ramírez 2006); RNA El Colibrí del Sol (Salaman *et al.* 2007a, XC 316462, G.M. Kirwan); RE San Sebastián La Castellana (XC 96100, D. Calderón-F.); RNA Loro Orejiamarillo (Salaman *et al.* 2007a); RMN El Mirador (Salaman *et al.* 2007a); RNA Giles-Fuertesi (XC 326761/62, A. Pinto-Gómez); RNSC La Sonadora (eBird: V. Marín); RNSC La Patasola and RNSC Alto Quindío Acaime (O.H. Marín-Gómez *in litt.* 2017); RNSC Herederos del Planeta El Encano (XC 179109, A. Mendoza): **G. n. obsoleta** RE Cotacachi Cayapas (XC 262134/35, O. Jahn); Reserva Geobotánica Pululahua (R.A. Gelis *in litt.* 2017); RBP Maquipucuna (J. Freile *in litt.* 2016); RP Yanacocha (Krabbe 1991); RP Verdecocha (AMNH); BP **Río Guajalito** (J. Freile *in litt.* 2017); RP Siempreverde (J. Freile *in litt.* 2017). **G. n. nuchalis Ecuador**: PN Podocarpus (Krabbe 1991, Poulsen 1993, Rassmussen *et al.* 1994, Ridgely & Tudor 1994); PN Llanganates (Benítez *et al.* 2001); PN Sangay (XC 250495 in background, N. Krabbe); PN Cayambe-Coca (AMNH 180253); PN Sumaco-Galeras (AMNH, ANSP); PN Yacurí (J. Freile *in litt.* 2017); RE Antisana (HFG); RE Cerro Plateado/BP Alto Nangaritza (XC 250884, N. Krabbe); RP Tapichalaca (Juiña *et al.* 2009, Greeney *et al.* 2010, Heinz 2010, Snow *et al.* 2015); RP Machay (L. Jost *in litt.* 2017); RP Estación Biológica Yanayacu (HFG); RP Guango Lodge (XC 217725, C. Mroczko); RP SierrAzul (ML: J.W. Wall); RP Guango Lodge (XC 217725, C. Mroczko). **Peru**: ACM

Bosque de Huamantanga (A. García-Bravo *in litt.* 2017). Chestnut-naped Antpitta almost certainly occurs within SN Tabaconas-Namballe.

OTHER NAMES *Grallaria ruficeps* (Sclater 1890, Sharpe 1901, Brabourne & Chubb 1912, Cory & Hellmayr 1924). **English** Chestnut-headed Antpitta [*ruficeps*] (Cory & Hellmayr 1924, Meyer de Schauensee 1952), Chestnut-naped Ant-Thrush [*nuchalis*] Brabourne & Chubb 1912), Chestnut-headed Ant-Thrush [*ruficeps*] Brabourne & Chubb 1912); Western Chestnut-naped Antpitta [*obsoleta*] (Cory & Hellmayr 1924); Russet-capped Antpitta [*ruficeps*] (Meyer de Schauensee 1966, Sibley & Monroe 1990). **Spanish** Tororoí Nuquicastaño (Krabbe & Schulenberg 2003); Tororoí chusquero [*ruficeps*] (Cardona *et al.* 2005, Salaman *et al.* 2007a); Tororoi Nuquicastaña [*nuchalis*] (Clements & Shany 2001); Gralaria nuquicastaña (Ridgely *et al.* 1998, Buitrón-Jurado 2008). **French** Grallaire à nuque rousse. **German** Rostnacken-Ameisenpitta (Krabbe & Schulenberg 2003).

Chestnut-naped Antpitta, adult (*nuchalis*), RP Tapichalaca, Zamora-Chinchipe, Ecuador, 7 December 2013 (*Dušan M. Brinkhuizen*).

Chestnut-naped Antpitta, mid-aged nestlings (*nuchalis*) removed from the nest for measuring, RP Tapichalaca, Zamora-Chinchipe, Ecuador, 17 November 2006 (*Mery E. Juiña J.*).

Chestnut-naped Antpitta, fledgling (*nuchalis*), RP Tapichalaca, Zamora-Chinchipe, Ecuador, 5 December 2006 (*Francisco Sornoza*).

Chestnut-naped Antpitta, adult, RN Río Blanco, Caldas, Colombia, 27 August 2012 (*Peter Hawrylyshyn*).

Chestnut-naped Antpitta, adult feeding juvenile (*ruficeps*), RNA Colibri del Sol, Antioquia, Colombia, 21 September 2009 (*Alonso Quevedo*).

PALE-BILLED ANTPITTA
Grallaria carrikeri   Plate 13

Grallaria carrikeri Schulenberg & Williams, 1982, Wilson Bulletin, vol. 94, p. 105, Cordillera de Colán, south-east of La Peca (05°34'S, 78°19'W), Dept. Amazonas, Peru, 2,450 m. The holotype, an adult male, is preserved as a skin and partial skeleton at the Louisiana State University collection (LSUMZ 88044) and was collected on 15 October 1978 by Morris D. Williams. All 11 designated paratypes are also housed there, five of these (three males, two females) being from the type locality. The remaining paratypes (three male, three female) are from Cumpang, above Utcubamba, on the trail to Ongón in Dept. La Libertad. In addition, the LSU collection includes two young nestlings collected by Wiedenfeld (1982), both fluid-preserved (Schulenberg & Williams 1982, Cardiff & Remsen 1994).

With its deep red eyes, bright pale bill and contrasting black hood, this large antpitta rivals its close relative, Jocotoco Antpitta, in being near the top of many a birder's to-see list (Schulenberg & Kirwan 2012e). Endemic to upper-middle elevations of the Andes in north-central Peru, it is known from just a handful of locations and is not yet known from any formally protected habitat (Schulenberg & Kirwan 2012e). Named for Mel Carriker, one of the most prolific collectors of Neotropical birds, who also significantly advanced our knowledge of their Mallophaga, Pale-billed Antpitta is among the least-studied *Grallaria*. Pale-billed Antpitta is undoubtedly closely related to both Chestnut-naped and Jocotoco Antpittas, with all three having generally similar plumage and heavy bills. The intraspecific relationships between them, however, remain unclear. Adults of both Pale-billed and Jocotoco Antpittas have bright red eyes and similarly proportioned bills, but similarities in immature plumage between Pale-billed Antpitta and the nominate race of Chestnut-naped Antpitta suggest a closer relationship

between them. Pale-billed Antpitta has a restricted range, much of which is inaccessible by road. Although one nest has been found, little else is known of its reproductive habits and the eggs are undescribed.

IDENTIFICATION 19–20cm. A large, rather simply patterned *Grallaria*, basically dark brown above (wings and tail more chestnut-brown), grey below, with a black head (tending to very dark brown on the crown). The most distinctive feature in the field is the heavy, strikingly pale, ivory-white bill. A good look through binoculars will also reveal its blood-red irides. The heavy pale bill alone should preclude confusion with other species of *Grallaria*. Furthermore, within its range, there are no other antpittas with grey underparts. Pale-billed Antpitta is similar in size and plumage to one of its putative sister-species, Chestnut-naped Antpitta, with which there is no range overlap. Chestnut-naped Antpitta lacks the pale bill, is more olive-brown above and the head is reddish rather than black.

DISTRIBUTION One of Peru's endemic antpittas, Pale-billed Antpitta is resident on the east slope of the northern Andes south of the Río Marañón from the Cordillera Colán, Dept. Amazonas (Davies *et al.* 1997), south to at least to Cumpang, east of Tayabamba, Dept. La Libertad (Schulenberg *et al.* 1982, Valqui 2004, T. Mark in Schulenberg & Kirwan 2012e). It has been registered at relatively few localities within its range, being known only from the original collecting sites until the mid-1990s (Cardiff & Remsen 1994). A few other locations have since been found (Begazo *et al.* 2001, Mark *et al.* 2008) but, despite these new records, it is somewhat surprising that the species has yet to be recorded in the Cordillera Carpish (Dept. Huánuco), despite increasing work in this area. Although Pale-billed Antpitta is considered likely to occur south of Cumpang, its true southern limit is unknown, due largely to the difficulties of accessing appropriate habitat in that part of the Andes (Schulenberg & Kirwan 2012e). North of the Río Marañón, it is replaced by Chestnut-naped Antpitta.

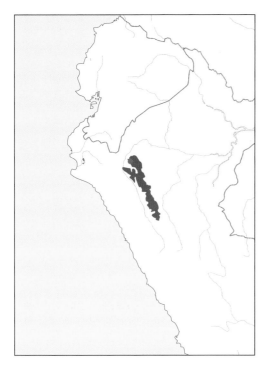

HABITAT Pale-billed Antpitta inhabits the understorey of humid montane forest, seemingly preferring tall, broken-canopy forest with heavy epiphyte loads (Schulenberg & Kirwan 2012e) and possibly favouring ridgetops (Davies *et al.* 1994). Like Chestnut-naped Antpitta, one of its putative relatives, Pale-billed Antpitta is thought to favour areas where *Chusquea* bamboo dominates the understorey (Schulenberg & Williams 1982, Schulenberg *et al.* 2007), and is even considered an indicator species of central Andean bamboo habitats by Stotz *et al.* (1996). Due to the natural heterogeneity of its montane habitat and the prevalence of bamboo throughout, it remains to be seen what the species' association is with bamboo. Some studies have, in fact, found it only in forests away from large patches of bamboo (Davies *et al.* 1994). The known altitudinal range is 2,350–2,900m (Fjeldså & Krabbe 1990, Schulenberg *et al.* 2007), somewhat narrower than other *Grallaria* (Graves 1985).

VOICE In general, the vocalisations are poorly documented. Fjeldså & Krabbe (1990) described the song as six 'staccato notes on the same pitch, lasting *c.*3s. There is a slight pause after the 1st and before the last note'. D.F. Lane (in Schulenberg *et al.* 2007) described the song as 'a series of low hoots, first a higher note, then a stuttered series of lower notes: *WHEE wur'KUK-KUK-KUK-KUK KUK*.' He also described the call as 'a series of high, metallic *teet* notes'. Mark *et al.* (2008) recently mentioned that Pale-billed Antpitta produces '4–13 high (5kHz) *tséep* notes delivered at *c.*1 per second and with intervals of 6–9 seconds'. They likened it to the call reportedly given in response to playback by Chestnut-naped Antpitta (*nuchalis*) in northern Peru (Parker *et al.* 1985). Mark *et al.* (2008) reported this vocalisation as being given by an unsexed individual in 'apparent response to a vocalising Rusty-tinged Antpitta', and was continued for *c.*5 minutes.

They did, however, hear similar calls after playback of the song of Pale-billed Antpitta on a separate occasion. They also recorded another vocalisation that they attributed to Pale-billed Antpitta and which they described 'as reminiscent of the better-known six-note 'staccato' call mentioned by Fjeldså & Krabbe (1990). It differed by containing additional notes (at least nine).

NATURAL HISTORY Other than one report of nesting (see below), the ecology and behaviour is virtually unknown. It is said to forage and behave like other *Grallaria* (Schulenberg & Kirwan 2012e), but some observers have suggested that it is less terrestrial than many others (Davies *et al.* 1994). It also appears that Pale-billed Antpitta is somewhat less vocal than other *Grallaria*, as suggested by surveys that detected the species visually but not its vocalisations (Davies *et al.* 1994). Wiedenfeld (1982), in the vicinity of an active nest, stated that adults vocalised frequently during mid-morning, but neither daily nor annual patterns of singing have been rigorously documented. While singing, Pale-billed Antpittas throw their head back and extend their neck and legs, flicking the tail and head with each note. With each progressive note they crouch slightly lower, gradually inflating their thoat and breast, and lifting their head more dramatically with each note (IBC 1319443, video K. Blomerley).

DIET The diet of Pale-billed Antpitta is poorly documented, but is probably similar to other large, montane *Grallaria*. At a nest where the pair was provisioning nestlings (Wiedenfeld 1982), earthworms were included in seven of 19 feedings, indicating their importance (at least to nestlings). A similar preference for earthworms as nestling food has been documented for several other *Grallaria* (Greeney *et al.* 2008a). Nevertheless, the author felt that earthworms might not make up a significant portion of adult diet, at least during the breeding season. From LSUMZ and MSB specimen labels, stomach contents of eight adults (one empty) included: four larval lepidopterans (12–34mm), three ants, one orthopteran, one earthworm and one spider. In addition, four stomachs contained coleopterans or their remains, and one 'insect eggs'. The stomach contents previously reported by Schulenberg & Williams (1982) are included among these data.

REPRODUCTION To date, only one nest has been described in the literature (Wiedenfeld 1982), and the following description is derived from this work. A single nest was found at 2,875m above Cumpang along the trail to Ongón (La Libertad). **NEST & BUILDING** The only described nest was just 5m from an active mule trail, which might suggest that Pale-billed Antpitta is fairly tolerant of human activity. It is possible that this was not accidental, rather the adults chose the site to facilitate foraging for prey in the wake of passing mules. The nest was 3m above the steeply sloping ground below, well supported by an angled, 0.5m-diameter tree trunk. It was well hidden among live mosses and ferns, and surrounded by humid leaf litter that had collected on top of the trunk. The walls of the nest included a few small sticks, but consisted mostly of a thick mass of decaying leaves similar to the naturally collected litter surrounding the nest (= mostly bamboo leaves and ferns; HFG examination of WFVZ 168915). 'The nest merged almost imperceptibly with detritus on the trunk, so that if the rootlets lining the cup had not been seen, the nest would have been difficult to distinguish.' Dimensions: outer diameter 20 × 19cm; inner diameter 14 × 12cm; inner depth 5cm. The nest was relatively shallow compared with other *Grallaria* nests (Greeney *et al.* 2008a), especially given the large size of Pale-billed Antpitta. Measurements of additional nests are needed before the nest of this species should be considered 'a shallow cup' (Krabbe & Schulenberg 2003). Measurement of the cup depth in this case may have been confounded by the angled nature of the substrate or have been a result of nestling activity stretching the nest from its original, pre-hatch dimensions. **EGG, LAYING & INCUBATION** No information. **NESTLING & PARENTAL CARE** The nestling description provided by Wiedenfeld (1982) was somewhat cursory. The two nestlings, which were estimated to be seven days old, still retained their egg-teeth and had not yet opened their eyes. They had 'sparse down', and bright orange bills and mouth linings. Based on this description, the mass of the nestlings (26–27g) and my own experience with the nestlings of Jocotoco Antpitta, I feel that an estimated age of seven days is fairly accurate. The nestlings were brooded by both adults for 60% of 11 hours of observation while the nestlings were *c.*7–8 days old. Brooding bouts never lasted longer than 41min, and 14 observed bouts lasted a mean 27min. This relatively high percentage of brooding further supports the assertion that they were around one week old. Although not observed directly, the adults were thought to be consuming faecal sacs at the nest. Food was also delivered by both adults, which brought multiple prey items per trip and apparently fed some to each of the nestlings on each visit. Provisioning rates averaged *c.*1 trip/nestling/h. Periods without feeding frequently lasted more than an hour, but were usually followed by several visits in quick succession. Apparently the adults were never heard vocalising around the nest, an observation in rather sharp contrast to experiences at the nests of other *Grallaria* (Greeney *et al.* 2008a). **SEASONALITY** The nest with young studied by Wiedenfeld (1982) was found on 14 October 1979. Using the 30-day incubation period of Jocotoco Antpitta (HFG) and an estimated hatch date of 7 October, I estimate that laying occurred during the first or second week of September. Fjeldså & Krabbe 1990 reported 'eggs' in September and October in Dept. La Libertad but the source of these data is unclear. These records taken together suggest that Pale-billed Antpitta breeds predominantly during the drier part of the year in its range. If true, this would make it one of many east Andean species that are at least facultatively associated with bamboo to display peak reproductive activity during the drier months (Greeney & Miller 2008). **BREEDING DATA** Nestlings, 14 October 1979, Cumpang (Wiedenfeld 1982); fledgling, skull 0% ossified, 24 July 2008, 4.5km N of Tullanya (MSB 32714; not examined); juvenile, skull <5% ossified, 17 July 2008, 4.5km N of Tullanya (MSB 32437); juvenile, skull 40% ossified, 21 October 1978, SE of La Peca (LSUMZ 88042; same record called 'juvenile' in Fjeldså & Krabbe 1990); juvenile, 17 October 2015, Trocha Cresta de San Lorenzo (ML 41082011, photo D. Beadle); subadult, 9 December 2015, Río Chido trail (C. Calle photos); subadult, 19 October 1978, SE of La Peca (LSUMZ 88043); 'juvenile', August, Amazonas (Fjeldså & Krabbe 1990).

TECHNICAL DESCRIPTION Sexes similar. *Adult* Most of the head is black, including the lores, forecrown, malar and base of ear-coverts. The crown and tips of ear-coverts are blackish-brown, grading to dark olive-brown on nape. The upperparts are generally dark olive-brown, the feathers narrowly tipped black. Remiges dusky-brown except outer

webs, which are dark chestnut. Upperwing-coverts dark chestnut and rectrices dusky-brown. The chin is black, grading to dark grey on throat. Remaining underparts light grey, darker on breast-sides, but becoming paler posteriorly. The central belly is nearly white. Feathers of the breast and upper belly are narrowly tipped black. Flanks and undertail-coverts buffy-olive. **Adult Bare Parts** *Iris* pale, reddish brown; *Bill* ivory or horn-coloured; *Tarsi & Toes* blue-grey. *Juvenile* My examination of a juvenile specimen (see Breeding Seasonality) that I consider roughly midway between fledgling and subadult plumages suggests the following. Forecrown black becoming charcoal-grey around mid-crown where the feathers show very faint, dark cinnamon barring, only visible on close inspection. This barring becomes ever more prominent rearwards until the nape and neck-sides are distinctly barred black and deep rufous-brown. Back largely similar to adults but with large, irregular patches of cinnamon (slightly paler than the nape) with black barring. Wings and tail similar to adults, but primary wing-coverts tipped cinnamon-buff and have a subterminal black band. Lores and face black, faintly washed grey on lores and dark cinnamon on ear-coverts. Chin and throat black, grading to dark charcoal-grey on lower throat and upper breast. Central breast is similarly patterned to back with sooty-grey adult feathers appearing, especially on sides and lower breast. The central barred pattern fades until the upper belly is cinnamon-brown, becoming white on the lower belly and dark grey mixed cinnamon-brown on the vent and flanks. **Juvenile Bare Parts** *Iris* 'straw' (= light brown?; MSB 32437 label, A. B. Johnson); *Bill* maxilla dark brownish=yellow, yellowish along tomia and at tip, culmen dark red-brown, mandible mostly dark yellow with wash of red-brown (dried skin; HFG), pale orange, darker along tomia and at base, becoming ivory as in adults; *Tarsi & Toes* dark yellowish with brownish wash (dried skin; HFG), lead-grey to bluish-grey (life; HFG). *Fledgling* Fjeldså & Krabbe (1990) and Schulenberg & Kirwan (2012e) described the 'juvenile' as having the head blackish-grey with black lores, the feathers of the nape tinged brown and tipped cinnamon. Hindcrown, back and breast rufous or cinnamon-brown and barred black. Primary wing-coverts tipped cinnamon-buff with a subterminal black band. Upper belly and flanks buffy with black barring, which becomes weaker on lower belly (creamy-buff on sides and white centrally). Fjeldså & Krabbe (1990) described the bill as 'black with orange flecks at base' and the tarsi as 'grey-flesh'. The colour of the iris is unknown but I suggest it will show a similar progression from brown to red, as in Jocotoco Antpitta. Based on the notably immature coloration of the bill (orange), I suggest that this description best applies to what I refer to as fledglings.

MORPHOMETRIC DATA The following measurements are from Schulenberg & Williams (1982). *Wing* mean 110.1mm, range 106.8–112.5mm ($n = 5$, ♂♂), mean 105.4mm, range 103.7–107.1mm ($n = 2$, ♀♀); *Tail* mean 64.6mm, range 62.6–66.0mm ($n = 4$, ♂♂), mean 63.4mm, range 61.4–65.7mm ($n = 4$, ♀♀); *Bill* total culmen mean 30.1mm, range 29.6–31.3mm ($n = 5$, ♂♂), mean 28.4mm, range 28.0–29.1mm ($n = 4$, ♀♀); *Tarsus* mean 59.7mm, range 59.0–60.6mm ($n = 5$, ♂♂), mean 57.7mm, range 56.4–58.9mm ($n = 4$, ♀♀). *Bill* from nares 16.8mm, exposed culmen 23.2mm, depth at front of nares 10.8mm, width at gape 13.2mm, width at front of nares 6.9mm, width at base of lower mandible 10.0mm; *Tarsus* 60.4mm (*n* = 1, ♂, MSB 32477). **Mass** mean 112g, 107g (*n* = 10, 5♂♂, 5♀♀, Schulenberg & Williams 1982); 119.2g(lt) (*n* = 1, ♂, MSB 32477); 116.2g(lt) (*n* = 1, ♂, juvenile MSB 32437); 97g(lt), 112g, 109g, 110g, 113g(m), 109g(h), 111g, 96g, 124g, 114g (*n* = 10; 6♀♀, 4♂♂, LSUMZ); 86g, 97.5g (*n* = 2, juvenile ♂, subadult ♂, LSUMZ 88042/43). **Total Length** 19–21cm (Fjeldså & Krabbe 1990, Clements & Shany 2001, Ridgely & Tudor 2009, Schulenberg *et al.* 2010).

DISTRIBUTION DATA Endemic to **PERU**: *Amazonas* SE of La Peca, *c.*05°33'S, 78°19'W (LSUMZ 88042–046); Campamento Bosque Quemado, 05°36'S, 78°15'W (F. Angulo P. *in litt.* 2017); Trocha Cresta de San Lorenzo, 05°48.5'S, 78°01.5'W (ML 41082011, photo D. Beadle); Río Chido trail above La Florida, 05°49.5'S, 78°01'W (Begazo *et al.* 2001, XC 7265/66, N. Athanas); 33km NE of Ingenio, *c.*05°54.5'S, 77°58'W (DMNH 60082); Atumpampa, 05°58.5'S, 77°535'W (F. Angulo P. *in litt.* 2017); *c.*4.5km N Tullanya, 06°05'S, 78°20'W (MSB). *San Martín* Laurel, 06°41.5'S, 77°41.5'W (FMNH 480824–826); Laguna de los Cóndores, 06°50'S 77°42'W (Mark *et al.* 2008). *La Libertad* Cumpang, 08°12'S, 77°10'W (LSUMZ, WVFZ, ML).

TAXONOMY AND VARIATION Monotypic. Although it has been suggested to be most closely allied to Chestnut-naped Antpitta by some authors (Sibley & Monroe 1990, Krabbe *et al.* 1999), Krabbe & Schulenberg (2003) suggested that the species may be sister to Jocotoco Antpitta. All three have unpatterned breasts, grey flanks and heavy bills. Both Pale-billed and Jocotoco Antpittas have contrasting dark heads and red irides. Phylogenetic analysis of mitochondrial DNA (cytochrome *b* and ND2) recovered a sister-species relationship between Chestnut-naped and Jocotoco Antpittas (Krabbe *et al.* 1999). Pale-billed Antpitta was not included in this analysis, however, and thus the interspecific relationships between these three species are unclear. In light of the immature

plumage descriptions provided here, the fledglings of the nominate race of Chestnut-naped Antpittas seem extremely similar to those of Pale-billed, including the pink to pinkish-grey tarsi (unlike blue-grey in Jocotoco Antpitta). Taken together, this evidence inclines me to agree with Krabbe *et al.* (1999), and to place the present species as closest to Chestnut-naped Antpitta.

STATUS Although Pale-billed Antpitta is a range-restricted species (Stattersfield *et al.* 1998), and is considered 'uncommon' (Schulenberg *et al.* 2007), the population trend was believed stable and, until recently, it was regarded as Least Concern (BirdLife International 2013). However, it has been uplisted to Near Threatened, now that an improved estimate of its distribution (BirdLife International 2017) has revealed just how small its range is (19,418km², Franke *et al.* 2007). As is true for many sensitive Andean bird species, Pale-billed Antpitta undoubtedly faces ongoing and increasing threats through habitat destruction, particularly clearance for crops and grazing (Schulenberg & Kirwan 2012e). This is particularly strong in key areas of the species' range, in Amazonas and San Martín, which have experienced a large influx of landless people from adjacent Cajamarca due to recent road improvements (Begazo *et al.* 2001). Despite this, Schulenberg & Kirwan (2012e) pointed out that there is still hope, as most of the Pale-billed Antpitta's range is remote with a relatively small human population. **Protected Populations** SN Cordillera de Colán (F. Angulo P. *in litt.* 2017); ACP Hierba Buena-Allpayacu (eBird: O. Janni); ACP Abra Patricia (eBird: M. Amershek).

OTHER NAMES Spanish Tororoí de Pico Pálido (Plenge 2017); Tororoí de Carriker (Krabbe & Schulenberg 2003); Tororoí Piquipálido (Clements & Shany 2001). **French** Grallaire de Carriker. **German** Blassschnabel-Ameisenpitta (Krabbe & Schulenberg 2003).

Pale-billed Antpitta, juvenile, San Lorenzo, Amazonas, Peru, 17 October 2015 (*David Beadle*).

Pale-billed Antpitta, subadult, Río Chido, Amazonas, Peru, 9 December 2015 (*Carlos Calle*).

Pale-billed Antpitta, adult, Río Chido, Amazonas, Peru, 2 December 2012 (*Carlos Calle*).

WHITE-THROATED ANTPITTA
Grallaria albigula   Plate 14

Grallaria albigula Chapman, 1923; American Museum Novitates, no. 86, p. 8; Inca Mine, Santo Domingo, 6000ft, Peru. [1,830m, 13°51'S, 69°41'W, Puno; Stephens & Traylor 1983.] The holotype, an adult male (AMNH 146167), was collected on 14 September 1916, by Harry Watkins (LeCroy & Sloss 2000).

White-throated Antpitta is among the least known of the montane antpittas. Described from Peru, it wasn't found in Bolivia until a decade later (Bond & Meyer de Schauensee 1942) and in Argentina not until almost half a century after its description (Olrog & Contino 1970). The southernmost ranging of the Andean *Grallaria*, it occurs discontinuously from south-east Peru to Bolivia and north-west Argentina (Parker *et al.* 1996, Ridgely & Tudor 2009). It forages almost exclusively on the ground for invertebrate prey, although few specifics have been published. Like its congeners, White-throated Antpitta is most likely to be initially detected by voice, an easily imitated, double-noted call. It is a medium-sized antpitta with a very simple plumage pattern, and is unlikely to be confused within its range. The taxonomic affinities of White-throated Antpitta are unresolved. Ridgely & Tudor (1994) suggested the nearly allopatric Red-and-white Antpitta as its likely sister species. Like its congeners, White-throated Antpitta is very difficult to see, even where fairly common. This, combined with its restricted range, has resulted in its being one of the most poorly studied *Grallaria*. Its diet is very poorly documented and its habitat requirements are unclear, possibly varying geographically. Even the validity of the only described subspecies demands investigation. In this account I provide the first description of subadult plumage, which in turn becomes the only reproductive information available.

IDENTIFICATION 18.5–20.0cm. White-throated Antpitta is aptly named for its bright white throat contrasting strongly with the chestnut-rufous head, brownish upperparts and grey underparts. Over most of its range, the white throat and prominent ocular ring make the species distinctive. In the northern Peruvian part of its range, it may occcur sympatrically with the similar Red-and-white Antpitta, albeit with only limited altitudinal overlap around 2,100m. Both White-throated Antpitta and Red-and-white Antpitta have a reddish-brown crown and a white throat, but the back and wings of the present species are olive-brown rather than chestnut-brown as in Red-and-white Antpitta. The breast and belly of White-throated are pale grey, the flanks lack distinct rufous markings, and the breast lacks a reddish band (Clements & Shany 2001, Schulenberg & Kirwan 2012c). Further south, Rufous-faced Antpitta also generally occurs at higher elevations and has a dark crown contrasting with a rufous face, unlike the all-rufous head of White-throated Antpitta (Ridgely & Tudor 1994).

DISTRIBUTION Endemic to the South Peruvian Andean subcentre of endemism (Cracraft 1985), inhabiting the east slope of the Andes of Peru south from the Cosñipata Valley of Cuzco and Puno (Walker *et al.* 2006), through the Andes of Santa Cruz, Cochabamba, Chuquisaca and Tarija in Bolivia (Bond & Meyer de Schauensee 1942, Remsen & Traylor 1989, Fjeldså & Maijer 1996, Schulenberg & Awbry

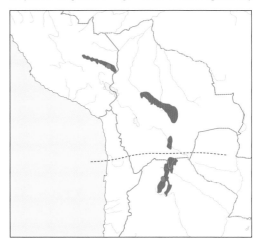

1997), to the Yungas of north-west Argentina in the upper Bermejo river basin of Jujuy and Salta (Narosky & Yzurieta 1993, Mazar Barnett & Pearman 2001, Malizia *et al.* 2005, Camperi *et al.* 2013).

HABITAT Occupies the understorey of humid montane forest (Parker *et al.* 1982, Parker & Rowlett 1984, Hennessey *et al.* 2003). In Peru, it occurs in tall humid forest (Schulenberg & Kirwan 2012c), but at some sites in Bolivia the species frequents shrubby areas, bamboo patches, second growth and scrub habitat (Fjeldså & Maijer 1996, Macleod *et al.* 2005). In Bolivia it is also found in semi-deciduous forest and alder (*Alnus*) woodland (Schulenberg & Aubry 1997, Krabbe & Schulenberg 2003), and in the Argentine portion of its range the species occupies humid temperate forests dominated by *Juglans* and *Cedrela*, with a rather shrubby understorey (Olrog & Contino 1970). Surveys of the drier inter-Andean valleys in Cochabamba failed to find this species, although it occurs at similar elevations on more humid slopes (Herzog *et al.* 1999). Across its entire range, White-throated Antpitta occurs at 600–2,700m (de la Peña 1988, Sibley & Monroe 1990, Fraga & Clark 1999, Hennessey *et al.* 2003, Ridgely & Tudor 2009), but seems to be most frequently recorded at 1,200–2,100m (Maijer *et al.* 2000, Walker *et al.* 2006, Schulenberg *et al.* 2007, Jankowski 2010). The species seems to have a greater altitudinal range, at least in some areas, near the southern end of its range (de la Peña 1988, Fjeldså & Maijer 1996), perhaps due to the lack of a higher elevation congener. The upper altitudinal limit of 3,200m cited by BirdLife International (2012) appears erroneous. Although the precise habitat requirements are unknown, competition with congeners may be a factor in some regions, as it is replaced by *G. erythroleuca* at higher elevations in the north of its range (Clements & Shany 2001) and by Rufous-faced Antpitta in parts of its southern range (Ridgely & Tudor 1994).

VOICE The two-note song is composed of mellow or hollow whistles, the second note slightly higher. It is described variously as: *hee-KEE* (D.F. Lane in Schulenberg *et al.* 2007), *Juú-juehh* (de la Peña 1988) or *hu-hooo* (Ridgely & Tudor 2009). Krabbe & Schulenberg (2003) provided a detailed description: the song lasts 1.0–1.2s, with the notes delivered at 1.1–1.3kHz and repeated at intervals of 8–14s for several minutes at a time. Overall, the song is similar to the two-noted song of Rufous-breasted Antthrush *Formicarius rufipectus*, but the two notes of White-throated Antpitta are usually considered to be on markedly different pitches (Schulenberg *et al.* 2007, Schulenberg & Kirwan 2012c). De la Peña & Rumboll (1998), however, described the song as an 'evenly-pitched *hooop hooop*' delivered slowly and deliberately. The call note is a single, short, descending whistle, *clew*, similar in pitch to the song, and repeated at 1–3s intervals, usually in series of 4–11 notes (Krabbe & Schulenberg 2003, Schulenberg *et al.* 2007). Ridgely & Tudor (2009) also reported a short trill, given in excitement.

NATURAL HISTORY Very few data have been published on the general behaviour of White-throated Antpitta. They forage on or near the ground, usually as alone, advancing in short hops while peering, head cocked to one side, at the leaf litter (Canevari *et al.* 1991, Krabbe & Schulenberg 2003). Such behaviour is typical of most *Grallaria*, and further details are certainly desired.

This species seems less likely to forage in open areas, preferring to remain in dense vegetation (Ridgely & Tudor 1994, Schulenberg *et al.* 2007).

DIET Almost no published data, with the contents of two stomachs including caterpillars (4.5cm long, one each in both stomachs), spiders, roaches and other unidentified insects (Krabbe & Schulenberg 2003, Schulenberg & Kirwan 2012c). Although earthworms are not known in their diet, the widespread consumption of earthworms within the genus, combined with the species' largely terrestrial habits, predicts that earthworms will be important prey.

REPRODUCTION Completely unknown. **SEASONALITY** R. Ramírez photographed a subadult *cinereiventris* on 2 April 2013, at Arroyo Negrito in PN Calilegua, Jujuy, Argentina (EcoReg 21925), and a subadult *albigula* collected 11 November 1980 at Abra de Maruncunca (LSUMZ 98366). With these records being the sum total of our knowledge of breeding for the species, inferences about seasonality are premature for now.

TECHNICAL DESCRIPTION Sexes similar. *Adult* Crown, nape and ear-coverts bright chestnut-rufous, deeper or darker on ear-coverts and forecrown, and becoming brighter posteriorly, especially on nape, where it has a somewhat olivaceous tinge as it grades into the coloration of the upper back. Orbital area whitish, forming a distinct ring, broadest postocularly. Lores and anterior malar greyish, dusky-white or sometimes white, slightly peppered black. The upperparts are olivaceous-brown, except the uppertail-coverts and tail, which are more reddish-brown. Wings are similar to tail and have the primaries edged paler tawny-brown on the leading edge, especially basally. The wing-coverts are brownish-olive, similar to back. The throat to the lower posterior edge of the ear-coverts is snowy white, contrasting with the pale grey to mid-grey breast and belly. The central breast and especially the belly is whiter, becoming darker grey laterally, and washed olive on the flanks. The undertail-coverts are pale grey washed olive-brown (Chapman 1923c, Krabbe & Schulenberg 2003). **Adult Bare Parts** *Iris* brown; *Orbital Skin* whitish to bluish-white; *Bill* maxilla blackish-grey, mandible blue-grey; *Tarsi & Toes* blue-grey (from Schulenberg & Kirwan 2012c). *Subadult* In his description of the type, Chapman (1923) stated that 'immature males' are similar to adult males but have the underparts washed brownish and the flanks and sides tinged with the same coloration as the back. He did not provide his basis, but presumably this description refers to juveniles nearly or wholly independent of their parents. Here, based on a photo taken by R. Ramírez and photos of LSUMZ 98366 (M. Brady), I describe the plumage of subadults in the very last stages of transition to adult plumage. The crown, nape and ear-coverts are bright chestnut-rufous as in adults, but the forecrown has dusky feathers mixed in, giving it a 'dishevelled' appearance. The bare skin around the eye is bluish-white as in adults and the lores and anterior malar region (base of bill and below eye) are dusky-white or blackish (overall darker and not as 'clean'-looking as in adults). The back appears to be the same olivaceous-brown as adult plumage, but I cannot describe the lower back, rump or uppertail-coverts. The underside of the tail and wings are dark, slightly reddish-brown and, as in adults, the leading edges of the primaries are paler tawny-brown.

The upperwing-coverts are slightly darker (duskier) than the flight feathers. They bear a fairly broad, darker blackish subterminal band and are thinly tipped buffy-chestnut. The throat to the lower posterior edge of the ear-coverts is white, irregularly washed with small areas of grey, unlike the clean, bright throat of adults. The contrasting line between the throat and underparts of adults is somewhat blurred. The remaining underparts appear similar to adults but the olive wash on the flanks is more extensive and there are scattered patches of olive or rusty-olive feathers on the belly and breast. The undertail-coverts were not clearly visible, but appeared darker and cleaner grey than adults. **Subadult Bare Parts** *Iris* dark; *Orbital Skin* pale bluish-white like adults; *Bill* maxilla blackish-grey, slightly paler distally and on the tomia, with the latter becoming more yellow-orange basally onto the only minimally inflated rictus, mandible blue-grey with the tomia of the basal third and the rictus as on the maxilla; *Tarsi & Toes* blue-grey as adults.

MORPHOMETRIC DATA Very few published data. Measurements refer to nominate subspecies unless indicated. *Wing* 106mm, 107mm, 101.5mm, 103mm ($n = 4$, 2♂♂ from Santa Cruz, 2♂♂ from Puno, Bond & Meyer de Schauensee 1942). *Wing* 99.5mm, 100.5mm ($n = 2$, ♀♀, Puno, Bond 1950). *Bill* from nares 15.2mm, 15.1mm; *Tarsus* 51.3mm, 53.2mm ($n = 2$, ♀,♂, AMNH 149898/99). *Wing* 109mm; *Tail* '86mm' [= 65mm?]; *Bill* exposed culmen 22mm; *Tarsus* 60mm ($n = 1$, ♀ holotype, *cinereiventris*, Olrog & Contino 1970). *Wing* 101mm; *Tail* 65mm; *Bill* [total?] culmen 28mm; *Tarsus* 55mm ($n = 1$, ♂ holotype, *albigula*, Chapman 1923c). **MASS** 95g ($n = 1$, ♀, *cinereiventris*, Jujuy, Salvador 1988, Dunning 2008). 88g(lt) ($n = 1$, ♂, *cinereiventris*, Jujuy, CUMV 52535); 84g(m), 92g(lt), 86g, 87g(lt), 90g(lt) ($n = 5$, 3♂♂, 2♀♀, LSUMZ, previously reported by Krabbe & Schulenberg 2003); 86.5g(lt) ($n = 1$, ♂ subadult, ♂, LSUMZ 98366); 116g(h) ($n = 1$, ♀, FMNH 398033). **TOTAL LENGTH** 18.5–20.3cm (Meyer de Schauensee 1970, de la Peña 1988, Narosky & Yzurieta 1989, Ridgely & Tudor 1994, de la Peña & Rumboll 1998, Krabbe & Schulenberg 2003).

TAXONOMY AND VARIATION The few molecular analyses of *Grallaria* did not include White-throated Antpitta (Krabbe *et al.* 1999, Rice 2005). Although Zimmer (1934) suggested that White-throated Antpitta is related to Chestnut-naped Antpitta, this seems rather unlikely given the strong differences in plumage, voice and morphology (Schulenberg & Williams 1982, Schulenberg & Kirwan 2012c). A seemingly more likely scenario is that White-throated Antpitta belongs to the superspecies formed by Red-and-white Antpitta, Bay Antpitta, Rusty-tinged Antpitta and White-bellied Antpitta (Lowery & O'Neill 1969, Krabbe & Schulenberg 2003). This relationship has been suggested by other authors (Chapman 1923c, Schulenberg & Williams 1982, Ridgely & Tudor 1994) and White-throated Antpitta is generally placed accordingly in linear arrangements (Clements 2007, Remsen *et al.* 2015). Two subspecies currently recognised (Dickinson 2003, Schulenberg & Kirwan 2012c) but the southern form (*cinereiventris*) is very poorly differentiated and White-throated Antpitta is perhaps best considered monotypic.

Grallaria albigula albigula Chapman, 1923. The nominate race occurs in south-east Peru and Bolivia. The population in Chuquisaca is separated by a gap in records of *c.*120km from the northernmost *cinereiventris* (see that subspecies).

A similar gap in records (*c.*160km) separates the Chuquisaca population from the southernmost records in Santa Cruz. I was unable to uncover sufficient evidence to conclusively assign Chuquisaca records to either race, but they are treated here as nominate *albigula*. In any case, these gaps, as well as the 330km gap between Calle and Piedras Blancas are likely artefacts of incomplete sampling, leaving no logical break between the two subspecies and further arguing against their separation (see below). **SPECIMENS & RECORDS PERU**: *Cuzco* Cock-of-the-Rock Lodge, 13°03.5'S, 71°32.5'W (XC 14537, R. Ahlman); Pensión Suecia, 13°06'S, 71°34'W (FMNH 398033/034); Cadena, 13°24'S, 70°43'W (FMNH 190087); Capiri, 13°25'S, 70°55'W (XC 22815, D. Geale). *Puno* Above San Juan del Oro, 14°14'S, 69°11.5'W (KUNHM); Abra de Maruncunca, 14°14'S, 69°17'W (Robbins *et al.* 2013, LSUMZ, ML; D.J. Lebbin). **BOLIVIA**: *La Paz* 2.75km S of Calle, 14°37'S 68°57'W (Hennessey & Gomez 2003, ML); Piedras Blancas, *c.*40km ENE of Inquisivi, 16°45.5'S, 66°47.5'W (MHNNKM 5268). *Cochabamba* Pampa Grande, 16°41'S, 66°29'W (S.K. Herzog *in litt.* 2017); Chapare road, *c.*17°24'S, 65°60'W (XC 223577–579, P. Boesman); 52km W of Villa Tunari, 17°09'S, 65°42'W (ML: T.A. Parker). *Santa Cruz* Las Quinas, 18°09'S, 63°52'W (UMMZ 222955); Samaipata, 18°11'S, 63°53'W (Bond & Meyer de Schauensee 1942, ANSP 140804); Abra Tabla, 18°36'S, 64°03'W (Maijer *et al.* 2000); La Yunga, 18°47'S, 63°52'W (XC 1852, S. Maijer). *Chuquisaca* Cerro Punto Lajas, 20°12'S 64°18'W (Fjeldså & Maijer 1996); Arcos, 20°15'S, 64°17'W (Fjeldså & Maijer 1996); 1.75km W of Lagunillas, 20°21'S, 64°15'W (Fjeldså & Maijer 1996); Cerro Campanarios, 20°47.5'S, 64°28'W (XC 1554, S. Maijer); Cerro Bufete, 20°50'S, 64°20'W (Fjeldså & Maijer 1996, Schulenberg & Awbry 1997); Campamiento, 20°54'S 64°31'W (Fjeldså & Maijer 1996).

Grallaria albigula cinereiventris Olrog & Contino, 1970, Neotrópica, vol. 47, p. 51, El Jordán, Valle Grande, cerro Calilegua, Jujuy, Argentina, 1,700m [*c.*23°29'S, 64°57'W from Camperi *et al.* 2013]. The holotype (FIML 13388) is an adult female collected 8 September 1968 by Francisco Contino. The authors mentioned a second adult female from the same locality, but did not provide its collection number. Apparently it was not assigned its own number at the time of description, but remains in the collection with a paratype label and the number FIML 13388b (S.B. Bertelli *in litt.* 2017). This subspecies is usually considered endemic to north-west Argentina in south-east Jujuy and northern Salta (Olrog 1979, de la Peña 1988). It is known from just a handful of specimens and its validity has been questioned in the past (Krabbe & Schulenberg 2003), an opinion that I share. Subspecies *cinereiventris* apparently differs from the nominate race in having the white underparts greyer, and having the flanks sepia-olive rather than grey (Olrog & Contino 1970). While a careful examination of additional specimens from Argentina is needed to ascertain the validity of *cinereiventris*, my description of what I consider a subadult (see Technical Description) is not dissimilar from that of *cinereiventris*. I provisionally include it until additional evidence comes to light and further include records just north of the Bolivian border (i.e. Pampa Grande) as pertaining to *cinereiventris*. Many of the following distribution points were extracted from Di Giacomo *et al.* (2007), and are somewhat general. The GPS provided here were adjusted slightly in some cases to account for a lower elevational limit of 600–700m. Specimens & Records BOLIVIA: *Tarija* Pampa Grande, 22°02'S, 64°26'W (XC 1704, S. Maijer). ARGENTINA: *Salta* Itiyuro-Tuyunti, 22°06'S, 63°40'W (Di Giacomo *et al.* 2007); Tartagal, 22°12'S, 63°57'W (XC 50497–498, B. Lopéz-Lanús); vicinity of Río Lipeo, *c.*22°26'S, 64°32'W (FIML 13483); PN Baritú, *c.*22°35'S, 64°37.5'W (XC 50499, B. Lopéz-Lanús); PP Laguna Pintascayoc, 22°45'S, 64°22'W (Di Giacomo *et al.* 2007); La Porcelana, 22°55'S, 64°12'W (Di Giacomo *et al.* 2007); Abra Grande, 23°00'S, 64°28.5'W (Di Giacomo *et al.* 2007); Fincas Santiago y San Andrés, 23°00'S, 64°53'W (Di Giacomo *et al.* 2007); Río Santa María, 23°17.5'S, 64°34.5'W (Di Giacomo *et al.* 2007); Valle Morado, 23°27'S, 64°36'W (Di Giacomo *et al.* 2007); PN El Rey, 24°43'S, 64°40'W (Di Giacomo *et al.* 2007). *Jujuy* Alto Calilegua, 23°23'S, 64°55'W (Di Giacomo *et al.* 2007); Valle Grande, *c.*23°29'S, 64°56'W (Salvador 1988); Vinalito, 23°40'S, 64°28'W (Di Giacomo *et al.* 2007, Camperi *et al.* 2013); Abra de Cañas, 23°40'S, 64°56'W (Camperi *et al.* 2013); 2km W of El Monolito, 23°40.5'S, 64°54'W (XC 345952–54, R. Dunn); below Aguada del Tigre, 23°42'S, 64°53'W (XC 139454–458; B. Lopéz-Lanús); Tres Cruces, 23°43.5'S, 64°51.5'W (XC 139332, B. Lopéz-Lanús); Arroyo Negrito, 23°41'S, 64°54'W (EcoReg 21925, R. Ramírez); Abra Colorada, 23°47'S, 65°01'W (Camperi *et al.* 2013, FIML 13358); 2km E of Ocloyas, 23°56.5'S, 65°13'W (CUMV 52535, KUNHM); Tiraxi, 23°59'S, 65°16'W (Di Giacomo *et al.* 2007, Camperi *et al.* 2013); Las Capillas, 24°03'S, 65°16'W (Di Giacomo *et al.* 2007, Camperi *et al.* 2013); Yala, 24°07'S, 65°23'W (Di Giacomo *et al.* 2007, Camperi *et al.* 2013); Lagunas San Miguel y El Sauce, 24°10'S, 64°36'W (Di Giacomo *et al.* 2007); Sierra de Zapala, 24°11'S, 65°04'W (Camperi *et al.* 2013); RNP Las Lancitas, 24°12'S, 64°22'W (Moschione & Segovia 2005, Di Giacomo *et al.* 2007); El Fuerte/ Santa Clara, 24°16'S, 64°25'W (Di Giacomo *et al.* 2007); La Cornisa, 24°25'S, 65°20'W (Di Giacomo *et al.* 2007).

STATUS Despite its geographic range being fairly small (Stattersfield *et al.* 1998), population trends are believed to be stable, and the species is currently considered Least Concern (BirdLife International 2015). Although known from relatively few localities and generally considered uncommon (Stotz *et al.* 1996, Krabbe & Schulenberg 2003) across its range, new records are continually being reported and the species can be fairly common (Clements & Shany 2001, Schulenberg *et al.* 2007). Like most Neotropical species, it would seem that the local effects of habitat destruction are of major concern. However, it inhabits a fairly broad spectrum of habitat types and has been regularly found in areas of moderate or high disturbance (Fjeldså & Maijer 1996, Macleod *et al.* 2005). Many known sites are currently afforded some degree of protection. Protected Populations *G. a. albigula* Peru: PN Manú (Merkord 2010). Bolivia: PN Amboró (Gemuseus & Sagot 1996, Hennessey *et al.* 2003b); PN Madidi (Hennessey & Gómez 2003); PN Carrasco (Hennessey *et al.* 2003b, XC). *G. a. cinereiventris* Bolivia: RNFF Tariquía (Hennessey *et al.* 2003b). Argentina: PN El Rey (Di Giacomo *et al.* 2007); PN Baritú (XC 50499, Bernabe Lopéz-Lanús); PN Calilegua (Narosky & Yzurieta 1989, Di Giacomo *et al.* 2007); PP Laguna Pintascayoc (Di Giacomo *et al.* 2007); RNP Las Lancitas (Moschione *et al.* 2005, Di Giacomo *et al.* 2007); Reserva de la Biosfera Serranía de Zapla (eBird: F. Cornell).

OTHER NAMES Spanish Tororoí Gorgiblanco (Krabbe & Schulenberg 2003); Chululú Pintado (Fraga & Narosky 1985); Chululu de Garganta Blanca (de la Peña 1988, de la Peña & Rumboll 1998); Chululú Mediano (Kempff-Mercado 1985, Canevari *et al.* 1991); Chululú Cabeza Rojiza (Narosky & Yzurieta 1989, Mazar Barnett & Pearman 2001); Tororoí gargantiblanco (Olgro 1979). French Grallaire à gorge blanche. German Grauflanken-Ameisenpitta (Krabbe & Schulenberg 2003).

White-throated Antpitta, adult, Manú Road, Cuzco, Peru, 03 November 2009 (*Kristian Svensson*).

White-throated Antpitta, subadult, Arroyo Negrito, PN Calilegua, Jujuy, Argentina, 2 April 2013 (*Ramíro Ramírez*).

White-throated Antpitta, adult, El Monolito, Jujuy, Argentina, 1 November 2015 (*Andrés Terán*).

White-throated Antpitta, adult, PN Calilegua, Jujuy, Argentina, 19 October 2009 (*Fabrice Schmitt/WINGS Birding Tours Worldwide*).

WHITE-BELLIED ANTPITTA
Grallaria hypoleuca   Plate 14

Grallaria hypoleuca P. L. Sclater, 1855, Proceedings of the Zoological Society of London, vol. 23, p. 88, 'Santa Fe de Bogotá', Colombia. Olivares (1969) suggested that the type locality is probably near Fusagasugá, Cundinamarca (04°21'N, 74°22'W). The exact location is unknown, as the type is one of the many trade skins collected and labelled with little thought for scientific accuracy. Sclater (1855b) stated that, at the time he noticed the specimen in Muséum national d'Histoire naturelle in Paris, the type was labelled as having been received in 1843 by M. Rieffer.

For more than a decade I lived with the present species singing from my 'front yard' at 2,050m on the east slope of the Andes in Ecuador's Napo province, just 20km from the type locality of subspecies *castanea* (Chapman 1923c). White-bellied Antpitta has long been considered closely related to Red-and-white, Bay, Rusty-tinged and Yellow-breasted Antpittas (Sibley & Monroe 1990, Ridgely & Tudor 1994, 2009, Krabbe & Schulenberg 2003, Schulenberg *et al.* 2013b). All of the aforementioned were sometimes placed into a single polytypic species, Bay-backed Antpitta *G. hypoleuca* (Meyer de Schauensee 1970) or, alternatively, maintained as separate species

with the exception of lumping Yellow-breasted and White-bellied (Peters 1951, Meyer de Schauensee 1966, Fjeldså & Krabbe 1990). Parker & O'Neill (1980), pointing to marked differences in songs within this 'Bay-backed Antpitta complex', proposed that all of these taxa be maintained as species, except *castanea*, which they maintained as a subspecies of White-bellied Antpitta. This suggestion has generally been followed by most subsequent authors (Krabbe & Schulenberg 2003, Ridgely & Tudor 2009). However, recent molecular analyses have shed new light on the 'traditional' view of these five species (Winger *et al.* 2015). Most notably, this study found Yellow-breasted Antpitta to be more closely related to Brown-banded Antpitta than to any of the 'bay-backed' species (see Brown-banded Antpitta for further discussion). Somewhat less surprising was the molecular support for adding Rufous-faced Antpitta to the complex. This species was recovered as sister to Red-and-white Antpitta, together with which there was weak support for a close relationship to White-bellied Antpitta (as here defined). Despite its relatively wide distribution and propensity to be fairly common in most of its range (Restall *et al.* 2006, Ridgely & Tudor 2009, Freile *et al.* 2010), little is known about the natural history of the present species. Its diet is all but unknown, only one nest has been described, and even published morphometric data are few. Its broad distribution and apparent tolerance of some disturbance may (to some degree) cushion White-bellied Antpitta from some of the immediate effects of habitat loss and alteration (Schulenberg *et al.* 2013b), and the general consensus suggests that White-bellied Antpitta 'may even benefit, locally, from human activities, such as a low level of clearing for subsistence agriculture' (Schulenberg *et al.* 2013b). As discussed below, however, I suggest that a careful evaluation of habitat and microhabitat use will be of great importance for making sound conservation decisions. This account describes a second nest and improves our knowledge of nestling appearance and parental care. In addition, it provides the first description of the plumages of fledglings, juveniles and subadults.

IDENTIFICATION 16–17cm. The relatively simple-plumaged White-bellied Antpitta is a medium-sized *Grallaria*. The nominate subspecies is brownish above, with the nape and neck-sides and face rufous-chestnut. The lores are dark grey and the underparts are off-white, cleaner white on the throat and central belly, washed grey across the breast and on the belly-sides. The flanks and breast-sides are a similar rufous-chestnut to the face. The southern race, *castanea*, is quite similar, but smaller and much more chestnut-brown above. Most similar in plumage, Yellow-breasted Antpitta has no range overlap with the present species, but can be separated by the yellow wash to the underparts. Within its range, there are really no candidates for confusion. Bicoloured and Chestnut-naped Antpittas are larger, considerably darker grey below, and generally occur at elevations above that of the present species.

DISTRIBUTION Found in the Central and East Andes of Colombia (Hilty & Brown 1986), and on the Amazonian slope in Ecuador (Ridgely & Greenfield 2001) and northern Peru (Schulenberg *et al.* 2007). Records of White-bellied Antpitta from western Ecuador (Pichincha Volcano) (Goodfellow 1902) may reflect errors in labelling (Freile 2000) or, quite possibly, refer to the similarly plumaged Yellow-breasted Antpitta. There is little

doubt some records do indeed fall into one of these two categories. However, Orton (1871) reported White-bellied Antpitta from the 'Valley of Quito', an area he made a point of clearly defining as Ecuador's inter-Andean valley from 01°N to 04°S. This work was published well before the description of Yellow-breasted Antpitta (Sclater 1877), making it possible he was referring to the latter, but failed to comment on its yellowish underparts.

HABITAT Inhabits the interior and borders of humid montane forests and secondary forests in advanced stages of regeneration (Hilty & Brown 1986, Ridgely & Greenfield 2001, Schulenberg *et al.* 2007). Many authors have suggested that the species prefers a dense understorey of tangled vegetation or bamboo (Olivares 1969, Freile 2000, Schulenberg *et al.* 2013b) or areas of naturally disturbed habitat such as treefalls and landslides (Goodfellow 1902, Ridgely & Tudor 1994, McMullan *et al.* 2010). Although my experience with this species is entirely limited to north-east Ecuador, I tend to disagree with these stated microhabitat preferences. In its steep, montane haunts, frequent landslides, flooding rivers and unpredictable treefalls are natural events that generate the spatially complex floral heterogeneity characteristic of these forests. Indeed, any organism in these forests is never likely to be far from a freshly disturbed area or one being recolonised by bamboo or other dense tangled growth. Both known nests, and the suspected location of a third, were in mature, high-canopy forest. Despite having spent a good amount of time crashing through bamboo and other tangled undergrowth, and having been followed there by the largely sympatric Chestnut-crowned Antpitta (see Natural History), I have never met White-bellied Antpitta in such habitat. Instead, all instances that I have interpreted as cases of White-bellied Antpitta being attracted to the noise of my passage and hoping to forage in my wake (Greeney 2012b) have been in relatively open-understorey, high-canopy, mossy forest. Like other observers, however, most of the individuals I locate are detected by voice, and are vocalising from dense understorey vegetation. Such locations are the logical choice for individuals seeking to announce their presence while remaining relatively safe from predators. My feeling is that mature forest is equally preferred habitat, if not more so, of the species for both foraging and breeding. I propose that it is a 'mosaic specialist', i.e. it would suffer from either the lack of *or* the predominance of regenerating habitat. There is little doubt that it can persist in heavily human-impacted areas such as by slash-and-burn agriculture (Fjeldså & Krabbe 1990), but we know little of how long the species remains in such areas, nor if it thrives there. In the region where I know this antpitta best, there are (thankfully) still large tracts of little or virtually unaltered habitat. Here, White-bellied Antpitta is rarely or never found in deforested or agricultural areas. I believe that careful, quantified studies of microhabitat use by White-bellied Antpitta will reveal interesting and complex patterns unlike those described in most works. This conservation-relevant point was also discussed by Freile *et al.* (2010), who noted that, while White-bellied Antpitta may be linked, to some degree, to 'naturally disturbed areas, especially dense secondary growth and bamboo stands, [they] are still reliant on 'good' forest coverage'. On a related note, Goodfellow (1902) suggested a preference for narrow montane riparian areas, but I have not found it to be especially tied to streams, at least in eastern Ecuador. Across its entire range, reported elevations for White-bellied Antpitta range from 1,400m to 2,700m. The highest reported localities are from its northernmost localities in Serranía de los Yariguíes, Santander, on the west slope of the East Andes, where it occurs at 1,500–2,700m (Donegan & Huertas 2005, Donegan *et al.* 2007). In Serranía de los Churumbelos, on the east slope of the East Andes in Huila, White-bellied Antpitta is found at 1,450–2,500m (Donegan & Salaman 1999). Most reported altitudinal ranges indicate a preference for elevations of 1,400–2,300m (Fjeldså & Krabbe 1990). Other given ranges include: 1,400–2,450m in Colombia (Hilty & Brown 1986, Salaman *et al.* 2002), 1,400–2,200m in Ecuador (Ridgely & Greenfield 2001) and 1,700–2,100m in Peru (Schulenberg *et al.* 2007).

VOICE The song is simple and fairly easily imitated. It consists of three clear whistles, the first longest and a half-tone lower than the following two, which are delivered after a slight pause and are usually nearly monotone. Following Krabbe & Schulenberg (2003), the song lasts 1.1–1.4s and is delivered at intervals of at least 7–15s. The first note is weakest, given at 2.0–2.2kHz and the final two are higher pitched, at 2.1–2.3kHz. It has been variously described as: *puuuh pü pü* (Hilty & Brown 1986), 'a fast *too, téw-téw*' (Ridgely & Greenfield 2001) and *hu HEW-HEW* (D.F. Lane in Schulenberg *et al.* 2007). Described calls include a series of soft *whee-whee-whee-whee* notes, delivered on one pitch and suggested to possibly be an alarm call (Parker *et al.* 1985, Fjeldså & Krabbe 1990), 'a strikingly pygmy-owl [*Glaucidium*]-like vocalization, *too, too, too, too* ...' (Ridgely & Greenfield 2001) and 'a series of rising whistled *clew'ee*? notes' (D.F. Lane in Schulenberg *et al.* 2007). There is apparently little documented variation between populations. The song of nominate *hypoleuca* near the northern extremity of the species' range was described by Donegan *et al.* (2007) as being nearly identical to that of *castanea* in Ecuador, 'a series of three slow whistles, the first lower (*c.*1.8–2.1kHz) than the others (*c.*1.9–2.2kHz), with a slightly longer interval between the first and second notes than the second and third'.

NATURAL HISTORY Essentially, the species' behaviour is undocumented, although it is generally credited as being similar to 'better-studied' members of the genus. Numerous authors have suggested that the species forages on or near the ground while remaining concealed in dense understorey vegetation, moving by running or hopping (Hilty & Brown 1986, Ridgely & Greenfield 2001, Krabbe & Schulenberg 2003, Schulenberg *et al.* 2007). I have observed White-bellied Antpittas foraging at raiding swarms of *Labidus* army ants, and have been followed by them on several occasions (Greeney 2012b). They are also among those *Grallaria* known to readily 'acclimatise' to worm-feeding stations (Woods *et al.* 2011). The importance or frequency of such commensal foraging relationships is unknown. Krabbe & Schulenberg (2003) reported that song perches are generally <2m above ground, with which my own data agree. On all but two of my 17 observations of singing adults, they were perched on horizontal branches less than 0.5m above ground. Once I found an adult singing from the ground and in the final instance the adult was 2.5m above ground in a dense tangle of vines and epiphytes against a tree trunk.

DIET Very few details of the diet are available. The contents of one stomach were described as 'insects' (Schulenberg *et al.* 2013b), data probably derived from the LSUMZ collection where the contents of four stomachs

are labelled the same. A photo that I examined, taken by L. Bushman in north-east Ecuador, shows an adult carrying a caterpillar of the (presumably) chemically defended butterfly, *Altinote dicaeus* (Nymphalidae). I have observed few other species of birds feeding on this caterpillar, despite its abundance in eastern Ecuador. Interestingly, most of these records are of immatures, but the significance of this is unknown.

REPRODUCTION The first and only previously described nest of White-bellied Antpitta (race *castanea*) was found in north-east Ecuador at Cabañas San Isidro (though not stated in the publication) (Price 2003). **NEST & BUILDING** The nest described by Price (2003) was 1.25m up, wedged in the V created by a split in a trunk. Externally, it was composed mainly of twigs and dead leaf matter (little moss), sparsely lined with flexible rootlets and dark fibres, and it was *c.*14.2cm in diameter (height not measured). Internally, the cup was 10.7cm wide by 6.4cm deep. **EGG, LAYING & INCUBATION** The only known clutch (Price 2003) was of two eggs. Price (2003) described them as 'light green with small light-brownish blotches fairly evenly distributed over the shell [but which] may simply be stains from the nest'. A fully shelled egg extracted from a female collected in Piura (*castanea*) was light blue and unmarked (Parker *et al.* 1985). Published egg weights: 9.17g, 7.85g (Price 2003), 8.3g (Parker *et al.* 1985). Published egg dimensions: 32 × 24mm, 30 × 23mm (Price 2003), 29.8 × 24.5mm (Schönwetter 1979). The eggs found by Price (2003) hatched 17 days after discovery, which is probably close to the full incubation period. At this nest, both adults participated in incubation with periods of attendance lasting 23–124min. Mean on-bout duration (± SD) from 07:00–13:30, 11 days prior to hatching, was 65 ± 26min (*n* = 5 bouts). At the same time of day, five days prior to hatching, mean on-bout duration was 99 ± 20min (*n* = 4 bouts). The eggs were incubated 88% of the 13 hours of observation (both days combined). It appeared that attendance was low during the early morning, but after 09:15 the eggs were covered almost continuously until at least 13:30 on both days. Exchanges between adults at the nest were brief and unceremonious, with the incubating adult seeming to anticipate the arrival of its mate. This observation suggests that there may be a type of soft vocalisation used between adults at the nest (but was not detected on video recordings). Of the seven changeovers observed by Price (2003), the arriving and departing adults traded places simultaneously twice, were present together at the nest for 1s on two additional occasions, and remained together for *c.*22s on three occasions. During these latter exchanges, the incubating adult remained covering the eggs while the second perched on the rim. After the incubating adult had departed, the new arrival generally remained on the rim of the nest for 19 ± 8s before settling over the eggs. On three of five well-observed arrivals, both adults brought a single thin fibre (lining material) which was dropped into the cup before starting to incubate. No mention was made of additional nest maintenance or sanitary behaviours common in other antpittas (Greeney *et al.* 2008). **NESTLING & PARENTAL CARE** The only previous, albeit cursory, description of *G. hypoleuca* nestlings is that of Price (2003) who described newly hatched young as 'covered with grey down [with] bright orange bills and gapes'. The duration of the nestling period is unknown, but is probably close to the 17-day period of often sympatric *G. ruficapilla* (Martin & Greeney

2006). **Seasonality** Apart from adults with developing gonads collected March–September in the Central (*hypoleuca*) and East (race?) Andes of Colombia (Hilty & Brown 1986), and one in June (*hypoleuca*) in Antioquia (Fjeldså & Krabbe 1990), clues as to the breeding season are scarce. **Breeding Data** *G. h. hypoleuca* Testes greatly enlarged, 19 June 1948, above Sevilla (FMNH 299492 = USNM 402473). *G. h. castanea* Female with shelled egg in oviduct, 7 July 1978, Playón (Parker *et al*. 1985, LSUMZ 88047); incubation, 19 May 2001, Cabañas San Isidro (Price 2003); nestling, Yanayacu Biological Station, 6 December 2011 (HFG); two records of fledglings, Cabañas San Isidro, 14 November 2013 (L. Bushman photos) and 15 July 2013 (G. Golumbeski photos); fledgling, Estación Biológica Yanayacu, 10 November 2008 (J. Simbaña *in litt*. 2008); juvenile, 13 February 2009, Cabañas San Isidro (R. Ahlman *in litt*. 2017); juvenile, Estación Biológica Yanayacu, 2 July 2003 (R.C. Dobbs *in litt*. 2012); subadult, Finca Roberto Aldaz, 7 September 2012 (HFG); subadult, NNE of San José de Lourdes, 20 August 1998 (LSUMZ 165301); subadult, Picorana, 28 March 1999 (LSUMZ 172181); subadult, 'Río Tigre' [= eastern Ecuador], 5 February 1936 (MLZ 17748). In addition, photos posted on Flickr provide four records of adults carrying worms away from feeders at Cabañas San Isidro (feeding nestlings or young out of the nest), all of which I feel represent separate breeding events. These are: 1 January 2010 (H. Mueller); 18 October 2010 (L. Kay); 5 November 2014 (H. Davis); 21 November 2015 (S. Riall).

TECHNICAL DESCRIPTION Sexes similar. *Adult* Crown dull rufescent-brown or greyish-brown, sometimes with inconspicuous black streaking on the forecrown. Remaining upperparts brown to reddish-brown, sometimes slightly brighter and paler across the nape. The head- and neck-sides are more rufous (less brown). The lores are somewhat variable, ranging from almost white, through various shades of grey, to almost black. The throat and central belly are white, while the breast and flanks are smoky grey, central belly white. The breast-sides are rufous-brown. Flanks, thighs and underwing-coverts are olive-brown or tawny-brown, and the undertail-coverts tawny (Sclater 1855b). **Adult Bare Parts** *Iris* brown; *Bill* black or almost so; *Tarsi & Toes* blue-grey. The immature plumages were previously undescribed (Schulenberg *et al.* 2013b). Via examination of photos of several individuals, I am able to give a reasonably complete description of the immature plumages of *G. h. castanea* at different post-fledge ages (see Reproduction Seasonality). *Subadult* As the juvenile (described below) was visiting the feeding station at Cabañas San Isidro, I was able to examine photos of what was (presumably) the same individual taken by L. Jönsson one month later. Its plumage was now mostly as in adults, with the forecrown still somewhat 'messy' and a few small patches of juvenile barred feathering on the nape. The bright chestnut tracts of adults were, on the whole, somewhat duller, especially on the face, head and upper back. The upperwing-coverts were still broadly barred dusky and somewhat darker, duller chestnut than the surrounding feathers. The bill was fairly similar to the description below, but now with the orange slightly reduced. The rictus was still slightly inflated. The greyish malar stripes were visible, but very faint, and the white underparts were still somewhat duller than in adults. *Juvenile* I examined photos of a juvenile, still being fed by an adult, photographed by G. Golumbeski. At this age the

juvenile had the forecrown dusky-brown with a peppering of paler feathers (compared to clean grey in adults). The crown had predominantly adult-type feathering, but generally duller and browner (less bright chestnut). The hindcrown, nape and neck-sides were dusky-blackish with fine buffy barring, broken by irregular patches of rufous to rufous-brown feathers similar to adults. The rest of the upperparts were similar to adults but with irregular blackish and buffy barring, especially on the scapulars, lower back and rump. The lores were whitish like adults but browner, less grey. The ocular area was not bare, but bore tiny greyish-brown feathers continuous with the lores. The postocular area was dull brown-chestnut, with the ear-coverts slightly brighter brownish-chestnut with fine pale shaft-streaks. The wings and tail were similar to adults, but the upperwing-coverts were broadly tipped with dusky blackish bars and not noticeably fringed chestnut. The underparts were, overall, similar to those of adults but the sides of the throat had a grey wash giving an indistinct moustached appearance, and the remaining white areas bore a faint greyish tinge, especially on the vent. Additionally, the central breast (and probably vent, though the latter was partly invisible) still bore a small patch of buffy- and black-barred immature plumage. **Juvenile Bare Parts** *Iris* brown; *Bill* blackish, maxilla pale yellowish at very tip, washed orange around nares and basally, where rictus orange and still slightly inflated, mandible with yellow tip and basal quarter orange; *Tarsi & Toes* similar to adults. *Fledgling* J. Simbaña shared photos of a very recent fledgling, unable to fly, probably 1–2 weeks old or less. Overall, its plumage was essentially like that of pre-fledging nestlings and was still covered in fluffy wool-like feathers, all dusky-blackish with indistinct pale spots and fine bars, the markings buffy-chestnut above, buffy on the upper breast and buffy-whitish on the lower breast and flanks. Barring on the lower breast, belly and flanks much broader, and overall these areas appeared more buff than dark, mixed white, buffy and dusky. The central belly and vent were white peppered dusky. The face and ocular area were weakly feathered but appeared dusky-brownish, except ear-coverts, which were washed brighter chestnut-buff and barred blackish, as were the neck-sides and scapulars. The ear-coverts additionally had narrow pale shaft-streaks. Upperwing-coverts warm brown with broad blackish subterminal bands, only narrowly fringed brownish-chestnut. **Fledgling Bare Parts** *Iris* light brown; *Bill* mostly bright orange, maxilla dusky in middle and along culmen, yellower near tip and still bearing an egg-tooth; *Tarsi & Toes* bluish-brown, with nails notably paler than toes and somewhat yellowish.

MORPHOMETRIC DATA *G. hypoleuca hypoleuca* Wing 92–93mm; *Bill* [total?] culmen 23.0–25.5mm; *Tail* 50–54mm; *Tarsus* 45.5–51.0mm (*n* = 3, ♂?, Chapman 1923c). *Wing* 90mm; *Tail* 53mm; *Bill* [total culmen?] 20mm (*n* = 1, ♂, holotype, Menegaux & Hellmayr 1906). *Bill* from nares 14.4mm, 14.5mm, 15.5mm, 15.6mm; *Tarsus* 47.2mm, 48.8mm, 48.6mm, 48.8mm (*n* = 4, ♂♀♂♂, vicinity of Fusagasugá, AMNH). *G. hypoleuca castanea* Wing 88–90mm; *Tail* 47–50mm; *Bill* [total?] culmen 22.5–23.0mm; *Tarsus* 43–45mm (*n* = 4, ♀?, Chapman 1923c). *Bill* from nares 12.6mm, 13.5mm, 14.0mm, 14.4mm, 14.6mm; *Tarsus* 44.2mm, 44.8mm, 45.4mm, 46.9mm, 47.3mm (*n* = 5, ♀,♂,♂,♀,♂, eastern Ecuador and Peru, MLZ, AMNH). *Bill* from nares 14.7mm; *Tarsus* 47.6mm (*n* = 1, ♂, La Candela, Huila, AMNH 116345). *Bill* from nares 13.3mm, exposed culmen 16.9mm, width at gape 11.8mm; *Tarsus* 44.4mm (*n* = 1, ♀, subadult, MLZ 17748). **Mass** 62–69g (*n* = 3, ♂, race?, Krabbe & Schulenberg 2003). 70.7g(lt), 60.5g(no), 64.5g(no), 58.5g(no), 46.0g(no) (n = 5, 4♂♂, 1♀ noted as being especially thin-looking, LSUMZ, M. Brady *in litt.* 2017). 61.9g(tr), 59.0g(no) (*n* = 2, ♂♂ subadults, LSUMZ). **Total Length** 13–18cm (Hilty & Brown 1986, Dunning 1987, Fjeldså & Krabbe 1990, Ridgely & Tudor 1994, Restall *et al.* 2006, McMullan *et al.* 2010).

TAXONOMY AND VARIATION Two subspecies recognised. Birds from the headwaters of the Río Magdalena, on the west slope of Colombia's East Andes in Huila are said to be 'intermediate' (Chapman 1923c, 1926a, Meyer de Schauensee 1952a) between nominate and *castanea* and, although sometimes assigned to the former (Chapman 1923c, Cory & Hellmayr 1924), are usually considered in *castanea* (Meyer de Schauensee 1945, 1950, 1964, Peters 1951, Fjeldså & Krabbe 1990, Krabbe & Schulenberg 2003). I provisionally include this population in *castanea* too. Overall, however, geographic variation in plumage (and vocalisations) is minimal. This lack of variation might suggest that the species is best considered monotypic, but Winger *et al.* (2015) did identify a fair amount of genetic structure in northern populations.

Grallaria hypoleuca hypoleuca P. L. Sclater, 1855 (Sclater 1855b). Nominate *hypoleuca* is endemic to Colombia, being found on both slopes of the Central Andes in Antioquia and on the west slope of the East Andes from Norte de Santander south to Cundinamarca. The northernmost records in the Central Andes of Antioquia are around Valdivia (07°17'N, 75°24'W) and it is apparently still fairly common south of there, near Anorí and Briceño, to Amalfi (06°54'N, 75°04'W). South of there, its range in the Central Andes is very poorly known, but is suggested to possibly extend further south (Schulenberg *et al.* 2013b). Several plausible but unsubstantiated records at RN Río Blanco near Manizales, Caldas (eBird) suggest this to be true. Nevertheless, it remains somewhat inexplicably unreported, if indeed it does occur south of Antioquia in the Central Andes. On the west slope of the East Andes the only records of specimens (AMNH, CUMV) from the general vicinity of the (proposed) type locality are at San Bernardo. The northernmost records for nominate *hypoleuca* (and the species as a whole) derive from recent recordings by T.M. Donegan near San Vicente de Chucuri. **Specimens & Records** *Norte de Santander* Agua de la Virgen, 08°13'N, 73°24'W (XC 40202, T.M. Donegan). *Santander* Near San Vicente de Chucuri, 06°51'N, 73°22.5'W (XC 102625/26, F. Schmitt); above Vereda Honduras, 06°37'N, 73°30'W (XC 34939; T.M. Donegan); La Luchata, 06°38'N, 73°19'W (XC 30808/09, T.M. Donegan); Virolín, 06°06.5'N, 73°12'W (USNM 373677). *Cundinamarca* Albán, 04°52.5'N, 74°26.5'W (MNHLS 39640); RN Chicaque, 04°36.5'N, 74°18'W (A.M. Cuervo *in litt.* 2017); La Aguadita, 04°23'N, 74°19.5'W (AMNH 121972/73, ICN 11804/805); Fusagasugá, 04°21'N, 74°22'W (AMNH 121970/71, CUMV 7213/14); San Bernardo, 04°12'N, 74°25'W (XC 93870–874; O. Cortes). *Antioquia* Above Sevilla, *c.*07°14'N, 75°29'W (FMNH 299492, USNM 402471/472); upper La Serrana, 07°06'N, 75°08'W (Cuervo *et al.* 2008); Ventanas, 07°04.5'N, 75°27'W (USNM 402472); RFP La Forzosa, 06°59'N, 75°08'W (Cadena *et al.* 2007, Cuervo *et al.* 2008); Bosque Guayabito, 06°52'N, 75°06'W (Cuervo *et al.* 2008); La Secreta, 06°49'N, 75°06'W (Cuervo *et al.* 2008). **Notes** EL Roble (AMNH 121974) is

placed by Paynter (1997) only a few km SE of La Aguadita, and together these two sites are less than 10km NE of Olivares' (1969) proposed type locality.

Grallaria hypoleuca castanea Chapman, 1923, American Museum Novitates, no. 86, p. 8, Baeza, 1,525m, Napo, Ecuador [= probably hills above the town, vicinity of 00°29'S, 77°55'W]. One of the many important specimens collected by Carlos Olalla and his sons, the adult male holotype (AMNH 176060) was collected on 25 November 1922 (LeCroy & Sloss 2000). Topotypical specimens are also held at MLZ and USNM. The known range of *castanea* extends from south-east Colombia, at the head of the Magdalena Valley (see above), along the east slope of the East Andes south to northern Peru, north and west of the Marañón (Ridgely & Greenfield 2001, Schulenberg *et al.* 2007). Records on the Amazonian slope in the upper Río Cusiana watershed in Boyacá are the northernmost records east of the Andes. Although they probably involve *castanea*, the race concerned requires confirmation. Between this region and the east slope just across the divide from the Huila population, north as far as Cordillera Los Picachos (Guacamaya area), there is a notable gap in observations, only partially explained by a lack of exploration. Southward, despite somewhat patchy records, *castanea* appears to occur fairly continuously as far as the mountains around 'Chaupe' and Puerto Libre (see Stephens & Traylor 1983 for locations below) in north-west Amazonas. It occurs on the Ecuadorian side of the Cordillera del Condor (Krabbe & Sornoza 1994) and presumably also in adjacent Peru (Loreto). A specimen (MLZ 17747) labelled *castanea* is quite similar to nominate *hypoleuca*, but is considered smaller and has the entire upperparts, including wings and tail, much more chestnut or reddish-brown compared with the duller brown nominate (Schulenberg *et al.* 2013b). The sides, especially flanks, are more strongly washed chestnut (Chapman 1923c). **Specimens & Records COLOMBIA:** *Boyacá* Río Cusiana watershed (05°26'N, 72°41.5'W) above Pajarito (Bohórquez 2002, Salaman *et al.* 2002). *Meta* RNSC Las Palmeras, 03°50.5'N, 73°54.5'W (J. Beckers *in litt.* 2017). *Caquetá* *c.*1.8km SSE of Guacamaya, 02°48'N, 74°51'W (long series in ML, M. Álvarez); old Florencia road, 01°49.5'N, 75°40'W (J. Beckers *in litt.* 2017). *Huila* Cañón de La Cristalina, 02°15.5'N, 75°27'W (ML 41473591, J. Muñoz García); Belén, 02°12'N, 76°03'W (USNM 446640); San Agustín, 01°53'N, 76°16'W (MCZ 81782, AMNH 116344); La Candela, 01°50'N, 76°20'W (ANSP 155848–851, AMNH 116345, USNM 446639); La Palma, 01°46'N, 76°22'W (AMNH-116356/364); El Dorón, 01°41'N, 76°14.5'W and Nabú, 01°36'N, 76°16'W (Donegan & Salaman 1999); Cueva de los Guácharos, 01°36.5'N, 76°06.5'W (ICN 27014, XC 65284, P. Florez); Quebrada La Quebradona, 01°35'N, 76°19'W (XC 304569, J.P. **López-Ordoñéz**). *Cauca* Valle de las Papas, 01°50'N, 76°35'W (AMNH 116358–363); *c.*4.5km S of Santa Rosa, 01°39'N, 76°34'W (eBird: F. Ayerbe-Quiñones); near San Juan de Villalobos, 01°33'N, 76°19'W (eBird: F. Ayerbe-Quiñones); Villa Iguana, 01°14'N, 76°31'W (Donegan & Salaman 1999). *Putumayo* 2.2km SSE of Vereda Patoyaco, 01°13'N, 76°48'W (XC 298525, J.A. Zuleta-Marín); Filo del Hambre, 01°08.5'N, 76°42'W (J. Beckers *in litt.* 2017); Mocoa–Pasto road, 01°07'N, 76°52'W (XC 7766, T. Mark). *Nariño* Río San Miguel [on Colombia–Ecuador border], 00°24.5'N, 77°24'W (ANSP 165175-77). **ECUADOR:** *Sucumbíos* *c.*10km S of Santa Bárbara, 00°33'N, 77°32'W (R. Ahlman *in litt.* 2017); *c.*15km NE of La Sofía, 00°26'N, 77°32'W (R.A. Gelis *in litt.* 2016).

Napo Cascada de San Rafael, 00°06'S, 77°35'W (A. Spencer *in litt.* 2015); Mirador, 00°14.5'S, 77°47.5'W (ANSP 185504); La Cabaña, 00°16'S, 77°46'W (J. Freile *in litt.* 2017); Río Oyacachi below El Chaco, 00°19.5'S, 77°48.5'W (AMNH, ANSP); Finca Roberto Aldaz, 00°35'S, 77°51'W (HFG); RP Cabañas San Isidro, 00°35'S, 77°53'W (Price 2003, XC); RP Estación Biológica Yanayacu, 00°36'S, 77°53.5'W (XC 2390, W. Halfwerk); Cocodrilos, 00°39'S, 77°48'8.40"W (HFG). *Orellana* Below Volcán Sumaco, *c.*00°28'S, 77°34'W (MCZ 138476, ANSP 83340). *Tungurahua* Machay, 01°24'S, 78°16'W (Taczanowski & Berlepsch 1885); Mapoto, *c.*01°25'S, 78°15'W (Taczanowski & Berlepsch 1885, ANSP 169714). *Cañar* Llavircay, 02°33.5'S, 78°37.5'W (P. Astudillo Webster *in litt.* 2017). **Morona-Santiago** *c.*7.5km E of Zúñac, 02°13'S, 78°18'W (R.A. Gelis *in litt.* 2017); Puente Río Abanico, 02°14'S, 78°13'W (R. Ahlman *in litt.* 2017); *c.*14km NE of Sucúa, 02°26'S, 78°03'W (XC 249775, N. Krabbe); Guarumales, 02°34'S, 78°30'W (J. Freile *in litt.* 2017); Cordillera de Cutucú *c.*18km SE of Logroño, 02°43'S, 78°05'W (ANSP 176863/64); Coangos, 03°28.5'S, 78°13.5'W (ML: T.A. Parker); Mirador Condor, 03°38.5'S, 78°24'W (MECN 8038). *Zamora-Chinchipe* RP El Zarza, 03°50.5'S, 78°35'W (eBird: M. Chelemer); lower Río San Francisco, 03°59'S, 79°05.5'W (MCZH 199127); upper Río San Francisco, 03°59.5'S, 79°05.5'W (XC 7173, N. Athanas); *c.*5.8km E of Chinapintza, 04°02.5'S, 78°35'W (R. Ahlman *in litt.* 2017); *c.*4.5km NNW of Valladolid, 04°31'S, 79°09'W (XC 250879/80, N. Krabbe); Cerro Panguri, 04°36'S, 78°58'W (ANSP 185505); RE Cerro Plateado, 04°37'S, 78°54'W (XC 343231, L. Ordóñez-Delgado); Reserva Comunitaria San Andrés, 04°47'S, 79°18'W (XC 101544, M. Sánchez N.); near La Chonta, 04°57'S, 79°06'W (XC 276039/40; J. Nilsson). **PERU:** *Amazonas* Ridge below Cerro Machinaza, 03°54'S, 78°25'W (ML 92826, T.S. Schulenburg). *Cajamarca* Picorana, 05°02'S, 78°51'W (LSUMZ-172180/81); *c.*3km NNE of San José de Lourdes, 05°04.5'S, 78°53'W (LSUMZ 165299–303, XC 132111/113, D.F. Lane); 3.5km W of Pueblo Libre, 05°06'S, 79°14'W (FMNH 480827); 'Chaupe', 05°10'S, 79°10'W (AMNH 181328–335); Montaña de Valle Primavera, 05°33.5'S, 79°06'W (ML 40353891, photo J.N. Cova). *Piura* Playón, *c.*05°02.5'S, 79°20'W (Parker *et al.* 1985, LSUMZ 88047); E slope of Cerro Chinguela, 05°07'S, 79°23'W (ML 21733/34, T.A. Parker). **Notes** I examined two adults labelled 'Río Pastaza, Andoas' and 'Río Tigre, Oriente' collected by Carlos Olalla (MLZ 17747/48). Despite some confusion as to the whereabouts of these localities (Paynter 1993), they are certainly mislabelled, as all hypothesised locations are far below the altitudinal range of White-bellied Antpitta. Similarly, specimens (AMNH 156295, AMNH 492211/12) from Papallacta (00°22.5'S, 78°08.5'W, *c.*3,150m) are almost certainly mislabelled, and they are more likely to originate from the steep slopes below 2,500m in the Quijos Valley, probably close to the type locality of Baeza.

STATUS White-bellied Antpitta is considered to be of Least Concern (BirdLife International 2016), based largely on its broad distribution and relative abundance (Stotz *et al.* 1996). It is considered 'uncommon to fairly common' in Ecuador (Ridgely & Greenfield 2001) and 'fairly common' in Peru (Schulenberg *et al.* 2007). Nevertheless, its estimated range has been diminished by *c.*24% in Ecuador due to habitat loss (Freile *et al.* 2010) and at the northern end of its range the nominate subspecies was found to have become locally extinct in some fragmented landscapes (Castaño-Villa & Patiño-Zabala 2008). **Protected Populations** *G. h. hypoleuca* PNN Serranía de los Yariguíes (Donegan *et al.* 2007, XC); RN Chicaque (A.M. Cuervo *in litt.* 2017); RFP La Forzosa (Cadena *et al.* 2007, Cuervo *et al.* 2008); RNA Arrierito Antioqueño and RNA Reinita Cielo Azul (Salaman *et al.* 2007a). Probably also occurs in SFF Iguaque (USNM 373677). *G. h. castanea* Colombia: PNN Cueva de los Guácharos (Ridgely & Gaulin 1980, Hilty & Brown 1986, ICN, XC); PNN Cordillera de los Picachos (ML: M. Álvarez); PNN Sumapaz (J. Beckers *in litt.* 2017); RFP Cuenca Alta Río Mocoa (D. Carantón *in litt.* 2017); RNSC Las Palmeras (J. Beckers *in litt.* 2017). **Ecuador**: PN Cordillera del Condor (ML: T.A. Parker); PN Sangay (R.A. Gelis *in litt.* 2017); PN Sumaco-Galeras (MCZ, ANSP, ML, XC); PN Podocarpus (XC 7173, N. Athanas); PN Cayambe-Coca (ANSP 185504); PN Yacurí (XC 101544, M.V. Sánchez N.); RE Antisana (HFG); RE Cerro Plateado (XC 343231, L. Ordóñez-Delgado); RP Cabañas San Isidro (Price 2003, many recordings XC); RP Estación Biológica Yanayacu (XC 2390, W. Halfwerk); RBP Estación Científica San Francisco, (J. Freile *in litt.* 2016); RP Tapichalaca (MECN 8216, XC 250879/80, N. Krabbe); RP El Zarza (eBird: M. Chelemer). **Peru**: SN Tabaconas-Namballe (FMNH 480827).

OTHER NAMES English Bay-backed Antpitta (Ridgely & Gaulin 1980, Hilty & Brown 1983, Rodner *et al.* 2000); Ecuadorian White-bellied Antpitta (*castanea*, Cory & Hellmayr 1924, Meyer de Schauensee 1950); White-bellied Ant-Thrush (Brabourne & Chubb 1912). **Spanish** Tororoí Ventriblanco (Krabbe & Schulenburg 2003); Tororoí Pechiblanco (Salaman *et al.* 2007a); Gralaria ventriblanca (Ortiz-Crespo *et al.* 1990); Comepán (J. Freile *in litt.* 2017); Gralaria dorsibaya (Valarezo-Delgado 1983). **French** Grallaire à ventre blanc. **German** Blassbauch-Ameisenpitta (Krabbe & Schulenburg 2003).

White-bellied Antpitta, adult (*castanea*), Cabañas San Isidro, Napo, Ecuador, 1 September 2013 (*Luis A. Salagaje M.*).

White-bellied Antpitta, fledgling (*castanea*), Estación Biológica Yanayacu, Napo, Ecuador, 10 November 2008 (*José Simbaña*).

White-bellied Antpitta, adult feeding juvenile (*castanea*), Cabañas San Isidro, Napo, Ecuador, 10 August 2013 (*Leif Jönsson*).

White-bellied Antpitta, two nestlings (*castanea*), begging in anticipation of adult arrival, Estación Biológica Yanayacu, Napo, Ecuador, 12 December 2011 (*Harold F. Greeney*).

White-bellied Antpitta, adult (*castanea*) at nest just after feeding two nestlings, Estación Biológica Yanayacu, Napo, Ecuador, 12 December 2011 (*Harold F. Greeney*).

RED-AND-WHITE ANTPITTA
Grallaria erythroleuca   **Plate 15**

Grallaria erythroleuca P. L. Sclater, 1873, Proceedings of the Zoological Society of London, vol. 41, p. 783, "Huasampilla", Cuzco, Peru (= Huaisampillo, 2,970m, 13°14'S, 71°26'W, Río Madre de Dios drainage; Stephens & Traylor 1983). The holotype (NHMUK 1889.9.20.623) is an unsexed adult collected by H. Whitely in March 1872 and housed in Tring (Warren & Harrison 1971).

The English name of this Peruvian endemic aptly mirrors its scientific name, taken from the Greek word *eruthroleukos*, meaning red-and-white (Jobling 2010). Its range covers only a small area of south-central Peru, in the Vilcabamba and Vilcanota mountains (Parker *et al.* 1982, Schulenberg & Kirwan 2012d). Although Red-and-white Antpitta is locally fairly common, it prefers dense understories and thickets and is very difficult to observe. If you are reading this book linearly, you will not be shocked

to learn that the behaviour, distribution, taxonomy and life history of this species are all poorly understood. Indeed, the population of Red-and-white Antpitta in the northern Vilcabamba Mountains has been suggested to merit taxonomic recognition based on its vocal distinctiveness and yellowish-washed underparts (Schulenberg & Kirwan 2012d). Very poorly known in general, this account provides the first description of subadult plumage.

IDENTIFICATION 17–18cm. Red-and-white Antpitta is a medium-sized *Grallaria*, overall reddish-brown above with contrasting white underparts, sometimes with a yellowish wash to the throat. The flanks are similar to the upperparts and the breast bears a somewhat variable broken band of red-brown blotches. In most of its range this antpitta is distinctive and unlikely to be confused with any other species. At the south end of its range in southern Cuzco, it may be sympatric with White-throated Antpitta (Schulenberg & Kirwan 2012d). There is probably only very limited elevational overlap, however, with White-throated Antpitta replacing the present species at lower elevations. Both species have a reddish-brown crown and a white throat, but the back and wings of White-throated Antpitta are olive-brown, not red-brown, and the breast and belly are pale grey, not white or pale yellow (Ridgely & Tudor 2009, Schulenberg & Kirwan 2012d; see Technical Description).

DISTRIBUTION Red-and-white Antpitta is endemic to the South Peruvian Andean subcentre of endemism (Cracraft 1985) on the east slope of the Andes in southern Peru (Schulenberg & Kirwan 2012d). For many years, it was thought to be restricted to the area around the type locality just east of the Río Urubamba at Huaisampillo and in the Occobamba Valley (Chapman 1921, Zimmer 1934, Meyer de Schauensee 1966), but more recently it has been found at several locations north and south of there in Ayacucho (Núñez *et al.* 2012) and the northern Cordillera Vilcabamba in Junín (Weske 1972, Schulenberg *et al.* 2001a). Currently, the southernmost known locality is the Maracapata Valley, in Cuzco (Schulenberg & Kirwan 2012d).

MOVEMENTS None has been documented and, despite seasonal changes in apparent abundance, the species is not thought to move altitudinally (Merkord 2010).

HABITAT Primarily a denizen of humid subtropical montane forest (Weske 1972, Parker *et al.* 1982), forest borders and tall secondary forest, Red-and-white Antpitta may also show affinities for disturbed areas such as regenerating landslides and patches of *Chusquea* bamboo (Parker & O'Neill 1980, Fjeldså & Krabbe 1990, Schulenberg *et al.* 2007). The altitudinal range is generally said to extend from 2,100m to 3,000m (Chapman 1921, Lane & Pequeño 2004, Walker *et al.* 2006, Schulenberg *et al.* 2007), although Sibley & Monroe (1990) gave the upper limit as 3,400m and some studies have reported it as low as 1,965m (Jankowski 2010) or 1,758m (Merkord 2010).

VOICE The song documented over most of the range is described as three loud, hollow whistles, with a brief pause following the first, and the final two slightly lower and monotone, *tew too-too* or *HEE hew-hew*, with the entire sequence lasting 1–2s and repeated every 5–10s (Fjeldså & Krabbe 1990, D.F. Lane in Schulenberg *et al.* 2007). In the Cordillera Vilcabamba (Schulenberg *et al.* 2001a), however, the song apparently contains four notes, best

transcribed as *heep hew-hew-hew* (D.F. Lane in Schulenberg *et al.* 2007, ML 92422, T.S. Schulenberg; headwaters Río Poyeni). These three- to four-note songs are most commonly heard at dawn and dusk (Fjeldså & Krabbe 1990). Calls include a single whistle *hye* (Fjeldså & Krabbe 1990) or 'a series of plaintive, descending *clew* notes' (D.F. Lane in Schulenberg *et al.* 2007).

NATURAL HISTORY Red-and-white Antpitta is among the most poorly known of the genus, especially with respect to breeding and general behaviour. Like most species, it is usually encountered alone but apparently is more frequently found in pairs than other *Grallaria* (Parker & O'Neill 1980, Walker & Fjeldså 2005). It is said to generally forage on the ground or below 1.5m in the dense understorey (Fjeldså & Krabbe 1990). While hopping across the ground, it periodically stops and cocks its head sideways in a 'listening' posture before continuing forward. On encountering prey, it lunges forward to pluck it from terrestrial substrates such as fallen leaves, mosses and mud (Parker & O'Neill 1980). The species is especially adept at remaining concealed in dense vegetation (Ridgely & Tudor 1994, Schulenberg *et al.* 2007), but when singing may climb to a horizontal branch up to 2m above ground (Fjeldså & Krabbe 1990, Walker & Fjeldså 2005). Parker & O'Neill (1980) reported at least nine individuals foraging within a small (50m²) patch of secondary forest, although it was unclear if they were aggregating for social reasons or exploiting an ephemerally available resource.

DIET The diet is all but undocumented, but probably consists largely of small invertebrates. Stomach contents reported by Parker & O'Neill (1980) included beetles, ants, spiders, and gravel and plant matter (possibly ingested incidentally while capturing terrestrial prey).

REPRODUCTION There are no published data on any aspect of the breeding biology. Based on hypothesised relationships within the genus, it is probable that the nest, eggs and nestlings are similar to those of White-bellied and Yellow-breasted Antpittas. **SEASONALITY** Two juveniles (USNM 273159, USNM 273160; see Juvenile Description), 31 July and 2 August 1915, Tocopoquen, Occobamba Valley (*c.*2,800m). These are presumably the specimens mentioned by Chapman (1921), although he failed to mention their immature plumage. Both were in the final stages of acquiring adult plumage, with the younger of the two collected 2 August (male) and the older, by perhaps 1–3 weeks, was a female. Without any additional knowledge of the duration or sequence of post-fledging moults, it is somewhat premature to estimate a breeding season from two data points. Based largely on my intuition, after examining juveniles of many *Grallaria*, I suggest that active nests, at least in this area, might be found in January–May. **BREEDING DATA** Juvenile transitioning, 10 May 1975, San Luís (FMNH 299606); subadult, 30 September 1949, Cadena (FMNH 208273); subadult, 14 August 1966, 14km E of Lusiana (AMNH 820070).

TECHNICAL DESCRIPTION Sexes similar. *Adult* Most of the upperparts bright rufous, somewhat browner on back. The throat and centre of the belly are white, with the bright rufous-red (sometimes slightly more rufous-brown) upperparts extending variably onto the sides, flanks and vent. This coloration is most extensive on the breast-sides, where it extends towards the centre of the breast to form an indistinct breast-band. Some of these rufous-red breast feathers are fringed white (Sclater 1873b), producing a spotted look and there is often an olive wash on the flanks (Fjeldså & Krabbe 1990, Ridgely & Tudor 1994, Krabbe & Schulenberg 2003). The breast-sides and flanks are rufous-brown, washed olive on the flanks. **Adult Bare Parts** (Parker & O'Neill 1980) *Iris* medium brown; *Bill* black; *Tarsi & Toes* slate-grey to blue-grey. *Juvenile* Previously undescribed. I examined two specimens at the Smithsonian (see Breeding Seasonality). One of these is a juvenile retaining signs of immature plumage and had possibly recently become independent. The upperparts are generally the same bright rufous as adults, but most of the hindcrown, nape and neck-sides are buffy-rufous with blackish barring, broken only by a few adult feathers. There are also a few smaller patches on the lower back and more on the rump. Like adults, the throat and central belly are bright white. The breast-sides and flanks are rufous-brown, similar to adults, some feathers on the flanks and vent being faintly barred grey. The irregular spotty breast-band is clearly apparent, similar to adults, perhaps slightly paler. The wings and tail are similar to the adults but the upperwing-coverts are slightly paler (more tawny-rufous) on the tips with dusky subterminal bars. **Juvenile Bare Parts** *Bill*, maxilla black with yellowish tip, mandible mostly yellowish, slightly duskier near base (in dried specimen). The other specimen is similar to that described above, but is perhaps 1–2 weeks older as the juvenile barring is reduced to just patches on the nape with a few on the lower back. Other than having less yellow on the mandible (lower third mostly dusky-blackish), the rest of the plumage is very like the first specimen.

MORPHOMETRIC DATA *Wing* 95mm; *Tail* 56mm; *Bill* 28mm; *Tarsus* 51mm (*n* = 1, unsexed holotype, Taczanowski 1884). *Bill* from nares 15.5mm, 14.0mm, 16.5mm, exposed culmen 20.2mm, 19.5mm, 22.9mm, depth at front of nares 8.1mm, 8.5mm, width at gape 11.5mm, width at front of nares 6.3mm, 6.5mm, 6.7mm, width at base of mandible 8.4mm, 8.9mm, 9.6mm; *Tarsus* 54.3mm, 47.9mm, 48.4mm (n = 3♂♂, MSB, AMNH). *Bill* from nares 15.6mm; *Tarsus* 51.2mm (*n* = 1, subadult ♀, AMNH 820070). *Bill* 12.9mm, 13.9mm; *Tarsus* 45.9mm, 48.8mm (*n* = 2, ♀♀, juveniles, USNM 273159–273160). **Mass** mean 76.8g, range 73–80g (*n* = 4♂♂, Krabbe & Schulenberg 2003); 80g (*n* = 1♂, Dunning 2008); mean 78g, range 76–79g (n = 3♀♀, Krabbe & Schulenberg 2003); 75.41g(tr), 75.7g (*n* = 2, ♂,♀, MSB 34488–34489). **Total Length** 17–18cm (Fjeldså & Krabbe 1990, Clements & Shany 2001, Walker & Fjeldså 2005, Ridgely & Tudor 2009, Schulenberg *et al.* 2010).

DISTRIBUTION DATA Endemic to **PERU**: *Junín* Río Poyeni headwaters, 11°33'S, 73°38'W (Schulenberg *et al.* 2001a, FMNH 390684–686, ML). *Cuzco* Tinkanari, 12°15.5'S, 72°05.5'W (Lane & Pequeño 2004); Tocopoqueyu, 12°32'S, 72°13'W (USNM 273159/60); Yanatili, 12°37'S, 72°14'W (MSB 34488/89); 14km E of Lusiana, 12°38'S, 73°36'W (AMNH 820070/277, USNM 512165); below San Luís, 13°04'S, 72°23'W (LSUMZ 78562–566); San Luís, 13°06'S, 72°25'W (FMNH 299606); Rocotal, *c*.13°06.5'S, 71°34'W (XC 35362, D. Geale; ML 65064961, photo N. Frade); Pillahuata, 13°08'S, 71°25'W (ML 91629, S. Connop); *c*.3km SW of Aguas Calientes, 13°10'S, 72°33'W (XC 47636/37, N. Krabbe); Estación Biológica Wayqecha, 13°10.5'S, 71°35.5'W (Walker 2009, XC); Paucartambo, 13°19'S, 71°36'W (ML 135308: B.J. O'Shea); Cadena, 13°24'S, 70°43'W (FMNH 208273); 'Ccachubamba' near Marcapata (Stephens & Traylor 1983), *c*.13°30'S, 70°55'W (FMNH 222148).

TAXONOMY AND VARIATION Monotypic. Zimmer (1934) proposed an interesting, but subsequently ignored relationship, suggesting that closely related to the present might be Bicoloured Antpitta. This suggestion was based on wing, tail, bill and tarsal measurements, the shape of the remiges, wing formula and 'texture' of the plumage. He concluded that these two antpittas were both 'members of the same ancient group of which only the two ends are now left [and] should stand next to each other in the check lists'. I mention this largely as a point interest, as I have been unable to locate any subsequent authority who has agreed with him (e.g. Meyer de Schauensee 1966, Howard & Moore 1984, Altman & Swift 1993, Clements 2000). Instead, as its simple plumage suggests, it probably forms a superspecies with Rusty-tinged, White-bellied and Bay Antpittas (Parker & O'Neill 1980, Krabbe & Schulenberg 2003), with which it has occasionally been treated as conspecific (Meyer de Schauensee 1970, Walters 1980). Despite Red-and-white Antpitta having such a limited range, there does appear to be some, as yet, incompletely documented variation. Most notably, birds in the Cordillera de Vilcabamba differ in having the underparts washed yellow, reminiscent of the White-bellied/Yellow-breasted Antpitta comparison. Combined with apparent differences in vocalisations (Schulenberg *et al.* 2001a, 2007, Schulenberg & Kirwan 2012d), it is possible that these birds represent an undescribed taxon, most likely subspecifically related to Red-and-white Antpitta. The song in other parts of the species' range typically comprises three notes (see Voice), but birds in the Cordillera Vilcabamba apparently deliver a four-noted song. In anticipation, both plumages are illustrated on Plate 15. The population of yellow-bellied individuals was discovered during rapid avifaunal inventories of the north-east end of the cordillera. There is an unfortunate paucity of comparative material from this range, but I have examined three specimens collected at its south-west end. None has yellow underparts and they appeared to be otherwise typical. It would seem that the yellow-bellied population is confined to one end of the Cordillera Vilcabamba.

STATUS Red-and-white Antpitta has a restricted geographic range (Stattersfield *et al.* 1998), but is considered uncommon to fairly common (Ridgely & Tudor 1994, Clements & Shany 2001) or locally fairly common (Fjeldså & Krabbe 1990, Walker 2005) and the population trend is believed to be stable, resulting in a current status of Least Concern (BirdLife International 2015). As noted by Herzog *et al.* (2012), the estimated range size used to assess its conservation status by BirdLife International (2015) was 32,700km^2, a figure approaching twice the size of the range (17,462km^2) modelled by Young *et al.* (2009). Needless to say, all estimates and models are in need of refinement with additional distributional and habitat-use data. Human activity has few direct short-term effects on Red-and-white Antpitta, other than habitat destruction (Schulenberg & Kirwan 2012d). Given that the species occupies regenerating habitats, it may even benefit, locally and in the short term, from human activities, such as low-level clearance for subsistence agriculture. In the longer term, habitat destruction could pose a significant threat, in view of the species' small geographic range, although currently much of this lacks roads and has a low human population density (Schulenberg & Kirwan 2012d). Several of the localities from which the species is known are, thankfully, under some form of protection. **Protected Populations** PN

Manú (Walker *et al.* 2006); SN Megantoni (Lane & Pequeño 2004); SH Machu Picchu (Walker & Fjeldså 2005); ÁCP Abra Málaga (ML 40168, T.S. Schulenberg); ACP San Luís (eBird: J.L. Avendaño Medina). Lane & Pequeño (2004) identified SN Megantoni as a key locality for Red-and-white Antpitta, which area is the only uninterrupted stretch of wilderness between Manú National Park and Machiguenga Community Reserve.

OTHER NAMES English Chestnut-brown Ant-Thrush (Brabourne & Chubb 1912); Chestnut-brown Antpitta (Cory & Hellmayr 1924, Howard & Moore 1984). **Spanish** Tororoí de Cuzco (Krabbe & Schulenberg 2003); Tororoí Rojiblanco (Clements & Shany 2001). **French** Grallaire de Cuzco. **German** Weißflecken-Ameisenpitta (Krabbe & Schulenberg 2003).

Red-and-white Antpitta, adult, Manu road, Cuzco Department, Peru, 20 October 2014 (*David Beadle*).

Red-and-white Antpitta, adult, Manu road, Cuzco, Peru, 1 November 2009 (*Kristian Svensson*).

Red-and-white Antpitta, adult singing from hidden perch, Manu road, Cuzco, Peru, 18 November 2009 (*Fabrice Schmitt/WINGS Birding Tours Worldwide*).

Red-and-white Antpitta, adult, Pillahuata, Cuzco, Peru, 12 October 2010 (*Daniel F. Lane*).

RUSTY-TINGED ANTPITTA
Grallaria przewalskii   Plate 15

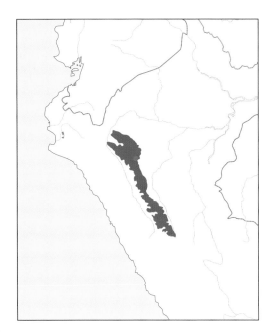

Grallaria przewalskii Taczanowski, 1882, Proceedings of the Zoological Society of London, vol. 50, p. 33, Ray-urmana, above Chirimoto, Huayabamba Valley, northern Peru, 8,000ft [see Distribution Data for approximate coordinates]. The original description is based on two adults, a male and a female, collected on 29 October 1880. The specific eponym honours the Russian explorer and geographer, Colonel Nikolay Mikhaylovich Przhevalsky (Przewalski is the Polish rendition of his surname), a well-known explorer of Asia, especially Mongolia and Tibet (Taczanowski 1882). Consequently, the scientific name of Rusty-tinged Antpitta should be pronounced *sha-val-skee-eye*.

Rusty-tinged Antpitta is endemic to humid montane forest in eastern Peru. The song is an easily imitated series of three notes, which usually betray its presence long before the species is detected visually. Although Rusty-tinged Antpitta was not considered globally threatened until recently, it is restricted to a single EBA and faces severe habitat alteration in the near future, leading to its categorisation of Vulnerable (BirdLife International 2017). There are almost no published data on any aspect of its natural history, ecology, behaviour or taxonomic affinities. To date, its breeding biology is among the least known of the genus. Of particular importance for conservation, given that it is not known how tolerant the species is of forest disturbance, future work on its habitat requirements is sorely needed.

IDENTIFICATION 16–17cm. Both sexes have a pale grey forecrown, darkening posteriorly. A thin whitish eye-ring, not always obvious in the field, is slightly broader postocularly. The upperparts, including lower nape, back and tail, are dark rufous-brown, becoming paler, rufous-chestnut on the neck-sides and upper back. The chin and belly are white, the throat dusky or sometimes yellowish-white, while the chest, lower neck-sides and face (below the eye) are rufous or rust-coloured, with this coloration extending across the upper breast to form a variable band, sometimes streaked white or broken into chestnut streaking. The flanks and vent are also rufous, but duskier, contrasting with the white belly. Within the limited range of Rusty-tinged Antpitta in northern Peru, it is unlikely to be confused. North of the Rio Marañón it is replaced by its putative sister species, White-bellied Antpitta, which has the rusty coloration much reduced on the flanks (greyish-white in White-bellied) and a cleaner white throat (without the dingy or yellowish appearance of Rusty-tinged).

DISTRIBUTION Endemic to Peru, in the Andes south and east of the Marañón within the East Peruvian Andean subcentre of endemism (Cracraft 1985). Within its limited range it can be fairly common, albeit somewhat patchily distributed, especially in Amazonas in the Cordillera de Colán and around Abra Patricia and Leimebamba (Mark *et al.* 2008, eBird). Note that Cory & Hellmayr (1924) place the type locality in Loreto, and several subsequent authors have given the range of Rusty-tinged Antpitta as extending to this department (Peters 1951, Meyer de Schauensee 1966, Sibley & Monroe 1990). Stephens & Traylor (1983), however, placed 'Ray-Urmana' in Dept. Amazonas, as did

Vaurie (1972) (as Ray-Hurmana Mountain). Most recent authors appear to have corrected this error, listing the departments of San Martín and La Libertad, in addition to Amazonas (Clements & Shany 2001, Schulenberg *et al.* 2007, Ridgely & Tudor 2009). To the south, Rusty-tinged Antpitta is replaced by Bay Antpitta, with a gap of *c.*200km. North, on the opposite bank of the Rio Marañón, with a gap of *c.*50km or less, Rusty-tinged Antpitta is replaced by White-bellied Antpitta. There is currently no known overlap in the ranges of these three closely related species.

HABITAT The habitat preferences are poorly documented, but most authors suggest that it prefers the understorey and floor of more intact forest (Fjeldså & Krabbe 1990, Davies *et al.* 1994, 1997, Schulenberg *et al.* 2007). As is typical of the forests within its known altitudinal range in eastern Peru, however, most habitat where the species is found has a relatively dense understorey dominated by *Chusquea* spp. bamboo. The extent to which bamboo is utilised is undocumented. Following comparatively recent fieldwork in Dept. Amazonas (Barnes *et al.* 1997) the altitudinal distribution is now known to be broader than once thought (Ridgely & Tudor 1994, Kirwan & Hornbuckle 1997), and is currently considered to be 1,700–2,750m (Schulenberg *et al.* 2007).

VOICE Following Krabbe & Schulenberg (2003), the song is short (1.2–1.5s) and given at 11–12s intervals. It consists of a series of three whistles at around 1.5–1.8kHz. The first note is usually weakest, followed by a short pause before the second and third notes. The second note may be similar in pitch to the first, or slightly higher, while the final note is generally higher pitched and louder than the other two and rises slightly. Overall, the song is similar to that of Chestnut-crowned Antpitta (Schulenberg *et al.* 2007), but does not drop between the first and second notes: *hip hew-HEE* (Schulenberg *et al.* 2007). The presumed call is a series of rising, interrogative, hollow *clew?* notes (Schulenberg *et al.* 2007).

NATURAL HISTORY The habits of Rusty-tinged Antpitta are poorly described, although they are probably similar to the species' better-known relatives such as White-bellied Antpitta. Like these, it forages on or very near the ground in the forest understorey, moving by running or hopping, with frequent, motionless pauses. It is described as always remaining concealed in dense vegetation where it is difficult to observe (Krabbe & Schulenberg 2003, Greeney & Kirwan 2012), a trait hardly separating it from the rest of its kin, and one which may be overstated due to the biases introduced by being watched by a repellent-soaked crowd of eager birdwatchers.

DIET Although there is little reason to suspect that this varies much from its congeners, the diet of Rusty-tinged Antpitta is largely unknown. Stomach contents indicate that arthropods (including a 4cm larval Lepidoptera) and 'small seeds' are taken (Schulenberg & Williams 1982, Krabbe & Schulenberg 2003, T.S. Schulenberg in Greeney & Kirwan 2012).

REPRODUCTION The nest, eggs, nestlings and breeding ecology are completely undocumented. SEASONALITY With respect to reproductive activity, there is also little to go on. Based on putative relationships within the genus, my guess is that Rusty-tinged Antpitta will be proven to breed during the wetter months of the year within its range. The only previously available information is the mention of a 'fledgling' in Amazonas in July (Fjeldså & Krabbe 1990). Based on the cursory description of this individual (see Description), I classify it as a middle-aged juvenile that probably hatched no earlier than early May. BREEDING DATA Fledgling, 16 August 1978, E of La Peca (LSUMZ 88049); subadult, 13 March 1925, La Lejía (AMNH 234692).

TECHNICAL DESCRIPTION Sexes similar. The following is based on Cory & Hellmayr's (1924) translation from French of Taczanowski (1882). *Adult* Upperparts rufescent-brown with crown more blackish. Cheeks bright ferruginous; broad but somewhat indistinct greyish eye-ring extends slightly rearwards in a faint line; chin whitish, throat pale buff, sometimes with yellowish tinge. Central breast and belly extensively washed grey, grading to brown on the vent; sides and flanks rufescent-brown but paler than back; undertail-coverts rufescent-brown; underwing-coverts rufous. The inner webs of the remiges are somewhat paler than the leading margins. **Adult Bare Parts** (coloration from Taczanowski 1882, Krabbe & Schulenberg 2003) *Iris* dark brown; *Bill* maxilla black, mandible blackish, sometimes slightly paler basally; *Tarsi & Toes* pale bluish-grey. Although the plumage maturation of Rusty-tinged Antpitta is undocumented, Fjeldså & Krabbe (1990) state that 'old males have buffy white throat, grey breast and sides, whitish belly, and often a few rufous-brown feathers on breast'. *Immature* The immature plumages have not been previously described in detail. Fjeldså & Krabbe (1990) stated that the throat is white, contrasting with a black breast while the remaining plumage is 'dark, finely barred with buff', while Remsen (1984) stated only that the immature plumage is darker than in adults and 'finely spotted'. These descriptions are of fledgling plumage, and were based on the same skin (LSUMZ 88049) (J. Fjeldså & J.V. Remsen *in litt.* 2016). *Fledgling* My own examination of photos of LSUMZ 88049 reveal the following. Fledglings are largely covered in dark, downy plumage, brownish-grey on upperparts, black across breast and dark sooty-grey mixed buff along flanks. The throat is white, the vent tawny and

central belly white washed pale buff, gradually suffused more tawny-buff, grey or blackish as it grades into the vent, flanks and breast, respectively. The feathers of the crown and nape each have a small buffy spot near the tip, creating a finely speckled look that approaches fine barring on the nape and upper back, as the spots align into rows depending on the arrangement of the feathers. Barring gradually becomes more distinct and coarser on the back and rump, and the buff markings tawnier. The face and lores are paler than the crown, pale brownish or greyish-white, with darker blackish and chestnut feathering behind the eye and on the ear-coverts. Flight feathers and (still emerging) tail feathers are similar in coloration to adults, but the tips of the tertials are tawny-rufous beyond a narrow black subterminal bar. The upperwing-coverts are dark brown to chestnut-brown and broadly tipped tawny-rufous. This rufous area is crossed by at least one thin black bar, with some of the secondary-coverts having an additional one or two black (but considerably less distinct) bars. **Fledgling Bare Parts** *Iris* brown; *Bill* mostly orange, blackish on culmen; *Tarsi & Toes* pinkish-grey (from LSUMZ 88049 label, ex. T.S. Schulenberg). *Juvenile* Undescribed. The fledgling described above was just beginning to acquire adult-like, rufescent-brown feathers in scattered patches above. This detail, along with what I have observed of subadult *przewalskii* and juveniles of other *Grallaria*, suggests that juveniles can retain fledgling plumage on the nape and variably on the crown. The underparts will probably approach the coloration of adults but are likely to be even more heavily suffused grey and and washed with variable amounts of chestnut on the throat and flanks. They likely have some fine dusky barring, at least on the upper breast, and probably on the flanks and vent. The wings and tails of juveniles probably resemble those of adults except by immature wing-coverts as described for fledglings. Bare-parts coloration probably now similar to adults but, at least in younger individuals, the gape and mandible are likely to have variable amounts of orange. *Subadult* The few subadult skins that I examined were largely indistinguishable from adults, except that (presumably) younger birds tend to have fewer chestnut feathers on the flanks, sides and upper breast. The youngest subadult I examined (LSUMZ 81991) had almost no chestnut wash on the underparts, which were heavily suffused grey. On the breast, it also retained a few buffy-white feathers crossed by 1–2 thin dusky bars. **Subadult Bare Parts** Not described but unlikely to differ from those of adults.

MORPHOMETRIC DATA There are almost no published measurements. *Wingspan* 325mm, 315mm; *Wing* 97mm; 100mm; *Tail* 65mm, 60mm; *Bill* [total culmen] 29mm, 29mm; *Tarsus* 51mm, 48mm; *Middle toe* 34mm, 33mm (*n* = 2, ♂,♀, Taczanowski 1882, 1884). *Bill* from nares 13.2mm, 13.5mm, 13.5mm, 13.7mm, 13.8mm; *Tarsus* 44.9mm, 46.6mm, 46.6mm, 48.1mm, 46.9mm (*n* = 5, ♀,♀,♀,♂,♂, AMNH). *Bill* from nares 13.5mm; *Tarsus* 48.0mm (*n* = 1, subadult ♂, AMNH 234692). *Bill* from nares 14.3mm, exposed culmen 21.1mm, depth at front of nares 7.9mm, width at gape 10.1mm, width at front of nares 6.4mm, width at base of lower mandible 8.3mm; *Tarsus* 47.0mm (*n* = 1, ♂, MSB 32685). **Mass** Data from labels at LSUMZ and MSB. ♂♂: 57g, 63g(×2), 63.5g, 64g, 66g, 66g(no), 67.1g(tr), 67.9g(m), 68g(lt)(×2), 68g(m), 69.2g(no), 70g, 71g(lt), 72g(lt), 74g(×2), 76g(tr), 77g, 83.5g. ♀♀: 62g, 66g, 69g, 69g(lt), 69.5g, 70g, 73g(no), 73g, 76g. ♀?: 60.5g. The fledgling discussed had mass 56g (♀;

LSUMZ 88049). See also Schulenberg & Williams (1982), Krabbe & Schulenberg (2003). **Total Length** 16–22cm (Taczanowski 1882, 1884, Meyer de Schauensee 1970, Fjeldså & Krabbe 1990, Ridgely & Tudor 1994, 2009, Krabbe & Schulenberg 2003, Schulenberg *et al.* 2007).

DISTRIBUTION DATA Endemic to **PERU**: *Amazonas* Campamento Palomino, 05°38.5'S, 78°15.5'W (F. Angulo P. *in litt.* 2017); García Trail (see discussion of locality in Ochre-fronted Antpitta account), 05°40'S, 77°46'W (ML 18072, J.P. O'Neill); Pampas de Copal, 05°47.5'S, 77°50'W (F. Angulo P. *in litt.* 2017); Río Chido Trail, 05°48.5'S, 78°01.5'W (XC 296850, R. Gallardy); 33km NE by road of Ingenio, 05°52'S, 77°57'W (LSUMZ 81990–992); ACP Bosque Berlín, 05°54.5'S, 78°25.5'W (XC 123530/792, J. Tiravanti); *c.*4.5km N of Tullanya, 06°05'S, 78°20'W (MSB); La Lejía, 06°10'S, 77°31'W (AMNH 234687–692); *c.*30km by road E of Florida de Pomacochas, 05°42'S, 77°49'W (Davis 1986, LSUMZ 116970–972); Cordillera Colán, E of La Peca, *c.*05°33'S, 78°19'W (Davies *et al.* 1994, 1997, LSUMZ 88048–053); La Esperanza, 05°40'S, 77°55'W (XC 83027, O. Janni); Campamento Quebrada Salas, 06°06.5'S, 77°26.5'W (A. García-Bravo *in litt.* 2017); Cerro Shucahuala, 06°14'S, 78°06.5'W (A. García- Bravo *in litt.* 2017); Cerro Montealegre, 06°35'S, 77°34'W (B.M. Winger *in litt.* 2017); Pillcopata, 06°45.5'S, 77°48'W (F. Angullo P. *in litt.* 2017). *San Martín c.*22km ENE of Florida de Pomacochas, 05°41'S, 77°45'W (LSUMZ 174010–012); Camp Utter Solitude, 05°43'S, 77°44.5'W (XC 132694, D.F. Lane); Laguna de los Cóndores, 06°51'S, 77°42'W (XC 167, W.-P. Vellinga); between Aguacate and Caanan, 06°54.5'S, 77°29'W (F. Schmitt *in litt.* 2017); Nuevo Bolívar, 07°19.5'S 77°27.5'W (eBird: P. Saboya del Castillo); Las Papayas, 07°27'S, 77°10'W (LSUMZ 104345); Puerta del Monte, 07°34'S, 77°09'W (LSUMZ 104487). *La Libertad* Cumpang, 08°12'S, 77°10'W (LSUMZ 92462–474, ML: long series, T.A. Parker). *Huánuco* Uchiza Valley, above San Pedro, 08°39'S 76°53.5'W (G. Seeholzer *in litt.* 2017); Bosque Zapatagocha, 09°40'S, 76°03'W (CM P153649); 28km NE of Huánuco, 09°43'S, 76°06'W (ML). **Notes** In choosing the most appropriate coordinates for the type locality, I slightly adjusted those given by Stephens & Traylor (1983), 06°28'S, 77°21'W, and corrected for elevation to place Ray-Urmana at 2,440m in the hills above Chirimoto, *c.*06°26.5'S, 77°19.5'W, on the Amazonas–San Martín border.

TAXONOMY AND VARIATION Monotypic. Based on general plumage and other morphological characteristics, the species clearly belongs to the subgenus *Oropezus* (Ridgway 1909, Lowery & O'Neill 1969), and is considered to form a superspecies with the similarly plumaged but allopatric Red-and-white Antpitta, White-bellied Antpitta and Bay Antpitta (Parker & O'Neill 1980, Krabbe & Schulenberg 2003). Indeed, some earlier authors (Meyer de Schauensee 1970) united all of these species within a polytypic *G. hypoleuca* (Bay-backed Antpitta) until Parker & O'Neill (1980) recommended their recognition as species. Chapman (1926) considered Bay Antpitta to be the closest relative of Rusty-tinged and many authors have placed them adjacent in linear arrangements (Clements 2000).

STATUS Rusty-tinged Antpitta is a range-restricted species (Stattersfield *et al.* 1998), with estimates of total range size varying from as tiny as 4,100km² (BirdLife International 2015) to 6,376km² (Franke *et al.* 2007, Young *et al.* 2009). Most of the known localities lack formal protection. The global population is unknown but, where present, it is described as 'fairly common' (Stotz *et al.* 1996). Based on predicted deforestation within its range (Soares-Filho *et al.* 2006) and its apparent sensitivity to fragmentation (Bird *et al.* 2012), BirdLife International (2016) predicted that the population will decline rapidly over the next three generations, and uplisted Rusty-tinged Antpitta to Vulnerable in 2012 (BirdLife International 2012). Apart from the assumption that this species will be detrimentally affected by habitat fragmentation, there are no specific threats to Rusty-tinged Antpitta (Greeney & Kirwan 2012). **Protected Populations** SN Cordillera de Colán (LSUMZ); BP Alto Mayo (Rosas 2003, XC); ACP Bosque Berlín (XC 123530/792, J. Tiravanti); ACP Hierba Buena-Allpayacu (eBird: O. Janni); ACP Abra Patricia-Alto Nieva (XC); ACP Huiquilla (eBird: A. Kratter).

OTHER NAMES *Grallaria* [*hypoleuca*] *przewalskii* (Sibley & Monroe 1990). **English** Przewalski's Ant-Thrush (Brabourne & Chubb 1912); Przewalski's Antpitta (Cory & Hellmayr 1924, Walters 1980); **Spanish** Tororoí Rojizo (Clements & Shany 2001, Bernis *et al.* 2003, Krabbe & Schulenberg 2003). **French** Grallaire de Przewalski. **German** Taczanowskiameisenpitta (Krabbe & Schulenberg 2003).

Rusty-tinged Antpitta, adult, Fundo Alto Nieva, San Martin, Peru (*Carlos Calle*).

Rusty-tinged Antpitta, adult, Fundo Alto Nieva, San Martin, Peru (*Carlos Calle*).

Rusty-tinged Antpitta, adult, Fundo Alto Nieva, San Martin, Peru (*Carlos Calle*).

BAY ANTPITTA
Grallaria capitalis   Plate 15

Grallaria capitalis Chapman, 1926, American Museum Novitates, no. 231, p. 2, Rumicruz, 9,700ft [2,960m], Dept. Junín, eastern Peru. Rumicruz is not in Junín, as reported in the type description, but lies in Pasco (Stephens & Traylor 1983). The holotype (adult male, AMNH 174089) is in New York, and was collected on 22 March 1922 by Harry Watkins (Chapman 1926, LeCroy & Sloss 2000). There are three topotypical specimens (all males, AMNH 174088/90/91), two of which are not in full adult plumage. See Distributional Data for further discussion as to the whereabouts of the type locality and suggested coordinates.

Bay Antpitta is a monotypic Peruvian endemic found east of the Andes in central Peru (Parker *et al.* 1982, 1996, Ridgely & Tudor 2009). It was for many years considered a subspecies of White-bellied Antpitta, along with other species with relatively 'simple' plumage. Despite being fairly common within its limited range (Schulenberg *et al.* 2007, Ridgely & Tudor 2009, Schulenberg & Kirwan 2012g), Bay Antpitta is one of the least-studied members of the genus, and almost nothing is known of its natural history. The nest and eggs are undescribed, but this account provides the first detailed description of immature plumage.

IDENTIFICATION 16–17cm. Bay Antpitta is a mid-sized *Grallaria* with relatively simple plumage: nearly uniform rufous-chestnut on the throat, breast and flanks, with a darker, brownish or sooty-black crown and slightly paler, rufous-buff belly that often looks vaguely streaked in the field (Fjeldså & Krabbe 1990, Ridgely & Tudor 1994, Schulenberg & Kirwan 2012g). It was described as a 'larger, stronger-billed version of Rufous and Chestnut Antpittas' by Clements & Shany (2001), and to resemble a 'saturated Rusty-tinged Antpitta' by Ridgely & Tudor

(1994). Bay Antpitta has no geographic overlap with the latter, but might be confused with one of several sympatric subspecies of Rufous Antpitta (which favour higher elevations). In general, Bay Antpitta is much larger and has brighter chestnut plumage compared to the various shades of duller brown exhibited by Rufous Antpitta in eastern Peru (Krabbe & Schulenberg 2003). Chestnut Antpitta, at generally lower elevations but with broad overlap (Schulenberg *et al.* 2007), is notably smaller, has dark barring on the belly, and a proportionately smaller bill that is black rather than bluish-grey (Ridgely & Tudor 1994, Schulenberg & Kirwan 2012g). Comparing Bay Antpitta with the closely related Rusty-tinged Antpitta (no range overlap), Chapman (1926) noted that Bay Antpitta has a much heavier bill and averages darker overall. Bay Antpitta is also largely pale rufous below but Rusty-tinged Antpitta is much greyer.

DISTRIBUTION Confined to the East Peruvian Andean subcentre of endemism on the east slope of the Andes in the departments of Junín, Pasco and Huánuco in central Peru (Cracraft 1985, Clements 2007). In Pasco, there is a population immediately south-south-east of Oxapampa on the west side of the Río Santa Cruz valley and another in the Cordillera Yanachaga east of Oxapampa (Schulenberg *et al.* 1984). In Huánuco, Bay Antpitta is fairly common in the Corillera Carpish (Parker & O'Neill 1976, Ridgely & Tudor 1994). Its northernmost known locality lies between Tingo María and Pucallpa in the Cordillera Divisoria, at the border between the departments of Huánuco and Ucayali (Parker & O'Neill 1981). At the northern extent of its known range there is a gap of *c.*200km between the southernmost known locality for Rusty-tinged Antpitta (Cumpang, La Libertad) and the Cordillera Carpish. The intervening region is relatively unexplored and it is possible that the ranges of the two species meet somewhere (Schulenberg & Kirwan 2012g). The southernmost records are fairly recent, and lie just north of the Río Apurimac in Ayacucho (Hosner *et al.* 2015).

HABITAT Bay Antpitta inhabits the dark understorey of humid subtropical forests (Parker *et al.* 1982, Schulenberg *et al.* 2007), primarily at elevations of 1,800–3,000m (Schulenberg & Kirwan 2012g) but occasionally as low as 1,500m in some areas (Parker & O'Neill 1981, Fjeldså & Krabbe 1990). It does, however, also venture from primary forest to forage in disturbed areas such as *Chusquea* bamboo-dominated landslips and treefalls, habitat regenerating following human disturbance (Fjeldså & Krabbe 1990, Ridgely & Tudor 1994, Clements & Shany 2001) and brushy Alder-dominated woodlands (LSUMZ labels).

VOICE As noted by Parker & O'Neill (1980), the song of Bay Antpitta differs distinctly from those of the closely related White-bellied Antpitta and Rusty-tinged Antpitta. Its easily imitated song is a four-note (occasionally five) hollow whistle: *HEEP hew-hew-hew* (D.F. Lane in Schulenberg *et al.* 2007). Alternatively, also described as a whistled *hy hyhyhy* (Fjeldså & Krabbe 1990) or *tew, too-too-too* (Ridgely & Tudor 1994). The quantitative description of Krabbe & Schulenberg (2003) stated that the song lasts 1s and is given at intervals of 6–10s, repeated for one minute or more. The first of the four notes is delivered at 1.9–2.0kHz. After a short pause, the last three notes are weaker, given at 1.7–1.8kHz, with the final note rising slightly.

NATURAL HISTORY Like most *Grallaria*, extremely secretive and usually encountered alone or in pairs foraging low in vegetation or on the ground (Krabbe & Schulenberg 2003), sometimes flicking aside leaves with its bill or probing soft earth (P. Hosner *in litt.* 2017). Fjeldså & Krabbe (1990) described it as foraging only 'sometimes on the ground' suggesting that the species may be less terrestrial than other *Grallaria*. While on the ground, Bay Antpitta moves by running or hopping but remaining well concealed in dense vegetation (Krabbe & Schulenberg 2003, Schulenberg *et al.* 2007).

DIET The diet, along with most details of this species' biology, remains virtually undocumented, but it presumably takes a variety of small- to medium-sized invertebrates. The only published data come from stomach

contents (LSUMZ labels) given by Schulenberg & Kirwan (2012g). These authors listed insect remains, including coleopterans, a larval lepidopteran *c.*3cm long, a 4cm grub (Coleoptera?) and small snails. Additional stomach contents from LSUMZ labels (M. Brady *in litt.* 2017) include small bones, possibly of a frog.

REPRODUCTION Nest, eggs and reproductive behaviour completely unknown. **SEASONALITY** A 'fledgling' was reported from Huánuco in December (Fjeldså & Krabbe 1990). In addition, I examined photos of a male (LSUMZ 80578) which is the 'juvenile' mentioned by Schulenberg & Kirwan (2012g). This specimen is, without a doubt, a fledgling that was probably still fairly dependent on its parents for care. Based on plumage, I would estimate this individual as no more than one month out of the nest, and probably closer to only three weeks. Birds on the Río Satipo reportedly had enlarged gonads and were highly vocal in October (Hosner *et al.* 2015, KUNHM). **BREEDING DATA** The following records are from Bosque Zapatogocha in Huánuco: fledgling, 18 June 1975 (LSUMZ 80578); fledgling transitioning, 3 December 1972 (FMNH 296702); juvenile, 11 February 1973 (FMNH 296701); juvenile transitioning, 18 June 1972 (FMNH 293400). It appears that the 'fledgling' reported from Huánuco in December (Fjeldså & Krabbe 1990) may be considered an additional record (J. Fjeldså *in litt.* 2015). Records of older immatures include: subadult, 20 June 1972, Micho (FMNH 293399); subadult transitioning, 15 July 1968, Bosque Taprag (FMNH 283700); two subadults transitioning, 19 and 25 March 1922, Rumicruz (AMNH 174088/90). Older immatures collected at Bosque Zapatogocha include: subadult, 1 September 1973 (USNM 582346); subadult transitioning, 2 March 1973 (FMNH 296707).

TECHNICAL DESCRIPTION Sexes similar. *Adult* Has a grey-brown or dark blackish-russet crown that becomes rufous-chestnut on the nape. The remaining upperparts, including the rump, wings and tail, are rufous-chestnut. Towards the front of the crown, the feathering becomes almost black and extends slightly onto the otherwise chestnut lores and to the base of the bill, where almost contiguous with the dark grey-brown or blackish feathering on the uppermost throat near the base of the bill. The underparts are, overall, brighter rufous than the upperparts, darker on the breast and paler on the belly. Centrally, the belly fades from pale rufous to whitish-buff and the vent is rufous. **Adult Bare Parts** *Iris* dark brown; *Bill* blue-grey with a darker, blackish culmen; *Tarsi & Toes* blue-grey. Chapman (1926) described the type specimen as follows, adding a few details to the above description. 'Upperparts deep chestnut-auburn; crown blackish extending on to the nape; cheeks and auriculars much like the back; anteorbital region paler; tail and wings externally like the back; underparts brighter than above, orange-rufous or Sanford's brown; the chin and center of the abdomen whitish; sides, flanks, and lower tail-coverts like the breast; feet blackish; bill blackish; mandible more plumbeous, its cutting edge and tip horn-color.' *Subadult* The subadult skins I examined (AMNH, USNM; see Reproduction) were in nearly complete adult plumage. Two retained only one or two immature upperwing-coverts that were slightly paler and browner than those of adults and bore thin, indistinct, darker subterminal bars (but were probably somewhat faded and worn just prior to being moulted). The USNM specimen is slightly younger, retaining all of its immature coverts. The tertials of all, however, still bore thin black subterminal bands. The tertials were broadly bordered black at the tips in one specimen and in another they were not. The tips of the feathers in the second individual, however, were heavily worn and had probably lost this dark fringe. The eye colour of the younger individual was dark brown. *Juvenile* Undescribed, but probably similar to the fledgling plumage described below, but lacking most or all of the barring below and appearing less 'fluffy'. Probably juveniles have similar bills to adults, but with some remnant orange near the base, possibly along the tomia and near the tip. *Fledgling* Fjeldså & Krabbe (1990) described the 'fledgling' plumage as 'blackish above, crown spotted and back barred with buff; throat and breast blackish, belly buffy, lower breast and sides barred buffy and blackish'. Remsen (1984) summarised the immature plumage as overall darker than the adult and finely spotted, probably referring to the same specimen I describe here (see Reproduction, Seasonality). From my examination of photos of this specimen (LSUMZ 80578), I describe it as follows. Head blackish, including lores, ear-coverts and cheeks, the latter with a trace of chestnut feathers; crown, nape and upper back with fine, buffy-chestnut spots; back blackish with traces of chestnut barring, especially towards rump, and a few irregular chestnut feathers. Wings chestnut with blackish inner webs. Tail chestnut, uppertail-coverts chestnut with black barring. Chin and sides of throat blackish, central and lower throat deep chestnut, upper breast blackish, including sides with a few chestnut feathers and black barring; lower breast, belly and flanks dark buffy-grey to dirty white, paler in centre and darker on flanks; some chestnut feathers in irregular patches across breast; flanks with traces of darker barring; undertail-coverts deep chestnut, brighter than those of adults. **Juvenile Bare Parts** (from dried specimen) *Iris* undescribed; *Bill* maxilla dark blackish along culmen, whitish near tip and horn-coloured on tomia, mandible horn-coloured, palest near base; *Tarsi & Toes* dark brownish.

MORPHOMETRIC DATA There are almost no previously published measurements for *G. capitalis*. I measured several skins: *Bill* from nares 14.0mm; *Tarsus* 46.2mm (*n* = 1, ♂, AMNH 174091, Pasco). *Bill* from nares 13.9mm, 13.9mm, 15.0mm; *Tarsus* 48.7mm, 46.9mm, 48.2mm (*n* = 3, ♂♂, subadults, USNM 582346, Huánuco, AMNH 174090, 174088, Pasco). **Mass** 72g(lt), 80g(lt) (*n* = 2, immature ♂, adult ♀, LSUMZ). For males, Schulenberg & Kirwan (2012g) provide a mean of 72.4g and range of 65–77g (*n* = 5; LSUMZ). Krabbe & Schulenberg (2003) gave a range of 72–77g for an unspecified number of ♂♂; they gave mass of a single ♀ as 72g. **Total Length** 14.5–17.0cm (Fjeldså & Krabbe 1990, Ridgely & Tudor 1994, Krabbe & Schulenberg 2003, Schulenberg *et al.* 2007).

DISTRIBUTION DATA The type locality of Rumicruz, is placed at 2,960m, *c*.10°44'S, 75°55'W by Vaurie (1972) and Stephens & Traylor (1983). These coordinates, however, are at an elevation of 4,150m in Google Earth, suggesting Rumicruz is inaccurately georeferenced, at least with respect to this species (AMNH 174088–891). Only 15km NE of Vaurie's (1972) Rumicruz are more appropriate elevations and habitats for Bay Antpitta, and I suggest 10°39'S, 75°48'W for the type locality. Records from the following list that likely belong to the somewhat distinctive 'southern form' (see Taxonomy & Variation) are marked[†]. Endemic to **PERU**: *Huánuco* Bosque Zapatogocha, 09°40'S,

76°03'W (FMNH, USNM, LSUMZ); Sendero Paty, 09°38'S, 76°08'W (XC 350400/02, G.M. Kirwan, ML 28775, A.B. van den Berg); below Carpish Tunnel, 09°43'S, 76°06'W (XC 100362/63, J. King); Bosque Taprag, 09°43'S, 76°04'W (FMNH 283700); Micho, 09°43'S, 76°00'W (FMNH 293399); Papayajo, 09°45'S, 76°03'W (FMNH 296705); Huanacaure, 09°46'S, 75°54.5'W (J. Barrio *in litt.* 2017); Cushi, 09°51'S, 75°37'W (LSUMZ 128532/33); Playa Pampa, 09°57'S, 75°42'W (LSUMZ). *Pasco* Sector Oso Playa, 10°19.5'S, 75°35'W (A. García-Bravo *in litt.* 2017); Cueva Blanca, 10°22'S, 75°36'W (XC 62987, W.-P. Vellinga); 5km SW of Esperanza, 10°34'S, 75°19'W (ML: T.S. Schulenberg); Ulcumano Ecolodge, 10°38.5'S, 75°26'W (XC 97941, R.P. Piana); Cumbre de Ollón (see discussion in Chestnut Antpitta account), 10°39'S, 75°17.5'W (LSUMZ 106083–085); Santa Cruz, 10°39'S, 75°22'W (LSUMZ 106086); Bosque Schóllet, 10°41'S, 75°19'W (XC 88115, A. Spencer). *Ucayali* 3km by road NE of Abra Divisoria, *c*.09°02.5'S, 75°41.5'W (LSUMZ 84952). *Junín* CC Bosque Puyu Sacha, *c*.11°05.5'S, 75°25.5'W (Gamarra-Toledo *et al.* 2012); Río Satipo, 11°30'S, 74°52'W (KUNHM)[†]; Concepción, 11°32'S, 74°49'W (XC 152920, H. van Oosten)[†]; Apalla–Andmarca road, 11°33'S, 74°49'W (XC 88112, A. Spencer)[†]. *Ayacucho* Tutumbaro, 12°44'S, 73°57'W (Hosner *et al.* 2015)[†]; below San Antonio, 12°58'S, 73°38'W (Hosner *et al.* 2015)[†]; below Rumichaca, 13°10'S, 73°35'W (Hosner *et al.* 2015)[†]; below Chupón, 13°15'S, 73°30'W (Hosner *et al.* 2015)[†].

TAXONOMY AND VARIATION Bay Antpitta was, for many years, considered a subspecies of White-bellied Antpitta (Lowery & O'Neill 1969, Meyer de Schauensee 1970, Walters 1980, Sibley & Monroe 1990). Along with other species that possess relatively simple plumage, Lowery & O'Neill (1969) considered Bay Antpitta a member of the subgenus *Oropezus*. It is currently considered to form a superspecies with Red-and-white, Rusty-tinged and White-bellied Antpittas (Parker & O'Neill 1980, Ridgely & Tudor 1994, 2009, Schulenberg & Kirwan 2012, Winger *et al.* 2015). Yellow-breasted Antpitta was typically included within this group, but is now thought to be more closely related to Brown-banded Antpitta (see these accounts). Chapman (1926) proposed a close relationship to Rusty-tinged Antpitta, although its closest relative is not known with certainty. Currently considered monotypic, a recently discovered population in southern Junín and northern Ayacucho is fairly distinctive in voice and plumage (Hosner *et al.* 2015). In plumage, it is overall similar to the description above (see Technical Description), especially the rufous-chestnut body and dark crown. The face, however, differs in that the dark feathers of the crown do not extend onto the lores or chin. The loral area is, instead, greyish-buff or dirty whitish. It appears to have a generally whiter (less buffy) central belly. The northern limit of this form appears to be somewhere around Apalla in southern Junín, while the southern limit of the 'northern form' appears to be the Puya Sacha area of northern Junín (P.A. Hosner & J. Barrio *in litt.* 2017).

STATUS A range-restricted species (Stattersfield *et al.* 1998), the total range is estimated to be 6,439–10,200km² (Franke *et al.* 2007, BirdLife International 2015). Its population appears to be stable, however, and it is generally considered fairly common in appropriate habitat (Fjeldså & Krabbe 1990, Stotz *et al.* 1996, Schulenberg *et al.* 2007). Consequently, Bay Antpitta is currently afforded a threat status of Least Concern (Stattersfield *et al.* 1998, BirdLife International 2015). The precise impacts of human activity have not been quantified but Schulenberg & Kirwan (2012) noted that, in light of its apparent occupation of regenerating habitats, Bay Antpitta may actually benefit from minor anthropogenic habitat alteration, at least locally and in the short term. The currently understood range falls largely within the Huánuco centre of the North-east Peruvian Cordilleras EBA (Fjeldså 1995, Stattersfield *et al.* 1998). Few localities lie within formally protected areas but, as predicted by Krabbe & Schulenberg (2003), it does occur in PN Yanachaga-Chemillén (XC 62987, W.-P. Vellinga). Fairly

recently it was reported from the privately protected CC Bosque Puyu Sacha in the Ríos Casca, Oxabamba and Palca watersheds, which protects montane forests between 1,500m and 2,500m (Gamarra-Toledo *et al.* 2012). 'Typical' Bay Antpittas will also probably be found in PN Tingo María, SN Pampa Hermosa and CC Maximo Flores, and perhaps in the highest parts of BP Bosque de Protección San Matías-San Carlos.

OTHER NAMES Spanish Tororoí Bayo (Clements & Shany 2001, Krabbe & Schulenberg 2003). **French** Grallaire châtaine. **German** Rostfarbene Ameisenpitta (Krabbe & Schulenberg 2003).

Bay Antpitta, adult, Satipo Road, Junín, Peru, 9 November 2015 (*Gunnar Engblom*).

Bay Antpitta, adult, Oxapampa, Pasco, Peru, 19 November 2017 (*Carlos Calle*).

RUFOUS-FACED ANTPITTA
Grallaria erythrotis Plate 16

Grallaria erythrotis P.L. Sclater & Salvin, 1876, Proceedings of the Zoological Society of London, vol. 1876, p. 357, 'Tilotilo, prov. Yungas, Bolivia' [= near Sandillani, 16°12'S 67°54'W, La Paz; see Paynter 1992]. The specific name is derived from the Greek words *eruthros* (= red) and *tis* (= eared) (Jobling 2010).

Named for its distinctive orange-rufous cheeks (Sclater 1877, 1890), Rufous-faced Antpitta is endemic to the humid montane forests of north and central Bolivia. Within this very small range, it could be confused only with the larger White-throated Antpitta, which in contrast has the entire head deep rufous and principally grey underparts. Rufous-faced Antpitta has a greyish crown and nape, and is largely buffy below, grading to whitish ventrally. The songs of the two species are also quite different, being triple-noted in Rufous-faced Antpitta, but only two notes in White-throated Antpitta. Rufous-faced Antpitta is not currently considered at risk (BirdLife International 2015). The scant data on breeding come from a single nest observed in captivity. Among the least known of the montane antpittas, this range-restricted species is in need of work on all aspects of its natural

history, ecology, behaviour, population size and taxonomic affinities. Only when we know its true distribution and habitat requirements, and understand why it is absent from some apparently suitable areas (Herzog *et al.* 1999), will we be able to develop effective conservation plans.

IDENTIFICATION 17–18cm. Among the more simple-plumaged *Grallaria*, with both sexes having dark olivaceous-grey to olivaceous-brown upperparts and tail. The lores, face and neck-sides are orange-rufous. The throat and belly are white, with the breast orange-rufous with faint white streaking. Within its small range, Rufous-faced Antpitta is unlikely to be confused except with White-throated Antpitta, which occurs at lower elevations (Ridgely & Tudor 1994). The more strongly patterned White-throated Antpitta differs by having the entire head (including crown) bright rufous. Although it shares a bright white throat with Rufous-faced Antpitta, White-throated Antpitta further differs in its duller, greyer belly. As noted by prior authors, the two species are most easily separated by the entirely rufous crown of White-throated Antpitta and by its lack of a bright rufous face (Sclater 1877, 1890).

DISTRIBUTION Endemic to Bolivia in the southern part of the South Peruvian Andean subcentre of endemism (Cracraft 1985) and the Upper Yungas EBA (Stattersfield *et al.* 1998) in the departments of La Paz

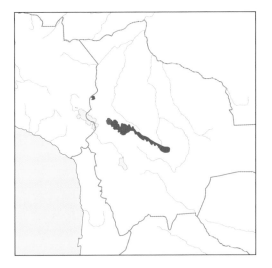

(Remsen & Parker 1995, Hennessey & Gomez 2003, Martínez & Rechberger 2007), Cochabamba (Bond & Meyer de Schauensee 1942, Fjeldså & Krabbe 1989, Macleod *et al.* 2005) and Santa Cruz (Parker & Rowlett 1984, Gemuseus & Sagot 1996).

HABITAT Appears to be partial to forest edges, road cuts and other areas of second growth, and is only infrequently reported from mature forest. Indeed, the one reference I found to an individual inside 'deep forest' was in a *Chusquea* sp. bamboo thicket (Remsen *et al.* 1982). Apart from this, little has been quantified concerning habitat use. Rufous-faced Antpitta appears to be most abundant at 2,000–3,000m (Remsen *et al.* 1982, Fjeldså & Krabbe 1990, Whitney *et al.* 1994, Macleod *et al.* 2005), but its full altitudinal range appears to be 1,700–3,300m (Ridgely & Tudor 2009, Macleod *et al.* 2005). I have been unable to find confirmed records from as low as 1,350m (Sibley & Monroe 1990).

VOICE The song resembles that of Tawny Antpitta, being a series of three, easily imitated whistles, either pure or with a double-note quality, described by Remsen *et al.* (1982) as *heelo-hee-hee*. The last two notes generally are higher pitched than the first, and may be even-pitched or the final note is slightly higher again. Song duration is short (1.2–1.4s), generally at 5–10s intervals. It ranges in frequency, across the entire song, from 0.2–0.3kHz. The short call of Rufous-faced Antpitta, usually given every 3–4 seconds, is a single *krie* produced at 2.2kHz and descending to 1.8kHz (Remsen *et al.* 1982, Fjeldså & Krabbe 1990, Krabbe & Schulenberg 2003).

NATURAL HISTORY Little has been published on the foraging behaviour, and the available information is limited to a few cursory accounts. These agree that, like other *Grallaria*, the species spends a good deal of time foraging on the ground, often running or hopping short distances, and frequently ventures into open areas at the edge of more dense habitat (Remsen *et al.* 1982, Ridgely & Tudor 1994).

DIET Almost undocumented, but almost certainly dominated by invertebrates. Remsen *et al.* (1982) reported the stomach contents of seven individuals as arthropods and a few small seeds. Arthropods included

a 15mm caterpillar (presumably Lepidoptera), beetles (Coleoptera), ants (Formicidae; Hymenoptera) and spiders (Araneae). Like other *Grallaria*, earthworms likely are important for adults and for young in the nest. To date, however, Rufous-faced Antpitta has not joined the ranks of species trained to visit worm-feeding stations (Woods *et al.* 2011).

REPRODUCTION Bell & Bruning (1976), in their somewhat obscure, but often-cited, paper, reported on the successful rearing of a single brood in the New York Zoological Park. To the best of my knowledge, this remains the only successful captive breeding event for any of the species treated herein. All of the following pertains to this nest, with no published data from the wild. **Nest & Building** Of necessity, the description that follows must be viewed with caution, given that it was constructed in an artificial enclosure. The nest was described as 'a rather ragged structure of small twigs, assorted leaves, moss and other plant parts' and measured: outer diameter 20cm; outer height 30cm; inner cup depth 14cm. **Egg, Laying & Incubation** This pair laid a total of six two-egg clutches, all of unmarked, pale blue eggs. Only one clutch hatched, *c.*15 days after clutch completion, but only because it was artificially incubated. **Nestling & Parental Care** Though not well described, based on a published photo, the newly hatched young were dark-skinned and bore dorsal tufts of long down feathers, similar to other *Grallaria* (Greeney 2012a). As the chicks were hand-reared, it is difficult to determine at what age they would have fledged, but it seems likely that fledging would have occurred no earlier than 20 days, at which point the chicks could stand and partially walk. **Seasonality** The only previously published breeding datum from the wild is that of Fjeldså & Krabbe (1990) who reported a 'juvenile' in November in Cochabamba. **Breeding Data** Four subadults, Incachaca, 22 May 1915 (AMNH 137174), 25 May 1927 (FMNH 110364), and 6 and 27 September 1927 (FMNH 180563/65); subadult, 24 March 1896, 'San Antonio' (AMNH 99206); subadult transitioning, 20 May 1926, Nequejahuira (AMNH 229209); subadult transitioning, 23 May 1915, Incachaca (AMNH 137173).

TECHNICAL DESCRIPTION Sexes similar. *Adult* With the exception of very faint white streaking on the orange-ochraceous breast, adult plumage is unstreaked. The crown, nape, back, wings and tail are dull ochraceous-grey or ochraceous-brown, with the face (including lores), ear-coverts and neck-sides a contrasting, bright orange-ochraceous. The face pattern connects to a distinct, somewhat narrow band across the upper breast, which contrasts strongly with the bright white throat. The lower breast and belly are bright white, becoming dull, pale ochraceous-grey on the flanks and vent. **Adult Bare Parts** *Iris* dark brown; *Bill* black with paler, horn-coloured tip; *Tarsi & Toes* grey or pinkish-grey (data from Remsen *et al.* 1982). *Juvenile* Fjeldså & Krabbe (1990) provided the only previous description of immature plumage. They described the 'juvenile' as grey-brown above, almost uniform on top of head. Back has cinnamon-buff spotting and rather inconspicuous tawny barring, the latter more apparent on the lower back. The chin is whitish and the rest of underparts 'very profusely barred' dusky-brown and cinnamon-buff. There may also be scattered pale spots on the breast. The belly is pale buff and the mandible orange. This description appears to match closely what I would consider a juvenile. Presumably, this description should be expanded to include immature wing-coverts similar to those described below for subadults. **Subadult** All of the immature skins I examined (see Breeding Data) were subadults and nearly indistinguishable from adults. One (AMNH 99206) had two immature feathers on the nape. These were cinnamon-rufous with a thick black bar crossing the distal end and pale shaft-streaks that were buffy basally but cinnamon-buff nearer the tip. Otherwise, all subadults appeared fully adult except their immature wing-coverts. These were brown with rich cinnamon-rufous fringes, broadest and most obvious at the tip. The cinnamon tips are fringed proximally by a single thin black bar. As in other species, there is a fair degree of variation in the brightness and clarity of these markings, most or all of which appeared to reflect varying amounts of wear. Comparing bare-part coloration to adult skins suggests that these differ little in life.

MORPHOMETRIC DATA No modern published measurements. *Wing* 89mm; *Tail* 51mm; *Tarsus* 51mm ($n = 1$, holotype, Sclater & Salvin 1876). The following data include both adults and subadults, and are from MCZ (J.C. Schmitt *in litt.* 2017), FMNH, AMNH, USNM (HFG): ♀♀ *Wing* 103mm, 103mm ($n = 2$); *Bill* from nares 13.9 ± 1.0mm ($n = 6$), exposed culmen 20.7 ± 0.7mm ($n = 3$), depth at nares 7.5 ± 0.2mm ($n = 3$), width at nares 6.0 ± 0.0mm ($n = 3$), width at base of mandible 8.6 ± 0.6mm ($n = 3$); *Tarsus* 51.3 ± 1.5mm ($n = 5$). ♂♂ *Wing* 99.2 ± 3.6mm ($n = 5$); *Bill* from nares 14.1 ± 0.3mm ($n = 9$), exposed culmen 19.2 ± 1.7mm ($n = 6$), depth at nares 7.6 ± 0.3mm ($n = 5$), width at nares 5.9 ± 0.2mm ($n = 5$), width at base of mandible 8.2 ± 0.3mm ($n = 5$); *Tarsus* 50.6 ± 1.3mm ($n = 9$). ♂? *Wing* 101.0mm ($n = 1$); *Bill* from nares 13.3 ± 0.4mm ($n = 4$), exposed culmen 20.0mm ($n = 1$), depth at nares 7.0mm ($n = 1$), width at nares 5.5mm ($n = 1$), width at base of mandible 8.3mm ($n = 1$); *Tarsus* 51.6 ± 1.3mm ($n = 3$). **MASS** 53g(no), 56g(lt), 60g(lt), 63g(lt), 66g(lt), 69.5g(no), 61g, 61g(lt) ($n = 8$, 6♂♂, 1♀, 1♂?, respectively; LSUMZ). These data presumably all included in summary statistics reported by Remsen *et al.* (1982) and Remsen (1985). **TOTAL LENGTH** 15.2–19.0cm (Sclater & Salvin 1876, Meyer de Schauensee 1970, Dunning 1987, Fjeldså & Krabbe 1990, Ridgely & Tudor 1994, Krabbe & Schulenberg 2003).

DISTRIBUTION DATA Endemic to **BOLIVIA**: **La Paz** Tokoaque, 14°37'S, 68°57'W (Hennessey & Gomez 2003, ML); Keara, 14°41.5'S, 69°03.5'W (XC 73598–600, J.A. Tobias); Piara, near Pelechuco, *c*.14°47'S, 69°00.5'W (AMNH 2506); Zongo Valley, *c*.16°06.5'S, 68°03'4W (ML 87679, A.B. Hennessey); Sayani, 16°08'S, 68°06'W (XC 47641, N. Krabbe); 'San Antonio', near Coroico, loc. uncertain (Paynter 1992), *c*.16°11.5'S, 67°43.5'W (AMNH 99206); Sandillani, 16°12'S 67°54'W (ANSP, AMNH, UMMZ); Sacramento Alto, 16°16'S, 67°47'W (LSUMZ 90732–736); 1km S of Chuspipata, 16°18'S, 67°50'W (LSUMZ 101398/102246); Nequejahuira, 16°20'S, 67°50'W (AMNH 229209); Bosque de Apa Apa, 16°21'S, 67°30'W (ML 132718, P.A. Hosner); 3km NW of Chacopampa, 16°39.5'S, 66°49'W (MHNNKM 5267); Inquisivi to Chulumani road, *c*.km180, 16°41.5'S, 67°13.5'W (A. Spencer *in litt.* 2017); Piedras Blancas, 16°45.5'S, 66°47.5'W (M.G. Harvey *in litt.* 2017). *Cochabamba* Pampa Grande, 16°41.5'S, 66°28.5'W (S.K. Herzog *in litt.* 2017); El Limbo, *c*.17°08.5'S, 65°37.5'W (LSUMZ 37763); 70km W of Villa Tunari, 17°09'S, 65°42'W (ML 33718/22, T.A. Parker); 3.5km W of Aguada, 17°10'S,

65°35'W (XC 2927–29, A.B. Hennessey); Tablas Monte, 17°10'S, 65°53.5'W (MHNNKM 4952, XC); Represa Corani, 17°13.5'S, 65°53.5'W (D.F. Lane *in litt.* 2017); Incachaca, 17°14'S, 65°49'W (ANSP, AMNH, CM, FMNH, MCZ, USNM, UMMZ); 'San Mateo', above La Fortaleza, *c*.17°43'S, 64°43.5'W (ANSP 170142). *Santa Cruz* Bosque Siberia, *c*.17°50.5'S, 64°41.5'W (MHNNKM 5265/69/70, ML, XC). **NOTES** Several specimens are labelled only 'Western Yungas' (AMNH 492207) or 'Chapare, Yungas' (FMNH 180561), referring to the eastern foothills of the Andes between Cochabamba and La Paz. Most of these were probably collected somewhere SW of Villa Tunari, *c*.17°10'S, 65°47'W (see Paynter 1992), but would require significant additional investigation into their precise origin, probably requiring a review of the itineraries and field notes of the collectors. For now, there is no clear indication that any of these would signficantly alter our understanding of the distribution and habitat of Rufous-faced Antpitta, and I have omitted them from the distributional data.

TAXONOMY AND VARIATION Monotypic. The relationships of *G. erythrotis* to congenerics are unresolved. Recent phylogenetic analyses of *Grallaria*, based on DNA sequence data, are incomplete and do not include Rufous-faced Antpitta (Krabbe *et al.* 1999, Rice 2005). Based on morphological characters, Rufous-faced Antpitta is a member of the subgenus *Oropezus* (Ridgway 1909, Lowery & O'Neill 1969). This group contains Bicoloured, Chestnut-naped, White-throated, Red-and-white, White-bellied, Yellow-breasted, Rusty-tinged, Bay Antpitta, Grey-naped, Rufous, Tawny and Brown-banded Antpittas. Chapman (1912, 1917) considered Rufous-faced Antpitta to be sister to Brown-banded, but at first glance its plumage suggests to me that Tawny Antpitta is a closer relative, as also suggested by Fjeldså (1992) and near to which it is often included in linear sequences (e.g., Howard & Moore 1984, Monroe & Sibley 1993, Clements 2007).

STATUS Based on an apparently stable population, no evidence of declining abundance and a designation of 'fairly common' within its limited range (Stotz *et al.* 1996), Rufous-faced Antpitta is considered Least Concern by Birdlife International (2017). A note of caution, however, that there is some debate over the actual range size. BirdLife International (2017) considered Rufous-faced Antpitta to have a range sufficiently large not to trigger any threat status, but estimates of range size (Young *et al.* 2009, Herzog *et al.* 2012, BirdLife International 2017) vary considerably. Recently, two studies suggested that the actual range of Rufous-faced Antpitta may be considerably smaller (20,391km²: Herzog *et al.* 2012, or 22,754km²: Franke *et al.* 2007) than the estimate used to evaluate its threat status (39,100km²: BirdLife International 2017). No threats to populations have been specifically documented, and the species occurs in a few formally protected areas, including PN Madidi (Hennessey & Gomez 2003), PN Carrasco (ML 87646/47/52, XC 2927–29, A.B. Hennessey), and PN Amboró (Gemuseus & Sagot 1996). It should be carefully monitored, however, despite its apparent tolerance of some anthropogenic disturbance.

OTHER NAMES **English** Rufous-cheeked Ant-Thrush (Brabourne & Chubb 1912). **Spanish** Tororoí Carirrufo. **French** Grallaire masquée. **German** Rostwangen-Ameisenpitta (Krabbe & Schulenberg 2003).

Rufous-faced Antpitta, adult, Manuel María Caballero, Santa Cruz, Bolivia, 7 November 2013 (*Paul B. Jones*).

Rufous-faced Antpitta, adult, Manuel María Caballero, Santa Cruz, Bolivia, 7 November 2013 (*Paul B. Jones*).

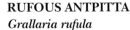

Fig. 3. **Rufous-faced Antpitta**, adult, Cueara, La Paz, Bolivia, 8 October 2006 (*Joseph Tobias*).

Fig. 4. **Rufous-faced Antpitta**, adult, Manuel Maria Caballero, Santa Cruz, Bolivia, 7 November 2013 (*Paul B. Jones*).

RUFOUS ANTPITTA
Grallaria rufula   Plate 16

Grallaria rufula Lafresnaye, 1843, Revue Zoologique par La Société Cuvierienne, 1843, p. 99, 'Bogotá, Colombia.

'The bird inhabits the floor of the thickets in the low temperate zone, where it is never an obtrusive resident. Of those I found, one seemed to materialize out of the shadows as my eyes became accustomed to the gloom on entering a shadowed ravine, and was quietly disappearing behind an adjacent tree when I stopped it with a shot. Another was in a sheltered pathway as I rounded a bend and, rail-like, made instant use of its long legs to carry it into concealment among the grasses.' – **John T. Zimmer, 1930, Peru**

Rufous Antpitta is found from western Venezuela south to central Bolivia. Some populations are less 'rufous' than others, but geographic variation in plumage is relatively minor, even when its presumed sister-species, Chestnut Antpitta, is considered. More noticeable is the variation in vocalisations, but the complexity of this has not been well explored. Without doubt, multiple species are involved, and the systematics of the '*Grallaria rufula* complex' are currently being tackled by a team of ornithologists who are attempting to place the complex regional variation within a sound phylogeographic context using both vocal characters and an extensive multi-gene molecular analysis (M. Isler *in litt.* 2017). For now, I have relied heavily on the excellent summary of vocal variation by Spencer (2012). Overall, however, the following account will require a good deal of re-evaluation once the many taxonomic questions have been resolved.

IDENTIFICATION 14.5–15.0cm. Over most of its range (nominate *rufula*), both sexes are rufous-brown above, less brown on the sides and upper breast, with most rufous saturation on the breast. The flanks are greyer rufous, fading to pale rufous on the lower breast and whitish on the lower belly and vent. The undertail-coverts vary from whitish through buffy-brown to dark brown (Fjeldså & Krabbe 1990). Eyes are brown, the bill blackish and tarsi greyish. In most of its range, Rufous Antpitta's small size, relatively narrow bill and nearly uniform rufous plumage make it unmistakable. Only in Peru, where similar Chestnut Antpitta occurs, is confusion likely. However, the latter is somewhat brighter rufous than sympatric or nearby populations of Rufous Antpitta, and also lacks an eye-ring, something that most Peruvian races of Rufous Antpitta possess. Finally, with good views, Rufous Antpitta lacks any barring on the belly, which character is somewhat variable in Chestnut Antpitta, being sometimes almost absent in the south of its range (Schulenberg *et al.* 2007).

DISTRIBUTION Inhabits middle to high elevations (Restall *et al.* 2006, Ridgely & Tudor 2009) from western Venezuela (Hilty 2003) through Colombia (McMullan *et al.* 2010), Ecuador (Butler 1979, Ridgely & Greenfield 2001) and Peru (Parker *et al.* 1982) to central Bolivia (Kempff-Mercado 1985, Remsen & Traylor 1989). For details see Taxonomy and Variation.

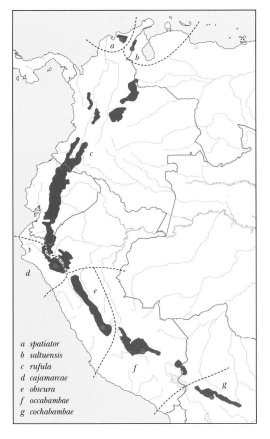

a *spatiator*
b *saltuensis*
c *rufula*
d *cajamarcae*
e *obscura*
f *occabambae*
g *cochabambae*

HABITAT Although some authors have mentioned a preference for dense bamboo or cane (Fjeldså & Krabbe 1990, Krabbe & Schulenberg 2003, Schulenberg *et al.* 2007), most authors mention humid, mossy, epiphyte-laden montane forests (Cresswell *et al.* 1999a, Eckhardt 2003, Arbeláez-Cortés *et al.* 2011a), forest borders and adjacent *páramo* (Meyer de Schauensee & Phelps 1978, Astudillo *et al.* 2015) as preferred habitats. It is often associated with dense, tangled undergrowth, within which it moves, ghost-like, appearing and reappearing as if by magic (Todd & Carriker 1922), but may also venture into the open, provided there is cover nearby, especially during cloudy or foggy weather (Poulsen 1993, HFG). It is also thought to have a predilection for boggy areas, seepage zones and damp riparian areas (Freile 2000, Krabbe & Schulenberg 2003). Across its range, the species is reported at 1,900–3,650m (Fjeldså & Krabbe 1990, Giner & Bosque 1998), but most records appear to be between 2,500m and 3,300m (Freile 2000, Ridgely & Greenfield 2001, Krabbe & Schulenberg 2003, Schulenberg *et al.* 2007, Ridgely & Tudor 2009). Within their very limited ranges, *saltuensis* and *spatiator* occur at *c.*2,850–3,100m (Meyer de Schauensee 1950) and 2,200–3,000m (Meyer de Schauensee 1950, Giner & Bosque 1998), respectively. In Peru, Rufous Antpitta is found primarily at 2,400–3,700m, locally to 2,000m in southern Peru (*occabambae*) (Schulenberg *et al.* 2007).

VOICE There is much geographic variation in the vocalisations of Rufous Antpitta across its broad range. As noted by Spencer (2012), in his excellent review of vocal variation using XC contributions, the distinction between songs and calls is somewhat subjective, especially given that their function has not been studied. In general terms, Rufous Antpitta vocalisations can be divided into short or long. Following Spencer (2012), in the descriptions below I generally refer to the shorter of the two as the 'call'. For Rufous Antpitta in particular, however, it appears that many of the shorter vocalisations can have song-like functions such as territorial defence (J.A. Tobias *in litt.* 2017). Vocal descriptions, incorporating the work of Spencer (2012) and Krabbe & Schulenberg (2003) are as follows. Race *spatiator* is described by Krabbe & Schulenberg (2003) as having a ringing, 3s trill of *c.*30 notes, evenly paced at 9.5–10.0 notes/s and falling gradually from 4.5–4.8kHz to 3.2–3.4kHz. Spencer (2012) noted that the song of *spatiator* in Santa Marta is the highest pitched of all of the described subspecies, a 3s series of short, high-pitched but descending trill, with the 'overall impression … more like an antwren than an antpitta'. What is presumed to be the call of *spatiator* is a short, high-pitched whistle that descends slightly, reminiscent of some *Grallaricula* antpittas. The song of the nominate subspecies in western Colombia and in Ecuador is what might be considered the typical vocalisation of the species. Spencer (2012) described the song as 'a short clear note followed by a rapid, mellow stutter, at a pitch a bit lower than that of the Eastern Andes form from Colombia [i.e. true *rufula*], in all lasting about a second. It is both lower pitched and much shorter than the song from the eastern Colombia cordillera, and shorter and made up of more notes than the birds from the Cajamarca area on the other side of the Marañón River.' The call of *rufula* is, essentially, a longer version of the second part of the song, described by Spencer (2012) as 'a rapid, mellow series of quick, bell-like notes at a medium pitch, lasting up to 6s and at times with short breaks between the note series. Unlike the song, this vocalisation is given at much longer intervals.' Krabbe & Schulenberg (2003) provided the acoustic details of this song: it lasts 5.6–7.3s with an introductory note at 2.3–2.9kHz, followed by accelerating series of notes lasting 2.7–3.9s and comprising 23–34 notes. These notes begin slightly lower than the introductory note, fall 0.1–0.2kHz and end with 2–6 progressively shorter bursts, each burst consisting of a loud note followed by 2–6 accelerated notes. They also described a frequently heard vocalisation, either a call or alternative song, which resembles one of the terminal 'bursts' described above. This vocalisation lasts 0.6–0.7s and is repeated every 2.5–4.0s for several minutes. It begins with a single loud note, then a short pause followed by 4–5 slightly lower-pitched, descending, accelerated notes. Finally, Krabbe & Schulenberg (2003) described an infrequently heard call as a ringing trill of 1.5–2.2s, with 16–24 notes at 2.5–3.0kHz, given a few times at intervals of 4–10s. They likened this call to the song of *G. blakei* except being higher pitched. Subspecies *saltuensis*, of the Perijá Mountains on the Colombia–Venezuela border, is one of the more distinctively plumaged races, but its vocalisations are poorly known (Spencer 2012). Spencer (2012) suggested that the population of *G. rufula* in the East Andes of Colombia around Bogota sounds very different from surrounding populations, with the song being 'made up of a series of long, clear, even whistles given at a fairly high pitch, with the whole series descending slightly'. He also noted that 'the call … is a very distinctive 'whistle-purr' a clear, fairly high-pitched whistle reminiscent of one note of the song, followed by a lower pitched 'purr,' an even, modulated note about the same length as the first'. Subspecies *cajamarcae* in northern Peru south of the Marañón River delivers a song comprising a slow series of medium-pitched, mellow notes, lasting 3–4 seconds, and consisting of 5–8 or more notes (Spencer 2012). Spencer (2012) likened it to songs from the population in the East Andes of Colombia, but described it as lower pitched and having shorter notes that are less whistled. The call of *cajamarcae* is a rapid series of short, loud notes, described by Spencer (2012) as having a 'distinctive upslur' near the end of the series. The entire call lasts only 2–4s and is similar in pitch to the song. While the two vocalisations of *cajamarcae* are designated song and call, Spencer (2012) pointed out that they are similar in duration, but the slower vocalisation is heard more frequently and given at shorter intervals, suggesting that it may function as a call rather than a song. The song of *obscura* in central Peru is described by Spencer (2012) as a buzzier version of the song of the nominate subspecies. It comprises a single, fairly clear note, followed by a fast, buzzing whistle. The latter portion is likened to a 'fast police whistle' by Spencer (2012). Krabbe & Schulenberg (2003) described the acoustic properties thus: it lasts 0.9s and is delivered at intervals of 3–6s. They noted that the introductory note is sometimes omitted but is followed by an accelerated series of *c*.30 notes all between 2.2 and 2.6kHz. They stated that this can be rather similar in duration and pitch to that of nominate *rufula*, but that *rufula* rarely delivers more than 16 notes. The call of *obscura* is a series of even-pitched, buzzy whistles, like the second note of the song but somewhat drier and shorter (Spencer 2012). This call is apparently delivered at fairly short intervals and lasts 2–6s. In most of the recordings available to Spencer (2012), this vocalisation speeds up and rises at the end, but he pointed out that along the

Satipo Road in Junín ('true *obscura*' if a split occurs) the song has a more even pitch and pace. In summary, although slight, there appear to be some vocal differences between populations of *obscura* immediately south of the Río Marañón, at least in the Leyemebamba / Abra Barra Negro area, and those in the centre of its range, at least near the type locality in Junín. Krabbe & Schulenberg (2003) described one call of *obscura* as an even-paced trill 3.5–3.8s long of 31–35 notes delivered at 2.2–2.4kHz, and a second as a 1–2s series of 4–7 notes like police whistles. The latter is delivered at 3kHz and repeated at intervals of 4–7s. Spencer (2012) felt that available recordings of calls of *obscura* showed much variation, but that there is less geographic variation in songs. Those of the southernmost subspecies, *occabambae* and *cochabambae*, are the least complex. Krabbe & Schulenberg (2003) described the song of *occabambae* as lasting 4.0–4.4s and consisting of an evenly paced trill of *c*.25 notes. The trill falls from 2.2 to 2.0kHz then evens out, rising only at the very end to 2.4kHz. They described the song of *cochabambae* as 2.5–3.3s long and consists of an even-paced trill of 25–35 notes at 2.6–2.7kHz with a slight fall in pitch at the start, but rising at the end. Krabbe & Schulenberg (2003) described the call of *occabambae* as two-noted and lasting only 0.3–0.4s. This call is given at intervals of 2–8s. They stated that calls given in Cuzco are at 2–2.2kHz, rather soft and of equal length, with the first note slightly weaker. Elsewhere in the range of *occabambae* this call is sharper, with the first note shorter, and in the La Paz population pitch ranges from 2.4 to 2.6kHz. They described the call of *cochabambae* as similar, two-noted rather like a sharp version of *occabambae* but at 2.0–2.2kHz. Overall, there seems to be much overlap and confusing variation within populations in the combined ranges of *occabambae* and *cochabambae* (Spencer 2012), and further detailed comparisons are needed before these two subspecies can be definitely considered vocally distinct. However, Spencer (2012) did suggest that, together at least, these two subspecies vary significantly from other races.

NATURAL HISTORY Although many accounts state that Rufous Antpitta is solitary or found in pairs, this small, quiet antpitta has been found with mixed-species flocks, perhaps more than many other *Grallaria* (Bloch *et al.* 1991, HFG). Like its larger congeners, the present species hops rapidly on the forest floor, pausing to flick aside leaves with its bill, frequently probing into soft soil or to reach under leaves on the ground. Krabbe & Schulenberg (2003) noted that the typical song of nominate *rufula* may be given only 1–3 times during an entire morning but, even when singing more frequently, its soft voice is easily overlooked. This quiet species usually flits and hops silently across the forest floor or through dense tangles, venturing only occasionally into the open on landslides, trails and mossy clearings at dawn and dusk, or in foggy weather (Goodfellow 1902, Peters & Griswold 1943). Adults sometimes perch in shrubs or brushy tangles, especially when disturbed and seeking a better vantage point to observe any disturbance (Krabbe & Schulenberg 2003). I have noticed this is particularly true during breeding (see Reproduction). The territorial behaviour of Rufous Antpitta has not been studied in detail, but it appears that territories may be either very large or, more likely, widely separated, as one study found almost 0.5km between vocalising individuals (mean 389 ± 159m SD; Harris & Haskell 2013).

DIET Rufous Antpittas are known to prey only on small invertebrates, with specific items reported to date including spiders, lepidopteran larvae, worms, grasshoppers, cockroaches and centipedes (Asociacíon Bogotana de Ornitología 2000, Krabbe & Schulenberg 2003). I identified 95 (of 131) prey items delivered to an older nestling near Papallacta (*rufula*). Prey included the following invertebrates (*n*): lepidopteran larvae (17; all hairless, Geometridae, Nymphalidae, unknown), Orthoptera (16; Acritidae, Tettigoniidae, unknown); Neuroptera (11, Chrysopidae, Hemerobiidae, Myrmeleontidae); Phasmida (11); spiders (9); earthworms (7); Tipulidae (7); lepidopteran adults (6; Tortricidae, Geometridae, unknown); Coleoptera (3, Scarabaeidae, unknown); Chilipoda (2); Diptera (3; unknown Brachycera); lepidopteran pupa (1, *Eretris* sp.); coleopteran pupa (1) and Isopoda (1). In addition, from photographs of adults feeding a mid-aged nestling at Wayqecha Biological Station (S. David *in litt.* 2017), I identified lepidopteran larvae (4, Geometridae, unknown), earthworms (2) and Hymenoptera (1).

REPRODUCTION The first description of the nest, eggs and young was based on a nest of nominate *rufula* in the vicinity of Volcán Nevado de Ruiz, Caldas, 3,620m (Colombia), which hatched around mid-July 1989 (de Fabrègues 1991). The following year, Whitney (1992) published the often-cited description of a nest and egg of nominate *rufula* found on 25 March 1990 along the Río Chalpi Grande below Papallacta Pass, Napo, north-east Ecuador, at 2,700m. Subsequently, a third nest of *rufula* was discovered, on 26 November 2003, at RP Tapichalaca, Zamora-Chinchipe, south-east Ecuador, at 2,500m (Greeney & Gelis 2005b). At the same location as this latter nest, Heinz (2002) described two nests believed to be of this species although they were not active at the time. As all aspects of the nests described in Heinz (2002) suggest that his supposition was correct, I have included these data below. To date, these remain the only published accounts of nests for *G. rufula*, with all but the nominate subspecies lacking published nest descriptions. **NEST & BUILDING** The nest is an open cup, as described for all members of the family. Nests are generally built fairly low above ground: 0.5m (de Fabrègues 1991), 0.75m (Whitney 1992), 2.0m (Greeney & Gelis 2005b), 0.85m and 0.85m (Heinz 2002). To these, I add data from three others nests: 1.1m, 0.8m and 1.4m (all *rufula*, eastern Ecuador). Nest substrates and the position of nests are all fairly similar, being well supported from below and all backed against a solid substrate (making them inaccessible from that direction). Nest substrates and positions are described as a: 'shallow hollow' on the side of a large, moss-covered, 1.5m-tall stump (Whitney 2002); a moss and epiphyte-laden tree supported below by a lateral growth on the trunk and by a sizeable quantity of moss apparently placed there by the adults (Greeney & Gelis 2005b); and one nest on a large clump of epiphytes against a tree trunk and a second wedged between a tree trunk and several mossy stems (Heinz 2002). Previously unreported nest locations include one built on a ledge on a small rocky cliff thickly covered in mosses. At this nest the thick moss and slight overhang above the nest made it appear as if built inside a shallow cavity. In fact, such a well-sheltered situation seems to be the preferred nest site for *G. rufula*. All of the nests mentioned above, except one of the two described by Heinz (2002) were in situations best described as 'tucked into' the substrate, and

were quite well protected both physically and visually from above. Nest dimensions (Whitney 1992, Greeney & Gelis 2005b, Heinz 2002, HFG; *n* = 1, 1, 2, 1, respectively) are as follows: outer diameter ?, 20cm, 18cm, 18cm, 17cm; outer height ?, 12cm, 13cm, 13cm, 14cm; inner diameter 9cm, 10cm, 9cm, 9.5cm, 9cm; inner cup depth 5cm, 6.5cm, 6cm, 5.5cm, 6.5cm. The materials used are described somewhat differently by various authors. In Colombia, de Fabrègues (1991) did not provide a description, but from the photo it is clear that the external portion of the nest, including the rim, as well as at least the upper half of the posterior part of the inner cup were all composed entirely of moss. Whitney (1992) stated that the rim of his nest was 'entirely of thin, yellowish grass stems' and the cup was lined with 'what appeared to be 30–40 yellowish mammal hairs'. The external part of this nest was not described, but apparently dark, damp leaves were visible below the sparse lining. A third nest was described as a large mossy cup lined with pale fibres and dark fungal rhizomorphs by Greeney & Gelis (2005b). Re-examination of my own photographs of this nest necessitates an amendment to this description. While the external portion was, undoubtedly, composed almost entirely of moss, the floor of the nest visible through the very sparse lining had the appearance of dark, damp organic material, including pieces of leaves. As neither I nor Whitney (1992) removed these nests for careful inspection, based on my subsequent observations of Rufous Antpitta I feel that it is likely that both of our nests were 'old' nests that had been built in a previous season and that the damp detritus in the base of the nests probably represented material that had collected there naturally between usage. Furthermore, on close inspection of photos of the 2005 nest, I can see only one piece of lining that is likely a fungal rhizomorph. The remaining dark fibres were actually unbranched, dark, flexible rootlets, perhaps around a dozen in number. The rest of the lining consists of relatively short (4–8cm), semi-flexible bamboo or grass stems. Like the nest described by Whitney (1992) the lining is very sparse and the base of the cup is clearly visible below it, with only the very bottom of the cup having any lining at all. The two nests described by Heinz (2002) were apparently also sparsely lined and included predominantly short, pale bamboo stems and a few dark fibres. It is perhaps noteworthy that, from my experience, the internal dimensions of the nest of Rufous Antpitta are (in proportion to the adult) much larger than any other *Grallaria* whose nest is described. This is readily apparent when one sees an incubating adult in the nest, as most or all of the bird is hidden below the rim of the cup (Fig. 7; Greeney & Gelis 2005b). **EGG, LAYING & INCUBATION** The full clutch at three nests of Rufous Antpitta apparently consisted of a single egg (de Fabrègues 1991, Whitney 1992, HFG; all *rufula*), and two nests were found with only a single nestling (HFG, *rufula*; S. David, *occabambae*). Nests with clutches of two eggs have been reported for *rufula* (*n* = 2; Greeney & Gelis 2005b, Greeney *et al.* 2010) and *occabambae* (*n* = 1; Londoño 2014). These data all suggest that the species regularly produce one-egg clutches and that mean clutch size is probably less than two. All described eggs were subelliptical and immaculate blue, pale blue or turquoise-blue (de Fabrègues 1991, Whitney 1992, Greeney & Gelis 2005b, Londoño 2014). The egg described by Heinz (2002) as 'dark green' seems unlikely, as I have now observed an additional five eggs of this species in Ecuador (*rufula*), all of which I would describe as immaculate turquoise-blue. Egg measurements from the literature are: 25.3 × 22.6mm, 25.4 × 22.5mm

(Greeney & Gelis 2005b). An additional two eggs, from separate clutches near Papallacta, measured 27.8 × 20.6mm and 26.0 × 21.20mm. There is little published on the specifics of adult behaviour during incubation. Whitney (1992) did not specifically note the distance at which the adult left the nest on his approach, but described the adult as sneaking away low to the ground and downslope from the nest, then perching motionless just *c.*6m away. These observations combined with my own experiences suggest that they are somewhat bolder than other *Grallaria*, with my few direct observations suggesting that incubating adults generally do not flush from the nest until an observer approaches to within 2m. The experiences of M.E. Juina J. are similar. Neither Whitney (1992) nor I have noted any vocalisations given, in alarm or otherwise, by adults at nests. The following data were not reported by Greeney & Gelis (2005b), who mentioned only that both adults partook in incubation duties and that they frequently added lining materials and probed the nest lining in the sharp and rapid manner described for other species (Greeney *et al.* 2008a). We videotaped the nest for a total of 12h 15min on the day after it was discovered with two eggs. Filming began just before sunrise, when there was no adult on the nest and the eggs were cold, suggesting that an adult had not spent the night there. No adult arrived for the first 4h 38min of daylight, at which point an adult visited briefly, added a small pale bamboo petiole, probed into the nest several times and left. The eggs were not incubated until 5h 49min after sunrise at which time an adult arrived, added a piece of lining and covered the egg for 11.5min before being flushed when the videotape had to be changed. Thereafter, adults did not return until 12:30, at which time they both arrived in relatively quick succession. Landing on the rim, dropping a piece of lining into the cup, probing a few times and then settling over the eggs, only to be replaced a few minutes later by the other adult who repeated the same procedure. Adults alternated incubating and bringing nest lining for 38min, covering the eggs for most of this time except during the brief switches, which occurred seven times during this period, the final arrival resulting in a 95min incubation bout before being replaced by the other adult. This adult sat for just 38min before being replaced, whereupon the other remained in the nest until after dark and presumably spent the night. In total, the eggs were covered for just 28% of daylight hours and adults brought material on all arrivals except the final visit of the day (9/10 arrivals). These observations strongly suggest that the eggs were laid no more than two days prior to these observations (*c.*25–26 November 2003), as most antpittas display limited incubation immediately following clutch completion (Greeney *et al.* 2008b). As reported by Greeney & Gelis (2005), the nest was filmed again two days later, when both eggs were predated by Turquoise Jays *Cyanolyca turcosa*. Unmentioned in their paper, however, was that after the video camera was set up at sunrise, no adult antpittas were seen at the nest until the jays arrived *c.*1h later. Interestingly, the first jay to arrive at the nest pecked into it several times but then flew away. It was not until half an hour later that the jays returned and consumed both eggs. Videotaping continued after they left and no antpittas had visited the nest by 11:00, when the camera was taken down. **NESTLING & PARENTAL CARE** I am aware of just one cursory description of young *G. rufula* nestlings. A recently hatched nestling was described by de Fabrègues (1991) as being covered in fairly dense black down. This author did not report any further details, but doubtless it was similar in natal down

coverage to other *Grallaria* on hatching (Greeney 2012a). Previously unpublished, I described a mid-aged nestling (*rufula*) near Papallacta, 3,565m, Napo, Ecuador. I first observed the nestling on the day prior to its primary pin feathers breaking their sheaths. The post-natal down feathers were ½–¾ emerged from their sheaths, coating the nestling in a dense, wool-like covering of sooty-black down. The upperparts were finely barred chestnut-brown, these markings paler and broken into small spots and streaks on the nape and crown. The down on the breast was also indistinctly marked, but so diffusely as to appear washed brownish. The belly and vent were unmarked and only slightly paler than other regions. The median coverts were unbroken while those of the secondaries had emerged from their sheaths 1–2mm. The skin was yellowish to orange-pink and the mostly open eyes were dark brown. The bill was mainly bright orange, duskier on the culmen with a distinct blackish area on the maxilla, just anterior of the nares. The pale yellowish egg-tooth was still attached. The rictal flanges were brighter orange and the mouth lining bright crimson-orange. The tarsi and toes were dusky-pinkish with an orange cast proximally. The nails were dull yellow. Four days later, the nestling's appearance was largely unchanged except for a few scattered feathers on the breast now with indistinct buffy-white shaft-streaks. The primary pin feathers had emerged 2–5mm from their sheaths, the secondaries 1–4mm. The median secondary and primary wing-coverts had emerged roughly halfway from their sheaths. All coverts and flight feathers were dark blackish-brown with the tips and leading edges reddish-chestnut. The median coverts were very finely margined black. The only nest watched during the (early) nestling period (de Fabrègues 1991) was not reported in detail. However, both adults attended and seemed to approach the nest from the same direction on most visits. They apparently had a favoured perch *c.*2m from the nest, where they paused before going straight to the rim to deliver food. During my presence at the above-mentioned Papallacta nest, on all seven occasions adults approached me quietly on the ground and low horizontal perches. They never dropped or consumed the prey they carried, despite my presence for nearly 20min on one occasion, and quickly fed the nestling within 2min of my departure. They frequently approached to within 2.5m, periodically flicking their wings but never vocalising. Based on the direction that adults arrived at the rim of the nest, it appears they also had favoured pre-arrival perches, either from the ground below or a horizontal log on the steep slope 2m above and 2m away from the nest. **SEASONALITY** Apart from the nests mentioned above, reproductive data is scarce. For the two northernmost races *spatiator* and *saltuensis*, Hilty & Brown (1986) reported birds of both subspecies collected in breeding condition between March and May. For the remaining subspecies, Whitney (1992) referred to breeding-condition birds collected in Carchi in mid-March 1992 (*rufula*) and to specimens from northern Peru in breeding condition in June–October, either *cajamarcae* or *obscura*. A nest found by F. Sornoza on 4 November 2001 at RP Tapichalaca apparently contained an abandoned (but not yet rotten) egg with a nearly fully formed chick inside (Heinz 2002). It should be noted that, although Whitney (1992) reported breeding-condition adults from Carchi in March, this evidence is based on 1–2mm ova in females. Given how little we know of the onset and pace of reproductive readiness in antpittas, and considering there was no reported evidence of active brood patches in these specimens (which should have been visible in both sexes), these records should be used cautiously. **Breeding Data G. r. rufula** Two records of adults carrying material, RP Yanacocha, 8 March 2014 (K. Miller *in litt.* 2015) and 20 December 2009 (M. Bauman); nest building, 22 September 2007, RP Tapichalaca (Greeney *et al.* 2010); incubation, 7 July 1989, Volcán del Ruiz (de Fabrègues 1991); two nests with one egg each, Río Chalpi Grande, 25 March 1990 (Whitney 1992) and 7 February 2016 (ML 24750661, photo D. Humple); two nests with eggs, RP Tapichalaca, 26 November 2003 (Greeney & Gelis 2005) and 22 September 2007 (Greeney *et al.* 2010); two nests with one egg each, one nest with a young nestling, c.4km N of Papallacta, fresh egg 2 March 2011, well-developed egg 16 November 2012, nestling 17 November 2015 (HFG); nestling, 4 November 2008, RP Tapichalaca (Greeney *et al.* 2010); nestling, 17 November 2015, c.4km N of Papallacta (HFG); adult carrying faecal sac, 14 April 2014, RP Yanacocha (F. Rowland *in litt.* 2017); two records of fledglings, RP Tapichalaca, 29 October 2007 (Greeney *et al.* 2010), 28 September 2010 (G.M. Kirwan *in litt.* 2017); fledgling, 10 September 2008, Cerro El Águila (A.M. Cuervo *in litt.* 2017); juvenile, 14 July 1911, Micay (AMNH 109634); juvenile, 31 May 1913, Volcán Pichincha (AMNH 124440); two juveniles, Verdecocha, 4–5 September 1922 (AMNH 173308/09); juvenile, 1 February 1899, Papallacta (AMNH 492264); two juveniles, Páramo de Frontino, 27 July 2010 (D. Calderón-F. *in litt.* 2017) and 20 August 1951 (USNM 436485); juvenile, 17 June 2011, RP Yanacocha (S. Olmstead *in litt.* 2015); subadult, 26 September 1923, Oyacachi (AMNH 180272); two subadults, RP Yanacocha, 16 July 1967 (MECN 2613, M.V. Sánchez N. *in litt.* 2017) and 14 July 2014 (IBC 1072364, D.M. Brinkhuizen); subadult, 1 December 1898, Volcán Pichincha (AMNH 492262); subadult transitioning, 15 March 1992, W slope of Cerro Mongus (MECN 6104, M.V. Sánchez N. *in litt.* 2017); 'fledgling', February, 'northeastern Ecuador' (Fjeldså & Krabbe 1990); 'juveniles/immatures' in July in Cauca, September in Quindío, December, May, June and September in NW Ecuador (Fjeldså & Krabbe 1990). **G. r. cajamarcae** subadult, 10 July 1026, Taulís (AMNH 235897). **G. r. obscura** juvenile, 14 June 2008, Ccano (KUNHM 112735); juvenile transitioning, 5 March 1922, Rumicruz (AMNH 174092); juvenile transitioning, 22 July 2012, Bosque Unchog (C. Calle photo); juvenile transitioning, 2 March 1973, Bosque Zapatogocha (FMNH 296698); 'juveniles/immatures,' March, 'Pasco' (Fjeldså & Krabbe 1990). **G. r. occabambae** Two records of nestlings, Wayqecha Cloud Forest Biological Station of nestlings, 10 December 2011 (S. David *in litt.* 2016) and a fledgling, 23 December 2011 (T. Grim & S. David *in litt.* 2016, see IBC 996417); two juveniles, 28 July and 1 August 1915, Tocopoqueyu (USNM 273162/164); juvenile transitioning, 3 August 1958, Marcapata (YPM 81713, S.S. Snow *in litt.* 2017); subadult, 26 July 1974, 14km NE of Abra Málaga (LSUMZ 78570).

TECHNICAL DESCRIPTION Sexes similar. **Adult** Nominate *rufula* is rufous-brown above, slightly brighter rufous on sides of head, throat and breast. Flanks dark greyish-rufous to tawny-rufous, with central belly paler rufous, sometimes with paler feather tips giving a slightly scaled appearance. Lower belly and vent buffy-white or white, with darker greyish feather bases sometimes giving mottled appearance. Undertail-coverts rather variable, from whitish to buffy or dark brown (Krabbe & Schulenberg 2003). **Adult Bare Parts** *Iris* dark brown; *Bill* maxilla blackish, mandible mostly blackish, sometimes paler near

base; *Tarsi & Toes* lead-grey to blue-grey. **Immature** Fjeldså & Krabbe (1990) described immature plumage (of unknown subspecific designation) as simply 'back barred buff, crown and underparts streaked whitish', a description that appears to best correspond to my definition of fledglings. Sclater (1877) published the first description of what I feel was likely a fledgling, stating: 'From the skin of an immature bird (from Bogotá) in my collection, the young plumage of [subspecies *rufula*] would appear to be of a blackish grey, with long white shaft-spots'. Here I describe two, apparently fairly different, fledgling plumages for two races of Rufous Antpitta (*rufula* and *occabambae*). It is unclear, however, the true extent of these differences, and to what degree the described differences are age related. This is, in part, because I have not yet seen young across the whole range of plumage maturation and because my descriptions are based on small sample sizes of differently aged young. The most striking difference between the fledglings described below are the heavily patterned underparts of *occabambae* versus the relatively unpatterned underparts of *rufula*. My examination of the mature nestlings of both subspecies reveals that these streaks appear in *occabambae* prior to fledging and almost certainly do not in *rufula*. Nevertheless, similar streaking is clearly present on juvenile *rufula*, suggesting that they may follow a similar ontogenetic progression, but may lag behind that of *occabambae* with respect to age. Based on the scant evidence, it appears that the upperwing-coverts are largely unpatterned at fledging, but are replaced by more 'typically patterned' immature coverts at some point post-fledging. These are seemingly retained for the rest of plumage maturation, apparently losing much of their pattern due to wear before they are replaced by adult coverts. Further investigation is certainly warranted before this progression can be confirmed. **Fledgling (rufula)** The following is based on my examination of an older nestling (NE Ecuador), probably only a day or two away from fledging. I saw no clear evidence that its appearance would change greatly prior to leaving the nest. It was entirely covered in thick, sooty-black down, darkest (black) on the crown and palest (dark grey) on the central belly and vent. The crown and nape were finely and indistinctly spotted and streaked buffy-brown, these markings forming fine barring (still indistinct) on the back and rump. Below the downy plumage was largely unmarked except sparse whitish or buffy streaking on the breast. The tail feathers were only just breaking the tips of their sheaths and it is unlikely that any part of the tail would have been visible beyond the down feathers at the time of fledging. The wing feathers were almost completely emerged from their sheaths, with flight feathers appearing similar to adults. The upperwing-coverts were dark grey edged on the leading margin with dull rufous-brown that extended over their tips. Primary wing-coverts were otherwise unmarked but secondary coverts bore very faint, thin, blackish subterminal bars. **Fledgling Bare Parts (rufula)** *Iris* dark brown; *Orbital Skin* dark, blackish-grey; *Bill* almost entirely bright orange, dusky on culmen and near tip of the mandible, rictal flanges inflated and a deeper, crimson-orange, tip of maxilla still with a pale yellowish egg-tooth, mouth lining deep crimson; *Tarsi & Toes* dusky-pinkish, yellow-orange on posterior margin of tarsi, nails dull yellow. **Fledgling (occabambae)** The following is based on images of very young fledglings, still incapable of sustaining an upright posture, photographed in Dept. Cuzco by T. Grim and S. David. They were superficially similar to young *rufula* described above in being covered in dark, fluffy down. On their upperparts, however, the fledglings seemed darker overall, nearer black. They also apparently lacked any barring, spotting or streaking on the crown and back. There were several small, irregular-sized and uneven patches of adult-like contour feathers disrupting the otherwise fluffy down. These were dull brown to olive-brown and most obvious on the upper back and nape. The ocular region was nearly fully feathered and similar in colour to the emerging back feathers. Below, the *occabambae* fledglings differed distinctly from young *rufula* in having bold, sparse, somewhat arrow-shaped streaking on the otherwise blackish breast. The streaks were formed by buffy-ochraceous shaft-streaks that broadened near the feather tips. The streaking on individual feathers was reduced, approaching spotting, on the breast-sides, but became pale buff and more diffuse on the lower breast and belly. These markings, along with buffy tips to the down plumes on the flanks, made most of the vent and lower belly appear largely buff or buffy-white. The feathers of the tail did not yet extend beyond the down. Their flight feathers were fairly similar to those of adults, brownish-chestnut on the leading web and dusky-brown on the trailing web. The upperwing-coverts were similar to those of *rufula*, but the chestnut-brown parts of the leading webs were not confined to narrow fringes, but instead covered most of the feather, leaving only a dark grey margin on the trailing edge. The tips of the coverts seemed heavily worn and were finely edged blackish, including the primary-coverts. It was unclear how they would have been patterned when fresh, but they probably bore thin subterminal bands as described for the secondary coverts of *rufula*. **Fledgling Bare Parts (*occabambae*)** As described for *rufula*, the ocular area is now largely feathered. ***Juvenile (*rufula*)*** An older juvenile *rufula* from north-east Ecuador (AMNH 492264) was in the process of moulting into subadult plumage when collected. The newly grown contour feathers, in all locations, appear to accurately approximate adult colour for nominate *rufula* from this region. Irregular patches of immature plumage are scattered across nearly all of the body, especially on the lower breast, central belly, and sides of the neck, nape and rump. Flight and tail feathers are fully emerged from their sheaths and appear fully adult in coloration. Wing-coverts are dark grey, broadly fringed rufous on the lateral margins. Each covert has a broad, poorly defined, pale rusty-buff subterminal bar. On some feathers this is reduced to a central, ill-defined spot. The coverts have blackish fringes, of somewhat variable breadth, seeming to vary based on wear, and almost absent in some (especially on secondary coverts). ***Subadult (*rufula*)*** As mentioned above (see Immature), the ontogenetic timing and frequency of upperwing-covert replacement is uncertain. I have examined several specimens of nominate *rufula* that were, with the exception of the coverts, in apparently full adult plumage. Pattern on the coverts varied individually, however, ranging from distinctly buffy-chestnut tips with clear black terminal bars, to almost unmarked coverts with little more than a thin black margin. It is clear to me that upperwing-covert markings are an indication of immaturity, but the degree to which the above-described variation corresponds to age or other factors remains to be seen.

MORPHOMETRIC DATA *G. r. spatiator* *Wing* 83.6mm; *Tail* 42mm; *Bill* exposed culmen 20mm; *Tarsus* 46mm ($n = 1$, ♂ holotype, Bangs 1898b). *Wing* 81–83mm / 77–79mm (♂/♀ range); *Tail* 40–42mm (♂+♀) ($n = 4$?, probably

includes holotype as authors mentioned only four known specimens; Cory & Hellmayr 1924). *Wing* 79mm, 77mm, 81mm; *Tail* 42mm, 40mm, 41mm; *Bill* [exposed? culmen] 19mm, 19mm, 19mm; *Tarsus* 44mm, 47mm, 46mm ($n = 3$, ♂,♀,♀, Todd & Carriker 1922). *G. r. saltuensis* *Wing* 80.2mm, 78.6mm, 79.3mm, 80.4mm; *Tail* 36.9mm, 39.9mm, 42.0mm, 42.5mm; *Bill* 'culmen from base' 17.8mm, 18.8mm, 19.1mm, 19.4mm; *Tarsus* 43.3mm, 44.5mm, 44.8mm, 45.8mm ($n = 4$, ♂,♂,♀,♀, including holotype, Wetmore 1946). *G. r. rufula* *Wing* 77mm; *Tail* 47.8mm; *Bill* from nares 11mm, exposed culmen 17.9mm, depth at front of nares 5.6mm, width at front of nares 4.7mm; *Tarsus* 41.6mm ($n = 1$ unsexed, Zamora-Chinchipe, Ecuador, B. Mila *in litt.* 2017). *Wing* 81.0mm, 81.0mm, 82.0mm; *Tail* 44.0mm, 43.0mm, 43.0mm; *Bill* from nares 11.5mm, 10.0mm; 10.0mm; *Tarsus* 44.0mm, 43.0mm, 42.0mm ($n = 3$, ♂,♀,♀, Tungurahua, Domaniewski & Sztolcman 1918). *Bill* from nares, 11.7mm, 11.2mm, 10.4mm; *Tarsus* 43.5mm, 44.5mm, 43mm ($n = 3$, ♂♂, Napo, WFVZ). *Tarsus* 44.3mm ($n = 1$, ♂, Morona-Santiago, MLZ 7278). *Bill* from nares 11.8mm, 10.9mm; *Tarsus* 45.1mm, 44.6mm ($n = 2$, ♀♀, Cauca, LACM 37383 Sucumbíos, LACM 40446). *Bill* from nares 10.1mm; *Tarsus* 42.2mm ($n = 1$, ♂, juvenile, Napo, AMNH 492264). *G. r. cajamarcae* *Wing* 83–88mm, 80mm; *Tail* 45mm, 44mm; *Bill* [exposed] culmen 22mm, 19.5mm; *Tarsus* 44–46mm, 44mm ($n = 4$, ♂,♂,♂,♀, Chapman 1927). *G. r. obscura* *Wing* 90mm; *Tail* 46.5mm; *Bill* [exposed] culmen 19.5mm; *Tarsus* 44.5mm ($n = 1$, ♀, holotype, Berlepsch & Sztolcman 1896). Graves (1987) provided mean ± SD (range) for 16 ♂♂/ 13 ♀♀. *Wing* 81.3 ± 1.7mm (78.8–84.3mm) / 81.6 ± 15mm (78.2–84.5mm); *Bill* from front of nares 10.9 ± 0.5mm (10.0–12.1mm) / 10.9 ± 0.6mm (9.9–12.0mm), width at front of nares 4.8 ± 0.2mm (4.2–5.2mm) / 4.9 ± 0.3mm (4.2–5.2mm); *Tarsus* length 41.7 ± 1.0mm (40.2–43.6mm) / 41.4 ± 1.3mm (39.3–43.2mm), width at narrowest medio-lateral point on tarsometatarsus 1.5 ± 0.1mm (1.4–1.6mm) / 1.5 ± 0.1mm (1.4–1.6mm); *Middle Toe* 22.5 ± 0.9mm (20.9–23.7mm) / 22.4 ± 0.5mm (21.5–23.3mm). *G. r. occabambae* *Bill* [exposed?] culmen 17–18mm, depth at base 5.5mm ($n = 2$, Chapman 1923a). *G. r. cochabambae* *Wing* 84.5mm, 79.5mm; *Tail* 45mm, 42mm; *Bill* [from nares] 13.5mm, 13mm; *Tarsus* 47mm, 44mm ($n = 2$, ♀,♂, including holotype, Bond & Meyer de Schauensee 1940). **Mass** 37.9g ($n = 1$ unsexed *rufula*, Zamora-Chinchipe, T. Smith *in litt.* 2017), 38.9g(lt), 38.4g(tr), 42.1g(lt) ($n = 3$, ♀,♀,♂, *rufula*; MSB, Chontalí, Cajamarca), 39g ($n = 1$ unsexed *rufula*, Rahbek *et al.* 1993), 41g ($n = 6$ *occabambae*, sexes combined, Parker & O'Neill 1980), mean 35.3g/34.7g ($n = 9$, 6♂♂/3♀♀, *cochabambae*, vicinity of Chusipata, Remsen 1985), mean 41.6g ($n = 15$, *obscura*, genders combined, Schulenberg & Williams 1982); male 35–46g, female 39–46g, unsexed mean 45.5g ($n = ?$, race?, Krabbe & Schulenberg 2003), 40.4g(no), 47.3g(no), 46.3g(no), 44.0g(lt), 44.0g(no), 45.0g(lt) ($n = 6$, 1♀, 5♂♂, *rufula*, Napo, WFVZ, MECN), 34.4g, 36.3g, 38.9g, 39.0g, 40.1g ($n = 5$, ♂,♂,♀,♂,♀, *cajamarcae*, Lembayque, MSB), 38.4g, 42.1g ($n = 2$, ♀,♂, *cajamarcae*, MSB, Cajamarca), 36.3g, 37.8g, 39.0g, 39.5g, 41.0g, 47.5g ($n = 6$, ♀,♀,♂,♂,♀,♂, *occabambae*, Cuzco, MSB), 36.3g, 38.2g ($n = 2$, ♂,♀, *rufula*, Pichincha, MSB), 38g ($n = 1$, ♀, Apurimac Valley, either *obscura* from Ayacucho or *occabambae* from Cuzco; Weske 1972), 33.8g ($n = 1$, ♂?, *occabambae*, Cuzco, Londoño *et al.* 2015). **Total Length** 13.0–15.5cm (Bangs 1898b, Meyer de Schauensee & Phelps 1978, Dunning 1982, 1987, Hilty & Brown 1986, Fjeldså & Krabbe 1990, Ridgely & Tudor 1994, Asociacíon Bogotana de Ornitología 2000, Clements & Shany 2001, Ridgely & Greenfield 2001, Schulenberg *et* *al.* 2007, McMullan *et al.* 2010). **OTHER** Londoño *et al.* (2015) reported the basal metabolic rate of a single bird (*occabambae*) at 0.52 watts.

TAXONOMY AND VARIATION Rufous Antpitta served as the type species for Ridgway's (1909) genus *Oropezus*, which is supposedly distinguished from other *Grallaria* in part by its proportionately longer legs and smaller bill, and unmarked upperparts. Seven subspecies have generally been recognised over the past 70 years (Peters 1951, Fjeldså & Krabbe 1990, Krabbe & Schulenberg 2003) but the species is currently the focus of a much-needed taxonomic revision (M. Isler & R.T. Chesser *in litt.* 2016). Domaniewski & Sztolcman (1918) described an eighth subspecies, *G. r. saturata*, on the basis of a female/male pair of syntypes (MZPW 33971/33976) and a topotypical female (MZPW 33969) collected by Sztolcman in March 1884 at 'S. Rafael' [= San Rafael, Tungurahua, *c.*01°22'S, 78°29'W, 2,750m, see Paynter 1993], eastern Ecuador (Taczanowski & von Berlepsch 1885, Mlíkovský 2009). This subspecies was synonymised with nominate *rufula* by Cory & Hellmayr (1924) and I have seen no evidence that it warrants recognition. Somewhat larger and distinctively less rufous than the other subspecies, *saltuensis* probably deserves species rank, a distinction already bestowed by some authors (Salaman *et al.* 2001, del Hoyo & Collar 2016).

Grallaria rufula spatiator Bangs, 1898, Proceedings of the Biological Society of Washington, vol. 12, p. 177, 'Macotama, 8,000ft, Santa Marta, Colombia' [= Páramo de Macotama, 2,440m, La Guajira, Sierra Nevada de Santa Marta, Colombia; 10°55'N, 73°30'W]. The holotype is an adult male collected 17 June 1898 by W.W. Brown, Jr. (Bangs 1898b, Allen 1900). This narrowly distributed subspecies is confined to the Sierra Nevada de Santa Marta of northern Colombia. Race *spatiator* resembles nominate *rufula*, but has a shorter bill, longer and more slender tarsi, is darker and duller brown overall, and the abdomen is dirty white instead of buffy (Bangs 1898b, Todd & Carriker 1922). In his original description, Bangs (1898b) provided the following details: above, including tail, 'about mummy brown' with many of the individual feathers washed reddish-olive; primaries dusky but edged reddish; chin whitish while the throat and breast are cinnamon-russet and flanks raw umber; the lower abdomen and vent appear dirty white, with the feathers 'somewhat marbled with raw umber and russet.' **SPECIMENS & RECORDS** *Magdalena* Estación Experimental San Lorenzo, 11°07'N, 74°03'W (XC 56827/28, B. Lopéz-Lanús); Cuchilla de San Lorenzo, 11°06'N, 74°04'W (CM P37917, XC 235761–766, N. Krabbe); Páramo de Macotama, 10°55'N, 73°30'W (MCZ 105683). *La Guajira* Cuchilla Caracas, 10°57'N, 73°34'W (CM P45155/223). *Cesar* Río Guatapurí, *c.*10°33'N, 73°28'W (USNM 387368/69). **SEE ALSO** Allen (1900), Todd & Carriker (1922), Strewe & Navarro (2004).

Grallaria rufula saltuensis Wetmore, 1946, Smithsonian Miscellaneous Collections, vol. 106, no. 16, p. 4, 'between 9,500ft and 10,000ft, south of the south Teta above Airoca, Sierra de Perijá, Depto. Magdalena, Colombia [Dept. Cesar, Municipio Becerril, above Eroca, *c.*5.75km SSW of Sabanas de San Genaro, 2,895–3,050m, 09°42'N, 73°05'W; GPS from Paynter 1997]. The holotype is an adult male housed at the Smithsonian (USNM 373673), collected 4 May 1942 by M.A. Carriker Jr. (Deignan 1961). Krabbe & Schulenberg (2003) suggested that *saltuensis* may, in fact, be better placed as a subspecies of *G. quitensis*. Even

at the time of its description, Wetmore (1946) somewhat reluctantly assigned *saltuensis* as a race of *rufula*, noting that it was fairly divergent in form and colour from both *spatiator* to the west and nominate *rufula* further south. His remarks included the possibility that it deserved species rank. Race *saltuensis* is endemic to the Perijá Mountains on the Venezuela (Zulia) / Colombia (César/La Guajira) border (Phelps & Phelps 1963, Meyer de Schauensee & Phelps 1978). As noted, *saltuensis* is fairly unique, perhaps most similar to *spatiator*, but much duller and paler overall. It is more olive and less rufescent above with slightly contrasting clay-coloured feather tips, while below it is generally paler and more whitish-grey, paler on the throat, belly, flanks and undertail-coverts (Wetmore 1946, Krabbe & Schulenberg 2003). In *saltuensis*, the upperparts are between buffy-brown and olive-brown, with the feathers of the forecrown somewhat whitish basally. The innermost primaries and secondaries have the outer margins 'snuff brown' or buffy grey-brown, while the outer edges of the outer primaries are more 'avellaneous' or light brownish-olive. The tail is buffy grey-brown. The 'bases of upper loral feathers [are] deep olive buff making an indistinct spot' and there is a narrow, indistinct whitish eye-ring. The ear-coverts are similar in colour to the upperparts, washed more brownish on their basal margins and fading to whitish on the throat. The remaining underparts, including the foreneck, breast, belly and undertail-coverts are dull white. The breast-sides, flanks and thighs are buffy-brown and the underwing-coverts light brownish-olive or brownish-grey (Wetmore 1946). **SPECIMENS & RECORDS COLOMBIA**: *La Guajira* 'Cerro Pintado', Laguna del Junco, 10°29'N, 72°55'W (USNM 373676). *Cesar* above Vereda El Cinco, 10°22'N, 72°57'W (XC 234286/89, A.M. Cuervo); Páramo de Sabana Rubia, 10°22'N, 72°56.5'W (XC 322051–053, D. Calderón-F.). **VENEZUELA**: *Zulia* Cerro Viruela, 10°25'N, 72°53'W (COP 74167–169); Serranía del Perijá at 'Poste Frontera 5' 10°23'N, 72°53'W (COP 72844/45); and 'Poste Frontera 2' on SE slope of Cerro Tetari, 10°05'N, 72°55'W (long series COP, AMNH).

Grallaria rufula rufula Lafresnaye, 1843, Revue Zoologique par La Société Cuvierienne, p. 99, 'Colombie' [= 'Bogotá'; Cory & Hellmayr 1924, Peters 1951]. The holotype of Rufous Antpitta (MCZ 76736) is, unfortunately, among the many 'Bogotá trade skins' of which there is ultimately no way of knowing their true provenance. Perhaps its collection locality might be better approximated through genotyping. Depending, however, on the taxonomic rearrangements proposed with respect to various Colombian populations, knowing where it was collected might prove to be of little importance. The nominate subspecies, as currently defined, is the most widely distributed race, with much individual and poorly understood variation (Graves 1987, M. Isler *in litt.* 2016). Nominate *rufula* ranges from Táchira and Apure in south-west Venezuela, south and west patchily through most of the Colombian Andes and Ecuador to northern Peru (Ridgely & Greenfield 2001, Schulenberg *et al.* 2007). Its range extends south through both chains of the Ecuadorian Andes, as well as some of the higher areas in the central valley (Krabbe *et al.* 1998). Southward, the range of *rufula* extends into northern Peru as far south as northern Piura (Parker *et al.* 1985, Vellinga *et al.* 2004) and northern Cajamarca (Fjeldså & Krabbe 1990). Specimens labelled as from near Gualea at 1,830m (i.e., AMNH 124437–439) are clearly mislabelled, as this location is closer to 1,550m, with no elevations appropriate for Rufous Antpitta in the vicinity. These specimens probably came from somewhere higher on the Andean slope closer to Quito. **SPECIMENS & RECORDS VENEZUELA:** *Táchira* Hacienda La Providencia, 07°33'N, 72°22'W (COP 62197–202); Buena Vista, 07°29'N, 72°26'W (USNM 505784); Páramo de Tamá, 07°24'N, 72°25'W (Phelps & Phelps 1950, COP 11104/05, FMNH 43600). *Apure* La Revancha, 07°26'N, 72°21'W (COP 73944/45); Paramito, 07°24'N, 72°23'W (J.E. Miranda T. *in litt.* 2017). **COLOMBIA:** *Norte de Santander* Páramo de Tamá, 07°23.5'N, 72°24'W (FMNH 4360001); trail to Cerro El Águila, 07°18'N, 72°22'W (A.M. Cuervo *in litt.* 2017). *Santander* Hacienda Las Vegas, 07°04'N, 72°56'W (USNM 411893/94); La Pica, 06°45'N,72°45'W (CM P59572). *Boyacá* Peña Blanca, 06°33'N, 72°30'W (CM P59689); *c.*5.5km SW of Soatá, 06°18'N, 72°43'W (XC 94519–521, O. Cortes-Herrera); SFF Iguaque, 05°54'N, 73°15'W (O. Cortes-Herrera *in litt.* 2017); *c.*3.7km W of San Pedro de Iguaque, 05°38'N, 73°29'W (XC 123355/56; D. Calderón-F.); PNN Pisba, 05°52'N, 72°34'W (D. Carantón *in litt.* 2017); RFP Nacimiento del Río Bogotá, 05°14'N, 73°31'W (XC 299523, K.C. Cubillos); Laguna Montejo, Páramo de Mamapacha, 05°13'N, 73°15'W (XC 293244, D.C. Macana). *Cundinamarca* Reserva BioAndina, 04°48'N, 73°47'W (XC 299364/66; D. Bradley); Aurora Alta, 04°47.5'N, 74°01'W (XC 91831, O. Cortes-Herrera); *c.*4.5km E of Buenavista, 04°45'N, 73°51'W (XC 102560, F. Schmitt); Vereda Chuscales, 04°44.5'N, 73°40'W (O. Cortes-Herrera *in litt.* 2017); *c.*2.3km S of Laguna Chingaza, 04°30'N, 73°45'W (XC 143646, D. Calderón-F.); Chipaque, 04°26.5'N, 74°02.5'W (Chapman 1917); El Peñon, 04°26'N, 74°18'W (Chapman 1917, Meyer de Schauensee 1950); 'Alturas de Quetame', *c.*04°21'N, 73°48.5'W (MCNLS 3969); Vereda El Calvario, 04°16'N, 73°48'W (O. Cortes-Herrera *in litt.* 2017). *Bogotá* Cerro de Monserrate, 04°39.5'N, 74°02.5'W (B.G. Freeman *in litt.* 2017); Quebrada La Vieja, 04°37'N, 74°03'W (XC 32225, N. Athanas); Laguna Chisacá, 04°17.5'N, 74°12.5'W (eBird: A. Pinto). *Meta c.*2km SSW of San Juanito, 04°26.5'N, 73°41'W (F. Rowland *in litt.* 2017). *Antioquia* Páramo de Frontino, 06°26'N, 76°05'W (Flórez *et al.* 2004, Krabbe *et al.* 2006, USNM, XC 27436–438, N. Krabbe); Farallones del Citará, 05°46'N, 76°04'W (Pulgarín-R. & Múnera-P. 2006); Cerro Ventanas road, 05°33'N, 75°48'W (XC 104251, D. Calderón-F.); RNA Loro Orejiamarillo, 05°32'N, 75°48'W (Salaman *et al.* 2007a, XC 16851/52, F. Lambert). *Chocó* La 'M', 06°01'N, 76°10'W (XC 143676, D. Calderón-F.). *Chocó* La 'M', 06°01'N, 76°10'W (XC 143676, D. Calderón-F.). *Risaralda* Near Santa Rosa de Cabal, 04°50'N, 75°33'W (XC 302280, J. Holmes); Cerro Montezuma road, 05°15'N, 76°06'W (XC 116284, D. Calderón-F.). *Caldas* Above Manizales, 05°00'N, 75°21'W (XC 101710, J. Minns); El Zancudo, 05°05'N, 75°30'W (Verhelst *et al.* 2001); Termales del Ruiz–Casa Teja road, 04°58'N, 75°23'W (D. Calderón-F. *in litt.* 2017). *Tolima* Finca La Martínica, 05°05'N, 75°20'W (XC 234299, A.M. Cuervo); Volcán del Ruiz, 04°53.5'N, 75°19.5'W (de Fabrègues 1991); Cajamarca, 04°29'N, 75°29'W (XC 127798, O.H. Marín-Gómez); 10km NW of Roncesvalles, 04°01'N, 75°38'W (XC 29517, N. Krabbe). *Quindío* Laguneta, 04°35'N, 75°30'W (Chapman 1917, AMNH 111997–2000); RMN El Mirador, 04°36'N, 75°44'W (XC 197058, O. Cortes-Herrera). *Cauca* Micay, 02°46'N, 76°55'W (loc. see Paynter 1997: 337, AMNH 109634/35); Cerro Munchique, 02°32'N, 76°57'W (FMNH 249750, LACM 37383); Páramo San Rafael, 02°25'N, 76°25'W (FMNH 255649/50); Coconuco, 02°20.5'N, 76°30'W (de Schuensee 1950, ANSP 142401); PNN Complejo Volcánico Doña Juana-Cascabel, 01°28'N 76°49'W (Ayerbe-Quiñones

2006). *Huila* PNN Puracé, 02°12'N, 76°21'W (XC 68422, F. Lambert). *Putumayo* Páramo Bordoncillo Santiago, 01°09'N, 77°04.5'W (XC 296512, B.C. Jaramillo). *Nariño* Corregimiento de Daza, 01°16'N, 77°15'W (XC 12707, O. Laverde); Volcán Cumbal, Miraflores, 01°02'N, 77°52'W (Strewe 2000); Puerres, 00°53'N, 77°30'W (Meyer de Schauensee 1951, ANSP 162063); Chorreado, 00°50'N, 77°25'W (Meyer de Schauensee 1951, ANSP 162062); Llorente, 00°49'N, 77°15'W (FMNH 292128/29); Aguas Hediondas, 00°49'N, 77°54'W (Krabbe 1991). **ECUADOR:** *Carchi* Tufiño–Maldonado road, 00°48'N, 78°00.5'W (R.A. Gelis *in litt.* 2016); Quebrada Agua Blanca, 00°45'N, 78°04'W (J. Freile *in litt.* 2017); 3–4km ENE of Páramo El Ángel, 00°39'N, 77°54'W (Krabbe 1991); Laguna Negra, 00°37'N, 77°40'W (Stotz & Valenzuela 2009); Loma Guagua, 00°30'N, 77°46.5'W (Buitrón-Jurado 2008, QCAZ); *c.*3km SE of Impuerán, 00°27'N, 77°52'W (Robbins *et al.* 1994, ANSP 184714–719, ML, MECN 6104). *Imbabura* Cayapachupa, 00°33'N, 78°28'W (XC 262136, O. Jahn); Apuela, near Laguna Cuicocha, 00°21.5'N, 78°30.5'W (MECN 2616); Intag, 00°20'N, 78°25'W (Poulsen 2002); Loma Taminanga, 00°17'N, 78°28'W (Krabbe 1991, XC 248415, N. Krabbe); BP Cerro Blanco, 00°13'N, 78°21'W (XC 349245, C. Vogt). *Pichincha* Laguna San Marcos, 00°06.5'N, 77°57.5'W (XC 109776, L. Ordóñez-Delgado); Alaspungo, 00°00'S, 78°36'W (MHNG, J. Freile *in litt.* 2017); Nono, 00°04'S, 78°34.5'W (NRM, J. Freile *in litt.* 2017); RP Verdecocha, 00°06.5'S, 78°37.5'W (ANSP 83346, AMNH 173308/09); RP Yanacocha, 00°07'S, 78°35'W (MECN 2612/13, XC, IBC); Loma Yanayacu, 00°08'S, 78°35'W (Krabbe 1991); slopes of Volcán Pichincha, *c.*00°10'S, 78°35'W (see Notes); Hacienda Rumiloma, 00°11'S, 78°32'W (R.A. Gelis *in litt.* 2017); Lloa, 00°15'S, 78°35'W (ANSP 163680); 7.3km W of Lloa, 00°15'S, 78°38'W (MVZB 160503/05); Volcán Pamba, 00°16.5'S, 78°36.5'W (MVZB 160504); Tablón, 00°22'S, 78°15'W (NR, J. Freile *in litt.* 2017); RVS Pasochoa, 00°28'S, 78°29'W (Fierro 1991); SW slope of Volcán El Corazón, *c.*00°33'S, 78°43'W (Salvadori & Festa 1899, Poulsen 2002, XC 248855/56, N. Krabbe). *Sucumbíos* 3km ENE of Santa Bárbara, 00°39'N, 77°30'W (XC 248355, N. Krabbe); Alto La Bonita, 00°29'N, 77°35'W (Stotz & Valenzuela 2009). *Napo* Oyacachi, *c.*00°13'S, 78°05'W (Baéz *et al.* 2000, Poulsen 2002, AMNH 180272, MCZ 138486, XC); Río Chalpi Grande, 00°22'S, 78°05'W (Whitney 1992); Papallacta and vicinity, *c.*00°23'S, 78°08.5'W (Salvadori & Festa 1899, Goodfellow 1902, AMNH 492263–266, MECN 2615, XC); 2km S of Pan de Azúcar, 00°27'S, 77°43'W (WFVZ 47677/78, MECN 2614); Cordillera de Guacamayos, 00°40'S, 77°51.5'W (Krabbe 1991, MECN 6494); 45km ENE of Salcedo, *c.*00°55'S, 78°12'W (ANSP 183354, WFVZ 49486); Río Ana Tenorio, 00°59'S, 78°20.5'W (Krabbe 1991, Benitez *et al.* 2001, Poulsen 2002, MECN, XC); 1km N of Aucacocha, 01°08'S, 78°19'W (MECN 6156). *Orellana* Upper Volcán Sumaco, 00°32'S, 77°37.5'W (AMNH 184368). *Cotopaxi* Caripero, 00°36'S, 79°00'W (J. Freile *in litt.* 2017); W slope of Volcán Iliniza, 00°42'S, 78°47.5'W (Krabbe 1991); W slope of Cerro Parcato, 00°44'S, 78°58'W (Krabbe 1991, XC 249414/15, N. Krabbe); above Chugchilán, *c.*00°48'S, 78°55'W and Pilaló, 00°57'S, 78°59.5'W (J. Freile *in litt.* 2017). *Tungurahua* 1km SE of Laguna del Tambo, 01°07'S, 78°22'W (XC 249491, N. Krabbe); Quebrada El Golpe, 01°08'S, 78°19'W (J. Freile *in litt.* 2017); Hacienda Pondoa, 01°11'S, 78°38'W (SBMNH 8614); El Triunfo, 01°15'S, 78°22'W (Benitez *et al.* 2001); Yunguilla, 01°23'S, 78°18'W (Berlioz 1932a); RP Machay, 01°23'S, 78°16'W (L. Jost *in litt.* 2017); near Ulba, 01°24'S,

78°23.5'W (FMNH 99577); NW slope of Volcán Tungurahua, 01°26'S, 78°28'W (XC 249515, N. Krabbe); RP Cerro Candelaria, 01°28.5'S, 78°19.5'W (L. Jost *in litt.* 2017). *Bolívar* 10km NW of Salinas, 01°21'S, 79°05'W (Poulsen 2002, XC 249507, N. Krabbe); BP Cashca Totoras, 01°45'S, 78°58'W (Bonaccoroso 2004). *Chimborazo* W slope of Volcán Tungurahua, 01°28'S, 78°28.5'W (Krabbe 1991); Riobamba, *c.*01°40'S, 78°40'W (RMNH, J. Freile *in litt.* 2017). *Cañar* Llavircay, 02°33'S, 78°37'W (P.X. Astudillo W. *in litt.* 2017); La Libertad, 02°36'S, 78°42'W (XC 34, B. Planqué). *Azuay* Palmas, 02°41'S, 78°35'W (Berlioz 1932b); Saucay, 02°45'S, 79°01'W (P.X. Astudillo W. *in litt.* 2017); Paredones de Molleturo, 02°47'S, 79°25'W (Krabbe 1991; Sural, 02°49'S, 79°28'W (Krabbe 1991, XC 250040, N. Krabbe); Laguna Llaviuco, 02°50.5'S, 79°08.5'W (XC 30302, S. Woods); Río Mazán, *c.*02°52'S, 79°07'W (King 1989, Poulsen 2002, XC); Yunga Totorillas, 03°00'S, 79°01'W (J. Freile *in litt.* 2017); Fasañán, 03°02'S, 78°42'W (J. Freile *in litt.* 2017); Tarqui, 03°06.5'S, 79°05'W (P.X. Astudillo W. *in litt.* 2017); San Fernando (Buza), 03°11.5'S, 79°17'W (P.X. Astudillo W. *in litt.* 2017); Oña Cappa, 03°34'S, 79°08'W (Bloch *et al.* 1991). *Morona-Santiago* Culebrillas de Sangay, 01°46'S, 78°17'W (MLZ 7278); Tambillo, 02°12'S, 78°28'W (R.A. Gelis *in litt.* 2017); Gualaceo–Limón road, *c.*03°02'S, 78°39'W (ZMUC); Matanga, 03°16'S, 78°54'W (Poulsen 2002); BP Cooperativa Jima Ltda, 03°17'S, 78°53'W (XC 132312/412, L. Ordóñez-Delgado); Achupallas, 03°27'S, 78°21'W (Parker 1997). *Loja* Cordillera de Chilla, 03°32'S, 79°22'W (Bloch *et al.* 1991); Lagunas de Maní, 03°34'S, 79°26'W (XC 154656, L. Ordóñez-Delgado); Bosque Comunal Angashcola, 04°35'S, 79°22'W (J. Freile *in litt.* 2017); Cerro Acanamá, 03°42'S, 79°13'W (Bloch *et al.* 1991, Krabbe 1991, XC); *c.*25km NNW of Loja, 03°50'S, 79°16'W (Krabbe 1991); 'farm of Dr David Espinosa', 03°58'S, 79°09'W (Bloch *et al.* 1991); Cajanuma, 04°06.5'S, 79°11'W (Krabbe 1991, ANSP 184720/21); Cerro Toledo, 04°23'S, 79°07'W (Poulsen 2002). *El Oro* Taraguacocha, 03°34'S, 79°35'W (Ridgely & Greenfield 2001, AMNH). *Zamora-Chinchipe* Estación Científica San Francisco, 03°58'S, 79°05.5'W (J. Freile *in litt.* 2017); Quebrada Honda, 04°29'S, 79°08'W (MECN 7215/16, ANSP); RP Tapichalaca, 04°29.5'S, 79°07.5'W (Heinz 2002, Greeney & Gelis 2005, Greeney *et al.* 2010); Cerro Panguri, 04°36.5'S, 78°58'W (MECN 7956); PN Yarucí, 04°45.5'S, 79°25'W (R. Ahlman *in litt.* 2017); 25km SSE of Jimbura, 04°47'S, 79°21'W (KUNHM, ANSP); Lagunillas, 04°47'S, 79°24'W (Poulsen 2002). **PERU**: *Piura* Cerro Chinguela, *c.*05°07'S, 79°23'W (LSUMZ, ML). *Cajamarca* Hito Jesús, 04°53.5'S, 78°53.5'W (XC 7014, T. Mark); Picorana, 05°02'S, 78°51'W (LSUMZ 169897); near Chaupe, 05°14.5'S, 79°07'W (A. García-Bravo *in litt.* 2017); Tabaconas, 05°19'S, 79°17'W (MUSA 3397); Cauaris, 05°19.5'S, 79°00'W (MSB 43402); near Laguna Azul, 05°32'S, 79°08'W (A. García-Bravo *in litt.* 2017); Agua Azul, 05°38'S, 79°04'W (MSB 42616/655); *c.*8km ESE of Sallique, 05°40'S, 79°14'W (LSUMZ 169894–897); ÁCM Bosque de Huamantanga, 05°40'S, 78°58'W (A. García-Bravo *in litt.* 2017); 'La Ashitas', near Jaén, *c.*05°43'S, 78°55'W (MUSA 4598). **NOTES** There are a few historical records of Rufous Antpitta among vague or imprecisely georeferenced historical collecting locations on the slopes of Volcán Pichincha (Paynter 1993), and specimens labelled simply 'Pichincha' (AMNH 173035/36, MCZ 138487) probably do fall into this category. The coordinates above are for a centrally located point on the volcano: see discussion under *G. q. quitensis* for further information. Some are given above and others include: Rucu Pichincha (AMNH 148097); Guagua Pichincha

(AMNH 148096); some variation on Mt. Pichincha (AMNH 124440/492261/62). **SEE ALSO** Allen (1889b), Sclater (1890), Lönnberg & Rendahl (1922), Chapman (1926), Moore (1934), Meyer de Schauensee (1950), Negret (1994), Cardona *et al.* (2005), Ayerbe-Quiñones *et al.* (2008), Arbeláez-Cortés *et al.* (2011a, 2011b), Calderón-Leytón *et al.* (2011).

Grallaria rufula cajamarcae (Chapman, 1927), American Museum Novitates, no. 250, p. 2, Chugur, 2,745m, 40 miles north-west of Cajamarca, Peru [= 06°40'S, 78°44'W]. Originally described as *Oropezus cajamarcae*, the adult male holotype (AMNH 229329) was collected by Harry Watkins on 30 April 1926 (LeCroy & Sloss 2000). This range-restricted subspecies is endemic to the western Andes of Peru, its range generally given as southern Cajamarca (Fjeldså & Krabbe 1990, Krabbe & Schulenberg 2003). The true range of *cajamarcae*, however, is not well understood. Observations (eBird) presumed to be of this race extend as far south as the city of Cajamarca, from there west as far as the area south of Niepos (06°55.5'S, 79°07.5'W) and as far east as the forests around Celendín. North of the type locality observations extend north to the mountains west of Pucará (eBird: A. García-Bravo, D. García Olaechea). Specimens are few, but birds collected 24km south-west of Pucará (MSB) in easternmost Lambayeque are also *cajamarcae* (R.T. Chesser *in litt.* 2017). The well-documented population in the Cruz Blanca area (*c.*05°20'S, 79°32'W, south-west of Huancabamba) was not racially identified by Parker *et al.* (1985) but recordings (XC 236362, D.F. Lane, also ML) and specimens (LSUMZ) are now available. The Cruz Blanca population appears isolated from nominate *rufula* in the Cerro Chinguela area by the Huancabamba Valley, and is clearly assignable to *cajamarcae*, greatly extending its range north from previous estimates. This race is also similar to *obscura*, but has whitish lower underparts and a darker reddish-brown back (rather than tawny cinnamon-brown). Chapman (1927) described *cajamarcae* as having the forecrown and face bright ochraceous-tawny, grading towards the nape into 'Dresden brown or Saccardo's umber', which continues onto the back and rump. The tail and wings are slightly more rufescent than the back, the throat and breast bright ochraceous-tawny, and the flanks more olivaceous. The feathers of the abdomen are slightly tipped whitish, the vent is whitish, and the undertail-coverts buffy-white. **SPECIMENS & RECORDS** *Piura* Abra Cruz Blanca, *c.*05°20'S, 79°32'W (LSUMZ 78576/77); Tambo, 05°21'S, 79°33'W (FMNH 222151); above Huarmaca, 05°30'S, 79°28'W (XC 83047, D. Geale). *Cajamarca* Cerro Mishuanga, 06°21.5'S, 79°15'W (XC 7811, T. Mark); Cerro Ilucan, 06°22'S, 78°50'W (A. García-Bravo *in litt.* 2017); Bosque Paja Blanca, 06°23.5'S, 79°07'W (XC: W.-P. Vellinga, B. Planqué); Taulís, 06°54'S, 79°03'W (AMNH 235897); Chacra Señor Filadelfio, 06°54.5'S, 78°39'W (A. García-Bravo *in litt.* 2017); above Miravalles, 06°59.5'S, 79°05.5'W (F. Angulo *in litt.* 2017); Paucal, 07°00'S, 79°10'W (Taczanowski 1884); Cruz Conga area on Cajamarca–Celendín road, *c.*07°02'S, 78°13'W (XC, many recordings); Celendín 06°54'S, 78°07'W (XC 158131, H. Matheve). *Lambayeque* Tres Lagunas, 06°14'S, 79°13.5'W (MSB).

Grallaria rufula obscura Berlepsch & Sztolcman, 1896, Proceedings of the Zoological Society of London, vol. 1896, p 385., 'Maraynioc,' 45 mi. north-east of Tarma, Dept. Junín, central Peru [= vicinity of Vitoc, 11°12.5'S, 75°20'W]. The holotype (MZPW 33975, originally MZBW

3085a; Mlíkovský 2009) is a female collected by Jan Kalinowski on 24 November 1891. I have not examined the type, but the only topotypical example that I am aware of (MCZ 266701) agrees well with the original description (photos J. Schmitt). Populations usually assigned to this Peruvian endemic race extend along the east Andean slope, south of the Marañon drainage, to Junín (Krabbe & Schulenberg 2003). Scattered records within its presumed range suggest a fairly continuous distribution, and include recordings and specimens from eastern La Libertad and west-central San Martín. Populations just south of the Río Marañón, from Cordillera de Colán (LSUMZ) south to around Leymebamba and Barra Negra perhaps represent a diagnosable taxon (Schulenberg *et al.* 2007) 09°42'N, 73°05'W. Hosner *et al.* (2015) suggested that birds in eastern Ayacucho, north of the Río Apurímac, are actually closer to Chestnut Antpitta, and may represent an undescribed taxon. Following their suggestion, these Ayacucho records have been included in Chestnut Antpitta (which see). The southernmost populations here referred to *obscura* 09°42'N, 73°05'W are probably in the mountains east of Huancayo (XC 20655, N. Athanas) north to Puerta del Monte (see above). Race *obscura* is most similar to nominate *rufula* in being nearly monochromatic, but differs in being less rufescent and more a deep, rich brown, slightly less brown below. It further differs from nominate by having a fairly distinct white postocular crescent like other southern subspecies. Peters & Griswold (1943) gave bare-parts coloration as: *Iris* dark brown; *Bill* black; *Tarsi & Toes* light blue-gray. SPECIMENS & RECORDS *Amazonas* NE of La Peca, 05°33'S, 78°19'W (LSUMZ 88058–069); Cerro Shucahuala, 06°14'S, 78°06.5'W (A. García-Bravo *in litt.* 2017); Cerro Montealegre, 06°35'S, 77°34'W (B.M. Winger *in litt.* 2017); near Leymebamba, *c*.06°43'S, 77°51'W (XC 40750, D. Beadle); Abra Barro Negro, 06°44'S, 77°53.5'W (XC 8284, H. van Oosten); *c*.3.7km E of Chuquibamba, 06°56'S, 77°49.5'W (A. García-Bravo *in litt.* 2017). **La Libertad** Mashua, 08°13'S, 77°12'W (LSUMZ 92475–477); Cumpang, 08°17'S, 77°18'W (Schulenberg & Williams 1982). **San Martín** 30km by road E of Florida de Pomacochas, 05°47'S, 77°44'W (Davis 1986); Quintecocha, 06°51.5'S, 77°42'W (XC 150, W.-P. Vellinga); Puerta del Monte, 07°34'S, 77°09'W (LSUMZ). **Huánuco** Mascarrón, 09°34'S, 76°07'W (ANSP 176468/69); Esperanza, 09°41'S, 76°03'W (FMNH 296697); Pan de Azúcar, 09°42'S, 75°59'W (FMNH 293401); Bosque Cutirragra, 09°42'S, 76°01'W (LSUMZ 74099–103); 2km NW of Punta de Saria, 09°43'S, 75°54'W (LSUMZ 128553); Bosque Unchog, 09°44.5'S, 76°10'W (LSUMZ 119013–015, XC); Achupampa, 09°44'S, 75°57.5'W (MUSA 4503); Bosque Zapatogocha, 09°44.5'S, 76°02.5'W (FMNH 296695/96, LSUMZ 74105/06); near Acomayo, 09°46'S, 76°05'W (LSUMZ 74104); 5km E of Chaglla, *c*.09°49'S, 75°51'W (LSUMZ 113578); Panao Mountains, 09°50'S, 76°02'W (Zimmer 1930, FMNH, LSUMZ); Millpo, 09°54'S, 75°44'W (long series; LSUMZ); 16km W of Panao, 09°54'S, 76°06'W (LSUMZ 113579); Pozuzo–Chaglla trail, 09°58'S, 75°46'W (LSUMZ 129902); Bosque Potrero, 09°59'S, 76°05'W (LSUMZ 113581/82, XC). **Junín** Maraynioc, 11°22'S, 75°24'W (MCZ 266701); Mariposa, 11°24'S, 74°45'W (XC 41551, J. Hornbuckle); below Toldopampa, near Puente Carrizales, *c*.11°29'S, 74°53.5'W (KUNHM 113873–875, XC 47027/28, A. Spencer); Chanchuleo, 11°31'S, 74°48'W (LSUMZ 127645–650); Chilifruta, 12°05.5'S, 74°54'W (XC 20655, N. Athanas). **Pasco** Santa Bárbara, 10°20.5'S, 75°38'W (XC 62928–932, W.-P. Vellinga); Cumbre de

Ollón, 10°34'S, 75°19'W (ML 35960, T.S. Schulenberg); Rumicruz, 10°42.5'S, 75°54'W (AMNH 174092).

Grallaria rufula occabambae (Chapman, 1923), American Museum Novitates, no. 67, p. 8, Occabamba Valley, 9,100ft [2,775m], Urubamba region, Peru [= Dept. Cuzco, vicinity of Urubamba, 13°18'S, 72°07'W]. Originally described as *Oropezus rufula occabambae* based on an adult male (AMNH 166533) collected 2 August 1915 by Edmund Heller (LeCroy & Sloss 2000). At the time, Chapman had just one other example, from nearby Machu Picchu (Cedrobamba; AMNH 166582) and had previously referred Urubamba specimens to *obscura* (Chapman 1921). The range of *occabambae* extends from southeast Peru, south of the Río Apurímac (including the Cordillera de Vilcabamba), to north-west Bolivia 09°42'N, 73°05'W. Records from northern Apurímac are scarce (e.g. Baiker 2011), but are here included as *occabambae*. Equally plausible is that Apurímac populations should be placed with those in Ayacucho. The south-easternmost population here included in *occabambae* is in eastern Puno and western La Paz (Hennessey & Gomez 2003) 09°42'N, 73°05'W, possibly as far south as Cholina. This population is separated by a *c*.180km gap in observations from the southernmost records in Dept. Cuzco (Marcapata region). Rufous Antpittas in eastern Apurímac (known from sight records alone) are tentatively considered to be *occabambae*, although they potentially might prove to represent the Ayacucho population. Probably closest to *obscura*, race *occabambae* is more olivaceous above and paler below. Chapman (1923a) regarded *occabambae* as most similar to *obscura*, but generally brighter. The upperparts are 'ochraceous tawny' rather than reddish-brown as in *obscura*. Compared to nominate *rufula*, *occabambae* is, in general, less rufescent, especially below, but the back is also more olivaceous (Krabbe & Schulenberg 2003). Its bill was suggested to be shorter and stouter than either *obscura* or *rufula* (Chapman 1923a), but there is considerable overlap. To date, however, assessment of any potential differences in bill morphology between subspecies is confounded by small sample sizes and variable measuring techniques. SPECIMENS & RECORDS **PERU** *Junín* Río Pomureni headwaters, 11°39'S, 73°40'W (Schulenberg *et al.* 2001a, FMNH 390659–662, ML). *Cuzco* Tocopoqueyu, 12°32'S, 72°12.5'W (USNM 273161–164); southern Cordillera de Vilcabamba area, *c*.12°37'S, 73°33'W (USNM, AMNH); Quebrada Honda/Abra Bella Vista, *c*.12°37'S, 72°15'W (MSB); Quillabamba, 12°52'S, 72°41.5'W (FMNH 324108); San Luís, 13°04'S, 13°04'S (Parker & O'Neill 1980, LSUMZ, FMNH); 14–15km NE of Abra Málaga, 13°04.5'S, 72°23.5'W (LSUMZ 78570–574, MSB); Cedrobamba, 13°05'S, 72°33'W (USNM 273165/66, AMNH 166582); Abra Málaga area on both slopes, *c*.13°08'S, 72°18'W (ML and many recordings XC); Pillahuata, 13°10'S, 71°35.5'W (FMNH 429996); Wayqecha Cloud Forest Biological Station, 13°10.5'S, 71°35'W (Londoño *et al.* 2015, XC); Acjanaco, 13°11'S, 71°37'W (XC 33922/102558, F. Schmitt); Huasampillo, 13°14'S, 71°26'W (Taczanowski 1884); Kurkur, 13°22.5'S, 73°08.5'W (P. Hosner *in litt.* 2017); Abuela, 13°25'S, 72°45'W (A. García-Bravo *in litt.* 2017); Marcapata 13°30'S, 70°53.5'W, and vicinity (AMNH, YPM, LSUMZ); Amacho, 13°35.5'S, 70°58.5'W (FMNH 222152). *Apurímac* *c*.1.9km NW of Cocha, 13°34'S, 72°50.5'W (J. Barrio *in litt.* 2017). *Puno* Valcón, 14°23'S, 69°22'W (long series LSUMZ); Sina, 14°29'S, 69°16'W (KUNHM). **BOLIVIA**: *La Paz* Tokoaque,

14°37'S, 68°57'W (Hennessey & Gomez 2003); Keara, 14°41'S, 69°06'W (XC 73601–605, J.A. Tobias); Cholina, 15°09'S, 68°58'W (S.K. Herzog *in litt.* 2017).

Grallaria rufula cochabambae Bond & Meyer de Schauensee, 1940, Notulae Naturae, vol 44, p. 3, Incachaca, 10,000ft, Cochabamba, Bolivia [3,050m, 17°14'S, 65°49'W; topotypical material held at AMNH, CM, SNOMNH, LSUMZ]. The type specimen is an adult male (ANSP 140348) collected 10 June 1937 by M. A Carriker Jr. It was described along with an adult female from the same location, these two specimens being the first record of Rufous Antpitta for Bolivia. For some time, *cochabambae* was considered endemic to its namesake Bolivian department. Subsequently discovered populations in central La Paz, in the Cordillera Real (east and north-east of Cerro Chacaltaya, Chuspipata area) were previously presumed to represent *cochabambae* (Krabbe & Schulenberg (2003), and are treated as such here. They are separated from the southernmost observations of *occabambae* by c.150km. Field observations in the Chusipata area (D.F. Lane *in litt.* 2017) suggest that, although their vocalisations are rather similar to the Cochabamba population, there appear to be consistent plumage differences that hint at the possibility of an unrecognised taxon. Race *cochabambae* is very similar to *occabambae*, but overall even duller. Bond & Meyer de Schauensee (1940b) described the holotype as having the upperparts deep olive-brown, crown slightly darker, flight feathers edged dull rufous and underparts dull olive-ochraceous (not rufescent). Specimens & Records *La Paz* Cotapata Trail, 16°17'S, 67°53'W (LSUMZ 96069–073; XC 1959/66, S. Mayer); vicinity of Chuspipata, c.16°17.5'S, 67°48'W (LSUMZ 102241–245, XC 124024, D.F. Lane). *Cochabamba* Pampa Grande, 16°42'S, 66°29'W (S.K. Herzog *in litt.* 2017); Cocapata (Pujyani), 16°47'S, 66°42'W (S.K. Herzog *in litt.* 2017); 5km ESE of Itimpampa, 17°08'S, 65°43'W (LSUMZ 36115–125); Cañadón, 17°12'S, 65°54'W (XC 145004, D.F. Lane); Presa Corani, 17°13.5'S, 65°54'W (XC 100391–397, J. King); Incachaca, 17°14'S, 65°49'W (ANSP, CM, LSUMZ, SNOMNH). *Santa Cruz* Bosque Siberia, 17°50'S, 64°40'W (XC 351780/352239, G.M. Kirwan); near Comarapa, 17°54.5'S, 64°32'W (XC 38865, F. Lambert). See Also Chapman (1921, 1923), Bond & Meyer de Schauensee (1942), Clements & Shany (2001), Walker *et al.* (2006).

STATUS Uncommon to fairly common across most of its range (Stotz *et al.* 1996) and not considered globally threatened (BirdLife International 2017). The two Colombian endemic subspecies are among the most range-restricted races, yet both are afforded some degree of protection in large reserves. In addition, *saltuensis* probably occurs in the Iroka and Sokorpa reserves in the Perijá Mountains (Krabbe & Schulenberg 2003). The widespread nominate race is well protected in many areas encompassing most subpopulations in Colombia and Ecuador. Despite its apparently well-protected status, however, Freile *et al.* (2010) estimated that, in Ecuador alone, nominate *rufula* has lost 38% of its potential habitat. The Peruvian races appear to have smaller percentages of their ranges covered by protected areas, and *cajamarcae* does not occur in any large, formal reserve. As one of the most range-restricted races, *cajamarcae* is probably at considerable risk. The Bolivian endemic race, *cochabambae*, is not known from any protected areas, but almost certainly occurs at upper elevations of PN Isiboro Sécure, PN Carrasco and perhaps PN Amboró. Protected Populations *G. r. spatiator* PNN Sierra Nevada de Santa Marta (XC 56827/28, B. Lopéz-Lanús); RNA El Dorado (Salaman *et al.* 2007a, XC). *G. r. saltuensis* RNA Chamicero del Perijá (ML 46892281, R. Gallardy, ML 38831091, photo D.M. Bell). *G. r. rufula* **Venezuela**: PN Tamá (COP). **Colombia**: PNN Puracé (Hilty & Silliman 1983); PNN Tatamá (Echeverry-Galvis & Córdoba-Córdoba 2007); PNN Complejo Volcánico Doña Juana-Cascabel (Ayerbe-Quiñones 2006); PNN El Tamá (FMNH 4360001); PNN Chingaza (XC 84005, A. Spencer); PNN Sumapaz (eBird: A. Pinto); PNN Munchique (Negret 1994, FMNH 249750, LACM 37383); PNN Los Nevados (Pfeifer *et al.* 2001); PNN Pisba (D. Carantón *in litt.* 2017); RN Chicaque (eBird: A. Bartels); SFF Iguaque (XC 123355/56, D. Calderón-F.); RNA Loro Orejiamarillo, RNA El Colibrí del Sol and RNA Mirabilis–Swarovski (Salaman *et al.* 2007a); RMN El Mirador (XC 197058, O. Cortes-Herrera); Reserva BioAndina (XC 315037/41, A. Pinto-Gómez); RFP Nacimiento del Río Bogotá (XC 299523, K.C. Cubillos); RFP Cárpatos (Andrade *et al.* 1994); RNSC Bosque Guajira (O. Cortes-Herrera *in litt.* 2017). **Ecuador**: PN Cayambe-Coca (Baéz *et al.* 2000, AMNH, MECN, MCZ, XC, ML); PN Sangay (R.A. Gelis *in litt.* 2017); PN Sumaco-Galeras (Krabbe 1991, MECN 6494, AMNH 184368); PN Llanganates (Krabbe 1991, Benítez *et al.* 2001, MECN); PN Yarucí (R. Ahlman *in litt.* 2017); PN Podocarpus (Bloch *et al.* 1991, Poulsen 1994, Rasmussen *et al.* 1994); PN Cajas (Tinoco & Astudillo 2007, Tinoco & Webster 2009, Latta *et al.* 2011); RE Cotacachi-Cayapas (XC 262136, O. Jahn); RE Antisana (HFG); RE Los Ilinizas (Krabbe 1991); RBP Guandera (Cresswell *et al.* 1999b); BP Mazán (Toral 1996, Astudillo *et al.* 2015); BP Cashca Totoras (Bonaccorso 2004); Reserva Geobotánico Pululahua (C. Hesse *in litt.* 2017); RVS Pasochoa (Fierro 1991); BP Cooperativa Jima Ltda (XC 132312/412, L. Ordóñez-Delgado); RP Verdecocha (ANSP 83346, AMNH 173308/308); RP Cerro Candelaria, RP Chamana and RP Machay (L. Jost *in litt.* 2017); RP Estación Científica San Francisco (J. Freile *in litt.* 2017); RP Tapichalaca (Greeney & Gelis 2005, Greeney *et al.* 2010); RP Yanacocha (MECN 2612/13, XC, IBC); RP Guango Lodge (Whitney 1992, ML). **Peru**: ÁCM Bosque de Huamantanga (A. García-Bravo *in litt.* 2017); Probably occurs in SN Tabaconas-Namballe. *G. r. cajamarcae* BP Pagaibamba (XC, W.-P. Vellinga). Probably occurs in RVS Bosques Nublados de Udima. *G. r. obscura* PN Río Abiseo (LSUMZ 104346–347); PN Yanachaga-Chemillén (XC 62928/29, W.-P. Vellinga); SN Cordillera de Colán (LSUMZ 88058–069); BP Alto Mayo (Davis 1986); ACP San Marcos (LSUMZ 113579/80). *G. r. occabambae* **Peru**: PN Otishi (FMNH 390659–662); SH Machu Picchu (Walker & Fjeldså 2005); ACP Abra Málaga (XC 277016, D.F. Lane). **Bolivia**: PN Madidi (Hennessey & Gomez 2003).

OTHER NAMES The following list of Latin synonomies is fairly complete, but will necessarily require carefully updating in light of upcoming taxonomic rearrangements. *G. r. rufula* Hypsibemon rufula (Gray 1869); *Hypsibemon rufulus* (Heine & Reichenow 1890); *Grallaria rufula saturata* (Domaniewski & Sztolcman 1918, Sztolcman 1926a); *Oropezus rufula* (Ridgway 1909, Chapman 1917, 1923, 1926, Moore 1934). *G. r. spatiator* Grallaria spiator (misspelling; Allen 1900); *Grallaria spatiator* (Bangs 1898b, Sharpe 1901, Brabourne & Chubb 1912, Ogilvie-Grant 1912); *Grallaria rufula spaitior* (misspelling; Rodner *et al.* 2000). *G. r. obscura* Grallaria obscura (Sharpe 1901, Brabourne & Chubb 1912, Ogilvie-Grant 1912); *Oropezus rufulus obscurus* (Chapman 1927). *G. r. cajamarcae* Oropezus cajamarcae (Chapman 1927); *Grallaria rufula cajamarcensis* (Vellinga *et al.* 2004). *G.

r. occabambae Oropezus rufula occabambae (Chapman 1923a); *Oropezus rufulus occabambae* (Chapman 1927); *Oropezus rufula obscura* (Chapman 1921). **English** Urubamba Rufous Antpitta (*occabambae*, Cory & Hellmayr 1924); Junin Rufous Antpitta (*obscura*, Cory & Hellmayr 1924); Perijá Ant-pitta/Antpitta (*saltuensis*, Meyer de Schauensee 1950, del Hoyo & Collar 2016); Little Rufous Ant-Thrush (*rufula*; Brabourne & Chubb 1912); Wandering Ant-Thrush (*spatiator*, Brabourne & Chubb 1912); Wandering Antpitta/ Ant-pitta (*spatiator*, Cory & Hellmayr 1924/Meyer de Schauensee 1950); Little Olivaceous Ant-Thrush (*obscura*, Brabourne & Chubb 1912). **Spanish** Tororoí Rufo (Krabbe & Schulenberg 2003; Asociacíon Bogotana de Ornitología 2000); Hormiguero Pichón Rufo (Phelps & Phelps 1963, Meyer de Schauensee & Phelps 1978, Hilty 2003); Chuncho balón, Huicundo café, Rundobalín (Valarezo-Delgado 1984), Gralaria rufa (Valarezo-Delgado 1984, Ortiz-Crespo *et al.* 1990, Rasmussen *et al.* 1994, Tinoco & Webster 2009); Tororoí flautista (Cardona *et al.* 2005, Salaman *et al.* 2007a); Pichón Rufo (Phelps & Phelps 1950); Tororoí del Perijá (del Hoyo & Collar 2016). **French** Grallaire rousse (Krabbe & Schulenberg 2003); Grallaire de la Perija (del Hoyo & Collar 2016). **German** Einfarb-Ameisenpitta (Krabbe & Schulenberg 2003); Perijáameisenpitta (del Hoyo & Collar 2016).

Rufous Antpitta, adult (*occabambae*), Wayquecha, Cuzco, Peru, 23 November 2011 (*Thomas Grim*).

Rufous Antpitta, adult (*rufula*), Papallacta, Napo, Ecuador, 17 November 2015 (*Harold F. Greeney*).

Rufous Antpitta, adult feeding nestlings (*occabambae*), Wayquecha, Cuzco, Peru, 11 December 2011 (*Santiago David*).

Rufous Antpitta, eggs in nest (*rufula*), RP Tapichalaca, Zamora-Chinchipe, Ecuador, 28 November 2003 (*Harold F. Greeney*).

Rufous Antpitta, adult incubating (*rufula*), RP Tapichalaca, Zamora-Chinchipe, Ecuador, 28 September 2007 (*Mery E. Juiña J.*).

Rufous Antpitta, adult (*spatiator*), RNA El Dorado, Magdalena, Colombia, 24 May 2009 (*Alonso Quevedo*).

Rufous Antpitta, fledgling (*occabambae*) Wayquecha, Cuzco, Peru, 23 November 2011 (*Thomas Grim*).

Rufous Antpitta, adult (*obscura*), Satipo Road, Junin, Peru, 29 September 2013 (*Carlos Calle*).

Rufous Antpitta, adult (*cajamarcae*), Cajamarca, Peru, 5 October 2011 (*Daniel F. Lane*).

Rufous Antpitta, adult (*cochabambae*), Keara, La Paz, Boliva (*Joseph Tobias*).

Rufous Antpitta, adult (*saltuensis*), Serrania de Perija, Cesar, Colombia (*Andrés Cuervo*).

CHESTNUT ANTPITTA
Grallaria blakei   **Plate 17**

Grallaria blakei Graves, 1987, Wilson Bulletin, vol. 99, p. 314, eastern slope of the Carpish Mountains, near the Carretera Central, *c.*2,400m, 09°40'S, 76°09'W, Huánuco, Peru. The holotype (LSUMZ 64228) is an adult female collected 17 August 1968 by John P. O'Neill (Graves 1987, Cardiff & Remsen 1994). The name honours Emmet Reid Blake's many contributions to the study of Neotropical birds.

Chestnut Antpitta was described from seven specimens taken in the Peruvian departments of Huánuco, Amazonas and Pasco (Cardiff & Remsen 1994). Currently, Chestnut Antpitta is considered monotypic, but birds from Pasco (in the south) have been suggested to merit subspecific rank based on plumage and vocal differences (Ridgely & Tudor 1994, Krabbe & Schulenberg 2003, Schulenberg & Kirwan 2012). Chestnut Antpitta is very similar to Rufous Antpitta, with which it has been considered closely allied (Krabbe & Schulenberg 2003). A close relationship between Rufous Antpitta and Chestnut Antpitta is also supported by mitochondrial DNA (cytochrome *b* and ND2) (Krabbe *et al.* 1999, Rice 2005). As currently defined, however, both species may consist of more than one taxon deserving full-species status, and more thorough sampling is required before drawing robust conclusions (Graves 1987, Isler & Whitney 2002, Krabbe & Schulenberg 2003). The fairly restricted distribution and the inaccessibility of its habitat have contributed to an extreme paucity of available data on ecology and natural history. The species' breeding and diet are completely undescribed, and an obvious priority for future studies. Additionally, the intra- and inter-specific relationships of Chestnut Antpitta beg thorough analysis. The late-stage immature plumage is described for the first time here.

IDENTIFICATION 14.5–15.0cm. One of the smaller *Grallaria*, with near-uniform rufous plumage. The belly, however, is whitish with variable but indistinct spotting and barring. As its range overlaps with those of two similar species, Bay and Rufous Antpittas, correct identification presents more of a challenge than most species. Altitudinally, the ranges of Chestnut and Bay Antpittas show considerable overlap, whereas the overlap with Rufous Antpitta is less. The present species is significantly smaller than Bay Antpitta, is browner overall, has a black (rather than blue-grey) bill, lacks black on the crown, and has a buff-coloured (not pure white) belly that is usually finely barred. Rufous Antpitta is very similar in size to Chestnut Antpitta and Peruvian populations of nominate Rufous Antpitta are extremely similar in plumage (Fjeldså & Krabbe 1990). This has led to much confusion between the two species (Graves 1987, Tallman 1974). Where they co-occur, Rufous is found at elevations above Chestnut Antpitta, whereas Bay Antpitta inhabits lower elevations (Schulenberg *et al.* 2007). Chestnut Antpitta is generally separable from the other three Peruvian subspecies of Rufous Antpitta (*cajamarcae, obscura* and *occabambae*) by being darker chestnut overall, especially by its darker red-brown head, by the lack of contrasting feather tips on the back and breast, and by its dark brown undertail-coverts (pale buff in Ayacucho), thicker bill and legs, and lack of the relatively distinct eye-ring of Rufous Antpitta (Graves 1987, Fjeldså & Krabbe 1990). Furthermore, the presence

of (albeit indistinct and geographically variable) barring on the belly may help to separate it from the unmarked Rufous Antpitta (Fjeldså & Krabbe 1990, Schulenberg & Kirwan 2012f), but certainly not everywhere in its range (P.A. Hosner *in litt.* 2017). These two species are easily distinguished by song.

DISTRIBUTION Chestnut Antpitta is endemic to Peru and reported from just a few sites, roughly separable into four populations, but is presumably more widespread than currently known (Schulenberg & Kirwan 2012f). The northernmost locality is the Cordillera Colán, south of the Río Marañón (Graves 1987, Hornbuckle 1999) in Amazonas, with most records from Abra Patricia south to Leimebamba. South of there, a *c.*360km gap in records separates the northern population from that in the Cordillera Carpish region of Huánuco. The population further south in the Cordillera de Yanachaga and Oxapampa area south to central Junín (east of Satipo) was postulated by Graves (1987) to represent an undescribed taxon, but is traditionally ascribed to *blakei* (Fjeldså & Krabbe 1990, Ridgely & Tudor 1994). In Distribution Data, localities that potentially pertain to the central population, or what is referred to as 'Oxapampa Antpitta', are denoted [†]. A skeletal specimen (LSUMZ 70234) was also collected at 2,485m in the Apurímac Valley in Ayacucho, suggesting to Graves (1987) the possibility of a third taxon of *blakei* (or *rufula*) much further south. This was further supported by recent work (Hosner *et al.* 2015). Records probably representing this population (marked [‡]) extend east and west of Tutumbaru (12°44'S, 73°57'W) to at least east of Churca (13°14'S, 73°34'W) (Hosner *et al.* 2015).

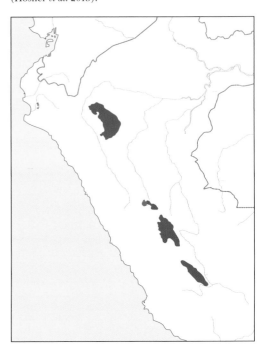

HABITAT A denizen of humid montane forests (Graves 1987, Davies *et al.* 1997), apparently preferring bamboo and other dense undergrowth (Ridgely & Tudor 1994, Krabbe & Schulenberg 2003), although it does not appear to be a 'bamboo specialist'. Altitudinal range is generally given as 1,700–2,500m, yet I am aware of only one record as low as 1,700m (Comboca, Cordillera Colán, Barnes *et al.* 1997, Kirwan & Hornbuckle 1997). There is apparently a record at 1,850m (below Tesoro, Cordillera Colán; Davies *et al.* 1997). Clements & Shany (2001), however, gave a lower limit of 1,650m. In the Cordillera Colán it ranges to 3,100m locally, with the altitudinal range reportedly lower in Huánuco and Pasco (Ridgely & Tudor 1994, 2009, Schulenberg *et al.* 2007). In Ayacucho, Chestnut Antpitta can be found as high as 3,800m (P.A. Hosner *in litt.* 2017).

VOICE The vocalisations are poorly known and the first recordings were not made until the late 20th century (Fjeldså & Krabbe 1990). From Amazonas south to Junín, the song is 'a rapid, monotone, slightly accelerating series of chiming notes: *chew'chu'u'u'u'u'u'u'u'u'u'u'u'* (Schulenberg *et al.* 2007). Krabbe & Schulenberg (2003) provided details: 3.2–4.6s long, given at 4–11s intervals, a ringing, even-pitched series of 38–52 notes at 2kHz given at even pace of 11–12/s'. The apparent song in Pasco and Junín (central Peru) is rather different. D.F. Lane (in Schulenberg *et al.* 2007) described the Pasco song as 'a single chiming note: *clew*' and Ridgely & Tudor (1994) as a 'steadily repeated single *toop* note continued several times at about 1s intervals.' What is presumably the alarm call lasts only 1.8–2.2s but is otherwise similar to song. The alarm is delivered at 1–3s intervals (Krabbe & Schulenberg 2003) and is described as 'a brief *weeoo*' (Barnes *et al.* 1997) or a 'single *clew* or short series of notes' (D.F. Lane in Schulenberg *et al.* 2007). A detailed record of geographic variation in song and calls is sorely needed.

NATURAL HISTORY The basic details of foraging behaviour and habitat use are virtually undocumented for this rarely seen species. Chestnut Antpitta usually forages below 1m, well inside dense understorey vegetation (Schulenberg & Kirwan 2012f), yet this hardly distinguishes it from most *Grallaria*. Davies *et al.* (1997) mentioned that Chestnut Antpitta is known to the residents of Comboca (Cordillera Colán) by 'its habit of hopping onto paths after mules have passed'. Interestingly, this is yet another piece of observational data supporting the proposed propensity for antpittas to regularly forage in the wake of large mammals (Greeney 2012b). Undoubtedly, individuals foraging in the path of passing mules will be richly rewarded with easily captured earthworms and other invertebrates exposed by the mules' hooves.

DIET No published data, although its diet is presumably similar to the closely related Rufous Antpitta. The stomach contents of a few specimens were dominated by insect larvae (P.A. Hosner *in litt.* 2017). Schulenberg & Kirwan (2012f) suggested that it may also eat small vertebrates, such as frogs.

REPRODUCTION The nest, eggs, nestlings and reproductive habits are undescribed. **SEASONALITY** Far too few data are available to accurately assess this. Graves (1987) reported juveniles with a few immature feathers in early August and late November in Huánuco. I examined these specimens, both collected near Cordillera Carpish, the first just beginning and the second just ending their transition to subadults (see below). These are presumably the records mentioned by Fjeldså & Krabbe (1990). The two juveniles photographed in the final transition to adult plumage (see below) would probably have been very

similar to adults within 1–2 weeks of these photographs being taken, and were likely similar in age to LSUMZ 75243). Graves (1987) reported that none of four adults (2 males, 10 August, 21 October; 2 females, 10–12 August) from the Cordillera Carpish were in breeding condition. Graves (1987) proposed a breeding season of December to April or May, an assertion that appears to be supported by the following data. **BREEDING DATA** Older juvenile, 2 August 1973, trail to Hacienda Paty (Graves 1987, LSUMZ 74098); juvenile transitioning, 27 June 2013, Carpish Tunnel (ML 61670731, photo L. Seitz); juvenile transitioning, 18 September 2013, Apalla–Andamarca road (N. Athanas photo, Fig. 1); juvenile transitioning, 14 June 2008, 2km S of Ccano (KUNHM 112735, photos P.A. Hosner); juvenile finishing transition, 22 November 1973, trail to Hacienda Paty (Graves 1987, LSUMZ 75243); enlarged ovary (10 × 8mm) with 3mm ova, 29 November 1983, E of Florida de Pomacochas (LSUMZ 116973); testes greatly enlarged, 21 October 1964, Cordillera Carpish (FMNH 275632).

TECHNICAL DESCRIPTION Sexes similar. The following is based on Graves (1987), Schulenberg & Kirwan (2012f) and *in situ* photos. **Adult** Upperparts, including crown, nape, scapulars, back and rump rich, uniform chestnut-brown, with wings and tail slightly more reddish-brown. Leading edge of outer primaries slightly paler. Lores, orbital area and face-sides slightly brighter chestnut, with a paler rufous throat and breast. The rufous fades to whitish-buff on belly, nearly white on the lower belly. The belly-sides are weakly spotted or barred dusky, due to the greyish outer edges to the feathers. Flanks and thighs more brown or olive-brown, the vent and undertail-coverts tawny-brown (except in Ayacucho, where they are pale buff; P.A. Hosner *in litt.* 2017). **Adult Bare Parts** *Iris* brown; *Bare Ocular Skin* purplish-grey; *Bill* black with paler, whitish gape; *Tarsi & Toes*; slate-grey to purplish- or bluish-grey. **Juvenile** Not properly described, but Graves (1987) described an immature moulting to adult plumage as having 'traces of streaks scattered on crown, nape, wing-coverts, throat and breast'. I have examined photos of an immature, in the final stages of acquiring subadult plumage, photographed by N. Athanas at Apalla (Fig. 1). Vegetation limited visibility, especially of the upper and lower underparts. Visible plumage included: lores blackish-chestnut, forecrown, crown and hindcrown chestnut, nape, upper back and neck-sides dark grey with indistinct buffy-chestnut spots and streaks, interspersed by irregular patches of adult-coloured chestnut feathers. Face chestnut, slightly buffier with faint buffy shaft-streaks on ear-coverts. Rest of upperparts, tail and wings mostly obscured. Chin whitish, throat pale chestnut. Breast similar to adult, but lower underparts not visible. Additionally, I examined photos (J.V. Remsen) of the younger juvenile at LSUMZ, more similar to the description in Fjeldså & Krabbe (1990) and probably at least 2–3 weeks younger than the previous individual. My description follows: lores blackish-chestnut; forecrown, crown and hindcrown chestnut, some feathers similar to adult coloration, others slightly darker; feathers of the nape and neck-sides dark grey, with a broad buffy or whitish stripe on the rachis, creating a streaked effect. The remaining upperparts, tail and wings are similar to adult plumage, duskier on rump and uppertail-coverts and still with immature upperwing-coverts that are similar to adult coverts but bear a thin blackish subterminal bar; long, plume-like feathers of rump and flanks grey basally, buffy olive-chestnut near fringes. Chin buffy-white; central throat pale chestnut bordered greyish with indistinct whitish-buff streaking; upper breast pale chestnut, darker on central breast, fading below to pale buffy, buffy-white in centre; lower belly and flanks pale buffy to buffy-white with indistinct dusky to brownish-grey vermiculations, more prominent and finer than in adult; undertail-coverts dark grey, thighs grey washed chestnut. Bare parts: *Bill* maxilla black, mandible brownish; *Tarsi & Toes* brownish. **Juvenile Bare Parts** *Iris* dark brown; *Bare Ocular Skin* purplish-grey; *Bill* black with faint hint of pale yellowish-orange gape; *Tarsi & Toes* dark grey to purplish-grey, similar to adults. **Fledgling** Undescribed.

MORPHOMETRIC DATA from Graves (1987). ♂♂ (*n* = 4) *Wing* mean 81.1mm ± 1.8mm, range 81.1–85.2mm; *Bill* from nares mean 11.6 ± 0.4mm, range 11.0–11.9mm, width at anterior margin of nares mean 5.5 ± 0.1mm, range 5.3–5.6mm; *Tarsus* mean 42.7 ± 0.3mm, range 42.7–43.4mm; the holotype measured: wing chord 78.5mm; *Bill* from front of nares 11.0mm; *Bill* width at front of nares 5.4mm; *Tarsus* 30.8mm, narrowest medio-lateral width of tarsus 1.6mm, middle toe (sum of outer two phlanges and claw) 21.8mm. ♀♀ (*n* = 4). *Wing* mean 81.6 ± 2.3mm, range 78.5–83.8mm; *Bill* from nares mean 10.8 ± 0.2mm, range 10.5mm–11.0mm, width at anterior margin of nares mean 5.5 ± 0.1mm, range 5.3mm–5.6mm; *Tarsus* mean 40.5 ± 1.8mm, range 38.0–42.2mm. *Wing* 79.8mm; *Bill* from nares 11.2mm; width at anterior margin of nares 5.2mm; *Tarsus* 39.9mm, narrowest medio-lateral width of tarsus 1.6mm; *Middle toe* 24.6mm (LSUMZ 106081, east of Oxapampa). **Mass** from Krabbe & Schulenberg (2003): 38g, 40g (*n* = 2, ♂♂), 39.5–47.0g (*n* = 4, ♀♀, mean 44.1g). **Total Length** 14.5–15.0cm (Krabbe & Schulenberg 2003, Schulenberg *et al.* 2007, Ridgely & Tudor 2009).

DISTRIBUTION DATA Records presumed to involve potentially unnamed taxa are marked † or ‡ (see Distribution). Endemic to **PERU**: *Amazonas* NE of La Peca, 05°33'S, 78°19'W (LSUMZ 88068); *c.*30km by road E of Florida de Pomacochas (near Trocha Graliria/Trocha Mono), 05°42'S, 77°49'W (LSUMZ 116973, XC, J. King); Omia, 06°28'S, 77°23.5'W (A. García-Bravo *in litt.* 2017). *San Martín* Leimebamba, *c.*06°42'S, 77°41'W (FMNH 480820–823). La Libertad **Huánuco** Playa Pampa, 09°57'S, 75°42'W (LSUMZ 170664); Pacoyán, 09°43'S, 76°05'W (XC 40510, D. Geale); trail to Hacienda Paty, 09°42'S, 76°05'W (LSUMZ, XC); Carpish Tunnel, *c.*09°40'S, 76°06'W (many recordings XC); Bosque Unchog, 09°44''S, 76°10'W (A. Farnsworth *in litt.* 2017). *Junín* Río Satipo†, 11°30.5'S, 74°50.5'W (KUNHM 14849/50) and 11°30.5'S, 74°51.5'W (KUNHM 113991, ML 171921, P.A. Hosner); Apalla–Andamarca road, 11°33'S, 74°49'W (XC 148512/13, N. Athanas)†; 14.5km SW of Satipo†, 11°22'S, 74°43'W (XC 350362–364; G.M. Kirwan). *Pasco* Oxapampa Antenna Road†, 10°38'S, 75°17'W (XC 334331, J. Beck); Cumbre de Ollón†, 10°39'S, 75°17.5'W (Schulenberg *et al.* 1984, LSUMZ 106081; see below). *Ayacucho* Punco‡, 12°43.5'S, 73°52'W (AMNH 820774/75); 2km S of Ccano‡, 12°47'S, 73°59.5'W (Hosner *et al.* 2015, KUNHM); Ccarapa Bridge‡, 12°56'S, 74°01'W (Graves 1987, LSUMZ 70234); Chupón‡, 13°15'S, 73°30'W (Hosner *et al.* 2015, ML 173860, M.B. Robbins). **Notes** Stephens & Traylor (1983) gave the coordinates of LSUMZ collecting locality, Cumbre de Ollón (2,500m), as 10°34'S, 75°10'W, while Schulenberg *et al.* (1984) gave 10°34'S, 75°19'W. The former placed it *c.*25km E of Oxapampa and the latter

*c.*8km E. Based on the map in Schulenberg *et al.* (1984), placing it SE of Oxapampa, along with the stated distance of *c.*12km from there, and a more modern estimation of its location (XC 41549/50, J. Hornbuckle), I have used the coordinates above as a closer approximation of its likely location.

TAXONOMY AND VARIATION Currently considered monotypic, much uncertainty surrounds the various populations and up to four taxa may be involved in *G. blakei*. Some of these may be referable to Rufous Antpitta (see Distribution). The presumed song of populations in Pasco, Junín and Ayacucho is very simple and strikingly different from that of birds further north (Schulenberg & Kirwan 2012f). At Cumbre de Ollón, Cordillera Yanachaga, for example (Schulenberg *et al.* 1984), one of the known populations vocalises differently, and Graves (1987) described the only specimen of *G. blakei* from here (2,500m, *c.*12km east of Oxapampa, Fjeldså & Krabbe 1990) as being darker above and lacking the ventral barring of northern populations. Based on this specimen, Graves (1987) proposed that this population, although closely allied to northern *G. blakei*, warrants a more detailed analysis and may deserve subspecific rank. Fjeldså & Krabbe (1990) mentioned a similar specimen from Ayacucho that may share the same differences and agree that they may warrant consideration as a separate taxon. They also suggested that Amazonas specimens of *G. blakei* have the brightest underparts and least distinct barring on belly. Contrary to these suggestions, a song similar to that of the Yanachaga population was recently recorded at another site in Pasco (Playa Pampa, 2,400m, Schulenberg & Kirwan 2012f). A specimen collected at Playa Pampa in response to playback, however, is virtually indistinguishable from those of *G. rufula* above Playa Pampa (T.S. Schulenberg *in litt.* 2016). Until further specimens and recordings are collected from this area, it is prudent to maintain the current arrangement, but the forthcoming revision of the '*rufula/blakei*' complex should clarify the affinities of the various populations.

STATUS Chestnut Antpitta has been ranked as Near Threatened for several decades (Collar *et al.* 1994, Stattersfield & Capper 2000, BirdLife International 2017). Its geographic range is small (*c.*3,832km²; Franke *et al.* 2007, Young *et al.* 2009) and numbers are probably declining due to ongoing habitat loss (Stattersfield & Capper 2000). However, habitat presumably suitable for *G. blakei* is not severely fragmented or restricted to a few locations within its known range. The species is present in the North-east Peruvian Cordilleras EBA (Stattersfield *et al.* 1998) at Abra Patricia, where it is afforded some degree of protection within the Alto Mayo Protected Forest (Rosas 2003). Populations in the Cordillera de Colán (Graves 1987), where extensive deforestation is ongoing (Davies *et al.* 1994), are thought to be rapidly declining (Schulenberg & Kirwan 2012f). More positively, the relatively narrow altitudinal range occupied by Chestnut Antpitta includes many large, remote and uninhabited (as well as unsurveyed) areas. This offers hope that populations may be larger than known. It is also at least presumed to occur in PN Río Abiseo and PN Yanachaga-Chemillén (Schulenberg & Kirwan 2012f). In areas of suitable habitat where Chestnut Antpitta occurs, it is considered uncommon (Parker *et al.* 1996, Ridgely & Tudor 2009) or even fairly common (Schulenberg *et al.* 2007). In summary, although much of the range of Chestnut

Antpitta is remote, and supports a relatively small human population, technology and improved roads continue to promote ever-more habitat destruction, primarily for agriculture and grazing (Schulenberg & Kirwan 2012f).

OTHER NAMES Spanish Tororoí Castaño (Clements & Shany 2001, Krabbe & Schulenberg 2003). **French** Grallaire de Blake. **German** Kastanienbraune Ameisenpitta (Krabbe & Schulenberg 2003).

Chestnut Antpitta, adult, south-west of Satipo on the Apalla–Andamarca road, Junín, Peru, 18 September 2013 (*Nick Athanas/ Tropical Birding*).

Chestnut Antpitta, juvenile transitioning to subadult, south-west of Satipo on the Apalla–Andamarca road, Junín, Peru, 18 September 2013 (*Nick Athanas/Tropical Birding*).

Chestnut Antpitta, adult, south-west of Satipo on the Apalla–Andamarca road, Junín, Peru, 18 September 2013 (*Nick Athanas/Tropical Birding*).

TAWNY ANTPITTA
Grallaria quitensis — Plate 17

Grallaria quitensis Lesson, 1844, Écho du Monde Savant, vol. 2, no. 49, 'vicinity of Quito,' Ecuador [*c*.00°16'S, 78°25'W, see Taxonomy and Variation].

'In fact it is almost impossible to force this humpty-dumpty thrush-like bird to open its wings, its long, robust legs enabling it to leap and jump and run with almost as much address as the famous long-tailed Paisano or Road-runner of Mexico. Strange is it not, that such diversely feathered birds should have such similar habits? Nothing can be more tiresome than the three-cornered '*Wu, weeo, weeou*' or whistled song of this constantly invisible bird. Especially does this apply to the feelings of the collector who has tried vainly from day to day to locate and secure the singer, which sits motionless in a low bush, or on the ground beneath, in such a way as to be completely obscured. The notes are ventriloquial, and you may actually walk away from it in endeavoring to get closer.' – **Samuel N. Rhoads, Volcán Pichincha, Ecuador, 1912**

Tawny Antpitta is one of only two species of antpittas that truly 'inhabits' the windswept, tundra-like Andes above the treeline (i.e., *páramo* and *puna* ecosystems). Rufous Antpitta may venture a short way from forest but Tawny Antpitta forages and nests in open *páramo*, capitalising on small gaps in the bunchgrass to dash about in search of prey while remaining concealed and safe from the cold winds that frequently sweep across the mountains bearing clouds, rain, sleet and sometimes snow. I will never forget, while reviewing behavioural video shot at a Tawny Antpitta nest, seeing a sudden snowstorm sweep through the area. As the sleet-like snow began to pile up I watched, fascinated, as the incubating adult steadfastly maintained its position tightly pressed over the eggs, while forced to lift its chin steadily higher to keep its bill above the snow. Just as it seemed the adult would be buried, the storm abated and the adult's mate arrived, pausing beside the nest to peer at the bill and eyes poking out of the snow, where only an hour ago it had left its nest. The exchange occurred so fast I had to watch it several times to record the details. The instant that its snow-covered partner pushed up and away to bound across the snow, the second dropped resolutely onto the nest to continue incubating their eggs. Watching a moment in nature such as this is what makes all the hours of tromping across uneven muddy train bent under the force of snow-bearing winds pushing across the *páramo* worth the effort. Here was an antpitta, the quintessential Neotropical bird, practically iconic of tropical montane forests, incubating in the snow as if it were a Grey Jay *Perisoreus canadensis* in Canada! Five years later, I found myself writing a book about antpittas. Coincidence? I think not. Despite antpittas' reputation for being impossible to see, the Tawny Antpitta is among the rare exceptions (Freile 2002, Ridgely & Tudor 2009, Greeney 2015e). Despite its relative abundance in appropriate habitat, the nest was described only a decade ago, based on nominate *quitensis* in Ecuador (Greeney & Martin 2005). Indeed, apart from the new information presented here, there are still no published data on the breeding of two of the rather different subspecies; *alticola*, endemic to Colombia, and *atuensis*, endemic to Peru. As I have noted (Greeney & Martin 2005), the ease with which one can find the nests of Tawny Antpitta makes it an excellent study organism for future investigations

into the biology of this poorly known group of birds. The present account includes the first complete description of post-fledging plumage development and the first published information for breeding seasonality and immature plumage of *alticola*. It also includes new nesting data on nominate *quitensis* from northern Ecuador.

IDENTIFICATION 16–18cm. The plumage shows some fairly clear differences between the races, but the sexes do not differ. The nominate race, across most of the species' range, is olivaceous above, usually with a distinct greyish-washed crown and back. The rump is dull rufous to rufous-brown. The area around the eye, extending to the lores, is whitish, while the rest of the head is a mix of olive-brown and blackish-brown. The underparts are pale rufous-brown to tawny-brown with whitish markings forming barely noticeable crescents and a whiter central belly. Race *alticola* in the north is similar but overall smaller, browner above, and has the underparts more pale rufous and distinctly marked with white crescents. Race *atuensis* is darker overall, also with more distinct white markings below, but looks mottled or spotted. Although Tawny Antpitta is a rather dull-plumaged bird without obvious field marks, it is much less skulking than other *Grallaria*. In fact, its behaviour alone, frequently running about and foraging in open habitats, is usually the best clue to its identity (Ridgely & Tudor 1994, Ridgely & Greenfield 2001). Rufous Antpitta is the only other *Grallaria* that overlaps extensively with Tawny Antpitta in altitudinal range, but it is much more uniformly rufous or rufescent. Plain-backed, Brown-backed and Urrao Antpittas are all similar to Tawny in being rather dull-plumaged, but all are essentially forest-dwelling species found at considerably lower elevations and are unlikely to be found syntopically with the present species. Finally, although easy to separate from Tawny Antpitta visually, in some areas where their ranges overlap (i.e. the inter-Andean valley in Ecuador), the three-noted song of Chestnut-crowned Antpitta might cause some auditory confusion.

DISTRIBUTION The range extends from the East and Central Andes of Colombia south to extreme northern Peru. The northern subspecies, *alticola*, is restricted to the East Andes of Colombia, while the southernmost race, *atuensis*, is confined to the high mountains of southern Amazonas, La Libertad and San Martín, south of the Rio Marañón. Most of its range is occupied by the nominate subspecies, in the West Andes of southern Colombia and Ecuador, somewhat patchily in the Central Andes of Colombia, and in the East Andes from southern Colombia to northern Peru (Hilty & Brown 1986, Fjeldså & Krabbe 1990, Rodner *et al.* 2000, Ridgely & Greenfield 2001, Ridgely & Tudor 2009, Altropico 2010, Schulenberg *et al.* 2010).

HABITAT In the main, Tawny Antpitta is thought of as a species of Andean *páramo* and humid montane forests with adjacent grass (Fjeldså & Krabbe 1990). It is especially tolerant of habitat disturbance and prone to use almost any type of open area at appropriate elevations (Sibley & Monroe 1990). Across its range natural habitats include barren cushion-plant and bunch-grass *páramo*, bunch-grass *páramo* with scattered trees and shrubs or frailejones (*Espeletia*), and the interior and edge of elfin and *Polylepis* forests (Freile 2000, Ridgely & Greenfield 2001). Much of the species' range has been heavily impacted by human civilisations since the Incas ruled the

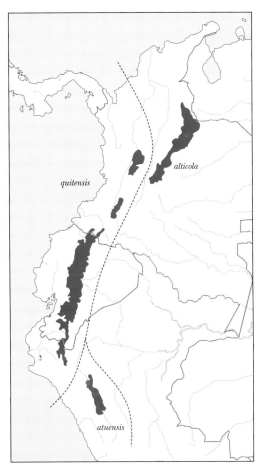

final two notes are rather uniform, with the whole song described by Schulenberg *et al.* (2007) as *WHEEP tu-tu*, by Ridgely & Greenfield (2001) as a slightly hollow *took, tu-tu*, and by Parker *et al.* (1985) as *uut—oo-oo*. Krabbe & Schulenberg (2003) also stated that *quitensis* may also give versions where the first note is lower than the others or where all three notes are somewhat lower pitched, at 1.4–1.5kHz. Although described as 'less common' than the three-note song above (Krabbe & Schulenberg 2003), the two-note variation, usually lower pitched than the full song, can be heard regularly in the *páramo* above Papallacta, Napo, Ecuador. This version was described by Fjeldså & Krabbe (1990) as *kyr kerk*. Descriptions of the calls of *quitensis*, which are usually given at 1–3s intervals, include: an 'explosive, single down-slurred note, beginning with first overtone loudest (at 4.5kHz) and ending with fundamental loudest (at 1.5kHz)' (Krabbe & Schulenberg 2003), a loud, sharp *lieu* (Fjeldså & Krabbe 1990), or a penetrating '*keeyurr!* recalling calls of the often sympatric Great Thrush *Turdus fuscater* (Ridgely & Greenfield 2001). Also, when 'excited' *quitensis* may deliver a somewhat whining or complaining *eenr* (Fjeldså & Krabbe 1990). Greeney & Martin (2005) mentioned soft one- or two-part 'pup' calls similar to the introductory notes of the full song. Other descriptions of the calls of *quitensis* include a squeaky *ts'EEW!* (Schulenberg *et al.* 2007) or a piercing *beeert* (Parker *et al.* 1985). Race *alticola* has a four-note (not three-note) song, with short pauses between the last three notes. Quantitatively, the notes tend to reach lower frequencies than *quitensis* and higher maximum frequencies than *atuensis* (del Hoyo *et al.* 2017d). Its song was described by Krabbe & Schulenberg (2003) as a lively *pit-wheer perwheedit* with the accent on the first syllable, or the final *perwheedit* is given alone. Olivares (1969) described the song of *alticola* (in Spanish) as *ú-u-ú, ú-u-ú, ú-u-ú*. Krabbe & Schulenberg (2003) provided acoustic details of the song of *atuensis*: slightly higher pitched than the song of *quitensis* (2.0–2.2kHz), the last note drawn out with strong upward inflection. Its song was described by Schulenberg *et al.* (2007) as *WHEEP tu-TUEE?* The call of *atuensis* is a harsh, rolling screech *TCHEE'ew* (Schulenberg *et al.* 2007), apparently quite distinct from other races. The call is qualitatively described as having a long, very burry start followed by an underslurred whistle (del Hoyo *et al.* 2017e) and quantitatively described as a long *tree-eh* at 1.7–3.0kHz, first rising with a treble quality, then falling at the end (Krabbe & Schulenberg 2003).

NATURAL HISTORY Perhaps because of the thin air at elevations where Tawny Antpitta is found, or perhaps because they are among the most easily observed of antpittas, the species seems to have inspired more than its share of humorous, but generally insightful quotes from early natural historians. As such, I cannot resist including the following here, one which I believe captures well the essence of this antpitta. 'The most persistent sound from the marshes was the loud, low-pitched call of the long-legged *Grallaria monticola*, Lafr. This bird dashed about in the runways between the tall clumps of *paja* and was very difficult to flush. On one occasion I had the opportunity of observing it calling when only a few feet from me. It raised its entire body as well as head and stretched its stilt-like legs to the full limit. When the call was completed, the bird collapsed like a jack-in-the-box.' (Moore 1934, Volcán Sangay, Ecuador). In general, Tawny Antpitta is found alone or, more rarely, in pairs. In

Andean highlands. Nevertheless, in some areas it is now apparently flourishing in hedgerows and small patches of second growth largely surrounded by agriculture, pine plantations and cattle pastures (Fjeldså & Krabbe 1990). I am unaware of any locations where it occurs within urban situations, but its absence may be a result of the devastating effects of cats, rather than its inability to occupy such areas. Adults often appear to favour lakeshores and swampy areas (Krabbe & Schulenberg 2003). Across its range, Tawny Antpitta is found at 2,200–4,500m, but usually is confined to above 2,800m (Freile 2000, Ridgely & Tudor 1994, Krabbe & Schulenberg 2003). In Ecuador it is reported mostly at 3,000–4,500m (Ridgely & Greenfield 2001). In Peru, Schulenberg *et al.* (2007) gave its preferred range as 2,850–3,400m. Chapman (1926) reported a lower altitudinal limit of 1,220m based on specimens from Huigra on the slopes of Volcán Pichincha. As noted by Paynter (1993), however, Huigra is actually in Chimborazo and the specimens to which Chapman (1926) referred are almost certainly mislabelled as to their provenance.

VOICE Following Krabbe & Schulenberg (2003) and Fjeldså & Krabbe (1990), the song of nominate *quitensis* lasts 0.9–1.3s and is repeated at intervals of 4–8s. It consists of three loud, piercing, somewhat doubled notes at 1.8–2.0kHz, with a slight pause after the first note, which is usually higher pitched than the others, *kyrkerk-kerk*. The northern Peru (*quitensis*), Parker *et al.* (1985) noted that Tawny Antpitta 'regularly foraged up to 30m from the cover of trees and shrubs in grassland, especially just after dawn or just before dusk, and throughout the day during foggy weather'. This is generally true throughout its range, but my own observations suggest that it will forage in the open on all but the sunniest of days. While in open areas, it forages by hopping on the ground, often pausing with its head tilted sideways, apparently searching the ground for prey. It favours the ground, whether in the open or in forested or shrubby areas, hopping rapidly and pausing to flick aside leaves with its bill, to probe soft mossy areas or damp ground, or to glean prey from low vegetation such as leaves, stems or cushion plants. Fjeldså & Krabbe (1990) stated that foraging adults will also sometimes 'flap [their] wings while clinging to a vertical stem and probing moss'. While moving, Tawny Antpitta frequently flicks its wings slightly up and bobs up and down (Fjeldså & Krabbe 1990, Ridgely & Greenfield 2001), with these movements apparently also associated with inter- or intra-specific conflict (Krabbe & Schulenberg 2003). Perhaps one of the most distinctive ways in which Tawny Antpitta deviates from the normal habits of *Grallaria* is its propensity to sing from partially to fully exposed locations (Fjeldså & Krabbe 1990, Ridgely & Greenfield 2001, pers. obs.). When singing, adults extend their long legs and hold their body vertically while throwing their head back and fluffing out the white feathers of the throat and neck (Ridgely & Greenfield 2001). As mentioned by Parker *et al.* (1985), however, songs are also frequently delivered from well-hidden perches at forest edges and in dense cover, generally *c.*1m above ground. The territorial behaviour has not been documented, but the species is thought to sing and maintain territories year-round (Krabbe & Schulenberg 2003). One curious behavioural observation, to the best of my knowledge documented previously only by Freile (2000), is the occasional regurgitation of a bolus of undigested food. As described by Freile (2000), I have also seen this twice while following Tawny Antpittas among tall bunch grass at Papallacta. On both occasions the bolus contained fragments of insect chiton and what appeared to be small amounts of fibrous plant matter (probably ingested incidentally during prey capture). Although undocumented in other species, this might be a regularly utilised method of dealing with hard-to-digest parts of chitinous insects and perhaps small vertebrate bones or incidentally ingested plant matter. Territory size has not been studied in detail, but Cresswell *et al.* (1999b) estimated densities of 0.1–0.9 individuals/ha.

DIET Like its congeners, Tawny Antpitta is thought to be a generalist on small to medium-sized invertebrates such as earthworms and insects. Several authors mention spiders, cockroaches (Blattodea), grasshoppers (Acrididae) and centipedes (Chilopoda) as known prey (Olivares 1973, Krabbe & Schulenberg 2003). Parker *et al.* (1985) documented caterpillars (Lepidoptera), beetles (Coleoptera), dipteran larvae, leeches and the bones of small vertebrates in the stomachs of adults (*quitensis*) from northern Peru (verified from LSUMZ specimen labels; M.L. Brady *in litt.* 2017). Calderón (2002) stated that frogs form part of their diet, and the bones examined by Parker *et al.* (1985) were also thought to belong to a frog. Apparently, one stomach contained 2mm-long seeds and another contained 'fruit' (Krabbe & Schulenberg 2003), but the extent to which vegetal matter is intentionally

consumed is unknown. The regurgitated pellets examined by J. Freile and myself contained beetle legs and elytra (including Carabidae, Scarabaeidae and Passalidae) and fragments of lepidopteran wings.

REPRODUCTION To date, only two studies have focused on the breeding biology of Tawny Antpitta, both of the same population in Napo, north-east Ecuador (*quitensis*). The first (Greeney & Martin 2005) provided a complete description of the nest and eggs, while the second (Greeney & Harms 2008) documented the specifics of incubation behaviour. Several other papers have provided cursory information on nests, all of nominate *quitensis* and all from Ecuador (Black-M. 1982, Greeney *et al.* 2011b). To date, no nests have been described for either of the other two subspecies, and the information below (except Seasonality) pertains entirely to nominate *quitensis*. **NEST & BUILDING** As noted by Greeney & Martin (2005), within any given pair's territory old nests and the currently active nest tend to be clustered, almost always associated with relatively isolated vegetation and usually on the leeward side of ridges where they receive most shelter from wind and weather. These authors mentioned *Gynoxys* [*acostae*], *Loricaria antisanensis* (both Asteraceae) and *Hypericum laricifolium* (Clusiaceae) as preferred nesting substrates in the Papallacta area (*quitensis*). With the exception of adding bunch-grass (*Festuca* sp., Asteraceae) as a favoured spot, my additional observations from this area support these data. Thirty-four nests, including the three nests described in Greeney & Martin (2005), were constructed in the following plants: *H. laricifolium* (*n* = 13), *G. acostae* (*n* = 13), *Festuca* sp. (*n* = 3), *L. antisanensis* (*n* = 2), *Pentacalia* sp., Asteraceae (*n* = 1), *Baccharis* sp., Asteraceae (*n* = 1), unknown (*n* = 1). Based on many hours searching for nests, I suggest that bunch-grass is more heavily utilised than the above data suggest, but they are more easily overlooked in this dense plant that frequently dominates the habitat in many areas. I expand here upon the 0.8–2.5m nest height of *G. quitensis* in Greeney & Martin (2005) with the above-mentioned sample of 34 nests (range 0.55–2.5m, mean ± SD of 1.2 ± 0.5m). Most nests are roughly centrally located in the substrate and often only partially concealed by sparse foliage (Greeney & Martin 2005). Previously, the nests of *G. quitensis* were described as bulky cups of moss, mud and small sticks, sparsely lined with pale grass stems (Greeney & Martin 2005). My additional observations suggest this is generally true of most nests. I believe, however, that the 'mud' often seen in the nest walls is frequently, if not always, as a result of repeated use of the same nests, resulting in the decomposition of previously added materials. There are no published observations of nest construction by *G. quitensis*, but if it proves that they actually bring mud to the nest, this would be the only known case of mud being used in the nest of any antpitta. Measurements of 34 nests, including the three measured by Greeney & Martin (2005) are: external diameter 19–30cm, mean 22.5 ± 2.9cm; external height 10–16cm, mean 13.0 ± 0.9cm; internal diameter 10–13cm, mean 11.2 ± 0.7cm; internal depth 6.0–7.5cm, mean 6.3 ± 1.0cm. It is not known, but it is presumed that both sexes are involved in nest-site selection and construction. **EGG, LAYING & INCUBATION** The eggs are blue to blue-green, subelliptical, and somewhat variably marked; completely unmarked, bear a few scattered cinnamon flecks and small spots, or display heavier (but still relatively sparse) cinnamon and lavender flecks and blotches scattered across the egg. Most eggs with markings tend to have the spots concentrated near the larger pole to some degree (Greeney & Martin 2005). Published measurements are as follows: 30.6 ± 0.3 × 24.8 ± 0.8mm, range 30.3–31.1 × 23.0–25.5mm (*n* = 8, Greeney & Martin 2005). These authors reported egg weights of 8.5g and 10.3g. To this I add measurements for nine additional eggs from four complete clutches (two eggs) and one incomplete clutch of nominate *quitensis* (clutches separated by semicolons): 31.4 × 25.0mm, 31.0 × 24.3mm; 31.6 × 24.6mm, 32.1 × 25.2mm; 30.8 × 22.9mm, 31.3 × 23.3mm; incomplete clutch, 29.5 × 24.4mm; 28.8 × 23.8mm, 30.8 × 24.1mm. The final clutch was from Yanacocha (Pichincha), the rest from Papallacta (Napo). Overall mean ± SD = 30.7 ± 0.8 × 24.5 ± 0.8mm (*n* = 17 eggs, all *quitensis*). Previously, I described an instance where I believed the first egg (of a two-egg clutch) had been deposited only minutes prior to my arrival, as it was still warm and sticky (Greeney & Harms 2008). Years of experience with the eggs of this and other *páramo*-nesting birds suggests that eggs remain slimy or sticky for several hours after being laid, especially on foggy days. They do not, however, retain their heat for very long at this elevation. At the nest in question, the fact that the egg was still warm even in the chilly *páramo* air is very strong evidence of its recent deposition. This discovery was made at *c*.16:30, and there is little doubt that this egg had been laid no more than 30min prior. At the same nest, *c*.48 hours passed before the second egg was laid, with our observations strongly suggesting it was also laid after 16:00 (Greeney & Harms 2008). It appeared that the adult did not spend the night on the nest the night after clutch completion, as we found the eggs icy cold and covered with dew at 06:00, only a few minutes after sunrise next day. Unfortunately, I have been unable to gather further information on time of egg-laying or commencement of incubation for this species. These two eggs hatched within 30min of each other (10:15 and 10:45) 21.5 days later, but this is the only record of incubation period for *G. quitensis*. Both pair members participate in incubation (Greeney & Martin 2005, Greeney & Harms 2008). By reviewing 252h of video recorded at this nest, Greeney & Harms (2008) documented limited and irregular incubation for the first three days after clutch completion. Across the entire incubation period, daylight coverage of the clutch was 44–97%. The lower percentage occurred on the day of clutch completion, with all subsequent days having over 64% attendance, with a mean 86% coverage across the incubation period if the first three days of irregular attendance are omitted. This is close to the 82% coverage documented by Greeney & Martin (2005) during only four hours of observation on a single day. Incubation bouts (*n* = 209) lasted a mean 55 ± 37min, while periods of absence (*n* = 164) lasted 12 ± 21min (Greeney & Harms 2008). Like other antpittas, for *G. quitensis* the longest period of absence in the day usually occurs around dawn (Londoño *et al.* 2004, Greeney 2006). Also similar to other antpittas (Greeney *et al.* 2008a), incubating *G. quitensis* observed by Greeney & Harms (2005) spent a measurable amount of time (3.9%) engaged in behaviours that likely reduce their ability to remain vigilant. Similar behaviours were also described at nests studied by Greeney & Martin (2005). Bouts of activity on the nest, when adults engaged in one of the various behaviours described below, may occur as frequently as 10–11 times/h or every 5–6min (Greeney & Harms 2008). During 27% of a total of 2,158 movement bouts, Greeney & Harms (2008) observed the

adults lean into the nest and probe at the nest lining, either with sharp pecks or with the sewing-machine-like movement described for other passerines (Haftorn 1994, Greeney 2004), including antpittas (Dobbs *et al.* 2003, Greeney & Gelis 2006). Twenty-eight percent of non-vigilant time on the nest was spent arranging material therein, while a relatively small (5%) amount of time was dedicated to preening or ruffling their feathers. Incubating adults sometimes close their eyes and appear to doze off, occasionally opening their bills widely but silently, presumably 'yawning' (Greeney & Martin 2005). Apparently, watching eggs develop is akin to watching paint dry. When returning to the nest, either to allow their partner to rest or get back onto the eggs, adults brought with them a piece of lining material which was rather unceremoniously dropped into the nest before settling down (28% of 229 arrivals). While incubating, adult *G. quitensis* vocalise frequently. Individual bouts of singing (with >30s between vocalizations) usually begin with soft, one or two-note versions of the normal song (see Vocalisations). Songs increase in volume slowly, and the bout of vocal activity usually ends with one or more typical, full-volume, three-note songs. Often the singing adult alternates between two- and three-note songs (Greeney & Martin 2005). The observed bouts range in duration from single songs to repeated songs (at 4–10s intervals) spanning four minutes or more and delivered at rates of up to 17 songs/min. At the nest studied by Greeney & Harms (2008), combining all time spent at the nest across the incubation period, adults sang at an overall rate of 4.5 songs/h. In addition to delivering two- or three-note songs, adults occasionally gave single-note calls, *keeyurr!*, usually in response to similar calls heard away from the nest. I presume that these calls, at least when given from the nest, function to inform mates of their location and status. They are certainly given in alarm (see Vocalisations) and may possess more than one function. While vocalising, adults strain their necks upwards, pointing their bills at an angle towards the sky as described for singing adults away from the nest (see Natural History), but incubating adults always remain in apparent contact with their eggs (Greeney & Martin 2005). While incubating, adults are reluctant to flush, usually only abandoning the nest when observers approach to within *c*.1m or less (Greeney & Martin 2005). With ten additional observations of flush distance at three additional nests (total *n* = 16 flushes at six nests), mean flush distance during incubation was 1.2 ± 0.4m (range 0.8–1.8m). When an approaching threat is detected, incubating adults raise their heads, fluff their throats, and hold their bills upward while keeping their bodies low to the nest (see *G. varia* for further comments on similar responses by other species). When leaving the nest in the presence of a threat, adults drop immediately to the ground and escape by running 10–20m. Broken-wing displays or other types of distraction behaviour have not been observed. Usually the displaced adult begins an incessant repetition of their harsh, piercing alarm call (see Vocalisations) and are quickly joined by their mate.

NESTLING & PARENTAL CARE There are no published studies on nestlings or parental care of *G. quitensis*. Previously unreported, the nestling period at one nest may have been 15 or 16 days, as I found the nest empty, with no signs of disturbance (pers. obs.). Compared to the longer nestling periods of many congeners (Greeney *et al.* 2008b), this seems rather short, especially considering the low temperatures faced by nestlings. Further observations are certainly warranted before a nestling period of 15–16 days is considered certain. **SEASONALITY** Across their entire range, there are observations suggesting breeding in nearly all months. At Papallacta in eastern Ecuador, however, breeding appears fairly seasonal (peak September–December). This coincides with generally milder weather and less precipitation in this area. This period is also marked by the mass emergence of scarab beetle adults ('*catzos*'), which are so numerous they are often harvested for food by humans (G. Buitrón-Jurado *in litt.* 2017) and possibly provide an important resource for Tawny Antpittas. Despite the plethora of records that follow, more data are needed for any region before seasonality can be accurately accessed in any portion of the range. **BREEDING DATA** *G. q. alticola* Fledgling, 14 October 2013, Laguna de Chisacá (ML 42994541, N.O. Díaz Martínez); two juveniles, 4 August and 25 July 1947, 6.4km E of Páramo Angostura (USNM 402477/78); juvenile, 16 December 2013, PNN Chingaza (J.D. Castillo Ramírez); subadult, 30 April 2017, Laguna de Chisacá (ML 56900721, photo T. Forrester); juvenile, *c.*27 December 1971, La Cueva (Olivares 1973); 'juvenile', 6 June 1965, Páramo de Sumapaz (Olivares 1967); 'juvenile/immature', June, Cundinamarca (Fjeldså & Krabbe 1990, possibly same record as Olivares 1967); adult carrying food, 10 June 2015, PNN Chingaza (Flickr: B. Caswell); ovary enlarged, 4 August 1947, E of Páramo Angostura (Hilty & Brown 1986, USNM 402476). *G. q. quitensis* **COLOMBIA**: Adult carrying food, 6 November 2015, RNA Loros Andinos (Flickr: ProAves); fledgling, 29 March 1912, Valle de las Papas (AMNH 116337); fledgling, 23 April 2016, entrance to Nevado del Ruíz (M. 27534421, photo D. López); fledgling, 31 May 2012, Nevado del Ruíz, Caldas/ Tolima border (Flickr: A. Bunting); fledgling transitioning, 15 September 1911, Páramo de Santa Isabel (AMNH 111980); fledgling transitioning, 21 April 2013, Villamaria (ML 43071661/71, photos N.O. Díaz M.); two juveniles, Valle de las Papas, 22 and 26 March 1912 (FMNH 51014/ AMNH 116332, AMNH 116335); juvenile, 31 May 2012, PNN Los Nevados (IBC 1214439, video M. Kennewell); juvenile, 15 September 1911, Páramo de Santa Isabel (AMNH 111981); juvenile, 3 March 1958, Páramo San Rafael (LACM 33122); juvenile, 25 January 1958, Malvasá (LACM 33120); juvenile, 3 February 2015, Casa de Nieves (ML 45289051, photo H. Cruz); juvenile transitioning, 22 March 1912, Valle de las Papas (AMNH 116333); juvenile transitioning, 21 April 1956, El Crucero (FMNH 226611); 'juvenile,' 12 June 2015, PNN Los Nevados (eBird: D. Avendaño); immature, 12 November 2011, Laguna Torreadora (eBird: P. Cantino); two subadults, 7 February 1952, Páramo Puracé (USNM 446644/45); subadult, Nevados del Ruiz, Caldas/Tolima border, 31 March 2014 (L.E. Urueña photos); subadult, 18 October 1958, El Crucero (FMNH 251325); 'fledglings' in March in Huila, September in Quindío (Fjeldså & Krabbe 1990); 'juvenile/ immature in August in Piura (Fjeldså & Krabbe 1990); 'juveniles,' March in Huila and September in Cauca and Quindio (Fjeldså & Krabbe 1990); adults in breeding condition, February, Cauca (Hilty & Brown 1986). **ECUADOR**: Five nests under construction, Papallacta Pass, 11 October 2006 (*n* = 2; HFG), 29 August 2006, 28 September 2005, 29 September 2005 (Greeney *et al.* 2011b); two reports of adults carrying material, 7 December 2003, Yanacocha (N. Athanas *in litt.* 2004) and 10 February 2016, Micacocha (eBird: D. Humple); two nests initiating laying, Papallacta Pass, 18 October 2006

(HFG), 11 October 2006 (hatch 24 October; Greeney 2012a); female with ruptured follicle [laying], 27 August 1991, *c.*21km NE of Salcedo (MECN 6017); 14 nests with eggs, Papallacta Pass, 26 September 2005 (×2), 27 October 2006 (Greeney *et al.* 2011b), 28 September 2005 (Greeney & Harms 2008), 16 and 22 September 2004, 17 October 2004 (×2), (Greeney & Martin 2005), 11 October 2006, 26 November 2011, 18 October 2012 (×2), 5 October 2014, 7 October 2014 (HFG); nest with eggs, RP Yanacocha, 2 February 2009 (Greeney *et al.* 2011b); three nests with nestlings, Yanacocha, 5 January 2010, 9 February 2009 (×2) (HFG); three nests with nestlings, Papallacta, 17 October and 22 October 2004, 7 October 2005 (Greeney *et al.* 2011b); two adults carrying food at Papallacta Pass, 7 February 2000 (R.C. Dobbs *in litt.* 2015) and 27 January 2016 (eBird: J. Tobin); adult carrying food, 17 October 2014, W of Papallacta Pass (eBird: M. Dettling); adult carrying food, 9 December 2010, Quebrada Chuquibantza (eBird: P.-Y. Henry); fledgling, 27 February 2014 (T. Friedel photos); fledgling, 8 February 2011, Cabañas San Jorge de Quito (G. Cruz photos); fledgling, 13 February 2011, Papallacta Pass (R. Ahlman *in litt.* 2017); fledgling (by measurements), 4 May 1883, Ceche (Domaniewski & Sztolcman 1918); fledgling transitioning, 22 November 1924, Volcán Antisana (AMNH 810739); fledgling transitioning (FMNH 99573) and juvenile (FMNH 99574), 1935, Cordillera de los Llanganates; juvenile, 21 February 2014, near Micacocha (ML 38791201, photo S. Malcom); juvenile, 17 February 1900, 7.3km W of Lloa (USNM 236464); juvenile, 29 January 1914, Guagua Pichincha (USNM 313769); juvenile, 11 June 1931, 'El Castillo' (MLZ 5195); two juveniles, 1 January 1899, near Quito (AMNH 492199/200); two juveniles, 7 and 8 April 1963, Volcán Pichincha (MECN 2622/23; M.V. Sánchez N. photos); two juveniles, Bestión, 5 January and 6 April 1921 (AMNH 167269/AMNH 167272); juvenile, 16 October 1987, Papallata Pass (WFVZ 42900); juvenile, 9 August 2015, RP Yanacocha (B. Al-Shahwany photo); juvenile, 6 June 2013, Yanacocha (D.M. Brinkhuizen photos); juvenile, 21 July 1929, Culebrillas de Sangay (MLZ 5166); juvenile, 23 February 1936, Páramo de Carbonsillos (MLZ 9749); juvenile, 6 April 1927, 'San Diego Chucho' (MLZ 7226); juvenile, 3 January 2009, PN Cajas (Y. Koleinsson photos); juvenile, 9 February 1900, Volcán Pichincha (USNM 236463); juvenile, 8 February 2011, Papallacta Pass (I. Maton photos); juvenile, 26 October 1922, near Papallacta (AMNH 176066); juvenile, 26 July 1937, Volcán Chimborazo (YPM 14770; S.S. Snow photos); juvenile, 20 April 2015, PN Cajas (A. Sorokin photos); juvenile, 18 September 2007, Páramo del Artezón (QCAZ 2990, G. Buitrón-Jurado photos); juvenile (by measurements), 4 May 1883, Ceche (Domaniewski & Sztolcman 1918); juvenile transitioning, 28 September 2015, Cabañas San Jorge de Quito (Flickr: J. Scarff); juvenile transitioning, 18 September 1923, Cerro Guamaní (AMNH 180262); three juveniles transitioning, Yanacocha, 20 April 1974 (MECN 2624, M.V. Sánchez N. photos), 8 June 2017 (IBC 1395943, K. Harvard) and 20 September 2013 (IBC 979826, D. Weaver); three juveniles transitioning, near Micacocha, 8 March 2014 (ML 39604801/11/21, photos M. Todd), 13 March 2016 (ML 53641321, photo E. Ocaña), 25 June 2016 (ML 37482291, photo M. Smith); juvenile transitioning, 7 April 1963, Volcán Pichincha (MECN 2625, M.V. Sánchez N. photos); juvenile transitioning, 5 March 1921, Bestión (AMNH 167271); juvenile transitioning, 14 June 1974, Laguna Cuicocha

(FMNH 373092); juvenile transitioning, 30 September 1923, Oyacachi (AMNH 180258); subadult, 7 July 2015, Cabañas San Jorge de Quito (G. Cruz *in litt.* 2015); subadult, 27 May 2007, Yanacocha (J.N. Rosenthal photo); subadult, 7 December 1991, Tablahuasi (WFVZ 49485); subadult, 5 July 1971, 'Cuyuja' (MECN 2618; M.V. Sánchez N. photos); two subadults, PN Cajas, 12 February 2009 (J. Hornbuckle photos), 17 May 2013 (Flickr: 'DM Pura Vida'); subadult, 31 August 1929, W of Mocha (MLZ 7213); two subadults, 22 August and 26 October 1931, Culebrillas de Sangay (MLZ 5175/83); subadult, 5 April 1927, 'Pacubamba' on Volcán Pichincha (MLZ 7227); four subadults, Papallacta Pass, 15 October 2014 (HFG), 18 July 2015 (Flickr: D. Betzler), 10 February 2016 (long photo series including ML 25761901/941/981, J. McGowan), 11 February 2016 (photo series including ML 25855131, J. McGowan); two 'fledglings' shot with an adult male in January near Quito, Pichincha (Goodfellow 1902); immature, 14 November 2012, Paja Blanca (eBird: J. Neill); 'juvenile/immatures,' January, in NW Ecuador and Azuay, also in July, October and November in NE Ecuador (Fjeldså & Krabbe 1990); 'juvenile' female collected last week of January near Cuenca, Azuay (Berlioz 1932); 'juvenile' in April in Pichincha (Salvadori & Festa 1899). **PERU:** Juvenile, 3 August 1980, Cruz Blanca (LSUMZ 97689); subadult, 15 July 1980, Cerro Chinguela (LSUMZ 97683). *G. q. atuensis* No previously published data for this subspecies. My only additions are two specimens from Mashua, a subadult collected 22 September 1979 (LSUMZ 92482) and a female with an unshelled egg (24 × 20mm) in the oviduct, 24 September 1979 (LSUMZ 92490). **ADDITIONAL COMMENTS** Olivares (1974) reported an adult male and a 'subadult' female (*alticola*) collected in late December in La Sierra Nevada del Cocuy in Colombia's West Andes (Dept. Boyacá). The male was in non-breeding condition but, confusingly, he gave the state of the female's ovaries as 'developing', despite describing her plumage as having the nape and neck-sides still barred with juvenile plumage. This observation of what is apparently a bird in immature plumage preparing to breed bears further consideration. We now know that immature plumage is retained for an extended period, at least in some antpitta species, but presumably in all. If, however, antpittas commence physiological preparations for breeding before moulting to adult plumage, this may explain, in part, the slow plumage replacement as a physiological trade-off in energy allocation between moult and breeding. To date, there are no observations of antpittas in immature plumage breeding (but see *G. ridgelyi*). In the future, however, researchers with the opportunity to closely examine birds attending nests should look for signs of immature feathering.

TECHNICAL DESCRIPTION Sexes similar. For detailed descriptions of the adults of the somewhat distinctly plumaged races of *G. quitensis* see Geographical Variation. **Adult Bare Parts** *Iris* brown to chestnut-brown; *Bill* black to blackish-grey, base of mandible sometimes brown (Parker *et al.* 1985); *Tarsi & Toes* pale to dark brownish (olive-grey: Restall *et al.* 2006; brownish-grey: Krabbe & Schulenberg 2003). Although Krabbe & Schulenberg (2003) were unaware of previous descriptions of immatures, I encountered several in the literature, which provide at least a partial idea of the appearance of young *G. quitensis*. Goodfellow (1902) described the 'young' of nominate *quitensis* as having the 'head and back speckled

with black arid brown, their breasts are darker than in the adult bird and are marked down the centre with black; gape yellowish red.' According to Fjeldså & Krabbe (1990), the 'juvenile' (subspecies not specified) is 'barred pale rufous and blackish brown almost throughout, belly mostly buffy-white'. The most complete description of the plumage of 'juveniles' previously available was by Olivares (1967) based on two individuals of *alticola*. He described the crown, nape, interscapular area and rump as blackish, strongly barred ochraceous. The greater wing-coverts are tipped ochraceous with a subterminal blackish band. The feathers of the throat and foreneck are edged blackish, while the chest and upper belly are barred black and pale ochraceous, with adult-coloured feathers scattered throughout. The thighs and undertail-coverts are yellowish. In dried skins Olivares (1967) noted that the maxilla was black and the mandible reddish with black splotches, while the tarsi and toes were reddish-black. A second specimen, described by Olivares (1967) as being older than the bird described above, was similar on the upperparts but had the underparts much like those of an adult but with scattered black blotches. He further noted that individuals in both plumages were already similar in size to adults. Below, I provide full descriptions of fledglings, juveniles and subadults of *alticola* for the first time, as well as these plumages for *quitensis* based on Greeney (2015e). The immature plumage of *atuensis* is undescribed. **Fledgling (*alticola*)** I examined photographs of a fledgling by N.O. Díaz Martínez. It was scarred on the forecrown and had a severely malformed maxilla making the appearance of this area difficult to describe accurately. From the crown to the nape, however, the plumage was grey, with each feather broadly striped buffy over the shaft, creating a distinctly streaked appearance. The face was nearly uniform dark tawny, slightly duskier on the ear-coverts and dusky mixed buffy-white on the malar. The throat was white, washed dusky in the centre and buffy mixed dusky on the lower part. The white of the throat extended rearwards giving a slightly moustached appearance. The upper back was already similarly coloured, with a few scattered feathers bearing indistinct tawny tips, these markings bordered above and below (on the feather margin) with variable amounts of dark grey. These feathers are otherwise coloured similarly to adults, and apparently belong to a plumage phase between fledgling and full adult plumage. The upper back still has a few patches of buffy-white or tawny-white feathers thickly barred black, with the lower back and rump mostly clothed in juvenile feathers. The neck-sides are finely barred and spotted tawny-buff and grey. The upper breast and sides are largely covered with adult-like tawny feathers boldly streaked white, already giving it the distinctive mottled appearance of race *alticola*. Two irregular, 'messy' bands of dusky immature plumage cross the breast. The lower breast and belly are wet, and somewhat difficult to determine their coloration, but they appear generally whitish, lightly washed buff. The flanks and vent look buffier. The tail is not visible, suggesting it is still even shorter than that of an adult. The wings are similar in coloration to an adult but the wing-coverts are distinctly tipped bright tawny with a thick subterminal black band. A fledgling described by Domaniewski & Sztolcman (1918) appears similar to this description. **Fledgling Bare Parts (*alticola*)** *Iris* dark brown; *Bill* maxilla dusky above, pale at very tip and orange on tomia, mandible entirely orange, paler distally, brighter orange on slightly inflated rictus (because of the bill deformity, it is difficult to determine the extent to which the rictus inflation has been effected); *Tarsi & Toes* dull bluish-lead, the nails appear somewhat yellowish, especially distally. **Fledgling (*quitensis*)** Precise sequence of post-fledgling plumage development is still not completely clear and appears somewhat variable with respect to when different tracts of the upperparts lose their downy, barred, fledgling appearance. In addition, Greeney (2015e) described what appears to be a fledgling at least a week or more older than other individuals, but which retains tufts of pale grey-white nestling (natal?) down on the crown- and nape-sides. In general, however, older fledglings have the forecrown to nape and upper back finely barred ochraceous or tawny-buff and blackish. The back and rump are coarsely barred tawny-ochraceous and black, but with irregular adult (olivaceous-brown) feathers, especially on the scapulars. In general coloration, the wings and tail are similar to adults, but the rectrices have distinctly buffy tips and the wing-coverts are slightly browner and finely edged buffy-ochraceous or tawny-rufous, with subterminal black bars. Additionally, the secondaries and tertials are tipped buffy with faint subterminal black bars or spots on the tertials. These black markings are either greatly reduced or lacking on the secondaries. The lores are whitish washed tawny to tawny-olivaceous. Fledglings at this stage already have an indistinct eye-ring, similar in colour to the lores in front of the eye and more whitish postocularly. The superciliary and head-sides behind the eyes are olivaceous-tawny or brownish-tawny. The ear-coverts and malar are similar in colour, but washed dusky, approaching blackish in some individuals, especially on the posterior ear-coverts and lower face-sides. The central throat is pale tawny bordered white, creating a fairly distinct submalar streak that may or may not extend rearward. The neck-sides are finely barred, continuous with the nape, becoming brighter tawny-buff towards throat and more thickly barred blackish. This barring continues across the breast just below the throat and onto the sides and the central breast. The immature plumage of the breast is broken by variable numbers of emerging adult-looking tawny feathers. The remaining underparts are usually still mostly in downy plumage, with a somewhat mottled tawny-buff and white appearance, with grey in patches. This fluffy plumage becomes buffier lower, and the vent and flanks are largely clean tawny-buff. **Fledgling Bare Parts (*quitensis*)** *Iris* dark brown; *Bill* dusky-orange, mandible blackish on culmen and more orange on tomia and at tip, mandible slightly duskier near tip than basally, overall mostly dusky-orange, rictus still somewhat expanded and bright orange; *Tarsi & Toes* dusky-pink to brownish-pink, nails distinctly yellow. **Juvenile (*alticola*)** J.D. Castillo Ramírez provided photos of a juvenile and I also examined two skins in similar plumage. At this age, the forecrown is tawny-olive with thin black shaft-streaks. The face is dark tawny with dusky feathers, especially on the ear-coverts and lower malar, with the cheeks and upper malar appearing slightly speckled. The crown, hindcrown, nape and upper back are blackish or dark grey and coarsely streaked (not spotted or barred as in *quitensis*) tawny-buff to pale buff. The back is now largely similar to adult's, but a few scattered feathers (otherwise similar to adult's) have tawny-buff tips and subterminal black bars. The wings are similar to adult's, but upperwing-coverts are dusky-olive with rusty-buff or tawny-rufous tips and thin subterminal black bars. The chin and throat are now clean white, with

the remaining underparts overall much like an adult *alticola*. Variable numbers of feathers across the lower breast have dusky spotting, and the flanks have a slight greyish wash. The undertail-coverts are buffy-whitish, the uppertail-coverts are warm orange-tawny, and the tail is warm tawny-brown with the tips of the rectrices edged pale tawny. **Juvenile Bare Parts (*alticola*)** *Iris* dark brown; *Bill* maxilla black with a paler tip and tomia, mandible black, dusky-orange on tomia and ridge, especially closer to base; *Tarsi & Toes* greyish to pale violet-grey, nails yellowish-grey. **Juvenile (*quitensis*)** G, Buitrón-Jurado supplied photos of a mid-aged juvenile. With these photos and examination of numerous museum specimens (see Reproduction, Seasonality), the descriptions of Greeney (2015e) can be summarised as follows. The crown is dusky-olivaceous with variable tawny-olive edges to the feathers, giving an indistinctly scaled appearance. The crown appears variably spotted in some individuals, probably a character that becomes less distinct with age. The forecrown is more tawny-olive and less dusky. The face is dark tawny with dusky feathers, especially on the ear-coverts and lower malar, with the cheeks and upper malar slightly speckled. Feathering is whitish around the eyes and on the lores. The hindcrown, nape and upper back are tawny-buff, finely barred (not streaked as in *alticola*) black, with a variable number of adult feathers throughout. The back is as in adults, sometimes with faint dusky scaling or barring. The wings are similar to adult's, but more tawny-olive, the primaries are distinctly tawny on the anterior margins and dark on the posterior webs. The upperwing-coverts are dusky-olive with tawny-buff tips (seemingly less tawny-rufous than *alticola*) and thin subterminal black bars. The secondaries have a similar pattern but are otherwise more brownish. The tail appears similar to adult's but has thin tawny-buff markings near the tips and traces of white or pale buff at the very tips. The uppertail-coverts are buffy-rufous to rusty-buff, the thighs tawny buff. The undertail-coverts are buffy, with the bases to the long fluffy rump and lower flank feathers dark grey. The chin and central throat are whitish, becoming tawny lower and on sides. The underparts are overall much like adult's but with variable, but usually irregular, patches of dusky black barring, especially on the breast-sides and flanks. A juvenile described by Domaniewski & Sztolcman (1918) is similar to this. **Juvenile Bare Parts (*quitensis*)** *Iris* dark brown; *Bill* maxilla black, with a dusky-white tip, dusky-orange on tomia, mandible black, dusky-orange on tomia and ridge, especially closer to base, and gape orange with only slightly inflated rictal flanges; (in dried specimens) maxilla black with a whitish tip, dusky-orange or paler dusky on tomia, mandible black, paler on tomia and centre of lower surface, especially near base; *Tarsi & Toes* greyish brown; (in dried specimens) brownish to reddish-brown. In life, slightly older juveniles have the bill mostly black, slightly more brownish on the tomia and paler at tip of the maxilla, but retain a bright orange rictus that is now only slightly enlarged. **Subadult (*quitensis*)** G. Cruz, D. Weaver and J.N. Rosenthal all sent me photos of individuals in almost complete adult plumage. Just prior to acquiring full-adult plumage, the crown is dusky-olivaceous, the forecrown more tawny-olive and less dusky. The face is dark tawny with dusky feathers, especially on the ear-coverts and lower malar, with the cheeks and upper malar appearing slightly speckled. Feathering is whitish around the eyes and on the lores. Hindcrown and nape is now mostly similar to adult's, but sometimes a few small patches of tawny-buff feathers, finely barred black, remain. The back is as adults but with faint dark scaling or barring created by dusky feather edges. The wings are tawny olive-brown, the primaries tawny, at least on the anterior margin, the upperwing-coverts are dusky-olive with buffy tips. These markings are somewhat variable, with primary-coverts replaced first, then the secondaries and tertials. The chin and central throat are whitish, becoming tawny lower and on sides. The underparts are overall like adults but younger individuals can retain a few small traces of dusky-black barring, especially on breast-sides. Overall, it appears that the last clearly juvenile plumage character to be lost is the traces of barring on the nape or neck-sides, followed by replacement of the upperwing-coverts. **Subadult Bare Parts (*quitensis*)** *Iris* as adults; *Bill* maxilla black with a white tip, mandible black, paler on tomia, gape orange with rictus no longer inflated and sometimes barely visible; *Tarsi & Toes* brownish-grey to dark grey, largely similar to adults.

MORPHOMETRIC DATA *G. q. alticola Wing* 93mm; *Tail* 50mm; *Bill* exposed culmen 21mm; *Tarsus* 46mm (*n* = 1, *alticola*, holotype, Todd 1919). Additional measurements (sexes combined) from FMNH and USNM (HFG). Adults: *Wing* 98.0.8 ± 7.1mm (*n* = 4); *Bill* from nares 13.9 ± 0.3mm (*n* = 5), exposed culmen 19.6 ± 0.6mm (*n* = 4), depth at nares 6.9 ± 0.2mm (*n* = 4), width at nares 5.4 ± 0.3mm (*n* = 4), width at base of mandible 7.9 ± 0.3mm (*n* = 4); *Tarsus* 44.2 ± 1.7mm (*n* = 4). From two juveniles (♀ USNM 402477, ♂ USNM 402478) *Bill* from nares 13.2mm, 13.2mm; *Tarsus* 45.1mm, 46.1mm. See also Olivares (1974). *G. q. quitensis* Measurements from FMNH, LACM, MLZ, AMNH, WFVZ (HFG), LSUMZ (M.L. Brady), MECN (M.V. Sánchez N.), QCAZ (G. Buitrón-Jurado). Adult ♀♀: *Wing* 96.8 ± 4.7mm (*n* = 8); *Tail* 46.6 ± 2.2mm (*n* = 5); *Bill* from nares 15.4 ± 0.7mm (*n* = 18), exposed culmen 22.6 ± 1.2mm (*n* = 18), total culmen 28.1 ± 1.5mm (*n* = 4), depth at nares 8.5 ± 0.4mm (*n* = 5), width at nares 6.8 ± 0.5mm (*n* = 6), width at base of mandible 9.1 ± 0.8mm (*n* = 6), width at gap 14.6 ± 0.5mm (*n* = 4); *Tarsus* 48.5 ± 2.1mm (*n* = 11). Juvenile ♀♀: *Wing* 97.4 ± 4.0mm (*n* = 5); *Tail* 45.7 ± 5.0 (*n* = 3); *Bill* from nares 14.6 ± 0.5mm (*n* = 9), exposed culmen 21.6 ± 1.3mm (*n* = 8), total culmen 26.6 ± 1.1mm (*n* = 3), depth at nares 7.7 ± 0.5mm (*n* = 3), width at nares 7.1 ± 0.6mm (*n* = 4), width at base of mandible 8.6 ± 0.6mm (*n* = 4), width at gap 13.4 ± 1.4mm (*n* = 6; *Tarsus* 47.9 ± 1.6mm (*n* = 9). Adult ♂♂: *Wing* 97.6 ± 5.1mm (*n* = 26); *Tail* 45.3 ± 3.3 (*n* = 7); *Bill* from nares 15.4 ± 0.7mm (*n* = 35), exposed culmen 21.9 ± 1.1mm (*n* = 29), total culmen 27.0 ± 1.0mm (*n* = 7), depth at nares 8.3 ± 0.7mm (*n* = 14), width at nares 6.7 ± 0.6mm (*n* = 15), width at base of mandible 8.8 ± 0.7mm (*n* = 14), width at gap 12.7 ± 2.4mm (*n* = 13); *Tarsus* 48.5 ± 2.4mm (*n* = 44). Juvenile ♂♂: *Wing* 100.6 ± 5.2mm (*n* = 5), *Tail* 52.2 ± 4.6 (*n* = 3); *Bill* from nares 14.7 ± 1.2mm (*n* = 9), exposed culmen 19.8 ± 2.1mm (*n* = 6), total culmen 25.2 (*n* = 1), depth at nares 7.9 ± 1.0mm (*n* = 5), width at nares 8.3 ± 3.9mm (*n* = 5), width at base of mandible 8.5 ± 0.3mm (*n* = 4), width at gape 13.0 ± 1.9mm (*n* = 2); *Tarsus* 48.9 ± 2.1mm (*n* = 9). Two fledglings transitioning (♀? FMNH 99573, ♀ FMNH 57239): *Wing* 103mm, 98mm; *Bill* from nares 11.0mm, 13.3mm, exposed culmen 15.1mm, 18.5mm, depth at nares n/m, 6.4mm, width at nares 6.1mm, 5.8mm, width at base of mandible 8.9mm, 8.0mm; *Tarsus* 49.5mm, 46.6mm. Previously unpublished tarsal measurements for nestlings (HFG) include: 17.9mm (five days old); 34.0mm,

35.1mm (one day pre-fledging). See also Domaniewski & Sztolcman (1918), Carriker (1933). *G. q. atuensis* Scant data in literature: *Wing* 87mm; *Tail* 52mm; *Tarsus* 40mm; *Middle toe* 28mm (*n* = 1, holotype, Carriker 1933). *Wing* 91mm; *Bill* exposed culmen 19mm; *Tarsus* 41mm (*n* = ?, Bond 1950). Data from LSUMZ (M.L. Brady), adults and subadults combined. Adult ♀♀: *Wing* 85.8 ± 2.2mm (*n* = 4); *Bill* from nares 13.4 ± 0.3mm (*n* = 6), exposed culmen 21.3 ± 0.8mm (*n* = 6); *Tarsus* 41.6 ± 1.3mm (*n* = 6). Adult ♂♂: *Wing* 87.9 ± 3.1mm (*n* = 11); *Bill* from nares 14.3 ± 0.4mm (*n* = 10), exposed culmen 21.5 ± 0.8mm (*n* = 10); *Tarsus* 40.9 ± 1.4mm (*n* = 11). **MASS** Without distinguishing subspecies, Krabbe & Schulenberg (2003) gave the following ranges: ♂♂ 62–78g, ♀♀ 58.5–81.2g. See also Parker *et al.* (1985).

G. q. quitensis Specimen label data from WFVZ, LSUMZ, MECN and QCAZ (adults and subadults combined). Adult ♀♀: 61g, 65g, 67g, 70g, 71.0g(lt), 75g(no), 81.2(m). Adult ♂♂: 66g(lt), 67.2g(m), 69g, 69.5g(no), 70g, 71g, 71g(no), 72g(tr), 72g(m), 72.3g(lt), 73g(x2), 73g(lt), 74g(x2), 74g(m), 78g(lt). Juvenile ♂♂: 68.6g, 70.8g(m), 75(tr). Juvenile ♀♀ 60g. Nestling 22.2g (five days old; HFG). *G. q. atuensis* All data from LSUMZ specimen labels. Adult ♀♀: 63g(lt), 62.5g, 60.0g, 62.5g, 58g. Adult ♂♂: 58.2g, 60g(lt), 62g(m), 69g(m), 59g, 63g(lt), 64g(lt), 60g, 57g, 66.5g, 60g. *G. q. alticola* No data. **TOTAL LENGTH** 16–18cm (Carriker 1933, Hilty & Brown 1986, Fjeldså & Krabbe 1990, Ridgely & Greenfield 2001, Schulenberg *et al.* 2007).

TAXONOMY AND VARIATION The taxonomic affinities have not been investigated in detail, but its inclusion in the subgenus *Oropezus* (Ridgway 1909, Lowery & O'Neil 1999) has never been disputed. Rufous Antpitta seems a likely sister species, a distinctive subspecies of which (*saltuensis*) may actually be better treated as a race of Tawny Antpitta (Krabbe & Schulenberg 2003; see discussion under Rufous Antpitta). Fjeldså (1992) suggested that 'a certain overall similarity' suggests a close relationship between the present species and Rufous-faced Antpitta. Krabbe & Schulenberg (2003) pointed out that the three races described below differ vocally, and may represent three separate species. Indeed, del Hoyo & Collar (2016) elevated each of the following taxa to species level based on somewhat minor, but seemingly consistent, plumage differences (see below), in conjunction with fairly marked vocal differences (Boesman 2016b; see Voice). Although I tend to agree with this split, I have maintained their subspecific status until a more rigorous analysis provides additional support for their recognition as full species.

Grallaria quitensis alticola Todd, 1919, Proceedings of the Biological Society of Washington, vol. 32, p. 115, Lagunillas, Boyacá, Colombia [06°15'N, 72°38'W]. The adult male holotype (CM P59904) was collected 17 March 1917 by M. A. Carriker Jr. Originally described as a species, this fairly distinctive race is restricted to the East Andes of northern Colombia, from Norte de Santander south to Cundinamarca. The following description is based on Todd (1919). Race *alticola* has the upperparts, including crown, deep sepia-brown, brighter brownish on uppertail-coverts. The wings and tail are similar in coloration, but the primaries tend to be slightly paler. The lores and head-sides are dull buffy, with the ear-coverts usually noticeably darker. The central throat is buffy-white, becoming pale tawny or yellow-orange below and on sides. Rest of underparts buffy-ochraceous, with whitish feather tips giving a mottled or scaled appearance. The underwing-coverts are bright ochraceous-tawny. Del Hoyo & Collar (2016) considered *alticola* to differ from nominate *quitensis* in its slightly smaller size, including bill, and browner rather than olive-brown upperparts. In my view, the overall more yellowish cast to the underparts and the mottled pattern of the breast make *alticola* especially distinctive. Distributional records, especially north of Bogotá, are fairly scarce. SPECIMENS & RECORDS *Santander* 6.4km E of Páramo Angostura, 07°23'N, 72°54'W (USNM 402476–478). *Boyacá* La Cueva, 06°25'N, 72°21'W (Olivares 1973, ICN 22973); PNN Pisba, *c*.05°51'N, 72°32.5'W (XC 215558, J. Zuluaga-Bonilla); Hacienda Baza, 05°18.5'N, 73°25.5'W (eBird: P. Kaestner). *Cundinamarca* Páramo del Palacio, 04°59'N, 74°26'W (ICN 11826/27); Lagunas de Siecha, 04°46'N, 73°51'W (IBC 977984, photo M. Rueda); Páramo Guasca, 04°43.5'N, 73°51.5'W (ICN 11836; XC 148116/18, F.G. Stiles); Carretera Chingaza, *c*.04°41'N, 73°47'W (ML 49892961, N. Komar); Lagunas Boca Grande, 04°18'N, 74°06'W (ICN 33381). *Bogotá* Laguna de Chisacá, 04°18'N, 74°12.5'W (ICN 11828–840, ANSP 168268/69); Páramo de Sumapaz, 04°14'N, 74°12'W (ICN 15561; XC 316173, G.M. Kirwan).

Grallaria quitensis quitensis Lesson, 1844, Écho du Monde Savant, vol. 2, no. 49, 'vicinity of Quito', Ecuador. Although for most species this vague type locality would be probably best interpreted as either on the slopes of Volcán Pichincha or the pass above Papallacta, for Tawny Antpitta it may be fairly accurate. J. Freile (*in litt.* 2017) informs me that Tawny Antpitta is sometimes found quite close to urban Quito, e.g. the Itchimbia and Puengasi hills, the Parque Metropolitano del Sur, Volcán Atacazo and on Volcán Ilaló (00°16'S, 78°25'W). I suggest that Ilaló, should the need to define a more precise type locality arise, could serve as a reasonably accurate estimate. The nominate race inhabits the Central Andes of Colombia in Caldas, Tolima, Cauca and Nariño, south from Carchi (Cresswell *et al.* 1999a), throughout the Andes of Ecuador, to extreme northern Peru in Cajamarca and Piura (Peters 1951, Fjeldså & Krabbe 1990). SPECIMENS & RECORDS COLOMBIA: *Caldas* El Zancudo, 05°05'N, 75°30'W (CM P70428); Villamaría, 05°02.5'N, 75°31'W (N.O. Díaz Martínez *in litt.* 2016); Nevado del Ruíz, 04°56'N, 75°21'W (XC 354120/21, R. Gallardy); entrance to Nevado del Ruíz, 04°53'N, 75°21.5'W (ML 27534421, photo D. López). *Tolima* La Leonera, 05°05'N, 75°20'W (ANSP, CM); Páramo del Ruiz, 04°54'N, 75°18'W (CM P70702/03); near Murillo, 04°53'N, 75°14'W (XC 188193–196, J. Poveda); Vereda La Estrella, 04°44.5'N, 75°14.5'W (XC 96374, Y.G. Molina-Martínez); Páramo de Romerales, 04°40'N, 75°19'W (ANSP 154010–12, UMMZ 222951); Finca el Vergel, 04°38.5'N, 75°25'W (XC 130336/38, O.H. Marín-Gómez); Casa de Nieves, 04°37'N, 75°17'W (ML 45289051, photo H. Cruz). *Risaralda* Páramo de Santa Isabel, 04°47'N, 75°28'W (FMNH, AMNH, CMNH); El Bosque, 04°44.5'N, 75°26'W (XC 53643/44, B. López-Lanús). *Quindío* Finca Buenos Aires, *c*.04°38.5'N, 75°29'W (XC 53660, B. López-Lanús); Finca La Judea, 04°09.5'N, 75°43'W (XC 129437/244778, O.H. Marín-Gómez); near Génova, 04°08.5'N, 75°44'W (XC 197056, O. Cortes-Herrera). *Cauca* Malvasá, 02°29'N, 76°18'W (LACM 33117–121, FMNH 249748); Páramo San Rafael, 02°25'N, 76°25'W (FMNH 255647/48, LACM 33122/23); Páramo Puracé, 02°24'N, 76°27'W (USNM); Mina de Azufre/El Crucero [Volcán Puracé; Paynter 1997], *c*.02°20'N, 76°23'W (FMNH, LACM, USNM); Paletará, 02°12.5'N,

76°30'W (CM, YPM); 'Valle de las Papas' 01°51'N, 76°36.5'W (Meyer de Schauensee 1950, AMNH, FMNH). *Huila* PNN Puracé, 02°12'N, 76°21'W (XC 68419, F. Lambert). *Nariño* Laguna Verde, 01°05.5'N, 77°43.5'W (eBird: J. Muñoz García); Lago Cumba, 00°57.5'N, 77°49'W (J.A. Zuleta-Marín *in litt.* 2017); E slope of Volcán Chiles, 00°52.5'N, 77°57.5'W (ANSP 149856); Mayasquer, 00°52'N, 78°00'W (Meyer de Schauensee 1950, ANSP 149857); Chorreado/Guanderal, *c*.00°50'N, 77°25'W (ANSP 162064–070; see Paynter 1997). *Putumayo* Páramo Bordoncillo, 01°09'N, 77°05.5'W (D. Carantón *in litt.* 2017). ECUADOR: *Carchi* Near Laguna Verde, 00°48'N, 77°55'W (LSUMZ 162111); Páramo del Artezón, 00°46.5'N, 77°54.5'W (QCAZ 2990); Quebrada Agua Blanca, 00°45'N, 78°04'W (J. Freile *in litt.* 2016); headwaters of Río Bobo, *c*.00°43'N, 77°47'W (N. Krabbe *in litt.* 2015); Loma El Voladero, 00°39.5'N, 77°53.5'W (J. Freile *in litt.* 2017); RBP Guandera, 00°36'N, 77°42'W (Cresswell *et al.* 1999a,b); Escudilla, 00°33'N, 77°46.5'W (ANSP 66852/53); W slope of Cerro Mongus, 00°27'N, 77°52'W (Robbins *et al.* 1994b, MECN 6106, ANSP, ML). *Sucumbíos* Alto La Bonita, 00°29.5'N, 77°35'W (Stotz & Valenzuela 2009); 2km SW of Cocha Seca, 00°39'N, 77°40'W (USNM 614860). *Imbabura* Pimantura, 00°30.5'N, 78°13.5'W (J. Freile *in litt.* 2017); Quebrada Chumachi, *c*.00°19.5'N, 78°21'W (LSUMZ 112589); Laguna Cuicocha, 00°18'N, 78°22'W (FMNH 373092); Otavalo, 00°14'N, 78°15.5'W (LACM 105723); Casa Mojanda, 00°10.5'N, 78°16.5'W (HFG). *Napo* Oyacachi, 00°13'S, 78°04.5'W (ANSP, AMNH, MCZ); Laguna de Loreto, 00°18'S, 78°08'W (J. Freile *in litt.* 2017); Papallacta Pass at Napo/Pichincha border, *c*.00°19.5'S, 78°11.5'W (Greeney & Martin 2005, Greeney & Harms 2008, WFVZ, MECN, USNM, MHNG); Papallacta and vicinity, *c*.00°23'S, 78°08'W (AMNH, SBMNH, MLZ, MCZ, LSUMZ); Volcán Antisana near Micacocha, 00°31'S, 78°12.5'W (AMNH, ANSP); confluence of Río Tambo and Río del Valle Vicioso, 00°40'S, 78°15'W (ML 39708/09, P. Coopmans). *Pichincha* Cerro Mojanda, 00°07.5'N, 78°16.5'W [loc. adjusted from Paynter 1983] (AMNH 810742/43); Laguna San Marcos, 00°06.5'N, 77°57.5'W (XC 109780, L. Ordóñez-Delgado); 'Tablahuasi', 00°04'N, 78°35'W (WFVZ 49485); Cayambe, 00°03'N, 78°09.5'W (ANSP, UMMZ); 'Cayambe Mts.' (near Volcán Cayambe), *c*.00°01.5'N, 77°59.5'W (AMNH 492202/03); Calacalí, 00°00'N, 78°31'W (NRM: J. Freile *in litt.* 2015); Alaspungo, *c*.00°00'S, 78°36'W (MECN 2620, MLZ 7216); Verdecocha, 00°06'S, 78°38'W (USNM 357203); RP Yanacocha, 00°06.5'S, 78°35'W (MECN, CAS, USNM, XC, ML, IBC); Cabañas San Jorge de Quito, 00°07'S, 78°31'W (G. Cruz photos); slopes and crater of Volcán Pichincha, *c*.00°10'S, 78°35'W (see Notes); Parque Metropolitano del Quito (and historical records for Quito), *c*.00°11'S, 78°28'W (Goodfellow 1902, Lévêque 1964; see below); 4km SE of Pifo, 00°16'S, 78°18'W (N. Athanas *in litt.* 2017); Loma de Huamanichupa, *c*.00°19'S, 78°13.5'W (WFVZ 47087); Cerro Guamaní, 00°20'S, 78°13'W (AMNH 180260–263); near Cerro Atacazo, *c*.00°21.5'S, 78°37.5'W (SBMNH 8623); RVS Pasochoa, *c*.00°26'S, 78°29'W (Fierro 1991); SW slope of Volcán Corazon, 00°33'S, 78°43'W (XC 248853, N. Krabbe); Laguna de Limpiopungo, 00°36.5'S, 78°28.5'W (MVZB 160506, XC); Iliniza, 00°39.5'S, 78°43'W (ANSP 169707); E slope of Volcán Cotopaxi, 00°41'S, 78°25'W (SBMNH 8615). *Cotopaxi* Paja Blanca, 00°38.5'S, 78°30.5'W (eBird: J. Neill); E entrance to PN Cotopaxi, *c*.00°43.5'S, 78°32'W (ML 55031, A.T. Chartier); *c*.5km W of Güingopana,

00°46.5'S, 78°49'W (S. Olmstead *in litt.* 2016); Volcán Quilotoa, 00°51.5'S, 78°54'W (MLZ 5186); *c.*21km NE of Salcedo, *c.*00°58'S, 78°25'W (WFVZ 49483, MECN 6017). ***Bolívar*** BP Cashca Totoras, 01°45'S, 78°58'W (J. Freile *in litt.* 2017). ***Cañar*** San José de Culebrillas, 02°25.5'S, 78°52'W and Cañar Zhuya, 02°34.5'S, 78°56.5'W (P.X. Astudillo W. *in litt.* 2017); La Libertad, 02°36'S, 78°42'W (XC 41/42, B. Planqué); Chanlud, 02°40'S, 79°02.5'W (P.X. Astudillo W. *in litt.* 2017). ***Tungurahua*** Laguna de Pisayambo, 01°05'S, 78°23'W (MECN 7685); Quebrada El Golpe, 01°08'S, 78°19'W (MECN 6151); Cordillera de los Llanganates, *c.*01°13'S, 78°15'W (FMNH 99573/74); Quebrada Chuquibantza, 01°13.5'S, 78°51.5'W (eBird: P.-Y. Henry); Llanganatillos, 01°16'S, 78°26.5'W (J. Freile *in litt.* 2017; QCAZ); Guayama Valley, W of Mocha, *c.*01°27'S, 78°44'W (MLZ 7212/13). This last is probably close to 'Mocha Cañon' (AMNH 145845). ***Chimborazo*** Volcán Chimborazo, *c.*01°28'S, 78°48'W (YPM 14770–772, AMNH 124435); Urbina, 01°30'S, 78°44'W (AMNH 145841–844); above Chambo (= Los Cubillines, MECN), *c.*01°44'S, 78°33'W (ANSP 59373–383, MECN); Alao, 01°54'S, 78°29'W (ML 18095, J.P. O'Neill); Ceche, 02°11'S, 78°51'W (Domaniewski & Sztolcman 1918); Cochaseca, *c.*02°11'S, 78°58'W (FMNH 57595); Laguna de Atillo, 02°11.5'S, 78°31'W (G. Buitrón-Jurado *in litt.* 2017). ***Azuay*** Laguna Torreadora, 02°46.5'S, 79°13.5'W (XC 17746/47, A. Spencer); Río Blanco, 02°49.5'S, 79°22'W (P.X. Astudillo W. *in litt.* 2017); PN Cajas, *c.*02°50'S, 79°12.5'W (XC 76276, I. Davies); Fasañán, 03°02'S, 78°42'W (J. Freile *in litt.* 2017); Portete, 03°06'S, 79°06'W (MLZ 5190); Carachulas-Shagli, 03°09'S, 79°22.5'W (P.X. Astudillo W. *in litt.* 2017); El Paso, 03°22'S, 79°05'W (AMNH 167275/76, MCZ 138474); Bestión, 03°29'S, 79°04'W (AMNH 167268–273); Páramos de Yacuambí, 03°35'S, 79°04.5'W (XC 196862, L. Ordóñez-Delgado). ***El Oro*** Taraguacocha (above), 03°34'S, 79°35'W (AMNH 167267). ***Morona-Santiago*** 'Culebrillas de Sangay', 01°46'S, 78°17'W (MLZ, CM); Páramo de Matanga, 03°16'S, 78°54'W (XC 250101, N. Krabbe). ***Loja*** Páramo de Carbonsillos, 03°33'S, 79°15'W (MLZ 9748–751); Lagunas de Mani, 03°34.5'S, 79°25.5'W (XC 153783/84, L. Ordóñez-Delgado); Cerro Acanamá, 03°42'S; 79°13'W (XC 76282, I. Davies); Cerro Toledo, 04°22.5'S, 79°06.5'W (R.A. Gelis *in litt.* 2017); Bosque Comunal Angashcola, 04°35'S, 79°22'W (J. Freile *in litt.* 2017); Páramo de Cordillera Las Lagunillas, *c.*04°43.5'S, 79°26'W (MECN 6493, ANSP 185502, XC 250405, N. Krabbe). ***Zamora-Chinchipe*** Besún, 03°26'S, 78°56'W (MCZ 138473); Estación Científica San Francisco, 03°58'S, 79°04'W (J. Freile *in litt.* 2017); Lagunas de Sabanillas, 04°30'S, 79°10'W (QCAZ); 25km by road SSE of Jimbura, *c.*04°45.5'S, 79°25'W (ANSP 186012, KUNHM 86825). **PERU:** *Piura* Cerro Chinguela, 05°07'S, 79°23'W (LSUMZ, ML, XC); Abra Cruz Blanca, 05°20'S, 79°32'W (LSUMZ 97687–689, ML). *Cajamarca* Tabaconas, 05°19'S, 79°17'W (MUSA 3337); Quebrada Lanchal, 05°41'S, 79°15'W (LSUMZ 169898). **NOTES** There are a large number of vague or imprecisely georeferenced historical collecting locations near the slopes and crater of Volcán Pichincha (Paynter 1993), and many specimens labelled simply 'Quito' were probably collected somewhere on the slopes of this volcano. The coordinates above are for a centrally located point between the two highest peaks, Guagua Pichincha (00°10.5'S, 78°36.5'W) and Rucu Pichincha (00°09.5'S, 78°34'W). Aside from the known or appoximate locations above, some of the most common 'mystery locations' include: San Diego Cucho (MLZ 7226);

Hacienda Garzón (ANSP 59367–372, UMMZ 222952); Guagua or Guagua forests (USNM 313768/69, ROM 30470, CM P102339/40); Pacubamba (MLZ 7227, MVZB 160508); Rinconada (SBMNH 8616/17); any variation on Mt. Pichincha (MCZ, AMNH 124431–34, ANSP 59365/66, SBMNH 8618, USNM 236463); Crater del Pichincha (MCZ 199121–126). The location 'El Castillo' was considered to be close to Mindo (Paynter 1993), but *quitensis* specimens (MLZ 5188/95, CM P142872) are probably from along the old road between Quito and Mindo, in the vicinity of Nono (00°03.5'S, 78°34.5'W). Several Ecuadorian specimens bear unlikely collecting localities, all of which are signficantly below appropriate elevations for Tawny Antpitta. The following were not included in the range map: Mindo, 00°03'S, 78°46.5'W (AMNH 492201); Cuyuja, 00°25'S, 78°01.5'W (MECN 2618); Nanegal, 00°08'N, 78°40.5'W (USNM 371393).

Grallaria quitensis atuensis Carriker, 1933, Proceedings Academy of Natural Sciences of Philadelphia, vol. 85, p. 22, Atuén, Amazonas, Peru, 12,000 ft [3,660m, 06°45'S, 77°52'W; Vaurie 1972]. This is the least known of the three taxa. For more than 50 years after its discovery, it was known only from the type specimen (ANSP 108127; Peters 1951, Parker *et al.* 1985). Race *atuensis* is endemic to the Central Andes of northern Peru, in southern Amazonas and eastern La Libertad (Krabbe & Schulenberg 2003). The following description follows Carriker (1933), Bond (1950) and del Hoyo *et al.* (2017e). Most similar to *quitensis*, but differs in the uniformly darker coloration, both of the upper- and underparts, especially the latter, which are dark olive-ochraceous, darker across the chest and buffy olive-brown on the flanks. The throat is clear cinnamon-ochraceous, instead of whitish-ochraceous or white. The white markings on the underparts consist of thin white tips to the feathers (not white on the median portion), creating a more spotted than scaled or streaked appearance. This pattern is also generally more conspicuous than in other races, at least compared to *quitensis*, and contrasts more with the darker ground colour. The lores, narrow ocular ring and broad postocular spot are deep cinnamon-ochraceous instead of buffy-white, while the ear-coverts are the same colour basally, but tipped blackish giving them a sooty look. The bill of *atuensis* is much smaller than in *quitensis*, and smaller than but closer in size to *alticola*. Finally, the legs and wings are somewhat shorter in *atuensis* than in either of the other races. **SPECIMENS & RECORDS** *Amazonas* La Muralla, 06°47'S, 77°46'W (XC 735, T. Mark); Chuquibamba, 06°54.5'S, 77°47'W (A. García-Bravo *in litt.* 2017). ***San Martín*** Río Lajasbamba, 06°50.5'S, 77°40.5'W (XC 166, W.-P. Vellinga); Puerta del Monte, 07°34'S, 77°09'W (LSUMZ). ***La Libertad*** Mashua, 08°12'S, 77°14'W (long series LSUMZ); E of Tayabamba, 08°17'S, 77°18'W (ML 17235/38, T.A. Parker).

STATUS Across its considerable range (27,179.8km² globally following Freile *et al.* 2010, and 16,272km² in Ecuador following Freile 2000), Tawny Antpitta is judged fairly common to common (Hilty & Brown 1986, Stotz *et al.* 1996, Ridgely & Greenfield 2001, Schulenberg *et al.* 2007). The nominate race is numerous in multiple national parks in Colombia and Ecuador, and occurs in at least one protected area in Peru. Even the range-restricted *alticola* is common in several Colombian national parks. Peruvian *atuensis* is perhaps the least common taxon, but is abundant locally and occurs in at least one national park. In addition to its wide range and apparently large population, Tawny Antpitta is relatively tolerant of disturbance and can be abundant in shrubby second growth around habitations and in agricultural areas (Fjeldsa & Krabbe 1990, Ridgely & Tudor 1994, Freile 2000, Ridgely & Greenfield 2001). It is among the first of the high-elevation species to re-colonise fire-damaged habitat (Koenen & Koenen 2000). Unsurprisingly, therefore, BirdLife International (2017) considered it Least Concern. Nevertheless, the *páramo* habitat of Tawny Antpitta is under increasing threat, especially from agricultural development and *Polylepis* wood-harvesting practices (HFG). One final word of caution for conservationists and decision-makers: although clearly more adaptable to anthropogenic changes than some species, Tawny Antpitta has almost certainly already lost more of its original habitat than most (Freile *et al.* 2010).

PROTECTED POPULATIONS *G. q. alticola* PNN Pisba (XC 215558, J. Zuluaga-Bonilla); PNN Chingaza (ICN, XC, IBC); PNN Sumapaz (del Hoyo *et al.* 2017d, XC 316173, G.M. Kirwan); PNN El Cocuy (Olivares 1973, del Hoyo *et al.* 2017d, ICN, XC). *G. q. quitensis* **Colombia**: PNN Los Nevados (Pfeifer *et al.* 2001, 2010); PNN Puracé (Hilty & Silliman 1983, USNM, FMNH, LACM, CM); RNA Loros Andinos (Flickr: ProAves); RMN El Mirador (Salaman *et al.* 2007a); RNSC Semillas de Agua (O. Cortes-Herrera *in litt.* 2017); ÁNÚ Volcán Azufral (eBird: J. Muñoz García). **Ecuador**: PN Podocarpus (Rasmussen *et al.* 1994); PN Cotopaxi (HFG); PN Sangay (MLZ, CM); PN Cayambe-Coca (Greeney & Martin 2005, Greeney & Harms 2008); PN Cajas (Tinoco & Astudillo 2007, Tinoco & Webster 2009, MECN, XC); PN Yaruci (ANSP 186012, KUNHM 86825); RE El Ángel (XC 248347/48, N. Krabbe); RE Antisana (AMNH, ANSP, XC, ML); BP Cashca Totoras (J. Freile *in litt.* 2017); BP Mazán (Astudillo *et al.* 2015); RP Yanacocha (MECN, XC, IBC); RP Verdecocha (USNM 357203); RBP Guandera (Cresswell *et al.* 1999a,b); RVS Pasochoa (Fierro 1991). **Peru**: SN Tabaconas-Namballe

(Amanzo 2003). *G. q. atuensis* PN Río Abiseo (Krabbe & Schulenberg 2003, del Hoyo *et al.* 2017e); ÁCP Llamapampa-La Jalca (eBird: A. Sagone).

OTHER NAMES Several years after Lesson (1844) described *G. quitensis*, de Lafresnaye (1847) described *Grallaria monticola* from 'Bolivia,' which, as noted by Cory & Hellmayr (1924), is clearly a synonym of nominate *quitensis*. To date, however, *G. quitensis* is not known to occur in Bolivia and the type locality is obviously erroneous (Chapman 1926). Prior to this, the name *G. monticola* was widely used in the literature (Sclater 1857c, 1858, 1860b, 1862, 1890, Hartert 1898, Salvadori & Festa 1899, Goodfellow 1902, Brabourne & Chubb 1912, Chapman 1917, Moore 1934), appearing sporadically even after publication of Cory & Hellmayr (1924). **English** Mountain Antpitta [*quitensis*], Eastern Mountain Antpitta [*alticola*] (Cory & Hellmayr 1924); Mountain Ant-Thrush (Brabourne & Chubb 1912); Quito Ant-Pitta [*quitensis*], Boyacá Ant-Pitta [*alticola*] (Meyer de Schauensee 1950); Western, Northern or Southern Tawny Antpitta (when races split; del Hoyo & Collar 2016). **Spanish** Tororoí Leonado (Clements & Shany 2001, Krabbe & Schulenberg 2003); Gralaria leonada (Valarezo-Delgado 1984, Ortiz-Crespo *et al.* 1990, Ortiz-Crespo & Carrión 1991, Ridgely *et al.* 1998; Ridgely & Greenfield 2001, Granizo 2009); Licuango (Delgado 1984, Carrión 2002, Altropico 2010), Güicundo, Huicundo (Valarezo-Delgado 1984, Carrión 1986, 2002); Vicoca (Olivares 1973); Tororoí leonado occidental, Tororoí leonado norteño, Tororoí leonado sureño (del Hoyo & Collar 2016). **French** Grallaire de Quito (Krabbe & Schulenberg 2003, del Hoyo & Collar 2016); Grallaire alticole, Grallaire d'Atuen (del Hoyo & Collar 2016). **German** Ockerwangen-Ameisenpitta (Krabbe & Schulenberg 2003, del Hoyo & Collar 2016); Grauwangen-Ameisenpitta, Dunkelameisenpitta (del Hoyo & Collar 2016).

Tawny Antpitta, adult (*alticola*), Parque Nacional Natural Chingaza, Cundinamarca, Colombia, 23 January 2012 (*Nick Athanas/ Tropical Birding*).

Tawny Antpitta, nest with eggs hidden in bunch grass (*quitensis*), Papallacta Pass, Napo, Ecuador, 10 October 2012 (*Harold F. Greeney*).

Tawny Antpitta, adult on nest (*quitensis*), Termas Papallacta, Napo, Ecuador, 20 October 2014 (*Harold F. Greeney*).

Tawny Antpitta, adult singing after snowstorm (*quitensis*), Papallacta Pass, Napo, Ecuador, 25 February 2011 (*Håkan Sandin*).

Tawny Antpitta, young nestlings begging (*quitensis*), Yanacocha, Pichincha, Ecuador, 2 February 2009 (*Harold F. Greeney*).

Tawny Antpitta, older fledgling (*quitensis*), Villamaría, Caldas, Colombia, 21 April 2013 (*Neil Orlando Díaz Martínez*).

URRAO ANTPITTA
Grallaria fenwickorum   **Plate 18**

Grallaria fenwickorum Barrera & Bartels, 2010, Conservación Colombiana, vol. 13, p. 9, Reserva Natural de las Aves Colibrí del Sol, Vereda El Chuscal, 06°25'53.1'N, 76°04'57.9'W, 3,130m, Páramo del Sol, Urrao, Antioquia, Colombia (Barrera *et al.* 2010).

= *Grallaria urraoensis* Carantón-Ayala & Certuche-Cubillos, 2010, Ornitología Colombiana, vol. 9, p. 58, Reserva Natural de las Aves Colibrí del Sol, Vereda El Chuscal, 06°26'N, 76°05'W, 2,850m, Páramo del Sol, Urrao, Antioquia, Colombia. The specific name used by Carantón-Ayala & Certuche-Cubillos (2010) refers to the municipality (Urrao) where *G. fenwickorum* was discovered and to which it may very well be confined. They designated a holotype (ICN 36689) and a paratype (ICN 36688), both males, collected 30 March and 20 February 2008, the latter an immature (skull ossification 30%). The word Urrao is of indigenous origin; these areas were originally occupied by several tribes, some currently represented by groups of the Embera-Katío culture (Carantón-Ayala & Certuche-Cubillos 2010).

Considerable controversy surrounds the description of Urrao Antpitta (Anon. 2010a,b, 2011, Cadena & Stiles 2010, González *et al.* 2011, Regalado 2011, Fundación ProAves 2013). Following the discovery of this antpitta, two publications by separate authors proposed different names: Fenwick's Antpitta *G. fenwickorum* (Barrera *et al.* 2010) and Urrao Antpitta *G. urraoensis* (Carantón-Ayala & Certuche-Cubillos 2010). In short, the controversy centres around the relative validity of type materials used in the two descriptions: feather samples and a photograph (Barrera *et al.* 2010) versus two skins (Carantón-Ayala & Certuche-Cubillos 2010). Following considerable discussion in the literature (see also Peterson 2013, Claramunt *et al.* 2014), based largely on the perceived inadequacy of type material used by, and technical flaws inherent in the description of, Barrera *et al.* (2010), some authors have used *G. urraoensis* (Greeney 2014e, Remsen *et al.* 2017), while others have used *G. fenwickorum* (McMullan *et al.* 2010, Dickinson & Christidis 2014, del Hoyo & Collar 2016). There has been some recent discussion in the literature concerning the use of, and need for, a dead individual as a type specimen (Smith *et al.* 1991, Timm *et al.* 2005, Dubois & Nemésio 2007, Donegan 2008b). Nevertheless, following the International Code of Zoological Nomenclature (ICZN), two aspects of the situation seem abundantly clear. First, the publication of the name *G. fenwickorum* (Barrera *et al.* 2010) preceeded the name *G. urraoensis* (Carantón-Ayala & Certuche-Cubillos 2010) by 37 days (Peterson 2013), making the latter name a junior synonym of the former. Secondly, there are several ways in which the description and 'holotype' of Barrera *et al.* (2010) failed to conform to the ICZN regulations (see Peterson 2013), potentially invalidating the description entirely (Claramunt *et al.* 2014). These two points led Peterson (2013) to consider *G. fenwickorum* a *nomen dubium* ['dubious name'] and to propose that the situation be rectified by designating the holotype of *G. urraoensis* as the 'neotype' (= lectotype) for *G. fenwickorum*. Irrespective of the outcome of this proposal, aside from the opinions of myself or my peers, and even irrespective of the validity of the *fenwickorum* description, strict adjerence to ICZN regulations requires the use of *fenwickorum* until such time as, and if, ICZN issues a contrary ruling (Dickinson & Christidis 2014). Therefore, somewhat deftly straddling the proverbial fence, I have followed previous authors (Fjeldså 2013, BirdLife International 2017) in referring to this species as Urrao Antpitta *Grallaria fenwickorum*. Urrao Antpitta is a Colombian endemic, known only from a tiny region at the northern tip of the West Andes. Nothing is known of its basic biology, other than what was published in the above-mentioned descriptions (Barrera *et al.* 2010, Carantón-Ayala & Certuche-Cubillos 2010). An inhabitant of seasonally dry, montane cloud forest, where it is partial to areas of *Chusquea* bamboo, Urrao Antpitta appears to breed during the drier months, but its nest and eggs are undescribed. Based on the distribution of seemingly suitable habitats, Carantón-Ayala & Certuche-Cubillos (2010) suggested that Urrao Antpitta may occur outside its currently known range in the northern sector of the Cordillera Occidental. The additional localities suggested, including the Nudo de Paramillo (Antioquia), Cerro Plateado, Farallones de Citará, and Cerro Caramanta (Antioquia/Chocó border), all are in dire need of surveys to ascertain the species' true distribution. In addition, as for most of its congeners, work should be undertaken to describe all aspects of the basic ecology such as diet, microhabitat requirements and breeding behaviour.

IDENTIFICATION 16–17cm (Fjeldså 2013). Urrao Antpitta is a rather dull-coloured, medium-sized antpitta. The head and upperparts, including wings, are olive-brown, darker and more brownish on the tail. The ear-coverts are slightly brighter olive-brown, tinged cinnamon-brown, while the lores and orbital area are pale grey. The throat and an indistinct malar stripe are light grey, with brownish feather tips. The underparts, including flanks, are pale grey with even paler tips giving a slightly scaled appearance in some lights. The breast-sides and thighs are tinged olive-grey, and the central belly is paler. The greater underwing-coverts are bright cinnamon, with the underwing otherwise dull brownish. Urrao Antpitta is most similar to Brown-banded Antpitta, but has more olivaceous upperparts and a brownish-olive rather than whitish throat. Urrao Antpitta also lacks the brown breast-band characteristic of Brown-banded Antpitta. Within its limited range, there are just four *Grallaria* sympatric with Urrao Antpitta (Barrera *et al.* 2010, Carantón-Ayala & Certuche-Cubillos 2010): Undulated, Rufous, Chestnut-naped and Chestnut-crowned Antpittas (Krabbe & Schulenberg 2003). All but Rufous Antpitta are strikingly patterned, and unlikely to be confused with the nearly uniformly Urrao Antpitta. The dull, olive-brown plumage of Urrao Antpitta immediately separates it from the bright rufescent Rufous Antpitta.

DISTRIBUTION Known only from the type locality, Colibrí del Sol Bird Reserve, on the south-east slope of the Páramo del Sol massif, at the northern tip of the West Andes of Colombia. The range of this species is so small that I have depicted it based on all available records, visual and otherwise, excluding areas with little or no available habitat. See below for a list of points used to determine the maximum extent of Urrao Antpitta's range.

HABITAT Apparently restricted to montane, epiphyte-laden cloud forest with a canopy height of 8–15m. Dominant tree species in this habitat are: *Quercus humboldtii* (Fagaceae), *Blakea longipes* (Melastomataceae), *Ocotea callophylla*, *Persea* sp. (Lauraceae), *Weinmannia* sp. (Cunnoniaceae) and *Podocarpus oleifolius* (Podocarpaceae) (Carantón-Ayala & Certuche-Cubillos 2010). Preferred microhabitat seems to be areas with heavy to moderate *Chusquea* bamboo in the understorey (Barrera *et al.* 2010, Carantón-Ayala & Certuche-Cubillos 2010) but, to date, no detailed study has focused on patterns of microhabitat use by Urrao Antpitta. It is possible that it equally utilises the mossy, wetter areas of its environment, and has been

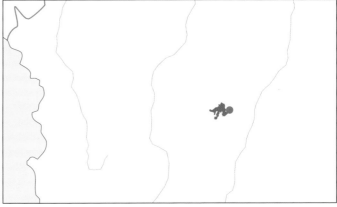

associated by researchers with bamboo largely due to the latter's ubiquitous nature at these elevations.

VOICE The following description is based on Carantón-Ayala & Certuche-Cubillos (2010). The song comprises three notes, increasing slightly in pitch and delivered in rapid succession, with a mean (± SD) 0.30 ± 0.03s between the first and second notes, and only 0.07 ± 0.02s between the second and third. The song's total duration is 0.9 ± 0.06s. The first and second notes are 0.13 ± 0.02s and 0.14 ± 0.02s in duration, respectively, whereas the third is slightly longer (0.18 ± 0.04s). The pitch of the first note is 2.93 ± 0.11kHz, the second 3.08 ± 0.10kHz and the third at 3.27 ± 0.10kHz. The call, apparently given after playback or in response to loud noises, is louder and higher pitched than the song, being a single abrupt note (duration, 0.31 ± 0.03s) that rises and falls.

NATURAL HISTORY As with almost all antpittas, Urrao Antpitta is inconspicuous and difficult to see, even when using playback. It is most often found alone or in pairs, foraging close to the ground, especially in dense bamboo and tangled understorey vegetation. While foraging, it moves leaf litter and soil with its feet. Prey are often beaten against the ground and broken into pieces before ingestion. There are no published data on territorial defence, maintenance or fidelity. Territory size is also poorly known, but Carantón-Ayala & Certuche-Cubillos (2010) found five males apparently defending territories within a 1.5km-long transect.

DIET The stomachs of specimens included arthropod fragments, of which beetles (Coleoptera) were the only identifiable items. Carantón-Ayala & Certuche-Cubillos (2010) also observed the species foraging on earthworms captured in the leaf litter and suggested that, like many *Grallaria* (Greeney et al. 2008), such prey may be important in the diet of adults and nestlings.

REPRODUCTION The nest, eggs and breeding behaviour are undescribed. **SEASONALITY** The two males collected by Carantón-Ayala & Certuche-Cubillos (2010) in February and March had enlarged gonads, suggesting breeding activity. They also observed more pronounced vocal activity during February to April and an adult male with a post-reproductive brood patch in June. A fledgling, fresh from the nest (see Description), was captured on 12 June 2008 (Barrera et al. 2010, Carantón-Ayala & Certuche-Cubillos 2010). Barrera et al. (2010) estimated this individual to have been c.18 days old and my estimate would be similar, 20–25 days old. From this, using a loosely calculated estimate of incubation period for Urrao Antpitta (18 days), I suggest a clutch-initiation date of c.1 July. The juvenile Urrao Antpitta (see Description) was captured and photographed on 20 August 2008 (Carantón-Ayala in litt. 2017), a fact omitted from the paper. Given how little we know of post-fledgling plumage development in most antpittas, estimating clutch initiation for this individual is nearly impossible. That said, placing Urrao Antpitta within the subgenus *Oropezus* as assigned to its putative relatives (Lowery & O'Neill 1969), within which the recently described Jocotoco Antpitta would be included, I can make a rough estimate based on what is known about plumage maturation in Jocotoco Antpitta. Scaling down for body mass, I would guess that this juvenile left the nest c.2.5–3.0 months prior to being captured, leading to a (very rough) clutch-initiaton date of 5–20 June. Climate in the species' known range is bimodal, with drier periods during the first months of the year (Velásquez-Ruiz

2005). Urrao Antpitta may breed predominantly during this period, thus the estimated breeding season of January through April or May (Carantón-Ayala & Certuche-Cubillos 2010) may be correct. However, my best guess is that most clutches are initiated in April–July, coinciding with the end of the drier period, and timing hatching and nestling care with the onset of rains and presumably increased insect availability and softer soil, facilitating earthworm capture. The observed increase in vocal activity early in the year may, in fact, represent increased territorial defence and courtship prior to breeding activities. Previously (Greeney 2014e) I suggested that Urrao Antpitta probably breeds during the drier months for similar reasons proposed for two other *Chusquea* bamboo-associated antpittas (Martin & Greeney 2006, Greeney & Miller 2008) and other montane bamboo 'specialist' passerines (Greeney et al. 2005, 2008b). However, although the few data for Urrao Antpitta agree that it is strongly associated with bamboo (Barrera et al. 2010, Carantón-Ayala & Certuche-Cubillos 2010, Carantón-Ayala et al. 2012), I am not yet convinced that this species is as closely associated as, for example, Chestnut-crowned Antpitta and Slate-crowned Antpitta, both of which appear to prefer nesting in bamboo as well as foraging in it (Martin & Greeney 2006, Greeney & Miller 2008, Greeney 2013k, Greeney & Kirwan 2013a). Conversely, the pale, tawny-buff coloration of the fledgling Urrao Antpitta described herein places it among a paraphyletic but ecologically similar group of antpittas, all of which appear to select nest sites with relatively high light levels and/or highly disturbed habitats, and which all share the (overall) pale nestling plumage of Urrao Antpitta. This line of reasoning would lead me to put my money on Urrao Antpitta being a bamboo-nester. For now, I prefer to await more definitive proof.

TECHNICAL DESCRIPTION Sexes similar. The following description of adults is paraphrased from the original descriptions given by Carantón-Ayala and Certuche-Cubillos (2010), using the colour nomenclature of Smithe (1975, 1981), and Barrera et al. (2010) using the colour nomenclature provided in Munsell Color (1977). *Adult* The forehead, crown, nape, back and rump are all dull olive-brown, with the wings and tail slightly darker and less olivaceous. The lores are creamy-whitish to greyish-buff, almost contiguous with an inconspicuous buffy eye-ring, with small black-tipped feathers creating an indistinct speckled look. The ear-coverts, neck-sides and malar region are warm brownish, with the mid-portion of individual feathers paler tawny-ochraceous below darker cinnamon-washed brown feather tips. The middle and upper secondary-coverts, and the external margin of the primary-coverts are rich brown, the internal margin of the primary-coverts dark grey. The underwing-coverts are warm cinnamon-brown. The primaries are brownish, washed dull brown-olive distally, while the secondaries are slightly darker. The feathers of the throat are light grey, slightly tipped dull brownish-olive, producing a mottled appearance, especially on the sides. Overall, the throat is greyish centrally, grading to brownish laterally. The breast and sides are pale grey, with the feathers narrowly edged whitish, especially on the upper breast, with the feathers of the sides irregularly and indistinctly washed olive-brown. The grey of breast fades to dull white on the centre of the abdomen. The flanks and undertail-coverts are grey, washed pale olive, the thighs olive-brown to dark olive-grey. **Adult Bare Parts** *Iris* dark brown; *Bill* maxilla black and mandible horn, with tomia and tip paler; *Tarsi & Toes* blue-grey (Carantón-Ayala & Certuche-Cubillos 2010), or *Iris* dark brown; *Bill* maxilla black, mandible dusky-horn or bluish-horn, both paler at tip; *Tarsi & Toes* grey to blue-grey (Barrera *et al.* 2010). *Juvenile* Carantón-Ayala & Certuche-Cubillos (2010) described the juvenile, an individual which I estimate as being at least one month from achieving full adult plumage, as having an 'overall scaled appearance, with patches of black down with chestnut tips in most of the crown, nape, and flanks, and in small patches in the scapular area, the rump and the breast. Its belly was buff-coloured, and the feathers on the sides of the throat were dark, giving it a blackish appearance. The maxilla was black proximally and orange distally, whereas the mandible was entirely orange; bill commissures were conspicuously red-orange.' Carantón-Ayala & Certuche-Cubillos (2010) provided the following measurements of this unsexed juvenile: *Wing* 90.0mm, *Tail* 60.5mm, *Bill* exposed culmen 18.8mm, total culmen 21.0mm, commissure width 15.3mm, bill height 7.1mm, *Tarsus* 44.1mm. *Fledgling* Carantón-Ayala & Certuche-Cubillos (2010) described a fledgling, quite obviously no more than a week out of the nest, as 'covered by a dense blackish down with brown edges on the upperparts; these edges were wider and more brightly coloured in the lower back and rump. The belly, flanks, and lower part of the chest were largely buff, its feet were coloured dark pink, and its bill was similar to that of the juvenile. Primaries and secondaries had emerged *c.*4mm from their sheaths, but rectrices had not yet started to appear.' Carantón-Ayala & Certuche-Cubillos (2010) provided the following measurements of this unsexed fledgling: *Wing* 51.6mm, *Bill* total culmen 16.0mm, commissure width 16.8mm, bill height 5.1mm, *Tarsus* 31.2mm; *Mass* 41g.

MORPHOMETRIC DATA From Carantón-Ayala & Certuche-Cubillos (2010) (*n* = 2, holotype, paratype, respectively) ♂: *Wing* 95.4mm, 96.5mm; *Tail* 63.0mm, 63.2mm; *Bill* exposed culmen 20.4mm, 18.9mm; total culmen 23.5mm, 22.2mm; width at gape 13.4mm, 13.0mm, depth at nares 7.0mm, 6.9mm; *Tarsus* 46.5mm, 44.6mm. These authors also measured three additional unsexed, live adults as follows: *Wing* 90mm, 89mm, 87mm; *Tail* 56.3mm, 58.4mm, 55.5mm; *Bill* exposed culmen 20.4mm, 20.5mm, 20.2mm, total culmen 23.0mm, 23.3mm, 23.3mm, width at gape 15.0mm, 14.6mm, 12.8mm; *Bill* height 6.8mm, 7.0mm, 7.2mm; *Tarsus* 46.3mm, 45.8mm, 44.7mm. The following is from Barrera *et al.* (2010)

(*n* = 1, ♀?) *Wing* flattened chord 99mm; *Tail* 57mm; *Bill* 'maxilla length' [total culmen] 22.9mm; *Tarsus* 49.9mm. **Mass** 56.4g(no), 57.4g(no) (*n* = 2, ♂♂, holotype, paratype, Carantón-Ayala & Certuche-Cubillos 2010); 53.5g (*n* = 1, ♀?, Barrera *et al.* 2010); 56.9g, 66.4g, 59.8g (*n* = 3, ♀?, Carantón-Ayala & Certuche-Cubillos 2010).

DISTRIBUTION DATA Known only from **Páramo de Frontino, Antioquia, Colombia**: 06°26'7.08'N, 76°04'54.12'W (XC 48972/73, K.C. Cubillos); 06°26'38.40'N, 76°06'46.80'W (XC 354207/08, R. Gallardy); 06°26'17.16'N, 76°05'18.24'W (XC 81791/92, A. Spencer); 06°20'55.85'N, 76°10'27.00'W (eBird: P. Kaestner); 06°24'13.47'N, 76°14'40.77'W (eBird: F. Murphy); 06°25'33.99'N, 76° 8'16.07'W (eBird: B. Crins).

TAXONOMY AND VARIATION Monotypic. Based on plumage and vocal characters, as well as range, Urrao Antpitta is probably closely related to Brown-banded Antpitta.

STATUS BirdLife International (2017) afforded Urrao Antpitta the conservation status Critically Endangered. Its known range is extremely small (*c.*80km²) and its global population is estimated to be fewer than 400 individuals. Although the overall range is suspected to be larger than is currently known, to date the species has been recorded only at the type locality and is in dire need of further conservation action. Although the known range of Urrao Antpitta is largely protected within RNA Colibrí del Sol, threats to its habitat, for cultivation and stock grazing, are ongoing. Land in the Páramo de Sol massif generally is not well protected, and faces increasing pressure due to timber extraction and mining efforts. There is hope, however, that the species also occurs just to the north, in PN Las Orquideas. Searching for new populations is certainly a priority, and Carantón-Ayala & Certuche-Cubillos (2010) recommended that such efforts are best directed at appropriate habitat north from Páramo de Frontino to the Nudo de Paramillo (Antioquia) and south to Cerro Tatamá (Risaralda), with particular attention to Farallones de Citará and Cerro Plateado.

OTHER NAMES Fenwick's Antpitta (Barrera *et al.* 2010, McMullan *et al.* 2010, Donegan *et al.* 2011, BirdLife International 2017). See Introduction. **Spanish** Tororoí de Urrao. **French** Grallaire d'Urrao. **German** Antioquia-Ameisenpitta (Fjeldså 2013, del Hoyo & Collar 2016).

Urrao Antpitta, fledgling, photographed at Reserva Natural de las Aves Colibrí del Sol, 12 June 2008 (*Diego Carantón-Ayala*).

Urrao Antpitta, juvenile, Reserva Natural de las Aves Colibrí del Sol, 20 August 2008 (*Diego Carantón-Ayala*).

Urrao Antpitta, adult, Reserva Natural de las Aves Colibrí del Sol, 3 September 2014 (*Dušan M. Brinkhuizen*).

Urrao Antpitta, adult, Reserva Natural de las Aves Colibrí del Sol, 3 September 2014 (*Dušan M. Brinkhuizen*).

BROWN-BANDED ANTPITTA
Grallaria milleri   Plate 18

Grallaria milleri Chapman, 1912, Bulletin of the American Museum of Natural History, vol. 31, p. 147, Laguneta, 10,300 ft, Central Andes, near Quindio Pass, Cauca, Colombia [= 3,140m, 04°35'N, 75°30'W, Dept. Quindío]. The type specimen (AMNH 111994) is an adult female collected 11 September 1911 by Arthur A. Allen & Leo E. Miller (LeCroy & Sloss 2000).

The relationships of this species to other members of the genus are not fully resolved. Based on general plumage and other morphological characteristics, Brown-banded Antpitta clearly forms part of the subgenus *Oropezus* (Ridgway 1909, Lowery & O'Neill 1969). Chapman (1912, 1917) considered it sister to Rufous-faced Antpitta, which seems unlikely. More plausibly, it has recently been suggested as a close relative of the newly described Urrao Antpitta (Carantón-Ayala *et al.* 2012), a hypothesis further supported by vocal similarities. Much remains to be learned of the natural history of this threatened and poorly understood antpitta. Its nest is undescribed and even its distribution, both historical and present, has not been fully elucidated (see Geographic Variation). After a *c.*40-year gap in observations, it is now recorded regularly at several localities, and as it is among those species now visiting worm-feeders, it should not prove too difficult to obtain such information. As aptly stated by Kattan *et al.* (2006), 'there is no quick substitute for long-term and detailed fieldwork'.

IDENTIFICATION 16.5–18.0cm. The plumage is uncomplicated and dull. Except in juveniles, it is unstreaked. The crown, nape, back, wings and tail are dull, dark brown, with the head-sides slightly paler. The lores are pale grey to whitish, as is the throat and belly. The whitish underparts are broken by a diffuse, but distinct, pale brown to ochraceous breast-band. The neck-sides are similar, and this brown-ochraceous coloration extends onto the flanks and, to a lesser degree, onto the face and ear-coverts. Brown-banded Antpitta is generally distinguished from other *Grallaria* in its range by its lack of obvious field marks. Tawny Antpitta, of higher elevations, lacks the distinctive band on the breast and is pale ochraceous (rather than brown) (Hilty & Brown 1986).

DISTRIBUTION A Colombian endemic, the species' range falls entirely within the North Andean subcentre of endemism (Cracraft 1985) and the Northern Central Andean EBA (Stattersfield *et al.* 1998). So far as is known, it is confined to the Central Andes, with records from the

west slope in Caldas to Quindío and on the east slope in Tolima (Verhelst *et al.* 2001, 2002, Losada-Prado *et al.* 2005, Parra-Hernández *et al.* 2007, Arbeláez-Cortés *et al.* 2011, ProAves 2014). Two 'Bogotá' specimens (AMNH 43555/59) are among the many trade skins of uncertain origin, and were presumably collected in the Central Andes, and not in Cundinamarca.

HABITAT As the majority of habitats within the species' range have been degraded, it is unlikely we will ever know the true habitat preferences of Brown-banded Antpitta. Although it may persist in relatively modified areas, its natural haunts would probably have been areas of regrowth following natural disturbances such as landslides. Areas dominated by *Chusquea* spp. bamboo or alder trees (*Alnus acuminata*), natural colonisers within the species' range, were probably favoured. Brown-banded Antpitta has been recorded at elevations of 1,800–3,150m (Hilty & Brown 1986, Fjeldså & Krabbe 1990, Ridgely & Tudor 1994, Kattan & Beltran 1999, 2002, López-Lanús *et al.* 2000). At any given locality, however, this may vary, perhaps affected by local patterns of habitat alteration or slight changes in microclimate.

VOICE Kattan & Beltrán (1997) rendered song as a soft, whistled *puuh, pü-pü*. Krabbe & Schulenberg (2003) described it as short (1.1s), generally given at intervals of 8–9s, and consisting of three clear whistles with a slight pause after the first. The second and third notes, especially the last, are higher pitched than the first. Overall it is similar to the three-note songs of species such as Tawny and Yellow-breasted Antpittas, but noticeably faster. The frequently heard call of Brown-banded Antpitta is a loud, whistled *wooee* (Kattan & Beltrán 1997).

NATURAL HISTORY Behaviour is poorly documented. It forages on or close to the ground, taking prey from the ground or low foliage, often running or hopping short distances (López-Lanús *et al.* 2000). The species generally remains in dense undergrowth, but may also forage for short periods in more open areas, and will visit feeders (Woods *et al.* 2011). As Brown-banded Antpitta has been observed following *Labidus* army ants (Nieto-R. & Ramírez 2006), the species may also forage in association with large terrestrial mammals, as suggested by Greeney (2012). While attending an ant swarm, Nieto-R. & Ramírez (2006) observed *G. milleri* spending most of its time foraging within the swarm, but only making swift forays into the centre. López-Lanús *et al.* (2000) observed a pair of adults feeding an older fledgling within a small, understorey mixed-species flock. During nearly an hour of observation, the family stayed within a 20m-diameter area of thick undergrowth. Brown-banded Antpitta is presumably territorial, with densities in acceptable habitat estimated at 1.3 individuals/ha, and territory sizes of 0.5–5.4ha (Kattan & Beltran 1999, 2002).

DIET The diet is essentially undocumented, but almost certainly is dominated by invertebrates. Like other *Grallaria*, earthworms probably are important for adults and for young in the nest; the species has been habituated to take earthworms at feeding stations (Woods *et al.* 2011).

REPRODUCTION The nest, eggs and nestlings of Brown-banded Antpitta are undescribed. **SEASONALITY** The only published account of breeding for this species is the observation of a 'juvenile' being fed by both parents mentioned above (López-Lanús *et al.* 2000). I infer from

their wording that this bird was an older fledgling or young juvenile. Chapman (1912) mentioned four immatures, with 'traces of juvenal plumage' from the type series (seven birds). His description of one of these, which he stated was a male collected 16 September 1911, is a very good match for the plumage of AMNH 11193 (a juvenile). Two others from this series (AMNH 111992 and NHMUK 1921.7.3.61), which I examined directly or from photos, respectively, are both subadults, or nearly so. As far as I can tell, none of the remaining four birds examined by Chapman (1912) (AMNH 111991/ 492208 and holotype examined directly, MCZ 81785 from photos) shows any indication of immaturity. Additionally, none was collected on 16 September, so Chapman's (1912) record of a juvenile presumably refers to AMNH 111993 (see below). **BREEDING DATA** Fledgling/juvenile, 17 June 1998, Volcán Tolima-Ruiz (López-Lanús *et al.* 2000); juvenile, 4 June 2014, RFP Río Blanco (F. Uribe photo); a juvenile and juvenile transitioning, 3 September 2011, Laguneta (AMNH 111993/992); juvenile transitioning; 19 August 2011, RFP Río Blanco (T. Friedel photos); two subadults, Laguneta, 31 August and 4 September 1911, (NHMUK 1921.7.3.61, MCZ 81785); subadult, 31 January 2011, RFP Río Blanco (Flickr: 'ferruge'); two records of adults prey-loading at feeders at RFP Río Blanco; 25 June 2011 (IBC 983652, D. Calderón-F.), 30 May 2012 (IBC 1048582, M. Flack).

TECHNICAL DESCRIPTION Sexes similar. The following description is based largely on Chapman's (1912) original. *Adult* Crown, nape and back deep, rich raw-umber; uppertail-coverts same as back, but rump slightly paler; tail slightly more greenish or, in some specimens, more rufescent than back; lores whitish, indistinctly speckled black; ear-coverts similar to upperparts but more ochraceous; wings overall similar to tail, with wing-coverts more similar to back coloration, the outermost primaries nearly uniform dusky-brown, with little if any rufous-brown on their distal margins; underwing-coverts tawny-ochraceous, proximal margins of secondaries narrowly ochraceous on basal half; throat greyish-white; sides of throat tawny-olive, matching the broad breast-band; sides and flanks more olivaceous, less tawny; most of remaining underparts dusky, but central belly creamy-white; undertail-coverts mixed grey and olivaceous; thighs sepia. My examination of skins and photos failed to reveal the long, thin, faint brown supercilium mentioned by Fjeldså & Krabbe (1990). **Adult Bare Parts** *Iris* dark brown; *Bare Ocular Skin* whitish to greyish, with more featherless skin exposed postocularly; *Bill* black at base with distal half to one-third pale horn-coloured; *Tarsi & toes* slate-grey to blackish. *Fledgling* López-Lanús *et al.* (2000) described a 'juvenile' still being fed by its parents as similar to juvenile *G. quitensis*, stating that it had dark barring on the crown and neck and 'the entire plumage flecked'. Its bill was mostly pinkish-red, blackish at the tip. I interpret this description as involving a fledgling nearing independence and approaching the juvenile stage of plumage development. Combining their observations with the description of immature plumage by Chapman (1912), it seems that *G. milleri* follows the same plumage progression as presumed relatives; first, developing adult-like underparts, then losing the barring on the crown (but retaining it on the nape). Most of the orange on the bill must be lost around the time that the barred feathers of the crown are moulted, presumably coinciding with full independence. *Juvenile* Chapman (1912) stated that 'black feathers on the sides of the throat appear to be the last evidences of immaturity'. He went on to describe a juvenile (as defined herein) as having the hindcrown and nape dark brown with ochraceous-buff shaft-streaks; the scapulars with a few similarly coloured but ochraceous-tipped feathers, while the flanks and abdomen still have several dusky feathers with broad ochraceous-buff bars, and there are black feathers at the sides of the throat'. Other than that, the plumage is similar to that of adults but overall duller, more olivaceous-brown. **Juvenile Bare Parts** Not described in life, but tarsi and toes probably similar to subadults (below), and the bill is likely intermediate between subadults and the fledgling. *Subadult* My examination of photos of several juveniles (see Reproduction) permits me to expand on previous descriptions. A juvenile just losing the very last of its immature plumage is nearly identical to an adult. Only a few patches of fine black barring remain on the nape. The chin is somewhat black-speckled, separated from the whitish, black-speckled lores only by the narrow gape-line. The ear-coverts and face are similar to adults but washed grey. Wings as in adults but, at least in photos of the individual I examined, basal third of second primary distinctly rufescent on outer margin (not 'little if any rufous-brown on their distal margin'; see above). Greater primary wing-coverts dark rufous with broad dusky edges, and greater secondary-coverts darker brown than wings, with very thin buffy margins and narrow black subterminal bars. Underparts similar to adult but lower throat washed smoky grey with scattered brown feathers on lower breast and belly (which becomes clean grey in mature individuals). Although I have not examined birds at the juvenile stage (as defined here), my work suggests that they will retain indistinct blackish barring on the nape and neck-sides, with markings on the hindcrown becoming less distinct and more spotted. Bare-parts coloration is also probably similar to adults, but tarsi and toes slightly paler. The most obvious feature that points to the immaturity of subadults is the distinctly bright orange gape (rictal flanges only marginally inflated). The black basal portion of the adult bill is paler and the tomia and distal third of the maxilla is dusky-orange. Almost the entire mandible is pale orange, only the very base being smoky-grey. A second subadult, photographed by F. Uribe, also appears to be in the final stages of acquiring adult plumage and is quite similar to the above description. There are, however, several differences. Only a few patches of fine black spotting (broken bars) remain on the crown, hindcrown and neck-sides (not nape). Primary wing-coverts are now like those of adults but it appears that the greater wing-coverts are still dark rufous with dusky edges (not very clear in the photo). The underparts are not visible. The bright orange gape of the other individual is now barely visible. It is still dull orange, but not inflated. The rest of the bill is as described for adults. Finally, a third individual, posted on Flickr by 'ferruge' is, judging by plumage alone, likely the oldest of the three. Only traces of immature plumage remain on the nape and neck-sides. Most other plumage characters are not clear from the photo. The bright orange gape is still fairly distinct, but is dull orange, and does not appear to be at all inflated.

MORPHOMETRIC DATA Few published measurements. *Wing* chord 89.6 ± 3.5mm; *Bill* culmen 19.3 ± 1.4mm, gape 14.0 ± 1.4mm; *Tarsus* 46.6 ± 1.9mm (*n* = 18, means ± SD, Kattan & Beltrán 1999). *Wing* 85mm; *Bill* culmen 19mm; *Tail* 53mm; *Tarsus* 42mm (*n* = 1, ♀ holotype, Chapman 1912). *Bill* from nares 12.8mm, exposed culmen 17.5mm, depth at front of nares 7.0mm, width at gape 8.9mm, width at front of nares 5.9mm, width at base of lower mandible

7.9mm; *Tarsus* 45.8mm (*n* = 1, ♂, AMNH 111996). *Bill* from nares 12.5mm, 14.2mm; *Tarsus* 44.1mm, 44.2mm (*n* = 2, ♂♂, juveniles, AMNH). **Mass** Mean 52.5 ± 3.2g (n = 18, unsexed; Kattan & Beltrán 1999). **Total Length** 13–18cm (Meyer de Schauensee 1964, 1982, Hilty & Brown 1986, Dunning 1987, Fjeldså & Krabbe 1990, Restall *et al.* 2006, McMullan *et al.* 2010).

DISTRIBUTION DATA Endemic to **COLOMBIA**: *Caldas* El Desquite, 05°06'N, 75°22.5'W (XC 117587–589, A.M. López); RFP Río Blanco, *c.*05°05'N, 75°25'W (Verhelst *et al.* 2001, Ocampo-T. 2002, Nieto-R. & Ramírez 2006); El Zancudo, 05°03.5'N, 75°26'W (CM P70234). *Risaralda* Finca San Miguel, 04°53'N, 75°30'W (Cadena *et al.* 2007); PRN Ucumarí, 04°39'N, 75°36'W (Kattan & Beltran 1997, 1999, 2002). *Quindío* Near Génova, 04°08'N, 75°44'W (XC 10721, N. Athanas); above Salento, 04°38'N, 75°34'W (Chapman 1917, Ospina-Duque & Granada-Castro 2012, AMNH, ANSP). *Tolima* *c.*3km N of Juntas, 04°35'N, 75°19'W (XC 48969, K.C. Cubillos); Cañon del Río Conbeima, 04°37'N, 75°17'W (Salaman *et al.* 2009b); Vereda San Julian, 04°29'N, 75°29'W (XC 128267/73, O.H. Marín-Gómez); Clarita Botero, 04°28'N, 75°14'W (Martinez 2014); SE flank of Volcán Tolima-Ruiz massif, 04°26'N, 75°22'W (López-Lanús *et al.* 2000).

TAXONOMY AND VARIATION Monotypic. A recently described subspecies of Brown-banded Antpitta (*Grallaria milleri gilesi*; Salaman *et al.* 2009b), which is known from a single specimen, was thought to be an immature *G. flavotincta* by Sclater (1890). This specimen (NHMUK 1889.7.10.875) is pictured in their description of *gilesi* (Salaman *et al.* 2009b), and I concur that it is not an immature *G. flavotincta*. On further reflection, I also disagree with my previous suggestion that it is an immature *milleri* (Greeney 2013l). Until the final process of moult to adult plumage is better understood, it seems prudent to refrain from accepting *gilesi* as a valid taxon based on this one specimen. Despite its apparently larger size, it seems indistinguishable by plumage from many specimens of nominate *milleri*. Furthermore, although the modelled distribution of Brown-banded Antpitta presented by Salaman *et al.* (2009b) suggests that suitable habitat extends as far north as the Medellín region, it is possible that the holotype of *gilesi* bears inaccurate locality information, a problem prevalent during the relevant era due to the commerciality of bird specimens. The range map here does not include Santa Elena in the known range of Brown-banded Antpitta. Without further evidence of northern populations bearing shared characters, we may never learn the true identity of NHMUK 1889.7.10.875.

STATUS Brown-banded Antpitta went unreported for 40 years following its original description, until two specimens were collected again at the type locality (Meyer de Schauensee 1945). After that, it was again 'lost to science' and was identified as a species of conservation concern ahead of many other antpittas (King 1981, Hilty 1985, IUCN 1986). Surveys within its range in the 1980s and early 1990s failed to find it (Collar *et al.* 1992, Kattan *et al.* 2006) and it was considered possibly extinct as recently as 1990 (Collar & Andrews 1988, Fjeldså & Krabbe 1990). However, the species was rediscovered in 1994 (Kattan & Beltran 1997). Much of what may have been its historical range is deforested and severely fragmented (Hilty 1985, Orejuela 1985, Collar *et al.* 1994, Stattersfield & Capper 2000) and BirdLife International (2012, 2013) reported a slow and ongoing degradation of habitat within the species' restricted range, but recently downgraded its conservation status from Endangered to Vulnerable (BirdLife International 2013), which status has been maintained since (BirdLife International 2017). At a national level, *G. milleri* is the focus of considerable conservation attention (Stiles 1997, Renjifo *et al.* 2000, Cardona *et al.* 2005), is currently considered Vulnerable (ProAves 2014) and, thankfully, is protected in several reserves. Apart from general destruction of habitat, no threats have been specified. Given that the species appears tolerant, to some degree, of modified habitat (Kattan & Beltran 1999), it may not be under extreme threat except where forest is completely cleared. Like all antpittas, the natural history and reproductive behaviour of Brown-banded Antpitta are almost undocumented. Studies that report territoriality, breeding and diet are especially needed. In addition, to better understand

the taxonomy and distribution, further searches for the presence of the proposed subspecies *gilesi* should be made (Salaman *et al.* 2009b). **PROTECTED POPULATIONS** PNN Los Nevados (Negret 2001, Pfeifer *et al.* 2001); RFPR Cañon del Quindío (Wege & Long 1995); PRN Ucumarí (Wege & Long 1995); RNA Giles–Fuertesi (O. Cortes-Herrera *in litt.* 2017); RNSC Patasola (O.H. Marín-Gómez *in litt.* 2017); RMN El Mirador (Salaman *et al.* 2007a); RFPR El Palmar (XC 48969, K.C. Cubillos); RNSC Alto Quindío Acaime (Wege & Long 1995).

OTHER NAMES English Miller's Antpitta (Cory & Hellmayr 1924) **Spanish** Tororoí Bandeado (Krabbe & Schulenberg 2003), Peon del Quindío (Negret 2001); Tororoí de Miller (Renjifo *et al.* 2001, Cardona *et al.* 2005, Salaman *et al.* 2007a). **French** Grallaire ceinturée. **German** Brustband-Ameisenpitta (Krabbe & Schulenberg 2003).

Brown-banded Antpitta, adult, Reserva Natural Río Blanco, Caldas, Colombia, 25 November 2016 (*Laval Roy*).

Brown-banded Antpitta, adult, Reserva Natural Río Blanco, Caldas, Colombia, 27 August 2012 (*Peter Hawrylyshyn*).

Brown-banded Antpitta, adult, Reserva Natural Río Blanco, Caldas, Colombia, 11 April 2014 (*Ramíro Ramírez Cardona*).

Brown-banded Antpitta, adult, Reserva Natural Río Blanco, Caldas, Colombia, 7 September 2009 (*Dušan M. Brinkhuizen*).

YELLOW-BREASTED ANTPITTA
Grallaria flavotincta **Plate 18**

Grallaria flavotincta P. L. Sclater, 1877, Ibis, 4th ser., vol. 1, p. 445, pl. 9, 'near Frontino', Antioquia, Colombia [*c*.06°47'N, 76°08'W]. The type locality lacks much precision, but may well lie within modern-day PN Las Orquídeas. The holotype is an unsexed adult (NHMUK 1889.9.20.622) collected on an unspecified date in 1876 by T. K. Salmon (Warren & Harrison 1971).

This rather simply, yet attractively, plumaged antpitta is uniform warm chestnut or rufous-brown above, with pale yellow underparts, from which its scientific and English names derive (Latin *flavus* = yellow, *tinctus* = coloured; Jobling 2010). Yellow-breasted Antpitta can appear bold and confident, but like most antpittas is not easily seen, even if using playback or whistled imitations of its song. Yellow-breasted Antpitta is confined to the Pacific slope of the West Andes from Colombia to north-west Ecuador, and appears to favour ravines and gullies in mature forest, but there have been no quantitative studies of its habitat preferences. Currently, it is not considered globally threatened, but its conservation status perhaps needs re-evaluation. Despite being among the first antpittas to visit worm-feeding stations, the foraging ecology and other aspects of its natural history are poorly studied. Its nest has recently been described, but details of other aspects of its breeding habits are unreported. Additionally, there remain several sizeable gaps in its range, despite it being generally assumed to be continuously distributed. This account provides novel reproductive information and the first description of juvenile plumage.

IDENTIFICATION 17–18cm. Adults are warm rufous-brown above with a dusky-grey loral area and pale bluish- or greyish-white ocular area. The underparts are mostly yellow, deepest on the throat and upper breast, fading gradually to white on the sides, flanks and lower belly. Within its range, there are no similarly coloured antpittas that are uniform chestnut above and yellowish below. Yellow-breasted Antpitta is most similar to the formerly conspecific White-bellied Antpitta, which occurs only east of the Andes and has a greyish (not yellow) wash to the underparts.

DISTRIBUTION Confined to the Pacific slope of the West Andes from Antioquia to north-west Ecuador (Hilty & Brown 1986, Ridgely & Greenfield 2001, Ayerbe-Quiñones *et al.* 2008, Freile *et al.* 2010, McMullan *et al.* 2010, Calderón-Leytón *et al.* 2011). There was some early confusion as to its range, and several authors (Cory & Hellmayr 1924, Meyer de Schauensee 1950, Peters 1951) included Colombia's Central Andes within its range. A.M. Cuervo (*in litt.* 2016) helped to resolve the source of this confusion, which seems to stem from Sclater & Salvin's (1879) inability to locate the 'Santa Elena' (of T. K. Salmon's collections) in Antioquia. At the time, they surmised it was on the west slope of the Central Andes. Soon afterwards, Sclater (1890) listed a specimen of *flavotincta* from Santa Elena. Peters (1951) did not question this and indicated the presence of Yellow-breasted Antpitta on the west slope of the Central Andes, while Meyer de Schauensee (1950) noted its presence on the '? eastern slope' of the Central Andes, probably because he knew the true location of Santa Elena (east of Medellín in the Valle de Aburrá; Paynter 1997). The true identity of this 'Santa

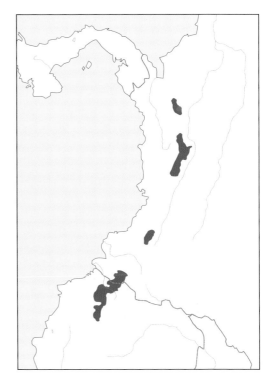

Elena specimen' is the source of some debate, and was used by Salaman *et al.* (2009b) to describe a new subspecies of Brown-banded Antpitta (which see). Irrespective of the true identity of this specimen, it is almost certainly not an immature *flavotincta*, as surmised by Sclater (1890). This confusion was then either resolved or ignored, as most (see Rodner *et al.* 2000) subsequent authors correctly stated that Yellow-breasted Antpitta is confined to the Pacific slope of the West Andes (Meyer de Schauensee 1970, Sibley & Monroe 1990, Clements 2000, 2007, Ridgely & Tudor 2009).

HABITAT Prefers the shady understorey of humid montane forests, especially on steep slopes and in riparian areas. Less frequently, they are found at forest edges, and the species appears relatively intolerant of disturbance (O'Dea & Whittaker 2007). However, Krabbe & Schulenberg (2003) noted that they sometimes are found in patches of *Chusquea* bamboo and at several locations J. Freile (*in litt.* 2017) has found them associated with naturally regenerating areas such as landslides and treefall gaps, but always continuous with (or within) mature forest. Yellow-breasted Antpitta is rarely common and seems to occur at lower densities than other similar-sized *Grallaria*. While using playback to estimate boundaries, J. Freile found that most territories do not immediately abut those of adjacent pairs, unlike the situation for other species (e.g., *G. rufula*). Across its latitudinal range, Yellow-breasted Antpitta is recorded at 1,170–2,350m (Meyer de Schauensee 1950, Hilty & Brown 1986, Stiles 1997, Ridgely & Greenfield 2001, Krabbe & Schulenberg 2003), but in north-west Ecuador seems more restricted to 1,600–2,300m (Greeney 2014i). In general, where its range overlaps Chestnut-crowned Antpitta (broadly in some areas), Yellow-breasted appears to favour less disturbed habitats (J. Freile *in litt.* 2017), mirroring the situation with White-bellied and Chestnut-crowned Antpittas in the East Andes (HFG).

VOICE The song is fairly short (1.1s) and consists of three easily whistled notes at 2.0–2.2kHz (Krabbe & Schulenberg 2003). The first note is weaker and the last lower pitched. Often the first note is so weak as to be inaudible at any distance (Ridgely & Tudor 2009) and sometimes has a hoarse or scratchy quality. Especially when heard from afar, the song recalls that of Rufous-breasted Antthrush *Formicarius rufipectus*, which is often sympatric. The three-note song is delivered at intervals of 13–22s, often for extended periods (5min or more). In response to playback, adults may respond with a drawn-out, shrill and piercing *eeeeeeeeee-yk* (M. Lysinger in Ridgely & Greenfield 2001).

NATURAL HISTORY Generally encountered alone, hopping on the ground or along fallen logs, pausing to probe the soil or leaf litter. They seem less prone than some of the other smaller *Grallaria* to climb far above ground. Yellow-breasted Antpitta is presumed to be monogamous, but there are no specific data on pair formation or courtship. Adults sing from concealed perches, usually on or very near the ground. Among the least-known aspects of most species' lives are their sleeping and roosting habits (Skutch 1989). J. Lyons reported a valuable observation of the sleeping habits of Yellow-breasted Antpitta made at Reserva Las Gralarias, Pichincha, Ecuador. On 13 March 2012, H. Imba found a sleeping adult perched on an arching palm frond 2–3m above a small forest stream *c*.3h after nightfall.

DIET There are no published records of prey taken by Yellow-breasted Antpitta. This species was, however, one of the first species habituated to visit worm-feeding stations by Sr. Angel Paz in north-west Ecuador (Woods *et al*. 2011). As with other *Grallaria*, worms are probably an important part of their diet.

REPRODUCTION Just one nest has been described (Greeney *et al*. 2009), found on 20 January 2009 at the private Las Gralarias Reserve in north-west Ecuador. **NEST & BUILDING** There are no published data concerning nest-site selection or construction, but it is likely that both adults participate. As with other *Grallaria* (Greeney *et al*. 2008), the nest was a bulky, relatively deep, open-cup structure. It was composed mostly of green moss and fern leaves intermixed with a few sticks, leaf rachises and vines. The nest was lined with dark rootlets and a few bare fern rachises. It was sited 1.5m above ground against the side of an epiphyte-laden tree trunk. These epiphytes partially shaded and provided most support for the nest. Externally, the cup was *c*.20cm in diameter and 15cm tall. The internal cup was 13cm wide by 12cm front to back, and 7.5cm deep. **EGG, LAYING & INCUBATION** When first discovered, the nest contained two turquoise eggs that may have been lightly spotted, but they were not examined closely. **NESTLING & PARENTAL CARE** Nestlings, estimated to be *c*.1 week old, weighed 29.8g and 32.7g. They were pale-skinned with patches of black down above. Their primaries had not yet broken their sheaths and their mouth linings and gape were bright crimson-orange. There are no detailed data on parental care or nesting success, but the Greeney *et al*. (2009) nest was thought to have been destroyed prior to fledging. **SEASONALITY** There are too few data to accurately define the breeding season. The nest described above contained eggs on 20 January, in the local wet season

(Greeney & Nunnery 2006). Also in north-west Ecuador, R. Ahlman photographed a young juvenile (still being fed by adults) at the worm-feeding station at Refugio Paz de las Aves on 24 September 2011 (Greeney 2014i). This juvenile was postulated to have left the nest sometime in April or May (Greeney 2014i). In light of my more recent observations, I would amend this estimation to May or June, still in accordance with wet-season breeding in north-west Ecuador. The only other published information on seasonality involves breeding-condition specimens collected in June in Colombia (Hilty & Brown 1986) and a pair showing 'precopulatory' behaviour in March in Cauca (Negret 1997). **BREEDING DATA** Nestlings, 14 February 2009, RP Las Gralarias (Greeney *et al*. 2009); mid-aged nestling, 15 December 2012, RP Las Gralarias (HFG); fledgling, 14 February 2006, Refugio Paz de las Aves (B. Herrera *in litt*. 2014); fledgling transitioning, 8 September 2011, Refugio Paz de las Aves (ML 39795711, A. Selin, also photos 24 September by R. Ahlman); juvenile, 11 September 2014, Cerro Montezuma (IBC 1069602, photo D.M. Brinkhuizen); juvenile transitioning, 14 September 2015, Refugio Paz de las Aves (Flickr: H. Singh); subadult, 15 September 2010, Refugio Paz de las Aves (D. Faulkner photos); subadult transitioning, 6 April 1958, near Ricaurte (YPM 58739, photos S.S. Snow).

TECHNICAL DESCRIPTION Sexes similar. *Adult* Uniform deep rufous-brown from the crown to the rump, including the wings. Lores and forecrown dusky-grey; the mostly bare ocular area is pale bluish- or greyish-white. Underparts white with a distinct yellow wash, strongest on the throat and upper breast, fading gradually to dusky-white on the sides, flanks and lower belly. The rufous-brown coloration of the upperparts extends irregularly onto the sides, flanks and vent, giving a variable, mottled appearance. **Adult Bare Parts** *Iris* dark brown; *Bare Orbital Skin* pale bluish- or greyish-white; *Bill* black; *Tarsi & Toes* bluish-grey. *Fledgling/Juvenile* Until recently, the immature plumage was undescribed. Greeney (2014) described a juvenile photographed by R. Ahlman, and the following refers to a fledgling/juvenile apparently still with the adults. Lores, forecrown, crown, hindcrown and nape dusky-grey, forecrown with a few yellowish feathers. Rest of upperparts similar rusty-brown to adults from lower nape to rump, with a few dusky feathers giving a slightly 'messy' appearance. Wings similar to back, upperwing-coverts less reddish, greyish-brown on leading edge and brownish-dusky on trailing webs, otherwise unmarked. Face dusky-olive mixed with yellowish and some whitish-yellow around eyes. Sides of neck and behind and below ear-coverts mixed brownish, indistinct yellowish malar separates face from dark grey throat, paler near base of bill. Upper breast with mix of yellow feathers and white-shafted grey ones, sides of breast and flanks mixed with a few chestnut feathers. From lower breast yellow, olive-washed laterally, becoming whiter on vent, with central breast and belly mixed grey and white. Thighs olive-brown. Rectrices dark, slightly edged chestnut, especially basally. This description closely matches a photo (ML 39795711, A. Selin) taken 16 days earlier of what is presumably the same individual (same feeding station). **Fledgling/ Juvenile Bare Parts:** *Iris* dark brown; *Bare Orbital Skin* yellowish to orange-yellow; *Bill* maxilla blackish, pale whitish at tip, tomia dusky-orange distally becoming bright orange proximally and nares distinctly orangey, gape partially expanded, bright orange-white, mandible bright orange, slightly duskier at tip and near base; *Tarsi & Toes* violet-greyish, some orange wash proximally and at posterior edge. *Juvenile/Subadult* The same individual described above in transition from fledgling to juvenile was photographed ten days later, when there was no detectable change in its appearance. Fifty-nine days later, however, subadult plumage was evident. Above, now essentially in full adult plumage, overall slightly duller brown, especially on crown. Lores lightly washed with yellow instead of clean grey. Ear-coverts slightly duskier. Below, also very similar to adult, yellow clean and bright, and no evidence of dusky wash except distinct central throat patch washed black. Bare skin around eye whitish, not bluish. Bill mostly black, maxilla dusky yellow-orange on distal third and inside nares, mandible dusky yellow-orange on distal half, no sign of inflated rictus. It appears that the last vestiges of immature plumage to disappear are the greyish feathers of the crown, slowly being replaced from the rear until only the greyish lores of adult plumage remain. Similarly, the underparts slowly become more yellow, first on the sides and belly, lastly on the central throat, with an overall progression towards a cleaner and unmarked breast and belly.

MORPHOMETRIC DATA Oniki & Willis (1991) provide the only previously published morphometrics for *G. flavotincta* based on one unsexed adult: *Wing* 96mm; *Tail* 45mm; *Bill* exposed culmen 25mm; *Tarsus* 50mm. Additional measurements: *Bill* from nares 15.7mm, 15.8mm, 15.4mm, 15.2mm, 14.8mm, 15.1mm; *Tarsus* 48.5mm, 47.1mm, 48.1mm, 48.8mm, 46.1mm, 48.1mm ($n = 6$, 1♀,5♂♂, LACM, AMNH). **Mass** 63g (Oniki & Willis 1991). Oniki & Willis (1991) also gave the cloacal temperature of the above-mentioned individual as 39.0°C. **Total Length** 17–18cm (Hilty & Brown 1986, Ridgely & Greenfield 2001, Restall *et al.* 2006, McMullan *et al.* 2010).

DISTRIBUTION DATA COLOMBIA: *Antioquia* Dabeiba, 07°05'N, 76°22'W (XC 297172, C. Olaciregui); Hacienda Potreros, 06°39'N, 76°09'W (USNM 426419–422); Ciudad Bolívar, 05°47'N, 76°05'W (XC 354022, S. Berrío); RNA Loro Orejiamarillo, 05°32.5'N, 75°48'W (O. Cortes-Herrera *in litt.* 2017). *Chocó* RNA Las Tangaras, 05°51'N, 76°13.5'W (ML 41022701, photo A. Spencer); La Selva, 04°55'N, 76°09'W (ANSP 157979/80); San José del Palmar, 04°52'N, 76°13'W (XC 96076–078; D. Calderón-F.); El Carmen de Atrato, 05°49'N, 76°11'W (XC 311843, A.E.H. Guarán). *Risaralda* RNSC Mesenia, 05°29'N, 75°54'W (Cadena *et al.* 2007, XC); 3.75km WSW of Mistrato, 05°17'N, 75°55'W (XC 325817, D. Orózco); Cerro Montezuma, *c.*05°15'N, 76°06'W (XC, IBC); Pueblo Rico, 05°14'N, 76°02'W (XC 43476/79, O. Cortes-Herrera); PRN Santa Emilia, 05°12'N, 75°54'W (Zuleta-Marín 2012); RFP AguaBonita, 05°11.5'N, 75°57.5'W (J.A. Zuleta-Marín *in litt.* 2017); Vereda Las Cumbres, 05°06.5'N, 76°00'W (XC 121572, O. Janni). *Valle del Cauca* Bitaco, 03°36'N, 76°36'W (CM P67338); Cerro El Inglés, 04°44'N, 76°18'W (XC 193169, O.H. Marín-Gómez); Vereda Balcanes Alto, 04°35'N, 76°19'W (XC 29845, K. Fierro-Calderón); Alto Galapagos, 04°50'N, 76°11'W (XC 59569–571, F. Lambert). *Cauca* Cerro Munchique, 02°32'N, 76°57'W (LACM 36432); Santander de Quilichao, 03°00'N, 76°30'W (ROM 118731); Brisas, 7km WNW, 02°17'N, 77°07'W (XC 14641, O. Cortes-Herrera); RNSC Tambito, 02°30'N, 77°00'W (Donegan & Dávalos 1999). *Nariño* Vicinity of Ricaurte, 01°13'N, 77°59'W, (LACM, ANSP, YPM, FMNH); RNSC Río Ñambí, 01°18'N, 78°05'W (XC 241331/32, O.H.

Marín-Gómez); RNSC La Planada, 01°05'N, 77°53'W (XC 12345/47, O. Laverde) and 01°09'N, 77°59'W (XC 68414, F. Lambert). **ECUADOR**: *Carchi* Maldonado, 00°54'N, 78°07'W (J. Freile *in litt.* 2017); Chilmá Alto, 00°52'N, 78°04'W (XC 262938/39, J. Nilsson); El Corazón, 00°50'N, 78°07'W (J. Freile *in litt.* 2017); Cerro Golondrinas, 00°49'N, 78°07'W (J. Freile *in litt.* 2017); Chical–Gualchán road, 00°50'N, 78°13'W (R.A. Gelis *in litt.* 2016). *Esmeraldas* Cordillera de Toisán, 00°25'N, 78°47'W (J. Freile *in litt.* 2017). *Imbabura* BP Los Cedros, 00°19'N, 78°47'W (HFG); above Cristal, 00°46'N, 78°27'W (J. Freile *in litt.* 2017). *Pichincha* RBP Maquipucuna, 00°05'N, 78°37'W (XC 248464, N. Krabbe); RP Santa Lucía, 00°07'N, 78°36.5'W (HFG); RP Las Gralarias, 00°01'S, 78°44'W (Greeney *et al.* 2009); Refugio Paz de las Aves, 00°00'N, 78°43'W (Collins 2006, XC, ML); BP Mindo-Nambillo, 00°02'S, 78°44'W (XC 3865/66, N. Athanas); Mindo, 00°03'S, 78°46.5'W (AMNH 173028). *Santo Domingo de las Tsáchilas* BP Río Guajalito, 00°14'S, 78°49'W (J. Freile *in litt.* 2017).

TAXONOMY AND VARIATION Monotypic. Yellow-breasted Antpitta was treated as a monotypic species by early authors (e.g. Cory & Hellmayr 1924), but subsequently as a subspecies of White-bellied Antpitta by many authorities (e.g. Meyer de Schauensee 1950, Peters 1951, 1970). Based on its different plumage, slightly different song and allopatric distribution, most modern treatments consider it a species (Hilty & Brown 1986, Ridgely & Tudor 1994, Ridgely & Greenfield 2001, Krabbe & Schulenberg 2003, McMullan *et al.* 2010). Due to its plumage similarities, Yellow-breasted Antpitta was considered most closely related to several other members of the *Oropezus* subgenus: Red-and-white, Bay, Rusty-tinged and White-bellied Antpittas (Lowery & O'Neill 1969, Parker & O'Neill 1980, Krabbe & Schulenberg 2003). Recently, however, mtDNA phylogenies have suggested a new hypothesis, i.e. that Yellow-breasted Antpitta is closely related to Brown-banded Antpitta (Winger *et al.* 2015). This relationship has yet to be tested with more rigorous genome-wide DNA sampling.

STATUS Yellow-breasted Antpitta is considered of Least Concern by BirdLife International (2015). However, it is a range-restricted Chocó endemic. In Ecuador, its range is estimated at just 1,572–2,880km², and has perhaps declined by >30% from its original size (Freile *et al.* 2010). Nevertheless, its global range is not tiny, and is estimated at 13,923–50,189km² (Graham *et al.* 2010). Due to its apparent intolerance of habitat disturbance, Krabbe & Schulenberg (2003) suggested that it could merit Near Threatened status. Yellow-breasted Antpitta was considered Endangered in Ecuador by Granizo *et al.* (2002), but its national status was suggested as Vulnerable by Freile *et al.* (2010), who further supported elevating its global status to Near Threatened. Although no human effects on the species have been specifically documented, logging and other habitat-modifying activities undoubtedly have adverse consequences (Greeney 2014i). Indeed, it is not presently known to occur in either of the two nationally protected areas in its Ecuadorian range (RE Cotacachi–Cayapas, Reserva Geobotánica Pululahua; Freile *et al.* 2010), but may be afforded some protection by the extremely rugged terrain in much of its range (Manne & Pimm 2001). **Protected Populations Colombia** PNN Las Orquídeas (USNM); PNN Tatamá (XC 121572, O. Janni); RFP AguaBonita (J.A.

Zuleta-Marín *in litt.* 2017); RNSC Tambito (Donegan & Dávalos 1999); RNSC La Planada (Willis & Schuchmann 1993); PNN Munchique (Negret 1997); PRN Santa Emilia (Zuleta-Marín 2012); RNA El Pangán and RNA Mirabilis Swarovski (Salaman *et al.* 2007a); RNA Las Tangaras (ML, XC, IBC 1211245, video M. Kennewell); RNA Loro Orejiamarillo (O. Cortes-Herrera *in litt.* 2017); RNSC Río Ñambí (XC 241331/32, O.H. Marín-Gómez); DMI Arrayanal (XC 325817, D. Orózco); DMI Cuchilla del San Juan (IBC 1004785, photo J.A. Zuleta-Marín); RNSC Mesenia (Cadena *et al.* 2007, XC); RNSC La Playita (J.A. Zuleta-Marín *in litt.* 2017); RNC Cerro El Inglés (XC 193169, O.H. Marín-Gómez); RNC Galápagos (XC 59569–571, F. Lambert). **Ecuador** BP Cerro Golondrinas, BP Río Guajalito (Freile 2000); BP Mindo-Nambillo (XC 3865/66, N. Athanas); RP Las Gralarias (Greeney *et al.* 2009); RBP Maquipucuna (Ridgely & Greenfield 2001).

OTHER NAMES For many years, Yellow-breasted Antpitta was treated as a race of White-bellied Antpitta and referred to as Bay-backed Antpitta, *G. hypoleuca flavotincta* (Peters 1951, Meyer de Schauensee 1970, 1982). *Grallaria hypoleuca* (Goodfellow 1902; specimens from 'Pichincha and Papallacta'); *Grallaria* ('*hypoleuca*') *flavotincta* (Krabbe 1991); *Grallaria* [*hypoleuca*] *flavotincta* (Sibley & Monroe 1990). **English** Yellow-breasted Ant-Thrush (Brabourne & Chubb 1912). **Spanish** Tororoí Pechiamarillo (Krabbe & Schulenberg 2003); Gralaria Pechiamarillenta (Ridgely *et al.* 1998, Granizo *et al.* 2002, Granizo 2009); Tororoí Rufoamarillo (Salaman *et al.* 2007a); Peón de pecho amarillo (Negret 1997); Comepán (locally in Ecuador; J. Freile *in litt.* 2017). **French** Grallaire à poitrine jaune. **German** Gelbbrust-Ameisenpitta (Krabbe & Schulenberg 2003).

Yellow-breasted Antpitta, adult, Refugio Paz de Las Aves, 19 May 2009 (*Roger Ahlman*).

Yellow-breasted Antpitta, juvenile, Refugio Paz de Las Aves, 24 September 2011 (*Roger Ahlman*).

Yellow-breasted Antpitta, nestlings, *c.*7 days old, Reserva Las Gralarias, 14 February 2009 (*Harold F. Greeney*).

Yellow-breasted Antpitta, adult on nest, Reserva Las Gralarias, 14 December 2012 (*Harold F. Greeney*).

Genus *Hylopezus*: mid-sized lowland antpittas

Hylopezus, **Ridgway, 1909**, Proceedings of the Biological Society of Washington, vol. 21, p. 71. Type, by original designation: *Hylopezus perspicillatus* (Lawrence, 1861).

Remarks on immature plumages of *Hylopezus* In the accounts that follow, it will become obvious that examination of photos and specimens revealed far fewer records of *Hylopezus* immatures than for *Grallaria* antpittas. I am left, therefore, with a rather incomplete picture of plumage development for the genus. Nevertheless, my observations of nestlings and fledglings of several species will most likely prove true for this rather morphologically conservative genus, and should questions arise about immature plumage for species not herein described, I suggest comparison with the descriptions of Streak-chested and Spotted Antpittas will prove informative. For now, I will provide what are *presumed* to be a few generalities concerning immature plumages, with the proviso that exceptions may occur. *Hylopezus* fledge still covered in fluffy, wool-like, rufescent or rusty-brown nestling down, at least over most of the upperparts. Below, there is probably some variation in colour, ranging from white to buffy, to pale rufous on the belly, flanks and vent. Their still-emerging tail feathers may be only just visible beyond the downy feathers of the rump. To date, no fledglings or nestlings are known to be spotted, streaked or otherwise marked. In very recently fledged individuals the face may still show fair amounts of bare skin, probably pinkish or bluish-grey. Individuals fresh from the nest have (in general) paler and shorter bills than adults, with obviously inflated rictal flanges that vary from orange to pale yellow. On the whole, however, bare-parts coloration is largely unknown for immature *Hylopezus*. Young at all stages have pale markings at the tips of the upperwing-coverts, irrespective of the presence of such markings in adults of the same species. These markings are whitish to rich buff but are strongly washed rusty-brown or chestnut, especially on the greater coverts. Immature coverts are further distinguished from those of adults by having bright rufous or reddish-chestnut margins at the tips and, sometimes, extending over the leading margin. Though not always obvious from photos or in the field, it appears that the tertials and inner secondaries are also marked at their tips, to some degree, either lightly washed or thinly fringed ochraceous or rusty-buff. Most covert spots are cleanly extended by a terminal rufous-chestnut fringe, which is never seen in full adults. These covert markings tend to fade with wear, becoming less obvious with age but, if the maturation process in *Hylopezus* mirrors that of *Grallaria*, it is likely that immature coverts are replaced with similarly patterned feathers at least once prior to a final moult to adult coverts. The process of transitioning to adult-like contour feathers is not yet clear, but appears to occur in a less haphazard or uneven manner than in *Grallaria*. Nevertheless, *Hylopezus* is similar to *Grallaria* in retaining evidence of downy fledgling plumage on the head for longer than any other part of the upperparts, although in *Hylopezus* the last down is on the crown rather than the nape. As immature *Hylopezus* approach fully adult-like plumage they frequently retain some rust-coloured feathers on the lower breast and central belly. This latter observation came to my attention while working with specimens, when I realised that what I was mistaking for pale belly feathers stained with blood during preparation were actually scattered dried blood-coloured feathers creating streaks on otherwise adult-like underparts. This late indicator of immaturity is shared with *Myrmothera*. Unlike most *Grallaria*, the first covering of adult-like contour feathers, including the hindcrown and nape, is patterned like, or very similar to, that of full adults. By this age the bill is probably close to adult size, but may retain slightly inflated rictal flanges. Interestingly, and unfortunately for the usefulness of this character in identifying immatures in the field, most *Hylopezus* retain a fairly obvious and slightly inflated rictus into adulthood. Although I believe that the upperwing-coverts will prove to be the last external indicator of immaturity, the (apparently) rather quick transition to adult contour feathers, with no spotting or scaling on the nape and crown (as in *Conopophaga* and *Grallaria*), combined with the retention into adulthood of spotted coverts in some species poses greater challenges for determining the age of immature *Hylopezus* than for other genera. Compounding my ability to delineate clear indicators of immaturity in nearly adult birds is my incomplete sampling of immatures for many species and my poor understanding of the effects of wear on covert coloration. It is likely, however that, like the thin black fringe on immature upperwing-coverts of *Conopophaga*, the rufous-chestnut fringes on the coverts of *Hylopezus* will consistently distinguish immatures from adults, even in species where these are strongly marked. Perhaps largely for continuity with other genera discussed herein, I tentatively define three principal categories of *Hylopezus* immatures. However, it should be noted that, within the framework of plumage ontogeny, the middle stage (juvenile) cannot be as clearly defined as I have done in other genera. This is perhaps due to a faster or more synchronised replacement of contour feathers, but certainly this aspect of plumage development deserves special attention before strong inferences can be made. For now, I define three phases of immature plumage. FLEDGLING Still covered in fluffy, wool-like, feathers, rictal flanges still obviously inflated and contrastingly coloured, still heavily dependent on adults for food and largely remains stationary while awaiting adult food delivery. JUVENILE Mostly covered in adult-like contour feathers, but may retain some rust-coloured patches of fledgling down on the crown, as well as scattered wisps of down on the nape and/or underparts; rictal flanges still brightly coloured but may be less obvious than in similarly aged individuals of other genera of antpittas. Behaviours at this age are largely unknown, but probably fairly mobile, following adults and actively soliciting food, possibly already almost independent. SUBADULT Very close or identical to adults in general coloration, only separable by immature wing-coverts, probably difficult to separate from adults in the field and almost certainly fully independent. Overall, plumage ontogeny appears to be more step-wise than in other genera of antpittas and gnateaters, but in a few cases I use the terms '**fledgling transitioning**' and '**juvenile transitioning**' for birds somewhere between the above-defined stages. The term '**subadult transitioning**' is reserved for individuals with heavily worn and faded upperwing-coverts (but with at least some indication of chestnut fringes) that are otherwise fully adult in appearance.

STREAK-CHESTED ANTPITTA
Hylopezus perspicillatus   **Plate 19**

Grallaria perspicillata Lawrence, 1861, Annals of the Lyceum of Natural History of New York, vol. 7, p. 3, 'Isthmus of Panama' (see discussion of type locality under Taxonomy & Variation).

'The call of the Streaked-chested Antpitta, seven or eight loud, mellow whistles delivered in a peculiar, hollow, melancholy tone is another woodland sound that I was long in tracing to its source. So secretive is the stout, long-legged, stubby-tailed bird that I never succeeded in seeing it as it called until I found a puzzling nest, quite different from any that I had seen. By concealing myself nearby, I was able to watch the antpitta approach; he hopped rapidly over the ground, now and then flicking aside fallen leaves with a vigorous sideways sweep of its bill. While sitting on the shallow bowl of dead leaves, warming the two mottled eggs, it delivered the unmistakable whistles that I had heard so many times before without associating them with any bird I knew, and another of the mysterious sounds of the forest ceased to be a mystery.' – **Alexander F. Skutch, 1971**, *A Naturalist in Costa Rica*

Streak-chested Antpitta was for many years known as Spectacled Antpitta, its English name being derived from its scientific name (*perspicillatus* = spectacled; Jobling 2010) which, in turn was derived from the bespectacled look imparted by its distinctive buffy eye-ring. Fairly widespread, latitudinally, from southern Honduras to north-west Ecuador. It is fairly common in Costa Rica and Panama, but notably rarer in Nicaragua and Colombia. It is a bird of wet, lowland, evergreen forests, spending the majority of its time on the forest floor of open understorey, closed-canopy forests. The species' habit of repeatedly fluffing then smoothing its contour feathers, recalling the inflating and deflating of a balloon, is an often observed but unexplained behaviour. Compared to most antpittas, breeding biology is fairly well known, with studies ranging from the beautifully detailed accounts of Alexander Skutch to evolutionarily and ecologically oriented works that use large numbers of nests to estimate predation rates and aspects of nest-site selection. Though not currently considered of conservation concern, the species appears relatively sensitive to habitat disturbance and is known to have been extirpated from several locations.

IDENTIFICATION 13–14cm. A relatively slender, mid-sized antpitta with a greyish crown and nape, a prominent buffy eye-ring, and heavy black streaking on the pale breast. The two rows of fairly large, buffy spots on the wing-coverts form diffuse wingbars. The upright posture, long legs, short tail and terrestrial habits make the species unlikely to be confused with anything other than its frequently sympatric congener, Thicket Antpitta. The latter, however, has a distinctly plainer face (no eye-ring), lacks spotted wing-coverts (when fully adult), is more extensively washed with colour below, and has much fainter breast streaking.

DISTRIBUTION Found on the Caribbean coast of southeast Honduras (Gracias a Dios) and most of Nicaragua, and then on both slopes from southern Nicaragua and northern Costa Rica through Panama. In South America its range extends east from southern Panama, across the northern end of the Colombian Andes to Santander, south along the west slope of the Andes to northern Ecuador (Stiles & Skutch 1989, Ridgely & Greenfield 2001, Restall *et al.* 2006, Ridgely & Tudor 2009, Angehr & Dean 2010, McMullan *et al.* 2010, 2013, Martínez-Sánchez *et al.* 2014). Though currently known only to range to southern Esmeraldas in Ecuador, historical records extend as far south as Los Ríos and northern Manabí. Records from El Oro (Ridgely & Tudor 1994) are erroneous.

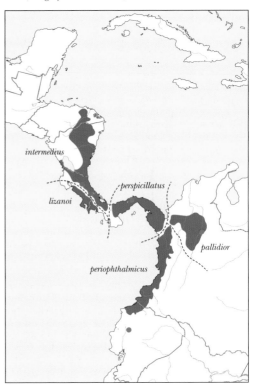

HABITAT Inhabits wet, lowland evergreen forest across its range, being absent from nearby dry forest (Slud 1964). Often credited with preferring more open understorey than congeners (Carriker 1910, Schemske & Brokaw 1981, Krabbe & Schulenberg 2003, Horsley *et al.* 2016), most common in mature forest with intact canopy, and rarely seen in second growth and small fragments (Slud 1960, Stiles 1985, Naranjo & de Ulloa 1997, Blake & Loiselle

2001). Occasional reports from disturbed habitats (Slud 1960) may involve wandering birds or opportunistic foragers, and it seems unlikely that they nest in such habitat. Across its range, the species is found from sea level to 1,250m (Slud 1964, Garrigues & Dean 2007, Angehr & Dean 2010), usually below 900m (Ridgely & Tudor 1994). In Costa Rica tends to be most numerous below 300m in the Caribbean lowlands (*intermedius*), apparently slightly overlapping with Thicket Antpitta, which replaces it at higher elevations (Carriker 1910). There appears to be a gradual shift in upper altitudes attained by the species, increasing south from maximum 500m in Nicaragua (Martínez-Sánchez *et al.* 2014) to 1,250m in southern Costa Rica and Panama, and 1,200m in Colombia (Hilty & Brown 1986, McMullan *et al.* 2010).

VOICE Both Skutch (1969) and Krabbe & Schulenberg (2003) stated that only males sing, but to the best of my knowledge this is not known with any certainty. The song lasts 2.3–2.4s and consists of 7–9 whistled notes, the second note louder and the first lowest (*c.*1.7kHz). The other notes are given at *c.*1.8kHz and gradually fall slightly in pitch and volume (Krabbe & Schulenberg 2003). What is considered an alarm was described by Krabbe & Schulenberg (2003) as a 2.5kHz whistled *keeuw*, followed by a decelerating rattle at *c.*2kHz and lasting 2–3s. In Costa Rica, Slud (1964) heard two principal vocalisations delivered at 'frequent if irregular intervals, but not interchangeably'. The more commonly heard 'song' is a somewhat deliberate series of up to ten similar *pew, pow* or *poo* notes that rise in pitch slightly in the faster middle portion. The call described by Slud (1964) was transcribed as *pee chyuchyiichyii-chyuichyuichyui*, with no real pause after the short, harsh, introductory *pee*. The sound weakens at the end, 'as though the bird were running out of breath'. In Panama, Eisenmann (1952) described as the song as a series of clear, melancholy whistles, at first rising a little in pitch, then falling off in descending couplets: *deh, dee, dee, dee, dee, dee, deé-eh, déhoh, dóh-a*. Presumably this same vocalisation was described by Willis & Eisenmann (1979) as a series of clear, melancholy whistles, *deh dee-dee-dee-dee-dce-dee, dew, dew, dew*, usually ending with three longer, lower notes. The call of Panamanian birds (Willis & Eisenmann 1979) was described as a resonant *pew, you-you-you-you*. Willis (1985) reported adults giving *kwirr* 'buzzes' while foraging at ant swarms on Barro Colorado Island. Skutch (1969) associated a 'loud rattle, dying away at the end' with adults (*lizanoi*) leaving the nest.

NATURAL HISTORY Usually encountered alone or in pairs (Slud 1964), and reports of this species travelling only with flocks (Jones 1977) probably reflect the difficulty with which they are found when foraging on their own. A recent radio-telemetry study in central Panama suggested that pairs spend >50% of their time together (H.S. Pollock in Horsley *et al.* 2016) and encounters with lone individuals may be the result of a failure to detect the other adult, or be when one adult is attending a nest. While stationary they repeatedly fluff out their contour feathers, with wings slightly spread, appearing to expand and deflate like a balloon (Slud 1964, Skutch 1969). Willis (1985) observed this behaviour by adults at ant swarms, with quick hops or dashes for prey interspersed with pauses accompanied by feather fluffing. While foraging, they generally keep to the ground or below 1m (Cody 2000). They move across the forest floor in rapid hops, but occasionally make quick dashes, pausing to peer at the

ground or to flick leaves aside with the bill (Slud 1964, Skutch 1969, 1971, Ridgely & Tudor 2009, Horsley *et al.* 2016). To sing, adults usually ascend to a slightly elevated perch, such as a log or low branch. Their posture while singing is somewhat theatrical, as they hold their head still while bobbing their body up and down in time with the notes. It is interesting to note that, while vocalising (IBC 1180799, video C. Beckman), Streak-chested Antpitta does not appear to inflate its throat to expose large areas of pink skin either side of its neck as frequently as many of its congeners (e.g. Amazoniana Antpitta). They do occasionally expand the throat sufficiently to expose bare skin, but when doing so the swelling appears to happen more centrally, creating a single pink patch below the chin (see ML 64483041, photo G.C. Benavides), as opposed to the pair of 'swellings' either side of the throat seen in other *Hylopezus*. It does, as described above, often fluff out the feathers of the belly and flanks, while extending its neck to sing, greatly altering its shape and appearing to take a deep breath (IBC 1342199, video J. del Hoyo). Sometimes, as also seen in this video, adults perform a very distinctive wing-flick, in which the wings are extended laterally, one-third to halfway, and then quickly snapped back into place. Although the adult in del Hoyo's video (IBC 1342199) performs both the 'belly-fluff' and the wing-flick while vocalising, both movements are also performed while not singing, perhaps in agitation (Ridgely & Tudor 2009), but their function is unclear. See ML 65382481 (photo R. Candee) for an example of an adult photographed flicking its wings. Heath's (1932) descriptive account suggested that Streak-chested Antpittas may vocalise throughout the day, even during periods of low vocal activity by other species. Although the species' population ecology has not been studied in detail, in central Panama Robinson *et al.* (2001b) recorded densities of *c.*21 pairs per 100ha. Little is known of its natural enemies, but the species is a known host of *Formicophagus* bird lice (Carriker 1957, Emerson 1981, Price & Clayton 1996).

DIET There have been no quantified studies of diet, but this presumably consists largely of arthropods. They are known, however, to at least occasionally consume small vertebrates such as frogs (Poulin *et al.* 2001, Lopes *et al.* 2005a).

REPRODUCTION The first description of the nest and eggs was based on three nests of subspecies *lizanoi* found by Skutch (1969). This, and his subsequent works (Skutch 1971, 1981), though limited to observations of only five nests, provided a wealth of detail about nesting biology during the incubation period. The nesting phase, however, is poorly known. Though subsequent studies have mentioned nests of Streak-chested Antpittas or used their nests in more theoretical studies of ecology and nesting success (Roper 1992, Robinson *et al.* 2000a, 2005), these all pertain to the same population of nominate *perspicillatus* in central Panama, and descriptive details of nest architecture and egg morphology are unreported for all subspecies but *lizanoi*. I was unable to find information of any type for subspecies *pallidior*. **NEST & BUILDING** The nest is a broad, shallow, open cup comprising an untidy arrangement of sticks, leaf petioles and large dead leaves, sparsely lined with dark, flexible fibres such as rootlets, fungal rhizomorphs or thin leaf petioles. They are generally sited below 1.5m (Robinson 2005), usually in situations such as at the base of a rosette of leaves from a fern, cyclanth, aeroid or palm tree, or other locations where detritus accumulates naturally. Nest height in central Panama (*perspicillatus*) averaged 0.72 ± 0.3m (*n* = 16; Robinson *et al.* 2005), five nests (*lizanoi*) found by Skutch (1969, 1981) were built 1.5m, 0.66m, 1m (×3) above ground, and five nests of *intermedius* in eastern Costa Rica were 0.4m, 0.3m, 1.1m, 0.7m and 0.5m up (HFG). Nests appear to be built in fairly exposed situations, and Robinson *et al.* (2005) reported average 'nest visibility' as 75 ± 4.4% at a distance of 1m, 25 ± 4.7% from 10m, 55 ± 8.1% from directly above, and 74 ± 8.7% from directly below. Nest dimensions (Skutch 1969, 1981): outer diameter 20cm, 15cm; 10cm, 20.3cm; outer height n/m, n/m, 5cm, 5cm; inner diameter 8.2cm, 8–9cm, n/m, n/m; inner depth 3.5cm, 2.5cm, n/m, n/m. **Egg, Laying & Incubation** The eggs are short and subellipitical, as is typical of the Grallariidae. They vary in ground colour from pale grey to greenish- or bluish-grey, and most appear to be fairly heavily marked with blotches and flecks of varying sizes and colour, ranging from dark brown to dull lavender and cinnamon. Both sexes participate in incubation, frequently in long sessions (5.0–6.5h), but also with occasional long absences (Skutch 1969, 1981). Exchanges at the nest are quick and unceremonious. Females spend the night on the nest. Skutch (1981, 1996) reported that both sexes sing while incubating, the male apparently more often. Both sexes of *Grallaria* antpittas are known to sing from the nest as well (Greeney *et al.* 2008a), but apparently much more frequently than reported for Streak-chested Antpitta. Clutch size = 2 (*n* = 28, Robinson *et al.* 2000a; *n* = 5, Skutch 1969; *n* = 3, WFVZ, NHMUK; *n* = 1, Willis & Eisenmann 1979). Incubation period = 22.6 ± 2.9 days (*n* = 4, Robinson *et al.* 2005); mean 22.5 days (*n* = 71, Brawn *et al.* 2011). Egg dimensions: 25.4 × 20.6mm, 25.4 × 20.6mm, 26.2 × 20.6mm, 26.2 × 20.6mm, 27.0 × 21.4mm (*n* = 5, *lizanoi*, Skutch 1969); mean of nine eggs, including the previous five, was 26.1 × 20.7mm. **Nestling & Parental Care** Dark-skinned and devoid of natal down at hatching. Bill dusky-orange, with bright ivory-white egg-tooth, yellow-orange rictal flanges and bright crimson-orange mouth lining. Within a few days of hatching they start to develop a dense coating of rusty-brown, wool-like down (HFG) similar to nestlings of *Myrmothera* and *Grallaricula*. Nestling periods 11.5 ± 1.0 days (*n* = 3; Robinson *et al.* 2000a); mean nestling period 11.5 days (*n* = 71, Brawn *et al.* 2011). **Nest Success** Robinson *et al.* (2000a) reported a calculated nest-success (Mayfield method) of 8.3% based on 20 nests of nominate *perspicillatus* in central Panama. Brawn *et al.* (2011) reported similarly poor nesting success in the same area: DSR 0.91385, range 0.89131–0.93208 (*n* = 71 nests; 668 observation days; cumulative rate of nest success 0.047 (SE = 0.074). **Breeding Data** *H. p. intermedius* Near-complete nest, 19 March 2015, Quebrada González (HFG); young nestlings, 16 June 2013, Quebrada González (I. Ausprey *in litt.* 2015); fledgling, 25 June 2012, Quebrada González (A. Spencer photos; Fig. 4); subadult transitioning, 21 May 1909, Peña Blanca (AMNH 103749). *H. p. lizanoi* Carrying nest material, 4 April 2013, Quebrada Bonita (ML 65957321, photo J. Straka); nest with eggs 18 June 1902, Pozo Azul de Pirrís (NHMUK 1904.7.15.77.78); two nests with eggs, 5 and 18 April 1971, vicinity of Rincón de Osa (WFVZ 58404/58259); five nests with eggs, 10 June and 30 April 1940, 29 August 1959, 9 June 1975 and May 1970, El General (Skutch 1969, 1981); subadult transitioning, January 1886, 'Las Trojas de Puntarenas' (USNM 119951, holotype). *H. p. perspicillatus* Nest with eggs, 4 July 1969, Pipeline Road (J. Karr in Willis & Eisenmann 1979);

fledgling, 12 July 1969, *c.*8km NE of Gamboa (Karr 1971b); subadult transitioning, 22 August 1926, Río Calovébora (AMNH 246763); subadult transitioning, 1 March 1912, Río Salaquí (AMNH 113349); subadult transitioning, 20 February 1915, Tapalisa (AMNH 135794); subadult transitioning, 25 March 1915, Cerro Tacarcuna (AMNH 135793). *H. p. periophthalmicus* Fledgling, 18 November 1900, Pulún (Hartert 1902, AMNH 492289); juvenile, 15 November 1900, Pulún (Hartert 1902, AMNH 492288); subadult, 1 January 1900, San Javier de Cachabí (AMNH 492293); subadult transitioning, 2 July 1912, Serranía de Baudó (AMNH 123352).

TECHNICAL DESCRIPTION Sexes similar. Nominate *perspicillatus*. **Adult** Grey crown, blending to olive-grey on nape and olive-brown over rest of upperparts. The interscapular area and central back have sparse but prominent buffy streaks. The lores and prominent ocular ring are dull buff. The throat is white bordered by a blackish malar stripe. Wings brown, the outermost primaries have the leading edge narrowly fringed rufescent-brown and pale bases that form a small, pale patch. The leading vane of the alula is also pale buff. Upperwing-coverts brown with contrasting buffy apical spots. Tail a warmer, rufescent-brown than back. Breast and flanks white, streaked black and washed ochraceous-buff across breast, and pale buff on flanks and vent. **Adult Bare Parts** *Iris* dark brown; *Bill* maxilla black, mandible pale, whitish or pinkish basally, darkening towards tip; *Tarsi & Toes* pale grey. **Fledgling** The youngest examples that I have seen were all of race *intermedius* from eastern Costa Rica, and the following is based largely on these. However, as fledglings, I have no reason to believe that there are any distinct differences between races. Young fledge covered predominantly in fluffy, dark rusty-brown down, typical of the genus. The down is slightly brighter reddish and washed buff along the flanks and vent, transitioning to pale ochraceous-buff on the lower breast and belly-sides, fading to pure white on the central belly. The face and throat are largely bare, exposing dark bluish-grey skin, but soon after fledging these areas start to develop pale feathering, whitish on the throat and buffy to buffy-olive on the face. The tail feathers are less than halfway emerged from their sheaths and probably not visible beyond the fluffy plumage for some time. Flight feathers are probably fully developed, but may retain small amounts of sheathing at the bases; it seems unlikely that they are capable of flight and are possibly able to run only in short, awkward bursts. Upperwing-coverts are similar to adult in having bright buffy-ochraceous tips, but differ in having the terminal spots variably suffused bright rufous and by not being triangular or diamond-shaped. Furthermore, the tips are distinctly fringed bright chestnut-rufous. Overall, they resemble 'blurry' rufous-washed versions of adult coverts. **Fledgling Bare Parts** *Iris* dark; *Bill* maxilla dusky, slightly yellowish or orange around nares, mandible largely pale, pinkish to orange-pink, rictal flanges inflated and white to yellow-white, slightly orange on their inner edges; *Tarsi & Toes* dusky-pink to vinaceous-grey. **Juvenile** I have not seen individuals in 'mid-juvenile' plumage, suggesting to me that the transition from fledgling to subadult may be rapid, reducing the amount of time I would consider them to be in juvenile plumage. Adult-like (i.e. not downy) feathering appears first on the face, throat and upper back/scapular region, followed by the upper breast, sides, lower back and nape. Though the degree of streaking on the upperparts has been used, to varing degrees, to distinguish between subspecies, the first adult-like contour feathers may be more streaked, especially on the upper back and scapulars, than in full adults. Otherwise, juvenile plumage is similar to that of adults, with the last tracts to lose their fluffy rust-coloured down being the crown, lower breast and vent. It appears that the upperwing-coverts are replaced by more adult-like feathers, the terminal spots now clearly defined and clean buff or ochraceous-buff. Juvenile coverts, however, still have bright reddish-chestnut margins. **Juvenile Bare Parts** *Iris* dark brown; *Bill* similar to adult but perhaps slightly whiter or more yellow-white at gape; *Tarsi & Toes* dark grey, similar to adult. *Subadult* Similar in all respects to adult, perhaps retaining 3–4 widely scattered rusty down feathers on the crown or central belly. Subadults have immature wing-coverts, as described for juveniles, for an unknown period, these gradually appearing more adult-like as the chestnut fringes fade and wear.

MORPHOMETRIC DATA *H. p. intermedius* Data from Ridgway (1911) [range (mean)]. Adult ♂♂: *Tail* 25.5–32.0mm (29.4mm) (*n* = 10); *Bill* [total] culmen 17–20mm (19mm) (*n* = 10); *Tarsus* 34–37mm (35.7mm) (*n* = 10), *Middle Toe* 17.0–18.5mm (17.3mm) (*n* = 10). Adult ♀♀: *Wing* 77–82mm (70.6) (*n* = 10); *Tail* 27–35mm (29mm); *Bill* [total] culmen 18–20mm (10.2mm); *Tarsus* 35.0–38.5mm (36.2mm); *Middle Toe* 16.5–18.0mm (17.2mm) (*n* = 10). Measurements from USNM, LACM, AMNH, FMNH, MZUCR (HFG). Adult ♂♂: *Bill* from nares 12.3 ± 0.6mm (*n* = 9); *Tarsus* 36.7 ± 1.0mm (*n* = 10). Adult ♀♀: *Bill* from nares 12.7mm, 12.5mm, 12.2mm; *Tarsus* 36.4mm, 36.2mm, 37.5mm. *H. p. lizanoi* Data from Ridgway (1911) [range (mean)]. Adult ♂♂: *Wing* 79.0–84.5mm (81.7mm) (*n* = 10); *Tail* 26.5–32.0mm (30.2mm) (*n* = 10); *Bill* [total] culmen 19.0–20.5mm (19.6mm) (*n* = 10); *Tarsus* 34.0–37.5mm (35.8mm) (*n* = 10); *Middle Toe* 16.5–18.5mm (17.3mm) (*n* = 10). Adult ♀♀: *Wing* 79.0–84.5mm (82.4mm) (*n* = 10); *Tail* 27.5–32.0mm (30.2mm) (*n* = 10); *Bill* [total] culmen 19–21mm (19.8mm) (*n* = 10); *Tarsus* 32.5–37.0mm (34.8mm) (*n* = 10); *Middle Toe* 16.5–18.0mm (17.5mm) (*n* = 10). Measurements from USNM, LACM, WFVZ, FMNH, MZUCR (HFG). Adult ♂♂: *Bill* from nares 12.4 ± 0.5mm (*n* = 8); *Tarsus* 36.9 ± 1.7mm (*n* = 8). Adult ♀♀: *Bill* from nares 12.8 ± 0.5mm (*n* = 5); *Tarsus* 35.7 ± 1.3mm (*n* = 7). *H. p. perspicillatus* Data from central Panama in Horsley *et al.* (2016) [range (mean)]. Adult ♂?: *Wing* 69.0–87.0mm (80.95 ± 3.83mm) (*n* = 20); *Tail* 26–34mm (30.13 ± 1.99mm) (*n* = 21); *Bill* total culmen 18.35–19.95mm (19.11 ± 0.61mm) (*n* = 6), width 4.73–5.52mm (5.11 ± 0.28mm) (*n* = 6), depth 5.93 ± 0.25mm, 5.44–6.13mm (*n* = 6); *Tarsus* 35.20–42.63mm (37.66 ± 2.67mm) (*n* = 6); *Kipp's distance* 10.12–15.38mm (13.16 ± 2.13mm) (*n* = 6). Data from Ridgway (1911) [range (mean)]. Adult ♂♂: *Wing* 77.5–82.5mm (79.9mm) (*n* = 5); *Tail* 26.5–30mm (28.7mm) (*n* = 5); *Bill* [total] culmen 18–20mm (18.8mm) (*n* = 5); *Tarsus* 34.0–37.5mm (35.2mm) (*n* = 5); *Middle Toe* 16.0–18.5mm (17.1mm) (*n* = 5). Adult ♀♀: *Wing* 78.0–79.5mm (78.7mm) (*n* = 4); *Tail* 28–30mm (29mm) (*n* = 4); *Bill* [total] culmen 19mm (no range or mean given) (*n* = 4?); *Tarsus* 34.0–36.5mm, (35.2mm) (*n* = 4); *Middle Toe* 16.5–18.0mm (17.4mm) (*n* = 4). Measurements from AMNH, FMNH (HFG). Adult ♂♂: *Bill* from nares 12.5 ± 1.0mm (*n* = 4); *Tarsus* 36.1 ± 1.1mm (*n* = 5). *H. p. pallidior* Measurements from USNM, LACM, WFVZ, FMNH (HFG). Adult ♂♂: *Bill* from nares 12.6 ± 0.5mm (*n* = 15);

Tarsus 36.6 ± 1.5mm (*n* = 18). Adult ♀♀: *Bill* from nares 12.7 ± 0.7mm (*n* = 17); *Tarsus* 37.1 ± 1.8mm (*n* = 17). Adult ♂?: *Bill* from nares n/m, 12.8mm; *Tarsus* 36.7mm, 35.1mm. *H. p. periophthalmicus* Measurements from FMNH, AMNH, USNM (HFG). Adult ♂♂: *Bill* from nares 12.1 ± 0.7mm (*n* = 18); *Tarsus* 35.2 ± 1.1mm (*n* = 19). Adult ♀♀: *Bill* from nares 11.9 ± 0.5mm (*n* = 8); *Tarsus* 35.5 ± 0.7mm (n = 6). See also Salvadori & Festa (1898a). **Mass** *H. p. intermedius* 44.4g(no), 49.5g(tr), 40.0g (*n* = 3, ♂♂, MZUCR). *H. p. perspicillatus* 47.71 ± 3.73g, range 39.5–54.0g (*n* = 23, ♂?, Horsley *et al.* 2016), 37g (*n* = 1, ♀?, Burton 1975). *H. p. periophthalmicus* 39g, 41g (*n* = 2, ♂,♀, Horsley *et al.* 2016). **Total Length** *H. p. intermedius* Data from Ridgway (1911) [range (mean)]. Adult ♂♂: 11.4–13.7cm (12.8cm). Adult ♀♀: 11.3–13.4cm (12.3mm) (*n* = 10). *H. p. lizanoi* Data from Ridgway (1911) [range (mean)]. Adult ♂♂: 12.0–13.5cm (12.6cm) (*n* = 10). 11.5–13.2cm (12.5cm) (*n* = 10). *H. p. perspicillatus* Data from Ridgway (1911) [range (mean)]. Adult ♂♂: 11.8–12.9cm (12.5cm) (*n* = 5). Adult ♀♀: 12.5–12.6cm (12.55cm) (*n* = 4). *H. p. periophthalmicus* Data from Salvadori & Festa (1898a). Adult ♂?: 13.0cm (*n* = 1). Most authors give a range of 12.5–14.0cm (Hilty & Brown 1986, Ridgely & Tudor 1994, Krabbe & Schulenberg 2003, Garrigues & Dean 2007, Angehr & Dean 2010, McMullan *et al.* 2010).

TAXONOMY AND VARIATION Five subspecies currently recognised, but vocal and plumage differences between some of them are not strong, and a careful revision is needed. In particular, separation of *periopthalmicus* from nominate *perspicillatus* is dubious, and perhaps clinal. In addition, there is no clear barrier separating these two populations. In general, following Horsley *et al.* (2016), *intermedius* has lores and eye-ring ochraceous, the upper back washed grey, and the edges of back with no (or few) thin streaks. The flanks and vent are pale ochraceous. Subspecies *lizanoi* and *periopthalmicus* are both very similar to nominate, but tend to have the back unstreaked or only weakly marked. Subspecies *pallidior* has crown and nape grey-brown and the outer primaries tawny-brown with a rich-buff wing patch. This summary suggests to me that perhaps only three taxa are involved. Following convention (Cory & Hellmayr 1924, Peters 1951, Krabbe & Shulenberg 2003, Horsley *et al.* 2016), however, all five subspecies are recognised here.

Hylopezus perspicillatus intermedius Ridgway, 1884, *Proceedings of the United States National Museum*, vol. 6, p. 406, 'Talamanca, eastern Costa Rica' [probably = Sipurio, *c.*09°32.5'N, 82°56'W; see Deignan 1961, Slud 1964]. The original description of *intermedius* was somewhat troublesome for two reasons: it was a footnote and no holotype was designated. In 1883, Charles C. Nutting began an article entitled 'On a collection of birds from Nicaragua' that was continued in April 1884, and concluded in the subsequent issue of the same year. Robert Ridgway was listed as the 'editor' and inserted numerous taxonomic comments throughout the manuscript. In the concluding part, Nutting mentioned a single specimen of Thicket Antpitta (USNM 91265) collected at Sábalos. Somewhat tangentially, in a footnote, Ridgway pointed out that several Costa Rican specimens, then labelled *Grallaria dives*, were somewhat intermediate between Thicket and Streak-chested Antpittas (nominate *perspicillatus*). Ridgway felt that specimens at USNM that belonged to the new taxon, for which he proposed the name *intermedia*, were from Angostura and Talamanca, Costa Rica, but failed to designate a holotype or even mention specimen numbers. At the time, there were only two specimens at USNM to which Ridgway could have been referring. One is an adult male (USNM 47484) from Angostura [Laguna Angostura?, *c.*99°51'N, 83°39'W, Cartago] and the other (USNM 64718) an unsexed adult from 'Talamanca'. At some point, Ridgway himself attached a red holotype label to the Angostura specimen and wrote '*Grallaria intermedia* Ridgw.' Nevertheless, subsequently, Ridgway (1911) officially listed the type locality simply as Talamanca (Cory & Hellmayr 1924), by inference making USNM 64718 the holotype. Deignan (1961) considered the two specimens as equivalent cotypes. Ridgway (in Nutting & Ridgway 1884) distinguished *intermedius* by the bright ochraceous, wholly unstreaked flanks (white in *perspicillatus*, ochraceous but streaked in *H. dives*). The back is slate-olive with few or no streaks. The breast is bright ruddy ochraceous with narrow, indistinct black streaks. The range of *intermedius* is currently considered to extend along the Caribbean slope of Central America, from north-east Honduras (Gracias a Dios) to western Panama (Boucard 1878a, Bangs 1901, Blake & Loiselle 2000, Anderson *et al.* 2004). SPECIMENS & RECORDS **HONDURAS**: *Colón c.*70km WNW of Wampusirpi, 15°22'N, 85°13.5'W (eBird: S. Schuette). *Gracias a Dios* Reserva de Biósfera Río Platano, 15°34'N, 84°57.5'W (XC 46041, R. Gallardo); Mairin Tighni, 15°30'N, 84°58'W (Marcus 1983); Río Tuskruhua, 15°25'N, 84°45'W (Marcus 1983, LSUMZ 99946); Montañas de Colón, 14°57'N, 84°52'W (Bonta & Anderson 2002); SW of Chile, 14°56.5'N, 84°32.5'W (ML 58981681, R. Rumm). **NICARAGUA**: *Jinotenga* Peña Blanca, 13°15'N, 85°40'W (AMNH 103748/49). *Río San Juan* Sábalos, 11°02.5'N, 84°28.5'W (AMNH 144044/45); confluence of Ríos Bartola and San Juan, 10°58'N, 84°09.5'W (Cody 2000). *Rivas* San Emilio, 11°31'N, 85°52.5'W (Cory & Hellmayr 1924, FMNH 22008). **COSTA RICA**: *Guanacaste* Hacienda Allemania, 11°07'N, 85°33'W (LACM 15057); Vocán Miravalles, 10°45'N, 85°09'W (YPM 56726); Estación Biológica San Gerardo, *c.*10°22'N, 84°47.5'W (HFG). *Alajuela* 9km NE of Dos Ríos de Upala, 10°57.5'N, 85°19.5'W (MZUCR 3607); Bijagua, 10°44'N, 85°03.5'W (MCZ 123383–386, USNM 211709/10); La Marina, 10°22'N, 84°23'W (UMMZ); Quesada, 10°19.5'N, 84°26'W (CUMV 7223); Texas Christian University Research Station, 10°15'N, 84°33'W (XC 92967/93315, T. Stevens); RB Alberto Manuel Brenes, 10°12.5'N, 84°37'W (eBird: J. Zook). *Heredia* Puerto Viejo de Sarapiquí, 10°27'N, 84°01'W (AMNH 492282); Estación Biológica La Selva, 10°25'N, 84°01'W (Visco *et al.* 2015, FLMNH, ML); 10km SSW of Puerto Viejo de Sarapiquí, 10°22'N, 84°03'W (MZUCR 2662). *San José* Quebrada González, 10°09.5'N, 83°56.5'W (HFG). *Cartago* Volcán Turrialba, *c.*10°01'N, 83°47.5'W (CM P23518). *Limón* Guápiles, 10°13'N, 83°47'W (CM); Guácimo and vicinity (i.e. El Hogar), *c.*10°13'N, 83°41'W (AMNH 390396/97, CM P27910/11); Jiménez, 10°10'N, 83°46.5'W (Cherrie 1891a, USNM 198342); Hacienda La Iberia, 10°05'N, 83°39'W (UCLA 21908/09, LACM 14554/55); 5km S of Puerto Vargas, 09°41'N, 82°49'W (MZUCR 4810); RB Hitoy Cerere 09°40.5'N, 83°01.5'W (eBird: J. Chaves); Cuabre, 09°36.5'N, 82°48.5'W (CM P23926). **PANAMA**: *Bocas del Toro* Laguna de Chiriquí, *c.*08°58'N, 82°09'W (Griscom 1935). *Ngäbe-Buglé* Cricamola, 08°45'N, 81°50'W (MCZ 141249).

Hylopezus perspicillatus lizanoi (Cherrie, 1891), Proceedings of the United States National Museum, vol. 14, p. 342, 'Trojas', Costa Rica. The label bears the locality 'Las

Trojas de Puntarenas' which according to the map in Slud (1964) is *c.*2km south of Capulín on the left bank of the Río Grande de Tárcoles, within present-day PN Carara (*c.*09°48'N, 84°35'W). The holotype (USNM 119951) is a male collected in January 1886 by Anastasio Alfaro (Diegnan 1961). The subspecific name honours the Costa Rican politician, Joaquín Lizano. As suggested in the original description, my examination of the holotype revealed that it is not in full adult plumage and appears to be in the process of acquiring adult wing-coverts. The lesser and median wing-coverts are already tipped with broad, somewhat triangular or diamond-shaped buffy spots characteristic of adult plumage. Most of the greater coverts, however, are broadly tipped rufescent-chestnut, buffier in the centre and finely fringed bright reddish-chestnut. As herein defined, the holotype is a subadult in transition to full adult. Subspecies *lizanoi* is confined to the Pacific slope of Costa Rica (Carriker 1910) including the Osa Peninsula (Wilson *et al.* 2011). At least historically, its range extended across the Panamanian border into western Chiriquí (Sclater 1870, Bangs 1901). I was unable to locate any evidence to suggest that it occurs north of the Río Grande de Tárcoles. Cherrie (1891b) described the underside of the wing as ochraceous and gave the bare-parts coloration as *Bill* mostly black with base of the mandible white; *Tarsi & Toes* 'very light plumbeous'. Bangs & Barbour (1922) and Hellmayr (Cory & Hellmayr 1924) felt that *lizanoi* was not clearly separable from the nominate race, an opinion I share. SPECIMENS & RECORDS **COSTA RICA**: *Puntarenas* Garabito/Quebrada Bonita, 09°46.5'N, 84°36.5'W (MZUCR 4809); Playa Agujas, 09°43'N, 84°39'W (LACM 16228/29); Pozo Azul de Pirrís, 09°35'N, 84°20'W (ROM, CM, MCZ, ANSP, NHMUK); Pozo Pital, 09°26.5'N, 84°06.5'W (Cherrie 1895, FMNH, USNM); Volcán de Osa, 09°12'N, 83°27.5'W (UCLA 13689); Buenos Aires, 09°10.5'N, 83°20'W (Cherrie 1893, AMNH, MCZ, ANSP); Térraba, 09°04'N, 83°17.5'W (Cherrie 1893); Paso Real, 09°00.5'N, 83°13'W (loc. from Slud 1964, MCZ, AMNH, USNM); Boruca area, *c.*09°00'N, 83°19.5'W (Bangs 1907, CM, ANSP, AMNH, FMNH, MCZ, USNM, UMMZ); Lagarto, 08°57'N, 83°18.5'W (Cherrie 1893, AMNH 492280); 'Pozo del Río Grande', *c.*08°56.5'N, 83°31.5'W (loc. from Slud 1964, CM, ANSP, MCZ); 4.8km NW of Piedras Blancas, 08°49'N, 83°16'W (UMMZ); Alto El Campo, 08°47.5'N, 82°58.5'W (eBird: J. Zook); Reserva Forestal Golfo Dulce, *c.*08°45.5'N, 83°29'W (FLMNH 14345); Cañas Gordas, 08°44.5'N, 82°55'W (AMNH 706137); vicinity of Rincón de Osa, 08°42'N, 83°30'W (Ricklefs 1996, WFVZ, UMMZ); *c.*2km WSW of Pueblo Bahía Drake, 08°41'N, 83°41'W (Wilson *et al.* 2011); La Gamba–Golfito road, 08°40.5'N, 83°12'W (C. Sánchez *in litt.* 2017); Los Planes (PN Corcovado), 08°38.5'N, 83°39.5'W (ML 64483061, photo G.C. Benavides); Puerto Jiménez, 08°32'N, 83°18.5'W (YPM, AMNH, CUMV, UMMZ, FMNH); Río Sirena, 08°29'N, 83°36'W (ML 54815451, photo O. Johnson). *San José* Macaw Lodge, 09°43.5'N, 84°31'W (J.D. Vargas-Jiménez *in litt.* 2017); San Isidro de Dota, 09°29.5'N, 83°58.5'W (eBird: B. Young); El General, 09°23'N, 83°39'W (Skutch 1969, 1981, MCZ 123379); Finca Los Cusingos, *c.*09°20.5'N, 83°37'W (FLMNH 5639); near Tinamaste, 09°17'N, 83°46.5'W (eBird: J. Zook). **PANAMA**: *Chiriquí* S slope of Volcán Chiriquí, 08°41.5'N, 82°33'W (Salvin 1870, Sclater 1890); Bugaba, 08°29'N, 82°37'W (AMNH 492285); Mina de Chorcha, 08°25'N, 82°16'W (Salvin 1870, Sclater 1890); Divalá, 08°24.5'N, 82°41'W (Bangs 1901, MCZ 107897–900).

Hylopezus perspicillatus perspicillatus (Lawrence, 1861), Annals of the Lyceum of Natural History of New York, vol. 7, p. 3, 'Isthmus of Panama'. George N. Lawrence began writing his 'Catalogue of a collection of birds made in New Grenada…' in 1861, describing a collection made by James McLeannan from the unfortunately broadly defined locality of the 'Atlantic side of the Isthmus of Panama, along the line of the Panama Railroad, from near the coast to about a central point between the two oceans. One part of this described *Hylopezus perspicillatus*, with no additional comments as to a more precise locality and without designating a holotype. As explained by LeCroy & Sloss (2000), Lawrence (1961) had only three specimens of *H. perspicillatus* at his disposal in this collection, an unsexed individual (AMNH 43556) collected by McLeannan, and a male (AMNH 43557) and female (AMNH 43558) collected by McLeannan and John R. Galbraith. These three, based on collecting itineraries, were certainly collected on the Caribbean slope, probably in the vicinity of Lion Hill, a station along the railway and a common base of operations at the time. This location is now an island, after the Gatun Reservoir was flooded. Indeed, Cory & Hellmayr (1924) used Lion Hill (09°13.5'N, 79°53.5'W) as the holotype location. As Lawrence (1861) described both male and female, all three of these birds were in his possession at the time of the original description. In addition, although Foster (1892) and Cory & Hellmayr (1924) gave 1862 as the description date, the issue of the Annals in which the description was published was issued in 1861, as noted by Wetmore (1972). SPECIMENS & RECORDS PANAMA: *Veraguas* Río Calovébora, 08°42'N, 81°13'W (AMNH 246759–760/762–763; loc. see Siegel & Olson 2008); Santa Fé, 08°30.5'N, 81°04.5'W (Sclater 1867, 1890, Angehr *et al.* 2008); Santiago, 08°06'N, 80°59'W (Sclater 1867). *Coclé* Cascajal, 08°43'N, 80°28'W (AMNH 492283); Tigre, 08°42'N, 80°36'W (USNM 476735); near El Copé, 08°40'N, 80°35.5'W (XC 3309, D. Bradley); Natá, 08°20'N 80°31'W (Cory & Hellmayr 1924, USNM 150803/04). *Colón* Cerro Bruja, 09°29'N, 79°34'W (USNM 229474/75); Fort Sherman, 09°20'N, 79°58'W (FLMNH 7824); Estación Hídro Salamanca, 09°17'N, 79°36'W (MCZ 171661); Limbo Hunt Club, 09°10'N, 79°45'W (Schemske & Brokaw 1981, FLMNH 8286, XC, ML); *c.*8km NE of Gamboa, 09°09.5'N, 79°44.5'W (Karr 1970, Karr *et al.* 1978, 1990, FMNH); Isla Barro Colorado (extirpated), 09°07.5'N, 79°51.5'W (Chapman 1929a, Heath 1932, Eisenmann 1952, Willis & Eisenmann 1979, Willis 1985a, AMNH 228951, ML 22665, E.S. Morton); Quebrada Juan Grande, 09°07.5'N, 79°43'W (ML 65382481, photo R. Candee). *Panamá* Playa Grande, 09°23'N 79°21'W (UWBM 76867/912); Estación Hídro El Peluca, 09°23'N, 79°33.5'W (USNM 474469); confluence of Ríos Chagresito and Chagres, 09°21.5'N, 79°19'W (UWBM 76892); Quebrada Carriaso, 09°14'N, 79°14'W (USNM 409507); Altos de Cerro Azul, 09°13'N, 79°24.5'W (P.A. Hosner *in litt.* 2017); San Antonio, 09°09'N, 79°03'W (YPM 54969); Pedro Miguel, 09°01.5'N, 79°36.5'W (USNM 459964); Capira, 08°45'N, 79°53'W (AMNH 492284); Altos del María, 08°40'N, 80°03.5'W (ML 34730991, photo A. Lin-Moore). *Guna Yala* Nusagandi, 09°20'N, 78°59'W (XC 101088; J. King); Permé, 08°45'N, 77°34'W (Griscom 1932a, MCZ 155553–558, YPM 54967/68); Puerto Obaldía, 08°40'N, 77°25'W (Griscom 1932a, MCZ 155559/60); Quebrada Venado/Armila, 08°40'N, 77°29'W (USNM 477778–780). *Emberá* Cerro Tacarcuna, 08°10'N, 77°18'W (AMNH 135792/93). *Darién* Quebrada Cauchero, 08°46'N,

78°32'W (USNM 533765–767); Esnápe, 08°05'N, 78°13'W (Bangs & Barbour 1922, MCZ 87365); Tacarcuna Village, 08°05'N, 77°17'W (USNM 484542); Capetí, 08°04'N, 77°33'W (AMNH); Rancho Frío, 08°01'N, 77°43'W (CUMV 55973, ML, XC); Casita, 08°01'N, 77°22'W (USNM 469749); Cituro, 08°00'N, 77°36'W (AMNH 136605–609); Jesusito, 07°59.5'N, 78°16.5'W (Bangs & Barbour 1922, MCZ 87364); Ríos Paya/Tuira confluence, 07°59.5'N, 77°33'W (USNM 470103); Tapalisa, 07°59'N, 77°26'W (AMNH 135794–796); Cana, 07°45'N, 77°42'W (USNM, FLMNH, XC, ML); Río Seteganti headwaters, 07°43'N, 77°44.5'W (USNM 474467/48); Río Piñas/Puerto Piña, 07°39'N, 78°12'W (Miller *et al.* 2011, UWBM 113313); Las Peñitas, 07°28'N, 77°59'W (USNM 389886–888); Ríos Imamadó/Jaqué confluence (= Chicao), 07°26.5'N, 77°58.5'W (USNM 389889–892). **COLOMBIA**: *Chocó* Río Salaquí, 07°27'N, 77°07'W (Chapman 1917, Meyer de Schauensee 1950a, AMNH 113349); Río Truandó, 07°26'N, 77°07'W (Haffer 1975, AMNH 787156); Juradó, 07°06.5'N, 77°45.5'W (Meyer de Schauensee 1950a). *Antioquia* Sautata, 07°50'N, 77°04'W (CM P64096).

Hylopezus perspicillatus pallidior Todd, 1919, Proceedings of the Biological Society of Washington, vol. 32, p. 115, El Tambor, Santander, Colombia (07°19'N, 73°16'W, 500m). The holotype is an adult male (CM 59199) collected 29 December 1916 by M. A. Carriker Jr. (Todd 1928). This race is endemic to northern Colombia, being confined to the valleys of the upper Río Sinú, the lower Río Cauca and the middle Río Magdalena (Ayerbe-Quiñones *et al.* 2008). Race *pallidior*, as its name implies, is paler in general coloration than nominate *perspicillatus*, but otherwise is similarly patterned. The crown is paler, duller grey and the back lighter olive-green. The wing markings and head-sides are paler buffy. SPECIMENS & RECORDS *Córdoba* Alto de Quimarí, 08°07'N, 76°22.5'W (Meyer de Schauensee 1950a,b, ANSP-160860–864); upper Quebrada Charrura, 08°00'N, 75°49'W (AMNH 787157); Cerro Murrucucú, 07°59'N, 76°00'W (Meyer de Schauensee 1950a,b, ANSP 160865); Socorré, 07°51'N, 76°17'W (USNM 411881–887); Quebrada Salvajín, 07°45'N, 76°16'W (USNM 411888–890). *Bolívar* Santa Rosa del Sur, 07°58'N, 74°03'W (USNM 398004–007). *César* San Alberto, 07°45.5'N, 73°23.5'W (LACM, WFVZ, YPM, FMNH, LSUMZ). *Antioquia* Nechí, 08°05.5'N, 74°46'W (FMNH 190881); El Real, 07°37.5'N, 74°47'W (USNM 402485); Tarazá, 07°35'N, 75°24'W (USNM 402486); Puerto Valdivia, 07°18'N, 75°23'W (Chapman 1917, Meyer de Schauensee 1950a); Hacienda Belén, 07°10'N, 74°43'W (USNM 402479–484); Finca La Banca, 06°36'N, 74°35'W (A.M. Cuervo *in litt.* 2017); *c.*3.3km E of Mulata, 05°56'N, 74°49.5'W (F. Rowland *in litt.* 2017). *Santander* Río Lebrija, 08°08'N, 73°47'W (Cory & Hellmayr 1924, Meyer de Schauensee 1950a); Hacienda Santana, 07°27'N, 73°08'W (USNM 411891/92); Cerro de La Paz, 06°59'N, 73°26'W (Donegan & Huertas 2005, XC); Meseta de Los Caballeros, *c.*06°56'N, 73°41'W (ICN 11824); Río Oponcito, *c.*06°39'N, 73°54'W (ICN 18825). *Boyacá* RNA El Paujil, 06°03'N, 74°15.5'W (Quevedo *et al.* 2006, Salaman *et al.* 2007a); Monte del Diablo, 05°38'N, 74°18'W (Stiles *et al.* 1999).

Hylopezus perspicillatus periophthalmicus (Salvadori & Festa, 1898), Bollettino di Musei di Zoologia ed Anatomia Comparata della Universita di Torino, vol. 13, no. 330, p. 2, Río Peripa, western Ecuador. Paynter (1993) described Río Peripa as a small tributary of the Río Daule, along which Salvadori was travelling when the holotype was collected. The Río Peripa apparently rises south-west of Santo Domingo de Los Colorados and enters the Río Daule *c.*28km north-west of Quevedo, at the border between Manabí and Guayas. This would place the type locality for *periophthalmicus* at *c.*00°53'S, 79°43'W, *c.*50km south-west of the only modern record for the species (Estación Científica Río Palenque) south of Esmeraldas province. The continued persistence of Streak-chested Antpitta at Río Palenque is unconfirmed, with apparently no records (historical or otherwise) between there and southern Esmeraldas (RP Río Canandé), a distance of at least 100km and currently the southernmost locality for Streak-chested Antpitta. The northern limit of *periophtalmicus* is not entirely clear, but I have followed Haffer (1967) in extending its range (west of the Andes) as far north as Mutatá and Pavarandocito in westernmost Antioquia. Most records, historical or otherwise, are from Esmeraldas, with scattered but fairly continuous records north of there, from Nariño (Calderón-Leytón *et al.* 2011) to central Chocó, where it is apparently fairly regular near the coastal town of Mutis. Subspecies *periophthalmicus* is said to differ from the nominate race in being deeper olive-brown with fewer buffy shaft-streaks on the upper back and in having a more ochraceous (less fulvous) face pattern (Hartert 1902, Chapman 1917). **SPECIMENS & RECORDS** *Antioquia* Pavarandocito, 07°16'N, 76°30'W (USNM 426432–435); Mutatá, 07°14.5'N, 76°26'W (Haffer 1967, 1975); Murindó, 06°59'N, 76°49.5'W (CM P64316); upper Río Murrí, 06°30'N, 76°29'W (Haffer 1975, AMNH 787158). *Chocó* *c.*5.8km SE of Mutis, 06°11'N, 77°22'W (B. Freeman *in litt.* 2017); Alto del Buey, 06°06'N, 77°13'W (ANSP 146302); Serranía de Baudó, *c.*06°00'N, 77°05'W (Chapman 1917, AMNH 123352/55); Río Jurubidá, 05°50'N, 77°16'W (USNM 443351–355); Río Nuquí, 05°40.5'N, 77°14'W (USNM 443346–350); Andagoya, 05°05.5'N, 76°41.5'W (CM P66325); Quebrada Sando, 05°03'N, 76°57'W (FMNH 255653); Potedo, 04°33'N, 77°00.5'W (CM P66533). *Cauca* Río Saija, *c.*02°52.5'N, 77°37'W (FMNH 255654); Guapí, 02°34.5'N, 77°53'W (ICN 11821–823). *Valle del Cauca* Río Tatabro, 03°42'N, 76°58'W (Naranjo & de Ulloa 1997). *Nariño* Barbacoas, 01°40.5'N, 78°08.5'W (Chapman 1917, AMNH 117882); Candelilla, 01°29'N, 78°43'W (FMNH 250851); La Guayacana, 01°26'N, 78°27'W (ANSP 157354/55, 159574–576). **ECUADOR**: *Carchi* Lita, 00°52.5'N, 78°28'W (Hartert 1902, AMNH 492291). *Esmeraldas* Humedal de Yalare, 01°09'N, 78°51'W (XC 276103, J. Nilsson); Carondelet, 01°06'N, 78°42'W (FMNH); Durango, 01°05'N, 78°41.5'W (XC 16463, R. Ahlman); Pulún, 01°05'N, 78°40'W (Hartert 1902, FMNH 57202–205, AMNH 492288/89, MCZ 94795); San Javier de Cachabí, 01°04'N, 78°47'W (Hartert 1902, AMNH 492292/93); 20km NNW of Alto Tambo by road, *c.*01°01'"N, 78°36'W (ANSP 182531); Cachabí, 00°58'N, 78°48'W (Hartert 1898, AMNH 492287); Playa del Oro, 00°51'N, 78°46'W (XC, ML); Río Zapallo Grande, 00°44'N, 78°56'W (USNM 331277); RE Cotacachi-Cayapas, 00°41.5'N, 78°54.5'W (XC 11867, N. Athanas); 30km S of Chontaduro, *c.*00°39.5'N, 79°25'W (ANSP 185510); RP Río Canandé, 00°31.5'N, 79°13'W (HFG). *Los Ríos* Estación Científica Río Palenque, 00°36.5'S, 79°21.5'W (Leck 1979).

STATUS Considered fairly common (Stotz *et al.* 1996) across most of its fairly large range (263,000km²), BirdLife International (2017) regards the species as Least Concern.

Neverthless, it is considered 'highly sensitive' to habitat alteration (Stotz *et al.* 1996) and has apparently already been extirpated from several areas where it was once common, such as Barro Colorado Island (Eisenmann 1952, Willis & Eisenmann 1979), where it was last seen in 1974 (Jones 1977: *contra* Robinson 1999), and at Estación Biológica La Selva, where race *intermedius* has declined dramatically (Slud 1960, Blake & Loiselle 2001, Sigel *et al.* 2006). The reasons for its disappearance from Barro Colorado are not entirely clear (Gillespie 2001, Sigel *et al.* 2010), but increased levels of nest predation, small population size compared with the nearby mainland, along with human activity and introduced predators, may have contributed and also led to its loss from other islands in Lago Gatún (Sieving 1992). **PROTECTED POPULATIONS** *H. p. lizanoi* PN Carara (Arevalo & Newhard 2011, MZUCR 4809); PN Corcovado (ML 73453, D.L. Ross Jr.); PN Los Quetzales (eBird: B. Young); Reserva Forestal Golfo Dulce (FLMNH 14345). *H. p. intermedius* **Honduras** Reserva de Biósfera Río Platano (Marcus 1983, LSUMZ 99946). **Nicaragua** RB Río Indio-Maiz (Cody 2000). **Costa Rica** PN Cahuita (eBird: C. Drysdale); PN Braulio Carrillo (Blake & Loiselle 2000, ML); PN Volcán Turrialba (CM P23518); RB Hitoy Cerere (eBird: J. Chaves); RB Alberto Manuel Brenes (eBird: J. Zook); RP Estación Biológica La Selva (Slud 1960, Blake & Loiselle 2001, Sigel *et al.* 2006, Boyle & Sigel 2015, Visco *et al.* 2015). **Panama** Probably occurs in BP Palo Seco. *H. p. perspicillatus* **Panama** PN Chagres (P.A. Hosner *in litt.* 2017); PN Soberanía (Robinson *et al.* 2000a, Poulin *et al.* 2001, Angehr *et al.* 2008); PN San Lorenzo (Angehr *et al.* 2008); PN General Omar Torrijos (XC 3309, D. Bradley); PN Darién (USNM, FLMNH). Probably in PN Santa Fé. *H. p. pallidior* RNA el Paují (Quevedo *et al.* 2006, Salaman *et al.* 2007a); RNA Pauxi pauxi (Salaman *et al.* 2009a). *H. p. periophthalmicus* **Colombia** PNN Ensenada de Utría (ANSP 146302); RNSC San Cipriano (eBird: D. Martínez Castaño). **Ecuador** RE Cotacachi-Cayapas (Horsley *et al.* 2016, XC); RP Río Canandé (HFG); RP Estación Científica Río Palenque (Leck 1979).

OTHER NAMES Streak-chested Antpitta was placed in the genus *Grallaria* for many years after its original description. Synonyms in the literature include: *H. p. perspicillatus Grallaria perspicillata* (Sclater 1870, Boucard 1878a, Hartert 1898, Clark 1971, Burton 1975); *Grallaria perspicillata perspicillata* (Cory & Hellmayr 1924, Griscom 1935, Meyer de Schauensee 1950a, Eisenmann 1952); *Hylopezus perspicillata perspicillata* (Chapman 1917, Bangs & Barbour 1922); *Grallaria perspicillata* (Sclater 1867, 1868, 1877, 1890, Dubois 1900, Sharpe 1901); *Hylopezus perspicillata* (Schemske & Brokaw 1981, Robinson 1999, Robinson *et al.* 2000a). *H. p. periophthalmicus Grallaria fulviventris barbacoae* (Olivares 1958, see Haffer 1959); *Grallaria perspicillata periophthalmica* (Hartert 1902, Cory & Hellmayr 1924, Meyer de Schauensee 1950a); *Grallaria periophthalmica* (Salvadori & Festa 1898a, Sharpe 1901, Dubois 1904, Brabourne & Chubb 1912); *Hylopezus perspicillata periophthalmica* (Hartert 1902, Chapman 1917); *Grallaria perspicillata* (Hartert 1898, Leck 1979); *Hylopezus perspicillatus periopthalmica* (Krabbe 1991); *Hylopezus perspicillata periopthalmica* (note misspelling of race, Chapman 1917, spelled correctly in Chapman 1926a). *H. p. lizanoi Grallaria perspicillata* (Salvin 1870, Bangs 1901, Skutch 1969); *Grallaria perspicillata lizanoi* (Cory & Hellmayr 1924, Slud 1964, Schönwetter 1979); *Grallaria lizanoi* (Cherrie 1891b, 1893, 1895, Dubois 1900, Sharpe 1901, Bangs 1907); *Grallaria intermedia*

(Zeledón 1885); *Hylopezus lizanoi* (Carriker 1910). *H. p. pallidior* Hylopezus perspicillata perspicillata (Chapman 1917); *Grallaria perspicillata pallidior* (Cory & Hellmayr 1924, Meyer de Schauensee 1950a,b). *H. p. intermedius* Grallaricula perspicillata (Sclater 1873a); *Grallaria intermedia* (Cherrie 1891a, Dubois 1900, Sharpe 1901); *Grallaricula perspicillata* (Sclater & Salvin 1873b); *Grallaria perspicillata* (Sclater 1877, 1890, Slud 1960); *Grallaria perspicillata intermedia* (Cory & Hellmayr 1924, Griscom 1935, Carriker 1957, Slud 1964); *Hylopezus intermedius* (Carriker 1910, Ridgway 1911). **English** Around two decades ago, Banks *et al.* (1997) formally changed the English name from Spectacled Antpitta to Streak-chested Antpitta, the former name being widespread in the literature for many years (e.g., Slud 1960, Davis 1972, Altman & Swift 1986, Ortiz-Crespo *et al.* 1990, Sibley & Monroe 1990, Principe 1991, Sieving 1992, Robinson 1999, Gillespie 2001); North Colombian Antpitta/Ant-pitta (*pallidior*; Cory & Hellmayr 1924, Meyer de Schauensee 1950a); Lawrence's

Antpitta/Ant-pitta (*perspicillata*; Ridgway 1911, Cory & Hellmayr 1924, Chapman 1929a, Heath 1932, Meyer de Schauensee 1950a); Lawrence's Antbird (*perspicillatus*; Stone 1918); Festa's Ant-pitta (*periophthalmicus*; Cory & Hellmayr 1924, Meyer de Schauensee 1950a) Talamanca Antpitta (*intermedius*; Ridgway 1911, Cory & Hellmayr 1924); Speckled Antpitta (*perspicillatus*; Jones 1977); Streak-cheeked Antpitta (Quevedo *et al.* 2006); Lizano's Antpitta (*lizanoi*; Ridgway 1911, Cory & Hellmayr 1924); Salvadori's Ant Thrush (*periophthalmicus*, Brabourne & Chubb 1912). **Spanish** Tororoí de Anteojos (Krabbe & Schulenberg 2003, Salaman *et al.* 2007a); Gallito montés pecho de rayas (Bonta & Anderson 2002); Gralaria pechilistada (Valarezo-Delgado 1984); Tororoí pechirrayado (Valarezo-Delgado 1984, Ridgely *et al.* 1998; Martínez-Sánchez *et al.* 2014); Tororoí de Anteojos (Ortiz-Crespo *et al.* 1990); Tororoí pechilistado (van Perlo 2006). **French** Grallaire à lunettes. **German** Orangewangen-Ameisenpitta (Krabbe & Schulenberg 2003).

Streak-chested Antpitta, adult (*lizanoi*), Piro, Osa, Costa Rica, 21 March 2015 (*Daniel Hernández*).

Streak-chested Antpitta, adult (*intermedius*) feeding nestlings, Quebrada González, San José, Costa Rica, 21 June 2013 (*Ian Ausprey*).

Streak-chested Antpitta, young nestlings (*intermedius*) in nest, Quebrada González, San José, Costa Rica, 15/16 June 2013 (*Deborah Visco*).

Streak-chested Antpitta, fledgling (*intermedius*) at nocturnal roost, Quebrada González, San José, Costa Rica, 25 June 2012 (*Adam Spencer*).

SPOTTED ANTPITTA
Hylopezus macularius   **Plate 19**

Pitta macularia Temminck, 1830, Nouveau recueil de
planches coloriées d'oiseaux, livr. 85, Genre Brève, 2nd
Section, composée des espèces du Nouveau-Monde.
Esp. 11, p. 4. Described from 'le Brésil', this was later
considered to be in error and the type locality was
amended by (Hellmayr (1910) to 'Cayenne' [= French
Guiana]. Coenraad Jacob Temminck described Spotted
Antpitta within a part of the vast monograph *Nouveau
recueil des planches coloriées d'oiseaux* (Temminck & Laugier
1820–39), co-authored by Meiffren Laugier de Chartrouse
and issued in 120 parts or 'livraisons' over the course
of *c.*20 years (Dickinson 2001). The species description
given above is that most frequently used, but there is
considerable confusion over the order and dates of issue
of the various parts (Dickinson 2001, 2011, Lebossé &
Bour 2011), and the validity of this citation bears further
investigation. The work was later compiled and re-
issued in four volumes (Temminck & Laugier 1850; see
Sayako & Dickinson 2001) available at the Biodiversity
Heritage Library. The species description for *macularius*
is in vol. 2, 'page 237' at http://biodiversitylibrary.org/
page/35246983.

Recent analysis of the 'Spotted Antpitta complex' indicates
that the species comprises just two subspecies separated
by the Rios Branco and Negro in northern Brazil. In
plumage they are very similar, but consistent vocal
differences suggest that they may eventually be recognised
as species. Found in the undergrowth of humid *terra
firme* forest, especially around treefalls and other areas of
dense vegetation, Spotted Antpitta ranges from western
Venezuela and the Guianas across northern Amazonia to
northern Peru. Diet and breeding biology are both very
incompletely known (just one nest has been found). As
noted by King (2013) there are important gaps in our
understanding of the species' vocalisations. Importantly,
more recordings of subspecies *dilutus*, especially from
southern Venezuela (Amazonas), will be particularly
useful to determine if it merits species rank. Similarly,
there are few records between southern Venezuela and
northern Peru, so visitors to south-east Colombia (Dept.
Amazonas) could make an important contribution simply
by confirming its presence there.

IDENTIFICATION 13.5–14.0cm. Sexes similar. Adults
have a grey (to olivaceous-grey) crown and nape
indistinctly streaked buff, buffy lores, an ochraceous
eye-ring, and dull ear-coverts streaked blackish and buffy.
Back, tail and wings olive-brown, the wing-coverts tipped
tawny-buff, forming faint wingbars. Primary-coverts black,
contrasting with small, well-defined tawny patch at base of
primaries (except two outermost). These wing markings
form a fairly distinctive wingbar visible mainly in flight
(Hilty 2003). Throat white, bordered by thin black malar
stripe. Breast white, washed buffy and has short (spot-
like) black streaks that become sparser lower, above a
clean white belly. Flanks and vent ochraceous-buff. Adults
have dark eyes, a black bill (paler at base of mandible),
and pale creamy or greyish legs. Thrush-like Antpitta is
similar in size and silhouette, but lacks the malar stripe and
prominent buffy eye-ring of Spotted Antpitta. Even more
heavily streaked individuals of Thrush-like have diffuse,
lightly marked underparts that are quite different from

the boldly marked underparts of Spotted. In Peru and
Colombia, White-lored Antpitta has only a weak eye-ring,
distinctly more diffuse streaking below and a (usually)
darker-looking crown.

DISTRIBUTION Ranges across much of the Guianan
Shield and northern Amazona, from eastern and southern
Venezuela, east through Guyana, Surinam and French
Guiana, and from there south and west across Brazil
(north of the Amazon) to south-east Colombia and north-
east Peru (Hilty 2003, Ridgely & Tudor 2009, McMullan *et
al.* 2010, Schulenberg *et al.* 2010).

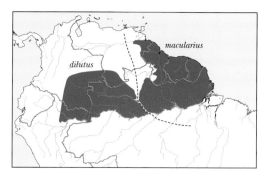

HABITAT Overall, a species of mature forest (Donahue
& Pierson 1982, Cohn-Haft *et al.* 1997, Braun *et al.* 2000,
Thiollay *et al.* 2001, Lentino *et al.* 2013a). Despite a
propensity to prefer disturbed areas, it appears to disappear
from small forest fragments (Bierregaard & Lovejoy 1989)
and to remain absent from anthropogenically modified
areas for many years even when the forest is left to regrow
(Powell 2013). It keeps largely to the floor, especially in
dense undergrowth around treefalls and along streams
in humid *terra firme* and gallery forest (Santos & da Silva
2007), less frequently foraging in more open understorey.
A species of the lowlands, it is found as high as 500m in
Venezuela (Giner & Bosque 1998).

VOICE The song of nominate *macularius*, following
Krabbe & Schulenberg (2003), is *c.*2.5s in duration,
delivered at intervals of 7–10s, and comprises an evenly
paced series of six notes, the first two and the fourth at
a frequency of *c.*1kHz, and the rest slightly higher. The
song of *dilutus* is also of six notes, but is shorter than
that of nominate *macularius* (*c.*1.8s) and is delivered at
intervals of 8–11s. The first four notes are on about the
same frequency (*c.*1kHz) and the final two are slightly
lower (*c.*0.8kHz) (Krabbe & Schulenberg 2003, Carniero
et al. 2012). Krabbe & Schulenberg (2003) transcribed the
songs of nominate and *dilutus* (respectively) as *whoa-whoa-
wok-whoa-wok-wok* and *hoor-hoor-hoor-hoor-ho-ho.*

NATURAL HISTORY Usually encountered alone
(Thiollay & Jullien 1998), Spotted Antpittas may
occasionally forage as pairs (Krabbe & Schulenberg 2003).
Like others in the genus, it hops or runs across the ground,
pausing to probe at prey or flick aside leaves with its bill
(Hilty 2003). Between bouts of movement they often
pause stiffly upright, sometimes fluffing out their feathers.
Singing occurs from the ground or a low perch, typically
a diagonal or horizontal branch less than 1.5m above
ground. With each song, the head is cocked back and the
throat swells slightly, exposing pink skin below the feathers
(Hilty 2003). There are no data concerning natural

enemies, but Tostain (1986) described watching as a White Hawk *Pseudastur albicollis* passed over an incubating adult. The bird remained motionless, remaining undetected by the low-flying raptor.

DIET One stomach held 'insects'. The stomach of a fledgling collected in Guyana (AMNH 586404, R.T. Brumfield; see Reproduction Seasonality) contained an ant and various insect remains, including beetles, as did an adult collected in Amapá (Aguilar & Junior 2008).

REPRODUCTION Only one nest and clutch of eggs has been described, of nominate *macularius* in French Guiana (Tostain 1986). The nest was discovered with a complete clutch of two eggs on 16 April 1983 at Saül (St. Laurent du Maroni) at an elevation of 325m. NEST & BUILDING The single nest was a shallow, very rudimentary cup of twigs and dead leaves, sparsely lined with a few rootlets and placed 75cm above ground atop a live horizontal leaf of a palm (*Astrocaryum paramaca*). Tostain (1986) likened the rather thin, shallow nest to that of a dove or cotinga. The maximum outer diameter of the nest, created by a few longer sticks that protruded from the main construction, was 30cm. The egg cup was 7cm in diameter and only 1.5cm deep. Nothing is known concerning site selection or building, but presumably these aspects are similar to other *Hylopezus*. Of described *Hylopezus* nests, the nest of *H. macularius*, as expected, appears most similar to that of *H. auricularis*. EGG, LAYING & INCUBATION Two eggs, with a mean mass of 5.7g, measured 25.4 × 19.5mm and 25.3 × 19.4mm. They were both pale greenish-cream with brown and beige speckling (Tostain 1986). NESTLING & PARENTAL CARE Unknown. SEASONALITY Very few records for this species. BREEDING DATA *H. m. macularius* Almost fully developed egg in oviduct, 11 June 1972, Sipaliwini (Haverschmidt & Mees 1994); incubation, 16 April 1983, Saül (Tostain 1986); fledgling transitioning, 28 September 1997, Baramita (USNM 586404); juvenile, 30 May 1922 Kartabu Point (AMNH 821520); 13 'juveniles' (precise age unclear) at the BDFFP field site, 6 August 1982, 19 September and 20 December 1983, 29 September, 2 October and 21 November 1984, 16 January and 2 March 1985, 26 February 1986, 7 December 1988 (*n* = 2), 5 October 1989, 20 November 1991 (BDFFP database; P. Stouffer *in litt.* 2015); subadult, 7 November 1960, Kayserberg Airstrip (FMNH 260494); subadult, 12 December 1917, Fleuve Oyapock (YPM 29657, S.S. Snow photos); enlarged ovary, 15 November 1960, Kayserberg Airstrip (Blake 1963, FMNH 260495); testes greatly enlarged, 10 December 1961, Wilhelmina Mountains (FMNH 264497). *H. m. dilutus* Subadult, 16 April 1929, El Merey (AMNH 432820).

TECHNICAL DESCRIPTION Sexes similar, the following description of nominate *macularius* is from Haverschmidt & Mees (1994). *Adult* Forehead, crown and nape dark grey; rest of upperparts, including the short tail, olive-brown; wings mostly brownish, similar to back, lesser and median coverts broadly edged orange-buff; primaries have orange-buff outer margins and bases, which contrast with the blackish-brown greater coverts; secondaries with broad olive-brown margins; underwing-coverts ochraceous. Lores and broad ring of small feathers around eyes orange-buff, ear-coverts olive-brown; there is a black streak below the ear-coverts and another from the chin back along the side of the throat (submalar). Underparts mostly pale ochraceous or buffy-white, the

breast heavily marked with short black streaks; sides, flanks and undertail-coverts orangey-buff. **Adult Bare Parts** *Iris* dark brown; *Bill* maxilla black, basal two-thirds of mandible pinkish-grey, tip black; *Tarsi & Toes* pinkish-brown to creamy-pink. *Immature* The following is based largely on my examination of a fledgling *macularius* transitioning to juvenile plumage (USNM 586404). Based on this specimen I have extrapolated somewhat forward and back. *Fledgling* Largely still covered in fluffy, wool-like down, above uniform rufescent-chestnut or dark, rusty-brown. The chest is similarly feathered, becoming paler, more ochraceous below and on the flanks, buffy to grey-white centrally and on vent. Wings similar to adult but buffy spots on upperwing-coverts not as well defined, paler (yellow-buff) and more extensive, covering most of the apical half of the feather in some of the lesser coverts, and broadly tipped or fringed bright rufous-chestnut. **Fledgling Soft Parts** *Iris* brown; *Bill* maxilla black, mandible pale pinkish, rictal flanges tinged orange; *Tarsi & Toes* pinkish-grey (USNM 586404, R.T. Brumfield). *Juvenile* Back and most of upperparts similar to adult, perhaps somewhat paler or cinereous. Facial feathering probably remains 'messy' for some time, with a mix of adult and immature feathers on the lores and ear-coverts. Eye-ring poorly developed, buffy-ochraceous in part, buffy-chestnut in others. Retains patches of fluffy rufescent-brown down on crown and nape, and in irregular patches on underparts, mostly across upper breast and central belly. Rest of underparts largely buffy, ochraceous-yellow, paler centrally, white on throat and irregularly spotted on breast, hinting at adult plumage. It appears that the submalar streaks are apparent early on. Wings as described for fledglings, with the number of chestnut tips likely reduced (from wear?) but possibly still present on coverts if they are replaced around this time. **Juvenile Soft Parts** Undescribed. Likely similar to fledglings but rictal flanges less apparent, paler yellow (little orange). *Subadult* Similar to adult, somewhat more cinereous, less olivaceous on back. In the few I have seen, the spotting is sparser below but this could be either age-related or individual variation. Ochraceous wash to underparts brighter and extends to central belly and lower breast, more than in adults. Eye-ring somewhat thinner and probably less apparent in the field than in adults. Upperwing-coverts are the most obviously different feature from adult, though perhaps not apparent in the field, being thinly margined rufous-chestnut, sometimes with the chestnut coloration 'bleeding' into the buffy-ochraceous spots. **Subadult Soft Parts** Not seen in life, but probably similar to adults.

MORPHOMETRIC DATA *H. m. dilutus* Data from AMNH and USNM (HFG). *Bill* from nares 12.4 ± 0.7mm (*n* = 8); *Tarsus* 35.5 ± 0.8mm (*n* = 9). Zimmer (1934) provided the following measurements for an adult ♂, his type specimen for '*diversus*'. *Wing* 86mm; *Tail* 33mm; *Bill* exposed culmen 18mm, total culmen 23mm; *Tarsus* 35mm. *H. m. macularius* From the BDFFP database (all captures, P. Stouffer *in litt.* 2015). *Wing* 82.5 ± 3.8mm (*n* = 39); *Tail* 34.1 ± 2.1mm (*n* = 37). Bierregaard (1988) reported the following measurements, sexes and ages combined, from the same project, presumably with some overlap in use of data. *Wing* 82.28 ± 2.61mm (*n* = 25); *Tail* 34.43 ± 1.95mm (*n* = 23). Data from AMNH, FMNH and USNM (HFG, sexes combined). Adults: *Wing* 83.8 ± 2.8mm (*n* = 4); *Tail* 34mm (*n* = 1); *Bill* from nares 12.1 ±

0.7mm (*n* = 9), exposed culmen 16.6 ± 0.2mm (*n* = 3), total culmen 18mm (*n* = 1), depth at nares 5.6 ± 0.3mm (*n* = 3), width at nares 5.1 ± 0.2mm (*n* = 3), width at base of mandible 6.7 ± 0.4mm (*n* = 3); *Tarsus* 34.9 ± 0.9mm (*n* = 12). Fledgling: *Bill* from nares 11.1mm; *Tarsus* 34.8mm (*n* = 1, ♂, USNM 586404). See also Chubb (1918b, 1921), Haverschmidt & Mees (1994). **Mass** Without indicating subspecies, Krabbe & Schulenberg (2003) gave a range of 47–53g for 3♀♀ and 43–47g for 2♂♂. The following weights are available for the nominate subspecies: 43g, 53g (*n* = 2, ♂,♀, Haverschmidt & Mees 1994), 48g, 43g (*n* = 2♂♂, Dick *et al.* 1984), 42g (*n* = ?, *macularius* Karr *et al.* 1990), 43.9g(tr) (*n* = 1♂, adult, USNM 625539); 40.2g(lt) (*n* = 1, ♂, fledgling, USNM 586404), 42.30 ± 3.59g (*n* = 34, sexes/ages combined, Bierregaard 1988). From the BDFFP database (all captures, P. Stouffer *in litt.* 2015): 42.8 ± 3.3g, range 34–50g (*n* = 45). **Total Length** 12.5–15.0cm (Lafresnaye 1842, Chubb 1918b, Hilty & Brown 1986, Clements & Shany 2001, Hilty 2003, Krabbe & Schulenberg 2003, Spaans *et al.* 2016).

TAXONOMY AND VARIATION Two subspecies currently recognised. Following the recent split of the *macularius* complex, only some authorities have followed Carneiro *et al.*'s (2012) recommendation to recognise *dilutus* as a monotypic species (e.g. WA). Carneiro *et al.* (2012) proposed Zimmer's Antpitta as a common name honouring J. T. Zimmer's early contribution to the taxonomy of the *H. macularius* complex (Zimmer 1934). Choosing a more conservative approach in their proposal to SACC, Carneiro & Aleixo (2013) suggested maintaining *dilutus* as a race of *macularius* until the relationship between them is further elucidated (see discussion under *dilutus*). This recommendation was followed by SACC and here. The taxonomic history of Spotted Antpitta and its relatives is rather complex. Two subspecific names are no longer considered valid (*macconnelli* and *diversus*, the latter having been widely used for many years (see Other Names). Furthermore, at one point, *H. auricularis*, *H. whittakeri* and *H. paraensis* were all considered subspecies of Spotted Antpitta. In light of these past complexities and the potential future split of *dilutus*, in the accounts below I have tried to be particularly diligent in documenting localities and have tried to point out all records where past taxonomic arrangements may present confusion.

Hylopezus macularius macularius (Temminck, 1830) The nominate race ranges from north-east Venezuela through the Guianas to north-east Brazil, north of the Amazon and east of the Rios Branco and Negro. In French Guiana, Thiollay (2002) found *macularius* at 19 of 20 sites across the entire country. **Specimens & Records VENEZUELA**: *Delta Amacuro* Campamento Río Grande, 08°03'N, 61°38'W (ML 40466, P. Coopmans); Altiplanicie de Nuria, 07°40'N, 61°20'W (COP 17127). *Bolívar* Near El Palmar, *c*.07°58'N, 61°53'W (ML, LSUMZ); E of Tumeremo, 06°54'N, 61°15'W (XC 11648, N. Pieplow); *c*.29km N of San Isidro, 06°25'N, 61°28'W (ML 44279, J.D. MacDonald); 5km W of San Isidro, 06°09'N, 61°28'W (XC 3577, N. Athanas); La Maloka, upper Río Cuyuní, 06°04'N, 61°28'W (Lentino *et al.* 2013b); 6.2km SW of La Laja, 06°00'N, 61°30'W (ML 60952, L. Macaulay). **GUYANA**: *Barima-Waini* Baramita area, *c*.07°21'N, 60°29'W (USNM, KUNHM); *Pomeroon-Supenaam* Along Ituribisi River, at *c*.07°05'N, 58°29'W (Zimmer 1934, AMNH 230196). *Cuyuni-Mazaruni* Cuyuni River near Aurora, 06°46'N, 59°45.5'W (ML 48535161/201, photos L. Moore); Kartabu

Point, 06°23'N, 58°41'W (AMNH); Kamarang River, Mission Paruima 05°54'N, 60°35'W (AMNH 492294/95, COP 4241); Merume Mountains, *c*.05°48'N, 60°06'W (USNM 90596). *Potaro-Siparuni* North Rupununi District, *c*.04°50'N, 59°20'W (Mistry *et al.* 2008); Iwokrama River Lodge, 04°40.5'N, 58°41'W (ML 23946961, photo L. Seitz); Kabocalli Landing, 04°17'N, 58°31'W (KUNHM, ANSP). *Upper Takutu-Upper Essequibo* Vicinity of Nappi village, 03°25'N, 59°34'W (ML 90671/92, D.J. Kerr); upper Rewa River, 02°58'N, 58°35'W (UNSM 637111); near Brian's Landing, 02°57'N, 58°58'W (ML 98686, D.W. Finch); Cacique Mountain, 03°11'N, 58°48'W (Finch *et al.* 2002); Parabara Savanna, 02°12'N, 59°22'W (Robbins *et al.* 2004, KUNHM 90849); Kassikaityu River, 01°49'N, 58°49'W (ML 117947, D.W. Finch); 10km SE of Gunn's Landing, 01°35'N, 58°34'W (USNM 625539); Kamoa River, 01°32'N, 58°50'W (Robinson *et al.* 2007, O'Shea 2013); north side of Acari Mountains, 01°23'N, 58°56'W (Robinson *et al.* 2007, O'Shea 2013, KUNHM 90356). *East Berbice-Corentyne* Itabu Creek Head, 01°33'N, 58°10'W (Blake 1950, FMNH 120238). **FRENCH GUIANA**: *St. Laurent du Maroni* km12 on Piste de la Crique Dardanelles, 05°21.5'N, 53°33'W (XC 122497/98, J. King); Tamanoir, 05°09'N, 53°45'W (CM, ANSP, YPM, UMMZ); Saül, 03°37'N, 53°12'W (Dick *et al.* 1984, Thiollay 1986, XC 213807, P. Ingremeau, ROM, KUNHM). *Cayenne* km30 on Piste de la Crique Dardanelles, 05°17.5'N, 53°29'W (XC 122492–494, J. King); Crique Nancibo, 04°37'N, 52°30.5'W (XC 128677, A. Renaudier); Montagne de Kaw, 04°34'N, 52°12'W (Thiollay 1999b); Reserve Naturelle des Nouragues, 04°05'N, 52°41'W (Thiollay 1994, 1999a, Jullien & Thiollay 1998, Thiollay & Jullien 1998, XC 44134, A. Renaudier); Fleuve Oyapock, *c*.03°53.5'N, 51°48.5'W (YPM 29657/58); Pied Saut, *c*.03°44'N, 51°56'W (long series CM); Pic du Croissant, 03°33'N, 52°15'W (Thiollay 1986); Mont Belvédère, 02°25'N, 53°06'W (Thiollay 1986). **SURINAM**: *Brokopondo* Brownsberg Natuurpark, *c*.04°57'N, 55°11'W (Donahue 1985, ML, XC). *Sipaliwini* Rudy Kappel Airstrip, 03°48'N, 56°09'W (Zyskowski *et al.* 2011); Wilhelmina Mountains, 03°26'N, 56°45'W (Blake 1963, FMNH 264497); Mozeskreek area, 04°51'N, 56°46'W (XC 272178, P. Boesman); Kayserberg Airstrip, 03°07'N, 56°27'W (Blake 1961, 1963, FMNH 260494–497); Palumeu, 03°21'N, 55°26'W (Haverschmidt & Mees 1994); SSE of Palumeu, 03°19'N, 55°26'W (XC 75015, D.F. Lane); confluence of Makrutu Creek/ Palumeu River, 02°47.5'N, 55°22'W (O'Shea & Ramcharan 2013a); Grensgebergte Rock, 02°28'N, 55°46'W (O'Shea & Ramcharan 2013a); Werehpai, 02°22'N, 56°42'W and Sipaliwini River, 02°17.5'N, 56°36.5'W and Kutari River, 02°10.5'N, 56°47'W (O'Shea & Ramcharan 2013b). **BRAZIL**: *Amapá* PN Montanhas do Tumucumaque, 02°11.5'N, 54°35.5'W (Coltro 2008); Rio Anacuí, 01°50.5'N, 52°44.5'W (Coltro 2008); Serra do Navio, 01°38'N, 52°17'W (MNRJ 29429, USNM 515628, WA 1224387, J.A. Alves); Calçoene, 01°25'N, 51°55.5'W (MPEG 21235); Cachoeira Belheira, 01°15.5'N, 49°56'W (MPEG 21172); Igarapé Capivara, 01°02'N, 51°44'W (MPEG 21181); Areia Vermelha 01°00.5'N, 51°39'W (Novaes 1974, MPEG 20427); Rio Falsino, 00°56'N, 51°35.5'W (Novaes 1974, MPEG 20428, LSUMZ 67361); Igarapé Novo, 00°30'N, 52°30'W (Novaes 1974, MPEG 29257); Cachoeira Amapá, 00°02'N, 51°55'W (Novaes 1974, MPEG 28744); RESEX Rio Cajari, 00°35'S, 52°16'W (Schunck *et al.* 2011). *Roraima* RE Xixuaú-Xipariná, 00°48'S, 61°33'W (Trolle & Walther 2004). *Pará* ESEC Grão-Pará Centre, 00°37'N, 55°43'W (Aleixo *et al.* 2011); Alenquer, 00°09'S, 55°11'W (Aleixo *et al.* 2011, MPEG

66053); *c.*140km N of Óbidos, 00°38'S, 55°43.5'W (MPEG 66675–678); Almeirim, 00°49'S, 53°55'W (Aleixo *et al.* 2011, MPEG 66340); FLOTA Trombetas, 00°57'S, 55°31'W (Aleixo *et al.* 2011); FLOTA Faro, 01°42'S, 57°12'W (MPEG 64739); **Óbidos,** 01°54'S, 55°31'W (Griscom & Greenway 1941, CM P83836). *Amazonas* Balbina Dam, 01°55'S, 59°28'W (Willis & Oniki 1988); Presidente Figueiredo, 02°01'S, 60°01'W (WA: A. Carvalho, J. Specian); BDFFP study plots, *c.*02°20'S, 60°00'W (Bierregaard 1988, Cohn-Haft *et al.* 1997, Stouffer *et al.* 2006, 2009); REBIO Cuieiras, *c.*02°34'S, 60°06'W (IBC 1193130, video J. del Hoyo); 11.5km SW of Dormida, 02°35'S, 60°12'W (ML 127371, C.A. Marantz); Ramal do Pau Rosa, 02°47.5'S, 60°02.5'W (ML 58167091, photo P. Diniz); Reserva Forestal Adolpho Ducke, *c.*02°58'S, 59°55.5'W (Willis 1977).

Hylopezus macularius dilutus (Hellmayr, 1910), Novitates Zoologicae, vol. 17, p. 370. Described as *Grallaria macularia diluta*, the holotype is an adult male (NMW 16440), collected 6 December 1830 by J. Natterer along the upper Rio Negro, 'below Thomar' (= Tomar). As Natterer apparently left Tomar and headed upstream to Santa Isabel do Rio Negro around this date (Stephens & Traylor 1993b), the type locality should probably be amended to above Tomar (= *c.*00°22.5'S, 64°04'W, Amazonas, Brazil). Carneiro *et al.* (2012) found strong support for the recognition of *dilutus* as a taxon separate from the rest of the *macularius* complex (including *H. paraensis* and *H. whittakeri*), but were unable to resolve its relationship to these other taxa. They recommended recognising it as a species, because it appeared at least as divergent from nominate *macularius* as from the *paraensis–whittakeri* clade. This subspecies is distributed north of the Amazon from the west banks of the Rios Negro and Branco in Amazonas (Brazil) through southern Venezuela (Amazonas), southern Colombia, eastern Ecuador and northern Peru west of the Río Ucayali (Hilty & Brown 1986, Ridgely & Tudor 2009, Schulenberg *et al.* 2010). Recent records from Colombia's Dept. Meta (Ramírez *et al.* in press), although not unexpected, represent an important range extension suggesting that Spotted Antpitta probably occurs throughout most of the Colombian Amazon, despite the lack of records. It is also worth noting that there are apparently no records in the Rio Jufari/Rio Branco interfluvium, north of where these rivers join the Rio Negro. Although its presence there would not be surprising, its absence from this area (western Roraima) is reflected in the map, provisionally showing the right (western) bank of the lower Rio Jufari as the north-eastern limit of the range of race *dilutus*. The single record of 'Spotted Antpitta' west of the Rio Madeira and south of the Amazon, from where there are no other records of this complex, on the upper Rio Urucu (Peres & Whittaker 1991, Whittaker *et al.* 2008) could either represent Alta Floresta Antpitta or the present taxon. Hellmayr (1910) described it as similar to the nominate race but with a slightly longer tail and much longer wings. The sides and flanks are overall paler than nominate *macularius*, being dull ochraceous-yellow with a slight olivaceous wash (not deep ochraceous). Carneiro *et al.* (2012) elaborated by noting the brownish rather than greenish-olivaceous back of *dilutus* with little or none of the pale shaft-streaks of *macularius*. They further showed that, comparatively, *dilutus* has shorter legs and a heavier bill (broader and deeper). Bartlett (1882) stated 'iris yellow', but it is unclear why this might have been true of the specimens he

received. **SPECIMENS & RECORDS VENEZUELA**: *Amazonas* Puerto Yapacana, 03°40'N, 66°52'W (COP 39363–374); Caño León, 03°25'N, 65°40'W (Zimmer 1934, AMNH 270916/19); El Merey, 03°04'N, 65°55'W (Zimmer 1934, AMNH 432820–825); Caño Duratamoní, 02°47'N, 65°59'W (AMNH 270915/18); mouth of Río Ocamo on Río Orinoco, 02°48'N, 65°13'W (Zimmer 1934, AMNH 432818/19); Rio Vaciva, below mouth on Brazo Casiquiare, 02°23'N, 66°25'W (Friedmann 1948, USNM 327112). **BRAZIL**: *Amazonas* Maraã, Lago Cumapi, 01°43.5'S, 65°53'W (FMNH 457233, MPEG 62966); Maguari, 01°50.5'S, 65°24'W (MPEG 42749/50); Terra Verde Lodge, 03°08'S, 60°28'W (ML 112792, C.A. Marantz). **COLOMBIA**: *Meta* Vereda Playa Nueva, 02°35.5'N, 72°39'W (Ramírez *et al.* in press). *Amazonas* Río Loretoyacu, NW of Leticia, *c.*03°44'S, 70°11'W (Bartlett 1882). **PERU**: *Loreto* 'Campamento Medio Algodón,' 02°35.5'S, 72°53.5'W (Stotz *et al.* 2016); Cocha Aguila, 02°42.5'S, 70°33'W (eBird: D.F. Stotz); 'Campamento Alto Cotuhé', 03°12'S, 70°54'W (Stotz & Alván 2011); ExplorNapo Lodge, 03°16'S, 72°56'W (XC 102595, F. Schmitt); Sabalillo, 03°22'S, 72°17'W (XC 20058, D. Edwards); Río Mazán, 03°28'S, 73°11'W (AMNH 407154); Puerto Indiana, 03°28'S, 73°03'W (Zimmer 1934); 'Middle Varillal campsite', 03°41'S, 74°14'W (S.J. Socolar *in litt.* 2017); Iquitos, 03°44'S, 73°15'W (Zimmer 1934); RN Allpahuayo-Mishana, *c.*03°55'S, 73°33'W (Alonso *et al.* 2012).

STATUS Not currently considered globally threatened (BirdLife International 2017). A bird of mature forest, Spotted Antpitta is one of many forest understorey insectivores that quickly disappears from small fragments (Bierregaard & Lovejoy 1989). It appears that a fair number of populations of both races occur within conservation units. **PROTECTED POPULATIONS** *H. m. macularius* **Venezuela** RF Imataca (Mason 1996, ML); Reserva El Dorado (Krabbe & Schulenberg 2003). **Guyana** Iwokrama Forest Reserve (KUNHM, ANSP, ML, XC). **Surinam** Brownsberg Natuurpark (Donahue 1985, ML, XC); Raleigh Vallen-Voltzberg Natuurpark (A. Farnsworth *in litt.* 2017). **French Guiana** Reserve Naturelle des Nouragues (Thiollay 1994, XC 44134, A. Renaudier). **Brazil** PN Montanhas do Tumucumaque (Coltro 2008); Reserva Forestal Adolpho Ducke (Willis 1977); RE Xixuaú-Xipariná (Trolle & Walther 2004); FLOTA Trombetas and FLOTA Faro (Aleixo *et al.* 2011); RESEX Rio Cajari (Schunck *et al.* 2011); ESEC Grão-Pará (Aleixo *et al.* 2011, MPEG 66053); REBIO Maicuru (Aleixo *et al.* 2011, MPEG 66340); REBIO Cuieiras (IBC 1193130, video J. del Hoyo). *H. m. dilutus* **Brazil** PN Jaú (Borges & Carvalhães 2000, Borges *et al.* 2001, Borges 2007). **Colombia** PNN Amacayacu (Krabbe & Schulenberg 2003). **Peru** RN Pacaya-Samiria (Krabbe & Schulenberg 2003); RN Allpahuayo-Mishana (Alonso *et al.* 2012); ÁCP Saballilo (S.J. Socolar *in litt.* 2017).

OTHER NAMES *Hylopezas macularia macconelli*, Chubb 1918, Bulletin of the British Ornithologists' Club, vol. 38, p. 86, Ituribisi River (Pomeroon-Supenaam, Guayana, *c.*07°05'N, 58°29'W following Stephens & Traylor 1985); note misspelling of *Hylopezus*. Chubb (1918b) considered the single specimen before him, named for its collector, to differ from nominate *macularius* in having the crown and nape paler grey, the back, innermost secondaries and tail olive-green (not ochraceous-brown), much paler flanks, and the undertail-coverts almost all white (versus ferruginous). The specimen (NHMUK 1922.3.5.2373) is in Tring (Warren & Harrison 1971). Subspecies *macconnelli* was soon synonymised with the nominate race by Cory & Hellmayr (1924). Other scientific synonyms: *Myioturdus macularius* (*macularius*, Lafresnaye 1838); *Calobathris macularia* (*macularius*, Schomburgk 1848); *Grallaria macularia* (*macularius*, Lafresnaye 1842, Lesson 1844, Burmeister 1856, Gray 1869, Dubois 1900, Willis 1977, Feduccia & Olson 1982); *Grallaria macularia* (*dilutus*, Bartlett 1882); *Grallaria macularia diluta* (Hellmayr 1910); *Pitta macularia* (*macularius*; Temminck & Laugier 1820–39, Lesson 1831); *Grallaria macularia diversa* (*dilutus*; Zimmer 1934, Dugand & Borrero 1946, Friedmann 1948, Phelps & Phelps 1950, Meyer de Schauensee 1950b); *Grallaria macularia macularia* (*macularius*; Zimmer 1934, Blake 1950, 1961, Phelps & Phelps 1950, Haverschmidt 1968, Novaes 1974); *Hylopezus macularius diversus* (*dilutus*; Naka 2010); *Hylopezus macularia macconnelli* (*macularius*; Chubb 1921; note the correct spelling of genus but different spelling of his own subspecific name!); *Hylopezus macularius paraensis* (*dilutus*, Pinto 1978). **English** Spot-breasted Ant-Thrush (both races, Brabourne & Chubb 1912); Zimmer's Antpitta (*dilutus*, Carniero *et al.* 2012); Western Spotted Ant-pitta (*dilutus*, Meyer de Schauensee 1950b); McConnell's Spotted-breasted Ant-Thrush (*macularius*, Chubb 1921), Napo Spotted Ant-pitta (*dilutus*; Friedmann 1984). **French** Grallaire Tachetée (*macularius*, Tostain 1986, Haverschmidt & Mees 1994; both races, Krabbe & Schulenberg 2003); Brève Moucheté (*macularius*, Temminck & Laugier 1820–39). **Spanish** Hormiguero Pichón Punteado (Phelps & Meyer de Schauensee 1979, 1994, Frisch & Frisch 1981, Hilty 2003, Verea *et al.* 2017); Tororoí Moteado (both races, Krabbe & Schulenberg 2003); Pichón Pechipunteado Guayanés (*macularius*), Pichón Pechipunteado Peruano (*diversus*) (Phelps & Phelps 1950). **Portuguese** torom-carijó (both races, Willis & Oniki 1991b, CBRO 2011); torom-torom-pintalgado (both races, Frisch & Frisch 1981); torom-do-imeri (*dilutus*, WA); Tovaquinha (Novaes 1974). **German** Östliche Brillenameisenpitta (both races, Krabbe & Schulenberg 2003). **Dutch** Roodflankmierpitta (*macularius*; Spaans *et al.* 2016).

Spotted Antpitta, adult (*macularius*), Manaus, Amazonas, Brazil, 7 September 2015 (*João Quental*).

Spotted Antpitta, adult (*macularius*), Manaus, Amazonas, Brazil, 1 August 2015 (*Robson Czaban*).

Spotted Antpitta, adult (*macularius*), Manaus, Amazonas, Brazil, 1 August 2015 (*Robson Czaban*).

ALTA FLORESTA ANTPITTA
Hylopezus whittakeri   Plate 19

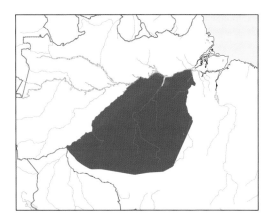

Hylopezus whittakeri Carneiro *et al.*, 2012, Auk, vol. 121, p. 348, Figueiredo in Belterra, FLONA do Tapajós, Sucupira base, km 117 of the BR 163 highway, 03°21'22"S, 54°56'57"W. The holotype of this newly recognised species is an adult male (MPEG 56099) collected 23 July 2002 by D. Davison, W. Figueiredo and L.W. Figueiredo, and is named in honour of Andrew Whittaker for his numerous contributions to Amazonian ornithology.

The distinctiveness of Alta Floresta Antpitta, especially from Spotted Antpitta north of the Amazon, has been recognised since at least since 1989 when Ted Parker first visited the Alta Floresta region (K. J. Zimmer in Carneiro & Aleixo 2013). Nevertheless, no formal attempt was made to describe this population, largely because it was assumed that it represented Snethlage's Antpitta, for which the song was not known until recently. More recently, while investigating the taxonomic complexity of the Spotted Antpitta group for his M.Sc., Carneiro (2009) recognised the distinctiveness of birds in the Rio Madeira–Xingu interfluvium, assigning the name *H. oreni* to these birds. This name was not published, however, and *H. whittakeri* became the official name for this population in 2012. Following Carneiro *et al.*'s (2012) thorough systematic revision, four main lineages of Spotted Antpitta were recovered, three of which corresponded to already named taxa (*dilutus*, *macularius* and *paraensis*), and the fourth to what is now known as Alta Floresta Antpitta. Details of the natural history and breeding biology of Alta Floresta Antpitta remain almost entirely unknown, and there remains much to learn concerning its vocal repertoire and distribution. Although it has not been evaluated from a conservation standpoint, its known range is not particularly small (being larger than that of Snethlage's Antpitta) and it does not face any well-defined threat not shared by other Amazonian birds.

IDENTIFICATION 14cm. Long-legged and very short-tailed, with typical upright posture of *Hylopezus*. Both sexes have dark grey crowns and dull greenish-brown upperparts and wings, the latter with a prominent dark rectangle on the leading edge near the top and two buffy wingbars broken into spots. The buffy-orange lores are separated from a similarly coloured and fairly prominent eye-ring by a narrow black crescent. The pale underparts, white centrally and buffy laterally, have distinctive black spotting, heaviest on the upper breast and becoming sparser and smaller lower down. Overall, very similar to Spotted and Snethlage's Antpittas. Probably not safely separated from either in the field except by range (no overlap) and by voice.

DISTRIBUTION Endemic to south-central Brazil (Amazonia), confined to the Rio Madeira–Rio Xingu interfluvium (Carneiro *et al.* 2012, Fjeldså 2017). At many of the following localities it was historically known as Spotted Antpitta. The single record of 'Spotted Antpitta' from west of the Rio Madeira and south of the Amazon, on the the upper Rio Urucu (Peres & Whittaker 1991, Whittaker *et al.* 2008) may or may not represent Alta Floresta Antpitta.

HABITAT Inhabits humid lowland forest, apparently most frequently in swampy or flooded areas in upland *terra firme* forest (Stotz *et al.* 1996), often around gaps such as treefalls (Wunderle *et al.* 2006). Also found in drier transitional forest at the southern limit of its range in northern Mato Grosso (Lees *et al.* 2008). Though not well studied, appears to show an affinity for treefall gaps and riparian areas, and rarely ventures into more disturbed areas (Carneiro *et al.* 2012).

VOICE Distinguished from all other members of the Spotted Antpitta complex by a loud-song normally composed of five (rarely four or six) whistles with identical tones, in which the second and third notes are separated by an unusually long interval (Carneiro *et al.* 2012). In particular, the song differs from Spotted Antpitta (race *dilutus*) in length of the second, third and fourth intervals.

NATURAL HISTORY Of the (relatively) many published observations on the behaviour of Spotted Antpitta, none has been specifically in reference to Alta Floresta Antpitta. Its natural history, however, can safely be assumed to be at least similar to that of other *Hylopezus*. Like related species, Alta Floresta Antpitta sings from the ground or from low perches, inflating its throat and exposing the pink skin below the feathers (ML 38268141, photo A. Van Norman).

DIET Nothing documented, but presumably similar to congeners, i.e. invertebrates and small vertebrates.

REPRODUCTION Nest and egg undescribed, along with all other aspects of the reproductive biology. **SEASONALITY** I found only two immatures, a subadult collected 26 July 1907 at Calama, and a subadult transitioning, 24 January 1920, at Vila Braga (YPM 29661; S.S. Snow photos).

TECHNICAL DESCRIPTION Sexes similar. *Adult* From Carneiro *et al.* (2012) and Fjeldså (2017). Top of head dark grey with sparse, poorly defined darker shaft-streaks, sides of crown with more prominent blackish shaft-streaks sharply separating grey cap from conspicuous buffy-yellow lores and broad yellow-ochre eye-ring. These buffy facial markings are bordered by a continuous narrow blackish crescent across lores and continuing as a long black moustachial stripe that contrasts with the whitish throat and centre of chin. Ear-coverts speckled blackish, greyish and pale buff, less speckled and with mainly pale ochre-buff feathers on the lower part. Mantle to uppertail olive-brown, sometimes washed slightly greyish; feathers of upper mantle and upper scapulars with narrow bright yellow-ochre shaft-streaks. Upperwing-coverts olive-brown with diffuse cinnamon-buff tips, broader on outer and reduced on inner ones, alula bicoloured, dark brownish on inner webs, orange-buff on outer webs, bend of wing buffy-orange. Primary-coverts dusky-brown contrasting with bright ochre or tawny patch at base of outer webs of primaries, otherwise primaries dark greyish on inner webs and olive-brown on outer webs, inner primaries and secondaries plainer olive-brown on outer webs, with minute orange-buff tips. Inner webs of tertials also plain olive-brown. Underwing-coverts tawny-buff to tawny-cinnamon. Underparts mostly creamy-white with buffy-cinnamon wash on sides and flanks, and blackish malar stripe. Feathers of breast fringed with black flecks, these markings at tips of feathers forming dense rows of inverted V-shaped markings, becoming much smaller and more spaced on lower breast, sides of belly and flanks. **Adult Bare Parts** *Iris* dark brown; *Bill* maxilla grey-black, mandible pinkish, darker greyish on distal third, rictal flanges orangish; *Tarsi & Toes* pale purple-pink with paler claws. **Subadult** Very similar to adult but fringes of upperwing-coverts retaining some rusty-rufous coloration. No further information.

MORPHOMETRIC DATA *Bill* from nares 14.2mm, 13.0mm, 12.8mm, 13.0mm; *Tarsus* 43.6mm, 34.4mm, 34.4mm, 36.2mm (*n* = 4, ♂♂♀♀, USNM, AMNH, LACM). *Wing* 89mm, n/m, 84mm; *Tail* 35mm, n/m, 36mm; *Bill* from nares 13.4mm, 11.8mm, 12.9mm, exposed culmen 19.6mm, 18.0mm, 19.2mm, depth at nares 5.6mm, 5.7mm, 5.6mm, width at nares 5.0mm, 5.0mm, 4.8mm, width base of mandible 8.2mm, 7.4mm, 7.9mm; *Tarsus* 37.8mm, 38.0mm, 36.8mm (*n* = 3, ♀?, FMNH). *Wing* 86mm; *Bill* from nares 11.5mm, exposed culmen 13.7mm; *Tarsus* 36.9mm (*n* = 1♂, MCZ 134957, J.C. Schmitt *in litt.* 2017). The following measurements are from Carneiro *et al.* (2012) [Mean (range)]: **Males** *Wing* 86.1mm (82.5–90.5mm); *Tail* 40.1mm (37.4–42.7mm); *Tarsus* 35.0mm (32.4–36.4mm); *Bill*, length 13.2mm (12.5–13.9mm), width 5.8mm (5.3–6.3mm), depth 5.8mm (5.5–5.9mm) (*n* = 10). **Females** *Wing* 85.2mm (83.9–86.9mm); *Tail* 40.8mm (36.6–44.6mm); *Tarsus* 34.4mm (33.8–34.7mm); *Bill* length 13.0mm (12.1–14.0mm), width 5.9mm (5.7–6.1mm), depth 5.8mm (5.7–5.9mm) (*n* = 3). **Sex unspecified** *Wing* 86.1mm (83.1–90.1mm); *Tail* 37.0mm (33.9–39.6mm); *Tarsus* 35.2mm (33.7–37.6mm); *Bill*, length 13.4mm (12.7–14.0mm), width 5.9mm (5.1–7.7mm), depth 6.2mm (5.6–6.7mm) (*n* = 10). **Mass Males** mean 44.1g (range 40–47 g) (n ≤ 10, Carneiro *et al.* 2012). **Females** mean 42.4g (*n* = 1, Carneiro *et al.* 2012). **Total Length** No data. Presumably similar to *H. macularius*.

DISTRIBUTION DATA Endemic to **BRAZIL**: *Pará* Cuçari, 01°53.5'S, 53°21'W (MPEG 4852); vicinity of Juruti, *c.*02°09'S, 56°05'W (Santos *et al.* 2011a); Limoãl, 02°14.5'S, 54°24.5'W (Zimmer 1934, AMNH 288617); Retiro, 02°23.5'S, 55°47'W (MPEG 66134); Santarém, 02°27'S, 54°42'W (Griscom & Greenway 1941, Lees *et al.* 2013a, MPEG, CM, UMMZ); Mojuí dos Campos, 02°42'S, 54°38'W (CM, YPM); Belterra, 03°04'S, 54°55.5'W (MPEG 47847); FLONA Tapajós, 03°17'S, 54°60'W (XC 90705, A.C. Lees); Lago Arauepá, 03°36'S, 55°20'W (LACM 31930); Fordlândia, 03°50'S, 55°30'W, (MZUSP 58838, paratype); Mirituba, 04°16'S, 55°54'W (Griscom & Greenway 1941, MCZ, CM); Vila Braga, 04°25'S, 56°17'W (Griscom & Greenway 1941, CM, YPM, USNM); Jacareacanga, 07°19'S, 57°28'W (WA 1010561, P. Cerqueira, WA 960624, recording G. Gonsioroski); confluence of Rios Jamanxim and Tapajós, 04°44'S, 56°25'W (Griscom & Greenway 1941); Jatobá, 05°04'S,

56°56'W (MPEG 75931); Cachoeira do Caí, 05°10'S, 56°27'W (Snethlage 1914a, MPEG 6476); Rio Ratão, 05°26'S, 56°54.5'W (MPEG 75919); near Rio Crepori/Rio Tapajós confluence, 05°46'S, 57°17.5'W (MPEG 76423); Comunidade São Martins, 06°06.5'S, 57°37'W (MPEG 75634); Rio São Benedito, 4km from confluence with Rio Teles Pires, 09°08'S, 57°03'W (MPEG 54706); 25km N of Alta Floresta, 09°24'S, 56°05'W (ML 48137/38, M.L. Isler). *Amazonas* Near Deus é Pai, 05°48'S, 59°15'W (ML 185519, G.F. Seeholzer); Humaitá, 07°33'S, 62°33'W (MPEG 58757, paratype); Pousada Rio Roosevelt, 08°29'S, 60°57'W (Whittaker 2009). *Rondônia* Calama, 08°01'S, 62°52'W (AMNH 492297); Machadinho d'Oeste, 08°45'S, 62°14.5'W (XC 167354, G. Leite); Jamarí, 08°45'S, 63°27'W (MNRJ 21895, paratype); Porto Velho area, 09°24'S, 64°26'W (XC 189546, L.G. Mazzoni); Cachoeira Nazaré, 09°44'S, 61°53'W (Stotz *et al.* 1997, FMNH, ML, MPEG 39819, paratype); Pedra Branca, 10°03'S, 62°07'W (Stotz *et al.* 1997, FMNH 344084); Pousada Ecológica Rancho Grande, 10°18'S, 62°52'W (R.C. Hoyer *in litt.* 2017); near Alvorada d'Oeste, *c.*11°24'S, 62°24'W (MPEG 38808, paratype). *Mato Grosso* Rio Teles Pires, *c.*72km NW of Alta Floresta, 09°24.5'S, 56°33.5'W (MPEG 69331); Fazenda Rio Paranaíta, 09°33.5'S, 56°42.5'W (MPEG 67501/502); Fazenda Aliança, 09°35'S, 56°43'W (MPEG 69333); RPPN Rio Cristalino, 09°52'S, 55°54'W (MPEG 51626, XC, ML); Alta Floresta area, *c.*09°53'S, 56°05'W (Zimmer *et al.* 1997); Rio Aripuanã, Cachoeira Dardanelos, 10°10'S, 58°37'W (MPEG 34420 paratype, MPEG 45606); Rondolândia, 10°27'S, 61°08'W (WA 2019207, recording G. Correa); Serra dos Caiabis, *c.*11°30'S, 56°30'W (Lees *et al.* 2008); Sinop, 11°51'S, 55°30'W (WA).

STATUS BirdLife International (2017) has not recognised Alta Floresta Antpitta as a species, and thus its conservation status has not been evaluated. Like other members of the Spotted Antpitta group, this species appears quite sensitive to habitat loss, fragmentation and perturbation (Lees & Peres 2006), given that it was found in only 25% of a sample of 31 variably sized (1.2–100,000 ha) forest patches in the Alta Floresta region, northern Mato Grosso, where the smallest occupied patch was 19ha (Lees & Peres 2010, A.C. Lees in Carneiro *et al.* 2012). It is also thought to be fairly sensitive to edges created by roads (Laurance 2004). The species' range is not particularly small and, thankfully, it occurs in several protected areas. **Protected Populations** FLONA Tapajós (Henriques *et al.* 2003, 2008, Wunderle *et al.* 2006, XC, ML); PN Campos Amazônicos (eBird: J. Dickens); FLOTA Rio Preto (XC 189672, L.G. Mazzoni); Pousada Ecológica Rancho Grande (R.C. Hoyer *in litt.* 2017); RPPN Rio Cristalino (MPEG 51626, XC, IBC, ML).

OTHER NAMES *Grallaria macularia paraensis* (Snethlage 1914a, Zimmer 1934, Pinto 1938, Griscom & Greenway 1941), Hellmayr (1929) used this name in reference to either or both *H. paraensis* and *H. whittakeri*. *Hylopezus macularius* (Zimmer *et al.* 1997, Henriques *et al.* 2003, 2008, Lees *et al.* 2013a), *Hylopezus macularius paraensis* (Pinto 1978, Clements 2007), *Hylopezus oreni* (Carniero 2009). The following references treat Spotted Antpitta (*sensu lato*, pre-recognition of *whittakeri*) and include distributional or other data pertaining to *H. whittakeri* (Meyer de Schauensee 1970, Howard & Moore 1984, Dunning 1987, Sibley & Monroe 1990, van Perlo 2009, Ridgely & Tudor 2009). **French** Grallaire de Whittaker. **German** Alta-Floresta-Brillenameisenpitta **Spanish** Tororoí de Alta Floresta. **Portuguese** torom-de-Alta-Floresta (WA).

Alta Floresta Antpitta, adult, Alta Floresta, Mato Grosso, Brazil, 13 August 2016 (*João Quental*).

Alta Floresta Antpitta, adult, Rio Cristallino, Matto Grosso, Brazil, 20 September 2011 (*Rich Hoyer*).

SNETHLAGE'S ANTPITTA
Hylopezus paraensis   Plate 19

Grallaria macularia paraensis Snethlage, 1910, Ornithologische Monatsberichte, vol. 18, p. 192, Ourém, Rio Guamá, Pará, Brazil [01°32'S, 47°07'W]. The holotype (MPEG 3272), an adult male, was collected 5 December 1903. Although Snethlage (1907b) initially described Snethlage's Antpitta as *Grallaria macularia berlepschi* in 1907, Hellmayr (1910) pointed out that *berlepschi* was a pre-occupied name within the genus, thus Snethlage (1910) assigned the new name *paraensis* to it, making 1910 the official date of description.

Snethlage's Antpitta, following the recent reorganisation of the Spotted Antpitta complex (Carneiro *et al.* 2012), is the most range-restricted of the recognised taxa, found only in north-east Pará and westernmost Maranhão. Like its relatives it is a poorly known, shy denizen of the forest understorey that is difficult to observe, even when detected by voice. Now that its true distinctiveness is recognised, studies are sorely needed of its biology and natural history.

IDENTIFICATION 13–14cm. Both sexes have a grey to olive-grey crown and nape, buffy-white lores and distinct ochraceous eye-ring. Rest of upperparts olive-brown with indistinct, sparse, thin pale streaking. Wings olive-brown with two rows of tawny-buff spots on upperwing-coverts and a small, well-defined tawny patch at base of primaries. Throat white with a thin black malar stripe. Rest of underparts white, washed buff across breast and on sides and flanks, with short black streaks on breast and sides. Bill dark, paler near base. Snethlage's Antpitta is probably most similar to western populations of Spotted Antpitta, but tends to be greyer above and slightly more streaked. Overall, however, separation of Snethlage's Antpitta from Alta Floresta and Spotted Antpittas is best done by vocalisation or range, as plumage differences are subtle and imprecisely defined.

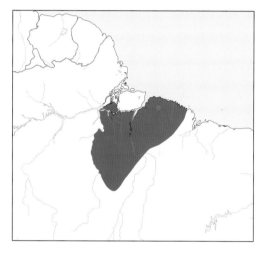

DISTRIBUTION Endemic to north-east Brazil, south of the Amazon from the Rio Xingu east in the state of Pará. Its range also extends slightly, but to an unknown degree, into western Maranhão (Oren 1990, 1991, Snethlage 1914a, Pacheco *et al.* 2007). The records and specimens reported below, under a variety of names, are all here considered to represent *H. paraensis*.

HABITAT Insufficiently well known to determine if this differs from related species. Like these, it inhabits the undergrowth of mature *terra firme* forest, probably preferring areas of dense understorey such as near treefalls and riparian areas. Apparently restricted to altitudes below 500m.

VOICE Krabbe & Schulenberg (2003) described the song as a decelerating series of 5–6 similar notes lasting about 3s, delivered at *c.*0.8kHz, sometimes falling slightly in pitch. It is usually repeated at intervals of 10–20s.

NATURAL HISTORY Due to past confusion in the literature with related species, there are no data specifically published concerning behaviour or other aspects of the species' natural history. It is undoubtedly similar to that of Spotted Antpitta. The pink skin below the feathers of the throat is often exposed while vocalising, as in other *Hylopezus* (see WA 1798450; A.C. Guerra).

DIET No data.

REPRODUCTION Virtually undocumented, and nest and eggs are undescribed. Undoubtedly its reproductive biology will prove similar to that of others in the 'macularius' species complex. Pinto (1953) reported a female with a well-formed egg in the oviduct collected 9 February 1929 at Belém (as 'Utinga' see Paynter & Traylor 1991a). No further information.

TECHNICAL DESCRIPTION Sexes alike. **Adult** Grey crown washed olive extends onto nape. Thin black line separates distinct buffy-ochraceous eye-ring from a buffy-white loral spot. This black line extends below eye to form a malar stripe separating white throat from neck-sides which, like the ear-coverts, are olive-brown indistinctly streaked and mottled black and buff. Upperparts grade to olive-brown from the olive-grey crown and, mostly in the scapulars, are sparsely marked with thin, pale buff streaks. Wings mostly olive-brown but primary wing-coverts blackish and contrast with tawny patch formed by pale bases to primaries, except two outermost. Lesser and median upperwing-coverts broadly tipped with tawny-buff spots, forming a pair of irregular wingbars. Throat white with a thin black submalar stripe below the malar stripe. Rest of underparts white, washed pale, rather dull buff or ochraceous-yellow across breast, slightly washed olive on sides and flanks, and marked with black spots forming short black streaks on breast, breaking into small spots on flanks. **Adult Bare Parts** *Iris* dark brown; *Bill* maxilla dark grey to blackish, mandible blackish at tip, pale grey to pinkish basally; *Tarsi & Toes* dusky-pink or pinkish-brown. **Immature** Undescribed, but probably very similar to *H. macularius*.

MORPHOMETRIC DATA *Wing* 95mm; *Tail* 34mm; *Bill* [exposed culmen?] 20mm; *Tarsus* 35mm (*n* = ?, Snethlage 1914a). *Wing* 87mm, 88mm, 88mm, 89mm, 90mm; *Tail* range 35–37mm; *Bill* 19–20mm (*n* = 5, ♀?, Cory & Hellmayr 1924). *Bill* from nares 13.6mm, n/m, n/m; *Tarsus* 36.6mm, 37.5mm, 37.5mm (*n* = 3♂♂, LACM). *Wing* 86.0mm; *Tail* 31.4mm; *Bill* total culmen 22.0mm; *Tarsus* 31.7mm (*n* = 1♂, Novaes & Lima 1998). *Wing* 83.6mm (81.1–86.8), 81.4mm (80.8–87.0), 83.9mm (81.1–87,0), [83.4mm]; *Tail* 38.5mm (34.1–39.2), 38.8mm (34.9–40.8), 36.3mm (34.1–38.6), [37.4mm]; *Bill* from nares 13.6mm (12.2–14.0), 12.3mm (12.3–14.2), 15.4mm (12.6–14.2), [14.6mm], width at front of nares 5.5mm (5.0–6.0), 5.2mm (5.2–6.4), 5.6mm (5.4–5.9), [5.5mm], depth at front of nares 5.3mm (5.1–5.8), 5.6mm (5.1–5.9), 5.8mm (5.7–6.0), [5.7mm] (*n* = 16, 4♂♂, 3♀♀, 9♂?, mean (range), [overall mean], Carniero *et al.* 2012). **Mass** Males, mean 43g, range 42–44g, females, mean 45.2g, range 42–48.4 (*n* = 2–4♂♂, 2–3♀♀; Carniero *et al.* 2012). **Total Length** No data. Presumably similar to *H. macularius*.

DISTRIBUTION DATA Endemic to **BRAZIL**: *Pará* Fazenda Santa Bárbara, 01°22'S, 46°02'W (MPEG 36922); Belém area, 01°26'S, 48°26'W (Pinto 1953); Alto Igarapé Pedral, 01°33'S, 47°06'W (MPEG 32407/08); Barcarena,

01°34'S, 48°30'W (MPEG 1670/71); Rio Acará, 01°40'S, 48°25'W (Snethlage 1914a, Griscom & Greenway 1941: GPS from Paynter & Traylor 1991 as general loc. for lower portion of river); Estação Cientifica Ferreira Penna, 01°44.5'S, 51°27.5'W (ML 113105/117/118, C.A. Marantz); Aurora do Pará, 02°12.5'S, 47°33.5'W (MPEG 15940); Senador José Porfírio, 02°35'S, 51°57'W (WA 2078011, recording T. Junqueira); Paragominas, 03°00'S, 47°21'W (Lees *et al.* 2012); Belo Monte, 03°05'S, 51°46'W (XC 18925, S. Dantas); Caracol, 03°27.5'S, 51°40.5'W (MPEG 64919/65326); Anapu, 03°28'S, 51°12'W (WA 1428572, V. Castro); Rio Xingu, right bank *c.*70km downstream from Altamira, 03°33'S, 51°44'W (MPEG 55691); Itapuama, 'área 1', 03°36.5'S, 52°20.5'W (MPEG 63448/449); Fazenda Rio Capim, 03°38.5'S, 48°31'W (MPEG 58982); Vila Temporária I, 03°45.5'S, 49°40.5'W (MPEG 36238); *c.*80km E of Goianésia do Pará, 03°49'S, 48°23'W (XC 86163, N. Moura); Goianésia do Pará, 03°50'S, 48°45.5'W (XC 155520, A. Lees); Pacajá, 03°51'S, 50°43'W (XC 212976, G. Leite); Dom Eliseu, 04°07.5'S, 47°33.5'W (MPEG 18127); Fazenda Sr. Zé Mário, 04°20'S, 51°11.5'W (MPEG 78312); Marabá, 05°23'S, 49°08'W (WA 01499996, K. Okada); Manganês, 06°09'S, 50°24'W (MPEG 37250); Parauapebas, 06°12'S, 49°58'W (XC 213250, G. Leite); Mina do Sossego, 06°26'S, 50°02.5'W (MPEG 72288); São Félix do Xingu, *c.*06°38'S, 51°59'W (Galvão & Gonzaga 2011, MPEG A05776); Rio Fresco, near Gorotire, 07°46.5'S, 51°07.5'W (MPEG 37123). *Maranhão* Carutapera, 01°21'S, 46°01'W (Galvão & Gonzaga 2011, MPEG 36922); REBIO Gurupi, 03°42'S, 46°45.5'W (eBird: J. Weckstein).

TAXONOMY AND VARIATION Monotypic. Carneiro *et al.* (2012) found two diagnostic loud-song variables that distinguish Snethlage's Antpitta from both subspecies of Spotted Antpitta and four variables distinguishing it from Alta Floresta Antpitta. Morphologically, however, the present species is difficult to separate from either of these species, except by slight differences in the colour of the underparts and bill measurements. For further details see the accounts for Spotted and Alta Floresta Antpittas.

STATUS The conservation status of Snethlage's Antpitta has not been formally evaluated since being separated from Alta Floresta Antpitta, but when considered a subspecies of Spotted Antpitta it was evaluated as Least Concern (BirdLife International 2017). As the most range-restricted of the newly recognised species within the Spotted Antpitta complex, however, it certainly merits a re-evaluation of its conservation status. **Protected Populations** FLONA Serra dos Carajás (Pacheco *et al.* 2007, IBC 1232987, video C. Albano); FLONA Caxiuanã (ML: C.A. Marantz); REBIO Gurupi (eBird: J. Weckstein).

OTHER NAMES *Grallaria macularia berlepschi* (Snethlage 1907b, 1909); *Grallaria macularia* (Snethlage 1907c); *Grallaria macularia diluta* (Hellmary 1910); *Grallaria macularia paraensis* (Snethlage 1910, 1914a, Cory & Hellmayr 1924); *Hylopezus macularius* (Pacheco *et al.* 2007, Galvão & Gonzaga 2011); Hellmayr (1929) used *Hylopezus macularius* to refer to either or both *H. paraensis* and *H. whittakeri*; *Grallaria macularia paraensis* (all new specimens discussed belong to *H. whittakeri*; Griscom & Greenway 1941); *Grallaria macularia paraensis* (includes *whittakeri* and *H. macularius dilutus*; Peters 1951); *Grallaria macularia paraensis* (= just *paraensis*; Hellmayr 1912, Pinto 1953).

Hylopezus macularius (includes all; Ridgway 1911). *Grallaria macularia paraensis* (includes *whittakeri* and *paraensis*; Snethlage 1914a); *Hylopezus macularius* (Lees *et al.* 2012); *Grallaria paraensis* (Brabourne & Chubb 1912). **English**

Para Ant-Thrush (Brabourne & Chubb 1912); Amazonian Antpitta (Cory & Hellmayr 1924). **Portuguese** torom-do-Pará (WikiAves); torom-carijó (Oren 1991). **Spanish** Tororoí de Snethlage.

Snethlage's Antpitta *Hylopezus paraensis*. Adult, FLONA Carajás, Pará, Brazil, 21 June 2015 (*Ciro Albano*).

MASKED ANTPITTA
Hylopezus auricularis   Plate 19

Grallaria auricularis Gyldenstolpe, 1941, Arkiv för Zoologi, vol. 13b, no. 13, p. 7, Victoria, at confluence of Madre de Dios and Beni rivers, Madre de Dios, Bolivia [*c.*125m, 10°59'S, 66°08'W].

Formerly treated as a subspecies of Spotted Antpitta (Gyldenstolpe 1945b, Peters 1951, Phelps & Phelps 1950, Meyer de Schauensee 1970), Masked Antpitta is a poorly known, and apparently genuinely rare, bird endemic to northern Bolivia. BirdLife International (2017) considers the species to be Vulnerable, as it is known from only a few localities in south-east Pando and northern Beni departments. It appears to favour low-stature, dense second growth, including thickets. The plumage is characterised by a grey crown, dark mask, brownish-olive upperparts, creamy-white breast streaked black, and warm buff undertail-coverts. Very little has been published on the behaviour and natural history of Masked Antpitta, but recent studies have provided the first detailed information, which suggests that its habits are similar to congeners. The species' precise distribution and ecological requirements are very poorly known and require more research. Field surveys of all suitable habitats in northern Bolivia should be undertaken, and more studies of its diet and reproductive ecology are warranted.

IDENTIFICATION 14cm. A typical *Hylopezus* and easily recognised as such by its posture, shape and behaviour. Crown and nape grey, lores white, and ear-coverts and

ocular ring red-brown, bordered below by black. Apart from the crown, upperparts and tail are olive-brown. Wings brown with primaries edged tawny-ochraceous. Throat white edged with a black or dusky-black lateral throat stripe. White breast washed buff at top, boldly streaked black and olive-brown. Belly clear white, flanks and vent buff (Krabbe & Schulenberg 2003). The blackish 'mask' and better-streaked breast of Masked Antpitta separate it from *H. macularius*, with which it was formerly considered conspecific. There is, however, no range overlap between them. Within its range, Masked Antpitta is unlikely to be confused with any other species, as it is not known to occur with any other *Hylopezus*. The range of Amazonian Antpitta may approach that of Masked but Amazonian lacks the blackish 'mask', has an olive-brown (not grey) crown, and a buffier breast.

DISTRIBUTION Endemic to northern Bolivia, known from only half a dozen localities in Beni and Pando departments, with most records near the town of Riberalta (Krabbe & Schulenberg 2003, Maillard 2009) and all records within an altitudinal range of *c.*100–200m (Hennessey *et al.* 2003b). The range map here is defined by all records, both current and historical. It should be noted however, that within this polygon are localities where Masked Antpitta has been searched for and not found. See Distribution Data for a list of localities, both recent and historical.

HABITAT Inhabits dense, tangled, wet second growth adjacent to open areas, with some affinity for low-lying muddy areas (Maijer 1998, 1999, Maillard 2012, Aponte *et al.* 2017).

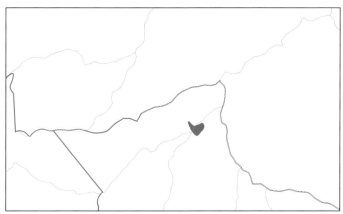

VOICE The song is a slow, trilled, slightly descending series of hollow and high-pitched *cu* notes. It lasts 3–4s and consists of *c*.40 short notes delivered at a steady pace and descending evenly from 1.5kHz to 1.2kHz. A call, often delivered at dusk or in alarm, is a quick succession of 2–3 melodious *fui* notes, followed by a short, lower-pitched *cuu*. Call notes are delivered at a frequency of *c*.1.5–2.0kHz (Maijer 1998, Stattersfield & Capper 2000, Krabbe & Schulenberg 2003). Another call, apparently uttered only at dusk, is a single note (Maijer 1998). Maijer (1998) observed that both calls were 'the last sounds of the day [ ] heard from the diurnal birds in the area'.

NATURAL HISTORY Occurs alone or in pairs, but habits very poorly known (Krabbe & Schulenberg 2003). Adults sing from low perches in tangles and forest edges (Maijer 1998, Maillard 2009). Preferred perches are horizontal vines and branches, roughly 0.1–3.0m above ground (Maillard 2012). They forage principally on the ground, probing leaf litter and soil, sometimes thrashing prey violently against the ground or a stem (Maillard 2012). Presumably monogamous (Krabbe & Schulenberg 2003).

DIET Nothing published, but the stomach contents of three specimens were insects (Maillard *et al.* 2008, Greeney 2014c, FMNH).

REPRODUCTION Nesting habits very poorly known, with the first available data published only very recently (Maillard *et al.* 2008) concerning an adult carrying a thin stick in late September, presumably for a nest. Maillard (2012) recently completed a detailed study of the species' reproductive habits, finding seven nests and making detailed behavioural observations at two of these. The following details are taken from the thesis of Maillard (2012), only recently amplified and formally published (Aponte *et al.* 2017). **NEST & BUILDING** The nest is a broad, relatively shallow, somewhat messy, open cup, placed on a loose platform of leaf strips and long, thick sticks (up to 40cm long). The cup is of smaller, more tightly woven sticks and a few dead leaves, and is lined with dark, flexible fibres such as rootlets and fungal rhizomorphs. Nests are placed in low vegetation, supported by stems, vines and small branches from below, not well attached to the substrate. Nest height ranged from 0.9–2.1m above ground (mean ± SD 1.26 ± 0.47 m, *n* = 6). Externally, nests were 5.5 ± 1.7cm tall (range 4–8cm) and 14.7 ± 3.6cm wide (range 11–20cm), with some sticks in the base extending out by as much as an additional 20cm. Internally, nests measured

7–9cm wide by 3.1 ± 0.9cm deep (range 1.8–4.0cm deep). **EGG, LAYING & INCUBATION** Based on three clutches (one found predated), the clutch consists of two subelliptical eggs. All six eggs, including shells from the predated nest, were pale brown, marked with irregular blotches of varying shades of brown and grey. Markings are distributed unevenly across the shell, but heaviest at the larger end. Four eggs measured 17.9 ± 2.3 × 22.1 ± 5.7mm (range 19.7–24.6mm × 21.3–34.7mm). Both sexes participated in nest construction and incubation. Mean daily coverage of the eggs during incubation was 85% of daylight hours. Absences from the nest ranged from 0.1–1.1h. **NESTLING & PARENTAL CARE** Nothing known, but presumably nestlings are born dark-skinned, without natal down, and with bright orange-red mouth linings. It is reasonable to suppose that both adults share the duties of provisioning and brooding the nestlings. **SEASONALITY** Nesting is thought to occur from September to January or late February. **NEST SUCCESS** Both of the two nests monitored by Maillard (2012) during incubation were predated, one by a Black-headed Squirrel Monkey *Saimiri boliviensis* and one by an unknown species of squirrel (*Sciurus* sp.).

TECHNICAL DESCRIPTION Sexes similar. The following is based on Gyldenstolpe (1945b). *Adult* Base of forecrown, ocular area, superciliary and ear-coverts brown. Feathers of the ear-coverts have pale shafts giving a thinly streaked appearance. Lores white. Crown and nape grey, becoming more olive posteriorly. Central crown feathers have dark centres. Upperparts pale brownish-olive, contrasting with grey nape. Uppertail-coverts paler than rest of upperparts with feathers tipped pale buff. Wing-coverts orange-buff, with primary-coverts dark brown basally and broadly tipped buffy-brown. Longest alula ochraceous at base of outer web, becoming paler and buffy-white on posterior portion of same web, with inner web blackish-brown. Anterior edge of wing brown, with primaries and secondaries darker brown and edged ochraceous-olive, brightest on inner primaries and secondaries. Tertials brownish-olive, all but innermost with cinnamon edges. Underwing-coverts tawny ochraceous-orange. Broad black lateral throat stripe, bordered above by whitish. Centre of chin and throat white, above creamy-white breast strongly marked with black, giving a heavily streaked appearance, which extends to flanks where feathers more yellow-buff and black streaking mixed with olive-brown contrasting with pale shafts. Undertail-coverts warm buff, becoming more ochraceous-buff on thighs. Central belly pure white. Rectrices rufescent-brown. **Adult Bare Parts** *Iris* dark brown; *Bill* maxilla black, mandible ivory to pinkish-white; *Tarsi & Toes* pale pinkish. **Immature** Undescribed.

MORPHOMETRIC DATA *Wing* 88mm, 92mm, 87mm, 88mm; *Tail* 38mm, 39mm, 35mm, 31mm; *Bill* [exposed?] culmen 21mm, 21mm, 20mm, 20mm; *Tarsus* 37mm, 37mm, 34mm, 35mm (*n* = 4, ♂ holotype, ♂,♀,♀ Gyldenstolpe 1941, 1945b). I measured two unsexed adults: *Wing* 90mm, 90mm; *Tail* 42mm, 40mm; *Bill* from nares 12.8mm, 14.3mm, exposed culmen 19.6mm, 20.6mm, depth at nares 5.6mm, 6.1mm, width at nares 5.2mm, 5.4mm, width at base of mandible 8.3mm, 8.8mm; *Tarsus* 38.6mm, n/m (FMNH 391157/158). **Mass** 42.8g(no), 43g(no), 38g(no), (*n* = 3, ♂,♂,♀, Maillard *et al.* 2008, also reported in Dunning 2008). **Total Length** 14cm (Stattersfield & Capper 2000, Krabbe & Schulenberg 2003, Ridgely & Tudor 2009).

DISTRIBUTION DATA Endemic to a small area of El Beni, **BOLIVIA**: I have divided the following localities into those recently confirmed by Aponte *et al.* (2017) and historical sites without confirmed reports in the last decade, with year of most recent observation in brackets. **Current** Puerto Remanso [2011], 10°56'S, 66°17.5'W (Aponte *et al.* 2017); Hamburgo [2006], 11°01'S, 66°05.5'W (ML 132761, P.A. Hosner); Hamburgo [2012], 11°01'S, 66°06'W (Maijer 1998, Aponte *et al.* 2017, MHNNK 1228/1604); San Vicente [2011], 11°02'S, 66°05.5'W (Aponte *et al.* 2017, FMNH 391156–158); Las Piedras [2001], 11°01'S, 66°07'W (Mailliard *et al.* 2009). **Historical** Hamburgo [1998], 11°02'S, 66°06'W (FMNH); Riberalta [1998], 10°59'0.24"S, 66°03'W (XC 2684, S. Maijer); Hamburgo [1998], 11°01'S, 66°06'W (XC 2719/20, S. Maijer); Tumichucua [1976], 11°08.5'S, 66°10'W (Maijer 1998).

TAXONOMY AND VARIATION Monotypic. Maijer (1998) provided evidence that, based largely on its distinctive vocalisations, *auricularis* deserves to be elevated again to species status. Recent molecular studies suggest that Masked Antpitta is more closely related to Speckle-breasted Antpitta than to the *macularius* group (Carneiro & Aleixo 2012, Carneiro *et al.* 2012).

STATUS The estimated range is very small, irrespective of the methodology used: 1,098km² (Young *et al.* 2009), 380–509km² (Herzog *et al.* 2012) or 210km² (Aponte *et al.* 2017). Currently, it is considered Vulnerable (BirdLife International 2017). Its apparent tolerance of, even preference for, heavily disturbed habitats, as well as the general lack of surveys over much of its range, however, led BirdLife International (2017) to suggest that it eventually may be downgraded to Near Threatened once more is known of its distribution and biology. Unfortunately, considering the most recent surveys (Aponte *et al.* 2017), this seems unlikely. Despite its (apparent) tolerance of disturbance, habitat within its known range is facing severe threats from many forms of human activities (Fjeldså & Rahbek 1998, Maillard 2009). Indeed, given all that we (don't) know, the species' global status could even be upgraded to Critically Endangered, as it has been at a national level (Maillard 2009), or at least to Endangered, as suggested by Aponte *et al.* (2017). Although a *c.*50,000ha Important Bird Area has been designated, covering Masked Antpitta's range (Soria & Hennessey 2005, Maillard *et al.* 2009), it is not protected within any reserves. Reducing threats and formally protecting habitat with the help of communities and local government authorities should be a priority, starting with known locations in dynamic floodplains (Aponte *et al.* 2017).

OTHER NAMES *Grallaria auricularis* (Gyldenstolpe 1941, Bond & Meyer de Schauensee 1942); *Grallaria macularius auricularis* (Zimmer & Mayr 1943, Gyldenstolpe 1945b, Peters 1951); *Grallaria macularia* (Meyer de Schauensee 1966); *Hylopezus macularius* (Meyer de Schauensee 1970); *Hylopezus macularius auricularis* (Howard & Moore 1984). **Spanish** Tororoí Enmascarado (Krabbe & Schulenberg 2003); Tilluche (Gyldenstolpe 1945b). **French** Grallaire oreillarde. **German** Südliche Brillenameisenpitta.

Masked Antpitta, adult, Hamburgo, Beni, Bolivia, 17 November 2006 (*Oswaldo Maillard*).

Masked Antpitta, adult, Riberalta, El Beni, Bolivia, 29 August 2013 (*Andrew Spencer*).

Masked Antpitta, nest and eggs, Hamburgo, Beni, Bolivia, 12 December 2006 (*Oswaldo Maillard*).

THICKET ANTPITTA
Hylopezus dives   Plate 19

Grallaria dives Salvin, 1865, Proceedings of the Zoological Society of London, vol. 32, p. 582, Tucurriquí, Costa Rica [= Tucurrique, 09°51'N, 83°43.5'W, 765m, Cartago].

Thicket Antpitta was long considered conspecific with White-lored Antpitta, and together they were known as Fulvous-bellied Antpitta. As currently defined, three subspecies are recognised, two confined to Central America and one endemic to the Chocó region of southern Panama and western Colombia. Morphologically, Thicket Antpitta is distinguished from other *Hylopezus* by the much less well defined and duller streaking below, and its predominantly buffy-rufous underparts. Thicket Antpitta inhabits lowlands below 900m and, as its name suggests, prefers dense tangles around streams and treefalls, also at forest edges and in adjacent plantations. The species is generally fairly common, but like all other Grallariidae, is more easily heard than seen. The reproductive biology, diet and behaviour of Thicket Antpitta are almost completely unknown. As is the case for most antpittas, very few quantified data have been gathered on any aspect of the species' behaviour and ecology. In particular, information on diet and breeding are sorely needed.

IDENTIFICATION 13–15cm. Sexes similar. Crown and nape dark grey, with buffy lores and a narrow eye-ring. Ear-coverts greyish-brown. Upperparts, including tail, dark olive-brown, with a few narrow, rather pale buffy shaft-streaks on upper back. Wings dark olive-brown with brighter edges to primaries. Throat and belly white, with the breast and sides washed ochraceous-buff and diffusely streaked blackish. Flanks and vent deep orange-rufous. Thicket Antpitta is distinguished from the similar (and previously conspecific) White-lored Antpitta by its smaller, buffier loral spot, and paler grey crown and ear-coverts. It also has slightly finer ventral streaking and deeper orange-rufous flanks, plus a distinctly different song (Krabbe & Schulenberg 2003). Thicket and White-lored Antpittas are allopatric, with White-lored only in north-west Amazonia. Streak-chested Antpitta, often sympatric, has a bold, buffy orbital area and buff-spotted wing-coverts,

which characters are lacking in Thicket Antpitta. It is also much whiter below, rather than having the buffy-rufous or fulvous wash to the underparts of Thicket Antpitta (Ridgely & Gwynne 1989).

DISTRIBUTION Resident on the Caribbean slope of Central America, from eastern Honduras south through Panama, including Bocas del Toro in the north-west, and into Colombia on the Pacific slope to western Nariño.

HABITAT Thicket Antpitta occurs in dense, tangled thickets and vegetation along streams and forest edges, inside mature forest only in disturbed areas such as riparian zones and treefalls (Stiles 1985, Hilty & Brown 1986, Garrigues & Dean 2007, Angehr & Dean 2010,

McMullan *et al.* 2010, Greeney 2014a), even cacao and banana plantations (Harvey & Villalobos 2007). Previously reported from sea level to 900m in Costa Rica (*dives*; Stiles & Skutch 1989), in some areas it may range as high as 1,100m (e.g., RB El Copal; L. Sandoval *in litt.* 2017).

VOICE The song is a rapid, ascending series of 10–13 clear whistles, lasting 1.6–2.0s, and increasing slightly in volume and speed. The first note is slightly doubled, the second has a slight stutter, and the final 3–4 notes are on the same pitch (Krabbe & Schulenberg 2003). It is faster than the song of Streak-chested Antpitta, which also differs by trailing off, rather than ending abruptly (Ridgely & Gwynne 1989, Angehr & Dean 2010). In Costa Rica, at least, the song is sometimes preceded by a single 'heavily whistled' *hoo* call (Slud 1964). The alarm is an accelerating roll, perhaps somewhat weaker than that of Streak-chested Antpitta (Krabbe & Schulenberg 2003). Greeney & Vargas-Jiménez (2017) described a rapid *ptiur-ur-ur-ur-ur-ur-ur-ur-u*, given at intervals of 8–10s by an adult in response to an observer's approach to the nest.

NATURAL HISTORY The general behaviour is very poorly known. It is, like most species of antpittas, secretive and difficult to see. Adults sing from low perches such as horizontal branches and logs, generally no more than 1.5m above ground. It forages close to, or on the ground, running or moving in short hops. This appears to be a particularly vocal species, often singing throughout the day (Howell 1957, Stiles & Skutch 1989). There is no information on territory size or maintenance, but pairs apparently hold territories year-round (Krabbe & Schulenberg 2003).

DIET Essentially unknown, it apparently feeds largely on arthropods (Krabbe & Schulenberg 2003), but perhaps also takes small vertebrates such as frogs, as does *H. perspicillatus* (Horsley *et al.* 2016). An adult male had several hard-shelled beetles in its stomach (MZUCR 1912) and another adult male had Hymenoptera and Orthoptera (MZUCR 2769, R. Delgado).

REPRODUCTION Only a single nest has been described (Greeney & Vargas-Jiménez 2017), found with two young (2–3 day-old) nestlings. Minimal observations were made, as the nest was apparently predated soon after discovery. **NEST & BUILDING** A shallow, saucer-like cup, 20cm above ground, placed loosely atop several thin branches within tangled vegetation. It was sited only 1m from the edge of an infrequently used (but well-maintained) forest trail in an area of dense understorey vegetation. The bulk of the cup was composed of small sticks that were loosely interlaced and mixed with a few dead leaves. The very sparse lining consisted of only a few flexible fibres, only slightly thinner than the twigs used in the outer nest. No measurements were taken but it was proportionally similar to nests of other *Hylopezus* I have seen. **EGG, LAYING & INCUBATION** Unknown. **NESTLING & PARENTAL CARE** The two young nestlings had their eyes slitted, but not open. Their primaries were still in pin, and were halfway emerged from the skin. The only pin feathers to have broken their sheaths were those in the scapular region. The emerging fluffy down-like feather tips appeared uniformly dull brownish, with little or no hint of reddish as in other *Hylopezus*. Their skin was blackish to dusky-lavender or dusky-pinkish. Their bills were largely dull, grey-orange, the maxilla brightest orange on the culmen and tomia, yellowish near the tip. Their whitish egg-teeth

remained attached. The inflated rictal flanges were bright orange to yellow-orange and their mouth linings were bright clear orange; not orange-crimson as described for most *Grallaria* (Greeney *et al.* 2008a). Interestingly, it appears that the nestlings begged silently, signalling their hunger by vigorously shaking their bright orange mouths when adults arrived with food. Only minimal behavioural observations were made, but these revealed that both adults provisioned the nestlings and consumed faecal sacs. Prey were usually delivered singly and consisted of small arthropods (0.5–1.5 times the length of the adults' bill). If one adult was brooding when the other arrived with food, it would stand and permit the nestlings to be fed. **SEASONALITY** Published data are few, but include four birds (*barbacoae*) collected in breeding condition in north-west Colombia in March–May (Hilty & Brown 1986). **BREEDING Data** *H. d. dives* Freshly built nest, 4 October 2012, Laguna de Arenal (Greeney & Vargas-Jiménez 2017); young nestlings, 19 September 2012, Laguna de Arenal (Greeney & Vargas-Jiménez 2017); juvenile transitioning, 6 August 1889, Jiménez (USNM 198343); subadult, 8 August 1926, La Iberia (YPM 56720, S.S. Snow photos); subadult, 5 March 2011, RB Hitoy Cerere (MZUCR 4820); subadult transitioning, 6 August 1926, La Iberia (YPM 56719, S.S. Snow photos). *H. d. barbacoae* Subadult, 14 March 1915, Cerro Tacarcuna (AMNH 135797).

TECHNICAL DESCRIPTION Sexes similar. The following refers to nominate *dives* (Ridgway 1911). **Adult** Crown and hindneck dull slate, indistinctly streaked blue-blackish. Rest of upperparts dull slate passing into olive posteriorly, the feathers (especially scapulars) with very narrow and mostly indistinct pale buffy shaft-streaks; uppertail-coverts and tail russet-brown. Wings generally deep olive-brown, outer webs of primaries paler and more rufescent-brown; outermost feather of alula edged buff or ochraceous-buff. Loral, orbital and suborbital regions buff, more or less flecked dusky, the lower ear-coverts deeper buffy; upper ear-coverts dull slate, more or less tinged olive; malar region, chin and throat white or buffy-white, the first more or less flecked dusky; breast and sides ochraceous, the feathers with median or central portion paler and edged black, producing a streaked effect. Sides, flanks, undertail-coverts and underwing-coverts plain ochraceous to rufous-tawny; inner webs of remiges become dull vinaceous-cinnamon on edges. Maxilla brownish, paler on tomia, darker (sometimes almost black) on culmen; mandible pale dull yellowish (in dried skins), usually tinged brownish laterally or terminally; legs and feet pale yellowish or yellowish-brown (in dried skins). **Adult Bare Parts** *Iris* dark brown; *Bill* blackish, basal half of mandible pale; *Tarsi & Toes* pale pinkish-grey. **Immature** I encountered relatively few immatures in collections (see Reproduction), the youngest was already beginning to transition from the juvenile plumage phase, but still retained enough of its downy fledgling plumage to suggest that the species is similar to other *Hylopezus* in having a dense coating of mostly rusty-chestnut down during the late nestling and early fledgling phases. I have not seen photographs of, or live, immatures but presumably bare-parts coloration is also similar to other species. *Juvenile* Probably replaces contour feathers with adult-like plumage in a sequence similar to other antpittas, starting with the central back and upper breast, completing replacement on most of the underparts slightly ahead of the back and rump, followed closely by the crown feathers from front to rear. The older juvenile I examined (USNM 198343) was in transition and retained several irregular patches of fledgling down on the nape and upper back, as well as scattered plumes mixed with adult-like feathering on the breast and flanks. Although there was already recognisable, if not somewhat indistinct, streaking on the breast, these scattered immature feathers gave the entire underparts a much more cinnamon-rufescent background coloration, as opposed to the brighter fulvescent wash of adults. In several areas, where two or three rusty down feathers remained, they created irregular dark streaking against the pale white or fulvous feathers; in the specimen at least these streaks might easily be dismissed as blood stains incurred during preparation. Small numbers of immature feathers were scattered irregularly across the crown, brow and face, and likely would have appeared wispy in the live bird. The upperwing-coverts appeared rather worn but each retained buffy-fulvous central patches near their tips. The spots had faded or worn almost completely on some. All, however, still had fairly broad terminal margins of rusty-chestnut. Probably these were much more apparent during the fledgling and early juvenile phases, with bright, diamond-shaped or triangular fulvous spots merging into bright chestnut fringes. **Juvenile Bare Parts** Not described in life, but juveniles probably retain orange or yellow rictal flanges, tomia and mandibular base. **Subadult** Overall similar to adult but somewhat 'scruffy' looking on the underparts and face with a mix of adult and immature feathers. The most significant difference from fully adult plumage is the narrow reddish-chestnut tips to the upperwing-coverts. It is likely that these markings slowly wear away or fade almost entirely prior to their replacement with adult wing-coverts. **Subadult Bare Parts** Unknown; likely very similar to adults.

MORPHOMETRIC DATA *H. d. dives* Bill from nares 11.9mm, exposed culmen 17.3mm, depth at front of nares 5.0mm, width at gape 8.9mm, width at front of nares 5.0mm, width at base of mandible 7.4mm; *Tarsus* 35.7mm (*n* = 1♀, AMNH 103387). *Bill* from nares 12.8mm; *Tarsus* 38.1mm (*n* = 1♂?, USNM 116346). *Bill* from nares 12.4mm, 12.6mm, 13.3mm, 12.8mm, 12.2mm; *Tarsus* 37.4mm, 37.4mm, 36.7mm, 37.8mm, 37.6mm (*n* = 5♂♂, USNM, WFVZ). *Wing* 75mm; *Tail* 30mm; *Bill* [exposed] culmen 18mm; *Tarsus* 36mm; *Middle toe* 19mm (*n* = 1♀, Ridgway 1911). *Wing* range 73.5–78.0mm, mean 75.8mm; *Tail* range 29–31mm, mean 30mm; *Bill* [total] culmen, range 19.0–19.5mm, mean 19.3mm; *Tarsus* 37mm (×3); *Middle toe* range 18.0–20.5mm (*n* = 3♂♂, Ridgway 1911). *Wing* range 75.5–81.6mm, mean 78.7mm); *Tail* range 28.5–31.4mm, mean 29.5mm; *Bill* culmen from base, range 21.0–21.2mm, mean 21.1mm; *Tarsus* range 36.8–37.5mm, mean 37.2mm (*n* = 4♂♂, Wetmore 1972). *Wing* 74.4mm; *Tail* 25.6mm; *Bill* culmen from base 20.6mm; *Tarsus* 37.4mm (*n* = 1♀, Wetmore 1972). *Bill* from nares 10.9mm; *Tarsus* 38.3mm (*n* = 1♂, juvenile, USNM 198343). *H. d. barbacoae* Wing range 74.1–79.5mm, mean 76.7mm; *Tail* range 28.5–33.5mm, mean 30.5mm; *Bill* culmen from base, range 20.0–22.6mm, mean 21.0mm; *Tarsus* range 35.3–40.0mm, mean 37.8mm (*n* = 10♂♂, Wetmore 1972). *Bill* from nares 11.9mm, 12.7mm, 12.5mm, 12.5mm, 12.8mm, 12.9mm, 13.4mm, 12.5mm, 12.9mm, 13.3mm, 14.1mm, 12.4mm, 12.0mm; *Tarsus* 35.4mm, 37.9mm, 39.6mm, 35.6mm, 36.9mm, 35.5mm, 38.7mm, 36.6mm, 35.6mm, 38.1mm, 37.5mm, 37.9mm, 36.4mm (*n* = 13, 3♀♀, 10♂♂, USNM, AMNH). *Bill* from nares 11.9m; *Tarsus* 37.0mm (*n* = 1♂, subadult, AMNH 135797). **Mass** 38g, 44g (*n* = 2, ♂♂, *barbacoae*, Robbins *et al.* 1985), 44g (*n* = ?♀?,

dives, Stiles & Skutch 1989), 41g (*n* = ?, ♀?, *dives*, Cody 2000), 40.3g(no), 47.5g(no), 40.4g(no), 40.3(vl), 47.5(lt), (*n* = 5♂♂, *dives*, WFVZ, MZUCR), 36.0g(tr) (*n* = 1♂, *dives*, subadult, MZUCR 4820). **Total Length** 11.8–12.9cm, mean 12.4cm (*n* = 3♂♂), 11.9cm (*n* = 1♀) (*dives*; Ridgway 1911). Generally given as 13.0–16.5cm (Hilty & Brown 1986, Ridgely & Gwynne 1989, Stiles & Skutch 1989, Garrigues & Dean 2007).

TAXONOMY AND VARIATION Described as *Grallaria dives* (Salvin 1865), Thicket Antpitta was maintained within the genus *Grallaria* (Cory & Hellmayr 1924, Peters 1951, Howell 1957, Monroe 1968) for many years, despite its assignment to *Hylopezus* by Ridgway (1911). Since the observations of Lowery & O'Neill (1969), however, most authors have considered *Hylopezus* a genus apart from *Grallaria* (Haffer 1975, Ridgely & Gwynne 1989, Monroe & Sibley 1993, Clements 2007). Thicket Antpitta was long considered a race of *H. fulviventris* (Cory & Hellmayr 1924, Peters 1951, Haffer 1975, Howard & Moore 1984), but considering the significant vocal differences noted by later authors (Ridgely & Gwynne 1989, Ridgely & Tudor 1994), it currently is considered a separate species, including *flammulatus* and *barbacoae* (Clements 2007, Remsen *et al.* 2017). There is little apparent plumage or vocal variation between the three currently recognised subspecies of *H. dives*. In fact, with respect to the putative subspecific plumage differences given below, there appears to be as much intra- as inter-population variation. Particularly with respect to the degree of breast streaking, this variation does not appear clinal. For example, some of the most heavily streaked (*cf. flammulatus*) individuals I have seen are from Colombia, while some of the least marked are from Limón province in south-eastn Costa Rica, scarcely 50km from the type locality of *flammulatus*. Although it does not fit with my general, admittedly vague, understanding of plumage development in *Hylopezus*, the observed variation in intensity and extent of breast streaking may, in fact, relate to age. Irrespective of their origin, this geographically inconsistent plumage variation, coupled with little or no consistent differences in vocalisations, suggest that a careful taxonomic review would result in *flammulatus* and *barbacoae* being considered invalid. Alternatively, and perhaps more likely given the apparent distributional gap in Panama, *flammulatus* might best be subsumed within *dives* and *barbacoae* kept as valid.

Hylopezus dives dives (Salvin, 1865). There are two designated syntypes in the British Museum, one of which is NHMUK 1889.7.10.903, collected by E. Arcé (Warren & Harrison 1971). The nominate subspecies occurs on the Caribbean slope of Central America from eastern Honduras south to Costa Rica. A significant portion of the range of nominate *dives* lies in Nicaragua and many important specimens were collected in the early 1900s by W. B. Richardson, who was notorious for his vague localities on mammalian and avian skins (Jones & Genoways 1970, Martínez-Sánchez & Will 2010). I have done my best to locate as many of these as possible, largely with the help of several key papers (Buchanan & Howell 1965, Howell 1971, Martínez-Sánchez & Will 2010). **Specimens & Records HONDURAS**: *Colón c.*43km SE of Bonito Oriental, 15°32.5'N, 85°23'W (XC 104731, M.M. Mejia). *Gracias a Dios* Las Marias, Río Plátano, 15°40.5'N, 84°53.5'W (Anderson *et al.* 1998, WFVZ 52988–990, LSUMZ 161380–382); *c.*40km WNW of Wampusirpi, 15°15'N, 84°58.5'W (ML 49482661, J. van Dort); *c.*38km S of Wampusirpi, 14°51'N,

84°34'W (XC 309678, A. Auerbach); Río Rus Rus bridge, 14°43'N, 84°27'W (ML 58901171, photo R. Rumm). *Olancho* *c.*13km E of Loma de Enmedio, 15°17'N, 85°19'W (XC 263332, C. Funes); 5km NW of Dulce Nombre de Culmí, 15°06'N, 85°36'W (XC 283061, A. Auerbach); *c.*23.5km SW of Walakitan, 14°25.5'N, 85°14.5'W (ML 49802141, A. Auerbach); Las Marías Abajo, 14°17'N, 85°37'W (XC 178774, O. Komar); Villa Nueva de Yamales, 14°07'N, 85°45'W (ML). **NICARAGUA**: *Nueva Segovia* El Rosario, 13°50'N, 85°52'W (eBird: J.P. Kjeldsen). *Atlántico Norte* 25km SSW of Waspan, *c.*14°31'N, 84°03'W (Howell 1971); Kum, 13°26'N, 84°56'W (UCLA 35598–600). *Jinotega* Río Labu, Bosawas area, 13°56.5'N, 85°08.5'W (A. Farnsworth *in litt.* 2017); 'Río Coco,' *c.*13°27'N, 85°55'W (AMNH 103387; loc. see Martínez-Sánchez & Will 2010); 'Peñas Blancas,' 13°15'N, 85°41'W (AMNH 103745–747). *Matagalpa* 'Río Tuma,' *c.*13°09'N, 84°49'W (AMNH 103601); 'Savala,' *c.*13°04.5'N, 85°27.5'W (AMNH 102551); 'Las Cañas,' 13°03'N, 85°45'W (AMNH 144042); 'Río Grande,' *c.*12°59'N, 84°44.5'W (AMNH). *Chontales* Refugio de Vida Silvestre Peñas Blancas, 12°16'N, 85°04.5'W (eBird: L. Diaz); 'Chontales' (Ridgway 1911). *Atlántico Sur* El Recreo, 12°10'N, 84°25'W (UCLA 35601); Río Escondido, 80.5km from Bluefields, *c.*12°09'N, 84°10.5'W (Richmond 1893, Ridgway 1911, USNM 128353). *Río San Juan* Sábalos, 11°03'N, 84°28.5'W (Ridgway 1911, AMNH 144043, USNM 91265); El Castillo, 11°01'N, 84°24'W (BRTC 7117); Greytown [= San Juan de Nicaragua], 10°56.5'N, 83°44'W (Lawrence 1865, Ridgway 1911, USNM 40430). **COSTA RICA**: *Guanacaste* Estación Biológica San Cristobal, 10°53'N, 85°23.5'W (HFG); Estación Biológica Pitilla, 10°40'N, 85°30'W (XC 6403, D. Bradley); Tilarán, 10°27'N, 84°58'W (DMNH 47550/51); Corobicí, 10°27'N, 85°07.5'W (eBird: R. Garrigues); Estación Biológica San Gerardo, 10°22'N, 84°47.5'W (J.B.C. Harris *in litt.* 2017). *Alajuela* Bijagua, 10°44'N, 85°03'W (Ridgway 1911, MCZ 123382); Lake Coter Eco Lodge, 10°35'N, 84°56'W (XC 7824, R. Carter); Arenal Observatory Lodge, 10°28.5'N, 84°39'W (XC 213921, K. Frisch); near Laguna de Arenal, 10°27'N, 84°44'W (Greeney & Vargas-Jiménez 2017); 8km SW of La Fortuna, 10°26'N, 84°42.5'W (XC 278629, O. Campbell); San Carlos, 10°22'N, 84°23'W (YPM 56722); Garabito entrance to PN Juan Castro Blanco, 10°21'N, 84°205'W (eBird: J. Zook); Quesada, 10°16'N, 84°26'W (YPM 56724/25); Hotel Valle Escondido, 10°15'N, 84°31'W (XC 1038, R. Carter); RB Alberto Manuel Brenes, 10°12'N, 84°37'W (XC 165204, O.R. Alán). *Heredia* Estación Biológica La Selva, 10°20'N, 84°00'W (Rangel-Salazar 1995, XC, FLMNH, ML); Rara Avis, 10°17'N, 84°03'W (ML: D.L. Ross Jr.). *Cartago* Rancho Naturalista, 09°50'N, 83°34'W (XC 5935, R.C. Hoyer); RB El Copal, 09°47'N, 83°45'W (L. Sandoval *in litt.* 2017). *Limón* Jiménez, 10°10'N, 83°46'W (Ridgway 1911, USNM 198343); Guápiles, 10°13'N, 83°47'W (YPM 56721); Guácimo, 10°13'N, 83°41'W (Ridgway 1911, CM P23594); La Iberia (= near Alegría, cantón Siquirres), *c.*10°02.5'N, 83°37'W (YPM 56718–21); RB Hitoy Cerere, 09°39'N, 83°00.5'W (MZUCR 4820); Volio, 09°38'N, 82°52'W (MZUCR 1912); Suretka, 09°36'N, 82°58'W (YPM 56723).

Hylopezus dives flammulatus Griscom, 1928. American Museum Novitates 293, p. 4, Almirante, Bocas del Toro, north-western Panama [09°18'N, 82°25'W]. The holotype (AMNH 233595) is an adult male collected 16 May 1927 by Rex R. Benson. There are a few topotypical specimens at USNM and LSUMZ, but overall this taxon is poorly collected. This subspecies is the most range-restricted

of the three, apparently being found only in the vicinity of the type locality, east to western Ngäbe-Buglé. This race is considered darker overall, to have more richly coloured wings, and to show heavier black flammulations on the breast (Griscom 1928). **SPECIMENS & RECORDS** *Bocas del Toro* Dos Bocas, 09°22.5'N, 82°32'''W (Angehr *et al.* 2008); Wekso Ecolodge, 09°21.5'N, 82°35'W (Angehr *et al.* 2008); Tierra Oscura, 09°11'N, 82°12'W (USNM); upper Río Changuinola, *c.*09°22'N, 82°37'W (LSUMZ 178099). *Ngäbe-Buglé* Near Río Cañaveral, 09°00'N, 81°42.5'W (V. Wilson recording; see eBird checklist S18487876). *Veraguas* Río Mulabá, *c.*7.5km WNW of Santa Fé, 08°31.5'N, 81°08'W (V. Wilson *in litt.* 2016).

Hylopezus dives barbacoae Chapman, 1914, Bulletin of the American Museum of Natural History, vol. 33, p. 617, Barbacoas, Nariño, Colombia [01°40'N, 78°08'W, *c.*35m]. The holotype (AMNH 117883) is an unsexed adult collected by W. B. Richardson on 8 September 1912. This race is found from eastern Panama (eastern Darién) south along the Pacific slope of the West Andes in Colombia to western Nariño. It resembles nominate *dives*, but has a darker crown and more olive back. The upper back also has fewer or less prominent streaks. From Chapman's original description (1914): 'Similar to *H. d. dives* Salv., but crown darker, its color extending little if any on to the back, which is dark olivaceous rather than slaty; back, as a rule, without fulvous shaft-streaks, exposed margins of the wing-quills averaging less cinnamomeus, dresden-brown rather than tawny. This is evidently an intermediate between *H. d. dives* and *fulviventris* from the eastern base of the Eastern Andes. It is based on four specimens from the type-locality and one from San José, W. Colombia, which have been compared with a single specimen of *fulviventris* from La Murelia and eight of *dives* from Nicaragua. In *fulviventris* the back is more purely olivaceous and it is furthermore, easily distinguished by whitish lores. There is no geographical reason why *dives* and *barbacoae* should not intergrade, but *fulviventris* is effectually isolated from the latter by the intervening Andes'. **SPECIMENS & RECORDS PANAMA**: *Guna Yala* Near Nusagandi, *c.*09°20'N, 78°59'W (Angehr *et al.* 2008). *Panamá* Upper Río Majé, 08°54'N, 78°40.5'W (Angehr *et al.* 2008). *Embera* Cerro Tacarcuna, 08°10'N, 77°18'W (AMNH 135797). *Darién* Río Sambú, 08°04'N, 78°14.5'W (Angehr *et al.* 2008); Cana, 07°46'N, 77°41'W (LSUMZ, MCZ, FLMNH, ML, XC). **COLOMBIA**: *Antioquia* Alto Bonito, 07°05'N, 76°30'W (AMNH, MCZ); Villa Arteaga, 07°23'N, 76°29'W (USNM 426429–431); *c.*11.5km WNW of Dabeiba, 07°02'N, 76°21'W (XC 265970, IBC 1096477, D.M. Brinkhuizen). *Chocó* 'Tacama' on Río Tolo, *c.*08°26.5'N, 77°16.5'W (Haffer 1959); vicinity of Bahía Solano and Nuquí, *c.*06°10'N, 77°22'W (FLMNH 16306, XC); Río Jurubidá, 06°06'N, 77°13'W (USNM 443356/57); El Valle road, 05°54'N, 77°18'W (XC 68492, F. Lambert); Andagoya, 05°06'N, 76°42'W (CM P66236); Río Capico/Río Docampadó confluence, 04°40'N, 77°15'W (FMNH 255651/52); Punto Muchimbo, 04°08'N, 77°04'W (USNM 443358). *Risaralda* Condoto, 05°21.5'N, 76°04.5'W (XC 319776, D. Orozco). *Córdoba* Socorré, 07°51'N, 76°17'W (USNM 411876); Bosque Carlos Bran, 07°50.5'N, 76°28.5'W (ML 63367511, F. Schmitt); Quebrada Salvajín, 07°45'N, 76°16'W (USNM). *Valle del Cauca* *c.*4.2km W of Córdoba, 03°52.5'N, 76°53.5'W (O. Cortes-Herrera *in litt.* 2017); San José, 03°51'N, 76°52'W (AMNH 107478); Río Saboletas, 03°51'N, 76°54'W (ML 83790, S.L. Hilty); Santa Bárbara, 03°44.5'N, 77°01.5'W (XC 325383, E. Fierro-

Calderón); Buenaventura, Bajo Anchicayá, 03°37'N, 76°55'W (J.A. Zamudio E. photo). *Nariño* RNA El Pangan, *c.*01°21'N, 78°12'W (Salaman *et al.* 2007a).

STATUS Not considered globally threatened (BirdLife International 2017), as it is a widespread species and is fairly common to locally common within most of its range (Slud 1960, Hilty & Brown 1986, Garrigues & Dean 2007, Angehr & Dean 2010, McMullan *et al.* 2010). As the common name suggests, the species appears very tolerant of anthropogenic disturbance, possibly even favoured by light habitat disturbance (Krabbe & Schulenberg 2003, Harvey & Villalobos 2007). No effects of human activity have been thoroughly documented, but Thicket Antpitta was found to be only slightly less common in banana and cacao plantations in one study (Harvey & Villalobos 2007), suggesting that at least some degree of human disturbance may benefit this species. Conversely, the species appears to be nearly completely extirpated from Estación Biológica La Selva, which seems undoubtedly to be a result of understorey disturbance due to inflated populations of Collared Peccaries *Tayassu tajacu* following the loss of large predators. The range-restricted subspecies, *flammulatus*, may not occur in any formally protected area, but both of the other subspecies occur in national parks and reserves across their ranges. **PROTECTED POPULATIONS** *H. d. dives* **Honduras** PN Sierra Río Tinto (XC 104731, M.M. Mejia); PN Patuca (XC 178774, O. Komar); Reserva de Biósfera Río Plátano

(LSUMZ, WFVZ). **Nicaragua** RN Macizos de Peñas Blancas (AMNH); RVS Peñas Blancas (eBird: L. Diaz). **Costa Rica** PN Volcán Arenal (Greeney & Vargas-Jiménez 2017); PN Tortuguero (Krabbe & Schulenberg 2003); Área de Conservación Guanacaste (XC 6403, D. Bradley); PN Juan Castro Blanco (eBird: J. Zook); RB El Copal (L. Sandoval *in litt.* 2017); RB Hitoy Cerere (MZUCR 4820); RB Alberto Manuel Brenes (XC 165204, O.R. Alán). Probably also occurs in PN Volcán Tenorio and PN Rincón de la Vieja. *H. d. flammulatus* **Panama** BP Palo Seco (LSUMZ 178099). Probably occurs in PN Santa Fé (V. Wilson *in litt.* 2017). *H. d. barbacoae* **Panama** PN Darién (LSUMZ, MCZ, FLMNH, XC, ML). **Colombia** PNN Los Katíos (Krabbe & Schulenberg 2003); PNN Ensenada de Utría (XC 148101, F.G. Stiles); PNN Farallones de Cali (J.A. Zamudio E. *in litt.* 2017); RNSC San Cipriano (XC 174797, J. Suárez); RNA El Pangan (Salaman *et al.* 2007a).

OTHER NAMES *Grallaria fulviventris barbacoae* (Haffer 1959); *Grallaria fulviventris* (*dives*, Howell 1971); *Hylopezus fulviventris* (Anderson *et al.* 1998, van Perlo 2006). **English** Fulvous-bellied Antpitta (Howell 1971, Hilty & Brown 1986); Dives Antpitta (Ridgway 1911), Flammulated Antpitta [*flammulatus*] (Krabbe & Schulenberg 2003). **Spanish** Tororoí Ventricanela Colombiano (Krabbe & Schulenberg 2003); Tororoí Carimanchado (Salaman *et al.* 2007a); Tororoí Pechicanelo (van Perlo 2006). **French** Grallaire buissonnière. **German** Orangeflanken-Ameisenpitta (Krabbe & Schulenberg 2003).

Thicket Antpitta, adult (*dives*), PN Volcán Arenal, Alajuela, Costa Rica, 19 September 2012 (*Juan Diego Vargas-Jiménez*).

Thicket Antpitta, adult (*barbacoae*), Buenaventura, Bajo Anchicayá, Valle del Cauca, Colombia, 17 April 2012 (*Jeisson Andres Zamudio E./Asociación Calidris*).

Thicket Antpitta, adult (*dives*), PN Volcán Arenal, Alajuela, Costa Rica, 17 September 2012 (*Juan Diego Vargas-Jiménez*).

Thicket Antpitta, adult preparing to brood two young nestlings (*dives*), PN Volcán Arenal, Alajuela, Costa Rica, 19 September 2012 (*Juan Diego Vargas-Jiménez*).

Thicket Antpitta, two young nestlings in nest (*dives*), PN Volcán Arenal, Alajuela, Costa Rica, 20 September 2012 (*Juan Diego Vargas-Jiménez*).

WHITE-LORED ANTPITTA
Hylopezus fulviventris   Plate 20

Grallaria fulviventris P. L. Sclater, 1858, Proceedings of the Zoological Society of London, vol. 26, pp. 68, 282, 'Rio Napo', Ecuador. The rather vague type locality probably refers to the area around Coca, Orellana province (00°28'S, 76°59'W), one the most accessible locations to access the Río Napo during the era when the type specimen was collected.

A poorly known inhabitant of the north-west Amazon Basin, from southern Colombia to northern Peru, and almost an Ecuadorian endemic. Two subspecies are recognised, with one known only from two adjacent localities in Colombia. The latter is, however, only doubtfully separable from the nominate race that occupies the rest of the species' range. Like many congeners, this antpitta favours dense undergrowth and tangles along rivers, around forest edges and treefall gaps, where it is difficult to see as it runs or hops across the forest floor and low logs, or sings from a concealed perch. Its somewhat trogon-like, short series of rough, hollow *coop* notes are ventriloquial such that even when singing from a few metres away, the species is a challenge to see. For many years, White-lored Antpitta was considered conspecific with Thicket Antpitta of western Colombia and Central America, but it has a distinctly different voice and plumage. Because of this, much of the old literature that potentially discusses White-lored Antpitta is, in fact, based largely or entirely on the better-known Thicket Antpitta. Natural history details, such as diet, breeding biology, and even range and habitat use are poorly documented.

IDENTIFICATION 14.5–15.0cm. Sexes similar. There is a prominent white loral spot and white triangular spot behind the eye. The white of the throat extends to the base of the ear-coverts. The crown, nape and head-sides are dark grey, contrasting with the dark olivaceous-brown of the remaining upperparts. The breast and sides are buffy-ochraceous, coarsely streaked dusky. The flanks and vent are rather bright fulvous. The similar

Spotted Antpitta (no overlap) has a distinctive buffy eye-ring, wingbars formed by buffy-orange bases to the primaries, and is distinctly spotted on the breast rather than streaked. Apparently there is possible overlap with the rather similar Amazonian Antpitta in northern Peru. The latter species lacks the dark grey crown and nape, and has a less distinctly patterned face (lores buffy, not white). The similar-sized Thrush-like Antpitta probably occurs in most locations where White-lored is found. Thrush-like Antpitta lacks the grey crown and strong face markings, and the fulvous wash to the underparts (instead washed dull brownish or olive-brown).

DISTRIBUTION Found in the North Amazon or Napo centre of endemism (Cracraft 1985), records extend from south-east Colombia (Caquetá, Putumayo, Nariño) over most of Amazonian Ecuador to northern Peru (Loreto, Amazonas) (Hilty & Brown 1983, Rodner *et al.* 2000, Ridgely & Greenfield 2001, Ridgely & Tudor 2009, McMullan *et al.* 2010, Schulenberg *et al.* 2010, Calderón-Leytón *et al.* 2011).

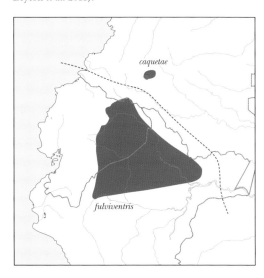

HABITAT Very dense undergrowth at forest edges (both *várzea* and *terra firme*), perhaps favouring damp riparian thickets. It also seems to use tangled growth bordering artificial clearings and large treefalls (Schulenberg *et al.* 2010). Generally found only below 600m, but may occur up to 750m in Napo (Ridgely & Greenfield 2001, Ridgely & Tudor 2009, WFVZ, ML), and has recently been found to be fairly common at 900m in the Serranía Sinangoe in north-east Ecuador (Schulenberg 2002).

VOICE The fairly short song lasts only 2–3s and consists of an evenly paced series of 4–6 similar downslurred *cuock* notes delivered at frequencies of 0.8–0.9kHz (Krabbe & Schulenberg 2003). Songs are repeated relatively frequently, usually every 5–10s. Krabbe & Schulenberg (2003) described the alarm call as a rapid, hollow, accelerating roll with a slightly higher-pitched introductory note, *e-o-o-o-o-o-o-oh*, presumably based on Willis (1988). The latter, however, spent most of his time in Colombia west of the Andes and his description probably derives from encounters with *H. dives barbacoae*. These authors liken it to the song of Cocha Antshrike *Thamnophilus praecox*, from which it differs in being considerably lower pitched.

NATURAL HISTORY Behaviour not reported, partially because most previous descriptions referred to the more commonly seen Thicket Antpitta. However, it probably behaves in a similar fashion to this and other congeners. Forages on the ground, walking, running or hopping, pausing occasionally to turn over leaves or thrust its bill into the litter. It is among the many species of anptitta that sing throughout the day, apparently from perches up to 2m above ground (Krabbe & Schulenberg 2003).

DIET Not documented.

REPRODUCTION No published data on any aspect of breeding biology. **Breeding Data** *H. f. fulviventris* Adult carrying food, 25 December 2006, Gareno Lodge (E.T. Miller *in litt.* 2007); juvenile, 29 June 1964, Puerto Asís (Olivares 1966); subadult, 17 November 1925, mouth of Río Curaráy (AMNH 255993); subadult, 9 February 1923, near Ávila (AMNH 179383). *H. f. caquetae* No data.

TECHNICAL DESCRIPTION Sexes similar. *Adult* Somewhat hooded, with entire crown and nape dark slate-grey, contiguous with the face and ear-coverts. The pale lores and a triangular white postocular spot both contrast fairly sharply with the dark grey hood. The rest of the upperparts are warm brown. The throat and malar area are white, separated by a thin black mesial stripe, often not very apparent. The throat may be tinged fulvous and black hair-like feather tips may give a slightly speckled appearance. Rest of underparts are white, washed pale fulvous on upper breast, the fulvous becoming richer, more ochraceous on sides and especially on the flanks and vent. Usually only the central belly and lower breast are clean white. Underparts are streaked dusky-brown or blackish, heaviest on upper breast and sides, becoming thinner on flanks and lower breast, sometimes extending slightly into white of upper belly. The markings are somewhat variable, usually heavy enough to give a scaled appearance on the upper breast but sometimes reduced to washed-out streaking similar to *H. dives*. **Adult Bare Parts** *Iris* dark brown; *Bill* blackish, base of mandible paler; *Tarsi & Toes* dusky-pinkish to fairly pink. *Juvenile/Subadult* Previously overlooked, a description of what is clearly a juvenile female was provided by Olivares (1966). He described

the bird as being in transitional plumage, indicating that the gape showed clear signs of immaturity, but failed to describe its coloration. The back was olivaceous, washed blackish. The face, crown and nape were dark greyish. The upperwing-coverts have subterminal ochraceous spots and black fringes. Below, largely as adult, but with scattered ochraceous (rusty-brown) immature feathers. Olivares described the tail as being olivaceous with a yellowish cast. There was no mention of immature feathers on the upperparts, suggesting this individual was in the last phase of transitioning from juvenile to subadult. Although Olivares (1966) did not describe the rictus in detail, the fact that he felt its condition was a clear sign of immaturity suggests that, even at this advanced stage of development, it may still have been associating with its parents.

MORPHOMETRIC DATA *H. f. caquetae* Tarsus 38.9mm (*n* = 1♂, USNM 446641). *Wing* 80mm; *Bill* [exposed] culmen 20mm; *Tarsus* 39mm (*n* = 1♂, Chapman 1917). *H. f. fulviventris* Bill from nares 13.3mm, 14.3mm, 13.8mm, exposed culmen 18.1mm, 19.6mm, 19.5mm, width at gape 9.9mm, 11.5mm, 10.0mm; *Tarsus* 37.6mm, 37.8mm, 40.0mm (*n* = 3, ♀,♂,♂, Napo, MLZ). *Bill* from nares 12.8mm, 13.5mm, 12.7mm; *Tarsus* 39.6mm, 39.1mm, 40.4mm (*n* = 3♂♂, Napo, AMNH, WFVZ): *Bill* from nares 13.4mm, n/m; *Tarsus* 39.2mm, 39.7mm (*n* = 2, ♂,♀, subadult, Napo, AMNH). **Mass** 47.5g, 57g, 43.7g (*n* = 3, ♂,♂,♀, Krabbe & Schulenberg 2003); 45.5g(tr) (*n* = 1♂, *fulviventris*, WFVZ 45744); 54g(light) (*n* = 1♂, *fulviventris*, LSUMZ 88072, Brooks *et al.* 2009). **Total Length** 13.5–15.0cm (Dunning 1987, Ridgely & Tudor 1994, Clements & Shany 2001, Restall *et al.* 2006, Schulenberg *et al.* 2010).

TAXONOMY AND VARIATION Two subspecies recognised, but *caquetae* is known from a very restricted area and very few specimens, of which I examined only the type and an additional adult male collected just north of the type locality (near Florencia) that was poorly prepared (USNM 446641). Based on this, and in the absence of any known vocal differences, it seems that *H. fulviventris* is probably best considered monotypic. I provisionally maintain them as distinct taxa for the purposes of this discussion.

Hylopezus fulviventris caquetae Chapman, 1923, American Museum Novitates, no. 96, p. 10. Described as *Hylopezus dives caquetae*. The holotype (AMNH 116350), an adult male, was collected 25 July 1912, by Leo E. Miller at 'La Morelia,' Caquetá, Colombia. The original label gives the locality as 'urelia', the M apparently not having printed correctly (LeCroy & Sloss 2000) but this, and the locality given by Chapman (1917, 1923a) are considered to refer to Morelia (01°29'N, 75°43'W, 240m) or a nearby locality in western Caquetá (Paynter & Traylor 1981, Paynter 1997) near the headwaters of the Río Bodoquero. Chapman (1923a) considered *caquetae* to differ from nominate *fulviventris* in having the upperparts 'brownish olive instead of dark greenish olive', buffy-olive lores, and by having a paler grey crown with less extensive grey plumage (not extending to back). **Specimens & Records** Only one other location is known, also in Caquetá, close enough to the type locality to almost be considered topotypical: Venecia, 01°34'N, 75°32'W (USNM 446641).

Hylopezus fulviventris fulviventris (P. L. Sclater, 1858), Proceedings of the Zoological Society of London, vol. 26, p. 68, 'Rio Napo', Ecuador. Generally said to range from north-east Ecuador south to the Río Marañón in northern Peru (Ridgely & Greenfield 2001, Krabbe & Schulenberg 2003, Schulenberg *et al.* 2010). Here I follow Meyer de Schauensee (1952a,b) in including populations in extreme southern Colombia, the only two records being >135km from the Caquetá records. **Specimens & Records COLOMBIA** *Putumayo* Puerto Asís, 00°30'N, 76°30'W (Olivares 1966). *Nariño* Río San Miguel, 00°23'N, 77°15'W (Meyer de Schauensee 1952b, ANSP 165179–181). **ECUADOR** *Sucumbíos* Río Verde, 00°14'N, 77°34.5'W (Stotz & Valenzuela 2009); Sinangoe, 00°11'N, 77°30'W (Schulenberg 2002); Cuccono, 00°08'N, 77°33.5'W (Schulenberg 2002); *c.*7.5km SW of Lumbaqui, 00°00.5'N, 77°23.5'W (R.A. Gelis *in litt.* 2016); RPF Cuyabeno, 00°01'S, 76°11.5'W (XC 135215, R. Ahlman); Pacayacu, 00°02'S, 76°35'W (ANSP 169719); Totoa Nai'qui, 00°02'S, 76°45'W (Stotz & Quenamá 2007); *c.*3km N of Limoncocha, 00°23'S, 76°38'W (ML 30302, A.B. van den Berg); Pañacocha, 00°25.5'S, 76°06'W (J. Freile *in litt.* 2017); Sani Isla, 00°26'S, 76°17'W (Hollamby 2012); Mandicocha, 00°29'S, 76°22'W (ML 86051, J.W. Wall). *Orellana* San José, Río Suno, 00°26'S, 77°20'W (Chapman 1923b, 1926a, AMNH, MLZ); Río Payamino, 00°27'S, 77°08'W (ANSP, ML); Isla Anaconda, Río Napo, 00°30'S, 76°24.5'W (ML 44547931, photo R. van Twest); Edén, 00°30'S, 76°05'W (MECN 2626); Añangucocha, 00°31.5'S, 76°26'W (HFG); Zancudococha, 00°36.5'S, 75°29.5'W (XC 234515, R.A. Gelis); near Ávila, *c.*00°38'S, 77°26'W (MLZ 7219, AMNH 179383); Concepción, 00°48'S, 77°25'W (MLZ 7217); Boanamo, 01°16'S, 76°23.5'W (Greeney *et al.* 2018). *Napo* Río Hollín, 00°41.5'S, 77°44'W (XC 258617, J.V. Moore); WildSumaco Lodge, 00°41.5'S, 77°36'W (XC 175016, R. Ahlman); 10km SE of Archidona, 00°58.5'S, 77°45'W (WFVZ 45744, MECN 3726); Gareno Lodge, 01°02'S, 77°24'W (E.T. Miller *in litt.* 2014); 8.3km SSW of Tena, 01°04'S, 77°48'W (Krabbe 1991, ML: P. Coopmans). *Pastaza* Shiripuno Research Center, 01°06'S, 76°44'W (XC 13460, N. Athanas); Puyo, 01°29.5'S, 78°00'W (MECN 2628); Territorio Achuar, 01°45'S, 76°30'W (J. Freile *in litt.* 2017); Río Tigre, 02°03'S, 76°04'W (J. Freile *in litt.* 2017); Kapawi Lodge, 02°32.5'S, 76°51'W (Ridgely & Greenfield 2001, XC 282719/720, C. Vogt); Río Bobonaza, 02°36'S, 76°38'W (Berlioz 1932a); Sarayacu, 01°44'S, 77°29'W (Sclater 1890, Cory & Hellmayr 1924, MECN 2627). **PERU** *Loreto* Panguana, 02°08'S, 75°09'W (Stotz & Alván 2007); mouth of Río Curaráy, 02°22'S, 74°06'W (AMNH 255993);

Isla Yagua on Río Napo, 02°59.5'S, 73°18.5'W (ML 29375, T.A. Parker); 2.2km W of Sucusari, 03°15'S, 72°55'W (ML 31726, T.A. Parker); Río Morona, W bank, 04°17'S, 77°14'W (LSUMZ 173005–007). *Amazonas* Vicinity of Huampami on Río Cenepa, 04°28'S, 78°09'W (Brooks *et al.* 2009, LSUMZ 88072).

STATUS Not considered globally threatened (BirdLife International 2016), being uncommon to locally common in Ecuador (Stotz *et al.* 1996, Ridgely & Greenfield 2001), within which most of its range falls. However, the species appears to be genuinely rare in Peru and Colombia. Race *caquetae* is not known from any protected area and its range is apparently very small. The following all refer to nominate *fulviventris*. **Protected Populations** PN Yasuní (Bass *et al.* 2010, XC); RPF Cuyabeno (XC 135215, R. Ahlman); RE Limoncocha (ML 30302, A.B. van den Berg); Reserva Privada El Para (WFVZ, MECN, XC); privately or community-protected forests of Napo Wildlife Center, Sani Lodge, Sacha Lodge, La Selva Lodge (all Río Napo), as well as ExplorNapo Lodge in Peru.

OTHER NAMES Subspecies *caquetae* *Grallaria fulviventris caquetae* (Cory & Hellmayr 1924, Meyer de Schauensee 1950b); *Hylopezus dives caquetae* (Chapman 1923a, LeCroy & Sloss 2000); *Hylopezus dives fulviventris* (Chapman 1917); Caquetá Antpitta (Cory & Hellmayr 1924); Caquetá Ant-pitta (Meyer de Schauensee 1950b). **Subspecies** *fulviventris* *Grallaria fulviventris fulviventris* (Berlioz 1932a, Peters 1951, Meyer de Schauensee 1952a,b); *Grallaria fulviventris* (Sclater 1862, 1890, Gray 1869, Sclater & Salvin 1873b, Dubois 1900, Brabourne & Chubb 1912). **Both subspecies** = *Grallaria fulviventris* (Meyer de Schauensee 1966, 1970, Gruson & Forster 1978). **Spanish** Tororoí Ventricanela Ecuatoriano (Krabbe & Schulenberg 2003); Tororoí Buchicanelo (Salaman *et al.* 2007a); Tororoí de Lorum Blanco (Schulenberg 2002); Tororoí Loriblanco (Clements & Shany 2001, Ridgely & Greenfield 2006, Granizo 2009); Gralaria ventrirrufa (Valarezo-Delgado 1984); Tororoí pechicanelo (Delgado 1984). **English** Fulvous-bellied Antpitta (Gruson & Forster 1978, Butler 1979, Meyer de Schauensee 1964, 1966, 1970, 1982, Parker *et al.* 1982, Howard & Moore 1984; Hilty & Brown 1986, Dunning 1987); Fulvous-bellied Ant-Thrush (Brabourne & Chubb 1912). **French** Grallaire à ventre fauve. **German** Weißwangen-Ameisenpitta (Krabbe & Schulenberg 2003).

White-lored Antpitta, adult (*fulviventris*), Sani Lodge, Sani Isla, Sucumbíos, Ecuador, 14 August 2012 (*Dušan M. Brinkhuizen*).

White-lored Antpitta, adult (*fulviventris*), Sani Lodge, Sani Isla, Sucumbíos, Ecuador, 30 May 2015 (*Nick Athanas/Tropical Birding*).

Fig. 3. **White-lored Antpitta**, adult (*fulviventris*), Sani Lodge, Sani Isla, Sucumbíos, Ecuador, 9 November 2011 (*Roger Ahlman*).

Fig. 4. **White-lored Antpitta**, adult (*fulviventris*), Limoncocha, Sucumbíos, Ecuador, 29 December 2016 (*George Cruz*).

AMAZONIAN ANTPITTA
Hylopezus berlepschi   Plate 20

Grallaria berlepschi Hellmayr, 1903, Verhandlungen der Kaiserlich-Königlichen Zoologisch-Botanischen Gesellschaft in Wien, vol. 53, p. 218. The type locality, given as Engenho do Gama, Rio Guaporé, Mato Grosso, is apparently somewhere in extreme south-west Mato Grosso (Paynter & Traylor 1991a), probably east of Comodoro (13°40'S, 59°47'W).

Amazonian Antpitta is widely distributed, albeit usually locally scarce, over most of southern Amazonia, in Brazil, eastern Peru and northern Bolivia. Like most congeners, it inhabits dense undergrowth of *terra firme* forest, usually below 500m. Above and on head, the plumage is medium brown, whereas the underparts are white with dark streaking on the breast and flanks, becoming warm buff on the posterior flanks. Like most antpittas, Amazonian Antpitta is far more frequently heard than seen, although its fairly simple song is not particularly 'ear-catching'. The usual song is a series of 3–5 somewhat croaked *cuock* notes given at regular intervals. Overall, with Screaming Pihas *Lipaugus vociferans* and toucans above, its monotonously repeated, somewhat trogon-like phrase might easily go unnoticed. Though present at several fairly well-studied research centres, almost nothing is known of its diet or behaviour, and its nest and breeding habits are undescribed. The two subspecies are not particularly separable by voice, plumage or distribution, and this species might be best considered monotypic. It has been suggested that the present species is only a subspecies of White-lored Antpitta, but its true taxonomic affinities remain uninvestigated.

IDENTIFICATION 14–15cm. Sexes similar. Fairly plain, with the upperparts, including nape and crown, wings and tail olive- or grey-brown. There is a buffy loral spot, an indistinct malar stripe and white on the throat and below the ear-coverts. Underparts white, washed on breast and flanks with buffy-ochraceous and diffusely streaked

dark brownish (Meyer de Schauensee 1970). Thrush-like Antpitta (extensive overlap) is similarly plain but usually has more diffuse breast streaking. White-lored Antpitta (limited overlap in northernmost Peru) is similar, but has a distinctly grey crown and nape, a more distinctly patterned face, and is somewhat paler overall below. Spotted Antpitta (limited overlap in Bolivia) has a bold ochraceous eye-ring and spotted, rather than streaked, breast.

DISTRIBUTION Amazonian Antpitta is just that, a bird of the Amazon Basin. It occurs from east-central Peru and immediately adjacent western Brazil, south to northern Bolivia. East, its range extends across the lower Brazilian Amazon through northern Mato Grosso to north-east Pará (Bond & Meyer de Schauensee 1942, Remsen *et al.* 1986, Remsen & Traylor 1989, Sibley & Monroe 1990, Parker & Bailey 1991, Altman & Swift 1993, Hennessey *et al.* 2003b, Clements 2007).

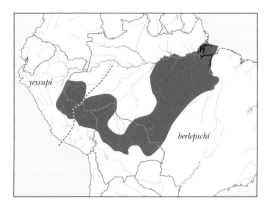

HABITAT Like most antpittas, it is found most often on or near the forest floor (O'Neill 1974, Dubs 1992). The species prefers forest edge (Henriques *et al.* 2003, Fávaro & Flores 2009), overgrown clearings (Robinson *et al.* 1990, Allen 1995, Lebbin 2007), riparian gallery forest within *cerrado* (Silva 1995, 1996), Amazonian alluvial forests dominated by *Guadua* bamboo, small palms (Aleixo & Guilherme 2010), *Gynerium* cane or *Cecropia* (Remsen & Parker 1983, Robinson & Terborgh 1997), but is also reported in *terra firme* (Brace *et al.* 1997, Pacheco *et al.* 2007). To 500m in north-east Peru (*yessupi*) (Clements & Shany 2001), locally to 600m (Walker *et al.* 2006) or occasionally 700m (Schulenberg *et al.* 2007).

VOICE The primary vocalisation is a hollow, guttural sequence of 3–5 notes, very similar to that of White-lored Antpitta but distinctly slower. Ridgely & Tudor (1994) likened it to the advertising call of Zigzag Heron *Zebrilus undulatus: kho, kho, kho, kho, kho, kho* or *coo!, coo!, coo!* (Schulenberg *et al.* 2007). Described in Brazil as 3–4 mid-to high, well-separated, hollow *Uh* notes (van Perlo 2009). Quantitatively, this vocalisation is 2–3s long, delivered at intervals of 5–10s and is composed of an even-paced series of 3–4 (rarely two) similar downslurred *cuock* notes at frequencies of 0.8–0.9kHz (Krabbe & Schulenberg 2003). The call note was described by Krabbe & Schulenberg (2003) as a 1.6s rattle of 14–16 notes at 0.8kHz.

NATURAL HISTORY The general behaviour of this elusive species is very poorly studied but is generally similar to other *Hylopezus*, probably most like White-lored Antpitta. It forages alone, occasionally in pairs. It walks, runs or hops on the ground or on low horizontal branches and logs (O'Neill 1974), lunging at potential prey. When singing, usually from a low perch, its throat expands greatly to expose bright pink skin below the feathers (WA: e.g., photos C. Albano, J. Quental, F.K. Ubaid). In Peru, Robinson & Terborgh (1997) reported densities of 7.5 territories per 100ha (*berlepschi*).

DIET Food consumed by Amazonian Antpitta is, essentially, undocumented, and specimen label data I was able to find listed little other than 'stomach contents insects'. Presumably, its diet includes a variety of arthropods and probably small vertebrates such as lizards and frogs.

REPRODUCTION The nest is undescribed. Probably it is a shallow, loosely woven, open cup composed externally of sticks with a sparse lining of flexible fibres. Atop vine tangles or at base of understorey palm leaf rosettes are likely places to look. **SEASONALITY** Very few data. Snethlage (1907a) reported a 'juvenile' male (*berlepschi*) collected 16 or 17 March 1904 at Bom Lugar on the Rio Purús, but gave no details to correctly age it. **BREEDING DATA *H. b. yessupi*** Subadult, 24 November 1927, Santa Rosa (AMNH 240344); subadult transitioning, 7 July 1964, San Pablo (FMNH 297946). ***H. b. berlepschi*** Skeletal specimen with 50% cranial ossification (juvenile?), 7 July 1981, Río Beni, N of Puerto Linares (LSUMZ 101400); subadult, 21 July 1015, Todos Santos (AMNH 137176); subadult, 5 February 1917, Astillero (AMNH 146168); subadult, 26 September 1959, Tapaina (FMNH 258117); subadult, 15 August 1986, 52km SSW of Altamira (USNM 572603); subadult transitioning, 28 July 1985, Hacienda Amazonia (FMNH 322344); female with enlarged ovary, 17 October 1958, Río Madre de Dios (FMNH 251971).

TECHNICAL DESCRIPTION Sexes similar. *Adult* The following description refers to the nominate race, but differences between the two subspecies are minimal. The upperparts are largely greyish- to olive-brown, including the tail. The head is simply marked with a buffy spot,

an indistinct blackish malar, and has the white of the throat extending to below the ear-coverts. The wings are similar to the back but outer webs of the primaries are paler, olive-buff. The central belly and lower chest are largely white, the chest washed pale fulvous and diffusely streaked brown. The flanks and undertail-coverts are deep ochraceous. **Adult Bare Parts** *Iris* brown; *Exposed Orbital Skin* pinkish-red; *Bill* maxilla pale brown, mandible dusky-pinkish, darker near tip; *Tarsi & Toes* pale pinkish-grey (Carriker 1930; Dubs 1992, Krabbe & Schulenberg 2003).

MORPHOMETRIC DATA *H. b. yessupi Wing* 82mm; *Tail* 36mm; *Bill* [total culmen?] 20mm; *Tarsus* 38mm ($n = 1$♂, holotype, Carriker 1930). *Bill* from nares 12.2mm, 12.5mm; *Tarsus* 36.0mm, 37.1mm ($n = 2$♀♀, AMNH). *Bill* from nares 12.1mm; *Tarsus* 36.7mm ($n = 1$♀, subadult, AMNH 240344). *H. b. berlepschi Wing* 83mm; *Tail* 36mm; *Bill* [total culmen?] 22mm; *Tarsus* 40mm ($n = 1$♂, Hellmayr 1903). *Wing* 87mm; *Tail* 33mm; *Bill* [exposed] culmen 17mm; *Tarsus* 34mm ($n = 1$♀, Gyldenstolpe 1945b). *Wing* 84mm; *Tail* 36mm; *Bill* [total culmen?] 21mm ($n = 1$♀, Cory & Hellmayr 1924). *Bill* from nares 12.6mm, n/m, n/m, 12.8mm, 13.0mm, n/m, 13.6mm; *Tarsus* 40.6mm, 38.7mm, 38.7mm, 38.9mm, 41.1mm, 39.4mm, 39.6mm ($n = 7$, 3♀♀,4♂♂, Pará, LACM). *Bill* from nares 12.2mm, 12.8mm; *Tarsus* 38.3mm, 42.9mm ($n = 2$♀♀, AMNH 286766 from Pará, AMNH 824073 from Madre de Dios). *Bill* from nares 13.3mm; *Tarsus* 37.7mm ($n = 1$♀, Pará, USNM 276962). *Wing* 86mm; *Tail* 42mm; *Bill* [total culmen?] 19mm; *Tarsus* 36mm ($n = ?$, ♀?, Snethlage 1914a, likely the source of measurements given by Ruschi 1979), *Wing* 83mm; *Tail* 35mm; *Bill* [total culmen?] 20mm ($n = ?$, ♀?, Dubs 1992). *Bill* from nares 12.6mm, 11.9mm, 11.8mm, 39.5mm, 40.7mm, 37.1mm ($n = 3$, ♀♀, subadults, USNM, AMNH). **Mass** 46g ($n = 1$♂, *yessupi*, O'Neill 1974); 48g ($n = 1$, ♀?, *berlepschi*, Tergorgh *et al.* 1990); range 46–54g, 36.0–48.9g ($n = ?$, ♂♂,♀♀, Krabbe & Schulenberg 2003); 39.0g(no) ($n = 1$♀, *berlepschi*, subadult, USNM 572603 presumably reported by Graves & Zusi 1990). **TOTAL LENGTH** 13–15cm (Ruschi 1979, Dunning 1987, Dubs 1992, Ridgely & Tudor 1994, Clements & Shany 2001, Krabbe & Schulenberg 2003, Schulenberg *et al.* 2007, van Perlo 2009).

TAXONOMY AND VARIATION The racial affiliation of some populations in south-east Peru and adjacent Brazil (see *berlepschi* below) is uncertain, but they are provisionally considered to belong to the nominate race, following Krabbe & Schulenberg (2003). Here I recognise two subspecies, but as differences between them are slight, and there is no apparent geographic barrier separating the two taxa, the species might best be considered monotypic.

Hylopezus berlepschi yessupi Carriker, 1930, Proceedings of the Academy of Natural Sciences of Philadelphia, vol. 82, p. 370. The type specimen (ANSP 91233) is an adult male collected at 'Puerto Yessup, Junin, Peru' [= Puerto de Yesup, Pasco, 10°26.5'S, 74°54.5'W] on 17 February 1930 by M.A. Carriker Jr. Race *yessupi* is found in eastern Peru, at least as far south as the type locality in Pasco, north through Ucayali to central Loreto (Schulenberg *et al.* 2001b) and immediately adjacent Brazil. Until further evidence presents itself, I consider *yessupi* to be largely restricted to the Río Ucayali drainage, including records in northern Cuzco and eastern Junín. Several plausible records from eBird in eastern Loreto and Ucayali are all in either the Rio Juruá or Río Javari drainages, and are also considered *yessupi*. Only slight further south, however, Peruvian and Brazilian records from the upper Rio Purús drainage are included in nominate (see below). According to Carriker (1930), *yessupi* differs from the nominate race in being slightly darker above, more brownish and less olivaceous, with the crown slightly darker than the back. The lores, head-sides and underparts except the central belly are rich fulvous or buffy-ochraceous, most richly coloured on the breast, flanks and undertail-coverts (unlike the largely white underparts of *berlepschi*) (Carriker 1930, Bond 1950). Entire breast, lower throat and sides irregularly marked with thin black streaks. Central belly a decidedly paler buffy-white. Upperwing-coverts are darker than the back, sooty-brown, with the median coverts tipped narrowly fulvous. The flight feathers are brown, broadly edged rufous-brown. Uppertail-coverts fulvous, distinct from the dull rufous tail. Underwing-coverts and inner margins of flight feathers are rich cinnamon-ochraceous. Maxilla black and mandible pinkish. Toes and tarsi are pinkish-lavender. **SPECIMENS & RECORDS PERU:** *Loreto* Actiamë, 06°19'S, 73°**09.5**'W (Stotz & Pequeño 2006). *Pasco* San Pablo, 10°27'S, 74°52'W (FMNH 297946). *Junín* Poyeni, 11°15'S, 73°40'W (A. García-Bravo *in litt.* 2017). *Cuzco* San Martín oil well, 11°47'S, 72°42'W (Angehr *et al.* 2001); 9km ESE of Puerto Mainiqui, 11°57'S, 72°57'W (eBird: R. Zeppilli T.). *Ucayali* Santa Rosa, 10°26'S, 73°31'W (AMNH 240344–344); Cerro Tahuayo, 08°08'S, 74°01'W (LSUMZ 156622); near Puntijao, 10°20'S, 73°58'W (ML 165674, M.G. Harvey); Pucani, 10°40'S, 73°31'W (UAM 30984); Quebrada Shicotsá, 10°26'S, 74°06'W (ML). **BRAZIL:** *Acre* Foz do Jurupari, 07°50.5'S, 70°04.5'W (Guilherme 2016); Porongaba, 08°40'S, 72°47'W (MPEG 49669); *c.*7km NNW of Marechal Thaumaturgo, 08°52.5'S, 72°47'W (MPEG 52916/17); Foz do Tejo, 08°59'S, 72°43'W (XC 90411, J. Minns); Foz do Breu, 09°24.5'S, 72°43'W (XC 90410/11, J. Minns). *Amazonas* Eirunepé, 06°50.5'S, 70°14.5'W (WA 63430, recording T.V.V. Costa).

Hylopezus berlepschi berlepschi (Hellmayr, 1903). Found only in Brazil, Peru and Bolivia. In Brazil it is known from a few localities in southern and western Amazonas, through most of Pará (Pacheco & Olmos 2005, Pacheco *et al.* 2007) and northern and north-west Mato Grosso (Naumburg 1930, Dubs 1992). In Pará, the easternmost records appear to be in the vicinity of São Geraldo do Araguaia and Canaã dos Carajás, and there is a notable absence of records from north-east Pará south of Belém, leading me to suggest the left bank of the lower Rio Araguaia as the eastern limit in Brazil. Populations in southern Acre in the upper drainage of the Rio Purús (Aleixo & Guilherme 2010) are of uncertain racial affiliation, but are here included in *berlepschi*. Populations near Balta, in extreme eastern Ucayali, are similarly worthy of evaluation, but are also in the Rio Purús drainage and are provisionally considered to represent *berlepschi*. In south-east Peru the nominate subspecies is reported from south-east Cuzco, through most of Madre de Dios (Lebbin 2007) and Puno, and south to north-central Bolivia in Pando, Cochabamba, El Beni, La Paz and Santa Cruz. **SPECIMENS & RECORDS BRAZIL:** *Amazonas* Bom Lugar, 08°43'S, 67°20'W (Gyldenstolpe 1951, MPEG 3651/52); Maués, 05°18'S, 58°11'W (Dantas *et al.* 2011). *Pará* Diamantina, 00°35'S, 49°06'W (USNM 276962); Santarém, 02°27'S, 54°42'W (Carriker 1930, Lees *et al.* 2013a, CM, FMNH, UMMZ, YPM); Maicá, 02°36'S, 54°23'W (ANSP 76556); Caxiricatuba, 02°50'S, 55°08'W (AMNH 286766, MCZ 175129); Tapaiuna, 02°54'S, 55°05'W (FMNH 258117); near Belo Monte, 03°04.5'S, 51°42'W (MPEG 82770); Tauari, 03°06'S, 55°07'W (MCZ 175130–

132); Lago Arauepá, 03°36'S, 55°20'W (LACM 31932–934); 52km SSW of Altamira, 03°39'S, 52°22'W (USNM 572603); Urucurituba, 03°47.5'S, 55°31.5'W (LACM 31931); *c.*8km west of Pacajá, 03°51'S, 50°43'W (XC 212975, G. Leite); Mirituba, 04°18'S, 55°57'W (Griscom & Greenway 1941, CM); Jardim Botánico (Parauapebas), 06°07'S, 49°50.5'W (ML 23728151, photo I. Thompson); São Geraldo do Araguaia, 06°12'S, 48°48'W (WA 1701310, recording P. Pacheco); Canaã dos Carajás, 06°32'S, 49°51'W (WA 874059, E. Lima); Rio Cururu, 07°12'S, 58°03'W (LACM 44314–316). *Mato Grosso* RPPN Rio Cristalino, 09°36'S, 55°56'W (Zimmer *et al.* 1997, ML); 40km S of Alta Floresta, 10°14'S, 56°05'W (ML 48054/59, P.R. Isler); Comodoro, 13°13'S, 59°39.5'W (WA 715312, V. Castro); Vila Bela da Santíssima Trindade area, *c.*15°00'S, 59°57'W (Silveira & d'Horta 2002); Tapirapuã, 14°51'S, 57°45'W (MNRJ 23694). *Acre* Santa Cruz Velha, 09°01'S, 69°32'W (Guilherme & Dantas 2011); RESEX Cazumbá-Iracema, 09°08'S, 68°57'W (eBird: T. Melo); Rio Branco, 09°58.5'S, 67°49.5'W (WA 1429223, J. Quental); PE Chandless, 10°01.5'S, 70°21.5'W (Buzzetti 2008); ESEC Rio Acre, *c.*10°57'S, 70°17'W (Aleixo & Guilherme 2010). **PERU:** *Ucayali* Río La Novia, 09°56'S, 70°42'W (Angulo *et al.* 2016); Balta, 10°06'S, 71°14'W (O'Neill 1974, 2003b, LSUMZ). *Cuzco* Pilcopata, 12°59'S, 71°17'W (ML 30051, T.A. Parker); Manu Road, 'Tangara corner', 13°02'S, 71°31.5'W (D. Beadle *in litt.* 2017). *Madre de Dios* Noaya, 11°07.5'S, 69°34'W (XC 146549, D.F. Lane); Oceania, 11°25.5'S, 69°32.5'W (ML 129506, P.A. Hosner); Cocha Cashu, 11°53'S, 71°24'W (Robinson & Terborgh 1997, AMNH 824073); Pakitza, 11°56'S, 71°15'W (Servat 1996); 5.5km NW of Iberia, 11°23'S, 69°32'W (ML: P.A. Hosner); Romero Rainforest Lodge, 12°13.5'S, 70°59'W (R.C. Hoyer *in litt.* 2017); Boca Manú, 12°17'S, 70°53.5'W (A. Farnsworth *in litt.* 2017); Manú Wildlife Center, 12°24'S, 70°42'W (XC 23232, P. Donahue); Centro de Investigación Río Los Amigos, 12°34'S, 70°06'W (ML, XC); Pantiacolla Lodge, 12°39.5'S, 71°14.5'W (XC 23375, D. Geal); 10km SW of Puerto Maldonado, 12°40'S, 69°15'W (ML); Maskoitania, 12°46.5'S, 71°23'W (FMNH 433523); Amazonia Lodge (= Hacienda Amazonia on labels), 12°52'S, 71°22.5'W (FMNH 322344–346, 315841, ML, XC); Collpa Guacamayo de Blanquillo, 13°08'S, 69°35'W (ML, C.A. Marantz). *Puno* Astillero, 13°22'S, 69°37'W (AMNH 146168). **BOLIVIA:** *Pando* Rutina, 11°25'S, 69°06'W (Schulenberg *et al.* 2000, ML); Área de Inmovilización Madre de Dios, *c.*11°34'S, 67°08'W (O'Shea *et al.* 2003). *La Paz* Río Beni, 20km N by river of Puerto Linares, 15°25'S, 67°29'W (LSUMZ). *El Beni* El Consuelo, 14°20'S, 67°15'W (Glydenstolpe 1945). *Santa Cruz* Flor de Oro, 13°22'S, 61°00.5'W (Bates *et al.* 1998); Los Fierros, 14°33.5'S, 60°56'W (Bates *et al.* 1998); Lago Caimán, 13°36'S, 60°55'W (Bates *et al.* 1998, XC, ML); Perseverancia, 14°44'S, 62°48'W (ML 90174, G. Cox); El Refugio, 14°45'S, 61°01.5'W (Bates *et al.* 1998); Estación Biológica Caparú, 14°48'S 61°10'W (Vidoz *et al.* 2010); Velasco, 14°50'S, 60°25'W (LSUMZ 153428); Puerto Peligro, 16°35'S, 64°44.5'W (MHNNKM 6021); Propiedad Nuevo Mundo, 17°23.5'S, 63°50'W (DMNH 65285); Buena Vista, 17°27.5'S, 63°40'W (LSUMZ 37768, CM P119862, FMNH 217711); 60km E of Asención, 17°51'S, 60°10'W (ML 90181/182, G. Cox). *Cochabamba* Sajta, 17°06'S, 64°46.5'W (XC 2553/62/75, S. Mayer); Todos Santos, 16°48'S, 65°08'W (AMNH 137176, ANSP 140806).

STATUS Not currently considered globally threatened (BirdLife International 2017). Overall, it is reported as uncommon to locally fairly common (Parker *et al.* 1994a,b,

Stotz *et al.* 1996), but seems genuinely rare in many parts of Peru (Parker *et al.* 1982, Terborgh *et al.* 1984, 1990, Servat 1996).The nominate subspecies is protected within a fair number of reserves. Race *yessupi* appears less protected, but its full range is much less well known and it may still be found in reserves within its known or potential range. Although sometimes considered vulnerable to disturbance (Bird *et al.* 2012), Amazonian Antpitta was among those species least affected by fragmentation in some studies (Lees & Peres 2006) and may, in fact, be favoured by mild habitat disturbance (Felton *et al.* 2007, Lees & Peres 2008, 2010). PROTECTED POPULATIONS *H. b. berlepschi* **Brazil** FLONA Pau-Rosa (Dantas *et al.* 2011); FLONA Humaitá (eBird: C. Gussoni); FLONA Tapajós (Forrester 1993, Oren & Parker 1997, Henriques *et al.* 2003); ESEC Rio Acre (Aleixo & Guilherme 2010); ESEC Terra do Meio (Fávaro & Flores 2009); RESEX Cazumbá-Iracema (eBird: T. Melo); PE Chandless (Buzzetti 2008); RPPN Rio Cristalino (XC, ML). **Bolivia** Reserva de la Biosféra Pilón Lajas (Hennessey *et al.* 2003a); PN Noel Kempff Mercado (Wheatley 1995, Bates *et al.* 1998); PN Madidi (Parker & Bailey 1991, Hennessey *et al.* 2003b); PN Amboró (Gemuseus & Sagot 1996); the 160,000ha Beni Biological Station (Brace *et al.* 1997); Reserva Nacional Amazónica de Vida Silvestre Manuripi (Hennessey *et al.* 2003b). **Peru** RN Tambopata and presumably the adjacent PN Bahuaja (Donahue 1994, Allen 1995); PN Manú (Terborgh *et al.* 1984, Walker *et al.* 2006); Reserva Cuzco Amazónico

(Davis *et al.* 1991); RP Amazonia Lodge (Wheatley 1995); RP Pantiacolla Lodge (Clements & Shany 2001); CC Río La Novia (Angulo *et al.* 2016). *H. b. yessupi* **Peru** RN Matsés (Stotz & Pequeño 2006). **Brazil** RESEX Alto Juruá (XC 90410/11, J. Minns). Although not recorded during surveys of Peru's Zona Reservada Sierra del Divisor (Schulenberg *et al.* 2006), *yessupi* probably occurs there, and in adjacent parts of Brazil's PN Serra do Divisor.

OTHER NAMES Subspecies *yessupi* Grallaria berlepschi *yessupi* (Carriker 1930, Bond 1950, Peters 1951). **Subspecies** *berlepschi* Grallaria berlepschi berlepschi (Bond & Meyer de Schauensee 1942, Gyldenstolpe 1951, Peters 1951); Grallaria fulviventris (von Pelzeln 1871), Grallaria berlepschi (von Ihering & von Ihering 1907, Brabourne & Chubb 1912, Snethlage 1914a, Cory & Hellmayr 1924, Naumburg 1930, Pinto 1938, Dubs 1992). **Both subspecies** = Grallaria berlepschi (Meyer de Schauensee 1966), Hylopezus [fulviventris] berlepschi (Sibley & Monroe 1990). **Portuguese** torom-torom (Snethlage 1914a, Pinto 1978, Frisch & Frisch 1981, Willis & Oniki 1991b, Dubs 1992, van Perlo 2009), torom-torom-garganta-branca (Ruschi 1979), trontrom (Pinto 1938). **Spanish** Tororoí Amazónico (Clements & Shany 2001, Krabbe & Schulenberg 2003, Plenge 2017). **English** Brazilian Ant-Thrush (Brabourne & Chubb 1912), Brazilian Antpitta (Cory & Hellmayr 1924), Berlepsch's Antpitta (Walters 1980). **French** Grallaire d'Amazonie (Krabbe & Schulenberg 2003). **German** Olivmantel-Ameisenpitta (Krabbe & Schulenberg 2003).

Amazonian Antpitta, adult singing (*berlepschi*), Rio Branco, Acre, Brazil, 20 August 2014 (*João Quental*).

Amazonian Antpitta, adult (*berlepschi*), Parauapebas, Pará, Brazil, 19 July 2014 (*Ciro Albano*).

Amazonian Antpitta, adult singing (*berlepschi*), Parauapebas, Pará, Brazil, 5 December 2013 (*Ciro Albano*).

367

Amazonian Antpitta, adult (*berlepschi*), Parauapebas, Pará, Brazil, 13 May 2017 (*Caio Brito*).

WHITE-BROWED ANTPITTA
Hylopezus ochroleucus   Plate 20

Myioturdus ochroleucus Wied, 1831, Beiträge zur Naturgeschichte von Breasilien, vol. 3, p. 1032., in der Nähe der Arrayal da Conquista, inneren Provinz Bahiá [= Vitória da Conquista, southern Bahia, Brazil, 926m, 14°51'S, 40°51'W]. The male holotype has apparently been lost (Allen 1889c, LeCroy & Sloss 2000) and to my knowledge a neotype has not been designated (Whitney *et al.* 1995).

Until recently, White-browed Antpitta was considered conspecific with Speckle-breasted Antpitta of the southern Atlantic Forest, but it is now separated on the basis of its strikingly different plumage, especially the white supercilium, different vocalisations and rather different habitat preferences. The two species are allopatric, with White-browed Antpitta inhabiting deciduous and semi-deciduous *caatinga* woodland of north-east Brazil. Although its behaviour appears largely similar to other *Hylopezus*, there have been no detailed studies on any aspect of its diet or foraging behaviour, and its nest has only recently been described. Ecological studies to determine the types of habitat required by this species, and to understand its level of tolerance to disturbance are of top priority. It will also be important to survey known localities and implement regular monitoring studies to determine population trends and rates of range contraction.

IDENTIFICATION 12–14cm. Adult characterised by white lores and eye-ring, a buffy-white postocular streak, and pale greyish-olive ear-coverts. Crown grey, nape olive-grey and rest of upperparts olive-brown or olive-greyish. In some, buffy wing-covert tips form two indistinct rows of spots, visible to varying degrees. Wings similar in coloration to back, with a white edge and roughly square-shaped patches of black and ochraceous at base of primaries. Throat white bordered by a dusky throat stripe, and upper breast white, with prominent black streaks

and spots. The sides and flanks are buffy-ochraceous. The upright Antpitta posture, bold black markings on a white breast and bright buffy rear underparts make White-browed Antpitta nearly unmistakable within its range. The somewhat similar, and formerly conspecific, Speckle-breasted Antpitta lacks the postocular stripe and has a distinctly more speckled breast. The range of White-browed Antpitta, however, does not overlap with Speckle-breasted, or with any other *Hylopezus*.

DISTRIBUTION Endemic to the *caatinga* of eastern Brazil from northern Ceará to northern Minas Gerais, as far south as the Serra do Espinhaço (Vasconcelos 2003a, Diniz *et al.* 2012).

HABITAT Inhabits the lower growth of both tall and shrubby *caatinga* (Santos 2004b, 2008), seeming to prefer dense tangles (van Perlo 2009, Albano 2010a). It is most common in lush, semi-deciduous woodland but can also be found in fully deciduous habitats (Olmos 1993). White-browed Antpitta generally occurs at elevations of 400–1,000m (Krabbe & Schulenberg 2003, Ridgely & Tudor 2009), in some areas reaching nearly 1,200m (Pereira *et al.* 2014).

VOICE The song is delivered in two or three parts, the first ascending, the second with doubled notes, and the final portion sometimes with triple-parted notes (van Perlo 2009, Ridgely & Tudor 2009). The song is 3–4s in duration and includes a total of 12–14 notes starting at *c.*2.2kHz and rising to *c.*2.6kHz. The first four notes alternate low-high-low-high, with the initial two notes slightly quieter and the next two somewhat transitional. The last 6–8 notes are even-paced, short, downslurred whistles, ending in a fairly weak note (Krabbe & Schulenberg 2003). This song may be transcribed as *whu-whú, whu-whú-whu-whu-wheú-wheú-wheú-wheú-wheú-wheú-wheú* (Ridgely & Tudor (2009), *piupiu-piupiu tetju tetju tetju tetjutu* (van Perlo 2009) or *uhu-uhu, hu-hu-hu-hu* (Gwynne *et al.* 2010).

NATURAL HISTORY No details of general behaviour are available for this poorly known species. Adults sing from somewhat elevated perches inside thickets and forage by walking and hopping on the forest floor, occasionally probing the leaf litter and soil (Krabbe & Schulenberg 2003, van Perlo 2009). Interestingly, as can be clearly seen in a video taken by C. Albano (IBC 1197313), White-browed Antpitta, at least occasionally, twitches its hindquarters back and forth in a manner very similar to *Grallaricula* antpittas. As far as I know, this is the only evidence of this stereotyped movement in a *Hylopezus*, and it would be interesting to see if this behaviour is more widespread in the genus. The species is a reported host of blood parasites (Haemoproteidae; Bennett *et al.* 1987), but nothing is known of its interactions with predators.

DIET The diet of White-browed Antpitta appears to be completely undocumented, but presumably includes a selection of arthropods, earthworms and small invertebrates, similar to that of other species of *Hylopezus* (Krabbe & Schulenberg 2003, Greeney 2014g, Horsley *et al.* 2016).

REPRODUCTION The reproductive biology was completely unknown until very recently. The following details are from a nest found on 25 March 2015 near Araripe (Greeney *et al.* 2016). **NEST & BUILDING** There are no published observations concerning nest-site selection or building behaviour, but it is likely that both adults partake in these activities. The nest was 0.5m above ground, in a forested area with typical deciduous *caatinga* and a relatively dense understorey of small dicots and tangled vines. The terrain around the nest was comparatively arid, uneven and hilly, more than 0.5km from the nearest riparian area. The frail, shallow cup was built of loosely woven sticks, leaf petioles and thin vines, described by the authors as dove-like, and supported from below by 5–6 thin, overlapping lianas and branches. It did not have a well-defined inner cup lining except for apparently a higher concentration of thin, flexible materials. To minimise disturbance to the nest, it was not measured directly. They provided the following estimated nest dimensions: outer diameter 12–15cm with some

twigs extending beyond the bulk of the nest an additional 4–5cm; external thickness 5–7cm; internal diameter 6–7cm; internal depth 3–4cm. **EGG, LAYING & INCUBATION** The nest was discovered when the authors followed an adult that was singing loudly and repeatedly in response to playback. They found that the adult was sat on the nest at the time, presumably in the final stages of incubation. It is impossible to know, from this single observation, how commonly the species sings from the nest, especially considering that in this instance it was in response to playback. It seems likely, however, that this is a normal part of their breeding biology as it is for *Grallaria* (Greeney *et al.* 2008a). The eggs are undescribed. **NESTLING & PARENTAL CARE** The authors were unable to make any detailed observations on the behaviour of adults while caring for young. They did, however, describe the newly hatched young. Nestlings estimated with a degree of accuracy to be 0–1 day old were completely naked. They had their eyes completely closed, dark greyish-pink skin, and bright orange bills with a yellow-white egg-tooth. Contour feather tracts were barely starting to develop, visible just below the skin. Three days later, the nestlings were well covered in dense, wool-like, rusty-red or reddish-brown down. The feathers of the capital and spinal tracts were dark rufescent-brown, those of the humeral tracts slightly paler, and the feathers of the femoral, pelvic spinal and ventral abdominal tracts pale rusty-buff. The unbroken flight-feather pins had emerged through the skin roughly 1.5–2.0cm, with the primary feather sheaths unbroken and those of the secondaries with 1–3mm of bright ochraceous-buff feather vanes exposed at their tips. Their eyes were just starting to open, their bills were still bright orange, but with the inflated rictal flanges pale yellow-orange. Brief observations made during two approaches to the nest provided what appears to be the first observation of a broken-wing display for any antpitta. When an adult was brooding the two hatchlings, it remained frozen until the observer approached to within 0.3m, then dropped from the nest and disappeared silently into the undergrowth. When the brooding adult was approached three days later, however, its response was quite different. It dropped from the nest when the observer was *c.*1m away, but instead of disappearing it lowered both wings to the ground and moved back and forth in small semi-circular movements, dragging its wings in an apparent distraction display. It remained within 1m of the observer, continuing to feign injury during the brief period at the nest. **SEASONALITY** The nestlings hatched on 4 or 5 April, which coincides with the middle of the wettest months in the relevant region.

TECHNICAL DESCRIPTION Sexes similar. *Adult* This plumage is similar in both sexes and is characterised by white or buffy-white lores and eye-ring, and a narrow white or buffy-white postocular streak. The latter is thinly bordered black and the lores often include some black feathers. Ear-coverts buffy-olive or greyish-olive with indistinct white shaft-streaks. Upperparts mostly olive-grey, more grey than olive on crown. Wing-coverts have light buff tips that sometimes are visible as two indistinct rows of spots or broken wingbars, but are not always prominent. Primary wing-coverts dark, contrasting with the ochraceous bases to the primaries, forming a distinct black and ochraceous wing patch. Alula bright white. Primaries and secondaries are similar to the lower back, somewhat more brownish, and the primaries have distinct ochraceous edges on the front margin. Throat and most of breast white except dusky lateral throat stripe and prominent streaks and spots on breast. From lower breast down, including sides and flanks, buffy-ochraceous, with some of the breast streaking sometimes extending into the buffy area, especially centrally. Central belly, undertail-coverts and thighs paler buff or whitish. **Adult Bare Parts** *Iris* dark brown; *Bill* black, with basal half of mandible paler; *Tarsi & Toes* pink to greyish-pink. **Immature** Undescribed.

MORPHOMETRIC DATA *Wing* 76.2mm; *Tail* 38.1mm; *Tarsus* 35.6mm (*n* = 1♀?, Sclater 1858). *Wing* 76mm; *Tail* 38mm; *Bill* [exposed?] culmen 18mm; *Tarsus* 35mm (*n* = 1♂, Pinto 1937a). *Bill* from nares 11.3mm; *Tarsus* 37.5mm (*n* = 1♂, Bahia, AMNH 243146). *Wing* 75mm, 78mm, 79mm; *Tail* 42mm, 43mm, 45mm; *Bill* [exposed culmen] 17mm, 18mm, 19mm (*n* = 3♂♂, FMNH 63518-20, Hellmayr 1929). I measured the same three ♂♂ measured by Hellmayr (1929): *Wing* 76mm, n/m, n/m; *Bill* from nares 11.4mm (×3), exposed culmen 17.0mm, 16.8mm, 16.6mm, depth at nares 4.5mm, 4.9mm, 4.5mm, width at nares 3.9mm, 4.1mm, 4.2mm, width at base of mandible 5.3mm, 5.9mm, 5.8mm; *Tarsus* 33.9mm, 38.3mm, 35.4mm. **Mass** 28g (*n* = ?♂?, Sick 1993). **Total Length** 12.5–14.0cm (Sclater 1858, Meyer de Schauensee 1970, Sick 1993, van Perlo 2009, Ridgely & Tudor 2009).

DISTRIBUTION DATA Endemic to **Brazil**: *Ceará* Tianguá, 03°44'S, 41°00'W (XC 202306, R. Silva Lopes Junior); São Paulo, Serra de Ibiapaba, 04°19.5'S, 40°48.5'W (Snethlage 1924, MPEG 7219/20, MNRJ 16255); Várzea Formosa, 04°45'S, 40°53'W (Hellmayr 1929, FMNH 63518–520); Crateús, 05°10.5'S, 40°40'W (WA 1488982, M. Holderbaum); Parambu, 06°17'S, 40°40.5'W (WA 1417174, J. Bob); Reserva Florestal Sítio Nascente, 07°13.5'S, 39°59'W (ML 35880341, photo F. Olmos); PNM Distrito de Brejinho, 07°13.5'S, 40°00'W (Greeney *et al.* 2016); FLONA Chapada do Araripe near Crato, 07°15'S, 39°35'W (XC 5031, N. Athanas); Barbalha, 07°18'S, 39°18'W (WA 1916972, G. Luiz). *Piauí* São Julião, 07°04'S, 40°49'W (WA 1796136, R.R. Rocha); Pavussu, 07°58'S, 43°13.5'W (WA 200643, C. Albano); 13km WSW of Eliseu Martins, 08°08'S, 43°47'W (XC 62916, C. Albano); PN Serra da Capivara, 08°30'S, 42°20'W (Olmos 1993); Projeto Cajugaia, 08°31'S, 43°17.5'W (MPEG 75511); Zabelê, 08°43'S, 42°33'W (MPEG 76761); *c*.15km NNW of São Raimundo Nonato, 08°53'S, 42°43.5'W (MPEG 75481); Baixão do Abílio, 08°58.5'S, 43°49'W (MPEG 76770); PN Serra das Confusões, 09°13'S, 43°29'W (Silveira & Santos 2012, MPEG 76083); near Morro Cabeça no Tempo, 09°39'S, 44°02.5'W (Santos *et al.* 2012, MPEG 68134); Curimatá, 09°41.5'S, 44°14.5'W (MPEG 68135). *Pernambuco* Araripina, 07°35'S, 40°30'W (WA 2038967, M. Holderbaum); Serrita, 07°45'S, 39°24'W (WA 171262, G. Malacco); Três Morros, 08°27.5'S, 37°15'W (Farias 2009); PN Vale do Catimbau, *c*.08°35'S, 37°11.5'W (Farias 2009, de Sousa *et al.* 2012); Floresta, 08°36'S, 38°34.5'W (WA 730352, recording M. Holderbaum); Buíque, *c*.08°37'S, 37°05'W (WA 2308551, F.G. Las-casas); REBIO Serra Negra, 08°40'S, 38°00'W (Coelho 1987, Roda & Carlos 2004). *Bahia* Sento Sé, 09°44'S, 41°52.5'W (WA 221703, M. Crozariol); Campo Formoso, 10°11.5'S, 40°48'W (WA 1620507, J. Anunciação); Bomfim, 10°27'S, 40°11'W (Pinto 1937a, 1938); Xique-Xique, 10°49'S, 42°43'W (WA 2098953, recording E. Felix); near Mirangaba, 10°57'S, 40°36'W (XC 265946, R. Silva e Silva); Formosa do Rio Preto, 11°02'S, 45°12'W (Whitney *et al.* 1995, MNRJ); Morro do Chapéu, 11°35'S, 41°07'W (WA 1978258, C. Silva);

Baixa Grande, 12°00'S, 40°11'W (WA 2283196, recording S. Sampaio); Brotas de Macaúbas, 12°08'S, 42°28.5'W (WA 759928, E. Legal); Penha, 12°12'S, 44°57'W (XC 169880, R.A. Souza); Gruta Lapa Doce, 12°20'S, 41°36'W (Parrini *et al.* 1999); Palmeiras, 12°30'S, 41°33'W (Whitney *et al.* 1995); *c*.12.5km NE of Ibiquera, 12°35'S, 40°50'W (MPEG 51162); Mucugê, 13°00.5'S, 41°22.5'W (WA 2207298, P. Ávila); Novo Acre (= Giguy on label, see Paynter & Traylor 1991b), 13°27'S, 41°06'W (Naumburg 1939, AMNH 243146); São Félix do Coribe, 13°30'S, 4°03.5'W (MNRJ MNA7262/7504, WA 2300937, recording M. Rezende); Contendas do Sincorá, 13°50'S, 41°05'W (WA 863673, K. Okada); Jequié, 13°51.5'S, 13°51.5'S (WA 2205667, recording C. Costa); Igaporã, 13°52.5'S, 42°40'W (WA 1064630/38, recordings P. Gouvêa); 5km S of Brejinho das Ametistas, 14°18'S, 42°32'W (Albano 2010b, XC 200963/64, N.P. Dreyer); PN Boa Nova, 14°20.5'S, 40°15'W (XC 215054, H. Matheve); *c*.7.5km S of Sebastião Laranjeiras, 14°38.5'S, 42°56.5'W (ML 91034, G. Bencke). *Minas Gerais c*.20km N of Manga, 14°34'S, 43°56'W (Whitney *et al.* 1995, Dornelas *et al.* 2012); PN Cavernas do Peruaçu, 15°02'S, 44°15'W (XC 80442/43, J. Minns); Itacarambi, 15°05'S, 44°07'W (Whitney *et al.* 1995); Januária, 15°29'S, 44°22'W (Mattos *et al.* 1991); Berizal, 15°43'S, 41°48'W (WA 1122086, P.R. Siqueira); Janaúba, 15°48'S, 43°19'W (Whitney *et al.* 1995, Vasconcelos & Neto 2007); Fazenda Lagoa das Pedras, 16°08'S, 43°45'W and Mata do Poção, 16°20'S, 43°48'W (Vasconcelos & Neto 2007); Rubelita, 16°21.5'S, 42°12'W (WA 1279630, R.S. Moreira); Fazenda Baixa da Lasca, 16°22'S, 43°33'W (Vasconcelos *et al.* 2006); Itinga, 16°36'S, 41°46'W (WA 645523, G. Malacco); Fazenda Maria das Neves, 16°36'S, 42°49'W (Vasconcelos *et al.* 2006, Vasconcelos & Neto 2007); Montes Claros, 16°44'S, 43°54'W and Boa Vista do Bananal, 16°44'S, 42°58'W (Vasconcelos & Neto 2007); Barragem de Juramento, 16°46'S, 43°39'W (Vasconcelos & Neto 2007); Mata do Lobo, 16°47'S, 43°01'W (Vasconcelos *et al.* 2006, Vasconcelos & Neto 2007); Araçuaí, 16°59'S, 42°00'W (WA 337552, E. Luiz); ESEC Acauã, 17°07'S, 42°46'W (Vasconcelos & Neto 2007); Campo Limpo, 17°12'S, 42°51'W (Vasconcelos & Neto 2007). **Notes** Paynter & Traylor (1991b) were unable to locate São Paulo (Ceará), but placed it somewhere in the Serra de Ibiapaba in NW Ceará at an elevation of *c*.1,000m. A search of Google Earth revealed three locations known as São Paulo in Ceará, but all are east of the Serra de Ibiapaba at or below 450m. The coordinates given above accompanied the two Snethlage specimens from MPEG and appear to be a good approximation of the true location of São Paulo (at *c*.850m).

TAXONOMY AND VARIATION Monotypic. White-browed Antpitta was, for many years, treated as conspecific with Speckle-breasted Antpitta (Peters 1951, Howard & Moore 1984), despite striking differences in voice, plumage, habitat and distribution (Whitney *et al.* 1995). Indeed, molecular studies suggest that Speckle-breasted Antpitta is more closely related to Masked Antpitta, and that White-browed Antpitta is sister to the Spotted Antpitta group (Carneiro & Aleixo 2012).

STATUS Currently ranked Near Threatened by BirdLife International (2017), White-browed Antpitta is usually considered a bird of only moderate sensitivity to human habitat alteration (Silva *et al.* 2003), able to persist in the face of low levels of habitat degradation (BirdLife International 2017). To date, however, no studies have focused on this aspect of its natural history, and its perception as a non-sensitive species may be largely based on its tendency to favour second-growth tangles, which does not necessarily mean the species can persist without mature habitat nearby. Habitat destruction for agriculture and colonisation, and degradation of understorey caused by overgrazing, are the principal threats within the species' geographic range. These threats have seriously increased since the 1970s when the Brazilian oil company, Petrobrás, began building roads in the core of its range, opening up new areas to settlers (Stattersfield *et al.* 1998, Volpe *et al.* 2003). **Protected Populations** PN Boa Nova (XC 215654, H. Matheve); FLONA Chapada do Araripe (Whitney *et al.* 1995, Nascimento *et al.* 2000, XC, IBC, WA); PNM Distrito de Brejinho (Greeney *et al.* 2016); PN Serra das Confusoes (Silveira & Santos 2012); PN Cavernas do Peruaçu (Kirwan *et al.* 2001, XC, ML); PN Serra da Capivara (Olmos 1993, Olmos & Albano 2012); PN Vale do Catimbau (de Farias 2009, Sousa *et al.* 2012); PE Veredas do Peruaçu (Benfica *et al.* 2012); PE Mata Seca (Dornelas *et al.* 2012); REBIO Serra Negra (Coelho 1987, Roda & Carlos 2004); ESEC Acauã (Vasconcelos & Neto 2007); Reserva Florestal Sítio Nascente (ML 35880341, photo F. Olmos).

OTHER NAMES *Grallaria ochroleuca* (Allen 1889c, Koenigswald 1896, Hagmann 1904, von Ihering & von Ihering 1907, Brabourne & Chubb 1912, Cory & Hellmayr 1924, Zimmer 1934, Pinto 1938); *Grallaria ochroleucus* (Lesson 1844); *Myioturdus ochroleucus* (zu Weid 1831, Ménétries 1835); *Grallaria martinsi* (Snethlage 1925, 1927, 1928, Hellmayr 1929); *Grallaria ochroleuca ochroleuca* (Peters 1951, Ruschi 1967); *Hylopezus ochroleucus ochroleucus* (Pinto 1978, Howard & Moore 1984); *Hylopezus ochroleucus martinsi* (Sick 1984, 1993); *Chamaezosa ochroleuca* (Burmeister 1856); *Turdus concretus* (as used by Lichtenstein cited in Lesson 1844). **English** Speckle-breasted Antpitta (Gruson & Forster 1978, Frisch & Frisch 1981, Meyer de Schauensee 1982, Howard & Moore 1984); Spotted-bellied Antpitta (Cory & Hellmayr 1924); Spot-bellied Ant-Thrush (Brabourne & Chubb 1912). **Portuguese** torom-do-nordeste (van Perlo 2009, Gwynne *et al.* 2010, CBRO 2011); torom-malhado (Willis & Oniki 1991b); pinta-do-mato (Sick 1984, 1993); torom-torom-do-Nordeste (Frisch & Frisch 1981); tovacuçú do norte (Ruschi 1967). **French** Grallaire teguy. **German** Weißbrauen-Ameisenpitta. **Spanish** Tororoí teguá (Krabbe & Schulenberg 2003).

White-browed Antpitta, adult singing, Tianguá, Ceará, Brazil, 21 December 2014, 21 December 2014 (*Francisco Valdicélio Ferreira*).

White-browed Antpitta, adult stretching, Araripe, Ceará, Brazil, 16 February 2015 (*Caio Brito*).

White-browed Antpitta, adult brooding young nestlings, Araripe, Ceará, Brazil, 5 April 2015 (*Jefferson Luis Gonçalves de Lima*).

White-browed Antpitta, nest with two mid-aged nestlings, Araripe, Ceará, Brazil, 11 April 2015 (*Jefferson Luis Gonçalves de Lima*).

White-browed Antpitta, adult, Crato, Chapada do Araripe, Ceará, Brazil, 14 August 2014 (*Ciro Albano*).

SPECKLE-BREASTED ANTPITTA
Hylopezus nattereri   **Plate 20**

Grallaria nattereri Pinto, 1937, Boletim Biologico (Nova Serie), vol. 3, no. 5, p. 7, Alto da Serra, São Paulo, Brazil [= Paranapiacaba, *c*.23°47'S, 46°19'W following Paynter & Traylor 1991a]. The holotype (MPUSP 4729) is an adult male collected on 27 August 1904 by João Lima.

Speckle-breasted Antpitta is endemic to the Atlantic Forest of south-east Brazil, eastern Paraguay and extreme north-east Argentina. Like most antpittas, it is usually first detected by virtue of its mildly 'haunting' song, a short, even-paced series of whistles lasting *c*.2.5s. It occupies the ground and low levels of humid and montane forests, and is the southern replacement of White-browed Antpitta, with which it was until recently considered conspecific. This long-standing taxonomic 'lump' led to a delayed acknowledgement that northern birds (White-browed) are likely to be considerably more threatened than the present species (Whitney *et al.* 1995). In Minas Gerais, the ranges of Speckle-breasted and White-browed Antpittas are separated by at least 300km, with the Serra dos Gerais and Serra do Cabral ranges, areas dominated by *cerrado* vegetation, in the intervening region (Whitney *et al.* 1995). Though we still have much to learn about the natural history of Speckle-breasted Antpitta, its nest and eggs were recently described (Bodrati & Di Sallo 2016), and the following account provides the first description of immature plumage.

IDENTIFICATION 13–14cm. Adult olive-brown above with pale buff lores, a partial eye-ring and a whitish moustachial. Wings somewhat browner than back with an ochraceous patch at base of primaries and dark primary-coverts. Underparts generally pale ochraceous-buff, whiter on throat and central belly, and paler buff on sides and flanks. There is a black or dusky-grey submalar streak and the breast-sides are slightly scalloped dusky, breaking up into profuse spots and speckles on breast, upper belly

and flanks. Speckle-breasted Antpitta is unmistakable within its range. The only other sympatric antpitta is the much larger, and very distinct, Variegated Antpitta. White-browed Antpitta (no known overlap) is distinctly greyer above (especially on crown), with a prominent white postocular stripe, overall has much less richly coloured plumage and tends to be less marked below.

DISTRIBUTION Endemic to the Atlantic Forest of south-east Brazil, adjacent easternmost Paraguay and extreme north-east Argentina (Ruschi 1979, Hayes 1995, Whitney *et al.* 1995, Mazar Barnett & Pearman 2001, Guyra Paraguay 2004, Ridgely & Tudor 2009, van Perlo 2009). It occurs in Minas Gerais, from Caeté and Ouro Preto south to extreme south-west Rio de Janeiro (Bauer & Pacheco 2000, Buzzetti 2000). From the Serra Geral of western

and central Paraná (Santos *et al.* 2004, Volpato *et al.* 2004, Straube 2008, Scherer-Neto *et al.* 2011) it occurs east to São Paulo (Silviera & Ueza 2011) and south to Rio Grande do Sul (Bencke *et al.* 2010). The southernmost localities are all north of Porto Alegre (30°02'S, 51°13'W). Unconfirmed reports (eBird) of *H. ochroleucus* from south-west Espírito Santo (just east of Pico de Bandeira) may refer to the present species, and would amplify the northern portion of its range east from the Belo Horizonte area. Outside Brazil, it is found only in extreme eastern Paraguay and north-east Argentina.

HABITAT Inhabits the ground and lower growth of humid and montane forest (Esquivel & Peris 2012), mature secondary woodland and borders; often in very dense tangles and bamboo (Whitney *et al.* 1995, dos Anjos *et al.* 2004, Volpato *et al.* 2006). Found mostly at 1,200–1,900m in the north of its range, but occurs much lower in some areas (Whitney *et al.* 1995) and to 2,450m in the Serra de Itatiaia (Mallet-Rodrigues *et al.* 2015). Overall, habitat use and altitudinal distribution seems to vary somewhat across its range. In hilly areas of Misiones, Argentina, Speckle-breasted Antpitta is most commonly found on steep, rocky hillsides at 300–700m (A. Bodrati *in litt.* 2017).

VOICE The song is an even-paced series of whistles that steadily increases in amplitude and, after the first few notes, rises in frequency. The individual notes are initially flat, but those towards the end of the song are downslurred (Whitney *et al.* 1995). Krabbe & Schulenberg (2003) gave the acoustic particulars as follows. The song is 2.0–2.5s in duration, consisting of 5–10 even-paced notes in the south of its range, but 8–14 whistles in the north. Across the song, the amplitude increases, with the first few notes slightly falling, but thereafter rising in pitch from *c.*2kHz to 2.5kHz. In the south, the final note is shorter (usually quieter) than the preceding one. According to van Perlo (2009), the song is a short, fast, fluted series of *c.*10 notes, increasing in volume and pitch. Ridgely & Tudor (1994) described the song as *teeu-teeu-teeu-teeu-teeu-tew-tew-tew-tew!* and it has also been described as a soft series of whistles, ascending slightly in pitch and volume, *Fü, fü, fü, fü, füihhh* (R. Straneck in de la Peña 2016). The call is a rattle, falling from around 2.7kHz to 2.2kHz, and consisting of seven short, sharply downslurred notes.

NATURAL HISTORY Whitney *et al.* (1995) described the species as primarily terrestrial, moving with short hops punctuated by abrupt stops, sometimes with head lowered and cocked towards the ground (Whitney *et al.* 1995) or hopping nimbly between low, horizontal perches (Canevari *et al.* 1991). Prey is captured with reaches to the leaf litter and occasional probes into curled leaves or possibly the soil. Individuals may occasionally probe at a single site for several seconds. They rarely fly, but when they do, it is short, swift and silent (Canevari *et al.* 1991). Singing adults usually perch in dense vegetation, on logs, on the ground or as high as 2m on smaller branches, but song perches are generally below 1m (Whitney *et al.* 1995). In response to playback, they more frequently sing from the ground and, apparently when agitated, will 'rock and roll the body from side-to-side without moving the head or legs' (Whitney *et al.* 1995). The latter authors likened this body-rocking motion to the rump-twitching movements of *Grallaricula*, and other authors have described it as nearly identical to the twitching of their smaller relatives (Canevari *et al.* 1991). A. Bodrati (*in litt.* 2017) described watching adults singing as they

move about, as opposed to from a perch. After walking a few metres the adult stops and sings from the ground, barely finishing its vocalisation before beginning to move. He further suggests that, in some cases, they may actually sing on the move, based on the changing locations of adults heard (but unseen) in the undergrowth.

DIET Undocumented. A. Bodrati (*in litt.* 2017) observed an adult feeding its fledgling a large (> than bill) black orthopteran. There is no reason to presume that the species' diet differs greatly from that of its congeners.

REPRODUCTION The breeding biology was, until very recently, virtually unknown. Canevari *et al.* (1991) stated that nests are cup-shaped and built in hollow logs ('huecos troncos'), generally 2m above ground. They did not provide a source for this information, but the suggested placement inside hollow trunks suggests that the description was based on misidentified nests. Subsequently, a nest was found, on 9 November 2009, at 1,720m in Itatiaia National Park, on the Rio de Janeiro/Minas Gerais border. Kirwan (2009) reported that an adult carried 'small items' to an empty nest and appeared to add them to the rim while perching briefly inside the nest. From these observations, Kirwan (2009) was unable to confirm if the nest was under construction at the time or was, perhaps, simply being visited in some sort of territorial or mating display. Not until recently was an active nest found and described, along with a clutch of eggs (Bodrati & Di Sallo 2016). Except where noted, the following data are from this paper. **NEST & BUILDING** The open-cup nest found by Kirwan (2009) was in a small patch of bamboo at the bottom of a narrow, shallow ditch, beside the Agulhas Negras road. It was *c.*2m above ground, and primarily supported by the base by a curved, narrow branch of an 18m-tall tree. Several thinner bamboo stems overlapped at this point and provided additional support at the base and sides of the nest. The external portion of the nest was constructed primarily of small twigs and dark vegetable fibres, with a few dead leaves mixed into the rim and dangling raggedly to 10cm below the nest, adding to its camouflage. Externally, the nest was *c.*7cm wide by 7cm tall, but it was not examined closely. The Argentine nest was built on a small slope above a small stream within fairly mature forest dominated by trees in the genera *Alchornea* (Euphorbiaceae), *Ocotea*, *Nectandra* (Lauraceae), *Holocalyx* (Leguminaceae), *Diatenopteryx* (Sapindaceae), *Cedrela* (Meliaceae) and *Araucaria* (Araucariaceae). A nearby stand of subcanopy-level tree-ferns (Cyatheaceae) shaded the nest in an otherwise relatively open understorey. This nest was only 31cm above ground, somewhat precariously supported from below by several angled stems and tree-fern fronds. It was a loose, open cup of dead sticks and dry leaves. The outside diameter, measured at perpendicular angles, was 13.1 × 14.0cm. It was 5cm tall with an inner cup measuring 6.5 × 6.6cm, and an internal depth of 2.9cm. The nest cup was lined by thin flexible materials such as leaf petioles, thin rootlets and fungal rhizomorphs (*Marasmius* spp.). On one side the external materials extended out slightly more than in other directions, providing a rudimentary platform. **EGG, LAYING & INCUBATION** The latter nest containd two subelliptical, bluish-green eggs heavily marked with irregularly shaped blotches of black, brown and grey. Dimensions of the eggs were: 24.5 × 19.1mm (4.4g) and 23.1 × 18.5mm (3.7g). Both adults were observed incubating, covering the eggs for 97% of observed daylight hours. The longest period of absence was 35min (around midday). Six complete on-bouts averaged 65 ±

35min (range 24–104min). Adults approached the nest by walking on the ground, sometimes pausing for up to 5min below the nest before hopping onto the rim. Adults were usually heard vocalising in the vicinity of the nest, always from dense vegetation, just prior to changeovers. The first exchange usually occurred within 2h of daybreak and the last usually around 1h before dusk. Once, the adult on the nest near nightfall emitted three short, sharp whistles, which resembled a lower-pitched version of the song of Solitary Tinamou *Tinamus solitarius*, but at lower volume. This appeared to attract its mate and, after switching places, the departing adult continued to vocalise in this manner as it walked away from the nest. **NESTLING & PARENTAL CARE** Unfortunately, the Argentine nest was found empty soon after the observations described above. The nestlings and parental care are unknown. **SEASONALITY** Three males (Rio Grande do Sul), collected in late September, late November and early December were described as having the testes 'moderately developed' by Belton (1994). In passing, Bodrati *et al.* (2010) mentioned a fledgling in Misiones, details of which were shared by A. Bodrati (*in litt.* 2017; PP Cruce Caballero record below). The AMNH subadult from Corvo (see below) had 'immature' on the label. Interestingly, however, apart from a single reddish-brown plume of nestling down among the feathers of the crown, I could not find any plumage characters that indicated immaturity and it was likely considered so based on incomplete skull ossification. This indicates that, at least in this species, birds may moult into adult plumage prior to complete skull ossification. **BREEDING DATA** Nest with eggs, 25 September 2015, PP Cruce Caballero (Bodrati & Di Sallo 2016); active nest, 2 September 2016, Reserva Forestal Morro Grande (IBC 1270901, photo R.Y. Castro); fledgling, 22 November 2010, PP Cruce Caballero (A. Bodrati *in litt.* 2017); two juveniles transitioning, Arroyo Urugua-í, 21 December 1957 (AMNH 771189), 14 January 1958 (AMNH 771184); juvenile transitioning, 28 December 2016, São Luiz do Paraitinga (WA 2428631, G. Balieiro); subadult, 2 July 1961, Piñalitos (YPM 66937); subadult transitioning, 19 February 1930, Corvo (AMNH 318523).

TECHNICAL DESCRIPTION Sexes similar. *Adult* Has an indistinct buffy or whitish moustachial and buffy to buffy-white lores that usually blend into a pale eye-ring, somewhat thicker posteriorly. Crown greyish-brown, grading to olive-brown on the back. Wings slightly browner than back, especially the primaries which at their bases form a small, ill-defined 'speculum' that contrasts with the dark primary-coverts. The indistinct pale moustachial streak extends into a duskier malar line that separates it from the white throat. Rest of underparts somewhat variable, both in coloration and marking. Individuals vary from almost entirely rich buffy-ochraceous underparts, usually fading to nearly white on central belly, to mostly pale buffy-white below, with only the sides and flanks washed ochraceous. The sides, especially of the breast, are scalloped blackish, the markings becoming duskier and reduced to spotting on the breast and along the flanks, but generally lacking from the lower breast, belly and vent. From examination of numerous specimens and hundreds of photographs on WA, it appears to me that there is a general trend in the underparts, from north to south, towards decreased richness and extent of the ochraceous wash, as well as an increase in the darkness, size and extent of speckling. Northern birds, especially in São Paulo, tend towards entirely ochracous underparts with the scalloping on sides sometimes extending well onto

the breast and only breaking into small, scattered spots on the central breast. **Adult Bare Parts** *Iris* dark brown; *Bare Orbital Skin* buffy-white; *Bill* maxilla black, mandible whitish or pinkish on basal half, dark near tip; *Tarsi & Toes* pinkish-grey. *Juvenile/Subadult* The following description is based on two older juveniles at AMNH (see Breeding). Both individuals appear to be in full adult plumage except for the back of head and wings. The face appears 'messy' and still retains a few rusty-brownish plumes, probably retained since leaving the nest. These give the lores a reddish or dull orange look, darker than adults. Asymmetrical patches of rusty immature feathers remain on the hindcrown, nape and neck-sides, the largest being *c.*1 × 1.5cm. The flight feathers appear fully adult but both the greater and lesser upperwing-coverts are still immature. The coverts are dull brown with very diffuse subterminal buffy spots and a broad fringe of dull rufous. These rufous edges are starting to wear off, to varying degrees, from the tips of the feathers and it seems likely that they were originally a much brighter rusty-red. From this, I surmise that subadults will be identifiable by the retained immature wing-coverts and perhaps some reddish wash to the lores and face. Juveniles probably slowly lose their rusty-red nestling plumage in similar order to other *Hylopezus*, with the crown, nape and neck-sides being the last areas to gain fully mature plumage. **Juvenile/Subadult Bare Parts** Bill, tarsi and toes of the two individuals I examined had dried to a very similar colour to those of adult specimens. Undoubtedly, however, at least until well into the juvenile stage, the pale base of the bill and rictus would have retained some orange or reddish coloration.

MORPHOMETRIC DATA Males *Wing* 78mm; *Tail* 36.5mm; *Bill* [exposed] culmen 17mm; *Tarsus* 39mm (*n* = 1, holotype, Pinto 1937a). *Wing* 81mm, 79mm; *Tail* 38mm, 36.5mm; *Bill* [exposed] culmen 15.5mm, 15.5mm; *Tarsus* 39.5mm, 40mm (*n* = 2, Naumburg 1939). *Wing* 82mm, 80mm, 80mm, 80mm; *Tail* 38mm, 37mm, 38mm, 38mm; *Bill* [exposed] culmen 17.5mm, 18.0mm, 17.0mm, 18.0mm; *Tarsus* 38mm, 40mm, 38mm, 39mm (*n* = 4, Partridge 1954). *Bill* from nares 11.2mm, 10.7mm, 10.3mm, 11.2mm, 10.5mm, n/m; *Tarsus* 39.9mm, 38.8mm, 40.0mm, 38.6mm, 39.1mm, 38.8mm (*n* = 6, AMNH, LACM). *Bill* from nares 10.0mm, 11.3mm, exposed culmen 15.7mm, 17.3mm, depth at front of nares 4.8mm, 4.3mm, width at front of nares 3.8mm, 4.3mm; *Tarsus* 39.3mm, 39.8mm (*n* = 2, AMNH). *Bill* from nares 10.5mm, 10.0mm; *Tarsus* 39.5mm, 40.2mm (*n* = 2, juvenile, subadult, AMNH). **Females** *Bill* from nares 10.9mm; *Tarsus* 39.8mm (*n* = 1, juvenile, AMNH 771189). **MASS** 31–33g (*n* = 3♂♂, Belton 1994 cited by Krabbe & Schulenberg 2003 (?) and then by Dunning 2008); mean 32g (*n* = 3, ♀?, Esquivel-M. & Peris 2012). **TOTAL LENGTH** 12.1–13.6cm (*n* = 3♂♂, Belton 1994); 13.5cm (Krabbe & Schulenberg 2003, Ridgely & Tudor 2009); 14cm (de la Peña 1988, van Perlo 2009).

DISTRIBUTION DATA BRAZIL: *Minas Gerais* Caeté, 19°54'S, 43°40'W (WA 164913, recording G. Malacco); Santa Bárbara, 19°58'S, 43°25'W (WA 2454782, recording D. Murta); Ouro Preto, 20°23'S, 43°31'W (WA 2345983, D. Murta); Bom Jardim de Minas, 21°57'S, 44°11'W (WA 2195724, L. Thiesen); Itamonte, 22°17'S, 44°52'W (WA 12894, J. Honkala); Serra do Juncal, *c.*22°45'S, 45°52'W (Vasconcelos & Neto 2009); Extrema, 22°51'S, 46°19'W (WA 584745, F. Cipriani). *Rio de Janeiro* PN Itatiaia, 22°26'S, 44°36'W (XC 11688, R. Gagliardi, ML); Angra dos Reis, 23°00'S, 44°19'W (WA 2398842/43, recordings

C.E. Blanco). **Paraná** PE Mata dos Godoy, 23°27'S, 51°14'W (Volpato *et al.* 2006, XC); Manoel Ribas, 24°31'S, 51°40.5'W (WA 2308913, A. Constantini); Vermelho, 24°55'S, 51°25'W (Sztolcman 1926b); Tunas do Paraná, 24°57'S, 48°55'W (WA 1950407, recording M. Arasaki); PE Vila Velha, 25°15'S, 49°55'W (dos Anjos & Baçon 1999); Estrada da Graciosa, 25°20'S, 48°54'W (XC 80494, J. Minns); Corvo, 25°21'S, 48°56'W (MPEG 64834, AMNH 318523); Irati, 25°28'S, 50°39'W (WA 964476, O. Slompo); Foz do Rio Jordão, 25°45'S, 52°15'W (Straube *et al.* 1996); Serra do Cabral, 25°45'S, 49°20'W (Kaminski & Cerrano 2004); Fazenda Concórdia, 25°45'S, 51°05'W (Sztolcman 1926b); Foz do Rio da Areia, 25°50'S, 51°15'W (Straube *et al.* 1996); Guaratuba, 25°53'S, 48°34.5'W (WA 962311, L.R. Deconto); UHE Segredo, 25°55'S, 52°05'W (Straube *et al.* 1996); Cruz Machado, 25°56'S, 51°13'W (WA 2401202, recording L.R. Deconto); Fazenda São Pedro, 26°22'S, 51°22'W (Straube *et al.* 1996). **São Paulo** Piquete, 22°36.5'S, 45°11'W (WA 684792, recording M. Cruz); PE Campos do Jordão, 22°38'S, 45°26'W (Willis & Oniki 1981); Bananal, 22°41'S, 44°19.5'W (MNRJ 23612, WA 2546824, J.F. Pacheco); Serra da Bocaina, 22°45'S, 44°45'W (ML); São Francisco Xavier, 22°54'S, 45°59.5'W (XC 146413, A. Silveira); São José dos Campos, 23°04'S, 45°56'W (XC 119066, G. Leite); São Luiz do Paraitinga, 23°13.5'S, 45°18.5'W (WA 2428631, G. Balieiro); Santa Virginia, 23°23.5'S, 45°09'W (XC 80440, J. Minns); Pico do Corcovado, 23°28'S, 45°12'W (Goerck 1999); Casa Grande, 23°37'S, 45°57'W (LSUMZ 60996/71412); Estação Biologica Boraceia, 23°39'S, 45°54'W (Cavarzere *et al.* 2010, ML); Ibiúna, 23°40.5'S, 47°13'W (WA 1320044, recording M. Merzvinskas); Reserva Forestal Morro Grande, 23°44'S, 46°58'W (IBC 1270901, photo R.Y. Castro); Juquitiba, 23°55.5'S, 47°04.5'W (WA 908543/44, recordings F.I. de Godoy); APA Municipal Capivari-Monos, 23°57.5'S, 46°39'W (XC 187632, M.A. Melo); Tapiraí, 23°58'S, 47°30.5'W (WA 2141908, C. Albano); PE Carlos Botelho, 24°04'S, 47°58'W (Willis & Oniki 1981); Itararé, 24°06.5'S, 49°22.5'W (XC 299151/52; F.I. de Godoy); PE Intervales, 24°16'S, 48°24.5'W (XC 224311, P. Boesman); Apiaí, 24°30.5'S, 48°50.5'W (WA 984542, recording A.E. Rupp). **Santa Catarina** Campo Alegre, 26°08.5'S, 49°14.5'W (WA 2435066, recording J. Ventura); Joinville, 26°18.5'S, 48°51'W (Costa 2015); Doutor Pedrinho, 26°43'S, 49°29'W (XC 333064, D. Meyer); Passos Maia, 26°49'S, 51°57.5'W (XC 5652/53, R.E.F. Santos); Taió, 27°07'S, 50°00'W (WA 328661, recording A.E. Rupp); Barra do Monte Alegre, Chapecó, 27°14'S, 52°35'W (XC 35990/42494, A.E. Rupp); Nova Trento, 27°17'S, 48°55'W (WA 174323, recording G. Kohler); Petrolândia, 27°33.5'S, 49°40.5'W (WA 237244, recording D. Meyer); Santo Amaro da Imperatriz, 27°41'S, 48°47'W (WA 694990, recording F. Farias); Campo Belo do Sul, 27°51'S, 50°48.5'W (WA 851659/1006487, recordings G. Willrich); Anitápolis, 27°54'S, 49°08'W (WA 1481853, P. Steinbach); Urupema, 27°57'S, 49°52'W (XC 118585, R. Silva e Silva); Siderópolis, 28°36'S, 49°26'W (WA 1986230, A. Bianco); Timbé do Sul, 28°50'S, 49°50.5'W (WA 1570011, recording J.G. Just). **Rio Grande do Sul** Derrubadas, 27°15'S, 53°53.5'W (WA 1147458, recording D. Meller); Tenente Portela, 27°22.5'S, 53°45.5'W (WA 1858928, C. Furini); Caxias do Sul, 29°10'S, 51°10.5'W (WA 493011, recording J.N. Martins); Pousada Estáncia do Cambará, 29°13'S, 50°11'W (XC 320790, B. Davis); Morro Gaúcho, *c.*29°21'S, 51°56'W (Lau 2004); Sinimbu, 29°24'S, 52°33'W (Oliveira & Köhler 2010); Morro Pelado, *c.*29°24.5'S, 50°46.5'W (MNRJ 32233, ML); São Francisco de Paula area, *c.*29°27'S, 50°35'W (MNRJ 32245, AMNH 314615, ML, XC); SE of

Itati, 29°30'S, 50°06'W (Bencke & Kindel 1999); Picada Verão, 29°33'S, 51°01'W (ML 88090/91, G. Bencke); Dois Irmãos, 29°35.5'S, 51°05'W (WA 1473842, A. Wittmann); Maquiné, 29°36.5'S, 50°13.5'W (WA 2507461, M. Sand). **ARGENTINA**: *Misiones* Paraje María Soledad, 25°51'S, 53°59'W (XC 8182, G.S. Cabanne); Arroyo Urugua-í, 25°54'S, 54°36'W (AMNH 771184–189); Piñalitos, 25°59'S, 53°54'W (YPM); Tobuna, 26°28'S, 53°54'W (LACM 48624/25, FIML 13398); PP Cruce Caballero, 26°31'S, 53°59.5'W (Bodrati *et al.* 2010); RP Yaguaroundi, 26°41.5'S, 54°16'W (XC 49064–067, R. Fraga); Área Experimental y Reserva Guaraní, 26°56'S 54°13'W (A. Bodrati *in litt.* 2017); Corpus Christi, 27°08'S, 55°28'W (Krauczuk 2008); *c.*4.5km N of El Soberbio, 27°15.5'S, 54°11.5'W (XC 154362–367; B. López-Lanús). **PARAGUAY**: *San Pedro* Yaguarete forest, 23°50'S, 56°09.5'W (GPDDB). *Caaguazú* Ka'aguy Rory, 24°44'S, 55°41'W (GPDDB). *Caazapá* PN Caaguazú, fracción Cristal, 26°05'S, 55°45'W (Madroño *et al.* 1997). *Itapúa* Santa Inés, 26°23'S, 55°46'W (XC 55422, M. Velazquez); PN San Rafael, 26°36.5'S, 55°40'W (Esquivel & Peris 2012); Estancia Nueva Gambach, 26°38'S, 55°40'W (Smith *et al.* 2006).

TAXONOMY AND VARIATION Monotypic, possibly with some clinal variation (see Description). Speckle-breasted and White-browed Antpittas were long considered conspecific (Peters 1951, Partridge 1954, Pinto 1978, Howard & Moore 1984, Sibley & Monroe 1990), despite their distinctly different plumage, vocalisations and habitat preferences (Zimmer & Mayr 1943, Whitney *et al.* 1995). In fact, the two are not particularly closely related and molecular evidence suggests that Speckle-breasted is probably sister to the southernmost member of the genus, Masked Antpitta (Carneiro & Aleixo 2012).

STATUS Endemic to the Atlantic Forest biome (Whitney *et al.* 1995), generally fairly common in appropriate habitat within both its Brazilian (Stotz *et al.* 1996) and Argentine ranges (A. Bodrati *in litt.* 2017), and appears to not be particularly threatened with habitat loss over most of its montane distribution. Forest loss in the southern, lower-elevation part of its range is, however, much more severe, and is ongoing (Whitney *et al.* 1995, Volpato *et al.* 2006). Cordeiro (2001) estimated a range of 255,888km², with around 57,475km² of potential habitat. Its distribution in the north includes a fair number of protected areas in Brazil. Having gone unreported in Paraguay for almost a century before its rediscovery there (Lowen *et al.* 1996), it appears to be genuinely rare in that part of its range (Guyra Paraguay 2004). As the species has a fairly large range and populations do not appear to be declining, BirdLife International (2017) currently ranks it as Least Concern. In Argentina, Speckle-breasted Antpitta is considered 'Vulnerable' (AA/AOP & SADS 2008), a national ranking roughly equivalent to the global category of Near Threatened. Nevertheless, *H. nattereri* may be fairly sensitive to forest fragmentation (dos Anjos *et al.* 2010, 2011) and populations should certainly be monitored carefully in the face of ever-encroaching deforestation. **PROTECTED POPULATIONS Brazil** PN Serra dos Órgãos (Mallet-Rodrigues *et al.* 2007); PN Serra do Itajaí (XC 30525, A.E. Rupp); PN Itatiaia (Pinto 1954b, Ridgely & Tudor 1994, Parker & Goerck 1997, MNRJ, XC, ML); PN Serra do Gandarela (eBird: H. Peixoto); PN Serra da Bocaína (Cavarzere *et al.* 2010); PN Aparados da Serra (ML: W. Belton); PN Araucarias (XC 5652/53, R.E.F. Santos); FLONA Canela (Franz *et al.* 2014); FLONA São Francisco de Paula (ML: G. Bencke); PE Carlos Botelho

(Willis & Oniki 1981); PE Itacolomi (WA: D. Murta); PE Serra do Mar (XC); PE Campos do Jordão (Willis & Oniki 1981); PE Vila Velha (dos Anjos & Baçon 1999); PNM Ronda (Franz *et al.* 2014); PE Pedra Selada (eBird: B. Rennó); PE Mata dos Godoy (Volpato *et al.* 2006); PE Intervales (XC 224311, P. Boesman); PE Serra do Tabuleiro (ML); REBIO Estadual Canela Preta (XC 46180, E. Legal); REBIO Estadual Sassafrás (XC 333064, D. Meyer); APA Municipal Capivari-Monos (XC 187632, M.A. Melo); APA Bacia do Rio Paraíba do Sul (XC 76036, M. Melo); RPPN Caraça (Vasconcelos & Junior 2001); RPPN Rio das Furnas (eBird: G. Carvalho); RPPN Salto Morato (Straube & Urben-Filho 2005a); RPPN Prima Luna (XC 44243, E. Legal); RPPN Bugerkopf (eBird: A.E. Rupp); Reserva Forestal Morro Grande (IBC 1270901, photo R.Y. Castro). **Argentina** PN Iguazú (Saibene *et al.* 1996); PP Cruce Caballero (Bodrati *et al.* 2010); PP Esmeralda (eBird: R. Ramírez); PP Moconá (eBird: G. Pugnali); PP Caa Yarí (eBird: C. Agulian); PP Piñalito (ML); PP Guardaparque Horacio Foerster (eBird: 'Aves del NEA'); PP Urugua-í

(XC 75359, F. Schmitt); Reserva Arroyo Alegría, Reserva San Jorge, Área Experimental y Reserva Guaraní, Reserva Cultural Papel Misionero, and Reserva Tangará (A. Bodrati *in litt.* 2017); RP Yaguaroundi (XC 49064–066, R. Fraga). **Paraguay** PN San Rafael and PN Caaguazú (Lowen *et al.* 1997, Madroño *et al.* 1997a,b, Esquivel-M. *et al.* 2007, Mattos 2010, Esquivel-M. & Peris 2011b, 2012).

OTHER NAMES *Grallaria nattereri* (Pinto 1937a); *Grallaria ochroleuca* (Sztolcman 1926b, Saibene *et al.* 1996, Chebez 1996), *Hylopezus ochroleucus* (de la Peña 1988, Canevari *et al.* 1991). **English** Natterer's Antpitta (Sibley & Monroe 1993). **Portuguese** torom-torom-malhado (Frisch & Frisch 1981). **Spanish** Tororoí de Natterer (Krabbe & Schulenberg 2003); Chululu'I (Guyra Paraguay 2004); Chululú chico (de la Peña 1988, Canevari *et al.* 1991, Mazar Barnett & Pearman 2001, Guyra Paraguay 2004, Esquivel-M. & Peris 2011b). **French** Grallaire de Natterer. **German** Fleckenbauch-Ameisenpitta (Krabbe & Schulenberg 2003).

Speckle-breasted Antpitta, adult, Tapiraí, São Paulo, Brazil, 19 September 2015 (*Marcelo Jordani*).

Speckle-breasted Antpitta, adult settling on two eggs, PP Cruce Caballero, Misiones, Argentina, 6 October 2015 (*Martjan Lammertink*).

Speckle-breasted Antpitta, juvenile transitioning into subadult plumage, São Luiz do Paraitinga, São Paulo, Brazil, 28 December 2016 (*Guto Balieiro*).

Speckle-breasted Antpitta, adult, Tapiraí, São Paulo, Brazil, 19 September 2015 (*Marcelo Jordani*).

Genus *Myrmothera*: mid-sized lowland antpittas

Myrmothera, **Sclater, 1858**, Analyse d'une nouvelle ornithologie élémentaire, p. 43. Type, by subsequent designation: *Myrmothera campanisona* (Hermann, 1783), by Sclater (1890).

Remarks on the immature plumages of *Myrmothera* Overall, I have examined relatively few immatures (none for *M. simplex*), but the sequence of plumage development in *Myrmothera* appears to be very similar to my understanding of that for *Hylopezus*. In particular, *Myrmothera* antpittas fledge when still covered in fluffy, wool-like, rufescent or rusty-brown nestling down, paler on the belly, flanks and vent. The tips of the upperwing-coverts are edged rufous-chestnut and may be washed to varying degrees with buff. These markings tend to fade with wear, becoming less obvious with age but, if the plumage maturation process in *Hylopezus* mirrors that of *Grallaria*, it is likely that immature coverts are replaced with similarly patterned feathers at least once prior to a final moult to adult coverts. The process of transitioning to adult-like contour feathers is incompletely known, but appears to mirror *Hylopezus* in timing and sequence. This includes the retention of scattered feathers that can appear blood-stained in specimens, as described under my remarks on immature plumages of *Hylopezus*. Unlike most *Grallaria*, but like *Hylopezus*, the first covering of adult-like contour feathers, including the hindcrown and nape, is patterned like or very similar to that of full adults. I define the three phases of immature plumage in *Myrmothera*, essentially as I did for *Hylopezus* (and also similar to *Grallaricula*). FLEDGLING Covered in fluffy, wool-like, feathers, rictal flanges still obviously inflated and pale orange or yellow-white, still heavily dependent on adults for food and largely remaining stationary while awaiting prey delivery. JUVENILE Mostly covered in adult-like contour feathers, but may retain some rust-coloured patches of fledgling down on crown, as well as scattered wisps of downy plumage on nape, rump and central underparts; rictal flanges probably still paler than adults but may not be distinctly immature in appearance. Behaviour at this age largely unknown, but probably fairly mobile, moving with adults and actively soliciting food, possibly already close to independence. SUBADULT Very close or identical to adults in general coloration, only separable by their immature wing-coverts, and likely difficult to separate in the field and almost certainly fully independent. As described in more detail for *Hylopezus*, plumage transition from fledgling to adult appears to be fairly quick, but in a few cases I use the terms '**fledgling transitioning**' and '**juvenile transitioning**' for birds that are somewhere between the above-defined stages. The term '**subadult transitioning**' is reserved for individuals with heavily worn and faded (but still at least some indication of chestnut margins to) upperwing-coverts that are otherwise fully adult in appearance.

THRUSH-LIKE ANTPITTA
Myrmothera campanisona   Plate 21

Myrmornim campanifonam Hermann, 1783, *Tabula affinitatum animalium olim academico specimine edita*, p. 189, 'Cayenne' [= unknown location in French Guiana].

"[The Thrush-like Antpitta] is one of the plainest of the interior antbirds, and is rarely seen, always hidden among the densest brush low to the ground. As it forages while surrounded by swarming ants, it jumps about on long legs resembling a tall hairy ball." **Penard brothers, 1910, Guyana** (translation from Dutch)

The Latin binomial above was first, by a few months, applied to Thrush-like Antpitta, based on the illustration published in Georges Louis Leclerc comte de Buffon's monumental treatise *Histoire naturelle des oiseaux* (Buffon 1771–86). In this work it received its first vernacular name, 'Le Grand Béfroi', or 'The Great Belfry'. Unaware of this formal naming, later the same year, Boddaert (1783) applied the name *Formucarius* [*sic*] *brevicauda*, basing on the painting of the 'Grand Béfroi, de Cayenne' printed between 1765 and 1783 as part of a long series of zoological illustrations by Edme-Louis Daubenton (pl. 706, fig. 1). Nearly simultaneous with the appearance of these two names, Gmelin (1789) christened Thrush-like Antpitta *Turdus tinniens*, also based on Daubenton's plate. It is from these early names that we take the scientific name that is currently used, *campanisona* (from Latin, *campana* = bell, *sonus* = sounding; Jobling 2010). Being one of the first-named

species treated in this book, this name game continued for some time, exacerbated by zoologists working in effective isolation in the pre-modern era. Thrush-like Antpitta is widely distributed across the Guianan Shield, the Amazon and in the Andean foothills. It is currently considered to comprise six subspecies, but morphological and plumage differences between them are subtle and weakly defined. Vocal differences between most populations, if they exist, are also poorly known, with the exception of the vocally distinct subspecies in east Amazonian Brazil (*subcanescens*). My preliminary conclusion (see Geographic Variation) is that Thrush-like Antpitta may comprise just two (possibly three) recognisable subspecies, but that one of these, *subcanescens*, might merit species status. In plumage, Thrush-like Antpitta is rather plain brownish above with lightly streaked, paler underparts and a pale mark behind the eye. It inhabits dense undergrowth in low-lying swampy areas in Amazonian *terra firme* forests and the Andean foothills (usually below 1,200m). Despite being broadly distributed, its natural history remains poorly studied. The nest has been described from several locations and is a broad, shallow cup, comprising predominantly sticks and leaves, placed less than 1m above ground. The clutch consists of two subelliptical, brown-spotted, blue-green eggs. In some areas this species can be quite abundant and, like previous authors (Ridgely *et al.* 2005), I suggest it would make an excellent candidate for natural history and ecological studies.

IDENTIFICATION 14.5–15.0cm. Sexes similar. A fairly slender antpitta, most similar in silhouette to a *Hylopezus* and, as its English name implies, sometimes likened to a tailless thrush (Hilty & Brown 1986). Very little plumage variation across its broad range, with little known concerning any intra-population plumage differences. Adult has somewhat buffy lores and a small white patch behind the eye. Malar white, mottled brown, while the head-sides and upperparts are dull rufous-brown. Ground colour of the underparts is predominantly white, with the sides and flanks often washed olive-brown or olive-grey. Breast and sides streaked brownish, usually leaving the central belly and vent pure white. Within most of its lowland and foothill range, Thrush-like Antpitta is unlikely to be confused, except in parts of its Amazonian range where it overlaps with one or more *Hylopezus*. In all cases, however, Thrush-like Antpitta is distinctly browner above and less marked below, giving a distinctive, overall faded appearance compared to the several sympatric *Hylopezus*. In northernmost portions of its range it may be sympatric with its only congener, Tepui Antpitta, but is separated by its streaked underparts.

DISTRIBUTION Broadly distributed across the Amazon, its range extending north-east through Suriname and the Guianas to eastern Venezuela, north-west into south-west Venezuela and eastern Colombia, and then south along the base of the Andes through eastern Ecuador and Peru to north-west Bolivia (Remsen & Traylor 1989, Rodner *et al.* 2000, Ridgely & Greenfield 2001, Hilty 2003, van Perlo 2009, Schulenberg *et al.* 2010, Spaans *et al.* 2016). An old sight record from Rio de Janeiro (Ménétriés 1835) is considered in error.

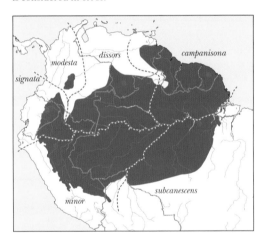

HABITAT An inhabitant of humid lowland rainforest and foothills, appearing to prefer dense understorey around treefalls, regrowing clearings and road edges (Hilty 2003, Schulenberg *et al.* 2010). It also has a predilection for low-lying, swampy areas and riparian zones, especially for nesting (Greeney 2017a,b), but appears to only infrequently use *Guadua* bamboo-dominated habitats (Lebbin 2007). In the foothills of eastern Ecuador, the tangled new growth at the edge of a light gap in a swampy depression is the first place I head when searching for the nests of these secretive birds. In at least one study, Thrush-like Antpitta was not found on the dense archipelago of river islands on the lower Rio Negro (Cintra *et al.* 2007) and the species is presumed absent from most Amazonian river island habitats. Across its range, it is generally found from sea level to 1,200m (Krabbe & Schulenberg 2003). Its

altitudinal limit in Venezuela appears to be closer to 800m (Meyer de Schauensee & Phelps 1978, Giner & Bosque 1998, Hilty 2003), perhaps because of the presence of its higher-elevation congener, Tepui Antpitta. Nevertheless, it is generally commonest below 700–800m in most of its range (Ridgely & Greenfield 2001, McMullan *et al.* 2010), but ranges as high as 1,300m in the Tepuyes de San Miguel de las Orquídeas of southern Ecuador (*signata*; Freile *et al.* 2013) and 1,500m in southern Peru (*minor*; Merkord 2010, Schulenberg *et al.* 2010), such as in the Villa Rica area of Pasco (XC: N. Athanas, F. Lambert). Although it prefers natural disturbance and can be found near roads, the species is probably sensitive to habitat degradation (Canaday 1997, Canaday & Rivadeneyra 2001, Lees & Peres 2006),and probably requires intact forest nearby (or at least a relatively well-structured canopy).

VOICE With apparently little variation between races in the north and west of its range, the song was described by Krabbe & Schulenberg (2003) as being 1.4–2.4s long and delivered at intervals of 6–14s. It consists of a slightly falling series of 5–8 similar-sounding, hollow whistles at frequencies of 0.7–0.8kHz. The first note is often slightly weaker and at a lower pitch than the rest, e.g. *wuhwuh- wuH-WUH-Wuh-wu* or *hoo* HOO HOO *hoo hoo* (Hilty & Brown 1986, Schulenberg *et al.* 2010). The call, at least sometimes given in alarm, is a low churr or musical chortle, lasting 1.0–1.6s and consisting of 14–24 notes. In frequency it ranges from 1.0–1.1kHz and may be repeated at intervals of 3–6s (often while flicking the wings; Ridgely & Greenfield 2001). Haverschmidt & Mees (1994) described the call of nominate *campanisona* as a low, rolling *tjooooorr-……..tjooooorrr*. The call of *subcanescens* has not been described as varying much from other populations, but its song is distinctly different (Krabbe & Schulenberg 2003). Specifically, it gradually increases in volume, rising steadily in pitch from 0.85 to 1.05kHz and is delivered at a faster rate. The song of *signata* is said to differ only slightly from that of *minor*, being higher-pitched and usually more monotone. Nemeth *et al.* (2001) provided a quantified description (mean ± SD) of the song of *dissors* from Venezuela as: lasting 2.06 ± 0.11s, with the lowest frequency of 0.846 ± 0.024kHz, the highest frequency of 0.898 ± 0.030kHz and the loudest frequency of 0.875 ± 0.010kHz.

NATURAL HISTORY Forages alone or, apparently infrequently, in pairs, hopping and walking across the forest floor, sometimes climbing onto logs and low branches (Hilty & Brown 1986). Although there are scattered, and usually rather oblique, references to the species foraging at army ant swarms (e.g., Penard & Penard 1910), there is apparently no concrete evidence that this represents regular behaviour. It is shy and retiring, generally remaining in cover of dense thickets and vine tangles, climbing onto low perches only infrequently, usually to sing (Sick 1984, 1993). Mean song-perch height in Venezuela (*dissors*) was 15cm (Nemeth *et al.* 2001). Thrush-like Antpitta, even when common, usually occurs at fairly low densities, although details of territory size are not well documented. Two neighbouring pairs nested 400m apart in French Guiana (Tostain & Dujardin 1988). Terborgh *et al.* (1990) estimated 4.5 pairs per 100ha, with mean territories of *c.*11ha in southern Peru (*minor*), English (1998) estimated four pairs per 100ha in Ecuador (*signata*) and Thiollay (1994) estimated 5.75–8.0 pairs/100ha in French Guiana. Territory size was estimated to be slightly smaller (9ha) just north of Manaus (*campanisona*) (Johnson *et al.* 2011). Although nothing has been published with respect to the species' predators, it is known to be at least occasionally infested with *Formicaphagus* chewing lice (Clayton *et al.* 1992) and avian malaria (Svensson-Coelho *et al.* 2013).

DIET Like other antpittas, the diet is poorly known. Specific prey include weevils (Curculionidae), leafcutter ants (Formicidae), grasshoppers (Acrididae) and millipedes (Diplopoda) (Haverschmidt 1968, Haverschmidt & Mees 1994). The stomachs of two birds in Amapá, Brazil (*campanisona*) held two beetles (Coleoptera) and 17 ants (Aguiar & Júnior 2008) and the stomach of an adult female in Venezuela (*dissors*) had small beetles (Williard *et al.* 1991).

REPRODUCTION The details of breeding biology are derived from observations on nests of half of the recognised subspecies, from studies in French Guiana (*campanisona*; Tostain & Dujardin 1988), Ecuador (*signata*; Greeney *et al.* 2005, Greeney 2017a) and Peru (*minor*; Londoño 2014). The first nest and clutch to be described was discovered in 1984 at Le Mont Belvédère, French Guiana, in the upper drainage of the Camopi River, at an elevation of 160m (Tostain & Dujardin 1988). The following details are derived largely from this and the nests found in eastern Ecuador. **NEST & BUILDING** All nests described to date were broad, somewhat shallow, open cups composed externally of long sticks (*c.*0.5–2.0cm diameter), internally of smaller sticks and dead, humid leaf matter, with sparse linings of long, dark, flexible rootlets and flexible, unbranched leaf petioles. All were in low-lying swampy areas or beside small, shady streams, generally in dense, tangled vegetation such as a treefall. All were rather precariously perched upon (not interwoven with) their supporting substrates. In most cases nests were supported by multiple thin (1–3cm diameter), crisscrossing supports such as the branches of living shrubs, aeroid leaf petioles, vines and dead branches. In one case, a thicker angled log, about half the width of the nest, provided most of the support. Two nests were supported by fairly dense clumps of grass and two by the rosettes of stem bases of a fern in one case and a palm in the other. Nest height: 29.6 ± 10.2cm (mean ± SD; *n* = 8, *signata*, Greeney 2017a); 15cm (*n* = 1, *signata*, Greeney *et al.* 2005); 25cm, 40cm, 60cm (*n* = 3, *campanisona*, Tostain & Dujardin 1988). Nest dimensions: inner diameter 9.1 ± 1.1cm; inner depth 4.3 ± 0.5cm; outer height (thickness) 9.5 ± 2.2cm; minimum outer diameter (where most material ends) 19.0 ± 4.6cm; maximum outer diameter (including longest sticks) 28.5 ± 2.7cm (mean ± SD; *n* = 8, *signata*, Greeney 2017a): inner diameter 9.5cm; inner depth 4cm; outer height *c.*10cm; minimum outer diameter 15cm; maximum outer diameter (including longest sticks) 23cm (*n* = 1, *signata*, Greeney *et al.* 2005): inner diameter 9cm, 8.5–9.0cm; inner depth 4cm, 3–4cm; outer height n/m, 10cm; outer diameter n/m, 17cm (*n* = 2, *campanisona*, Tostain & Dujardin 1988). Brief observations at a nest in Ecuador (Greeney 2017a), found late in the construction phase, revealed that both sexes build, as in other antpitta genera (Greeney *et al.* 2008b). At this nest, observed during all daylight hours five days prior to laying the first egg, an adult visited the nest only once, at *c.*10.00. The adult walked onto the rim of the nest and dropped a long flexible rootlet into it, remaining there for ten minutes, alternating between tucking loose material into the nest, shaping the nest, and

singing. The adult grasped loose rootlets with its bill, tucking them into the rim or nest lining with the same gentle, but rapid, movement of the bill as described below for egg-rolling. It shaped the nest by lowering itself into the cup, crossing its wings partially over its back, pressing its breast downward, and vibrating slightly, a movement used by many species for nest-shaping (Hoyt 1961, Nolan 1978). Most tucking or shaping movements were separated by a 90° shift in orientation, the adult completing several complete circles during its time at the nest. During its ten-minute stay at the nest, the adult vocalised 25 times, producing shorter (2–5 note) versions of the typical song. Brief observations at this nest, closer to the day of clutch initiation, also revealed a similar lack of activity and provided further evidence that laying may be delayed by at least five days after the nest is completed. During the few additionally observed visits to the nest, singing from the nest continued, varying from 1.6–2.5 songs/min. **EGG, LAYING & INCUBATION** Reported clutch size is always two eggs: *campanisona* (*n* = 2; Tostain & Dujardin 1988); *signata* (*n* = 4; Greeney *et al.* 2005b, Greeney 2017a); *minor* (*n* = 2; Londoño 2014). Eggs are pale blue to blue-green, somewhat variably marked with irregular-shaped brownish and lavender splotches or fine spotting, sometimes forming a wreath at the larger end (Greeney 2017b). Two eggs from a clutch in the Nangaritza Valley of south-east Ecuador (*signata*; Greeney 2017a) differed fairly significantly from those in French Guiana (*campanisona*; Tostain & Dujardin 1988), as well as from other eggs of *signata* I have observed elsewhere in Ecuador (Greeney *et al.* 2005b, Greeney 2017a) in being paler blue with much finer and paler speckling (rather than blotching), but the significance of this is unknown. Overall, however, the shade of blue and the type of markings appear to vary greatly in Thrush-like Antpitta, even within populations, as is also apparent from my inspection of photographs from southern Peru (*minor*; Londoño 2014, G.A. Londoño *in litt.* 2016). Egg-laying has been documented twice, both at nests in eastern Ecuador. At one, the hour of egg deposition was in the afternoon, with roughly 48 hours separating the two eggs. At another, the precise hour of egg-laying was determined by video surveillance. At this nest (Greeney 2017a) the strong abdominal contractions of the female were clearly visible on the video as she laid the first egg at 16.30. Soon after laying the egg she stood, peered into the nest and gently rolled the egg by tucking her bill under it and using a series of rapid, sewing-machine-like bill movements. This rapid probing movement was nearly identical, though perhaps more gentle, than the proposed nest-cleaning or parasite removal behaviour described for other antpittas (Greeney *et al.* 2008), but similar to that used by some *Grallaricula* to, at least occasionally, roll their eggs (see Slate-crowned Antpitta). This movement was, in form and force, similar to those used in nest construction, described above. At this nest, the second egg was laid at 16.25, almost precisely 48 hours after the first. The adults spent very little time at the nest in the period between eggs, and the egg was not attended at night until the evening after clutch completion. Both adults occasionally sat on the incomplete clutch, but only for brief periods during which they spent most of time preening, adjusting material in the nest, or singing. Thus, egg development is almost certainly delayed until the clutch is complete, suggesting that hatching will be synchronous in this species (though undocumented). Four hours of video observation (midday) at a nest somewhere midway through incubation (Greeney *et al.* 2005b) revealed that both adults shared incubation duties and that the eggs were covered nearly continuously. The incubating adult appeared to anticipate the arrival of the second, walking off the nest only a few seconds prior to the arrival of its mate. One complete incubation bout measured at this nest lasted 139 minutes. No further details have been published concerning adult behaviour during incubation, and incubation period is unknown. Egg dimensions (clutches separated by /): 27.0×21.5mm, 28.0×21.5mm / 27.3×20.9mm, 27.0×20.9mm ($n = 4$, *campanisona*, Tostain & Dujardin 1988); 24.9×19.9mm, 22.5×19.0mm (Greeney *et al.* 2005). 22.6×19.4mm, 23.4×19.7mm / 25×20mm, 24×19mm / 25.6×19.9mm, 25.1×20.2mm ($n = 6$, *signata*, Greeney 2017). Greeney (2017a) gave the fresh mass of four eggs of *signata*: 4.67g, 4.90g / 5.57g, 5.50g. **NESTLING & PARENTAL CARE** No information. **SEASONALITY** Overall, reproductive seasonality is poorly documented. With such a broad range, it likely varies greatly by region, but few data are available from most areas. So far as is known, it breeds during the wet season (December–January) in French Guiana (Tostain *et al.* 1992). Robbins *et al.* (2004) found evidence of reproductive activity in southern Guyana in March and April (*campanisona*). Laying in eastern Ecuador appears to begin in late December or early January, and continues until at least early May, also suggesting that breeding is timed to the wetter season in this part of its range (Greeney 2017b). **BREEDING RECORDS** *M. c. modesta* Subadult transitioning, 6 March 1957, Río Guapaya (FMNH 248948). *M. c. dissors* Juvenile, 23 January 1929, Cerro Duida (AMNH 270917); two subadults, Salto do Hua, 20 and 25 November 1930 (USNM 326446/449, Friedmann 1948 only reported one of these as immature); subadult transitioning ('immature' in Friedmann 1948), 22 February 1931, below Caño Caripo (USNM 327110). *M. c. campanisona* Incubation, 6 December 1986, beside Piste de St.-Elie (Tostain & Dujardin 1988); incubation, 5 December 1984, Le Mont Belvédère (Tostain & Dujardin 1988); juvenile transitioning 11 March 1887, Takutu River (AMNH 492333); subadult, 19 October 1920, Kartabu Point (AMNH 805766); two ♂♂ with testes enlarged, 25 December 1960 and 5 February 1961, Kayser Gebergte Airstrip (Blake 1963, FMNH 260490/491); ♂ with testes greatly enlarged, 24 January 1962, West River, Wilhelmina Mountains (Blake 1963, FMNH 264496). *M. c. signata* Nest under construction, 23 February 2012, Cabañas Yankuam (Greeney 2017a); nearly complete nest, 22 December 2010, Gareno Lodge (Greeney 2017a); clutch initiation, 5 January 2007, Shiripuno Research Center (Greeney 2017a); clutch initiation, 22 February 2012, Cabañas Yankuam (Greeney 2017a); incubation, 2 May 2009, Tiputini Biodiversity Station (R.A. Gelis in Greeney 2017a); nest with eggs, 20 May 2003, *c.*7.5km ESE of Pilche (Orellana, not Sucumbíos as reported; Greeney *et al.* 2005b); fledgling transitioning, 20 May 1996, Carretera Pompeya (MECN 7090; M.V. Sánchez N. photos); subadult, 27 February 2004, Noytzabaja (MECN 8169, M.V. Sánchez N. photos). *M. c. minor* Juvenile transitioning, 2 July 1930, Lago Sampaio (AMNH 282114); subadult, 9 November 1922, 1km NNW of Arimã (YPM 29645); subadult, 29 September 2001, Moskitania (FMNH 433462); enlarged ovary, 1 October 1958, mouth of Río Inambari (FMNH 251970). *M. c. subcanescens* Juvenile transitioning, 7 May 1930, Rosainho (AMNH 282119); 'immature' ♂ 23 July 1907 (AMNH?, specimen not

located) and a 'juvenile' ♀ 3 August 1907 (AMNH 492336, not examined), Calama (Hellmayr 1910); subadult, 18 June 1931, 'Igarapé Brabo', Rio Tapajós (AMNH 286765).

TECHNICAL DESCRIPTION Sexes similar. *Adult* This species shows little plumage variation except for some inter- and intra-population variation in the warmth of the upperparts and breast streaking, and the degree to which the pale underparts are washed greenish or greyish. The following refers to nominate *campanisona*, following Chubb (1921). Above generally uniform rufous-brown including the head, back, wings and tail. Inner webs of primaries and secondaries dark brown with paler margins; ear-coverts similar to back; throat and underparts white, edged greyish-brown on breast and more broadly on breast-sides and body-sides; underwing-coverts cinnamon-rufous, quill-linings also tinged this colour, apical portion of the quills brown like the underside of the tail. **Adult Bare Parts** *Iris* dark brown; *Bill* maxilla black, mandible whitish or pinkish-yellow with dark tip and tomia; *Tarsi & Toes* pinkish-grey. *Immature* Zimmer (1934) described an immature female (*dissors*) as having 'the mantle, sides of breast and sides of head (including lores and superciliary region), chin and throat, and remiges and rectrices apparently adult. The whole top of the head (from forehead to nape is rich Bay; the center of the breast is Bay; the sides of the upper belly are Hazel; the remainder of the belly is white, tinged with Hazel; the flanks are Bay; the rump is Bay; the upperwing-coverts are brownish like the back but with dull Auburn tips; the scapulars are dull Bay.' This description appears to refer to what I consider a juvenile, specifically AMNH 270917, which I examined and used for the juvenile description below. *Fledgling* I have not examined a fledgling fresh from the nest, but based on an older fledgling *subcanescens* (MECN 7090) and photographs of older nestlings of *minor* provided by G.A. Londoño, the following description is probably fairly accurate. There is, to date, no reason to suspect that fledglings of the different races vary significantly. They leave the nest while still covered almost entirely in fluffy, dark rusty-brown down (= 'bay' from Zimmer 1934). The throat and face are largely bare, exposing the dark pinkish-grey skin, but are soon covered in pale whitish or olivaceous-grey feathering, respectively. The down on the flanks and belly tends to buffy-brown, becoming buffy-white on the central belly. The tail is not fully grown and in life is probably not yet visible beyond the downy rump. Their wings are similar to adults, probably close to fully developed, but the upperwing-coverts are lightly washed pale buff at the tips and distinctly edged bright rufous-chestnut. The amount of buffy at the tips of the coverts appears to vary, but it is unclear if this represents individual or racial variation. On the whole, inner coverts appear to have more buff, outer coverts sometimes lack buff markings altogether. **Fledgling Bare Parts** *Iris* dark; *Bill* dusky-orange overall, maxilla largely blackish with more orange or yellow on tomia, mandible more orange, tending towards yellowish on tomia, rictal flanges inflated and yellow-white, orange on inner surfaces; *Tarsi & Toes* dusky-pinkish, washed orange, especially on posterior edge. *Juvenile* The first covering of non-downy feathers are already similar in coloration to those of adults. These grow in, replacing the rusty down of the fledgling stage, first on the head-sides, upper breast, upper back and scapulars. As described by Zimmer (1934), the last areas to retain significant amounts of fledgling plumage are the crown, rump, flanks and central breast. Juveniles still retain upperwing-coverts distinctly fringed bright reddish-chestnut, sometimes still with a hint of buff near the tips. I have not seen juveniles in life, but bare parts are presumably now similar to adults, with the bill probably retaining some orange or yellow tones, especially near the base of the mandible and the rictal flanges. *Subadult* Appears nearly indistinguishable from adult in overall coloration, but occasionally retains scattered downy feathers in the centre of the breast or belly, on the rump or crown. Those individuals I have seen with retained immature feathers on the head tend to have them on the edges of the crown, in one case almost forming an indistinct superciliary. Subadults retain chestnut fringes at the tips of the upperwing-coverts. Bare parts are presumably now indistinguishable from those of adults.

MORPHOMETRIC DATA *M. c. dissors Wing* range 79–82mm, mean 79.8mm; *Tail* range 31–37mm, mean 33mm; *Bill* culmen from base range 19.0–21.75mm, mean 20.6mm; *Tarsus* range 38–42mm, mean 40.7mm (*n* = ?, ♀♀, Zimmer 1934). *Wing* range 80–85mm, mean 82.3mm; *Tail* 32–36mm, mean 34.5, culmen from base 20–23mm, mean 24.6mm; *Tarsus* 39–43mm, mean 40.6mm (*n* = ?, ♂♂, Zimmer 1934). *Bill* from nares 11.9mm, 10.9mm, 11.5mm, 12.7mm; *Tarsus* 41.8mm, 41.3mm, 42.0mm, 45.7mm (*n* = 4, ♀,♂,♂,♂, USNM). *Bill* from nares 11.6mm, 11.3mm; *Tarsus* 43.2mm, 40.0mm (*n* = 2, ♂,♀, subadults, USNM). *Wing* 77mm, 79mm; *Tail* 33.5mm, 34mm; *Bill* total culmen 21mm, 20mm; *Tarsus* 34.5mm, 38mm (*n* = 2, ♂,♀, Olivares 1964). *M. c. modesta Bill* from nares 11.0mm; *Tarsus* 43.0mm (*n* = 1, ♀?, USNM 372265). *M. c. campanisona Wing* 81mm; *Tail* 37mm; *Bill* [exposed] culmen 19mm; *Tarsus* 45.5mm (*n* = 1, ♂, Chapman 1917). *Wing* 80.5mm; *Tail* 35mm; *Bill* [exposed] culmen 30.75mm; *Tarsus* 44.5mm (*n* = 1, ♂, von Berlepsch 1908). *Wing* 80mm; *Tail* 30mm; *Bill* exposed culmen 20mm; *Tarsus* 40mm (*n* = 1, ♂, Chubb 1921). *Wing* 82mm, 88mm (*n* = 2♂♂, Haverschmidt & Mees 1994). *Wing* range 80–89mm, mean 84.6mm (*n* = 8, ♀♀, Haverschmidt & Mees 1994). *Wing* 80–81mm (*n* = ?, ♂?, Haverschmidt 1968). *Wing* mean 80.0mm; *Tail* 37.0mm (*n* = 2, ♂?, Bierregaard 1988). *M. c. signata Wing* range 76.5–80.5mm, mean 79mm; *Tail* 29–36mm, mean 33.2mm; *Bill* culmen from base 20–21mm, mean 20.6mm; *Tarsus* 39–41mm, mean 40mm (*n* = 5♀♀, Zimmer 1934). *Wing* range 77–84mm, mean 80mm; *Tail* 28–34.5mm, mean 32.5mm; *Bill* culmen from base, 20.5–21.5mm, mean 21mm; *Tarsus* 37.0–41.5mm, mean 39.9mm (*n* = 6♂♂, Zimmer 1934). *Bill* from nares 11.1mm; *Tarsus* 41.6mm (*n* = 1♀, WFVZ 44888). *Bill* from nares 10.8mm, 11.8mm, n/m, 11.9mm, 12.8mm, exposed culmen 15.2mm, 18.3mm, n/m, 16.9mm, 18.0mm, width at gape 9.3mm, 10.8mm, 9.6mm, 10.6mm, 10.7mm; *Tarsus* 40.7mm, 39.5mm, 39.0mm, 42.4mm, 39.8mm (*n* = 5, ♀,♀,♂,♂,♂, MLZ). *Wing* 76mm, 77mm; *Tail* 35mm, 34mm; *Bill* [exposed] culmen 18mm, 18mm; *Tarsus* 36.5mm, 40.5mm (n = 2♂♂, as *minor*, Chapman 1917). *M. c. minor Wing* range 83–88mm, mean 85.6mm; *Tail* 31.5–38mm, mean 34.6mm; *Bill* culmen from base 21–25mm, mean 23.2mm; *Tarsus* 42–44mm, mean 42.9mm (*n* = 8, ♂♂, Zimmer 1934). *Wing* range 78–86.5mm, mean 82.1mm; *Tail* range 29.5–39mm, mean 33.4mm; *Bill* culmen from base 20.0–23.5mm, mean 23.2mm; *Tarsus* range 41–45mm, mean 42.2mm (*n* = 4♀♀, Zimmer 1934). *Wing* 88mm, 85mm; *Tail* 38mm, 35mm; *Bill* [exposed] culmen 18mm, 19mm; *Tarsus* 31mm, 42mm (*n* = 2, ♂,♀, Gyldenstolpe

1945). *M. c. subcanescens Wing* 85mm, 85mm, 89mm; *Tail* 38mm, 41mm, 40mm; *Bill* [exposed culmen] 19mm, 19mm, 20.5mm (*n* = 3, ♀ juv.', ♂ 'imm.', ♂, Hellmayr 1910). *Wing* 86mm; *Tail* 40mm; *Bill* [exposed culmen?] 20mm; *Tarsus* 44mm (*n* = 1♂, holotype, Todd 1927). **Mass** 45.0g (*n* = 1, ♀, *dissors*, USNM 586403), 46g (n = 1♀, *dissors*, Williard *et al.* 1991), 39.5g, 54g (*n* = 2, ♂♂, *campanisona*, Haverschmidt & Mees 1994), range 42–64g, mean 49.6g (*n* = 6♀♀, *campanisona*, Haverschmidt & Mees 1994), 43.8g (*n* = 1, ♂, *signata*, WFVZ 44888), 45.8g (*n* = 1♀, *minor*, O'Neill 1974), 47g (*n* = ?, ♂?, Hilty 2003), mean 47.7 ± 2.1g (*n* = 3, ♂?, *campanisona*, Bierregaard 1988). Several studies have used 47g or 48g as the mass of this species for calculations or analyses (Karr *et al.* 1990, Nemeth *et al.* 2001). Though the precise source of these data are not *always* clear, they are usually taken from Dunning (1992, 2008), with or without an unspecified inclusion of their own data. **Total Length** 12.9–15.0cm (Chubb 1921, Meyer de Schauensee & Phelps 1978, Ruschi 1979, Hilty & Brown 1986, Iafrancesco *et al.* 1987, Sick 1993, Ridgely & Tudor 1994, Ridgely & Greenfield 2001, Krabbe & Schulenberg 2003, Hilty 2003, Restall *et al.* 2006, McMullan *et al.* 2010, Schulenberg *et al.* 2010, Spaans *et al.* 2016).

TAXONOMY AND VARIATION Despite seemingly little variation in plumage across its rather broad Amazonian range, six subspecies of *M. campanisona* are currently recognised (Clements 2007). More than a decade ago, A. Whittaker and K.J. Zimmer (Whittaker 2004) remarked on the unique loud-song and calls of *subcanescens*, but despite continued agreement as to its vocal distinctiveness (Krabbe & Schulenberg 2003), the possibility that it deserves species rank remains unexplored. During my research on subspecific ranges, I listened to calls and songs from across the species' range and examined hundreds of skins and photos. My own conclusion is that *subcanescens* may be the only taxon discussed here that deserves recognition as separate from nominate *campanisona*, based largely on consistent differences in song. Its status as a subspecies or full species remains to be seen. Race *subcanescens* is also the only taxon to have clearly defined dispersal barriers isolating it from other populations: the Amazon from nominate *campanisona*; Rio Madeira from *minor*. Of the remaining subspecies, there appear to be no consistent plumage characters to separate them, and their genetic isolation from each other is questionable. The lower Rios Negro and Branco likely isolate nominate *campanisona* from *dissors* and there is an apparent gap in distribution in the northern portion of the Rio Negro/Branco interfluvium, suggesting possible isolation of *campanisona* from other races (from *subcanescens* by the Amazon). Similarly, *dissors* north of the Amazon are likely isolated from populations of *minor* on the south bank. As hypothesised here, *dissors* is isolated from *signata* by the lower Río Napo and by an apparent gap in distribution in the central Colombian Amazon, a range gap that also isolates *dissors* from *modesta* in Colombia. Though this gap in records is unlikely, there are several potential dispersal barriers in this region (e.g., Ríos Vaupés, Japurá and Putumayo). Nevertheless, it seems quite possible that populations at the base of the Colombian Andes (*signata*, *modesta*) are contiguous with Colombian and Brazilian *dissors*. The remaining races (*modesta*, *signata*, *minor*) are all found up to at least 1,000m in most parts of their ranges in the East Andean foothills, making it more challenging to identify potential dispersal barriers between them. In sum, although there is probably some degree of genetic isolation between populations in central Amazonia, the species' near-continuous distribution around the edges of the Amazon Basin makes it likely that there are few restrictions to gene flow between populations currently considered as *minor*, *signata*, *modesta*, *dissors* and *campanisona*. Zimmer (1934) appears to be the last person to have published their ideas on this matter. At that time there were only four subspecies recognised: *modesta*, *campanisona*, *subcanescens* and *minor*. The latter was considered to include all birds from Ecuador and south-east Colombia (now *signata*; see Other Names). This review led Zimmer (1934) to describe both *dissors* and *signata*, but it is clear from his discussion that the lines between races are far from clear. In particular, he referred several times to a confusing overlap in characters that occurs in populations in several locations at the eastern base of the Andes and clearly states that the races are frequently inseparable except in adult plumage. Other authors (i.e., Chapman 1926) have also questioned the validity of *modesta*. Nevertheless, Zimmer's (1934) arrangement, both taxonomic and geographic, has been maintained, but in an (understandably) rather vague fashion, leaving some uncertainty as to the true ranges of currently recognised taxa. These are discussed under the subspecific accounts below.

Myrmothera campanisona modesta (P. L. Sclater, 1855), Proceedings of the Zoological Society of London, vol. 23, p. 89, plate 94, 'Santa Fé de Bogota', Colombia. The holotype, described as *Grallaria modesta*, is an unsexed adult at Tring (NHMUK 1854.1.25.72). The specimen was apparently part of a collection purchased from S. Stevens (Warren & Harrison 1971). Its provenance is unknown, although it is fairly likely that it was collected somewhere along the Amazonian slope of the Colombian East Andes near Villavicencio. Note that the citation (Sclater 1877) provided by Warren & Harrison (1971) is incorrect. The most range-restricted of the currently recognised subspecies, *modesta* is apparently confined to the base of the East Andes of Colombia south from the vicinity of Cabuyaro in northern Meta to the Serranía de la Macarena (Hilty & Brown 1986). Race *modesta* is said to be more olive-brown, less rufescent, than nominate *campanisona* (Krabbe & Schulenberg 2003). Sclater (1855b) described *modesta* as 'a rather uniformly-coloured species, of which the British Museum contains a single specimen. There are indications of darker marginations to the feathers of the nape and back. The breast feathers are medially yellowish-white, broadly margined with olivaceous.' SPECIMENS & RECORDS *Meta* Cabuyaro, 04°16'N, 73°03'W (XC 148243/48, F.G. Stiles); Villavicencio, 04°09'N, 73°38'W (Chapman 1917, Iafrancesco *et al.* 1987, CUMV, USNM, AMNH, ICN); Los Micos, *c.*03°17'N, 73°53'W (MHNG 1128023); Río Güejar, 02°54.5'N, 73°14'W (ICN 13714); Río Duda, *c.*02°45'N, 73°55'W (Olivares 1962, Hilty & Brown 1986, AMNH 460422/23, ICN 11664); La Uribe, 02°45'N, 74°25'W (XC 157702; J.A. Suárez); Centro de Investigaciones Ecológicas Macarena, 02°40'N, 74°10'W (Cadena *et al.* 2000, ML 81022, I. Jiménez); Río Guapaya, 02°28'N, 73°48'W (FMNH 248947–950); Finca El Porvenir, 02°23'N, 74°37.5'W (ML 63743521, J. Muñoz García).

Myrmothera campanisona dissors J. T. Zimmer, 1934, American Museum Novitates, no. 703, p. 11, Río Cassiquiare, Venezuela right bank, opposite El Merey [= 03°05'N, 65°55'W, 100m, state of Amazonas]. The holotype (AMNH 417393) is an adult male collected 20

April 1929 by the Olalla brothers, and several topotypical skins are held in New York (AMNH 417394, 423644/45). The range of *dissors* is generally described as extending across the north-west Amazon from eastern Colombia to southern Venezuela and north-west Brazil. Hilty & Brown (1986) followed Zimmer (1934) in considering populations in Caquetá and western Putumayo to represent this taxon. Based on what we now know of the distribution of Thrush-like Antpitta, however, these populations are almost certainly contiguous with those in north-east Ecuador (at least north of the Río Napo) and I have provisionally assigned them to *signata* (see below). For the Colombian range of this subspecies, I include only records from Amazonas, Vaupés and eastern Caquetá. There are no reports of the species from Guaviare, Vichada or Guainía. Collections at San Fernando de Atabapo and recordings from Pintado (both Amazonas, Venezuela), at the borders of these latter two departments, do suggest its presence in this region of Colombia. From Venezuela's western border, *dissors* is reported from scattered localities across Amazonas (Venezuela) to the headwaters of the Río Siapa, at the western border with Brazil. South of the Venezuelan border, however, reports are scarce, predominantly in the upper Rio Negro watershed (e.g., São Gabriel da Cachoeira), sites on the north bank of the Amazon, and the west bank of the lower Rio Negro. Recently recorded populations (Borges *et al.* 2014) on the left bank of the Rio Negro (Rio Negro/Branco interfluvium) and several scattered records west of the Rio Branco in north-west Roraima are assigned to this race following Naka *et al.* (2012). Presumably based on Zimmer's (1934) somewhat reluctant assignment of skins from the left bank of the lower Rio Madeira (Rosainho) to *dissors*, its range has been considered to cross the Amazon somewhere in this region (e.g., Krabbe & Schulenberg 2003). This assumption is somewhat puzzling, considering all other populations south of Rio Amazonas, both east and west of there, are usually referred to other subspecies. I examined two skins in New York (AMNH 282114/119) from this region, presumably the same 'adult and immature' considered by Zimmer (1934). I was unable to reach the same conclusion, and indeed found it difficult to (clearly) assign these specimens to any race. Based on biogeographical considerations, however, I have excluded this region (south of the Amazon) from the range of *dissors*. I will further muddy the situation by adding the following observations. Rather confusingly, there are two localities called 'Rosainho' on the middle/lower Rio Madeira (Paynter & Traylor 1991b), one just north of Lago Sampaio, on the left bank (i.e., *minor*), and one further south on the right bank (i.e. *subcanescens*). The two skins in question were both collected by the Oallas, two months apart. One label gives the locality as 'Rosainho', the other as 'Rosainho-Lago Sampaio'. Following the distributions herein hypothesised, this suggests to me that, *in theory*, one of these two skins assigned to *dissors* by Zimmer (1934) was *subcanescens* and the other *minor*, which is the arrangement adopted here. This somewhat 'minor' point is, for me, a perfect illustration of the need for a taxonomic revision, but both are immature (*contra* Zimmer 1934; see Reproduction), one a juvenile with obvious tufts of nestling plumage and one a subadult still with obviously patterned upperwing-coverts. Hilty & Brown (1986) were unsure of the subspecific affinities of Thrush-like Antpitta in Amazonas (Colombia) and

Krabbe & Schulenberg (2003) assigned all Peruvian birds south of the Colombian border and 'north of the Rio Amazonas' to *signata*. However, I have provisionally considered the range of *dissors* to include this region, extending north-west along the left bank of the Río Napo to at least the area around Nueva Florencia (02°11'S, 74°06'W). Vouchered records from this region are scarce, partially in agreement with Zimmer (1934) who also considered birds from the Río Napo/Putumayo interfluvium (e.g., Pebas) to belong to *signata*, but further speculated that they might better be assigned to *dissors* (as they are here). Race *dissors* is supposedly more olive-brown above (less rufescent) than nominate or *subcanescens* (but less olive than *modesta*), slightly paler rufescent than *minor* and much paler than *signata* (Zimmer 1934). Bare-parts coloration of *dissors* is described as: *Iris* dark brown; *Bill* black, mandible pink basally; *Tarsi & Toes* pale grey (Willard *et al.* 1991). SPECIMENS & RECORDS VENEZUELA: *Amazonas* 9km ENE of Pintado, 05°28'N, 67°30'W (ML: C.A. Marantz); San Fernando de Atabapo, 04°03'N, 67°42'W (Hilty 2003, COP 21864/65); localities at south end of Cerro Duida [i.e., Caño León, Playa del Río, La Laja] *c.*03°18'N, 65°44'W (Zimmer 1934, Hilty 2003, AMNH 272999–3003, MCZ 199729); below Caño Caripo, 03°07'N, 65°51'W (Friedmann 1948, USNM 327110); Pica Yavita-Pimichín, 02°54'N, 67°25'W (COP 34413–423); Chapazón, 02°03'N, 67°05'W (USNM 327111); Río Siapa, headwaters, 02°02'N, 63°56'W (COP 34885); Río Casiquiare/Guainía confluence, 02°01'N, 67°06'W (Zimmer 1934, AMNH 423646); El Carmen, 01°16.5'N, 66°52.5'W (COP 41340–341); base of Cerro de la Neblina, *c.*00°50'N, 66°10'W (Williard *et al.* 1991, FMNH 319335). COLOMBIA: *Vaupés* Mitú, 01°15.5'N, 70°14'W (XC 81768, A. Spencer); Caño Cubigú, 01°02'N, 70°12'W (Olivares 1964, ICN 9905/06); Comunidad Teresita, *c.*00°33.5'N, 70°03'W (ICN 35109); Bocas del Pirá, 00°26'S, 70°14'W (Stiles 2010); Jotabeyá, 00°37'S, 70°11'W (Stiles 2010). *Caquetá* Río Cuñaré, 00°31'N, 72°37'W (Álvarez *et al.* 2003); Río Cuñaré/Río Amú confluence, 00°13'N, 72°25'W (Álvarez-R. *et al.* 2003); Puerto Abeja, 00°04'N, 72°27'W (Álvarez *et al.* 2003); Río Mesay, Bombonal, 00°04'N, 72°12.5'W (ICN 32913). *Amazonas* Jirijirimo, 00°03'S, 70°57'W (Stiles 2010); MiritíParaná, Puerto Rastrojo, 01°12'S, 69°48.5'W (ICN 28751); La Pedrera, 01°19'S, 69°35'W (D. Cantarón *in litt.* 2017); Estación Ecológica Omé, *c.*03°33'S, 69°56'W (XC 58410, G.N. Forero); PNN Amacaycaú, 03°52'S, 70°09'W (XC 163576, F. Lambert); Las Malokas, 04°11'S, 69°56'W (eBird: N. Ocampo-Peñuela); Río Amazonas, Isla Ronda, 04°09'S, 70°00'W (A.M. Cuervo *in litt.* 2016). **BRAZIL**: *Roraima* ESEC Maracá, 03°25'N, 61°40'W (Moskovits *et al.* 1985); Colonia Apiaú, 02°38'N, 61°12'W (FMNH 344085). *Amazonas* Surumoni, 03°10'N, 65°40'W (Nemeth *et al.* 2001); Rio Toototobi, near Barcelos, 01°52'N, 63°25.5'W (MPEG 37522/23); Marabitanas, 00°56'N, 66°49'W (von Pelzeln 1871, Zimmer 1934); Salto do Hua, 00°40'N, 66°08'W (Friedmann 1948, USNM, ANSP); São Gabriel da Cachoeira, 00°08'S, 67°05'W (MPEG 77461, AMNH 276105/06); Umarituba, 00°04'N, 67°15'W (Zimmer 1934); Lagoa Comprida, 01°53'S, 61°44'W (ML 117014, C.A. Marantz); Iaunari, 00°31'S, 64°50'W (Zimmer 1934); Tonantins, 02°52'S, 67°48'W (CM); Lago Acajatuba, 03°06'S, 60°29.5'W (MPEG 12469); Manacapuru, 03°14'S, 60°41'W (Pinto 1837b, 1938, CM); Cacau Pereira, 03°10'S, 60°10'W (Zimmer 1934). In addition there are several reported localities within the Rio Negro/Branco

interfluvium, 00°19'N, 62°59'W, 00°18'S, 62°44'W, 00°17'N, 62°45'W (Borges *et al.* 2014). **PERU**: *Loreto* Middle Río Campuya, 01°31'S, 73°49'W, Río Ere headwaters, 01°41'S, 73°43'W, and lower Río Ere, 02°01'S, 73°15'W (Stotz & Inzunza 2013); Quebrada Bufeo, 02°20'S, 71°36.5'W (Stotz *et al.* 2016); 'Campamento Medio Algodón', 02°35.5'S, 72°53.5'W (Stotz *et al.* 2016); 'Campamento Choro', 02°36.5'S, 71°29'W (Stotz & Alván 2011); 'Campamento Cachimbo', 02°43'S, 70°31.5'W (Stotz & Alván 2011); 'Campamento Piedras', Río Algodoncillo, 02°47.5'S, 72°55'W (Stotz & Alván 2010); 'Campamento Yaguas', 02°52'S, 71°25'W (Stotz & Alván 2011); Río Yanayacu, 'Campamento Corupa', 02°53'S, 73°01'W (Stotz & Alván 2010); Río Yanayacu, *c.*90km N of Iquitos, 03°05'S, 73°08'W (LSUMZ 115591/92); 'Campamento Alto Cotuhé', 03°12'S, 70°54'W (Stotz & Alván 2011); Explornapo Lodge, 03°14'S, 72°55'W (XC 28131, D. Geale); Sucusari and vicinity, *c.*03°15'S, 72°54'W (LSUMZ 110246/47, ML); Pebas, 03°19'S, 71°51.5'W (Sclater & Salvin 1867, Zimmer 1934); ACP Sabalillo, 03°22'S, 72°17.5'W (XC 20059, D. Edwards); Orosa, 03°26'S, 72°08'W (as *minor*; Zimmer 1924, AMNH 231936/37); Otorongo Lodge, 03°28'S, 72°30.5'W (S.J. Socolar *in litt.* 2017). NOTES The locality São Gabriel da Cachoeira, on the left bank of the Rio Negro, has been variously called 'San Gabriel' or 'São Gabriel' on specimen labels and by various authors. Paynter & Traylor (1991b) referred to it as Uaupés.

Myrmothera campanisona campanisona (Hermann, 1783), *Tabula affinitatum animalium olim academico specimine edita*, p. 189, 'Cayenne' [= unknown location in French Guiana]. The range of nominate *campanisona* extends from eastern Venezuela through the Guianas and Suriname (Thiollay 2002, Braun *et al.* 2007) and south to the north bank of the Amazon. In north Brazil, there are records from eastern Roraima, Amapá, northern Pará, west to the region north of Manaus and east of Rio Negro. Following Naka *et al.* (2012), I have used the left bank of the Rio Branco as the western limit of *campanisona* in this region, and included north-west Roraima records at ESEC Maracá and Colonia Apiaú (both west of Rio Branco) within the range of *dissors*. SPECIMENS & RECORDS VENEZUELA: *Delta Amacuro* Altiplanicie de Nuria, 07°40'N, 61°20'W (COP 17126). *Bolívar* Sierra de Imataca, Río Grande–El Palmar area (see Notes), *c.*08°01.5'N, 61°46'W (Mason 1996, LSUMZ 68556, XC, ML); Sierra de Lema, km67, 06°21'N, 61°22'W (D. Beadle *in litt.* 2017); Caño El Buey, 06°11'N, 61°32'W (COP 46156/57); La Maloka, upper Río Cuyuní, 06°04'N, 61°28'W (Lentino *et al.* 2013b); Sierra de Lema, km109, 06°01'N, 61°23'W (COP 64726/27). GUYANA: *Barima-Waini* Baramita, 07°22'N, 60°29'W (KUNHM 89050, USNM 586403). *Cuyuni-Mazaruni* Arimu Mine, 06°28'N, 59°09'W (AMNH 805763); Bartica and vicinity (*c.*06°24'N, 58°37'W), e.g., Kartabu Point, Penal Settlement (Salvin 1885, Davis 1953, AMNH 805764–766, USNM); near Marshall Falls, *c.*06°23'N, 58°43.5'W (ML 84950, D.W. Finch); Merume Mountains [= Ourumee on labels], *c.*05°48'N, 60°06'W (ANSP 51071). *Potaro-Siparuni* Mount Ayangganna, 05°18'N, 59°50.5'W (Milensky *et al.* 2016); Tumatumari, 05°16'N, 59°09'W (Zimmer 1934, Snyder 1966, AMNH 125716); Iwokrama Forest Reserve, *c.*04°14.5'N, 58°54'W (Ridgely *et al.* 2005, ANSP 188776, XC). *Upper Takutu-Upper Essequibo* Maipaima Creek, 03°23'N, 59°30'W (Parker *et al.* 1993, ML); Cacique Mountain, 03°11'N, 58°49'W (Finch *et al.* 2002); Takutu

River, 02°54'N, 58°56'W (AMNH 492332/33); SE of Aishalton, *c.*02°30.5'N, 59°12'W (ML 72396, D.W. Finch); *c.*6km WSW of Parabara, 02°12'N, 59°22'W (Robbins *et al.* 2004, KUNHM 90848); 5.5km SW of Parabara, 02°11'N, 59°20'W (O'Shea *et al.* 2017); Kuyuwini River, *c.*02°06'N, 59°14'W (ML 106321, D.W. Finch); Gunn's Strip, 01°39'N, 58°37'W (Robinson *et al.* 2007); upper Essequibo River, 01°35'N, 58°38'W (Robinson *et al.* 2007, USNM 625540); Kamoa River, 01°32'N, 58°49.5'W (Robinson *et al.* 2007, O'Shea 2013); Sipu River, Acary Mountains, 01°24'N, 58°56.5'W (Robinson *et al.* 2007, O'Shea 2013, ML). *Upper Demerara-Berbice* Rockstone, 05°59'N, 58°31'W (Snyder 1966, FMNH 108350); S of Linden, *c.*05°59'N, 58°18.5'W (ML 134777, B.J. O'Shea); Kamakabra River, 05°28'N, 58°22'W (Snyder 1966). *East Berbice-Corentyne* Tiger Creek, 05°41'N, 57°10'W (Snyder 1966); headwaters of Itabu Creek, 01°33'N, 58°10'W (Blake 1950, FMNH 120239). **SURINAME**: **Commewijne** Near Meerzorg, 05°48.5'N, 55°05.5'W (eBird: T. Tromblee). *Para* Powakka, 05°27'N, 55°05'W (CM A1471); Jodensavanne, 05°26'N, 54°58.5'W (eBird: B.J. O'Shea); Natuurreservaat Boven-Coesewijne, 05°22'N, 55°36.5'W (eBird: B.J. O'Shea); Shotel Weg, 05°20.5'N, 55°12'W (MCZ 144902). *Brokopondo* Brownsberg Natuurpark, *c.*04°57'N, 55°10'W (ML 134306, B.J. O'Shea, XC 114042, J. King). *Sipaliwini* Mozes Kreek area, 04°51'N, 56°46.5'W (XC 272042/43; P. Boesman); Foengoe Eiland, 04°43.5'N, 56°12.5'W (ML 2131/36/50, T.H. Davis); Raleigh Vallen-Voltzberg Natuurpark, 04°40'N, 56°10'W (ML 21015, T. Pepper); Bakhuis Gebergte, 04°21'N, 56°45'W (ML 134545, B.J. O'Shea); Lely Gebergte, 04°16'N, 54°44.5'W (O'Shea 2007); Grace Falls, *c.*03°54'N, 56°09.5'W (Zyskowski *et al.* 2011, YPM 139432); Rudy Kappel Airstrip, 03°48´N, 56°09'W (Zyskowski *et al.* 2011); Poti Hill, 03°27.5'N, 55°22.5'W (XC 74996; D.F. Lane); West River, Wilhelmina Mountains, 03°26'N, 56°45'W (Blake 1963, FMNH 264496); Kayser Gebergte Airstrip, 03°07'N, 56°27'W (Blake 1963, FMNH 260490–492); 'Kasikasima camp', 02°58.5'N, 55°23'W (O'Shea & Ramcharan 2013a); confluence of Makrutu Creek–Palumeu River, 02°47.5'N, 55°22'W (O'Shea & Ramcharan 2013a); 'Juuru Camp', upper Palumeu River, 02°28.5'N, 55°38'W (O'Shea & Ramcharan 2013a); Grensgebergte Rock, 02°28'N, 55°46'W (O'Shea & Ramcharan 2013a); Werehpai, 02°22'N, 56°42'W (O'Shea & Ramcharan 2013b); Sipaliwini River, 02°17.5'N, 56°36.5'W (O'Shea & Ramcharan 2013b); Kutari River, 02°10.5'N, 56°47'W (O'Shea & Ramcharan 2013b). *Marowijne* Wane Kreek, 05°36'N, 54°14'W (Haverschmidt 1955). **FRENCH GUIANA**: *Saint Laurent Du Maroni* Tamanoir, 05°09'N, 53°45'W (CM, AMNH, YPM); Camp Aya, Reserve Naturelle La Trinité, 04°36'N, 53°24.5'W (XC 316503, O. Claessens); Saül, 03°37'N, 53°12'W (XC 214776, P. Ingremeau). *Cayenne* km25 on Piste Crique Dardanelles, 05°17.5'N, 53°30.5'W (XC 122505/06, J. King); 90km W of Cayenne, 04°55'N, 53°09'W (Thiollay 1992, 1997); 2km E of Cacao, 04°33'N, 52°28.5'W (XC 59209, A. Renaudier); Montagne de Kaw, *c.*04°28'N, 52°00'W (XC 57113, P. Ingremeau); Guisanbourg, 04°25'N, 51°55.5'W (AMNH 492334); Crique Ipoucin, 04°09'N, 52°24'W (von Berlepsch 1908); Nourages Field Station, 04°05'N, 52°41'W (Thiollay 1992, 1994, Thiollay & Jullien 1998, Thiollay *et al.* 2001); Oyapock, 03°53.5'N, 51°48.5'W (YPM 29638); Pied Saut, 03°43.5'N, 51°54.5'W (CM); Le Pic du Croissant, 03°33'N, 52°15'W (Thiollay 1986); Camopi, *c.*03°10'N, 52°20.5'W (Ménégaux 1904); Mont Belvédère, 02°25'N, 53°06'W (Thiollay 1986, Tostain & Dujardin

1988). **BRAZIL**: *Roraima* PN do Viruá, 01°29'N, 61°01'W (XC 10119, R.A. de By); Posto de Apoio BR-174, 01°29'N, 61°00'W (MPEG 56420). *Amapá* Oiapoque, 03°45.5'N, 51°54'W (Novaes 1974, MPEG 15188); Rio Anotaie, 03°30'N, 52°18'W (Coltro 2008); PN Montanhas do Tumucumaque, 02°12'N, 54°35'W (Coltro 2008); Rio Anacuí, 01°50.5'N, 52°44.5'W (Coltro 2008); Calçoene, 01°25'N, 51°55.5'W (MPEG 21232); Rio Mutum, 01°23'N, 51°55.5'W (Coltro 2008); Cachoeira Belheira, 01°15.5'N, 49°56'W (MPEG 21216/17); Laranjal do Jari, 01°06'N, 53°13'W (WA 736592, K. Okada); Igarapé Capivara, 01°02'N, 51°44'W (MPEG 21168); near Ferreira Gomes, 00°56'N, 51°35.5'W (MPEG 20426); Rio Amapari, Serra do Navio, 00°54'N, 52°01'W (MNRJ 29428); Pedra Branca do Amapari, 00°43'N, 51°55'W (MPEG 20425); Porto Grande (= Macapá), 00°43'N, 51°32'W (LSUMZ 67358, MPEG 20424); Cachoeira Itaboca, 00°05'N, 51°54'W (MPEG 28746); Cachoeira Amapá, 00°02'N, 51°55'W (MPEG 28745); Cachoeira Pancada, 00°10'S, 51°50'W (MPEG 20423); Igarapé Novo, 00°30'S, 52°30'W (MPEG); Igarapé Rio Branco, 00°33.5'S, 51°35.5'W (MPEG 16276/78); RESEX Rio Cajari, 00°35'S, 52°16'W (Schunck *et al.* 2011). *Pará* Alenquer, 00°09'S, 55°11'W (Aleixo *et al.* 2011, MPEG 65466/67); Colônia do Veado, 01°49.5'S, 55°41'W (MPEG 9518); Óbidos and vicinity, *c.*01°54'S, 55°31'W (Griscom & Greenway 1941, CM, YPM, MPEG, XC); Faro, 01°13'S, 57°44'W (Zimmer 1934); FLOTA Faro, 01°42'S, 57°12'W (Aleixo *et al.* 2011, MPEG 64781); *c.*25km NE of Nova Vida, *c.*00°43'S, 53°34'W (Barlow *et al.* 2007); The following records by Aleixo *et al.* (2011) document the occurrence of nominate *campanisona* across most of northern Pará, ESEC Grão-Pará Norte (01°17'N, 58°41'W); ESEC Grão-Pará Centre (00°37'N, 55°43'W); REBIO Maicuru (00°49'N, 53°55'W), FLOTA Trombetas (00°57'S, 55°31'W); FLOTA Paru (00°56'S, 53°14'W). *Amazonas* Presidente Figueiredo, 01°27'S, 60°02'W (WA 2007322, W. Coppede); Balbina Dam, 01°56'S, 59°28'W (Willis & Oniki 1988); Rio Paratucu/Nhamundá confluence, 01°59'S, 56°58'W (AMNH 283955); 30km NNW of Rio Preto da Eva, 02°26'S, 59°46'W (ML 113138, C.A. Marantz); 80km N of Manaus (BDFFP study area), 02°20'S, 60°00'W (Bierregaard 1988, Stotz & Bierregaard 1989, Cohn-Haft *et al.* 1997, Borges & Stouffer 1999, Laurance 2004); *c.*40km W of Manaus, 02°32'S, 60°49'W (Cintra *et al.* 2007); Ramal Pau de Rosa, 02°51'S, 60°04'W (XC 346010, J. Honkala). **Notes** There are are numerous recordings in ML (P.A. Schwartz, C.D. Duncan, L.R. Macaulay) and XC (A.T. Chartier, C. Parrish, N. Athanas) that are accompanied by a variety of coordinates, most of which appear to be rather general but all of which appear to refer to the region around El Palmar (08°01'N, 61°54.5'W) and Río Grande (08°03'N, 61°38'W, Bolívar/Delta Amacuro border). The coordinates provided above represent an arbitrary point halfway between the two locations.

Myrmothera campanisona signata J. T. Zimmer, 1934, American Museum Novitates, no. 703, p. 10, 'below San José', eastern Ecuador [*c.*00°26'S, 77°20'W, Orellana province]. The holotype (AMNH 184361) is an adult male collected 8 April 1924 by the indefatigable Carlos Olalla and sons. This subspecies is distributed from the base of the East Andes in southern Colombia south to the left bank of the Río Marañón and then east through northern Peru, north of the Río Amazonas to the right bank of the Río Napo. Birds at the base of the Andes in south-east

Colombia (western Putumayo, eastern Cauca, western Caquetá), though apparently contiguous with populations in eastern Nariño, were considered to represent *dissors* by Hilty & Brown (1986). This is undoubtedly based on Zimmer's (1934) suggestion that skins from this region (Florencia and La Morelia) belonged with this race. Even Zimmer (1934), however, noted that they were 'approaching' *signata* in overall coloration, and I provisionally include them in *signata* here. This is based largely on geography, as I can find no consistent plumage or vocal characters to separate these two taxa. Ridgely & Greenfield (2001) assigned all Ecuadorian Thrush-like Antpittas to *signata* and, indeed, birds both north and south of the upper Río Napo are fairly uniform in vocal and plumage characters. Along the lower Río Napo, however, the situation is unclear, with the possibility that the range of *dissors* may extend slightly into eastern Ecuador, north of the Río Napo. Race *signata* is said to be more olive-brown, less rufescent, than nominate *campanisona* and distinctly darker and more rufous than *minor* (Zimmer 1934, Krabbe & Schulenberg 2003). In summary, the subspecific affinities of birds in southern Colombia (extreme eastern Caquetá and Putumayo), extreme eastern Ecuador (eastern Orellana and Sucumbíos, north of the Río Napo) and northern Peru (northern Loreto and Amazonas) are unclear, but are provisionally included in *signata* here. In the hypothesised distribution of *signata* mapped, it is replaced south of the Río Marañón by *minor* and north (east) of the lower Río Napo by *dissors*. **Specimens & Records COLOMBIA**: *Caquetá* Mirador de los Tucanes, 01°49'N, 75°40'W (ML 59846251, J. Muñoz García); Florencia, 01°36.5'N, 75°36.5'W (AMNH 116346–347); Morelia, 01°29'N, 75°43.5'W (AMNH 116348–349); Hacienda Villa María, 01°15'N, 75°30'W (Iafrancesco *et al.* 1987, MLS 3957); Puerto Bello, 01°08'N, 76°16'W (Salaman *et al.* 2007); Laguna Pelegrino, 00°03.5'N, 74°36'W (J. Beckers *in litt.* 2017). *Cauca* Alto Río Hornoyaco, 01°14'N, 76°32'W (Salaman & Donegan 1998, Donegan & Salaman 1999, Salaman *et al.* 1999); Río Guayuyaco, 01°00'N, 76°22'W (ROM); Río Indiyaco, 01°06'N, 76°34'W (D. Carantón *in litt.* 2017). *Nariño* Río Churoyacu, 00°24'N, 77°06'W (Meyer de Schauensee 1952, ANSP 165170); Río Rumiyacu, 00°24'N, 77°11'W (Meyer de Schauensee 1952, ANSP 165171–174, UMMZ 222949). *Putumayo* Orito, 00°44'N, 76°51'W (ROM 103992); Bosque Nuevo Mundo, 00°44'N, 77°06'W (XC 305688, B.C. Jaramillo); Estación de Bombeo Guamués, 00°40'N, 77°00'W (FMNH 292936); San Miguel, 00°21'N, 76°55'W (ANSP 165168, ROM 98982); Puerto Umbría, 00°52'N, 76°35'W (ANSP 160036). **ECUADOR**: *Sucumbíos* Bermejo, 00°15'N, 77°23'W (Schulenberg 2002); Río Verde, 00°14'N, 77°34.5'W (Stotz & Valenzuela 2009); Río San Miguel, 00°35'N, 77°35'W (UMMZ 222948); Sinangoe, 00°11'N, 77°30'W (Schulenberg 2002); Cuccono, 00°08'N, 77°33.5'W (Schulenberg 2002); Baboroé, 00°02'N, 76°44.5'W (Stotz & Quenamá 2007); Pisorié Setsa'cco, 00°00.5'N, 76°40'W (Stotz & Quenamá 2007); 6km S of Marian, 00°05'S, 76°20'W (Canaday 1997); Tarapoa, 00°07.5'S, 76°21'W (Canaday & Rivadeneyra 2001); Güeppicillo, 00°11'S, 75°41'W (Stotz & Valenzuela 2008); Limoncocha, 00°24'S, 76°37'W (Pearson *et al.* 1972, LSUMZ, MCZ); Pañacocha, 00°25.5'S, 76°06'W (J. Freile *in litt.* 2017); Sacha Lodge, 00°28.5'S, 76°27.5'W (XC 10029–031, A.T. Chartier); Redondococha, 00°33.5'S, 75°15'W (eBird: D.F. Stotz). *Orellana* 12km N of Lumbaqui, 00°10'S, 77°18'W (ML

74784, M.B. Robbins); Yarina Lodge, 00°28'S, 76°50'W (J. Freile *in litt.* 2017); *c.*7.5km ESE of Pilche, S of Río Napo, 00°30'S, 76°20.5'W (Greeney *et al.* 2005b); Napo Wildlife Center, 00°31.5'S, 76°26.5'W (XC 276260, J. Nilsson); RBP Río Bigal, 00°32'S, 77°25.5'W (XC 175009, R. Ahlman); Yuturi Lodge, 00°33'S, 76°02.5'W (XC 50380/81, B. López-Lanús); Zancudococha, 00°36'S, 75°29'W (ANSP 183351, MECN 4038); Río Suno near Ávila, *c.*00°38'S, 77°26'W (Chapman 1926, AMNH 179379–381, MECN 2629, SBMNH 8626); Tiputini Biodiversity Station, 00°38'S, 76°09'W (Svensson-Coelho *et al.* 2013, Blake & Loiselle 2015, Greeney 2017a); km37 on Maxus road, 00°38.5'S, 76°27.5'W (XC 249102/03, N. Krabbe); PetroEcuador Block 16, 00°39'S, 76°27'W (English 1998); Loreto, 00°41.5'S, 77°18.5'W (ANSP 163655/66); Río Suno, below Loreto, 00°42'S, 77°15'W (MLZ); Pasohurco, 00°44'S, 77°23'W (ANSP 185495); Guaticocha, 00°45'S, 77°24'W (MCZ 299194); Concepción, 00°48'S, 77°25'W (ANSP 169693); Carretera Pompeya, *c.*100km SE of Coca, 00°59'S, 76°14'W (MECN 7090). *Napo* Río Pucuno, 00°48'S, 77°16'W (MCZ 299192/93, LACM 40447); near Mushullacta, 00°50'S, 77°34'W (XC 291673, N. Krabbe); Yachana Lodge, 00°52.5'S, 77°16'W (HFG); RP Hakuna Matata, 00°55.5'S, 77°49.5'W (R.A. Gelis *in litt.* 2003); RP El Pará, 00°57'S, 77°44'W (XC: N. Athanas, P. Coopmans); Gareno Lodge, 01°02'S, 77°24'W (Greeney 2017a; XC); RBP Jatun Sacha, 01°04'S, 77°37'W (D. Beadle *in litt.* 2017); Lisanyacu, 01°07.5'S, 77°57'W (R.A. Gelis *in litt.* 2016). *Pastaza* Shiripuno Research Center, 01°06.5'S, 76°44'W (Greeney 2017a); Churunalpi, 01°32.5'S, 77°45'W (XC 249575/76, N. Krabbe); Sarayacu, 01°44'S, 77°29'W (MECN 3152); Territorio Achuar, *c.*01°45'S, 76°30'W (J. Freile *in litt.* 2017); Río Tigre, 02°03'S, 76°04.5'W (J. Freile *in litt.* 2017); Montalvo, 02°04'S, 76°58'W (ANSP 163657); Kapawi Lodge, 02°32.5'S, 76°51.5'W (XC 260836, M. Lysinger). *Morona-Santiago* RP Nantar, 02°14'S, 78°02.5'W (XC 116460, S. Olmstead); 5km SW of Taisha, 02°20'S, 77°27.5'W (ANSP 182529/30, MECN 2671, ML); Unnsuants, 02°33'S, 77°54'W (XC 249952/53, N. Krabbe); Miazal, 02°38'S, 77°47'W (J. Freile *in litt.* 2017); *c.*3.5km SW of Puerto Morona, 02°54.5'S, 77°42.5'W (R.A. Gelis *in litt.* 2017); Noytzabaja, 03°00'S, 78°18'W (MECN 8169); Santiago, 03°03'S, 78°00.5'W (ANSP 181676); ACM Riberas del Zamora, 03°05.5'S, 78°23.5'W (XC 166792, L. Ordóñez-Delgado); Warientza, 03°12'S, 78°17'W (WFVZ 44888). *Zamora-Chinchipe* *c.*12km N of Guayzimi, 03°56.5'S, 78°41.5'W (R. Ahlman *in litt.* 2017); RP Maicu, 04°12.5'S, 78°38.5'W (XC 194447–49, L. Ordóñez-Delgado); Cabañas Yankuam, 04°15'S, 78°39.5'W (Freile *et al.* 2013b, Greeney 2017a); Quebrada Río Shaime, 04°22'S, 78°44'W (Balchin & Toyne 1998). **PERU**: *Loreto* Panguana, 02°08'S, 75°09'W (Stotz & Alván 2007); mouth of Río Curaráy, 02°22'S, 74°06'W (Zimmer 1934, AMNH 255991/92); Alto Nanay, 02°48.5'S, 74°49.5'W (Stotz & Alván 2007); *c.*110km W of Iquitos, 03°40'S, 74°13.5'W (S.J. Socolar *in litt.* 2017); RN Allpahuayo-Mishana, *c.*03°55'S 73°33'W (Alonso *et al.* 2012, XC); *c.*8km SW of Varillal, 03°57'S, 73°24.5'W (S.J. Socolar *in litt.* 2017); Rio Tigre, *c.*11km SW of Nueva York, 04°23'S, 74°17.5'W (J. Socolar *in litt.* 2017). *Amazonas* Quebrada Katerpiza, 04°01'S, 77°35'W (Inzunza *et al.* 2012); Quebrada Kampankis, 04°02.5'S, 77°32.5'W (Inzunza *et al.* 2012); several localities along Río Cenepa, N of Cháves Valdivia, *c.*04°22'S, 78°16'W (LSUMZ 84948–950, ML 17494, MVZB 175640); 'Río Santiago', N of Río Marañón, *c.*04°25.5'S, 77°38.5'W (see below) (AMNH 407155); Quebrada

Ajachim, 04°39.5'S, 78°05.5'W (A. García-Bravo *in litt.*
2017). **Notes** My examination of the Río Santiago
specimen listed above (AMNH 407155) failed to provide
me with clear characters to convince me of its racial
designation. The Río Santiago is a north-bank tributary
of the Río Marañón and, although the coordinates for this
location given by Stephens & Traylor (1983) place it on
the south bank, I have provisionally referred the specimen
to *signata* and used coordinates to place its origin north
of the Río Marañón.

Myrmothera campanisona minor (Taczanowski, 1882),
Proceedings of the Zoological Society of London, vol. 50,
p. 33, Yurimaguas [= 05°54'S, 76°05'W, 182m, left bank of
Río Huallanga, Dept. Loreto]. The holotype, a male
described as *Grallaria minor*, was originally housed in the
Warsaw Museum before it was destroyed (Zimmer 1934).
This south-western subspecies occurs throughout the
Peruvian Amazon (south of the Río Marañón), including
most or all of the outlying ridges such as Cordillera
Vilcabamba and Cerros del Sira (Terborgh & Weske 1975),
and east into adjacent western Brazil to at least the Rio
Purus, and south to north-west Bolivia (Hennessey *et al.*
2003). There is a notable absence of records from south
of the Brazilian border in eastern Pando, but should
populations be discovered there they would likely represent
minor, with its southern limit in this region perhaps being
the Río Madre de Dios. Birds north and east of this region,
in southern Acre (Aleixo & Guilherme 2010), clearly
represent *minor* based on vocal similarities (WA recordings
from Porto Acre (09°38'S, 67°43'W: song, WA1378207, E.
Kaseker; call, WA969705, T. Melo). These recordings differ
markedly from vocalisations recorded just south of this
region, on the opposite side of the Rio Madeira in northern
Rondônia (see below). Subspecies *minor* is considered to
be duller above, more olive-brown, less rufescent, than
nominate (Todd 1927). **Specimens & Records PERU:** *Loreto*
Quebrada Vainilla, 03°32.5'S, 72°44.5'W (Caparella 1987,
Robbins *et al.* 1991, LSUMZ 114463/64); Río Yanayacu,
03°55'S, 73°05'W (Caparella 1987); Río Yavarí, 04°21'S,
70°02'W (Zimmer 1934); Quebrada Limera, 04°31'S,
71°54'W (Lane *et al.* 2003); Quebrada Buenavista, 04°50'S,
72°23.5'W (Lane *et al.* 2003); Quebrada Curacinha,
05°03'S, 72°43.5'W (Lane *et al.* 2003); Estación Biológica
Pithecia, 05°05'S, 74°35'W (ANSP 177840/841, ML);
Chamicuros, 05°30'S, 75°30'W (Sclater & Salvin 1873a,
Zimmer 1934); Chonco, 05°33.5'S, 73°36.5'W (Stotz &
Pequeño 2006); 'Campamento Alto Cahuapanas', 05°40'S,
76°50.5'W (Stotz *et al.* 2014); Wiswincho, 05°49'S, 73°52'W
(O'Shea *et al.* 2015); 'Campamento Alto Cachiyacu',
05°51.5'S, 76°43'W (Stotz *et al.* 2014); Itia Tëbu, 05°51.5'S,
73°45.5'W (Stotz & Pequeño 2006); Anguila, 06°16'S,
73°55'W (O'Shea *et al.* 2015); Actiamë, 06°19'S, 73°09.5'W
(Stotz & Pequeño 2006); 79km WNW of Contamana,
07°08'S, 75°41'W (LSUMZ 161804); Sierra de Divisor,
*c.*07°12.5'S, 73°53'W (Schulenberg *et al.* 2006); upper Río
Tapiche, 07°12.5'S, 73°56'W (Schulenberg *et al.* 2006);
Cerro Azul, 07°13'S, 74°38'W (Traylor 1958, FMNH
187620). *Amazonas* Peña Blanca, 04°45'S, 78°08'W (D.
Beadle *in litt.* 2017); Pamau Nain, 'Campamento 1', 05°14'S,
78°10.5'W (A. García-Bravo *in litt.* 2017); Pomará, 05°16'S,
78°26'W (Zimmer 1934). *San Martín* 'Río Negro', *c.*23km
NW of Moyobamba, 05°56'S, 77°09'W (Zimmer 1934,
AMNH 234693; loc. see Stepehens & Traylor 1983);
Shapaja, 06°35'S, 76°1.5'W (Bond 1950, ANSP 117470);
Lago Lindo, 06°44'S, 76°14'W (XC 180696/97, M. Nelson);

Puerto Bermúdez, 07°32'S, 76°26'W (Zimmer 1930, FMNH
66251). *Huánuco* Hacienda Santa Elena, 09°11'S, 75°48'W
(LSUMZ 72514); PN Tingo María, 09°22'S, 76°00'W
(Angulo *et al.* 2015); Cerros del Sira, 09°25'S, 74°44'W (Mee
et al. 2002); 'Campamento 3 de Mayo', 09°25'S, 76°00'W
(F. Angulo P. *in litt.* 2017); above Yuyapichis, 09°29'S
74°47'W (Socolar *et al.* 2013). *Pasco* Villa Rica, 10°47'S,
75°17'W (XC 20692, N. Athanas, XC 38987/88, F.
Lambert); Chuchurras, 10°06S, 75°09'W, (Zimmer 1934,
AMNH 492337/338). *Ucayali* SE slope of Cerro Tahuayo,
08°08'S, 74°01'W (LSUMZ 156623); Río La Novia, RC
Purús, 09°56'S, 70°42'W (Angulo *et al.* 2016); Balta,
10°06'S, 71°14'W (LSUMZ); Santa Rosa, 10°26'S, 73°31'W
(Zimmer 1934, AMNH 240339/340); Lagarto, 10°40'S,
73°54'W (Zimmer 1934, AMNH 239270–274); Oventini,
10°45'S, 74°13'W (Harvey *et al.* 2011); Monte Tambor,
10°53'S, 74°11'W (Harvey *et al.* 2011). *Junín* Chanchamayo,
11°03'S, 75°19'W (FMNH 123301); Cerro Quitchungari,
11°03'S, 74°11'W (Harvey *et al.* 2011); above Llaylla, 11°26'S,
74°39'W (G.F. Seeholzer *in litt.* 2017). *Ayacucho* Hacienda
Luisiana, 12°39'S, 73°44'W (Terborgh & Weske 1969, Weske
1972, AMNH 788334/819711). *Cuzco* 50km NW of Pangoa,
11°47'S, 72°42'W (Angehr *et al.* 2001); Pagoreni Well site,
11°42'S, 72°54'W (Angehr *et al.* 2001, MUSA 3512); 2km
W of Tangoshiari, 11°47'S, 73°20.5'W (Schulenberg *et al.*
2001a); Kapiromashi, 12°09.5'S, 72°34.5'W (Lane &
Pequeño 2004); 'Balseadero', Río Nusiniscato, *c.*13°15'S,
70°34'W (FMNH 208272). *Madre de Dios* Cocha Cashu,
11°54'S, 71°18'W (Terborgh *et al.* 1984, 1990, Robinson *et
al.* 1990, XC); Tipishca Lodge, 12°14'S, 69°19.5'W (XC
65580, D. Geale); Amazon Manú Lodge, 12°20'S, 70°43.5'W
(XC 22997, D. Geale); Reserva Cuzco Amazónico, 12°33'S,
69°03'W (Davis *et al.* 1991); Centro de Investigación y
Capacitación Río Los Amigos, 12°34'S, 70°06'W (ML
128947, D.J. Lebbin); Río Palotoa, 12km from mouth,
*c.*12°34.5'S, 71°24.5'W (FMNH 322348/49); Pantiacolla
Lodge, 12°39.5'S, 71°14.5'W (XC 14487, R. Ahlman);
mouth of Río Inambari, 12°43'S, 69°44'W (FMNH 251970);
Moskitania, 12°46'S, 71°23'W (FMNH 433462–466); lower
Río Heath, 12°48.5'S, 68°49.5'W (Parker *et al.* 1994b); Río
La Torre/Tambopata confluence (Explorer's Inn), 12°50'S,
69°17.5'W (Parker *et al.* 1994c, LSUMZ 86540, ML);
Amazonía Lodge, 12°52'S, 71°22'W (XC: D. Geale, F.
Schmitt); 30km S of Puerto Maldonado, *c.*12°52'S, 69°12'W
(ML, M. Palmer); Primavera Baja, 12°56'S, 70°08.5'W (XC
22996, D. Geale); RC Amarakaeri, 12°59.5'S, 71°00.5'W
(eBird: M. Harvey); Collpa Guacamayo de Blanquillo,
13°09'S, 69°36'W (Parker *et al.* 1994a, ML 75520–522, C.A.
Marantz); lower Río Chocolatillo, 13°13.5'S, 70°14'W (XC
153935, R.P. Piana). *Puno* Cerros del Távara, 13°30'S,
69°41'W (Parker & Wust 1994); Huacamayo, 13°31'S,
69°38'W (ANSP 103263–265). **BRAZIL:** *Acre* Cruzeiro do
Sul, 07°37.5'S, 72°40.5'W (Novaes 1957, MPEG 24717); Foz
do Jurupari, 07°50.5'S, 70°04.5'W (Guilherme 2016);
Porongaba, 08°40'S, 72°47'W (MPEG 49668); *c.*7km NNE
of Marechal Thaumaturgo, 08°52.5'S, 72°47'W (MPEG
52918); Foz do Tejo, 08°59'S, 72°43'W (FMNH 395576, XC
90421, J. Minns); Manoel Urbano, 09°07'S, 69°45'W (XC
198521, E. Guilherme); RESEX Chico Mendes, 10°22'S,
68°43'W (Mestre *et al.* 2010); ESEC Rio Acre, *c.*10°57'S,
70°23'W (Aleixo & Guilherme 2010); near Jarinal,
*c.*09°54.5'S, 68°28.5'W (Guilherme & Santos 2009, MPEG
61438). *Rondônia* Igarapé São Lourenço, 09°25'S, 64°57'W
(Olmos *et al.* 2011); Porto Velho, left (N) bank of Rio
Madeira, 09°35'S, 65°03'W (C. Brito *in litt.* 2017). *Amazonas*
São Paulo de Olivença, 03°39'S, 69°06'W (CM, YPM); Lago

Sampaio, *c*.03°46'S, 59°08'W (AMNH 282114/20); Tupana Lodge, 04°04'S, 60°40'W (XC 38704, A. Renaudier); Carauari, 04°53'S, 66°54'W (WA 1772479, G. Leite); near Bauana, 05°26'S, 67°17'W (XC 88030, A.C. Lees); Jaburú, 05°36'S, 64°03'W (Gyldenstolpe 1945, 1951); Rio Purus near Arimã, 05°47'S, 63°38'W (CM, YPM 29645); 165km N of Pauini, 06°14'S, 66°54'W (ML 113151/52, C.A. Marantz); Igarapé do Gordão, 06°30'S, 69°50'W (Gyldenstolpe 1945); Eirunepé, 06°40'S, 69°52'W (Gyldenstolpe 1945); Huitanaã, 07°41'S, 65°45'W (CM). **BOLIVIA**: *Pando* 'Campamento Manoa', 09°41.5'S, 65°24.5'W (Stotz *et al.* 2003); 'Campamento Piedritas', 09°57'S, 65°20.5'W (Stotz *et al.* 2003); 'Campamento Caimán', 10°13.5'S, 65°22.5'W (Stotz *et al.* 2003); several localities *c*.17km SW of Cobija, near Mucden, 11°07'S, 68°55'W (Parker & Remsen 1987, LSUMZ 133061/62, ML 38929/930, T.A. Parker). *La Paz* 80km SE of Puerto Maldonado, 13°10'S, 68°45'W (ML 52374/388, T.A. Parker); Alto Madidi, 13°38'S, 68°44.5'W (Parker 1991); NE of San José de Uchupiamones, 14°12.5'S, 67°57'W (Hosner *et al.* 2009); Chalalan Ecolodge, 14°25.5'S, 67°55'W (XC 64605, S. Hampton); 17.5km SW of San Buenaventura, 14°33'S, 67°43'W (ML 110717, A.B. Hennessey); Tuichi-Hondo, 14°34'S, 67°44'W (XC 3203, A.B. Hennessey). **Notes** The collecting locality 'mouth of Río Urubamba' (Zimmer 1934) is probably somewhere within the range of *minor*, but is apparently considered to be a fraudulent locality label, made up by the collectors (Stephens & Traylor 1983).

Myrmothera campanisona subcanescens Todd, 1927, Proceedings of the Biological Society of Washington, vol. 40, p. 176, Colonia do Mojuy, Santarem, Brazil [= Mujuí dos Campos, 02°41'S, 54°39'W, Pará]. The holotype (CM 74906), an adult male, was collected on 6 November 1919 by Samuel M. Klages. Described as *Myrmothera campanisoma* [*sic*] *subcanescens*, this race is endemic to Brazil, so far as is known, and is reported from central Amazonas (south of the Amazon and east of the Madeira) south to Rondônia, southern Pará and the Alta Floresta area of northern Mato Grosso. Though Krabbe & Schulenberg (2003) indicated it was found only as far east as the right bank of the lower Rio Tapajós, recent records (XC, WA) extend its range as far east as the lower Rio Xingu. I am unable to find any records of Thrush-like Antpitta east of the Rio Xingu in south-central Pará. The precise eastern limit of its range in southern Pará is unclear, with no records available east of the Alta Floresta region. There is a similar dearth of records from north-east Mato Grosso and southern Rondônia, but vocalisations recorded in northern Rondônia clearly represent the vocally distinct *subcanescens* (e.g. XC 90425, 25km north of Ariquemes, J. Minns). A report of *M. campanisona* from just north-east of Chapada dos Guimarães in southern Mato Grosso (eBird: C. Williams) requires confirmation, but would significantly extend the southern limit of *subcanescens*. Todd (1927) considered *subcanescens* to be fairly similar to nominate *campanisona*, but larger, with the upperparts more brownish (less rufescent), and the streaking of the underparts tending towards greyish rather than brown. The underwing-coverts are paler, buffy rather than ochraceous. Compared to *minor*, *subcanescens* is brighter brown above, less olivaceous (Todd 1927). **Specimens & Records BRAZIL**: *Pará* vicinity of Juruti, *c*.02°09'S, 56°05'W (Santos *et al.* 2011a); Santarém, 02°27'S, 54°42'W (CM, YPM); Platô Capiranga, trilha 196, 02°30'S,

56°11.5'W (MPEG 60975); Caxiricatuba, 02°33'S, 54°57'W (MCZ 175122–124); Rio Arapiuns, *c*.02°34'S, 55°22'W (Pinto 1947); Aramanaí, 02°40.5'S, 54°59'W (Zimmer 1934, AMNH 286754–759, MPEG 13958); Mujuí dos Campos, 02°41'S, 54°38.5'W (CM P74655–656); Vitória do Xingu, 02°53'S, 52°01'W (WA 1634579, T. Junqueira); Tauarí, 03°05.5'S, 55°07.5'W (Zimmer 1934, AMNH 286750–753); Belterra, Bacia 357, 03°22.5'S, 54°51'W (XC 94889, A.C. Lees); UHE Belo Monte, left bank Rio Xingu, 03°22.5'S, 51°56'W (MPEG 55488/89); FLONA Tapajós, *c*.03°29'S, 54°55'W (Oren & Parker 1997, Wunderle *et al.* 2006, Henriques *et al.* 2003, 2008, XC); Lago Cuitena, 03°36'S, 55°20'W (MCZ 175121); 20km N of Rurópolis, 03°54'S, 54°55'W (ML 47908, P.R. Isler); Miritituba, 04°16'S, 55°54'W (CM, MCZ, YPM); Itaituba, 04°16'S, 55°59.5'W (YPM 29651, CM, XC); along BR-230 at Rio Itapacurazinho (km25), 04°22.5'S, 55°50.5'W (MPEG 34421/47846); Vila Braga, 04°25'S, 56°17'W (CM, YPM, FMNH); PN Amazônia, 04°31'S, 56°18'W (XC: J. Minns); 22km E of Trairão, 04°40'S, 55°37'W (Pacheco & Olmos 2005); Bella Vista, 04°43'S, 56°18'W (MCZ 134949, CM P76031/32); Rio Novo, *c*.04°58'S, 53°26'W (Fávaro & Flores 2009); Rio Jamanxim, Santa Helena, 05°13.5'S, 56°17.5'W (MPEG 6477); Comunidade Penedo, 05°34'S, 57°07.5'W (MPEG 76458); Vila Mamãe-anã, 05°45.5'S, 57°25'W (MPEG 75730); FLONA Altamira, *c*.06°04'S, 55°19'W (MPEG 63897); Mina de Ouro Palito, 06°18.5'S, 55°47'W (MPEG 77866); 30km SSE of Novo Progresso, 07°10'S, 55°06'W (Pacheco & Olmos 2005); Fazenda São Francisco, 09°18'S, 54°50'W (Santos *et al.* 2011c). *Amazonas* Parintins, 02°39'S, 56°43'W (Zimmer 1934, AMNH 277991); FLONA Pau-Rosa, *c*.03°54'S, 58°24'W (Dantas *et al.* 2011); Borba, 04°39'S, 59°28'W (Gyldenstolpe 1945); Manicoré, 05°49'S, 61°17'W (WA 1578172; recording A. Bianco); Rosarinho, 05°53.5'S, 61°26'W (AMNH 282119); Novo Aripuanã, 06°18'S, 60°02'W (WA 1147142, M. Padua); Pousada Rio Roosevelt, 08°29'S, 60°57'W (Whittaker 2009); *c*.11.5km NNE of Jatuarana, 08°41'S, 61°24'W (Aleixo & Poletto 2007). *Mato Grosso* Alta Floresta area, 09°52'S, 56°05'W (Zimmer *et al.* 1997, Lees & Peres 2006); Cristalino Jungle Lodge, 09°35'S, 55°56'W (many recordings XC, ML). *Rondônia* Calama, 08°01'S, 62°52'W (Hellmayr 1910, AMNH 492335/336); Machadinho d'Oeste, 08°55'S, 62°05'W (XC 167314, G. Leite); Fazenda Rio Candeias, 08°57'S, 63°41.5'W (MPEG 35156); Jaci-Paraná, 09°27'S, 64°22.5'W (XC 189661, L.G. Mazzoni); Cachoeira Nazaré, 09°44'S, 61°55'W (FMNH 389881–886, MPEG 39823–827, ML); 25km N of Ariquemes, 09°54'S, 63°02'W (XC 90425, J. Minns); Terra Indígena Igarapé Lourdes, 10°25.5'S, 61°39.5'W (Santos *et al.* 2011b); REBIO Traçadal, 11°23'S, 64°50'W (Olmos *et al.* 2011); Igarapé Tiradentes, 11°48'S, 64°15'W (Olmos *et al.* 2011). **Notes** The coordinates given for Pinhy (MCZ 175125–128) and Patauá (Gyldenstolpe 1945) by Paynter & Traylor (1991b) are very close to the town of Tauarí. Paynter & Traylor (1991a) were unable to locate Limoãl, Pará (AMNH 288618), but Haffer (1974) placed it south of the Amazon on the Rio Tapajós. The locality 'Igarapé Brabo, Rio Tapajós' (AMNH 286760–765) was also not located by Paynter & Traylor (1991a), but its location along the Rio Tapajós would fall within the range of *subcanescens* (as here defined).

STATUS In parts of its range, Thrush-like Antpitta can be fairly common (Stotz *et al.* 1996), although it is said to be less common in Peru (Parker *et al.* 1982, Schulenberg *et al.* 2010). As it occurs in several protected areas, is fairly common and has a relatively large range (estimated at 4,560,000km²), it is not considered of conservation concern (BirdLife International 2016). Despite its preference for tangles and edges created by disturbance, and its apparent ability to persist in isolated forest fragments (Quintela 1986, Stratford & Stouffer 1999, 2013), Thrush-like Antpitta is a forest-floor insectivore and it seems unlikely that populations will thrive for long in human-altered landscapes. **PROTECTED POPULATIONS** *M. c. campanisona* **Venezuela** PN Canaima (Lentino *et al.* 2013b); RF Imataca (Mason 1996). **Guyana** Iwokrama Forest Reserve (Ridgely *et al.* 2005, ANSP). **Suriname** Natuurreservaat Boven-Coesewijne (eBird: B.J. O'Shea); Raleigh Vallen-Voltzberg Natuurpark (ML 21015, T. Pepper); Brownsberg Natuurpark (ML 134306, B.J. O'Shea, XC 114042, J. King). **French Guiana** Reserve Naturelle La Trinité (XC 316503, O. Claessens). **Brazil** PN Montanhas do Tumucumaque (Bernard 2008); PN Viruá (MPEG 56420, XC 10119, R.A. de By); RESEX Rio Cajari (Schunck *et al.* 2011); Reserva Forestal Adolpho Ducke (Willis 1977, Bueno *et al.* 2012); records by Aleixo *et al.* (2011) include the following: ESEC Grão-Pará; REBIO Maicuru; FLOTA Trombetas; FLOTA Faro; FLOTA Paru. *M. c. dissors* **Brazil** PN Jaú (Borges & Carvalhães 2000, Borges *et al.* 2001, Borges 2006, Borges & Almeida 2011); ESEC Maracá (Moskovits *et al.* 1985). **Colombia** PNN Serranía de Chiribiquete (Álvarez *et al.* 2003); PNN Amacayacú (XC 163576, F. Lambert); Estación Ecológica Omé (XC 58410, G.N. Forero). **Peru** ACP Saballilo (XC 20059, D. Edwards). *M. c. modesta* PNN Tinigua (Cadena *et al.* 2000); PNN Cordillera de los Picachos (XC 157702, J.A. Suárez). *M. c. signata* **Colombia** PNN Serranía de Chiribiquete (Álvarez *et al.* 2003). **Ecuador** PN Yasuní (English 1998, Canaday & Rivadeneyra 2001, MECN, XC); PN Sumaco-Galeras (XC 251228/29, XC 291673, N. Krabbe); RPF Cuyabeno (Canaday 1997, Canaday & Rivadeneyra 2001); RE Limoncocha (Pearson *et al.* 1972, LSUMZ); BP Alto Nangaritza (Freile *et al.* 2013b, Greeney 2017a); ACM Riberas del Zamora (XC 166792, L. Ordóñez-Delgado); RBP Jatun Sacha (D. Beadle *in litt.* 2017); RBP Río Bigal (XC 175009, R. Ahlman); RP Maicu (XC 194447–449, L. Ordóñez-Delgado); RP El Pará (XC, N. Athanas, P. Coopmans); RP Cabañas Hakuna Matata (R.A. Gelis *in litt.* 2003); RP Nantar (XC 116460, S. Olmstead); RP Yachana Lodge (HFG). **Peru** RN Allpahuayo-Mishana (Alonso *et al.* 2012, XC). *M. c. subcanescens* PN Amazônia (XC); PN Campos Amazônicos (B. Rennó *in litt.* 2017); FLONA Altamira (MPEG 63897); RPPN Rio Cristalino (XC, ML); FLONA Pau-Rosa (Dantas *et al.* 2011); FLONA Tapajós (Oren & Parker 1997, Henriques *et al.* 2003, 2008, Wunderle *et al.* 2006, XC); FLOTA Rio Preto (XC 167314, G. Leite). *M. c. minor* **Peru** PN Tingo María (Angulo *et al.* 2015); PN Otishi (eBird: D. García Olaechea); PN Cordillera Azul (eBird: D.F. Stotz); PN Bahuaja-Sonene (Parker *et al.* 1994c, ANSP 103263–265); PN Manú (Terborgh *et al.* 1984, 1990); PN Yanachaga-Chemillen (A. García-Bravo *in litt.* 2017); PN Sierra del Divisor (Schulenberg *et al.* 2006); RN Tambopata (ML 23817; T.A. Parker); RN Matsés (Stotz & Pequeño 2006); SN Megantoni (Lane & Pequeño 2004); Reserva Cuzco Amazónico (Davis *et al.* 1991); ACR Cordillera Escalera (D. Beadle/R. Ahlman *in litt.* 2017); Estación Biológica BIOLAT (Servat 1996); RC Amarakaeri (eBird: M. Harvey); PN Río Avisado (AMNH 234693); CC Río La Novia (Angulo *et al.* 2016); Centro de Investigación y Capacitación Río Los Amigos (ML

128947, D.J. Lebbin). **Brazil** PN Mapinguari (C. Brito *in litt.* 2017); FLONA Humaitá (eBird: C. Gussoni); PE Chandless (Buzzetti 2008); FLOTA Rio Vermelho (eBird: B. Rennó); ESEC Rio Acre (Aleixo & Guilherme 2010); RESEX Alto Juruá (FMNH 395576); RESEX Cazumbá-Iracema (eBird: B. Rennó); RESEX Chico Mendes (Mestre *et al.* 2010). **Bolivia** PN Madidi (Hennessey *et al.* 2003, Hosner *et al.* 2009).

OTHER NAMES Zimmer (1934) applied the names *dissors* and *signata* to populations that, up to that point, had been referred to in the literature as *minor*. Both of these subspecies, therefore, are referred to in much of the older literature under *minor*. For some reason, the Latin name easily holds the honour of being the most frequently misspelled species name treated in this book. For those performing digital searches, it is important to note that the specific name has been misspelled with an 'm' repeatedly throughout the literature. The history of scientific synonyms is so complex for this species, I have listed the following synonyms based on the currently recognised subspecies that they refer to: *M. c. campanisona* *Formucarius brevicauda* (Boddaert 1783); *Myrmornis campanisona* (Hermann 1783); *Turdus tinniens* (Gmelin 1789, Latham 1790, Cuvier 1817, 1829, Cuvier & d'Orbigny 1836); *Turdus tinnicus* (Griffith *et al.* 1829); *Grallaria tinniens* (Lesson 1831, Sundeval 1835, Lafresnaye 1843, Müller 1847, Bonaparte 1850, Burmeister 1856); *Colobathris tinniens* (Schomburgk 1848); *Pitta tiniens* (Temminck & Laugier 1820–39); *Grallaria brevicauda* (Salvin 1885, Hagmann 1904, Ménégaux 1904, von Berlepsch 1908, Penard & Penard 1910, Beebe 1917, Chubb 1921; also Snethlage 1914a from Óbidos); *Grallaria brevicauda brevicauda* (Chapman 1917); *Myrmothera campanisoma* (Donahue & Pierson 1982, Thiollay 1986, 1992, 1997, 2002, Thiollay & Jullien 1998, Barlow *et al.* 2007); *Myrmothera companisona* (Blake 1950, Cohn-Haft *et al.* 1997). *M. c. dissors* *Grallaria brevicauda* (Sclater & Salvin 1867); *Myrmothera campanisa* (Andrade & Rubio-Torgler 1994). *M. c. minor* *Grallaria tiniens* (von Tschudi 1845); *Grallaria brevicauda* (Sclater & Salvin 1873a, von Ihering 1905); *Grallaria minor* (Taczanowski 1882); *Grallaria brevicauda minor* (Taczanowski 1884); *Myrmothera campanisoma minor* (Traylor 1958); *Myrmothera campanisoma* (Kempff-Mercado 1985, Servat 1996); *Mermothera campanisona* (Terborgh *et al.* 1990). *M. c. subcanescens* *Grallaria brevicauda* (Hellmayr 1910, Snethlage 1914a from Santa Helena); *Myrmothera campanisoma subcanescens* (Todd 1927, 1928). *M. c. signata* *Grallaria brevicauda minor* (Chapman 1917); *Myrmothera campanisona minor* (Chapman 1926); *Myrmothera campanisoma* (Blake & Loiselle 2015); *Myrmothera campanisona discors* (Iafrancesco & Pineda 1984, Iafrancesco *et al.* 1987); *Myromothera campanisona* (Butler 1979). *M. c. modesta* *Grallaria modesta* (Sclater 1855b, 1877, 1890, Sclater & Salvin 1873b, Dubois 1900, Sharpe 1901, Chapman 1917). **Various subspecies referred to with same name** *Grallaria brevicauda* (Sclater 1858, 1862, 1890, Sclater & Salvin 1873b, Dubois 1900, Sharpe 1901, von Ihering & von Ihering 1907, Brabourne & Chubb 1912). **Subspecies unknown** erroneous report from Rio de Janeiro as *Myioturdus tinniens* (Ménétriés 1835) and *Grallaria tinniens* (Burmeister 1856). **English** Ecuadorian Ant-pitta [*signata*] (Meyer de Schauensee 1952b); Little Antpitta (Cory & Hellmayr 1924); Little Ant-Thrush (Brabourne & Chubb 1912, Beebe 1917, Chubb 1921); Sclater's Colombian Ant-Thrush (Brabourne &

Chubb 1912); Sclater's Colombian Antpitta [*modesta*] (Cory & Hellmayr 1924); Sclater's Colombian Ant-pitta [*modesta*] (Meyer de Schauensee 1950); Western Little Antpitta [*minor*] (Cory & Hellmayr 1924); Venezuelan Little Ant-pitta (Meyer de Schauensee 1950); Zimmer's Ant-pitta [*dissors*] (Friedmann 1948); Brown-backed Ground-bird [*campanisona*] (Penard & Penard 1910); Alarum Thrush (Latham 1783). **Portuguese** tovaquinha (Frisch & Frisch 1981, Moskovits *et al.* 1985); tovaca-patinho (Sick 1993, van Perlo 2009, CBRO 2011, Borges *et al.* 2014); torom-patinho (Willis & Oniki 1991, Cintra *et al.* 2007); tovaquinha-olivácea (Ruschi 1979); patinho (Sick 1993). **Spanish** Tororoí Campanero (Ridgely *et al.* 1998, Krabbe & Schulenberg 2003, Ridgely & Greenfield 2006,

Alonso *et al.* 2012, Plenge 2017); Chululú Campanero (Greeney *et al.* 2005); Tororoí Campanera (Ortiz-Crespo *et al.* 1990); Gralaria Campanera (Delgado 1984); Hormiguero Campanero (Phelps & Meyer de Schauensee 1979, 1994, Frisch & Frisch 1981, Valarezo-Delgado 1984, Kempff-Mercado 1985, Hilty 2003); Hormiguero Campanero de Cayena [*campanisona*], Hormiguero Campanero del Casiquiare [*dissors*] (Phelps & Phelps 1950). **French** Grallaire grand-beffroi (Haverschmidt & Mees 1994, Krabbe & Schulenberg 2003); Brève Beffroi (Temminck & Laugier 1820–39). **German** Strichelbrust-Ameisenpitta (Krabbe & Schulenberg 2003). **Dutch** Bruinrug Grondmierenvogel (Penard & Penard 1910); Lijstermierpitta (Spaans *et al.* 2016).

Thrush-like Antpitta, adult (*signata*), Limoncocha, Napo, Ecuador, 3 July 2014 (*George Cruz*).

Thrush-like Antpitta, eggs (*signata*) and nest, Añangu, Orellana, Ecuador, 20 May 2003 (*Harold F. Greeney*).

Thrush-like Antpitta, young nestlings (*minor*) in nest, Pantiacolla, Madre de Dios, Peru, 7 October 2013 (*Manuel Andrés Sánchez*).

Thrush-like Antpitta, adult (*campanisona*), Manaus, Amazonas, Brazil, 23 August 2008 (*Robson Czaban*).

TEPUI ANTPITTA
Myrmothera simplex   **Plate 21**

Grallaria simplex Salvin & Godman, 1884, Ibis, vol. 26, p. 451, 'Mt. Roraima, 5,000ft, British Guiana' [= *c.*05°08'N, 60°46'W, Bolívar, Venezuela, 1,525m]. Although the type locality was originally cited as being in British Guiana, it was later realised to be in Venezuela (Phelps 1938, Phelps & Phelps 1950). The type series was collected by H. Whitely in August and September 1883 (Chubb 1921).

"After 10 minutes in the densest part of the 1850-meter forest, during which time I had kept up a soft sobbing 'squeak,' I was greeted by a hollow whistle, emitted about 50 yards away. I imitated the whistle and induced *Myrmothera* [*simplex simplex*] to approach. Its call became loud and piercing and finally the long-legged bird revealed itself by hopping '*Grallaria*-like' onto a low log." **E. Thomas Gilliard, 1941, Mt. Auyán-Tepui, Venezuela**

Tepui Antpitta, as its name implies, is a Pantepui endemic (Mayr & Phelps 1967). Known to many as Brown-breasted Antpitta, the overall plumage is dark chestnut-brown above, with a pale postocular spot, white throat and generally greyish, brownish or olivaceous underparts. Despite the translation of its generic name meaning 'ant-hunter' (Jobling 2010), Tepui Antpitta has apparently never been reported with army ant swarms unlike many other antpittas (Greeney 2012b). Overall, this is a rather poorly studied species, but it appears to run and bound across the ground and along low logs in search of prey in the same manner as most of its relatives. Although Cory & Hellmayr (1924) suggested that Tepui Antpitta might be conspecific with its only congener, Thrush-like Antpitta, it appears that the present species represents a derived lineage that may be only distantly related to Thrush-like Antpitta (Mayr & Phelps 1967). The four described subspecies of *M. simplex* are rather poorly defined, in both plumage and vocalisations, and a careful taxonomic revision is certainly in order. In addition, the diet is all but unknown and just one nest has been described. This was collected during incubation and the nestlings plus details of parental care remain a mystery. Indeed, there is a notable dearth of reproductive data in general, with only three studies providing secondary evidence of such activity.

IDENTIFICATION 15–16cm. This rather plain species has upperparts, wings and tail chestnut-brown, grading to ash-grey, olivaceous-grey or olive-brown on breast and sides. Throat, chin and central belly are white and most birds have a small but prominent white postocular spot with variable grey on the lores. Vent tawny-brown. Races (and individuals?) show much variation in the wash on the breast, from little to sufficient to form a fairly distinct band. The only other similar-sized antpitta in the region, apart from the distinctly different (and larger) Scaled Antpitta, is Thrush-like Antpitta. This congener, however, generally occurs at lower elevations (with no known areas of syntopy). The two are quickly distinguished, however, by the strong ventral streaking on the underparts of Thrush-like and the clean underparts (often with a breast-band) of Tepui Antpitta.

DISTRIBUTION Inhabits the slopes of most tepuis in south and south-east Venezuela and adjacent Guyana (Snyder 1966, Ridgely & Tudor 1994, Rodner *et al.* 2000, Barnett *et al.* 2002, Clements 2007). Its range is frequently

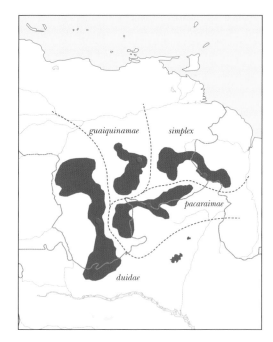

considered to extend into immediately adjacent northern Brazil (Pinto 1978, Ruschi 1979, Altman & Swift 1986, Naka *et al.* 2006, van Perlo 2009). Although this is almost certainly the case, there are no definite records from the border regions of Brazil (CBRO 2011). Nevertheless, Tepui Antpitta has very recently been confirmed to occur in Brazil (M. Cohn-Haft *in litt.* 2016), but from two fairly widely separated mountain ranges in central Roraima, more than 175km from the Venezuelan border and >200km from the nearest previously known confirmed records (see Taxonomy and Variation for details).

HABITAT Dense, humid forests in the upper tropical, subtropical and temperate zones, generally preferring areas with dense undergrowth (Meyer de Schauensee 1966, Dickerman & Phelps 1982, Restall *et al.* 2006) such as the melastome-dominated second growth on slopes of tepuis (Hilty 2003). On their summits, this species is also found in somewhat more open habitats and stunted forest (Ridgely & Tudor 1994). Across its limited range, the species has been found at 600–2,400m but is generally more common above 1,200m (Ridgely & Tudor 1994, Giner & Bosque 1998, Hilty 2003). Phelps & Phelps (1963) gave the altitudinal range of nominate *simplex* as 1,000–2,400m, *duidae* as 600–2,250m and *guaiquinimae* as 900–2,200m. Where *M. simplex* overlaps geographically with *M. campanisona* there appears to be little or no altitudinal overlap (Barnett *et al.* 2002). On the north slope of Mt. Roraima, Barber & Robbins (2002) recorded *M. campanisona* no higher than 500m and *M. simplex* no lower than 700m. In areas where *M. simplex* does not occur, however, the same authors found *M. campanisona* ranging as high as 1,000m (Acari Mountains of southern Guyana). Similarly, Ridgely & Tudor (1994) reported that the upper altitudinal limit of *M. campanisona* in the Andean foothills (where *M. simplex* does not occur), can reach 1,200m. This implies strong competition between them, which perhaps constrains habitat use and altitudinal distributions. Overall, *M. simplex* is found at 600–2,400m

(Meyer de Schauensee & Phelps 1978, Willard *et al.* 1991, Giner & Bosque 1998, Braun *et al.* 2003, Hilty 2003), a rather broad range amongst antpittas.

VOICE Despite the number of subspecies, song varies little between populations (Krabbe & Schulenberg 2003). Ridgely & Tudor (1994) described it as a 'series of 7–10 hollow, whistled notes on the same pitch, *wh-wh-wh-whoh-whoh-whoh-whoh-whoh-whoh*, gradually building to a crescendo.' Hilty (2003) described song as consisting of 6–7 (occasionally more) low-pitched, deliberately paced, hollow notes, *whu-whu-whu-WU-hu-hu*, with the last notes rising slightly in pitch and increasing in volume, and the last 2–3 notes on the same pitch. Likening the song to *M. campanisona*, Krabbe & Schulenberg (2003) described it as lasting 3–3.5s, given at intervals of 9–14s. The rising series of 10–11 notes are similar, evenly paced, with the volume gradually increasing and the whistles at 1.0–1.2kHz. The first 3–4 notes sometimes fall in pitch. Hilty (2003) also described what is presumably a call as a 'low, hollow, rattle-trill'.

NATURAL HISTORY This shy and retiring species appears, like most antpittas, usually to forage alone (Restall *et al.* 2006), but some authors have more frequently found it in pairs (Barnett *et al.* 2002). Almost entirely terrestrial, like its only congener, Tepui Antpitta bounds and runs on the ground (Hilty 2003), apparently appearing curious, at times boldly venturing into the open, unlike the more skittish *M. campanisona* (Krabbe & Schulenberg 2003, Restall *et al.* 2006). Several authors have suggested that it has a predilection for tangled undergrowth around fallen branches and treefall gaps (Hilty 2003, Restall *et al.* 2006). Singing adults tend to perch below 3m (Hilty 2003) but can still be frustratingly difficult to locate. Tepui Antpitta is said to vocalise more frequently in mornings and evenings, delivering both its typical song and also 'low, chuckling alarm calls' similar to those of *M. campanisona* (Braun *et al.* 2003). The latter authors also observed that adults respond 'vigorously' to playback, dashing about in short spurts covering 2–3m before freezing.

DIET No available information, although Barnett *et al.* (2002) observed a pair of nominate *simplex* that 'pecked at' fallen gesneriad flowers (*Alloplectus savannarum*), possibly taking insects.

REPRODUCTION The reproductive biology is virtually unknown with the exception of a single paper describing observations at a nest (race *simplex*) found 24 March 2001 on the northern slope of Mt. Roraima at 700m, in Guyana (Barber & Robbins 2002). The female was flushed from the nest and collected (KUNHM 92366) and brief observations were made on the behaviour of the male at the nest before he too was collected (USNM 622748) along with the nest and eggs (KUNHM 92366). **NEST & BUILDING** The nest found by Barber & Robbins (2002) was in a relatively level section of an otherwise steeply sloping region of extensive undisturbed forest. The cup-shaped nest was sturdily built and placed *c.*0.6m above ground, built into a clump of leaf bases and petioles of *Philodendron linnaei* (Araceae). The materials used in construction included small sticks and dead leaves, the latter apparently forming the base and the former presumably giving shape to the nest. The inner cup was lined with smaller 'sticks and rootlets'. Nest dimensions were: outer height 14.5cm, inner diameter roughly 10.1cm by 9.0cm, inner

cup depth 5.6cm. The outer diameter was not reported. **EGG, LAYING & INCUBATION** Eggs were described by Barber & Robbins (2002) as oval with a light blue ground colour 'paler than that of the American Robin, *Turdus migratorius*'. Both had sepia-coloured spots concentrated at the larger end. Photos of these eggs permits me to amplify this description somewhat, and make it more comparable to other grallariid eggs. My description would concur with their assessment of ground colour, which appears fairly similar to the blue of other species, perhaps slightly paler and more 'sky-blue' than the turquoise or greenish-blue of some species. The spotting, concentrated in a small 'cap' at the larger pole, is a mix of small spots, flecks and a few larger blotches of sepia (or cinnamon) and pale lavender. These eggs have the markings more concentrated at one end than I have observed in any other species, which instead have the markings more evenly dispersed across the shell. Egg dimensions were 26 × 20mm and 27 × 20mm. Unfortunately, Barber & Robbins (2002) were not able to observe a complete on-bout by the male, rather a single bout of incubation that lasted for 57min before the bird was flushed. They noted that, during this time it stood once, peered into the cup, and appeared to shift the eggs with its bill. Apart from this, he remained relatively still with only occasional vigilant movements of his head. These limited observations do confirm that both sexes participate in incubation and that, like other grallariids, periods of nest attendance can be fairly long (Greeney *et al.* 2008). **NESTLING & PARENTAL CARE** No data available on parental care or nestling appearance. **SEASONALITY** Apart from the above-described nest, there are almost no available data to estimate reproductive period. Barber & Robbins (2002) provided data on the breeding condition of the adults at their nest, which will be useful for comparisons with gonads of specimens of unknown reproductive status. They reported that the male's testes measured 7mm and 4mm, and that the female's ovary was 7mm × 5mm, while her oviduct was 3mm in diameter and convoluted, with light body fat. Chubb (1921) reported a 'young male' collected in August in Guyana (*simplex*). Finally, Barrowclough *et al.* (1995) collected a male (*diudae*) in the Serrania de Tapirapecó (Amazonas, Venezuela) that had moderately enlarged testes and was simultaneously in body moult some time between late January and mid-March, leading them to suggest that breeding occurs during the drier months in this region (November–March). Just south of there, on Cerro de la Neblina, Willard *et al.* (1991) reported the following adults with 'large' gonads: ♂ and ♀, January; ♂, February; ♂, March; ♂, December. A female in February had a moderately enlarged ovary and was moulting the contour feathers (degree of moult not reported).

TECHNICAL DESCRIPTION Sexes similar. *Adult* The following applies to the nominate subspecies. Upperparts chestnut-brown including the crown, entire back, wings and tail. Inner webs of flight feathers darker and more blackish-brown than leading webs. Base of lores and short feathers around eye whitish, with black bristly tips. Throat and central underparts dull white, while the breast and body-sides are dusky-grey. Undertail-coverts and thighs chestnut-brown. Underwing-coverts more chocolate-brown, and underside of flight feathers chestnut- or rust-brown, paler on their inner edges (Chubb 1921). **Adult Bare Parts** *Iris* brown; *Bill* black, paler near base of mandible; *Tarsi & Toes* grey. *Immature* There is, apparently, some variation in the plumage of immatures, which may be geographic. Restall *et al.* (2006) described the 'juvenile' plumage of nominate *simplex* as being similar to adults but with a 'white belly heavily flammulated rufous' and rufous undertail-coverts. They described the 'juvenile' of *guaiquinimae* as 'almost wholly deep rufous-brown with [a] whitish chin' and the base of the mandible as yellow-horn. In the former case, their 'juvenile' would appear to best fit the subadult category used herein, while in the latter they appear to be describing a fledgling. Chubb (1921) described a 'young male' as differing from the adult by being 'darker and almost entirely grey on the breast and abdomen with the remains of a few bright rufous feathers on the latter, which points to the fact that the first plumage is much brighter'. Certainly, the details of immature plumage, especially variation with ontogeny and geography, are in need of detailed study before any conclusions can be drawn.

MORPHOMETRIC DATA *M. s. guaiquinimae* Wing 86mm; *Tail* 45mm; *Bill* exposed culmen 18mm, culmen from base 24.5mm; *Tarsus* 43mm (*n* = 1♂, holotype, Zimmer & Phelps 1946). *Wing* range 83–86mm, mean 84.6mm; *Tail* range 41–46mm, mean 44.3mm; *Bill* culmen from base range 23–25mm, mean 23.9mm (*n* = 6♂♂, Zimmer & Phelps 1946). *Wing* range 80–86mm, mean 83.5mm; *Tail* range 40–46mm, mean 43.5mm; *Bill* culmen from base range 22.0–22.5mm, mean 22.2mm (*n* = 6♀♀, Zimmer & Phelps 1946); *Tarsus* 42.8mm (*n* = 1♀, USNM 383248). *M. s. duidae* *Tarsus* 45mm (*n* = 1♂, holotype, Chapman 1929b). *Bill* from nares 12.3mm; *Tarsus* 46.3mm (*n* = 1♀, USNM 323087). *M. s. simplex* *Wing* 85mm; *Tail* 43mm; *Bill* exposed culmen 19mm; *Tarsus* 43mm (n = ?♂, Chubb 1921). *Wing* 78mm (*n* = ? ♀, Chubb 1921); *Bill* from nares 11.1mm, 12.7mm, 12.2mm, 11.9mm, 11.4mm; *Tarsus* 42.1mm, 46.0mm, 42.7mm, 43.9mm, 43.4mm (*n* = 5♂♂, USNM). Mass 54g (Hilty 2003); 52g (*n* = 1♂, holotype, *pacaraimae*, Phelps & Dickerman 1980); 46.8g(lt) (*n* = 1♀ during active incubation, *simplex*, Barber & Robbins 2002); 51.8g, 53.0g, 53.0g, 51.8g, 53.0g (*n* = 5, ♂,♂,♂,♀,♀, *duidae*, Willard *et al.* 1991); 45.3g (*n* = 1♂, (USNM 626774). Total Length 13.0–16.5cm (Meyer de Schauensee & Phelps 1978, Ruschi 1979, Dunning 1987, Ridgely & Tudor 1994, Hilty 2003, Krabbe & Schulenberg 2003, Restall *et al.* 2006).

TAXONOMY AND VARIATION The four currently recognised subspecies appear to differ rather little in plumage and apparently even less in vocalisations (Krabbe & Schulenberg 2003). Indeed, from the subspecific descriptions of various authors it is clear that there is sufficient variation within subspecies, as well as between localities, to blur the lines between taxa considerably. Phelps (1977) felt that birds from Meseta del Cerro de Jáua show a combination of characters found in *pacaraimae*, *guaiquinimae* and *duidae*. 'Most are as pale dorsally as *pacaraimae*, yet ventrally have a well-developed brownish breast band similar to *duidae*. Two are most like *guaiquinimae* dorsally; one is darker and more like *duidae*, whereas the rest are as pale as *pacaraimae*.' The population at Meseta del Cerro de Jáua is traditionally (including here) treated as *guaiquinimae* (Dickerman & Phelps 1982, Phelps 1977, Hilty 2003). Nevertheless, some authors have been certain of the distinctness of at least some subspecies. Dickerman & Phelps (1982) in relation to nominate *simplex* stated that it was consistently distinguishable from *duidae* and *guaiquinimae* by its

dark, 'near chocolate-brown' upperparts and by having a reduced or very pale breast-band. They further stated that the nominate race is the 'most uniform of any of the named subspecies', following examination of the very large series in the Phelps collection. The conviction of previous authors notwithstanding, my own examination of variation leads me to suggest that *M. simplex* might best be considered monotypic, especially given the relative lack of vocal variation. Irrespective of the most appropriate taxonomy, following currently accepted subspecific divisions the following is a useful summary of differences between the four races of *M. simplex* by Restall *et al.* (2006). Adults of nominate *simplex* have the head and upperparts dark brown while the underparts are white with an olive-grey wash to the breast-sides and flanks, which extends across the breast as a broad but diffuse band. The white throat of *simplex* is finely streaked grey, and *duidae* is most similar to *guaiquinimae* but has more white on the belly, which is somewhat flecked darker. Race *guaiquinimae* is dark umber above and has the white throat finely streaked cinnamon above a broad cinnamon or reddish-brown breast-band. The breast-sides, flanks and undertail-coverts are also washed cinnamon, rather than brown as in *simplex*. Race *pacaraimae* is similar in coloration to *simplex* but overall paler. As mentioned under Distribution, only recently has Tepui Antpitta been confirmed to occur in Brazil (M. Cohn-Haft *in litt.* 2016). Specimens, recordings and photographs have now confirmed its presence in two ranges in central Roraima: Serra da Mocidade and Serra do Apiaú. From the few specimens I have seen and published descriptions (e.g. Phelps & Phelps 1963, Phelps 1977), I feel that these newly discovered populations are closest in plumage to race *duidae*, and they are tentatively included therein below.

Myrmothera simplex duidae Chapman, 1929, American Museum Novitates, no. 380, p. 17, Mt. Duida, Central Camp, Amazonas, Venezuela 4,800ft [= 1,465m, 03°23'N, 65°35'W]. The holotype (AMNH 245926) is an adult male collected on 28 December by George H.H. Tate (LeCroy & Sloss 2000). Race *duidae* occurs in southern Venezuela (Amazonas) and presumably in adjacent Amazonas, Brazil (Peters 1951, Pinto 1978), as it is fairly common around 1,300m just north of Cerro Tamacuarí in the Serranía de Tapirapecó and Sierra de Unturán (Barrowclough & Escalante-Pliego 1990, Barrowclough *et al.* 1995). The recent localities from Roraima given below (and discussed above) are provisionally considered to represent *duidae*. Chapman (1929b) described *duidae* quite simply, stating that it is similar to nominate *simplex* but with the 'wash on breast, sides and flanks more rufescent, less olivaceous and heavier'. Similarly, Dickerman & Phelps (1982) considered *duidae* to be more ochraceous dorsally, less brown, and to have a stronger wash across the breast. Bare-parts coloration was given by Willard *et al.* (1991) as: *Iris* dark brown; *Bill* black, mandible with pinkish base; *Tarsi & Toes* medium grey. Specimens & Records VENEZUELA: *Amazonas* Cerro Yaví, 05°43'N, 65°53'W (COP 37700–702); Cerro Paraque, 05°06'N, 67°27'W (COP 33584–589); Caño Ceje, 05°05'N, 67°28'W (COP 33590/91); Cerro Cuculito, 04°40'N, 65°40'W, and Serranía Parú, 04°25'N, 65°50'W (Phelps & Phelps 1963); Cerro Marahuaca, 03°36'N, 65°22'W (COP 74985); headwaters of Caño Culebra, 40km NNW of Esmeralda, c.03°31'N, 65°27'W (USNM 505783/650468); several sites on Cerro Duida, c.03°20'N, 65°34'W (COP, MCZ, AMNH);

Sierra de Unturán, 01°33'N, 65°14'W (Barrowclough & Escalante-Pliego 1990); Cerro Tamacuarí, 01°13'N, 64°42'W (Barrowclough *et al.* 1995, AMNH 4270); Cerro Jimé, 01°00'N, 65°58'W (COP 60022–034); several sites on Cerro de la Neblina, *c.*00°50'N, 66°05'W (Phelps 1973, USNM, FMNH, AMNH, COP). **BRAZIL**: *Roraima* Serra do Apiaú, *c.*55km W of Mucajaí, 02°26'N, 61°25'W (M. Cohn-Haft *in litt.* 2016, INPA); Serra da Mocidade, *c.*75km W of Caracaraí, 01°42'N, 61°47'W (M. Cohn-Haft *in litt.* 2016, INPA, WA 2015695, A. d'Affonseca); Serra da Mocidade, SW portion, 01°34.5'N, 62°05.5'W (eBird: G. Leite).

Myrmothera simplex guaiquinimae J. T. Zimmer & Phelps Sr., 1946, American Museum Novitates, no. 1312, p. 9, Mt. Guaiquinima, Paragua River, Bolívar, 1,540m [= NW slope of Cerro Guaiquinima, 05°50'N, 63°39'W]. The holotype (COP 28825) is an adult male collected 26 January 1945 by F. Benedetti. Originally in the Phelps collection, it is now on deposit in New York. The range of *guaiquinimae* is restricted to south-east Venezuela in central and south-east Bolívar (Phelps & Phelps 1963). Note that Phelps & Phelps (1963) erroneously listed Pauraitepui as within the range of *guaiquinimae*, but populations there belong to *pacaraimae*. Race *guaiquinimae* is most similar to *duidae*, but differs by having greener, less ochraceous-brown sides, flanks and breast-band. It is separated from *simplex* by having brighter upperparts, the brown tending towards yellow-brown, and by having a stronger olivaceous wash on the breast (Zimmer & Phelps 1946). Following the latter authors' description of the holotype, adults are amber- or yellowish-brown above, paler on the uppertail-coverts. The lores are whitish and there is a prominent postocular white crescent. There is an indistinct, dusky malar and dusky-olive ear-coverts. The chin and throat are white with fine, blackish, hair-like tips that give these areas a finely streaked look. The breast is washed reddish-brown or cinnamon, forming a band on the breast that is paler near the centre and darker laterally. The sides, flanks and undertail-coverts are warm yellowish-brown, contrasting somewhat with the grey belly that becomes mixed with white centrally. The primaries and secondaries are blackish with all but the outermost primaries having brown leading margins. The tertials and upperwing-coverts are blackish-brown. The underwing-coverts are dusky-olive and the underside of the inner primaries and secondaries is similar but paler brown basally. The tail is similar to the back but somewhat duller. The iris is dark, the bill mostly black, greyish-white basally, and the toes and tarsi are slate-grey. **SPECIMENS & RECORDS** *Bolívar* Meseta del Cerro de Jáua, 04°48'N, 64°26'W (Phelps 1977, COP 72422–435); Cerro Tabaro, Río Nichare, 06°30'N, 64°45'W (Phelps & Dickerman 1980, COP 62732–735); Cerro Sarisariñama, 04°30'N, 64°14'W (Phelps 1977, COP 72431).

Myrmothera simplex simplex (Salvin & Godman, 1884). The nominate subspecies is endemic to the Sierra de Lema highlands and tepuis surrounding the Gran Sabana in south-east Bolívar (Phelps & Phelps 1950, Peters 1951, Dickerman & Phelps 1982) and immediately adjacent Guyana (Barnett *et al.* 2002). **SPECIMENS & RECORDS GUYANA**: *Cuyuni-Mazaruni* Mt. Holitipu, 05°59'N, 61°03'W (USNM 639140); Mt. Ayanganna, 05°23'N, 59°59.5'W (Zykowski *et al.* 2011, Milensky *et al.* 2016); N slope of Cerro Roraima, *c.*05°16'N, 60°44'W (Braun *et al.* 2003, AMNH, KUNHM, USNM, ML). *Potaro-Siparuni* Mt. Kopinang, 04°58'N, 59°54'W (Milensky *et al.* 2016);

Mt. Kowa, 04°52'N, 59°41.5'W (Barnett *et al.* 2002). **VENEZUELA**: *Bolívar* Sierra de Lema, km92, 06°09.5'N, 61°26'W (eBird: J. VanderGaast); La Escalera, 06°05'N, 61°21.5'W (ML 30418/34427, T.A. Parker); Sierra de Lema, km122–125, *c.*05°52'N, 61°27'W (COP 64728, LSUMZ 64932, AMNH 11994); Uaipán-tepui, 05°47'N, 62°37'W (COP, FMNH); Ptari-tepui, 05°46'N, 61°46'W (long series, COP); Auyán-tepui, 05°44'N, 62°31'W (Gilliard 1941, COP, AMNH); Aprada-tepui, 05°26'N, 62°25'W (COP); Chimantá-tepui, 05°17'N, 62°15'W (long series, COP); Cerro Cuquenán, 05°13'N, 60°51'W (COP); W slope of Cerro Roraima, 05°12'N, 60°49'W (COP-50268–280); Acopán-tepui, 05°10'N, 62°02'W (COP 42100/01); N slope of Uei-tepui, 05°01'N, 60°37'W (Phelps & Phelps 1962, COP 44656-59, FMNH).

Myrmothera simplex pacaraimae W.H. Phelps Jr. & Dickerman, 1980, Boletín de la Sociedad Venezolana de Ciencias Naturales, vol. 33, p. 144, Cerro Urutaní, Sierra Pacaraima, at the head of the Río Paragua, Bolívar, Venezuela, 1,280m [*c.*03°46'N, 63°03'W]. The holotype (COP 73643) is an adult male collected by M. Castro on 23 February 1977. Originally in the Phelps collection in Caracas, it is on deposit at AMNH in New York (Phelps & Dickerman 1980). There is a long series of topotypical material still held in Caracas (COP 73643–652), as well as several skins in Brazil (MPEG 33161/62). Race *pacaraimae* is endemic to the mountain range after which it is named, the Pacaraima Mountains in Bolívar. Within this range it is found from Cerro Paurai-tepui west to the Alto Río Ocamo, south-east Amazonas (Venezuela). The following description of *pacaraimae* and comparisons with other subspecies is taken largely from Phelps & Dickerman (1980) and Dickerman & Phelps (1982). Like other subspecies, *pacaraimae* differs little but is overall most similar to *simplex*. The upperparts are generally paler than *simplex*, *duidae* and most populations of *guaiquinimae* except those on Meseta de Jáua. The underparts of *pacaraimae* are most similar to those of *simplex* but the dark breast-band is either paler and not as broad, or lacking altogether. The somewhat reduced pectoral band also separates *pacaraimae* from *guaiquinimae* and all populations of *duidae* except those on Cerro Yaví and Cerro Paraque. **SPECIMENS & RECORDS VENEZUELA** *Amazonas* Alto Río Ocamo, 02°46'N, 64°27'W (COP 71093/94); La Faisca, Pauraí-tepui, 04°40'N, 61°25'W (COP 32586).

STATUS Tepui Antpitta is a restricted-range species that, as its English name implies, is found in the Tepuis EBA (BirdLife International 2017). Within its fairly small range, however, it is generally considered uncommon to fairly common in Venezuela (Stotz *et al.* 1996, Hilty 2003), but perhaps less numerous in Guyana (Restall *et al.* 2006). It is considered Least Concern (BirdLife International 2017). Although found in several areas under formal protection, Tepui Antpitta is predicted to lose 5.8–11.7% of suitable habitat within its distribution before the year 2020 (Soares-Filho *et al.* 2006, Bird *et al.* 2012). Ultimately, the extremely rugged terrain and mostly inaccessible areas where it lives is probably the species' best protection. **PROTECTED POPULATIONS** *M. s. simplex* PN Canaima (Krabbe & Schulenberg 2003, COP). *M. s. duidae* PN Duida–Marahuaca (COP, AMNH, MCZ); PN Parima-Tapirapecó (Barrowclough & Escalante-Pliego 1990, Hilty 2003, COP); PN Serranía de la Neblina (COP, AMNH, FMNH). *M. s. guaiquinimae* PN Jáua-Sarisariñama (Hilty 2003, COP). *M. s. pacaraimae* PN Parima-Tapirapecó (COP).

OTHER NAMES For many years, *M. simplex* was known as Brown-breasted Antpitta (Cory & Hellmayr 1924, Meyer de Schauensee 1966, Snyder 1966, Phelps 1973, Meyer de Schauensee & Phelps 1978, Howard & Moore 1980, Altman & Swift 1986, Willard *et al.* 1991, Sibley & Monroe 1993). Ridgely & Tudor (1994), however, pointed out that this name is somewhat misleading given that just one race has a brown breast, with grey being predominant in most. They further noted that this name is confusingly similar to that used for *G. milleri* (Brown-banded Antpitta), a species far more deserving of such a name. Following this, the vernacular name of *M. simplex* was changed to Tepui Antpitta. As the only species of antpitta restricted to the tepui region, this is indeed more fitting. Scientific synonyms are few: *Grallaria simplex* (Salvin 1885, Sclater 1890, Sharpe (1901), Penard & Penard 1910, Brabourne & Chubb 1912, Chubb 1921). **Spanish** Tororoí Flautista (Krabbe & Schulenberg 2003); Hormiguero Flautista (Phelps 1973, Meyer de Schauensee & Phelps 1978, Phelps & Meyer de Schauensee 1979, 1994); Hormiguero Flautista del Roraima [*simplex*], Hormiguero Flautista del Duida [*duidae*], Hormiguero Flautista del Guaiquinima [*guaiquinimae*] (Phelps & Phelps 1950). **Portuguese** tovaquinha-castanha (Ruschi 1979); torom-de-peito-pardo (Willis & Oniki 1991, van Perlo 2009, CBRO 2011). **English** Brown-breasted Antbird (Dickerman & Phelps 1982); Salvin's Ground-bird (Penard & Penard 1910); Grey-breasted Ant-Thrush (Brabourne & Chubb 1912, Chubb 1921). **Dutch** Salvin's Grondmierenvogel (Penard & Penard 1910). **French** Grallaire sobre. **German** Olivbauch-Ameisenpitta (Krabbe & Schulenberg 2003).

Tepui Antpitta, adult (*simplex*), La Escalera, Bolívar, Venezuela, 21 October 2005 (*Nick Athanas/Tropical Birding*).

Tepui Antpitta, adult (*duidae*), Caracaraí, Roraima, Brazil, 29 January 2001 (*Anselmo d'Affonseca*).

Tepui Antpitta, adult (*simplex*), slopes of Cerro Roraima, Bolívar, Venezuela, 29 October 2015 (*Marco Guedes*).

Genus *Grallaricula*: small montane antpittas

Grallaricula, P. L. Sclater, 1858, Proceedings of the Zoological Society of London, vol. 26, p. 283. Type, by subsequent designation, *Grallaricula flavirostris* (Sclater, 1858), by Sclater (1890).

Remarks on the immature plumages of *Grallaricula* As will be apparent in the following accounts, I can say with some certainty that the fledglings of *Grallaricula* are remarkably uniform in their appearance, being round, fluffy balls of wool-like rufescent or rusty-brown down ('vinaceous pink' down; Fjeldså & Krabbe 1990). The details of timing and sequence of post-fledging plumage development in *Grallaricula* appear to be very similar to my understanding of development for *Hylopezus* and *Myrmothera*, in particular, with respect to the colour of this first covering of downy plumage. Like these two genera, transition to more adult-like plumage in *Grallaricula* is poorly known, perhaps for the reason that I have suggested for *Myrmothera* and *Hylopezus*; the transition is faster than for *Grallaria*. Unlike most species of *Grallaria*, the first covering of adult-like contour feathers, including the hindcrown and nape, is patterned very similar to that of full adults. Of the four antpitta genera, *Grallaricula* species appear to have the least obviously marked immature wing-coverts. Nevertheless, the presence of pale buffy or chestnut edgings to the upperwing-coverts is unquestionably a sign of immaturity in most or all species. These markings tend to fade fairly quickly with wear and they are usually not as immediately evident as in other genera. Indeed, several taxa (e.g. *G. flavirostris*) have inadvertently been described from specimens in subadult plumage. It is still unclear but, unlike my suspicions for other genera, it appears that these first wing-coverts are retained for most or all of their plumage development, as opposed to being replaced at least once during the process. To approximate the stage of immaturity of *Grallaricula* I use the following definitions, in most ways similar to my definitions for *Myrmothera* plumages. FLEDGLING Covered in fluffy, wool-like, feathers, rictal flanges still obviously inflated and pale orange or yellow-white, tail may not be visible beyond the layer of down, flight feathers may not be completely emerged, still heavily dependent on adults for food and mostly remain stationary while awaiting food, probably incapable of competent flight and may still appear awkward even when walking. JUVENILE Mostly covered in adult-like contour feathers, but may retain some rust-coloured patches of fledgling down on crown and breast, as well as scattered wisps of downy plumage on nape, rump and central belly; rictal flanges probably still paler than in adults but may not be distinctly immature in appearance, behaviour largely unknown, but probably fairly mobile, moving behind adults and soliciting food. SUBADULT Very close or identical to adults in general coloration, perhaps only separable by their immature wing-coverts, although these may not be strongly marked, making them difficult to separate from adults in the field, may sometimes also retain scattered patches (1–2 feathers) of wispy rust-coloured down on crown. As described in more detail for *Hylopezus*, and seems also the case for *Myrmothera*, plumage transition from fledgling to adult appears fairly quick, but in a few cases I use the terms '**fledgling transitioning**' and '**juvenile transitioning**' for birds somewhere between the above-defined stages. The term '**subadult transitioning**' is reserved for individuals with heavily worn and faded upperwing-coverts that are otherwise fully adult in appearance.

OCHRE-BREASTED ANTPITTA
Grallaricula flavirostris   Plate 22

Grallaria flavirostris P. L. Sclater, 1858, Proceedings of the Zoological Society of London, vol. 26, p. 68, 'Río Napo, Ecuador'. The holotype (NHMUK 1889.9.20.647; Warren & Harrison 1971) is an unsexed adult collected by Manuel Villavicencio and brought to Europe by J. P. Verreaux as part of a large collection of birds from north-east Ecuador. There is no way to know the precise origin of most of Villavicencio's collections, although he was based in Puerto Napo (near Tena) during this time. Although Sclater (1890) gave 'Río Napo, Ecuador' as the type locality, in the original publication he indicates a somewhat more precise region in the opening text of the article: 'upper Río Napo, in the province of Quixos [Cantón Quijos], in the eastern part of the republic of Ecuador'. This would certainly be a more appropriate type locality for Ochre-breasted Antpitta as there are few, if any, locations along the Río Napo that could serve as appropriate habitat. Almost certainly the true origin of the specimen is in the East Andean foothills, between Baeza, Loreto, and Archidona, Napo province, *c*.00°40'S, 77°45'W.

Ochre-breasted Antpitta inhabits the undergrowth of humid montane forests from Costa Rica to central Bolivia, and has the most extensive range of any species of the genus. Although nests have now been described from locations spanning nearly its entire range, this diminutive antpitta remains very poorly known in terms of its ecology, behaviour, distribution, vocalisations, taxonomy and geographic variation. Up to nine subspecies have been recognised but intra-population plumage variation is considerable, and subspecific limits are badly in need of revision. Our understanding of the situation, however, remains hindered by a lack of vocal data as well as poorly understood distributional limits and habitat preferences. These particular information gaps are closely correlated: the quiet, simple vocalisations are difficult to detect and hard to identify, undoubtedly resulting in its being overlooked in many parts of its range.

IDENTIFICATION 10.0–10.2cm. Sexes similar. There is much geographic variation in plumage and bare-part colours, details of which are not well understood. In most of its range, adults are olive-brown above, greyer on the crown, with an ochraceous eye-ring and loral streak, and varying amounts of cinnamon-ochraceous wash across the sides of the face and ear-coverts. The throat, breast and flanks are ochraceous, sometimes fading to whitish on the belly. Feathers of the breast and flanks are variably margined blackish, creating dark streaks or scallops. Rusty-breasted and Slate-crowned Antpittas are probably the species most similar to Ochre-breasted Antpitta and with which there is the greatest range overlap. These species are, however, generally found at higher elevations and have little or no streaking on the underparts, whose ground colour varies from rusty-orange to orange-buff (not ochraceous). Most races of Rusty-breasted Antpitta are also distinguished by their distinct pale eye-ring or postocular crescent. Confusion with Slate-crowned Antpitta is especially likely in areas such as south-east Ecuador, where this species occasionally shows a fair amount of dark edging to the breast feathers (see Slate-crowned Antpitta account). Two other species that may overlap in range with Ochre-breasted Antpittta and have more similar altitudinal preferences are Peruvian and Ochre-fronted Antpittas. Both of these species have distinct breast scalloping but are immediately distinguished by their largely reddish-chestnut to orange-chestnut heads and lack of ochraceous on the underparts. In most cases of potential confusion, no other species of *Grallaricula* is likely to have as much yellow on the bill as Ochre-breasted Antpitta in that region. See also Taxonomy and Variation for a discussion of racial variation.

DISTRIBUTION The most widely distributed species of the genus, from northern Costa Rica to central Bolivia. Overall, we still lack a clear picture of its precise range and new populations are still being discovered (see Taxonomy and Variation). The most northerly locations are in the humid montane forests of Costa Rica's Volcán Arenal, south and east through the mountains to eastern Panama, with an apparently isolated population in the Serranía Darien. The next northernmost populations,

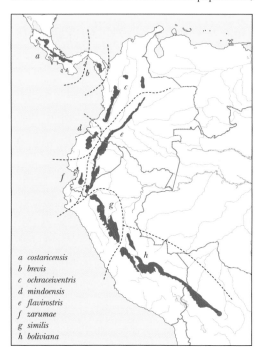

a *costaricensis*
b *brevis*
c *ochraceiventris*
d *mindoensis*
e *flavirostris*
f *zarumae*
g *similis*
h *boliviana*

also apparently isolated, are at the north ends of the Central and Eastern Andes in Colombia. From central Colombia, with several notable gaps, this antpitta is found on the Pacific slope of the West Andes south to southern Ecuador and on the Amazonian slope of the East Andes south to central Bolivia. **See Also** Hilty & Brown (1986), Stiles & Skutch (1989), Fjeldså & Krabbe (1990), Rodner *et al.* (2000), Ridgely & Greenfield (2001), Hennessey *et al.* (2003), Krabbe & Schulenberg (2003), Restall *et al.* (2006), Garrigues & Dean (2007), Ridgely & Tudor (2009), Angehr & Dean (2010), McMullan *et al.* (2010), Schulenberg *et al.* (2010).

HABITAT Ochre-breasted Antpitta inhabits the undergrowth of humid and wet montane forest, and adjacent (mature) second growth (Stiles & Skutch 1989, Fjeldså & Krabbe 1990). In appropriate habitat, this shade-loving antpitta seems to favour low-lying and riparian areas, avoiding more exposed ridgetops and forest edges. Across its substantial range, the species is known from elevations of 300–2,750m, but is most frequently found at 900–2,200m with a mean somewhere around 1,550m (Graves 1985). On the whole, it is found within a narrower range and at lower elevations in the north, and is commonest at higher elevations but present over a broader range in the south. In Costa Rica, records extend from 700 to 1,500m, at 900–1,850m in the south-west (Garrigues & Dean 2007, Stiles & Skutch 1989), and in Panama it is found at 900–1,500m (Ridgely & Gwynne 1989). Records extend as low as 500m in Colombia and 300m in Ecuador (*zarumae*; San Luís). An old record from 75m at Puebloviejo, Los Ríos (MLZ) seems unlikely, but is less than 20km from the base of the West Andes where much appropriate habitat is its more likely provenance. In Peru most populations are found at 1,400–2,200m (*similis/boliviana*; Weske 1972, Terborgh & Weske 1975, Terborgh 1985, Mekord 2010) but also as low as 800m and as high as 2,300m (Walker *et al.* 2006, Schulenberg *et al.* 2010). The southernmost records (*boliviana*) suggest an altitudinal range of roughly 1,600–2,500m in Bolivia (Hennessey *et al.* 2003). The suggested upper limit of 2,750m given by Fjeldså & Krabbe (1990) seems slightly high.

VOICE The species' vocalisations remain poorly known, seemingly in part due to its relatively non-vocal nature. Krabbe & Schulenberg (2003) quantitatively described the 'song' as *c*.7s in duration and consisting of an evenly paced series of *c*.30 notes that first rises in volume and pitch (3.0–3.4kHz), levelling out over the second half. This vocalisation is seemingly heard more often in the north of its range, and described as a high trill with a rattling quality by Garrigues & Dean (2007). The more frequently heard, and more geographically widespread call-like vocalisation is a much simpler, single, emphatic whistle, produced every 10–20s. Krabbe & Schulenberg (2003) provided a description: 0.3s long, delivered at 3.0–3.4kHz, repeated at intervals of 8–14s. Other variations include a weak piping *weeeu*, a descending *tew*, a slightly longer *teew* or a rising-falling *wheew* (Ridgely & Tudor 2009, Schulenberg *et al.* 2010).

NATURAL HISTORY Ochre-breasted Antpitta is generally considered to forage alone or in pairs (Krabbe & Schulenberg 2003, Ridgely & Tudor 2009), but its non-vocal and unobtrusive nature often lead to its being overlooked in the 'heat of the moment' when cacophonous and chaotic mixed-species flocks are chanced upon by observers. It has been reported in ant-following flocks (Greeney 2012b) and is among those species of antpittas trained to receive food at the now-famous worm-feeding stations (Woods *et al.* 2011), good evidence that there is considerably more to learn about its foraging behaviour. In further support of its suggested inclination to forage in association with large mammals (Greeney 2012b), J. Freile (*in litt.* 2017) recounted being followed, on two occasions, by a silent but apparently fearless Ochre-breasted Antpitta that hopped from perch to perch, and occasionally to the ground. While searching for food the species hops through dense undergrowth, close to the ground on thin horizontal branches, often clinging sideways to mossy trunks and vines. It makes brief sally-strikes to glean small prey from foliage, trunks or, less frequently, the ground (Hilty & Brown 1986, Stiles & Skutch 1989). Frequently, when perched, it twitches its lower body back and forth, while holding its head still, in typical *Grallaricula* fashion (see IBC 1222421, video D. Allen). The reason for this odd movement is unknown, but adults may also move thus at the nest (IBC 1150281, video HFG). It is among the least vocal of antpittas, but is said to occasionally sing from perches within dense thickets, 1–3m above ground. Delgado-V. (2002) described what appears to be a fairly elaborate courtship behaviour and nuptial feeding that commenced with one adult (*mindoensis*) approaching another perched on a low limb. The first adult carried a white lepidopteran in its bill and made a 'sharp and spaced call'. While continuing to vocalise, the adult with food hopped from perch to perch around the other, which remained stationary but vibrated its wings in what sounds like a typical begging posture. Given the possibility that the observer had overlooked the subtle cues that this begging individual was an immature, I would have been unconvinced of the courtship nature of this behaviour had the adult with food not continued to circle the other for *c.*15min. At the end of this period the lepidopteran was passed to the begging bird, which quickly consumed it and then followed the first into the undergrowth. No aspects of territoriality have been investigated, but tightly clumped nests in Costa Rica (Holley *et al.* 2001) and possible re-use of a site in south-west Ecuador (Greeney *et al.* 2012) suggest that territories may be held across multiple breeding attempts.

DIET The diet is very poorly known. Robbins *et al.* (1985) mentioned that the contents of two stomachs were unidentified insects (*similis*), and various invertebrates almost certainly form the bulk of their diet. The stomach of an incubating female (*boliviana*) contained larval lepidopterans and other arthropods (Maillaird-Z. & Vogel 2003). Large lepidopteran larvae, adult Lepidoptera and small flies (Nematocera) have all been reported as nestling food (*flavirostris*; Greeney *et al.* 2012). At a nest with *c.*3- or 4-day-old nestlings, the same authors documented the delivery of a *c.*2cm-long (SVL) *cf. Eleutherodactylus* sp. frog.

REPRODUCTION Although the general breeding biology and behaviour are not as well described as that of other *Grallaricula*, Ochre-breasted Antpitta holds the honour of having nest descriptions published from across the most widely scattered populations. Nests have been described for five of the eight subspecies, including from both extremes of the species' range. The egg was first described from western Ecuador (*mindoensis*; Robbins & Ridgely 1990). Holley *et al.* (2001) provided the first description

of the nest, with observations made at three nests in Costa Rica (*costaricensis*). Shortly thereafter, a nest from Bolivia (*boliviana*) was described (Maillard-Z. & Vogel 2003). Most recently, Greeney *et al.* (2012) presented data on nests in both eastern and western Ecuador, adding information for an additional three of the recognised subspecies (*flavirostris, mindoensis* and *zarumae*). For some embarrassing reason, for nominate *flavirostris* nests were not described by Greeney *et al.* (2012), and the following data accredited to this source refer to *mindoensis* and *zarumae* (in that order). Data for four nests of nominate *flavirostris* (HFG) include information for nests discussed by Greeney *et al.* (2012). Eggs of *costaricensis* were only qualitatively described by Holley *et al.* (2001) and egg dimensions and additional nest data given below for this subspecies are from my own examination of, and label data on, a Costa Rican nest and clutch collected by E.M. Carman (MZUCR H119). **NEST & BUILDING** Nests are relatively broad, shallow, open mossy cups, variously composed externally of green moss and dead plant materials such as twigs, leaf petioles, ferns and dead leaves. Internally they are dark-coloured and neatly lined with fine fibres, rootlets and fungal rhizomorphs. Though not explicitly stated in most descriptions, I believe most comprise a sparse stick and petiole base supporting a firmer, well-formed mossy cup, as is the case for nests I have observed directly. All known nests were located in areas with near-complete canopy cover and a fairly open understorey (= mature forest). Most appear to have been sited in poorly concealed locations, but their overall resemblance to a natural clump of detritus made them quite cryptic. Nests are built on a variety of substrates and in different situations but, overall, I consider most to be in fairly flimsy, poorly supported situations (compared to the logs, stumps and thick branches used by some *Grallaria*). Supporting structures: multiple forks of slim saplings (*n* = 5; Holley *et al.* 2001, HFG); base of small branches or epiphytic ferns sprouting from trunk of a sapling (*n* = 4; Holley *et al.* 2001, Greeney *et al.* 2012, HFG); leaves of a *Tillandsia* bromeliad on a dead branch (Maillard-Z. & Vogel 2003); base of epiphytes on side of large trunk (*n* = 1, Greeney *et al.* 2012); vine tangle (*n* = 1, Greeney *et al.* 2012); thin branch of a sapling (MZUCR H119). Nest heights: 0.46m (Maillard-Z. & Vogel 2003), 3.5m, 2.3m, 3.5m (Holley *et al.* 2001), 0.8m, 0.45m, 1.2m (Greeney *et al.* 2012); 1.6m, 1.4m, 1.7m, 1.4m (HFG), 1m (MZUCR H119). Dimensions: outer diameter 8cm (Maillard-Z. & Vogel 2003), 10.4cm, 9.5cm, 11.0cm (Holley *et al.* 2001), 7.5cm, 11.7cm, 11.0cm (Greeney *et al.* 2012), 9.5cm, 11.0cm, 12.0cm, 11.5cm (HFG); outer height 5.5cm (Maillard-Z. & Vogel 2003), 13.0cm, 9.0cm, 15.5cm (Greeney *et al.* 2012), 9.5cm, 9.0cm, 10.0cm, 9.0cm (HFG), 11.5cm (MZUCR H119); inner diameter 6.5cm, 6.5cm, 7.0cm, 6.0cm (HFG), 6.7cm (MZUCR H119); inner depth 6.8cm (Maillard-Z. & Vogel 2003), 4.0cm, 3.2cm, 4.5cm (Holley *et al.* 2001), 3.8cm, 3.8cm, 4.7cm (Greeney *et al.* 2012), 4.5cm, 4.0cm, 3.5cm, 3.5cm (HFG), 4.5cm (MZUCR H119). **EGG, LAYING & INCUBATION** The incubation period is not known with any certainty, but Holley *et al.* (2001) fairly estimated a period of 17–21 days. Incubation is by both adults (HFG). Three nests of *costaricensis* held two eggs, and one a single egg, but only one was watched from clutch initiation to completion (Holley *et al.* 2001; MZUCR H119). The single *boliviana* nest held two eggs (Maillard-Z. & Vogel 2003) and in Ecuador three nests held two eggs, one had one egg, one a single nestling, and one had two nestlings (Greeney *et al.* 2012, HFG). Of the Ecuadorian nests, only one was followed through laying (HFG). It thus appears that normal clutch size is two, possibly with occasional one-egg clutches. The eggs are subelliptical to short-subelliptical. Their ground colour appears to vary slightly, from pale brown or light coffee-brown (*flavirostris*, *boliviana*) to pale reddish-brown (*costaricensis*) or pale brown with a greenish cast (*mindoensis*). Eggs from all regions, however, are variously blotched and flecked brown, cinnamon, black and lavender, concentrated at the larger end to varying degrees. The white ground colour of the egg described by Robbins & Ridgely (1990) from a shelled egg removed from the oviduct (*mindoensis*) either represents further variation or was perhaps incompletely pigmented. Egg dimensions: *costaricensis* 20.3 × 16.1mm (2.9g), 19.8 × 15.7mm (2.7g) (MZUCR H119); *flavirostris* 20.9 × 16.6mm, 20.7 × 16.6mm (Greeney *et al.* 2012); *mindoensis* 22 × 16mm (Robbins & Ridgely 1990), 20.5 × 16.1mm, 20.4 × 15.0mm, 21.0 × 16.9mm (Greeney *et al.* 2012); *boliviana* 20.3 × 16.8mm, 20.8 × 16.6mm (Maillard-Z. & Vogel 2003). Nothing is known about egg laying and the only data on incubation behaviour comes from Holley *et al.* (2001), who reported brief observations on an adult's hesitant movements at the nest when observers were present. NESTLING & PARENTAL CARE Holley *et al.* (2001) estimated a nestling period of 14–16 days, which I feel is likely an under-estimation. Eggs hatched synchronously at one nest in Ecuador (within the same hour) and nestlings still not fully out of their shells were capable of begging vigorously by silently waving their bright orange gapes (Greeney *et al.* 2012). The following nestling descriptions are based on data and photos in Greeney *et al.* (2012) and my own observations (see also Greeney 2012b). Newly hatched young weigh about 3g, are dark-skinned and devoid of natal down. They have dusky-orange bills, legs and cloacas, and their mouth linings are bright orange, highlighted by pale creamy-yellow rictal flanges. At *c.*4 days of age nestlings weigh *c.*7g and are still dark-skinned and bare. They retain their bright white egg tooth and their eyes remain closed. Contour and flight feather pins are just beginning to break the skin's surface. By 10–11 days of age nestlings are almost completely covered in thick rusty down and weigh *c.*13.5g. At this age their legs are pale greyish-pink to dusky-orange. Their bills are still as described for hatchlings, largely orange but have darkened to blackish on the culmen. Their primary pin feathers are emerged 1.0–1.5mm from their sheaths, while the inner flight feathers have emerged slightly further, suggesting that primary pin feathers break around day 9–10. Greeney *et al.* (2012) reported details of nestling care based on video recordings made at two nests (*flavirostris*), confirming that both genders participate in all aspects of nestling care. On the day of hatching, all food brought by adults appeared to be tiny (<5mm) invertebrates. Faecal sacs were consumed immediately by the attending adult and the young were brooded for 73% of the 3.75h observation period, in bouts lasting 12.1 ± 8.7min. While brooding, adults infrequently stood and rapidly probed the nest lining. At a nest with *c.*4- or 5-day-old nestlings, adults brooded for 77% of daylight hours the nest was filmed. Once an adult arrived with a 4–5cm, hairless, green lepidopteran larva that proved too big to feed to the nestlings, despite several attempts, and was finally consumed by the adult. About half of faecal sacs produced were eaten by adults at the nest and half were carried from the nest (see IBC 1150281, video HFG). While brooding, adults occasionally stood and pecked or rapidly probed

into the nest, presumably removing parasites or debris (Haftorn 1994, Greeney 2004). Two nestlings, 10–11 days old, were fed small invertebrates, mostly <10mm, but were only brooded for a very small fraction of daylight hours. SEASONALITY With so few records available, spread too thinly across the species' range, any assessment of seasonality is premature. BREEDING DATA *G. f. costaricensis* Building, 13 May 2000, Las Tablas (Holley *et al.* 2001); adult carrying nest material, 30 May 2014, PN Tapantí (eBird: K. Borgmann); lightly incubated eggs, 25 April 2001, Taus (MZUCR H119, ex. E.M. Carman); two nests with eggs, 17 April 1999 and 27 April 2000, Las Tablas (Holley *et al.* 2001). *G. f. brevis* Subadult, 1 May 1912, Río Limón headwaters (USNM 238069, holotype). *G. f. ochraceiventris* Adults feeding dependent young [= fledgling/juvenile], 30 August 2015, Cerro Montezuma (eBird: K. Borgmann); juvenile transitioning, 28 July 2017, RNA Las Tangaras (ML 65136771, T. Brooks). I also examined the following undated specimen: AMNH 492301, subadult, Nóvita. *G. f. flavirostris* Two near-complete nests, 25 April 2005, Mushullacta (HFG) and 19 November 2015, WildSumaco Lodge (HFG); incubation, 5 December 2002, N slope of Volcán Sumaco (Greeney *et al.* 2012); incubation, 26 August 2012, WildSumaco Lodge (R. Ahlman *in litt.* 2017); mid-aged nestlings, 3 December 2002, N slope of Volcán Sumaco (Greeney *et al.* 2012); juvenile, 8 March 2002, San Carlos de Limón (MECN 8091); subadult, 24 July 1992, Panguri (MECN 6387); subadult, 19 August 2013, WildSumaco Lodge (ML 44551131, photos R. van Twest); subadult transitioning, 8 December 1970, El Carmen (FMNH 399877). *G. f. mindoensis* Laying, 4 March 2007, RP Intillacta (Greeney *et al.* 2012); shelled egg in oviduct, 8 August 1988, 4km SSW of Chical (Robbins & Ridgely 1990); shelled egg in oviduct, between 11 Feburary and 14 March 1989, RNSC La Planada (Oniki & Willis 1991); incubation, 2 September 2006, RP Intillacta (Greeney *et al.* 2012); *c.*2-day-old nestling, 10 April 2010, RP San Jorge de Milpe (G. Cruz photo); fledgling, 5 August 1988, 4km SSW of Chical (ANSP 181207); 'fledgling' [= juvenile], 30 July 2004, Tandayapa Bird Lodge (Solano-Ugalde *et al.* 2007); adults feeding fledgling transitioning, 14 August 2014, Refugio Paz de Las Aves (L. Vining photos); immature with orange mandible [juvenile], 3 August 2005, RP Las Tangaras (D. Becker *in litt.* 2015); subadult, 8 June 1929, Puebloviejo (MLZ 7307); subadult transitioning, 7 August 1988, 4km SSW of Chical (MECN 4089). I also examined the following undated specimens: MECN 3729, subadult, 'northwest Ecuador'. *G. f. zarumae* Nestling, 6 March 2007, RP Buenaventura (Greeney *et al.* 2012); female with vascularised brood patch, 15 April 1991, W of Piñas (MECN 5649); subadult, 14 April 1991, W of Piñas (MECN 5650); subadult, 28 January 1991, San Luís (MECN 5648). *G. f. similis* Subadult, 11 October 1973, Sariapampa (FMNH 296692); subadult, 15 June 1985, Playa Pampa (LSUMZ 128557); subadult transitioning, 30 June 1010, Quebrada Huancabamba (FMNH 474181). *G. f. boliviana* Fresh eggs, 25 November 2001, Río Palcabamba (Maillard-Z. & Vogel 2003); immature, skull 50% ossified [juvenile?], 10 May 1971, Huanhuachayo (LSUMZ 70233, skeleton); subadult, 16 May 1915, Incachaca (AMNH 137177, holotype); two subadults, Cordillera Vilcabamba, 17 August 1967 (AMNH 820258) and 29 June 1968 (AMNH 820259); subadult, 26 November 1980, Abra de Maruncunca (LSUMZ 99377); old brood patch between 14 and 28 July 1998, Llactahuaman (Pequeno *et al.* 2001b).

TECHNICAL DESCRIPTION Sexes indistinguishable based on plumage and there is considerable individual and geographic variation. The following description best fits a general description of nominate *flavirostris*. **Adults** Dull brownish above, the crown having a slight grey wash. The brownish face, often indistinctly washed ochraceous-rufous, has a loral crescent and an eye-ring of variable width, both ochraceous. The face is separated from the throat by an often indistinct black malar stripe. The wings and tail are brown, the flight feathers edged rufescent and the base of the primaries slightly paler brown. The outer web of the alula is distinctly buffy and the underwing-coverts are fulvous or ochraceous. The throat and breast are variably ochraceous, becoming slightly paler, tawny or fulvous on the flanks. The feathers of the breast and often the sides are margined olive-brown or blackish, creating distinct streaking or scalloping, tending towards scallops in the centre and streaks on the sides and sometimes flanks. The central belly and vent are white. **Adult Bare Parts** *Iris* dark brown; *Bill* maxilla largely or entirely brownish or black, mandible mostly yellow; *Tarsi & Toes* pinkish-grey. **Immature** The post-fledging plumage development was previously largely undescribed. Fjeldså & Krabbe (1990) described 'juveniles' as 'uniform vinaceous pink; wing coverts tipped rufous', unquestionably in reference to the fledgling plumage described below. As with other *Grallaricula*, the presence of markings on the upperwing-coverts is the aspect of immature plumage that is retained longest, something that was previously overlooked and led to several taxa being described from specimens in subadult plumage (see Taxonomy and Variation). It appears that these first wing-coverts are retained for most or all of their development and that they gradually fade from chestnut to pale buff. This colour shift may be facilitated by the wearing away of the feather fringes. *Fledgling* Probably leave the nest before their flight feathers are full and before they are capable of effective flight. They are almost entirely covered in a dense, dark, rusty- or rufescent-brown wool-like down that is somewhat shorter on the face, with dusky-pink skin visible. The central belly and vent are pale greyish or buffy-white. The upperwing-coverts are brownish like the flight feathers, but are broadly tipped dark chestnut to buffy-chestnut, slightly paler or duller near the middle and gradually brightening towards the margins, which may be bright reddish-chestnut in fresh plumage. The tail is probably not even visible beyond the fluffy down feathers until they have begun transitioning to juvenile plumage, but is already coloured as in more mature birds. **Fledgling Bare Parts** *Iris* brown; *Bill* bright orange, dusky on culmen and near tip, which probably still bears a bright white egg-tooth for some time after fledging, rictal flanges inflated and pale yellowish-white; *Tarsi & Toes* dark greyish on the anterior edge, fading to dull yellow or yellow-orange posteriorly. *Juvenile* Has largely acquired adult plumage on most of upperparts, retaining varying amounts of fluffy rufescent feathering on hindcrown, nape, upper back and sometimes in scattered patches near rump. The face pattern of adults is now evident but may not be as cleanly delimited as adults; the loral crescent seemingly becomes evident before the ocular ring. Below they may appear largely in adult plumage but usually show a messy mixture of fully adult feathers interspersed by irregular tufts of fluffy immature feathers. This mixture usually results in an overall decrease in the degree of dark streaking and scaling, and a reduction in the intensity of the ochraceous parts. The last few immature feathers may

create a rusty wash to the flanks. Upperwing-coverts as described for fledglings. **Juvenile Bare Parts** *Iris* brown; *Bill* mandible dusky or brownish-yellow, similar to adults but orange-yellow around nares, at tip and along tomia, mandible dusky yellow-orange, becoming yellower with age, rictal flanges probably still somewhat inflated early on, gradually becoming less evident with age, more yellow than white before they disappear; *Tarsi & Toes* similar to adults. **Subadult** Differs from adults largely by retaining their immature wing-coverts. Prior to being replaced by adult coverts, many or all of the thin rufescent-chestnut margins may wear off and the centres are well faded, leaving only a dull buffy crescent of variable thickness. **Subadult Bare Parts** As described for adults.

MORPHOMETRIC DATA *G. f. costaricensis* Wing 66mm; *Tail* 20mm; *Bill* [exposed] culmen 14.5mm; *Tarsus* 19.5mm (approximated due to damage) (*n* = 1♀, holotype of *vegata*, Bangs 1902). Wingspan 197mm, length 66.7mm; *Tail* 27mm; *Bill* 'from front' 12.7mm, 'from rictus' 19mm; *Tarsus* 20.6mm (*n* = 1♂, holotype, Lawrence 1866). *Bill* from nares 8.1mm, 8.7mm, 8.7mm, 7.3mm, 7.1mm, 8.8mm, 8.9mm; *Tarsus* 21.9mm, 22.7mm, 22.0mm, 21.9mm, 23.0mm, 21.6mm, 21.7mm (*n* = 7, 3♀♀, 4♂♂, MNCR, USNM, last measurements are of holotype). *Bill* from nares 8.3mm, 9.1mm; *Tarsus* 22.2mm, 22.6mm (*n* = 2, ♀? , MNCR, USNM). *Wing* 66mm, 62.9mm, 61.5mm, 64mm; *Tail* 26mm, 22.6mm, 18.0mm, 24.5mm; *Bill* [exposed] culmen 13.5mm, 14.0mm, 13.5mm, 14.5mm; *Tarsus* 21.5mm, 20.6mm, 20.5mm, 21.0mm, middle toe 13.5mm, 13.6mm, 13.5mm, 14.0mm (*n* = 4, ♂,♀,♀,♀, Ridgway 1911). *G. f. brevis* Wing 64mm; *Tail* 26mm; *Bill* culmen 16.5mm; *Tarsus* 24mm (*n* = 1♂, subadult, holotype USNM 238069, Nelson 1912). *Bill* from nares 8.9mm, 9.4mm, 9.2mm; *Tarsus* 23.0mm, 21.7mm, 22.5mm (*n* = 3, ♀,♀,♂, USNM). *Bill* from nares 9.5mm, exposed culmen 14.0mm, depth at front of nares 5.0mm, width at gape 9.1mm, width at front of nares 5.7mm, width at base of mandible 7.7mm; *Tarsus* 23.8mm (*n* = 1♂, subadult holotype USNM 238069). *G. f. ochraceiventris* Wing 67mm, 68mm; *Tail* 25mm, 24.5mm; *Bill* exposed culmen 15.5mm, 15mm; *Tarsus* 23mm, 21mm (*n* = 2, ♀,♂ holotype, Chapman 1922). *G. f. flavirostris* Wing 67mm; *Tail* 28mm; *Bill* [exposed] culmen 12mm, width at nares 7mm; *Tarsus* 22mm (*n* = 1, ♀?, 'Bogotá', Chapman 1926). *G. f. mindoensis* Wing 67mm, 72mm, 67mm; *Tail* 25mm, 28mm, 27mm; *Bill* exposed culmen 13mm, 15mm, 14mm; *Tarsus* 23mm, 25mm, 23mm (*n* = 3, ♀,♂,♀, Oniki & Willis 1991). *G. f. zarumae* Wing 65mm, 65mm; *Tail* 25mm, 25mm; *Bill* exposed culmen 14.5mm, 15mm; *Tarsus* 22mm, 23mm (*n* = 2, ♀,♂ holotype, Chapman 1922). *G. f. similis* Wing 70mm, 70mm; *Tail* 30mm, 32mm; *Bill* 15.5mm, 16.5mm (*n* = 2♂♂, including type, Carriker 1933). Wing 70mm; *Tail* 27mm; *Bill* 15mm (*n* = 1♀, Carriker 1933). Wing 63.3mm, 65.7mm, 63.0mm, 64.2mm, 63.0mm; *Tail* 26.3mm, 27.6mm, 24.4mm, 27.4mm, 26.2mm; *Bill* total culmen 15.1mm, 16.3mm, 15.4mm, 16.1mm, 15.6mm; *Tarsus* 21.9mm, 23.0mm, 22.3mm, 24.4mm, 23.2mm (*n* = 5, 2♂,3♀, Graves *et al.* 1983). Wing range 70–72mm (*n* = 3, ♂,♂,♀, including type, Bond 1950). *Bill* from nares 9.3mm; *Tarsus* 24.3mm (*n* = 1♂, MSB 36160). *Bill* from nares 8.1mm, exposed culmen 11.9mm, depth at front of nares 4.9mm, width at gape 9.5mm, width at front of nares 5.1mm, width at base of mandible 6.4mm; *Tarsus* 22.0mm (*n* = 1♂, MSB 36157). *G. f. boliviana* Wing 60/61mm; *Tail* 28/27mm; *Bill* [exposed culmen/from nares] 12.5/9.5mm, width at nares ?/5mm; *Tarsus* 21/21mm (*n* = 1♂, holotype, respectively, Chapman 1919/1926). Note for the same specimen, Carriker (1933) gave *Wing* 61mm; *Tail* 25mm; *Bill* [exposed] culmen 13mm. *Wing* 64.5–67.0mm (*n* = 5, ♀?, including type, Bond 1950). **Mass** 17.0g (*n* = 1, ♀?, *costaricensis*, Karr *et al.* 1978), 18g (*n* = ?, Stiles & Skutch 1989), 14–17g (*n* = 3♂, *brevis*, Robbins *et al.* 1985), 14.5mm, 15.2mm, 15.6mm, 17.7mm, 17.1mm (*n* = 5, 2♂,3♀, Graves *et al.* 1983), mean 17.13 ± 2.16g (*n* = 9, ♀?, *similis/boliviana*, both slopes of Apurímac Valley, Weske 1972), 13.8g, 18.0g, 15.3g (*n* = 3, ♀,♀,♂, *costaricensis*, MNCR), 20.1g, 20.2g, 24.0g (*n* = 3, ♀?, *flavirostris*, eastern Ecuador, MCZ), 16.6g, 17.5g, 15.5g, 17.7g (*n* = 4♂, *similis*, MSB), 18.0g, 19.5g, 21.5g (*n* = 3, ♀,♂,♀, mindoensis, Oniki & Willis 1991). **Total Length** 9.5–11.4cm (Davis 1972, Hilty & Brown 1986, Ridgely & Gwynne 1989, Stiles & Skutch 1989, Fjeldså & Krabbe 1990, Clements & Shany 2001, Restall *et al.* 2006, Garrigues & Dean 2007, Schulenberg *et al.* 2007, Ridgely & Tudor 2009).

TAXONOMY AND VARIATION A fair amount of plumage variation, especially in the colour and pattern of the underparts, and the coloration of the bill, has led to the description of nine subspecies. There is considerable confusion, however, as to how much of this variation coincides with current subspecific limits and how much is age-related. Birds from the east slope (*flavirostris*), however, can have entirely dark bills and most birds from south-west Ecuador (*zarumae*) have all-yellow bills (Robbins & Ridgely 1990, Ridgely & Tudor 2009). Most birds in south-west Ecuador (*zarumae*) are uniform ochraceous on throat and breast with little or no streaking. Birds from central Peru and Bolivia (*similis*, *boliviana*) have much heavier scalloping below and a distinctly blackish malar stripe (Ridgely & Tudor 2009). In the past, these two subspecies, both found south of the Río Marañón, have been treated as a separate species (Carriker 1933), and Krabbe & Schulenberg (2003) suggested this arrangement might warrant reconsideration. Together, these two races are fairly distinctive, with overall darker and more heavily streaked underparts. The precise extent of their respective ranges and intra- and inter-specific plumage variation is poorly understood, however, and there may be 1–3 taxa within Peruvian and Bolivian populations (see further discussion below). On the Pacific slope, from central Colombia to southern Ecuador, there are currently three recognised subspecies (*ochraceiventris*, *mindoensis*, *zarumae*) that have been distinguished entirely on plumage. Despite apparently sizeable gaps in distribution between each of these taxa, all of these populations show considerable, broadly overlapping plumage variation, leading Krabbe & Schulenberg (2003) to suggest they may best be treated as one taxon. This suggestion is further supported by a lack of any clear geographic barriers separating them. The confusing nature of individual vs. geographic variation in plumage is well illustrated by Chapman's (1925) statement: 'I am at loss to explain the close resemblance of this Mindo bird to the form in eastern Ecuador [nominate *flavirostris*] and its marked distinctness from the forms of western Colombia (*ochraceiventris*) and south-western Ecuador (*zarumae*). Although the known ranges of the two birds last-named are apparently separated by that of *mindoensis*, they are so much like each other and so unlike *mindoensis* that it is difficult to imagine them intergrading through that form.' It should be noted, however, that a similarly convoluted statement might well be applied to comparison of any combination of the currently described

races, pointing to the need for a thorough taxonomic revision based on more than just plumage. Unfortunately, the species' vocalisations are incompletely known, making it impossible to draw any firm conclusions as to whether more than one species is involved or too many subspecies are recognised. Taking all of the available information gathered during the writing of this book, I suggest both statements may be true. It seems likely to me that three species are involved: (1) *G. costaricensis*, including *brevis*, *ochraceiventris*, *mindoensis* and *zarumae* as subspecies; (2) *G. flavirostris* (monotypic); and (3) *G. boliviana*, either monotypic or with up to three subspecies. For now, following Krabbe & Schulenberg (2003), I recognise eight subspecies. Bangs (1902) described *Grallaricula vegata* from the Caribbean slope of Volcán de Chiriquí based on an adult female (MCZ 108552) collected 12 June 1901 by W. W. Brown. Subspecies *vegata* was, quite correctly, quickly synonymised with *costaricensis* (Cory & Hellmayr 1924).

Grallaricula flavirostris costaricensis Lawrence, 1866, Annals of the Lyceum of Natural History of New York, vol. 8, p. 346, 'Costa Rica, Barranca'. The holotype (USNM 41433) is a male collected 10 April 1865 by F. Carmiol. The locality most likely refers to San Antonio de Barranca (10°10'N, 84°25'W), 3.5km south-west of Zarcero, Alajuela province, at *c*.1,600m on the west slope of Volcán Poás, north-west of San José. In his description, Lawrence (1866) noted that the 'larger wing coverts [are] dark rich or Vandyke brown, narrowly edged with dark rufous', a character that almost certainly indicates it was not in full adult plumage. Of the other specimens of *costaricensis* I examined, and that were accompanied with clear indications of adulthood (i.e., gonads, skull ossification), none showed evidence of rufous edges to the upperwing-coverts. On examining the holotype myself, I was unable to find any other plumage character clearly indicating immaturity, but confirmed the presence of subadult wing-coverts. The present taxon shows a somewhat puzzling distribution, being found on both slopes of the Costa Rican Cordillera Central, but is then restricted to the Pacific slope of the Cordillera de Talamanca, and Fila Cruces above San Vito (Stiles & Skutch 1989, Lindell *et al.* 2004, Garrigues & Dean 2007). At this point, it apparently 'reappears' on the Caribbean slope on the Panamanian side of Cerro Echandi (Galindo & Sousa 1966, USNM), with records on the Caribbean slope extending as far east as the mountains around La Verrugosa (08°47'N, 82°11'W) in northern Ngäbe-Buglé. Despite the fact that some of these records are modern, given the notoriously inaccuracy of historical collecting labels and the relatively few modern records (without specimens), I suggest that the species' presence on the Caribbean slope of Panama should be considered hypothetical. Beyond the Panamanian border on the Pacific slope, there are numerous historical and modern records to suggest its presence east at least as far as the Reserva Forestal de Fortuna (Garcés 2007) and historical records extend its range as far as Chitra and Calobre. The latter site is probably too low for Ochre-breasted Antpitta (100m), but is clearly on the Pacific slope (Siegel & Olson 2008) and references to this locality probably refer to the mountains north of there. Somewhat more troublesome are numerous references (Salvin 1870, Sclater 1890, Salvin & Godman 1904) to specimen(s?) from Calovébora (08°47'N, 81°12.5'W) on the Caribbean coast. This location is clearly in error, but it is unclear from which side of the continental divide the specimen/s came from. The collector, Enrique Arcé, ranged widely across Veraguas, on both slopes and sometimes as far west as Chiriquí, and collected prodigious numbers of vaguely labelled specimens (Salvin 1870, Siegel & Olson 2008). Regardless, the oft-repeated description of the range as extending 'east to Veraguas', also demands careful modern scrutiny. With a fair amount of variation, *costaricensis* appears to be among the darker of the northern races, with a fair number of specimens approaching the heavy streaking and degree of ochraceous saturation on the underparts of the southernmost races (*similis, boliviana*). Upperparts brownish-olive, duller on the head, with a rufous tinge on the forecrown. The underwing-coverts, lores and ocular ring are pale rufous. The ear-coverts are brownish-rufous and there is a narrow black crescent just anterior of the eye. The throat, breast and sides are dull orange-rufous, duskier on the sides and fading centrally to buffy-white on the central belly and undertail-coverts. The breast and sides have sparse, narrow black streaking. SPECIMENS & RECORDS COSTA RICA: *Guanacaste* Estación Biológica San Gerardo, 10°22'N, 84°47.5'W (HFG). *Alajuela* Sierra de Tilarán, 10°18'N, 84°45'W (Young *et al.* 1998); Cariblanco de Sarapiquí, 10°16'N, 84°11'W (Carriker 1910, Chapman 1922, AMNH 492302–305, MCZ 117062); RB Alberto Manuel Brenes, 10°13'N, 84°36'W (MNCR 24636); Bosque de Paz, 10°12.5'N, 84°19'W (F. Rowland *in litt.* 2017); Los Ángeles Norte, 10°10'N, 84°31.5'W (ML 53993501, J.G. Campos). *Heredia* La Paz Waterfall Gardens, 10°12'N, 84°09.5'W (ML 26464611, photo J. Zook). *Limón* c.15km SW of Guápiles (El Jilguero), 10°06.5'N, 83°52'W (J. Zook *in litt.* 2017); Pacuare Ecolodge, 09°59.5'N, 83°33'W (eBird: D. Quesada). *Puntarenas* Above (E of) Monteverde, 10°18'N, 84°47.5'W (F. Rowland *in litt.* 2017); Buena Vista S slope of Cerro de la Muerte (see Slud 1964), c.09°30.5'N, 83°39'W (Chapman 1922); Bosque del Tolomuco, 09°28.5'N, 83°42'W (eBird, R. Garrigues); Altamira de Biolley, c.09°02'N, 83°00.5'W (J. Zook *in litt.* 2017); Las Alturas de Cotón/Las Tablas, 08°57'N, 82°50'W (Holley *et al.* 2001, Lindell & Smith 2003); Estación Biológica Las Cruces, 08°47'N, 82°58.5'W (Karr *et al.* 1978, Daily *et al.* 2001). *San José* PN Braulio Carrillo, c.10°09.5'N, 83°58.5'W (MZUCR 2415); San Joaquín de Dota, 09°35'N, 83°59'W (Griscom 1933, MCZ, YPM). *Cartago* N slope of Volcán Turrialba, c.10°04'N, 83°47.5'W (Carriker 1910, CM P13585); Rancho Naturalista, 09°50'N, 83°34'W (J. VanderGaast *in litt.* 2017); Cachí, 09°49.5'N, 83°48'W (Cory & Hellmayr 1924); Navarro, 09°48.5'N, 83°52.5'W (Ridgway 1884, Carriker 1910, USNM 91842); El Muñeco, 09°47.5'N, 83°54.5'W (YPM 56711/12); Taus, 09°47'N, 83°43'W (MZUCR H119); RB El Copal, 09°47'N, 83°45'W (MZUCR 4187/4777); Quebrada Segunda, 09°45.5'N, 83°47'W (XC 84820, P. O'Donnell, MNCR 27164); RVS La Marta, 09°44.5'N, 83°40'W (ML 35914981, photo K. Reyes). PANAMA: *Bocas del Toro* Río Changuena, c.09°02.5'N, 82°41.5'W (Galindo & Sousa 1966, USNM 476082). *Ngäbe-Buglé* Camino de la Divisoria Continental, 08°47.5'N, 82°12.5'W (ML 41372941, photo E. Campos); RF La Fortuna, 08°44'N, 82°14.5'W (Garcés 2007). *Chiriquí* Caribbean slope of Volcán de Chiriquí, 08°48.5'N, 82°31.5'W (MCZ 108552, type of *G. vegata*); Volcán, 08°46'N, 82°38'W (USNM 456231); Cordillera de Tolé, 08°17'N, 81°40'W (Salvin 1867, Sclater 1890, Salvin & Godman 1904, Wetmore 1972). *Veraguas* Chitra, 08°32'N, 80°55'W (Salvin 1870, Chapman 1925); Calobre, 08°19'N, 80°51'W (Salvin & Godman 1904, Cory & Hellmayr 1924).

Grallaricula flavirostris brevis Nelson 1912, Smithsonian Miscellaneous Collections, no. 60, p. 12. The type specimen (USNM 238069) is a male collected on 1 May 1912 on 'Mount Pirri, 4,500ft, near head of Río Limon, eastern Panama' [Cerro Pirre, headwaters of the Río Limon, Darién, 1,370m, c.07°46'N, 77°44'W, see Fairchild & Handley 1966, Siegel & Olson 2008]. Although not mentioned in the original description, my examination of the type revealed that it is, in fact, a subadult. The upperwing-coverts are clearly margined dull rufous. The coverts of the other three USNM specimens I examined (also available to Nelson) were uniform brownish. So far as is known, *brevis* is endemic to the Serranía de Pirre around the type locality (Krabbe & Schulenberg 2003, Angehr & Dean 2010). Nelson (1912) described *brevis* as 'generally similar to typical *flavirostris* but smaller with much larger bill; upper parts with less brownish suffusion, the crown more olive greyish; back nearly plain olive and outside of wings darker and more olive brown; tawny ochraceous of underside of neck, breast and sides of body about the same, but with black edgings to feathers usually well marked but narrower and less numerous on both throat and breast than in typical *flavirostris*'. Nelson's (1912) description was based on four specimens, and he noted that three of them were 'much alike' but that the fourth lacked the 'dark edges to feathers of neck and breast as in *costaricensis*', and the ochraceous feathers of these parts were 'lighter and yellower'. There are still relatively few specimens, but it appears that *similis* is probably no less variable than other races, even with such a restricted distribution. Robbins *et al.* (1985) gave the bare-part colours, described from live birds, as: *Iris* dark; *Bill* yellow with black culmen; *Tarsi & Toes* greyish-green. SPECIMENS & RECORDS *Darién* 9km NW of Cana, c.07°48.5'N, 77°45.5'W (LSUMZ 104696–701); Cerro Pirre, 07°46'N, 77°44'W (ANSP 131902/03); Altos de Quía, 07°35'N, 77°31'W (Ridgely & Gwynne 1989).

Grallaricula flavirostris ochraceiventris Chapman 1922, American Museum Novitates, no. 31, p. 6, Cocal, 4,000ft, Western Andes, Colombia [Cocal, Dept. Cauca, 02°31'N, 77°09'W, 1,220m]. The adult male holotype (AMNH 109636) was collected 13 June 1911 by W. B. Richardson. This Colombian endemic is found in the western foothills of the West Andes, from south-west Antioquia to central Cauca. Two specimens reportedly from Nóvita (c.70m, 04°57.5'N, 76°36.5'W; AMNH 492300/301) were considered erroneous by Cory & Hellmayr (1924). Although the precise locality is undoubtedly incorrect, the outlying mountains less than 15km east of there would appear to provide appropriate habitat and I have provisionally extended the range of *ochraceiventris* west from the main Andean chain to include these mountains. Two apparently isolated populations, one on the east slope of the West Andes in central Antioquia, and the more recently discovered (Donegan *et al.* 2007, 2008) population in the Serranía de los Yariguíes, on the west slope of the East Andes in Santander, are provisionally assigned to *ochraceiventris*. The true taxonomic affinities of these populations, especially the latter, are certainly in need of investigation. Chapman (1922) described *ochraceiventris* as most similar to *costaricensis* but with a longer wing and stouter, longer bill. He derived the Latin name from the more extensive amount of ochraceous on the underparts than most races (but see *zarumae*). From nominate *flavirostris* he separated *ochraceiventris* based on the more olive-washed upperparts, more

intense and extensive ochraceous plumage below, and lighter and less extensive streaking. He commented on its similarities to *zarumae*, feeling that the two should be considered distinct, based largely on the presence of the heavily streaked, white-bellied birds (*mindoensis*) found in the area separating *ochraceiventris* from *zarumae*.
SPECIMENS & RECORDS *Santander* El Talisman, 06°51'N, 73°22'W (Donegan *et al.* 2007); RNA Reinita Cielo Azul, 06°50'N, 73°22.5'W (Salaman *et al.* 2007a); La Luchata, 06°38'N, 73°19'W (Donegan *et al.* 2007); Alto Honduras, 06°37'N, 73°30'W (Donegan *et al.* 2007); Suaita, 06°06 N, 73°26.5 W (Donegan *et al.* 2007); Vereda El Taladro, 05°59'N, 73°13'W (eBird: J. Castaño). *Antioquia* Alto La Serrana, 07°06'N,75°08'W (Cuervo *et al.* 2008); Yarumal El Rosario, 07°04'N, 75°24'W (eBird: E. Munera); RNA Arrierito Antioqueño, 06°59.5'N, 75°06.5'W (Salaman *et al.* 2007a); RFP La Forzosa, 06°59'N, 75°08'W (Delgado V. 2002, Cuervo *et al.* 2008); Bosque de El Abuelo, 06°56'N, 75°00'W (Cuervo *et al.* 2008); La Secreta, 06°49'N,75°06'W (Cuervo *et al.* 2008); *c.*2.2km N of La Siberia, 06°32'N, 76°14.5'W (eBird: E.H. Guarín); RNA Loro Orejiamarillo, 05°32.5'N, 75°48'W (F. Schmitt *in litt.* 2017). *Chocó* RNA Las Tangaras, 05°49.5'N, 76°13'W (XC 108940, O. Cortes-Herrera); Cerro Tatamá, 05°01.5'N, 76°06.5'W (Hilty & Brown 1986); La Selva, 04°55'N, 76°09'W (Meyer de Schauensee 1950, ANSP 157981). *Risaralda* Cerro Montezuma road, 05°14.5'N, 76°06'W (XC 102517, J. Minns). *Valle del Cauca* Calima, 04°01'N 76°30'W (eBird: J. Zamudio); El Danubio, 03°36.5'N 76°52.5'W (eBird: J. Luna Solarte); Alto Anchicayá, 03°32'N, 76°45'W, (ML 56118801, photo J.M. García); San Antonio, 03°30'N, 76°38'W (Chapman 1917, 1925, Meyer de Schauensee 1950). *Cauca* Río Mechengue, 02°40'N, 77°12'W (Meyer de Schauensee 1950, ANSP); Cerro Munchique, 02°32'N, 76°57'W (Meyer de Schauensee 1950, ANSP 144678); RNSC Tambito, *c.*02°30'N, 77°00'W (Donegan & Dávalos 1999, Casas & López-Ordóñez 2006a).

Grallaricula flavirostris mindoensis Chapman 1925, American Museum Novitates, no. 205, p. 5, near Mindo, Ecuador [*c.*00°03'S, 78°46.5'W, 1,675m]. The holotype is an adult male (AMNH 173037) collected 13 October 1915. Ochre-breasted Antpitta is fairly common in the foothills surrounding the type locality (Kirwan & Marlow 1996) and a fair number of topotypical skins are available (Lönnberg & Rendahl 1922; USNM, AMNH, ANSP) to help with much-needed taxonomic revisions (hint, hint). Race *mindoensis* is not, as suggested by Krabbe & Schulenberg's (2003) wording, endemic to north-west Ecuador, as records extend north across the Colombian border into southern Nariño. The apparently isolated population in the Cordillera Mache-Chindul (at *c.*500m; Carrasco *et al.* 2008) is provisionally included here, although its taxonomic affinities have not been investigated. Chapman (1925a) described *mindoensis* as similar to nominate *flavirostris* but with more olivaceous upperparts and the markings of the loral, orbital and ear-covert regions much more orange-ochraceous, with this coloration reaching onto the chin and throat (less pure white than *flavirostris*). He differentiated *mindoensis* from *zarumae* by its cleaner white belly and lower breast (not ochraceous or buffy) and by its blackish-brown maxilla (not yellowish as in *zarumae*). The latter character, however, is unlikely to prove informative under close scrutiny, largely due to age-related variation in the character (see Robbins & Ridgely 1990). **SPECIMENS & RECORDS COLOMBIA**: *Nariño*

RNSC Río Ñambí, 01°18'N, 78°05'W (XC 241274, O.H. Marín-Gómez); Ricaurte, 01°13'N, 77°59'W (Calderon-Leyton *et al.* 2011, LACM, ANSP); RNSC La Planada, 01°09'N, 77°59.5'W (Gartner *et al.* 1982, Oniki & Willis 1991, Willis & Schuchmann 1993, Restrepo & Gómez 1998). **ECUADOR**: *Carchi* 4km SSW of Chical, 00°54'N, 78°12'W (Robbins & Ridgely 1990, ANSP, MECN, ML); Chilma, 00°52'N, 78°04'W (XC 276050, J. Nilsson); *c.*11.5km N of Guadal, 00°50'N, 78°13.5'W (R.A. Gelis *in litt.* 2017); El Corazón, *c.*00°50'N, 78°07'W (J. Freile *in litt.* 2016). *Esmeraldas* El Cristal Alto, 00°50'N, 78°32'W (Jahn & Valenzuela 2006, XC 262142–144, O. Jahn); RBP Bilsa, 00°21.5'N, 79°42'W (Carrasco *et al.* 2008). *Imbabura* BP Los Cedros, 00°19'N, 78°47'W (HFG). *Pichincha* Palmitopamba, 00°10'N, 78°40'W (J. Freile *in litt.* 2017); Mashpi, 00°09'N, 78°50'W (D. Becker *in litt.* 2015); RP Santa Lucía, 00°07'N, 78°36.5'W (HFG); Gualea, 00°05'N, 78°44.5'W (ANSP 169687/88); Nanegalito, 00°04'N, 78°41'W (MLZ 7206); RP San Jorge de Milpe, 00°03.5'N, 78°54.5'W (G. Cruz photo); RP Intillacta, 00°03'N, 78°42'W (Greeney *et al.* 2012); Milpe, 00°02'N, 78°52'W (MECN 2631, MLZ 7614); RP Refugio Paz de las Aves, 00°00.5'N, 78°42.5'W (HFG); Tandayapa Bird Lodge, 00°00'N, 78°40.5'W (Solano-Ugalde *et al.* 2007); RP Las Tangaras, 00°05'S, 78°46'W (D. Becker *in litt.* 2017). *Santo Domingo de los Tsáchilas* 'Las Palmeras', near Chiriboga, 00°13.5'S, 78°46.5'W (WFVZ 47088); Estación Científica Guajalito, 00°14'S, 78°48'W (J. Freile *in litt.* 2017); Tinalandia, 00°18.5'S, 79°03'W (HFG). *Los Ríos* Puebloviejo, 01°33'S, 79°32'W (MLZ 7307).

Grallaricula flavirostris zarumae Chapman 1922, American Museum Novitates, no. 31, p. 7, near Zaruma, El Oro proince, 6,000ft [03°41'S, 79°36.5'W, 1,825m]. The holotype is an adult male (AMNH 129758) collected 5 October 1913 by W. B. Richardson. So far as is known, this race is confined to more humid areas of south-west Ecuador, being absent from drier regions, and reports from the West Andes in Piura (Clements & Shany 2001) are apparently in error. Once its distribution is more completely known, the complex geographic mosaic of moisture gradients in this region will likely result in a true distribution polygon that is shaped something akin to an octopus and a squid in a fist fight! Chapman (1922) felt that *zarumae* was most similar to *ochraceiventris* but differed in having all of the ochraceous areas yellower, less orange-ochraceous, making the ochraceous markings of the forehead, lores and ocular region less pronounced but more strongly contrasting with the black crescent in front of the eye. He described the ear-coverts as more olivaceous than *ochraceiventris* and the maxilla yellow or olive-yellow, rather than dark brown. He separated *zarumae* from nominate *flavirostris* by its pale olive-brown back (rather than clear brown), unstreaked breast, ochraceous (not white) abdomen, and by its even yellower maxilla. **SPECIMENS & RECORDS** *Azuay* San Luís, 02°53.5'S, 79°34'W (ANSP 183350, MECN 5648). *El Oro* 'La Chonta' [= Satayán], 03°35'S, 79°53'W (Chapman 1922 as 'Santa Rosa-Zaruma Trail'; AMNH 171409); El Chiral, 03°38'S, 79°41'W (AMNH 167279–283); RP Buenaventura, 03°38.5'S, 79°45.5'W (Greeney *et al.* 2012, ML 124477, L.R Macaulay); 9.5km W of Pinas, *c.*03°41'S, 79°36'W (MECN 5649/50, ANSP 177604/05). *Loja* Salvias, 03°47'S, 79°21'W (AMNH 167284). **NOTES** A specimen labelled 'El Porotillo S. José Loja' (MLZ 10905) appears to me to represent nominate *flavirostris*. Paynter (1993) was unable to locate this Olalla collecting locality, but suggested it was somewhere south of Loja, placing it at one of several locations on the west slope of the East Andes. Of the suggested locations, only two appear to possess suitable habitat (04°05'S, 79°12'W; 04°14'S, 79°15'W). If either of these locations are the true origin of the specimen, it would either be the only record of nominate *flavirostris* on the west slope or a 40–50km south-east range extension for *zarumae*. Probably the location is incorrect or the specimen is mislabelled.

Grallaricula flavirostris flavirostris (Sclater 1858). The nominate race is the most broadly distributed, found along the Amazonian slope from central Colombia perhaps to northern Peru, but with numerous gaps in confirmed records. Its known range extends from eastern Colombia to south-east Ecuador. It is often reported to 'probably' occur in northern Peru (Krabbe & Schulenberg 2003), but despite a fair amount of collecting there, its presence there has not been proven (Schulenberg *et al.* 2010). SPECIMENS & RECORDS COLOMBIA: *Boyacá* Campamento Las Moyas, 04°53.5'N 73°17.5'W (eBird: A. Pinto). *Meta* Bosque Bavaria, 04°11'N, 73°39'W (B.G. Freeman *in litt.* 2017); Caño Gramalote/Villavicencio area, *c.*04°10'N, 73°37'W (Niceforo 1945, Friedmann 1947, Meyer de Schauensee 1950); RNSC Las Palmeras, 03°50.5'N, 73°54.5'W (J. Beckers *in litt.* 2017). *Caquetá* Cerro La Mica, 02°44'N, 74°51'W (Salaman *et al.* 1999, 2002, Borhoquez 2002). *Cauca* Villa Iguana, 01°14'N, 76°31'W (Salaman & Donegan 1998, Donegan & Salaman 1999, Salaman *et al.* 1999, 2007). *Nariño* El Carmen, 00°40'N, 77°10'W (FMNH 399877/78). *Putumayo* Vereda San Martín, 01°14'N, 76°43'W (eBird: E. Rosero Chates); San Miguel, 00°20.5'N, 76°54.5'W (Meyer de Schauensee 1952a,b, ANSP 165167). **ECUADOR**: *Orellana* RBP Río Bigal, 00°38'S, 77°19'W (J. Freile *in litt.* 2017); Mushullacta, 00°50.5'S, 77°33.5'W (HFG). *Napo* N slope of Volcán Sumaco, 00°31'S, 77°38'W (Greeney 2012b, Greeney *et al.* 2012, IBC); RP WildSumaco, 00°40.5'S, 77°36'W (XC, ML, IBC). *Tunugurahua* Machay, 01°23'S, 78°17'W (Benitez *et al.* 2001); La Guatemala, 01°24'S, 78°09.5'W (ML 20146131, I. Davies). *Pastaza* RP Río Zuñac, 01°22.5'S, 78°08.5'W (L. Jost *in litt.* 2017); Sarayacu (see discussion above). *Morona-Santiago* Nueva Alianza, 02°05'S, 78°09'W (J. Freile *in litt.* 2017); Kichikentza, 02°11'S, 77°51'W (J. Freile *in litt.* 2017); 'Colimba', *c.*02°17.5'S, 78°13.5'W (AMNH 408358/59); San Luís de Ininkis, 02°24'S, 78°05'W (J. Freile *in litt.* 2017); San Carlos de Limón, 03°12.5'S, 78°26'W (MECN 8091). *Zamora-Chinchipe* RP El Zarza, 03°50.5'S 78°35'W (eBird: M. Chelemer); Cordillera del Condor, *c.*03°52'S, 78°31'W (MECN 8092); Río Bombuscaro, 04°08'S, 78°58'W (Rasmussen *et al.* 1996); Miazi Alto, 04°15'S, 78°37'W (Freile *et al.* 2013b); Cerro Panguri, 04°36'S, 78°58'W (Rasmussen *et al.* 1996, ANSP, MECN); near Panguri, 04°39'S, 78°55'W (XC 250905, N. Krabbe).

Grallaricula flavirostris similis Carriker 1933, Proceedings of the Academy of Natural Sciences of Philadelphia 85, p. 21, Río Jelashte, Dept. San Martin, Peru, 5,000ft [06°48'S, 77°12'W, 1,525m]. Carriker (1933) described *similis* as a race of Chapman's (1919) *Grallaricula boliviana*, based on an adult male (ANSP 108128) collected 20 August 1932 by its describer. Race *similis* is endemic to Peru, and its range is generally stated to extend from northern Peru, south and east of the Río Marañón to Pasco (Krabbe & Schulenberg 2003). With more recent records of Ochre-breasted Antpitta that bridge the gap

between *boliviana* in Cuzco and *similis* in Pasco, but the distributions of these two taxa are uncertain. On the range map I have shown the population north of the Río Apurímac/Río Ucayali valley (from eastern Ayacucho to southern Pasco) as a separate polygon and the Cordillera El Sira population in Huanúco and Ucayali as a fourth polygon. Based on the specimens and other records available, I was unable to satisfactorily assign birds from these regions to either *similis* or *boliviana*. I provisionally include both populations in *boliviana* (which see), but this decision would certainly be challenged by someone with access to further data. Race *similis* is most similar in plumage to *boliviana*. Carriker (1933) considered *similis* to have a significantly paler ochraceous throat and chest, with heavier black streaks, paler and narrower malar streaks, and darker ear-coverts. He further distinguished *similis* by its larger size, conspicuous cinnamon-ochraceous eye-ring (absent in *boliviana*), and by the feathers of the crown and nape having the fringes washed dark olive. Nevertheless, these differences are, in my opinion, rather slight, and I found separation of specimens of these two taxa most challenging, suggesting they are synonymous. SPECIMENS & RECORDS *Amazonas* Alto Wawas, 05°19'S, 78°20'W (Dauphiné 2008); La Peca Nueva, 05°35'S, 78°22'W (Graves *et al.* 1983); Quebrada Huacabamba, 06°35.5'S, 77°33'W (FMNH 474181–184). *San Martín* 15km by road NE of Abra Patricia, 05°40'S, 77°45'W (LSUMZ 81993/94); Afluente, 05°44'S, 77°31'W (Parker & Parker 1982, LSUMZ 84963–965); Río Jelache, 06°48'S, 77°12'W (Bond 1950, ANSP 108128–130); Nuevo Bolívar, 07°19.5'S 77°27.5'W (eBird; P. Saboya del Castillo); 2.7km S of Plataforma, 07°25'S, 76°17.5'W (MSB). *Ucayali* 3km by road NE of Abra Divisoria, 09°03'S, 75°43'W (LSUMZ); Divisoria en Cordillera Azul, 09°05'S, 75°46'W (LSUMZ 62307): *Huánuco* Sariapampa, 09°43'S, 75°54'W (FMNH 296692/93); Cushi, 09°51'S, 75°37'W (LSUMZ 128564/65); Playa Pampa, 09°57'S, 75°43'W (LSUMZ 128554–563). **NOTES** The location '*c.*24km ENE of Florida' (LSUMZ 174013–016) is less than 1km from the LSUMZ locality NE of Abra Patricia given above.

Grallaricula flavirostris boliviana Chapman, 1919, Proceedings of the Biological Society of Washington 32, p. 257, Incachaca, 7,700ft, Prov. Cochabamba, Bolivia [17°14'S, 65°48'W, 2,350m]. The type specimen (AMNH 137177) is an adult male collected 16 May 1915 by 'Miller & Boyle'. As Incachaca was a regular collecting locality, there is a fair amount of topotypical material held at ANSP, CM and UMMZ. At the time of its discovery, no *Grallaricula* was known from Peru, and the description of *boliviana* extended the range of the genus by nearly 1,600km south (Chapman 1919). Race *boliviana* was long considered a Bolivian endemic, although more recent treatments usually extend its range north to Puno. As mentioned above (under *similis*), recent records of Ochre-breasted Antpitta north of Río Apurímac are of uncertain taxonomic affinity. Pending further data I have included here, as *boliviana*, records from north-east Cuzco (AMNH, USNM, Pequeño *et al.* 2001a, 2001b) and north-east along the Cordillera de Vilcabamba as far east as the headwaters of the Río Poyeni in eastern Junín (Schulenberg *et al.* 2001a). I also include records in the Cordillera El Sira that extend from the northern end (east of Puerto Inca, Huánuco; 09°22'S, 74°57'W; Terborgh & Weske 1975, Mee *et al.* 2002, Socolar *et al.* 2013) to the range's southern extreme (east of Oventeni, Uyacali). I follow the tradition of considering birds as far north as

Pasco in *boliviana* but have separated records north of the Rio Apurímac to emphasise their uncertain taxonomic affinity (see also discussion under *similis*). Other than *similis*, this race is probably most similar to *costaricensis*, but Chapman (1919) considered it unique enough that he suggested it was 'not closely related to any known species'. The throat is ochraceous streaked black, the upper breast white and lower breast ochraceous. The breast feathers are heavily fringed black, giving a scaled appearance, likened by Chapman (1919) to the breast markings of a *Premnoplex*. Chapman (1919) described the holotype as follows: 'upper parts between brownish olive and light brownish olive, the crown with darker centers, sides of the forehead basally ochraceous; lores and a narrow eye-ring ochraceous, minutely tipped with black; tail fuscous, the outer feathers externally margined with olive-brown; wings fuscous, externally margined with olive-brown, the inner margins of all but the outer quills ochraceous-buff; under wing-coverts, bend of the wing and outer margin of alula deeper; ear-coverts tinged with cinnamon-brown; throat ochraceous, loosely streaked with black; a narrow white breast-patch; a broad ochraceous band across the lower breast, the feathers sharply bordered with black; flanks more olivaceous, with some buffy feathers obscurely margined with blackish; center of the abdomen white.' Records from the Cordillera El Sira are indicated [†] and those from the Ayacucho-Pasco region marked [‡]. **Specimens & Records** *Huánuco* N end of Cerros del Sira[†], 09°25'S, 74°44'W (Socolar *et al.* 2013, AMNH 820965/1021). *Pasco* Santa Cruz[‡], 10°37'S, 75°20'W (Schulenberg *et al.* 1984, LSUMZ 106087). *Ucayali* Two sites at S end of Cerros del Sira: upper Santeni Valley[†], 10°42'S, 74°09.5'W, and upper Shaani Valley[†] 10°42'S, 74°07'W (Harvey *et al.* 2011). *Junín* Río Poyeni headwaters, 11°33.5'S, 73°38.5'W (Schulenberg *et al.* 2001a; ML); Yurinaqui Alto[‡], 10°46'S, 75°05'W (FMNH 296694). *Ayacucho* Tutumbaro[‡], 12°43.5'S, 73°57'W (Williams 2010, KUNHM); Huanhuachayo[‡], 12°44'S, 73°47'W (LSUMZ, AMNH). *Cuzco* Katarompanaki, 12°11.5'S, 72°28'W (Lane & Pequeño 2004); Tinkanari, 12°15.5'S, 72°05.5'W (Lane & Pequeño 2004); Cordillera Vilcabamba, 12°38'S, 73°39'W (AMNH, USNM); 9km E of Luisiana, 12°39.5'S, 73°39'W (Weske 1972, AMNH 819995); Wayrapata, 12°50'S, 73°29.5'W (Pequeno *et al.* 2001); Llactahuaman, 12°52'S, 73°31'W (Pequeño *et al.* 2001b); Hacienda Huyro, 12°58'S, 72°36'W (LSUMZ 78543); Bosque Aputinye, 13°00'S, 72°32'W (LSUMZ); Pensión Suecia, 13°06'S, 71°34'W (FMNH 398035–039); El Rocotal, 13°07'S, 71°34'W (MUSA 1681, XC). *Puno* c.7km NNW of Iparo, 14°02.5'S, 69°12'W (ML 66308451, D.F. Lane); Abra de Maruncunca, 14°18'S, 69°14'W (long series LSUMZ, MUSA, XC). **BOLIVIA:** *La Paz* Pata-Virgen, 14°36'S, 68°42'W (ML 120995, A.B. Hennessey); Tokoaque, 14°37'S, 68°57'W (Hennessey & Gomez 2003); Inciensal Sauce, 14°39'S, 68°37'W (ML 120961/73/75, A.B. Hennessey); Río Palcabamba, 14°49.5'S, 68°56.5'W (Maillard-Z. & Vogel 2003, AMNH); Sandillani, 16°12'S, 67°54'W (Bond & Meyer de Schauensee 1942, ANSP 120359–361); Estación Biológica Tunquini, 16°12'S, 67°52'W (XC 4738, S.K. Herzog); c.3km NNE of Apa Apa, 16°21'S, 67°30'W (XC 38976–978, F. Lambert). *Cochabamba* Pampa Grande, 16°41'S, 66°29'W (Macleod *et al.* 2005); San José, 17°06'S, 65°47'W (CM P85281); 'Alto Palmar', c.17°06'S, 65°29'W (LSUMZ 51271); old Cochabamba road, 17°10'S, 65°35'W (XC 3424/32, S.K. Herzog).

STATUS Although Ochre-breasted Antpitta has a fairly large range and can be locally fairly common (Stotz *et al.* 1996, Krabbe & Schulenberg 2003) it is scarce and poorly known over most of its distribution. As the largest portion of its range is on the Amazonian slope of the Andes and its primary habitat is largely mid-elevation and foothill forests, but it is under severe, ongoing threat from intense logging and clearing for agricultural development, and will possibly lose close to 20% of suitable habitat within the next three generations (Soares-Filho *et al.* 2006, Bird *et al.* 2012). Although it may be tolerant of edge habitat (O'Dea & Whittaker 2007), it probably requires extensive primary forest nearby, and is thus susceptible to predicted fragmentation of its habitat and may lose up to 30% of its population over three generations (BirdLife International 2017). Because of these predictions, Ochre-breasted Antpitta was upgraded to Near Threatened in 2012 (BirdLife International 2012, 2017). Across its broad range, most currently recognised subspecies occur in at least one protected area, but not as many as might be predicted from range size alone. It is worth noting that, should multiple species taxa be involved, each will deserve careful re-evaluation of its threat level based on revised genetic and distributional limits, especially those populations with potentially very small ranges such as *brevis* and *zarumae* (Freile *et al.* 2010). **Protected Populations** *G. f. costaricensis* **Costa Rica**: PN Tapantí (Sánchez 2002, MNCR, MZUCR, XC, ML); PN Braulio Carillo (Blake & Loiselle 2000, MZUCR); PN Los Quetzales (Griscom 1933, MCZ, YPM); PN La Amistad (Holley *et al.* 2001, Lindell & Smith 2003); PN Volcán Turrialba (Carriker 1910, CM P13585); RVS La Marta (ML 35914981, photo K. Reyes); RB Alberto Manuel Brenes (MNCR 24636); RB El Copal (MZUCR 4187/4777); RB Las Quebradas (eBird, N. Ureña); Monteverde Cloud Forest Reserve (eBird). Should be looked for in PN Volcán Arenal, PN Juan Castro Blanco, PN Volcán Poás and PN Chirripó. **Panama**: RF La Fortuna (Garcés 2007). *G. f. brevis* Almost all specimens and records of this race are from PN Darién. *G. f. ochraceiventris* PNN Tatamá (XC, ML); PNN Las Orquídeas (eBird; E.H. Guarín); PNN Serranía de los Yariguíes (Donegan *et al.* 2007); RNSC Tambito (Donegan & Dávalos 1999, Casas & López-Ordóñez 2006a); RNA Las Tangaras (XC, ML); RNA Arrierito Antioqueño and RNA Mirabilis–Swarovski (Salaman *et al.* 2007a); PNN Munchique (Negret 1994); RNA Loro Orejiamarillo (F. Schmitt *in litt.* 2017); RFP La Forzosa (Delgado V. 2002, Cuervo *et al.* 2008); RNA Reinita Cielo Azul (Salaman *et al.* 2007a); SFF Iguaque (eBird: J. Castaño): *G. f. flavirostris* **Colombia**: PNN Cordillera de Los Picachos (Salaman *et al.* 1999, 2002, Borhoquez 2002); RNA Halcón Colorado (Salaman *et al.* 2009a); RNSC Las Palmeras (J. Beckers *in litt.* 2017). **Ecuador**: PN Sumaco-Galeras (Greeney *et al.* 2012); PN Llanganates (Benitez *et al.* 2001); PN Podocarpus (Rasmussen *et al.* 1994); RP Río Zuñac (L. Jost *in litt.* 2017); RP WildSumaco (XC, ML, IBC); RP El Zarza (eBird: M. Chelemer). *G. f. mindoensis* **Colombia**: RNA El Pangán (Salaman *et al.* 2007a); RNSC Río Ñambí (XC 241274, O.H. Marín-Gómez); RNSC La Planada, (Oniki & Willis 1991, Willis & Schuchmann 1993). **Ecuador**: RE Cotocachi-Cayapas (Jahn & Valenzuela 2006); RE Mache-Chindul (Carrasco *et al.* 2008); RBP Maquipucuna (Mordecai *et al.* 2009). *G. f. zarumae* RP Buenaventura (Greeney *et al.* 2012). *G. f. similis* PN Cordillera Azul (LSUMZ); BP Alto Mayo (LSUMZ); ACP Abra Patricia-Alto Nieva (eBird); occurs very close to, and likely in, SN Cordillera de Colán (Graves *et al.* 1983, Dauphiné 2008); should be searched for in PN

Río Abiseo. *G. f. boliviana* **Peru**: PN Otishi (Schulenberg *et al.* 2001a); PN Manú (Walker *et al.* 2006, Jankowski 2010); SN Megantoni (Lane & Pequeño 2004). **Bolivia**: PN Madidi (Hennessey & Gomez 2003, ML); PN Carrasco (Hennessey *et al.* 2003, XC); PN y ÁNMIN Cotapata (Hennessey *et al.* 2003, XC); PN y ÁNMIN Apolobamba (Maillard-Z. & Vogel 2003, AMNH).

OTHER NAMES Because the northernmost and southernmost races were sometimes considered full species, a few synonmys were used in the older literature. *G. f. costaricensis* *Grallaricula costaricensis costaricensis* (Chapman 1925); *Grallaricula costaricensis* (von Frantzius 1869, Salvin 1867, 1870, Zeledón 1885, Ridgway 1884, 1911, Bangs 1902, Sclater 1868, Davis 1972); *Grallaricula vegata* (Bangs 1902). *G. f. ochraceiventris* *Grallaricula costaricensis ochraceiventris* (Chapman 1925); *Grallaricula costaricensis* (Chapman 1917). *G. f. mindoensis* *Grallaricula costaricensis* (Lönnberg & Rendahl 1922). *G. f. similis* *Grallaricula boliviana similis* (Carriker 1933). *G. f. boliviana* *Grallaricula boliviana boliviana* (Carriker 1933); *Grallaricula boliviana* (Chapman 1922, 1925). **English** Ochraceous Pygmy Antpitta/Ant-pitta (Slud 1964, Wetmore 1972); Ochraceous Antpitta/Ant-pitta (Walters 1980, Lodge 1991); Darien Grallaricula (*brevis*, Nelson 1912); Ochreous-breasted Grallaricula [*ochraceiventris*] (Cory & Hellmayr 1924); Mount Pirri Grallaricula [*brevis*] (Cory & Hellmayr 1924); Yellow-billed Grallaricula [*flavirostris*] (Cory & Hellmayr 1924, Meyer de Schauensee 1950); Costa Rica/Rican Grallaricula [*costaricensis*] (Ridgway 1911, Cory & Hellmayr 1924); Ochraceous-breasted Grallaricula [*ochraceiventris*] (Meyer de Schauensee 1950); Costa Rican Antpitta (Davis 1972); Bolivian Grallaricula [*boliviana*] (Cory & Hellmayr 1924); Zaruma Grallaricula [*zarumae*] (Cory & Hellmayr 1924). **Spanish** Ponchito ocráceo (Krabbe & Schulenberg 2003); Gralaria ocrácea (Valarezo-Delgado 1984); Gralarita ocrácea (Ortiz-Crespo *et al.* 1990, Ridgely & Greenfield 2001, Granizo 2009); Ponchito Ocroso (Wetmore 1972); Gralarita Pechiocrácea (*mindoensis*, Carrasco *et al.* 2008), Tororoí Pechiocráceo (Clements & Shany 2001), Tororoí ocráceo (*costaricensis*, Garcés 2007) Tororoí Piquigualdo (Valarezo-Delgado 1984, Stiles & Skutch 1989, van Perlo 2006, Salaman *et al.* 2007a); Hormiguero Pecho Ocráceo (Maillard-Z. & Vogel 2003); Güicundo (Valarezo-Delgado 1984); Güicundo chico (MLZ 10905 label; C. Olalla). **French** Grallaire ocrée. **German** Ockerbrust-Ameisenpitta (Krabbe & Schulenberg 2003).

Ochre-breasted Antpitta, adult (*ochraceiventris*), RNA Las Tangaras, Chocó, Colombia, 19 July 2016 (*Fabrice Schmitt/Wings Birding Tours Worldwide*).

Ochre-breasted Antpitta, adult (*mindoensis*), Refugio Paz de las Aves, Pichincha, Ecuador, 4 July 2014 (*Harold F. Greeney*).

Ochre-breasted Antpitta, nestlings in the process of hatching (*flavirostris*), Volcán Sumaco, Napo, Ecuador, 10 December 2002 (*Harold F. Greeney*).

Ochre-breasted Antpitta, adult (*costaricensis*), Orosi, Cartago, Costa Rica, 16 August 2016 (*Jorge Obando Gutierrez*).

Ochre-breasted Antpitta, adult (*flavirostris*), WildSumaco Lodge, Napo, Ecuador, 21 March 2014 (*Christian Hagenlocher*).

Ochre-breasted Antpitta, adult (*boliviana*), Apa Apa, La Paz, Bolivia, 12 March 2005 (*Joseph Tobias*).

SCALLOP-BREASTED ANTPITTA
Grallaricula loricata   Plate 22

Grallaria loricata P. L. Sclater, 1857, Proceedings of the Zoological Society of London, vol. 25, p. 129, 'Venezuela, in *vicin. urbis* Caraccas'. In the original description, Sclater mentioned only a single specimen in the Paris museum, with no more precise data than near Caracas. Subsequently, however, he appeared to designate the two specimens now in NHMUK as types, one from San Esteban and one labeled only 'Venezuela'. Warren & Harrison (1971) noted this discrepancy, but apparently the issue has not been resolved. Similarly unclear is the precise collecting location of the Paris specimen. There are numerous references in the literature to Silla de Caracas as the easternmost limit of this species' distribution (Hellmayr & von Seilern 1912a, Phelps & Phelps 1963), but I can find no direct evidence of Scallop-breasted Antpitta having been collected there. References to its occurrence there appear to have begun between Ménégaux & Hellmayr (1906) and Hellmayr & von Seilern (1912a). Nevertheless, it has appeared to have become the *de facto* eastern limit of the species and my guess is that this is due to it being the presumed type locality. The Silla de Caracas is a highly visible saddle between two peaks looming over the city of Caracas. The low point of the saddle (10°32'N, 66°51'W) is at the border between Miranda and Distrito Federal, between the city and the coast, and has attracted the attention of naturalists since the late 1700s (e.g., Alejandro von Humboldt, Aimé Bonpland, Andrés Bello) making it a likely location for Leveraud on his 1856 expedition and may well be where the type was collected.

This strikingly patterned, diminutive antpitta is clearly a *Grallaricula*, the genus to which it was transferred soon after its description (Sclater 1858c). Although not specified by Sclater (1857a), it appears that Scallop-breasted Antpitta was named for the strong scalloping on the breast, giving it the appearance of wearing a breastplate of chainmail armour (Jobling 2010). Endemic to the north coastal mountains of Venezuela, this attractive and boldly patterned antpitta is infrequently seen and is poorly studied. It inhabits the understorey of humid montane forest, usually above 1,400m, but occurs locally down to 800m. There is some geographical overlap with Rusty-breasted Antpitta and Slate-crowned Antpitta, but these species generally occur at lower and higher elevations, respectively. Adult males have a largely orange-rufous head, becoming paler over the face and throat, brown upperparts and very pale yellow underparts, with a black malar stripe and heavy black scalloping over the breast and flanks. The breeding biology of Scallop-breasted Antpitta remains poorly studied, as indeed are most aspects of its behaviour and ecology. BirdLife International (2017) recommended additional ecological studies to determine the species' true habitat requirements, as the extent of its tolerance of secondary habitats and fragmentation is still unknown. I further suggest that the species should be searched for outside of its presently known range, especially in the coastal mountains of Vargas and northern Miranda, and in PNN Cueva del Guácharo and environs (see Distribution).

IDENTIFICATION 10–11cm. Adults have an olive forecrown, laterally edged black, a rufous crown and a buffy eye-ring. The upperparts are olive-brown except the wings and tail, which are less olivaceous. The throat is pale buff with a strong black malar. Below they are maize-yellow, with the breast and flank heavily scalloped black. Within its restricted geographic range, Scallop-breasted Antpitta is unique and unlikely to be confused with any other species. Peruvian Antpitta is very similar (but there is no range overlap), with no clear, consistent plumage differences in males, apart from the generally brighter rufous-orange crown of Scallop-breasted. Female Peruvian Antpitta has a distinctly orange-rufous crown (as opposed to the brownish crown of female Scallop-breasted) and buffy lores. Ochre-fronted Antpitta (also no range overlap) lacks rufous on the crown and has a paler, buffy face. Both sexes of Ochre-fronted Antpitta are distinctly washed buff on the breast and flanks, similar to Scallop-breasted Antpitta.

DISTRIBUTION Endemic to the Venezuelan Montane centre of endemism (Cracraft 1985) in the coastal mountains of northern Venezuela, patchily from north-east Falcón to Distrito Federal, but its range may be incompletely known. Its presence east of Paso de Choroní has been documented only by unvouchered records, as far east as PN El Ávila, in the mountains west of Boquerón (*c.*10°33'N, 66°52'W) in the Distrito Federal. There are four specimens at AMNH with the locality given as Caripé

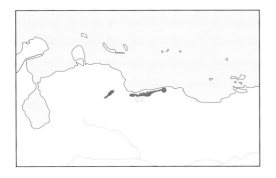

(Monagas, 800 m, 10°12'N, 63°29'W). These specimens have been, along with others from the Albert Mocquerys collection, the subject of much debate over their locality (Hellmayr 1912, Zimmer & Phelps 1954, Phelps & Phelps 1963), which has been accepted for some specimens, but rejected outright for others (Zimmer & Phelps 1954). Although the species' presence in this area has never been confirmed, I do not reject it entirely. It is worth noting that a similar species, Peruvian Antpitta, managed to go undetected for many years at a location subject to fairly intense birdwatching activity. To date, then, the easternmost extent of Scallop-breasted Antpitta's range appears to be Silla de Caracas (Phelps & Phelps 1963; see above). Its presence in the mountains east of there (at least as far as the eastern border between Vargas and Miranda), seems highly probable but is undocumented. The continued existence of the outlying population in the Sierra de San Luís, documented only by two specimens from Cerro La Danta, has not been confirmed despite fairly intensive surveys by observers familiar with the species (J.E. Miranda T. *in litt.* 2017).

HABITAT Found in humid, montane, floristically diverse cloud forests, frequenting the dense or semi-open understorey (Verea 2004). Infrequently, the species can be found in nearby second growth (Krabbe & Schulenberg 2003) and apparently shows some affinity for riparian areas. They appear to be replaced at lower elevations and in drier areas by Rusty-breasted Antpitta (López *et al.* 2012). The elevational range has a breadth of *c.*1,400m, with the range of elevations provided by various authors being 800–2,200m (Wetmore 1939, Phelps & Phelps 1963, Schäfer 1969, Giner & Bosque 1998, Hilty 2003, Verea 2004, Restall *et al.* 2006).

VOICE The range of vocalisations given by Scallop-breasted Antpitta is not well known. The most frequently heard is a melancholy *shiiiuu* note, repeated three to five times or separated by 3s intervals (Verea 2004), similar to that of Peruvian Antpitta (Greeney *et al.* 2004b, Greeney 2009). Males (alone?) produce a second vocalisation, a drawn-out, descending and similarly melancholy *shiiiiiiiiuuuuu*, which has been suggested to be used in territory defence (Verea 2004). Adults apparently also produce soft, nasal contact calls while foraging in pairs (López *et al.* 2012). D. Ascanio (*in litt.* 2016) likened the most commonly heard vocalisation to the song of Venezuelan Tyrannulet *Zimmerius improbus*, but suggested that it is lower in pitch near the end.

NATURAL HISTORY Found alone or in pairs, always near the ground in the understorey of mature, humid, montane forest, often in areas with more open and diverse vegetation (Krabbe & Schulenberg 2003, Verea 2004, Ridgely & Tudor 2009, López *et al.* 2012). Individuals search for food on branches, including in the mosses that grow on them (Verea 2004), pausing occasionally to remain motionless except for typical '*Grallaricula* twitching' in which the lower half of the body is rhythmically moved sideways in a staccato fashion while the head and upper breast remain frozen (Greeney 2009, López *et al.* 2012). Occasionally perches sideways on small vertical stems (Hilty 2003) in a manner recalling many typical antbirds (Thamnophilidae). There are few published data on territory defence, maintenance or fidelity, but the vocalisations of adult males, especially in response to playback, suggest territorial defence (Verea 2004, López *et al.* 2012). At La Cumbre de Rancho Grande 4–5 individuals (four adults, one juvenile) have been seen occupying a relatively small area (400m²), suggesting some intraspecific tolerance, but perhaps only during non-breeding periods (C. Verea *in litt.* 2017). Fieldwork (mist-netting) in the cloud forests of Henri Pittier National Park (C. Verea) found 2–3 pairs every 100 m, suggesting territories of *c.*8,000m². López *et al.* (2012), however, found only four pairs along a 400m transect (Sierra de Aroa), providing an estimated territory size of closer to 10,000m². Territory size, therefore, possibly ranges from 8,000–10,000m². Further observations are needed, however, and it seems likely that territories are probably at the smaller end of this range, and the species occurs in sufficiently low densities that territories frequently do not abut. Scallop-breasted Antpitta is probably monogamous (Verea 2004), but data are few. There are no published data concerning social or interspecific behaviours, although López *et al.* (2012) reported that adults react strongly to playback, suggesting that territorial defence may be strong. In response to playback, males raise their head, flare the feathers of the crown and peer in the direction of the vocalisation (Verea 2004).

DIET Poorly documented, but reported prey items include small arthropods such as crustaceans (Isopoda), arachnids (Aranae), caterpillars (Lepidoptera), beetles (Coleoptera), wasps (Hymenoptera) and true bugs (Hemiptera) (Schäfer 1969, Verea 2004).

REPRODUCTION The nest, eggs and nestlings have only recently been described, and the following is based on two active nests and five inactive nests reported by Miranda *et al.* (in review). **NEST & BUILDING** All seven nests were architecturally very similar, being shallow, open cups composed externally of moss, and lined with interwoven and coiled dark fibres and rootlets. Below these well-constructed cups were sparse platforms of twigs or leaf petioles, arranged to provide a supporting platform for the main nest, much as described for Peruvian Anptitta (Greeney 2009) and Ochre-breasted Antpitta (Greeney *et al.* 2012). Nest measurements for the two active nests were: external diameter, 111 × 108mm; 115 × 110mm; external height, 53–55mm, *c.*65mm; internal diameter, 78 × 70mm, 70 × 70mm; internal depth, 38mm, 40mm. All nests were in the understorey of mossy, closed-canopy forest. The nests were 55–119cm above ground, on the whole in fairly poorly supported positions such as on small branches or epiphyte clumps on the sides of small saplings or vine tangles. Such precarious positioning of nests is something that is certainly shared with Peruvian Antpitta (which see). In being largely composed of mossy materials externally, the nest of Scallop-breasted Antpitta is most similar to Peruvian (Greeney 2009), Ochre-breasted

(Holley *et al.* 2001, Maillard-Z. & Vogel 2003, Greeney *et al.* 2012) and Crescent-faced Antpittas (Greeney & Jipa 2012). The nests of these four species differ from the leaf, stick and petiole composition of Slate-crowned Antpitta (Greeney & Sornoza 2005) and Rusty-breasted Antpitta (Schwartz 1957, Niklison *et al.* 2008). Although Miranda *et al.* (in prep.) did not observe nest construction, my own examination of their photos and their description of nest architecture both suggest that Scallop-breasted Antpitta initiates construction by forming a loose platform of twigs that then helps to support the well-formed nest cup. Even without descriptions for the nests of all *Grallaricula* species, it seems very likely that this architectural detail will prove to be a synapomorphy for the genus. **Egg, Laying & Incubation** One nest held two nestlings and the complete clutch of one nest consisted of two eggs, suggesting two-egg clutches may be most frequent. The eggs were short subelliptical, with a pale, creamy-brown to off-white ground colour and sparsely marked with variably sized flecks and splotches of brown, cinnamon and lavender. **Nestling & Parental Care** The two mid-aged nestlings described by Miranda *et al.* (in prep.) were dark-skinned, but largely largely covered in dense, wool-like, rufescent or rusty-brown down, likely acquired after hatching. Their bills were bright orange, yellower near the expanded rictal flanges, and they had deep orange-crimson mouth linings. Overall, they showed little deviation from what would be expected based on the descriptions of other *Grallaricula* nestlings (Greeney 2012a). They reported the following measurements from the two nestlings: wing 13.6mm, 13.5mm; bill depth at nares 13.4mm, 13.1mm; bill width at nares 14.4mm, 14.8mm; bill length from front of nares 14.8mm, 14.6mm; exposed culmen 18.8mm, 18.3mm; tarsus 20.9mm, 20.5mm. **Seasonality** Early authors indicated a reproductive period of March to May, coinciding with the onset of the rainy season, without offering any further details or evidence (Schäfer & Phelps 1954, Schäfer 1969). Verea (2004) reported the breeding condition and moult stage of adults captured in mist-nets and, based on these data, concluded that Scallop-breasted Antpitta may, in fact, breed during most of the year. Further sampling is needed to confirm this, but it is possible that the species has two reproductive peaks during the year as suggested by records of the allied Peruvian Antpitta and as in other montane passerines (Greeney 2010). Moult in Scallop-breasted Antpitta occurs from August to December (Verea 2004, Verea *et al.* 2009), although the process and extent of moults have not been described. There appears to be some overlap of moulting with breeding, although most moult probably occurs directly following reproduction (Verea *et al.* 2009). Verea's (2004) data also suggested that at least some individuals may repeat their reproductive cycle immediately upon completion of the first one (Verea 2004). **Breeding Data** The only two described active nests were in the El Silencio portion of Sierra de Aroa National Park, near Pico El Tigre (Miranda *et al.* in prep.). The first contained nestlings on 26 May 2013, and the second completed laying on 1 June 2013. In addition I examined two birds in subadult plumage: a male collected 7 November 1937 at Rancho Grande (USNM 351952) and a female collected 7 January 1910 at Cumbre de Valencia (AMNH 492326).

TECHNICAL DESCRIPTION Sexes similar. The following translation of Verea's (2007) description uses the colour designations provided by Ridgway (1912). *Adult*

Forehead olivaceous infused with fuscous, giving it a dirty aspect; crown and nape chestnut-rufous, with black rachis, creating sparse streaking; supercilium chestnut-rufous; ear-coverts medal bronze; lores medal bronze, sometimes infused olivaceous or fuscous from forehead; chin and throat buff-yellow; subauricular and malar stripe fuscous, extending from base of moustache-like markings on sides of throat, and bordering the malar area is buff-yellow; these stripes widen lower and can sometimes join at base of throat; eye-ring buff-yellow, wider anteriorly and posteriorly to eye; back feathers, scapulars, rump, uppertail-coverts and rectrices olivaceous; greater coverts medal bronze, lightly edged rufous; primary-coverts rufous; primaries fuscous, medal bronze-edged; secondaries medal bronze; primaries and secondaries have black rachis. Anterior vane in alula and tenth primary strongly edged buff-yellow. Underside of primaries and secondaries fuscous, with posterior vanes strongly edged buff-yellow. These fringes increase in width and length from the ninth primary to the tenth secondary. Underwing-coverts buff-yellow, lightly edged fuscous. Breast feathers maize-yellow, losing intensity below, becoming cream-coloured on belly. The breast feathers are thickly edged black near tip and bright olive near base, creating a scaled appearance. Some rictal bristles are equal in length to the bill. **Adult Bare Parts** *Iris* brown; *Bill* maxilla black, mandible yellowish with a black tip; *Tarsi & Toes* pinkish-grey. **Subadult** The following is modified from Verea's (2007) 'juvenile' description, based on my own observations of specimens. Overall, rather similar to plumage of full adults. Throat tends to be whiter than in adults, becoming more washed with yellow or orange with age. The subauricular and malar stripes are paler, more like dull fuscous than black. The fringes of the breast feathers are pale olive, rather than black, overall appearing less heavily scalloped. Wings like adults but upperwing-coverts narrowly margined rufescent-chestnut. **Subadult Bare Parts** *Iris* brown; *Bill* mostly yellow, duskier on culmen, with no black tip to mandible, rictal flanges yellow; *Tarsi & Toes* similar to adults.

MORPHOMETRIC DATA *Wing* 70mm; *Tail* 27mm; *Bill* [total culmen] 14mm (*n* = 1, holotype, Ménégaux & Hellmayr 1912). *Wing* 67–71mm; *Tail* 27–30mm; *Bill* [total culmen] 13.0–14.5mm (*n* = 6♂♂, Hellmayr & von Seilern 1912a). *Wing* 67–70mm; *Tail* 27–30mm; *Bill* [total] culmen 14mm (*n* = 4♀♀, Hellmayr & von Seilern 1912a). *Wing* 68.6 ± 1.8mm, range 65–71mm (*n* = 15); *Tail* 26.3 ± 2.1mm, range 24–30mm (*n* = 15); *Bill* [total culmen] 12.7 ± 0.5mm, range 11.8–13.5mm (*n* = 11); *Tarsus* 24.3 ± 1.0mm, range 23.0–26.1mm (*n* = 11) (♀♀, Verea 2007). *Wing* 70.0mm, 68.0mm; *Tail* 28.0mm, 25.0mm; *Bill* [total] culmen 12.0mm, 12.6mm; *Tarsus* 22.0mm, 25.1mm (*n* = 2♂♂, Verea 2007). *Wing* mean 68.4 ± 1.8mm, range 66–71mm (*n* = 6); *Tail* mean 22.8 ± 2.3mm, range 20–25mm (*n* = 6); *Bill* [total] culmen mean 13.3 ± 1.2mm, range 12–14mm (*n* = 3); *Tarsus* 23.5mm, 24.0mm (*n* = 2) (♂?, Verea 2007). *Bill* from nares 9.0mm, 9.1mm, 9.2mm, 9.4mm, 8.2mm, 8.6mm, 8.9mm, 9.0mm; *Tarsus* 25.3mm, 23.5mm, 25.1mm, 24.9mm, 24.8mm, 24.5mm, n/m, 25.2mm (*n* = 8, 3♂♂, 1♀, 4♂?, USNM, AMNH). *Bill* width at base of mandible 5.9mm; *Tarsus* 25.4mm (*n* = 1♀, FMNH 254873). *Wing* 67mm; *Bill* from nares 8.3mm, exposed culmen 8.7mm; *Tarsus* 23.9mm (*n* = 1♀, MCZ 134948, J.C. Schmitt *in litt.* 2016). *Wing* 68mm, 69mm, 68mm, 70mm; *Tail* 27mm, 30mm, 26mm, 28mm; *Bill* [total culmen] 13mm, 14mm,

13mm, 13.5mm (*n* = 4, ♂,♂,♀,♀, 'juveniles', Hellmayr & von Seilern 1912a). *Wing* 69.5mm; *Tail* 24.0mm; *Tarsus* 25.0mm (*n* = 1♂, 'juvenile', Verea 2007). *Bill* from nares 8.6mm, 9.2mm; *Tarsus* 25.1mm, 24.2mm (*n* = 2, ♂,♀, subadults, USNM 351952, AMNH 492326). **Mass** 22.2g(NO) (*n* = 1, ♂, USNM 575269 label), 22.2g, 20.0g (*n* = 2 ♂♂), 20.9 ± 2.0g, range 18.5–24.5g (*n* = 11♀♀), 20.2 ± 1.7g, range 18.4–22.5g (*n* = 6, ♀?), 21.0g (*n* = 1, 'juvenile' ♂) (Verea 2007). Verea *et al.* (1999) presented the following for ten unsexed individuals, all of which are included in the previous weights: 20.8 ± 2.1g, range 18.4–23.0g (cited by Dunning 2008). **Total Length** 9.5–12.5cm, mean 10.4 ± 0.7cm (*n* = 15♀♀, Verea 2007), 10cm (Meyer de Schauensee & Phelps 1978, Ridgely & Tudor 1994, Krabbe & Schulenberg 2003), 10–11cm, mean 10.6 ± 0.4cm (*n* = 6, ♀?, Verea 2007), 10.2cm (*n* = 1 unsexed, Sclater 1857a), 10.7cm (Meyer de Schauensee 1970, Hilty 2003), 10.5cm (*n* = 1♂, Verea 2007), 10.8cm (Restall *et al.* 2006), 11cm (*n* = 1, 'juvenile' ♂, Verea 2007, also Dunning 1987), 11.5cm (*n* = 1♂, Verea 2007), 12.0cm (*n* = 1♀, MCZ 134948 label), 12.5cm (*n* = 1♀, FMNH 254873 label).

DISTRIBUTION DATA Endemic to **VENEZUELA**: *Falcón* Cerro La Danta, *c.*11°13'N, 69°33'W (Verea 2004). *Yaracuy* Pico El Tigre, 10°25'N, 68°47'W (J.E. Miranda T. & K. López *in litt.* 2017); PN Yurubí, Sector El Silencio, 10°24.5'N, 68°48.5'W (Miranda *et al.* in prep.); Cerro La Trampa del Tigre, 10°24'N, 68°47'W (Phelps & Phelps 1963, COP 64001); *c.*1km NNW of El Rosario, 10°20'N, 68°50'W (XC 223752, P. Boesman); Las Antenas, 10°22'N, 68°50'W (COP 83995/96); Sierra de Aroa, La Soledad, 10°14.5'N, 68°58.5'W (J.E. Miranda T. & K. López *in litt.* 2017). *Carabobo* [above] San Esteban (adjusted for elevation), 10°24'N, 67°59.5'W (Sclater 1890); Portachuelo, *c.*10°21'N, 67°41'W (Wetmore 1939b); Cumbre de Valencia, 10°20'N, 68°00'W (Sclater & Salvin 1868, Sclater 1890, Phelps & Phelps 1963, ANSP, CM, FMNH, MCZ, AMNH, NHMUK); Hacienda Santa Clara, 10°19'N, 67°44'W (Phelps & Phelps 1963, COP 24828). *Aragua* Pico La Mesa, 10°24'N, 67°33'W (C. Verea *in litt.* 2017, MEBRG); Pico La Florida, *c.*10°23.5'N, 67°18'W (Verea 2004); Cumbre de Rancho Grande, 10°22'N, 67°39'W (Verea 2004); Cerro Piedra de Turca, 10°22'N, 67°38'W (Fernández-Badillo 1997); trail to Pico Guacamayo, 10°22'N, 67°40.5'W, (XC 14300, D. Ascanio); Pico Guacamayo, 10°22'N, 67°39.5'W (Verea

2004, Verea & Solórzano 2011, COP 80777/78); Paso de Choroní, 10°21.5'N, 67°35'W (XC 66097, D.F. Lane). *Distrito Federal* Boca de Tigre, 10°33'N, 66°54.5'W (J.E. Miranda T. *in litt.* 2017).

TAXONOMY AND VARIATION Monotypic. Though no formal taxonomic work has been undertaken, Scallop-breasted Antpitta is probably closely related to Peruvian and Ochre-fronted Antpittas, which two are considered to form a superspecies (Graves *et al.* 1983). In fact, Meyer de Schauensee (1966) actually suggested that Peruvian Antpitta might be a subspecies of Scallop-breasted Antpitta. The plumage of all three also suggest a possible affinity with Hooded Antpitta.

STATUS Scallop-breasted Antpitta is a restricted-range species (Cracraft 1985) present in the Cordillera de la Costa Central EBA. In parts of its range it can be fairly common, being recorded in 80–83% of surveys (Verea 2004). Despite this, however, the species has a small range and numbers are likely to be declining due to habitat loss, leading to a threat status of Near Threatened (Collar *et al.* 1994, BirdLife International 2017). Although extensive forest cover remains in most parts of its limited range, deforestation has been severe around Caracas, and many other areas within its range have been severely affected (Stattersfield *et al.* 1998). **Protected Populations** PN Henri Pittier (e.g., Wetmore 1939a,b, Schäfer & Phelps 1954, Visbal *et al.* 1996, Verea 2004, Verea & Solórzano 2011); PN Yurubí (López *et al.* 2012); PN El Ávila (J. Miranda *in litt.* 2017); PN San Estéban (BirdLife International 2017); PN Sierra de San Luís (Verea 2004); MN Pico Codazzi (Krabbe *et al.* 2017). Though there are no reports I am aware of, Scallop-breasted Antpitta should be looked for in PN Macarao.

OTHER NAMES *Grallaria loricata* (Sclater 1857a). **English** Scale-breasted Antpitta (Phelps & Phelps 1963); Levraud's Grallaricula (Cory & Hellmayr 1924). **Spanish** Ponchito lorigado (Krabbe & Schulenberg 2003); Ponchito Pechiescamado (Phelps & Phelps 1963, Meyer de Schauensee & Phelps 1978, Phelps & Meyer de Schauensee 1979, 1994, Ponchito de Levraud (Phelps & Phelps 1950, Schäfer & Phelps 1954). **French** Grallaire maillée. **German** Schuppenameisenpitta (Krabbe & Schulenberg 2003).

Scallop-breasted Antpitta, adult, PN Henri Pittier, Aragua, Venezuela, 9 June 2005 (*Ron Hoff*).

Scallop-breasted Antpitta, PN Henri Pittier, Aragua, Venezuela, 9 June 2005 (*Ron Hoff*).

Scallop-breasted Antpitta, PN Henri Pittier, Aragua, Venezuela, 12 September 2012 (*Carlos Verea*).

Scallop-breasted Antpitta, PN Henri Pittier, Aragua, Venezuela, 12 September 2012 (*Carlos Verea*).

HOODED ANTPITTA
Grallaricula cucullata   **Plate 22**

Conopophaga cucullata P. L. Sclater, 1856, Proceedings of the Zoological Society of London, vol. 1856, p. 29, 'Bogotá', Colombia.

Immediately separated from all other *Grallaricula* within its range by the bright orange-rufous head and throat, Hooded Antpitta is nearly endemic to Colombia but for several records in westernost Venezuela (Táchira). Its Colombian range encompasses all three Andean chains, where it inhabits the dense understorey of mossy, humid montane forests at 1,500–2,800m. The known range of Hooded Antpitta, however, is best described by points rather than polygons, and more surveys of potentially viable habitat are needed to fill in distribution gaps and understand potential threats. Almost nothing is known about the species' breeding ecology, general behaviour or even its vocalisations, and it is currently afforded a conservation status of Vulnerable. Studies are urgently needed of all aspects of its distribution, taxonomy, behaviour and natural history. Although an egg was described more than 100 years ago, no nest has been reported, and no aspect of the breeding biology has been studied. The taxonomic affinities of Hooded Antpitta are unclear, although in linear classifications it is usually placed close to other species of *Grallaricula* with prominent orange-rufous markings on the head, such as Peruvian and Ochre-fronted Antpittas (Cory & Hellmayr 1924, Peters 1951, Sibley & Monroe 1990, Remsen *et al.* 2017). This account provides the first description of immature plumage for this poorly known species.

IDENTIFICATION 10–12cm. The simple but striking plumage consists of its characteristic orange-rufous hood that extends to the lower nape and throat, an olive-brown back, and pale grey underparts. It has yellowish legs and a bright yellowish bill. The bright hood, unmarked underparts and pale bill immediately separate this species from all other *Grallaricula*. As noted by Ridgely & Tudor (2009), Hooded Antpitta is, in parts of its range, sympatric with the similarly sized and similarly patterned Rufous-headed Pygmy-Tyrant *Pseudotriccus ruficeps*, which has shorter, yellow legs, and darker underparts.

Even with only a quick look, however, the characteristic rhythmic twitching of the rear by Hooded Antpitta will likely give it away.

DISTRIBUTION The range remains poorly understood, with new localities still being discovered. Currently it is known from a few scattered localities in all three ranges of the Colombian Andes and just across the border in western Venezuela. The nominate race occurs in Colombia on both slopes of the West Andes and the western slope of the Central Andes.

HABITAT The specific habitat requirements are poorly known, but the species is generally found in the undergrowth of humid montane forests. In at least one study, it was searched for and not found in tree plantations adjacent to otherwise (presumably) acceptable habitat (Durán & Kattan 2005) and it may not be tolerant of anthropogenic habitat modification. Reported altitudinal records include elevations of 1,500–2,800m (Hilty & Brown 1986, Fjeldså & Krabbe 1990, Stiles 1997, Hilty 2003, Ridgely & Tudor 2009, McMullan *et al.* 2010), with most records at 1,800–2,150m (Ridgely & Tudor 1994, Krabbe & Schulenberg 2003).

VOICE Like Peruvian Antpitta, the vocalisations are almost unknown and have not been formally described (Krabbe & Schulenberg 2003, Restall *et al.* 2006). Downing & Hickman (2004) reported an unsexed adult 'giving a di- or trisyllabic, high-pitched but quite liquid call, repeated 2–3 times' but there are no further descriptions available. Presumably it is less vocal than other members of the genus. The most frequently recorded vocalisation is a single, drawn-out note sounding very similar to that of Peruvian Antpitta (e.g., XC 265017, D.M. Brinkhuizen).

NATURAL HISTORY The behaviour is poorly documented. It apparently forages low to the ground, rarely or never on the ground, much as described for congeners (Krabbe & Schulenberg 2003, Downing & Hickman 2004). It is reported to make the rocking motion of its lower body characteristic of other species of *Grallaricula* (Gertler 1977). The intra- and inter-specific behaviours of Hooded Antpittas are undocumented, although it apparently joins mixed-species flocks only rarely (Ridgely & Tudor 1994).

DIET Essentially, the diet is completely unknown. A female collected in Antioquia, Colombia, had insects in its stomach (Sclater & Salvin 1879a), which could have been easily predicted based on our knowledge of other *Grallaricula*.

REPRODUCTION The nest remains undescribed, as are aspects of the species' breeding behaviour and ecology. Sclater & Salvin (1879a) reported a fully shelled egg removed from a specimen collected in September in Antioquia. The egg (NHMUK 1962.1.472) is subelliptical, slightly glossy, buffy-tan or pale coffee-coloured, and marked with large blotches and scrawls of cinnamon-brown at the large end, and with smaller lavender and sparse rufous spots and flecks elsewhere (photos courtesy of N. Heming). The measurements of this single egg, provided by three separate authors, vary somewhat: 20.3 × 16.5mm (Sclater & Salvin 1879a); 20.3 × 16.0mm (Oates & Reid 1903); 20.4 × 16.4mm (Schönwetter 1979). From the descriptions of these authors and my own examination of photos, the egg appears very similar to those of Slate-crowned, Peruvian, Scallop-breasted and Ochre-breasted Antpittas. **SEASONALITY** None of the ten individuals captured by Marín-Gómez *et al.* (2015) between June and February 2010 were in breeding condition. With so few data, it is somewhat premature to describe seasonality. If I were to search for its nest, however, I might begin during the rainier periods in its range, based largely on my own intuition, its mid-elevation distribution and apparent affinity for mossy (rather than *Chusquea* bamboo-dominated) habitats. A juvenile photographed on 19 December 2016 at Santuario de Flora y Fauna Otún-Quimbaya was older, but still begging from its parents (J. Flórez photos). One of the specimens I examined from La Candela (AMNH 116352), collected 17 May 1912, was in subadult plumage.

TECHNICAL DESCRIPTION Sexes similar. *Adult* A distinct bright orange-rufous hood covers the face, throat, nape and sides of the neck. The upperparts are uniform cinereous olive-brown while the wings and tail are similar but tend to be somewhat less olive, more warm brown. The underwing-coverts are fulvous-brown, the upperwing-coverts slightly darker brown. Below the rufous hood, the upper breast bears an indistinct, narrow white crescent that separates the rufous from the grey breast and flanks. The central belly and lower breast are unmarked white, sometimes washed brownish or greyish near the vent. **Adult Bare Parts** *Iris* dark brown; *Bill* bright yellow; *Tarsi & Toes* light yellow-brown or greyish-yellow. *Juvenile* The following description is based on photos taken by J. Flórez of an immature of the nominate subspecies, which was still begging food from its parents. Overall, it had the appearance of a 'messy' adult. The margins of the orange-rufous hood and the throat were distinctly yellowish or orange-buff. The upperparts were already similar to adults, but the underparts were distinctly mottled with small patches of dark rusty-brown nestling down, as described in other *Grallaricula* juveniles. The photos did not clearly show the upperwing-coverts, but they were presumably as described for subadults. **Juvenile Bare Parts** *Iris* brown; *Bill* maxilla blackish, orange at the nares, base, along the tomia and at the tip, mandible mostly orange, rictal flanges slightly inflated and yellow-orange; *Tarsi & Toes* dusky-yellow. **Subadult** So far as I can tell, distinguishable from adults only by their upperwing-coverts, which are indistinctly tipped with pale buff spots and a thin rusty-red margin. It appears that some may also have a slightly paler throat and a less distinct upper breast crescent.

MORPHOMETRIC DATA *G. c. venezuelana Wing* 66mm; *Tail* 31mm; *Bill* exposed culmen 13mm, culmen from base 17mm; *Tarsus* 25mm (*n* = 1, holotype, Phelps & Phelps 1956). *Wing* 66.5mm; *Tail* 30mm; *Bill* culmen from base 16mm (*n* = 1♂, Phelps & Phelps 1956). *G. c. cucullata Wing* 67–69mm; *Tail* 23–27mm; *Bill* culmen from base 16.0–16.5mm (*n* = 2♂♂, Phelps & Phelps 1956). *Bill* from nares 8.4mm, 9.2mm; *Tarsus* 23.0mm, 24.4mm (*n* = 2♂♂, USNM 446637, AMNH 492488). *Wing* 67mm; *Tail* 29mm; *Bill* culmen from base 15mm (*n* = 1♀, Phelps & Phelps 1956). *Bill* from nares 7.6mm; *Tarsus* 23.4mm (*n* = 1♀, AMNH 116351). *Wing* 69mm; *Tail* 28mm; *Bill* culmen from base 15.5mm (*n* = 1♂?, Phelps & Phelps 1956). *Bill* from nares 8.7mm, 8.2mm; *Tarsus* 24.8mm, 24.8mm (*n*=2, ♂,?, subadults, AMNH). The following means (range) are of unsexed adults in the field (*n* = 10, O.H. Marín-Gómez *in litt.* 2016): *Wing* 66.2 ± 4.5mm (55–71mm); *Tail* 30.9 ± 1.4mm (29–34mm); *Bill* exposed culmen 11.6 ± 0.6mm (10.8–12.5mm), total culmen 16.0 ± 1.5mm (15.1–18.2mm), width at nares 6.4 ±0.6mm (5.1–6.9mm), depth at nares 5.0 ± 0.2mm (4.8–5.3mm); *Tarsus* 21.0 ± 1.7mm (18.5–23.5mm). **Mass** 21g, 14.3g, 19.0g, 19.8g, 19.4g, 20.6g, 18.4g, 18.4g, 19.0g, 20.0g (*n*=10, ♀?, *cucullata*, O.H. Marín-Gómez *in litt.* 2016). **Total Length** 10–12.1cm (Meyer de Schauensee 1964, Meyer de Schauensee & Phelps 1978, Hilty & Brown 1986, Dunning 1987, Fjeldså & Krabbe 1990, Ridgely & Tudor 1994, Krabbe & Schulenberg 2003, Restall *et al.* 2006, McMullan *et al.* 2010).

TAXONOMY AND VARIATION Two subspecies are currently recognised, but differences between the two are subtle and Hooded Antpitta might best be considered monotypic. Records from the east slope of the East Andes in Caquetá and Cundinamarca have not yet been definitively assigned to subspecies, but are included within the range of nominate *cucullata* here (*contra* Krabbe & Schulenberg 2003).

Grallaricula cucullata venezuelana Phelps & Phelps, 1956, Río Chiquito, Hacienda La Providencia, Táchira, Venezuela, 1,800m [07°33'N, 72°22'W]. This subspecies has the more restricted range of the two. It occurs in western Venezuela in south-west Táchira and adjacent western Apure. Race *venezuelana* is rather similar to nominate *cucullata* but differs in having a strong olivaceous wash to the grey breast, making it appear darker overall. Additionally, the lower abdomen and undertail-coverts are pale cream-coloured or yellowish (instead of whitish). The following detailed description of the holotype, a male, is taken from Phelps & Phelps (1956). 'Top of head and lores Amber Brown with olivaceous tint towards nape; sides of head Sudan Brown; nape narrowly olivaceous; back and rump yellower than Saccardo's Olive; upper tail-coverts tinted with Dresden Brown. Throat Ochraceous-Tawny; a white patch between throat and breast; breast Buffy Citrine merging into the greyish olive of sides and flanks; abdomen whitish, slightly creamish posteriorly; thighs Ochraceous-Tawny; undertail coverts pale buffy. Wings Bone Brown; outermost primaries margined with greyish; other remiges margined with brownish olive, more rufous on tertials; inner vanes of remiges margined internally with pale Ochraceous-Salmon; wing-coverts lightly margined with brownish olive; bend of wing, under wing-coverts and axillaries mixed Ochraceous-Tawny and dusky. Tail dusky olivaceous, the under surface paler; outer vanes of rectrices Medal Bronze.' SPECIMENS & RECORDS *Táchira* 1,800m, Río Chiquita, Hacienda La Providencia, 07°33'N, 72°22'W (Phelps & Phelps 1956, COP 61222/23). *Apure* Upper Río Oirá, c.07°25'N, 72°19'W (Hilty 2003).

Grallaricula cucullata cucullata (Sclater, 1856), 'Bogotá, central Colombia'. The holotype (NHMUK 1889.9.20.654; Warren & Harrison 1971) and a topotypical specimen (NHMUK 1889.9.20.655) are two of four specimens held in Tring; both are unsexed and are Bogotá trade skins, making their precise collecting localities unknown. It is worth noting that the holotype is unquestionably an immature, as indicated by the rusty-brown wash to the belly and the rusty-red margins to the upperwing-coverts. These characters were described (in Latin) by Sclater (1856) but also subsequently repeated (Sclater & Salvin 1879a) without, apparently, recognition of their immature nature. Endemic to Colombia, the nominate subspecies is known from west of Cali in the West Andes, on the west slope of the Central Andes of Antioquia, Risaralda and Valle del Cauca, and on the west slope of the East Andes in Huila at the head of the Magdalena Valley. SPECIMENS & RECORDS *Antioquia* Santa Elena, 06°12.5'N, 75°30'W (Sclater & Salvin 1879a, NHMUK 1889.9.10.932/9.20.59); Paramillo, 05°29'N, 75°53.5'W (eBird: J. Castaño). *Chocó* San José del Palmar, 04°58.5'N, 76°13.5'W (S.M. Durán *in litt.* 2017). *Risaralda* RFP AguaBonita, 05°11.5'N, 75°57.5'W (J.A. Zuleta-Marín *in litt.* 2017); PNN Tatamá, 05°06.5'N, 76°04'W (XC 187298, J.A. Zuleta-Marín); La Bananera, 04°46.5'N, 75°37.5'W (XC 317141, G.M. Kirwan); El Cedral, 04°42'N, 75°32'W (XC 301993/94, J. Holmes); La Pastora, 04°41.5'N, 75°30'W (Downing 2005); SFF Otún-Quimbaya, 04°44'N, 75°35'W (Durán & Kattan 2005). *Caldas* Río Blanco, 05°05.5'N, 75°27'W (S.M. Durán *in litt.* 2017). *Tolima* El Agrado, 04°53'N, 75°06.5'W (Molina-Martínez *et al.* 2008). *Quindío* Vereda Boquía, 04°41'9.96"N, 75°32'49.92"W (Marín-Gómez *et*

al. 2015); Alto Quindío, 04°38.5'N, 75°27'W (Arbeláez-Cortés *et al.* 2011b); Río Santo Domingo, Calarcá, 04°31.5'N, 75°39'W (Arbeláez-Cortés *et al.* 2011b); El Corazón, 04°11.5'N, 75°44.5'W (eBird: D. Duque). *Valle del Cauca* Vereda Cristales, 04°08'N, 75°52.5'W (S.M. Durán *in litt.* 2017); near La Cristalina, 04°00.5'N, 76°28.5'W (eBird: E. Vallejo); km27 on Cali–Buenaventura highway, 03°34'N, 76°36.5'W (Downing & Hickman 2004); Ríolima, 03°30'N, 76°38'W (Collar *et al.* 1992, AMNH 492488). *Cundinamarca* Cerro El Reitro, near Ubalá, c.04°44.5'N, 73°32'W (Salaman *et al.* 2002). *Cauca* RN los Yalcones, 01°36'N, 76°39'W (S.M. Durán *in litt.* 2017); Río Mandiyaco, 01°22'N, 76°24'W (eBird: J.P. López-Ordoñez). *Caquetá* Cerro La Mica, 02°44'N, 74°51'W (Bohórquez 2002, Salaman *et al.* 2002b); La Esmeralda–Río Yurayaco, 01°21'N, 76°06'W (IAvH recordings, M. Álvarez). *Huila* La Candela, 01°50'N, 76°20'W (Chapman 1917, Collar *et al.* 1992, AMNH, ANSP, USNM); Pesebre, 01°43.5'N, 76°08'W (Marín-Gómez *et al.* 2015); Nabú (Finca Playón), 01°36'N, 76°16'W (Salaman *et al.* 1999, 2002b, 2007); PNN Cueva de los Guácharos, c.01°33.5'N, 76°08.5'W (Gertler 1977).

STATUS A range-restricted species found in both the Colombian East Andean EBA and the Colombian Inter-Andean EBA (Collar *et al.* 1992, Stattersfield *et al.* 1998, Krabbe & Schulenberg 2003). In the late 1980s, the poorly known Hooded Antpitta was considered Data Deficient (IUCN 1988) but was soon upgraded to Vulnerable status (Collar *et al.* 1992, Groombridge 1993). In Colombia it was originally considered Vulnerable (Renjifo *et al.* 2000) but, as new records expanded its known distribution, it was downgraded to Near Threatened (Renjifo *et al.* 2002). The more poorly known subspecies (*venezuelana*) is considered Vulnerable at a national level (Sharpe 2008). Although Hooded Antpitta is still considered Vulnerable at a global level by BirdLife International (2017), it suggested that, if additional records continue to surface, the status may warrant a downgrade to Near Threatened. The primary threat is anthropenic habitat destruction for lumber extraction and agriculture (Collar *et al.* 1992, Stattersfield *et al.* 1998). Although, when found, it is generally reported to be uncommon or fairly common (Gertler 1977, Marín-Gómez *et al.* 2015), the species' apparently non-vocal nature probably causes it to be overlooked or under-recorded in some areas. PROTECTED POPULATIONS *G. c. venezuelana* PN Tamá (Sharpe 2008, COP): *G. c. cucullata* PNN Cueva de los Guácharos (Gertler 1977), PRN Ucumarí (Downing 2005, Marín-Gómez *et al.* 2015); PNN Munchique (Negret 1997); RFP Río Blanco (Marín-Gómez *et al.* 2015); SFF Otún-Quimbaya (Durán & Kattan 2005); PNN Tatamá (XC 187298, J.A. Zuleta M.); RFP AguaBonita (J.A. Zuleta M. *in litt.* 2017). Although still unreported from there, the species likely occurs in PN Los Farallones de Cali (S.M. Durán *in litt.* 2017).

OTHER NAMES *Conopophaga cucullata* (Sclater 1856, 1858c, 1862). **English** Hooded Ant-Thrush (Brabourne & Chubb 1912); Hooded Grallaricula (Cory & Hellmayr 1924). **Spanish** Ponchito encapuchado (Krabbe & Schulenberg 2003); Tororoí Cabecirrufo (Salaman *et al.* 2007a); Ponchito Cabecicastaño (Phelps & Phelps 1963, Hilty 2003, Sharpe 2008); Peoncito de cabeza naranja (Negret 1997); Peoncito de cabeza roja (Negret 2001). **French** Grallaire à capuchin. **German** Rotkopf-Ameisenpitta (Krabbe & Schulenberg 2003).

Hooded Antpitta, adult, El Cedral, SFF Otún-Quimbaya, Risaralda, Colombia, 17 April 2016 (*Fabrice Schmitt/Wings Birding Tours Worldwide*).

Hooded Antpitta, juvenile, Pereira, SFF Otún-Quimbaya, Risaralda, Colombia, 19 December 2016 (*Johan Flórez*).

Hooded Antpitta, adult, El Cedral, SFF Otún-Quimbaya, Risaralda, Colombia, 8 April 2016 (*Peter Hawrylyshyn*).

Hooded Antpitta, adult, El Cedral, SFF Otún-Quimbaya, Risaralda, Colombia, 8 April 2016 (*Peter Hawrylyshyn*).

PERUVIAN ANTPITTA
Grallaricula peruviana   Plate 23

Grallaricula peruviana Chapman, 1923, American Museum Novitates, no. 96, p. 11, Chaupe, Cajamarca, Peru, 1,860m [05°10'S, 79°10'W].

Peruvian Antpitta is a strikingly patterned antpitta that, like others in the genus, frequently twitches its hindquarters in a slow, rhythmic fashion when perching otherwise motionless in the understorey. It was originally described (Chapman 1923b) from a single male specimen collected at Chaupe in north-east Peru, north of the Río Marañón, and close to the Ecuadorian border. This extremely elusive species was not found again until 1978, when it was rediscovered in northern Peru, not far from the type locality (Parker *et al.* 1985). Peruvian Antpitta was first discovered in Ecuador a few years later (Fjeldså & Krabbe 1986). It is a poorly known inhabitant of Andean forests under intense human pressure. So far as is known,

Peruvian Antpitta is restricted to the undergrowth of mature forest, often near riparian areas. Little is known of any aspect of its biology. Ironically, however, considering the level of knowledge for most Neotropical understorey passerines, most of our knowledge of this species concerns its breeding ecology. No studies have investigated its foraging or habitat requirements, and it remains uncertain what vocalisation, if any, constitutes its song. Much habitat within the known range of Peruvian Antpitta is under immediate threat from habitat conversion, i.e., for milk production, trout farming, timber and agriculture in north-east Ecuador (HFG), and for open-pit mining in south-east Ecuador (J. Freile *in litt.* 2017). Precise data on its habitat requirements, including altitudinal range, are sorely needed. In particular, more data on its reproductive and foraging ecology are needed to identify specific habitat requirements for this poorly known species. For use in future surveys, the identification and description of its song will be a most welcome contribution.

413

IDENTIFICATION 10–11cm. The rich rufous crown is duller brown in females and brighter rufous in males, with a distinct malar and pale pre-loral markings. The upperparts are uniform brown to olive-brown, and the bright white breast bears strong black scalloping. Within its range, Peruvian Antpitta is nearly unmistakable. While it is rather similar in colour and pattern to Ochre-fronted Antpitta of northern Peru, so far as is known the two species are wholly allopatric. Peruvian Antpitta is somewhat similar in pattern, but not in coloration, to the parapatric Crescent-faced Antpitta, which occurs at significantly higher elevations. There is currently no confirmed region of altitudinal overlap with Ochre-breasted Antpitta, but it is quite possible that their ranges overlap in Ecuador at several locations, particularly on the slopes of Volcán Sumaco and the Cordillera Cutucú. Both species have their breasts scalloped with black but Peruvian Antpitta is immediately distinguished by the bright white (rather than buffy-ochre) underparts.

DISTRIBUTION Appears to be restricted to the east slope of the Andes, from north-east Ecuador (Napo) south to northernmost Peru (eastern Piura and northern Cajamarca). As currently understood, its range is very patchy and its presence has been confirmed at just a handful of localities in Ecuador in Napo, Morona-Santiago and Zamora-Chinchipe (see below). In Peru it is known from the vicinity of the type locality in Cajamarca (Chapman 1923b, Krabbe & Schulenberg 2003) and in north-east Piura (Ridgely & Tudor 1994, Clements & Shany 2001). As the species is easily overlooked, and its altitudinal range and habitat requirements are poorly understood, the creation of robust distribution models are challenging. Thus, despite the model generated by Freile *et al.* (2010b), it is possible that its range is more continuous than currently known. It seems unlikely, however, that its range extends south of the Rio Marañón, where it is probably replaced by Ochre-fronted Antpitta. It seems likely, however, that it will eventually be found at least as far north as the East Andes in Nariño and Putumayo (Greeney 2009).

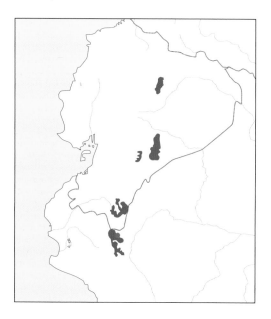

HABITAT As the species' ecology is so poorly known, it is difficult to quantify habitat requirements. Recent studies in north-east Ecuador (Greeney *et al.* 2004a,b, Greeney 2006) suggest that riparian areas may be important, at least for breeding. So far as is known, Peruvian Antpitta is confined to intact, humid, moss- and epiphyte-laden, mid-elevation forests. Within these, it appears to avoid areas with an understorey dominated by bamboo or clogged with brush and vine tangles. Instead the species prefers a more open understorey or at least dominated by herbaceous plants and small saplings which presumably provide suitable foraging perches (see below). Overall, its total altitudinal range is uncertain, with discontinuous records east of the Andes at 1,750–2,100m in Ecuador and 1,650–2,100m in Peru (Clements & Shany 2001)

VOICE The vocalisations are very poorly known and it is unclear if the most frequently heard vocalisation is a simple contact call or constitutes true song. The most frequently heard vocalisation is a single, rising–falling, high-pitched but muted *peeeu*, *wheeu?* or *wheee* note, repeated at regular intervals of <15s (Greeney *et al.* 2004b, Schulenberg *et al.* 2007, Ridgely & Tudor 2009, Krabbe *et al.* 2012). When agitated, may give a slightly slurred or 'hoarse-sounding' version of this note (e.g. second note in XC118112; D.M. Brinkhuizen). This vocalisation is heard both around and away from nests, and is rather similar (but slightly sharper) to a one-note call made by Ochre-breasted Antpitta around their nests (Greeney 2009). Greeney *et al.* (2004b) reported a male near a nest giving a piercing *seeee-UP!* alarm note. After falling prematurely from the nest, a nestling repeatedly made a loud, scratchy distress call (Greeney 2009).

NATURAL HISTORY Although very secretive and apparently easily overlooked, Peruvian Antpitta is not particularly shy and may continue foraging while only several metres from an observer (Greeney 2009). Adults rarely fly distances more than 5m, preferring instead to hop between close perches low above ground in dense vegetation, only occasionally making short flights. Adults most frequently forage from thin, horizontal perches, often remaining on one for several minutes. While perched they remain fairly still or, commonly, twitch their lower body in typical *Grallaricula* fashion, peering to locate prey on surrounding foliage. They also occasionally perch on near-vertical thin stems or vines, similar to *Conopophaga* spp., leaning to peck small insects on leaves and stems. They occasionally make short upward strikes to glean arthropods from the base of leaves. Although they have been trained to feed on worms in eastern Ecuador at Cabañas San Isidro (Krabbe *et al.* 2012), where they readily hop on logs and the ground to retrieve the proffered snacks, in the 'wild' they are rarely or never seen on the ground (Greeney 2009). It is suggested that Peruvian Antpitta's apparent boldness around humans and willingness to approach feeding stations are both indications that foraging in the wake of large mammals may be a regular (though not an obligatory) foraging tactic (Greeney 2012b). There are no firm data on territoriality, but the spatial distribution of nests suggests that pairs frequently build in the vicinity of previous nests and that the 'nesting areas' of different pairs are *c.*0.4km apart (Greeney 2009).

DIET There are few published data on diet. Although the species is known to visit worm-feeding stations (Krabbe *et al.* 2012), it is unknown if earthworms are a normal constituent of diet. Parker *et al.* (1985) reported 'beetles, other insects, fruit' in the stomachs of specimens collected in northern Peru. Nestlings are fed a diet of predominantly invertebrates, many of them aquatic insects, and also small vertebrates such as frogs (Greeney *et al.* 2004b, Greeney 2009). The aquatic invertebrates included dragonfly and damselfly nymphs and adults, as well as partially emerged imagos, suggesting that these larvae were gathered from riparian areas as nymphs crawled from the water to moult to adulthood (Greeney 2009).

REPRODUCTION Considering its rarity, the breeding behaviour is fairly well known, but all available data is from a single population in the private reserves of Cabañas San Isidro and Yanayacu Biological Station. **NEST & BUILDING** Five nests described: they were somewhat fragile, shallow mossy cups built atop a sparse platform of horizontally crossed sticks, internally lined with dark rootlets and flexible fibres (Greeney *et al.* 2004a,b, Greeney 2009). One nest was 6cm tall overall, had an external diameter of 13.5cm, an internal diameter of 8.5cm, and a cup depth of 4.5cm (Greeney *et al.* 2004b). All five nests were well inside primary, moss-laden forests, usually in riparian areas within 30m of small streams. Peruvian Antpittas appear to prefer rather precarious sites for nest placement, on small saplings with multiple small fern petioles, thin branches or epiphytic leaves as supports. Nests are placed on the substrate, rather than affixed to them, with no apparent attempt to interweave nest materials with surrounding structures. Five nests were 0.8–2.4m above ground (Greeney 2009). A sixth nest (L.A. Salagaje *in litt.* 2014) was 1.3m above ground and placed in a near-identical situation to that described by Greeney (2006). Few data are available on details of behaviour during nest construction. Greeney (2006) reported both adults bringing and arranging material. An additional nest, found during construction, was also built by both adults (L.A. Salagaje *in litt.* 2014). When the nest was approached, the male was very shy, disappearing almost immediately, while the female remained nearby, silently observing Salagaje's activity. Nest construction also may continue throughout the incubation period, with adults occasionally arriving at the nest with small fibres to add to the lining of the nest (see below). **EGG, LAYING & INCUBATION** Only one egg has been described (Greeney *et al.* 2004a). It was pale brown with heavy brown, black, lavender and white markings, and measured 21.4 × 17.6mm. Confirmed clutch size was one egg at two nests (Greeney 2006, L.A. Salagaje & E.M. Quipo *in litt.* 2014), and another nest contained a single nestling when found (Greeney *et al.* 2004b). These observations suggest that one-egg clutches are the norm. During incubation, one egg lost mass at a mean rate of 0.9% of its original mass/day, with an estimated loss of 18% prior to hatching (Greeney 2006). Incubation period was 20 days at one nest (Greeney *et al.* 2004a). The following description of adult behaviour during incubation is taken from a nest studied by Greeney (2006). Both sexes incubate, covering the egg for 78% of daylight hours between laying and hatching. When replacing each other at the nest during incubation, adults are rarely present simultaneously. Most often one adult leaves and its replacement arrives 1–2 minutes later. The duration of incubation bouts, both sexes combined, ranged from <1min to 3h, but were frequently interrupted briefly by a short foray from the nest to capture small prey

items. Mean duration of incubation bouts was 37 ± 34 minutes (*n* = 167, brief absences to capture food omitted). With brief change-overs omitted, the egg was covered for mean periods of 66 ± 61 minutes. Regular incubation rhythms, however, with daily attendance being more than 70%, did not begin until four days after the egg was laid, with only 50% attendance during the first two days. During the final eight days prior to hatching the egg was incubated for >90% of daylight hours. The male was never observed spending the night at the nest, a duty apparently reserved for the female. Most mornings she left the nest just prior to sunrise, returning for the night up to 30min after sunset. On most days the male replaced the female shortly after her morning departure and incubation time at the nest was fairly evenly divided, with the male investing only slightly more (58%). Mean periods of attendance by the male lasted 45 ± 40 minutes, while the female's visits averaged 29 ± 25 minutes. While sitting on the egg, both adults frequently stood suddenly, peered into the nest, and then leaned forward to sharply peck at the nest lining or rim, or to thrust their bill in and out of the nest lining in a sewing-machine-like fashion. Both behaviours are presumed to be forms of ectoparasite removal (Haftorn 1994, Greeney 2004). Incredibly, this frequent standing up in the nest resulted in a near catastrophe for the egg. On one occasion, caught on video, the female's sudden movement pushed the egg backwards onto the rim of the nest. Oblivious, she probed vigorously into the nest, with each sharp movement pushing the egg slightly closer to the edge. Still unaware of the precarious position of her egg, she sat back down in the nest but appeared restless. After a few minutes she turned around and discovered the egg, now held on the rim by little more than a few scraps of moss. She sat back down and peered at the egg, pecking it gently 5–6 times before remaining in the incubating position, facing the egg, for >20 minutes. When she next left the nest, by hopping to the rim and dropping over the side, her movement shook the nest sufficiently to almost send the egg tumbling to the ground. Soon after, on my approach, I noticed the egg and returned it to the nest cup prior to her next arrival, which would certainly have dislodged it completely. Other behaviours while incubating included preening and arranging stray pieces of nest material. Most of the time adults remained still, occasionally glancing about with sharp head movements. Both sexes continued to add material to the nest, generally rootlets for the lining, until the day prior to hatching. Once, when relieving her mate, the female fed him a small insect. The male, however, was never observed reciprocating. Only the female vocalised at the nest, several times making the one-note call at regular intervals for up to 5min. **NESTLING & PARENTAL CARE** Nestlings (Greeney 2012a) are born naked with dark skin. The legs are pale orange, as is the cloaca. The mouth lining throughout the nestling period is brilliant crimson-orange, outlined by paler, yellow-white rictal flanges. Older nestlings are covered with dense red-brown down. When approached, older nestlings will often expose their bright mouth lining and silently spread their wings (Fig. 4). Effort during nestling care is shared relatively equally between the sexes (Greeney 2009) and food is delivered as large, single items (IBC 1150678, video HFG). **SEASONALITY** We are far from understanding patterns (if any) of seasonality in breeding. The records below suggest either a bimodal pattern or perhaps year-round reproduction, with the suggestion of a peak from August to November (dry months). The following records (except where noted) are all from the Yanayacu/San Isidro area. **BREEDING DATA** Building, 20 April 2003 (Greeney 2009); building, almost finished, 5 October 2013, Cordillera Guacamayos (L.A. Salagaje *in litt.* 2014); incubation, 7 August 2011, fledging 29 August (L.A. Salagaje & E.M. Quipo *in litt.* 2014); nestling, 29 September 2002 (Greeney *et al.* 2004a,b). **NEST SUCCESS** The small sample size of monitored nests hinders a meaningful estimation of fledging success. It is interesting to point out, however, that one nest failed after a falling epiphyte left the nest still attached to the substrate branch, but tilted sideways. The nestling died when it fell from the nest and drowned in a shallow pool of water on the swampy ground below the nest (Greeney *et al.* 2004b, IBC 1150678, video HFG). A second nest, although successful, probably would have failed had I not intervened (see above).

TECHNICAL DESCRIPTION Sexes similar but distinguishable in the field. The following description is from Greeney *et al.* (2004b) and Greeney (2009). *Adult male* Crown is rich orange-rufous, with this colour extending just onto upper nape, and in some lights appears separated from the brown cheeks by a narrow, buffy stripe. The front and back of the eyes are rimmed by broad, incomplete whitish eye-rings. Cheeks brown, similar to mantle, and faint rufous-cream highlights on the pre-loral area extend either side of the bill base to the forecrown, giving an indistinct tufted effect. A white malar stripe extends to the upper chest with a blackish-brown submalar stripe starting at the bill base. The submalar markings gradually broaden, terminating on sides of throat well above upper chest. Central throat white, broadly bordered by submalar stripes. Breast and belly white, bearing upward-curving crescents on upper chest and upwards to below ear-coverts. This scalloping cleanly divides the throat and malar marks from chest scallops. Most of the breast to the mid/lower belly is blackish, narrowly scalloped white, with the only exception being the central breast where there is a *c.*1.5cm diameter patch that is whiter with narrow black scalloping. This white-on-black scalloping becomes finer on the belly and abruptly gives way to cleaner white, with very little black below. Entire back and wings even rich brown with no fringes to the wing-coverts or remiges. *Adult female* Crown rufous-brown, this colour extending to hindcrown. Eye-ring crescents distinctly buffy, with cheeks and nape brown and similar to mantle coloration. Pre-loral 'tufts' as extensive as in male, but buffy, similar in coloration to eye-ring crescents. Throat's clean white malar stripe connects with a broad white crescent on upper chest. A slightly broader, blackish submalar stripe terminates in a white upper chest crescent. Breast and upper belly bear broad, white, upward-curving crescents that cover the upper chest and extend to below ear-coverts. A blackish band crosses the upper breast and is thickly scalloped white. More diffuse scalloping continues well onto flanks and sides of breast and belly, quickly giving way to a clean white belly. Mid-breast also white. Entire back and wings medium brown with no fringes to coverts or remiges. **Adult Bare Parts** *Iris* dark chestnut; *Bill* blackish or dusky at base, becoming dull yellow towards tip; *Tarsi & Toes* pale pinkish-grey. *Fledgling* Like other *Grallaricula*, the fledglings leave the nest before their flight feathers are full and before they are capable of flight. They are entirely covered in a dense, dark, rusty-rufescent wool-like down. Only on the lower, central belly and vent is the plumage paler, white

or buffy-white. The upperwing-coverts are not complete when they first leave the nest, but the narrow chestnut edges at their tips are already visible. **Fledgling Bare Parts** *Iris* brown; *Bill* bright orange, dusky on culmen and near tip, which sometimes still bears a bright white egg-tooth, rictal flanges inflated and pale yellowish-white, mouth lining bright crimson-orange; *Tarsi & Toes* dark greyish on anterior edge, fading to bright orange posteriorly, toes dusky-orange, nails pale yellowish. *Subadult* Not well described. Fjeldså & Krabbe (1986), however, described an 'immature' female mist-netted in southern Ecuador as being very similar to adult females but having narrow rufous tips to the greater wing-coverts.

MORPHOMETRIC DATA Very few published data. *Wing* 70.0mm, 70.8mm, 69.6mm, 70.6mm, 70.7mm; *Tail* 30.6mm, 30.7mm, 31.1mm, 31.5mm, 31.6mm; *Bill* culmen from base 16.6mm, 16.6mm, 15.8mm, 16.5mm, 16.6mm; *Tarsus* 23.5mm, 24.0mm, 23.7mm, 24.4mm, 24.5mm ($n = 5$, ♂,♂,♀,♀,♀, Graves *et al.* 1983). *Bill* at nares 9.0mm, exposed culmen 14.0mm; *Tarsus* 25.4mm ($n = 1$♂, holotype, AMNH 178388). **MASS** 17.0g, 17.5g, 19.5g, 20.1g, 21.0g ($n = 5$, ♂,♂,♀,♀,♀, Peru, Graves *et al.* 1983, Parker *et al.* 1985); 18g ($n = 1$♀, Ecuador, Rahbek *et al.* 1993). **TOTAL LENGTH** 10.0–10.5cm (Clements & Shany 2001, Ridgely & Greenfield 2001, Krabbe & Schulenberg 2003, Restall *et al.* 2006, Ridgely & Tudor 2009, Schulenberg *et al.* 2010).

DISTRIBUTION DATA ECUADOR: *Napo* RP Estación Biológica Yanayacu and RP Cabañas San Isidro, *c.*00°36'S, 77°53'W (Greeney *et al.* 2004a, 2004b, Greeney 2006, XC, IBC); Cordillera Guacamayos, 00°38'S, 77°50.5'W (L.A. Salagaje photos). *Morona-Santiago* Cordillera de Cutucú, 02°42'S, 78°03'W (Fjeldså & Krabbe 1986, Robbins *et al.* 1987); Gualaceo–Limón Road, 03°01'S, 78°32'W (Ridgely & Greenfield 2001, Freile *et al.* 2010a). *Zamora-Chinchipe* Quebrada Avioneta, 04°17'S, 78°56'W (Rahbek *et al.* 1993, Rasmussen *et al.* 1996, ANSP, MECN); Quebrada Honda, *c.*04°29'S, 79°08'W (Hosner 2010). **PERU**: *Piura* Playón, 05°01'S, 79°19'W (LSUMZ 88091); Lucuma, 05°03'S, 79°22'W (LSUMZ 97708/09); Machete, 05°05'S, 79°24'W (LSUMZ 97706/07). *Cajamarca* 3.5km W of Pueblo Libre, 05°06'S, 79°14'W (FMNH 480828–830); Quebrada Las Palmas, 05°40'S, 79°12'W (LSUMZ 165304/05); ÁCM Bosque de Huamantanga, 05°40'S, 78°56.5'W (A. García-Bravo *in litt.* 2017).

TAXONOMY AND VARIATION Monotypic. The taxonomic affinities are unknown, but the species probably forms a superspecies with Ochre-fronted Antpitta and Scallop-breasted Antpitta (Graves *et al.* 1983, Krabbe & Schulenberg 2003). Some authors have considered the possibility that it is best treated as a race of Scallop-breasted Antpitta (Meyer de Schauensee 1966).

STATUS Historically, and currently, considered a range-restricted, globally Near Threatened species (Collar *et al.* 1994, Stattersfield & Capper 2000, Birdlife International 2016). It occurs in the Ecuador-Peru East Andes EBA (Stattersfield *et al.* 1998) and all aspects of its natural history and distribution are poorly known. The habitat requirements are also poorly understood. It is rarely observed, even where it is known to occur, but available data all point to its need for intact montane forest. With increasing logging and mining threats to foothill and cloud forest habitats within its relatively small altitudinal range and geographic distribution (Stattersfield *et al.* 1998, Stattersfield & Capper 2000), I suggest it may deserve

Endangered status, a sentiment echoed by other authors (Freile *et al.* 2010b). The few nesting records suggest that riparian habitats and aquatic insects may be important for reproduction, and any activities that threaten such habitat (e.g., timber extraction, trout farming) within its range are likely threats. No deleterious effects of human activity have been specifically documented, but this species has been overlooked in at least one heavily birded area. Additionally, as recent pressures on East Andes foothill forests have increased, it is likely that available habitat for Peruvian Antpitta has been, or will be, severely impacted. **PROTECTED POPULATIONS Ecuador**: PN Podocarpus (Rahbek *et al.* 1993, Rasmussen *et al.* 1996); Quebrada Honda portion of RP Tapichalaca

(Hosner 2010), RP Estación Biológica Yanayacu (HFG); RP Cabañas San Isidro (Greeney *et al.* 2004a,b, Greeney 2006); PN Sumaco-Galeras (L.A. Salagaje photos). **Peru**: SN Tabaconas-Namballe (FMNH 480828–830; in or very near); ÁCM Bosque de Huamantanga (A. García-Bravo *in litt.* 2017).

OTHER NAMES No known scientific synonyms. **Spanish** Tororoí Peruano (Clements & Shany 2001); Gralarita Peruana (Rasmussen *et al.* 1994); Ponchito Peruano (Krabbe & Schulenberg 2003); Tororito Peruano (Freile *et al.* 2017). **English** Peruvian Grallaricula (Cory & Hellmayr 1924). **French** Grallaire du Pérou. **German** Südliche SchmuckAmeisenpitta (Krabbe & Schulenberg 2003).

Peruvian Antpitta, adult female, Cordillera Guacamayos, Napo, Ecuador, 5 October 2013 (*Luis A. Salagaje M.*).

Peruvian Antpitta, hatching young in nest, Cabañas San Isidro, Napo, Ecuador, 15 May 2003 (*Harold F. Greeney*).

Peruvian Antpitta, adult female, Cabañas San Isidro, Napo, Ecuador, 4 October 2010 (*Roger Ahlman*).

Peruvian Antpitta, mid-aged nestling in nest, Cabañas San Isidro, Napo, Ecuador, 25 May 2003 (*Harold F. Greeney*).

Peruvian Antpitta, threat display of mid-aged nestling in nest, Cabañas San Isidro, Napo, Ecuador, 8 October 2002 (*Harold F. Greeney*).

OCHRE-FRONTED ANTPITTA
Grallaricula ochraceifrons   **Plate 23**

Grallaricula ochraceifrons Graves, O'Neill & Parker, 1983, Wilson Bulletin, vol. 95, issue 1, p. 1, 10km (by road) below (NE) of Abra Patricia, *c.*1,890m; Dpto. San Martín, Peru. The holotype (LSUMZ 81998) is an adult male collected 30 August 1976 by J.P. O'Neill. Four adult paratypes, three males and a female, are also held at LSUMZ (Cardiff & Remsen 1994). It should be noted that the coordinates given for the type locality (05°46'S, 77°41'W) by Graves *et al.* (1983) are in error. More appropriate coordinates would be 05°40'S, 77°46'W. This location is at the border between the departments of Amazonas and San Martín.

The strikingly patterned Ochre-fronted Antpitta was discovered in 1976. It is extremely range-restricted, being known only from isolated ridges of the East Andes, in the departments of Amazonas and San Martín, just south of the Marañón Valley. Currently considered Endangered, it inhabits the dense undergrowth of epiphyte-laden, humid cloud forests. I agree with Schulenberg & Kirwan (2012a) that research into the species' precise range and habitat preferences are of the most urgent conservation priority. Though some work has been undertaken (Pratolongo 2012), more surveys are needed, along with dietary information, population densities and reproductive strategies, if its conservation needs are to be properly addressed in the face of ongoing habitat destruction within its narrow range.

IDENTIFICATION 11.5cm. Sexes different. Both sexes are similarly patterned below, with ochre sides heavily streaked black, white or buffy-white in the centre. Above, both sexes are largely olive-brown. The female has a brown crown and only slightly ochraceous forecrown, while the male is bright ochre-rufous on the forecrown. Overall, Ochre-fronted Antpitta is distinctive within its limited range. Ochre-breasted Antpitta, with which there is some range overlap at lower elevations, is less olivaceous and paler tawny on the upperparts, buffier below, and has a bold buffy eye-ring. Ochre-breasted Antpitta also lacks the distinctive ochraceous-buff forecrown of the present species. Ochre-

fronted Antpitta is rather similar in colour pattern to Peruvian Antpitta, but there is no known range overlap, with the present species found only north of the Marañón River and Ochre-fronted Antpitta only to the south. Both sexes of Peruvian Antpitta have a prominent buffy loral spot, which is lacking in Ochre-fronted. In addition, the crown and nape of male Peruvian Antpitta are rufous as opposed to rufous-olive in Ochre-fronted Antpitta.

DISTRIBUTION Endemic to Peru, has been found only in an extremely small area on the Amazonian slopes of the Andes just south of the Río Marañón (Parker *et al.* 1982, Schulenberg *et al.* 2010). In the Cordillera Colán, Ochre-fronted is sympatric with Rusty-breasted Antpitta, this site once having been thought to be the only location with sympatric *Grallaricula* (Graves *et al.* 1983, Krabbe & Schulenberg 2003). Though poorly documented in either of these two species, it seems likely that they

segregate to some degree by habitat, as Peruvian and Slate-crowned Antpittas appear to do in north-east Ecuador. Interestingly, as previously noted (Angulo *et al.* 2008, Schulenberg & Kirwan 2012a), the very limited range of Ochre-fronted Antpitta is nearly identical to that of Long-whiskered Owlet *Xenoglaux loweryi* (O'Neill & Graves 1977, Schulenberg & Harvey 2012), yet it remains unclear why both species' ranges are, at least apparently, so restricted.

HABITAT Known only from the understorey of epiphyte-laden, humid, stunted montane forest (Graves *et al.* 1983, Davies *et al.* 1994, Hornbuckle 1999, Schulenberg & Kirwan 2012a). The known altitudinal range is 1,850–2,400m (Schulenberg *et al.* 2007), but if the species is a latitudinal replacement of Peruvian Antpitta, possibly it will eventually be found at lower elevations.

VOICE The vocalisations are virtually unknown. Schulenberg *et al.* (2007) described the 'song' as a single whistled note, given every 6–15s, a rising-falling *wheeu?* This is similar to the most commonly heard vocalisations of Ochre-breasted and Peruvian Antpittas, especially the latter (Schulenberg *et al.* 2007, Schulenberg & Kirwan 2012a). Apparently in aggressive response to playback, it may also deliver a 'chatter' (D.F. Lane in Schulenberg *et al.* 2010).

NATURAL HISTORY Based on information in the literature, one would think that Ochre-fronted Antpitta was virtually unknown in life, as was the case for Peruvian Antpitta a little more than a decade ago (Greeney *et al.* 2004b). The summation of published information is: a secretive bird of the forest understorey (Krabbe & Schulenberg 2003, Schulenberg & Kirwan 2012a). Given that it is now known from several well-birded tourist locations, it is high time that someone rectified this situation. Let this officially serve as a call to arms.

DIET I have been unable to find any published data on the diet or foraging behaviour of this poorly studied species. Presumably, however, both are similar to putative relatives with similar altitudinal distributions; *peruviana* and *loricata*.

REPRODUCTION The nest and eggs are undescribed and there are no available data to suggest seasonality. The few specimens were all collected in July and August; none were in breeding condition (Graves *et al.* 1983). It is probable that the reproductive habits resemble those of Peruvian Antpitta.

TECHNICAL DESCRIPTION Sexes similar, with a few notable differences. The following description is based on Graves *et al.* (1983) and Schulenberg & Kirwan (2012a). *Adult male* Forecrown, lores and wide eye-ring ochraceous-buff. Rear crown olive-brown. Upperparts olive-brown. Throat white and malar white or buffy-white, with a black lateral throat-stripe. Underparts white, tinged buff on sides and flanks. Breast streaked black, flanks indistinctly streaked olive-brown. *Adult female* Similar to male, but lacks ochraceous-buff on head; crown olive-brown with only a faint suffusion of ochraceous at the front. Remaining upperparts generally darker brown than in male. *Adult Bare Parts Iris* dark brown; *Bill* maxilla black, distal half of mandible black, base light pink; *Tarsi & Toes* greyish-pink. *Immature* Undescribed, but probably similar to those of Peruvian Antpitta.

MORPHOMETRIC DATA There are very few published data available for this species. The following measurements are all from Graves *et al.* (1983). ♂♂ (*n* = 4) *Wing* mean 73.2mm, range 73.2–75.3mm; *Tail* mean 32.6, range 31.7–33.5mm; *Bill* culmen from base mean 17.1mm, range 16.7–17.3mm; *Tarsus* mean 25.9mm, range 24.8–26.7mm. ♀♀ (*n* = 1) *Wing* 71.2mm; *Tail* 31.9mm; *Bill* culmen from base 16.9mm; *Tarsus* 27.0mm. **MASS** 23.75g(lt), 22.5g(lt) (LSUMZ: these are the only specimens in the type series with weights, indicating the data from Graves *et al.* 1983 were derived from two, not four, specimens). **TOTAL LENGTH** 10.5–12.0cm (Clements & Shany 2001, Ridgely & Tudor 2009, Schulenberg *et al.* 2010).

DISTRIBUTION DATA Endemic to **PERU:** *Amazonas* trail 20km E of La Peca, 05°36.5'S, 78°21'W (LSUMZ 88092/93); García Trail, 05°40'S, 77°46'W (XC 296999, ML 22634731, recording and photo, R. Gallardy); Fundo Alto Nieva, 05°40.5'S, 77°47'W (R. Ahlman *in litt.* 2017); ÁCP Abra Patricia–Alto Nieva area, *c.*05°42'S, 77°49'W (XC); 6km WNW of Yambrasbamba, 05°44'S, 77°57'W (ML 66317811, photo F. Angulo P.); Pampas de Copal, 05°47.5'S, 77°50'W (F. Angulo P. *in litt.* 2017). *San Martín c.*22km ENE of Florida, 05°41'S, 77°45'W (LSUMZ 174018–022); *c.*26.5km WNW of Naranjos, 05°41.5'S, 77°44'W (XC 122211, E. DeFonso); 'Camp Utter Solitude', 05°43'S, 77°44.5'W (XC 132704, D.F. Lane). **NOTES** The location 'García' (05°50'S, 77°46.5'W) given by Hornbuckle (1999) is apparently in error (J. Hornbuckle *in litt.* 2016) and was meant to refer to the García Trail location listed above.

TAXONOMY AND VARIATION Monotypic. Although phylogenetic relationships in *Grallaricula* have not been investigated in detail, Ochre-fronted Antpitta is similar in appearance, elevational distribution and habitat preference to Peruvian Antpitta. These similarities strongly suggest a sister-species relationship (Graves *et al.* 1983, Schulenberg & Kirwan 2012a).

STATUS Known only from the Andean Ridge-top Forests EBA (047), one of the most threatened and least protected regions of conservation concern (Wege & Long 1995, Stattersfield *et al.* 1998). Originally considered Near Threatened (Collar *et al.* 1994, Stattersfield *et al.* 1998), Davies *et al.* (1997) encouraged an upgrade to Vulnerable. Following this, in 2000 the species was listed as Endangered, a classification maintained to date (BirdLife International 2017). This decision was largely based on its very small range (2,300km²) and continuously declining habitat, especially in the Cordillera de Colán (Davies *et al.* 1994, 1997), although the Abra Patrica area is also under continued threat (BirdLife International 2017). Ochre-fronted Antpitta has previously been described as 'uncommon or rare' (Krabbe & Schulenberg 2003) or 'uncommon (and local?)' (Schulenberg *et al.* 2010), but genuinely rare is perhaps the best descriptor. Outside two specimens from SN Cordillera de Colán (LSUMZ 88092/93) and the occasional eBird record it remains virtually undocumented away from a few key localities in BP Alto Mayo and ACP Abra Patricia–Alto Nieva (Hornbuckle 1999, Rosas 2003, Angulo *et al.* 2008). It is clear that proper protection of upper-elevation forest in Alto Mayo is essential and protection of habitat in the Cordillera de Colán is of urgent priority (Angulo *et al.* 2008, Schulenberg & Kirwan 2012a).

OTHER NAMES Spanish Ponchito Frentiocre (Krabbe & Schulenberg 2003); Tororoí Frentiocrácea (Clements & Shany 2001). **French** Grallaire à front ocre. **German** Ockerstirn-Ameisenpitta (Krabbe & Schulenberg 2003).

Ochre-fronted Antpitta, adult female, Alto Nieva, San Martín, Peru, 28 October 2015 (*David Beadle*).

Ochre-fronted Antpitta, adult male, Fundo Alto Nieva, San Martín, Peru, 27 May 2013 (*Carlos Calle*).

Ochre-fronted Antpitta, adult female, Alto Nieva, San Martín, Peru, 28 October 2015 (*David Beadle*).

Ochre-fronted Antpitta, adult female, Alto Nieva, San Martín, Peru, 28 October 2015 (*David Beadle*).

Ochre-fronted Antpitta, adult female, Abra Patricia, Amazonas, Peru, 11 September 2014 (*Nick Athanas/Tropical Birding*).

Ochre-fronted Antpitta, adult male, Fundo Alto Nieva, San Martín, Peru, 27 May 2013 (*Carlos Calle*).

RUSTY-BREASTED ANTPITTA
Grallaricula ferrugineipectus   Plate 24

Grallaria ferrugineipectus P. L. Sclater, 1857, Proceedings of the Zoological Society of London, vol. 25, p. 129, 'near Caracas', Venezuela.

Rusty-breasted Antpitta is distributed discontinuously through the Andes from northern and western Venezuela south to western Bolivia. Like other members of the genus *Grallaricula*, this species is difficult to observe, even with playback, as these tiny birds can be incredibly difficult to spot while perched low and inconspicuously in the undergrowth. The plumage is mainly brown above and on the head, becoming warmer and more rufous over much of the face and underparts, although the belly is largely white. There is some degree of association between this species and bamboo stands, at least locally, but in general Rusty-breasted Antpitta inhabits humid or semi-humid montane forest. Compared to other *Grallaricula*, the nesting biology is fairly well documented, apparently differing by having a grey-green (rather than buffy-brown) ground coloration to the eggs. There is no aspect of the biology and natural history unworthy of further investigation. Of particular interest, however, would be a taxonomic revision of the group, helping to delimit subspecific ranges and to evaluate the taxonomic status of the southernmost population (*leymebambae*).

IDENTIFICATION 10–12cm. Sexes similar. Plain brown above, with a poorly differentiated brown or rufous-brown crown, a large buffy or white loral spot, and a prominent clean white or whitish-buff arc behind the eye. Throat and breast rufous-buff in the two northern subspecies, in Venezuela and Colombia, but more tawny-buff in the southern subspecies. Central belly white. In general terms, similar to Slate-crowned Antpitta, but the latter's crown is distinctly slate-grey. Slate-crowned Antpitta also has darker brown (or olive-brown) upperparts. The underparts of the somewhat similar Ochre-breasted Antpitta are scalloped with dusky or blackish markings, and are often more extensively whitish below. Where their ranges overlap, Ochre-breasted tends to occur at lower elevations than Rusty-breasted, and Slate-crowned occurs at higher elevations.

DISTRIBUTION Resident in montane areas from the coastal mountains of Venezuela south to Bolivia. To date, however, records are very scattered and its known range is highly fragmented, apparently into many rather isolated populations. Future records from additional populations may reveal greater connectivity, but aside from fairly evenly distributed populations in coastal and Andean Venezuela (*ferrugineipectus*) and eastern Peru and north-east Bolivia (*leymebambae*), there appear to be isolated populations in the Sierra Nevada de Santa Marta (*ferrugineipectus*), Sierra de Perijá (*rara*), Central Colombian Andes (*rara*), East Andes in Cundinamarca (*rara*), West Andes in north-west Ecuador (*leymebambae*) and the Central/West Andes in south-west Ecuador and northern Peru (*leymebambae*). Additionally there is apparently a poorly understood population in Norte de Santander that is provisionally considered to represent *rara* (Phelps & Phelps 1963, Schulenberg & Remsen 1982, Hilty & Brown 1983, Ridgely & Greenfield 2001, Hennessey *et al.* 2003b, Hilty 2003, Schulenberg *et al.* 2007, McMullan *et al.* 2010, Robbins *et al.* 2013).

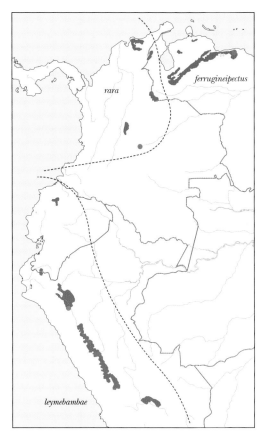

HABITAT Rusty-breasted Antpitta, in general, inhabits the undergrowth of humid and semi-humid montane forests, especially near thickets and vine tangles (Meyer de Schauensee 1964, Hilty & Brown 1986, Hilty 2003, Walker & Fjeldså 2005) and the edge of *Chusquea* bamboo-covered landslides (Fjeldså & Krabbe 1990, J. Freile *in litt.* 2017). The elevational distribution varies across its broad latitudinal range, apparently increasing in the south: 250–2,200m (Venezuela; Phelps & Phelps 1963, Giner & Bosque 1998); 700–2,000m (Colombia; McMullan *et al.* 2010); 1,750–3,350m but mostly 2,500–3,200m (Peru; Weske 1972, Walker & Fjeldså 2005, Walker *et al.* 2006; Schulenberg *et al.* 2007, Ridgley & Tudor 2009).

VOICE The song of the nominate subspecies (Sierra Nevada de Santa Marta, Colombia, and the Andes and coastal cordilleras of Venezuela) lasts 2.0–2.2s, is repeated at intervals of 7–9s, and consists of a series of 15–16 chipping notes in three parts. The first part is of 5–6 notes delivered with increasing volume, with pitch at 2.2kHz or rising from 2kHz to 2.2kHz. The second part comprises eight notes that rise from 2kHz to 2.4kHz and the last part is of 1–2 descending notes delivered at *c.*2kHz (Krabbe & Schulenberg 2003). The song is transcribed as *twa-twa-twa-twa-twa-twa-twa-cwi-cwi-cwi-cwi-cwi-cwi-cwi, cu-cu*, with the *cwi* notes higher and louder, and the last two notes lower (Schwartz 1957), or as *twa-twa-twa-twa-twa-twa-qwe-qwe-qwe-qwe-qwe-qwe-qwi-qua-qua* (Hilty 2003). The song of race *leymebambae* of Peru and Bolivia is a series of 15–18 notes, given at a pace of 6.3–6.7 notes, with a pitch of 2.6–3.0kHz (male?) or 3.0–3.2kHz (female?) (Krabbe & Schulenberg 2003). This song is a moderate to fast-paced (6–9 notes/s), even-pitched or slightly rising-falling, pure-toned series of whistled notes: *hee-hee-hee-hee-hee-hee-hee-hee* (D.F. Lane in Schulenberg *et al.* 2007). Krabbe & Schulenberg (2003) described a slower version of this song, 2.3–2.6s long and consisting of an even-paced (3.7 notes/s) series of 9–10 soft notes at 3.1kHz, with the first 2–3 notes lower and gradually rising from 2.9kHz. Calls include 'a sad, liquid *quierk* or doubled *quiu quiu* alarm call' (*ferrugineipectus*, Hilty 2003) and (*leymebambae*) a descending *tew* and *tew tip* (the number of *tip* notes variable), as well as a moderately paced (3–5 notes/s) slightly descending series of hollow, descending whistles: *chew-chew-chew-chew-chew* (D.F. Lane in Schulenberg *et al.* 2007).

NATURAL HISTORY Rusty-breasted Antpittas forage alone or in pairs, up to 2–3m above ground in the forest understorey, but rarely on the ground (Meyer de Schauensee & Phelps 1978, Fjeldså & Krabbe 1990, Ridgely & Tudor 1994, Hilty 2003). They forage predominantly by hopping on mossy stems and thin horizontal branches, occasionally flicking their wings and tail, and sometimes making short fly-catching sallies (Walker & Fjeldså 2006). Schwartz (1957) and subsequent authors have observed that, once encountered, individuals may be relatively confiding, and will often pause for long periods on a nearby branch, constantly twitching their rump in the rhythmic manner stereotypical of the genus. They perch on thin twigs, either horizontal or angled (but rarely vertical), probing into mossy branches, under leaves, on bark, and occasionally into leaf litter in search of prey. They sometimes make short sallies or upward strikes to glean prey from nearby vegetation, but rarely make long flights. No specific data have been published on territory defence, maintenance or fidelity, but the repeated inter-annual visits of Schwartz (1957) to the same area suggested that territories may be held for at least several years, presumably by the same pair. In the same study Schwartz estimated that territories were around 3.5ha in size.

DIET The diet is not as well documented as other aspects of its biology. Schwartz (1957) recorded small spiders, moths, cockroaches, beetles, crickets and other invertebrates being brought to feed nestlings. Parker *et al.* (1985) described the stomach contents of Peruvian specimens (*leymebambae*) as 'beetles and other insects'. Fierro-Calderón *et al.* (2006) reported the stomach contents of one unsexed individual (*rara*) as hymenopterans, beetles, orthopterans and hemipterans.

REPRODUCTION The breeding biology is one of the best documented in the genus. Todd & Carriker (1922) provided the first published information on breeding biology based on the cursory examination of a single nest with eggs at Cincinnati in the Sierra Nevada de Santa Marta (*ferrugineipectus*). The first detailed studies came from Schwartz (1957) who studied the nominate race at a site 2km south of Petare (850–900m), Miranda, Venezuela. Apart from the details provided below, Schwartz (1957) noted that both adults, during his observations of the nestling period, occasionally lean into the nest and probe into its lining after feeding the nestlings. He interpreted this as a form of nest cleaning, noting that often they appeared to ingest small items after probing rapidly several times. This behaviour is well documented and widespread in other antpittas (Greeney *et al.* 2008a) and these authors

similarly interpreted this behaviour (see discussion under Slate-crowned Antpitta). In 1969, Schäfer described nests of the nominate race as being 0.5–1.0m above ground in the undergrowth, and containing grey-green eggs marked brown. No citation was provided, but it is safe to assume that these generalisations were largely based on Schwartz (1957), which was the first detailed study of the species. No further breeding data were published until recently, when Niklison *et al.* (2008) published an excellent and detailed report on the nesting biology of the nominate race based on observations at 40 nests in PN Yacambú, Lara, Venezuela. This study provides us with the only quantified information on nesting success. All of the nesting failures observed by Niklison *et al.* (2008) were thought to have been caused by predation, and overall Mayfield (1961) calculated daily survival probability (±SE) was 0.94 ± 0.01, which gives a predicted success rate of just 15%. Below, I have supplemented observations from the aforementioned work with my own data from the north-west Ecuadorian population of uncertain racial affinity (see Geographical Variation). **NEST & BUILDING** Most described nests have been in areas of natural or human-modified second growth, generally in microhabitats with abundant vine tangles or stands of *Chusquea* sp. bamboo. Nests are frail structures, consisting of a slightly concave platform of unbranched twigs and leave petioles, on top of which dark (black or red-brown) rootlets, fungal rhizomorphs and thin, flexible leaf rachides are woven into a shallow cup (Schwartz 1957, Niklison *et al.* 2008). Nests contain little or no moss, similar to those of Slate-crowned Antpitta. Schwartz (1957) found nests at heights of 0.6–1.2m, with three active nests at 0.6m, 0.75m and 0.9m above ground. He described them as being supported by thin vine tangles and small bushes. Most nests found by Niklison *et al.* (2008) were supported by slender vines or placed on top of palm leaves at heights of 0.5–2.0m (*n* = 17; mean 1.30 ± 0.14m). Published nest dimensions are: outer diameter *c.*13–17cm (but see below*), outer height 6.0cm, 4.5cm, 3.5cm, inner diameter 5.50cm, 5.75cm, 5.75cm (but see below**), inner depth 2.5cm, 3.0cm, 2.2cm (*n* = 3; Schwartz 1957). Niklison *et al.* (2008) provided the following nest measurements (*n* = 20, mean ± SD); outer diameter 10.5 ± 0.3cm, outer height 4.4 ± 0.3cm, inner diameter 6.0 ± 0.2cm, inner depth 2.2 ± 0.2cm. (*) Schwartz (1957) provided two measurements of outer diameter, taken at perpendicular angles, and appears to have measured to the ends of the longest twigs protruding from the base of the nest, perhaps explaining the differences in external diameters provided by Schwartz and Niklison *et al.* (2008). Above I have given a rough approximation of what I interpret to be the span of the 'bulk' of the nest platform. The actual measurements given by Schwartz for the diameter (apparently including loose sticks in the base) of the three nests he measured were: 18.0 × 8.5cm, 16.0 × 12.0cm, and 20.0 × 13.5cm. He also provided the following measurements (respectively) for the outer diameter of the more tightly woven 'inner nest' or egg cup: 6.0 × 5.0cm, 6.5 × 6.0cm, 7.0 × 7.0cm. Furthermore, with respect to the above nest measurements (**) Schwartz (1957) also measured the internal diameter of nests at perpendicular angles. For simplicity, above I provided the mean of these two measurements. His actual data for three nests were: 6.0 × 5.0cm, 6.0 × 5.5cm, 6.0 × 5.5cm. **EGG, LAYING & INCUBATION** Eggs are short and subelliptical. Todd & Carriker (1922) reported pale greenish-white eggs marked with burnt umber, mainly at the larger end. Schwartz (1957) described their coloration as 'very pale emerald' or 'pale gray-green', or 'pearly-green' (translated from Spanish) and noted that eggs appeared greener as incubation progressed. He described eggs as flecked and spotted with red-brown, sepia and lavender markings of various shades, and relatively dispersed across the entire shell. Niklison *et al.* (2008) described eggs similarly, and their figure of an egg in the nest confirms that the grey-green ground colour of Rusty-breasted Antpitta eggs differs from those described for other species of *Grallaricula*, all of which have a pale brown ground colour (Greeney *et al.* 2008a). Fresh egg weight ranges from 2.40g to 3.02g (*n* = 15, mean 2.78 ± 0.05g, Niklison *et al.* 2008). Mean dimensions in one study were 18.8 ± 1.3mm by 17.3 ± 1.5mm (*n* = 3, Niklison *et al.* 2008), while those of another were 19.9 × 15.4mm (*n* = 6; Schwartz 1957). The normal clutch size appears to be two eggs. Todd & Carriker (1922) reported a clutch of two eggs, Schwartz (1957) a clutch size of two eggs at three nests, and Niklison *et al.* (2008) observed two eggs or two nestlings in 37 occupied nests. Schwartz (1957) estimated incubation at 16–17 days at one nest, but was unable to determine if both sexes incubated. Schwartz (1957) did not quantify incubation rhythms in detail, but noted that, on 15 visits to nests during incubation at various times of day, only once (in early morning) was the nest unattended, suggesting fairly constant coverage during daylight hours. In the Lara study, both adults participated in incubation, covering the eggs for 87–99% of daylight hours (Niklison *et al.* 2008). Niklison *et al.* (2008) observed a mean (±SE) incubation period of 17.0 ± 0.12 days (*n* = 4, range 16.0–17.5 days).

NESTLING & PARENTAL CARE The following description of nestling development is based on my translation of Schwartz (1957). Rusty-breasted Antpittas are naked at birth, with 'orange-flesh' skin that darkens to greyish or blackish (especially dorsally) by day four. Rictal flanges are pale orange and the mouth lining is bright orange or scarlet-orange, as described for other members of the genus (Greeney 2012a). By day four the eyes are still closed and pin feathers on the wings have emerged from the skin 2–3mm. The dense covering of wool-like feathers has begun to emerge from the skin, with down on the head forming a distinct cap. By four days of age nestlings respond to an observer bumping the nest by begging vigorously for food. By ten days the nestlings are well covered in 'wool-like' down feathers, which are predominantly red-brown, becoming more orange-brown below and whitish near the vent. This description closely matches photos of other mid-aged *Grallaricula* nestlings in Greeney (2012a). At ten days the pins of the inner primaries have emerged from the skin 17–20mm and have begun to break their sheaths 2–5mm. At 13 days the inner primaries measure 27–30mm and have emerged 16–18mm from their sheaths. Niklison *et al.* (2008) reported that the eighth primary feather begins to emerge from its sheath on day eight of the nestling period. Schwartz (1957) noted that both sexes fed and brooded the nestlings with relatively equal effort, and estimated the nestling period as 14–16 days. Subsequently, Niklison *et al.* (2008) reported a nestling period of 13.37 ± 0.37 days, suggesting that Schwartz's (1957) observations may have over-estimated or that there is some geographical variation in nestling development. Combined with the incubation period given above, this is the shortest nesting cycle described for the genus *Grallaricula* (Greeney *et al.* 2008a). The percentage of daylight hours that adults brooded the nestlings decreased with nestling age in the study by Niklison *et al.* (2008), who reported that nestlings grew slowly (using the *k* growth constant of Ricklefs 1967), with

respect to both tarsus length (*k* = 0.24) and mass (*k* = 0.41). Niklison *et al.* (2008) confirmed that both sexes brought small arthropods to the nest and recorded the following details of nestling provisioning behaviour at nests with two young: 3.77 trips/hr (*n* = 1 nest) on the first day of the nestling period; 3.91 ± 1.06 trips/h (*n* = 3 nests) on day two; 4.48±0.07 trips/h (*n* = 2 nests) on day six; 9.08 trips/h (*n* = 1 nest) on day eight. **SEASONALITY** Nesting in Venezuela begins with the rainy season and Schwartz (1957) suggested that the species is especially sensitive to weather with regards to the initiation of breeding. Most clutches observed by Niklison *et al.* (2008) in Lara were initiated in April, and the distribution of approximate laying dates suggested a single annual reproductive peak in this area. In Peru, in addition to the records provided below, Fjeldså & Krabbe (1990) observed fledglings in May and June in Huánuco. They also recorded a fledgling from Bolivia (La Paz) in January. **BREEDING DATA** *G. f. ferrugineipectus* Three clutches completed 25–30 May 1956, 2km S of Petare (Schwartz 1957); partially incubated eggs, 10 October 1916, Cincinnati (Todd & Carriker 1922). *G. f. leymebambae* Fledgling, 8 December 1983, E of Florida de Pomacochas (Fjeldså & Krabbe 1990, LSUMZ 116975); fledgling, 16 July 1973, Huailaspampa (Fjeldså & Krabbe 1990, LSUMZ 74090); juvenile transitioning, 18 February 1973, Huailaspampa (FMNH 296686); subadult, 20 March 1973, Huailaspampa (FMNH 296690); subadult transitioning, 30 November 2008, 5km N of Rodríguez de Mendoza (FMNH 473755); two subadults transitioning at Lonya Grande, 7 and 17 July 2008 (MSB 32068/32441); subadult transitioning, 20 May 1972, Bosque Taprag (FMNH 293398); enlarged testes, 5 October 1967, Chuquisyunca (FMNH 282599).

TECHNICAL DESCRIPTION Sexes similar. The following description is from Greeney (2013e) and is based on Bangs (1899) and Krabbe & Schulenberg (2003). It refers to nominate *ferrugineipectus* (see also Geographic Variation). *Adult* Upperparts, including head-sides, pale to dark brown with a slight olivaceous wash. Wings and tail tawny-brown with primaries, secondaries and tertials edged dull olive-cinnamon. The face bears a large (but somewhat indistinct) buff or whitish loral spot and a prominent buffy or whitish postocular crescent. Underparts bright ochraceous or tawny-rufous, with a semi-concealed white crescent on the lower throat. Central belly also white. **Adult Bare Parts** (from Krabbe & Schulenberg 2003): *Iris* brown; *Bill* black except base of mandible, which is white or pinkish; *Tarsi & Toes* pink or dusky grey-pinkish. *Fledgling* Described by Krabbe & Schulenberg (2003) as 'covered with vinaceous-pink fluffy down (least so on head), wing-coverts tipped rufous'. I examined photos (J.V. Remsen) of two fledglings in LSUMZ (see Breeding Data) that match this description. Both are fledglings that are only just beginning to show scattered adult-like contour feathers on the upper back. They are otherwise almost entirely covered in rusty-brown (i.e. 'vinaceous-pink') down that is slightly washed buff on the flanks and vent, approaching white on central belly. Chin whitish, washed ochraceous-buff and partially separated from a buff-washed white upper-chest crescent by an indistinct rusty-brown 'collar' of down feathers. The Amazonas specimen (LSUMZ 116975) appears slightly younger, with less white on chin and the white chest crescent strongly washed ochraceous-buff and fully separated from the chin patch by a band of rusty-brown on throat. As described by Krabbe & Schulenberg (2003), the upperwing-coverts are tipped bright rufous. Though not described by them, Walker & Fjeldså (2005) illustrated a 'juvenile' that closely matches my description. **Fledgling Bare Parts** (LSUMZ 74090 label; D. Tallman) *Iris* dark brown; *Bill* black; *Tarsi & Toes* brown. Though not noted on the label, the pale coloration to which the bill has dried strongly suggests that it was not all black, and almost certainly was partially orange or yellowish near the tip of the maxilla and at least basally on the mandible. *Juvenile/ Subadult* The next youngest skin I examined (FMNH 296686) was almost in full adult plumage. However, it had scattered rusty-brown down plumes on the hindcrown and bright rufous-chestnut tips to the upperwing-coverts. These coverts appeared quite fresh and it seems likely they had been acquired at some point post-fledging (i.e., a second set of immature coverts).

MORPHOMETRIC DATA *G. f. ferrugineipectus Wing* 66.0mm; *Tail* 20.3mm; *Tarsus* 40.6mm (*n* = 1♂, holotype, Sclater 1857a). *Wing* 61mm, 62mm; *Tail* 29.0mm, 29.4mm; *Bill* exposed culmen 13.0mm, 12.3mm; *Tarsus* 23.2mm, 23.6mm (*n* = 2, ♂,♀, Bangs 1899). *Wing* 61–66mm, mean 64mm; *Tail* range 30–32mm, mean 31mm (*n* = 3♂♂, Carriker 1933). *Wing* mean 63.9 ± 1.6mm, range 61.0–67.0mm (*n* = 14); *Tail* mean 30.7 ± 1.6mm, range 29.0–34.0mm (*n* = 13); *Bill* [total] culmen mean 15.8 ± 0.5mm, range 15.0–16.5mm (*n* = 13), width mean 5.3 ± 0.4mm, range 5.0–5.6mm (*n* = 2); *Tarsus* mean 22.5 ± 0.9mm, range 21.0–23.0mm (*n* = 14♂?, Donegan 2008). *G. f. rara Wing* 69mm; *Tail* 32mm; *Bill* 14mm (*n* = 1♂?, holotype, Hellmayr & Madarász 1914). *Wing* mean 63.0 ± 1.7mm (*n* = 13, range 60.0–66.0mm); *Tail* mean 30.6 ± 1.3mm (*n* = 14, range 29.0–33.0mm); *Bill* total culmen, mean 16.0 ± 0.7mm (*n* = 13, range 15.0–17.0mm), width 5.0mm (*n* = 1); *Tarsus* mean 23.0 ± 0.9mm (*n* = 14, range 21.0–24.0mm) (Donegan 2008). *Wing* 68mm, 68mm; *Tail* 34mm, 32mm; *Bill* exposed culmen 14mm, 14mm; *Tarsus* 23mm, 23mm (*n* = 2, ♂,♀, Olivares 1969). *G. f. leymebambae Wing* 71–72mm; *Tail* 43–44mm (*n* = 2♂♂, Carriker 1933). *Wing* mean 72.0mm (range 69.5–74.9mm); *Tail* mean 39.9mm (range 38.0–41.6mm); *Bill* culmen from base, mean 16.7mm (range 15.4–17.5mm); *Tarsus* mean 26.2 (range 25.0–27.2mm) (*n* = 7♂♂, Graves et al. 1983). *Wing* mean 67.7mm (range 66.8–68.4mm); *Tail* mean 37.6mm (range 36.1–38.6mm); *Bill* total culmen mean 15.9mm (range 15.5–16.1mm); *Tarsus* mean 26.2 (range 25.9–26.4mm) (*n* = 3♀♀, Graves et al. 1983). **Mass** Mean 17.5g, range 15–21g (*n* = 23♂♂, *leymebambae*, Parker et al. 1985), mean 16.8g, range 13–18g (*n* = 20♀♀, *leymebambae*, Parker et al. 1985), mean 16.6g, range 15.5–18.5g (*n* = 7♂♂, *leymebambae*, Graves et al. 1983). Mean 16.7g, range 16.5–17.0 g (*n* = 3♀♀, *leymebambae*, Graves et al. 1983). Mean 13.9g (*n* = 2♂?, *ferrugineipectus*, Faaborg 1975), 15.0g (*n* = 1♂, *ferrugineipectus*, Thomas 1982), 18.0g (*n* = 1♂?, *leymebambae*, Weske 1972), 16.48 ± 0.43g (*n* = 10♂?, *ferrugineipectus*, Niklison et al. 2008). **Total Length** 9.7–12.1cm (Sclater 1857a, Meyer de Schauensee 1964, 1970, Meyer de Schaeuensee & Phelps 1978, Dunning 1982, Hilty & Brown 1986, Fjeldså & Krabbe 1990, Clements & Shany 2001, Hilty 2003, Krabbe & Schulenberg 2003, Walker & Fjeldså 2005, Ridgely & Tudor 2009, McMullan et al. 2010).

TAXONOMY AND VARIATION Although the species' phylogenetic relationships have not been investigated, Parker *et al.* (1985) remarked that it might form a superspecies with Slate-crowned Antpitta, a suggestion reflected by its usual position in taxonomic arrangements (Dickinson 2003, Ridgely & Tudor 2009, Dickinson &

Christidis 2014, del Hoyo & Collar 2016). Three subspecies are recognised, but following Krabbe & Schulenberg's (2003) suggestion that subspecies *leymebambae* 'probably' warranted recognition as a separate species based on vocalisations, recent authors (Ridgely & Tudor 2009, del Hoyo & Collar 2016, Krabbe *et al.* 2017b) have elevated the latter taxon to species rank. This recommendation seems justified, but given the complex, fragmented distribution of known populations, further research is warranted to clearly draw taxonomic boundaries. Taxon *rara* also was described as a separate species, and is the most divergent in terms of plumage coloration, although differences in vocalisations are poorly known. There is also a fairly recently discovered population in north-west Ecuador (here included in *leymebambae*) that may represent an undescribed taxon (Krabbe *et al.* 2017b). Following Krabbe *et al.* (2017b), I have provisionally included records from Norte de Santander (Sinai, El Diamante) as *rara*, but their racial affinity has not been carefully evaluated. In addition, I have provisionally included records from southern Santander and the east slope of the West Andes (Caldas) with *rara*. Even more tentatively, records from the west slope of the West Andes in Valle del Cauca are here treated as *leymebambae*. If the north-west Ecuadorian population proves to warrant separation as a new taxon, these nearby Colombian records seem likely to be closely allied. Clearly, this puzzle deserves closer inspection.

Grallaricula ferrugineipectus ferrugineipectus (Sclater, 1857), Proceedings of the Zoological Society of London, vol. 25, p. 129, 'near Caracas', Distrito Federal, Venezuela. The nominate subspecies was originally described as *Grallaria ferrugineipectus* and its precise collecting locality is unknown. It may or may not have come from the Caracas area, as it appears to have been a trade skin, but was, most likely, collected somewhere in the Venezuelan coastal mountains. This subspecies occurs in Colombia and Venezuela. In Colombia it is found only in the Sierra Nevada de Santa Marta, in the north (Meyer de Schauensee 1950b), but in Venezuela it is distributed along the Andes from Mérida and Barinas north to Trujillo and Lara. It is also known from the mountains of Lara and Yaracuy, and in the coastal cordilleras east to the Distrito Federal and Miranda (Phelps & Phelps 1963, Hilty 2003). **Specimens & Records VENEZUELA**: *Vargas* Picacho de Galipán, 10°34'N, 66°55'W (AMNH, MCZ); *c.*1km E of La Macanilla, 10°30'N, 67°09'W (eBird; J. Kvarnback). *Miranda* La Tahona, 10°26'N, 66°51'W (J.E. Miranda T. *in litt.* 2017); 2km S of Petare, *c.*10°26'N, 66°48'W (Schwartz 1957, ML 7089–092, P.A. Schwartz); Bosques de La Lagunita, 10°24.5'N, 66°45'W (J.E. Miranda T. *in litt.* 2017); near San Antonio de Los Altos, 10°22'N, 66°58'W (J.E. Miranda T. *in litt.* 2017); Club Monteclaro, 10°21.5'N, 66°53.5'W (photo; D. Ascanio); Santa Lucía, 10°18'N, 66°40'W (COP); Barlovento region, no GPS (Cordero-Rodriguez 1987). *Distrito Federal* San Antonio de Galipán/San Isidro, 10°33'N, 66°53'W (COP 62592–594); La Montañita, 10°33'N, 66°55'W (COP 3347); Silla de Caracas, 10°32.5'N, 66°51'W (ML 22084861, photo J. McGowan); Los Venados, 10°32'N 66°54'W (ML 38408751/771, J. McGowan); 1km SW of Fuerte Tiuna, 10°26'N, 66°55'W (ML 62198/99, P.A. Schwartz). *Aragua* Colonia Tovar–Limón Road, 10°25'N, 67°17'W (ML 62400; P.A. Schwartz); Quebrada Honda, 10°22'N, 67°09'W (J.E. Miranda T. *in litt.* 2017); 6km E of Cansamacho, 10°22'N, 67°41'W (ML: P.A. Schwartz); Cerro Golfo Triste, 10°02'N, 66°55'W (COP 19220–224).

Carabobo San Estéban, 10°25'N, 68°00'W (CM P35266); Sierra de Carabobo, 10°22'N, 68°01'W (CM, UMMZ); Rancho Grande, 10°21.5'N, 67°40'W (CM); near Palmichal, 10°21'N, 68°12'W (XC 223750, P. Boesman); Sierra de Montalban La Rosa, 10°15'N, 68°18'W (XC 223743, P. Boesman). *Cojedes* *c.*7km S of Bejuma, 10°07'N, 68°16'W (ML 105919, C.K. Hanks). *Falcón* Serranía de San Luis, 11°12'N, 69°42'W (Boesman 1998, XC); Cerro Galicia, 11°11'N, 69°42'W (COP 81330); Mirimire, 11°09'N, 68°43'W (COP 63653); Cerro El Cogollal, 10°24'N, 70°44'W (COP 18517–524). *Yaracuy* Bucaral, 10°23'N, 68°49'W (COP 26764–767); Cerro El Candelo, 10°23'N, 68°50'W (COP 64000); near Rabo Frito, *c.*10°13'N, 68°29'W (J.E. Miranda T. *in litt.* 2017); Macizo de Nirgua, 10°02'N, 68°26'W (J.E. Miranda T. *in litt.* 2017). *Lara* Cerro El Cerrón, *c.*10°19'N, 70°38'W (COP 18525–528); 35km S of Cabudare, 09°55'N, 69°16'W (COP 72053–055); El Blanquito, 09°42.5'N, 69°34.5'W (J.E. Miranda T. *in litt.* 2017). *Portuguesa* *c.*9km SW of Biscucuy, 09°18.5'N, 70°03.5'W (J.E. Miranda T. *in litt.* 2017). *Mérida* Posada Monteverde, 08°45.5'N, 71°29'W (J. Beckers *in litt.* 2017); La Azulita, 08°43'N, 71°26.5'W (FMNH 289080–082, CM P90352); La Escasez, 08°39'N, 70°47.5'W (eBird; J. Kvarnback); Ejido, 08°33'N, 71°14'W (Hellmayr 1913, AMNH 492306). *Barinas* San Isidro Quarry, 08°50'N, 70°35'W (XC 65917, D.F. Lane); 16km NNW of Barinitas, 08°53'N, 70°29'W (XC 147596, F. Deroussen). **COLOMBIA**: *Magdalena* Las Vegas, 11°12'N, 73°53'W (Todd & Carriker 1922, CM, FMNH 254872); Las Nubes, 11°10'N, 73°56'W (Allen 1900, AMNH 71110); Minca, 11°08.5'N, 74°07'W (MHNG 1219056); *c.*2.2km SE of Minca, 11°07.5'N, 74°06'W (ML 40592071, photo K. Hansen); RNA El Dorado, 11°06.5'N, 74°04'W (ML 39684391, photo A. Van Norman); Cincinnati, 11°06'N, 74°05'W (Todd & Carriker 1922, USNM 387377–378, CM P42592); *c.*4km SSW of Palomino, 11°00'N, 73°40'W (XC 223746, P. Boesman); RNSC El Congo, 10°59'N, 74°04'W (Strewe & Navarro 2004). *Cesar* Nabusimaque, *c.*10°34'N, 73°36.5'W (J. Beckers *in litt.* 2017); Pueblo Viejo, 10°32'N, 73°25'W (Todd & Carriker 1922, ANSP 63088/89, MCZ 105581). *La Guajira* Río San Salvador, *c.*11°08'N, 73°32'W (Strewe & Navarro 2003); El Pueblito, 10°59'N, 73°27'W (Todd & Carriker 1922, CM); Páramo de Chirigua, 10°56'N, 73°22'W (AMNH 131004); Cherua, 10°52'N, 73°23'W (MCZ 106177–180, USNM 170556).

Grallaricula ferrugineipectus rara Hellmayr & Madarász, 1914, Annales Historico-Naturales Musei Nationalis Hungarici, vol. 12, p. 88, 'Dpt. Cundinamarca, Medina: Lanos'. The type locality clearly refers to the town of Medina (04°30'N, 73°21'W) on the east slope of the East Andes, *c.*54km north-east of Villavicencio, at an elevation of 575m in the Colombian *llanos* (Paynter 1997). This is at the low end of records for the species, so the true collecting locality is probably somewhere just west of there, closer to the foothills. Hellmayr, however, questioned the validity of this location (Cory & Hellmayr 1924) and most subsequent authors have 'moved' the type locality to an unspecified location on the west slope of the East Andes (Meyer de Schauensee 1950b, Olivares 1969). Given the relatively unexplored nature of this area, and the ghost-like ability of this and other antpittas to remain undetected, even in better-surveyed locations, it seems hasty to rule out the validity of Medina (or nearby foothills) as the true type locality. Originally described as *Grallaricula rara*, this subspecies is distributed in western Venezuela and the Andes of eastern Colombia. In Venezuela it occurs

in the Sierra de Perijá, where its range interrupts that of nominate *ferrugineipectus* (which occurs to the west in the Sierra Nevada de Santa Marta, and to the east in the Venezuelan Andes). As far as I can ascertain, however, there are no records from the Colombian part of the Serranía de Perijá (López-O. *et al.* 2014). In Colombia *rara* occurs on the east slope of the East Andes in Norte de Santander, and on the west slope of the East Andes in Cundinamarca (Meyer de Schauensee 1950b, Hilty & Brown 1986). Hellmayr & Madarász (1914) described the type (unsexed) as follows: 'Above reddish-brown, more reddish on the front and face, and washed with olive on the rump; a patch before the eye blackish; chin, throat and side of body rusty-brown; middle of belly and undertail-coverts pure white; underwing-coverts rusty-brown. Bill dark horn-brown, the base of mandible and feet yellowish.'

Specimens & Records VENEZUELA: *Zulia* Cerro Frontera, 10°05'N, 72°45'W (COP 54880); Cerro Pejochaina, 10°02'N, 72°49'W (long series, COP); 'Campamento Avispa', 10°16'N, 72°42'W (COP 57608); 5.5km SW of Canoquapa, 09°51'N, 72°56'W (COP 72837). **COLOMBIA**: *Norte de Santander* Pamplona, 07°23'N, 72°39'W (ANSP 167608); Sinai, 07°06'N, 72°14'W (IAvH); El Diamante, 07°34'N, 72°38'W (MLS 3946). *Santander* Suaita, 06°06'N, 73°26'W (ICN). *Caldas* Hacienda Tintiná, 05°15'N, 75°41'W (Alvarez *et al.* 2002, IAvH). *Cundinamarca* Vereda Guadualito, Yacopí, 05°29'N, 74°21'W (ICN 32526/27); Laguna de Tabacal, 05°01.5'N, 74°19.5'W (ICN 16547, photo D. Uribe; Fig. 1); Vereda El Peñon, 05°01'N, 74°14.5'W (XC 93042–044, O. Cortes-Herrera); Finca El Encanto, 05°00'N, 74°21'W (ICN 16282); Finca Agualinda, 04°59'N, 74°17.5'W (J. Beckers *in litt.* 2017); Sasaima, 04°58'N, 74°26'W (Niceforo 1947, Meyer de Schauensee 1950, Iafrancesco-V. *et al.* 1987, MLS 3945); Albán, 04°52.5'N, 74°26.5'W (Iafrancesco-V. *et al.* 1987, MLS 3944).

Grallaricula ferrugineipectus leymebambae Carriker, 1933, Proceedings of the Academy of Natural Sciences of Philadelphia, vol. 85, p. 21, Leymebamba, Dept. Amazonas, Peru, alt. 7,000 feet [= 2,134m, *c.*06°42.5'S, 77°48'W]. The holotype (ANSP 108131) is an adult male collected 30 August 1932, by M. A. Carriker. Bond (1950) referred to a paratype, ANSP 108132. This subspecies is found in extreme south-west Ecuador (Loja province) and north-west Peru (Dept. Piura). A disjunct population is also known from eastern Peru south of the Río Marañón along the east slope of the Andes from Amazonas south to western Bolivia (Dept. La Paz). Race *leymebambae* is very similar to nominate *ferrugineipectus*, but is considerably larger. Carriker (1933) in comparing the holotype male to the nominate subspecies, noted that it was similar 'in having the jugular patch and middle of abdomen white, but differs as follows: throat, breast and flanks rich cinnamon ochraceous, paler on throat and around border of the white abdominal patch, and with the chin and sides of throat indistinctly streaked with blackish and the chest mottled with sooty olive (no dusky mottling on throat or chest in *ferrugineipectus*, the underparts being clear cinnamon ochraceous, immaculate, and paler than *leymebambae*); lores and narrow front buffy ochraceous basally (white in *ferrugineipectus*), and with a blackish ring extending from the front of the auriculars forward around the eye and ending at the upper anterior side (absent in *ferrugineipectus*); pileum, back, and scapulars uniform rich olive brown, slightly dusky on the occiput and nape, while in *ferrugineipectus* these parts are light brown (darker in Santa Marta birds) and entirely without olive tinge. The lesser wing coverts are the colour of the back, but the median and greater ones are dark chestnut brown on the outer webs; the remiges are broadly edged with seal brown and the under wing coverts and edges of remiges are rich cinnamon ochraceous, more cinnamomeus on remiges and much darker and richer in colour than in *ferrugineipectus*. The bill is black, with the base of the mandible flesh, about the same colour as in *ferrugineipectus*'. Specimens from Bolivia differ slightly from those in central Peru by having the crown greyer, the centre of the back more olive, and have a reduced loral spot (Schulenberg & Remsen 1982), although the differences are possibly clinal (del Hoyo *et al.* 2017c). **SPECIMENS & RECORDS COLOMBIA**: *Valle del Cauca* Río Blanco, *c*.03°36'N, 76°48.5'W (UV 1656/1747). **ECUADOR**: *Imbabura* Cordillera de Toisán, 00°25'N, 78°47'W (O. Rodríguez, J. Freile *in litt.* 2016); *c*.6.6km ENE of Apuela, 00°22.5'N, 78°27'W (R.A. Gelis *in litt.* 2017). *Pichincha* Reserva Geobotánica Pululahua, 00°04.5'N, 78°31.5'W (MECN 7694–696, XC). *Loja* RP Bosque de Hanne, 04°21.5'S, 79°43'W (XC 262941/276049, J. Nilsson); RP Utuana, 04°22'S, 79°42'W (MECN 7700/01, ML). **PERU**: *Piura* Bosque de Cuyas, 04°36'S, 79°44'W (Vellinga *et al.* 2004); 3km N of Ayabaca, 04°36'S, 79°43'W (XC 264645/46, P. Coopmans); near Canchaque, *c*.05°21'S, 79°32.5'W (Schulenberg & Parker 1981, LSUMZ 78547, MHNSM 4331). *Lambayeque* El Porongo, 06°16.5'S, 79°28.5'W (Angulo *et al.* 2012, XC 66383/85, F. Angulo P.). *Amazonas* Cordillera de Colán, 05°18'S, 78°20'W (Dauphine 2008); trail 20km E of La Peca, *c*.05°36.5'S, 78°21'W (LSUMZ 88073–077); 'Campamento Palomino', 05°38.5'S, 78°15.5'W (F. Angulo P. *in litt.* 2017); *c*.30km by road E of Florida de Pomacochas (= Trocha Mono), 05°42'S, 77°49'W (LSUMZ 116974/75, XC 100405–407, J. King); *c*.33km by road NE of Ingenio (= Pedro Ruíz Gallo), 05°52'S, 77°57'W (LSUMZ 78548); 2km W of Abra Patricia, 05°47'S, 77°44'W (Davis 1986); Río Chido Trail, 05°48.5'S, 78°01.5'W (XC 296849, R. Gallardy); Lonya Grande, 06°05'S, 78°20'W (MSB); 'Camp Lowbrow', 06°05.5'S, 78°20.5'W (ML 58867261, D.F. Lane); 5km N of Rodríguez de Mendoza, 06°20'S, 77°26.5'W (FMNH 473754/55); Cerro Montealegre, 06°35'S, 77°33.5'W (FMNH 474185/86); *c*.6km W of Leimebamba, 06°42.5'S 77°52'W (ML 41867201, photo J. Sims); Abra Barro Negro, 06°44'S, 77°53.5'W (XC 8570/9271, H. van Oosten). *San Martín* *c*.22km ENE of Florida de Pomacochas, 05°41'S, 77°45'W (LSUMZ 174017); between Aguacate and Caanan, 06°54.5'S, 77°29'W (F. Schmitt *in litt.* 2017); Puerta del Monte, 07°34'S, 77°09'W (LSUMZ 104493). *La Libertad* Cumpang, 08°12'S, 77°10'W (LSUMZ 92494–501). *Huánuco* Uchiza Valley, above San Pedro, 08°39'S 76°53.5'W (ML 191430, M.G. Harvey); Bosque Zapatagocha, 09°40'S, 76°03'W (FMNH, LSUMZ); Huailaspampa, 09°42'S, 76°02'W (FMNH 296686–690, LSUMZ 74090); NE of Carpish Pass, 09°42'S, 76°05'W (long series, LSUMZ); Bosque Taprag, 09°44'S, 76°03'W (FMNH, LSUMZ); Playa Pampa, 09°57'S, 75°42'W (LSUMZ 128566); 4km SE of Cushi, 10°00'S, 75°40'W (ML, T.S. Schulenberg). *Pasco* Ulcumano Ecolodge, 10°38.5'S, 75°26'W and Bosque Schóllet, 10°40.5'S, 75°19'W (F. Schmitt *in litt.* 2017). *Junín* CC Puyu Sacha, 11°05.5'S, 75°25.5'W (MUSA 4263); Chuquisyunca, 11°15'S, 75°32'W (FMNH 282599); upper Satipo Road, 11°30.5'S, 74°51.5'W (P.A. Hosner *in litt.* 2017); Chucho Acha, 11°47.5'S, 74°46.5'W (XC 23640/41, F. Schmitt); Chilifruta, 12°00.5'S, 74°54'W (F. Schmitt *in litt.* 2017). *Cuzco* Cordillera Vilcabamba, *c*.12°38'S, 73°39'W (AMMH, USNM); ÁPC Abra Málaga, 13°07'S, 72°21'W (R. Ahlman *in litt.* 2017); Estación Biológica Wayqecha,

13°10.5'S, 71°35.5'W (XC 70150, M. Dehling). *Puno* Abra Maruncunca, *c*.14°14'S, 69°10'W (Robbins *et al.* 2013). **BOLIVIA**: *La Paz* Cotapata Track, 16°16'S, 67°51'W (P.A. Hosner *in litt.* 2017); Mina Copacabana, 16°17'S, 67°50'W (Martínez & Rechberger 2007); 1km S of Chuspipata, 16°18'S, 67°50'W (LSUMZ 102249–253).

STATUS Rusty-breasted Antpitta has a very large range and does not approach the threshold for Vulnerable under the range size criterion. In addition, population trends appear stable, and for these reasons the species has been evaluated as of Least Concern (BirdLife International 2017). It appears that the species is fairly tolerant of some degree of forest disturbance, presumably because its preferred habitat is one incorporating natural colonisers of disturbed habitats like vines and *Chusquea* bamboo. Nevertheless, even considering its apparent tendency to be easily overlooked, the species' known range is extremely patchy and there may be very specific habitat requirements that are, as yet, undocumented, making it much more of a habitat specialist than currently understood. Additionally, with apparently highly disjunct populations, and the presumed genetic diversity that accompanies such a range, this is a species which may be best evaluated region by region with respect to its conservation needs. Work is urgently needed to understand the true habitat requirements, as well as its true status and distribution. Once these aspects are better understood, its conservation needs can be evaluated more accurately (Hilty 1985). Most of the races appear to occur in at least one protected area, but it is unclear if the population of *rara* in the Sierra de Perijá occurs in the Iroka and Sokorpa reserves. **PROTECTED POPULATIONS** *G. f. ferrugineipectus* **Venezuela**: PN El Ávila (COP, AMNH, MCZ, ML); PN Yurubí (COP 26764–767); PN Sierra de la Culata (FMNH 289080–082, CM P90352, XC, ML); PN Terepaima (COP); PN Yacambú (XC 223742, P. Boesman); PN San Estéban (CM); PN Macarao (J. Miranda *in litt.* 2017); PN Sierra de San Luís (Boesman 1998, XC); PN Henri Pittier (Schäfer & Phelps 1954, CM, LSUMZ, XC); PN Tirgúa (eBird: C. Afán de Rivera). **Colombia**: PNN Sierra Nevada de Santa Marta (CM, MCZ, AMNH); RNA El Dorado (Salaman *et al.* 2007a, ML); RNSC El Congo (Strewe & Navarro 2004). *G. f. rara* **Venezuela**: PN Perijá (COP) and probably PN Tamá. **Colombia**: PNN Tamá (IAvH). *G. f. leymebambae* **Ecuador**: Reserva Geobotánica Pululahua (MECN); RE Cotacachi-Cayapas (Ridgely & Greenfield 2006); RP Bosque de Hanne (XC 262941/276049; J. Nilsson); RP Utuana (MECN). **Peru**: SN Cordillera de Colán (Dauphine 2008); RVS Laquipampa (Pratolongo *et al.* 2012); BP Alto Mayo (Davis 1986); ACP Abra Patricia–Alto Nieva (XC 100405–407, J. King); ACP Abra Málaga (R. Ahlman *in litt.* 2017); SH Machu Picchu (Walker & Fjeldså 2002, 2005); Zona Reservada Cordillera de Colán (Dauphine 2008). **Bolivia**: PN y ANMI Cotapata (Martínez & Rechberger 2007).

OTHER NAMES *Conopophaga browni* (Bangs 1899, Ogilvie-Grant 1912) described from Cherua (Cesar Dept.) is synonymous with nominate *ferrugineipectus* (Ridgeway 1911, Hellmayr 1913, Cory & Hellmayr 1924). *Grallaricula rara* (Cory & Hellmayr 1924); *Conopophaga* sp. (*ferrugineipectus*; Bangs 1898a). **Spanish** Tororoí Ferruginoso (Alvarez *et al.* 2002, Salaman *et al.* 2007a); Tororoí Pechirrojizo (Clements & Shany 2001); Ponchito Pechicastaño (Meyer de Schauensee & Phelps 1978, Phelps & Meyer de Schauensee 1979, 1994, Hilty 2003, Krabbe &

Schulenberg 2003, Verea *et al.* 2017); Ponchito Caraqueño Pechicastaño [*ferrugineipectus*], Ponchito Pechicastaño de Cundinamarca [*rara*] (Phelps & Phelps 1950); Ponchito pechicastaño norteño [*leymebambae*] (del Hoyo *et al.* 2017c). **English** Rusty-breasted Ant-Thrush (Brabourne & Chubb 1912); Rusty-breasted Grallaricula [*ferrugineipectus*] (Cory & Hellmayr 1924, Meyer de Schauensee 1950); Rufous-breasted Grallaricula [*rara*] (Cory & Hellmayr 1924,

Meyer de Schauensee 1950); Rufous-breasted Antpitta [*leymebambae*] (del Hoyo *et al.* 2017c); Leymebamba Antpitta [*leymebambae*] (Gill & Wright 2006, Ridgely & Tudor 2009). **French** Grallaire à poitrine rousse (Krabbe & Schulenberg 2003); Grallaire de Leymebamba [*leymebambae*] (del Hoyo *et al.* 2017c). **German** Rostbrust-Ameisenpitta (Krabbe & Schulenberg 2003); Orangebrust-Ameisenpitta [*leymebambae*] (del Hoyo *et al.* 2017c).

Rusty-breasted Antpitta, adult (*rara*), Laguna de Tabacal, Cundinamarca, Colombia, 16 November 2014 (*Daniel Uribe*).

Rusty-breasted Antpitta, adult (*leymebambae*), Reserva Geobotánico Pululahua, Pichincha, Ecuador, 1 June 2013 (*Nick Athanas/Tropical Birding*).

Rusty-breasted Antpitta, adult, (*ferrugineipectus*), RN El Dorado, Magdalena, Colombia, 25 May 2009 (*Alonso Quevedo*).

Rusty-breasted Antpitta, adult, (*ferrugineipectus*), RN El Dorado, Magdalena, Colombia, 25 May 2009 (*Alonso Quevedo*).

Rusty-breasted Antpitta, empty nest (*leymebambae*), Reserva Geobotánico Pululahua, Pichincha, Ecuador, 27 August 2012 and 17 March 2013 (*Harold F. Greeney*).

Rusty-breasted Antpitta, empty nest (*leymebambae*), Reserva Geobotánico Pululahua, Pichincha, Ecuador, 27 August 2012 and 17 March 2013 (*Harold F. Greeney*).

SLATE-CROWNED ANTPITTA
Grallaricula nana   Plate 24

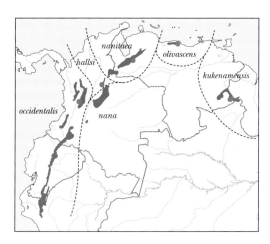

Grallaria nana Lafresnaye, 1842, Revue Zoologique par La Société Cuvierienne, 'vol.' 1842, p. 334, 'Colombie'.

Slate-crowned Antpitta is one of the most widespread species of *Grallaricula* and, in appropriate habitat, can be quite numerous. It occurs in the mountains from north coastal Venezuela south and west, somewhat discontinuously, to northernmost Peru, as well as disjunctly in south-east Venezuela and adjacent western Guyana. The plumage is rather plain; uniform brownish on the back and wings, with a notably greyer crown, and principally unmarked rufous underparts. Like others of its genus, Slate-crowned Antpitta is usually shy and difficult to observe, although it can be rather confiding. It occurs in the undergrowth of humid forest, almost exclusively associated with montane bamboos of the genus *Chusquea*. Across its broad geographical range, the species occupies a broad altitudinal band that spans at least 700–3,500m, although in most of its range the species is most abundant at 2,000–3,000m. Up to eight subspecies have been recognised, but recent authors have suggested that Slate-crowned Antpitta is better treated as three separate species, based on both vocalisations and morphology. Most authorities to date have partially followed this hypothesis by treating Sucre Antpitta of north-east Venezuela as a separate species, but have not yet elected to recognise the Tepui taxon, *G. n. kukenamensis*, specifically. The natural history and reproductive biology of this widespread species are incompletely known, but a few studies have reported data on nest architecture, eggs and nestling development, all from eastern Ecuador. Apart from requiring a careful evaluation of its taxonomic status and subspecific range limits, studies of its diet, foraging behaviour, territoriality and demography would be especially welcome.

IDENTIFICATION 10–12cm. Slate-crowned Antpitta is simply patterned, with a dark brown to olive-brown back contrasting with a grey crown. Below it is uniform ochraceous-orange or rusty-buff with the centre of the belly white. Some races have a white crescent on the upper breast or lower throat, but this character is somewhat variable. Most subspecies have some evidence of a pale orange-brown ocular ring and/or pre-loral spot and in some the breast feathers are variably edged dusky giving a somewhat scaled appearance (Plate 24, Fig. f; Fig. 2, page 436). Slate-crowned Antpitta is similar to Rusty-breasted Antpitta, but the slate-grey crown for which it is named contrasts with the brown back, unlike the more uniformly coloured upperparts of Rusty-breasted Antpitta. There is apparently no known range overlap between them, although they may come close to sympatry in parts of northern Peru. In several regions (i.e., Venezuela, East Andes of Colombia; T.M. Donegan *in litt.* 2017) Ochre-breasted Antpitta has extensive scalloping below, and compared to Slate-crowned might be confused with more heavily patterned individuals of the latter, although they apparently replace each other altitudinally in areas of potential overlap. Slate-crowned Antpitta is also similar to Sucre Antpitta of north-east Venezuela, with which it was long considered conspecific, but the two are allopatric.

DISTRIBUTION Occurs in montane areas from Venezuela south to Peru, being notably absent from the west slope of the Andes in most of Ecuador and in Peru. It occurs in the coastal cordillera of northern Venezuela, in Aragua and the Distrito Federal, and disjunctly in the Venezuelan Andes from Lara and Trujillo to Táchira, and into adjacent Colombia in Norte de Santander and Santander. The Colombian portion of the range is somewhat discontinuous, with several widely separated populations and no intervening records. This is especially true in the West Andes, with the recent addition of a third (apparently) isolated population in north-west Ecuador (J. Freile *in litt.* 2017). In the East Andes, Slate-crowned Antpitta is known much more continuously from the head of the Magdalena Valley to central Cajamarca, but not south of the Río Marañón (Phelps & Phelps 1950, Meyer de Schauensee 1950b, Snyder 1966, Parker *et al.* 1978, Hilty & Brown 1986, Ridgely & Greenfield 2001, Braun *et al.* 2003, Hilty 2003, Ayerbe-Quiñones *et al.* 2008, Schulenberg *et al.* 2010, Arbeláez-Cortés *et al.* 2011b, Calderón-Leytón *et al.* 2011). Finally, there is a highly disjunct population in the Tepui region of south-east Venezuela and adjacent Guyana that might well deserve recognition as a separate species (Donegan 2008).

HABITAT Dense thickets of *Chusquea* spp. bamboo are unquestionably the preferred habitat (Krabbe & Schulenberg 2003, Greeney 2013k). Where large areas of forest are mixed with patches of bamboo, the species is rarely seen far from dense tangles of *Chusquea*. Probably due to its affinity for naturally disturbed habitat, Slate-crowned Antpitta appears fairly tolerant of disturbance, and has been found in heavily degraded fragments of secondary forest (Delgado-V. 2002), but it is unknown how long the species persists in such habitats. Across the broad geographic range, it can be found as low as 700m, and up to the treeline, as high as 3,500m. Specific altitudinal ranges by country are: Venezuela 700–3,000m (Meyer de Schauensee & Phelps 1978, Giner & Bosque 1998, Hilty 2003); Colombia 1,300–3,500m (Hilty & Brown 1986, Verhelst *et al.* 2001, Losada-Prado *et al.* 2005, Krabbe *et al.* 2006, McMullan *et al.* 2010); Ecuador 2,000–2,900m (Freile 2000, Ridgely & Greenfield 2001); Peru 2,000–2,900m (Clements & Shany 2001, Schulenberg *et al.* 2010). In Colombia, and over most of its range south of there, records below 1,800m should be questioned (T.M. Donegan *in litt.* 2017).

VOICE Generally speaking, the song can be described as a rapid series of fife-like or piping notes that run slightly up the scale, then continue in a long arcing descent, *we* *'e'e'ti'ti'ti'ti'ti'ti'ti'ti'ti'te'te'te'e'e* ... fairly high-pitched and ventriloquial, making the bird hard to locate (Venezuela; Hilty 2003); 'a loud fast trill, ascending, then dropping (3s), *we'ti'ti'ti'ti'ti'te'tee'too*' (Colombia; Hilty & Brown 1986); 'a short, pretty, rather high-pitched series of 14–18 notes that fade and descend slightly, e.g., *we-e-e-e-e-e-e-e-e-e-e-ew*' (Ecuador; Ridgely & Greenfield 2001b); and 'a moderately paced (about 7 notes/s), rising-falling series of plaintive whistled notes: *pee-pee-PEE-pee-pee-pee-pee-pee*' (Peru; D.F. Lane in Schulenberg *et al.* 2007). There is subtle geographic variation in song throughout the range, which was documented thoroughly by Donegan (2008, 2010). Populations in the West Andes of Colombia (currently assigned to race *occidentalis*) differ from others (including other populations of *occidentalis*) by having a greater number of notes (mean 20.54 ± 2.54 notes, range 15–23), a greater frequency variation (mean 0.71 ± 0.14kHz, range 0.51–0.97) and a greater change of pace (mean 2.38 ± 0.50, range 1.83–3.50) (Donegan 2008). The song is more homogeneous across the rest of the range of *occidentalis* (from the Central Andes of Colombia south to northernmost Peru); comparable representative statistics (from songs in Ecuador and Peru) are 16.77 ± 1.97 notes/s (range 13–22), frequency variation of 0.57 ± 0.18kHz (range 0.31–1.10) and a change of speed of 1.63 ± 0.27 (range 1.06–2.18). Songs of *nana* in the East Andes of Colombia are slightly lower in frequency compared to other populations, whereas songs of *nanitaea* on the Colombia/Venezuela border are slightly higher (Donegan 2008). The call is described as 'a short, sharply descending *chew* at *c.*3.5kHz' (Krabbe & Schulenberg 2003), 'a short and abrupt *tchew*' (Ecuador; Ridgely & Greenfield 2001b) and 'a descending *tew* note' (Peru; D.F. Lane in Schulenberg *et al.* 2007). Donegan (2010) noted that sample sizes are too small to make strong inferences but that the call of *kukenamensis* appears to differ from other subspecies in note shape, comprising a fast rasp numbering three notes.

NATURAL HISTORY Like other species of *Grallaricula*, it forages alone or in pairs, rarely far above ground and almost always in dense vegetation (Ridgely & Greenfield 2001b, Delgado-V. 2002). Parker *et al.* (1985) reported the species as making short forward sallies to foliage and stems, and occasionally attempting to hawk insects from the air within a few centimetres of a perch. Parker *et al.* (1985) also observed these antpittas descending to the ground and foraging like a small thrush (*Catharus*), hopping several times then stopping abruptly to capture small insects and arthropods on the forest floor, but this behaviour has not been reported by other observers. Like other *Grallaricula*, it frequently twitches both wings simultaneously. It appears that Slate-crowned Antpitta rarely joins mixed-species flocks, but at least occasionally it attends swarms of highland army ants (*Labidus* sp.) (Greeney 2012b). Nothing has been published on the territorial habits, but my casual observations suggest that territorial disputes can be fierce and, at least occasionally, resolved via physical altercations. On two separate occasions I witnessed two adults grappling as they tumbled to the ground. On one of these the two adults remained locked in an 'embrace' as they tumbled down a steep, bamboo-covered slope for >5m until they were lost from view. I did not note any distinct vocalisations during these interactions, although they might easily have been missed due to the violent fluttering of wings.

DIET The limited information on the diet indicates that the species is, as expected, insectivorous. The stomach contents of an adult female (holotype, *hallsi*) included the remains of unidentified beetles (Coleoptera) and other arthropods (Donegan 2008). Delgado-V. (2002) reported fragments of beetle elytra in a faecal sample from an unsexed adult (*occidentalis*). Prey items delivered to a nest in north-east Ecuador were generally small (>1cm) and delivered singly. Most were unidentifiable, but at least 10% were lepidopteran larvae and 7% earthworms (Greeney & Miller 2008). A pair of adults at a nest in south-east Ecuador delivered 36 food items to a nestling comprising 21 small unidentified items, nine worms (Oligochaeta), three Lepidoptera larvae, two adult beetles (Coleoptera) and one wasp (Hymenoptera) (Greeney *et al.* 2010). Fierro-Calderón *et al.* (2006) examined the contents of two stomachs of adults (*nana*) finding beetles, moths and harvestmen (Opiliones).

REPRODUCTION Information on the reproductive biology has only recently been published, with the nest and egg described for the first time not much more than a decade ago (Greeney & Sornoza 2005). To date, just seven nests and four eggs have been described, all of race *occidentalis* (based on subspecific ranges as defined here; see Donegan 2008, Greeney 2013k). The first description of the nest and egg derived from observations at three nests found at RP Tapichalaca (Greeney & Sornoza 2005). Subsequently, Greeney & Miller (2008) described three additional nests, two from the same locality and one at Estación Biológica Yanayacu, providing the first nestling description. Most recently, Greeney *et al.* (2010) published the description of a seventh nest, also in south-east Ecuador, supplementing our knowledge with information on nestling development. Note that, because of the recent splitting of taxa *cumanensis* and *pariae* into a species separate from Slate-crowned Antpitta (Sucre Antpitta; Donegan 2008, Remsen *et al.* 2016), Schönwetter's (1979) often-cited description of the eggs of Slate-crowned Antpitta (Wiedenfeld 1982, Greeney & Sornoza 2005, Greeney *et al.* 2008a) in fact refers to *G. c. cumanensis*.
NEST & BUILDING The nests are shallow, frail cups of leaves and leaf strips supported by a sparse platform of sticks and leaf petioles, and sparsely lined with fine, flexible fibres and rootlets (Greeney & Sornoza 2005, Greeney & Miller 2008). They are placed on obviously fragile and/or unstable substrates such as rosettes of *Chusquea* leaf petiole bases, thin branches or small vine tangles. Nest heights reported in the literature are: 2.0m, 2.4m (Greeney & Sornoza 2005). Mean (±SD) dimensions of five nests (Greeney & Miller 2008) are: external diameter 10.7 ± 1.0cm; external height 5.6 ± 1.6cm; internal diameter 6.7 ± 0.5cm; internal depth 2.5 ± 0.6cm. Greeney & Miller (2008) also collected, dried and described the materials used in four nests, observing that nest linings were so sparse that we were unable to reliably separate them from the rest of the structure. Three nests collected in south-east Ecuador had total dry weights of 11.0g, 15.4g and 10.3g, and one from north-east Ecuador weighed 10.4g. Nests comprised sticks, thin dark fibres, thin pale fibres, *Chusquea* bamboo leaves, dicot leaves and fern leaves. They also contained a small amount of moss, which was almost certainly brought in incidentally, probably attached to sticks. Overall, nest architecture and composition is nearly indistinguishable from those of Rusty-breasted Antpitta (Schwartz 1957, Greeney 2013e).

To date, all described nests have been found within, or immediately adjacent to, dense stands of *Chusquea* spp. bamboo. No quantified studies have documented habitat preferences for nesting, but given this antpitta's predilection for foraging in bamboo, it seems likely that nesting is also somewhat dependent on such habitat. **Egg, Laying & Incubation** The eggs are pale brown or beige, with heavy brown, dark cinnamon and lavender spots and flecks distributed fairly evenly across the shell, but usually slightly thicker at the large end (Greeney & Sornoza 2005, Greeney *et al.* 2010). Egg measurements are 22.6 × 18.9mm and 23.7 × 16.6mm (Greeney & Sornoza 2005). Reported clutches (*n* = 4) all consisted of one egg and two additional nests contained a single nestling, strongly suggesting that the normal clutch size is one (Greeney & Sornoza 2005, Greeney & Miller 2008, Greeney *et al.* 2010). Both sexes participate in incubation. A single-egg clutch in south-east Ecuador (Greeney & Sornoza 2005) was covered by adults for 14.8h (65%) of the 22.7h observation period. The duration of on-bouts (*n* = 17; ±SD) was 53.8 ± 44.3min and mean periods of adult absence (*n* = 17) were 10.6 ± 15.7min. The longest observed off-bout was 2.3h. While covering the eggs, both adults frequently stood and peered into the nest (*n* = 120 times in *c*.15h of observation), usually engaging in one or several of the following behaviours. On 58 occasions the adult leaned forward and rapidly thrust its bill in and out of the nest lining, vibrating both its entire body and the nest. This was interpreted by the authors as a form of parasite searching/ removal, and has been documented in other antpittas (Greeney *et al.* 2008a). It may also have functioned to roll the egg. Adults also occasionally thrust their bill quickly into the nest, appearing to eat something small upon standing up, supporting the belief that parasites were being removed. They also performed nest maintenance by arranging stray fibres or sticks during 11 of the 120 observations. Interestingly, adults also occasionally leaned forward, pressed their breast into the cup with their wings slightly raised back, and shuffled or vibrated their body in a manner suggesting that they were shaping the nest cup. In total, adults spent 2.4% of their time moving in the nest, and sitting quietly and peering about with sharp movements of the head during the rest of their time covering the eggs. **Nestling & Parental Care** Slate-crowned Antpittas are dark grey-pink or blackish at hatching, and lack natal down (Greeney 2013k), as are other species of *Grallaricula* (Greeney 2012a). Within a few days they quickly develop a thick coat of dense, wool-like down, rusty-red or rufous-brown in coloration (Greeney & Miller 2008). Four days prior to fledging, one nestling weighed 19.6g, filling the nest cup. At this age the maxilla was almost completely dark, while the mandible was mainly dusky-orange. The cloaca was orangish, contrasting with the surrounding skin which was blackish (Greeney & Miller 2008). The feathers of the wings were well emerged from their sheaths and generally dark grey. The secondaries and inner primaries were edged rufous, with similar but stronger markings on the wing-coverts. On hatching, and throughout the nestling period, the young have creamy-yellow rictal flanges and pale whitish mouth linings (Greeney & Miller 2008, Greeney *et al.* 2010). This character is rather surprising, given that mouth linings of all other *Grallaria*, *Grallaricula* and *Hylopezus* nestlings have striking bright orange or orange-red mouth linings (Greeney *et al.* 2008a). The significance of this is still unknown. Greeney

et al. (2010) measured a single nestling 4–12 days after hatching, providing the scant data available on nestling growth rates. Mass, wing and tarsal measurements, respectively, were as follows: day 4: 8.5g, 10mm, 13mm; day 5: 9g, 11mm, 14mm; day 6: 10g, 12mm, 15mm; day 7: 11.5g, 12.5mm, 16mm; day 8: 12.5g, 14mm, ?; day 9: 12.5g, 16mm, 18mm; day 10: 14g, 16mm, 20mm; day 11: 15.5g, 17mm, 21mm; day 12: 17g, 19mm, 22mm. Using these observations of mass gain and weight of a nestling four days prior to fledging (Greeney & Miller 2008), I estimate the nestling period to be around 17 days, but there are no direct observations of complete nestling periods. Both sexes care for nestlings. In north-east Ecuador, Greeney & Miller (2008) made the following observations during the three days prior to fledging. On average, the nestling was fed 5.4 times/h and produced one faecal sac/h. All faecal sacs were removed by the adults and carried away from the nest. During these final days the nestling was brooded for <4% of the observation period, and then only during a brief period of rain. At another nest in south-east Ecuador, during observations of a nestling that was 7–10 days old, Greeney *et al.* (2010) reported that adults brought food 2.2 times/h and the nestling produced one faecal sac/h. Unlike the nest observed later in the nestling period by Greeney & Miller (2008), all faecal sacs were consumed by the adults at the nest. While at the nest, generally just after feeding the nestling, adults performed the rapid-probing movements described above (see Incubation). **Seasonality** Nests in Zamora-Chinchipe province, south-east Ecuador, were found with complete clutches on 8 September 2003, 25 November 2003 and 28 October 2006 (Greeney & Sornoza 2005, Greeney *et al.* 2010). The egg in the last-mentioned nest hatched on 29 October. These records collectively suggest that most breeding occurs in the dry season in eastern Ecuador (*occidentalis*), in agreement with observed patterns of breeding seasonality for most bamboo-nesting birds in this region (Greeney *et al.* 2005a, 2008b, 2011a). **Breeding Data G. n. olivascens** Very young fledgling, 7 March 2013, Instituto Venezolano de Investigaciones Científicas (IVIC) at San Antonio de Los Altos (J.E. Miranda T. *in litt.* 2017). **G. n. kukenamensis** Two juveniles transitioning, 11 April 2001, Cerro Roraima (USNM 626865/938); subadult, 3 April 2001, Cerro Roraima (USNM 626961). **G. n. nana** Two 'juveniles', 18 July and 21 August 1989, Estación Biológica Carpanta (Andrade & Lozano 1997). **G. n. occidentalis** Two nearly complete nests, RP Estación Biológica Yanayacu, 19 November 2007 (J. Simbaña *in litt.* 2008), 5 July 2011 (HFG); four nests with eggs, 25 November 2003, 14 October 2003, 19 November 2004, 28 October 2006, RP Tapichalaca (Greeney & Sornoza 2005, Greeney & Miller 2008, Greeney *et al.* 2010); nestling, 17 November 2006, RP Estación Biológica Yanayacu (Greeney & Miller 2008); fledgling, 12 February 1924, Volcán Sumaco (AMNH 184358); fledgling, 27 January 2010, RP Tapichalaca (HFG); juvenile, 22 June 1951, Páramo de Sonsón (USNM 436475); juvenile, 5 September 1957, Cerro Munchique (LACM 30859); juveniles, 24–29 February 2004, 25km NW of Santuario (Echeverry-Galvis & Córdoba-Córdoba 2007); 'fledgling', February, NE Ecuador (Fjeldså & Krabbe 1990); 'juvenile', 18 December 2014, RFP Río Blanco (D. Uribe *in litt.* 2017); two subadults, Cerro Munchique, 20 August 1952 (USNM 446636) and 31 August 1957 (FMNH 249746); subadult transitioning, 26 June 1970, Llorente (FMNH 292125); subadult transitioning, 3 September 1957, Cerro

Munchique (FMNH 249745). In addition I examined the following poorly labelled specimen: juvenile (*nanitaea*, AMNH 492318), 'Mérida'.

TECHNICAL DESCRIPTION Sexes similar. The following description of nominate *nana* is based on Greeney (2013k). *Adult* Slate-grey crown only weakly contrasts (especially under field conditions) with the dark olive-brown plumage of the remaining upperparts. Wings dark tawny-brown. Face marked with a large loral spot that is contiguous with a broad eye-ring, both of which are orange-rufous to buffy-orange. Underparts variable, especially in colour saturation, even within small geographic areas, ranging from orange-rufous to yellow-buff. Lower throat and central belly white. Feathers of the breast sometimes have thin dusky margins that can create a scalloped appearance (see Taxonomy and Variation). **Adult Bare Parts** *Iris* brown; *Bill* black, base of mandible white or pinkish; *Tarsi & Toes* greyish or bluish-slate. *Immature* Ontogeny of plumage development not fully known. Donegan (2008) described an immature as having patchy dark rufous feathers on the crown and breast, and more prominent blackish marks on the throat and breast, but otherwise like adults, a description that I consider to refer to an older juvenile. Having now seen a few older immatures (photos, specimens), I can provide fairly complete descriptions for all three stages of development. *Fledgling* Young leave the nest still covered in their red-brown nestling plumage and, at least sometimes, without their flight feathers fully emerged. During the first few days post-fledging, they probably differ little from nestlings (see description under Reproduction). It appears that the first adult feathers begin to appear on the throat and back, but the timing of this transition to juvenile plumage is uncertain. **Fledgling Bare Parts** As described for nestlings. *Juvenile* No evidence, at present, to suggest that the juvenile plumage of the various races varies greatly from that of *nanitaea* (Plate 24, Fig. g). Juveniles retain scattered patches of reddish nestling/fledgling down, especially on crown, rump and breast. Upper back, scapulars, throat and face already beginning to appear like adults. Feathers of breast replaced by adult-like feathers during this stage, and the underparts as a whole are likely to appear as a mix of adult and immature feathers. **Juvenile Bare Parts** *Iris* brown; *Bill* maxilla dusky, yellowish near tip and base, orange-yellow on tomia and at gape, mandible still mostly yellow to yellow-orange, dusky near tip; *Tarsi & Toes* similar to adult but somewhat more yellowish on posterior edge of tarsi and on toes. *Subadult* At this age appear similar in most ways to adults, but may retain 'scraps' of reddish down poking through adult-like feathering, particularly on nape and hindcrown. Younger individuals may also retain scattered immature feathers on lower breast or belly, occasionally the rump. The most persistant difference from adult plumage is the upperwing-coverts, which are narrowly margined dull chestnut to buffy-brown at their tips. **Subadult Bare Parts** As described for adult.

MORPHOMETRIC DATA (mean ± SD, ranges) *G. n. olivascens* Donegan (2008) provided measurements from specimens collected in the Coastal Cordillera of Venezuela: *Wing* 65.3 ± 2.4mm, 61.0–70.0mm (*n* = 22); *Tail* 34.0 ± 1.4mm, 31.5–37.0mm (*n* = 20); *Bill* total culmen 16.0 ± 0.9mm, 14.5–17.5mm (*n* = 20), width (at midpoint of nares) 3.6 ± 0.1mm, 3.5–3.7mm (*n* = 5); *Tarsus* 29.9

± 1.1mm, 28.0–32.0mm (*n* = 21). *Wing* 68–71mm (♂), 67–70mm (♀); *Tail* 33–38mm (both); *Bill* [total? culmen] 13.5–15.0mm (both) (*n* = 11, 6♂♂, 5♀♀, Cory & Hellmayr 1924). *G. n. nanitaea Wing* 66.0mm; *Tail* 31.0mm; *Bill* total culmen 16.5mm, width at nares, 4.0mm; *Tarsus* 31.0mm (*n* = 1♂, holotype, Donegan 2008). *Wing* 66.9 ± 3.2mm, 61.0–73.0mm (*n* = 11); *Tail* 33.5 ± 0.9mm, 32.0–35.0mm (*n* = 10); *Bill* total culmen 15.7 ± 0.8mm, 15.0–17.0mm (*n* = 9), width (at midpoint of nares) 4.8 ± 0.5mm, 4.2–5.1mm (*n* = 3); *Tarsus* 29.2 ± 0.9mm, 27.5–30.5mm (*n* = 10, range) (Donegan 2008). *G. n. hallsi Wing* (in life) 78.0mm, (from skin) 74.0mm; *Tail* 33.0mm; *Bill* total culmen 15.0mm, depth at nares 4.8mm, width at nares 5.2mm; *Tarsus* 32.0mm (*n* = 1♀, holotype, Donegan 2008). *Wing* 70.0 ± 2.6mm, 68.0–73.0mm (*n* = 3, from Appendix 2 in Donegan 2008, although they conflict with the measurement given earlier [Donegan 2008: 162] for maximum wing length); *Tail* 32.2 ± 0.6mm, 32.0–33.0mm (*n* = 3); *Bill* total culmen 15.8 ± 0.8mm, 15.0–16.5mm (*n* = 3), width (at midpoint of nares), 5.0 ± 0.2mm, 4.9–5.2mm (*n* = 3); *Tarsus* 31.0 ± 0.9mm, 30.5–32.0mm (*n* = 3). *G. n. nana* From Donegan (2008): *Wing* 62.7 ± 1.5mm, 61.0–64.0mm (*n* = 3); *Tail* 32.5 ± 0.7mm, 32.0–33.0mm (*n* = 2); *Bill* total culmen 15.7 ± 0.3mm, 15.5–16.0mm (*n* = 3); *Tarsus* 30.3 ± 0.3mm, 30.0–30.5mm (*n* = 3) (Sierra de los Picachos). Also from Donegan (2008): *Wing* 68.7 ± 1.2mm, 68.0–71.0mm (*n* = 6); *Tail* 31.3 ± 1.2mm, 30.0–32.0mm (*n* = 3); *Bill* total culmen 15.9 ± 0.2mm, 15.5–16.0mm (*n* = 5), width at midpoint of nares, 4.4 ± 0.4mm, 3.9–5.0mm (*n*= 6); *Tarsus* 29.8 ± 1.1mm, 28.0–31.0mm (*n* = 5) (Colombian East Andes). *G. n. occidentalis* Following from Donegan (2008): *Wing* chord 68.9 ± 1.7mm, 66.0–72.0mm (*n* = 37); *Tail* 34.3 ± 3.0mm, 28.0–38.8mm (*n* = 30); *Bill* total culmen, 16.8 ± 1.7mm, 13.8–18.8mm (*n* = 29), width at midpoint of nares, 4.5 ± 0.2mm, 4.2–4.6mm (*n* = 4); *Tarsus* 30.8 ± 1.2mm, 27.5–33.4mm (*n* = 30, Colombian West Andes). *Wing* 68.0 ± 2.3mm, 63.5–71.0mm (*n* = 15); *Tail* 33.8 ± 3.0mm, 30.0–38.0mm (*n* = 10); *Bill* total culmen, 16.0 ± 0.5mm, 14.9–16.8mm (*n* = 11), width at midpoint of nares, 4.4 ± 0.4mm, 4.0–5.0mm (*n* = 8); *Tarsus* 29.8 ± 1.1mm, 27.9–31.4mm (*n* = 12, Colombian Central Andes). *Wing* chord 68.8 ± 2.3mm, 66.0–71.0mm (*n* = 5); *Tail* 33.2 ± 0.8mm, 32.0–34.0mm (*n* = 5); *Bill* total culmen 16.0 ± 0.6mm, 15.5–16.5mm (*n* = 4), width at midpoint of nares, 4.0 ± 0.1mm, 3.9–4.0mm (*n* = 4); *Tarsus* 29.4 ± 0.5mm, 29.0–30.0mm (*n* = 5, eastern Ecuador). *G. n. kukenamensis* Donegan (2008) provided the greatest number of morphological data for this race: *Wing* chord mean 64.5 ± 2.0mm, range 60.0–68.0mm (*n* = 21); *Tail* 28.5 ± 1.7mm, 25.0–31.5mm (*n* = 21); *Bill* total culmen, 16.2 ± 0.4mm, 15.5–17.0mm (*n* = 19), width at midpoint of nares, 5.2 ± 0.3mm, 4.8–5.5mm (*n* = 17); *Tarsus* 23.7 ± 0.7mm, 23.0–25.0mm (*n* = 21, specimens from tepui region of Venezuela and Guyana). Additional data include: *Wing* 62mm; *Tail* 28mm; *Bill* exposed culmen 13mm; *Tarsus* 25mm (*n* = 1♀, holotype, Chubb 1918b). **Mass** Donegan (2008) provided the following: 20.4 ± 2.4g, 18.4–23.0g (*n* = 3♀?, *hallsi*), 19.4 ± 2.1g, 18.0–23.0g (*n* = 5♀?, *nanitaea*), 20.2 ± 0.9g, 18.5–22.0g (*n* = 34♀?, *occidentalis* from Colombian West Andes), 20.7 ± 0.8g, 19.9–22.0g (*n* = 6♀?, *occidentalis* from Colombian Central Andes). Additional published data include: mean 19.5g, range 18.5–20.5g (*n* = 4♂♂, *occidentalis* from northern Peru, Parker *et al.* 1985), mean 19.8g, range 17.5–21.5g (*n* = 4♀♀, *occidentalis* from northern Peru, Parker *et al.* 1985), 21g (*n* = 1♂, *occidentalis*,

Zamora-Chinchipe, Rahbek *et al.* 1993): 19.3g, 20.0g, 21.0g (*n* = 3♀♀, *nana*, Echeverry-Galvis *et al.* 2006), mean 19.4 ± 0.9g, range 18.3–20.3g (*n* = 5♂♂, *nana*, Echeverry-Galvis *et al.* 2006). **TOTAL LENGTH** 10.0–12.1cm (Lafresnaye 1842, Chubb 1918b, Meyer de Schauensee 1964, 1970, Meyer de Schauensee & Phelps 1978, Dunning 1982, Dunning 1987, Clements & Shany 2001).

TAXONOMY AND VARIATION The taxonomic history of Slate-crowned Antpitta is somewhat confusing, with anywhere between three and eight subspecies having been recognised. Combining the taxonomic suggestions of Remsen *et al.* (2016), Krabbe & Schulenberg (2003) and Donegan (2008), two taxa traditionally classified as subspecies, *G. n. cumanensis* and *G. n. pariae*, are together considered a separate species, Sucre Antpitta. Prior to Donegan's (2008) revision, Sucre Antpitta was considered subspecifically related to Slate-crowned Antpitta (Cory & Hellmayr 1924, Meyer de Schauensee 1982, Altman & Swift 1993, Clements 2000), which remains its most likely sister species. Rusty-breasted Antpitta is another potential relative, having similar plumage, voice and nesting habits (Parker *et al.* 1985). Five subspecies of Slate-crowned Antpitta are recognised here, two of them recently described (Donegan 2008, Donegan *et al.* 2009b). Race *occidentalis*, described from the West and Central Andes of Colombia, has sometimes been synonymised with nominate *nana*, based on the assertion that similar birds occur in northernmost Peru (Krabbe & Schulenberg 2003, Clements 2007). This was, however, prior to nominate *nana* being restricted by Donegan (2008) to refer only to very rufous birds in the main part of the East Andes of Colombia. Birds in Peru, referred here to *occidentalis*, are quite variable in plumage (Fjeldså & Krabbe 1986), as are those in extreme south-east Ecuador (IBC 1267100, video A. Carrasco), with these populations showing considerable variation in the amount of dark edging to the breast and upper belly feathers. The overall effect, at least in some individuals, is suggestive of an Ochre-breasted Antpitta (Plate 22, Fig. f; Fig. 2, page 436). My observations suggest that these 'scallop-breasted' individuals are fairly unique to this region and the possibility of a cryptic species warrants further investigation, despite no apparent differences in vocalisations. The songs of populations from Ecuador, Peru and the Central Andes of Colombia are virtually indistinguishable, even using average differences (Donegan 2010), and I have followed Donegan's (2008) suggestion by including all of these populations within *occidentalis*. This includes somewhat variable populations at the head of the Magdalena Valley which come fairly close to bridging the distributional gap between West/Central Andean populations and those of the East Andes. Race *olivascens* of coastal Venezuela is consistently darker and has a somewhat faster song, with more notes, than most other populations, and is apparently fully diagnosable by voice from all other subspecies by more than three variables, except that most proximate, *nanitaea*, which differs by fewer than three vocal variables (Donegan 2008, 2010). With the arrangement adopted here, the nominate subspecies, which was described based on 'Bogotá' specimens, is confined to the East Andes of Colombia, replaced to the north by newly described *nanitaea* and *hallsi*, and to the south and west by *occidentalis*.

Grallaricula nana olivascens Hellmayr, 1917, Verhandlungen der Ornithologische Gesellschaft in Bayern, vol. 13, p. 117, 2,000m, Galián, Cerro de Ávila outside of Caracas

in north-east Venezuela. The holotype (ZSM 151700) is an adult male collected 15 December 1913 by S.M. Klages. This race is confined to the coastal mountains of Venezuela in Aragua and the Distrito Federal, where it is allopatric with (until recently conspecific) Sucre Antpitta. Race *olivascens* is similar to nominate *nana*, but is generally paler above, including the wings and crown, with the back more greenish olive. Following Donegan (2008), an extralimital sight record of *G. n. olivascens* at Cerro Tucusito, at the Miranda/Anzoátegui border (10°00'N, 65°39'W, 550m), (Hilty 2003) is not included pending confirmation. It quite possibly refers to a misidentified Rusty-breasted Antpitta, which is known from specimens taken nearby. **SPECIMENS & RECORDS *Vargas*** Picacho de Galipán, 10°34'N, 66°54.5'W (CM, AMNH, MCZ); El Junquito, 10°28'N, 67°05'W (COP); El Limón, 10°28'N, 67°17'W (CM P104302/03); 2km S of La Guaira, 10°35'N, 66°56'W (ML 40338/39, P. Coopmans). ***Distrito Federal*** San Antonio de Galipán, 10°33'N, 66°53'W (COP 62591). ***Miranda*** Ño León, 10°26'N, 67°09'W (COP 58454); San Antonio de Los Altos, 10°24'N, 66°59'W (J.E. Miranda T. *in litt.* 2017); Laguneta de la Montaña, 10°21'N, 67°07'W (J.E. Miranda T. *in litt.* 2017). ***Aragua*** Near Colonia Tovar, *c.*10°25'N, 67°17'W (COP, AMNH, ML, XC); MN Pico Codazzi 10°24.5'N, 67°20'W (ML 171308, J.W. McGowan); Fundo Jeremba, 10°26'N, 67°15'W (COP 75776).

Grallaricula nana nanitaea Donegan, 2008, Bulletin of the British Ornithologists' Club, vol. 128, p. 164, 3,000m, La Culata, near the city of Mérida, Mérida state, Venezuela [08°54'N, 70°38'W]. The holotype (AMNH 146661) is an adult male collected 24 January 2011 by Briceño, S. B. Gabaldón and sons. Two paratypes were also designated by Donegan (2008), both from the same type locality (AMNH 96305/ 492317). Race *nanitaea* is found throughout the entire Cordillera de Mérida in Venezuela, from north-east Trujillo at least to central Táchira. Although there appears to be a distributional discontinuity at the Táchira Depression (T.M. Donegan *in litt.* 2017), a low area that separates the Cordillera de Merida from the northern East Andes in Venezuela and Colombia, records from north-east Colombia, north of the Río Chicamocha (right bank), to extreme southern Táchira are also included in *nanitaea*. Race *nanitaea* is replaced to the west in the Coastal Cordillera of Venezuela by *olivascens* and south in the East Andes by nominate *nana* south of the Río Chicamocha in Colombia. In his type description, Donegan (2008) noted that there is notable variation in the extent of the white breast collar, which is usually reduced in females. He further indicated that this collar can be difficult to detect in live birds, except in singing males (when the head is thrown back and the breast expanded). Race *nanitaea* differs from *hallsi* in having a higher frequency song. Donegan (2008) additionally noted that the song of the population of *nanitaea* near the city of Mérida is faster, has a greater acoustic frequency variation, and more notes. *G. n. nanitaea* can be separated from nominate *nana* by having a more olivaceous back, a paler breast, and a higher acoustic frequency in both songs and calls. Race *occidentalis* is less olivaceous on the mantle, and *olivascens* has a more olivaceous mantle and a faster-paced song than *nanitaea*. Donegan (2008) pointed out that the illustrations of Slate-crowned Antpitta in Hilty (2003) and Restall *et al.* (2006), both labelled as race *nana*, actually refer to *nanitaea*. Also noted by Donegan (2008), the type specimen's label states that the irides were blue, but this somewhat implausible assertion is made less credible by my observation that a large percentage of other antpittas and antthrushes collected by Briceño and Galbadón are also described as having blue eyes (specimen labels; USNM, AMNH). The iris of *nanitaea*, like other races, is dark brown (Donegan 2008). SPECIMENS & RECORDS VENEZUELA: *Trujillo* Páramo Las Rosas, *c.*09°42'N, 70°07'W (CM); Páramo de Cendé, 09°33'N, 70°07'W (COP 19961/62); Páramo Misisí, 09°20'N, 70°18'W (COP 4958); PN Guaramacal, *c.*09°10'N, 70°11'W (XC 223754, P. Boesman). *Mérida* Páramo El Escorial, 08°49'N, 70°58.5'W (AMNH 492314–316, CUMV 7192, ROM 33.9.1.590); La Azulita, 08°43'N, 71°23'W (COP 65392); Quintero, 08°41'N, 71°06'W (COP 14524); Valle, 08°40'N, 71°06'W (AMNH, USNM); Pico Humboldt, 08°37.5'N, 71°02'W (XC 6858, N. Athanas); La Mucuy, 08°36.5'N, 71°03.5'W (XC 43323–326, J. Klaiber); 6.5km SSE of El Muerto, 08°17'N, 71°09'W (XC 223756, P. Boesman). Táchira Boca de Monte, 08°04'N, 71°50'W (COP 24544–546); Páramos El Batallón y La Negra, 08°01'N, 71°58'W (COP 84126–128); Hacienda La Providencia, 07°33.6'N, 72°22'W (COP 62203–206); Páramo El Zumbador, 08°00'N, 72°05'W (FMNH 288322, CM A1831, ML). *Apure* Cerro El Retiro, 07°26'N, 72°21'W (COP 73941–943); Páramo de Tamá, 07°25'N, 72°24'W (FMNH 43603). **COLOMBIA**: *Norte de Santander* Páramo de Tamá, 07°23.5'N, 72°24'W (ICN 33933, FMNH 43602). *Santander* Suratá, 07°22'N, 73°00'W (Donegan *et al.* 2007, Donegan 2008, ICN 33933/36125); Finca Jericó, 07°01'N, 72°58'W (eBird: J. Avendaño).

Grallaricula nana hallsi Donegan, 2008, Bulletin of the British Ornithologists' Club, vol. 128, p. 161, 2,900m, Lepipuerto in the upper Río Chimera drainage, west slope of Serranía de los Yariguíes, in Simacota or El Carmen municipality, Dept. Santander, Colombia, 06°28'N, 73°28'W. The holotype (ICN 35195) is an adult female collected by T.M. Donegan on 10 January 2005. DNA from the holotype is on deposit at IAvH's molecular laboratory in Cali and vocalisations while in the hand are available online (XC 21507–508). This race is considered endemic to the Serranía de los Yariguíes in north-east Colombia, where it occurs on both slopes of the range. To date, *hallsi* has been found at 2,450–2,900m, but suitable habitat occurs to 3200m (the treeline in parts of Yariguíes). *G. n. hallsi* differs from nominate *nana* in having paler (more orange, less ferruginous) underparts and being more olivaceous dorsally. While quite similar in plumage to *nanitaea*, it differs in vocal characters by the lower acoustic frequency of its song and call. Race *hallsi* can be separated from *occidentalis* by its more olivaceous upperparts. Additionally, it differs from East Andean populations of *occidentalis* by the lower frequency song, and from West Andean populations of *occidentalis* by having a shorter song with less variation in frequency. SPECIMENS & RECORDS *Santander* San Vicente de Chucurí, 06°49'N, 73°22'W (Donegan *et al.* 2007); La Aurora, 06°38'N, 73°24'W (Donegan *et al.* 2007, XC 25508/18, T.M. Donegan, ICN 35555, paratype).

Grallaricula nana nana (Lafresnaye, 1842). Originally described as *G*[*rallaria*] *nana*, the type locality was given as just 'Colombie'. Todd (1927) in his description of race *occidentalis* pointed out that the holotype of *nana* is darker-breasted than Central Andean *occidentalis*, indicating that the holotype of *nana* probably came from the East Andes. While it is conceivable that the type of *nana* came from near the head of the Magdalena Valley or from a location in the West Andes based on its plumage (see Donegan 2008), for now I treat only birds from the East Andes of Colombia, in southern Santander, Boyacá and Cundinamarca, as belonging to the nominate subspecies, separate from paler-breasted *occidentalis* of the West and Central Andes of Colombia and the variable populations east of the Andes south from the head of the Magdalena Valley. Records from extreme north-west Caquetá are provisionally also included with *nana*, the apparent gap between these records and southern Cundinamarca probably reflects poor sampling and genuine rarity in that region. SPECIMENS & RECORDS COLOMBIA: *Santander* Vereda Las Minas, 06°04'N, 73°00'W (Donegan 2008; IAvH). *Boyacá c.*5km N of San Jeronimo, 05°50'N, 73°20'W (ML 80952, M. Álvarez); Cañon del Río Pómeca, 05°48'N, 73°28'W (Álvarez-R. 2000, IAvH); Rogitama, 05°47'N, 73°31'W (J. Zuluaga in Donegan 2008). *Cundinamarca* Bosques de la Falla del Tequendama, 04°43'N, 74°22'W (Donegan 2008, IAvH), Finca San Cayetano, 04°37'N, 74°18'W (O. Laverde in Donegan 2008); RFP Ríos Blanco y Negro, 04°42'N, 73°51'W (Stiles & Rosselli 1998); RB Carpanta, 04°35'N, 73°43'W (Andrade *et al.* 1994, Andrade & Lozano 1997, ICN 31322); near Monterredondo, 04°17'N, 73°47'W (XC 315044, A. Pinto-Gómez). *Caquetá* San Vicente del Caguán, 02°48'N, 74°51'W (M. Álvarez in Donegan 2008); Finca Andalucía, Cerro La Mica, 02°44'N, 74°53'W (Bohórquez 2002, Donegan 2008, IAvH-10253–55).

Grallaricula nana occidentalis Todd, 1927, Proceedings of the Biological Society of Washington, vol. 40, p.176, 'Sancudo', Caldas, Colombia [= El Zancudo, 05°05'N, 75°30'W]. The holotype (CM P70434) is an adult male collected 2 September 1918 by M.A. Carriker. Although Todd (1927) described the type locality as being in the West Andes of Colombia, El Zancudo is in the Central Andes (Paynter & Traylor 1981). Todd (1927) also mentioned examining two additional specimens at Carnegie Museum also from the 'Western Andes'. He did not provide any specifics, but the only other specimens currently in the Carnegie collection are all Carriker specimens from El Zancudo (CM P70299/423), confirming that *occidentalis* refers to Central Andes populations. Krabbe & Schulenberg (2003) did not recognise *occidentalis*, synonymising it with nominate *nana* based on its similarity to paler-breasted birds in eastern Ecuador and northern Peru. Indeed, Central Andean birds (*occidentalis*) are indistinguishable from Ecuadorian and Peruvian populations by voice or plumage. However, these populations are separated from East Andean populations in northern Colombia (nominate *nana*) in having paler breasts, and songs and calls at higher frequency. Therefore, following Donegan (2008) but contrary to other literature (Chapman 1926a, Clements 2000, Ridgely & Greenfield 2001), I consider *occidentalis* to range from north-west Colombia in the West and Central Andes, south through the Central Andes, along the east slope of the Andes in Ecuador, to northern Peru. The taxonomy and subspecific limits of Slate-crowned Antpitta are far from resolved. Although somewhat clarified by Donegan (2008), he also noted that West Andean *occidentalis* exhibits a few small plumage and average vocal differences from Central Andean *occidentalis*, but no diagnostic vocal variable has been identified. He pointed out the need for further comparisons to test the consistency of any plumage characters that might separate West Andes *occidentalis* from

Central Andes/Ecuador/Peru populations, especially in light of the high level of endemism known from the West Andes (Salaman *et al.* 2003). Furthermore, Donegan (2008) noted the presence of apparent intermediates around Nudo de Pasto in southern Colombia and northern Ecuador. Very recently, J. Freile, J.M. Loaiza and P. Molina made the first records of Slate-crowned Antpitta in the West Andes of Ecuador. Until more information becomes available, these records are also included in *occidentalis*. Following Todd (1927) and Donegan (2008), *occidentalis* is separable from nominate *nana* principally by its paler underparts (ochraceous to yellow-buff, rather than *ferruginous*). Previously unremarked upon in the literature, Slate-crowned Antpittas in south-east Ecuador occasionally show thin dusky or blackish margins to the feathers of the upper breast and sides, creating a scaled look that bears a good resemblance to Ochre-breasted Antpitta (Plate 22). I have seen photos and examined specimens of four individuals from Zamora-Chinchipe that are scaled thus (see ML 36591401, photo, L. Seitz, for a good example). One bird near Baeza also had scaling on the upper breast, but I did not see it well enough to judge its similarity to Zamora-Chinchipe birds. Another individual with fairly dark scaling (albeit confined to the upper breast) was recorded at Cajanuma (IBC 1267100, video A. Carrasco). **SPECIMENS & RECORDS COLOMBIA**: *Antioquia* Páramo Belmira, 06°37'N, 75°38'W (XC 77165, P. Flórez); Vereda La Lana, 06°28'N, 75°36'W (Donegan *et al.* 2009b); Páramo de Frontino, 06°26'N, 76°05'W (Flórez *et al.* 2004, Krabbe *et al.* 2006, XC); Vereda Ovejas, 06°22'N, 75°39'W (Donegan *et al.* 2009b); Parque Regional Ecoturistico Arví, *c.*06°16'N, 75°30'W (Castaño-Villa & Patiño-Zabala 2008); Alto San Luis, 06°06'N, 75°33'W (Delgado-V. 2002); Páramo de Sonsón, 05°42.5'N, 75°06.5'W (USNM 436475/76, XC); Vereda La Linda, 05°38'N, 75°48'W (Cuervo *et al.* 2003); RNSC La Mesenia, 05°29'N, 75°54'W (XC 119719–720, D. Edwards). *Risaralda* PNN Tatamá, 05°12'N, 76°06'W (IAvH 13387/410); 25km NW of Santuario, 05°08'N, 76°02'W (Echeverry-Galvis & Córdoba-Córdoba 2007); Vereda Las Cumbres, 05°06.5'N, 76°00'W (XC 146186, J.A. Zuleta-Marín); La Pastora, 04°46'N, 75°37'W (UV 6165); DMI Campoalegre, 04°52'N, 75°31'W (Donegan 2008, IAvH). *Caldas* RFP Río Blanco, *c.*05°07.5'N, 75°28'W (XC, many recordings, ML, many photos). *Quindío* Vereda Boquía, 04°41'N, 75°33'W (XC 214952, O.H. Marín-Gómez); Salento, 04°38'N, 75°34'W (Chapman 1917, UV 4802–04, AMNH 112010); Alto Quindío, 04°38'N, 75°27'W (Arbeláez-Cortés *et al.* 2011b, XC); Laguneta, 04°35'N, 75°30'W (Chapman 1917, AMNH 112011, ANSP 155176); RMN El Mirador, *c.*04°10.5'N, 75°43.5'W (Salaman *et al.* 2007a). *Tolima* Nevado del Ruiz road, *c.*05°00'N, 75°19.5'W (D. Uribe *in litt.* 2017); Cuenca del Río Toche, 04°36'N, 75°24'W (Donegan 2008, IAvH); Cajamarca, 04°29'N, 75°29'W (XC 128350, O.H. Marín-Gómez); near Ibagué, 04°26'N, 75°14'W (XC 12421, O. Laverde); RNA Loro Coroniazul, 04°17.5'N, 75°32.5'W (O. Cortes-Herrera *in litt.* 2017). *Valle del Cauca* Corea, Farallones de Cali, 03°21.5'N, 76°53'W (ICN 25920/21). *Cauca* Cerro Munchique, 02°32'N, 76°57'W (Bond & Meyer de Schauensee 1940a, LACM, FMNH, MHNG); San Antonio, 02°37'N, 76°54'W (Bond & Meyer de Schauensee 1940a, ANSP 142395); Quebradón, Moscopán, 02°14'N, 76°10'W (IAvH 2489); Tatauí, 01°37'N, 76°16'W (Salaman *et al.* 2007b). *Huila* La Candela [16km SW of San Augustín], 01°48'N, 76°17.5'W (ANSP 155852); Serranía de las Minas, 02°10'N, 76°11'W

(Donegan 2008, IAvH); PNN Cueva de los Guácharos, 01°36'N, 76°08'W (Donegan 2008, IAvH). *Caquetá* La Esmeralda, 01°21'N, 76°06'W (XC 11554–45 A.M. Umaña). *Putumayo* El Silencio, 01°05'N, 76°49'W (ML 23476001, B.C. Jaramillo). *Nariño* Lago Cumbal, 00°57.5'N, 77°49'W (J.A. Zuleta-Marín *in litt.* 2017); Llorente, 00°49'N, 77°15'W (FMNH 292120–125); La Victoria, 00°35'N, 77°10'W (FMNH 292119, MHNG 1179045). **ECUADOR**: *Carchi* c.8.3km ESE of El Gaaltal, 00°49'N, 78°05.5''W (J. Freile *in litt.* 2017). *Sucumbíos* 3km E of Santa Bárbara, 00°39'N, 77°30'W (XC 248354, N. Krabbe); La Sofía road, 00°27'N, 77°35'W (XC 78769, R. Ahlman). *Napo* Oyacachi, 00°13'S, 78°04.5'W (J. Freile *in litt.* 2017); 12km NNE of El Chaco, 00°14'S, 77°47.5'W (ANSP 185499, MECN 3715); Río Chalpi Grande, 00°21'S, 78°05'W (HFG); 2km S of Pan de Azúcar, 00°27'S, 77°43'W (MECN 2632, WFVZ 47679); Río Bermejo, 00°31'S, 77°53'W (XC 250908, N. Krabbe); RP Estación Biológica Yanayacu, 00°36'S, 77°54'W (Greeney & Miller 2008); Cordillera Guacamayos, 00°37.5'S, 77°50.5'W (WFVZ, MECN, ANSP, XC, ML); RP SierrAzul (Hacienda Aragón), 00°41'S, 77°55'W (MECN 6670, KUNHM 87156); Río Anatenorio, 00°59'S, 78°17'W (J. Freile *in litt.* 2017). *Pastaza* Comunidad Ingaru, 01°16.5'S, 78°04.5'W (QCAZ, J. Freile *in litt.* 2016). *Orellana* Volcán Sumaco, 00°32'S, 77°38'W (Chapman 1926a, AMNH 184358, MCZ 138488, ANSP 83348). *Chimborazo* Arenales, 02°34.5'S, 78°34.5'W (XC 250020, N. Krabbe). *Morona-Santiago* Guabisai, cantón Sucúa, 02°23.5'S, 78°18.5'W (MECN 8812); Cordillera Zapote Najda, 03°01'S, 78°31'W (Fjeldså & Krabbe 1986); 13.7km SE of San Carlos de Limón, 03°15.5'S, 78°19'W (MECN 8107/08); 24km ESE of Gualaquiza, 03°27'S, 78°21'W (ML 79686/70, T.A. Parker). *Zamora-Chinchipe* Estación Científica San Francisco, 03°58'S, 79°04'W (J. Freile *in litt.* 2016); Quebrada Honda, 04°29'S, 79°08'W (Rahbek *et al.* 1993, MECN, ANSP); RP Tapichalaca, 04°30'S, 79°08'W (Greeney & Sornoza 2005, XC, ML); above Río Blanco, 04°35'S, 79°17'W (XC 250907, N. Krabbe); Cerro Panguri, 04°39'S, 78°58'W (MECN 7954/59); 6km NW of San Andres, 04°46'S, 79°19'W (ANSP 185500). **PERU**: *Piura* Cerro Chinguela, 05°07'S, 79°23'W (ML: T.A. Parker). *Cajamarca* Hito Jesús, 04°54'S, 78°54'W (XC 7013, T. Mark); Picorana, 05°02'S, 78°51'W (LSUMZ 172185–188); Miraflores, Tabaconas, *c.*05°19'S, 79°17'W (MUSA 3183); Agua Azul, 05°35'S, 79°10'W (MSB); ÁCM Bosque de Huamantanga, 05°40'S, 78°56.5'W (A. García-Bravo *in litt.* 2017); Quebrada Lanchal, 05°41'S, 79°15'W (LSUMZ 169907–910); Bosque Paja Blanca, 06°23'S, 79°07'W (XC 37724, B. Planqué); BP Pagaibamba, 06°26'S, 79°04'W (XC 23696, W.-P. Vellinga); Montaña Negra, 05°04'S, 78°44'W (XC 10434, T. Mark).

Grallaricula nana kukenamensis Chubb, 1918b, Bulletin of the British Ornithologists' Club, vol. 38, p. 86, '5,000ft, Kukenam Mountains, British Guiana' [= 05°13'N, 60°51'W, 1,524m, Cerro Cuquenán, Bolívar, Venezuela: see Peters 1951, Paynter 1982]. The holotype is an adult female collected 31 August 1883 by Henry Whitely. This race is confined to the south-east tepui portion of the Gran Sabana, in Venezuela, and the area around Cerro Roraima in western Guyana. Morphologically diagnosable *kukenamensis* is also the most geographically isolated taxon considered here, being separated by at least 350km of unsuitable lowland habitat from other populations. It can be reliably separated from other races by its shorter tarsus, and from the two geographically closest subspecies, *olivascens* and *nanitaea*, in having a broader bill and shorter tail. Unfortunately, the vocalisations of *kukenamensis* are poorly known, with only one recording of a song and call note (Donegan 2008), together with a separate description of the call note unsupported by a recording (Braun *et al.* 2003). Donegan (2008) compared plumage, biometrics and the limited vocal data for *kukenamensis* with both subspecies of Sucre Antpitta and with the other races of Slate-crowned Antpitta described here, and suggested that, along with the elevation of Sucre Antpitta (races *cumanensis* and *pariae*) to species rank, *kukenamensis* should also be considered a species apart. This recommendation has not yet been fully accepted, pending the collection of further molecular and/or vocal evidence (Remsen *et al.* 2017). For now, I include *kukenamensis* as a race of Slate-crowned Antpitta. The situation, however, is far from clear as, with respect to biometrics, it is closer to *cumanensis*, but in plumage it is closer to nominate *nana* and its voice resembles neither (see Donegan 2008). Race *kukenamensis* differs from nominate *nana* in being paler overall, particularly the ash-grey crown (not slate-grey) and the ochraceous-brown back, wings and tail (not rich brown) (Chubb 1921). **SPECIMENS & RECORDS VENEZUELA**: *Bolivar* c.11.8km SE of San Isidro, 06°05'N, 61°22'W (AMNH 12024/38, FMNH 395496, COP 79001); Soldier's Monument area, 05°54'N, 61°26'W (XC 223759, P. Boesman); SW flank of Ptari-tepui, 05°46'N, 61°46'W (COP 27095–099); N flank of Sororopán-tepui, 05°42'N, 61°45'W (COP 27087–094); W peak of Chimantá-tepui, 05°17'N, 62°15'W (COP 35779–081); Cerro Cuquenán, 05°13'N, 60°51'W (Mayr & Phelps 1967); Acopán-tepui, 05°10'N, 62°02'W (COP 42094/95); Cerro Roraima, 'Phillip Camp', 05°09'N, 60°47'W (Chapman 1931, AMNH 236691/92); Arabupu, 05°06'N, 60°40'W (COP 4240). **GUYANA**: *Cuyuni-Mazaruni* Mount Ayanganna, 05°22.5'N 59°58.5'W (Milensky *et al.* 2016, IBC 1079044, photo J.R. Saucier); N slope of Mount Roraima, 05°16'N, 60°44'W (Braun *et al.* 2003, KUNHM, USNM, ML).

STATUS Birdlife International (2017) and other recent evaluations of the threat status of Slate-crowned Antpitta (Freile *et al.* 2010b) agree that it should be ranked as Least Concern. No threats have been specifically documented, but apparently the species can tolerate some degree of human modification of its habitat, especially the presence of bamboo-dominated regrowth. **PROTECTED POPULATIONS** *G. n. olivascens* PN El Ávila (COP); PN Macarao (J. Miranda *in litt.* 2017); MN Pico Codazzi (ML 171308, J.W. McGowan). *G. n. nanitae* **Venezuela**: PN Guaramacal (Boesman 1998); PN Tamá (COP, FMNH 43603); PN Sierra Nevada (USNM, FMNH, XC); PN Páramos de Batallón y La Negra (COP 84126–128, FMNH 288322, CM A1831); PN Tapo-Caparo (XC 223756, P. Boesman). **Colombia**: PNN Tamá (IaVH). *G. n. hallsi* PNN Serranía de los Yariguíes (Donegan *et al.* 2007). *G. n. nana* PNN Cordillera de los Picachos (Bohírquez 2002, Donegan 2008, IAvH); SFF Iguaque (ML 80952, M. Álvarez); RFP Ríos Blanco y Negro (Stiles & Rosselli 1998); RB Carpanta (Andrade *et al.* 1994, Andrade & Lozano 1997, ICN). *G. n. occidentalis* **Colombia**: PNN Farallones de Cali (ICN 25920/21); PNN Munchique (Negret 1994); PNN Tatamá (Echeverry-Galvis & Córdoba-Córdoba 2007); PNN Puracé (Hilty & Silliman 1983); PNN Los Nevados (Pfeifer *et al.* 2001); PRN Ucumarí (Kattan & Beltrán 1997); Parque Regional Arví Ecoturistico (Castaño-Villa & Patiño-

Zabala 2008); SFF Otún-Quimbaya (XC 54156, B. López-Lanús); RMN El Mirador (Salaman *et al.* 2007a); RFP La Montaña (Arbeláez-Cortés *et al.* 2011b, XC); RNA Ranita Dorada (Salaman *et al.* 2009a); RNA Loro Orejiamarillo (IBC 994882, N. Athanas); RNA El Colibrí del Sol (ML 45284721, photo N. Voaden); RNA Loro Coroniazul (O. Cortes-Herrera *in litt.* 2017); RNA Mirabilis–Swarovski (Salaman *et al.* 2007a); DMI Campoalegre (Donegan 2008, IAvH). **Ecuador**: PN Cayambe-Coca and RE Antisana (HFG); PN Sangay (XC 250020, N. Krabbe); PN Sumaco-Galeras (HFG); PN Llanganates (J. Freile *in litt.* 2017); PN Podocarpus (Ridgely & Greenfield 2001, IBC); RP Tapichalaca (Greeney & Sornoza 2005, Greeney *et al.* 2010), RP Guango, RP Cabañas San Isidro and RP Estación Biológica Yanayacu (HFG); RP Ankaku (J. Freile *in litt.* 2016). **Peru**: ACM Bosque de Huamantanga (A. García-Bravo *in litt.* 2017). *G. n. kukenamensis* PN Canaima (Mayr & Phelps 1967, Hilty 2003, COP, AMNH).

OTHER NAMES *Conopophaga nana* (Sclater 1855c); *Grallaricula nana nana* (= *occidentalis*; Bond & Meyer de Schauensee 1940a). **English** Guianan Antpitta [*kukenamensis*] (Donegan 2010) Red-breasted Ant-Thrush (Brabourne & Chubb 1912); Kukenam Ant-Thrush (Chubb 1921); Slate-crowned Grallaricula [*nana*] (Cory & Hellmayr 1924, Meyer de Schauensee 1950b); Caracas Grallaricula [*olivascens*] (Cory & Hellmayr 1924); Kukenaam Grallaricula [*kukenamensis*] (Cory & Hellmayr 1924); Western Slate-crowned Grallaricula [*occidentalis*] (Meyer de Schauensee 1950b). **Spanish** Ponchito Enano (Meyer de Schauensee & Phelps 1978, Phelps & Meyer de Schauensee 1979, 1994, Valarezo-Delgado 1984, Krabbe & Schulenberg 2003, Verea *et al.* 2017); Tororoí de Corona Pizarrosa (Plenge 2017); Tororoí enano (Cardona *et al.* 2005); Tororoí Coronipizarrosa (Clements & Shany 2001); Gralaria coronipizarra (Valarezo-Delgado 1984); Ponchito Enano Andino [*nana*], Ponchito Enano del Kukenam [*kukenamensis*], Ponchito Enano Avileño [*olivascens*] (Phelps & Phelps 1950). **French** Grallaire naine. **German** Graukappen-Ameisenpitta (Krabbe & Schulenberg 2003).

Slate-crowned Antpitta, adult (*occidentalis*), RNA Río Blanco, Manizales, Caldas, Colombia, 20 February 2013 (*Daniel Uribe*).

Slate-crowned Antpitta, adult (*occidentalis*) with scaled breast, RP Tapichalaca, Zamora-Chinchipe, Ecuador, 19 April 2011 (*Luke Seitz*).

Slate-crowned Antpitta, egg in nest (*occidentalis*), RP Tapichalaca, Zamora-Chinchipe, Ecuador, 25 November 2003 (*Harold F. Greeney*).

Slate-crowned Antpitta, older nestling (*occidentalis*) in nest, RP Estación Biológica Yanayacu, Napo, Ecuador, 17 November 2006 (*Eliot T. Miller*).

Slate-crowned Antpitta, adult (*occidentalis*), RNA Río Blanco, Manizales, Caldas, Colombia, 9 April 2014 (*Fabrice Schmitt/WINGS Birding Tours Worldwide*).

SUCRE ANTPITTA
Grallaricula cumanensis   **Plate 24**

Grallaricula cumanensis Hartert, 1900, Bulletin of the British Ornithologists' Club, vol. 11, p. 37, Las Palmales and Rincón de San Antonio, Cumana, Venezuela.

Prior to a recent morphological and vocal study, Sucre Antpitta was universally treated as a subspecies of the wide-ranging Slate-crowned Antpitta. Sucre Antpitta is endemic to north-east Venezuela in the region of the Paria Peninsula, where it is the only species of *Grallaricula*. The plumage is largely rufous-red below and dark brown above, with a slate-grey crown and nape, but getting a good view of this small antpitta often requires considerable fortune. It inhabits the undergrowth of humid forest above 700m elevation, and is almost certainly already threatened with extinction; it is currently treated as Vulnerable. The eggs were described more than 45 years ago, but this remains the only published information on its reproductive habits. Like its congeners, and indeed most antpittas, very little is known about any aspect of the natural history, habitat requirements and behaviour of Sucre Antpitta

IDENTIFICATION 11.0–11.5cm. Sexes similar. Adults have a slate-grey crown, a large orange-rufous pre-loral spot and a buffy rufous eye-ring. The upperparts are uniformly dark olive-brown. Below, they are bright rufous on the throat, chest and flanks, with a white crescent on the lower throat and upper chest, and the central and lower belly are bright white. Sucre Antpitta is the only *Grallaricula* within its range, and is unlikely to be confused with any sympatric species. Slate-crowned Antpitta has paler or duller underparts, less white below, and a different song. Despite Slate-crowned occurring in the coastal cordilleras of north-central Venezuela, its range does overlap with Sucre Antpitta to the east. Additionally, allopatric Rusty-breasted Antpitta lacks the grey crown and is paler below, with a strikingly different song.

DISTRIBUTION Endemic to the Paria Montane centre of endemism (Cracraft 1985) in the coastal mountains of Venezuela (Krabbe & Schulenberg 2003). It occurs on the Paria Peninsula, in the state of Sucre (*pariae*), and disjunctly on the Turimiquire Massif (both the Serranía de Turimiquire west of the San Antonio valley, and the Cordillera de Caripe to the east) at the borders of the states of Sucre, Anzoategui and Monagas (*cumanensis*).

HABITAT Sucre Antpitta inhabits the understorey of humid, epiphyte-laden, montane forests at elevations of 600–1,850m (Phelps & Phelps 1949, 1963). In the Serranía de Turimiquire, it is known from altitudes of 1,000–1,850m. On Cerro Humo, it occurs at 1,100–1,200m, and on Cerro El Olvido at 600–885m (BirdLife International 2017). No specific data exist concerning habitat requirements, but it probably favours areas of *Chusquea* bamboo as *G. nana*.

VOICE The song (*pariae*) lasts 1.6–2.0s and consists of 29–35 (*n* = 12, mean = 32.8 ± 1.6 notes) evenly paced notes beginning at *c.*2.2kHz, rising gradually to around 2.7kHz (*n* = 12; mean = 2.71 ± 0.05kHz) near the middle of the series, then falling gradually to finish at *c.*2.2kHz. This trill is delivered at a speed of 17.73 ± 0.31 notes/s (*n* = 12; range = 17.30-18.33 notes/s). The volume increases during the first few notes, then holds steady (Krabbe & Schulenberg 2003, Donegan 2008). The song of the nominate race is similar, but seemingly slightly more variable between individuals. Its song includes 33.7 ± 4.8 notes (*n* = 11; range = 25–38 notes), lasts 1.90 ± 0.27s (*n* = 13; range = 1.49–2.15s), is delivered at a speed of 17.78 ± 0.56 notes/s (*n* = 11; range = 16.82–18.78 notes/s), and reaches a maximum frequency of 2.84 ± 0.09kHz (*n* = 13; range = 2.68–2.98kHz) (Donegan 2008).

NATURAL HISTORY There are no published descriptions of the species' behaviour. Undoubtedly, however, it forages low to the ground, often in dense vegetation, in a manner similar to Slate-crowned Antpitta (Ridgely & Tudor 1994, Krabbe & Schulenberg 2003).

DIET The diet is unknown, although almost certainly it feeds predominantly on small invertebrates.

REPRODUCTION Nest and breeding behaviour undescribed. **Egg, Laying & Incubation** Kreuger (1968) gave the mean measurements of six eggs as 19.98 × 16.16mm and described the eggs as having a creamy-white ground colour with small brown or dark brown blotches and speckles. He did not, however, provide any information on clutch size, collection date or locality. Schönwetter (1979) expanded on this by providing the range of linear measurements for these eggs as 19.3–20.8 × 15.9–16.1mm. Confusingly, Schönwetter (1979) then provided mean egg measurements as 20.0 – 16.3mm (note that mean diameter is greater than the upper range of measurements provided).

Schönwetter (1979) also gave the average weight of the (empty) eggs as 0.17g (range 0.163–0.183g) and calculated the weight of fresh eggs as 2.8g. **Seasonality** Phelps & Phelps (1949) reported that 'May specimens have breeding gonads'. Apart from that, I found only two immature specimens in collections I examined. *G. c. cumanensis* Subadult transitioning, 14 March 1932, Cerro Turumiquire (FMNH 92099). *G. c. pariae* Subadult trasitioning, 28 May 1948, Cerro Azul (FMNH 189771).

TECHNICAL DESCRIPTION Sexes similar. The following is based on (Greeney 2013i). *Adult* Sucre Antpitta has a slate-grey forecrown, crown and nape, with a prominent orange-rufous loral spot and a buffy to buffy-rufous eye-ring, sometimes reduced to a postocular crescent. The upperparts are dark olive-brown, with the wings slightly browner and the primaries narrowly edged buff. The ear-coverts, neck-sides and throat are bright rufous-orange. The lower throat and upper chest are variably white or washed rufous, forming a partial collar delimited below by a band of rufous-orange on the chest. This band may be complete or broken by white in the centre that connects the white collar to the white central and lower belly. The breast-sides and flanks are bright rufous-orange, contrasting with the bright white belly. **Adult Bare Parts** *Iris* brown; *Bill* black, base of mandible whitish; *Tarsi & Toes* grey. *Subadult* Not well known. Phelps & Phelps (1949) described 'immature plumage' as similar to the adult but with the central forehead, crown and nape dark brown. They stated that white on the abdomen is lacking, presumably implying that the belly is entirely rufous-orange like the flanks. The only two subadults I examined had very worn wing-coverts, but these still bore the remains of buffy feather tips (mostly on the leading vane) and narrow rusty feather margins.

MORPHOMETRIC DATA *G. c. cumanensis* *Wing* 70mm, 67mm; *Tail* 31mm, 33mm; *Tarsus* 25mm, 25mm (*n* = 2, ♂, ♀, Hartert 1900). *Wing* mean 66.6mm, range 64–69mm; *Tail* 27mm; *Bill* culmen from base, mean 17.5mm, range 17–18mm (*n* = 5♂♂, Phelps & Phelps 1949). *Wing* mean 64.3mm, range 62–66mm; *Tail* mean 26.3mm (range 26–27mm); *Bill* culmen from base, mean 17.7mm, range 17–18mm (*n* = 3♀♀, Phelps & Phelps 1949). *Wing* 68mm, 64mm, 67mm; *Tail* 31mm, 28mm, 28mm; *Bill* 15mm (×3) (*n* = 3, ♂ holotype, ♀, ♀, Hellmayr 1917, see Hartert 1900 measurements). From Donegan 2008 (sex not given): *Wing* mean 63.0 ± 2.3mm, range 60.0–65.0mm (*n*=6); *Tail* mean 26.9 ± 1.3mm, range 25.0–28.5mm (*n*=6); *Bill* total culmen mean 16.4 ± 0.8mm, range 15.5–17.0mm (*n* = 4), width at nares, mean 5.4 ± 0.2mm, range 5.3–5.7mm (*n* = 4); *Tarsus* mean 25.4 ± 1.4mm, range 23.0–26.5mm (*n* = 5). *G. c. pariae* *Wing* 66mm; *Tail* 28mm; *Bill* exposed culmen 15mm, culmen from base 18mm; *Tarsus* 25mm (*n* = 1♂ holotype, Phelps & Phelps 1949). *Wing* mean 66mm, range 64–70mm; *Tail* mean 28.8mm, range 28–30mm; *Bill* culmen from base 18mm (*n*=5♂♂, Phelps & Phelps 1949). *Wing* mean 65mm, range 64–66mm; *Tail* mean 27.2mm, range 26–28mm; *Bill* culmen from base mean 18.4mm, range 18.0–19.0mm (*n*=5♀♀, Phelps & Phelps 1949), Donegan (2008) provided the following additional measurements. *Wing* mean 63.1 ± 2.0mm, range 60.0–67.0mm (*n* = 26); *Tail* mean 27.6 ± 0.9mm, range 26.0–29.5mm (*n* = 25); *Bill* culmen mean 17.2 ± 0.6mm, range 16.0–18.5mm (*n* = 25), width mean 6.0 ± 0.2mm, range 5.6–6.3mm (*n* = 23); *Tarsus* mean 24.6 ± 1.0mm, range 22.5–26.5mm (*n* = 25). **Mass** There are no published data, but the species is probably similar in weight to Slate-crowned Antpitta (e.g., 17.5–21.5g, Dunning 2008). **Total Length** There are no total length values published that are specific to Sucre Antpitta, but it is similar in size to Slate-crowned Antpitta.

TAXONOMY AND VARIATION Long considered a race of the more widely distributed Slate-crowned Antpitta (Cory & Hellmayr 1924, Phelps & Phelps 1963, Howard & Moore 1984, Fjeldså & Krabbe 1990, Sibley & Monroe 1990, Hilty 2003, Krabbe & Schulenberg 2003, Clements 2007, Ridgely & Tudor 2009). Ridgely & Tudor (1994) suggested that the two subspecies of Sucre Antpitta should be elevated to species status under the vernacular name Paria Antpitta. Following a revision of the various races of Slate-crowned Antpitta, Donegan (2008) recommended elevating *G. n. cumanensis* and *G. n. pariae* to species rank and suggested the name Sucre Antpitta. This suggestion was followed by Remsen *et al.* (2017). Parker *et al.* (1985) suggested that Sucre Antpitta (included within Slate-crowned Antpitta) forms a superspecies with Rusty-breasted Antpitta, and these three species are probably closely related.

Grallaricula cumanensis cumanensis Hartert, 1900, Las Palmales [10°17'N, 63°45'W] and Rincón de San Antonio [10°16'N, 63°43'W], Cumana [Sucre], Venezuela. In the original description, Hartert (1900) listed male and female syntypes, later designating as lectotype (AMNH 492320) a male collected 17 February 1898 at Las Palmales by Henry Caracciolo (Hartert 1922). The paralectotype, also collected by Caracciolo, is a female collected 15 March 1898 at Rincón de San Antonio (AMNH 492322) (LeCroy & Sloss 2000). Hartert (1900) provided the following description of nominate *cumanensis*: 'Lores, sides of head, throat, wide breast-band and sides of body bright dark ferruginous; middle of abdomen and jugular patch white; top of head and hindneck dark grey; remainder of upper surface olive-brown; remiges blackish brown, externally edged with the colour of the back, with the extreme bases and inner webs edged with pale ferruginous; primary-coverts black; tail brown; under wing-coverts ferruginous.' **Specimens & Records** *Sucre* Cumanacoa, 10°15'N, 63°55'W (CM P106886); Latal, 10°10'N, 63°55'W (Chapman 1925b); Cerro Turumiquire, 10°07'N, 63°53'W (FMNH 92099); Cerro Peonía, 10°05'N, 64°18'W (COP 15537). *Monagas* Cerro Negro, 10°11'N, 63°33'W (COP 81287/88, COP 23057, XC 223735, P. Boesman); Cerro Piedra de Moler, 10°05.5'N, 63°49'W (COP 81269).

Grallaricula cumanensis pariae Phelps & Phelps, 1949, Proceedings of the Biological Society of Washington, vol. 62, p. 36, Cerro Azul, Cristóbal Colón, Península de Paria, Sucre, Venezuela [*c.*10°42'N, 61°57'W]. The holotype (COP 44025), is an adult male collected 28 May 1948 by Ramón Urbano and is on deposit at AMNH. Originally described as *G. nana pariae*, it is similar to the nominate race, but is a cleaner olivaceous-brown on the upperparts, without any of the yellow-brown tones of the latter (Phelps & Phelps 1949). However, Donegan (2008) failed to find any consistent plumage differences between them, suggesting that a re-evaluation of taxon limits is necessary. Phelps & Phelps (1949) described the type specimen, a male, as follows: 'Forehead, except in the center, Orange-Buff Cadmium Yellow; center line of forehead, crown and nape Deep Mouse Gray; back, rump and upper tail-coverts Medal Bronze, the long back feathers covering the basal half of tail subterminally white and tipped with pale buff forming a band; orbital ring anteriorly buffy, posteriorly

whitish; a dusky preocular lunule. Below Ochraceous-Orange, darkest on breast, sides and flanks and paler on chin and throat; a prominent white spot on anterior breast; entire abdomen white; shanks and undertail coverts buffy. Remiges Fuscous, outermost edged with pale buff, the others with brownish, most extensively on tertials; remiges edged with buff internally, the outer ones only basally; greater and median upper wing coverts tipped with dark buffy forming two indistinct bands; primary coverts entirely blackish partly covering a buffy speculum; bend of wing buffy; axillaries and under wing coverts Cadmium-Orange. Tail darker, browner, than the back.' **Specimens & Records** *Sucre* Cerro El Olvido, 10°42'N, 61°59'W (BirdLife International 2017); Cerro Humo, 10°41'N, 62°37'W (long series at COP, XC); Ensenada Yacua, 10°40.5'N, 61°59'W (USNM 344248); Las Melenas, 10°39.5'N, 62°30'W (J.E. Miranda T. *in litt.* 2017); 'Marino', *c.*10°41'N, 62°27.5'W (AMNH 388107). **Notes** Mariño is a municipality in Sucre, the centre (and capital) of which is San Antonio de Irapa, at *c.*100m. Just north of there, however, the mountains reach more appropriate elevations for Sucre Antpitta, although no nearby localities appear to achieve the 1,100m on the label of AMNH 388107. The above GPS places the most likely collecting locality for this specimen at around 1,000m, *c.*8km NE of San Antonio de Irapa.

STATUS Sucre Antpitta has a very small range and is restricted to just two areas in northern Venezuela. Largely due to this extremely limited distribution, it is considered Vulnerable (BirdLife International 2017). The regions in which it occurs, the Turimiquire Massif and Paria Peninsula, are both experiencing widespread forest clearance for coffee, mango, banana and citrus plantations (BirdLife International 2017). The extent to which Sucre Antpitta tolerates disturbance is unknown, but large-scale clearing and burning is undoubtedly detrimental. Indeed, were it not known from at least seven localities, the species would qualify as Endangered (BirdLife International 2017).

OTHER NAMES *Grallaricula nana cumanensis/pariae* (Rodner *et al.* 2000, Restall *et al.* 2006, Clements 2007). **English** Bermudez Grallaricula [*cumanensis*] (Cory & Hellmayr 1924). **Spanish** Ponchito de Sucre (Krabbe & Schulenberg 2003); Ponchito Enano Oriental [*cumanensis*]; Ponchito Enano de Paria [*pariae*] (Phelps & Phelps 1950). **French** Grallaire de Cumana. **German** Sucreameisenpitta (Krabbe & Schulenberg 2003).

Sucre Antpitta, adult (*pariae*), Cerro Humo, Sucre, Venezuela, 29 October 2005 (*Nick Athanas/Tropical Birding*).

Sucre Antpitta, adult (*pariae*), Cerro Humo, Sucre, Venezuela, 29 October 2005 (*Nick Athanas/Tropical Birding*).

Sucre Antpitta, adult (*pariae*), Las Malenas, Sucre, Venezuela, 28 May 2007 (*Joseph Tobias*).

CRESCENT-FACED ANTPITTA
Grallaricula lineifrons   Plate 23

Apocryptornis lineifrons Chapman, 1924, American Museum Novitates, vol. 123, p. 5, Oyacachi, upper Papallacta River, Ecuador [00°13'S, 78°05'W, Napo Prov.]. The type (AMNH 180279), an adult female, was procured on 24 September 1923, by the prolific collector Carlos Olalla and his sons (LeCroy & Sloss 2000).

Crescent-faced Antpitta is one of the most strikingly plumaged members of the genus, perhaps even of the entire family. I know it was among the first birds on my 'must study' list when I first became interested in avian biology. There is no significant plumage variation, despite statements to the contrary (Lehmann *et al.* 1977). Adults are dark brown above with a slate-grey crown and nape, a prominent white crescent in front of the eye, and buffy-orange to white underparts boldly streaked black. Although the type specimen was collected in north-east Ecuador in 1923, Crescent-faced Antpitta remained unreported in life for 53 years (Peters 1951, Meyer de Schauensee 1966, 1970), until Lehmann *et al.* (1977) provided the first observations on its behaviour and reported it from Colombia. Today, based on regular reports and sightings by tourists and researchers, it is known to be more widespread than previously thought, inhabiting the undergrowth of humid forests at elevations around 3,000m. Its distribution appears somewhat patchy, extending from southern Colombia to southern Ecuador on the east slope of the Andes. Only recently, 90 years after its formal description, was breeding reported in the literature, based on a single nest found very close to the type locality in north-east Ecuador (Greeney & Jipa 2012). BirdLife International (2017) has currently assigned the species Near Threatened status. Proposed surveys of suitable habitats within its known range, particularly at locations where it has yet to be recorded, and studies to determine the species' true level of tolerance of secondary habitats are conservation priorities. Apart from such surveys, future research should focus on discovering and studying other aspects of its natural history. In particular, more information on breeding biology, diet and post-fledging care and dispersal would be useful. An additional line of investigation might be a careful comparison of habitat use to explain its apparently wholly allopatric distribution with Slate-crowned Antpitta, despite the two species overlapping in altitudinal range but replacing each other in adjacent valleys (Krabbe *et al.* 1997, Krabbe 2008).

IDENTIFICATION 11.5–12.0cm. Boldly patterned, with crown, nape and sides of head a dark slate-grey, broken by conspicuous orange-buff or ochraceous triangular spots on sides of neck and by bold white crescents that curve above and in front of the eye and blend into a 'moustache' just ahead of the anterior edge of the ear-coverts. Central throat and upper chin white, contrasting with black lower chin and throat-sides. Lower throat buffy, breaking into distinct streaking on the breast and belly, created by black feather edges. Sides and flanks olive-brown and vent pale buff (Fjeldså & Krabbe 1990, Ridgely & Tudor 1994, Restall *et al.* 2006). The bold white facial crescent, large ochraceous spot on the neck-side and obviously streaked underparts immediately distinguish Crescent-faced Antpitta from other species within its limited range. It is unlikely to occur sympatrically with any other *Grallaricula* except Slate-crowned Antpitta, which lacks bold head and face markings, and never shows breast streaking as pronounced as that of Crescent-faced.

DISTRIBUTION Confined to the North Andean centre of endemism (Cracraft 1985), from northern Colombia (Caldas) to southern Ecuador (Loja) (Rodner *et al.* 2000, Clements 2007). Similar to several other *Grallaricula*, the known range has expanded greatly in the past few decades, following its discovery at quite a few new locations. In fact, for nearly 50 years, Crescent-faced Antpitta was

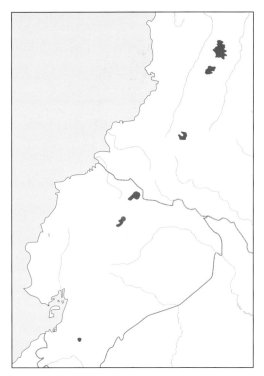

known only from the type locality in north-east Ecuador (Chapman 1924). In the mid-1970s, however, it was discovered in Colombia's Puracé National Park (Lehmann *et al.* 1977) providing a significant northward range extension. In Ecuador, following an absence in records for *c.*70 years, it was relocated in 1991 (Ridgely & Greenfield 2001) and has since been found at several places, thereby extending its known range both south and north. Most likely, its range is more contiguous than currently known, but I have restricted the range map to known localities, to facilitate the reporting of new ones. In Colombia it is now reported from scattered localities between Caldas and Nariño (Ayerbe-Quiñones *et al.* 2008, McMullan *et al.* 2010, Arbeláez-Cortes *et al.* 2011, Calderón-Leytón *et al.* 2011). In Ecuador, its distribution is similarly patchy, with most records pertaining to very specific locations from Carchi to Loja (Ridgely & Greenfield 2001, Buitrón-Jurado 2008). Currently, hotspots for finding the species (from eBird) are mostly in Ecuador, and include the Cerro Acanama area, the vicinity of Papallacta, and around Cerro Mongus and Guandera.

HABITAT Crescent-faced Antpitta inhabits the undergrowth of humid, montane cloud and elfin forests at, and just below, the treeline (Hilty & Brown 1986, Ridgely & Tudor 2009), but is not found in *páramo* habitats above the treeline as suggested by Restall *et al.* (2006). Published accounts, and my own observations, suggest that steep slopes may be preferred (Fjeldså & Krabbe 1990, Freile 2000), but it might also be noted that 'steep slopes' describes most of the terrain within its known altitudinal range. Overall, it appears Crescent-faced Antpitta uses *Chusquea* bamboo habitats far less than the only other *Grallaricula* of similar altitudinal distribution, Slate-crowned Antpitta. However, the two species are not

known to be sympatric at any location, with Crescent-faced Antpitta replacing Slate-crowned Antpitta above 2,900m (Krabbe 2008). Indeed, although Krabbe *et al.* (1997) found the two species at the same elevation in northern Ecuador, these locations were in adjacent valleys, each harbouring only one of the two species. Krabbe (2008), in his discussion of the importance of arid valleys as dispersal barriers in the Andes, went on to suggest that the two species provide an excellent example of competitive exclusion. In eastern Ecuador the Río Zamora is probably an important dispersal barrier for several species of Andean birds, including Crescent-faced Antpitta. Slate-crowned Antpitta, which ranges south of the Río Zamora, ascends all the way to treeline, strongly suggestive of a case of competitive release in the absence of Crescent-faced Antpitta. Previously published altitudinal ranges generally are within the band 2,900–3,400m (Parker *et al.* 1996, Krabbe *et al.* 1997, Ridgely & Greenfield 2001, Graham *et al.* 2010). The lower limit of 1,800m given by McMullan *et al.* (2010) is undoubtedly an error, and I was unable to find records at the lower limit of 2,500m reported by Fjeldså & Krabbe (1990). Recent records, however, extend the upper limit to at least 3,700m in Colombia (XC 90240, D. Calderón-F.) and to at least 3,500m in Ecuador (XC 41654, D.M. Brinkhuizen).

VOICE The most frequently described vocalisation, believed to be the song, is a rapid series of 14–20 notes delivered with steadily increasing pitch and volume, levelling at the end (Krabbe & Schulenberg 2003), or an ascending trill ending with several shrill pips, e.g., *pu-pu-pu-pe-pe-pee-pee-pi-pi-pi*? (Ridgely & Greenfield 2001, Restall *et al.* 2006). Lehmann *et al.* (1977) also mentioned a series of hard notes *clip clip clip*, apparently given in alarm. Robbins *et al.* (1994a) described the song of adult males (collected and sexed) in northern Ecuador as a slightly ascending series of closely spaced notes. One spontaneously vocalising male consistently gave 21 notes per song, whereas males responding to playback usually gave 13–15 notes per song. These authors gave a song duration of around 3.5s, with a frequency range of 2.6–3.9kHz. They reported an adult female (collected) giving this same vocalisation after her (presumed) mate was collected. Both sexes made single-noted, slightly downslurred whistles, likened by Robbins *et al.* (1994a) to the well-known call of Ochre-breasted Antpitta. This vocalisation was also given by both adults in the vicinity of their active nest (Greeney & Jipa 2012), and was likened to the call of Peruvian Antpitta given near its nest (Greeney *et al.* 2004b). At Cerro Mongus, Carchi, N. Krabbe and F. Sornoza recorded a juvenile male (ANSP 184703) giving a song that was described by Robbins *et al.* (1994a) as being 'similar to the adult's …, except that the notes were more raspy and much lower in frequency.'

NATURAL HISTORY Behaviour and ecology are poorly documented. Most often encountered as solitary individuals or in pairs, this species apparently only rarely joins mixed-species flocks (Ridgely & Tudor 1994, Greeney 2012c). Birds move about the understorey by clinging to small twigs, often upright ones, and making short sally-gleans to surrounding foliage. They are rarely seen on the ground, but may hop on fallen logs and stumps (Greeney 2012c). Like other *Grallaricula*, they frequently pause for long periods and remain motionless except for a steady, rhythmic rump-twitch. Also similar to congeners, Crescent-faced Antpittas frequently flick their wings in an agitated manner, especially in response to playback or when disturbed by an observer's presence (Lehmann *et al.* 1977, Robbins *et al.* 1994a). There are no concrete data concerning territorial behaviour. What are presumably the same pairs, however, are consistently found in the same areas in north-east Ecuador (Greeney 2012c). Lehmann *et al.* (1977) reported that an adult in Colombia approached in response to *spissh* noises. When it came into view it perched about 1m above ground, well hidden in dense vegetation, and scolded the offending 'spissher' with several sharp notes (see Voice).

DIET Like its congeners, presumed to be largely or entirely insectivorous, but few data are available. Reported stomach contents for several individuals included fragments of small spiders, adult and larval Coleoptera, and other unidentified arthropods (Robbins *et al.* 1994a), plus 'well-ground insect chitin' (Lehmann *et al.* 1977). Adults foraging near an active nest searched in displaced moss and vegetation on the ground that had been overturned by observers, suggesting that the species may also have a propensity to forage in association with large, montane herbivores such as tapirs or bears (Greeney 2012b). Further support for this hypothesis may be the somewhat cryptic statement of Fjeldså & Krabbe (1990) that Crescent-faced Antpitta is 'attracted to strange sounds'.

REPRODUCTION Other than a young collected in northern Ecuador (Robbins *et al.* 1994a), nothing was published on breeding until very recently. Greeney & Jipa (2012) described the nest and nestling for the first time based on observations of a nest found 10 February 2012 near Papallacta (Napo) in north-east Ecuador, very close to the type locality. **NEST & BUILDING** The nest described by Greeney & Jipa (2012) was 3.6m above ground and was an open, mossy cup built into a loose tangle of vines that hung 1m below a small, horizontal tree branch. The nest comprised a loose mixture of green moss, short sticks and leaf petioles that formed a 10.5cm-tall platform onto which the nest itself was built. The accumulation of material below the nest was clearly placed there by the adults and gave the entire structure a globular appearance, rather than the fairly shallow, open-cup nest that it was. The outer portion of the nest itself was composed of similar materials to those used for the base, but the moss was more tightly packed, allowing the nest itself to be easily separated from the support. Internally, the nest was neatly lined with a 0.5cm-thick layer of crisscrossed, dark rootlets and flexible fibres, with a few pale grass stems intermixed. I have since found two additional nests composed of nearly identical materials, both within the territory occupied by the pair studied by Greeney & Jipa (2012). The most recent of these was seemingly complete when I found it, and was almost identical to the published description, including the thick base of material below the nest. This nest was 3.4m above ground, and supported by many small, interlaced branches of an unknown Ericaceae. The supporting plant was growing from a portion of the slope that was extremely steep, and was only 1.2m horizontal distance from the ground where the supporting shrub sprouted, allowing me to access it for photos and measurements. I found a third nest with half of a recently-consumed egg beside the rim (see below) and large numbers of fleas that swarmed up my arms as I examined it. Just 1.8m above ground, its size and composition were nearly identical to the other two nests but, unlike them, did not include a thick pile of loose base material. It was, instead, built atop an abandoned, globular, oven-shaped nest of White-browed Spinetail *Hellmayrea gularis* (Greeney & Zyskowski 2008) that was suspended in a loose 'web' of vines below a heavy mossy branch. Only the neat, tightly packed upper portion of the nest rim and inner cup were built by the antpittas, yet the pile of material below this, giving the other two nests their overall globular shape, was replaced by the spinetail nest, producing the same effect. Having spent many hours searching for nests in the moss-covered understorey of Andean forests at this elevation, I examined this clump of moss under the assumption that I had found the nest of White-browed Spinetail, or perhaps Rufous-headed Pygmy-Tyrant *Pseudotriccus ruficeps*, both of whose nests are more apparent and easily found. While searching for the lateral entrance to the nest, only by accident did I discover the shallow cup nest of the antpitta built on top. My own blunder leads me to suggest that the purpose of the copious amounts of material below the other two nests may serve to disguise their outline and perhaps distract potential predators from the true whereabouts of their precious contents. I will be very interested to find more nests of Crescent-faced Antpitta to determine if the piling of material below the cup (far more than is needed to support it) represents normal architecture, and to pursue further support for my hypothesis of nest crypsis. I was able to visit this nest irregularly during the year after I found it, and watched the cup slowly fill with leaves and detritus from the canopy above. Almost one year later, however, the cup had been cleaned and fresh rootlets added to the interior, suggesting Crescent-faced Antpitta is rather faithful to nesting locations, perhaps even refurbishing nests from previous seasons. The process of nest construction and site selection is undocumented. However, Greeney & Jipa (2012) observed both sexes carrying material, suggesting that building duties are shared. **EGG, LAYING & INCUBATION** The egg remains beside the third nest described above cannot be attributed to Crescent-faced Antpitta with certainty, but the remnants strongly suggested that it was bluntly subelliptical in shape, like those of other antpittas, and I feel strongly that it was an egg of this species. There were unconsumed portions of flesh within the shell, suggesting that it had been well developed when destroyed. The unconsumed portion of shell was pale greenish to greenish blue and irregularly blotched and speckled with sparse, brown and dark cinnamon markings. If the nest and egg were, indeed, those of Crescent-faced Antpitta, this would be only the second *Grallaricula* to have the ground colour of its eggs deviate from the 'normal' beige or pale coffee, the other being Rusty-breasted Antpitta. There are no available data concerning incubation behaviour and, obviously, the above description of egg coloration demands confirmation. Presumably both adults partake in incubation duties as in other *Grallaricula* (Greeney *et al.* 2008a). **NESTLING & PARENTAL CARE** The single nestling was not observed closely, but from photos was described by Greeney & Jipa (2012) as being covered in a dense coat of rusty-brown downy plumage and having its primary feathers at least halfway emerged from their sheaths. The mouth lining was bright crimson-orange, the mandible dusky-orange, maxilla dark grey and rictal flanges yellow-white. This description closely matches descriptions for other *Grallaricula* (Greeney 2012a). Both adults arrived at the nest via a circuitous route, making short hops between small vines and branches, gradually approaching the nest from below and then pausing to peer about while perched

20–50cm away at the level of the nest, before hopping to the rim. All prey items I observed (*n* = 5) were small, ≤10mm, and delivered singly (see Food and Feeding). After one prey delivery, the nestling hunkered low in the nest and raised its rump while shaking it slightly from side to side. The adult responded by stretching up and peering over the nestling. At the moment the faecal sac began to emerge, the adult hopped quickly sideways on the rim, leaned forward and grasped the sac as it emerged. With the faecal sac in its bill, the adult dropped swiftly from the nest, descending in a slow arc to perch on a thin vine 1m above ground and *c.*10m from the nest. It dropped the faeces into the leaf litter before disappearing into thick undergrowth. On all provisioning visits, I could detect no begging vocalisations from my position *c.*6m away. **Seasonality** A pair was observed carrying material in early February in Napo (Greeney & Jipa 2012) and a fledgling was photographed nearby on 27 March 2011 (E.M. Quipo photos). A juvenile male (ANSP 184703; see Description) was collected on 22 March 1992, at Cerro Mongus, Carchi (Robbins *et al.* 1994a). At Cerro Mongus, Crescent-faced Antpitta was noted as especially vocal in March (early to mid-wet season) and less so in July (end of rains) (Robbins *et al.* 1994b). The same seasonal pattern of vocal activity has been noted at Guandera (J. Freile *in litt.* 2017). The nest with a nestling, found not far south of this (Greeney & Jipa 2012), was set to fledge near the start of the rainy season (mid-February 2012), the nest found under construction (discussed above) was found on 22 November 2015 (mid-dry season) and the recently predated nest full of fleas (also discussed above) on 7 November 2012. The latter was apparently freshly refurbished on 18 December 2013. These observations suggest that, at least in northern Ecuador, reproduction peaks at the end of the dry season or early wet season, probably extending from at least November to April (laying December–February). This is close to the pre-March breeding period predicted by Robbins *et al.* (1994a) and is also aligned with the general peak in avian breeding at these altitudes in northern Ecuador (Greeney *et al.* 2011b).

TECHNICAL DESCRIPTION Sexes similar. *Adult* Following Chapman (1924): the head and nape are dark sooty-grey, approaching black, the ear-coverts, forehead and region below eye and above anterior part of eye pure black. Bright white loral area extends upwards in a broad arching crescent to crown, offsetting a small but conspicuous white postocular spot. Back brownish-olive, tail and wings browner, with leading web of outermost primary and alula light yellow-buff. Lesser underwing-coverts brownish-olive and greater coverts white tipped black or dark grey. Conspicuous spots on neck-sides and the posterior part of the broad malar stripe ochraceous-buff. Latter fades to white anteriorly and extends rearward as a band passing behind the ear-coverts to the edge of the nape. Central chin and upper throat white edged in black, forming an inverted V (its base at the chin). Underparts white, strongly washed ochraceous-buff, especially on breast, leaving central abdomen pure white. Most feathers of underparts have broad, conspicuous black lateral margins, broadest on upper breast, narrower below abdomen. Sides and flanks dull brownish-olive, brightened with a wash of ochraceous-buff and indistinctly striped dark grey. Undertail-coverts ochraceous-buff. **Adult Bare Parts** (Lehmann *et al.* 1977, Robbins *et al.* 1994a) *Iris* dark brown; *Bill* black; *Tarsi &*

Toes vinaceous-grey to blue-grey (the only *Grallaricula* with dark legs *fide* Rice 2005a). *Fledgling* I am unable to provide a full description of the fledgling, but it seems certain they leave the nest appearing much like nestlings. From photographs of a recent fledgling (E.M. Quipo), it appears that the first areas to acquire adult-like feathers are the face and throat. This individual was still largely covered in dense red-brown nestling down but appeared 'masked' due to emerging blackish feathers on the face, lores and throat. Several feathers at the base of the malar and on the chin were white to pale buff, already hinting at adult plumage. The downy plumage of the lower breast or upper belly gradually faded to buffy chestnut and to buff on the flanks and lower belly. No further details available from the photographs. **Fledgling Bare Parts** *Iris* dark brown; *Bill* maxilla mostly blackish, suffused orange near base, mandible largely pale yellowish-orange, brighter orange basally, rictal flanges greatly inflated and bright orange, slightly more yellowish on their inner margins; *Tarsi & Toes* unknown. *Juvenile* Although not mentioned by previous authors (Krabbe *et al.* 2013), Robbins *et al.* (1994a) described a juvenile male (see Reproduction Seasonality) as follows: 'hindcrown and nape feathers are a fluffy, dull reddish-brown, [...] the olive-green back and the streaked underparts also are mixed with a few of the dull reddish-brown feathers. The abdomen is lightly washed with buff.' After examining photos of the specimen on which this description was made, I can add only that the upperwing-coverts are narrowly edged pale chestnut-buff (perhaps more strongly marked in fresher specimens). **Juvenile Bare Parts** Undescribed in life. At this age, the bill is probably close to adult size but retains some hint of orange, especially near the base, around the nares and on the tomia. The tarsi and toes are likely similar in colour to those of adults, perhaps slightly more vinaceous (see Plate 23; Fig. b).

MORPHOMETRIC DATA There are few specimens in collections and even fewer published measurements. *Wing* 75mm; *Bill* culmen 16mm, depth at base of nostril 5mm, breadth at base of nostril 4.3mm; *Tail* 33mm; *Tarsus* 29mm; *Middle Toe* without claw 19mm, with claw 23mm, 'greatest length of crown feathers' 14mm (*n* = 1♀ holotype, Chapman 1924). *Wing* 76mm; *Bill* exposed culmen 11.6mm; *Tail* 41.9mm; *Tarsus* 29.1mm (*n* = 1, unsexed, T. Smith). *Wing* flattened chord 80mm, 76mm; *Bill* culmen 14mm, 17mm; *Tarsus* 27mm, 30mm (*n* = 2 unsexed adults, Rivera-Pedroza & Ramírez 2005). Robbins *et al.* (1994a) reported no significant difference in measurements or mass between the sexes, with the following means for nine individuals (± SD): *Bill* culmen from base 16.2 ± 0.5mm; *Wing* chord 77.1 ± 1.7mm; *Tail* central rectrix 37.7 ± 1.3mm; *Tarsus* 27.8 ± 0.4mm. **Mass** mean 21.1g (*n* = 9 both sexes, Robbins *et al.* 1994a, Dunning 2008); range, ♂♂ 20.5–22.5 g (*n* = 6); ♀♀ 17–22 g (*n* = 3) (Krabbe & Schulenberg 2003); 22.1g (*n* = 1, unsexed, T. Smith). **Total Length** 11.0–12.2cm (Hilty & Brown 1986, Dunning 1987, Fjeldså & Krabbe 1990, Ridgely & Tudor 1994, 2009, Ridgely & Greenfield 2001, Restall *et al.* 2006, McMullan *et al.* 2010).

DISTRIBUTION DATA COLOMBIA: *Caldas* Cuenca de Gallinazo, 05°01'N, 75°26'W (D. Uribe *in litt.* 2017); La Gruta, 04°59'N, 75°23.5'W (ML 63372511, photo D. Monsalve S.); 1.5km W of El Apriso, 04°58'N, 75°23'W (Uribe 2011). *Quindío* Reserva Municipal El Mirador, 04°10'N, 75°43'W (D. Cantarón *in litt.* 2017), RN Valle Lindo, 04°10'N, 75°42'W and Hacienda Río Azul, 04°16'N,

75°39'W (Rivera-Pedroza & Ramírez 2005). *Tolima* Nevado del Ruiz, 04°55'N, 75°19'W (XC 273469, P. Boesman); *c.*3.5km E of Arenales La Azulera, 04°46'N, 75°22.5'W (XC 86230, O. Cortes-Herrera); Laguna Las Mellisas, 04°12'N, 75°35'W (ML 28900731, photo J. Sanabria-Mejía). *Cauca* Quebrada Tierra Adentro, 02°40'N, 76°50'W (Lehmann *et al.* 1977, AMNH 819382); Termales de San Juan, *c.*02°20.5'N, 76°19'W (ML 35175921, J. Muñoz García, XC 332882, O. Janni). *Huila* 9.2km ENE of Ullacos, 02°12'N, 76°21'W (XC 203511, G.M. Kirwan). *Nariño* Volcán Galeras, 01°12.5'N, 77°21.5'W (Kattan & Renjifo 2002, ICN 33814). **ECUADOR**: *Carchi* Cerro Mongus, 00°27'N, 77°52'W (Robbins *et al.* 1994a, Ridgely & Greenfield 2001, MECN, ANSP, KUNHM, XC, ML); RBP Guandera, 00°36'N, 77°42'W (Cresswell *et al.* 1999a). *Sucumbíos* Alto La Bonita, 00°29'N, 77°35'W (Stotz & Valenzuela 2009). *Napo* 4.5km below Oyacachi, 00°13'S, 78°02'W (Krabbe *et al.* 1997); *c.*3.9km N of Papallacta, 00°20'S, 78°09'W (Greeney & Jipa 2012); 7.5km WNW of Papallacta, 00°21'S, 78°12'W (XC 41654; D.M. Brinkhuizen); Río Mulatos, 00°58'S, 78°13'W (Benítez *et al.* 2001). *Tungurahua* Near El Triunfo, 01°15'S, 78°22'W (Benítez *et al.* 2001). *Cañar* 'Hacienda La Libertad', N of Taday, 02°34'S, 78°43W (Pearman 1994, Robbins *et al.* 1994a, Ridgely & Greenfield 2001, ANSP 184002). *Azuay* Patacocha, *c.*02°55.5'S, 78°43'W (C. Dingle *in litt.* 2015). *Morona-Santiago* Páramos de Matanga, 03°16'S, 78°54'W (Krabbe *et al.* 1997, Poulsen 2002, XC 250111/12, N. Krabbe). *Loja* Antenas Militares de Saraguro, 03°38'S, 79°13'W (L. Seitz *in litt.* 2017); Cerro Acanamá, 03°41'S, 79°15'W (XC 250226, ZMUC 104301, N. Krabbe); Cordillera de Cordoncillo, 03°41'S, 79°13'W (Robbins *et al.* 1994a, ANSP 184707/08).

TAXONOMY AND VARIATION Monotypic. As noted by Robbins *et al.* (1994a), the unique, bold markings, along with its nearly unique (within *Grallaricula*) high-altitude range, make the species' taxonomic affinities uncertain. It has been suggested to be closely related to Slate-crowned Antpitta (Krabbe 2008), to which it shows very little plumage resemblance but does share similar ecological and distributional characters. The strongly marked breast and bold facial markings suggest closer affinities to Peruvian, Ochre-fronted and Scallop-breasted Antpittas (Graves *et al.* 1983). Really, however, the plumage of Crescent-faced Antpitta is rather unique, so much so that it was originally described in a separate genus, *Apocryptornis* (Chapman 1924), and only later transferred to *Grallaricula* (Peters 1951), where it clearly belongs.

STATUS With the total extent of its Ecuadorian range estimated at 1,145km² (Freile *et al.* 2010b) and its global range at 8,109km² (Graham *et al.* 2010), Crescent-faced Antpitta is frequently considered of conservation concern due to its limited distribution (Orejuela 1985, Negret 2001). Its conservation status was first evaluated in 1990 (IUCN 1990), leading to its assessment of Near Threatened (Collar *et al.* 1994). It was first considered Near Threatened in Ecuador (Granizo *et al.* 1997), but subsequently Vulnerable in both Ecuador and Colombia (Granizo *et al.* 2002, Kattan & Renjifo 2002). Nevertheless, high-Andean, *Polylepis* forests often inhabited by Crescent-faced Antpitta are under ongoing and severe threat due to agricultural expansion and timber extraction for charcoal (Stattersfield *et al.* 1998, Sierra *et al.* 1999, Stattersfield & Capper 2000, Freile & Santander 2005, Freile *et al.* 2010a). Contrary to the insensitivity to habitat degradation suggested by Ridgely & Greenfield (2001),

the species does not appear to persist in human-modified areas (Cresswell *et al.* 1999a, Greeney 2012b) and, as noted by Freile *et al.* (2010b), its current status approaches the threshold for Vulnerable. Its geographical range falls within the Northern Central Andes EBA and Southern Central Andes EBA (Krabbe & Schulenberg 2003). In Colombia, it is fairly well protected (Wege & Long 1995) in several national parks and other conservation units, but somewhat less so in Ecuador. In particular in Ecuador, lowering of the treeline by cutting and burning for pasture, even in national parks, is a major problem (Granizo *et al.* 2002, Freile & Santander 2005). **PROTECTED POPULATIONS Colombia**: PNN Puracé (Lehmann *et al.* 1977, Hilty & Silliman 1983, Negret 2001, XC, ML); PNN Munchique (Lehmann *et al.* 1977, AMNH 819382); PNN Los Nevados (Uribe 2011, XC, ML); RMN El Mirador (D. Carantón *in litt.* 2017); RN Valle Lindo (Rivera-Pedroza & Ramírez 2005); SFF Volcán Galeras (Renjifo *et al.* 2002, ICN 33814). **Ecuador**: PN Cayambe-Coca (Baez *et al.* 2000, Greeney & Jipa 2012, XC); PN Llanganates (Benítez *et al.* 2001); BP Huashapamba (R. Ahlman *in litt.* 2017); RBP Guandera (Cresswell *et al.* 1999a, 1999b), presumably in PN Sumaco-Galeras and PN Sangay (Freile 2000, Granizo *et al.* 2002, Krabbe & Schulenberg 2003).

OTHER NAMES *Apocryptornis lineifrons* (Chapman 1924, 1926), *Aptocryptornis lineifrons* (misspelling; Chapman 1926). **Spanish** Ponchito medialuna (Krabbe & Schulenberg 2003); Gralarita Frentilunada (Buitrón-Jurado 2008); Gralarita Carilunada (Valarezo-Delgado 1984, Ortiz-Crespo *et al.* 1990, Granizo *et al.* 2002, Ridgely & Greenfield 2006, Granizo 2009, Stotz & Valenzuela 2009); Tororoí Medialuna (Renjifo *et al.* 2000, Rivera-Pedroza & Ramírez 2005, Uribe 2011); peón (Negret 2001). **French** Grallaire demi-lune. **German** Halbmond-Ameisenpitta (Krabbe & Schulenberg 2003).

Crescent-faced Antpitta, adult feeding older nestling, PN Cayambe-Coca, Napo, Ecuador, 12 February 2012 (*Murray Cooper*).

Crescent-faced Antpitta, adult, above Salento, Quindío, Colombia, 7 May 2017 (*Daniel Uribe*).

Crescent-faced Antpitta, adult, Cuenca de Gallinazo, Caldas, Colombia, 23 July 2011 (*Daniel Uribe*).

Crescent-faced Antpitta, older nestling begging as an adult approaches from below, PN Cayambe-Coca, Napo, Ecuador, 12 February 2012 (*Murray Cooper*).

Crescent-faced Antpitta, young fledgling, PN Cayambe-Coca, Napo, Ecuador, 10 February 2012 (*Edison Marcelo Quipo*).

Crescent-faced Antpitta, adult feeding older nestling, PN Cayambe-Coca, Napo, Ecuador, 10 February 2012 (*Harold F. Greeney*).

BIBLIOGRAPHY

Aben, J., Dorenbosch, M., Herzog, S.K., Smolders, A.J.P. & Van Der Velde, G. 2008. Human disturbance affects a deciduous forest bird community in the Andean foothills of central Bolivia. *Bird Conserv. Intern.* 18: 363–380.

Accordi, I.A. 2003. Contribuição ao conhecimento ornitológico da Campanha Gaúcha. *Atualidades Orn.* 112: 12.

Accordi, I.A. & Barcellos, A. 2006. Composição da avifauna em oito áreas úmidas da bacia hidrográfica do Lago Guaíba, Rio Grande do Sul. *Rev. Bras. Orn.* 14: 101–115.

Adams, M.P., Cooper, J.H. & Collar, N.J. 2003. Extinct and endangered ('E&E') birds: a proposed list for collection catalogues. *Bull. Brit. Orn. Club* 123A: 338–354.

Aguiar, K.M.O. & Júnior, L.A.C. 2008. Dietas de algumas espécies de aves das famílias Thamnophilidae, Grallariidae e Formicariidae do Amapá. *Rev. Bras. Orn.* 16: 376–379.

Albano, C. 2010a. Birding in north-east Brazil, part 1: Ceará, Pernambuco, Alagoas and northern Bahia. *Neotrop. Birding* 6: 56–63.

Albano, C. 2010b. Birding in north-east Brazil, part 2: the vast state of Bahia. *Neotrop. Birding* 7: 49–61.

Albano, C. & Girão, W. 2008. Aves das matas úmidas das serras de Aratanha, Baturité e Maranguape, Ceará. *Rev. Bras. Orn.* 16: 142–154.

Alderete, C. & Capllonch, P. 2010. El peso de aves en familias pertenecientes al orden Suboscines en la Argentina. *Nótulas Faunísticas (Segunda Ser.)* 58: 1–5.

Aleixo, A. 1999. Effects of selective logging on a bird community in the Brazilian Atlantic Forest. *Condor* 101: 537–548.

Aleixo, A. & Galetti, M. 1997. The conservation of the avifauna in a lowland Atlantic forest in south-east Brazil. *Bird Conserv. Intern.* 7: 235–261.

Aleixo, A. & Guilherme, E. 2010. Avifauna da Estação Ecológica do Rio Acre, estado do Acre, na fronteira Brasil/Peru: composição, distribuição ecológica e registros relevantes. *Bol. Mus. Para. E. Goeldi* 5: 279–309.

Aleixo, A. & Poletto, F. 2007. Birds of an open vegetation enclave in southern Brazilian Amazonia. *Wilson J. Orn.* 119: 610–630.

Aleixo, A., Poletto, F., Lima, M.F.C., Castro, M., Portes, E. & Miranda, L.S. 2011. Notes on the Vertebrates of northern Pará, Brazil: A forgotten part of the Guianan Region, II. Avifauna. *Bol. Mus. Para. E. Goeldi* 6: 11–65.

Aleixo, A. & Vielliard, J.M.E. 1995. Composição da avifauna da Mata de Santa Genebra, Campinas, São Paulo, Brasil. *Rev. Bras. Zool.* 12: 493–511.

Alexandrino, E.R., Bovo, A.A.A., Luz, D.T.A., Costa, J.C., Betini, G.S., Ferraz, K.M.P.M.B. & Couto, H.T.Z. 2013. Aves do campus "Luiz de Queiroz" Piracicaba, SP da Universidade de São Paulo: mais de 10 anos de observações neste ambiente antrópico. *Atualidades Orn.* 173: 40–52.

Allen, J.A. 1889a. List of the birds collected in Bolivia by Dr. H. H. Busby, with field notes by the collector. *Bull. Amer. Mus. Nat. Hist.* 2: 77–112.

Allen, J.A. 1889b. Notes on a collection of birds from Quito, Ecuador. *Bull. Amer. Mus. Nat. Hist.* 2: 69–76.

Allen, J.A. 1889c. On the Maximilian types of South American birds in the American Museum of Natural History. *Bull. Amer. Mus. Nat. Hist.* 2: 209–276.

Allen, J.A. 1900. List of birds collected in the district of Santa Marta, Colombia, by Mr. Herbert H. Smith. *Bull. Amer. Mus. Nat. Hist.* 13: 117–184.

Allen, S. 1995. *A Birder's Guide to Explorer's Inn, Peru.* Published privately, Lima.

Almeida, A., Couto, H.T.Z. & Almeida, A.F. 2003. Diversidade beta de aves em hábitats secundários da pré-Amazônia maranhense e interação com modelos nulos. *Ararajuba* 11: 157–171.

Almeida, M.E.C., Vielliard, J.M.E. & Dias, M.M. 1999. Composição da avifauna em duas matas ciliares do Rio Jacaré-Pepira, São Paulo, Brasil. *Rev. Bras. Zool.* 16: 1087–1098.

Alonso, J.Á., Alván, J.D. & Shany, N. 2012. Avifauna de la Reserva Nacional Allpahuayo Mishana, Loreto, Perú. *Cotinga* 34: 61–84.

Altman, A. & Swift, B. 1986. *Checklist of the Birds of South America* (1st edn.). Published privately, Great Barrington, MA.

Altman, A. & Swift, B. 1989. *Checklist of the Birds of South America* (2nd edn.). Published pivately, Great Barrington, MA.

Altman, A. & Swift, B. 1993. *Checklist of the Birds of South America* (3rd edn.). Published privately, Great Barrington, MA.

Álvarez-R., M. 2000. *Cantos de Aves de la Cordillera Oriental de Colombia.* Instituto de Investigación de Recursos Biologicos Alexander von Humboldt, Villa de Leyva, Boyacá.

Álvarez-R., M., Umaña, A.M., Córdoba, S. & Estela, F. 2002. *Inventario de la Avifauna Presente en las Cuencas de los Ríos Tapias-Tareas y Aferentes Directos al Cauca Zona Sur, Departamento de Caldas, Colombia.* Instituto de Investigación de Recurso Biológicos Alexander von Humboldt, Bogotá.

Álvarez-R., M., Umaña, A.M., Mejía, G.D., Cajiao, J., von Hildebrand, P. & Gast, F. 2003. Aves del Parque Nacional Natural Serranía de Chiribiquete, Amazonia - Provincia de la Guiania, Colombia. *Biota Colombiana* 4: 49–63.

Alves, M.A.S. & Duarte, M.F. 1996. Táticas de forrageamento de *Conopophaga melanops* (Passeriformes: Formicariidae) na area de Mata Atlântica da Ilha Grande, Estado do Rio de Janeiro. *Ararajuba* 4: 110–112.

Alves, M.A.S. & Vecchi, M.B. 2009. Birds, Ilha Grande, state of Rio de Janeiro, Southeastern Brazil. *Check List* 5: 300–313.

Alves, M.A.S., Jenkins, C.N., Pimm, S.L., Stornia, A., Raposo, M.A., Brooke, M.L., Harris, G. & Foster, J. 2009. Birds, montane forest, state of Rio de Janeiro, southeastern Brazil. *Check List* 5: 289–299.

Alves, M.A.S., Rocha, C.F., Van Sluys, M. & Vecchi, M.B. 2001. Incubação e alimentação de ninhegos em *Conopophaga melanops* (Passeriformes: Conopophagidae) em área de Mata Atlântica do Rio de Janeiro. P. 3 *in* Straube, F.C. (ed.) *Livro de Resumos do IX Congresso Brasileiro de Ornitologia: Ornitologia Sem Fronteras.* Curitiba.

Alves, M.A.S., Rocha, C.F., Van Sluys, M. & Vecchi, M.B. 2002. Nest, eggs and effort partitioning in incubation and rearing by a pair of the Black-cheeked Gnateaters, *Conopophaga melanops* (Passeriformes,

Conopophagidae), in an Atlantic rainforest area of Rio de Janeiro, Brazil. *Ararajuba* 10: 67–71.

Amanzo, J. 2003. *Evaluación Biológica Rápida del Santuario Nacional Tabaconas-Namballe y Zonas Aledañas. Informe WWF-OPP: QM-91.* World Wildlife Fund, Lima.

Amaral, F.Q., Fernandes, A.M. & Rodrigues, M. 2003. Aves de um fragmento florestal do Vale do Rio Doce. P. 161 in Machado, C.G. (ed.) *Livro de Resumos do XI Congresso Brasileiro de Ornitologia.* Feira de Santana.

Ames, P.L. 1971. The morphology of the syrinx in passerine birds. *Bull. Peabody Mus. Nat. Hist.* 37: 1–194.

Ames, P.L., Heimerdinger, D.L. & Warter, S.L. 1968. The anatomy and systematic position of the antpipits, *Conopophaga* and *Corythopis. Postilla* 114: 1–32.

Anciães, M. & Marini, M.Â. 2000. The effects of fragmentation on fluctuating asymmetry in passerine birds of Brazilian tropical forests. *J. Appl. Ecol.* 37: 1013–1028.

Anderson, D.L., Bonta, M. & Thorn, P. 1998. New and noteworthy bird records from Honduras. *Bull. Brit. Orn. Club* 118: 178–183.

Anderson, D.L., Wiedenfeld, D.A., Bechard, M.J. & Novak, S.J. 2004. Avian diversity in the Moskitia region of Honduras. *Orn. Neotrop.* 15: 447–482.

Andrade, G.I. & Lozano, I.E. 1997. Ocurrencia del Hormiguero de Corona Pizarra *Grallaricula nana* en la Reserva Biológoca Capanta, macizo de Chingaza, Cordillera Oriental Colombiana. *Cotinga* 7: 37–38.

Andrade, G.I., Rosas, M.L. & Repizzo, A. 1994. Notas preliminares sobre la avifauna y la integridad biologica de Carapanta. Pp. 207–228 in Andrade, G.I. (ed.) *Carpanta. Selva Nublada y Páramo.* Fundación Natura, Bogotá.

Andrade, G.I. & Rubio-Torgler, H. 1994. Sustainable use of the tropical rain forest: evidence from the avifauna in a shifting-cultivation habitat mosaic in the Colombian Amazon. *Conserv. Biol.* 8: 545–554.

Andrle, R.F. 1967. Birds of the Sierra de Tuxtla in Veracruz, Mexico. *Wilson Bull.* 79: 163–187.

Andrle, R.F. 1968. A Biogeographical Investigation of the Sierra de Tuxtla in Veracruz, Mexico. Ph.D. thesis. Louisiana State Univ., Baton Rouge, LA.

Angehr, G.R. 2003. *Directory of Important Bird Areas in Panama.* Sociedad Audubon de Panama, Balboa.

Angehr, G.R. & Christian, D.G. 2000. Distributional records from the highlands of the Serranía de Majé, an isolated mountain range in eastern Panama. *Bull. Brit. Orn. Club* 120: 173–178.

Angehr, G.R. & Dean, R. 2010. *The Birds of Panama.* Cornell Univ. Press, Ithaca, NY.

Angehr, G.R., Aucca, C., Christian, D.G., Pequeño, T. & Siegel, J. 2001. Structure and composition of the bird communities of the Lower Urubamba Region, Peru. Pp. 151–170 in Dallmeier, F., Alonso, A. & Campbell, P. (eds.) *Biodiversity of the Lower Urubamba Region, Peru. SIMAB Series 7.* Smithsonian Institution/MAB Biodiversity, Washington DC.

Angehr, G.R., Christian, D.G. & Aparicio, K.M. 2004. A survey of the Serranía de Jungurudó, an isolated mountain range in eastern Panama. *Bull. Brit. Orn. Club* 124: 51–62.

Angehr, G.R., Engleman, D. & Engleman, L. 2008. *A Bird-finding Guide to Panama.* Cornell Univ. Press, Ithaca, NY.

Angulo P., F. 2009. Informe Ornitológico de la Visita a la Cuenca Alta de los Ríos Zaña y Chancay. Unpubl. report to BirdLife International, Lima.

Angulo P., F., Novoa, J. & Lazo, I. 2016. Aves del río La Novia en la cuenca del Purús, Ucayali, Perú. Pp. 122–128, 287–308 in Mena, J.L. & Germaná, C. (eds.) *Diversidad Biológica del Sudeste de la Amazonía Peruana: Avances en la Investigación.* Consorcio Purús-Manu: WWF, CARE Perú, ProNaturaleza, ProPurús, Sociedad Zoológica de Fráncfort & ORAU, Lima.

Angulo P., F., Palomino, W., Arnal, H., Aucca, C. & Uchofen, O. 2008. *Corredor de Conservación de Aves Marañón-Alto Mayo: Análisis de Distribución de Aves de Alta Prioridad de Conservación e Identificación de Propuestas de Áreas para su Conservación.* Asociación Ecosistemas Andinos & American Bird Conservancy, Cusco.

dos Anjos, L. 2001. Bird communities in five Atlantic forest fragments in southern Brazil. *Orn. Neotrop.* 12: 11–27.

dos Anjos, L. 2002. Forest bird communities in the Tibagi River hydrographic basin, southern Brazil. *Ecotropica* 8: 67–79.

dos Anjos, L. & Boçon, R. 1999. Bird communities in natural forest patches in southern Brazil. *Wilson Bull.* 111: 397–414.

dos Anjos, L. & Graf, V. 1993. Riqueza de aves da Fazenda Santa Rita, região dos Campos Gerais, Palmeira, Paraná, Brasil. *Rev. Bras. Zool.* 10: 673–693.

dos Anjos, L. & Schuchmann, K.-L. 1997. Biogeographical affinities of the avifauna of the Tibagi River Basin, Paraná drainage system, southern Brazil. *Ecotropica* 3: 43–65.

dos Anjos, L., Bochio, G.M., Campos, J.V., McCrate, G.B. & Palomino, F. 2009. Sobre o uso de níveis de sensibilidade de aves à fragmentação florestal na avaliação da integridade biótica: um estudo de caso no norte do Estado do Paraná, sul do Brasil. *Rev. Bras. Orn.* 17: 28–36.

dos Anjos, L., Schuchmann, K.-L. & Berndt, R. 1997. Avifaunal composition, species richness, and status in the Tibagi river basin, Paraná State, southern Brazil. *Orn. Neotrop.* 8: 145–173.

dos Anjos, L., Zanette, L. & Lopes, E.V. 2004. Effects of fragmentation on the bird guilds of the Atlantic forest in north Paraná, southern Brazil. *Orn. Neotrop.* 15: 137–144.

dos Anjos, L., Volpato, G.H., Lopes, E.V., Serafini, P.P., Poletto, F. & Aleixo, A. 2007. The importance of riparian forest for the maintenance of bird species richness in an Atlantic Forest remnant, southern Brazil. *Rev. Bras. Zool.* 24: 1078–1086.

dos Anjos, L., Holt, R.D. & Robinson, S.K. 2010. Position in the distributional range and sensitivity to forest fragmentation in birds: a case history from the Atlantic forest, Brazil. *Bird Conserv. Intern.* 20: 392–399.

dos Anjos, L., Collins, C.D., Holt, R.D., Volpato, G.H., Mendonca, L.B., Lopes, E.V., Bocon, R., Bisheimer, M.V., Serafini, P.P. & Carvalho, J. 2011. Bird species abundance-occupancy patterns and sensitivity to forest fragmentation: implications for conservation in the Brazilian Atlantic forest. *Biol. Conserv.* 144: 2213–2222.

Anon. 2010a. Science notebook: introducing...... *Science News* 177: 4.

Anon. 2010b. Editorial sobre la descripción de una nueva especie de *Grallaria. Conserv. Colombiana* 13: 4–7.

Anon. 2011. Response to *Ornitología Colombiana* editorial by Cadena and Stiles concerning *Grallaria fenwickorum. Conserv. Colombiana* 4: 161–168.

Antunes, A.Z. 2005. Alterações na composição da comunidade de aves ao longo do tempo em um fragmento florestal no sudeste do Brasil. *Ararajuba* 13: 47–61.

Antunes, A.Z. 2007. Riqueza e dinâmica de aves endêmicas da Mata Atlântica em um fragmento de floresta estacional semidecidual no sudeste do Brasil. *Rev. Bras. Orn.* 15: 61–68.

Antunes, A.Z. 2008. Diurnal and seasonal variability in bird counts in a forest fragment in southeastern Brazil. *Rev. Bras. Zool.* 25: 228–237.

Aponte, M.A., Maillard-Z., O., Hennessey, B.A., Lebbin, D.J. & Greeney, H.F. 2017. Distribution, behaviour and conservation of Masked Antpitta *Hylopezus auricularis*. *Cotinga* 39: 44–52.

Arbeláez-Cortés, E., Marín-Gómez, O.H., Baena-Tovar, O. & Ospina-González, J.C. 2011a. Aves, Finca Estrella de Agua - Páramo de Frontino, municipality of Salento, Quindío, Colombia. *Check List* 7: 64–70.

Arbeláez-Cortés, E., Marín-Gómez, O.H., Duque-Montoya, D., Cardona-Camacho, P.J., Renjifo, L.M. & Gómez, H.F. 2011b. Birds, Quindío department, Central Andes of Colombia. *Check List* 7: 227–247.

Arévalo, J. & Newhard, K. 2011. Traffic noise affects forest bird species in a protected tropical forest. *Rev. Biol. Trop.* 59: 969–980.

Arribalzaga, E.L. 1926. Nombres vulgares argentinos de las aves silvestres de La Republica. *Hornero* 3: 349–363.

Arzua, M. & Barros-Battesti, D. 1999. Parasitism of *Ixodes* (Multidentatus) *auritulus* Neumann (Acari: Ixodidae) on birds from the city of Curitiba, state of Paraná, southern Brazil. *Mem. Inst. Oswaldo Cruz* 94: 597–603.

Arzua, M., Silva, M.N., Famadas, K.M., Beati, L. & Barros-Battesti, D.M. 2003. *Amblyomma aureolatum* and *Ixodes auritulus* (Acari: Ixodidae) on birds in southern Brazil, with notes on their ecology. *Exp. Appl. Acarol.* 31: 283–286.

Asociación Bogotana de Ornitología. 2000. *Aves de la Sabana de Bogotá: Guía de Campo*. Asociación Bogotana de Ornitología, Bogotá.

Astudillo, P.X., Tinoco, B.A. & Siddons, D.C. 2015. The avifauna of Cajas National Park and Mazán Reserve, southern Ecuador, with notes on new records. *Cotinga* 37: 1–11.

Avanzo, V.C. & Sanfilippo, L.F. 2000. Levantamento preliminar da avifauna de Ilha Comprida, São Paulo. *Bol. Cent. Estud. Orn. São Paulo* 14: 10–14.

Aves Argentinas/Asociación Ornitológica de la Plata & Secretaría de Ambiente y Desarrollo Sustentable. 2008. *Categorización de las Aves de la Argentina según su Estado de Conservación*. Aves Argentinas, Asociación Ornitológica del Plata & Secretaría de Ambiente y Desarrollo Sustentable, Buenos Aires.

Ayerbe-Quiñones, F. 2006. *Avifauna del Complejo Volcánico Doña Juana-Cascabel: Riqueza, Endemismos y Especies Amenazadas. Informe Técnico Convenio Marco UAESPNN-WWF.* WWF-CORPODOÑAJUANA, La Cruz, Nariño.

Ayerbe-Quiñones, F., Bernal, L.G.G., Ordóñez, J.P.L., Burbano, M.B.R., Sierra, J.V.S. & Rojas, M.F.G. 2009. Avifauna de Popayán y municipios aledaños. *Nov. Colombianas* 9: 1–27.

Ayerbe-Quiñones, F., López, J., González, M.F., Estela, F., Ramírez, M.B., Sandoval, J.V. & Gómez, L.G. 2008. Aves del departamento del Cauca - Colombia. *Biota Colombiana* 9: 77–132.

Azvedo, M.A.G. 2006. Contribuição de estudos para licenciamento ambiental ao conhecimento da avifauna de Santa Catarina, Sul do Brasil. *Biotemas* 19: 93–106.

Baiker, J. 2011. *Guía Ecoturística: Mancomunidad Saywite-Choquequirao-Ampay (Apurímac, Perú). Con Especial Referencia a la Identificación de Fauna, Flora, Hongos y Líquenes en el Departamento de Apurímac y Sitios Adyacentes en el Departamento de Cusco. Serie Investigación y Sistematización No. 15.* Programa Regional ECOBONA-INTERCOOPERATION, Lima.

Balchin, C.S. & Toyne, E.P. 1998. The avifauna and conservation status of the Río Nangaritza valley, southern Ecuador. *Bird Conserv. Intern.* 8: 237–253.

Balderrama, J.A. 2009. Aves. Pp. 305–418 *in* Aguirre, L.F., Aguayo, R., Balderrama, J.A., Cortez-F., C. & Tarifa, T. (eds.) *Libro Rojo de la Fauna Silvestre de Vertebrados de Bolivia*. Ministerio de Medio Ambiente y Agua, Viceministerio de Medio Ambiente, Biodiversidad y Cambios Climáticos, La Paz.

Bangs, O. 1898a. On some birds from Pueblo Viejo, Colombia. *Proc. Biol. Soc. Wash.* 12: 157–160.

Bangs, O. 1898b. On some birds from the Sierra Nevada de Santa Marta, Colombia. *Proc. Biol. Soc. Wash.* 12: 171–182.

Bangs, O. 1899. On some new or rare birds from the Sierra Nevada de Santa Marta, Colombia. *Proc. Biol. Soc. Wash.* 13: 91–108.

Bangs, O. 1901. On a collection of birds made by W. W. Brown, Jr., at David and Divala, Chiriqui. *Auk* 18: 355–370.

Bangs, O. 1902. On a second collection of birds made in Chiriqui, by W. W. Brown, Jr. *Proc. New England Zool. Club* 3: 16–70.

Bangs, O. 1907. On a collection of birds from western Costa Rica. *Auk* 24: 287–312.

Bangs, O. & Barbour, T. 1922. Birds from Darien. *Bull. Mus. Comp. Zool.* 65: 191–229.

Bangs, O. & Noble, G.K. 1916. List of birds collected on the Harvard Peruvian expedition of 1916. *Auk* 35: 442–463.

Banks, R.C., Fitzpatrick, J.W., Howell, T.R., Johnson, N.K., Monroe, B.L., Ouellet, H., Remsen, J.V. & Storer, R.W. 1997. Forty-first supplement to the American Ornithologists' Union *Check-List of North American Birds*. *Auk* 114: 542–552.

Banks-Leite, C., Ewers, R.M., Pimentel, R.G. & Metzger, J.P. 2012. Decisions on temporal sampling protocol influence the detection of ecological patterns. *Biotropica* 44: 378–385.

Barber, B.R. & Robbins, M.B. 2003. Nest and eggs of the Tepui Antpitta (*Myrmothera simplex*). *Wilson Bull.* 114: 287–288.

Barlow, J., Mestre, L.A.M., Gardner, T.A. & Peres, C.A. 2007. The value of primary, secondary and plantation forests for Amazonian birds. *Biol. Conserv.* 136: 212–231.

Barlow, J.C. & Dick, J.A. 1969. Ochre-striped Antpitta in Colombia. *Auk* 86: 759.

Barnes, R., Butchart, S.H.M., Davies, C.W.N., Fernández, M. & Seddon, N. 1997. New distributional information on eight bird species from northern Peru. *Bull. Brit. Orn. Club* 117: 69–74.

Barnett, A., Shapley, R., Benjamin, P., Henry, E. & McGarrell, M. 2002. Birds of the Potaro Plateau, with eight new species for Guyana. *Cotinga* 18: 19–36.

Barrera, L.F. & Bartels, A. 2010. A new species of antpitta (family Grallariidae) from the Colibrí del Sol Bird Reserve, Colombia. *Conserv. Colombiana* 13: 8–24.

Barrio, J., García-Olaechea, D. & More, A. 2015. The avifauna of El Angolo Hunting Reserve, north-west Peru: natural history notes. *Bull. Brit. Orn. Club* 135: 6–20.

Barrowclough, G.F. & Escalante-Pliego, P. 1990. Notes on

the birds of the Sierra de Unturán, southern Venezuela. *Bull. Brit. Orn. Club* 110: 167–169.

Barrowclough, G.F., Escalante-Pliego, P., Aveledo-Hostos, R. & Perez-Chinchilla, L.A. 1995. An annotated list of the birds of the Cerro Tamacuarí region, Serranía de Tapirapecó, Federal Territory of Amazonas, Venezuela. *Bull. Brit. Orn. Club* 115: 211–219.

Bartlett, E. 1882. On some mammals and birds collected by Mr. J. Hauxwell in eastern Peru. *Proc. Zool. Soc. Lond.* 50: 373–375.

Bass, M.S., Finer, M., Jenkins, C.N., Kreft, H., Cisneros-Heredia, D.F., McCracken, S.F., Pitman, N.C.A., English, P.H., Swing, K., Villa, G., Di Fiore, A., Voigt, C.C. & Kunz, T.H. 2010. Global conservation significance of Ecuador's Yasuni National Park. *PLoS ONE* 5: e8767.

Batalha-Filho, H., Pessoa, R.O., Fabre, P.-H., Fjeldså, J., Irestedt, M., Ericson, P.G.P., Silveira, L.F. & Miyaki, C.Y. 2014. Phylogeny and historical biogeography of gnateaters (Passeriformes, Conopophagidae) in the South America forests. *Mol. Phyl. & Evol.* 79: 422–432.

Bates, J.M., Stotz, D.F. & Schulenberg, T.S. 1998. The avifauna of Parque Nacional Noel Kempff Mercado and surrounding areas. Pp. 317–340 *in* Killeen, T.J. & Schulenberg, T.S. (eds.) *A Biological Assessment of Parque Nacional Noel Kempff Mercado, Bolivia. RAP Working Paper No. 10.* Conservation International, Washington DC.

Bauer, C. & Pacheco, J.F. 2000. Lista das aves da região de Visconde de Mauá, Serra da Mantiqueira, no limite dos Estados do Rio de Janeiro e Minas Gerais. *Atualidades Orn.* 97: 7.

Becker, C.D., Loughin, T.M. & Santander, T. 2008. Identifying forest-obligate birds in tropical moist cloud forest of Andean Ecuador. *J. Field Orn.* 79: 229–244.

Becker, L., Mongelos, R. & Dall'Agnol, L.S. 2007. Monitoramento da avifauna atropelada na RS 020, Rio Grande do Sul, Brasil. P. 214 *in* Fontana, C.S. (ed.) *Livro de Resumos do XV Congresso Brasileiro de Ornitologia.* Porto Alegre.

Beebe, C.W. 1917. List of the birds of Bartica District. Pp. 127–137 *in* Beebe, C.W., Hartley, G.I. & Howes, P.G. (eds.) *Tropical Wild Life in British Guiana.* New York Zoological Society, New York.

Beebe, C.W. 1952. Introduction to the ecology of the Arima Valley, Trinidad, B.W.I. *Zoologica* 37: 157–183.

Begazo, A.J. & Valqui, T.H. 1998. Birds of Pacaa-Samiria National Reserve with a new population (*Myrmotherula longicauda*) and new record for Peru (*Hylophilus semicinereus*). *Bull. Brit. Orn. Club* 118: 159–166.

Begazo, A.J., Valqui, T.H., Sokol, M. & Langlois, E. 2001. Notes on some birds from central and northern Peru. *Cotinga* 15: 81–87.

Beja, P., Santos, C., Santana, J., Pereira, M., Marques, J., Queiroz, H. & Palmeirim, J. 2010. Seasonal patterns of spatial variation in understory bird assemblages across a mosaic of flooded and unflooded Amazonian forests. *Biodiver. & Conserv.* 19: 129–152.

Belcher, C. & Smooker, G.D. 1936. Birds of the colony of Trinidad and Tobago, Part 4. *Ibis* (8)6: 792–813.

Bell, J. & Bruning, D. 1976. Hatching and handrearing the Rufous-faced Antpitta. Notes on antbirds kept at the New York Zoological Park. *Avicult. Mag.* 82: 119–122.

Belton, W. 1985. Birds of Rio Grande do Sul, Brazil. Part 2. Formicaridae through Corvidae. *Bull. Amer. Mus. Nat. Hist.* 180: 1–242.

Belton, W. 1994. *Aves do Rio Grande do Sul: Distribuição e*

Biologia. Ed. UNISINOS, São Leopoldo.

Bencke, G.A. 1996. Annotated list of birds of Monte Alverne, Central Rio Grande do Sul. *Acta Biol. Leopoldensia* 18: 17–42.

Bencke, G.A. 2001. *Lista de Referência das Aves do Rio Grande do Sul.* Fundação Zoobotânica do Rio Grande do Sul, Porto Alegre.

Bencke, G.A. & Kindel, A. 1999. Bird counts along an altitudinal gradient of Atlantic Forest in northeastern Rio Grande do Sul. *Ararajuba* 7: 91–107.

Bencke, G.A., Dias, R.A., Bugoni, L., Agne, C.E., Fontana, C.S., Mauricio, G.N. & Machado, D.B. 2010. Revisão e atualização da lista das aves do Rio Grande do Sul, Brasil. *Iheringia, Sér. Zool.* 100: 519–556.

Benfica, C.E., Mazzoni, L.G., Neto, S.D.A. & Canuto, M. 2012. Atlantic Forest, Cerrado or Caatinga? The influence of each biome on avifaunal composition in Parque Estadual Veredas do Peracu. P. 568 *in* More, A., Olaechea, D.G. & Olaechea, A.G. (eds.) *Libro de Resúmenes del IX Congreso de Ornitología Neotropical y el VIII Congreso Peruano de Ornitología.* Cuzco.

Benítez, V., Sánchez, D. & Larrea, M. 2001. Evaluación ecológica rápida de la avifauna en el Parque Nacional Llanganates. Pp. 67–107 *in* Vázquez, M., Suárez, L. & Larrea, M. (eds.) *Biodiversidad en el Parque Nacional Llanganates: Reporte de las Evaluaciones Ecológicas y Socioeconómicas Rápidas.* Ecociencia, Herbario Nacional del Ecuador, Museo Ecuatoriano de Ciencias Naturales, Instituto Internacional de Reconstrucción Rural & Ministerio de Medio Ambiente, Quito.

Bennett, G.F., Caines, J.R. & Woodworth-Lynas, C.B. 1987. Avian Haemoproteidae. 24. The hemoproteids of the New World passeriform families Formicariidae, Furnariidae, Mimidae, and Vireonidae. *Canadian J. Zool.* 65: 317–321.

van den Berg, A.B. & Bosman, C.A.W. 1984. Range extensions and rare records of birds in Ecuador. *Bull. Brit. Orn. Club* 104: 152–154.

Bergmann, F.B., Amaral, H.L.C., Santos, P.R.S., Gomes, G.C. & Mauricio, G.N. 2015. Avifauna de dois remanescentes florestais da Serra dos Tapes, Rio Grande do Sul, Brasil. *Atualidades Orn.* 186: 33–40.

Berla, H.F. 1944. Lista das aves colecionadas em Pedra Branca, município de Parati, Estado do Rio de Janeiro, com algumas notas sobre sua biologia. *Bol. Mus. Nac., N. Sér. (Zool.)* 18: 1–21.

Berla, H.F. 1946. Lista das aves colecionadas em Pernambuco, com descrição de uma subespécie n., de um alótipo e notas de campo. *Bol. Mus. Nac. Rio de Janeiro* 65: 1–35.

Berlanga, H., Kennedy, J.A., Rich, T.D., Arizmendi, M.C., Beardmore, C.J., Blancher, P.J., Butcher, G.S., Couturier, A.R., Dayer, A.A., Demarest, D.W., Easton, W.E., Gustafson, M., Iñigo-Elias, E., Krebs, E.A., Panjabi, A.O., Rodriguez-Contreras, V., Rosenberg, K.V., Ruth, J.M., Santana-Castellón, E., Vidal, R.M. & Will, T.C. 2010. *Saving our Shared Birds: Partners in Flight Tri-national Vision for Landbird Conservation.* Cornell Lab of Ornithology, Ithaca, NY.

von Berlepsch, H. 1893. Diagnosen neuer südamerikanischer Vogelarten. *Orn. Monatsb.* 1: 11–12.

von Berlepsch, H. 1908. On the birds of Cayenne. Part 1. *Novit. Zool.* 15: 103–164.

von Berlepsch, H. 1912. Beschreibung neuer Vogelformen aus dem Gebiete des unteren Amazonas. *Orn. Monatsb.* 20: 17–21.

von Berlepsch, H. & von Ihering, H. 1885. Die Vogel der Umgegend von Taquara do mundo novo, Prov. Rio Grande do Sul. *Zeit. Ges. Orn.* 2: 97–184.

von Berlepsch, H. & Leverkühn, P. 1890. Studien über einige südamerikanische Vögel nebst Beschreibungen neuer Arten. *Ornis* 6: 1–32.

von Berlepsch, H. & Sztolcman, J. 1896. On the ornithological researches of M. Jean Kalinowski in central Peru. *Proc. Zool. Soc. Lond.* 1896: 322–388.

von Berlepsch, H. & Sztolcman, J. 1901. Descriptions d'oiseaux noveaux du Pérou central recueillis par le voyageur polonaise Jean Kalinowski. *Ornis* 11: 191–195.

von Berlepsch, H. & Sztolcman, J. 1906. Rapport sur les nouvelles collections ornithologiques faites au Pérou par M. Jean Kalinowski. *Ornis* 13: 63–133.

von Berlepsch, H. & Taczanowski, L. 1884. Deuxieme list des oiseaux recueillis dans l'Ecuadeur Occidental par MM. Stolzmann et Simiradski. *Proc. Zool. Soc. Lond.* 1884: 281–313.

Berlioz, J. 1927 Étude d'une collection d'oiseaux de l'Equateur donnée au Muséum par M. Clavery *Bull. Mus. Natl. d'Hist. Nat. (Paris), 2ème sér.* 33: 486–493.

Berlioz, J. 1932a. Contribution a l'étude des oiseaux de l'Ecuador. *Bull. Mus. Natl. d'Hist. Nat. (Paris), 2ème sér.* 4: 228–242.

Berlioz, J. 1932b. Nouvelle contribution a l'étude des oiseaux de l'Ecuador. *Bull. Mus. Natl. d'Hist. Nat. (Paris), 2ème sér.* 4: 620–628.

Berlioz, J. 1937. Notes sur quelques oiseaux rares ou peu connus de l'Equateur. *Bull. Mus. Natl. d'Hist. Nat. (Paris), 2ème sér.* 9: 114–118.

Berlioz, J. 1939. Étude d'une collection d'oiseaux du Chiapas (Mexique). *Bull. Mus. Natl. d'Hist. Nat. (Paris), 2ème sér.* 11: 360–377.

Berlioz, J. 1962. Étude d'une collection d'oiseaux de Guyane française. *Bull. Mus. Natl. d'Hist. Nat. (Paris), 2ème sér.* 34: 131–143.

Bernard, E. 2008. Inventários biológicos rápidos no Parque Nacional Montanhas do Tumucumaque, Amapá, Brasil. *RAP Bull. Biol. Assess.* 48: 1–149.

Bernis, F., de Juana, E., del Hoyo, J., Fernández-Cruz, M., Ferrer, X., Sáez-Royuela, R. & Sargatal, J. 2003. Nombres en castellano de las aves del mundo recomendados por la Sociedad Española de Ornitología (octava parte: Orden Passeriformes, familias Eurylaimidae a Rhinocryptidae). *Ardeola* 50: 103–110.

Bertoni, A.W. 1901. Aves nuevas del Paraguay, continuación á Azara. *An. Cienc. Paraguayos* 1: 1–216.

Bertoni, A.W. 1914. *Fauna Paraguaya: Catálogos Sistemáticos de los Vertebrados del Paraguay.* Gráfico M. Brossa, Asunción.

Bertoni, A.W. 1939. Catálogos sistemáticos de los vertebrados del Paraguay. *Rev. Soc. Cienc. Paraguay* 4: 1–59.

Best, B.J. & Clarke, C.T. 1991. *The Threatened Birds of the Sozoranga Region, Southwest Ecuador.* International Council for Bird Preservation, Cambridge, UK.

Best, B.J., Broom, A.L., Checker, M. & Thewlis, R. 1992. An ornithological survey of El Oro and western Loja province, south-west Ecuador, January-March 1991. Pp. 137–210 *in* Best, B.J. (eds.) *The Threatened Forests of South-west Ecuador.* Privately published, Leeds.

Best, B.J., Checker, M., Thewlis, R.M., Best, A.L. & Duckworth, W. 1996. New bird breeding data from southwestern Ecuador. *Orn. Neotrop.* 7: 69–73.

Bierregaard, R.O. 1988. Morphological data from understory birds in terra firme forest in the central Amazonian basin. *Rev. Bras. Biol.* 48: 169–178.

Bierregaard, R.O. & Lovejoy, T.E. 1989. Effects of forest fragmentation on Amazonian understory bird communities. *Acta Amazonica* 19: 215–241.

Binford, L.C. 1989. A distributional survey of the birds of the Mexican state of Oaxaca. *Orn. Monogr.* 43: 1–418.

Bird, J.P., Buchanan, G.M., Lees, A.C., Clay, R.P., Develey, P.F., Yepez, I. & Butchart, S.H.M. 2012. Integrating spatially explicit habitat projections into extinction risk assessments: a reassessment of Amazonian avifauna incorporating projected deforestation. *Divers. & Distrib.* 18: 273–281.

BirdLife International. 2008. *Critically Endangered Birds: A Global Audit.* BirdLife International, Cambridge, UK.

BirdLife International. 2012a. *Grallaria eludens.* http://dx.doi.org/10.2305/IUCN.UK.2012-1.RLTS.T22703265A39180695.en.

BirdLife International. 2012b. *Grallaria kaestneri.* http://dx.doi.org/10.2305/IUCN.UK.2012-1.RLTS.T22703268A38083999.en.

BirdLife International. 2012c. *Grallaria przewalskii.* http://dx.doi.org/10.2305/IUCN.UK.2012-1.RLTS.T22703311A39186367.en.

BirdLife International. 2013a. *Grallaria carrikeri.* http://dx.doi.org/10.2305/IUCN.UK.2013-2.RLTS.T22703299A49265637.en.

BirdLife International. 2013b. *Grallaria watkinsi.* http://dx.doi.org/10.2305/IUCN.UK.2013-2.RLTS.T22703277A50414618.en.

BirdLife International. 2013c. Species factsheet: Jocotoco Antpitta *Grallaria ridgelyi.* http://www.birdlife.org.

BirdLife International. 2016. Species factsheet: Undulated Antpitta *Grallaria squamigera.* http://www.birdlife.org.

BirdLife International 2017. *IUCN Red List for Birds.* BirdLife International, Cambridge, UK.

Bisbal, F. 1983. *Inventario Preliminar de Fauna de la isla de Margarita.* Privately published, Caracas.

Bispo, A.A. & Scherer-Neto, P. 2010. Taxocenose de aves em um remanescente da Floresta com Araucária no sudeste do Paraná, Brasil. *Biota Neotropica* 10: 121–130.

Black-M., J. 1982. Los páramos del Antisana. *Rev. Geográfica* 17: 25–51.

Blake, C.H. 1957. Design quantities of some Chiapas birds. *Bull. Mus. Comp. Zool.* 116: 286–289.

Blake, E.R. 1950. Birds of the Acary Mountains of southern British Guiana. *Fieldiana Zool.* 32: 419–474.

Blake, E.R. 1953. *Birds of Mexico.* Univ. of Chicago Press, Chicago.

Blake, E.R. 1958. Birds of Volcán de Chiriquí. *Fieldiana Zool.* 36: 499–577.

Blake, E.R. 1961. New bird records from Surinam. *Ardea* 49: 178–183.

Blake, E.R. 1963. The birds of southern Surinam. *Ardea* 51: 53–72.

Blake, J.G. 2007. Neotropical forest bird communities: a comparison of species richness and composition at local and regional scales. *Condor* 109: 237–255.

Blake, J.G. & Loiselle, B.A. 2000. Diversity of birds along an elevational gradient in the Cordillera Central, Costa Rica. *Auk* 117: 663–686.

Blake, J.G. & Loiselle, B.A. 2001. Bird assemblages in second-growth and old-growth forests, Costa Rica: perspectives from mist nets and point counts. *Auk* 118: 304–326.

Blake, J.G. & Loiselle, B.A. 2015. Enigmatic declines in bird numbers in lowland forest of eastern Ecuador may be a consequence of climate change. *PeerJ* 3: e1177.

Boddaert, P. 1783. *Table des Planches Enluminéez d'histoire Naturelle de M. Daubenton.* Utrecht.

Bodrati, A., Cockle, K.L., Segovia, J.M., Roesler, I., Areta, J.I. & Jordan, E. 2010. La avifauna del Parque Provincial Cruce Caballero, Provincia de Misiones, Argentina. *Cotinga* 32: 41–64.

Bodrati, A. & Di Sallo, F.G. 2016. Primera descripción del nido, huevos y comportamiento de incubación del chululú chico (*Hylopezus nattereri*) en la Selva Atlántica de Argentina. *Orn. Neotrop.* 27: 197–201.

Boesman, P. 1998. Some new information on the distribution of Venezuelan birds. *Cotinga* 9: 27–39.

Boesman, P. 2016a. Notes on the vocalizations of Rufous Gnateater (*Conopophaga lineata*). HBW Alive Ornithological Note 66. *In* del Hoyo, J., Elliott, A., Sargatal, J., Christie, D.A. & de Juana, E. (eds.) *Handbook of the Birds of the World Alive.* Lynx Edicions, Barcelona (retrieved from http://www.hbw.com/node/931950 on 9 June 2017).

Boesman, P. 2016b. Notes on the vocalizations of Tawny Antpitta (*Grallaria quitensis*). HBW Alive Ornithological Note 72. *In* del Hoyo, J., Elliott, A., Sargatal, J., Christie, D.A. & de Juana, E. (eds.) *Handbook of the Birds of the World Alive.* Lynx Edicions, Barcelona.

Bohórquez, C.I. 2002. La avifauna de la vertiente oriental de los Andes de Colombia. Tres evaluaciones en elevación subtropical. *Rev. Acad. Colombiana Cienc. Exactas, Fisicas y Nat.* 26: 419–442.

Bonaccorso, E. 2004 Avifauna of a high Andean forest: Bosque Protector Cashca Totoras, Bolivar Province, Ecuador. *Orn. Neotrop.* 15: 483–492.

Bonaccorso, E., Santander, T., Freile, J.F., Tinoco, B. & Rodas, F. 2007. Avifauna and conservation of the Cerro Negro-Cazaderos area, Tumbesian Ecuador. *Cotinga* 27: 61–66.

Bonaparte, C.L. 1850. *Conspectus Generum Avium.* E.J. Brill, Brittenburg.

Bond, J. 1950. Notes on Peruvian Formicariidae. *Proc. Acad. Nat. Sci. Phil.* 102: 1–26.

Bond, J. 1953. Notes on Peruvian Dendrocolaptidae, Conopophagidae and Rhinocryptidae. *Not. Naturae* 248: 1–7.

Bond, J. & Meyer de Schauensee, R. 1940a. On some birds from southern Colombia. *Proc. Acad. Nat. Sci. Phil.* 92: 153–169.

Bond, J. & Meyer de Schauensee, R. 1940b. Descriptions of new birds from Bolivia. Part III - Mesomydi. *Not. Naturae* 44: 1–4.

Bond, J. & Meyer de Schauensee, R. 1942. The birds of Bolivia, Part 1. *Proc. Acad. Nat. Sci. Phil.* 94: 307–391.

Bonta, M. & Anderson, D.L. 2002. *Birding Honduras: A Checklist and Guide.* EcoArtes, Tegucigalpa.

Borges, S.H. 2006. Rarity of birds in the Jaú National Park, Brazilian Amazon. *Anim. Biodiv. Conserv.* 29: 179–189.

Borges, S.H. 2007. Análise biogeográfica da avifauna da região oeste do baixo Rio Negro, Amazônia Brasileira. *Rev. Bras. Zool.* 24: 919–940.

Borges, S.H. & Almeida, R.A.M. 2011. Birds of the Jau National Park and adjacent areas, Brazilian Amazon: new species records with reanalysis of a previous checklist. *Rev. Bras. Orn.* 19: 108–133.

Borges, S.H. & Carvalhães, A. 2000. Bird species of black water inundation forests in the Jaú National Park

(Amazonas state, Brazil): their contribution to regional species richness. *Biodiver. & Conserv.* 9: 201–214.

Borges, S.H. & Stouffer, P.C. 1999. Bird communities in two types of anthropogenic successional vegetation in Central Amazonia. *Condor* 101: 529–536.

Borges, S.H., Cohn-Haft, M., Carvalhães, A.M.P., Henriques, L.M., Pacheco, J.F. & Whittaker, A. 2001. Birds of Jaú National Park, Brazilian Amazon: species check-list, biogeography and conservation. *Orn. Neotrop.* 12: 109–140.

Borges, S.H., Whittaker, A. & Almeida, R.A.M. 2014. Bird diversity in the Serra do Aracá region, northwestern Brazilian Amazon: preliminary check-list with considerations on biogeography and conservation. *Zoologia* 31: 343–360.

Bornschein, M.R. & Reinert, B.L. 2000. Aves de três remanescentes florestais do norte do Estado do Paraná, sul do Brasil, com sugestões para a conservação e manejo. *Rev. Bras. Zool.* 17: 615–636.

Botero, J.E., Lentijo, G.M., Lopez, A.M., Castellanos, O., Aristizabal, C., Franco, N. & Arbelaez, D. 2005. Adiciones a la lista de aves del municipio de Manizales. *Bol. Soc. Antioqueña Orn.* 15: 69–88.

Boucard, A. 1878a. On birds collected in Costa Rica. *Proc. Zool. Soc. Lond.* 1878: 37–71.

Boucard, A. 1878b. Liste des oiseaux recoltes au Guatemala en 1877. *Ann. Soc. Linn. Lyon* 25: 1–47.

Boyle, W.A. & Sigel, B.J. 2015. Ongoing changes in the avifauna of La Selva Biological Station, Costa Rica: Twenty-three years of Christmas Bird Counts. *Biol. Conserv.* 188: 11–21.

Brabourne, L. & Chubb, C. 1912. *The Birds of South America*, vol. 1. Taylor & Francis, London.

Brace, R.C., Hornbuckle, J. & Pearce-Higgins, J.W. 1997. The avifauna of the Beni Biological Station, Bolivia. *Bird Conserv. Intern.* 7: 117–159.

Brandt, C.S., Hasenack, H., Laps, R.R. & Hartz, S.M. 2009. Composition of mixed-species bird flocks in forest fragments of southern Brazil. *Zoologia* 26: 488–498.

Braun, M.J., Finch, D.W., Robbins, M.B. & Schmidt, B.K. 2000. *A Field Checklist of the Birds of Guyana.* Smithsonian Institution, Washington DC.

Braun, M.J., Robbins, M.B., Milensky, C.M., O'Shea, B.F., Barber, B.R., Hinds, W. & Prince, W.S. 2003. New birds for Guyana from Mts Roraima and Ayanganna. *Bull. Brit. Orn. Club* 123: 24–33.

Braun, M.J., Finch, D.W., Robbins, M.B. & Schmidt, B.K. 2007. *A Field Checklist of the Birds of Guyana*, 2nd edn. Smithsonian Institution, Washington DC.

Brawn, J.D., Angehr, G., Davros, N., Robinson, W.D., Styrsky, J.N. & Tarwater, C.E. 2011. Sources of variation in the nesting success of understory tropical birds. *J. Avian Biol.* 42: 61–68.

Braz, V.S. & Cavalcanti, R.B. 2001. A representatividade de áreas protegidas do Distrito Federal na conservação da avifauna do Cerrado. *Ararajuba* 9: 61–69.

Brodkorb, P. 1943. Birds from the Gulf lowlands of southern Mexico. *Misc. Publ. Mus. Zool. Univ. Mich.* 55: 1–88.

Brooks, D.M., O'Neill, J.P., Foster, M.S., Mark, T., Dauphine, N. & Franke, I.J. 2009. Avifauna of the Pongos Basin, Amazonas Department, Peru. *Wilson J. Orn.* 121: 54–74.

Brooks, T., Tobias, J. & Balmford, A. 1999. Deforestation and bird extinctions in the Atlantic Forest. *Anim. Conserv.* 2: 211–222.

Brooks, T.M., Barnes, R., Batrina, L., Butchart, S.H.M., Clay,

R.P., Esquivel, E.Z., Etcheverry, N.I., C., L.J. & Fincent, J.P. 1993. *Bird Surveys and Conservation in the Paraguayan Atlantic Forest Project CANOPY 92' Final Report*. BirdLife Study Report 57. BirdLife International, Cambridge, UK.

Brown, K.S. & Brown, G.G. 1992. Habitat alteration and species loss in Brazilian forests. Pp. 119–142 *in* Whitmore, T.C. & Sayer, J.A. (eds.) *Tropical Forest and Extinction*. Chapman & Hall, London.

Brumfield, R.T. & Maillard, O. 2007. Birds of the central rio Paracti Valley, a humid montane forest in departamento Cochabamba, Bolivia. *Orn. Neotrop.* 18: 321–337.

Buchanan, M.O. & Howell, T.R. 1965. Observations on the natural history of the thick-spined rat, *Hoplomys gymnurus*, in Nicaragua. *Ann. & Mag. Nat. Hist.* 8: 549–559.

Buckingham, M.A. 2011. *The Avian Communities of Iguazu National Park*. M.Sc. dissertation. Stephen F. Austin State Univ., Nacogdoches.

Bueno, A.S., Bruno, R.S., Pimentel, T.P., Sanaiotti, T.M. & Magnusson, W.E. 2012. The width of riparian habitats for understory birds in an Amazonian forest. *Ecol. Appl.* 22: 722–734.

Buffon, G.-L.L. 1771–86. *Histoire Naturelles des Oiseaux* (10 vols.). Paris.

Bugoni, L., Mohr, L.V., Scherer, A., Efe, M.A. & Scherer, S.B. 2002. Biometry, molt and brood patch parameters of birds in southern Brazil. *Ararajuba* 10: 85–94.

Buitrón-Jurado, G. 2008. Composición y diversidad de la avifauna en cuatro localidades de la Provincia del Carchi dentro del área de intervención del Proyect GISRENA. Pp. 33–65 *in* Boada, C. & Campaña, J. (eds.) *Composición y Diversidad de la Flora y la Fauna en Cuatro Localidades en la provincia del Carchi. Un Reporte de las Evaluaciones Ecológicas Rápidas*. EcoCiencia y GPC, Quito.

Burmeister, H. 1856. *Systematische Übersicht der Thiere Brasiliens, welche während einer Reise durch die Provinzen von Rio de Janeiro und Minas Geraes gesammelt und beobachtet wurden. Zweiter Theil. Vögel (Aves)*. Georg Reimer, Berlin.

Burton, P.J.K. 1975 Passerine bird weights from Panama and Colombia with some notes on 'soft-part' colours. *Bull. Brit. Orn. Club* 95: 82–86.

Buskirk, W.H. 1972. *Foraging Ecology of Bird Flocks in a Tropical Forest*. Ph.D. thesis. Univ. of California, Davis.

Butler, T.Y. 1979. *The Birds of Ecuador and the Galapagos Archipelago*. Ramphastos Agency, Portsmouth, NH.

Buzzetti, D.R.C. 2000. Distribuição altitudinal de aves em Angra dos Reis e Parati, sul do Estado do Rio de Janeiro, Brasil. Pp. 131–148 *in* Alves, M.A.S., Silva, J.M.C., van Sluys, M., Bergallo, H.G. & Rocha, C.F.D. (eds.) *A Ornitologia no Brasil: Pesquiza Atual e Perspectivas*. Univ. do Estado do Rio de Janeiro, Rio de Janeiro.

Buzzetti, D. 2008. *Plano de Manejo do Parque Estadual Chandless: Avaliação Ecológica Rápida - Relatório do Grupo Aves*. SOS Amazônia, Secretaria de Estado do Meio Ambiente e Recursos Naturais do Estado do Acre, Rio Branco.

Buzzetti, D.R.C. & Silva, S. 2008. *Berços da Vida, Ninhos de Aves Brasileiras*. Editora Terceiro Nome, São Paulo.

Byers, K.A. 2013. *Correlations in Morphology Between the Sexes in Feather Mites (Acari: Astigmata): Precopulatory Guarding and Reproductive Morphologies*. M.Sc. dissertation. Univ. of Alberta.

Cabanis, J. 1873. Neue Vögel des Berliner Museums, von C. Jelski in Peru entdeckt. *J. Orn.* 21: 315–320.

Cabanis, J. & Heine, F. 1859. *Museum Heineanum: Verzeichniss der ornithologischen Sammlung des Oberamtmann Ferdinand Heine*, vol. 2. R. Frantz, Halberstadt.

Cadena, C.D. & Stiles, F.G. 2010. El costo de la prioridad. *Orn. Colombiana* 9: 1–10.

Cadena, C.D., Alvarez, M., Parra, J.L., Jiménz, I., Mejía, C.A., Santamaría, M., Franco, A.M., Botero, C.A., Mejía, G.D., Umaña, A.M., Calixto, A., Aldana, J. & Londoño, G.A. 2000. The birds of CIEM, Tinigua National Park, Colombia: an overview of thirteen years of ornithological research. *Cotinga* 13: 46–54.

Cadena, C.D., López-Lanús, B., Bates, J.M., Krabbe, N., Rice, N.H., Stiles, F.G., Palacio, J.D. & Salaman, P. 2007. A rare case of interspecific hybridization in the tracheophone suboscines: Chestnut-naped Antpitta *Grallaria nuchalis* × Chestnut-crowned Antpitta *G. ruficapilla* in a fragmented Andean landscape. *Ibis* 149: 814–825.

Calderón-Leytón, J.J. 2002. *Aves de la Laguna de la Cocha*. Asociación para el Desarrollo Campesino ADC, San Juan de Pasto.

Calderón-Leytón, J.J., Paí, C.F., Cabrera-Finley, A. & Mora, Y.R. 2011. Aves del departamento de Nariño, Colombia. *Biota Colombiana* 12: 31–116.

Camperi, A.R., Darrieu, C.A., Grilli, P.G. & Burgos, F. 2013. Avifauna de la provincia de Jujuy, Argentina: lista de especies (Passeriformes). *Acta Zool. Lilloana* 57: 72–129.

Canaday, C. 1997. Loss of insectivorous birds along a gradient of human impact in Amazonia. *Biol. Conserv.* 77: 63–77.

Canaday, C. & Rivadeneyra, J. 2001. Initial effects of a petroleum operation on Amazonian birds: Terrestrial insectivores retreat. *Biodiver. & Conserv.* 10: 567–595.

Candia-Gallardo, C.E. 2005. Cubatão vinte anos depois: comunidade de aves de uma floresta de encosta na Serra do Mar. P. 107 *in Livro de Resumos do XIII Congresso Brasileira de Ornitologia*. Belém.

Cândido, J.F. 2000. The edge effect in a forest bird community in Rio Claro, São Paulo state, Brazil. *Ararajuba* 8: 9–16.

Canevari, M., Canevari, P., Carrizo, G., Harris, R.G., Rodriguez-Mata, J. & Straneck, R.J. 1991. *Nueva guía de las aves Argentinas* (2 vols.). Fundación Acindar, Buenos Aires.

Capllonch, P., Lobo, R., Ortiz, D. & Ovejero, R. 2005. La avifauna de la selva de galería en el noreste de Corrientes, Argentina: biodiversidad, patrones de distribución y migración. *Insugeo, Miscelánea* 14: 483–498.

Capparella, A.P. 1987. *Effects of Riverine Barriers on Genetic Differentiation of Amazonian Forest Undergrowth Birds*. Ph.D. thesis. Louisiana State Univ., Baton Rouge, LA.

Carantón-Ayala, D. & Certuche-Cubillos, K. 2010. A new species of antpitta (Grallariidae: *Grallaria*) from the northern sector of the western Andes of Colombia. *Orn. Colombiana* 9: 56–70.

Carantón-Ayala, D., Certuche-Cubillos, K., Cadena, C.D. & Gómez-Martínez, M.J. 2012. Ecologia y relaciones filogeneticas de cuatro especies del genero *Grallaria* (Grallariidae) endemicas de Colombia. P. 148 *in* More, A., Olaechea, D.G. & Olaechea, A.G. (eds.) *Libro de Resúmenes del IX Congreso de Ornitología Neotropical y el VIII Congreso Peruano de Ornitología*. Cuzco.

Cardiff, S.W. 1983. Three new bird species for Peru, with other distributional records from northern

Departamento de Loreto. *Gerfaut* 73: 185–192.

Cardiff, S.W. & Remsen, J.V. 1981. Three bird species new to Bolivia. *Bull. Brit. Orn. Club* 101: 304–305.

Cardiff, S.W. & Remsen, J.V. 1994. Type specimens in the Museum of Natural Science, Louisiana State University. *Occ. Pap. Mus. Nat. Sci. Louisiana State Univ.* 68: 1–33.

Cardona, C.A., Cárdenas, E.A., Giraldo, L.M., Castaño, D.F., Obando, J.C., Salazar, Á.M. & Fernández, Y. 2005. Caracterización de avifauna e identificación y priorización de objetos de conservación de la vereda la Antioqueña, resguardo nuestra Señora de la Candelaria de la Montaña Ríosucio – Caldas. *Bol. Cient. Mus. Hist. Nat.* 9: 85–109.

Carneiro, L.S. 2009. Variação Morfológica, Vocal e Molecular em *Hylopezus macularius* (Temminck, 1830) (Aves, Grallariidae). M.Sc. dissertation. Univ. Federal do Pará, Belém.

Carneiro, L.S. & Aleixo, A. 2012. Molecular phylogenetics and chronology of the genus *Hylopezus* (Grallariidae). P. 368 *in* More, A., Olaechea, D.G. & Olaechea, A.G. (eds.) *Libro de Resúmenes del IX Congreso de Ornitología Neotropical y el VIII Congreso Peruano de Ornitología*. Cuzco.

Carneiro, L.S. & Aleixo, A. 2013. Recognize newly described *Hylopezus whittakeri* and split *Hylopezus macularius* into two species. Proposal (#622) to South American Classification Committee. http://www.museum.lsu.edu/~Remsen/SACCprop622.htm.

Carneiro, L.S., Gonzaga, L.P., Rego, P.S., Sampaio, I., Schneider, H. & Aleixo, A. 2012. Systematic revision of the Spotted Antpitta (Grallariidae: *Hylopezus macularius*), with description of a cryptic new species from Brazilian Amazonia. *Auk* 129: 338–351.

Carnevalli, N., Machado, R.B., Brandt, A., Lamas, I.R., Lins, L.V., Barros, L.P. & Souza, A.L.T. 1989. *Estudo qualitativo da avifauna da Estação de Pesquisa e Desenvolvimento Ambiental de Peti - EPDA-PETI*. Universidade Federal de Minas Gerais & Companhia Energética de Minas Gerais, Belo Horizonte.

Carrasco, L., Cook, A. & Karubian, J. 2008. Extensión del rango de distribución de ocho especies de aves en las montañas de Mache-Chindul, Ecuador. *Cotinga* 29: 72–76.

Carriker, M.A. 1910. An annotated list of the birds of Costa Rica, including Cocos Island. *Ann. Carnegie Mus.* 6: 314–915.

Carriker, M.A. 1930. Descriptions of new birds from Peru and Ecuador. *Proc. Acad. Nat. Sci. Phil.* 82: 367–376.

Carriker, M.A. 1931. Descriptions of new birds from Peru and Bolivia. *Proc. Acad. Nat. Sci. Phil.* 83: 455–467.

Carriker, M.A. 1933. Descriptions of new birds from Peru, with notes on other little-known species. *Proc. Acad. Nat. Sci. Phil.* 85: 1–38.

Carriker, M.A. 1935. Descriptions of new birds from Peru and Ecuador, with critical notes on other little-known species. *Proc. Acad. Nat. Sci. Phil.* 87: 343–359.

Carriker, M.A. 1955. Notes on the occurrence and distribution of certain species of Colombian birds. *Nov. Colombianas* 1: 48–64.

Carriker, M.A. 1957. Studies in Neotropical Mallophaga, XVI: Bird lice of the suborder Ischnocera. *Proc. US Natl. Mus.* 106: 409–439.

Carriker, M.A. & Meyer de Schauensee, R. 1935. An annotated list of two collections of Guatemalan birds in the Academy of Natural Sciences of Philadelphia. *Proc. Acad. Nat. Sci. Phil.* 87: 411–455.

Carrión, J.M. 1986. *Aves del Valle de Quito y sus Alrededores: Una Guía Ilustrada para Reconocer las Especies Comunes*. Fundación Natura, Quito.

Carrión, J.M. 2002. *Aves de Quito, Retratos y Encuentros*. Simbioe, Quito.

Carvalho, B.H.G., Bichinski, T.A.T., Foerster, N.E., Bazilio, S. & Cochak, C. 2016. Avifauna da Floresta Nacional de Piraí do Sul (Paraná, sul do Brasil). *Atualidades Orn.* 192: 41–49.

Casas, C. & López-Ordóñez, J.P. 2006a. Caracterización y diversidad de la avifauna en la Reserva Natural Tambito, Cauca, Andes Occidentales de Colombia. Informe técnico proyecto corredor de conservación biológico y multicultural sector Munchique-Pinche, Cordillera Occidental Colombiana. Unpubl. report for Fundación Proselva, Popayán.

Cassin, J. 1860. Catalogue of birds collected during a survey of a route for a ship Canal across the Isthmus of Darien, by order of the Government of the United States, made by Lieut. N. Michler, of the U.S. Topographical Engineers, with notes and descriptions of new species. *Proc. Acad. Nat. Sci. Phil.* 12: 188–197.

Cassin, J. 1864. Notes on some species of birds from South America. *Proc. Acad. Nat. Sci. Phil.* 16: 286–288.

Castaño-Villa, G.J. & Patiño-Zabala, J.C. 2000. Cambios en la composición de la avifauna en Santa Helena durante el siglo XX. *Crónica Forest. Med. Amb.* 15: 139–162.

Castaño-Villa, G.J. & Patiño-Zabala, J.C. 2007. Composición de la comunidad de aves en bosques fragmentados en la región de Santa Elena, Andes Centrales Colombianos. *Bol. Cien. Centro Mus., Mus. Hist. Nat. Univ. Caldas* 11: 47–64.

Castaño-Villa, G.J. & Patiño-Zabala, J.C. 2008. Extinciones locales de aves en fragmentos del bosque en la región de Santa Elena, Andes centrales, Colombia. *Hornero* 23: 23–34.

Cavarzere, V. & Arantes, F. 2017. Birds of a habitat mosaic in the threatened Cerrado of central São Paulo, Brazil. *Cotinga* 39: 27–39.

Cavarzere, V., Moraes, G.P. & Silveira, L.F. 2010. Boracéia Biological Station: an ornithological review. *Pap. Avuls. Dept. Zool., São Paulo* 50: 189–201.

Cavarzere, V., Moraes, G.P., Dalbeto, A.C., Maciel, F.G. & Donatelli, R.J. 2011. Birds from cerradão woodland, an overlooked forest of the Cerrado region, Brazil *Pap. Avuls. Dept. Zool., São Paulo* 51 259–273.

Cavarzere, V., Marcondes, R.S., Moraes, G.P. & Donatelli, R.J. 2012. Comparação quantitativa da comunidade de aves de um fragmento de floresta semidecidual do interior do Estado de São Paulo em intervalo de 30 anos. *Iheringia, Sér. Zool.* 102: 384–393.

Cendrero, L. 1972. *Zoologia Hispanoamericana*. Editorial Porrúa, Ciudad de México.

Chapman, F.M. 1912. Diagnoses of apparently new Colombian birds. *Bull. Amer. Mus. Nat. Hist.* 31: 139–166.

Chapman, F.M. 1914. Diagnoses of apparently new Colombian birds. 3. *Bull. Amer. Mus. Nat. Hist.* 33: 603–637.

Chapman, F.M. 1915. Diagnoses of apparently new Colombian birds. 4. *Bull. Amer. Mus. Nat. Hist.* 34: 635–662.

Chapman, F.M. 1917. The distribution of bird-life in Colombia: a contribution to the biological survey of South America. *Bull. Amer. Mus. Nat. Hist.* 36: 3–729.

Chapman, F.M. 1919. Descriptions of proposed new birds

from Peru, Bolivia, Brazil, and Colombia. *Proc. Biol. Soc. Wash.* 32: 253–268.

Chapman, F.M. 1921. The distribution of bird life in the Urubamba Valley of Peru. *Bull. US Nat. Mus.* 117: 1–138.

Chapman, F.M. 1922. Descriptions of apparently new birds from Colombia, Ecuador, and Argentina. *Amer. Mus. Novit.* 31: 1–8.

Chapman, F.M. 1923a. Descriptions of proposed new birds from Panama, Venezuela, Ecuador, Peru and Bolivia. *Amer. Mus. Novit.* 67: 1–12.

Chapman, F.M. 1923b. Descriptions of proposed new birds from Venezuela, Colombia, Ecuador, Peru, and Chile. *Amer. Mus. Novit.* 96: 1–12.

Chapman, F.M. 1923c. Descriptions of proposed new Formicariidae and Dendrocolaptidae. *Am. Mus. Novit.* 86: 1–20.

Chapman, F.M. 1924. Descriptions of new genera and species of Tracheophonae from Panama, Ecuador, Peru, and Bolivia. *Am. Mus. Novit.* 123: 1–9.

Chapman, F.M. 1925a. Descriptions of one new genus and of species of birds from Peru and Ecuador. *Am. Mus. Novit.* 205: 1–11.

Chapman, F.M. 1925b. Remarks on the life zones of northeastern Venezuela with descriptions of new species of birds. *Am. Mus. Novit.* 191: 1–15.

Chapman, F.M. 1926a. The distribution of bird-life in Ecuador. *Bull. Amer. Mus. Nat. Hist.* 40: 1–784.

Chapman, F.M. 1926b. Descriptions of new birds from Bolivia, Peru, Ecuador, and Brazil. *Amer. Mus. Novit.* 231: 1–7.

Chapman, F.M. 1927. Descriptions of new birds from northwestern Peru and western Colombia. *Am. Mus. Novit.* 250: 1–7.

Chapman, F.M. 1929a. *My Tropical Air Castle.* D. Appleton & Co., New York.

Chapman, F.M. 1929b. Descriptions of new birds from Mt. Duida, Venezuela. *Amer. Mus. Novit.* 380: 1–27.

Chapman, F.M. 1931. The upper zonal bird-life of Mts. Roraima and Duida. *Bull. Amer. Mus. Nat. Hist.* 63: 1–135.

Chapman, G., Cleese, J., Idle, E., Gilliam, T., Jones, T. & Palin, M. 1975. *Monty Python and the Holy Grail.* Directed by Gilliam, T. & Jones, T. (directors). EMI Films, London.

Chaves, J.A. & Freile, J. 2005. *Aves Comunes de Otonga y los Bosques Nublados Noroccidentales del Ecuador.* Imprenta Mariscal, Quito.

Chebez, J.C. 1996. *Fauna Misionera. Catálogo Sistemático y Zoogeográfico de los Vertebrados de la provincia de Misiones (Argentina).* LOLA, Buenos Aires.

Cherrie, G.K. 1891a. Descriptions of new genera, species, and subspecies of birds from Costa Rica. *Proc. US Natl. Mus.* 14: 337–346.

Cherrie, G.K. 1891b. Notes on Costa Rican birds. *Proc. US Natl. Mus.* 14: 517–537.

Cherrie, G.K. 1893. *Exploraciones Zoológicas Efectuadas en la Parte meridional de Costa Rica por los Años de 1891-92.* Museo Nacional de Costa Rica, San José.

Cherrie, G.K. 1895. *Exploraciones Zoológicas Efectuadas en el Valle del Rio Naranjo en al Año de 1893: Aves.* Institudo Físico-Geográico Nacional de Costa Rica, San José.

Christiansen, M.B. & Pitter, E. 1997. Species loss in a forest bird community near Lagoa Santa in southeastern Brazil. *Biol. Conserv.* 80: 23–32.

Chrostowski, T. 1921. Sur les types d'oiseaux néotropicaux du Musée Zoologique de l'Académie des Sciences de Pétrograde. *Ann. Zool. Mus. Polon. Hist. Nat.* 1: 9–30.

Chubb, C. 1910. On the birds of Paraguay. Part III. *Ibis* 52: 517–534.

Chubb, C. 1916. [Exhibition and description of new birds from Ecuador]. *Bull. Brit. Orn. Club* 36: 46–48.

Chubb, C. 1917. [Descriptions of new forms of South American birds]. *Bull. Brit. Orn. Club* 38: 29–34.

Chubb, C. 1918a. [Descriptions of new forms from South America]. *Bull. Brit. Orn. Club* 38: 47–48.

Chubb, C. 1918b. [Descriptions of new forms from South America]. *Bull. Brit. Orn. Club* 38: 83–87.

Chubb, C. 1918c. Descriptions of new genera and a new subspecies of South American birds. *Ann. & Mag. Nat. Hist.* (9)2: 122–124.

Chubb, C. 1921. *Birds of British Guiana,* vol. 2. Bernard Quarritch, London.

Cicchino, A.C. & Valim, M.P. 2008. Three new species of *Formicaphagus* Carriker, 1957 (Phthiraptera, Ischnocera, Philopteridae), parasitic on Thamnophilidae and Conopophagidae (Aves, Passeriformes). *Zootaxa* 1949: 37–50.

Cintra, R. & Yamashita, C. 1990. Habitats, abundância e ocorrência das espécies de aves do Pantanal de Poconé, Mato Grosso, Brasil. *Pap. Avuls. Dept. Zool., São Paulo* 37: 1–21.

Cintra, R., Sanaiotti, T.M. & Cohn-Haft, M. 2007. Spatial distribution and habitat of the Anavilhanas archipelago bird community in the Brazilian Amazon. *Biodiver. & Conserv.* 16: 313–336.

Claramunt, S.J., Cuervo, A.M., Piacentini, V.Q., Bravo, G.A. & Remsen, J.V. 2014. Comment on *Grallaria fenwickorum* Barrera & Bartels, 2010 (Aves, Grallariidae): proposed replacement of an indeterminate holotype by a neotype. *Bull. Zool. Nomencl.* 71: 40–43.

Clark, G.A. 1971. The occurrence of bill-sweeping in the terrestrial foraging of birds. *Wilson Bull.* 83: 66–73.

Clayton, D.H. & Walther, B.A. 2001. Influence of host ecology and morphology on the diversity of Neotropical bird lice. *Oikos* 94: 455–467.

Clayton, D.H., Gregory, R.D. & Price, R.D. 1992. Comparative ecology of Neotropical bird lice (Insecta: Phthiraptera). *J. Anim. Ecol.* 61: 781–795.

Clements, J.F. 2000. *Birds of the World: A Checklist.* Pica Press, Robertsbridge.

Clements, J.F. 2007. *The Clements Checklist of Birds of the World.* Cornell Univ. Press, Ithaca, NY.

Clements, J.F. & Shany, N. 2001. *A Field Guide to the Birds of Peru.* Ibis Publishing, Temecula, CA.

Coates-Estrada, R. & Estrada, A. 1989. Avian attendance and foraging at army-ant swarms in the tropical rain forest of Los Tuxtlas, Veracruz, Mexico. *J. Trop. Ecol.* 5: 281–292.

Cockle, K.L. 2003. The Bird Community of Shade-grown Yerba Mate and Adjacent Atlantic Forest in Canindeyu, Paraguay. M.Sc. dissertation. Dalhousie Univ., Halifax, Nova Scotia.

Cockle, K.L., Leonard, M.L. & Bodrati, A.A. 2005. Presence and abundance of birds in an Atlantic forest reserve and adjacent plantation of shade-grown yerba mate, in Paraguay. *Biodiver. & Conserv.* 14: 3265–3288.

Cody, M.L. 2000. Antbird guilds in the lowland Caribbean rainforest of southeast Nicaragua. *Condor* 102: 784–794.

Coelho, A.G.M. 1987. Aves da Reserva Biológica de Serra Negra (Floresta-PE), lista preliminar. *Publ. Avuls. Univ. Fed. Pernambuco* 2: 1–8.

Coelho, I.P., Kindel, A. & Coelho, A.V.P. 2008. Roadkills of vertebrate species on two highways through the Atlantic

Forest Biosphere Reserve, southern Brazil. *Eur. J. Wildl. Res.* 54: 689–699.

Cohn-Haft, M., Whittaker, A. & Stouffer, P.C. 1997. A new look at the "species-poor" central Amazon: the avifauna north of Manaus, Brazil. *Orn. Monogr.* 48: 205–235.

Collar, N.J. & Andrew, P. 1988. *Birds to Watch. The ICBP World Checklist of Threatened Birds.* International Council for Bird Preservation, Cambridge, UK.

Collar, N.J., Gonzaga, L.P., Krabbe, N., Madroño Nieto, A., Naranjo, L.G., Parker, T.A. & Wege, D.C. 1992. *Threatened Birds of the Americas: The ICBP/IUCN Red Data Book.* International Council for Bird Preservation, Cambridge, UK.

Collar, N.J., Crosby, M.J. & Stattersfield, A.J. 1994. *Birds to Watch 2: The World List of Threatened Birds.* BirdLife International, Cambridge, UK.

Collar, N.J., Wege, D.C. & Long, A.J. 1997. Patterns and causes of endangerment in the New World avifauna. *Orn. Monogr.* 48: 237–260.

Collins, C. 2006. Antpitta paradise. *Neotrop. Birding* 1: 68–70.

Coltro, L.A. 2008. A avifauna do Parque Nacional Montanhas do Tumucumaque registrada durante o projeto de inventários biológicos rápidos. Pp. 33–37; 94–102 in Bernard, E. (ed.) *Inventários Biológicos Rápidos no Parque Nacional Montanhas do Tumucumaque, Amapá, Brasil. RAP Bull. Biol. Assess.* 48. Conservacion International, Arlington, VA.

Comitê Brasileiro de Registros Ornitológicos. 2011. Lista das aves do Brasil, 10ª ed. (versão 25 Jan 2011). http://www.cbro.org.br/CBRO/listabr.htm (accessed 10 August 2015).

Contreras, J.R. 1983. Notas sobre el peso de aves argentinas II. *Historia Natural* 3: 39–40.

Contreras, J.R. 1987. Lista preliminar de la avifauna correntina II. Passeriformes. *Historia Natural* 7: 61–70.

Cook, A.G. 1996. Avifauna of North-western Peru Biosphere Reserve and its environs. *Bird Conserv. Intern.* 6: 139–165.

Cordeiro, P.H.C. 2001. Areografia dos Passeriformes endêmicos da Mata Atlântica. *Ararajuba* 9: 125–137.

Cordero-Rodriguez, G.A. 1987 Composición y diversidad de la fauna de vertebrados terrestres de Barlovento, estado Miranda, Venezuela. *Acta Cienc. Venez.* 38: 234–258.

Corrêa, L., Bazílio, S., Woldan, D. & Boesing, A.L. 2008. Avifauna da Floresta Nacional de Três Barras (Santa Catarina, Brasil). *Atualidades Orn.* 143: 38–41.

Cortes-Herrera, F., Villagran-Chavarro, D.X. & Cortes-Herrera, O. 2012. Uso de habitat del hormiguero (*Grallaria kaestneri*) en bosque andino de Cundinamarca, Colombia. Pp. 568–569 in More, A., Olaechea, D.G. & Olaechea, A.G. (eds.) *Libro de Resúmenes del IX Congreso de Ornitología Neotropical y el VIII Congreso Peruano de Ornitología.* Cuzco.

Cory, C.B. 1916. Descriptions of apparently new South American birds, with notes on some little known species. *Publ. Field Mus. Nat. Hist. (Orn. Ser.)* 1: 337–346.

Cory, C.B. & Hellmayr, C.E. 1924. Catalogue of birds of the Americas. Part III. Pteroptochidae, Conopophagidae, Formicariidae. *Publ. Field Mus. Nat. Hist. (Zool. Ser.)* 13: 1–369.

Costa, L.S. 2015. Contribuição ao conhecimento da ornitofauna do município de Joinville, Santa Catarina, Brasil. *Saúde Meio Ambiente* 4: 16–31.

Cracraft, J. 1985. Historical biogeography and patterns of differentiation within the South American avifauna: areas of endemism. *Orn. Monogr.* 36: 49–84.

Cresswell, W., Hughes, M., Mellanby, R., Bright, S., Catry, P., Chaves, J., Freile, J., Gabela, A., Martineau, H. & Macleod, R. 1999a. Densities and habitat preferences of Andean cloud-forest birds in pristine and degraded habitats in north-eastern Ecuador. *Bird Conserv. Intern.* 9: 129–146.

Cresswell, W., Mellanby, R., Bright, S., Catry, P., Chaves, J., Freile, J.F., Gabela, G., Hughes, M., Martineau, H., MacLeod, R., McPhee, F., Anderson, N., Holt, S., Barabas, S., Chapel, C. & Sanchez, T. 1999b. Birds of the Guandera Biological Reserve, Carchi province, north-east Ecuador. *Cotinga* 11: 55–63.

Crowley, P. 1883. *A List of Birds' Eggs in the Collection of Philip Crowley.* Waddon House, Surrey.

Cuadros, T. 1988. Aspectos ecológicos de la comunidad de aves en un bosque nativo en la Cordillera Central en Antioquia (Colombia). *Hornero* 13: 8–20.

Cubillos, K.C. & Carantón-Ayala, D.A. 2012. Ecologia del tororoi de Santa Marta (*Grallaria bangsi*) en la Sierra Nevada de Santa Marta, Magdalena, Colombia. Pp. 396–397 in More, A., Olaechea, D.G. & Olaechea, A.G. (eds.) *Libro de Resúmenes del IX Congreso de Ornitología Neotropical y el VIII Congreso Peruano de Ornitología.* Cuzco.

Cuervo, A.M., Pulgarín, P.C., Calderón-F., D., Ochoa-Quintero, J.M., Delgado, C.A., Palacio, A., Botero, J.M. & Múnera, W.A. 2008. Avifauna of the northern Cordillera Central of the Andes, Colombia. *Orn. Neotrop.* 19: 495–515.

Cuervo, A.M., Stiles, F.G., Cadena, C.D., Toro, J.L. & Londoño, G.A. 2003. New and noteworthy bird records from the northern sector of the Western Andes of Colombia. *Bull. Brit. Orn. Club* 123: 7–24.

Cuvier, G. 1817. *Le Règne Animal Distribué d'après son Organisation, pour servir de base à l'Histoire Naturelle des Animaux et d'Introduction à l'Anatomie Comparée*, vol. 1. Chez Déterville, Paris.

Cuvier, G. 1829. *Le Règne Animal Distribué d'après son Organisation, pour servir de base à l'Histoire Naturelle des Animaux et d'Introduction à l'Anatomie Comparée. Nouvelle Édition, Revue et Augmentée*, vol. 1. Chez Déterville, Paris.

Cuvier, G. & d'Orbigny, A. 1836. *Le Règne Animal Distribué d'après son Organisation, pour servir de base à l'Histoire Naturelle des Animaux et d'Introduction à l'Anatomie Comparée. Troisième Édition, Avec Figures Dessinées d'après Nature*, vol. 1. Louis Hauman, Brussels.

Dabbene, R. 1919. Especies de aves poco comunes o nuevas para la Republica Argentina. *Hornero* 1: 259–266.

Daily, G.C., Ehrlich, P.R. & Sánchez-Azofeifa, G.A. 2001. Countryside biogeography: Use of human-dominated habitats by the avifauna of southern Costa Rica. *Ecol. Appl.* 11: 1–13.

Dantas, G.P.M., Santos, F.R. & Marini, M.Â. 2007. Genetic variability of *Conopophaga lineata* (Conopophagidae) (Wied-Neuwied, 1831) in Atlantic Forest fragments. *Braz. J. Biol.* 67: 859–865.

Dantas, G.P.M., Santos, F.R. & Marini, M.Â. 2009. Sex ratio and morphological characteristics of Rufous Gnateaters, *Conopophaga lineata* (Aves, Passeriformes) in Atlantic forest fragments. *Iheringia, Sér. Zool.* 99: 115–119.

Dantas, S.M., Faccio, M.S. & Lima, M.F. 2011. Avifaunal inventory of the Floresta Nacional de Pau-Rosa, Maués, state of Amazonas, Brazil. *Rev. Bras. Orn.* 19: 154–166.

Dario, F.R. & de Vincenzo, M.C.V. 2011. Avian diversity and relative abundance in a restinga forest of Sao Paulo, Brazil. *Trop. Ecol.* 52: 25–33.

Dauphiné, N.S. 2008. Bird Ecology, Conservation, and Community Responses to Logging in the Northern Peruvian Amazon. Ph.D. thesis. Univ. of Georgia, Athens.

Davies, C.W.N., Barnes, R., Butchart, S.H.M., Fernández, M. & Seddon, N. 1994. *The Conservation Status of the Cordillera de Colan: A Report Based on Bird and Mammal Surveys in 1994.* Published privately, Cambridge, UK.

Davies, C.W.N., Barnes, R., Butchart, S.H.M., Fernández, M. & Seddon, N. 1997. The conservation status of birds on the Cordillera de Colan, Peru. *Bird Conserv. Intern.* 7: 181–195.

Davis, D.E. 1945. The annual cycle in plants, mosquitoes, birds, and mammals in two Brazilian forests. *Ecol. Monogr.* 15: 243–295.

Davis, L.I. 1972. *A Field Guide to the Birds of Mexico and Central America.* Univ. of Texas Press, Austin.

Davis, T.A.W. 1953. An outline of the ecology and breeding seasons of birds of the lowland forest region of British Guiana. *Ibis* 95: 450–467.

Davis, T.J. 1986. Distribution and natural history of some birds from the departments of San Martin and Amazonas, northern Peru. *Condor* 88: 50–56.

Davis, T.J., Fox, C., Salinas, L., Ballon, G. & Arana, C. 1991. Annotated checklist of the birds of Cuzco Amazonico, Peru. *Occ. Pap. Mus. Nat. Hist. Univ. Kansas* 144: 1–19.

Deignan, H.G. 1961. Type specimens of birds in the United States National Museum *Bull. US Natl. Mus.* 221: 1–718.

Delgado-V., C.A. 2002. Observations of the Ochre-breasted (*Grallaricula flavirostris*) and Slate-crowned (*G. nana*) Antpittas in Colombia. *Orn. Neotrop.* 13: 423–425.

Delhey, K. 2017. Gloger's Rule. *Current Biol.* 27: R689–R691.

Develey, P.F. & Martensen, A.C. 2006. As aves da Reserva Florestal do Morro Grande (Cotia, SP). *Biota Neotropica* 6: 1–16.

Develey, P.F. & Metzger, J.P. 2006. Emerging threats to birds in Brazilian Atlantic Forest: the roles of forest loss and configuration in a severely fragmented ecosystem. Pp. 269–290 *in* Peres, C.A. & Laurance, W.F. (eds.) *Emerging Threats to Tropical Forests.* Univ. of Chicago Press, Chicago.

Dick, J.A., McGilllivray, W.B. & Brooks, D.J. 1984. List of birds and their weights from Saul, French Guiana. *Wilson Bull.* 96: 347–365.

Dickerman, R.W. 1990. The Scaled Antpitta, *Grallaria guatimalensis*, in Mexico. *Southwest. Natur.* 35: 460–463.

Dickerman, R.W. & Phelps, W.H., Jr. 1982. An annotated list of the birds of Cerro Urutaní on the border of Estado Bolívar, Venezuela and Territorio Roraima, Brazil. *Amer. Mus. Novit.* 2732: 1–20.

Dickey, D.R. & van Rossem, A.J. 1938. The birds of El Salvador. *Publ. Field Mus. Nat. Hist. (Zool. Ser.)* 23: 1–609.

Dickinson, E.C. 2001. Systematic notes on Asian birds. 9. The "Nouveau recueil de planches coloriées" of Temminck & Laugier (1820-1839). *Zool. Verhand.* 335: 7–54, figs. 51–55.

Dickinson, E.C. (ed.) 2003. *The Howard & Moore Complete Checklist of the Birds of the World* (3rd. edn.). Christopher Helm, London.

Dickinson, E.C. 2011. The first twenty livraisons of "Les Planches Coloriées d'Oiseaux" of Temminck & Laugier (1820-1839): II. Issues of authorship, nomenclature and taxonomy. *Zool. Bibliography* 1: 151–166.

Dickinson, E.C. 2017. A study of d'Orbigny's "Voyage dans l'Amerique Meridionale" I. The contents of the parts of the volumes on natural history. *Zool. Bibliography* 5: 1–12.

Dickinson, E.C. & Christidis, L. (eds.) 2014. *The Howard and Moore Complete Checklist of the Birds of the World*, vol. 2. (4th edn.) Aves Press, Eastbourne.

Dickinson, E.C. & Lebossé, A. 2017. A study of d'Orbigny's "Voyage dans l'Amerique Meridionale" II. On the composition of the 1837 and 1838 volumes of the Magasin de Zoologie. *Zool. Bibliography* 5: 13–37.

Dietsch, T.V. 2005. Seasonal variation of infestation by ectoparasitic chigger mite larvae (Acarina: Trombiculidae) on resident and migratory birds in coffee agroecosystems of Chiapas, Mexico. *J. Parasitol.* 91: 1294–1303.

Diniz, M.G., Mazzoni, L.G., Neto, S.D., Vasconcelos, M.F., Perillo, A. & Benedicto, G.A. 2012. Historical synthesis of the avifauna from the Rio São Francisco basin in Minas Gerais, Brazil. *Rev. Bras. Orn.* 20: 329–349.

Dobbs, R.C., Martin, P.R., Batista, C., Montag, H. & Greeney, H.F. 2003. Notes on egg-laying, incubation, and nestling care in Scaled Antpitta *Grallaria guatimalensis*. *Cotinga* 19: 65–70.

Dobbs, R.C., Martin, P.R. & Kuehn, M.J. 2001. On the nest, eggs, nestlings, and parental care of the Scaled Antpitta (*Grallaria guatimalensis*). *Orn. Neotrop.* 12: 225–233.

Domaniewski, J. & Sztolcman, J. 1918. Przyczynek do znajomo ci form rodzaju *Grallaria* Vieill. *Sprawozdania z Posiedzie Towarzystwa Naukowego Warszawskiego* 3: 474–484.

Donahue, P.K. 1985. Notes on some little known or previously unrecorded birds of Suriname. *Amer. Birds* 39: 229–230.

Donahue, P.K. 1994. *Birds of Tambopata. A Checklist.* Tambopata Reserve Society, London.

Donahue, P.K. & Pierson, J.E. 1982. *Birds of Suriname: An Annotated Checklist.* South Harpswell, ME.

Donatelli, R.J. & Ferreira, C.D. 2009. Aves da Estação Ecológica de Caetetus, Gália, SP. *Atualidades Orn.* 148: 55–57.

Donatelli, R.J., Costa, T.V.V. & Ferreira, C.D. 2004. Dinâmica da avifauna em fragmento da mata na Fazenda Rio Claro, Lençóis Paulista, São Paulo. *Rev. Bras. Zool.* 21: 97–114.

Donatelli, R.J., Ferreira, C.D., Dalbeto, A.C. & Posso, S.R. 2007. Análise comparativa da assembléia de aves em dois remanescentes florestais no interior do Estado de São Paulo, Brasil. *Rev. Bras. Zool.* 24: 362–375.

Donegan, T.M. 2008a. Geographical variation in Slate-crowned Antpitta *Grallaricula nana*, with two new subspecies, from Colombia and Venezuela. *Bull. Brit. Orn. Club* 128: 150–178.

Donegan, T.M. 2008b. New species and subspecies descriptions do not and should not always require a dead type specimen. *Zootaxa* 1761: 37–48.

Donegan, T.M. 2010. *Grallaricula nana*: subspecies and species limits. http://www.xeno-canto.org/feature-view.php?blognr=34.

Donegan, T.M. & Dávalos, L.M. 1999. Ornithological observations from Reserva Natural Tambito, Cauca, south-west Colombia. *Cotinga* 12: 48–55.

Donegan, T.M. & Huertas, B.C. 2005. *Threatened Species of Serranía de los Yariguíes: Final report. Colombian EBA Project Report Series No. 5.*

Donegan, T.M. & Salaman, P.G.W. 1999. *Rapid Biodiversity Assessments and Conservation Evaluations in the Colombian Andes: northeast Antioquia & highlands of Serranía de los Churumbelos. Colombian EBA Project Report Series No. 2.* Fundación Proaves, Bogotá.

Donegan, T.M., Avendaño-C., J.E., Briceño-L., E.R. & Huertas, B.C. 2007. Range extensions, taxonomic and ecological notes from Serranía de los Yariguíes, Colombia's new national park. *Bull. Brit. Orn. Club* 127: 172–213.

Donegan, T.M., Huertas, B.C., Avendaño-C., J.E., Briceño, E. & Donegan, M. 2008. The first biological explorations of Serrania de los Yariguies, Colombia. *Neotrop. Birding* 3: 38–43.

Donegan, T.M., Avendaño-C., J.E., Briceño-L., E.R., Luna, J.C., Roa, C., Parra, R., Turner, C., Sharp, M. & Huertas, B.C. 2010. Aves de la Serranía de los Yariguíes y tierras bajas circundantes, Santander, Colombia. *Cotinga* 32: 23–40.

Donegan, T.M., Avendaño-C., J.E., Huertas, B.C. & Flórez, B. 2009a. Avifauna de San Pedro de los Milagros, Antioquia: una comparación entre colecciones antiguas y evaluaciones rápidas. *Bol. Cien. Mus. Hist. Nat.* 13: 63–72.

Donegan, T.M., Salaman, P.G.W. & Caro, D. 2009b. Revision of the status of various bird species occurring or reported in Colombia. *Conserv. Colombiana* 8: 80–86.

Donegan, T.M., Quevedo, A., McMullan, M. & Salaman, P.G.W. 2011. Revision of the status of bird species occurring or reported in Colombia 2011. *Conserv. Colombiana* 15: 4–21.

Dornelas, A.A.F., Paula, D.C., Santo, M.M.E., Sánchez-Azofeifa, G.A. & Leite, L.O. 2012. Avifauna do Parque Estadual da Mata Seca, norte de Minas Gerais. *Rev. Bras. Orn.* 20: 378.

Dorst, J. 1956. Étude d'une collection d'oiseaux rapportée du Pérou central. *Bull. Mus. Natl. d'Hist. Nat. (Paris), 2ème sér.* 28: 265–272.

Dorst, J. 1957. La vie dur les hauts plateau andins du Pérou. *Rev. d'Ecol. (Terre et Vie)* 104: 3–50.

Downing, C. 2005. New distributional information for some Colombian birds, with a new species for South America. *Cotinga* 24: 13–15.

Downing, C. & Hickman, J. 2004. New record of Hooded Antpitta *Grallaricula cucullata* in the western Cordillera of Colombia. *Cotinga* 21: 76.

Dreyer, N.P. 2002. A nest of Ash-throated Gnateater *Conopophaga peruviana* in Amazonian Ecuador. *Cotinga* 17: 79.

Dubois, A. 1900. Synopsis avium, Fasciculus ii, iii, iv, v. Pp. 81–368 in Dubois, A. (ed.) *Nouveau Manuel d'Ornithologie.* H. Lambertin, Brussels.

Dubois, A. 1904. *Synopsis Avium: Nouveau Manuel d'Ornithologie.* H. Lamertin, Brussels.

Dubois, A. & Nemésio, A. 2007. Does nomenclatural availability of nomina of new species or subspecies require the deposition of vouchers in collections? *Zootaxa* 1409: 1–22.

Dubs, B. 1992. *Birds of Southwestern Brazil.* Betrona Verlag, Küsnacht.

Dugand, A. & Borrero H., J.I. 1946. Aves de la ribera colombiana del Amazonas. *Caldasia* 4: 131–167.

Dunning, J.B. 1993. *CRC Handbook of Avian Body Masses,* 1st edn. CRC Press, Boca Raton, FL.

Dunning, J.B. 2008. *CRC Handbook of Avian Body Masses,* 2nd edn. CRC Press, Boca Raton, FL.

Dunning, J.S. 1982. *South American Land Birds: A Photographic Aid to Identification,* 1st edn. Harrowood Books, Newtown Square, PA.

Dunning, J.S. 1987. *South American Land Birds: A Photographic Aid to Identification,* 2nd edn. Harrowood Books, Newtown Square, PA.

Durães, R. & Marini, M.Â. 2003. An evaluation of the use of tartar emetic in the study of bird diets in the Atlantic Forest of southeastern Brazil. *J. Field Orn.* 74: 270–280.

Durães, R. & Marini, M.Â. 2005. A quantitative assessment of bird diets in the Brazilian Atlantic Forest, with recommendations for future diet studies. *Orn. Neotrop.* 16: 65–83.

Durán, S.M. & Kattan, G.H. 2005. A test of the utility of exotic tree plantations for understory birds and food resources in the Colombian Andes. *Biotropica* 37: 129–135.

Echeverry-Galvis, M.Á. 2001. Patrones Reproductivos y Procesos de Muda en Aves de Bosque alto Andino del Flanco Sur Occidental de la Sabana de Bogotá. B.Sc. thesis. Pontificia Universidad Javeriana, Bogotá.

Echeverry-Galvis, M.Á. & Córdoba-Córdoba, S. 2007. New distributional and other bird records from Tatamá Massif, West Andes, Colombia. *Bull. Brit. Orn. Club* 127: 213–224.

Echeverry-Galvis, M.Á. & Morales-Rozo, A. 2007. Lista anotada de algunas especies de la vereda "Cerca de Piedra", Chia, Colombia. *Bol. Soc. Antioqueña Orn.* 17: 87–93.

Echeverry-Galvis, M.Á., Córdoba-Córdoba, S., Peraza, C.A., Baptiste, M.P. & Ahumada, J.A. 2006. Body weights of 98 species of Andean cloud-forest birds. *Bull. Brit. Orn. Club* 126: 291–298.

Eckhardt, K. 2003. Evaluación de la diversidad biológica de aves del Santuario Nacional Tabaconas-Namballe. Pp. 81–93 in Amanzo, J. (ed.) *Evaluación Biológica Rapida del Santuario Nacional Tabaconas-Namballe y zonas aledañas. Informe WWF-OPP: QM-91.* World Wildlife Fund, Cuzco.

Edwards, E.P. 1967. Nests of the Common Bush-Tanager and the Scaled Antpitta. *Condor* 69: 605.

Edwards, E.P. 1968. *Finding Birds in Mexico.* Privately published, Sweet Briar, VA.

Edwards, E.P. 1972. *A Field Guide to the Birds of Mexico.* Privately published, Sweet Briar, VA.

Edwards, E.P. 1989. *A Field Guide to the Birds of Mexico,* 2nd edn. Privately published, Sweet Briar, VA.

Edwards, E.P. & Lea, R.B. 1955. Birds of the Monserrate area, Chiapas, Mexico. *Condor* 57: 31–54.

Efe, M.A., Mohr, L.V., Bugoni, L., Scherer, A. & Scherer, S.B. 2001. Inventário e distribuição da avifauna do Parque Saint Hilaire, Viamão, Rio Grande do Sul, Brasil. *Tangara* 1: 12–25.

Eisenmann, E. 1952. Annotated list of the birds of Barro Colorado Island, Panama Canal Zone. *Smithsonian Misc. Coll.* 117: 1–62.

Eisermann, K. & Avendaño, C. 2007. *Lista Comentada de las Aves de Guatemala.* Lynx Edicions, Barcelona.

Eisermann, K., Avendaño, C. & Tanimoto, P. 2013. Birds of the Cerro El Amay Important Bird Area, Quiché, Guatemala. *Cotinga* 35: 81–93.

Ellery, T. 2015. The Serranía del Perijá—an exciting new destination in Colombia. *Neotrop. Birding* 17: 58–67.

Emerson, K.C. 1981. Status of five species of Mallophaga described by M. A. Carriker, Jr. *Proc. Ent. Soc. Wash.* 83: 137–139.

English, P.H. 1998. Ecology of Mixed-species Understory Flocks in Amazonian Ecuador. Ph.D. thesis. Univ. of Texas, Austin.

Érard, C. 1982. Le nid et la ponte de *Lipaugus vociferans,*

Cotingidé, et de *Grallaria varia*, Formicariidé. *Alauda* 50: 311–313.

Erickson, H.T. & Mumford, R.E. 1976. Notes on birds of the Viçosa, Brazil region. *Bull. Agricult. Exp. Sta. Perdue Univ.* 131: 1–29.

Esquivel-M., A. 2010. Comunidades de Aves del Bosque Atlántico del Paraguay. Ph.D. thesis. Univ. de Salamanca, Salamanca.

Esquivel-M., A. & Peris, S.J. 2008. Influence of time of day, duration and number of counts in point count sampling of birds in an Atlantic Forest of Paraguay. *Orn. Neotrop.* 19: 229–242.

Esquivel-M., A. & Peris, S.J. 2011a. Comunidades de aves del Bosque Atlántico del Paraguay. *Ardeola* 58: 436–437.

Esquivel-M., A. & Peris, S.J. 2011b. *Aves de San Rafael.* Universidad de Salamanca/Asociación ProCosara, Asunción.

Esquivel-M., A. & Peris, S.J. 2012. Estructura y organización de una comunidad de aves del Bosque Atlántico de San Rafael, Paraguay. *Orn. Neotrop.* 23: 569–584.

Esquivel-M., A., Velázquez, M.C., Bodrati, A., Fraga, R., del Castillo, H., Klavins, J., Clay, R.P., Madroño, A. & Peris, S.J. 2007. Status of the avifauna of San Rafael National Park, one of the last large fragments of Atlantic Forest in Paraguay. *Bird Conserv. Intern.* 17: 301–317.

Euler, C. 1900. Descripção de ninhos e ovos das aves do Brazil. *Rev. Mus. Paulista* 4: 9–148.

Faaborg, J. 1975. Patterns in the Structure of West Indian bird Communities. Ph.D. thesis. Princeton Univ., Princeton, NJ.

de Fabrègues, F.P. 1991. Die Einfarb-Ameisen Pitta *Grallaria rufula. Trochilus* 12: 67–69.

Fairchild, G.B. & Handley, C.O. 1966. Gazetteer of collecting localities in Panama. Pp. 9–20 *in* Wenzel, R.L. & Tipton, V.J. (eds.) *Ectoparasites of Panama.* Field Museum of Natural History, Chicago, IL.

Faria, C.M.A. & Rodrigues, M. 2005. Aves seguidoras de formigas-de-correicao em um fragmento de Mata Atlantica do Medio Rio Doce, leste de Minas Gerais. Pp. 183 *in Livro de Resumos do XIII Congresso Brasileira de Ornitologia.* Belém, Pará, Brasil.

Faria, C.M.A. & Rodrigues, M. 2009. Birds and army ants in a fragment of the Atlantic Forest of Brazil. *J. Field Orn.* 80: 328–335.

Faria, C.M.A., Rodrigues, M., Amaral, F.Q., Módena, É. & Fernandes, A.M. 2006. Aves de um fragmento de Mata Atlântica no alto Rio Doce, Minas Gerais: colonização e extinção. *Rev. Bras. Zool.* 23: 1217–1230.

Faria, I.P. & de Paula, W.S. 2008. Body masses of birds from Atlantic Forest region, southeastern Brazil. *Orn. Neotrop.* 19: 599–606.

Farias, G.B. 2009. Aves do Parque Nacional do Catimbau, Buíque, Pernambuco, Brasil. *Atualidades Orn.* 147: 36–39.

Fávaro, F.L. & Flores, J.M. 2009. Aves da Estação Ecológica Terra do Meio, Pará, Brasil: resultados preliminares. *Ornithologia* 3: 115–131.

Feduccia, J.A. & Olson, S.L. 1982. Morphological similarities between the Menurae and the Rhinocryptidae, relict passerine birds of the Southern Hemisphere. *Smithsonian Contrib. Zool.* 366: 1–22.

Felton, A., Hennessey, B.A., Felton, A.M. & Lindenmayer, D.B. 2007. Birds surveyed in the harvested and unharvested areas of a reduced-impact logged forestry concession, located in the lowland subtropical humid forests of the Department of Santa Cruz, Bolivia. *Check List* 3: 43–50.

Fernández-Badillo, A. 1997. *El Parque Nacional Henri Pittier. Tomo II: Los Vertebrados.* Trabajo de Ascenso, Facultad de Agronomía, Universidad Central de Venezuela, Maracay.

Ferraro, C. & Lentino, M. 1992. *Venezuela, Paraíso de Aves.* Gráficas Armitano, Caracas.

Ferraz, G., Nichols, J.D., Hines, J.E., Stouffer, P.C., Bierregaard, R.O. & Lovejoy, T.E. 2007. A large-scale deforestation experiment: effects of patch area and isolation on Amazon birds. *Science* 315: 238–241.

Ferreira, J.D., Costa, L.M. & Rodrigues, M. 2009. Aves de um remanescente florestal do Quadrilátero Ferrífero, Minas Gerais. *Biota Neotropical* 9: 39–54.

ffrench, R. 1973. *A Guide to the Birds of Trinidad and Tobago* (1st edn.). Livingston, Wynnewood, PA.

ffrench, R. 1977. Some interesting bird records from Trinidad and Tobago. *Living World, J. Trinidad & Tobago Field Nat. Club* 1977–1978: 9–10.

ffrench, R. 1985. Additional notes on the birds of Trinidad and Tobago. *Living World, J. Trinidad & Tobago Field Nat. Club* 1985–1986: 9–11.

ffrench, R. 1991. *Birds of Trinidad and Tobago* (2nd edn.). Cornell Univ. Press, Ithaca, NY.

ffrench, R. 2012. *Birds of Trinidad and Tobago* (3rd edn.). Cornell Univ. Press, Ithaca, NY.

Fierro, C.A. 1991. *Una Guía de las Aves para el Bosque Protector Pasochoa.* Fundación Natura, Quito.

Fierro-Calderón, K., Estela, F.A. & Chacón-Ulloa, P. 2006. Observaciones sobre las dietas de algunas aves de la Cordillera Oriental de Colombia a partir del análisis de contenidos estomacales. *Orn. Colombiana* 4: 6–15.

Finch, D.W., Hinds, W., Sanderson, J. & Missa, O. 2002. Avifauna of the eastern edge of the Eastern Kanuku Mountains, Lower Kwitaro River, Guyana. Pp. 43–46 *in* Montambault, J.R. & Missa, O. (eds.) *A Biodiversity Assessment of the Eastern Kanuku Mountains, Lower Kwitaro River, Guyana: RAP Bull. Biol. Assess.* 26. Conservation International, Washington DC.

Fjeldså, J. 1992. Biogeographic patterns and evolution of the avifauna of relict high-altitude woodlands of the Andes. *Steenstrupia* 18: 9–62.

Fjeldså, J. 1993. The avifauna of the *Polylepis* woodlands of the Andean highlands: the efficiency of basing conservation priorities on patterns of endemism. *Bird Conserv. Intern.* 3: 37–55.

Fjeldså, J. 1995. Geographical patterns of neoendemic and older relict species of Andean highlands: the significance of ecological stability areas. Pp. 89–102 *in* Churchill, S., Balslev, H., Forero, E. & Luteyn, J.L. (eds.) *Biodiversity and Conservation of Neotropical Montane Forests.* New York Botanical Garden, New York.

Fjeldså, J. 2002 Key areas for conserving the avifauna of *Polylepis* forests. *Ecotropica* 8 125–131.

Fjeldså, J. 2013. Urrao Antpitta (*Grallaria fenwickorum*). P. 213 in del Hoyo, J., Elliott, A., Sargatal, J. & Christie, D.A. (eds.) *Handbook of the Birds of the World. Special volume: new species and global index.* Lynx Edicions, Barcelona.

Fjeldså, J. 2017. Alta Floresta Antpitta (*Hylopezus whittakeri*). *In* del Hoyo, J., Elliott, A., Sargatal, J., Christie, D.A. & de Juana, E. (eds.) *Handbook of the Birds of the World Alive.* Lynx Edicions, Barcelona (retrieved from http://www.hbw.com/node/204348 on 3 June 2017).

Fjeldså, J. & Kessler, M. 1996. *Conserving the Biological Diversity of Polylepis Woodlands of the Highland of Peru and Bolivia.*

A Contribution to Sustainable Natural Resource Management in the Andes. NORDECO, Copenhagen.

Fjeldså, J. & Krabbe, N. 1986. Some range extensions and other unusual records of Andean birds. *Bull. Brit. Orn. Club* 106: 115–124.

Fjeldså, J. & Krabbe, N. 1989. An unpublished major collection of birds from the Bolivian highlands. *Zool. Scripta* 18: 321–329.

Fjeldså, J. & Krabbe, N. 1990. *The Birds of the High Andes.* Zool. Mus., Univ. of Copenhagen & Apollo Books, Svendborg.

Fjeldså, J. & Rahbek, C. 1998. Priorities for conservation in Bolivia, illustrated by a continent-wide analysis of bird distributions. Pp. 313–327 *in* Barthlott, W. & Winiger, M. (eds.) *Biodiversity—A Challenge for Development, Research and Policy.* Springer-Verlag, Berlin.

Fjeldså, J., Krabbe, N. & Parker, T.A. 1987 Rediscovery of *Cinclodes excelsior aricomae* and notes on the nominate race. *Bull. Brit. Orn. Club* 107: 112–114.

Flórez, P., Krabbe, N., Castaño, J., Suárez, G. & Arango, J.D. 2004. *Evaluación Avifauna del Páramo de Frontino, Antioquia, Agosto 2004. Colombian EBA Project Report Series No. 6.* Fundación ProAves, Bogotá.

Fontana, C.S., Burger, M.I. & Magnusson, W.E. 2011. Bird diversity in a subtropical South-American city: effects of noise levels, arborisation and human population density. *Urban Ecosyst.* 14: 341–360.

Forero-Medina, G., Terborgh, J., Socolar, S.J. & Pimm, S.L. 2011. Elevational ranges of birds on a tropical montane gradient lag behind warming temperatures. *PLoS ONE* 6: e28535.

Forrester, B.C. 1993. *Birding Brazil. A Check-list and Site Guide.* Published privately, Rankinston.

Fraga, R.M. & Clark, R. 1999. Notes on the avifauna of the upper Bermejo River (Argentina and Bolivia) with a new species for Argentina. *Cotinga* 12: 77–78.

Fraga, R.M. & Narosky, S. 1985. *Nidificación de las Aves Argentinas (Formicariidae a Cinclidae).* Asociación Ornitológica del Plata, Buenos Aires.

Franke, I., Hernandez, P.A., Herzog, S.K., Paniagua, L., Soto, A., Tovar, C., Valqui, T. & Young, B.E. 2007. Birds. Pp. 46–53 *in* Young, B.E. (ed.) *Endemic Species Distributions on the East Slope of the Andes in Peru and Bolivia.* NatureServe, Arlington, VA.

von Frantzius, A. 1869. Ueber die geographische Verbreitung der Vögel Costaricas und deren Lebensweise. *J. Orn.* 17: 289–318.

Franz, I., Barros, M.P., Cappelatti, L., Dala-CorteII, R.B. & Ott, P.H. 2014. Birds of two protected areas in the southern range of the Brazilian Araucaria forest. *Pap. Avuls. Dept. Zool., São Paulo* 54: 111–127.

Fraser, K.C. 2010. Seasonal Migration and Diet Use in a Neotropical Community of Birds and Bats. PhD thesis. Univ. of New Brunswick, New Brunswick.

Fraser, K.C., McKinnon, E.A., Diamond, A.W. & Chavarría, L. 2011. The influence of microhabitat, moisture and diet on stable-hydrogen isotope variation in a Neotropical avian food web. *J. Trop. Ecol.* 27: 563–572.

Freile, J.F. 2000. Patrones de Distribucion y sus Implicaciones en la Conservación de los Generos *Grallaria* y *Grallaricula* (Aves: Formicariidae) en el Ecuador. Tesis de Licenciatura. Pontificia Universidad Catolica del Ecuador, Quito.

Freile, J.F. 2002. The case of the antpittas of Ecuador. *Bird Conserv. Intern.* 19: 8.

Freile, J.F. & Chaves, J.A. 2004. Interesting distributional records and notes on the biology of bird species from a cloud forest reserve in north-west Ecuador. *Bull. Brit. Orn. Club* 124: 6–16.

Freile, J.F. & Renjifo, L.M. 2003. First nesting records of the Moustached Antpitta (*Grallaria alleni*). *Wilson Bull.* 115: 11–15.

Freile, J.F. & Rodas, F. 2008. Conservación de aves en Ecuador: ¿cómo estamos y qué necesitamos hacer? *Cotinga* 29: 48–55.

Freile, J.F. & Santander, T. 2005. *Áreas Importantes para la Conservación de las Aves en Ecuador.* Aves & Conservación, Quito.

Freile, J.F., Cisneros-Heredia, D.F., Santander, T., Boyla, K.A. & Diaz, D. 2010a. Important bird areas of the Neotropics: Ecuador. *Neotrop. Birding* 7: 4–14.

Freile, J.F., Parra, J.L. & Graham, C. 2010b. Distribution and conservation of *Grallaria* and *Grallaricula* antpittas (Grallariidae) in Ecuador. *Bird Conserv. Intern.* 20: 410–431.

Freile, J.F., Piedrahita, P., Buitrón-Jurado, G., Rodríguez, C.A. & Bonaccorso, E. 2013a. Aves de los tepuyes de la cuenca alta del Río Nangaritza, Cordillera del Cóndor. Pp. 63–75 *in* Guayasamin, J.M. & Bonaccorso, E. (eds.) *Evaluación Ecológica Rápida de la Biodiversidad de los Tepuyes de la Cuenca Alta del Río Nangaritza, Cordillera del Cóndor, Ecuador.* Conservation International, Washington DC.

Freile, J.F., Piedrahita, P., Buitrón-Jurado, G., Rodríguez, C.A. & Bonaccorso, E. 2013b. Lista de las aves de los tepuyes de la cuenca alta del Río Nangaritza, Cordillera del Cóndor. Pp. 119–123 *in* Guayasamin, J.M. & Bonaccorso, E. (eds.) *Evaluación Ecológica Rápida de los Tepuyes de la Cuenca Alta del Río Nangaritza, Cordillera del Cóndor, Ecuador.* Conservation International, Washington DC.

Freile, J.F., Krabbe, N., Piedrahita, P., Buitrón-Jurado, G., Rodríguez-Saltos, C.A., Ahlman, F., Brinkhuizen, D.M. & Bonaccorso, E. 2014. Birds, Nangaritza River Valley, Zamora Chinchipe Province, southeast Ecuador: Update and revision. *Check List* 10: 54–71.

Freile, J.F., Mouret, V. & Siol, M. 2015. Amidst a crowd of birds: birding Río Bigal, Ecuador. *Neotrop. Birding* 17: 47–55.

Freile, J.F., Brinkhuizen, D.M., Greenfield, P.J., Lysinger, M., Navarrete, L., Nilsson, J., Ridgely, R.S., Solano-Ugalde, A., Ahlman, R. & Boyla, K.A. 2017. Lista de las Aves del Ecuador / Checklist of the Birds of Ecuador. Comité Ecuatoriano de Registros Ornitológicos, Quito. https://ceroecuador.wordpress.com/.

Freitas, G.H.S. & Costa, L.M. 2007. Avifauna de "Cabeça de Boi", vertente leste da Serra do Cipo, Minas Gerais. P. 85 in Fontana, C.S. (ed.) *Livro de Resumos do XV Congresso Brasileiro de Ornitologia.* Porto Alegre.

Freitas, M.A. 2011. Avifauna do município de Mata de São João, Bahia, Brasil. *Atualidades Orn.* 163: 48–56.

Freitas, M.A., Silva, T.F.S. & Silva, C.S. 2007. Levantamento e monitoramento da avifauna da Fazenda Palmeiras, Itapebi, Bahia. *Atualidades Orn.* 138: 43–47.

Friedmann, H. 1947. Colombian birds collected by Brother Nicéforo. *Caldasia* 4: 471–494.

Friedmann, H. 1948. Birds collected by the National Geographic Society's Expeditions to northern Brazil and southern Venezuela. *Proc. US Natl. Mus.* 97: 373–569.

Frisch, J.D. & Frisch, S. 1964. *Aves Brasileiras.* Editora Dalgas-Ecoltec Ecologia Técnica e Comércio, São Paulo.

Frisch, J.D. & Frisch, S. 1981. *Aves Brasileiras*, vol. 1. Editora

Dalgas-Ecoltec Ecologia Técnica e Comércio, São Paulo.

Fundação Biodiversitas. 2008. *Lista de Espécies Ameaçadas de Extinção da Fauna do Estado de Minas Gerais.* Fundação Biodiversitas, Belo Horizonte.

Fundación ProAves. 2013. Comment on *Grallaria fenwickorum* Barrera & Bartels, 2010 (Aves, Grallariidae): proposed replacement of an indeterminate holotype by a neotype. *Bull. Zool. Nomencl.* 70: 256–269.

Fundación ProAves. 2014. El estado de las aves en Colombia 2014: prioridades de conservacion de la avifauna colombiana. *Conserv. Colombiana* 20: 4–42.

Galindo, P. & Sousa, O.E. 1966. Blood parasites of birds from Almirante, Panama, with ecological notes on the hosts. *Rev. Biol. Trop.* 14: 27–46.

Galvão, A. & Gonzaga, L.A.P. 2011. Morphological support for placement of the Wing-banded Antbird *Myrmornis torquata* in the Thamnophilidae (Passeriformes: Furnariides). *Zootaxa* 3122: 37–67.

Gamarra-Toledo, V., Ugarte-Lewis, M. & Flores, L.M. 2012. Notas sobre la comunidad de aves del bosque Puya Sacha, un bosque montano de la vertiente oriental de los Andes en Peru. Pp. 537–538 *in* More, A., Olaechea, D.G. & Olaechea, A.G. (eds.) *Libro de Resúmenes del IX Congreso de Ornitología Neotropical y el VIII Congreso Peruano de Ornitología.* Cuzco.

Garcés, P.A. 2007. Analisis de la avifauna reportada en la Reserva Forestal del Proyecto Hidroelectrico Fortuna, Provincia de Chiriqui. *Technociencia* 9: 133–150.

Gardner, T.A., Ferreira, J., Barlow, J., Lees, A.C., Parry, L., Vieira, I.C.G., Berenguer, E., Abramovay, R., Aleixo, A., Andretti, C., Aragão, L.E.O.C., Araújo, I., Avila, W.S., Bardgett, R.D., Batistella, M., Begotti, R.A., Beldini, T., Blas, D.E., Braga, R.F., Braga, D.L., Brito, J.G., Camargo, P.B., Campos dos Santos, F., Oliveira, V.C., Cordeiro, A.C.N., Cardoso, T.M., Carvalho, D.R., Castelani, S.A., Chaul, J.C.M., Cerri, C.E., Costa, F.A., Costa, C.D.F., Coudel, E., Coutinho, A.C., Cunha, D., D'Antona, Á., Dezincourt, J., Dias-Silva, K., Durigan, M., Esquerdo, J.C.D.M., Feres, J., Ferraz, S.F.B., Ferreira, A.E.M., Fiorini, A.C., Silva, L.V.F., Frazão, F.S., Garrett, R., Gomes, A.S., Gonçalves, K.S., Guerrero, J.B., Hamada, N., Hughes, R.M., Igliori, D.C., Jesus, E.C., Juen, L., Junior, M., Junior, J.M.B.O., Junior, R.C.O., Junior, C.S., Kaufmann, P., Korasaki, V., Leal, C.G., Leitão, R., Lima, N., Almeida, M.F.L., Lourival, R., Louzada, J., Nally, R.M., Marchand, S., Maués, M.M., Moreira, F.M.S., Morsello, C., Moura, N., Nessimian, J., Nunes, S., Oliveira, V.H.F., Pardini, R., Pereira, H.C., Pompeu, P.S., Ribas, C.R., Rossetti, F., Schmidt, F.A., Silva, R., Silva, R.C.V.M., Silva, T.F.M.R., Silveira, J., Siqueira, J.V., Carvalho, T.S., Solar, R.R.C., Tancredi, N.S.H., Thomson, J.R., Torres, P.C., Vaz-de-Mello, F.Z., Veiga, R.C.S., Venturieri, A., Viana, C., Weinhold, D., Zanetti, R. & Zuanon, J. 2013. A social and ecological assessment of tropical land uses at multiple scales: the Sustainable Amazon Network. *Phil. Trans. Roy. Soc. B: Biol. Sci.* 368: 20120166.

Garrigues, R. & Dean, R. 2007. *The Birds of Costa Rica.* Cornell Univ. Press, Ithaca, NY.

Garrod, A.H. 1877a. Notes on the anatomy of passerine birds. Part II. *Proc. Zool. Soc. Lond.* 1877: 447–452.

Garrod, A.H. 1877b. Notes on the anatomy of passerine birds. Part III. *Proc. Zool. Soc. Lond.* 1877: 523–526.

Gartner, J.E.O., Figueroa, G.C. & Alberico, M.S. 1982. Estudio de dos comunidades de aves y mamiferos en Nariño, Colombia. *Cespedesia* 41–42: 41–67.

Geffen, E. & Yom-Tov, Y. 2000. Are incubation and fledging periods longer in the tropics? *J. Anim. Ecol.* 69: 59–73.

Gemuseus, R.C. & Sagot, F. 1996 *A Guide to the World's Best Bird-watching Place: Armonia Santa Cruz, Bolivia.* Armonia/BirdLife International, Santa Cruz de la Sierra.

Gertler, P.E. 1977. Hooded Antpitta (*Grallaricula cucullata*) in eastern Andes of Colombia. *Condor* 79: 389.

Ghizoni, I.R., Farias, F.B., Vieira, B.P., Willrich, G., Silva, E.S., Mendonça, E.N., Albuquerque, J.L.B., Gass, D.A., Ternes, M.H., Nascimento, C.E., Roos, A.L., Couto, C.C.M., Serrão, M., Serafini, P.P., Dias, D., Fantacini, F.M., Santi, S., Souza, M.C.R., Silva, M.S., Barcellos, A., Albuquerque, C. & Espínola, C.R.R. 2013. Checklist da avifauna da Ilha de Santa Catarina, sul do Brasil. *Atualidades Orn.* 171: 50–75.

Gill, F.B. & Wright, M.T. 2006. *Birds of the World: Recommended English Names.* Princeton Univ. Press, Princeton, NJ.

Gillespie, T.W. 2001. Application of extinction and conservation theories for forest birds in Nicaragua. *Conserv. Biol.* 15: 699–709.

Gilliard, E.T. 1939. A new race of *Grallaria excelsa* from Venezuela. *Amer. Mus. Novit.* 1016: 1–3.

Gilliard, E.T. 1941. The birds of Mt. Auyan-tepui, Venezuela. *Bull. Amer. Mus. Nat. Hist.* 77: 439–508.

Giner, S. & Bosque, C. 1998. Distribución altitudinal de las subfamilias Grallarinae, Formicariinae y Thamnophilinae (Aves, Formicariidae) en Venezuela. *Bol. Soc. Biol. Concepción, Chile* 69: 115–121.

Girão, W. & Albano, C. 2008. *Conopophaga lineata cearae* (Cory, 1916). Pp. 505–506 *in* Machado, A.B.M., Drummond, G.M. & Paglia, A.P. (eds.) *Livro Vermelho da Fauna Brasileira Ameaçada de Extinção.* Ministério do Meio Ambiente & Fundação Biodiversitas, Brasília & Belo Horizonte.

Girão, W., Albano, C., Pinto, T. & Silveira, L.F. 2007. Avifauna da Serra de Baturité: dos naturalistas à atualidade. Pp. 187–224 *in* Oliveira, T.S. & Araújo, F.S. (eds.) *Biodiversidade e Conservação da Biota na serra de Baturité, Ceará.* Edições UFC, Coelce, Fortaleza.

Giraudo, A.R., Baldo, J.L. & Abramson, R.R. 1993. Aves observadas en el sudoeste, centro y este de Misiones (Republica Argentina), con la mención de especies nuevas o poco conocidas para la provincia. *Nótulas Faunísticas (Primera Ser.)* 49: 1–13.

Giraudo, A.R., Matteucci, S., Alonso, J., Herrera, J. & Abramson, R. 2008. Comparing bird assemblages in large and small fragments of the Atlantic Forest hotspots. *Biodiver. & Conserv.* 17: 1251–1265.

Gloger, C.L. 1833. *Das Abändern der Vögel durch Einfluss des Klima's.* August Schulz & Co., Breslau.

Glowska, E. & Schmidt, B.K. 2014. New taxa of the subfamily Picobiinae (Cheyletoidea: Syringophilidae) parasitizing antbirds and gnateaters (Passeriformes: Thamnophilidae, Conopophagidae) in Guyana. *Zootaxa* 3861: 193–200.

Gmelin, J.-F. 1789. *Systema Naturae.* 13th edn. Tome 1, pars II. Impensis Georg. Emanuel Beer, Leipzig.

Goeldi, E.A. 1901–06. *Album de Aves Amazonicas.* Museu Goeldi (Museu Paraense) de Historia Natural e Ethnographia, Rio de Janeiro.

Goerck, J.M. 1999. Distribution of birds along an elevational gradient in the Atlantic forest of Brazil: implications for the conservation of endemic and endangered species. *Bird Conserv. Intern.* 9: 235–253.

Gonzaga, L.A.P., Pacheco, J.F., Bauer, C. & Castiglioni, G.D.A. 1995. An avifaunal survey of the vanishing montane Atlanta forest of southern Bahia, Brazil. *Bird Conserv. Intern.* 5: 279–290.

Gonzaga, L.A.P., Scott, D.A. & Collar, N.J. 1987. The status and birds of some forest fragments in eastern Brazil: report on a survey supported by CVRD, October 1986. Companhia Vale do Rio Doce (CVRD/VALE). Unpubl. report.

González, J., Proctor, G.R. & Bruno, E. 2011. The nomenclatural availability of and priority between two recently described names for the same new antpitta species from Colombia. *Conserv. Colombiana* 15: 45–54.

González-Acuña, D., Venzal, J.M., Keirans, J.E., Robbins, R.G., Ippi, S. & Guglielmone, A.A. 2005. New host and locality records for the *Ixodes auritulus* (Acari: Ixodidae) species group, with a review of host relationships and distribution in the Neotropical zoogeographic region. *Exp. Appl. Acarol.* 37: 147–156.

Goodfellow, W. 1901. Results of an ornithological journey through Colombia and Ecuador - Part III. *Ibis* 43: 699–715.

Goodfellow, W. 1902. Results of an ornithological journey through Colombia and Ecuador. *Ibis* 44: 59–67.

Graham, C.H., Silva, N. & Velasquez-Tibata, J. 2010. Evaluating the potential causes of range limits of birds of the Colombian Andes. *J. Biogeogr.* 37: 1863–1875.

Granizo, T. 2009. *Etimología de los Nombres Científicos de las Aves del Ecuador.* Simbioe, Quito.

Granizo, T., Guerrero, M., Pacheco, C., Phillips, R., Ribadeneira, M.B. & Suárez, L. 1997. *Lista de Aves Amenazadas de Extinción en el Ecuador.* UICN-Sur, CECIA, INEFAN, EcoCiencia, & BirdLife International, Quito.

Granizo, T., Pacheco, C., Ribadeneira, M.B., Guerrero, M. & Suárez, L. 2002. *Libro Rojo de las Aves de Ecuador.* SIMBIOE, Conservation International, EcoCiencia, Ministerio del Ambiente & IUCN, Quito.

Graves, G.R. 1985. Elevational correlates of speciation and intraspecific geographic variation in plumage in Andean forest birds. *Auk* 102: 556–579.

Graves, G.R. 1987. A cryptic new species of antpitta (Formicariidae, *Grallaria*) from the Peruvian Andes. *Wilson Bull.* 99: 313–321.

Graves, G.R. & Zusi, R.L. 1990. Avian body weights from the lower Rio Xingu, Brazil. *Bull. Brit. Orn. Club* 110: 20–25.

Graves, G.R., O'Neill, J.P. & Parker, T.A. 1983. *Grallaricula ochraceifrons*, a new species of antpitta from northern Peru. *Wilson Bull.* 95: 1–6.

Gray, G.R. 1840. *A List of the Genera of Birds, With an Indication of the Typical Species of Each Genus.* R. & J.E. Taylor, London.

Gray, G.R. 1869. *Hand-list of the Genera and Species of Birds, Distinguishing those Contained in the British Museum.* Trustees of the British Museum, London.

Greeney, H.F. 2004. Breeding behavior of the Bicolored Antvireo (*Dysithamnus occidentalis*). *Orn. Neotrop.* 15: 349–356.

Greeney, H.F. 2006. Incubation behavior of the Peruvian Antpitta (*Grallaricula peruviana*). *Orn. Neotrop.* 17: 461–466.

Greeney, H.F. 2009. Peruvian Antpitta (*Grallaricula peruviana*). *In* Schulenberg, T.S. (ed.) *Neotropical Birds Online.* Cornell Lab of Ornithology, Ithaca, NY (https://doi.org/10.2173/nb.perant1.01).

Greeney, H.F. 2010. Bimodal breeding seasonality of an understory bird, *Premnoplex brunnescens*, in an Ecuadorian cloud forest. *J. Trop. Ecol.* 26: 547–549.

Greeney, H.F. 2012a. The natal plumages of antpittas (Grallariidae). *Orn. Colombiana* 12: 65–68.

Greeney, H.F. 2012b. Antpittas and worm-feeders: a match made by evolution? Evidence for possible commensal foraging relationships between antpittas (Grallariidae) and mammals. *Neotrop. Biol. & Conserv.* 7: 140–143.

Greeney, H.F. 2012c. Crescent-faced Antpitta (*Grallaricula lineifrons*). *In* Schulenberg, T.S. (ed.) *Neotropical Birds Online.* Cornell Lab of Ornithology, Ithaca, NY (https://doi.org/10.2173/nb.crfant1.01).

Greeney, H.F. 2012d. The nest, egg, and nestling of Stripe-headed Antpitta (*Grallaria andicolus*) in southern Peru. *Orn. Neotrop.* 23: 367–374.

Greeney, H.F. 2013a. Rufous-faced Antpitta (*Grallaria erythrotis*). *In* Schulenberg, T.S. (ed.) *Neotropical Birds Online.* Cornell Lab of Ornithology, Ithaca, NY (https://doi.org/10.2173/nb.rufant2.01).

Greeney, H.F. 2013b. Ochre-striped Antpitta (*Grallaria dignissima*). *In* Schulenberg, T.S. (ed.) *Neotropical Birds Online.* Cornell Lab of Ornithology, Ithaca, NY (https://doi.org/10.2173/nb.ocsant1.01).

Greeney, H.F. 2013c. Black-bellied Gnateater (*Conopophaga melanogaster*). *In* Schulenberg, T.S. (ed.) *Neotropical Birds Online.* Cornell Lab of Ornithology, Ithaca, NY (https://doi.org/10.2173/nb.blbgna1.01).

Greeney, H.F. 2013d. Elusive Antpitta (*Grallaria eludens*). *In* Schulenberg, T.S. (ed.) *Neotropical Birds Online.* Cornell Lab of Ornithology, Ithaca, NY (https://doi.org/10.2173/nb.eluant1.01).

Greeney, H.F. 2013e. Rusty-breasted Antpitta (*Grallaricula ferrugineipectus*). *In* Schulenberg, T.S. (ed.) *Neotropical Birds Online.* Cornell Lab of Ornithology, Ithaca, NY (https://doi.org/10.2173/nb.rubant5.01).

Greeney, H.F. 2013f. Jocotoco Antpitta (*Grallaria ridgelyi*). *In* Schulenberg, T.S. (ed.) *Neotropical Birds Online.* Cornell Lab of Ornithology, Ithaca, NY (https://doi.org/10.2173/nb.jocant1.01).

Greeney, H.F. 2013g. Hooded Gnateater (*Conopophaga roberti*). *In* Schulenberg, T.S. (ed.) *Neotropical Birds Online.* Cornell Lab of Ornithology, Ithaca, NY (https://doi.org/10.2173/nb.hoogna1.01).

Greeney, H.F. 2013h. Hooded Antpitta (*Grallaricula cucullata*). *In* Schulenberg, T.S. (ed.) *Neotropical Birds Online.* Cornell Lab of Ornithology, Ithaca, NY (https://doi.org/10.2173/nb.hooant1.01).

Greeney, H.F. 2013i. Sucre Antpitta (*Grallaricula cumanensis*). *In* Schulenberg, T.S. (ed.) *Neotropical Birds Online.* Cornell Lab of Ornithology, Ithaca, NY (https://doi.org/10.2173/nb.slcant5.01).

Greeney, H.F. 2013j. Rufous-crowned Antpitta (*Pittasoma rufopileatum*). *In* Schulenberg, T.S. (ed.) *Neotropical Birds Online.* Cornell Lab of Ornithology, Ithaca, NY (https://doi.org/10.2173/nb.rucant3.01).

Greeney, H.F. 2013k. Slate-crowned Antpitta (*Grallaricula nana*). *In* Schulenberg, T.S. (ed.) *Neotropical Birds Online.* Cornell Lab of Ornithology, Ithaca, NY (https://doi.org/10.2173/nb.slcant2.01).

Greeney, H.F. 2013l. Brown-banded Antpitta (*Grallaria milleri*). *In* Schulenberg, T.S. (ed.) *Neotropical Birds Online.* Cornell Lab of Ornithology, Ithaca, NY (https://doi.org/10.2173/nb.brbant2171.2101).

Greeney, H.F. 2014a. Thicket Antpitta (*Hylopezus dives*). *In* Schulenberg, T.S. (ed.) *Neotropical Birds Online.*

Cornell Lab of Ornithology, Ithaca, NY (https://doi.org/10.2173/nb.thiant1.01).

Greeney, H.F. 2014b. Chestnut-belted Gnateater (*Conopophaga aurita*). *In* Schulenberg, T.S. (ed.) *Neotropical Birds Online*. Cornell Lab of Ornithology, Ithaca, NY (*https://doi.org/10.2173/nb.chbgna1.01*).

Greeney, H.F. 2014c. Masked Antpitta (*Hylopezus auricularis*). *In* Schulenberg, T.S. (ed.) *Neotropical Birds Online*. Cornell Lab of Ornithology, Ithaca, NY (https://doi.org/10.2173/nb.masant1.01).

Greeney, H.F. 2014d. Cundinamarca Antpitta (*Grallaria kaestneri*). *In* Schulenberg, T.S. (ed.) *Neotropical Birds Online*. Cornell Lab of Ornithology, Ithaca, NY (https://doi.org/10.2173/nb.cunant1.01).

Greeney, H.F. 2014e. Urrao Antpitta (*Grallaria urraoensis*). *In* Schulenberg, T.S. (ed.) *Neotropical Birds Online*. Cornell Lab of Ornithology, Ithaca, NY (https://doi.org/10.2173/nb.antant1.01).

Greeney, H.F. 2014f. Gray-naped Antpitta (*Grallaria griseonucha*). *In* Schulenberg, T.S. (ed/) *Neotropical Birds Online*. Cornell Lab of Ornithology, Ithaca, NY (https://doi.org/10.2173/nb.gynant1.01).

Greeney, H.F. 2014g. White-browed Antpitta (*Hylopezus ochroleucus*). *In* Schulenberg, T.S. (ed.) *Neotropical Birds Online*. Cornell Lab of Ornithology, Ithaca, NY (https://doi.org/10.2173/nb.whbant7.01).

Greeney, H.F. 2014h. Bicolored Antpitta (*Grallaria rufocinerea*). *In* Schulenberg, T.S. (ed.) *Neotropical Birds Online*. Cornell Lab of Ornithology, Ithaca, NY (https://doi.org/10.2173/nb.bicant3.01).

Greeney, H.F. 2014i. Yellow-breasted Antpitta (*Grallaria flavotincta*). *In* Schulenberg, T.S. (ed.) *Neotropical Birds Online*. Cornell Lab of Ornithology, Ithaca, NY (https://doi.org/10.2173/nb.yebant1.01).

Greeney, H.F. 2015a. Variegated Antpitta (*Grallaria varia*). *In* Schulenberg, T.S. (ed.) *Neotropical Birds Online*. Cornell Lab of Ornithology, Ithaca, NY (https://doi.org/10.2173/nb.varant2.01).

Greeney, H.F. 2015b. Giant Antpitta (*Grallaria gigantea*). *In* Schulenberg, T.S. (ed.) *Neotropical Birds Online*. Cornell Lab of Ornithology, Ithaca, NY (https://doi.org/10.2173/nb.giaant1.01).

Greeney, H.F. 2015c. Santa Marta Antpitta (*Grallaria bangsi*). *In* Schulenberg, T.S. (ed.) *Neotropical Birds Online*. Cornell Lab of Ornithology, Ithaca, NY (https://doi.org/10.2173/nb.samant1.01).

Greeney, H.F. 2015d. Great Antpitta (*Grallaria excelsa*). *In* Schulenberg, T.S. (ed.) *Neotropical Birds Online*. Cornell Lab of Ornithology, Ithaca, NY (https://doi.org/10.2173/nb.great2.01).

Greeney, H.F. 2015e. Tawny Antpitta (*Grallaria quitensis*). *In* Schulenberg, T.S. (ed.) *Neotropical Birds Online*. Cornell Lab of Ornithology, Ithaca, NY (https://doi.org/10.2173/nb.tawant1.01).

Greeney, H.F. 2017a. Observations on the reproductive biology of Thrush-like Antpitta *Myrmothera campanisona*. *Cotinga* 39: 84–87.

Greeney, H.F. 2017b. Thrush-like Antpitta (*Myrmothera campanisona*). *In* Schulenberg, T.S. (ed.) *Neotropical Birds Online*. Cornell Lab of Ornithology, Ithaca, NY (https://doi.org/10.2173/nb.thlant1.01).

Greeney, H.F. & Gelis, R.A. 2005a. Juvenile plumage and vocalization of the Jocotoco Antpitta *Grallaria ridgelyi*. *Cotinga* 23: 79–81.

Greeney, H.F. & Gelis, R.A. 2005b. A nest of Rufous Antpitta

Grallaria rufula depredated by a Turquoise Jay *Cyanolyca turcosa*. *Cotinga* 24: 110–111.

Greeney, H.F. & Gelis, R.A. 2006. Observations on parental care of the Moustached Antpitta (*Grallaria alleni*) in northwestern Ecuador. *Orn. Neotrop.* 17: 313–316.

Greeney, H.F. & Harms, I. 2008. Behavior of the Tawny Antpitta (*Grallaria quitensis*) in northern Ecuador. *Orn. Neotrop.* 19: 143–147.

Greeney, H.F. & Jipa, M. 2012. The nest of Crescent-faced Antpitta *Grallaricula lineifrons* in north-east Ecuador. *Bull. Brit. Orn. Club* 132: 217–220.

Greeney, H.F. & Juiña-J., M.E. 2010. First description of the nest of Jocotoco Antpitta (*Grallaria ridgelyi*). *Wilson J. Orn.* 122: 392–395.

Greeney, H.F. & Juiña-J., M.E. 2011. First description of the nest of Undulated Antpitta *Grallaria squamigera*, from south-west Ecuador. *Bull. Brit. Orn. Club* 131: 67–69.

Greeney, H.F. & Kirwan, G.M. 2012. Rusty-tinged Antpitta (*Grallaria przewalskii*). *In* Schulenberg, T.S. (ed.) *Neotropical Birds Online*. Cornell Lab of Ornithology, Ithaca, NY (https://doi.org/10.2173/nb.rutant5.01).

Greeney, H.F. & Kirwan, G.M. 2013. Chestnut-crowned Antpitta (*Grallaria ruficapilla*). *In* Schulenberg, T.S. (ed.) *Neotropical Birds Online*. Cornell Lab of Ornithology, Ithaca, NY (https://doi.org/10.2173/nb.chcant2.01).

Greeney, H.F. & Kirwan, G.M. 2014. Tachira Antpitta (*Grallaria chthonia*). *In* Schulenberg, T.S. (ed.) *Neotropical Birds Online*. Cornell Lab of Ornithology, Ithaca, NY (https://doi.org/10.2173/nb.tacant1.01).

Greeney, H.F. & Martin, P.R. 2005. High in the Ecuadorian Andes: the nest and eggs of the Tawny Antpitta (*Grallaria quitensis*). *Orn. Neotrop.* 16: 567–571.

Greeney, H.F. & Miller, E.T. 2008. The nestling and parental care of the Slate-crowned Antpitta (*Grallaricula nana*) in northeastern Ecuador. *Orn. Neotrop.* 19: 457–461.

Greeney, H.F. & Nunnery, T. 2006. Notes on the breeding of north-west Ecuadorian birds. *Bull. Brit. Orn. Club* 126: 38–45.

Greeney, H.F. & Sornoza, F. 2005. The nest and egg of the Slate-crowned Antpitta (*Grallaricula nana*), with observations on incubation behavior in southern Ecuador. *Orn. Neotrop.* 16: 137–140.

Greeney, H.F. & Valencia-Herverth, R. 2016. Two 'abnormal' clutches of Scaled Antpitta *Grallaria guatimalensis*. *Cotinga* 38: 98–101.

Greeney, H.F. & Vargas-Jiménez, J.D. 2017. First description of the nest and nestlings of Thicket Antpitta *Hylopezus dives*. *Orn. Neotrop.* 28: 181–185.

Greeney, H.F. & Zyskowski, K. 2008. A novel nest architecture within the Furnariidae: first nests of the White-browed Spinetail. *Condor* 110: 584–588.

Greeney, H.F., Gelis, R.A., Hannelly, E.C. & DeVries, P.J. 2004a. The egg and incubation period of the Peruvian Antpitta (*Grallaricula peruviana*). *Orn. Neotrop.* 15: 403–406.

Greeney, H.F., Hannelly, E.C. & Lysinger, M. 2004b. First description of the nest and vocalisations of the Peruvian Antpitta *Grallaricula peruviana* with a northward range extension. *Cotinga* 21: 14–17.

Greeney, H.F., Dobbs, R.C., Martin, P.R., Haupkla, K. & Gelis, R.A. 2005a. Nesting and foraging ecology of the Rufous-crowned Tody-Flycatcher (*Poecilotriccus ruficeps*) in eastern Ecuador. *Orn. Neotrop.* 16: 427–432.

Greeney, H.F., Hualinga-L., J.B.L., Branstein, J.F.V. & Bixenmann, R. 2005b. Observations at a nest of Thrush-like Antpitta *Myrmothera campanisona* in eastern Ecuador. *Cotinga* 23: 60–62.

Greeney, H.F., Gelis, R.A., Dingle, C., Vaca B., F.J., Krabbe, N. & Tidwell, M. 2006. The nest and eggs of the Plain-backed Antpitta (*Grallaria haplonota*) from Eastern Ecuador. *Orn. Neotrop.* 17: 601–604.

Greeney, H.F., Dobbs, R.C., Martin, P.R. & Gelis, R.A. 2008a. The breeding biology of *Grallaria* and *Grallaricula* antpittas. *J. Field Orn.* 79: 113–129.

Greeney, H.F., Jaffe, D.F. & Manzaba B., O.G. 2008b. Incubation behavior of the Yellow-billed Cacique (*Amblycercus holosericeus*) in eastern Ecuador. *Orn. Colombiana* 7: 83–87.

Greeney, H.F., Juiña-J., M.E., Lliquin, S.M.I. & Lyons, J.A. 2009a. First nest description of the Yellow-breasted Antpitta *Grallaria flavotincta* in north-west Ecuador. *Bull. Brit. Orn. Club* 129: 256–258.

Greeney, H.F., Miller, E.T. & Gelis, R.A. 2009b. Observations on parental care and fledging of Watkins's Antpitta (*Grallaria watkinsi*). *Orn. Neotrop.* 20: 619–622.

Greeney, H.F., Juiña-J., M.E., Harris, J.B.C., Wickens, M.T., Winger, B., Gelis, R.A., Miller, E.T. & Solano-Ugalde, A. 2010. Observations on the breeding biology of birds in south-east Ecuador. *Bull. Brit. Orn. Club* 130: 61–68.

Greeney, H.F., Jipa, M., Gibson, W.S., Gordon, P.A., Suson, B., Miller, E.T., Gordon, C.E. & Gelis, R.A. 2011a. The nest and eggs of Black-capped Hemispingus (*Hemispingus atropileus*) in eastern Ecuador. *Kempffiana* 7: 34–38.

Greeney, H.F., Martin, P.R., Gelis, R.A., Solano-Ugalde, A., Bonier, F., Freeman, B.G. & Miller, E.T. 2011b. Notes on the breeding of high-Andean birds in northern Ecuador. *Bull. Brit. Orn. Club* 131: 24–31.

Greeney, H.F., Solano-Ugalde, A., Juiña-J., M.E. & Gelis, R.A. 2012. Observations on the breeding of Ochre-breasted Antpitta (*Grallaricula flavirostris*) in Ecuador. *Orn. Colombiana* 12: 4–9.

Greeney, H.F., Rivera-Ortíz, A., Rodríguez-Flores, C., Soberanes-González, C. & Arizmendi, M.C. 2013. Scaled Antpitta (*Grallaria guatimalensis*). *In* Schulenberg, T.S. (ed.) *Neotropical Birds Online*. Cornell Lab of Ornithology, Ithaca, NY (https://doi.org/10.2173/nb.scaant1.01).

Greeney, H.F., Lima, J.L.G. & Silva, T.T. 2016. First description of the nest of White-browed Antpitta *Hylopezus ochroleucus*. *Rev. Bras. Orn.* 24: 213–216.

Greeney, H.F., Gualingua, D., Read, M., Medina, D., Puertas, C., Evans, L., Baihua, O. & Killackey, R. P. 2018. Rapid inventory, preliminary annotated checklist, and breeding records of the birds (Aves) of the Boanamo indigenous community, Orellana Province, Ecuador. *Neotrop. Biodiv.* 4.

Greeney, H. F., Angulo P., F., Dobbs, R. C., Crespo, S., Miller, E. T., Caceres, D., Gelis, R. A., Angulo, B. & Salagaje, L. in review. Notes on the breeding biology of the Tumbesian avifana in southwest Ecuador and northwest Peru. Rev. Ecuatoriana Orn.

Griffith, E., Pidgeon, E. & Gray, J.E. 1829. *The Animal Kingdom Arranged, in Conformity with its Organization, by the Baron Cuvier. Volume 6, the Class Aves*, vol. 1. G.B. Whittaker, London.

Griscom, L. 1928. New birds from Mexico and Panama. *Amer. Mus. Novit.* 293: 1–6.

Griscom, L. 1929. A collection of birds from Cana, Darien. *Bull. Mus. Comp. Zool.* 69: 147–190.

Griscom, L. 1932a. The ornithology of the Caribbean coast of extreme eastern Panama. *Bull. Mus. Comp. Zool.* 72: 303–372.

Griscom, L. 1932b. The distribution of bird-life in Guatemala. *Bull. Amer. Mus. Nat. Hist.* 64: 1–439.

Griscom, L. 1933. Notes on the Havemeyer collection of Central American birds. *Auk* 50: 297–308.

Griscom, L. 1934. The ornithology of Guerrero, Mexico. *Bull. Mus. Comp. Zool.* 75: 367–422.

Griscom, L. 1935. The ornithology of the Republic of Panama. *Bull. Mus. Comp. Zool.* 78: 261–382.

Griscom, L. 1937. A collection of birds from Omilteme, Guerrero. *Auk* 54: 192–199.

Griscom, L. 1950. Distribution and origin of the birds of Mexico. *Bull. Mus. Comp. Zool.* 103: 341–382.

Griscom, L. 1957. Family Formicariidae, antbirds. Pp. 55–58 *in* Miller, A.H., Friedmann, H., Griscom, L. & Moore, R.T. (eds.) *A Distributional Checklist of the Birds of Mexico. Part II*. Cooper Ornithological Society, Berkeley, CA.

Griscom, L. & Greenway, J.C. 1941. Birds of lower Amazonia. *Bull. Mus. Comp. Zool.* 88: 83–344.

Groombridge, B. 1993. *1994 IUCN Red List of Threatened Animals*. IUCN, Gland.

Gruson, E.S. & Forster, R.A. 1976. *Checklist of the World's Birds: A Complete List of the Species, with Names, Authorities, and Areas of Distribution*. Quadrangle/New York Times Book Co., New York, NY.

Guerrero, F.T. 1996. *Aves del Bosque de Mazán*. Empresa Pública Municipal de Teléfonos, Cuenca.

Guevara, E.A., Santander, T., Guevara, J.E., Gualotuna, R. & Ortiz, V. 2010. Birds, lower Sangay National Park, Morona-Santiago, Ecuador. *Check List* 6: 319–325.

Guilherme, E. 2009. Avifauna do Estado do Acre: Composição, Distribuição Geográfica e Conservação. Ph.D. thesis. Univ. Federal do Pará/Museu Paraense Emílio Goeldi, Belém.

Guilherme, E. 2012. Birds of the Brazilian state of Acre: diversity, zoogeography, and conservation. *Rev. Bras. Orn.* 20: 393–442.

Guilherme, E. 2016. *Aves do Acre*. Editora da Univ. Federal do Acre (Edufac), Rio Branco.

Guilherme, E. & Borges, S.H. 2011. Ornithological records from a campina/campinarana enclave on the upper Juruá River, Acre, Brazil. *Wilson J. Orn.* 123: 24–32.

Guilherme, E. & Dantas, S.M. 2011. Avifauna of the upper Purus River, state of Acre, Brazil. *Rev. Bras. Orn.* 19: 185–199.

Guilherme, E. & Santos, M.P.D. 2009. Birds associated with bamboo forests in eastern Acre, Brazil. *Bull. Brit. Orn. Club* 129: 229–240.

Gussoni, C.O.A. 2007. Avifauna de cinco localidades no município de Rio Claro, Estado de São Paulo, Brasil. *Atualidades Orn.* 136: 1–7.

Guyra Paraguay. 2004. *Lista Comentada de las Aves de Paraguay*. Guyra Paraguay, Asunción.

Gwynne, J.A., Ridgely, R.S., Tudor, G. & Argel, M. 2010. *Birds of Brazil, Vol. 1: The Pantanal & Cerrado of Central Brazil*. Cornell Univ. Press, Ithaca, NY.

Gyldenstolpe, N. 1941. Preliminary diagnoses of some new birds from Bolivia. *Ark. Zool.* 33B: 1–10.

Gyldenstolpe, N. 1945a. The bird fauna of the rio Juruá in western Brazil. *K. Sven. Vetensk. Akad. Zool.* 22: 1–338.

Gyldenstolpe, N. 1945b. A contribution to the ornithology of northern Bolivia. *K. Sven. Vetensk. Akad. Zool.* 23: 1–300.

Gyldenstolpe, N. 1951. The ornithology of the rio Purus region in western Brazil. *Ark. Zool.* (2)2: 1–320.

Haffer, J. 1959. Notas sobre las aves de la region de Uraba. *Lozania* 12: 1–49.

Haffer, J. 1967. On birds from the northern Chocó region, NW-Colombia. *Veröffentlichungen derZoologischen Staatssammlung München* 11: 123–149.

Haffer, J. 1974. Avian speciation in tropical South America, with a systematic survey of the toucans (Ramphastidae) and jacamars (Galbulidae). *Publ. Nuttall Orn. Club* 14.

Haffer, J. 1975. Avifauna of northwestern Colombia, South America. *Bonn Zool. Monogr.* 7.

Haftorn, S. 1994. The act of tremble-thrusting in tit nests, performance and possible functions. *Fauna Norvegica Ser. C. Cinclus* 17: 55–74.

Hagmann, G. 1904. As aves brasilicas mencionadas e descriptas nas obras de Spix (1825), de Wied (1830-1833), Burmeister (1854) e Pelzeln (1874) na sua nomenclatura scientífica actual. *Bol. Mus. Para. E. Goeldi, Sér. Hist. Nat. & Etnogr.* 4: 198–308.

Harris, J.B.C. & Haskell, D.G. 2013. Simulated birdwatchers' playback affects the behavior of two tropical birds. *PLoS ONE* 8: e77902.

Hartert, E. 1898. On a collection of birds from northwestern Ecuador, collected by W. F. H. Rosenberg. *Novit. Zool.* 5: 477–505.

Hartert, E. 1900. [Exhibition of new South-American birds]. *Bull. Brit. Orn. Club* 11: 37–40.

Hartert, E. 1902. Some further notes on the birds of north-west Ecuador. *Novit. Zool.* 9: 599–617.

Hartert, E. 1922. Types of birds in the Tring Museum. B. Types in the general collection. *Novit. Zool.* 29: 365–412.

Harvey, C.A. & Villalobos, J.A.G. 2007. Agroforestry systems conserve species-rich but modified assemblages of tropical birds and bats. *Biodiver. & Conserv.* 16: 2257–2292.

Harvey, M.G., Winger, B.M., Seeholzer, G.F. & Cáceres A, D. 2011. Avifauna of the Gran Pajonal and southern Cerros del Sira, Peru. *Wilson J. Orn.* 123: 289–315.

Haugaasen, T., Barlow, J. & Peres, C.A. 2003. Effects of surface fires on understorey insectivorous birds and terrestrial arthropods in central Brazilian Amazonia. *Anim. Conserv.* 6: 299–306.

Haverschmidt, F. 1955. *List of the birds of Surinam.* Publications Foundation for Scientific Research Surinam, Netherlands Antilles,

Haverschmidt, F. 1968. *Birds of Surinam.* Oliver & Boyd, London.

Haverschmidt, F. & Mees, G.F. 1994. *Birds of Suriname.* Vaco Uitgeversmaatschappij, Paramaribo.

Hayes, F.E. 1995. *Status, Distribution and Biogeography of the Birds of Paraguay.* American Birding Association, Albany, NY.

Hayes, F.E. 2014. Breeding season and clutch size of birds at Sapucái, Departamento Paraguarí, Paraguay. *Bol. Mus. Nat. Hist. Nat. Paraguay* 18: 77–97.

Hayes, F.E. & Scharf, P.A. 1995. The birds of Parque Nacional Ybycuí, Paraguay. *Cotinga* 4: 14–19.

Heath, R.E. 1932. Notes on some birds of Barro Colorado Island, Canal Zone. *Ibis* 74: 480–486.

Heine, F. & Reichenow, A. 1890. *Nomenclator Musei Heineani Ornithologici.* R. Friedländer & Sohn, Berlin.

Heinz, M. 2002. Ecology, Habitat, and Distribution of the Jocotoco Antpitta (Formicariidae: *Grallaria ridgelyi*) in south Ecuador. B.Sc. thesis. Institute of Landscape Ecology, Univ. of Münster, Münster.

Heinz, M., Schmidt, V. & Schaefer, M. 2005. New distributional record for the Jocotoco Antpitta *Grallaria ridgelyi* in south Ecuador. *Cotinga* 23: 24–26.

Hellmayr, C.E. 1903. Über neue und wenig bekannte südamerikanische Vögel. *Verh. Ges. Wien* 53: 199–223.

Hellmayr, C.E. 1905a. [Exhibition of 4 new species of South American birds]. *Bull. Brit. Orn. Club* 15: 54–57.

Hellmayr, C.E. 1905b. Notes on a collection of birds, made by Mons. A. Robert in the district of Pará, Brazil. *Novit. Zool.* 12: 269–305.

Hellmayr, C.E. 1906. Notes on a second collection of birds from the district of Pará, Brazil. *Novit. Zool.* 13: 353–385.

Hellmayr, C.E. 1907. Another contribution to the ornithology of the lower Amazons. *Novit. Zool.* 14: 1–39.

Hellmayr, C.E. 1908. An account of the birds collected by Mons. G. A. Baer in the State of Goyaz, Brazil. *Novit. Zool.* 15: 13–102.

Hellmayr, C.E. 1910. The birds of the Rio Madeira. *Novit. Zool.* 17: 257–428.

Hellmayr, C.E. 1911a. A contribution to the ornithology of western Colombia. *Proc. Zool. Soc. Lond.* 1911: 1084–1213.

Hellmayr, C.E. 1912. Zoologische Ergebnisse einer Reise in das Mündungsgebiet des Amazonas. II. Vögel. *Abh. Math. Phys. Klasse König. Bayer. Akad. Wissen. Berlin* 26: 1–142.

Hellmayr, C.E. 1913. Critical notes on the types of little-known species of Neotropical birds, 2. *Novit. Zool.* 20: 227–256.

Hellmayr, C.E. 1917. Beschreibung von sechs neuen neotropischen Vovelformen, nebst einer Bemerkung uber *Ampelion cinctus* (Tsch.). *Verh. Orn. Ges. Bayern* 13: 106–119.

Hellmayr, C.E. 1921. Review of the birds collected by Alcide d'Orbigny in South America, Part 1. *Novit. Zool.* 28: 171–212.

Hellmayr, C.E. 1929. A contribution to the ornithology of northeastern Brazil. *Publ. Field Mus. Nat. Hist., Zool. Ser.* 12: 235–501.

Hellmayr, C.E. & Madarász, J. 1914. Description of a new Formicarian-bird from Colombia. *Ann. Hist. Nat. Mus. Nation. Hungarici* 12: 88.

Hellmayr, C.E. & von Seilern, J. 1912a. Beiträge zur Ornithologie von Venezuela. *Archiv f. Naturgeschichte* 5: 34–166.

Hellmayr, C.E. & von Seilern, J. 1912b. [Descriptions of two new subspecies of birds from the Island of Trinidad]. *Bull. Brit. Orn. Club* 31: 13–14.

Hellmayr, C.E. & von Seilern, J. 1914. Neue Vögel aus dem tropischen Amerika. *Verh. Orn. Ges. Bayern* 12: 87–92.

Hennessey, A.B. & Gómez, M.I. 2003. Four bird species new to Bolivia: an ornithological survey of the Yungas site Tokoaque, Madidi National Park. *Cotinga* 19: 25–33.

Hennessey, A.B., Herzog, S.A., Kessler, M. & Robison, D. 2003a. Avifauna of the Pilón Lajas Biosphere Reserve and communal lands, Bolivia. *Bird Conserv. Intern.* 13: 319–349.

Hennessey, A.B., Herzog, S.K. & Sagot, F. 2003b. *Lista Anotada de las Aves de Bolivia.* Armonía/BirdLife International, Santa Cruz de la Sierra.

Henriques, L.M.P., Wunderle, J.M., Oren, D.C. & Willig, M.R. 2008. Efeitos da exploração madeireira de baixo impacto sobre uma comunidade de aves de sub-bosque na Floresta Nacional do Tapajós, Pará, Brasil. *Acta Amazonica* 38: 267–289.

Henriques, L.M.P., Wunderle, J.M. & Willig, M.R. 2003. Birds of the Tapajós National Forest, Brazilian Amazon: A preliminary assessment. *Orn. Neotrop.* 14: 307–338.

Herklots, G.A.C. 1961. *The Birds of Trinidad and Tobago.* Collins, London.

Hermann, J. 1783. *Tabula Affinitatum Animalium Olim Academico Specimine Edita: Nunc Uberiore Commentario Illustrata cum Annotationibus ad Historiam Naturalem Animalium Augendam Facientibus.* Impensis Joh. Georgii Treuttel, Strasbourg.

Hernandez, P., Franke, I., Herzog, S., Pacheco, V., Paniagua, L., Quintana, H., Soto, A., Swenson, J., Tovar, C., Valqui, T., Vargas, J. & Young, B. 2008. Predicting species distributions in poorly-studied landscapes. *Biodiver. & Conserv.* 17: 1353–1366.

Hernández-Baños, B.E., Peterson, A.T., Navarro-Sigüenza, A.G. & Escalante-Pliego, B.P. 1995. Bird faunas of the humid montane forests of Mesoamerica: biogeographic patterns and priorities for conservation. *Bird Conserv. Intern.* 5: 251–277.

Hernández-Camacho, J.I. & Rodríguez-M., J.V. 1979. Dos nuevos taxa del genero *Grallaria* (Aves: Formicariidae) del alto valle del Magdalena (Colombia). *Caldasia* 12: 573–580.

Herzog, S.K., Fjeldså, J., Kessler, M. & Balderrama, J.A. 1999. Ornithological surveys in the Cordillera Cocapata, depto. Cochabamba, Bolivia, a transitional zone between humid and dry intermontane Andean habitats. *Bull. Brit. Orn. Club* 119: 162–177.

Herzog, S.K., Maillard Z., O., Embert, D., Caballero, P. & Quiroga, D. 2012. Range size estimates of Bolivian endemic bird species revisited: the importance of environmental data and national expert knowledge. *J. Orn.* 153: 1189–1202.

Hillman, S.W. & Hogan, D.R. 2002. First nest record of the Ash-throated Gnateater (*Conopophaga peruviana*). *Orn. Neotrop.* 13: 293–295.

Hilty, S.L. 1975. Notes on a nest and behavior of the Chestnut-crowned Gnateater. *Condor* 77: 513–514.

Hilty, S.L. 1985. Distributional changes in the Colombian avifauna: a preliminary blue list. *Orn. Monogr.* 36: 1000–1012.

Hilty, S.L. 1997. Seasonal distribution of birds at a cloud-forest locality, the Anchicayá Valley, in western Colombia. *Orn. Monogr.* 321–343.

Hilty, S.L. 1999. Three bird species new to Venezuela and notes on the behaviour and distribution of other poorly known species. *Bull. Brit. Orn. Club* 119: 220–235.

Hilty, S.L. 2003. *Birds of Venezuela.* Princeton Univ. Press, Princeton, NJ.

Hilty, S.L. & Brown, W.L. 1983. Range extensions of Colombian birds as indicated by the M. A. Carriker Jr. collection at the National Museum of Natural History, Smithsonian Institution. *Bull. Brit. Orn. Club* 103: 5–17.

Hilty, S.L. & Brown, W.L. 1986. *A Guide to the Birds of Colombia.* Princeton Univ. Press, Princeton, NJ.

Hilty, S.L. & Silliman, J.R. 1983. Puracé National Park, Colombia. *Amer. Birds* 37: 247–256.

Hinkelmann, C. & Fiebig, J. 2001. An early contribution to the avifauna of Paraná, Brazil. The Arkady Fiedler expedition of 1928/29. *Bull. Brit. Orn. Club* 121: 116–127.

Hoffmann, D. & Geller-Grimm, F. 2013. A catalog of bird specimens associated with Prince Maximilian of Wied-Neuwied and potential type material in the natural history collection in Wiesbaden. *Zookeys* 353: 81–93.

Höfling, E. & Lencioni, F. 1992. Avifauna da floresta atlântica, região de Salesópolis, Estado de São Paulo. *Rev. Bras. Biol.* 52: 361–378.

Hollamby, N. 2012. Sani Lodge: the best-kept secret in the Ecuadorian Amazon—until now. *Neotrop. Birding* 10: 69–76.

Holley, D.R., Lindell, C.A., Roberts, M.A. & Biancucci, L. 2001. First description of the nest, nest site, and eggs of the Ochre-breasted Antpitta. *Wilson Bull.* 113: 435–438.

Holt, E.G. 1928. An ornithological survey of the Serra do Itatiaya, Brazil. *Bull. Amer. Mus. Nat. Hist.* 57: 251–326.

Hornbuckle, J. 1999. The birds of Abra Patricia and the upper Río Martín, north Peru. *Cotinga* 12: 11–28.

Horsley, N.P., Eddy, D.K., Maguire, C. & Pollock, H.S. 2016. Streak-chested Antpitta (*Hylopezus perspicillatus*). *In* Schulenberg, T.S. (ed.) *Neotropical Birds Online.* Cornell Lab of Ornithology, Ithaca, NY (http://neotropical. birds.cornell.edu/portal/species/overview?p_p_ spp=406441).

Hosner, P.A. 2010. Birding off the beaten path: visiting and documenting the birdlife of poorly surveyed areas. *Neotrop. Birding* 7: 42–47.

Hosner, P.A., Behrens, K.D. & Hennessey, A.B. 2009. Birds (Aves), Serrania Sadiri, Parque Nacional Madidi, Depto. La Paz, Bolivia. *Check List* 5: 222–237.

Hosner, P.A., Andersen, M.J., Robbins, M.B., Urbay-Tello, A., Cueto-Aparicio, L., Verde-Guerra, K., Sánchez-Goncález, L.A., Navarro-Sigüenza, A.G., Boyd, R.L., Núñez, J., Tiravanti, J., Combe, M., Owens, H.L. & Peterson, A.T. 2015. Avifaunal surveys of the Upper Apurímac River Valley, Ayacucho and Cuzco departments, Peru: new distributional records and biogeographic, taxonomic, and conservation implications. *Wilson J. Orn.* 127: 563–581.

Howard, R. & Moore, A. 1980. *A Complete Checklist of the Birds of the World.* Oxford Univ. Press, New York.

Howard, R. & Moore, A. 1984. *A Complete Checklist of the Birds of the World*, revised edn. Macmillan, London.

Howard, R. & Moore, A. 1991. *A Complete Checklist of the Birds of the World*, 2nd edn. Academic Press, London.

Howell, S.N.G. & Webb, S. 1992a. New and noteworthy bird records from Guatemala and Honduras. *Bull. Brit. Orn. Club* 112: 42–49.

Howell, S.N.G. & Webb, S. 1992b. A little-known cloud forest in Hidalgo, Mexico. *Euphonia* 1: 7–11.

Howell, S.N.G. & Webb, S. 1995. *A Guide to the Birds of Mexico and Northern Central America.* Oxford Univ. Press, Oxford.

Howell, T.R. 1957. Birds of a second-growth rain forest area of Nicaragua. *Condor* 59: 73–111.

Howell, T.R. 1971 An ecological study of the birds of the lowland pine savanna and adjacent rain forest in northeastern Nicaragua. *Living Bird* 10: 185–242.

del Hoyo, J. & Collar, N.J. 2016. *HBW and BirdLife International Illustrated Checklist of Birds of the World*, vol. 2. Lynx Edicions, Barcelona.

del Hoyo, J., Collar, N.J. & Kirwan, G.M. 2017a. Black-breasted Gnateater (*Conopophaga snethlageae*). *In* del Hoyo, J., Elliott, A., Sargatal, J., Christie, D.A. & de Juana, E. (eds) *Handbook of the Birds of the World Alive.* Lynx Edicions, Barcelona (retrieved from http://www. hbw.com/node/1343613 on 14 April 2017).

del Hoyo, J., Collar, N.J. & Kirwan, G.M. 2017b. Ceara Gnateater (*Conopophaga cearae*). *In* del Hoyo, J., Elliott, A., Sargatal, J., Christie, D.A. & de Juana, E. (eds) *Handbook of the Birds of the World Alive.* Lynx Edicions, Barcelona (retrieved from http://www.hbw.com/ node/1343614 on 11 June 2017).

del Hoyo, J., Collar, N.J. & Kirwan, G.M. 2017c. Rufous-breasted Antpitta (*Grallaricula leymebambae*). *In* del Hoyo, J., Elliott, A., Sargatal, J., Christie, D.A. & de Juana, E. (eds) *Handbook of the Birds of the World Alive*. Lynx Edicions, Barcelona (retrieved from http://www.hbw.com/node/1343618 on 18 August 2017).

del Hoyo, J., Collar, N.J. & Kirwan, G.M. 2017d. Northern Tawny Antpitta (*Grallaria alticola*). *In* del Hoyo, J., Elliott, A., Sargatal, J., Christie, D.A. & de Juana, E. (eds.) *Handbook of the Birds of the World Alive*. Lynx Edicions, Barcelona (retrieved from http://www.hbw.com/node/1343616 on 26 September 2017).

del Hoyo, J., Collar, N.J. & Kirwan, G.M. 2017e. Southern Tawny Antpitta (*Grallaria atuensis*). *In* del Hoyo, J., Elliott, A., Sargatal, J., Christie, D.A. & de Juana, E. (eds) *Handbook of the Birds of the World Alive*. Lynx Edicions, Barcelona (retrieved from http://www.hbw.com/node/1343617 on 26 September 2017).

Hoyt, S.F. 1961. Nest-building movements performed by juvenal Song Sparrow. *Wilson Bull.* 73: 386–387.

Huber, O. & Alarcón, C. 1988. *Mapa de la Vegetación de Venezuela*. Ministério del Ambiente y de los Recursos Naturales Renovables (División de Vegetación) and Nature Conservancy, Caracas.

Iafrancesco-V., G.M. & Pineda, C.L.M. 1984. Contribución al estudio de la ornitología de Colombia, entrega 1. Dendrocoláptidos del Museo de Ciencias Naturales de la Universidad de La Salle. *Bol. Cienc. Univ. La Salle* 1: 55–75.

Iafrancesco-V., G.M., Pineda, C.L.M. & Plaza, G.O. 1987. Contribución al estudio de los Passeriformes Formicariidos de Colombia, entrega 3. Formicariidos del Museo de Ciencias Naturales de la Universidad de La Salle. *Bol. Cienc. Univ. La Salle* 2: 63–144.

von Ihering, H. 1900. Catálogo crítico-comparativo dos ninhos e ovos das aves do Brazil. *Rev. Mus. Paulista* 4: 191–300.

von Ihering, H. 1902. Contribuições para o conhecimento da ornitologia de São Paulo. *Rev. Mus. Paulista* 5: 261–329.

von Ihering, H. 1905. O Rio Juruá. *Rev. Mus. Paulista* 6: 385–460.

von Ihering, H. & von Ihering, R. 1907. *Catalogos da fauna Brazileira*, vol. 1. Museu Paulista, São Paulo, Brasil.

Inzunza, E.R., Tizón, R.Z. & F., S.D. 2012. Aves / Birds. Pp. 271–279, 366–385 *in* Pitman, N., Inzunza, E.R., Alvira, D., Vriesendorp, C., Moskovits, D.K., del Campo, Á., Wachter, T., Stotz, D.F., Sesén, S.N., Cerrón, E.T. & Smith, R.C. (eds.) *Rapid Biological and Social Inventories 24: Perú: Cerros de Kampankis*. The Field Museum, Chicago.

Isler, P.R. & Whitney, B.M. 2002. *Songs of the Antbirds* (CDs). Cornell Lab. of Ornithology, Ithaca, NY.

IUCN. 1986. *1986 IUCN Red List of Threatened Animals*. International Union for Conservation of Nature & Natural Resources, Gland.

IUCN. 1988. *1988 IUCN Red List of Threatened Animals*. International Union for Conservation of Nature & Natural Resources, Gland.

IUCN. 1990. *1990 IUCN Red List of Threatened Animals*. International Union for Conservation of Nature & Natural Resources, Cambridge, UK.

Jacoboski, L.I., Mendonça-Lima, A. & Hartz, S.M. 2016. Structure of bird communities in eucalyptus plantations: nestedness as a pattern of species distribution. *Braz. J. Biol.* 76: 583–591.

Jahn, O. 2011. Bird communities of the Ecuadorian Chocó: a case study in conservation. *Bonn Zool. Monogr.* 56: 1–514.

Jahn, O. & Valenzuela, P.M. 2002. Pitasoma coronirrufa (*Pittasoma rufopileatum*). Pp. 295–296 *in* Granizo, T., Pacheco, C., Ribadeneira, M.B., Guerrero, M. & Suárez, L. (eds.) *Libro Rojo de las Aves de Ecuador*. SIMBIOE, Conservation International, EcoCiencia, Ministerio del Ambiente & IUCN, Quito.

Jahn, O. & Valenzuela, P.M. 2006. *Status and Ecology of the Cerulean Warbler* Dendroica cerulea *in Northwestern Ecuador*. Nature Conservancy, Quito.

Jankowski, J.E. 2010. Distributional Ecology and Diversity Patterns of Tropical Montane Birds. Ph.D. thesis. Univ. of Florida, Gainesville.

Jenkins, C.N., Alves, M.A.S. & Pimm, S.L. 2010. Avian conservation priorities in a top-ranked biodiversity hotspot. *Biol. Conserv.* 143: 992–998.

Jenkinson, M.A. & Tuttle, M.D. 1976. *Accipiter poliogaster* from Peru, and remarks on two collecting localities named "Sarayacu". *Auk* 93: 187–189.

Jobling, J.A. 2010. *The Helm Dictionary of Scientific Bird Names from Aalge to Zusii*. Christopher Helm, London.

Johns, A.D. 1991. Responses of Amazonian rain forest birds to habitat modification. *J. Trop. Ecol.* 7: 417–437.

Johnson, E.I., Stouffer, P.C. & Vargas, C.F. 2011. Diversity, biomass, and trophic structure of a central Amazonian rainforest bird community. *Rev. Bras. Orn.* 19: 1–16.

Jones, J.K. & Genoways, H.H. 1970. Harvest mice (genus *Reithrodontomys*) of Nicaragua. *Occ. Pap. West. Found. Vert. Zool.* 2: 1–16.

Jones, S.E. 1977. Coexistence in mixed species antwren flocks. *Oikos* 29: 366–375.

Juiña J., M.E., Harris, J.B.C. & Greeney, H.F. 2009. Description of the nest and parental care of the Chestnut-naped Antpitta (*Grallaria nuchalis*) from southern Ecuador. *Orn. Neotrop.* 20: 305–310.

Jullien, M. & Thiollay, J.-M. 1998. Multi-species territoriality and dynamics of Neotropical forest understory bird flocks. *J. Anim. Ecol.* 67: 227–252.

Junge, G.C.A. & Mees, G.F. 1961. The avifauna of Trinidad and Tobago. *Zool. Verh.* 37: 1–172.

Junior, T.M.O., França, B.R.A., Nascimento, E.P.G., Neto, M.R., Silva, M. & Pichorim, M. 2008. Aves de quarto fragmentos florestais no sul do Rio Grande do Norte, Brasil. P. 301 *in* Dornas, T. & Barbosa, C.E. (eds.) *Livro de Resumos do XVI Congresso Brasileiro de Ornitologia*. Palmas.

Just, J.P.G., Romagna, R.S., Rosoni, J.R.R. & Zocche, J.J. 2015. Avifauna na região dos contrafortes da Serra Geral, Mata Atlântica do sul de Santa Catarina, Brasil. *Atualidades Orn.* 187: 33–54.

Kaminski, N. & Carrano, E. 2004. Comunidade de aves em um ecótone (floresta ombrófila densa e f.o. mista) na Serra do Cabral, município de Tijucas do Sul, Paraná. P. 252 *in* Testoni, A.F. & Althoff, S.L. (eds) *Livro de Resumos do XII Congresso Brasileiro de Ornitologia*. Blumenau.

Karr, J.R. 1970. A Comparative Study of the Structure of Avian Communities in Selected Tropical and Temperate Habitats. Ph.D. thesis. Univ. of Illinois at Urbana-Champaign, Urbana.

Karr, J.R. 1971a. Ecological, behavioral, and distributional notes on some Central Panama birds. *Condor* 73: 107–111.

Karr, J.R. 1971b. Structure of avian communities in selected Panama and Illinois habitats. *Ecol. Monogr.* 41: 207–233.

Karr, J.R. 1977. Ecological correlates of rarity in a tropical forest bird community. *Auk* 94: 240–247.

Karr, J.R., Willson, F. & Moriarty, D.J. 1978. Weights of some Central American birds. *Brenesia* 14–15: 249–257.

Karr, J.R., Robinson, S.K., Blake, J.G. & Bierregaard, R.O. 1990. Birds of four neotropical forests. Pp. 237–269 *in* Gentry, A.H. (ed.) *Four Neotropical Rainforests.* Yale Univ. Press, New Haven, CT.

Kattan, G.H., Alvarez-Lopez, H. & Giraldo, M. 1994. Forest fragmentation and bird extinctions: San Antonio 80 years later. *Conserv. Biol.* 8: 138–146.

Kattan, G.H. & Beltrán, J.W. 1997. Rediscovery and status of the Brown-banded Antpitta *Grallaria milleri* in the central Andes of Colombia. *Bird Conserv. Intern.* 7: 367–371.

Kattan, G.H. & Beltrán, J.W. 1999. Altitudinal distribution, habitat use, and abundance of *Grallaria* antpittas in the Central Andes of Colombia. *Bird Conserv. Intern.* 9: 271–281.

Kattan, G.H. & Beltrán, J.W. 2002. Rarity in antpittas: territory size and population density of five *Grallaria* spp. in a regenerating habitat mosaic in the Andes of Colombia. *Bird Conserv. Intern.* 12: 231–240.

Kattan, G.H., Franco, P., Saavedra-Rodríguez, C.A., Valderrama, C., Rojas, V., Osorio, D. & Martínez, J. 2006. Spatial components of bird diversity in the Andes of Colombia: Implications for designing a regional reserve system. *Conserv. Biol.* 20: 1203–1211.

Kattan, G.H. & Renjifo, L.M. 2002. *Grallaricula lineifrons*. Pp. 329–330 *in* Renjifo, L.M., Franco-Maya, A.M., Amaya-Espinel, J.D., Kattan, G.H. & López-Lanús, B. (eds.) *Libro Rojo de Aves de Colombia. Serie Libros Rojos de Especies Amenazadas de Colombia.* Instituto de Investigación de Recursos Biológicos Alexander von Humboldt y Ministerio del Medio Ambiente, Bogotá.

Kempff-Mercado, N. 1985. *Aves de Bolivia.* Editorial Gisbert, La Paz.

Kenefick, M. 2012. Report of the Trinidad and Tobago Rare Birds Committee: rare birds in 2008–10. *Cotinga* 34: 100–105.

Kenefick, M., Restall, R.L. & Hayes, F.E. 2007. *Birds of Trinidad & Tobago.* Christopher Helm, London.

Kikuchi, D.W. 2009. Terrestrial and understorey insectivorous birds of a Peruvian cloud forest: species richness, abundance, density, territory size and biomass. *J. Trop. Ecol.* 25: 523–529.

King, J.R. 1989. Notes on the birds of the Rio Mazan Valley, Azuay Province, Ecuador, with special reference to *Leptosittaca branickii, Hapalopsittaca amazonina pyrrhops* and *Metallura baroni. Bull. Brit. Orn. Club* 109: 140–147.

King, J. 2013. Spotted Antpitta splits into three (or four). http://www.xeno-canto.org/feature-view.php?blognr=143.

King, W.B. 1981. *Endangered Birds of the World, the ICBP Bird Red Data Book.* Smithsonian Institution Press, Washington DC.

Kirwan, G.M. 2009. Notes on the breeding biology and seasonality of some Brazilian birds. *Rev. Bras. Orn.* 17: 121–136.

Kirwan, G.M. & Hornbuckle, J. 1997. Neotropical notebook. *Cotinga* 7: 75–82.

Kirwan, G.M. & Marlow, T. 1996. A review of avifaunal records from Mindo, Pichincha province, north-western Ecuador. *Cotinga* 6: 47–57.

Kirwan, G.M., Mazar Barnett, J. & Minns, J. 2001. Significant ornithological observations from the Rio São Francisco

Valley, Minas Gerais, Brazil, with notes on conservation and biogeography. *Ararajuba* 9: 145–161.

Knowlton, J.L. 2010. Breeding records of birds from Tumbesian region of Ecuador. *Orn. Neotrop.* 21: 109–129.

Knox, A.G. & Walters, M.P. 1994. *Extinct and Endangered Birds in the Collections of the Natural History Museum.* British Ornithologists' Club, Tring.

Koenen, M.T. & Koenen, S.G. 2000. Effects of fire on birds in páramo habitat of northern Ecuador. *Orn. Neotrop.* 11: 155–163.

Koenigswald, G. 1896. Ornithologia Paulista. *J. Orn.* 44: 332–398.

Koepcke, M. 1961. Birds of the western slope of the Andes of Peru. *Amer. Mus. Novit.* 2028: 1–32.

Koepcke, M. 1970. *The Birds of the Department of Lima, Peru.* Livingston Publishing Co., Wynnewood, PA.

Kofoed, E.M. & Auer, S.K. 2004. First description of the nest, eggs, young, and breeding behavior of the Great Antpitta (*Grallaria excelsa*). *Wilson Bull.* 116: 105–108.

Komar, O. 1998. Avian diversity in El Salvador. *Wilson Bull.* 110: 511–533.

Komar, O. 2002a. Birds of Montecristo National Park, El Salvador. *Orn. Neotrop.* 13: 167–193.

Komar, O. 2002b. Priority conservation areas for birds in El Salvador. *Anim. Conserv.* 5: 173–183.

Komar, O. & Herrera, N. 1995. Avian inventory of El Imposible National Park, San Benito and Rio Guayapa Sectors. Pp. 6–32 *in* Komar, O. & Herrera, N. (eds.) *Avian Diversity at El Imposible National Park and San Marcelino Wildlife Refuge, El Salvador.* Wildlife Conservation Society, Bronx, NY.

Krabbe, N. 1991. *Avifauna of the Temperate Zone of the Ecuadorian Andes.* Zool. Mus., Univ. of Copenhagen, Copenhagen.

Krabbe, N. 1992. Notes on distribution and natural history of some poorly known Ecuadorean birds. *Bull. Brit. Orn. Club* 112: 169–174.

Krabbe, N. 2007. Birds collected by P. W. Lund and J. T. Reinhardt in south-eastern Brazil between 1825 and 1855, with notes on P. W. Lund's travels in Rio de Janeiro. *Rev. Bras. Orn.* 15: 331–357.

Krabbe, N. 2008. Arid valleys as dispersal barriers to high-Andean forest birds in Ecuador. *Cotinga* 29: 28–30.

Krabbe, N. & Coopmans, P. 2000. Rediscovery of *Grallaria alleni* (Formicariidae) with notes on its range, song and identification. *Ibis* 142: 183–187.

Krabbe, N. & Schulenberg, T.S. 2003. Family Formicariidae (ground-antbirds). Pp. 682–731 *in* del Hoyo, J., Elliott, A. & Christie, D.A. (eds.) *Handbook of the Birds of the World,* vol. 8. Lynx Edicions, Barcelona.

Krabbe, N. & Sornoza-M., F. 1994. Avifaunistic results of a subtropical camp in the Cordillera del Condor, southeastern Ecuador. *Bull. Brit. Orn. Club* 14: 55–61.

Krabbe, N., DeSmet, G., Greenfield, P.J., Jacome, M., Matheus, J.C. & Sornoza, F. 1994. Giant Antpitta *Grallaria gigantea. Cotinga* 2: 32–34.

Krabbe, N., Poulsen, B.O., Frølander, A. & Barahona, O.R. 1997. Range extensions of cloud forest birds from the high Andes of Ecuador: new sites for rare or little-recorded species. *Bull. Brit. Orn. Club* 117: 248–256.

Krabbe, N., Skov, F., Fjeldså, J. & Petersen, I.K. 1998. *Avian Diversity in the Ecuadorian Andes: An Atlas of Distribution of Andean Forest Birds and Conservation Priorities.* DIVA Technical Report No. 4. Centre for Research on the Cultural and Biological Diversity of

Andean Rainforests (DIVA), Rønde.

Krabbe, N., Agro, D.J., Rice, N.H., Jacome, M., Navarrete, L. & Sornoza-M., F. 1999. A new species of antpitta (Formicariidae: *Grallaria*) from the southern Ecuadorian Andes. *Auk* 116: 882–890.

Krabbe, N., Flórez, P., Suárez, G., Castaño, J., Arango, J.D. & Duque, A. 2006. The birds of Páramo de Frontino, Western Andes of Colombia. *Orn. Colombiana* 4: 39–50.

Krabbe, N., Schulenberg, T.S., Bonan, A. & Boesman, P. 2012. Peruvian Antpitta (*Grallaricula peruviana*). *In* del Hoyo, J., Elliott, A., Sargatal, J., Christie, D.A. & de Juana, E. (eds.) *Handbook of the Birds of the World Alive.* Lynx Edicions, Barcelona (retrieved from http://www.hbw.com/node/56916 on 25 October 2014).

Krabbe, N., Schulenberg, T.S. & de Juana, E. 2013. Crescent-faced Antpitta (*Grallaricula lineifrons*). *In* del Hoyo, J., Elliott, A., Sargatal, J., Christie, D.A. & de Juana, E. (eds.) *Handbook of the Birds of the World Alive.* Lynx Edicions, Barcelona (retrieved from http://www.hbw.com/node/56920 on 1 September 2014).

Krabbe, N., Schulenberg, T.S. & Bonan, A. 2017a. Black-crowned Pittasoma (*Pittasoma michleri*). *In* del Hoyo, J., Elliott, A., Sargatal, J., Christie, D.A. & de Juana, E. (eds.) *Handbook of the Birds of the World Alive.* Lynx Edicions, Barcelona (retrieved from http://www.hbw.com/node/56921 on 5 June 2017).

Krabbe, N., Schulenberg, T.S. & Kirwan, G.M. 2017b. Rusty-breasted Antpitta (*Grallaricula ferrugineipectus*). in del Hoyo, J., Elliott, A., Sargatal, J., Christie, D.A. & de Juana, E. (eds.) *Handbook of the Birds of the World Alive.* Lynx Edicions, Barcelona (retrieved from http://www.hbw.com/node/56918 on 5 August 2017).

Krauczuk, E.R. 2008. Riqueza específica, frecuencia, abundancia y ambiente de las aves de la cuenca baja y media del arroyo Santo Pipó y del municipio de Corpus, San Ignacio, Misiones, Argentina. *Lundiana* 9: 29–39.

Kreuger, R. 1968. Some notes on the oology of members of the family Formicariidae (antbirds); family Conopophagidae (gnat-eaters or antpits); family Rhinocryptidae (Tapaculos); family Cotingidae (Cotingas); family Pipridae (Manakins). *Oölogists' Rec.* 42: 9–15.

Krügel, M.M. & dos Anjos, L. 2000. Bird communities in forest remnants in the city of Maringá, Paraná state, southern Brazil. *Orn. Neotrop.* 11: 315–330.

Lafresnaye, M. 1842. Oiseaux nouveaux de Colombie. *Rev. Zool. Soc. Cuvierienne* 1842: 333–336.

Lafresnaye, M. 1843. Quelques nouvelles espèces d'Oiseaux. *Rev. Zool. Soc. Cuvierienne* 1843: 97–99.

Lamm, D.W. 1948. Notes on the birds of the states of Pernambuco and Paraiba, Brazil. *Auk* 65: 261–283.

Land, H.C. 1970. *Birds of Guatemala.* Livingston Publishing, Narberth, PA.

Lane, D.F. & Pequeño, T. 2004. Aves / Birds. P. 299 *in* Vriesendorp, C., Chávez, L.R., Moskovits, D. & Shopland, J. (eds.) *Rapid Biological and Social Inventories 15: Perú: Megantoni.* The Field Museum & CEDIA, Chicago.

Lane, D.F., Pequeño, T. & Villar, J.F. 2003. Aves / Birds. Pp. 67–73, 150–156, 254–267 *in* Pitman, N., Vriesendorp, C. & Moskovits, D.K. (eds.) *Rapid Biological and Social Inventories 11: Peru: Yavarí.* The Field Museum, Chicago.

Latham, J. 1783. *A General Synopsis of Birds*, vol. 2. Leigh & Sotheby, London.

Latham, J. 1785. *A General Synopsis of Birds*, vol. 3. Leigh & Sotheby, London.

Latham, J. 1790. *Index Ornithologicus*, vol. 1. Leigh & Sotheby, London.

Latta, S.C., Tinoco, B.A., Astudillo, P.X. & Graham, C.H. 2011. Patterns and magnitude of temporal change in avian communities in the Ecuadorian Andes. *Condor* 113: 24–40.

Lau, R. 2004. Listagem preliminar da avifauna do Morro Gaucho e arredores do Vale do Taquari, Rio Grande do Sul, Brasil. P. 261 *in* Testoni, A.F. & Althoff, S.L. (eds.) *Livro de Resumos do XII Congresso Brasileiro de Ornitologia.* Blumenau.

Laubmann, A. 1940. *Die Vögel von Paraguay*, vol. 2. Strecker & Schröder, Stuttgart.

Laurance, S.G.W. 2004. Responses of understory rain forest birds to road edges in central Amazonia. *Ecol. Appl.* 14: 1344–1357.

Lawrence, G.N. 1861. Descriptions of three new species of birds. *Ann. Lyceum Nat. Hist. New York* 7: 303–305.

Lawrence, G.N. 1862a. Catalogue of a collection of birds, made in New Granada, by James McLeannan, Esq., of New York, with notes and descriptions of new species. Part I. *Ann. Lyceum Nat. Hist. New York* 7: 288–302.

Lawrence, G.N. 1862b. Catalogue of a collection of birds, made in New Granada, by James McLeannan, Esq., of New York, with notes and descriptions of new species. Part II. *Ann. Lyceum Nat. Hist. New York* 7: 315–334.

Lawrence, G.N. 1865. Catalogue of a collection of birds in the Museum of the Smithsonian Institution, made by Mr. H. E. Holland at Greytown, Nicaragua, with descriptions of new species. *Ann. Lyceum Nat. Hist. New York* 8: 178–184.

Lawrence, G.N. 1866. Characters of seven new species of birds from Central and South America, with a note on *Thaumatias chionurus*, Gould. *Ann. Lyceum Nat. Hist. New York* 8: 344–350.

Lawrence, G.N. 1868. A catalogue of the birds found in Costa Rica. *Ann. Lyceum Nat. Hist. New York* 9: 86–149.

Lebbin, D.J. 2007. *Habitat Specialization Among Amazonian Birds: Why Are There so Many Guadua Bamboo Specialists?* Ph.D. thesis. Cornell Univ., Ithaca, NY.

Lebossé, A. & Bour, R. 2011. The first twenty livraisons of "Les Planches Coloriées d'Oiseaux" by Temminck & Laugier (1820-1839): I. The ten wrappers now known. *Zool. Bibliography* 1: 141–150.

Lebossé, A. & Dickinson, E.C. 2015. Fresh information relevant to the make-up of the livraisons of the "Galerie des Oiseaux" by Vieillot (1748-1831) & Oudart (1796-1860). *Zool. Bibliography* 3: 25–58.

Leck, C.F. 1979. Avian extinctions in an isolated tropical wet-forest preserve, Ecuador. *Auk* 96: 343–352.

Leck, C.F., Ortiz-Crespo, F.I. & Webster, R. 1980. Las aves del Centro Científico Río Palenque. *Rev. Univ. Católica Ecuador* 8: 75–90.

LeCroy, M. & Sloss, R. 2000. Type specimens of birds in the American Museum of Natural History. Part 3, Passeriformes: Eurylaimidae, Dendrocolaptidae, Furnariidae, Formicariidae, Conopophagidae, and Rhinocryptidae. *Bull. Amer. Mus. Nat. Hist.* 257: 1–88.

Lees, A.C. & Peres, C.A. 2006. Rapid avifaunal collapse along the Amazonian deforestation frontier. *Biol. Conserv.* 133: 198–211.

Lees, A.C. & Peres, C.A. 2008. Conservation value of remnant riparian forest corridors of varying quality for Amazonian birds and mammals. *Conserv. Biol.* 22: 439–449.

Lees, A.C. & Peres, C.A. 2009. Gap-crossing movements predict species occupancy in Amazonian forest fragments. *Oikos* 118: 280–290.

Lees, A.C. & Peres, C.A. 2010. Habitat and life history determinants of antbird occurrence in variable-sized Amazonian forest fragments. *Biotropica* 42: 614–621.

Lees, A.C., Davis, B.J.W., Oliveira, A.V.G. & Peres, C.A. 2008. Avifauna of a structurally heterogeneous forest landscape in the Serra dos Caiabis, Mato Grosso, Brazil: a preliminary assessment. *Cotinga* 29: 149–159.

Lees, A.C., Moura, N.G., Santana, A., Aleixo, A., Barlow, J., Berenguer, E., Ferreira, J. & Gardner, T.A. 2012. Paragominas: a quantitative baseline inventory of an eastern Amazonian avifauna. *Rev. Bras. Orn.* 20: 93–118.

Lees, A.C., Moura, N.G., Andretti, C.B., Davis, B.J.W., Lopes, E.V., Henriques, L.M.P., Aleixo, A., Barlow, J., Ferreira, J. & Gardner, T.A. 2013a. One hundred and thirty-five years of avifaunal surveys around Santarém, central Brazilian Amazon. *Rev. Bras. Orn.* 21: 16–57.

Lees, A.C., Zimmer, K.J., Marantz, C.A., Whittaker, A., Davis, B.J.W. & Whitney, B.M. 2013b. Alta Floresta revisited: an updated review of the avifauna of the most intensively surveyed locality in south-central Amazonia. *Bull. Brit. Orn. Club* 133: 178–239.

Lees, A.C., Thompson, I. & Moura, N.G. 2014. Salgado Paraense: an inventory of a forgotten coastal Amazonian avifauna. *Bol. Mus. Para. E. Goeldi, Sér. Cienc. Nat.* 9: 135–168.

Lehmann-V., F.C. 1957. Contribuciones al estudio de la fauna de Colombia XII. *Nov. Colombianas* 1: 101–156.

Lehmann-V., F.C., Silliman, J.R. & Eisenmann, E. 1977. Rediscovery of the Crescent-faced Antpitta in Colombia. *Condor* 79: 387–388.

Leite, G.A., Gomes, F.B.R. & MacDonald, D.B. 2012. Description of the nest, nestling and broken-wing behavior of *Conopophaga aurita* (Passeriformes: Conopophagidae) *Rev. Bras. Orn.* 20: 128–131.

Lentijo, G.M. & Botero, J.E. 2013. La avifauna de localidades cafeteras de los Municipios de Manizales y Palestina, Departamento de Caldas, Colombia. *Bol. Cien. Hist. Nat.* 17: 111–128.

Lentijo, G.M. & Kattan, G.H. 2005. Estratificacion vertical de las aves en una plantacion monoespecifica y en bosque nativo en la Cordillera Central de Colombia. *Orn. Colombiana* 3: 51–61.

Lentino, M., Salcedo, M. & Ascanio, D. 2013a. Aves de la cuenca alta del Río Cuyuní, estado Bolívar: resultados del RAP Alto Cuyuní 2008. Pp. 156–163 *in* Lasso, C.A., Señaris, J.C., Rial, A. & Flores, A.L. (eds.) *Evaluación Rápida de la Biodiversidad de los Ecosistemas Acuáticos de la Cuenca Alta del Río Cuyuní, Guayana Venezolana.* Conservation International, Washington DC.

Lentino, M., Salcedo, M. & Ascanio, D. 2013b. Listado de aves registradas durante el RAP Alto Cuyuní 2008 en las diferentes localidades estudiadas. Pp. 217–224 in Lasso, C.A., Señaris, J.C., Rial, A. & Flores, A.L. (eds.) *Evaluación Rápida de la Biodiversidad de los Ecosistemas Acuáticos de la Cuenca Alta del Río Cuyuní, Guayana Venezolana.* Conservation International, Washington DC.

Lesson, R.P. 1831. *Traité d'ornithologie, ou, Tableau méthodique des ordres, sous-ordres, familles, tribus, genres, sous-genres et races d'oiseaux.* Chez F.G. Levrault, Brussels.

Lesson, R.P. 1844. Révision du genere *Grallaria. L'Echo du Monde Savant* 11, 2me semestre (49): column: 847–848.

Lévêque, R. 1964. Notes on Ecuadorian birds. *Ibis* 106: 52–62.

Lichtenstein, M.H.K. 1823. *Verzeichniss der Doubletten des Zoologischen Museums der Königl. Universität zu Berlin: nebst Beschreibung vieler bisher unbekannter Arten von Säugethieren, Vögeln, Amphibien und Fischen.* T. Trautwein, Berlin.

Lima, A.L.C., Manhães, M.A. & Piratelli, A.J. 2011. Ecologia trófica de *Conopophaga lineata* (Conopophagidae) em uma área de mata secundária no sudeste do Brasil. *Rev. Bras. Orn.* 19: 315–322.

Lima, A.M.X. 2007. Sitios de nidificacao de *Conopophaga melanops* (Conopophagidae) na Reserva Natural Salto Morato, Guaraquecaba, PR. Pp. 109–110 *in* Fontana, C.S. (ed.) *Livro de Resumos do XV Congresso Brasileiro de Ornitologia.* Porto Alegre.

Lima, A.M.X. 2008. *Dinâmica pobulacional de aves de sub-bosque na Floresta Atlântica do Paraná.* Universidade Federal do Paraná, Curitiba.

Lima, A.M.X. & Roper, J.J. 2007. Dinamica populacional das aves de sub-bosque de Floresta Atlantica no Paraná. P. 245 *in* Fontana, C.S. (ed.) *Livro de Resumos do XV Congresso Brasileiro de Ornitologia.* Porto Alegre.

Lima, A.M.X. & Roper, J.J. 2009a. Population dynamics of the Black-cheeked Gnateater (*Conopophaga melanops,* Conopophagidae) in southern Brazil. *J. Trop. Ecol.* 25: 605–613.

Lima, A.M.X. & Roper, J.J. 2009b. The use of playbacks can influence encounters with birds: an experiment. *Rev. Bras. Orn.* 17: 37–40.

Lima, A.M.X. & Roper, J.J. 2013. Early singing onset in the Black-cheeked Gnateater (*Conopophaga melanops*). *Rev. Bras. Orn.* 21: 5–9.

Lima, B. 2011. Lista de aves de Tingão e Dedo de Deus-Peruíbe-SP, BR. www.aultimaarcadenoe.com.br (accessed 11 June 2017).

Lima, C.A., Siqueira, P.R., Goncalves, R.M.M., Vasconcelos, M.F. & Leite, L.O. 2010. Dieta de aves da Mata Atlantica: uma abordagem baseada em conteudos estomacais. *Orn. Neotrop.* 21: 425–438.

Lima, P.C. 2006. Aves do Litoral Norte da Bahia. *Atualidades Orn.* 134: 1–661. (http://www.ao.com.br/).

Lima, P.C. & Grantsau, R. 2005. *Conopophaga melanops nigrifrons,* Pinto, 1943. Nova ocorrência para a Bahia. *Atualidades Orn.* 127: 5.

Lindell, C.A. & Smith, M. 2003. Nesting bird species in sun coffee, pasture, and understory forest in southern Costa Rica. *Biodiver. & Conserv.* 12: 423–440.

Lindell, C.A., Chomentowski, W.H. & Zook, J.R. 2004. Characteristics of bird species using forest and agricultural land covers in southern Costa Rica. *Biodiver. & Conserv.* 13: 2419–2441.

Lizarazo-B., J. & Londoño, G.A. in review. Nesting behavior of the Chestnut-crowned Gnateater (*Conopophaga castaneiceps*). *Wilson J. Orn.*

Lloyd, H. 2004. Habitat and population estimates of some threatened lowland forest bird species in Tambopata, south-east Peru. *Bird Conserv. Intern.* 14: 261–277.

Lloyd, H. 2008. Abundance and patterns of rarity of *Polylepis* birds in the Cordillera Vilcanota, southern Perú: implications for habitat management strategies. *Bird Conserv. Intern.* 18: 164–180.

Lobo-Araújo, L.W., Sugliano, G.O.S., Lima, G.S.T., Macario, P. & Santos, J.G. 2008. Levantamento das aves endêmicas e ameaçadas em remanescente de Floresta Atlântica no Município de Campo Alegre, Alagoas. P. 364 in

Dornas, T. & Barbosa, M.O. (eds.) *Livro de Resumos do XVI Congresso Brasileiro de Ornitologia*. Palmas.

Lobo-Araújo, L.W., Toledo, M.T.F., Efe, M.A., Malhado, A.C.M., Vital, M.V.C., Toledo-Lima, G.S., Macario, P., Santos, J.G. & Ladle, R.J. 2013. Bird communities in three forest types in the Pernambuco centre of endemism, Alagoas, Brazil. *Iheringia, Sér. Zool.* 103: 85–96.

Lodge, W. 1991. *Birds. Alternative Names: A World Checklist*. Sterling Publishing, New York.

Lombardi, V.T., Vasconcelos, M.F. & Neto, S.D.A. 2007. Novos registros ornitológicos para o centro-sul de Minas Gerais (alto Rio Grande): municípios de Lavras, São João Del Rei e adjacências, com a listagem revisada da região. *Atualidades Orn.* 139: 33–42.

Lombardi, V.T., Santos, K.K., Neto, S.D.A., Mazzoni, G., Renno, B., Faetti, R.G., Epifanio, A.D. & Miguel, M. 2012. Registros notaveis de aves para o Sul do Estado de Minas Gerais, Brasil. *Cotinga* 34: 32–45.

Londoño, G.A. 2014. *Anidación de Aves en un Gradiente Altitudinal*. Rapid Color Guide 514, versión 1. The Field Museum, Chicago.

Londoño, G.A., Saavedra-R., C.A., Osorio, D. & Martínez, J. 2004. Notas sobre la anidación del Tororoi Bigotudo (*Grallaria alleni*) en la Cordillera Central de Colombia. *Orn. Colombiana* 2: 19–24.

Londoño, G.A., Chappell, M.A., Castaneda, M.R., Jankowski, J.E. & Robinson, S.K. 2015. Basal metabolism in tropical birds: latitude, altitude, and the 'pace of life'. *Functional Ecol.* 29: 338–346.

Lönnberg, E. & Rendahl, H. 1922. A contribution to the ornithology of Ecuador. *Ark. Zool.* 14: 1–87.

Lopes, E.V., dos Anjos, L., Loures-Ribeiro, A., Gimenes, M.R., Mendonça, L.B., Volpato, G.H. & Silva, R.J. 2001. Efeito da fragmentação florestal sobre aves da família Formicariidae na região de Londrina, norte do Paraná. Resumo 117 *in* Straube, F.C. (ed.) *IX Congresso Brasileiro de Ornitologia, Resumos: Ornitologia Sem Fronteras*. Curitiba.

Lopes, L.E., Fernandes, A.M. & Marini, M.Â. 2005a. Predation on vertebrates by Neotropical passerine birds. *Lundiana* 6: 57–66.

Lopes, L.E., Fernandes, A.M. & Marini, M.Â. 2005b. Diet of some Atlantic Forest birds. *Ararajuba* 13: 95–103.

Lopes, L.E., Malacco, G.B., Vasconcelos, M.F., Carvalho, C.E.D., Duca, C., Fernandes, A.M., Neto, S.D. & Marini, M.A. 2008. Aves da região de Unaí e Cabeceira Grande, noroeste de Minas Gerais, Brasil. *Rev. Bras. Orn.* 16: 193–206.

Lopes, L.E., Peixoto, H.J.C. & Nogueira, W. 2012. Aves da Serra Azul, sul da Cadeia do Espinhaço, Minas Gerais, Brasil. *Atualidades Orn.* 169: 41–53.

Lopes, L.E., Peixoto, H.J.C. & Hoffmann, D. 2013. Notas sobre a biologia reprodutiva de aves brasileiras. *Atualidades Orn.* 171: 33–49.

Lopes, O.S., Sacchetta, L.A. & Dente, E. 1980. Longevity of wild birds obtained during a banding program in São Paulo, Brasil. *J. Field Orn.* 51: 144–148.

López, K., Miranda, J., Espinoza, F. & Machado, M. 2012. Observaciones sobre *Grallaricula loricata*, una especie endemica y amenazada de la cordillera de la costa en Venezuela. Pp. 381–382 *in* More, A., Olaechea, D.G. & Olaechea, A.G. (eds.) *Libro de Resúmenes del IX Congreso de Ornitología Neotropical y el VIII Congreso Peruano de Ornitología*. Cuzco.

López-Lanús, B., Salaman, P.G.W., Cowley, T.P., Arango, S. & Renjifo, L.M. 2000. The threatened birds of the Río

Toche, Cordillera Central, Colombia. *Cotinga* 14: 17–23.

López-Ordóñez, J.P., Cortés-Herrera, J.O., Paez-Ortíz, C.A. & González-Rojas, M.F. 2013. Nuevos registros y comentarios sobre la distribución de algunas especies de aves en los Andes Occidentales de Colombia. *Orn. Colombiana* 13: 21–36.

López-O., J.P., Avendaño, J.E., Gutiérrez-Pinto, N. & Cuervo, A.M. 2014. The birds of Serranía de Perijá: the northernmost avifauna of the Andes. *Orn. Colombiana* 14: 62–93.

Losada-Prado, S., Carvajal-Lozano, A.M. & Molina-Martínez, Y.G. 2005. Listado de especies de aves de la cuenca del río Coello (Tolima, Colombia). *Biota Colombiana* 6: 101–116.

Lotta, I.A., Gonzalez, A.D., Pacheco, M.A., Escalante, A.A., Valkiunas, G., Moncada, L.I. & Matta, N.E. 2015. *Leucocytozoon pterotenuis* sp. nov. (Haemosporida, Leucocytozoidae): description of the morphologically unique species from the Grallariidae birds, with remarks on the distribution of *Leucocytozoon* parasites in the Neotropics. *Parisitol. Res.* 114: 1031–1044.

Lowen, J.C., Bartrina, L., Brooks, T.M., Clay, R.P. & Tobias, J.A. 1996. Project Yucutinga '95: bird surveys and conservation priorities in eastern Paraguay. *Cotinga* 5: 14–19.

Lowen, J.C., Clay, R.P., Mazar Barnett, J., Madroño Nieto, A., Pearman, M., López-Lanús, B., Tobias, J.A., Liley, D.C., Esquivel, E.Z. & Reid, J.M. 1997. New and noteworthy observations on the avifauna of Paraguay. *Bull. Brit. Orn. Club* 117: 275–293.

Lowen, J.C., Clay, R.P., Brooks, T.M., Esquivel, E.Z., Bartrina, L., Barnes, R., Butchart, S.H.M. & Etcheverry, N.I. 1995. Bird conservation in the Paraguayan Atlantic Forest. *Cotinga* 4: 58–64.

Lowery, G.H., Jr. & Dalquest, W.W. 1951. Birds from the state of Veracruz, Mexico. *Univ. Kansas Publ. Mus. Nat. Hist.* 3: 531–649.

Lowery, G.H. & O'Neill, J.P. 1969. A new species of antpitta from Peru and a revision of the subfamily Grallariinae. *Auk* 86: 1–12.

Lunardi, V.O. 2004. Análise genética molecular (RAPD) de *Conopophaga melanops*, Vieillot 1818 (Aves, Conopophagidae), em escala fina de Mata Atlântica e sua implicação para a conservação da espécie. M.Sc. dissertation. Univ. Federal de São Carlos, São Carlos.

Lunardi, V.O., Francisco, M.R. & Galleti, P.M. 2007. Population structuring of the endemic Black-cheeked Gnateater, *Conopophaga melanops melanops* (Vieillot 1818) (Aves, Conopophagidae), in the Brazilian Atlantic Forest. *Braz. J. Biol.* 67: 867–872.

Lüthi, H. 2011. Birdwatching in Peru: 1963-2006. *Rev. Peru. Biol.* 18: 27–90.

Lyra-Neves, R.M., Dias, M.M., Azevedo, S.M., Júnior, W.R.T. & Larrazábal, M.E.L. 2004. Comunidade de aves da Reserva Estadual de Gurjaú, Pernambuco, Brasil. *Rev. Bras. Zool.* 21: 581–592.

Machado, A.B.M., Drummond, G.N. & Paglia, A.P. 2008. *Livro Vermelho da Fauna Brasileira Ameaçada de Extinção*. Ministério do Meio Ambiente & Ministério da Educação, Brasília.

Machado, R.B. & Fonseca, G.A.B. 2000. The avifauna of Rio Doce valley, southeastern Brazil, a highly fragmented area. *Biotropica* 32: 914–924.

MacLeod, R., Ewing, S.K., Herzog, S.K., Bryce, R., Evans, K.L. & MacCormick, A. 2005. First ornithological inventory and conservation assessment for the yungas forests of the Cordilleras Cocapata and Mosetenes, Cochabamba, Bolivia. *Bird Conserv. Intern.* 15: 361–382.

Madroño Nieto, A., Clay, R.P., Robbins, M.B., Rice, N.H., Faucett, R.C. & Lowen, J.C. 1997a. An avifaunal survey of the vanishing interior Atlantic forest of San Rafael National Park, Departments Itapúa/Caasapá, Paraguay. *Cotinga* 7: 45–53.

Madroño Nieto, A., Robbins, M.B. & Zyskowski, K. 1997b. Contribución al conocimiento ornitológico del Bosque Atlantico interior del Paraguay: Parque Nacional Caaguazú, Caazapá. *Cotinga* 7: 54–60.

Magalhães, V.S., Júnior, S.M.A., Lyra-Neves, R.M., Telino-Júnior, W.R. & de Souza, D.P. 2007. Biologia de aves capturadas em um fragmento de Mata Atlântica, Igarassu, Pernambuco, Brasil. *Rev. Bras. Zool.* 24: 950–964.

Maia-Gouvea, E.R., Gouvea, E. & Piratelli, A.J. 2005. Comunidade de aves de sub-bosque em uma área de entorno do Parque Nacional do Itatiaia, Rio de Janeiro, Brasil. *Rev. Bras. Zool.* 22: 859–866.

Maijer, S. 1998. Rediscovery of *Hylopezus* (*macularius*) *auricularis*: distinctive song and habitat indicate species rank. *Auk* 115: 1072–1073.

Maijer, S. 1999. Bolivian Spinetail *Cranioleuca henricae* and Masked Antpitta *Hylopezus auricularis*. *Cotinga* 11: 71–73.

Maijer, S., Christiansen, M.B. & Pitter, E. 2000. Birds observed along the road Vallegrande - Masicurí, depot. Santa Cruz, Bolivia, in 1991–1993. Unpublished Report.

Maillard-Z., O. 2009. *Hylopezus auricularis* (Gyldenstolpe, 1941). Pp. 326–327 *in* Aguirre, L.F., Aguayo, R., Balderrama, J., Cortez, C. & Tarifa, T. (eds.) *Libro Rojo de la Fauna Silvestre de Vertebrados de Bolivia*. Ministerio de Medio Ambiente y Agua, Viceministerio de Medio Ambiente, Biodiversidad y Cambios Climáticos, La Paz.

Maillard-Z., O. 2012. Hábitat, comportamiento y nidificación del Tororoí Enmascarado (*Hylopezus auricularis*), en el norte de la amazonia boliviana. B.Sc. dissertation. Universidad Autónoma "Gabriel Rene Moreno", Santa Cruz de la Sierra.

Maillard-Z., O. & Vogel, C.J. 2003. First description of nest and eggs of the Ochre-breasted Antpitta (*Grallaricula flavirostris*). *Orn. Neotrop.* 14: 129–131.

Maillard-Z., O., Vidoz, J.Q. & Herrera, M. 2008. Registros significativos de aves para el Departamento de Beni, Bolivia: parte 2. *Kempffiana* 4: 8–12.

Maillard-Z., O., Davis, S.E. & Hennessey, A.B. 2009. Bolivia. Pp. 91–98 *in* Devenish, C., Fernández, D.F.D., Clay, R.P., Davidson, I. & Zabala, I.Y. (eds.) *Important Bird Areas Americas - Priority Sites for Biodiversity Conservation*. BirdLife International, Quito.

Maldonado-Coelho, M. & Marini, M.Â. 2003. Composição de bandos mistos de aves em fragmentos de mata atlântica no sudeste do Brasil. *Pap. Avuls. Dept. Zool., São Paulo* 43: 31–54.

Malizia, L.R., Blendinger, P.G., Alvarez, M.E., Rivera, L.O., Politi, N. & Nicolossi, G. 2005. Bird communities in Andean premontane forests of northwestern Argentina. *Orn. Neotrop.* 16: 231–251.

Mallet-Rodrigues, F. 2005. Molt-breeding cycle in passerines from a foothill forest in southeastern Brazil. *Rev. Bras. Orn.* 13: 155–160.

Mallet-Rodrigues, F. & Noronha, M.L.M. 2003. The avifauna of low elevations in the Serra dos Órgãos, Rio de Janeiro state, south-east Brazil. *Cotinga* 20: 51–56.

Mallet-Rodrigues, F. & Noronha, M.L.M. 2009. Birds in the Parque Estadual dos Três Picos, Rio de Janeiro state, south-east Brazil. *Cotinga* 31: 96–107.

Mallet-Rodrigues, F. & Pacheco, J.F. 2003. O registro supostamente brasileiro de *Grallaria guatimalensis roraimae* Chubb, 1921. *Ararajuba* 11: 269–270.

Mallet-Rodrigues, F., Parrini, R. & Pacheco, J.F. 2007. Birds of the Serra dos Órgãos, state of Rio de Janeiro, southeastern Brazil: a review. *Rev. Bras. Orn.* 15: 5–35.

Mallet-Rodrigues, F., Parrini, R., Pimentel, L.M.S. & Bessa, R. 2010. Altitudinal distribution of birds in a mountainous region in southeastern Brazil. *Zoologia* 27: 503–522.

Mallet-Rodrigues, F., Parrini, R. & Rennó, B. 2015. Bird species richness and composition along three elevational gradients in southeastern Brazil. *Atualidades Orn.* 188: 39–58.

Manhães, M.A. 2007. Ecologia trófica de aves de sub-bosque em duas áreas de Mata Atlântica secundária no sudeste do Brasil. Ph.D. thesis. Univ. Federal de São Carlos, São Carlos.

Manhães, M.A. & Dias, M.M. 2011. Spatial dynamics of understorey insectivorous birds and arthropods in a southeastern Brazilian Atlantic woodlot. *Brazil. J. Biol.* 71: 1–7.

Manhães, M.A. & Loures-Ribeiro, A. 2011. Avifauna da Reserva Biológica Municipal Poço D'Anta, Juiz de Fora, MG. *Biota Neotropica* 11: 275–286.

Manhães, M.A., Loures-Ribeiro, A. & Dias, M.M. 2010. Diet of understorey birds in two Atlantic Forest areas of Southeast Brazil. *J. Nat. Hist.* 44: 469–489.

Manne, L.L. & Pimm, S.L. 2001. Beyond eight forms of rarity: which species are threatened and which will be next? *Anim. Conserv.* 4: 221–229.

Marchant, S. 1959. The breeding season in S.W. Ecuador. *Ibis* 101: 137–152.

Marchant, S. 1960a. The breeding of some S.W. Ecuadorian birds. *Ibis* 102: 349–382.

Marchant, S. 1960b. The breeding of some S.W. Ecuadorian birds (cont.). *Ibis* 102: 584–599.

Marcus, M.J. 1983. Additions to the avifauna of Honduras. *Auk* 100: 621–629.

Marenzi, R., Zimmermann, C.E. & Marenzi, A.W.C. 2006. Bird community in an Atlantic rain forest fragment - Praia Vermelha, Santa Catarina, Brazil. *J. Coastal Res.* 1789–1792.

Marín-Gómez, O.H. 2005. Avifauna del campus de la Universidad del Quindío. *Bol. Soc. Antioqueña Orn.* 15: 42–60.

Marín-Gómez, O.H. 2012. Inventario de la avifauna de la reserva natural "La montaña del ocaso". *Rev. Asoc. Colombiana Cienc. Biol.* 24: 129–142.

Marín-Gómez, O.H., Polanco, J.M., Arango-Giraldo, D. & Ospina-Duque, A. 2015. A new population of the Hooded Antpitta (*Grallaricula cucullata*: Grallaridae) for the Colombian Central Andes. *Acta Biol. Colombiana* 20: 229–232.

Marini, M.Â. 2001. Effects of forest fragmentation on birds of the cerrado region, Brazil. *Bird Conserv. Intern.* 11: 13–25.

Marini, M.Â. 2010. Bird movement in a fragmented Atlantic Forest landscape. *Stud. Neotrop. Fauna Environ.* 45: 1–10.

Marini, M.Â., Aguilar, T.M., Andrade, R.D., Leite, L.O., Anciães, M., Carvalho, C.E.A., Duca, C., Maldonado-Coelho, M., Sebaio, F. & Gonçalves, J. 2007. Biologia da nidificação de aves do sudeste de Minas Gerais, Brasil.

Rev. Bras. Orn. 15: 367–376.

Marini, M.Â., Motta-Junior, J.C., Vasconcellos, L.A.S. & Cavalcanti, R.B. 1997. Avian body masses from the cerrado region of Central Brazil. *Orn. Neotrop.* 8: 93–99.

Marini, M.Â., Reinert, B.L., Bornschein, M.R., Pinto, J.C. & Pichorim, M.A. 1996. Ecological correlates of ectoparasitism on Atlantic Forest birds, Brazil. *Ararajuba* 4: 93–102.

Mark, T., Augustine, L., Barrio, J., Flanagan, J. & Vellinga, W.-P. 2008. New records of birds from the northern Cordillera Central of Peru in a historical perspective. *Cotinga* 29: 108–125.

Marques, F.C. & dos Anjos, L. 2014. Sensitivity to fragmentation and spatial distribution of birds in forest fragments of northern Paraná. *Biota Neotropica* 14: 1–8.

Martínez-Sánchez, J.C. & Will, T. 2010. Thomas R. Howell's *Check-list of the Birds of Nicaragua* as of 1993. *Orn. Monogr.* 68: 1–108.

Marsden, S., Whiffin, M., Sadgrove, L. & Guimarães, P.R. 2003. Bird community composition and species abundance on two inshore islands in the Atlantic forest region of Brazil. *Ararajuba* 11: 181–187.

Martensen, A.C., Ribeiro, M.C., Banks-Leite, C., Prado, P.I. & Metzger, J.P. 2012. Associations of forest cover, fragment area, and connectivity with neotropical understory bird species richness and abundance. *Conserv. Biol.* 26: 1100–1111.

Martin, P.R. & Dobbs, R.C. 2004. Description of the nest, egg and nestling of Watkins's Antpitta *Grallaria watkinsi* *Cotinga* 21: 35–37.

Martin, P.R. & Greeney, H.F. 2006. Description of the nest, eggs and nesting period of the Chestnut-crowned Antpitta *Grallaria ruficapilla* from the eastern Ecuadorian Andes. *Cotinga* 25: 47–49.

Martínez, O. & Rechberger, J. 2007. Características de la avifauna en un gradiente altitudinal de un bosque nublado andino en La Paz, Bolivia. *Rev. Peru. Biol.* 14: 225–236.

Martinez, Y.G.M. 2014. Birds of the Totare River Basin, Colombia. *Check List* 10: 269–286.

Martínez-Sánchez, J.C., Chavarría-Duriaux, L. & Muñoz, F.J. 2014. *A Guide to the Birds of Nicaragua.* VerlagsKG Wolf, Magdeburg.

Mason, D. 1996. Responses of Venezuelan understory birds to selective logging, enrichment strips, and vine cutting. *Biotropica* 28: 296–309.

Mattos, G.T., Andrade, M.A. & Freitas, M.V. 1991. Levantamento de aves silvestres na região noroeste de Minas Gerais. *Rev. Soc. Orn. Mineira* 39: 26–29.

Maurício, G.N. & Dias, R.A. 2000. New distributional information for birds in southern Rio Grande do Sul, Brazil, and the first record of the Rufous Gnateater *Conopophaga lineata* for Uruguay. *Bull. Brit. Orn. Club* 120: 230–236.

Maurício, G.N., Bencke, G.A., Repenning, M., Machado, D.B., Dias, R.A. & Bugoni, L. 2013. Review of the breeding status of birds in Rio Grande do Sul, Brazil. *Iheringia, Sér. Zool.* 103: 163–184.

Mayfield, H. 1961. Nesting success calculated for exposure. *Wilson Bull.* 73: 255–261.

Mayr, E. & Phelps, W.H., Jr. 1967. The origin of the bird fauna of the south Venezuelan highlands. *Bull. Amer. Mus. Nat. Hist.* 13: 269–328.

Mazar Barnett, J., Carlos, C.J. & Roda, S.A. 2005. Renewed hope for the threatened avian endemics of northeastern Brazil. *Biodiver. & Conserv.* 14: 2265–2274.

Mazar Barnett, J. & Kirwan, G.M. 1999. Neotropical notebook. *Cotinga* 11: 96–105.

Mazar Barnett, J. & Pearman, M. 2001. *Lista Comentada de las Aves Argentinas.* Lynx Edicions, Barcelona.

McMullan, W.M. & Navarrete, L. 2013. *Fieldbook of the Birds of Ecuador including the Galapagos Islands.* Jocotoco Foundation, Quito.

McMullan, W.M., Donegan, T.M. & Quevedo, A. 2010. *Field Guide to the Birds of Colombia.* Fundación ProAves, Bogotá.

McNish M., T. 2007. *Las Aves de los Llanos de la Orinoquía.* M & B, Bogotá.

Mee, A., Ohlson, J., Stewart, I., Wilson, M., Örn, P. & Ferreyra, J.D. 2002. The Cerros del Sira revisted: birds of submontane and montane forest. *Cotinga* 18: 46–57.

Ménégaux, A. 1904. Catalogue des oiseaux rapportés par M. Geay de la Guyane française et du Contesté Franco-Bresilien. *Bull. Mus. Natl. d'Hist. Nat. (Paris)* 10: 174–186.

Ménégaux, A. 1910. Étude d'une collection d'oiseaux du Pérou. *Bull. Mus. Natl. d'Hist. Nat. (Paris)* 16: 359–367.

Ménégaux, A. 1911. *Mission du Service Geographique de l'Armée pour la mesure d'un arc de méridien équatorial en Amérique du Sud sous le contrôle scientifique de l'Académie des Sciences, 1899-1906: Étude des oiseaux de L'Équateur rapportés par le Dr Rivet.* Gauthier-Villars, Imprimeur Libraire du Bureau des Longitudes, de l'Ecole Polytechnique, Paris.

Ménégaux, A. & Hellmayr, C.E. 1906. Etude des espèces critiques et des types du groupe des Passereaux Trachéophones de l'Amérique tropicale. *Bull. Soc. Phil. Paris* (9)8: 24–58.

Ménétries, E. 1835. Monographie de la famille des Myiotherinae. *Mem. Acad. Imperiale Sci. St. Petersbourgh* (6)3: 443–543.

Merkord, C.L. 2010. Seasonality and Elevational Migration in an Andean Bird Community. Ph.D. thesis. Univ. of Missouri, Columbia, MO.

Merkord, C.L., Mark, T., Susanibar, D., Johnson, A. & Witt, C.C. 2009. Avifaunal survey of the Río Chipaota Valley in the Cordillera Azul region, San Martín, Peru. *Orn. Neotrop.* 20: 535–552.

Mestre, L.A.M. 2004. Avifauna da Fazenda Arapongas - Floresta com Araucaria-Lapa-Paraná. P. 301 *in* Testoni, A.F. & Althoff, S.L. (eds.) *Livro de Resumos do XII Congresso Brasileiro de Ornitologia.* Blumenau.

Mestre, L.A.M., Thom, G., Cochrane, M.A. & Barlow, J. 2010. The birds of Reserva Extrativista Chico Mendes, south Acre, Brazil. *Bol. Mus. Para. E. Goeldi, Sér. Cienc. Nat.* 5: 311–333.

Meyer, D. 2016. Avifauna do município de Salete, Santa Catarina. *Atualidades Orn.* 193: 65–77.

Meyer de Schauensee, R. 1945. Notes on Colombian antbirds, ovenbirds and woodhewers, with the description of a new form from Peru. *Not. Naturae* 153: 1–15.

Meyer de Schauensee, R. 1950a. Colombian Zoological Survey. Part VII: a collection of birds from Bolívar, Colombia. *Proc. Acad. Nat. Sci. Phil.* 102: 111–139.

Meyer de Schauensee, R. 1950b. The birds of the Republic of Colombia. *Caldasia* 5: 645–747.

Meyer de Schauensee, R. 1951. Colombian zoological survey. Part VIII. On birds from Nariño, Colombia, with the description of four new subspecies. *Not. Naturae* 232: 1–6.

Meyer de Schauensee, R. 1952a. The birds of the Republic of Colombia (addenda and corrigenda). *Caldasia* 5: 1115–1223.

Meyer de Schauensee, R. 1952b. Colombian zoological survey. Part X. A collection of birds from southeastern Nariño, Colombia. *Proc. Acad. Nat. Sci. Phil.* 104: 1–33.

Meyer de Schauensee, R. 1964. *The Birds of Colombia and Adjacent Areas of South and Central America.* Livingston Publishing, Narberth, PA.

Meyer de Schauensee, R. 1966. *The Species of Birds of South America and Their Distribution.* Livingston Publishing, Narberth, PA.

Meyer de Schauensee, R. 1970. *A Guide to the Birds of South America.* Livingston Publishing Co., Wynnewood, PA.

Meyer de Schauensee, R. 1982. *A Guide to the Birds of South America.* Academy of Natural Science, Philadelphia, PA.

Meyer de Schauensee, R. & Phelps, W.H. 1978. *A Guide to the Birds of Venezuela.* Princeton Univ. Press, Princeton, NJ.

Milensky, C.M., Robbins, M.B., Saucier, J.R., O'Shea, B.J., Radosavljevic, A., Davis, T.J. & Pierre, M. 2016. Notes on breeding birds from the Guyana highlands with new records from a recent inventory of Mount Ayanganna. *Cotinga* 38: 64–78.

Miller, A.H. 1963. Seasonal activity and ecology of the avifauna of an American equatorial cloud forest. *Univ. Cal. Publ. Zool.* 66: 1–78.

Miller, E.T., Wagner, S.K., Klavins, J., Brush, T. & Greeney, H.F. 2015. Striking courtship displays in the becard clade *Platypsaris. Wilson J. Orn.* 127: 123–126.

Miller, L.E. 1918. *In the Wilds of South America; Six Years of Exploration in Colombia, Venezuela, British Guiana, Peru, Bolivia, Argentina, Paraguay, and Brazil.* Scribner, New York.

Miller, M.J., Weir, J.T., Angehr, G.R., Guitton-M., P. & Bermingham, E. 2011. An ornithological survey of Piñas Bay, a site on the Pacific coast of Darién Province, Panama. *Bol. Soc. Antioqueña Orn.* 20: 29–38.

Mironov, S. & Hernandes, F.A. 2014. Two new species of the feather mite genus *Analloptes* (Trouessart, 1885) (Acariformes: Astigmata: Xolalgidae) from passerines (Aves: Passeriformes) in Brazil. *Zootaxa* 3889: 589–600.

Mischler, T. 2012. Status, abundance, seasonality, breeding evidence and an updated list of the birds of Cerro Blanco, Guayaquil, Ecuador. *Cotinga* 34: 60–72.

Mischler, T.C. & Sheets, D.R. 2007. *Catálogo Diagnóstico de las Aves del Bosque Protector Cerro Blanco* (2nd edn.). BirdLife International, Fundación Pro-Bosque & Universidad Católica de Santiago de Guayaquil, Guayaquil.

Mistry, J., Berardi, A. & Simpson, M. 2008. Birds as indicators of wetland status and change in the North Rupununi, Guyana. *Biodiver. & Conserv.* 17: 2383–2409.

Mlíkovský, J. 2007. Types of birds in the collections of the Museum and Institute of Zoology, Polish Academy of Sciences, Warszawa, Poland. Part 1: introduction and European birds. *J. Natl. Mus. (Prague), Nat. Hist. Ser.* 176: 15–31.

Mlíkovský, J. 2009. Types of birds in the collections of the Museum and Institute of Zoology, Polish Academy of Sciences, Warszawa, Poland. Part 3: South American birds. *J. Natl. Mus. (Prague), Nat. Hist. Ser.* 178: 17–180.

Mlíkovský, J. & Frahnert, S. 2009. Type specimens and type localities of Peruvian birds described by Jean Cabanis on the basis of Konstanty Jelski's collections. *Zootaxa* 2171: 29–47.

Molina-Martínez, Y.G., Diaz, H.M. & Gómez, C. 2008. Aves. Pp. 200–229 *in* Reinoso-Florez, G., Villa-Navarro, F.A., Esquivel, H.E., Garcia-Melo, J.E. & Vejarano-Delgado, M.A. (eds.) *Biodiversidad Faunística y Florística de la Cuenca del río Lagunillas - Biodiversidad Regional Fase IV.* Grupo de Investigación en Zoología, Universidad del Tolima, Ibagué.

Monroe, B.L. 1968. A distributional survey of the birds of Honduras. *Orn. Monogr.* 7: 1–458.

Monroe, B.L. & Sibley, C.G. 1993. *A World Checklist of Birds.* Yale Univ. Press, New Haven, CT.

Montambault, J.R. 2002. Informes de las evaluaciones biológicas Pampas del Heath, Perú, Alto Madidi, Bolivia y Pando, Bolivia. *RAP Bull. Biol. Assess.* 24: 1–125.

Moore, R.T. 1934. The Mt. Sangay labyrinth and its fauna. *Auk* 51: 141–156.

Mordecai, R.S., Cooper, R.J. & Justicia, R. 2009. A threshold response to habitat disturbance by forest birds in the Choco Andean corridor, northwest Ecuador. *Biodiver. & Conserv.* 18: 2421–2431.

Moritz, C. 2002. Strategies to protect biological diversity and the evolutionary processes that sustain it. *Syst. Biol.* 51: 238–254.

Morony, J.J., Bock, W.J. & Farrand, J. 1975. *Reference List of the Birds of the World.* American Museum of Natural History, New York.

Morrison, A. 1939. The birds of the Department of Huancavelica, Peru. *Ibis* 81: 453–486.

Morrison, A. 1940. Las aves del Departamento de Huancavelica. *Bol. Mus. Hist. Nat. Javier Prado* 4: 242–246.

Morrison, A. 1948. Notes on the birds of the Pampas River valley, south Peru. *Ibis* 90: 119–126.

Moschione, F., Segovia, J. & Burgos, F. 2005. Reserva Natural Las Lancitas. Pp. 219–220 *in* Di Giacomo, A.S. (ed.) *Áreas Importantes para la Conservación de las Aves en Argentina. Sitios Prioritarios para la Conservación de la Biodiversidad.* Aves Argentinas/Asociación Ornitológica del Plata, Buenos Aires.

Moskovits, D.K., Fitzpatrick, J.W. & Willard, D.E. 1985. Lista preliminar das aves de Estação Ecológica de Maracá, Território de Roraima, Brasil, e áreas adjacentes. *Pap. Avuls. Dept. Zool., São Paulo* 36: 51–68.

Motta-Junior, J.C. 1990. Estrutura trófica e composição das avifaunas de tres habitats terrestres na região central do estado de São Paulo. *Ararajuba* 1: 65–71.

Mountfort, G. & Arlott, N. 1988. *Rare Birds of the World.* Collins, London.

Moura, A.S., Corrêa, B.S. & Machado, F.S. 2015. Riqueza, composição e similaridade da avifauna em remanescente florestal e áreas antropizadas no sul de Minas Gerais *Rev. Agrogeoambiental* 7: 41–52.

Moura, N.G., Lees, A.C., Aleixo, A., Barlow, J., Berenguer, E., Ferreira, J., MacNally, R., Thomson, J.R. & Gardner, T.A. 2016. Idiosyncratic responses of Amazonian birds to primary forest disturbance. *Oecologia* 180: 903–916.

Moyle, R.G., Chesser, R.T., Brumfield, R.T., Tello, J.G., Marchese, D.J. & Cracraft, J. 2009. Phylogeny and phylogenetic classification of the antbirds, ovenbirds, woodcreepers, and allies (Aves: Passeriformes: infraorder Furnariides). *Cladistics* 25: 386–405.

Müller, J. 1847. Über die bisher unbekannten typischen Verschiedenheiten der Stimmorgane der Passerinen. *Abh. K. Akad. Wissen. Berlin* 1845: 321–391, 405–406.

Müller, J.A., Scherer-Neto, P., Carrano, E. & Andreiv, J. 2001. Avifauna do Parque Natural Municipal São Francisco de Assis, Blumenau, Santa Catarina. Resumo 137 *in* Straube, F.C. (ed.) *Livro de Resumos do IX Congresso Brasileiro de Ornitologia: Ornitologia Sem Fronteiras.* Curitiba.

Müller, J.A., Scherer-Neto, P., Carrano, E., Andreiv, J. &

Zimmermann, A.F. 2003. A diversidade de aves de uma Reserva Florestal Particular no Município de Blumenau, Santa Catarina. P. 105 *in* Machado, C.G. (ed.) *Livro de Resumos do XI Congresso Brasileiro de Ornitologia*. Feira de Santana.

Munsell Color. 1977. *Munsell® Book of Color*. Munsell Color, Macbeth Division of Kollmorgen Corp., Baltimore, MD.

Munves, J. 1975. Birds of a highland clearing in Cundinamarca, Colombia. *Auk* 92: 307–321.

Naka, L.N. 2010. The Role of Physical and Ecological Barriers in the Diversification Process of Birds in the Guiana Shield, Northern Amazonia. Ph.D. dissertation. Louisiana State Univ., Baton Rouge, LA.

Naka, L.N., Rodrigues, M., Roos, A.L. & Azevedo, M.A.G. 2002. Bird conservation on Santa Catarina Island, southern Brazil. *Bird Conserv. Intern.* 12: 123–150.

Naka, L.N., Cohn-Haft, M., Mallet-Rodrigues, F., Santos, M.P.D. & Torres, M.F. 2006. The avifauna of the Brazilian state of Roraima: bird distribution and biogeography in the Rio Branco basin. *Rev. Bras. Orn.* 14: 197–238.

Naka, L.N., Bechtoldt, C.L., Henriques, L.M.P. & Brumfield, R.T. 2012. The role of physical barriers in the location of avian suture zones in the Guiana Shield, northern Amazonia. *Amer. Natur.* 179: E115–E132.

Naranjo, L.G. 1994. Composicion y estructura de la avifauna del Parque Regional Natural Ucumari. Pp. 305–325 *in* Rangel, J.O. (ed.) *Ucumari: Un Caso Típico de la Diversidad Biotica Andina*. Corporation Autonoma Regional de Risaralda, Pereira.

Naranjo, L.G. & de Ulloa, P.C. 1997. Diversity of understory insects and insectivorous birds in disturbed tropical rainforest habitats. *Caldasia* 19: 507–520.

Narosky, T. & Yzurieta, D. 1987. *Guía para la Identificación de las Aves de Argentina y Uruguay*. Asociación Ornitológica del Plata, Buenos Aires.

Narosky, T. & Yzurieta, D. 1989. *Birds of Argentina and Uruguay: A Field Guide* (1st edn.). Asociación Ornitológica del Plata, Buenos Aires.

Narosky, T. & Yzurieta, D. 1993. *Birds of Argentina and Uruguay: A Field Guide* (2nd edn.). Asociación Ornitológica del Plata, Buenos Aires.

Nascimento, A.M.A., Cursino, L., Goncalves-Dornelas, H., Reis, A., Chartone-Souza, E. & Marini, M.Â. 2003. Antibiotic-resistant gram-negative bacteria in birds from the Brazilian Atlantic Forest. *Condor* 105: 358–361.

Nascimento, J.L.X., Nascimento, I.L.S. & Júnior, S.M.A. 2000. Aves da Chapada do Araripe (Brasil): biologia e conservação. *Ararajuba* 8: 115–125.

Naumburg, E.M.B. 1930. The birds of Matto Grosso, Brazil: a report on the birds secured by the Roosevelt-Rondon Expedition. *Bull. Amer. Mus. Nat. Hist.* 60: 1–432.

Naumburg, E.M.B. 1937. Studies of birds from eastern Brazil and Paraguay, based on a collection made by Emil Kaempfer. Conopophagidae, Rhinocryptidae, Formicariidae (part). *Bull. Amer. Mus. Nat. Hist.* 74: 139–205.

Naumburg, E.M.B. 1939. Studies of birds from eastern Brazil and Paraguay, based on a collection made by Emil Kaempfer. Formicariidae. *Bull. Amer. Mus. Nat. Hist.* 76: 231–276.

Navarro-Sigüenza, A.G. 1992. Altitudinal distribution of birds in the Sierra Madre del Sur, Guerrero, Mexico. *Condor* 94: 29–39.

Navarro-Sigüenza, A.G., Chavez, M.G.T. & Pliego, B.P.E. 1991. *Catalogo de Aves (Vertebrata: Aves) del Museo de Zoologia "Alfonso L. Herrera"*. Universidad Nacional Autonoma de Mexico, Mexico City.

Negret, A.J. 1994. Lista de aves registradas en el Parque Nacional Munchique, Cauca. *Nov. Colombianas, Nueva Época* 6: 69–83.

Negret, A.J. 1997. Adiciones a la avifauna del Parque Nacional Munchique, Cauca. *Nov. Colombianas, Nueva Época* 7: 8.

Negret, A.J. 2001. *Aves en Colombia Amenazadas de Extinción*. Editorial Universidad del Cauca, Popayán.

Nehrkorn, A. 1899. *Katalog der Eiersammlung nebst Beschreibungen der Aussereuropäschen Eier*. Harald Bruhn, Braunschweig.

Nehrkorn, A. 1910. *Katalog der Eiersammlung nebst Beschreibungen der aussereuropäschen Eier. 2 Auflage*. R. Friedländer & Sohn, Berlin.

Nehrkorn, A. 1914. *Nachträge zu Nehrkorn's Eierkatalog*. R. Friedländer & Sohn, Berlin.

Nelson, E.W. 1898. Description of new birds from Mexico, with a revision of the genus *Dactylortyx*. *Proc. Biol. Soc. Wash.* 12: 57–68.

Nelson, E.W. 1912. Descriptions of new genera, species and subspecies of birds from Panama, Colombia, and Ecuador. *Smithsonian Misc. Coll.* 60: 1–25.

Nemeth, E., Winkler, H. & Dabelsteen, T. 2001. Differential degradation of antbird songs in a Neotropical rainforest: adaptation to perch height? *J. Acoustic. Soc. Amer.* 110: 3263–3274.

Neumann, L.G. 1931. Neue Unterarten südamerikanischer Vögel. *Mitt. Mus. Berlin* 17: 441–445.

Niceforo M., H. 1945. Notas sobre aves de Colombia I. *Caldasia* 3: 367–395.

Niceforo M., H. 1947. Notas sobre aves de Colombia II. *Caldasia* 4: 317–377.

Nieto-R., M. & Ramírez, J.D. 2006. Notas sobre aves de tierras áltas que siguen marchas de hormigas arrieras para su alimentación, en la Reserva Natural Río Blanco, Manizales, Caldas. *Bol. Soc. Antioqueña Orn.* 16: 59–66.

Niklison, A.M., Areta, J.I., Ruggera, R.A., Decker, K.L., Bosque, C. & Martin, T.E. 2008. Natural history and breeding biology of the Rusty-breasted Antpitta (*Grallaricula ferrugineipectus*). *Wilson J. Orn.* 120: 345–352.

Nilsson, J., Jonsson, R. & Krabbe, N. 2001. First record of Bicoloured Antpitta *Grallaria rufocinerea* from Ecuador, with notes on the species' vocalisations. *Cotinga* 16: 105–106.

Nogueira, D.M., Freitas, A.A.R., Silva, C.P. & Souza, L.M. 2005. Estudio de la avifauna y sus ectoparasitos en un fragmento de Bosque Atlantico en la ciudad de Rio de Janeiro, Brasil. *Bol. Soc. Antioqueña Orn.* 15: 26–36.

Nolan, V. 1978. Ecology and behavior of the Prairie Warbler, *Dendroica discolor*. *Orn. Monogr.* 26: 1–595.

Nores, M., Cerana, M.M. & Serra, D.A. 2005. Dispersal of forest birds and trees along the Uruguay River in southern South America. *Divers. & Distrib.* 11: 205–217.

Novaes, F.C. 1947. Notas sôbre os Conopophagidae do Museu Nacional: Passeriformes, Aves. *Summa Bras. Biol.* 1: 243–250.

Novaes, F.C. 1957. Contribuição a ornitologia do noroeste do Acre. *Bol. Mus. Para. E. Goeldi, N. Sér. (Zool.)* 9: 1–30.

Novaes, F.C. 1958. As aves e as comunidades bióticas no alto Rio Juruá, territorio do Acre. *Bol. Mus. Para. E. Goeldi, N. Sér. (Zool.)* 14: 1–13.

Novaes, F.C. 1970. Distribuição ecológica e abundância das aves em um trecho da mata do baixo Rio Guamá (Estado do Pará). *Bol. Mus. Para. E. Goeldi, N. Sér. (Zool.)* 71: 1–54.

Novaes, F.C. 1974. Ornitologia do território do Amapá I. *Publ. Avuls. Mus. Para. E. Goeldi* 25: 1–121.

Novaes, F.C. 1976. As aves do rio Aripuanã, Estados de Mato Grosso e Amazonas. *Acta Amazonica* 6: 61–85.

Novaes, F.C. & Lima, M.F.C. 1998. *Aves da Grande Belém: Municípios de Belém e Ananindeua, Pará.* Museu Paraense Emílio Goeldi, Belém.

Núñez, J.U., Trujillo, J.V., Lewis, M.U., Toledo, V.G. & Olaechea, D.G. 2012. Cubriendo los vacios: 74 registros para la Cordillera de Vilcabamba. Pp. 322–323 *in* More, A., Olaechea, D.G. & Olaechea, A.G. (eds.) *Libro de Resúmenes del IX Congreso de Ornitología Neotropical y el VIII Congreso Peruano de Ornitología.* Cuzco.

Nutting, C.C. & Ridgway, R. 1884. On a collection of birds from Nicaragua. *Proc. US Natl. Mus.* 6: 372–410.

Oates, E.W. & Reid, S.G. 1903. *Catalogue of the Collection of Birds' Eggs in the British Museum (Natural History),* vol. 3. Trustees of the British Museum, London.

Obando-Calderón, G., Chaves-Campos, J., Garrigues, R., Montoya, M., Ramírez, O., Sandoval, L. & Zook, J. 2009. Lista oficial de las aves de Costa Rica 2009. *Zeledonia* 13: 1–34.

Ocampo-T., S. 2002. Río Blanco. Reserva hidrogeográfica, forestal y parque ecológico. Paraíso de las aves. *Bol. Soc. Antioqueña Orn.* 13: 48–61.

O'Dea, N. & Whittaker, R.J. 2007. How resilient are Andean montane forest bird communities to habitat degradation? *Vert. Conserv. & Biodiver.* 16: 1131–1159.

O'Donnell, P. 2014. Black-crowned Antpitta (*Pittasoma michleri*). *In* Schulenberg, T.S. (ed.) *Neotropical Birds Online.* Cornell Lab of Ornithology, Ithaca, NY (http://neotropical.birds.cornell.edu/portal/species/overview?p_p_spp=410441).

Ogrzewalska, M., Pacheco, R.C., Uezu, A., Ferreira, F. & Labruna, M.B. 2008. Ticks (Acari: Ixodidae) infesting wild birds in an Atlantic forest area in the state of São Paulo, Brazil, with isolation of *Rickettsia* from the tick *Amblyomma longirostre*. *J. Med. Ent.* 45: 770–774.

Ogrzewalska, M., Pacheco, R.C., Uezu, A., Richtzenhain, L.J., Ferreira, F. & Labruna, M.B. 2009. Ticks (Acari: Ixodidae) infesting birds in an Atlantic Rain Forest region of Brazil. *J. Med. Ent.* 46: 1225–1229.

Olivares, A. 1958. Aves de la Costa del Pacífico, Municipio del Guapí, Cauca Colombia, III. *Caldasia* 8: 217–251.

Olivares, A. 1962. Aves de la región sur de la Sierra de la Macarena, Meta, Colombia. *Rev. Acad. Colombiana Cienc. Exactas, Fisicas, y Nat.* 11: 305–344.

Olivares, A. 1964. Adiciones a las aves de la Comisaria del Vaupes (Colombia) II. *Caldasia* 9: 151–184.

Olivares, A. 1966. Algunas aves de Puerto Asis, Comisaria del Putumayo, Colombia. *Caldasia* 9: 379–393.

Olivares, A. 1967 Avifaunae colombiensis. Notulae II. Seis nuevas aves para Colombia y apuntaciones sobre sesenta especies y subespecies registradas antiormente. *Caldasia* 10: 39–58.

Olivares, A. 1969. *Aves de Cundinamarca.* Univ. Nacional de Colombia, Bogotá.

Olivares, A. 1973. Aves de la Sierra Nevada de Cocuy, Colombia. *Rev. Acad. Colombiana Cienc. Exactas, Fisicas, y Nat.* 14: 39–48.

Oliveira, P.R.R., Alberts, C.C. & Francisco, M.R. 2011. Impact of road clearings on the movements of three understory insectivorous bird species in the Brazilian Atlantic Forest. *Biotropica* 43: 628–632.

Oliveira, P.R.R., Alberts, C.C. & Francisco, M.R. 2012. Impact of road clearings on the movements of three understory insectivorous bird species in the Brazilian Atlantic Forest. P. 270 *in* More, A., Olaechea, D.G. & Olaechea, A.G. (eds.) *Libro de Resúmenes del IX Congreso de Ornitología Neotropical y el VIII Congreso Peruano de Ornitología.* Cuzco.

Oliveira, S.L. & Köhler, A. 2010. Avifauna da RPPN da UNISC, Sinimbu, Rio Grand do Sul, Brasil. *Biotemas* 23: 93–103.

Olmos, F. 1993. Birds of Serra da Capivara National Park, in the "caatinga" of north-eastern Brazil. *Bird Conserv. Intern.* 3: 21–36.

Olmos, F. 1996. Missing species in São Sebastião Island, southeastern Brazil. *Pap. Avuls. Dept. Zool., São Paulo* 39: 329–349.

Olmos, F. 2003. Birds of Mata Estrela private reserve, Rio Grande do Norte, Brazil. *Cotinga* 20: 26–30.

Olmos, F. & Albano, C. 2012. As aves da região do Parque Nacional Serra da Capivara (Piauí, Brasil). *Rev. Bras. Orn.* 20: 173–187.

Olmos, F. & Brito, G.R.R. 2007. Aves da região da Barragem de Boa Esperança, médio rio Parnaíba, Brasil. *Rev. Bras. Orn.* 15: 37–52.

Olmos, F., Silveira, L.F. & Benedicto, G.A. 2011. A contribution to the ornithology of Rondônia, southwest of the Brazilian Amazon. *Rev. Bras. Orn.* 19: 200–229.

Olrog, C.C. 1963. Lista y distribución de las aves argentinas. *Opera Lilloana* 9: 1–377.

Olrog, C.C. 1979. Notas ornitológicas sobre la colección del Instituto Miguel Lillo (Tucumán): XI. *Acta Zool. Lilloana* 33: 1–7.

Olrog, C.C. 1984. Frecuencia de especies: individuos en un sotobosque de las yungas, Bolivia (Aves). *Historia Natural* 4: 105–109.

Olrog, C.C. & Contino, F. 1970. Una nueva subespecie de *Grallaria albigula* (Chapman) (Aves, Formicariidae). *Neotrópica* 16: 51–52.

O'Neill, J.P. 1974. The birds of Balta, a Peruvian dry tropical forest locality, with an analysis of their origins and ecological relationships. Ph.D. dissertation. Louisiana State Univ., Baton Rouge, LA.

O'Neill, J.P. 2003a. Avifauna de la región de Balta, un poblado cashinahua en el río Curanja. Pp. 97–106 *in* Pitman, R.L., Pitman, N. & Álvarez, P. (eds.) *Alto Purús: Biodiversidad, Conservación y Manejo.* Center for Tropical Conservation, Lima.

O'Neill, J.P. 2003b. Appendix 4: Lista de aves registradas por John O'Neill en Balta de 1963 a 1972. Pp. 318–329 *in* Pitman, R.L., Pitman, N. & Álvarez, P. (eds.) *Alto Purús: Biodiversidad, Conservación y Manejo.* Center for Tropical Conservation, Lima.

O'Neill, J.P. 2006. Museum expedition to northern Perú. *Mus. Quar., LSU Mus. Nat. Sci.* 24: 8–10.

O'Neill, J.P. & Graves, G.R. 1977. A new genus and species of owl (Aves: Strigidae) from Peru. *Auk* 94: 409–416.

Oniki, Y. 1974. Some temperatures of birds of Belém, Brazil. *Acta Amazonica* 4: 63–68.

Oniki, Y. 1981a. Individual recognition of nestlings. *J. Field Orn.* 52: 147–148.

Oniki, Y. 1981b. Weights, cloacal temperatures, plumage and molt condition of birds in the state of São Paulo. *Rev. Bras. Biol.* 41: 451–460.

Oniki, Y. & Willis, E.O. 1972. Studies of ant-following birds north of the eastern Amazon. *Acta Amazonica* 2: 127–151.

Oniki, Y. & Willis, E.O. 1982. Breeding records of birds from Manaus, Brazil: III. Formicariidae to Pipridae. *Rev. Bras. Biol.* 42: 563–569.

Oniki, Y. & Willis, E.O. 1991. Morphometrics, molt, cloacal temperatures and ectoparasites in Colombian birds. *Caldasia* 16: 519–524.

Oniki, Y., Kinsella, J.M. & Willis, E.O. 2002. *Pelecitus helicinus* Railliet & Henry, 1910 (Filarioidea, Dirofilariinae) and other nematode parasites of Brazilian birds. *Mem. Inst. Oswaldo Cruz* 97: 597–598.

d'Orbigny, A. 1838. *Voyage dans l'Amérique Méridionale, exécuté pendant les années 1826, 1827, 1828, 1829, 1830, 1831, 1832, et 1833. Tome Quatrième, 3. Partie: Oiseaux.* Libraire de la Société de géologique de France, Strasbourg.

d'Orbigny, A. 1843. *Voyage dans l'Amérique Méridionale, exécuté pendant les années 1826, 1827, 1828, 1829, 1830, 1831, 1832, et 1833. Tome Deuxième, Partie Historique.* Libraire de la Société de géologique de France, Strasbourg.

d'Orbigny, A. & Lafresnaye, M. 1837. Synopsis avium ab Alcide d'Orbigny in ejus per Americam meridionalem itinere, collectarum et ab ipso viatore necnon. *Mag. Zool., Anat. Comp. & Palaeont.* 7 (2ᵉ Classe): 1–88.

Orcés-V., G. 1974. Notas sobre la distribución geográfica de algunas aves del Ecuador. *Ciencia y Naturaleza* 15: 8–11.

Ordóñez-Delgado, L., Reyes-Bueno, F., Orihuela-Torres, A. & Armijos-Ojeda, D. 2016. Registros inusuales de aves en la hoya de Loja, Andes sur del Ecuador. *Avan. Cienc. Ing.* 8: 26–36.

Orejuela, J.E. 1985. Tropical forest birds of Colombia: a survey of problems and a plan for their conservation. Pp. 95–114 *in* Diamond, A.W. & Lovejoy, T.E. (eds.) *Conservation of Tropical Forest Birds.* International Council for Bird Preservation, Cambridge, UK.

Oren, D.C. 1990. New and reconfirmed bird records from the state of Maranhão, Brazil. *Goeldiana Zool.* 4: 1–13.

Oren, D.C. 1991. Aves do estado do Maranhão, Brasil. *Goeldiana Zool.* 9: 1–55.

Oren, D.C. & Parker, T.A. 1997. Avifauna of the Tapajós National Park and vicinity, Amazonian Brazil. *Orn. Monogr.* 48: 493–525.

Orians, G.H. & Paulson, D.R. 1969. Notes on Costa Rican birds. *Condor* 71: 426–431.

Ortiz-Crespo, F.I. & Carrión, J.M. 1991. *Introducción a las Aves del Ecuador.* FECODES, Quito.

Ortiz-Crespo, F.I., Greenfield, P.J. & Matheus, J.C. 1990. *Aves del Ecuador: Continente y Archipielago de Galapagos.* Fundación Ecuatoriana de Promoción Turística/Corporación Ornitológica del Ecuador, Quito.

Orton, J. 1871. Contributions to the natural history of the valley of Quito. No. 1. *Amer. Natur.* 10: 619–626.

O'Shea, B.J. 2007. Birds of Lely Gebergte, Suriname. Pp. 104–106, 238–241 *in* Alonso, L.E. & Mol, J.H. (eds.) *A Rapid Biological Assessment of the Lely and Nassau Plateaus, Suriname (With Additional Information on the Brownsberg Plateau).* Conservation International, Washington DC.

O'Shea, B.J. 2013. Preliminary bird species checklist of the Konashen COCA, southern Guyana. Pp. 78–86 *in* Alonso, L.E., McCullough, J., Naskrecki, P., Alexander, E. & Wright, H.E. (eds.) *A Rapid Biological Assessment of the Konashen Community Owned Conservation Area, Southern Guyana.* Conservation International, Washington DC.

O'Shea, B.J. & Ramcharan, S. 2013a. A rapid assessment of the avifauna of the Upper Palumeu watershed, southeastern Suriname. Pp. 145–160 *in* Alonso, L.E. & Larsen, T.H. (eds.) *A Rapid Biological Assessment of the Upper Palumeu River Watershed (Grensgebergte and Kasikasima) of Southeastern Suriname.* Conservation International, Washington DC.

O'Shea, B.J. & Ramcharan, S. 2013b. Avifauna of the Kwamalasamutu region, Suriname. Pp. 131–143 *in* O'Shea, B.J., Alonso, L.E. & Larsen, T.H. (eds.) *A Rapid Biological Assessment of the Kwamalasamutu Region, Southwestern Suriname.* Conservation International, Washington DC.

O'Shea, B.J., Condori, J. & Moskovits, D. 2003. Aves / Birds. Pp. 73–75, 95–101 *in* Alverson, W.S. (ed.) *Rapid Biological Inventories 05: Bolivia, Pando, Madre de Dios.* The Field Museum, Chicago.

O'Shea, B.J., Milensky, C.M., Claramunt, S., Schmidt, B.K., Gebhard, C.A., Schmitt, C.G. & Erskine, K.T. 2007. New records for Guyana, with description of the voice of Roraiman Nightjar *Caprimulgus whitelyi. Bull. Brit. Orn. Club* 127: 118–128.

O'Shea, B.J., Stotz, D.F., del Castillo, P.S. & Inzunza, E.R. 2015. Aves / Birds. Pp. 305–320, 446–471 *in* Pitman, N., Vriesendorp, C., Chávez, L.R., Wachter, T., Reyes, D.A., del Campo, Á., Gagliardi-Urrutia, G., González, D.R., Trevejo, L., González, D.R. & Heilpern, S. (eds.) *Rapid Biological and Social Inventories 27: Perú: Tapiche-Blanco.* The Field Museum, Chicago.

O'Shea, B.J., Wilson, A. & Wrights, J.K. 2017. Additions to the avifauna of two localities in the southern Rupununi region, Guyana. *Check List* 13: 113–120.

Ospina-Duque, A. & Granada-Castro, J.S. 2012. Diversidad de la avifauna en un bosque montano en Salento, Quindío. *Orn. Colombiana* 12: 87.

Pacheco, J.F. 2004. Pílulas históricas VI. Sabará ou Cuiabá? O problema das localidades de Ménétriès. *Atualidades Orn.* 117: 4–5.

Pacheco, [J.]F. & Maciel, E. 2005. Inventario preliminar da avifauna do Parque Estadual da Pedra Branca (PEPB), Rio de Janeiro. P. 135 *in Livro de Resumos do XIII Congresso Brasileira de Ornitologia.* Belém.

Pacheco, J.F. & Olmos, F. 2005. Birds of a latitudinal transect in the Tapajós-Xingu interfluvium, eastern Brazilian Amazonia. *Ararajuba* 13: 29–46.

Pacheco, J.F. & Olmos, F. 2006. As aves do Tocantins 1: região sudeste. *Rev. Bras. Orn.* 14: 85–100.

Pacheco, J.F., Parrini, R., Fonseca, P.S.M., Whitney, B.M. & Maciel, N.C. 1996. Novos registros de aves para o Estado do Rio de Janeiro: região norte. *Atualidades Orn.* 72: 10–12.

Pacheco, J.F., Parrini, R., Whitney, B.M., Bauer, C. & Fonseca, P.S.M. 1997a. Novos registros de aves para o Estado do Rio de Janeiro: região sul do vale do rio Paraiba do Sul. *Atualidades Orn.* 79: 4–5.

Pacheco, J.F., Parrini, R., Whitney, B.M., Bauer, C. & Fonseca, P.S.M. 1997b. Novos registros de aves para o Estado do Rio de Janeiro: Costa Verde. *Atualidades Orn.* 78: 4–5.

Pacheco, J.F., Kirwan, G.M., Aleixo, A., Whitney, B.M., Minns, J., Zimmer, K.J., Whittaker, A., Fonseca, P.S.M., Lima, M.F.C. & Oren, D.C. 2007. An avifaunal inventory of the CVRD Serra dos Carajás project, Pará, Brazil. *Cotinga* 27: 15–30.

Pacheco, J.F., Parrini, R., Lopes, L.E. & Vasconcelos, M.F. 2008. A avifauna do Parque Estadual do Ibitipoca e áreas adjacentes, Minas Gerais, Brasil, com uma revisão crítica dos registros prévios e comentarios sobre biogeografica e conservacão. *Cotinga* 30: 16–32.

Pacheco, J.F., Parrini, R., Kirwan, G.M. & Serpa, G.A. 2014. Birds of Vale da Taquaras region, Nova Friburgo, Rio de Janeiro state, Brazil: checklist with historical and trophic approach. *Cotinga* 36: 74–102.

Parker, T.A. 1982. Observations of some unusual rain-forest and marsh birds in southeastern Peru. *Wilson Bull.* 94: 477–493.

Parker, T.A. & Bailey, B. 1991. *A Biological Assessment of the Alto Madidi Region and Adjacent Areas of Northwest Bolivia, May 18-June 15, 1990. RAP Working Paper 1.* Conservation International, Washington DC.

Parker, T.A. & Goerck, J.M. 1997. The importance of national parks and biological reserves to bird conservation in the Atlantic Forest region of Brazil. *Orn. Monogr.* 48: 527–541.

Parker, T.A. & O'Neill, J.P. 1976a. Introduction to bird-finding in Peru. Part I. *Birding* 8: 140–144.

Parker, T.A. & O'Neill, J.P. 1976b. Introduction to bird-finding in Peru. Part II. *Birding* 8: 205–216.

Parker, T.A. & O'Neill, J.P. 1980. Notes on little known birds of the upper Urubamba Valley, southern Peru. *Auk* 97: 167–176.

Parker, T.A. & O'Neill, J.P. 1981. Introduction to bird-finding in Peru. Part III. *Birding* 12: 100–106.

Parker, T.A. & Parker, S.A. 1982. Behavioural and distributional notes on some unusual birds of a lower montane cloud forest in Peru. *Bull. Brit. Orn. Club* 102: 63–70.

Parker, T.A. & Remsen, J.V. 1987. Fifty-two Amazonian bird species new to Bolivia. *Bull. Brit. Orn. Club* 107: 94–106.

Parker, T.A. & Rowlett, R.A. 1984. Some noteworthy records of birds from Bolivia. *Bull. Brit. Orn. Club* 104: 110–113.

Parker, T.A. & Wust, W. 1994. Birds of the Cerros del Távara (300-900 m). Pp. 83–90 *in* Foster, R.B., Carr, J.L. & Forsyth, A.B. (eds.) *The Tambopata-Candamo Reserved Zone of Southeastern Peru: A Biological Assessment. Rapid Assessment Program Working Papers 6.* Conservation International, Washington DC.

Parker, T.A., Castillo V., A., Gell-Mann, M. & Rocha-O., O. 1991. Records of new and unusual birds from northern Bolivia. *Bull. Brit. Orn. Club* 111: 120–138.

Parker, T.A., Hilty, S.L. & Robbins, M.B. 1976. Birds of El Triunfo cloud forest, Mexico, with notes on the Horned Guan and other species. *Amer. Birds* 30: 779–782.

Parker, T.A., Parker, S.A. & Plenge, M.A. 1978. *A Checklist of Peruvian Birds.* Privately published, Tucson, AZ.

Parker, T.A., Parker, S.A. & Plenge, M.A. 1982. *An Annotated Checklist of Peruvian Birds.* Buteo Books, Vermillion, SD.

Parker, T.A., Schulenberg, T.S., Graves, G.R. & Braun, M.J. 1985. The avifauna of the Huancabamba region, northern Peru. *Orn. Monogr.* 36: 169–197.

Parker, T.A., Foster, R.B., Emmons, L.H., Freed, P., Forsyth, A.B., Hoffman, B. & Gill, B.D. 1993. *A Biological Assessment of the Kanuku Mountain Region of Southwestern Guyana Rapid Assessment Program Working Papers 5.* Conservation International, Washington DC.

Parker, T.A., Donahue, P.K. & Schulenberg, T.S. 1994a. Birds of the Tambopata Reserve (Explorer's Inn Reserve). Pp. 106–124 *in* Foster, R.B., Carr, J.L. & Forsyth, A.B. (eds.) *The Tambopata-Candamo Reserved Zone of Southeastern Peru: A Biological Assessment. Rapid Assessment Program Working Papers 6.* Conservation International, Washington DC.

Parker, T.A., Kratter, A.W. & Wust, W. 1994b. Birds of the Ccolpa de Guacamayos, Madre de Dios. Pp. 91–105 *in* Foster, R.B., Carr, J.L. & Forsyth, A.B. (eds.) *The Tambopata-Candamo Reserved Zone of Southeastern Peru: A Biological Assessment. Rapid Assessment Program Working Papers 6.* Conservation International, Washington DC.

Parker, T.A., Schulenberg, T.S. & Wust, W. 1994c. Birds of the Lower Río Heath, Including the Pampas del Heath, Bolivia/Perú. Pp. 125–139 *in* Foster, R.B., Carr, J.L. & Forsyth, A.B. (eds.) *The Tambopata-Candamo Reserved Zone of Southeastern Peru: A Biological Assessment. Rapid Assessment Program Working Papers 6.* Conservation International, Washington DC.

Parker, T.A., Schulenberg, T.S., Kessler, M. & Wust, W. 1995. Natural history and conservation of the endemic avifauna of north-west Peru. *Bird Conserv. Intern.* 5: 201–232.

Parker, T.A., Stotz, D.F. & Fitzpatrick, J.W. 1996. Database B: distribution of Neotropical bird species by country. Pp. 292–377 *in* Stotz, D.F., Fitzpatrick, J.W., Parker, T.A. & Moskovits, D.K. (eds.) *Neotropical Birds: Ecology and Conservation.* Univ. of Chicago Press, Chicago.

Parra-Hernández, R.M., Carantón-Ayala, D.A., Sanabria-Mejía, J.S., Barrera-Rodríguez, L.F., Sierra-Sierra, A.M., Moreno-Palacios, M.C., Yate-Molina, W.S., Figueroa-Martinez, W.E., Díaz-Jaramillo, C., Florez-Delgado, V.T., Certuche-Cubillos, J.K., Loaiza-Hernández, H.N. & Florido-Cuellar, B.A. 2007. Aves del municipio de Ibagué - Tolima, Colombia. *Biota Colombiana* 8 199–220.

Parrini, R., Raposo, M.A., Pacheco, J.F., Carvalhaes, A.M.P., Melo Júnior, T.A., Fonseca, P.S.M. & Minns, J.C. 1999. Birds of the Chapada Diamantina, Bahia, Brazil. *Cotinga* 11: 86–95.

Partridge, W.H. 1954. Estudio preliminar sobre una colección de aves de Misiones. *Rev. Mus. Argentino Cienc. Nat. Zool.* 3: 87–153.

Patten, M.A., de Silva, H.G., Ibarra, A.C. & Smith-Patten, B.D. 2011. An annotated list of the avifauna of Palenque, Chiapas. *Rev. Mex. Biodiver.* 82: 515–537.

Paynter, R.A. 1957. Birds of Laguna Ocotal. *Bull. Mus. Comp. Zool.* 116: 249–285.

Paynter, R.A. 1982. *An Ornithological Gazetteer of Venezuela.* Museum of Comparative Zoology, Harvard Univ., Cambridge, MA.

Paynter, R.A. 1989. *An Ornithological Gazetteer of Paraguay.* Museum of Comparative Zoology, Harvard Univ., Cambridge, MA.

Paynter, R.A. 1992. *An Ornithological Gazetteer of Bolivia.* Museum of Comparative Zoology, Harvard Univ., Cambridge, MA.

Paynter, R.A. 1993. *An Ornithological Gazetteer of Ecuador.* Museum of Comparative Zoology, Harvard Univ., Cambridge, MA.

Paynter, R.A. 1997. *An Ornithological Gazetteer of Colombia,* 2nd edn. Museum of Comparative Zoology, Harvard Univ., Cambridge, MA.

Paynter, R.A. & Traylor, M.A. 1981. *Ornithological Gazetteer of Colombia,* 1st edn. Museum of Comparative Zooloogy, Harvard Univ., Cambridge, MA.

Paynter, R.A. & Traylor, M.A. 1991a. *An Ornithological Gazetteer of Brazil. A–L Bibliography.* Museum of Comparative Zoology, Harvard Univ., Cambridge, MA.

Paynter, R.A. & Traylor, M.A. 1991b. *An ornithological gazetteer of Brazil. M–Z Bibliography.* Museum of Comparative Zoology, Harvard Univ., Cambridge, MA.

Pearman, M. 1993. Some range extensions and five species new to Colombia, with notes on some scarce or little known species. *Bull. Brit. Orn. Club* 113: 66–75.

Pearman, M. 1994. Neotropical notebook. *Cotinga* 1: 26–29.

Pearson, D.L., Tallman, D.A. & Tallman, E.J. 1972. *The Birds of Limoncocha, Napo Province, Ecuador.* Instituto Linguistico de Verano, Quito.

Pearson, D.L., Anderson, C.D., Mitchell, B.R., Rosenberg, M.S., Navarrete, R. & Coopmans, P. 2010. Testing hypotheses of bird extinctions at Rio Palenque, Ecuador, with informal species lists. *Conserv. Biol.* 24: 500–510.

Pelayo, R.C. & Soriano, P.J. 2010. Diagnóstico ornitológico del estado de conservación de tres cuencas altoandinas venezolanas. *Ecotrópicos* 23: 79–99.

von Pelzeln, A. 1871. *Zur Ornithologie Brasiliens: Resultate von Johann Natterers Reisen in den Jahren 1817 bis 1835.* A. Pichler's Witwe & Sohn, Wien.

de la Peña, M.R. 1988. *Guia de Aves Argentinas.* Literature of Latin America, Buenos Aires.

de la Peña, M.R. & Rumboll, M. 1998. *Birds of Southern South America and Antarctica.* HarperCollins, London.

Penard, F.P. & Penard, A.P. 1910. *De Vogels van Guyana*, vol. 2. Published privately, Paramaribo.

Pequeño, T., Salazar, E. & Aucca, C. 2001a. Bird species observed at Wayrapata (2445 m), southern Cordillera de Vilcabamba, Peru. Pp. 252–254 *in* Alonso, L.E., Alonso, A., Schulenberg, T.S. & Dallmeier, F. (eds.) *Biological and Social Assessments of the Cordillera Vilcabamba, Peru. RAP Working Papers 12.* Conservation International, Washington DC.

Pequeño, T., Salazar, E. & Aucca, C. 2001b. Bird species observed at Llactahuaman (1710 m), Southern Cordillera de Vilcabamba, Peru. Pp. 249–251 *in* Alonso, L.E., Alonso, A., Schulenberg, T.S. & Dallmeier, F. (eds.) *Biological and Social Assessments of the Cordillera Vilcabamba, Peru. RAP Working Papers 12.* Conservation International, Washington DC.

Peraza, C.A. 2011. Aves, bosque oriental de Bogotá Protective Forest Reserve, Bogotá, D.C., Colombia. *Check List* 7: 57–63.

Pereira, G.A., Medcraft, J., Santos, S.S. & Neto, F.P.F. 2014. Riqueza e conservação de aves em cinco áreas de caatinga no nordeste do Brasil. *Cotinga* 36: 17–27.

Pereira, G.A., Araujo, H.F.P. & Azevedo, S.M. 2016. Distribution and conservation of three important bird groups of the Atlantic Forest in north-east Brazil. *Braz. J. Biol.* 76: 1004–1020.

Peres, C.A. & Whittaker, A. 1991. Annotated checklist of the bird species of the upper Rio Urucu, Amazonas, Brazil. *Bull. Brit. Orn. Club* 111: 156–171.

van Perlo, B. 2006. *Birds of Mexico and Central America.* Princeton Univ. Press, Princeton, NJ.

van Perlo, B. 2009. *A Field Guide to the Birds of Brazil.* Oxford Univ. Press, New York.

Perry, A., Kessler, M. & Helme, N. 1997. Birds of the central Río Tuichi Valley, with emphasis on dry forest, Parque Nacional Madidi, Dept. La Paz, Bolivia. *Orn. Monogr.* 48: 557–576.

Pessoa, R.O. 2001. Variação Geográfica em *Conopophaga melanops* (Aves: Conopophagidae), uma Espécie Endêmica da Mata Atlântica. M.Sc. dissertation. Univ. Federal da Paraíba, João Pessoa.

Pessoa, R.O. & Silva, J.M.C. 2003. Variação geográfica em *Conopophaga melanops* (Conopophagidae), uma espécie endêmica da Mata Atlântica. P. 215 *in* Machado, C.G. (ed.) *Livro de Resumos do XI Congresso Brasileiro de Ornitologia.* Feira de Santana.

Pessoa, R.O., Cabanne, G.S., Silveira, L.F. & Miyaki, C.Y. 2004. Estudo da estrutura genetica populacional de *Conopophaga lineata* (Conopophagidae) no sudeste da floresta Atlantica. P. 328 *in* Testoni, A.F. & Althoff, S.L. (eds.) *Livro de Resumos do XII Congresso Brasileiro de Ornitologia.* Blumenau.

Pessoa, R.O., Cabanne, G.S., Silveira, L.F. & Miyaki, C.Y. 2005. Filogeografia de *Conopophaga lineata* (Passeriformes: Conopophagidae): Fragmentacao historica no sudeste na Mata Atlantica. P. 219 *in Livro de Resumos do XIII Congresso Brasileira de Ornitologia.* Belém.

Pessoa, R.O., Cabanne, G.S., Sari, E.H., Santos, F.R. & Miyaki, C.Y. 2006. Comparative phylogeography of the Rufous Gnateater (Conopophagidae) and Lesser Woodcreeper (Dendrocolaptidae): congruent history of two passerines from the South American Atlantic forest. *J. Orn.* 147(suppl.): 227–228.

Peters, J.L. 1931. Additional notes on the birds of the Almirante Bay Region of Panama. *Bull. Mus. Comp. Zool.* 71: 293–345.

Peters, J.L. 1951. *Check-list of Birds of the World*, vol. 7. Museum of Comparative Zoology, Harvard, Cambridge, MA.

Peters, J.L. & Griswold, J.A. 1943. Birds of the Harvard Peruvian Expedition. *Bull. Mus. Comp. Zool.* 92: 281–328.

Peterson, A.T. 2013. Case 3623. *Grallaria fenwickorum* Barrera *et al.* 2011 (Aves, Formicariidae): proposed replacement of an indeterminate holotype by a neotype. *Bull. Zool. Nomencl.* 70: 99–102.

Peterson, A.T. & Navarro-Siguenza, A.G. 2006. Hundred-year changes in the avifauna of the Valley of Mexico, Distrito Federal, Mexico. *Huitzil* 7: 4–14.

Peterson, R.T. & Chalif, E.L. 1973. *A Field Guide to Mexican Birds.* Houghton Mifflin Co., Boston, MA.

Pfeifer, A.M., Verhelst, J.C. & Botero, J.E. 2001. Estado de conservación de las especies de aves en el Parque Nacional Los Nevados, Colombia. *Bol. Soc. Antioqueña Orn.* 12: 21–41.

Phelps, W.H. 1938. The geographical status of the birds collected at Mount Roraima. *Bol. Soc. Venez. Cienc. Nat.* 36: 83–95.

Phelps, W.H., Jr. 1966. Contribución al análisis de los elementos que componen la avifauna subtropical de las cordilleras de la costa norte de Venezuela. *Bol. Acad. Cienc. Fis. Mat. Nat. Caracas* 25: 14–34.

Phelps, W.H., Jr. 1973. Adiciones a la lista de aves de Sur América, Brasil y Venezuela y notas sobre aves venezolanas. *Bol. Soc. Venez. Cienc. Nat.* 30: 23–40.

Phelps, W.H., Jr. 1977. Aves colectadas en las mesetas de Sarisariñama y Jáua durante tres expediciones al macizo de Jáua, estado Bolívar. Descripciones de dos nuevas subespecies. *Bol. Soc. Venez. Cienc. Nat.* 33: 15–42.

Phelps, W.H., Jr. & Dickerman, R.W. 1980. Cuatro subespecies nuevas de aves (Furnariidae, Formicariidae) de la región de Pantepui, estado Bolívar y Territorio Amazonas, Venezuela. *Bol. Soc. Venez. Cienc. Nat.* 33: 139–147.

Phelps, W.H. & Gilliard, E.T. 1940. Six new birds from the Perijá Mountains of Venezuela. *Amer. Mus. Novit.* 1100: 1–8.

Phelps, W.H. & Gilliard, E.T. 1941. Seventeen new birds from Venezuela. *Amer. Mus. Novit.* 1153: 1–17.

Phelps, W.H., Jr. & Meyer de Schauensee, R. 1979. *Una Guía de las Aves de Venezuela.* Gráficas Armitano, C.A., Caracas.

Phelps, W.H., Jr. & Meyer de Schauensee, R. 1994. *Una Guía de las Aves de Venezuela*, 2nd edn. ExLibris, Caracas.

Phelps, W.H. & Phelps, W.H., Jr. 1947. Descripción de seis

aves nuevas de Venezuela y notas sobre veinticuatro adiciones a la avifauna de Brasil. *Bol. Soc. Venez. Cienc. Nat.* 11: 53–74.

Phelps, W.H. & Phelps, W.H., Jr. 1949. Eight new birds from the subtropical zone of the Paria Peninsula, Venezuela. *Proc. Biol. Soc. Wash.* 62: 33–44.

Phelps, W.H. & Phelps, W.H., Jr. 1950. Lista de las aves de Venezuela con su distribución. Tomo I, Parte II. Passeriformes, primera edición. *Bol. Soc. Venez. Cienc. Nat.* 12: 1–427.

Phelps, W.H. & Phelps, W.H., Jr. 1953a. Eight new birds and thirty-three extensions of ranges to Venezuela. *Proc. Biol. Soc. Wash.* 66: 125–144.

Phelps, W.H. & Phelps, W.H., Jr. 1953b. Eight new subspecies of birds from the Perija Mountains, Venezuela. *Proc. Biol. Soc. Wash.* 66: 1–12.

Phelps, W.H. & Phelps, W.H., Jr. 1956. Five new birds from Río Chiquito, Táchira, Venezuela and two extensions of ranges from Colombia. *Proc. Biol. Soc. Wash.* 68: 47–58.

Phelps, W.H. & Phelps, W.H., Jr. 1962. Cuarentinueve aves nuevas para la avifauna brasileña del Cerro Uei - Tepui (Cerro del Sol). *Bol. Soc. Venez. Cienc. Nat.* 23: 32–39.

Phelps, W.H. & Phelps, W.H., Jr. 1963. Lista de las aves de Venezuela con su distribución. Tomo I, Parte II. Passeriformes, segunda edición. *Bol. Soc. Venez. Cienc. Nat.* 24: 1–479.

Pichorim, M. & Bóçon, R. 1996. Estudo da composição avifaunística dos municípios de Rio Azul e Mallet, Paraná, Brasil. *Acta Biol. Leopoldensia* 18: 129–144.

Pimentel, L. & Olmos, F. 2011. The birds of Reserva Ecologica Guapiacu (REGUA), Rio de Janeiro, Brazil. *Cotinga* 33: 8–24.

Pinto, O.M.O. 1936. Contribuição à ornitologia de Goyaz. *Rev. Mus. Paulista* 8: 1–172.

Pinto, O.M.O. 1937a. *Grallaria ochroleuca* Pelzeln prova ser ave diversa do *Myioturdus ochroleucus* Wied. *Bol. Biol. (N. Ser.)* 3: 6–7.

Pinto, O.M.O. 1937b. Nova contribução a omithologia amazonica. *Rev. Mus. Paulista* 23: 495–604.

Pinto, O.M.O. 1938. Catalogo das aves do Brasil, 1ª Parte. *Rev. Mus. Paulista* 22: 1–566.

Pinto, O.M.O. 1940. Aves de Pernambuco. *Arq. Zool. Estado São Paulo* 1: 219–282.

Pinto, O.M.O. 1944. Sôbre as aves do distrito de Monte Alegre, município de Amparo (São Paulo, Brasil). *Pap. Avuls. Dept. Zool., São Paulo* 4: 117–149.

Pinto, O.M.O. 1947. Contribuição à ornitologia do baixo Amazonas. *Arq. Zool. Estado São Paulo* 5: 311–482.

Pinto, O.M.O. 1953. Sobre a coleção Carlos Estevão de peles, ninhos, e ovos de aves de Belém (Pará). *Pap. Avuls. Dept. Zool., São Paulo* 11: 111–222.

Pinto, O.M.O. 1954a. Resultados orniológicos de duas viagens científicas ao estado de Alagoas. *Pap. Avuls. Dept. Zool., São Paulo* 11: 1–198.

Pinto, O.M.O. 1954b. Aves do Itatiaia: lista remissiva e novas achegas à avifauna da região. *Bol. Parque Nacional do Itatáia* 3: 1–87.

Pinto, O.M.O. 1978. *Novo Catalogo das Aves do Brasil. 1ª Parte.* Empresa Grafica Rev. Tribunais, São Paulo.

Pinto, O.M.O. & Camargo, E.A. 1955. Lista anotada de aves colecionadas nos limites ocidentais do Estado do Paraná. *Pap. Avuls. Dept. Zool., São Paulo* 12: 215–234.

Piratelli, A., Oliveira, E.F., Moraes, S.M.M. & Andrade, V.A. 2003. Aves em região serrana do Rio de Janeiro: Município de Miguel Pereira. P. 192 *in* Machado,

C.G. (ed.) *Livro de Resumos do XI Congresso Brasileiro de Ornitologia.* Feira de Santana.

Pizo, M.A. & Melo, A.S. 2010. Attendance and co-occurrence of birds following army ants in the Atlantic Rain Forest. *Condor* 112: 571–578.

Plenge, M.A. 2014. Bibliographic references of the birds of Peru. Lima, Peru. III - Passeriformes Suboscines. Available at: https://sites.google.com/site/boletinunop/bibliographic-references (accessed 22 November 2014). Unión de Ornitólogos del Perú, Lima.

Plenge, M.A. 2017. Lista de las aves de Perú. https://sites.google.com/site/boletinunop/checklist (accessed 15 March 2017).

Plenge, M.A., Parker, T.A., Hughes, R.A. & O'Neill, J.P. 1989. Additional notes on the distribution of birds in west-central Peru. *Gerfaut* 79: 55–68.

Pollock, H.S., Cheviron, Z.A., Agin, T.J. & Brawn, J.D. 2015. Absence of microclimate selectivity in insectivorous birds of the Neotropical forest understory. *Biol. Conserv.* 188: 116–125.

Portes, C.E.B., Carneiro, L.S., Schunck, F., Silva, M.S., Zimmer, K.J., Whittaker, A., Poletto, F., Silveira, L.F. & Aleixo, A. 2011. Annotated checklist of birds recorded between 1998 and 2009 at nine areas in the Belem area of endemism, with notes on some range extensions and the conservation status of endangered species. *Rev. Bras. Orn.* 19: 167–184.

Poulin, B., Lefebvre, G., Ibáñez, R., Jaramillo, C., Hernández, C. & Rand, A.S. 2001. Avian predation upon lizards and frogs in a neotropical forest understorey. *J. Trop. Ecol.* 17: 21–40.

Poulsen, B.O. 1993. Change in mobility among crepuscular ground-living birds in an Ecuadorian cloud forest during overcast and rainy weather. *Orn. Neotrop.* 4: 103–105.

Poulsen, B.O. 1994. Mist-netting as a census method for determining species richness and abundances in an Andean cloud forest bird community. *Gerfaut* 84: 39–49.

Poulsen, B.O. 1996. Structure, dynamics, home range and activity pattern of mixed-species bird flocks in a montane alder-dominated secondary forest in Ecuador. *J. Trop. Ecol.* 12: 333–343.

Poulsen, B.O. 2002. A comparison of bird richness, abundance and trophic organization in forests of Ecuador and Denmark: are high-altitude Andean forests temperate or tropical? *J. Trop. Ecol.* 18: 615–636.

Poulsen, B.O. & Krabbe, N. 1997a. Avian rarity in ten cloud forest communities in the Andes of Ecuador: implications for conservation. *Biodiver. & Conserv.* 6: 1365–1375.

Poulsen, B.O. & Krabbe, N. 1997b. Avian diversity: the birds of Oyacachi. Pp. 80–85; Appendix 83 *in* Baez, S., Fjeldså, J., Krabbe, N., Morales, P., Navarrete, H., Poulson, B.O., Resl, R., Schjellerup, I., Skov, F., Ståhl, B. & Øllgaard, B. (eds.) *Oyacachi - People and Biodiversity: DIVA Technical Report 2.* Center for Research on Cultural and Biological Diversity of Andean Rainforests, Rønde.

Poulsen, B.O. & Krabbe, N. 1998. Avifaunal diversity of five high-altitude cloud forests on the Andean western slope of Ecuador: testing a rapid assessment method. *J. Biogeogr.* 25: 83–93.

Powell, L.L. 2013. Recovery of Understory Bird Movement in Post-pasture Amazonia. Ph.D. thesis. Louisiana State Univ., Baton Rouge, LA.

Powell, L.L., Wolfe, J.D., Johnson, E.I. & Stouffer, P.C. 2016. Forest recovery in post-pasture Amazonia: testing a conceptual model of space use by insectivorous understory birds. *Biol. Conserv.* 194: 22–30.

Pozza, D.D. & Pires, J.S.R. 2003. Bird communities in two fragments of semideciduous forest in rural São Paulo state. *Braz. J. Biol.* 63: 307–319.

Pratolongo, F.A. 2012. Evaluación de la distribución, estado de conservación y amenazas de cuatro especies amenazadas en el norte del Peru. P. 299 *in* More, A., Olaechea, D.G. & Olaechea, A.G. (eds.) *Libro de Resúmenes del IX Congreso de Ornitología Neotropical y el VIII Congreso Peruano de Ornitología.* Cuzco.

Pratolongo, F.A., Flanagan, J.N.M., Vellinga, W.-P. & Durand, N. 2012. Notes on the birds of Laquipampa Wildlife Refuge, Lambayeque, Peru. *Bull. Brit. Orn. Club* 132: 162–174.

Prévost, F. & Des Murs, O. 1846. Oiseaux. Pp. 177–284 *in Voyage Autour du Monde sur la Frégate la Vénus Commandée par Abel de Petit-Thouars. Zoologie: Mammifères, Oiseaux, Reptiles, et Poissons.* Gide & J. Baudry, Paris.

Price, E.R. 2003. First description of the nest, eggs, hatchlings, and incubation behavior of the White-bellied Antpitta (*Grallaria hypoleuca*). *Orn. Neotrop.* 14: 535–539.

Price, R.D. & Clayton, D.H. 1996. Revision of the chewing louse genus *Formicaphagus* (Phthiraptera: Philopteridae) from Neoptropical antbirds and gnateaters (Aves: Passeriformes). *J. Kansas Ent. Soc.* 69: 346–356.

Principe, B. 1991. *A Check List of the Birds of Costa Rica.* Bird Processing Electronic Publishers, Flintridge, CA.

Protomastro, J.J. 2000. Notes on the nesting of Variegated Antpitta *Grallaria varia. Cotinga* 14: 39–41.

Puebla-Olivares, F., Rodríguez-Ayala, E., Hernández-Baños, B.E. & Navarro S., A.G. 2002. Status and conservation of the avifauna of the Yaxchilán Natural Monument, Chiapas, Mexico. *Orn. Neotrop.* 13: 381–396.

Pulgarín-R., P.C. & Múnera-P., W.A. 2006. New bird records from Farallones del Citará, Colombian Western Cordillera. *Bol. Soc. Antioqueña Orn.* 16: 44–53.

Pulido-Santacruz, P. & Renjifo, L.M. 2011. Live fences as tools for biodiversity conservation: a study case with birds and plants. *Agroforestry Syst.* 81: 15–30.

Quevedo, A., Salaman, P. & Donegan, T. 2006. Serranía de las Quinchas: establishment of a first protected area in the Magdalena Valley of Colombia. *Cotinga* 25: 24–32.

Quintela, C.E. 1986. Forest Fragmentation and Differential Use of Natural and Man-made Edges by Understory birds in Central Amazonia. M.Sc. dissertation. Univ. of Illinois, Chicago.

Quintela, C.E. 1987. First report of the nest and young of the Variegated Antpitta (*Grallaria varia*). *Wilson Bull.* 99: 499–500.

Rahbek, C., Bloch, H., Poulsen, M.K. & Rasmussen, J.F. 1993. Avian body weights from southern Ecuador. *Bull. Brit. Orn. Club* 113: 103–108.

Ramírez, J.D. 2006. Redescubrimiento de *Grallaria rufocinerea* (Formicariidae) en el Valle de Aburrá, Antioquia, Colombia. *Bol. Soc. Antioqueña Orn.* 16: 17–23.

Ramírez, W.A., Chaparro-Herrera, S., López, R.C., Arredondo, C. & Sua-Becerra, A. in press. Ampliación del rango de distribución del Tororoí Carimanchado (*Hylopezus macularius*) en Colombia. *Cotinga* 40.

Ramírez-Albores, J.E. 2010. Avifauna de sitios asociados a la selva tropical en la depresión central de Chiapas, México. *Acta Zool. Mex., N. Ser.* 26: 539–562.

Ramos, M.A. 1985. Endangered tropical birds in Mexico and northern Central America. Pp. 305–318 *in* Diamond,

A.W. & Lovejoy, T.E. (eds.) *Conservation of Tropical Forest Birds.* International Council for Bird Preservation, Cambridge, UK.

Rand, A.L. & Traylor, M.A. 1954. *Manual de las Aves de El Salvador.* Univ. de El Salvador, San Salvador.

Rangel-Salazar, J.L. 1995. Diversidad de Aves de Sotobosque Asociadas a Plantaciones de Banano en Costa Rica. M.Sc. dissertation. Univ. Nacional Autónoma de Costa Rica, Heredi.

Rangel-Salazar, J.L., Enríquez, P.L. & Sántiz-López, E.C. 2009. Variación de la diversidad de aves de sotobosque en el Parque Nacional Lagos de Montebello, Chiapas, México. *Acta Zool. Mex., N. Ser.* 25: 479–495.

Rasmussen, J.F., Rahbek, C., Horstman, E., Poulsen, M.K. & Bloch, H. 1994. *Birds of Podocarpus National Park, An Annotated Checklist.* Fundación Aage V. Jensen, Quito.

Rasmussen, J.F., Rahbek, C., Poulsen, B.O., Poulsen, M.K. & Bloch, H. 1996. Distributional records and natural history notes on threatened and little known birds of southern Ecuador. *Bull. Brit. Orn. Club* 116: 26–45.

Raton, R. & Gomes, Y.M. 2015. Biodiversidade da avifauna de um fragmento antropizado na região serrana do Espírito Santo. *Atualidades Orn.* 187: 61–67.

Regalado, A. 2011. Feathers are flying over Colombian bird name flap. *Science* 331: 1123–1124.

Reichenow, A. 1884. *Bericht über die Leistungen in der Naturgeschichte der Vögel,* Bd. 3. Nicolaische Verlags-Buchhandlung, Berlin.

Reinert, B.L., Pinto, J.C., Bornschein, M.R., Pichorim, M. & Marini, M.Â. 1996. Body masses and measurements of birds from southern Atlantic forest, Brazil. *Rev. Bras. Zool.* 13: 815–820.

Remsen, J.V. 1984. Natural history notes on some poorly known Bolivian birds. Part 2. *Gerfaut* 74: 163–179.

Remsen, J.V. 1985. Community organization and ecology of birds of high elevation humid cloud forest of the Bolivian Andes. *Orn. Monogr.* 36: 733–756.

Remsen, J.V. 2015. Elevate *Conopophaga lineata cearae* to species rank. Proposal (#684) to South American Classification Committee. http://www.museum.lsu.edu/~Remsen/SACCprop684.htm.

Remsen, J.V. & Parker, T.A. 1983. Contribution of river-created habitats to bird species richness in Amazonia. *Biotropica* 15: 223–231.

Remsen, J.V. & Parker, T.A. 1995. Bolivia has the opportunity to create the planet's richest park for terrestrial biota. *Bird Conserv. Intern.* 5: 181–199.

Remsen, J.V. & Traylor, M.A. 1989. *An Annotated List of the Birds of Bolivia.* Buteo Books, Vermillion, SD.

Remsen, J.V., Parker, T.A. & Ridgely, R.S. 1982. Natural history notes on some poorly known Bolivian birds. *Gerfaut* 72: 77–87.

Remsen, J.V., Traylor, M.A. & Parkes, K.C. 1986. Range extensions for some Bolivian birds, 2 (Columbidae to Rhinocryptidae). *Bull. Brit. Orn. Club* 106: 22–32.

Renaudier, A. 2009. Birding French Guiana. *Neotrop. Birding* 5: 39–47.

Renjifo, L.M. 1998. Effect of the Landscape Matrix on the Composition and Conservation of Forest Bird Communities. Ph.D. dissertation. Univ. of Missouri Saint Louis, Saint Louis.

Renjifo, L.M. 1999. Composition changes in a Subandean avifauna after long-term forest fragmentation. *Conserv. Biol.* 13: 1124–1139.

Renjifo, L.M. 2001. Effect of natural and anthropogenic

landscape matrices on the abundance of Subandean bird species. *Ecol. Appl.* 11: 14–31.

Renjifo, L.M. & Andrade, G.I. 1987. Estudio comparativo de la avifauna entre un area de bosque andino primario y un crecimiento secundario en el Quindio, Colombia. Pp. 121–127 *in* Alvarez-Lopez, H., Kattan, G. & Murcia, C. (eds.) *Memorias del Tercer Congreso de Ornitologia Neotropical.* Sociedad Vallecaucana de Ornitologia & Universidad del Valle, Cali.

Renjifo, L.M., Servat, G.P., Goerck, J.M., Loiselle, B.A. & Blake, J.G. 1997. Patterns of species composition and endemism in the northern Neotropics: a case for conservation of montane avifauna. *Orn. Monogr.* 48: 577–594.

Renjifo, L.M., Franco-Maya, A., Álvarez-López, H., Álvarez, M., Borja, R., Botero, J.E., Córdoba, S., de la Zerda, S., Didier, G., Estela, F., Kattan, G., Londoño, E., Márquez, C., Montenegro, M.I., Murcia, C., Rodríguez, J.V., Samper, C. & Weber, W.H. 2000. *Estrategia Nacional para la Conservación de las Aves de Colombia.* Instituto Alexander von Humboldt, Bogotá.

Renjifo, L.M., Franco-Maya, A., Amaya-Espinel, G.H., Kattan, G.H. & López-Lanús, B. 2002. *Libro Rojo de Aves de Colombia.* Instituto Alexander von Humboldt & Ministerio del Medio Ambiente, Bogotá.

Renjifo, L.M., Gómez, M.F., Velásquez-Tibatá, J., Amaya-Villarreal, A.M., Kattan, G.H., Amaya-Espinel, J.D. & Burbano-Girón, J. 2014. *Libro Rojo de Aves de Colombia,* vol. 1. Editorial Pontificia Univ. Javeriana & Instituto Alexander von Humboldt, Bogotá.

Renner, S.C., Waltert, M. & Mühlenberg, M. 2006. Comparison of bird communities in primary vs. young secondary tropical montane cloud forest in Guatemala. *Forest Diver. & Manag.* 2: 485–515.

Repenning, M. & Fontana, C.S. 2011. Seasonality of breeding, moult and fat deposition of birds in subtropical lowlands of southern Brazil. *Emu* 111: 268–280.

Restall, R., Rodner, C. & Lentino, M. 2006. *Birds of Northern South America. An Identification Guide.* Christopher Helm, London.

Restrepo, C. & Gomez, N. 1998. Responses of understory birds to anthropogenic edges in a Neotropical montane forest. *Ecol. Appl.* 8: 170–183.

Restrepo, M.P. 1997. Ornitofauna presente en una zona de construccion carretera Florencia - Altamira. Interventoria integral S.A. *Bol. Soc. Antioqueña Orn.* 8: 30–45.

Reynaud, P.A. 1998. Changes in understory avifauna along the Sinnamary River (French Guyana, South America). *Orn. Neotrop.* 9: 51–70.

Rhoads, S.N. 1912. Birds of the paramo of central Ecuador. *Auk* 29: 141–149.

Ribeiro, S.F., Sebaio, F., Branquinho, F.C.S., Marini, M.Â., Vago, A.R. & Braga, E.M. 2005. Avian malaria in Brazilian passerine birds: parasitism detected by nested PCR using DNA from stained blood smears. *Parasitology* 130: 261–267.

Ribeiro, V., Guedes, L.B.S., Cavalcante, W.S. & Caparroz, R. 2013. Prevalence of *Chlamydia* in free-living birds in Distrito Federal, Brazil. *Rev. Bras. Orn.* 21: 114–119.

Ribon, R. 2000. Lista preliminar da avifauna do município de Ijaci, Minas Gerais. *Rev. Ceres* 47: 665–682.

Ribon, R., Neves, N.R. & Nieto, E.C.A. 2001. Abundância de aves em um fragmento de Mata Atlântica da Zona da Mata de Minas Gerais através do método de contagem por pontos. Resumen 173 *in* Straube, F.C. (ed.) *Livro de Resumos do IX Congresso Brasileiro de Ornitologia: Ornitologia Sem Fronteras.* Curitiba.

Ribon, R., Simon, J.E. & Mattos, G.T. 2003. Bird extinctions in Atlantic Forest fragments of the Viçosa region, southeastern Brazil. *Conserv. Biol.* 17: 1827–1839.

Ribon, R., Lamas, I.R. & Gomes, H.B. 2004. Avifauna da Zona da Mata de Minas Gerais: municípios de Goianá e Rio Novo, com alguns registros para Coronel Pacheco e Juiz de Fora. *Rev. Árvore* 28: 291–305.

Rice, N.H. 2000. Phylogenetic Relationships of the Ground Antbirds and their Relatives (Aves: Formicariidae). Ph.D. dissertation. Univ. of Kansas, Lawrence.

Rice, N.H. 2005a. Phylogenetic relationships of antpitta genera (Passeriformes: Formicariidae). *Auk* 122: 673–683.

Rice, N.H. 2005b. Further evidence for paraphyly of the Formicariidae (Passeriformes). *Condor* 107: 910–915.

Richmond, C.W. 1893. Notes on a collection of birds from eastern Nicaragua and the Río Frio, Costa Rica, with a description of a supposed new trogon. *Proc. US Natl. Mus.* 16: 479–532.

Ricklefs, R.E. 1996. Morphometry of the digestive tracts of some passerine birds. *Condor* 98: 279–292.

Ridgely, R.S. 2012. Discovering the Jocotoco. *Neotrop. Birding* 10: 4–8.

Ridgely, R.S. & Gaulin, S.J.C. 1980. The birds of Finca Merenberg, Huila Department, Colombia. *Condor* 82: 379–391.

Ridgely, R.S. & Greenfield, P.J. 2001. *Birds of Ecuador.* Cornell Univ. Press, Ithaca, NY.

Ridgely, R.S. & Greenfield, P.J. 2006. *Aves del Ecuador.* Cornell Univ. Press, Ithaca, NY.

Ridgely, R.S. & Gwynne, J.A. 1989. *A Guide to the Birds of Panama with Costa Rica, Nicaragua, and Honduras.* Princeton Univ. Press, Princeton, NJ.

Ridgely, R.S. & Tudor, G. 1994. *The Birds of South America,* vol. 2. Univ. of Texas Press, Austin.

Ridgely, R.S. & Tudor, G. 2009. *Field Guide to the Songbirds of South America: The Passerines.* Univ. of Texas Press, Austin.

Ridgely, R.S., Greenfield, P.J. & Guerrero, M. 1998. *Una Lista Anotada de las Aves del Ecuador Continental.* CECIA, Quito.

Ridgely, R.S., Agro, D. & Joseph, L. 2005. Birds of the Iwokrama Forest. *Proc. Acad. Nat. Sci. Phil.* 154: 109–112.

Ridgely, R.S., Gwynne, J.A., Tudor, G. & Argel, M. 2016. *Birds of Brazil, Vol. 2: The Atlantic Forest of Southeast Brazil.* Cornell Univ. Press, Ithaca, NY.

Ridgway, R. 1884. On some Costa Rican birds, with descriptions of several supposed new species. *Proc. US Natl. Mus.* 6: 410–415.

Ridgway, R. 1891. Notes on some birds from the interior of Honduras. *Proc. US Natl. Mus.* 14: 467–471.

Ridgway, R. 1909. New genera, species and subspecies of Formicariidae, Furnariidae, and Dendrocolaptidae. *Proc. Biol. Soc. Wash.* 22: 69–74.

Ridgway, R. 1911. *The Birds of North and Middle America,* part V. Smithsonian Institution, Washington DC.

Ridgway, R. 1912. *Color Standards and Color Nomenclature.* A. Hoen & Co., Washington DC.

Rivera-Pedroza, L.F. & Ramírez, M.P. 2005. Una extensión de la distribución del Tororoi Medialuna *Grallaricula lineifrons* (Fomicariidae) en Colombia. *Orn. Colombiana* 3: 81–83.

Robbins, M.B. & Ridgely, R.S. 1986. A new race of *Grallaria haplonota* (Formicariidae) from Ecuador. *Bull. Brit. Orn. Club* 106: 101–103.

Robbins, M.B. & Ridgely, R.S. 1990. The avifauna of an upper tropical cloud forest in southwestern Ecuador. *Proc. Acad. Nat. Sci. Phil.* 142: 59–71.

Robbins, M.B., Parker, T.A. & Allen, S.E. 1985. The avifauna of Cerro Pirre, Darién, eastern Panama. *Orn. Monogr.* 36: 198–232.

Robbins, M.B., Ridgely, R.S., Schulenberg, T.S. & Gill, F.B. 1987. The avifauna of the Cordillera de Cutucú, Ecuador, with comparisons to other Andean localities *Proc. Acad. Nat. Sci. Phil.* 139: 243–259.

Robbins, M.B., Capparella, A.P., Ridgely, R.S. & Cardiff, S.W. 1991. Avifauna of the Río Manití and Quebrada Vainilla, Peru. *Proc. Acad. Nat. Sci. Phil.* 143: 145–159.

Robbins, M.B., Krabbe, N., Ridgely, R.S. & Molina, F.S. 1994a. Notes on the natural history of the Crescent-faced Antpitta. *Wilson Bull.* 106: 169–173.

Robbins, M.B., Krabbe, N., Rosenberg, G.H. & Molina, F.S. 1994b. The tree line avifauna at Cerro Mongus, Prov. Carchi, Northeastern Ecuador. *Proc. Acad. Nat. Sci. Phil.* 145: 209–216.

Robbins, M.B., Braun, M.J. & Finch, D.W. 2004. Avifauna of the Guyana southern Rupununi, with comparisons to other savannas of northern South America. *Orn. Neotrop.* 15: 173–200.

Robbins, M.B., Braun, M.J., Milensky, C.M., Schmidt, B.K., Prince, W., Rice, N.H., Finch, D.W. & O'Shea, B.J. 2007. Avifauna of the upper Essequibo River and Acary Mountains, southern Guyana. *Orn. Neotrop.* 18: 339–368.

Robbins, M.B., Geale, D., Walker, B., Davis, T.J., Combe, M., Eaton, M.D. & Kennedy, K.P. 2011. Foothill avifauna of the upper Urubamba Valley, dpto. Cusco, Peru. *Cotinga* 33: 34–45.

Robbins, M.B., Schulenberg, T.S., Lane, D.F., Cuervo, A.M., Binford, L.C., Nyári, A.S., Combe, M., Arbeláez-Cortes, E., Wehtje, W. & Lira-Noriega, A. 2013. Abra Maruncunca, dpto. Puno, Peru, revisited: vegetation cover and avifauna changes over a 30-year period. *Bull. Brit. Orn. Club* 133: 31–51.

Roberts, H.R. 1934. List of Trinidad birds with field notes. *Trop. Agricult. (Trinidad)* 11: 87–99.

Robinson, S.K. & Terborgh, J. 1997. Bird community dynamics along primary successional gradients of an Amazonian whitewater river. *Orn. Monogr.* 48: 641–672.

Robinson, S.K., Terborgh, J. & Munn, C.A. 1990. Lowland tropical forest bird communities of a site in western Amazonia. Pp. 229–258 *in* Keast, A. (ed.) *Biogeography and Ecology of Forest Bird Communities.* SPB Academy Publications, The Hague.

Robinson, W.D. 1999. Long-term changes in the avifauna of Barro Colorado Island, Panama, a tropical forest isolate. *Conserv. Biol.* 13: 85–97.

Robinson, W.D., Robinson, T.R., Robinson, S.K. & Brawn, J.D. 2000a. Nesting success of understory forest birds in central Panama. *J. Avian Biol.* 31: 151–164.

Robinson, W.D., Brawn, J.D. & Robinson, S.K. 2000b. Forest bird community structure in central Panama: influence of spatial scale and biogeography. *Ecol. Monogr.* 70: 209–235.

Robinson, W.D., Angehr, G.R., Robinson, T.R., Petit, L.J., Petit, D.R. & Brawn, J.D. 2004. Distribution of bird diversity in a vulnerable neotropical landscape. *Conserv. Biol.* 18: 510–518.

Robinson, W.D., Styrsky, J.N. & Brawn, J.D. 2005. Are artificial bird nests effective surrogates for estimating predation on real bird nests? A test with tropical birds. *Auk* 122: 843–852.

Rocha, O.O. & Quiroga O., C. 1996. Aves. Pp. 95–164 *in* Ergueta, P.S. & de Morales, C. (eds.) *Libro Rojo de los Vertebrados de Bolivia.* Centro de Datos para la Conservación, La Paz.

Roda, S.A. 2008. *Conopophaga melanops nigrifrons* Pinto, 1954. Pp. 507–508 *in* Machado, A.B.M., Drummond, G.M. & Paglia, A.P. (eds.) *Livro Vermelho da Fauna Brasileira Ameaçada de Extinção.* Ministério do Meio Ambiente & Fundação Biodiversitas, Brasília & Belo Horizonte.

Roda, S.A. & Carlos, C.J. 2004. Composição e sensitividade da avifauna dos brejos de altitude do estado de Pernambuco. Pp. 211–228 *in* Porto, K.C., Cabral, J.J.P. & Tabarelli, M. (eds.) *Brejos de altitude em Pernambuco e Paraíba: História Natural, Ecologia e Conservação.* Ministério do Meio Ambiente & Universidade Federal de Pernambuco, Brasília.

Roda, S.A., Carlos, C.J. & Rodrigues, R.C. 2003. New and noteworthy records for some endemic and threatened birds of the Atlantic forest of north-eastern Brazil. *Bull. Brit. Orn. Club* 123: 227–236.

Rodner, C., Lentino R., M. & Restall, R.L. 2000. *Checklist of the Birds of Northern South America.* Yale Univ. Press, New Haven, CT.

Rodrigues, R.C., do Amaral, A.C.A. & Sales, L.G. 2003. Inventário da avifauna na Área de Proteção Ambiental do Maciço do Baturité, CE. P. 119 *in* Machado, C.G. (ed.) *Livro de Resumos do XI Congresso Brasileiro de Ornitologia.* Feira de Santana.

Rodríguez, J.P. & Rojas-Suárez, F. 1995. *Libro Rojo de la Fauna Venezolana, Primera Edición.* Provita y Shell Venezuela, Caracas.

Rodríguez, J.P., Rojas-Suárez, F. & Sharpe, C.J. 2004. Setting priorities for the conservation of Venezuela's threatened birds. *Oryx* 38: 373–382.

Rojas, R., Marini, M.Â. & Coutinho, M.T.Z. 1999. Wild birds as hosts of *Amblyomma cajennense* (Fabricius, 1787) (Acari: Ixodidae). *Mem. Inst. Oswaldo Cruz* 94: 315–322.

Roper, J.J. 1992. Nest predation experiments with quail eggs: too much to swallow? *Oikos* 65: 528–530.

do Rosário, L.A. 1996. *As Aves em Santa Catarina: Distribuição Geográfica e Meio Ambiente.* Fundação do Meio Ambiente, Florianópolis.

Rosas, C.V. 2003. *Reporte de los Trabajos Realizados y los Registros Existentes para la Flora y Fauna del Bosque de Protección Alto Mayo.* Parkswatch Peru: http://www.parkswatch.org/parkprofiles/slide-shows/ampf/informe.pdf.

Rosselli, L., Stiles, F.G. & Camargo, P.A. 2017. Changes in the avifauna in a high Andean cloud forest in Colombia over a 24-year period. *J. Field Orn.* 88: 211–228.

Rother, D.C., Alves, K.J.F. & Pizo, M.A. 2013. Avian assemblages in bamboo and non-bamboo habitats in a tropical rainforest. *Emu* 113: 52–61.

Rothschild, W. 1900. [Exhibition of a new species of *Grallaria*]. *Bull. Brit. Orn. Club* 11: 36–37.

Rowley, J.S. 1966. Breeding records of birds of the Sierra Madre del Sur, Oaxaca, Mexico. *Proc. West. Found. Vert. Zool.* 1: 107–204.

Rowley, J.S. 1984. Breeding records of land birds in Oaxaca, Mexico. *Proc. West. Found. Vert. Zool.* 2: 73–224.

Rudge, A.C., Corrêa, J.S., Sousa, S.D., Piratelli, A. & Piña-Rodrigues, F. 2005. Dieta de aves e dispersao de sementes na regiao de Teresopolis, RJ. P. 66 *in Livro de Resumos do*

XIII Congresso Brasileira de Ornitologia. Belém.

Ruschi, A. 1953. Lista das aves do estado do Espírito Santo. *Bol. Mus. Biol. Prof. Mello Leitão, Sér. Zool.* 11: 1–21.

Ruschi, A. 1965. As aves do recinto da sede do Mus. Biol. M. Leitão em S. Teresa, observadas durante os anos 1931-1951. *Bol. Mus. Biol. Prof. Mello Leitão, Sér. Proteção à Natureza* 26A: 1–13.

Ruschi, A. 1967. Lista atual das aves do Estado do Esp. Santo. *Bol. Mus. Biol. Prof. Mello Leitão, Sér. Zool.* 28A: 1–23.

Ruschi, A. 1969. As aves do recinto da sede do Museu de Biologia Profesor Mello Leitão, na cidade de Santa Teresa, observadas durante o ano de 1968-1969, e a influência das áreas circunvizeinhas. *Bol. Mus. Biol. Prof. Mello Leitão, Sér. Proteção à Natureza* 31: 1–14.

Ruschi, A. 1977. A ornitofauna da estação biológica do Museu Nacional. *Bol. Mus. Biol. Prof. Mello Leitão, Sér. Zool.* 88: 1–10.

Ruschi, A. 1979. *Aves do Brasil*, vol. 1. Editora Rios Ltda., São Paulo.

Ryder, T.B. & Wolfe, J.D. 2009. The current state of knowledge on molt and plumage sequences in selected Neotropical bird families: a review. *Orn. Neotrop.* 20: 1–18.

Sada, A.M., Phillips, A.R. & Ramos, M.A. 1987. *Nombres en Castellano para las Aves Mexicanas.* Instituto Nacional de Investigaciones sobre Recursos Bióticos, Xalapa, Veracruz.

Saibene, C.A., Castelino, M.A., Rey, N.R., Herrera, J. & Calo, J. 1996. *Inventario de las Aves del Parque Nacional Iguazú, Misiones, Argentina.* Literature of Latin America, Buenos Aires.

Salaman, P.G.W. 1994. *Surveys and Conservation Biodiversity in the Chocó, South-west Colombia.* BirdLife International, Cambridge, UK.

Salaman, P.G.W. 2001. *The study of an understory avifauna community in an Andean premontane pluvial forest.* Ph.D. thesis. Univ. of Oxford, Oxford.

Salaman, P.G.W. & Donegan, T.M. 1998. *Colombia '98 Expedition to Serranía de los Churumbelos: Preliminary Report. Colombian EBA Project Report Series No. 1.* Fundación ProAves, Bogotá. www.proaves.org.

Salaman, P.G.W. & Donegan, T.M. 2001. *Presenting the First Biological Assessment of Serranía de San Lucas. Colombian EBA Project Report Series No. 3.* Fundación ProAves, Bogotá. www.proaves.org.

Salaman, P.G.W., Donegan, T.M. & Cuervo, A.M. 1999. Ornithological surveys in Serranía de los Churumbelos, southern Colombia. *Cotinga* 12: 29–39.

Salaman, P.G.W., Cuadros, T., Jaramillo, J.G. & Weber, W.H. 2001. *Lista de Chequeo de las Aves de Colombia.* Sociedad Antioqueña de Ornitología, Medellín.

Salaman, P.G.W., Donegan, T.M. & Cuervo, A.M. 2002a. New distributional bird records from Serrania de San Lucas and adjacent Central Cordillera of Colombia. *Bull. Brit. Orn. Club* 122: 285–304.

Salaman, P.G.W., Stiles, F.G., Bohórquez, C.I., Álvarez-R., M., Umaña, A.M., Donegan, T.M. & Cuervo, A.M. 2002b. New and noteworthy bird records from the east slope of the Andes in Colombia. *Caldasia* 24: 157–189.

Salaman, P.G.W., Coopmans, P., Donegan, T.M., Mulligan, M., Cortés, A., Hilty, S.L. & Ortega, L.A. 2003. A new species of wood-wren (Troglodytidae: *Henicorhina*) from the western Andes of Colombia. *Orn. Colombiana* 1: 4–21.

Salaman, P.G.W., Donegan, T.M. & Caro, D. 2007a. Listado de avifauna Colombiana 2007 - Checklist of Colombian avifauna 2007. *Conserv. Colombiana* 2 (suplemento): 1–85.

Salaman, P.G.W., Donegan, T.M., Davison, D. & Ochoa, J.M. 2007b. Birds of Serranía de los Churumbelos, their conservation and elevational distribution. *Conserv. Colombiana* 3: 29–58.

Salaman, P.G.W., Donegan, T.M. & Caro, D. 2009a. Checklist to the birds of Colombia 2009. *Conserv. Colombiana* 8: 1–89.

Salaman, P.G.W., Donegan, T.M. & Prys-Jones, R. 2009b. A new subspecies of Brown-banded Antpitta *Grallaria milleri* from Antioquia, Colombia. *Bull. Brit. Orn. Club* 129: 5–17.

Salvador, S.A. 1988. Datos de peso de aves argentinas. *Hornero* 13: 78–83.

Salvador, S.A. & Di Giacomo, A.G. 2014. Datos de pesos de aves argentinas. Parte 3. *Historia Natural (Tercera Ser.)* 4: 63–88.

Salvadori, T. & Festa, E. 1898a. Viaggio del Dott. E. Festa nella Repubblica dell'Ecuador e regioni vicine XIII, descrizione di tre nuove specie di uccelli. *Bol. Mus. Zool. Anat. Comp. Univ. Torino* 13: 1–2.

Salvadori, T. & Festa, E. 1898b. Viaggio del dott. E. Festa nel Darien e regioni vicine. *Bol. Mus. Zool. Anat. Comp. Univ. Torino* 15: 1–13.

Salvadori, T. & Festa, E. 1899. Viaggio del Dr. Enrico Festa nell' Ecuador. Uccelli. Part II. Passeres Clamatores. *Bol. Mus. Zool. Anat. Comp. Univ. Torino* 15: 1–34.

Salvin, O. 1865. Descriptions of seventeen new species of birds from Costa Rica. *Proc. Zool. Soc. Lond.* 1865: 579–586.

Salvin, O. 1867. On some collections of birds from Veragua. *Proc. Zool. Soc. Lond.* 1867: 129–161.

Salvin, O. 1870. On some collections of birds from Veragua - Part II. *Proc. Zool. Soc. Lond.* 1870: 175–219.

Salvin, O. 1882. *A Catalogue of the Collection of Birds Formed by the late Hugh Edwin Strickland.* Cambridge Univ. Press, Cambridge, UK.

Salvin, O. 1885. A list of birds obtained by Mr. Henry Whitely in British Guiana (Part 3). *Ibis* 27: 418–439.

Salvin, O. 1895. On birds collected in Peru by Mr. O. T. Baron. *Novit. Zool.* 2: 1–22.

Salvin, O. & Godman, F.D. 1884. Notes on birds from British Guiana, Part III. *Ibis* 26: 443–452.

Salvin, O. & Godman, F.D. 1892. *Biologia Centrali-Americana: Zoology, Botany and Archaeology*, vol. 1 (Aves). R. H. Porter, London.

Salvin, O. & Godman, F.D. 1904. *Biologia Centrali-Americana: Zoology, Botany and Archaeology*, vol. 4 (Aves). R. H. Porter, London.

Sánchez, G. & Aponte, M.A. 2006. Primera descripción del nido y huevos de *Conopophaga ardesiaca. Kempffiana* 2: 102–105.

Sánchez, J.E. 2002. *Aves del Parque Nacional Tapantí, Costa Rica.* Instituto Nacional de Biodiversidad, Santo Domingo.

Sánchez-Martínez, M.A. & Londoño, G.A. 2016a. Nesting behavior of male and female Undulated Antshrikes (*Frederickena unduliger*). *J. Field Orn.* 87: 21–28.

Sánchez-Martínez, M.A. & Londoño, G.A. 2016b. Nesting behavior of three species of *Chlorospingus* (*C. flavigularis, C. flavopectus*, and *C. parvirostris*) in southeastern Peru. *Wilson J. Orn.* 128: 784–793.

Santana, C.R. & dos Anjos, L. 2010. Associação de aves a agrupamentos de bambu na porção Sul da Mata Atlântica, Londrina, Estado do Paraná, Brasil. *Biota Neotropica* 10: 39–44.

Santana-Castellón, E. 2000. Dynamics of Understory Birds along a Cloud Forest Successional Gradient. Ph.D. thesis. Univ. of Wisconsin, Madison.

Santos, A.M.R. 2004. Comunidades de em remanescentes florestais secundários de uma área rural no sudeste do Brasil. *Ararajuba* 12: 41–49.

Santos, M.F.B. & Cademartori, C.V. 2010. Estudo comparativo da avifauna em áreas verdes urbanas da região metropolitana de Porto Alegre, sul do Brasil. *Biotemas* 23: 181–195.

Santos, M.P.D. 2003. Aves do Parque Nacional do Viruá, Estado de Roraima. P. 121 *in* Machado, C.G. (ed.) *Livro de Resumos do XI Congresso Brasileiro de Ornitologia.* Feira de Santana.

Santos, M.P.D. 2004a. New records of birds from the Brazilian state of Roraima. *Bull. Brit. Orn. Club* 124: 223–226.

Santos, M.P.D. 2004b. As comunidades de aves em duas fisionomias da vegetação de caatinga no estado do Piauí, Brasil. *Ararajuba* 12: 113–123.

Santos, M.P.D. 2008. Bird community distribution in a Cerrado-Caatinga transition area, Piauí, Brazil. *Rev. Bras. Orn.* 16: 323–338.

Santos, M.P.D. & Silva, G.C. 2005. Inventario da Avifauna da Terra Indigena Nove de Janeiro Humaita, Amazonas, Brasil. P. 94 *in Livro de Resumos do XIII Congresso Brasileira de Ornitologia.* Belém.

Santos, M.P.D. & Silva, J.M.C. 2007. As aves das savanas de Roraima. *Rev. Bras. Orn.* 15: 189–207.

Santos, M.P.D., Aleixo, A., d'Horta, F.M. & Portes, C.E.B. 2011a. Avifauna of the Juruti Region, Pará, Brazil. *Rev. Bras. Orn.* 19: 134–153.

Santos, M.P.D., Silva, G.C. & Reis, A.L. 2011b. Birds of the Igarapé Lourdes Indigenous Territory, Jí-Paraná, Rondônia, Brazil. *Rev. Bras. Orn.* 19: 230–243.

Santos, M.P.D., Silveira, L.F. & Silva, J.M.C. 2011c. Birds of Serra do Cachimbo, Pará state, Brazil. *Rev. Bras. Orn.* 19: 244–259.

Santos, M.P.D., Santana, A., Soares, L.M.S. & Sousa, S.A. 2012. Avifauna of Serra Vermelha, southern Piauí, Brazil. *Rev. Bras. Orn.* 20: 188.

Santos, M.P.D., Soares, L.M.S., Lopes, F.M., Carvalho, S.T., Silva, M.S. & Santos, D.D. 2013. Birds of Sete Cidades National Park, Brazil: ecotonal patterns and habitat use. *Cotinga* 35: 48–60.

Santos, R.E.F., Patrial, E.W. & Carrano, E. 2004. Composição, estrutura e conservação da avifauna do Distrito do Bugre, Balsa Nova, Paraná, Brasil. P. 361 *in* Testoni, A.F. & Althoff, S.L. (eds.) *Livro de Resumos do XII Congresso Brasileiro de Ornitologia.* Blumenau.

Sanz, V., Oviol, L., Medina, Á. & Moncada, R. 2010. Avifauna del estado Nueva Esparta, Venezuela: recuento histórico y lista actual con nuevos registros de especies y reproducción. *Interciencia* 35: 329–339.

Sari, E.H.R. & Santos, F.R. 2004. Diversidade genetica e filogeografia de populações de *Conopophaga lineata* (Passeriformes: Conopophagidae) em Minas Gerais. P. 363 *in* Testoni, A.F. & Althoff, S.L. (eds) *Livro de Resumos do XII Congresso Brasileiro de Ornitologia.* Blumenau.

Sari, E.H.R., Lacerda, D.R., Vilaca, S.T., Marini, M.Â. & Santos, F.R. 2006. Congruent phylogeography of the Rufous Gnateater and Variable Antshrike (Passeriformes) reveals secondary contact between lineages. *J. Orn.* 147: 106–107.

Sari, E.H.R., Vilaça, S.T., Bertelli, M.Q., Lacerda, D.R.,

Marini, M.Â. & Santos, F.R. 2005. Filogeografia comparada de dois Passeriformes da Mata Atlantica: *Conopophaga lineata* e *Trichothraupis melanops*. P. 223 *in Livro de Resumos do XIII Congresso Brasileira de Ornitologia.* Belém.

Sayako, N. & Dickinson, E.C. 2001. Systematic notes on Asian birds. 10. The "Nouveau recueil de planches coloriées" of Temminck & Laugier (1820-1839): the little known impression of 1850. *Zool. Verh.* 335: 55–60.

Schäfer, E. 1969. Lebensweise und Ökologie der im Nationalpark von Rancho Grande (Nord-Venezuela) nachgewiesenen Ameisenvogelarten (Formicariidae). *Bonn Zool. Beitr.* 20: 99–109.

Schäfer, E. 2002. *Die Vogelwelt Venezuelas und ihre Ökologischen Bedingungen*, Bd. 3. Wirtemberg-Verlag Lang-Jeutter & Jeutter, Berglen.

Schäfer, E. & Phelps, W.H. 1954. Aves del Parque Nacional "Henri Pittier" (Rancho Grande) y sus funciones ecológicas. *Bol. Soc. Venez. Cienc. Nat.* 16: 3–167.

Schaldach, W.J., Escalante, B.P. & Winker, K. 1997. Further notes on the avifauna of Oaxaca, Mexico. *An. Inst. Biol. Univ. Nat. Autónomo Mex., Ser. Bot.* 68: 91–135.

Schemske, D.W. & Brokaw, N. 1981. Treefalls and the distribution of understory birds in a tropical forest. *Ecology* 62: 938–945.

Scherer-Neto, P. & Bispo, A.A. 2011. Avifauna do Parque Estadual de Vila Rica do Espírito Santo, Fênix, Paraná *Biota Neotropica* 11: 317–329.

Scherer-Neto, P. & Toledo, M.C.B. 2012. Bird community in an *Araucaria* forest fragment in relation to changes in the surrounding landscape in Southern Brazil. *Iheringia, Sér. Zool.* 102: 412–422.

Scherer-Neto, P., dos Anjos, L. & Straube, F.C. 1994. Avifauna do parque estadual de Vila Velha, estado do Paraná. *Arq. Biol. Tecn. Curitiba* 37: 223–229.

Scherer-Neto, P., Straube, F.C., Carrano, E. & Urben-Filho, A. 2011. *Lista das Aves do Paraná.* Hori Consultoria Ambiental, Curitiba.

Schmitt, C.J., Schmitt, D.C., Tiravanti C., J., Angulo P., F., Franke, I., Vallejos, L.M., Pollack, L. & Witt, C.C. 2013. Avifauna of a relict *Podocarpus* forest in the Cachil Valley, north-west Peru. *Cotinga* 35: 17–25.

Schomburgk, R.H. 1848. *Reisen in Britisch-Guiana in den Jahren 1840-1884.* J. J. Weber, Leipzig.

Schönwetter, M. 1967. *Handbuch der Oölogie*, Bd. 1. Akademie-Verlag, Berlin.

Schönwetter, M. 1979. *Handbuch der Oölogie.* Akademie-Verlag, Berlin.

Schönwetter, M. & Meise, W. 1988. *Handbuch der Oölogie*, Lieferung 44 & 45. Akademie-Verlag, Berlin.

Schubart, O., Aguirre, A.C. & Sick, H. 1965. Contribução para o conhocimento da alimentação das aves brasileiras. *Arq. Zool. Estado São Paulo* 12: 95–249.

Schulenberg, T.S. 1987. New records of birds from western Peru. *Bull. Brit. Orn. Club* 107: 184–189.

Schulenberg, T.S. 2002. Aves / Birds. Pp. 141–148 *in* Pitman, N., Alverson, W.S., Moskovits, D.K. & Borman A., R. (eds.) *Rapid Biological Inventories 03: Ecuador, Serranías Cofán-Bermejo, Sinangoe.* The Field Museum, Chicago.

Schulenberg, T.S. & Awbrey, K. 1997. *A Rapid Assessment of the Humid Forests of South Central Chuquisaca, Bolivia. RAP Working Paper 8.* Conservation International, Washington DC.

Schulenberg, T.S. & Harvey, M. 2012. Long-whiskered Owlet (*Xenoglaux loweryi*). *In* Schulenberg, T.S. (ed.) *Neotropical*

Birds Online. Cornell Lab of Ornithology, Ithaca, NY (https://doi.org/10.2173/nb.lowowl1.01).

Schulenberg, T.S. & Kirwan, G.M. 2011. Stripe-headed Antpitta (*Grallaria andicolus*). *In* Schulenberg, T.S. (ed.) *Neotropical Birds Online.* Cornell Lab of Ornithology, Ithaca, NY (https://doi.org/10.2173/nb.sthant2.01).

Schulenberg, T.S. & Kirwan, G.M. 2012a. Ochre-fronted Antpitta (*Grallaricula ochraceifrons*). *In* Schulenberg, T.S. (ed.) *Neotropical Birds Online.* Cornell Lab of Ornithology, Ithaca, NY (https://doi.org/10.2173/nb.ocfant1.01).

Schulenberg, T.S. & Kirwan, G.M. 2012b. Moustached Antpitta (*Grallaria alleni*). *In* Schulenberg, T.S. (ed.) *Neotropical Birds Online.* Cornell Lab of Ornithology, Ithaca, NY (https://doi.org/10.2173/nb.mouant1.01).

Schulenberg, T.S. & Kirwan, G.M. 2012c. White-throated Antpitta (*Grallaria albigula*). *In* Schulenberg, T.S. (ed.) *Neotropical Birds Online.* Cornell Lab of Ornithology, Ithaca, NY (https://doi.org/10.2173/nb.whtant2.01).

Schulenberg, T.S. & Kirwan, G.M. 2012d. Red-and-white Antpitta (*Grallaria erythroleuca*). *In* Schulenberg, T.S. (ed.) *Neotropical Birds Online.* Cornell Lab of Ornithology, Ithaca, NY (https://doi.org/10.2173/nb.rawant1.01).

Schulenberg, T.S. & Kirwan, G.M. 2012e. Pale-billed Antpitta (*Grallaria carrikeri*). *In* Schulenberg, T.S. (ed.) *Neotropical Birds Online.* Cornell Lab of Ornithology, Ithaca, NY (https://doi.org/10.2173/nb.pabant1.01).

Schulenberg, T.S. & Kirwan, G.M. 2012f. Chestnut Antpitta (*Grallaria blakei*). *In* Schulenberg, T.S. (ed.) *Neotropical Birds Online.* Cornell Lab of Ornithology, Ithaca, NY (https://doi.org/10.2173/nb.cheant2.01).

Schulenberg, T.S. & Kirwan, G.M. 2012g. Bay Antpitta (*Grallaria capitalis*). *In* Schulenberg, T.S. (ed.) *Neotropical Birds Online.* Cornell Lab of Ornithology, Ithaca, NY (https://doi.org/10.2173/nb.bayant1.01).

Schulenberg, T.S. & Parker, T.A. 1981. Status and distribution of some northwest Peruvian birds. *Condor* 83: 209–216.

Schulenberg, T.S. & Remsen, J.V. 1982. Eleven bird species new to Bolivia. *Bull. Brit. Orn. Club* 102: 52–57.

Schulenberg, T.S. & Williams, M.D. 1982. A new species of antpitta (*Grallaria*) from northern Peru. *Wilson Bull.* 94: 105–113.

Schulenberg, T.S. & Wust, W.H. 1997. Birds of the upper Río Comainas, Cordillera del Cóndor. Pp. 180–187 *in* Schulenberg, T.S. & Awbrey, K. (eds) *The Cordillera del Cóndor region of Ecuador and Peru: A Biological Assessment. RAP Working Papers 7.* Conservation International, Wshington DC.

Schulenberg, T.S., Allen, S.E., Stotz, D.F. & Wiedenfeld, D.A. 1984. Distributional records from the Cordillera Yanachaga, central Peru. *Gerfaut* 74: 57–70.

Schulenberg, T.S., Parker, T.A. & Wust, W.H. 1997. Birds of the Cordillera del Cóndor. Pp. 63–64 *in* Schulenberg, T.S. & Awbrey, K. (eds.) *The Cordillera del Cóndor Region of Ecuador and Peru: A Biological Assessment. RAP Working Papers 7.* Conservation International, Washington DC.

Schulenberg, T.S., Quiroga O., C., Jammes, L. & Moskovits, D.K. 2000. Aves / Birds. Pp. 36–39 *in* Alverson, W.S., Moskovits, D.K. & Shopland, J.M. (eds.) *Rapid Biological Inventories 01: Bolivia, Pando, Río Tahuamanu.* The Field Museum, Chicago.

Schulenberg, T.S., López, L., Servat, G. & Valdes, A. 2001a. Preliminary list of the birds at three sites in the Northern Cordillera de Vilcabamba, Peru. Pp. 241–248 *in* Alonso, L.E., Alonso, A., Schulenberg, T.S. & Dallmeier, F. (eds.) *Biological and Social Assessments of the Cordillera Vilcabamba, Peru. RAP Working Papers 12.* Conservation International, Washington DC.

Schulenberg, T.S., O'Neill, J.P., Lane, D.F., Valqui, T. & Albújar, C. 2001b. Aves / Birds. Pp. 146–155 *in* Alverson, W.S., Rodríguez, L.O. & Moskovits, D.K. (eds.) *Rapid Biological Inventories 02: Perú, Biabo Cordillera Azul.* The Field Museum, Chicago.

Schulenberg, T.S., Albujar, C. & Rojas, J.I. 2006. Aves / Birds. Pp. 186–195, 263–273 *in* Vriesendorp, C., Schulenberg, T.S., Alverson, W.S., Moskovits, D.K. & Rojas, J.I. (eds.) *Rapid Biological and Social Inventories 17: Perú: Sierra del Divisor.* The Field Museum, Chicago.

Schulenberg, T.S., Stotz, D.F., Lane, D.F., O'Neill, J.P. & Parker, T.A. 2007. *Birds of Peru.* Princeton Univ. Press, Princeton, NJ.

Schulenberg, T.S., Stotz, D.F., Lane, D.F., O'Neill, J.P. & Parker, T.A. 2010. *Birds of Peru: Revised and updated edition.* Princeton Univ. Press, Princeton, NJ.

Schulenberg, T.S., Greeney, H.F. & Terrill, R.S. 2013a. Watkins's Antpitta (*Grallaria watkinsi*). *In* Schulenberg, T.S. (ed.) *Neotropical Birds Online.* Cornell Lab of Ornithology, Ithaca, NY (https://doi.org/10.2173/nb.watant1.01).

Schulenberg, T.S., Price, E.R. & Kirwan, G.M. 2013b. White-bellied Antpitta (*Grallaria hypoleuca*). *In* Schulenberg, T.S. (ed.) *Neotropical Birds Online* Cornell Lab of Ornithology, Ithaca, NY (https://doi.org/10.2173/nb.whbant3.01).

Schunck, F., De Luca, A.C., Piacentini, V.D., Rego, M.A., Rennó, B. & Corrêa, A.H. 2011. Avifauna of two localities in the south of Amapá, Brazil, with comments on the distribution and taxonomy of some species. *Rev. Bras. Orn.* 19: 93–107.

Schwartz, P. 1957. Observaciones sobre *Grallaricula ferrugineipectus. Bol. Soc. Venez. Cienc. Nat.* 18: 42–62.

Sclater, P.L. 1854. List of a collection of birds received by Mr. Gould from the Province of Quijos, in the Republic of Ecuador. *Proc. Zool. Soc. Lond.* 1854: 109–115.

Sclater, P.L. 1855b. Descriptions of some new species of ant-thrushes (Formicariinae) from Santa Fé di Bogota. *Proc. Zool. Soc. Lond.* 1855: 88–90.

Sclater, P.L. 1855c. On the birds received in collections from Santa Fé di Bogota. *Proc. Zool. Soc. Lond.* 1855: 131–164.

Sclater, P.L. 1856. On some additional species of birds received in collections from Bogota. *Proc. Zool. Soc. Lond.* 1856: 25–31.

Sclater, P.L. 1857a. Descriptions of twelve new or little-known species of the South American family Formicariidæ. *Proc. Zool. Soc. Lond.* 1857: 129–133.

Sclater, P.L. 1857b. Characters of some apparently new species of American ant-thrushes. *Proc. Zool. Soc. Lond.* 1857: 46–48.

Sclater, P.L. 1857c. Further additions to the list of birds received in collections from Bogota. *Proc. Zool. Soc. Lond.* 1857: 15–20.

Sclater, P.L. 1857d. Catologue of the birds collected by M. Auguste Sallé in southern Mexico, with descriptions of new species. *Proc. Zool. Soc. Lond.* 1857: 283–311.

Sclater, P.L. 1858a. Notes on a collection of birds received by M. Verreaux, of Paris, from the Rio Napo in the Republic of Ecuador. *Proc. Zool. Soc. Lond.* 1858: 59–77.

Sclater, P.L. 1858b. On the birds collected by Mr. Fraser in the vicinity of Riobamba, in the Republic of Ecuador. *Proc. Zool. Soc. Lond.* 1858: 549–556.

Sclater, P.L. 1858c. Synopsis of the American ant-birds (Formicariidae), Part III, containing the third subfamily Formicariinae, or ant-thrushes. *Proc. Zool. Soc. Lond.* 1858: 272–289.

Sclater, P.L. 1859a. On some new or little-known birds from the Rio Napo. *Proc. Zool. Soc. Lond.* 1859: 440–441.

Sclater, P.L. 1859b. On a series of birds collected in the vicinity of Jalapa, in southern Mexico. *Proc. Zool. Soc. Lond.* 1859: 362–369.

Sclater, P.L. 1859c. List of birds collected by M. A. Boucard in the state of Oaxaca in south-western Mexico, with descriptions of new species. *Proc. Zool. Soc. Lond.* 1859: 369–393.

Sclater, P.L. 1860a. List of additional species of birds collected by Mr. Louis Fraser at Pallatanga, Ecuador, with notes and descriptions of new species. *Proc. Zool. Soc. Lond.* 1860: 63–73.

Sclater, P.L. 1860b. List of birds collected by Mr. Fraser in the vicinity of Quito and during excursions to Pichincha and Chimborazo with notes and descriptions of new species. *Proc. Zool. Soc. Lond.* 1860: 73–83.

Sclater, P.L. 1860c. List of birds collected by Mr. Fraser in Ecuador at Nanegal, Calacali, Perucho, and Puellaro, with notes and descriptions of new species. *Proc. Zool. Soc. Lond.* 1860: 83–97.

Sclater, P.L. 1860d. On two new birds from the Rio Napo. *Ann. & Mag. Nat. Hist.* (3)5: 498.

Sclater, P.L. 1861. Descriptions of twelve new species of American birds of the families Dendrocolaptidae, Formicariidae, and Tyrannidae. *Proc. Zool. Soc. Lond.* 1861: 377–383.

Sclater, P.L. 1862. *Catalogue of a Collection of American Birds Belonging to Philip Lutley Sclater.* N. Trübner & Co., London.

Sclater, P.L. 1868. Descriptions of some new or little-known species of formicarians. *Proc. Zool. Soc. Lond.* 1868: 571–575.

Sclater, P.L. 1873a. Additions to the list of birds of Nicaragua. *Ibis* 15: 372–373.

Sclater, P.L. 1873b. On Peruvian birds collected by Mr. Whitely - Part VII. *Proc. Zool. Soc. Lond.* 1873: 779–784.

Sclater, P.L. 1874. [Exhibition of two new species of birds from Antioquia]. *Proc. Zool. Soc. Lond.* 1873: 728–729.

Sclater, P.L. 1877. Description of two new ant-birds of the genus *Grallaria*, with a list of the known species of the genus. *Ibis* 19: 437–451.

Sclater, P.L. 1890. *Catalogue of the Passeriformes, or Perching Birds, in the Collection of the British Museum. Tracheophonae, or the families Dendrocolaptidae, Formicariidae, Conopophagidae, and Pteroptochidae.* British Museum of Natural History, London.

Sclater, P.L. & Salvin, O. 1859. On the ornithology of Central America, Part II. *Ibis* 1: 117–138.

Sclater, P.L. & Salvin, O. 1864. Notes on a collection of birds form the Isthmus of Panama. *Proc. Zool. Soc. Lond.* 1864: 342–373.

Sclater, P.L. & Salvin, O. 1867. List of birds collected at Pebas, upper Amazons, by Mr. John Hauxwell, with notes and descriptions of new species. *Proc. Zool. Soc. Lond.* 1867: 977–981.

Sclater, P.L. & Salvin, O. 1868. On Venezuelan birds collected by Mr. A. Goering, Part II. *Proc. Zool. Soc. Lond.* 1868: 626–632.

Sclater, P.L. & Salvin, O. 1869. Descriptions of six new species of American birds of the families Tanagridae, Dendrocolaptidae, Formicariidae, Tyrannidae, and Scolopacidae. *Proc. Zool. Soc. Lond.* 1869: 416–420.

Sclater, P.L. & Salvin, O. 1870. On Venezuelan birds collected by Mr. A. Goering - Part IV. *Proc. Zool. Soc. Lond.* 1870: 779–788.

Sclater, P.L. & Salvin, O. 1873a. On the birds of eastern Peru. *Proc. Zool. Soc. Lond.* 1873: 252–311.

Sclater, P.L. & Salvin, O. 1873b. *Nomenclator Avium Neotropicalium.* Sumptibus Auctorum, London.

Sclater, P.L. & Salvin, O. 1874. On Peruvian birds collected by Mr. Whitely - Part VIII. *Proc. Zool. Soc. Lond.* 1874: 677–670.

Sclater, P.L. & Salvin, O. 1876. On new species of Bolivian birds. *Proc. Zool. Soc. Lond.* 1876: 352–358.

Sclater, P.L. & Salvin, O. 1879a. On the birds collected by the late Mr. T. K. Salmon in the State of Antioquia, United States of Colombia. *Proc. Zool. Soc. Lond.* 1879: 486–550.

Sclater, P.L. & Salvin, O. 1879b. On the birds collected in Bolivia by Mr. C. Buckley. *Proc. Zool. Soc. Lond.* 1879: 589–645.

Sclater, P.L. & Salvin, O. 1880. On new birds collected by Mr. C. Buckley in eastern Ecuador. *Proc. Zool. Soc. Lond.* 1880: 155–161.

Scott, D.A. & Brooke, M.L. 1985. The endangered avifauna of southeastern Brazil: a report on the BOU/WWF expeditions of 1980/81 and 1981/82. Pp. 115–139 *in* Diamond, A.W. & Lovejoy, T.E. (eds.) *Conservation of Tropical Forest Birds.* International Council for Bird Preservation, Cambridge, UK.

Sebaio, F., Braga, E.M., Branquinho, F., Fecchio, A. & Marini, M.Â. 2012. Blood parasites in passerine birds from the Brazilian Atlantic Forest. *Rev. Bras. Parasit. Vet.* 21: 7–15.

Servat, G.P. 1996. An annotated list of birds of the BIOLAT Biological Station at Pakitza, Peru. Pp. 555–575 *in* Wilson, D.E. & Sandoval, A. (eds.) *Manu: The Biodiversity of Southeastern Peru.* Smithsonian Institution Press, Lima.

Sharpe, C.J. 2008. Ponchito cabecicastaño *Grallaricula cucullata* Sclater 1856. P. 146 *in* Rodríguez, J.P. & Rojas-Suárez, F. (eds.) *Libro Rojo de la Fauna Venezolana*, 3rd edn. Provita y Shell Venezuela, Caracas.

Sharpe, C.J. & Ascanio, D. 2008. Hormiguero tororoi excelso *Grallaria excelsa* Berlepsch 1893. P. 145 *in* Rodríguez, J.P. & Rojas-Suárez, F. (eds.) *Libro Rojo de la Fauna Venezolana*, 3rd ed. Provita y Shell Venezuela, Caracas.

Sharpe, C.J. & Lentino, M. 2008. Hormiguero tororoi tachirense *Grallaria chtonia* Wetmore & Phelps 1956. P. 144 *in* Rodríguez, J.P. & Rojas-Suárez, F. (eds.) *Libro Rojo de la Fauna Venezolana*, 3rd edn. Provita y Shell Venezuela, Caracas.

Sherborn, C.D. & Griffin, F.J. 1934. On the dates of publication of the natural history portions of Alcide d'Orbigny's 'Voyage Amérique méridionale'. *Ann. & Mag. Nat. Hist.* 13: 130–134.

Sherborn, C.D. & Woodward, B.B. 1901. Notes on the dates of publication of the natural history portions of some French voyages - Part I. 'Amérique méridionale'; 'Indes orientales'; 'Pôle Sud' ('Astrolabe' and 'Zélée'); 'La Bonite'; 'La Coquille'; and 'L'Uranie et Physicienne'. *Ann. & Mag. Nat. Hist.* 7: 388–390.

Sibley, C.G. & Monroe, B.L. 1990. *Distribution and Taxonomy of Birds of the World.* Yale Univ. Press, New Haven, CT.

Sibley, C.G. & Monroe, B.L. 1993. *A Supplement to Distribution and Taxonomy of Birds of the World.* Yale Univ. Press, New Haven, CT.

Sick, H. 1957. Roßhaarpilze als Nestbau-Material brasilianischer Vögel. *J. Orn.* 98: 421–431.

Sick, H. 1965. Sons emitidos pelas aves independentemente do órgão vocal; caso de *Conopophaga lineata* (Wied). *An. Acad. Bras. Ciênc.* 37: 131–146.

Sick, H. 1984. *Ornitologia Brasileira, uma Introdução*, vol. 1. Univ. de Brasília e Linha Grafica Ed., Brasília.

Sick, H. 1993. *Birds in Brazil: A Natural History*. Princeton Univ. Press, Princeton, NJ.

Sick, H. 1997. *Ornitologia Brasileira*. Editora Nova Fronteira, Rio de Janeiro.

Sick, H. & Pabst, L.F. 1963. As aves do Rio de Janeiro (Guanabara). *Arq. Mus. Nac. Rio de Janeiro* 53: 99–160.

Siegel, D.C. & Olson, S.L. 2008. *The Birds of the Republic of Panama. Part 5. Gazetteer and Bibliography*. Buteo Books, Shipman, VA.

Sierra M., R., Campos, F. & Chamberlin, J. 1999. *Areas Prioritarias para la Conservación de la Biodiversidad en el Ecuador Continental: Un Estudio Basado en la Biodiversidad de Ecosistemas y su Ornitofauna*. Ministerio de Medio Ambiente, Proyecto INEFAN/GEF-BIRF, EcoCiencia & Wildlife Conservation Society, Quito.

Sieving, K.E. 1992. Nest predation and differential insular extinction among selected forest birds of central Panama. *Ecology* 73: 2310–2328.

Sigel, B.J., Sherry, T.W. & Young, B.E. 2006. Avian community response to lowland tropical rainforest isolation: 40 years of change at La Selva Biological Station, Costa Rica. *Conserv. Biol.* 20: 111–121.

Sigel, B.J., Robinson, W.D. & Sherry, T.W. 2010. Comparing bird community responses to forest fragmentation in two lowland Central American reserves. *Biol. Conserv.* 143: 340–350.

Silva, E.S., Espínola, C.R.R., Schmitt, D.A.M. & Albuquerque, J.L.B. 2008. Estudo da avifauna florestal no morro da Lagoa da Conceição, Ilha de Santa Catarina. P. 234 in Dornas, T. & Barbosa, M.O. (eds.) *Livro de Resumos do XVI Congresso Brasileiro de Ornitologia*. Palmas.

Silva, H.G. 2005. The nesting season: Mexico. *North Amer. Birds* 59: 659–661.

Silva, H.G., González-García, F. & Casillas-Trejo, M.P. 1999. Birds of the upper cloud forest of El Triunfo, Chiapas, Mexico. *Orn. Neotrop.* 10: 1–26.

Silva, J.M.C. 1995. Birds of the Cerrado region, South America. *Steenstrupia* 21: 69–92.

Silva, J.M.C. 1996. Distribution of Amazonian and Atlantic birds in gallery forests of the Cerrado region, South America. *Orn. Neotrop.* 7: 1–18.

Silva, J.M.C., Lima, M.F.C. & Marceliano, M.L.V. 1990. Pesos de aves de duas localidades na Amazonia Oriental. *Ararajuba* 1: 99–104.

Silva, J.M.C., Souza, M.A., Bieber, A.G.D. & Carlos, C.J. 2003. Aves da caatinga: status, uso do habitat e sensitividade. Pp. 237–273 in Leal, I.R., Tabarelli, M. & da Silva, J.M.C. (eds.) *Ecologia e conservacao da caatinga*. Univ. Federal de Pernambuco, Recife.

Silva, M.S., Rodrigues, E.B., Carvalho, S.T., Sousa, D.D.S. & Santos, M.P.D. 2004. Analise ecologica das comunidades de aves na area do eco resort Nazareth, Municipio de Jose de Freitas, Piaui. P. 380 in Testoni, A.F. & Althoff, S.L. (eds.) *Livro de Resumos do XII Congresso Brasileiro de Ornitologia*. Blumenau.

Silva, R.R.V. 2001. Ornitofauna encontrada no Complexo Esportivo do SESI em Caxias do Sul, Rio Grande do Sul. Resumo 225 in Straube, F.C. (ed.) *IX congresso brasileiro de ornitologia, resumos: Ornitologia Sem Fronteras*. Curitiba.

Silveira, L.F. 1998. The birds of Serra da Canastra National Park and adjacent areas, Minas Gerais, Brazil. *Cotinga* 10: 55–63.

Silveira, L.F. 2008. *Grallaria varia intercedens* Berlepsch & Leverkühn, 1890. Pp. 557–558 in Machado, A.B.M., Drummond, G.M. & Paglia, A.P. (eds.) *Livro Vermelho da Fauna Brasileira Ameaçada de Extinção*. Ministério do Meio Ambiente & Fundação Biodiversitas, Brasília & Belo Horizonte.

Silveira, L.F. & d'Horta, F.M. 2002. A avifauna da regiao de Vila Bela da Santissima Trindade, Mato Grosso. *Pap. Avuls. Dept. Zool., São Paulo* 42: 265–286.

Silveira, L.F. & Santos, M.P.D. 2012. Bird richness in Serra das Confusões National Park, Brazil: how many species may be found in an undisturbed caatinga? *Rev. Bras. Orn.* 20: 173.

Silveira, L.F. & Uezu, A. 2011. Checklist das aves do estado de São Paulo, Brasil. *Biota Neotropica* 11: 83–110.

Silveira, L.F., Olmos, F. & Long, A.L. 2003. Birds in Atlantic Forest fragments in north-east Brazil. *Cotinga* 20: 32–46.

Simon, J.E. 2000. Composição da avifauna da Estação Biológica de Santa Lúcia, Santa Teresa - ES. *Bol. Mus. Biol. Prof. Mello Leitão, N. Sér.* 11/12: 149–170.

Simon, J.E., Ribon, R., Mattos, G.T. & Abreu, C.R.M. 1999. A avifauna do Parque Estadual da Serra do Brigadeiro, Minas Gerais. *Rev. Árvore* 23: 33–48.

Simon, J.E., Ruschi, P.A. & Peres, J. 2008. Levantamento preliminar da avifauna na Serra das Torres, sul do estado do Espirito Santo. P. 368 in Dornas, T. & Barbosa, M.O. (eds.) *Livro de Resumos do XVI Congresso Brasileiro de Ornitologia*. Tocantins.

Skutch, A.F. 1940. Some aspects of Central American bird-life. II. Plumage, reproduction and song. *Scientific Monthly* 51: 500–511.

Skutch, A.F. 1969. *Life histories of Central American birds III*. Pacific Coast Avifauna 35. Cooper Ornithological Society, Berkeley, CA.

Skutch, A.F. 1971. *A Naturalist in Costa Rica*. Univ. of Florida Press, Gainesville.

Skutch, A.F. 1981. *New studies of tropical American birds*. Publ. Nuttall Orn. Club 19. Nuttall Ornithological Club, Cambridge, MA.

Skutch, A.F. 1996. *Antbirds and Ovenbirds: Their Lives and Homes*. Univ. of Texas Press, Austin.

Slud, P. 1960. The birds of finca "La Selva," Costa Rica: a tropical wet forest locality. *Bull. Amer. Mus. Nat. Hist.* 121: 53–148.

Slud, P. 1964. The birds of Costa Rica: distribution and ecology. *Bull. Amer. Mus. Nat. Hist.* 128: 1–430.

Smith, E.F.G., Arctander, P., Fjeldså, J. & Amir, O.G. 1991. A new species of shrike (Laniidae: *Laniarius*) from Somalia, verified by DNA sequence data from the only known individual. *Ibis* 133: 227–235.

Smith, P., del Castillo, H., Batjes, H., Montiel, M. & Wainwright, B. 2005. An avifaunal inventory of Estancia Laguna Blanca, Departamento San Pedro, north-eastern Paraguay. *FAUNA Paraguay Tech. Publ.* 2: 1–16.

Smith, P., del Castillo, H., Bankovics, A., Hansen, L. & Wainwright, B. 2006. An avifaunal inventory of San Rafel "National Park", Departamento Itapúa, southern Paraguay. *FAUNA Paraguay Tech. Publ.* 3: 1–13.

Smithe, F.B. 1975. *The naturalist's Color Guide. Parts I and II*. American Museum of Natural History, New York.

Smithe, F.B. 1981. *The naturalist's Color Guide. Part III*.

American Museum of Natural History, New York.

Snethlage, E. 1906. Über brasilianische Vögel. *Orn. Monatsb.* 14: 9–10.

Snethlage, E. 1907a. Sobre uma collecção de aves do Rio Purús. *Bol. Mus. Para. E. Goeldi, Sér. Zool.* 5: 43–76.

Snethlage, E. 1907b. Neue Vogelarten aus Südamerika (2). *Orn. Monatsb.* 15: 193–196.

Snethlage, E. 1907c. Über unteramazonische Vögel [continued]. *J. Orn.* 55: 283–299.

Snethlage, E. 1909. Novas especies de aves amazonicas das collecções do Museu Goeldi. *Bol. Mus. Para. E. Goeldi, Sér. Zool.* 5: 437–448.

Snethlage, E. 1910. Berichtigung. *Orn. Monatsb.* 18: 192.

Snethlage, E. 1914a. Catalogo das aves Amazonicas: contendo todas as especies descriptas e mencionadas até 1913. *Bol. Mus. Para. E. Goeldi* 8: 1–530.

Snethlage, E. 1914b. Neue Vogelarten aus Amazonien. *Orn. Monatsb.* 22: 39–44.

Snethlage, E. 1925. Novas especies de aves do N. E. do Brasil. *Bol. Mus. Nac. Rio de Janeiro* 1: 407–412.

Snethlage, E. 1935. Beiträge zur Fortpflanzungsbiologie brasilianischer Vögel. *J. Orn.* 83: 532–562.

Snethlage, H. 1927. Meine Reise durch Nordostbrasilien. *J. Orn.* 75: 453–484.

Snethlage, H. 1928. Meine Reise durch Nordbrasilien. II. Biologische Beobachtungen. *J. Orn.* 76: 503–581.

Snow, D.W. 1985. Affinities and recent history of the avifauna of Trinidad and Tobago. *Orn. Monogr.* 36: 238–246.

Snow, S.S., Field, D.J. & Musser, J.M. 2015. Interspecific competition in *Grallaria* antpittas: observations at a feeder. *Bull. Peabody Mus. Nat. Hist.* 56: 89–93.

Snyder, D.E. 1966. *The birds of Guyana.* Peabody Museum, Salem, MA.

Soares-Filho, B.S., Nepstad, D.C., Curran, L.M., Cerqueira, G.C., Garcia, R.A., Ramos, C.A., Voll, E., McDonald, A., Lefebvre, P. & Schlesinger, P. 2006. Modelling conservation in the Amazon basin. *Nature* 440: 520–523.

Socolar, S.J., González, Ó. & Forero-Medina, G. 2013. Noteworthy bird records from the northern Cerros del Sira, Peru. *Cotinga* 35: 24–36.

Solano-Ugalde, A., Arcos-Torres, A. & Greeney, H.F. 2007. Additional breeding records for selected avian species in northwest Ecuador. *Bol. Soc. Antioqueña Orn.* 17: 17–25.

Solano-Ugalde, A., Paz, Á. & Paz, W. 2009. First description of the nest, nest site, egg, and young of the Giant Antpitta (*Grallaria gigantea*). *Orn. Neotrop.* 20: 633–638.

Somenzari, M., Silveira, L.F., Piacentini, V.Q., Rego, M.A., Schunck, F. & Cavarzere, V. 2011. Birds of an Amazonia-Cerrado ecotone in southern Pará, Brazil, and the efficiency of associating multiple methods in avifaunal inventories. *Rev. Bras. Orn.* 19: 260–275.

Sornoza M., F. 2000. Fundación Jocotoco: conservation in action in Ecuador. *World Birdwatch* 22: 14–17.

Sousa, A.E.B.A., Lima, D.M. & Lyra-Neves, R.M. 2012. Avifauna of the Catimbau National Park in the Brazilian state of Pernambuco, Brazil: species richness and spatiotemporal variation. *Rev. Bras. Orn.* 20: 215.

Sousa, S.D., Corrêa, J.S., Rudge, A.C. & Piratelli, A. 2004. Avifauna de fragmentos florestais em areas agricolas na região de Teresopolis, RJ. P. 394 *in* Testoni, A.F. & Althoff, S.L. (eds.) *Livro de Resumos do XII Congresso Brasileiro de Ornitologia.* Blumenau.

Soye, Y., Schuchmann, K.-L. & Matheus, J.C. 1997. Field notes on the Giant Antpitta *Grallaria gigantea. Cotinga* 7: 35–36.

Spaans, A.L., Ottema, O.H. & Ribot, J.H.J.M. 2016. *Field Guide to the Birds of Suriname.* Koninklijke Brill, Leiden.

Spencer, A. 2012. Rufous Antpitta vocal variation. http://www.xeno-canto.org/feature-view.php?blognr=130.

Spencer, A. 2014. Northern Colombia: January 18–26, 2014 (Tropical Birding Trip Report).

Stark, J. 1828. *Elements of Natural History, Adapted to the Present State of the Science, Containing the Generic Characters of Nearly the Whole Animal Kingdom, and the Descriptions of the Principal Species.* W. Blackwood, Edinburgh.

Stattersfield, A.J. & Capper, D.R. 2000. *Threatened Birds of the World.* Lynx Edicions, Barcelona.

Stattersfield, A.J., Crosby, M.J., Long, A.J. & Wege, D.C. 1997. *A Global Directory of Endemic Bird Areas.* BirdLife International, Cambridge, UK.

Stattersfield, A.J., Crosby, M.J., Long, A.J. & Wege, D.C. 1998. *Endemic Bird Areas of the World. Priorities for Biodiversity Conservation.* BirdLife International, Cambridge, UK.

Stenzel, R.N. & de Souza, P.B. 2014. Biologia reprodutiva do cuspidor-de-máscara-preta, *Conopophaga melanops* (Passeriformes: Conopophagidae), no Jardim Botanico do Rio de Janeiro, Brasil. *Atualidades Orn.* 177: 26–27.

Stephens, L. & Traylor, M.A. 1983. *An Ornithological Gazetteer of Peru.* Museum of Comparative Zoology, Harvard Univ., Cambridge, MA.

Stephens, L. & Traylor, M.A. 1985. *An Ornithological Gazetteer of Guiana.* Museum of Comparative Zoology, Harvard Univ., Cambridge, MA.

Stiles, F.G. 1984. Inventario preliminar de las aves de las selvas nubladas de Monte Zerpa y La Mucuy, Merida, Venezuela. *Bol. Soc. Venez. Cienc. Nat.* 39: 11–24.

Stiles, F.G. 1985. Conservation of forest birds in Costa Rica: problems and perspectives. Pp. 141–168 *in* Diamond, A.W. & Lovejoy, T.E. (eds.) *Conservation of Tropical Forest Birds.* International Council for Bird Preservation, Cambridge, UK.

Stiles, F.G. 1992. A new species of antpitta (Formicariidae: *Grallaria*) from the eastern Andes of Colombia. *Wilson Bull.* 104: 389–399.

Stiles, F.G. 1997. Aves endémicas de Colombia. Pp. 378–385, 428–432 *in* Cháves, M.E. & Arango, N. (eds.) *Informe Nacional Sobre el Estado de la Biodiversidad en Colombia,* vol. 1. Instituto de Investigación de Recursos Biológicos Alexander von Humboldt, PNUMA & Ministerio de Medio Ambiente, Bogotá.

Stiles, F.G. 2010. La avifauna de la parte media del Río Apaporis, Departamentos de Vaupés y Amazonas, Colombia *Rev. Acad. Colombiana Cien. Exactas, Fisicas, y Nat.* 34 381–396.

Stiles, F.G. & Alvarez-López, H. 1995. La situación del tororoi pechicanela (*Grallaria haplonota,* Formicariidae) en Colombia. *Caldasia* 17: 607–610.

Stiles, F.G. & Bohórquez, C.I. 2000. Evaluando el estado de la biodiversidad: el caso de la avifauna de la Serranía de las Quinchas, Boyacá, Colombia. *Caldasia* 22: 61–92.

Stiles, F.G. & Rosselli, L. 1998. Inventario de las aves de un bosque altoandino: comparación de métodos. *Caldasia* 20: 29–43.

Stiles, F.G. & Skutch, A.F. 1989. *A guide to the birds of Costa Rica.* Cornell Univ. Press, Ithaca, NY.

Stiles, F.G., Rosselli, L. & Bohórquez, C.I. 1999. New and noteworthy records of birds from the middle Magdalena valley of Colombia. *Bull. Brit. Orn. Club* 119: 113–129.

Stone, W. 1918. Birds of the Panama Canal Zone with special reference to a collection made by Lindsey L. Jewel. *Proc.*

Acad. Nat. Sci. Phil. 70: 239–280.

Stone, W. 1928. On a collection of birds from the Para region, eastern Brazil. *Proc. Acad. Nat. Sci. Phil.* 80: 149–176.

Stone, W. 1932. The birds of Honduras with special reference to a collection made in 1930 by John T. Emlen, Jr., and C. Brooke Worth. *Proc. Acad. Nat. Sci. Phil.* 84: 291–342.

Storer, R.W. 1989. Notes on Paraguayan birds. *Occ. Pap. Mus. Zool. Univ. Mich.* 719: 1–21.

Storni, A., Alves, M.A.S. & Amorim, M. 2001. Passeriformes como hospedeiros de *Amblyomma longirostre* (Acari: Ixodidae) na Praia do Sul e do Aventureiro, Ilha Grande, Rio de Janeiro. Resumo 213 *in* Straube, F.C. (ed.) *IX Congresso Brasileiro de Ornitologia, Resumos: Ornitologia Sem Fronteras*. Curitiba.

Stotz, D.F. 1993. Geographic variation in species composition of mixed species flocks in lowland humid forests in Brazil. *Pap. Avuls. Dept. Zool., São Paulo* 38: 61–75.

Stotz, D.F. & Alván, J.D. 2007. Aves / Birds. Pp. 67–73, 134–140, Apéndice/Appendix 135 *in* Vriesendorp, C., Álvarez, J.A., Barbagelata, N., Alverson, W.S. & Moskovits, D.K. (eds.) *Rapid Biological and Social Inventories 18: Perú, Nanay-Mazán-Arabela.* The Field Museum, Chicago.

Stotz, D.F. & Alván, J.D. 2010. Aves / Birds. Pp. 81–90, 197–205, 288–310 *in* Gilmore, M.P., Vriesendorp, C., Alverson, W.S., del Campo, Á., von May, R., Wong, C.L. & Ochoa, S.R. (eds.) *Rapid Biological and Social Inventories 22: Perú: Maijuna.* The Field Museum, Chicago.

Stotz, D.F. & Alván, J.D. 2011. Aves / Birds. Pp. 237–245, 336–355 *in* Pitman, N., Vriesendorp, C., Moskovits, D.K., von May, R., Alvira, D., Wachter, T., Stotz, D.F. & del Campo, A. (eds.) *Rapid Biological and Social Inventories 23: Perú: Yaguas-Cotuhé.* The Field Museum, Chicago.

Stotz, D.F. & Bierregaard, R.O. 1989. The birds of the fazendas Porto Alegre, Esteio and Dimona north of Manaus, Amazonas, Brazil. *Rev. Bras. Biol.* 49: 861–872.

Stotz, D.F. & Inzunza, E.R. 2013. Aves / Birds. Pp. 257–262, 362–373 *in* Pitman, N., Inzunza, E.R., Vriesendorp, C., Stotz, D.F., Wachter, T., del Campo, A., Alvira, D., Rodríguez-Grández, B., Smith, R.C., Sáenz-Rodríguez, A.R. & Soria-Ruiz, P. (eds.) *Rapid Biological and Social Inventories 25: Perú: Ere-Campuya-Algodón.* The Field Museum, Chicago.

Stotz, D.F. & Pequeño, T. 2006. Aves / Birds. Pp. 88–98, 197–205 *in* Vriesendorp, C., Pitman, N., Rojas-M., J.I., Pawlak, B.A., Rivera, L., Calixto-M., L.C., Vela-C., L.M. & Fasabi-R., P. (eds.) *Rapid Biological and Social Inventories 16: Perú: Matsés.* The Field Museum, Chicago.

Stotz, D.F. & Quenamá, F.Q. 2007. Aves / Birds. Pp. 99–102, 160–169 *in* Borman, R., Vriesendorp, C., Alverson, W.S., Moskovits, D.K., Stotz, D.F. & del Campo, Á. (eds) *Rapid Biological and Social Inventories 19: Ecuador, Territorio Cofan Dureno.* The Field Museum, Chicago.

Stotz, D.F. & Valenzuela, P.M. 2008. Aves / Birds. Pp. 222–229, 324–351 *in* Alverson, W.S., Vriesendorp, C., del Campo, Á., Moskovits, D.K., Stotz, D.F., García D., M. & Borbor, L.A. (eds) *Rapid Biological and Social Inventories 20: Ecuador-Perú, Cuyabeno-Güeppí.* The Field Museum, Chicago.

Stotz, D.F. & Valenzuela, P.M. 2009. Aves / Birds. Pp. 199–209, 288–307 *in* Vriesendorp, C., Alverson, W.S., del Campo, A., Stotz, D.F., Moskovits, D.K., Fuentes C., S., Coronel T., B. & Anderson, E.P. (eds.) *Rapid Biological*

and Social Inventories 21: Ecuador, Cabeceras Cofanes-Chingual. The Field Museum, Chicago.

Stotz, D.F., Fitzpatrick, J.W., Parker, T.A. & Moskovits, D.K. 1996. *Neotropical Birds, Ecology and Conservation.* Univ. of Chicago Press, Chicago.

Stotz, D.F., Lanyon, S.M., Schulenberg, T.S., Willard, D.E., Peterson, A.T. & Fitzpatrick, J.W. 1997. An avifaunal survey of two tropical forest localities on the middle Rio Jiparaná, Rondônia, Brazil. *Orn. Monogr.* 48: 763–781.

Stotz, D.F., O'Shea, B., Miserendino, R., Condori, J. & Moskovits, D. 2003. Aves / Birds. Pp. 92–96, 125–135 *in* Alverson, W.S., Moskovits, D.K. & Halm, I.C. (eds.) *Rapid Biological Inventories 06: Bolivia, Pando, Federico Román. Rapid Biological Inventories 6.* The Field Museum, Chicago.

Stotz, D.F., del Castillo, P.S. & Ruelas-Inzunza, E. 2014. Aves / Birds. Pp. 138–154, 330–344, 482–503 *in* Pitman, N., Vriesendorp, C., Alvira, D., Markel, J.A., Johnston, M., Ruelas Inzunza, E., Lancha-Pizango, A., Sarmiento-Valenzuela, G., Álvarez-Loayza, P., Homan, J., Wachter, T., del Campo, Á., Stotz, D.F. & S. Heilpern, S. (eds.) *Rapid Biological and Social Inventories 26: Perú: Cordillera Escalera-Loreto.* The Field Museum, Chicago.

Stotz, D.F., del Castillo, P.S. & Laverde-R., O. 2016. Aves / Birds. Pp. 131–140, 311–319, 466–493 *in* Pitman, N., Bravo, A., Claramunt, S., Vriesendorp, C., Alvira-Reyes, D., Ravikumar, A., del Campo, Á., Stotz, D.F., Wachter, T., Heilpern, S., Rodríguez-Grández, B., Sáenz-Rodríguez, A.R. & Smith, R.C. (eds.) *Rapid Biological and Social Inventories 28: Perú: Medio Putumayo-Algodón.* The Field Museum, Chicago.

Stouffer, P.C. 2007. Density, territory size, and long-term spatial dynamics of a guild of terrestrial insectivorous birds near Manaus, Brazil. *Auk* 124: 291–306.

Stouffer, P.C. & Bierregaard, R.O. 1995. Use of Amazonian forest fragments by understory insectivorous birds. *Ecology* 78: 2429–2445.

Stouffer, P.C. & Bierregaard, R.O. 2007. Recovery potential of understory bird communities in Amazonian rainforest fragments. *Rev. Bras. Orn.* 15: 219–229.

Stouffer, P.C., Bierregaard, R.O., Strong, C. & Lovejoy, T.E. 2006. Long-term landscape change and bird abundance in Amazonian rainforest fragments. *Conserv. Biol.* 20: 1212–1223.

Stouffer, P.C., Strong, C. & Naka, L.N. 2009. Twenty years of understorey bird extinctions from Amazonian rain forest fragments: consistent trends and landscape-mediated dynamics. *Divers. & Distrib.* 15: 88–97.

Stratford, J.A. & Stouffer, P.C. 1999. Local extinction of terrestrial insectivorous birds in a fragmented landscape near Manaus, Brazil. *Conserv. Biol.* 13: 1416–1423.

Stratford, J.A. & Stouffer, P.C. 2013. Microhabitat associations of terrestrial insectivorous birds in Amazonian rainforest and second-growth forests. *J. Field Orn.* 84: 1–12.

Straube, F.C. 1988. Contribuição ao conhecimento da avifauna da região sudoeste do Estado do Paraná. *Biotemas* 1: 63–75.

Straube, F.C. 1989 Notas bionômicas sobre *Conopophaga melanops* (Vieillot, 1818) no Estado do Paraná. *Biotemas* 2: 91–95.

Straube, F.C. 2003. Avifauna da Área Especial de Interesse Turístico do Marumbi (Paraná, Brasil). *Atualidades Orn.* 113: 12.

Straube, F.C. 2008. Avifauna da Fazenda Barra Mansa

(Arapoti, Paraná), com anotações sobre a ocupação de monoculturas de essências arbóreas. *Atualidades Orn.* 142: 46–50.

Straube, F.C. & Urben-Filho, A. 2004. Uma revisão crítica sobre o grau de conhecimento da avifauna do Parque Nacional do Iguaçu (Pará, Brasil) e áreas adjacentes. *Atualidades Orn.* 118: 6–32.

Straube, F.C. & Urben-Filho, A. 2005a. Avifauna da Reserva Natural Salto Morato (Guaraqueçaba, Paraná). *Atualidades Orn.* 124: 1–21.

Straube, F.C. & Urben-Filho, A. 2005b. Observações sobre a avifauna de pequenos remanescentes florestais na região noroeste do Paraná (Brasil). *Atualidades Orn.* 123: 1–14.

Straube, F.C. & Urben-Filho, A. 2008. Notas sobre a avifauna de nove localidades na Bacia do Rio Piquiri (Região Oeste do Paraná, Brasil). *Atualidades Orn.* 141: 33–37.

Straube, F.C., Bornschein, M.R. & Scherer-Neto, P. 1996. Coletânea da avifauna da região noroeste do Estado do Paraná e áreas limítrofes (Brasil). *Arq. Biol. Tecn. Curitiba* 39: 193–214.

Straube, F.C., Willis, E.O. & Oniki, Y. 2002. Aves colecionadas na localidade de Fazenda Caiuá (Paraná, Brasil) por Adolph Hempel, com discussão sobre a sua localização exata. *Ararajuba* 10: 167–172.

Straube, F.C., Krul, R. & Carrano, E. 2005. Coletânea da avifauna da região sul do estado do Paraná (Brasil). *Atualidades Orn.* 125: 1–10.

Straube, F.C., Accordi, I.A. & Argel, M. 2007. Nomes populares de aves brasileiras coletados por Johann Natterer (1817-1835). *Atualidades Orn.* 136: 1–6.

Strewe, R. 2000a. Las aves y la importancia de la conservacion de la reserva natural El Pangán, Nariño, en el suroeste de Colombia. *Bol. Soc. Antioqueña Orn.* 11: 56–73.

Strewe, R. 2000b. New distributional sightings of 28 species of birds from Dpto. Nariño, SW Colombia. *Bull. Brit. Orn. Club* 120: 189–195.

Strewe, R. & Navarro, C. 2003. New distributional records and conservation importance of the San Salvador Valley, Sierra Nevada de Santa Marta, Colombia. *Orn. Colombiana* 1: 29–41.

Strewe, R. & Navarro, C. 2004. The threatened birds of the río Frío Valley, Sierra Nevada de Santa Marta, Colombia. *Cotinga* 22: 47–55.

Sturgis, B.B. 1928. *Field Book of Birds of the Panama Canal Zone.* G.P Putnam's Sons, New York.

Sumichrast, F. 1869. The geographical distribution of the native birds of the Department of Vera Cruz, with a list of the migratory species. *Mem. Boston Soc. Nat. Hist.* 1: 542–563.

Svensson-Coelho, M., Blake, J.G., Loiselle, B.A., Penrose, A.S., Parker, P.G. & Ricklefs, R.E. 2013. Diversity, prevalence, and host specificity of avian *Plasmodium* and *Haemoproteus* in a western Amazon assemblage. *Orn. Monogr.* 76: 1–47.

Swainson, W. 1841. *A Selection of the Birds of Brazil and Mexico: The Drawings.* H.G. Bohn, London.

Sztolcman, J. 1926a. Revision des oiseaux néotropicaux de la collection du Musée Polonais d'Historie Naturelle à Varsovie. I. *Ann. Zool. Mus. Polon. Hist. Nat. Warsaw* 5: 197–235.

Sztolcman, J. 1926b. Étude des collections ornithologiques de Paraná. *Ann. Zool. Mus. Polon. Hist. Nat. Warsaw* 5: 107–196.

Sztolcman, J. & Domaniewski, J. 1927. Typy opisowe ptaków w Polskiem Pa stwowem Muzeum Przyrodniczem. *Ann.*

Zool. Mus. Polon. Hist. Nat. Warsaw 6:

Taczanowski, L. 1874. Liste des oiseaux recuellis par M. Constantin Jel dans la partie centraleski du Perou occidental. *Proc. Zool. Soc. Lond.* 1874: 501–565.

Taczanowski, L. 1879. Liste des oiseaux recuellis au Nord de Pérou occidental par M. Stolzmann en 1878. *Proc. Zool. Soc. Lond.* 1879: 220–245.

Taczanowski, L. 1880. Liste des oiseaux recuellis au Nord de Pérou occidental par M. Stolzmann pendant les derniers mois de 1878 et dans la première moitié de 1879. *Proc. Zool. Soc. Lond.* 1880: 189–219.

Taczanowski, L. 1882. Liste des oiseaux recuellis par M. Stolzmann au Pérou nord-oriental *Proc. Zool. Soc. Lond.* 1882: 2–49.

Taczanowski, L. 1884. *Ornithologie du Pérou*, vol. 2. R. Friedländer & Sohn, Berlin.

Taczanowski, L. 1886. *Ornithologie du Pérou*, vol. 3. R. Friedländer & Sohn, Berlin.

Taczanowski, L. & von Berlepsch, H. 1885. Troisieme liste des oiseaux recuellis par M. Stolzmann dans l'Ecuadeur. *Proc. Zool. Soc. Lond.* 1885: 67–124.

Tallman, D.A. 1974. Colonization of a Semi-isolated Temperate Cloud Forest: Preliminary Interpretation of Distributional Patterns of Birds in the Carpish Region of the Department of Huánuco, Peru. M.Sc. dissertation. Louisiana State Univ., Baton Rouge, LA.

Tallman, E.J. & Tallman, D.A. 1994. The trematode fauna of an Amazonian antbird community. *Auk* 111: 1006–1013.

Telino-Júnior, W.R., Dias, M.M., Júnior, S.M.A., Lyra-Neves, R.M. & Larrazabal, M.E.L. 2005. Trophic structure of bird community of Reserva Estadual de Gurjau, Zona da Mata Sul, Pernambuco State, Brazil. *Rev. Bras. Zool.* 22: 962–973.

Telles, M. & Dias, M.M. 2010. Bird communities in two fragments of Cerrado in Itirapina, Brazil. *Braz. J. Biol.* 70: 537–550.

Temminck, C.J. & Laugier, M. 1820–39. *Nouveau Recueil de Planches Coloriées d'Oiseaux*, 102 livraisons (600 plates). Levrault, Paris.

Temminck, C.J. & Laugier, M. 1850. *Nouveau Recueil de Planches Coloriées d'Oiseaux*, 'vol. 2'. J.-B. Baillière, Livraire de L'Academie de Medecine, Paris.

Temple, S.A. 2002. Extinction-prone birds of Trinidad and Tobago: making predictions from theory. Pp. 180–193 in Hayes, F.E. & Temple, S.A. (eds.) *Studies in Trinidad and Tobago Ornithology Honouring Richard ffrench.* Dept. of Life Sciences, Univ. of the West Indies, St. Augustine, Trinidad.

Terborgh, J. 1985. The role of ecotones in the distribution of Andean birds. *Ecology* 66: 1237–1246.

Terborgh, J. & Weske, J.S. 1969. Colonization of secondary habitats by Peruvian birds. *Ecology* 50: 765–782.

Terborgh, J. & Weske, J.S. 1975. The role of competition in the distribution of Andean birds. *Ecology* 56: 562–576.

Terborgh, J., Fitzpatrick, J.W. & Emmons, L.H. 1984. Annotated checklist of bird and mammal species of Cocha Cashu Bioloigcal Station, Manu National Park. *Fieldiana Zool.* 21: 1–29.

Terborgh, J., Robinson, S.K., Parker, T.A., Munn, C.A. & Pierpont, N. 1990. Structure and organization of an Amazonian forest bird community. *Ecol. Monogr.* 60: 213–238.

Thiollay, J.-M. 1986. Structure comparée du peuplement avien dans trois sites de foret primaire en Guyane. *Rev. Ecol. (La Terre et la Vie)* 41: 59–105.

Thiollay, J.-M. 1992. Influence of selective logging on bird species diversity in a Guianan rain forest. *Conserv. Biol.* 6: 47–63.

Thiollay, J.-M. 1994. Structure, density and rarity in an Amazonian rainforest bird community. *J. Trop. Ecol.* 10: 449–481.

Thiollay, J.-M. 1997. Disturbance, selective logging and bird diversity: A Neotropical forest study. *Biodiver. & Conserv.* 6: 1155–1173.

Thiollay, J.-M. 1999a. Frequency of mixed species flocking in tropical forest birds and correlates of predation risk: An intertropical comparison. *J. Avian Biol.* 30: 282–294.

Thiollay, J.-M. 1999b. Responses of an avian community to rain forest degradation. *Biodiver. & Conserv.* 8: 513–534.

Thiollay, J.-M. 2002. Avian diversity and distribution in French Guiana: patterns across a large forest landscape. *J. Trop. Ecol.* 18: 471–498.

Thiollay, J.-M. & Jullien, M. 1998. Flocking behaviour of foraging birds in a neotropical rain forest and the antipredator defence hypothesis. *Ibis* 140: 382–394.

Thiollay, J.-M., Jullien, M., Théry, M. & Erard, C. 2001. Appendix 4: Bird species recorded in the Nouragues area (Guyane) (from Nouragues inselbergs to Arataye River). Pp. 357–370 in Bongers, F., Charles-Dominique, P., Forget, P.M. & Théry, M. (eds.) *Nouragues: Dynamics and Plant-Animal Interactions in a Neotropical Rainforest.* Kluwer Academic Publishers, Dordrecht.

Thomas, B.T. 1982. Weights of some Venezuelan birds. *Bull. Brit. Orn. Club* 102: 48–52.

Thurber, W.A., Serrano, J.F., Sermeño, A. & Benítez, M. 1987. Status of uncommon and previously unreported birds of El Salvador. *Proc. West. Found. Vert. Zool.* 3: 109–293.

Timm, R.M., Ramey R. R., I. & Nomenclatural Committee of the American Society of Mammalogists 2005. What constitutes a proper description? *Science* 309: 2163–2164.

Tinoco, B.A. & Astudillo, P. 2007. *Guía de Campo de las Aves del Parque Nacional Cajas.* ETAPA, Cuenca.

Tinoco, B.A. & Webster, P.A. 2009. *Guía de Campo para Observación de Aves del Parque Nacional Cajas.* ETAPA, Cuenca.

Tobias, J.A. 2009. *Antbirds of Peru and Bolivia.* The Field Museum, Chicago.

Todd, W.E.C. 1915. Preliminary diagnoses of apparently new South American birds. *Proc. Biol. Soc. Wash.* 28: 79–82.

Todd, W.E.C. 1919. Descriptions of apparently new Colombian birds. *Proc. Biol. Soc. Wash.* 32: 113–117.

Todd, W.E.C. 1927. New gnateaters and antbirds from tropical America, with a revision of the genus *Myrmeciza* and its allies. *Proc. Biol. Soc. Wash.* 40: 149–178.

Todd, W.E.C. 1928. List of types of birds in the collection of the Carnegie Museum on May 1, 1928. *Ann. Carnegie Mus.* 18: 329–364.

Todd, W.E.C. 1932. Seven apparently new South American birds. *Proc. Biol. Soc. Wash.* 45: 215–220.

Todd, W.E.C. & Carriker, M.A. 1922. The birds of the Santa Marta region of Colombia: a study in altitudinal distribution. *Ann. Carnegie Mus.* 14: 1–611.

Toral, F. 1996. Variación en la Composición de las Comunidades de Aves en Diferentes Tipos de Vegetación en el Bosque Protector de Mazán. B.Sc. dissertation. Univ. del Azuay, Cuenca.

del Toro, M.Á. 1958. Lista de las especies de aves que habítan en Chiapas. *Rev. Soc. Mex. Hist. Nat.* 19: 73–113.

del Toro, M.Á. 1964. *Lista de las aves de Chiapas: Endémicas, Emigrantes y de Paso.* Instituto de Ciencias y Artes de Chiapas, Tuxtla Gutiérrez.

del Toro, M.Á. 1971. *Las Aves de Chiapas.* Univ. Autonoma de Chiapas, Tuxtla Gutierrez.

Tostain, O. 1986. Description du nid et de la ponte de deux formicariidés guyanais: *Hylopezus macularius* et *Thamnophilus nigrocinereus. Alauda* 54: 170–176.

Tostain, O. & Dujardin, J.-L. 1988. Nesting of the Wing-banded Antbird and the Thrush-like Antpitta in French Guiana. *Condor* 90: 236–239.

Tostain, O., Dujardin, J.-L., Érard, C.H. & Thiollay, J.-M. 1992. *Oiseaux de Guyane.* Société d'Etudes Ornithologiques, Brunoy.

Traylor, M.A. 1958. Birds of northeastern Peru. *Fieldiana Zool.* 35: 87–141.

Tristram, H.B. 1889. *Catalogue of a Collection of Birds Belonging to H. B. Tristram.* Advertiser Office, Durham.

Trolle, M. & Walther, B.A. 2004. Preliminary bird observations in the rio Jauaperí region, rio Negro basin, Amazonia, Brazil. *Cotinga* 22: 81–85.

von Tschudi, J.J. 1844. Avium conspectus, quae in Republica Peruana reperiuntur et pleraeque observatae vel collectae sunt in itinere a Dr. J. J. de Tschudi. *Arch. Naturges.* 10: 262–317.

von Tschudi, J.J. 1845. *Untersuchungen über die Fauna Peruana. Ornithologie.* Scheitlin & Zollikofer, St. Gallen.

Tubelis, D.P. & Tomas, W.M. 2003. Bird species of the Pantanal wetland, Brazil. *Ararajuba* 11: 5–37.

Uribe, D. 2011. Confirmada la presencia del Tororoi Medialuna en Caldas. *Merganetta* 48: 1–2.

Valadão, R.M. 2012. As aves da Estação Ecológica Serra das Araras, Mato Grosso, Brasil. *Biota Neotropica* 12: 263–281.

Valarezo-Delgado, S. 1984. Aves del Ecuador: sus nombres vulgares. Tomo II. *Publ. Mus. Ecuatoriano Cien. Nat., Ser. Monogr.* 1: 222–528.

Valdez-Juarez, S.O. & Londoño, G.A. 2011. Nesting of the Pectoral Sparrow (*Arremon taciturnus*) in southeastern Peru. *Wilson J. Orn.* 123: 808–813.

Valdez-Juarez, S.O. & Londoño, G.A. 2016. Nesting biology of Carmiol's Tanager (*Chlorothraupis carmioli frenata*) in southeastern Peru. *Wilson J. Orn.* 128: 794–803.

Valencia-Herverth, R., Valencia-Herverth, J., Calderón, R.P., Nochebuena, M.O., Hernández, R.H. & Nochebuena, M.O. 2012. Información adicional sobre la avifauna de Hidalgo, México. *Huitzil* 13: 95–103.

Vallejo, E.A. 2003. *Modernizando la Naturaleza: Desarrollo Sostenible y Conservación de la Naturaleza en la Amazonia Ecuatoriana.* Simbioe, Quito.

Valqui, T. 2004. *Where to Watch Birds in Peru.* Grafica Ñañez, Lima.

Vasconcelos, M.F. 2003a. Padrões de distribuição geográfica da avifauna na região da Cadeia do Espinhaço de Minas Gerais. P. 261 in Machado, C.G. (ed.) *Livro de Resumos do XI Congresso Brasileiro de Ornitologia.* Feira de Santana.

Vasconcelos, M.F. 2003b. A avifauna dos campos de altitude da Serra do Caparaó, estados de Minas Gerais e Espirito Santo, Brasil. *Cotinga* 19: 40–48.

Vasconcelos, M.F. & Melo-Júnior, T.A. 2001. An ornithological survey of Serra do Caraça, Minas Gerais, Brazil. *Cotinga* 15: 21–31.

Vasconcelos, M.F. & Neto, S.D. 2007. Padrões de distribuição e conservação da avifauna na região central da Cadeia do Espinhaço e áreas adjacentes, Minas Gerais, Brasil. *Cotinga* 28: 27–44.

Vasconcelos, M.F. & Neto, S.D. 2009. First assessment of the avifauna of Araucaria forests and other habitats from extreme southern Minas Gerais, Serra da Mantiqueira, Brazil, with notes on biogeography and conservation. *Pap. Avuls. Dept. Zool., São Paulo* 49: 49–71.

Vasconcelos, M.F., Neto, S.D., Brandt, L.F.S., Venturin, N., Oliveira-Filho, A.T. & Costa, F.A.F. 2002. Avifauna de Lavras e municípios adjacentes, sul de Minas Gerais, e comentários sobre sua conservaço. *Unimontes Científica* 4: 153–164.

Vasconcelos, M.F., Neto, S.D., Kirwan, G.M., Bornschein, M.R., Diniz, M.G. & Silva, J.F. 2006. Important ornithological records from Minas Gerais state, Brazil. *Bull. Brit. Orn. Club* 126: 212–238.

Vaurie, C. 1972. An ornithological gazetteer of Peru (based on information compiled by J. T. Zimmer). *Amer. Mus. Novit.* 2491 1–36.

Vecchi, M.B. & Alves, M.A.S. 2015. Bird assemblage mist-netted in an Atlantic Forest area: A comparison between vertically-mobile and ground-level nets. *Brazil. J. Biol.* 75: 742–751.

Velásquez-Ruiz, C.A. 2005. *Paleoecología de Alta Resolución del Holoceno Tardío en el Páramo de Frontino Antioquia*. Univ. Nacional de Colombia, Medellín.

Velho, P.P.P. 1932. Descripço de alguns ovos de aves do Brasil existentes nas coleções do Museu. *Bol. Mus. Nac. Rio de Janeiro* 8: 49–60.

Vellinga, W.-P., Flanagan, J.N.M. & Mark, T.R. 2004. New and interesting records of birds from Ayabaca province, Piura, north-west Peru. *Bull. Brit. Orn. Club* 124: 124–142.

Venturini, A.C., Rehen, M.P., de Paz, P.R. & do Carmo, L.P. 2001. Contribuição ao conhecimento das aves da região centro serrana do Espírito Santo: municípios de Santa Maria do Jetibá e Itarana (parte 2). *Atualidades Orn.* 99: 12.

Verea, C. 2004. Contribución al conocimiento del Ponchito Pechiescamado (*Grallaricula loricata*) (Formicariidae) de los bosques nublados del Parque Nacional Henri Pittier, norte de Venezuela. *Orn. Neotrop.* 15: 225–235.

Verea, C. 2007. Algunas notas sobre las características externas del Ponchito Pechiescamado (*Grallaricula loricata*), una especie endémica del norte de Venezuela. *Orn. Neotrop.* 18: 1–9.

Verea, C. & Greeney, H.F. 2014. Scallop-breasted Antpitta (*Grallaricula loricata*). *In* Schulenberg, T.S. (ed.) *Neotropical Birds Online*. Cornell Lab of Ornithology, Ithaca, NY (http://neotropical.birds.cornell.edu/portal/species/overview?p_p_spp=408201).

Verea, C. & Solórzano, A. 2011. Avifauna asociada al sotobosque musgoso del Pico Guacamaya, Parque Nacional Henri Pittier, Venezuela. *Interciencia* 36: 324–330.

Verea, C., Solórzano, A. & Fernández-Badillo, A. 1999. Pesos y distribución de aves del sotobosque del Parque Nacional Henri Pittier al norte de Venezuela. *Orn. Neotrop.* 10: 217–231.

Verea, C., Solórzano, A., Díaz, M., Parra, L., Araujo, M.A., Antón, F., Navas, O., Ruíz, O.J.L. & Fernández-Badillo, A. 2009. Registros de actividad reproductora y muda en algunas aves del norte de Venezuela. *Orn. Neotrop.* 20: 181–201.

Verea, C., Rodríguez, G.A., Ascanio, D., Solórzano, A., Sainz-Borgo, C., Alcocer, D. & González-Bruzual, L.G. 2017. *Los Nombres Comunes de las Aves de Venezuela*, 4ta edición. Comité de Nomenclatura Común de las Aves de Venezuela; Unión Venezolana de Ornitólogos (UVO), Caracas.

Verhelst, J.C., Rodríguez, J.C., Orrego, O., Botero, J.E., López, J.A., Franco, V.M. & Pfeifer, A.M. 2001. Aves del Municipio de Manizales, Caldas, Colombia. *Biota Colombiana* 2: 265–284.

Vidoz, J.Q., Jahn, A.E. & Mamani, A.M. 2010. The avifauna of Estación Biológica Caparú, Bolivia: natural history and range extensions of some poorly known species. *Cotinga* 32: 5–22.

Vieillot, L.P. 1816. *Analyse d'une Nouvelle Ornithologie Élémentaire*. Deterville, Paris.

Vieillot, L.P. 1818. *Nouveau Dictionnaire d'Histoire Naturelle*. Deterville, Paris.

Vieillot, L.P. 1823. *La Galerie des Oiseaux: Livraison 25*. De Rignoux, Paris.

Vieillot, L.P. 1824. *La Galerie des Oiseaux: Livraison 45*. Carpentier-Méricourt, Paris.

Vieillot, L.P. 1834. *La Galerie des Oiseaux*, vol. 1. Carpentier-Méricourt, Paris. [Reprinted compilation of original, see Lebossé & Dickinson 2015].

Visbal, R., Manzanilla, J. & Fernández, A. 1996. *Importancia de los Vertebrados del Parque Nacional Henri Pittier y Consideraciones para su Conservación*. Univ. Central de Venezuela, Maracay.

Visco, D.M., Michel, N.L., Boyle, W.A., Sigel, B.J., Woltmann, S. & Sherry, T.W. 2015. Patterns and causes of understory bird declines in human-disturbed tropical forest landscapes: a case study from Central America. *Biol. Conserv.* 191: 117–129.

Vogt, C. 2007. Range extensions and noteworthy records for mainland Ecuador. *Bull. Brit. Orn. Club* 127: 228–233.

Volpato, G.H., dos Anjos, L., Lopes, E.V., Fávaro, F.L. & Mendonça, L.B. 2004. Abundancia e distribuição, da familia Formicariidae na Floresta Atlantica da bacia hidrografica do Rio Tibagi, Paraná, Brasil. P. 413 *in* Testoni, A.F. & Althoff, S.L. (eds.) *Livro de Resumos do XII Congresso Brasileiro de Ornitologia*. Blumenau.

Volpato, G.H., dos Anjos, L., Poletto, F., Serafini, P.P., Lopes, E.V. & Fávaro, F.L. 2006. Terrestrial passerines in an Atlantic forest remnant of southern Brazil. *Braz. J. Biol.* 66: 473–478.

Volpe, M.M., Neto, J.B.F., Pereira, A.R., Diaz, D., Pedro, Lobão, P.S.P. & Meira, L.P.C. 2003. Avaliação da avifauna na área de influência da Linha de Transmissão LT 230 kV, Poções–Brumado (Bahia - Brasil). P. 131 *in* Machado, C.G. (ed.) *Livro de Resumos do XI Congresso Brasileiro de Ornitologia*. Feira de Santana.

Walker, B. 2002. Observations from the Tumbes Reserved Zone, dpto. Tumbes, with notes on some new taxa for Peru and a checklist of the area. *Cotinga* 18: 37–43.

Walker, B. 2009. Birding the Manu Biosphere Reserve, Peru. *Neotrop. Birding* 5: 49–58.

Walker, B. & Fjeldså, J. 2002. *Field Guide to the Birds of Machu Picchu, Peru*. Fondo Nacional para Areas Naturales Protegidas por el Estado (PROFONANPE) y Programa Machu Picchu, Lima.

Walker, B. & Fjeldså, J. 2005. *The Birds of Machu Picchu and the Cusco Region*. Nuevas Imagenes, S. A., Lima.

Walker, B., Stotz, D.F., Pequeño, T. & Fitzpatrick, J.W. 2006. Birds of the Manu Biosphere Reserve Pp. 23–49 *in* Patterson, B.D., Stotz, D.F. & Solari, S. (eds.) Mammals and Birds of the Manu Biosphere Reserve, Peru. *Fieldiana Zool.* 110.

Walters, M. 1980. *The Complete Birds of the World*. David & Charles, Newton Abbot.

Warren, R.L.M. & Harrison, C.J.O. 1971. *Type-specimens of Birds in the British Museum (Natural History)*, vol. 2. British Museum of Natural History, London.

Wege, D.C. & Long, A.J. 1995. *Key Areas for Threatened Birds in the Neotropics*. BirdLife International, Cambridge, UK.

Welford, M.R. 2000. The importance of early successional habitats to rare, restricted-range, and endangered birds in the Ecuadorian Andes. *Bird Conserv. Intern.* 10: 351–359.

Welford, M.R. & Nunnery, T. 2001. Behaviour and use of a human trail by Giant Antpitta *Grallaria gigantea*. *Cotinga* 16: 67–68.

Weske, J.S. 1972. *The Distribution of the Avifauna in the Apurimac Valley of Peru with respect to Environmental Gradients, Habitat, and Related Species*. Ph.D. thesis. Univ. of Oklahoma, Norman.

Wetmore, A. 1939a. Observations on the birds of northern Venezuela. *Proc. US Natl. Mus.* 87: 173–260.

Wetmore, A. 1939b. Lista parcial de los pájaros del Parque Nacional de Venezuela. *Bol. Soc. Venez. Cienc. Nat.* 5: 269–298.

Wetmore, A. 1943. The birds of southern Veracruz, Mexico. *Proc. US Natl. Mus.* 93: 215–340.

Wetmore, A. 1945. A review of the Giant Antpitta *Grallaria gigantea*. *Proc. Biol. Soc. Wash.* 58: 17–20.

Wetmore, A. 1946. New birds from Colombia. *Smithsonian Misc. Coll.* 106: 1–14.

Wetmore, A. 1972. *Birds of the Republic of Panama. Part III. Dendrocolaptidae (Woodcreepers) to Oxyruncidae (Sharpbills)*. Smithsonian Institution Press, Washington DC.

Wetmore, A. & Phelps, W.H., Jr. 1956. Further additions to the list of birds of Venezuela. *Proc. Biol. Soc. Wash.* 69: 1–10.

Wheatley, N. 1995. *Where to Watch Birds in South America*. Princeton Univ. Press, Princeton, NJ.

Whitney, B.M. 1992. A nest and egg of the Rufous Antpitta in Ecuador. *Wilson Bull.* 104: 759–760.

Whitney, B.M. 1997. Birding the Alta Floresta region, northern Mato Grosso, Brazil. *Cotinga* 7: 64–68.

Whitney, B.M. 2003. Family Conopophagidae (gnateaters). Pp. 732–747 *in* del Hoyo, J., Elliott, A. & Christie, D.A. (eds.) *Handbook of the Birds of the World*, vol. 8. Lynx Edicions, Barcelona.

Whitney, B.M., Oren, D.C. & Pimentel-Neto, D.C. 1997. Avaliação ecológica rápida da avifauna do Parque Nacional da Serra do Divisor, Acre, Brasil, com comentários sobre mamíferos, a conservação, o manejo e o potencial de ecoturismo no PNSD. Unpubl. report.

Whitney, B.M., Rowlett, J.L. & Rowlett, R.A. 1994. Distributional and other noteworthy records for some Bolivian birds. *Bull. Brit. Orn. Club* 114: 149–162.

Whitney, B.M., Pacheco, J.F., Isler, P.R. & Isler, M.L. 1995. *Hylopezus nattereri* (Pinto, 1937) is a valid species (Passeriformes: Formicariidae). *Ararajuba* 3: 37–42.

Whittaker, A. 2004. Noteworthy ornithological records from Rondônia, Brazil, including a first country record, comments on austral migration, life history, taxonomy and distribution, with relevant data from neighbouring states, and a first record for Bolivia. *Bull. Brit. Orn. Club* 124: 239–271.

Whittaker, A. 2009. Pousada Rio Roosevelt: a provisional avifaunal inventory in south-western Amazonian Brazil, with information on life history, new distributional data and comments on taxonomy. *Cotinga* 31: 23–46.

Whittaker, A. & Oren, D.C. 1999. Important ornithological records from the Rio Juruá, western Amazonia, including twelve additions to the Brazilian avifauna. *Bull. Brit. Orn. Club* 119: 235–260.

Whittaker, A., Oren, D.C., Pacheco, J.F., Parrini, R. & Minns, J. 2002. Aves registradas na reserva extrativista do Alto Juruá. Pp. 81–99 *in* Cunha, M.C. & Almeida, M.B. (eds.) *Enciclopédia da Floresta: O Alto Juruá: Práticas e Conhecimentos das Populações*. Companhia das Letras, São Paulo.

Whittaker, A., Aleixo, A. & Poletto, F. 2008. Corrections and additions to an annotated checklist of birds of the upper Rio Urucu, Amazonas, Brazil. *Bull. Brit. Orn. Club* 128: 114–125.

Wied, M. 1831. *Beiträge zur Naturgeschichte von Brasilien, Vögel* 3(2). Landes-Industrie-Comptoirs, Weimar.

Wiedenfeld, D.A. 1982. A nest of the Pale-billed Antpitta (*Grallaria carrikeri*) with comparative remarks on antpitta nests. *Wilson Bull.* 94: 580–582.

Wiedenfeld, D.A., Schulenberg, T.S. & Robbins, M.B. 1985. Birds of a tropical deciduous forest in extreme northwestern Peru. *Orn. Monogr.* 305–315.

Willard, D.E., Foster, M.S., Barrowclough, G.F., Dickerman, R.W., Cannell, P.F., Coats, S.L., Cracraft, J.L. & O'Neill, J.P. 1991. The birds of Cerro de la Neblina, Territorio Federal Amazonas, Venezuela. *Fieldiana Zool.* 65: 1–80.

Williams, R. 1995. Neotropical notebook: Brazil. *Cotinga* 4: 66–67.

Williams, R.A.J. 2010. *Ecology and Geography of Avian Viruses Using Niche Models and Wild Bird Surveillance*. Ph.D. thesis, Univ. of Kansas, Lawrence.

Willis, E.O. 1972. The behavior of Spotted Antbirds. *Orn. Monogr.* 10: 1–162.

Willis, E.O. 1977. Lista preliminar das aves da parte noroeste e áreas vizinhas da Reserva Ducke, Amazonas, Brazil. *Rev. Bras. Biol.* 37: 585–601.

Willis, E.O. 1979. The composition of avian communities in remanescent woodlots in southern Brazil. *Pap. Avuls. Dept. Zool., São Paulo* 33: 1–25.

Willis, E.O. 1985a. Antthrushes, antpittas, and gnateaters (Aves, Formicariidae) as army ant followers. *Rev. Bras. Zool.* 2: 443–448.

Willis, E.O. 1985b. *Myrmeciza* and related antbirds (Aves, Formicariidae) as army ant followers. *Rev. Bras. Zool.* 2: 433–442.

Willis, E.O. 1987. Primeiros registros de *Geotrygon saphirina* (Aves: Columbidae) e *Grallaria* sp. cf. *eludens* (Aves, Formicariidae) no oeste do Brasil. P. 153 *in Livro de Resumos do XIV Congresso Brasileira de Zoologia*. Juiz de Fora.

Willis, E.O. 1988. Behavioral notes, breeding records, and range extensions for Colombian birds. *Rev. Acad. Colombiana Cien. Exactas, Fisicas, y Nat.* 16: 137–150.

Willis, E.O. 2003a. Birds of a neotropical woodlot after fire. *Orn. Neotrop.* 14: 233–246.

Willis, E.O. 2003b. Birds of a eucalyptus woodlot in interior São Paulo. *Braz. J. Biol.* 63: 141–158.

Willis, E.O. 2004. Birds of a habitat spectrum in the Itirapina Savanna, São Paulo, Brazil (1982-2003). *Braz. J. Biol.* 64: 901–910.

Willis, E.O. 2006. Protected Cerrado fragments grow up and lose even metapopulational birds in central São Paulo, Brazil. *Braz. J. Biol.* 66: 829–837.

Willis, E.O. & Eisenmann, E. 1979. A revised list of birds of Barro Colorado Island, Panama. *Smithsonian Contrib. Zool.* 1–39.

Willis, E.O. & Oniki, Y. 1972. Studies of ant-following birds north of the eastern Amazon. *Acta Amazonica* 2: 127–151.

Willis, E.O. & Oniki, Y. 1978. Birds and army ants. *Ann. Rev. Ecol. & Syst.* 9: 243–263.

Willis, E.O. & Oniki, Y. 1981. Levantamento preliminar de aves em treze áreas do Estado de São Paulo. *Rev. Bras. Biol.* 41: 121–135.

Willis, E.O. & Oniki, Y. 1988. Aves observadas em Balbina, Amazonas e os prováveis efeitos da barragem. *Ciência e Cultura* 40: 280–284.

Willis, E.O. & Oniki, Y. 1991a. Avifaunal transects across the open zones of northern Minas Gerais, Brazil. *Ararajuba* 2: 41–58.

Willis, E.O. & Oniki, Y. 1991b. *Nomes Gerais para as Aves Brasileiras.* Gráfica da Região, Américo Brasiliense, São Paulo.

Willis, E.O. & Oniki, Y. 2002a. Birds of Santa Teresa, Espírito Santo, Brazil: do humans add or subtract species? *Pap. Avuls. Dept. Zool., São Paulo* 42: 193–264.

Willis, E.O. & Oniki, Y. 2002b. Birds of a central São Paulo woodlot: 1. Censuses 1982-2000. *Braz. J. Biol.* 62: 197–210.

Willis, E.O. & Schuchmann, K.-L. 1993. Comparison of cloud-forest avifaunas in southeastern Brazil and western Colombia. *Orn. Neotrop.* 4: 55–63.

Willis, E.O., Oniki, Y. & Silva, W.R. 1983. On the behavior of Rufous Gnateaters (*Conopophaga lineata*, Formicariidae). *Naturalia* 8: 67–83.

Wilson, S., Collister, D.M. & Wilson, A.G. 2011. Community composition and annual survival of lowland tropical forest birds on the Osa Peninsula, Costa Rica. *Orn. Neotrop.* 22: 421–436.

Winger, B.M., Hosner, P.A., Bravo, G.A., Cuervo, A.M., Aristizabal, N., Cueto, L.E. & Bates, J.M. 2015. Inferring speciation history in the Andes with reduced-representation sequence data: an example in the bay-backed antpittas (Aves; Grallariidae; *Grallaria hypoleuca s. l.*). *Mol. Ecol.* 24: 6256–6277.

Woods, S., Athanas, N. & Olmstead, S. 2011. Antpitta paradise: a 2010 update. *Neotrop. Birding* 8: 5–10.

Woodworth-Lynas, C.R., Caines, J.R. & Bennett, G.F. 1989. Prevalence of avian haematozoa in São Paulo State, Brazil. *Mem. Inst. Oswaldo Cruz* 84: 515–526.

Wunderle, J.M., Henriques, L.M.P. & Willig, M.R. 2006. Shortterm responses of birds to forest gaps and understory: an assessment of reduced-impact logging in a lowland Amazon forest. *Biotropica* 38: 235–255.

Young, B.E., DeRosier, D. & Powell, G.V.N. 1998. Diversity and conservation of understory birds in the Tilarán Mountains, Costa Rica. *Auk* 115: 998–1016.

Young, B.E., Franke, I., Hernandez, P.A., Herzog, S.K., Paniagua, L., Tovar, C. & Valqui, T. 2009. Using spatial models to predict areas of endemism and gaps in the protection of Andean slope birds. *Auk* 126: 554–565.

Zeledón, J.C. 1885. Catalogue of the birds of Costa Rica, indicating those species of which the United States National Museum possesses specimens from that country. *Proc. US Natl. Mus.* 8: 104–118.

Zimmer, J.T. 1930. Birds of the Marshall Field Peruvian Expedition, 1922-1923. *Field Mus. Nat. Hist. (Zool. Ser.)* 282.

Zimmer, J.T. 1931. Studies of Peruvian birds 1, New and other birds from Peru, Ecuador, and Brazil. *Am. Mus. Novit.* 500: 1–23.

Zimmer, J.T. 1934. Studies of Peruvian birds 12, notes on *Hylophylax, Myrmothera,* and *Grallaria. Am. Mus. Novit.* 703: 1–21.

Zimmer, J.T. 1937. Studies of Peruvian birds 25, Notes on the genera *Thamnophilus, Thamnocharis, Gymnopithys,* and *Ramphocaenus. Am. Mus. Novit.* 917: 1–16.

Zimmer, J.T. & Mayr, E. 1943. New species of birds described from 1938 to 1941. *Auk* 60: 249–262.

Zimmer, J.T. & Phelps, W.H. 1945 New species and subspecies of birds from Venezuela 2. *Amer. Mus. Novit.* 1274: 1–9.

Zimmer, J.T. & Phelps, W.H. 1946. Twenty-three new subspecies of birds from Venezuela and Brazil. *Amer. Mus. Novit.* 1312: 1–23.

Zimmer, J.T. & Phelps, W.H. 1954. A new flycatcher from Venezuela, with remarks on the Mocquerys Collection and the piculet, *Picumnus squamulatus. Amer. Mus. Novit.* 1657: 1–7.

Zimmer, K.J., Parker, T.A., Isler, M.L. & Isler, P.R. 1997. Survey of a southern Amazonian avifauna, the Alta Floresta region, Mato Grosso, Brazil. *Orn. Monogr.* 48: 887–918.

Zotta, A.R. 1944. *Lista Sistemática de las Aves Argentinas: Comprende la Enumeración de Todas las Especies y Subespecies Conocidas en la Argentina, Ordenadas Taxonómicamente Según la Moderna Nomenclatura Ornitológica, con sus Nombres Técnicos y Vulgares y su Distribución Geográfica.* Talleres Gráficos Tomás Palumbo, Buenos Aires.

Zuleta-Marín, J.A. 2012. *Inventarios Preliminares de Biodiversidad del Parque Regional Natural Santa Emilia.* Corporación Autónoma Regional de Risaralda, Belén de Umbría, Risaralda.

Zyskowski, K., Mittermeier, J.C., Ottema, O., Rakovic, M., O'Shea, B.J., Lai, J.E., Hochgraf, S.B., de León, J. & Au, K. 2011. Avifauna of the easternmost Tepui, Tafelberg in central Suriname. *Bull. Peabody Mus. Nat. Hist.* 52: 153–180.

INDEX

Numbers in **bold** refer to plate numbers, page numbers in *italic* refer to the caption text in the plate section. Other page numbers refer to the first occurrence of species and subspecies names in the relevant species account. Some subspecies have detailed sub-entries later in the species account.